Table of Atomic Masses*

	Symbol	Atomic Number	Atomic Mass		Symbol	Atomic Number	Atomic Mass
Actinium	Ac	89	(227)†	Mercury	Hg	80	200.6
Aluminum	Al	13	26.98	Molybdenum	Mo	42	95.94
Americium	Am	95	(243)	Neodymium	Nd	60	144.2
Antimony	Sb	51	121.8	Neon	Ne	10	20.18
Argon	Ar	18	39.95	Neptunium	Np	93	(237)
Arsenic	As	33	74.92	Nickel	Ni	28	58.70
Astatine	At	85	(210)	Niobium	Nb	41	92.91
Barium	Ba	56	137.3	Nitrogen	N	7	14.01
Berkelium	Bk	97	(247)	Nobelium	No	102	(259)
Beryllium	Be	4	9.012	Osmium	Os	76	190.2
Bismuth	Bi	83	209.0	Oxygen	O	8	16.00
Boron	B	5	10.81	Palladium	Pd	46	106.4
Bromine	Br	35	79.90	Phosphorus	P	15	30.97
Cadmium	Cd	48	112.4	Platinum	Pt	78	195.1
Calcium	Ca	20	40.08	Plutonium	Pu	94	(244)
Californium	Cf	98	(251)	Polonium	Po	84	(209)
Carbon	C	6	12.01	Potassium	K	19	39.10
Cerium	Ce	58	140.1	Praseodymium	Pr	59	140.9
Cesium	Cs	55	132.9	Promethium	Pm	61	(145)
Chlorine	Cl	17	35.45	Protactinium	Pa	91	(231)
Chromium	Cr	24	52.00	Radium	Ra	88	226.0
Cobalt	Co	27	58.93	Radon	Rn	86	(222)
Copper	Cu	29	63.55	Rhenium	Re	75	186.2
Curium	Cm	96	(247)	Rhodium	Rh	45	102.9
Dysprosium	Dy	66	162.5	Rubidium	Rb	37	85.47
Einsteinium	Es	99	(252)	Ruthenium	Ru	44	101.1
Erbium	Er	68	167.3	Samarium	Sm	62	150.4
Europium	Eu	63	152.0	Scandium	Sc	21	44.96
Fermium	Fm	100	(257)	Selenium	Se	34	78.96
Fluorine	F	9	19.00	Silicon	Si	14	28.09
Francium	Fr	87	(223)	Silver	Ag	47	107.9
Gadolinium	Gd	64	157.3	Sodium	Na	11	22.99
Gallium	Ga	31	69.72	Strontium	Sr	38	87.62
Germanium	Ge	32	72.59	Sulfur	S	16	32.06
Gold	Au	79	197.0	Tantalum	Ta	73	180.9
Hafnium	Hf	72	178.5	Technetium	Tc	43	(98)
Helium	He	2	4.003	Tellurium	Te	52	127.6
Holmium	Ho	67	164.9	Terbium	Tb	65	158.9
Hydrogen	H	1	1.008	Thallium	Tl	81	204.4
Indium	In	49	114.8	Thorium	Th	90	232.0
Iodine	I	53	126.9	Thulium	Tm	69	168.9
Iridium	Ir	77	192.2	Tin	Sn	50	118.7
Iron	Fe	26	55.85	Titanium	Ti	22	47.90
Krypton	Kr	36	83.80	Tungsten	W	74	183.9
Lanthanum	La	57	138.9	Uranium	U	92	238.0
Lawrencium	Lr	103	(260)	Vanadium	V	23	50.94
Lead	Pb	82	207.2	Xenon	Xe	54	131.3
Lithium	Li	3	6.941	Ytterbium	Yb	70	173.0
Lutetium	Lu	71	175.0	Yttrium	Y	39	88.91
Magnesium	Mg	12	24.31	Zinc	Zn	30	65.38
Manganese	Mn	25	54.94	Zirconium	Zr	40	91.22
Mendelevium	Md	101	(258)				

*The values given here are to four significant figures. A table of more accurate atomic masses is given in Appendix F.

†A value given in parentheses denotes the mass of the longest-lived isotope.

GENERAL CHEMISTRY

GENERAL CHEMISTRY

SECOND EDITION

DONALD A. McQUARRIE

UNIVERSITY OF CALIFORNIA, DAVIS

PETER A. ROCK

UNIVERSITY OF CALIFORNIA, DAVIS

W. H. FREEMAN AND COMPANY · NEW YORK

COVER IMAGE: An oxy-hydrogen torch (Fig. D-1).

Photograph by Chip Clark.

Library of Congress Cataloging-in-Publication Data

McQuarrie, Donald A. (Donald Allan)
General chemistry.

Includes index.
1. Chemistry. I. Rock, Peter A., 1939–
II. Title.
QD31.2.M388 1987 540 86-18371
ISBN 0-7167-1806-5

Printed in the United States of America

2 3 4 5 6 7 8 9 0 KP 5 4 3 2 1 0 8 9 8 7

CONTENTS IN BRIEF

CONTENTS

INTERCHAPTER A
SEPARATION OF MIXTURES 36

CHAPTER 2
CHEMICAL REACTIONS AND THE PERIODIC TABLE 43

INTERCHAPTER B
THE ALKALI METALS 82

CHAPTER 3
CHEMICAL CALCULATIONS 88

INTERCHAPTER C
THE MAIN GROUP METALS 119

CHAPTER 4
THE PROPERTIES OF GASES 126

INTERCHAPTER D
HYDROGEN AND OXYGEN 155

CHAPTER 5
THERMOCHEMISTRY 163

CHAPTER 20 ELECTROCHEMISTRY 623

INTERCHAPTER M THE HALOGENS 658

CHAPTER 21 NUCLEAR AND RADIOCHEMISTRY 670

INTERCHAPTER N THE TRANSITION METALS 700

CHAPTER 22 TRANSITION METAL COMPLEXES 713

CHAPTER 23 ORGANIC CHEMISTRY 741

CHAPTER 24 SYNTHETIC AND NATURAL POLYMERS 769

APPENDIX A A MATHEMATICAL REVIEW 791

APPENDIX B

APPENDIX C

APPENDIX D

APPENDIX E

APPENDIX F

PREFACE

The experience of preparing this second edition of our text was quite unlike that of writing the first. We approached the first edition with the courage of our convictions, believing that what worked for us in the classroom would work for others. We approached the second edition with a different kind of confidence: the confidence of knowing that the first edition had been widely and continuously used, and that we could draw upon the experiences of the text's many users. The enthusiastic reception of our approach was most gratifying, and it became clear that the second edition should be written in the spirit of the first.

We take what we call the "experimental" approach to chemistry. We believe students should be exposed to the science as it is practiced and applied. In most cases we introduce and discuss experimental data before developing the theory to tie these data together. We make a great effort to use actual chemical compounds and real data in formulating in-chapter examples and end-of-chapter problems. We avoid discussions of unspecified substances A and B or reactions such as A + B → C; rather, we use actual chemical species to make the presentation real and vivid.

Two integral features contribute to the success of our approach. The first is our early introduction of descriptive chemistry and the second is our presentation of substances and reactions in color throughout the text.

INTERCHAPTERS PROVIDE AN INVITING AND FLEXIBLE WAY TO INCORPORATE DESCRIPTIVE CHEMISTRY

We retain the format of *interchapters,* each briefly focusing on the chemistry of a particular element or group of elements. These present chemistry in an everyday context whenever possible. They also explore various industrial applications that most students find especially interesting and relevant.

In this edition we have made a greater effort to integrate interchapters with preceding text and to reiterate relevant principles, but, as before, the interchapters contain no new principles and may be covered in any order or even omitted entirely. They are brief and focused enough to be given as assigned reading. At the University of California, Davis, we cover the interchapters successfully in discussion sections.

New to this edition are Interchapters D, on oxygen and hydrogen, and E, on energy utilization.

COLOR IS USED TO TEACH CHEMISTRY

In keeping with the experimental approach, ours was the first general chemistry text to show chemistry as the colorful subject that it is. We've greatly expanded the use of color throughout the book in this edition. However, we've also made a determined and sustained effort to avoid using color frivolously. With the help of Travis Amos, our superb photo researcher, and Chip Clark, a scientific photographer for the Smithsonian Institution, we assembled a collection of striking and unique photographs that illustrate the beauty and excitement of chemistry. Thus, we are able to show many elements, compounds, reactions, and other chemical events that can be appreciated fully only when seen in color.

THERE ARE SEVERAL NOTEWORTHY CHANGES IN THIS EDITION

Length

Many of you who used our first edition will be pleased to note that this volume is shorter and lighter, and you may be wondering what was left out. While it's true that we condensed and combined some topics and sought to eliminate unnecessary detail, no economies of space were achieved at the expense of content. In fact, several new topics and a wealth of new illustrations have been added. Subtle but significant changes in design and layout are largely responsible for the smaller size of this volume.

Organization

The enthusiastic reception of our first edition dictates that we limit our organizational changes. Accordingly, they are few.

We begin as before by introducing many of the fundamental concepts of chemistry in Chapter 1, "Atoms and Molecules." Next, we shorten and combine the chapters on the elements and the periodic table and on chemical reactions. In the new Chapter 2, "Chemical Reactions and the Periodic Table," we develop the group properties of the chemical elements as a lead-in to our discussion of the periodic table. We then discuss several types of chemical reactions, which leads in a natural way to chemical nomenclature, to the relative reactivities of metals, and to an introduction to the chemistry of acids and bases.

This early introduction to descriptive chemistry allows us to integrate it throughout the rest of the text. When a compound is used to illustrate a principle or a type of calculation, we often comment on one or more of its chemical or physical properties.

Other basic organizational changes include the following:

—Our chapter on chemical kinetics (Chapter 13) now precedes our chapter on chemical equilibrium (Chapter 14). The relative order of these two chapters enjoys no consensus, there being about the same degree of preference for either approach. To this end, we introduce the basic ideas of both rate and equilibrium (particularly the concept of dynamic equilibrium) in our discussion of vapor pressure in Chapter 11, "Liquids and Solids," and of the dissolution of a solid in a liquid in Chapter 12, "Properties of Solutions."

—We now cover thermodynamics (Chapter 19) before electrochemistry (Chapter 20). As many of our readers suggest, this is a more logical order of presentation, even though the study of electrochemical cells aided in the development of thermodynamics.

Treatment of Advanced Topics

In this edition we include a few advanced topics such as the Clapeyron-Clausius equation, the van't Hoff equation, the titration of weak acids with strong bases, and hybrid orbitals involving *d* orbitals; we also expand the treatment of the van der Waals equation. In most cases, however, we place advanced material in the final section of a chapter, where it may be passed over without interrupting the discussion of a primary topic. In fact, we purposefully isolate topics of optional interest as often as possible to better accommodate a wide range of course levels and emphases.

Chapter-ending problems

In keeping with the format of the first edition, most of the chapter-ending problems are arranged in matched pairs, each member dealing with the same principle or operation. This gives students who have difficulty with a particular problem a chance to work a similar but new problem in order to test their understanding. The problems are grouped by topic in the order covered in the chapter. For each chapter we now include a section of about 20 additional problems that are not identified by topic, that are not paired, and that are often more challenging than the rest of the problems. The intent of this change is, again, to accommodate the varied needs of our users.

As before, interchapters end with a set of questions based solely on the material presented in each interchapter. In addition, we now include a set of problems at the end of each interchapter. While these are not necessarily computation problems, they do require students to use the chemical principles developed in the previous chapters.

As a checkpoint for students, the answers to odd-numbered problems appear in Appendix F at the end of the text.

WELL-RECEIVED FEATURES HAVE BEEN RETAINED

The following are special, noteworthy emphases of our text:

—We feel that first-year students should become proficient in writing Lewis formulas, and we've taken special care in developing Chapter 8 to teach this skill. Rules for writing Lewis formulas are especially clearcut in this edition. This chapter can be covered in two lectures.

—Chapter 9, "VSEPR Theory" (formerly entitled "The Shapes of Molecules") develops the valence-shell electron-pair repulsion theory. We feel VSEPR theory is easy to understand, easy to apply, and amazingly reliable. It both reinforces the writing of Lewis formulas and introduces first-year students to a large number of compounds. In our experience, students enjoy VSEPR theory because of its simplicity and predictive power. For the instructor who does not wish to present VSEPR, most, if not all, of the chapter can be omitted.

—Our section headings take the form of declarative sentences rather than brief terms or phrases. These headings focus on the underlying principle or primary objective of each section. Simply reading the section headings gives the student a good overview of any chapter.

—We order the material with the companion laboratory course in mind. We introduce stoichiometry, solutions, and the more elementary properties of acids and bases early to accommodate a variety of laboratory schedules.

—We use SI units almost exclusively. Authors of textbooks today face a dilemma with regard to units. Although SI units are endorsed by numerous organizations and journals, many instructors are reluctant to change to SI units, or may even be hostile to the idea. Neither of us was a strong advocate of SI units before writing this text, but in the process we found that we became comfortable with them with little effort. Thus, we use joules instead of calories and picometers instead of angstroms. One SI unit that we could not readily adjust to, however, is the pascal, the SI unit of pressure. We generally express pressure in units of atmospheres or torr (mmHg), although we include a separate section of gas-law problems involving pascals for instructors who have made a complete transition to SI units.

—We use the Guggenheim slash notation to label headings in tables and graph axes in figures. This notation is endorsed by the International Union of Pure and Applied Chemistry (IUPAC) and is explained in Section 1-18. Although not yet widely used by American authors, this notation is so much more convenient and less ambiguous than other notations that its use is expanding rapidly.

—Each chapter ends with a summary, a list of terms you should know (cross-referenced to the page numbers on which the terms are introduced), and a list of equations you should know how to use.

A COMPLETE INSTRUCTIONAL PACKAGE IS AVAILABLE

We stated in the preface to our first edition that we firmly believed the *Study Guide/Solutions Manual* that accompanied the text to be of real benefit to the student. The overwhelming general acceptance of the *Study Guide/Solutions Manual* bore us out. Consequently, except for the fact that the answers to the even-numbered problems are no longer included, the format of the *Study Guide/Solutions Manual* remains unchanged. We continue to be sensitive to the difficulty many students have with numerical problems, and thus we give considerable emphasis to problem-solving skills.

For each chapter in the text, the *Study Guide/Solutions Manual* provides

—an outline of the chapter (section headings and short descriptive sentences)

—a self-test (about 50 fill-in-the-blank questions; no computational problems)

—a list of calculations you should know how to do

—solutions to the odd-numbered problems (detailed solutions are unquestionably the most valuable feature to the student)

—the answers to the self-test

There is also a glossary at the end of the *Study Guide/Solutions Manual,* which is cross-referenced to the text.

Our accompanying lab manual, *General Chemistry in the Laboratory,* second edition, by Julian Roberts, J. Leland Hollenberg, and James Postma, is derived from the popular Frantz-Malm series and provides suggestions for using the 42 experiments in conjunction with our text. The accompanying Instructor's Manual contains filled-in report forms for all the experiments in the manual.

For the instructor, our Test Bank, by Robert J. Balahura, allows selection from among more than 1300 questions and includes both multiple-choice items and a variety of subjective items—short-answer questions, crossword puzzles, and brain teasers. A computerized version of the Test Bank is available for use with IBM or Apple personal computers. Our Instructor's Manual contains detailed solutions to all even-numbered problems in the text and answers to all interchapter questions and problems. The carefully selected set of 120 Overhead Transparencies in two colors provides a useful lecture aid.

ACKNOWLEDGMENTS

We begin by thanking the 150 teachers who either answered our survey or were interviewed by phone and the scores of others who sent in unsolicited comments: your encouragement and suggestions very much influenced the course we took with this edition. In addition, we also wish to remember by name the more than 40 reviewers of the first edition, whose counsel not only led to the success of that work but has had an ongoing effect on our writing. David L. Adams, North Shore Community College; Robert C. Atkins, James Madison University; Robert J. Balahura, University of Guelph; Otto T. Benfey, Guilford College; Larry E. Bennett, San Diego State University; David W. Brooks, University of Nebraska-Lincoln; Bruce W. Brown, Portland State University; George Brubaker, Illinois Institute of Technology; Ian S. Butler, McGill University; Harvey F. Carroll, Kingsborough Community College, CUNY; Ronald J. Clark, Florida State University; John M. D'Auria, Simon Fraser University; Derek A. Davenport, Purdue University; Daniel R. Decious, California State University, Sacramento; Robert Desiderato, North Texas State University; Timothy C. Donnelly, University of California, Davis; Frank J. Gomba, United States Naval Academy; Charles G. Haas, Jr., Pennsylvania State University; Edward D. Harris, Texas A&M University; Henry M. Hellman, New York University; Forrest C. Hentz, Jr., North Carolina State University; Earl S. Huyser, University of Kansas; Joseph E. Ledbetter, Contra Costa College; Edward C. Lingafelter, University of Washington; William M. Litchman, University of New Mexico; Saundra Y. McGuire, University of Tennessee; Arlene M. McPherson, Tulane University; John M. Newey, American River College; Dennis G. Peters, Indiana University; Grace S. Petrie, Nassau Community College; Henry Po, California State University, Long Beach; James M. Postma, California State University, Chico; W. H. Reinmuth, Columbia University; Randall J. Remmel, University of Alabama in Birmingham; Don Roach, Miami-Dade Community College; Charles B. Rose, University of Nevada; Barbara Sawrey, San Diego State University; William M. Scovell, Bowling Green State University; Donald Showalter, University of Wisconsin; R. T. Smedberg, American River College; James C. Thompson, University of Toronto; Russell F. Trimble, Southern Illinois University; Carl Trindle, University of Virginia; Carl A. von Frankenberg, University of Delaware; E. J. Wells, Simon Frazer University; Helmut Wieser, University of Calgary.

Many individuals deserve special recognition here for their detailed reviews of our present work: Edwin H. Abbott, Montana State University; Ed Acheson, Millikin University; Hugh Akers, Lamar University; Robert J. Balahura, University of Guelph; Larry E. Bennett, San Diego State University; Muriel B. Bishop, Clemson University; Weldon Burnham, Richland College; Kevin Cadogan, California State University, Hayward; Gordon J. Ewing, New Mexico State University; Marcia L. Gillette, Indiana University, Kokomo; Clarence Josephson, Millikin

University; Mike Lamb, Cambrian College; Joseph Ledbetter, Contra Costa College; Gilbert J. Mains, Oklahoma State University; Betty Moser, Millikin University; Charles P. Nash, University of California, Davis; Richard Pizer, Brooklyn College of the City University of New York; Ron T. Smedberg, American River College; Nicholas E. Takach, University of Tulsa; Richard J. Wittebort, University of Louisville.

Finally, there are several people at W. H. Freeman and Company to whom we give special thanks: Linda Chaput, President, for always giving us her enthusiastic support and encouragement; Carol Pritchard-Martinez for directing the development of the text; Barbara Brooks for overseeing the project, including development of the ancillaries; Stephen Wagley for coordinating the editing; Ellen Cash for directing the production of the work; and Mike Suh, for his talented handling of the design and layout. We also thank Travis Amos for his outstanding photo research and numerous excellent suggestions, Chip Clark for his brilliant photography, Carole McQuarrie and Joseph Ledbetter for their generous and invaluable scientific help, Betsy Galbraith for her meticulous copy editing, and Elaine Rock for her prompt and accurate typing of this manuscript.

Donald A. McQuarrie
Peter A. Rock

GENERAL CHEMISTRY

ATOMS AND MOLECULES

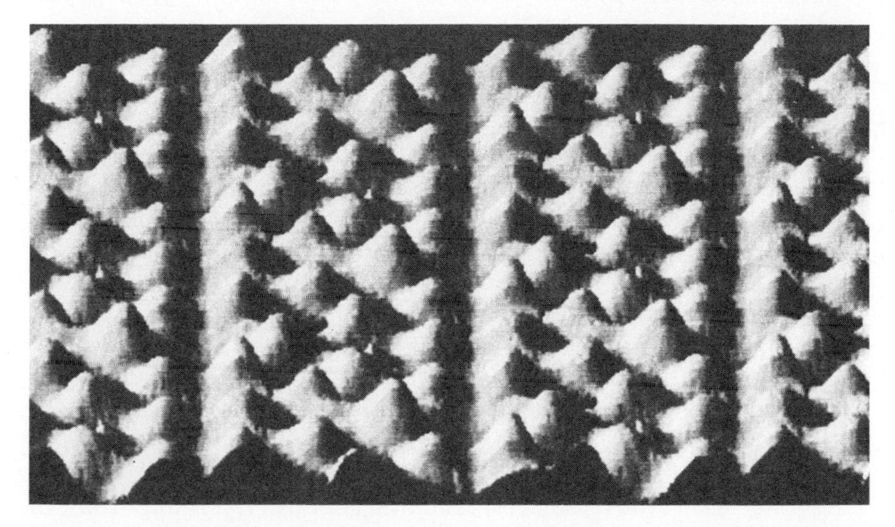

A microscopic view of the surface of a silicon crystal, in which the locations of the silicon atoms correspond to the peaks in the photo.

Chemistry is the study of the properties of various substances and of how they react with each other. To understand the nature of these substances better, we must begin by examining the atomic theory, one of the fundamental theories of chemistry. The atomic theory pictures all substances as consisting either of atoms or of groups of atoms called molecules. Proposed in the early 1800s by an English schoolteacher named John Dalton, the atomic theory gives a simple picture of chemical reactions and provides explanations for many chemical observations. We begin with a picture of the atom as a solid, structureless sphere and conclude the chapter with the nuclear model of the atom as a small, relatively massive nucleus containing protons and neutrons with electrons surrounding the nucleus.

1-1. WHY SHOULD YOU STUDY CHEMISTRY?

You and about 400,000 other students in the United States and Canada are about to begin your first college course in chemistry. Although most of you do not plan to become professional chemists (only about 10,000 students graduate each year in the United States and Canada with a bachelor's degree in chemistry), your proposed major field of study probably requires at least one year of college chemistry. A knowledge of elementary chemistry is necessary in so many fields that general chemistry is one of the courses with the largest enrollment at most colleges.

Chemistry plays a pervasive role in all our lives. Hundreds of materials that you and your family use directly and indirectly every day are products of chemical research. The development of fertilizers, which is one of the major areas of the chemical industry, has profoundly affected agricultural production in developed countries. Another major area of the chemical industry is the pharmaceutical, or drug, industry. Who among us has not used an antibiotic to cure an infection or various drugs to alleviate the pain associated with dental work, accidents, or surgery? Modern medicine, which rests firmly upon chemistry, has increased our life expectancy by about 15 years since the 1920s. It is hard to believe that little over a century ago many people actually died from simple infections and other diseases that we seldom hear of anymore.

Perhaps the chemical products most familiar to all of us are plastics. The annual production of synthetic fibers in the United States exceeds 10 billion pounds, or over 50 pounds per person. About 50 percent of industrial chemists are involved with the development or production of plastics. There is hardly an activity in your daily life that does not include some plastic product. Names such as nylon, polyethylene, Formica, Saran, Teflon, Hollofil, Gore-Tex, polyester, and silicone are familiar to most of us. Chemistry also plays a major role in materials science—from the manufacture of computer chips to paper and wood products to structural metals such as steel and lightweight titanium and aluminum alloys for ships and aircraft.

Regardless of your reasons for studying chemistry, it is important to remember that it is a requirement for your major because the people who work in your field consider it necessary and useful. We are confident that with a reasonable effort on your part, you will find chemistry both interesting and enjoyable.

1-2. CHEMISTRY HAS AN EXPERIMENTAL BASIS

Chemistry is an experimental science based on the **scientific method.** The essence of the scientific method is the use of carefully controlled experiments to answer scientific questions.

As an example, consider the statement "Hot water freezes faster than cold water." This statement seems obviously incorrect to many people because, as they argue, hot water first has to get cold before it can freeze, and therefore hot water must take longer to freeze than cold water. However, an argument is not an experiment. We can develop just as reasonable an argument for the original statement from some observed properties of water. Water contains dissolved air, which is expelled during boiling and also during freezing. (This explains why boiled water freezes to clear ice, whereas unboiled tap water freezes to opaque ice. The opaqueness is a consequence of the trapping of expelled air between ice crystals as they form.) Also, water usually freezes from the top down, and some of the expelled air from unboiled water accumulates in a layer between the ice already formed and the remaining liquid. This air layer can interfere with the freezing process by impeding the transfer of water from the liquid to the solid. Thus it can be argued that boiled water, which contains virtually no dissolved air, will freeze completely more rapidly than tap water, even though its initial temperature is higher when it is placed in the freezer.

For a simple experimental test of the statement, take two identical ice-cube trays and fill one with freshly boiled water and the other with cold tap water. Then place the two trays side by side in the freezer and observe them periodically. We leave the discovery of the result to the reader, as an exercise in the application of the scientific method.

Experiments and observations yield the results that form the factual basis of science. A **scientific law,** or a **natural law,** is a concise summary of a large body of experimental results. However, a law is not an explanation of the observations. For example, a scientific law called the first law of thermodynamics states that energy is always conserved in any process. The first law of thermodynamics is a significant and pervasive scientific law, but the statement that energy is conserved is not an explanation of the facts but is a summary statement of them.

The role of **theory** in science is to provide explanations for the laws of science and to aid in making predictions that lead to new knowledge. A theory should be subjected to experimental tests that are designed to disprove the theory. A theory can never be proved correct by experiment. Experimental results can provide additional supporting data for a theory, but no matter how many experiments are found to yield results consistent with a theory, there always remains the possibility that additional experiments will demonstrate a flaw in the theory. A theory is not a fact.

Scientific theories are subject to ongoing revision, and most theories in use have known limitations. An imperfect theory is often useful because of its predictive value, even though we cannot have complete confidence in the theoretical predictions. Because scientific theories produce a unification of ideas, imperfect theories generally are not abandoned until a better theory is developed. A **hypothesis** is a proposition put forth as the possible explanation for an observation or a phenomenon, which serves as a guide to further investigation. Hypotheses are the seeds of scientific theories, in that hypotheses evolve into scientific theories if they are supported by a number of experimental observations.

Definitions play an important role in scientific communication. When we state a value for the density of a substance at a particular temperature and pressure, we know what that value means, because we have agreed on the definition of density (d) as the mass per unit volume ($d = m/V$) of a substance. Definitions are not facts; they are chosen arbitrarily in the interests of convenience and utility. When you encounter a new definition in this text, you should commit it to memory so that you can understand the language of chemistry.

We conclude this section with a short story that illustrates the importance of having carefully defined quantities in science.

"A certain retired sea captain made his home in a secluded spot on the island of Zanzibar. As a sentimental reminder of his seafaring career he still had his ship's chronometer and religiously kept it wound and in good operating condition. Every day exactly at noon, as indicated on his chronometer, he observed the ritual of firing off a volley from a small cannon. On one rare occasion he received a visit from an old friend who inquired how the captain verified the correctness of his chronometer. 'Oh,' he replied, 'there is a horologist over there in the town of Zanzibar where I go whenever I lay in supplies. He has very reliable time and as I have fairly frequent occasion to go that way I almost always walk past his window and check my time against his.' After his visit was over the visitor dropped into the horologist's shop and inquired how the horologist checked his time. 'Oh,' replied he, 'there's an old sea captain over on the other end of the island who, I am told, is quite a fanatic about accurate time and who shoots off a gun every day exactly at noon, so I always check my time and correct it by his.' "*

1-3. ELEMENTS ARE THE SIMPLEST SUBSTANCES

Almost all the millions of different chemicals known today can be broken down into simpler substances. Any substance that cannot be broken down into simpler substances is called an **element.** This is strictly an operational definition, but it does lend itself to experimental testing. Pure substances that can be broken down into simpler substances are called **compounds.** Before the early 1800s, many substances were incorrectly classified as elements because methods to break them down had not yet been developed, but these errors have been rectified over the years. Although our definition of an element is a satisfactory working definition, we shall learn later that the modern definition is that an element is a substance that consists of only atoms with the same nuclear charge.

There are 108 known chemical elements. Some of the elements are very rare; fewer than half of them constitute 99.99 percent of all substances. Table 1-1 lists the most common elements found in the earth's crust, the oceans, and the atmosphere. Note that only ten elements make up over 99 percent of the total. Oxygen and silicon are the most common elements because they are the major constituents of sand, soil, and rocks. Oxygen also occurs as a free element in the atmosphere and in combination with hydrogen in water. Table 1-2 lists the most common elements found in the human body. Note that only ten elements constitute over 99.8 percent of the total mass of the human body. The high-percentage-by-mass data for hydrogen and oxygen result from the fact that about 70 percent of the mass of the human body is water.

Table 1-1 Elemental composition of the earth's surface, which includes the crust, oceans, and the atmosphere

Element	Percent by mass
oxygen	49.1
silicon	26.1
aluminum	7.5
iron	4.7
calcium	3.4
sodium	2.6
potassium	2.4
magnesium	1.9
hydrogen	0.88
titanium	0.58
chlorine	0.19
carbon	0.09
all others	0.56

*This story comes from E. R. Cohen, K. M. Crowe, J. W. M. Drummond, *Fundamental Constants of Physics* (John Wiley Interscience, 1957); they attribute the story to Professor George Harrison. Used by permission of John Wiley Interscience.

1-4. ABOUT THREE FOURTHS OF THE ELEMENTS ARE METALS

One broad classification of the elements is into **metals** and **nonmetals.** We are all familiar with the properties of metals. They have a characteristic luster; can be rolled or hammered into sheets, drawn into wires, melted, and cast into various shapes; and are usually good conductors of electricity and heat. About three fourths of the elements are metals. All the metals except mercury are solids at room temperature (about 20°C). Mercury is a shiny, silver-colored liquid at room temperature and used to be called quicksilver.

Table 1-3 lists some metals and their **chemical symbols.** Chemical symbols are abbreviations used to designate the elements and are usually the first one or two letters in the name of the element. Some chemical symbols do not seem to correspond at all to the names because the symbols are derived from the Latin names of those elements (Table 1-4). It is necessary to memorize the chemical symbols of the more common elements because we shall be using them throughout this book.

Unlike metals, nonmetals vary greatly in their appearance. Over half of the nonmetals are gases at room temperature, and the others are solids except for bromine, which is a red-brown, corrosive liquid. In contrast to metals, nonmetals are poor conductors of electricity and heat, cannot be rolled into sheets or drawn into wires, and do not have a characteristic luster. Table 1-5 lists several common nonmetals, their

Starting at the upper left and going in a clockwise direction, we have the elements sulfur, copper, bromine, mercury, and iodine.

Table 1-2 Elemental composition of the human body

Element	Percent by mass
oxygen	64.6
carbon	18.0
hydrogen	10.0
nitrogen	3.1
calcium	1.9
phosphorus	1.1
chlorine	0.40
potassium	0.36
sulfur	0.25
sodium	0.11
magnesium	0.03
iron	0.005
zinc	0.002
copper	0.0004
tin	0.0001
manganese	0.0001
iodine	0.0001

Table 1-3 Some common metals and their chemical symbols

Element	Symbol	Element	Symbol
aluminum	Al	mercury	Hg
barium	Ba	nickel	Ni
cadmium	Cd	platinum	Pt
calcium	Ca	potassium	K
chromium	Cr	silver	Ag
cobalt	Co	sodium	Na
copper	Cu	strontium	Sr
gold	Au	tin	Sn
iron	Fe	titanium	Ti
lead	Pb	tungsten	W
lithium	Li	uranium	U
magnesium	Mg	zinc	Zn
manganese	Mn		

Table 1-4 Elements whose symbol corresponds to the Latin name

Element	Symbol	Latin name
antimony	Sb	stibium
copper	Cu	cuprum
gold	Au	aurum
iron	Fe	ferrum
lead	Pb	plumbum
mercury	Hg	hydrargyrum
potassium	K	kalium
silver	Ag	argentum
sodium	Na	natrium
tin	Sn	stannum

chemical symbols, and their appearances. Note that several of the symbols of the nonmetallic elements in Table 1-5 have a 2 subscript. This indicates that these elements—hydrogen (H_2), nitrogen (N_2), oxygen (O_2), fluorine (F_2), chlorine (Cl_2), bromine (Br_2), and iodine (I_2)—exist in nature as two atoms joined together. A unit of two or more atoms that are joined together is called a **molecule,** and a molecule consisting of two atoms is called a **diatomic molecule.** Scale models of some diatomic molecules are shown in Figure 1-1.

1-5. ANTOINE LAVOISIER WAS THE FOUNDER OF MODERN CHEMISTRY

Although chemistry was beginning to develop as a science by the eighteenth century, it still lacked one ingredient essential for becoming a modern science. That ingredient was **quantitative measurement.** A

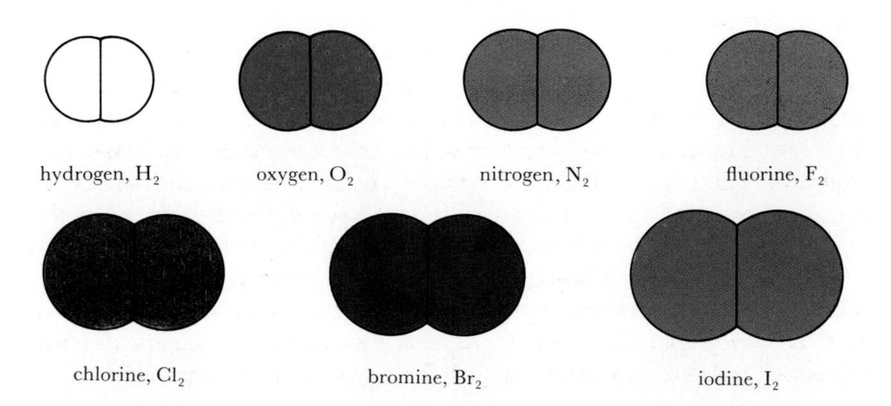

Figure 1-1 Scale models of molecules of hydrogen, oxygen, nitrogen, fluorine, chlorine, bromine, and iodine. These substances exist as diatomic molecules in their natural state but are still classified as elements, because the molecules consist of identical atoms. We shall discuss atoms in Section 1-7.

Table 1-5 Some common nonmetals and their appearances at room temperature

Element	Symbol*	Appearance
Gases		
hydrogen	H_2	colorless
helium	He	colorless
nitrogen	N_2	colorless
oxygen	O_2	colorless
fluorine	F_2	pale yellow
neon	Ne	colorless
chlorine	Cl_2	green-yellow
argon	Ar	colorless
krypton	Kr	colorless
xenon	Xe	colorless
Liquids		
bromine	Br_2	red-brown
Solids		
carbon	C	black (in the form of coal or graphite)
phosphorus	P	pale yellow or red
sulfur	S	lemon yellow
iodine	I_2	violet-black

*The subscript 2 tells us that, at room temperature, the element exists as a diatomic molecule, that is, a molecule consisting of two atoms.

quantitative measurement is one in which the result is expressed as a number. For example, the determination that the mass of 1.00 cm^3 (cubic centimeter) of gold is 19.3 g (grams) or that 1.25 g of calcium reacts with 1.00 g of sulfur is a quantitative measurement. Compare these statements with **qualitative observations,** where we note general characteristics, such as color and odor. An example of a qualitative statement is that lead is much denser than aluminum; the quantitative statement is that the mass of 1.00 cm^3 of lead is 11.3 g while that of 1.00 cm^3 of aluminum is 2.70 g.

It was the French scientist Antoine Lavoisier (Figure 1-2) who first appreciated the importance of carrying out quantitative measurements. Lavoisier designed special balances that were more accurate than ever before and discovered the **law of conservation of mass:** in a chemical reaction, the total mass of the reacting substances is equal to the total mass of the products formed. Lavoisier's influence on the development of chemistry as a modern science cannot be overstated. In 1789, he published his *Elementary Treatise on Chemistry,* in which he presented a unified picture of chemical knowledge. The *Elementary Treatise on Chemistry* (Figure 1-3) was translated into many languages and was the first textbook of chemistry based on quantitative experiments.

Figure 1-2 The French chemist Antoine Lavoisier and his wife and colleague, Marie-Anne Pierrette. Marie-Anne assisted Antoine in much of his work and illustrated and helped write his famous book, *Elementary Treatise on Chemistry.* Because of his financial connection with a much-hated tax-collecting firm, Lavoisier was denounced, arrested, and guillotined in 1794 by supporters of the French Revolution.

1-6. THE LAW OF CONSTANT COMPOSITION STATES THAT THE RELATIVE AMOUNT OF EACH ELEMENT IN A COMPOUND IS ALWAYS THE SAME

The quantitative approach pioneered by Lavoisier was used in the chemical analysis of compounds. The quantitative chemical analysis of a great many compounds led to the **law of constant composition:** the relative amount of each element in a particular compound is always the same, regardless of the source of the compound or how the compound is prepared. For example, if calcium metal is heated with sulfur, the compound called calcium sulfide is formed. We can specify the relative amounts of calcium and sulfur in calcium sulfide by the mass percentage of each element. The mass percentages of calcium and sulfur in calcium sulfide are defined as

$$\begin{pmatrix}\text{mass percentage of calcium}\\ \text{in calcium sulfide}\end{pmatrix} = \frac{\text{mass of calcium}}{\text{mass of calcium sulfide}} \times 100$$

$$\begin{pmatrix}\text{mass percentage of sulfur}\\ \text{in calcium sulfide}\end{pmatrix} = \frac{\text{mass of sulfur}}{\text{mass of calcium sulfide}} \times 100$$

Suppose we analyze 1.630 g of calcium sulfide and find that it consists of 0.906 g of calcium and 0.724 g of sulfur. Then the mass percentages of calcium and sulfur in calcium sulfide are

$$\begin{pmatrix}\text{mass percentage of calcium}\\ \text{in calcium sulfide}\end{pmatrix} = \frac{\text{mass of calcium}}{\text{mass of calcium sulfide}} \times 100$$

$$= \frac{0.906\ \text{g}}{1.630\ \text{g}} \times 100 = 55.6\%$$

$$\begin{pmatrix}\text{mass percentage of sulfur}\\ \text{in calcium sulfide}\end{pmatrix} = \frac{\text{mass of sulfur}}{\text{mass of calcium sulfide}} \times 100$$

$$= \frac{0.724\ \text{g}}{1.630\ \text{g}} \times 100 = 44.4\%$$

The law of constant composition says that the mass percentage of calcium in calcium sulfide is 55.6 percent whether the calcium sulfide is prepared by heating a large amount of calcium with a small amount of sulfur or by heating a small amount of calcium with a large amount of sulfur. The mass percentage of calcium in calcium sulfide is always 55.6 percent, and the mass percentage of sulfur in calcium sulfide is always 44.4 percent.

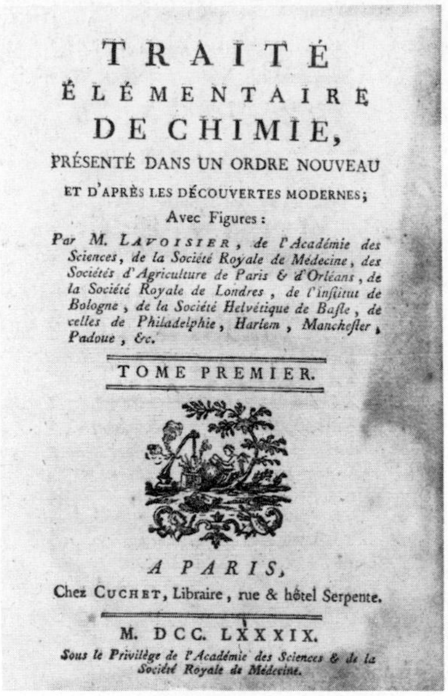

TRAITÉ
ÉLÉMENTAIRE
DE CHIMIE,
PRÉSENTÉ DANS UN ORDRE NOUVEAU
ET D'APRÈS LES DÉCOUVERTES MODERNES;
Avec Figures :
Par M. LAVOISIER, de l'Académie des Sciences, de la Société Royale de Médecine, des Sociétés d'Agriculture de Paris & d'Orléans, de la Société Royale de Londres, de l'Institut de Bologne, de la Société Helvétique de Basle, de celles de Philadelphie, Harlem, Manchester, Padoue, &c.
TOME PREMIER.
A PARIS,
Chez CUCHET, Libraire, rue & hôtel Serpente.
M. DCC. LXXXIX.
Sous le Privilège de l'Académie des Sciences & de la Société Royale de Médecine.

Figure 1-3 The title page to Lavoisier's textbook of chemistry.

▪ Although many people use the terms mass and weight interchangeably, these terms are not the same. Mass is the inherent amount of material of an object, whereas weight is the force of attraction of the object to a large body such as the earth or the moon. An object on the moon weighs about one-sixth as much as it does on earth, but its mass is the same in both places. We shall use the term mass throughout the book.

Example 1-1: Suppose we analyze 2.83 g of a compound of lead and sulfur and find that it consists of 2.45 g of lead and 0.380 g of sulfur. Calculate the mass percentages of lead and sulfur in the compound, called lead sulfide.

Solution: The mass percentage of lead in lead sulfide is

$$\begin{pmatrix}\text{mass percentage of lead}\\ \text{in lead sulfide}\end{pmatrix} = \frac{\text{mass of lead}}{\text{mass of lead sulfide}} \times 100$$

$$= \frac{2.45\ \text{g}}{2.83\ \text{g}} \times 100 = 86.6\%$$

The mass percentage of sulfur in lead sulfide is

$$\begin{pmatrix}\text{mass percentage of sulfur}\\ \text{in lead sulfide}\end{pmatrix} = \frac{\text{mass of sulfur}}{\text{mass of lead sulfide}} \times 100$$

$$= \frac{0.380\ \text{g}}{2.83\ \text{g}} \times 100 = 13.4\%$$

The law of constant composition assures us that the mass percentage of lead in lead sulfide is independent of the source of the lead sulfide. The principal source of lead sulfide is the ore galena (Figure 1-4).

Figure 1-4 Many metal sulfides are valuable ores of the respective metals. Shown here is galena, the principal ore of lead.

1-7. DALTON'S ATOMIC THEORY EXPLAINS THE LAW OF CONSTANT COMPOSITION

By the end of the eighteenth century, many compounds had been analyzed and a large amount of experimental data had been accumulated. A theory was needed to bring all these data into a single framework. In 1803, John Dalton, an English schoolteacher, proposed an **atomic theory** that provided a simple and beautiful explanation of the law of constant composition. We can express the postulates of Dalton's atomic theory in modern terms as follows:

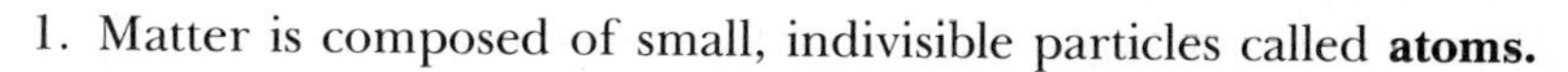

1. Matter is composed of small, indivisible particles called **atoms.**
2. The atoms of a given element all have the same mass and are identical in all respects, including chemical behavior.
3. The atoms of different elements differ in mass and in chemical behavior.
4. Chemical compounds are composed of two or more different atoms joined together. The particle that results when two or more atoms join together is called a **molecule.** The atoms in a molecule do not necessarily have to be different. If the atoms are the same, it is a molecule of an element. If the atoms are different, it is a molecule of a compound.
5. In a chemical reaction, the atoms involved are rearranged to form different molecules; no atoms are created or destroyed.

As we shall see, some of these postulates were later modified, but the main features of Dalton's atomic theory still are accepted today.

The law of constant composition follows nicely from Dalton's atomic theory. Consider calcium sulfide, which we know consists of 55.6 percent calcium and 44.4 percent sulfur by mass. Suppose that there is one calcium atom for each sulfur atom in calcium sulfide. Because we know that the relative masses of a calcium atom and a sulfur atom are 55.6 and 44.4, we know that the ratio of the mass of a calcium atom to that of a sulfur atom is

$$\frac{\text{mass of a calcium atom}}{\text{mass of a sulfur atom}} = \frac{55.6}{44.4} = 1.25$$

or

$$\text{mass of a calcium atom} = 1.25 \times \text{mass of a sulfur atom}$$

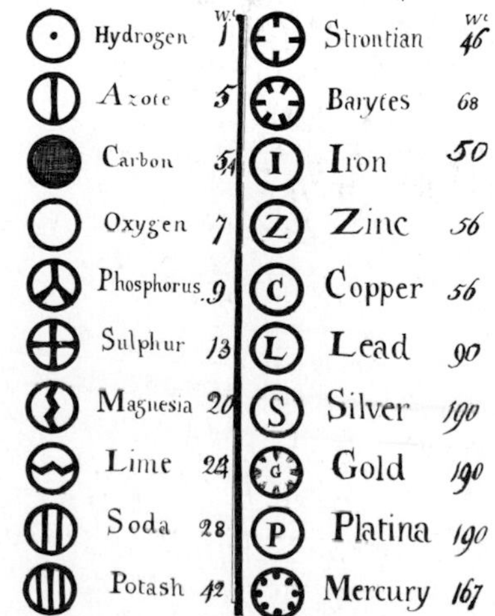

Dalton's symbols for chemical elements. Some of these are now known to be compounds, not elements.

Thus, even though we cannot easily determine the mass of any individual atom, we can use the quantitative results of chemical analysis to determine the *relative* masses of atoms. Of course, our result for calcium and sulfur is based on the assumption that there is one atom of calcium for each atom of sulfur in calcium sulfide.

Let's consider another compound, hydrogen chloride. Quantitative chemical analysis shows that the mass percentages of hydrogen and chlorine in hydrogen chloride are 2.76 percent and 97.24 percent, respectively. Once again, assuming that one atom of hydrogen is combined with one atom of chlorine, we find that

$$\frac{\text{mass of a chlorine atom}}{\text{mass of a hydrogen atom}} = \frac{97.24}{2.76} = 35.2$$

or

$$\text{mass of a chlorine atom} = 35.2 \times \text{mass of a hydrogen atom}$$

By continuing in this manner with other compounds, it is possible to build up a table of relative atomic masses. We define a quantity called **atomic mass,** which is the ratio of the mass of a given atom to the mass of some particular atom. At one time the mass of hydrogen, the lightest atom, was arbitrarily given the value of exactly 1 and used as the standard by which all other atomic masses were expressed. As discussed later in this chapter, however, a form of carbon is now used as the standard. Thus, today the atomic mass of hydrogen is 1.008 instead of exactly 1. The atomic masses of the elements are given on the inside front cover.

▪ Atomic masses are relative masses.

Being relative quantities, atomic masses have no units. Nevertheless, it is often convenient to assign a unit called an **atomic mass unit** (amu) to atomic masses. Thus, for example, we can say that the atomic mass of carbon is 12.01 or 12.01 amu; both statements are correct.

1-8. MOLECULES ARE GROUPS OF ATOMS JOINED TOGETHER

The original statement of Dalton's atomic theory postulated that an element is a substance that consists of identical atoms and that a compound is a substance that consists of identical molecules. Although Dalton did not realize it at the time, some of the elements consist of molecules containing the same kind of atoms. As noted in Table 1-5, the elements hydrogen, nitrogen, oxygen, fluorine, chlorine, bromine, and iodine exist as diatomic molecules of the same kind of atoms (see Figure 1-1). Consequently, these substances are classified as elements. Compounds, on the other hand, are made up of molecules containing different kinds of atoms. Some examples of such molecules are:

▪ These formulas indicate how the atoms are joined together in the molecules. We shall learn how to write such formulas in Chapter 8.

H—Cl	Cl—F	O H ╱ ╲ H	H—N—H │ H	H │ H—C—O—H │ H	H │ H—C—H │ H
hydrogen chloride, HCl	chlorine fluoride, ClF	water, H_2O	ammonia, NH_3	methyl alcohol, CH_3OH (wood alcohol)	methane, CH_4 (principal constituent of natural gas.)

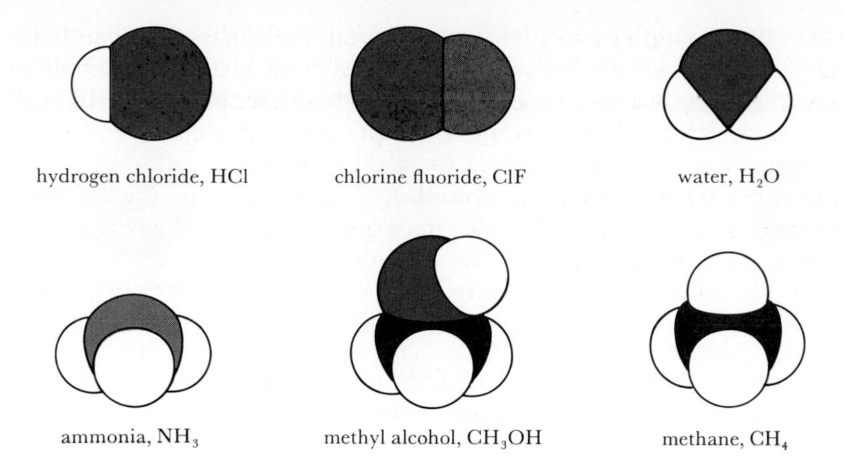

Figure 1-5 Scale molecular models of hydrogen chloride, chlorine fluoride, water, ammonia, methyl alcohol, and methane.

Molecular models of these molecules are shown above in Figure 1-5.

Dalton's atomic theory provides a nice pictorial view of chemical reactions. Recall that Dalton proposed that, in a chemical reaction, the atoms in the reactant molecules are separated and then rearranged into product molecules. According to this view, the chemical reaction between hydrogen and oxygen to form water involves the following rearrangement:

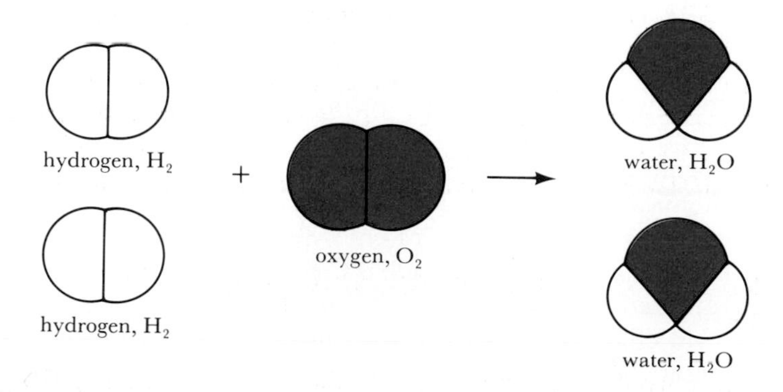

Note that completely different molecules and hence completely different substances are formed in a chemical reaction. Hydrogen and oxygen are gases, whereas water is a liquid.

As another example, consider the burning of carbon in oxygen to form carbon dioxide:

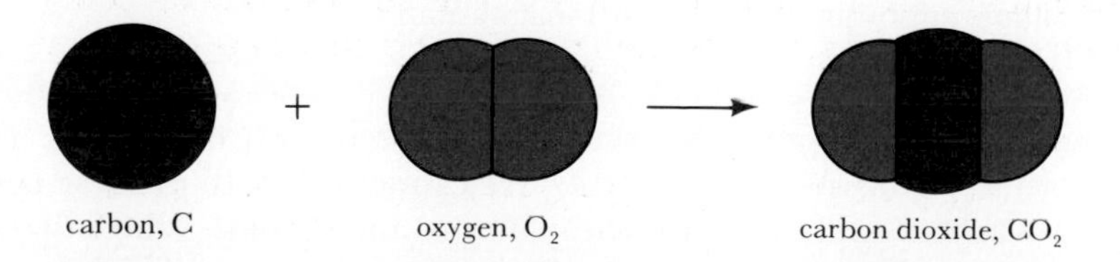

Once again, note that a completely new substance is formed. Carbon is a black solid; the product, carbon dioxide, is a colorless gas.

As a final example, consider the reaction between steam (hot gaseous water) and red-hot carbon to form hydrogen and carbon monoxide:

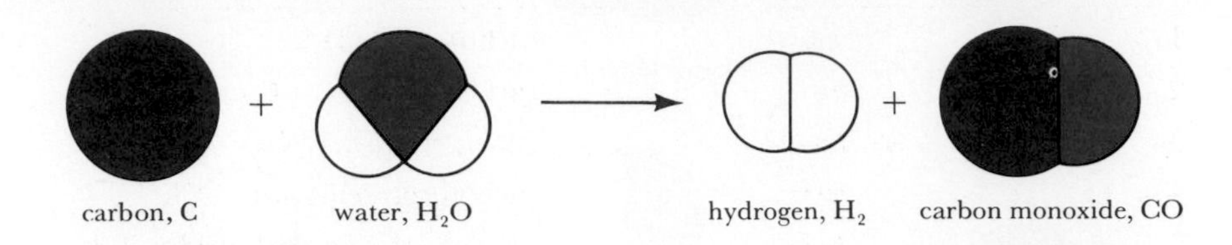

Notice that in each of the three reactions pictured here, the numbers of each kind of atom do not change. Atoms are neither created nor destroyed in chemical reactions; they are simply rearranged into new molecules.

1-9. COMPOUNDS ARE NAMED BY AN ORDERLY SYSTEM OF CHEMICAL NOMENCLATURE

The assignment of names to compounds is called **chemical nomenclature.** At this point we discuss the system of naming compounds that consist of just two elements, or **binary compounds.** When the two elements that make up a binary compound are a metal and a nonmetal (Tables 1-3 through 1-5), the compound is named by first naming the metal and then the nonmetal, with the ending of the name of the nonmetal changed to *-ide*. For example, the name of the compound formed between calcium and sulfur is calcium sulf*ide*. Because calcium sulfide consists of one atom of calcium for each atom of sulfur, we write the **chemical formula** of calcium sulfide as CaS; in other words, we simply join the chemical symbols of the two elements. In a different case, calcium combines with *two* atoms of chlorine to form calcium chlor*ide;* thus, the formula of calcium chloride is $CaCl_2$. Note that the number of atoms is indicated by a subscript. The subscript 2 in $CaCl_2$ means that there are two chlorine atoms per calcium atom in calcium chloride. Table 1-6 lists the *-ide* nomenclature for some common nonmetals.

Table 1-6 The *-ide* nomenclature of some common nonmetals

Element	*-ide* nomenclature
oxygen	oxide
nitrogen	nitride
sulfur	sulfide
fluorine	fluoride
chlorine	chloride
bromine	bromide
iodine	iodide
phosphorus	phosphide
hydrogen	hydride
carbon	carbide

Example 1-2: Name the following compounds:

(a) K_2O (b) $AlBr_3$ (c) $CdCl_2$ (d) MgH_2

Solution: Use Table 1-6 for the correct *-ide* nomenclature.

(a) potassium oxide
(b) aluminum bromide
(c) cadmium chloride
(d) magnesium hydride

Many binary compounds are combinations of two nonmetals. For example, let's consider CO and CO_2. We cannot call both of these compounds carbon oxide because the name is ambiguous. When two or more compounds can result from the same two nonmetallic elements, we distinguish among them by means of Greek numerical prefixes:

CO carbon *mono*xide CO_2 carbon *di*oxide

Table 1-7 Greek prefixes used to indicate the number of atoms of a given type in a molecule

Number	Prefix	Example
1	*mono-*	carbon monoxide, CO
2	*di-*	carbon dioxide, CO_2
3	*tri-*	sulfur trioxide, SO_3
4	*tetra-*	carbon tetrachloride, CCl_4
5	*penta-*	phosphorus pentachloride, PCl_5
6	*hexa-*	sulfur hexafluoride, SF_6

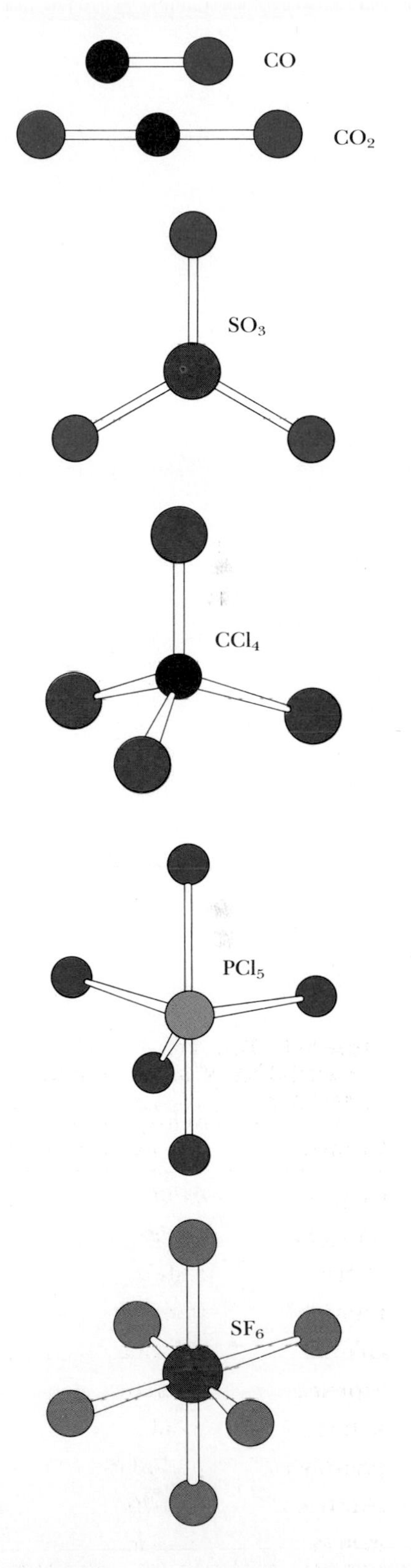

Some other examples are

SO_2	sulfur *di*oxide	SO_3	sulfur *tri*oxide
SF_4	sulfur *tetra*fluoride	SF_6	sulfur *hexa*fluoride
PCl_3	phosphorus *tri*chloride	PCl_5	phosphorus *penta*chloride

The Greek prefixes used are summarized in Table 1-7; the examples are illustrated in the margin at right.

Example 1-3: Name the following compounds:

(a) BrF_3 and BrF_5

(b) XeF_2 and XeF_4

(c) N_2O, NO, N_2O_3, NO_2, and N_2O_5

Solution:

(a) Because there is more than one compound formed by bromine and fluorine, we must distinguish between them by using Greek prefixes. Bromine is written first in the formulas, and so we name them

BrF_3 bromine trifluoride BrF_5 bromine pentafluoride

(b) Xenon is written first in the formulas, and so we have

XeF_2 xenon difluoride XeF_4 xenon tetrafluoride

(c) This series of compounds represents various oxides of nitrogen. Following the rules we have developed, we assign them the names

N_2O	dinitrogen monoxide
NO	nitrogen oxide (the prefix mono- is often dropped)
N_2O_3	dinitrogen trioxide
NO_2	nitrogen dioxide
N_2O_5	dinitrogen pentoxide

Dinitrogen oxide is commonly called by its less systematic name, nitrous oxide. Nitrous oxide was the first known general anesthetic (laughing gas) and is used as a foam propellant for whipped cream and shaving cream. Except for N_2O_5, which is a solid, all the oxides of nitrogen are gases at room temperature.

At this point you should understand how to name binary compounds if you are given the formula. In the next chapter, we shall learn how to write a correct formula from the name.

1-10. MOLECULAR MASS IS THE SUM OF THE ATOMIC MASSES OF THE ATOMS IN A MOLECULE

The sum of the atomic masses of the atoms in a molecule is called the **molecular mass** of the substance. For example, a water molecule, H_2O, consists of two atoms of hydrogen and one atom of oxygen. Using the table of atomic masses given on the inside front cover, we see that the molecular mass of water is

$$\begin{aligned}\begin{pmatrix}\text{molecular mass}\\ \text{of } H_2O\end{pmatrix} &= 2(\text{atomic mass of H}) + (\text{atomic mass of O})\\ &= 2(1.008) + (16.00)\\ &= 18.02\end{aligned}$$

Using the table of atomic masses given on the inside front cover, we see that the molecular mass of dinitrogen pentoxide, N_2O_5, is

$$\begin{aligned}\begin{pmatrix}\text{molecular mass}\\ \text{of } N_2O_5\end{pmatrix} &= 2(\text{atomic mass of N}) + 5(\text{atomic mass of O})\\ &= 2(14.01) + 5(16.00)\\ &= 108.02\end{aligned}$$

The following example shows how to use atomic and molecular masses to calculate the mass percentage composition of compounds.

Example 1-4: Using the fact that the atomic mass of lead is 207.2 and that of sulfur is 32.06, calculate the mass percentages of lead and sulfur in the compound lead sulfide, PbS.

Solution: As the formula PbS indicates, lead sulfide consists of one atom of lead for each atom of sulfur. The molecular mass of lead sulfide is

$$\begin{aligned}\begin{pmatrix}\text{molecular mass}\\ \text{of lead sulfide}\end{pmatrix} &= \text{atomic mass of lead} + \text{atomic mass of sulfur}\\ &= 207.2 + 32.06\\ &= 239.3\end{aligned}$$

■ The two mass percentages in Example 1-4 do not add up to 100.00 percent because of a slight round-off error.

The mass percentages of lead and sulfur in lead sulfide are

$$\begin{aligned}\text{mass percentage of lead} &= \frac{\text{atomic mass of lead}}{\text{molecular mass of lead sulfide}} \times 100\\ &= \frac{207.2}{239.3} \times 100 = 86.59\%\end{aligned}$$

$$\begin{aligned}\text{mass percentage of sulfur} &= \frac{\text{atomic mass of sulfur}}{\text{molecular mass of lead sulfide}} \times 100\\ &= \frac{32.06}{239.3} \times 100 = 13.40\%\end{aligned}$$

Note that this is the same result we got in Example 1-1. The table of atomic masses must be consistent with experimental values of mass percentages.

One of the great attractions of Dalton's atomic theory was that he was able to use it to devise a table of atomic masses that could then be used in chemical calculations like those in Example 1-4.

1-11. MOST OF THE MASS OF AN ATOM IS CONCENTRATED IN ITS NUCLEUS

For most of the nineteenth century, atoms were considered to be indivisible, stable particles. Toward the end of the century, however, new experiments indicated that the atom is composed of even smaller, **subatomic particles.** One of the first of these experiments was carried out by the English physicist J. J. Thomson in 1897. Some years earlier, it had been discovered that an electric discharge (glowing current) flows between metallic electrodes that are sealed in a partially evacuated glass tube, as shown in Figure 1-6. Using an apparatus of the type shown in Figure 1-6, Thomson deflected the electric discharge with electric and magnetic fields and showed that it was actually a stream of identical, negatively charged particles and that the mass of each particle was only 1/1837 that of a hydrogen atom. Because the hydrogen atom is the lightest atom, he correctly reasoned that these particles, which are now called **electrons,** are constituents of atoms. The electron was the first subatomic particle to be discovered.

If an atom contains electrons, which are negatively charged particles, then it also must contain positively charged particles because atoms are electrically neutral. The total amount of negative charge in a neutral atom must be balanced by an equal amount of positive charge. The question is, how are the positively charged particles and electrons arranged within an atom? The first person to answer this question was a New Zealand–born physicist, Ernest Rutherford.

To gain a sense of Rutherford's experiment, we must first mention another discovery of the 1890s, radioactivity. About the same time that Thomson discovered the electron, the French scientist Henri Becquerel discovered **radioactivity,** the process by which certain atoms spontaneously break apart. Becquerel showed that uranium atoms are **radioactive.** Shortly after Becquerel's discovery, Marie and Pierre Curie, working in Paris, discovered other radioactive elements such as radium (so named because it emits rays) and polonium (named for Poland,

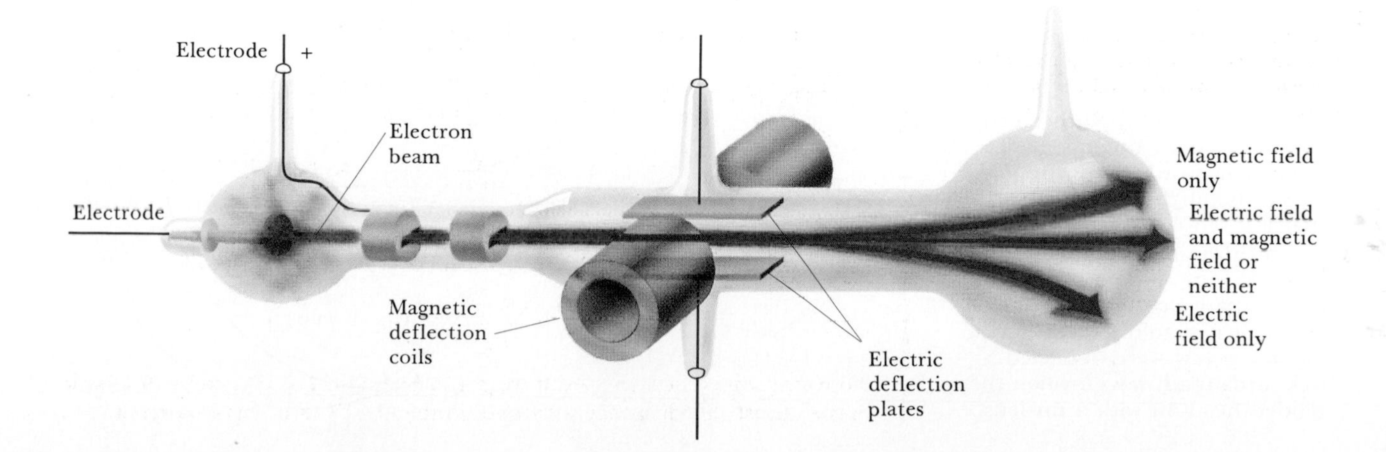

Figure 1-6 An apparatus like that Thomson used to discover the electron. When a voltage is applied across electrodes that are sealed in a partially evacuated glass tube, the space between the electrodes glows. Thomson showed that this glow discharge consisted of a stream of identical negatively charged particles, now called electrons. He was able to do this by deflecting the stream of particles in electric and magnetic fields.

Table 1-8 Properties of the three radioactive emissions discovered by Rutherford

Original name	Modern name	Mass*	Charge
α-ray	α-particle	4.00	+2
β-ray	β-particle (electron)	5.49×10^{-4}	−1
γ-ray	γ-ray	0	0

*In atomic mass units.

Marie Curie's native country). It was discovered that the radiation emitted by radioactive substances consists of three types, which are now called **α-particles** (alpha particles), **β-particles** (beta particles), and **γ-rays** (gamma rays). Experiments by a number of researchers showed that α-particles have a charge equal to that of two electrons but of opposite sign and a mass equal to the mass of a helium atom (4.00 amu); that β-particles are just electrons that result from radioactive disintegrations; and that γ-rays are very similar to X-rays. Table 1-8 summarizes the properties of these three common radioactive emissions.

▪ Radioactivity is discussed in Chapter 21.

Rutherford became intrigued with the idea of using α-particles as subatomic projectiles. In a now-famous experiment, one of Rutherford's students, Ernest Marsden, took a piece of gold and rolled it into an extremely thin foil (gold is very **malleable,** meaning that it can easily be rolled into a thin foil). He then directed a beam of α-particles at the gold foil and observed the paths of the particles by watching them strike a fluorescent screen surrounding the foil (Figure 1-7). Contrary to expectations, most of the particles passed right through the foil, but a few were deflected through large angles (pathway c in Figure 1-7). Rutherford interpreted this unexpected result by saying that an atom is mostly empty space and that all the positive charge and essentially all the mass of an atom are concentrated in a very small volume in the center of the atom, which he called the **nucleus.** Most α-particles passed through the gold foil; the few that were deflected through large angles were the result of collisions of the α-particles with gold nuclei.

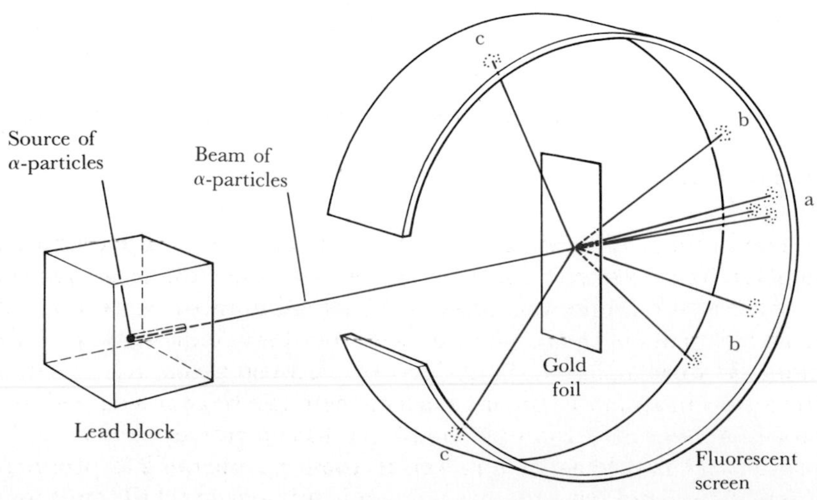

Figure 1-7 In 1911, Rutherford and Marsden set up an experiment in which a thin gold foil was bombarded with α-particles. Most of the particles passed through the foil (pathway a). Some were deflected only slightly (pathway b) when they passed near a gold nucleus in the foil, and a few were deflected backwards (pathway c), when they collided head-on with a nucleus.

Those α-particles that were deflected through intermediate angles (pathway b in Figure 1-7) had passed near gold nuclei and were repelled by their positive charge. Because α-particles are positively charged, they would be expected to be repelled by a positively charged nucleus because like charges repel each other. By counting the numbers of α-particles deflected in various directions, Rutherford was able to show that the diameter of a nucleus is about 1/100,000 times the diameter of an atom.

Rutherford subsequently discovered that the positive charge in an atom is due to **protons,** which are subatomic particles that have a positive charge equal in magnitude to that of an electron but opposite in sign. The mass of a proton is almost the same as the mass of a hydrogen atom, about 1836 times the mass of an electron.

The size of the nucleus relative to the size of the whole atom can be grasped from the following analogy. If an atom could be enlarged so that its nucleus were the size of a pea, then the entire atom would be about the size of Yankee Stadium. The electrons in an atom are located throughout the space surrounding the nucleus (Figure 1-8). Just how the electrons are arranged in an atom is taken up in later chapters.

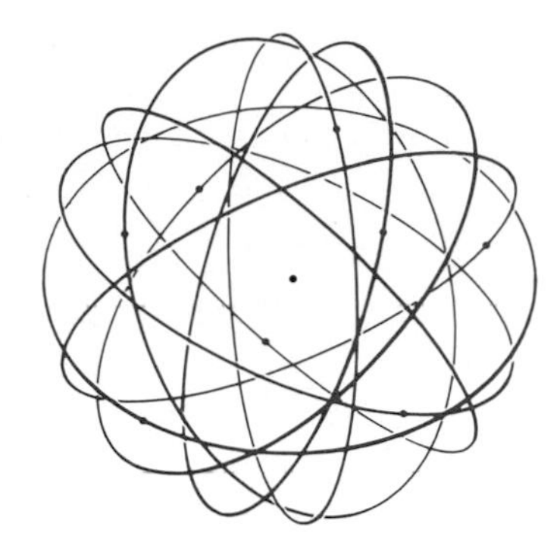

Figure 1-8 The nuclear model of the atom. The nucleus is very small and located at the center. The electrons are located in the space around the nucleus. We shall see in Chapter 6 that electrons do not travel around a nucleus in well-defined orbits, as shown in this commonly seen but rather fanciful picture of an atom.

1-12. ATOMS CONSIST OF PROTONS, NEUTRONS, AND ELECTRONS

Our picture of the atom is not yet complete. Later experiments suggested that the mass of a nucleus could not be attributed to the protons alone. It was hypothesized in the 1920s, and experimentally verified by James Chadwick in 1932, that there is another particle in the nucleus. This particle has essentially the same mass as a proton and is called a **neutron** because it is electrically neutral.

▪ Rutherford proposed the existence of neutrons in 1920.

The modern picture of an atom, then, consists of three types of particles—electrons, protons, and neutrons. The properties of these three subatomic particles are

Particle	**Charge***	**Mass†**	**Where located**
proton	+1	1.0073	in nucleus
neutron	0	1.0087	in nucleus
electron	−1	5.49×10^{-4}	outside nucleus

*Relative to the charge on a proton. The actual charge on a proton is 1.602×10^{-19} coulomb.

†In atomic mass units.

The number of protons in an atom is called the **atomic number,** denoted by Z, of that atom. In a neutral atom, the number of electrons is equal to the number of protons. The total number of protons and neutrons in an atom is called the **mass number,** denoted by A, of that atom. As we shall learn, the differences between atoms are a result of the different atomic numbers. Each element can be characterized by its atomic number. For example, hydrogen has an atomic number of 1 (1 proton in the nucleus), helium has an atomic number of 2 (2 protons in the nucleus), and uranium has an atomic number of 92 (92 protons in the nucleus). The table of the elements given on the inside front cover of this book lists the atomic numbers and atomic masses of all the known elements.

1-13. MOST ELEMENTS OCCUR IN NATURE AS MIXTURES OF ISOTOPES

Because nuclei are made up of protons and neutrons, each of which has a mass of approximately 1 amu, you might expect that atomic masses should be approximately equal to whole numbers. Although many atomic masses are approximately whole numbers (for example, the atomic mass of carbon is 12.0 and the atomic mass of fluorine is 19.0), many others are not. Chlorine ($Z = 17$) has an atomic mass of 35.45, magnesium ($Z = 12$) has an atomic mass of 24.31, and copper ($Z = 29$) has an atomic mass of 63.55. The explanation lies in the fact that many elements consist of two or more **isotopes,** which are atoms that contain the same number of protons but different numbers of neutrons. Realize that it is the number of *protons* that characterizes a particular element, and so nuclei of the same element may have varying numbers of neutrons. For example, the most common isotope of the simplest element, hydrogen, contains one proton and one electron. There is another, less common isotope of hydrogen that contains one proton, one neutron, and one electron. These two isotopes are identical chemically; they both undergo the same chemical reactions. The heavier isotope is called heavy hydrogen or, more commonly, **deuterium,** and is denoted by the symbol D. Water that is made from deuterium is called **heavy water** and is often denoted by D_2O.

▪ Heavy water is used in some nuclear reactors.

An isotope is specified by its atomic number and its mass number. The notation used to designate isotopes is the chemical symbol of the element written with its atomic number as a left subscript and its mass number as a left superscript:

$$\text{mass number} \rightarrow {}^{A}_{Z}\text{X} \leftarrow \text{chemical symbol}$$
$$\text{atomic number} \rightarrow Z$$

For example, an ordinary hydrogen atom is denoted ${}^{1}_{1}\text{H}$ and a deuterium atom is denoted ${}^{2}_{1}\text{H}$.

▪ The number of neutrons, N, in an atom is equal to $A - Z$.

Example 1-5: Fill in the blanks:

	Symbol	Atomic number	Number of neutrons	Mass number
(a)	____	22	____	48
(b)	____	____	110	184
(c)	${}^{?}_{?}\text{Co}$	____	____	60

Solution:

(a) The number of neutrons is the mass number minus the atomic number, or $48 - 22 = 26$ neutrons. The element with atomic number 22 is titanium (see inside front cover), and so the symbol of this isotope is ${}^{48}_{22}\text{Ti}$, called titanium-48.

(b) The number of protons is the mass number minus the number of neutrons, or $184 - 110 = 74$ protons. The element with atomic number 74 is tungsten, and so the symbol is $^{184}_{74}W$.

(c) According to the symbol, the element is cobalt, whose atomic number is 27. The symbol of the particular isotope is $^{60}_{27}Co$, and the isotope has $60 - 27 = 33$ neutrons. Cobalt-60 is used as a γ-radiation source for the treatment of cancer.

Although one of the postulates of Dalton's atomic theory was that all the atoms of a given element have the same mass, we see that this is not so. Isotopes of the same element have different masses. Several common natural isotopes and their corresponding masses are given in Table 1-9. Of particular note is that the isotopic mass of carbon-12 is exactly 12. This is the convention on which the modern atomic mass scale is based. All atomic masses are given relative to the mass of carbon-12, which is defined by international convention to be exactly 12.

Table 1-9 also indicates that helium has two isotopes, helium-3 and helium-4. The atomic number of helium is 2, which means that a helium nucleus has two protons and a nuclear charge of +2. A helium-4 nucleus has a charge of +2 and an atomic mass of 4, the same as an α-particle (Table 1-8). In fact, an α-particle is simply the nucleus of a helium-4 isotope.

As Table 1-9 implies, many elements occur in nature as a mixture of isotopes with a **natural abundance** for each isotope. Although naturally

▪ Data like these in Table 1-9 are available for all the naturally occurring elements.

Table 1-9 Naturally occurring isotopes of some common elements

Element	Isotope	Natural abundance/%	Isotopic mass	Protons	Neutrons	Mass number
hydrogen	$^{1}_{1}H$	99.985	1.0078	1	0	1
	$^{2}_{1}H$	0.015	2.0141	1	1	2
	$^{3}_{1}H$	trace	3.0160	1	2	3
helium	$^{3}_{2}He$	1.4×10^{-4}	3.0160	2	1	3
	$^{4}_{2}He$	99.99986	4.0026	2	2	4
carbon	$^{12}_{6}C$	98.89	12.0000	6	6	12
	$^{13}_{6}C$	1.11	13.0034	6	7	13
	$^{14}_{6}C$	trace	14.0032	6	8	14
oxygen	$^{16}_{8}O$	99.758	15.9949	8	8	16
	$^{17}_{8}O$	0.038	16.9991	8	9	17
	$^{18}_{8}O$	0.204	17.9992	8	10	18
fluorine	$^{19}_{9}F$	100.0	18.9984	9	10	19
magnesium	$^{24}_{12}Mg$	78.99	23.9850	12	12	24
	$^{25}_{12}Mg$	10.00	24.9858	12	13	25
	$^{26}_{12}Mg$	11.01	25.9826	12	14	26
chlorine	$^{35}_{17}Cl$	75.77	34.9689	17	18	35
	$^{37}_{17}Cl$	24.23	36.9659	17	20	37

occurring chlorine consists of two isotopes, they occur as 75.77 percent $^{35}_{17}Cl$ and 24.23 percent $^{37}_{17}Cl$, independent of any natural source of the chlorine. Chlorine obtained from, say, salt deposits in Africa or Australia has essentially the same isotopic composition as that given in Table 1-9. The atomic mass of chlorine is the sum of the masses of each isotope, each weighted by its natural abundance. If we use the isotopic masses and natural abundances of chlorine given in Table 1-9, then we obtain

$$\begin{pmatrix}\text{atomic mass}\\ \text{of chlorine}\end{pmatrix} = (34.97)\left(\frac{75.77}{100}\right) + (36.97)\left(\frac{24.23}{100}\right) = 35.45$$

This is the atomic mass of chlorine given on the inside front cover of the book. The factors 75.77/100 and 24.23/100 must be included in order to take into account the relative natural abundance of each isotope.

Example 1-6: Naturally occurring chromium is a mixture of four isotopes with the following isotopic masses and natural abundances:

Mass number	Isotopic mass	Natural abundance/%
50	49.946	4.35
52	51.941	83.79
53	52.941	9.50
54	53.939	2.36

Calculate the atomic mass of chromium.

Solution: The atomic mass is the sum of the masses of the four isotopes weighted by their abundances:

$$\begin{pmatrix}\text{atomic mass}\\ \text{of chromium}\end{pmatrix} = (49.946)\left(\frac{4.35}{100}\right) + (51.941)\left(\frac{83.79}{100}\right) + (52.941)\left(\frac{9.50}{100}\right) + (53.939)\left(\frac{2.36}{100}\right)$$

$$= 52.00$$

Small variations in natural abundances of the isotopic composition of the elements limit the precision with which atomic masses can be specified. The masses of individual isotopes are known much more accurately than atomic masses given in the periodic table. In the next section, we discuss how isotopic masses are determined.

1-14. ISOTOPIC MASSES ARE DETERMINED WITH A MASS SPECTROMETER

The mass and percentage of each isotope of an element can be determined by using a **mass spectrometer.** When a gas is bombarded with electrons from an external source, electrons are knocked out of the neutral atoms in the gas, producing positively charged, atomic-sized

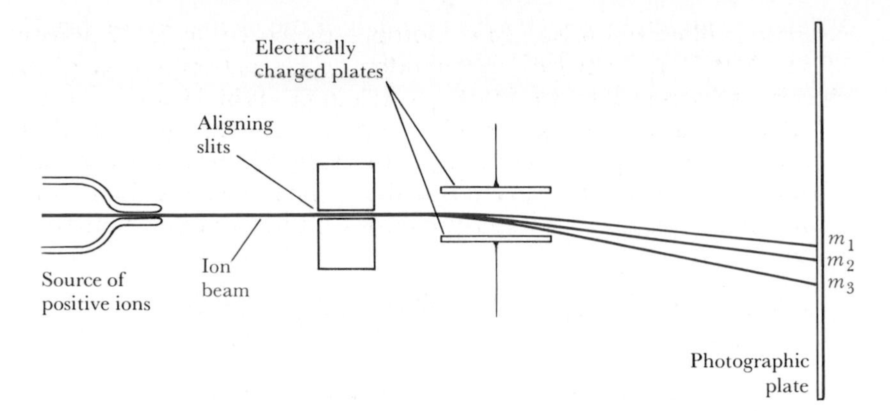

Figure 1-9 Schematic diagram of a mass spectrometer. A gas is bombarded with electrons that knock electrons out of the gas atoms or molecules and produce positively charged ions. These ions are then accelerated by an electric field and form a sharply focused beam as they pass through the two aligning slits. The beam is then allowed to pass through an electric or a magnetic field. The ions of different masses (m) are deflected to different extents, and the beam is split according to the masses of the various ions in the beam. The particles of different masses strike a detector, such as a photographic plate, at different places, and the amount of plate exposure at various places is proportional to the number of particles having a particular mass.

particles called **ions.** Ions are atoms that have either a deficiency of electrons (in which case the ions are positively charged) or extra electrons (in which case the ions are negatively charged). The positively charged gas ions produced by electron bombardment are accelerated by an electric field and passed through slits to form a narrow, well-focused beam (Figure 1-9). The beam of ions is then passed through an electric field, which deflects each ion according to its mass. Thus, the original beam of ions is split into several separate beams, one for each isotope of the gas. The intensities of the separated ion beams, which can be determined experimentally, are a direct measure of the number of ions in each beam. In this manner, it is possible to determine not only the mass of each isotope of any element (by the amount of deflection of each beam) but also the percentage of each isotope (by the intensity of each beam).

We shall encounter ions throughout our study of chemistry, and so we introduce a notation for them here. An atom that has lost one electron has a net charge of $+1$; one that has lost two electrons has a charge of $+2$; one that has gained an electron has a charge of -1; and so on.

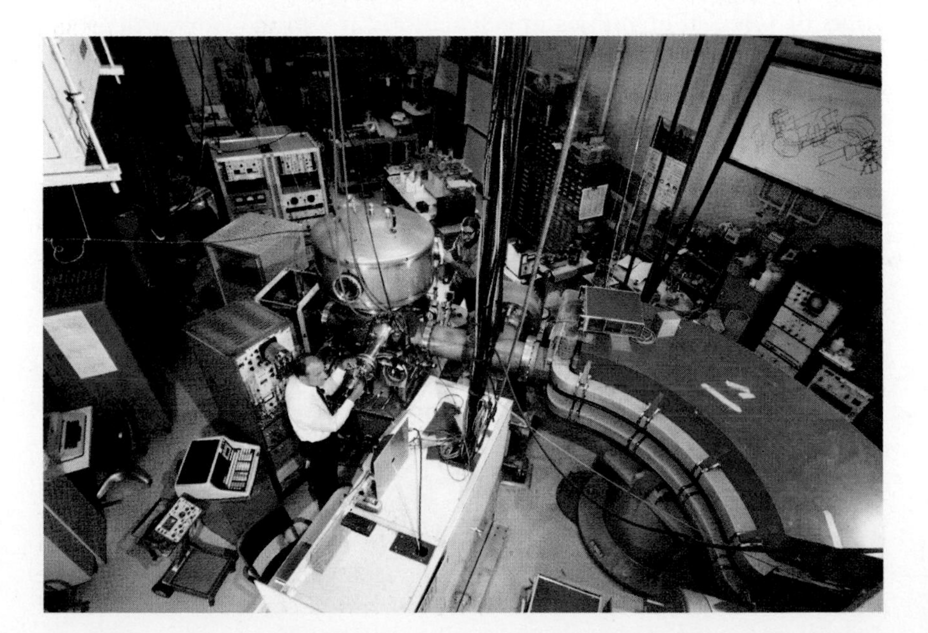

Figure 1-10 A mass spectrometer like the one shown here can be used to separate the isotopes of an element by exploiting the difference in their masses.

We denote an ion by the chemical symbol of the element with a right-hand superscript to indicate its charge:

Na^{+}	singly charged sodium ion
Mg^{2+}	doubly charged magnesium ion
Cl^{-}	singly charged chloride ion
S^{2-}	doubly charged sulfide ion

Positive ions are called **cations** and negative ions are called **anions.** Note from these examples that the names of anions have the -ide ending characteristic of the second element in binary compounds. We shall see in Chapter 2 that some compounds consist of ions rather than molecules. Such compounds are called **ionic compounds.**

Example 1-7: How many electrons are there in K^{+} and Cl^{-}?

Solution: From the table on the inside front cover, we see that the atomic number of potassium is 19. The K^{+} ion is a potassium atom that is lacking one electron, and so K^{+} has 18 electrons.

The atomic number of chlorine is 17. The Cl^{-} ion has one electron more than a chlorine atom, and so Cl^{-} has 18 electrons.

Notice that both K^{+} and Cl^{-} have 18 electrons. Species that contain the same number of electrons are said to be **isoelectronic.**

1-15. THE PRECISION OF A MEASURED QUANTITY IS INDICATED BY THE NUMBER OF SIGNIFICANT FIGURES

Figure 1-11 A modern analytical balance, used to determine mass. The mass determination is made by turning a dial to find the known mass (using a set of standard masses enclosed within the balance) that balances the unknown mass. When the two masses just balance, they are equal within the accuracy of the balance.

Before we leave this chapter, we discuss a matter of technical importance regarding chemical calculations. To carry out scientific calculations, you must know how to handle significant figures, units of measurement, and unit conversions. The determination of the number of significant figures in a calculated quantity and the manipulation of the units of physical quantities in calculations are straightforward procedures. If you learn a few simple rules on significant figures and always include the units of various quantities, then you will find that your calculations will go smoothly.

The counting of objects is the only type of experiment that can be carried out with complete accuracy, that is, without any inherent error. Let's consider the problem of determining how many pennies there are in a jar. It is possible to determine the exact number of pennies simply by counting them. Suppose that there are 1542 pennies in the jar. The number 1542 is exact; there is no uncertainty associated with it. It is a different matter, however, when we determine the mass of 1542 pennies with a balance (Figure 1-11). Suppose that the balance we use is capable of measuring the mass of an object to the nearest one tenth of a gram and that the experimentally determined mass of the 1542 pennies is 4776.2 ± 0.1 g. The ± 0.1 denotes the experimental uncertainty in the measured mass of the pennies. The actual mass lies somewhere between 4776.1 and 4776.3 g. We cannot tell from the result 4776.2 ± 0.1 g if the mass is, say, 4776.15 g, or 4776.24 g, or some other mass in the range 4776.1 to 4776.3. We would need a more sensitive balance to determine the mass of the pennies to the nearest hundredth of a gram ($\pm$0.01 g), and a still more sensitive balance to determine the mass to the nearest milligram (mg, $\pm$0.001 g).

Denoting the uncertainty of a measured quantity by the $\pm$ notation is desirable in precise work, but such a notation is too cumbersome for our purposes. We indicate the precision of measured quantities by the number of **significant figures** used to express a result. All digits in a numerical result are significant if only the last digit has some uncertainty. A result expressed as 4776.2 g means that only the 2 has some uncertainty and that there are five significant figures in the result.

Zeros in a measured quantity require special consideration. Zeros are not taken as significant figures if they serve only to position the decimal point, as is illustrated in the following examples:

Result	**Number of significant figures**	**Comment**
(a) 0.0056	2	The three zeros are used only to position the decimal point and are thus not significant figures.
(b) 5.6×10^{-3}	2	Compare with (a); the 10^{-3} simply positions the decimal point.
(c) 38.70	4	The zero is not necessary to position the decimal point and thus must be regarded as a significant figure.
(d) 100.0	4	All the zeros are significant in this case.

In certain cases, it is not clear just how many significant figures are implied. Consider the number 100. With the number presented as 100, we might mean that the value of the number is *exactly* 100. On the other hand, the two zeros might not be significant and we might mean that the value is *approximately* 100, say 100 $\pm$ 10, for example. The number of significant figures in such a case is uncertain. Usually the number of significant figures can be deduced from the statement of a problem. It is the writer's obligation to indicate clearly the number of significant figures.

Example 1-8: State the number of significant figures in each of the following numbers:

(a) 0.0312 (b) 0.03120 (c) 3120

Solution:

(a) 0.0312 has three significant figures: 3, 1, and 2.

(b) 0.03120 has four significant figures: 3, 1, 2, and 0.

(c) 3120 has four significant figures if we mean *exactly* 3120 and three significant figures if we mean 3120 $\pm$ 10.

1-16. CALCULATED NUMBERS SHOULD SHOW THE CORRECT NUMBER OF SIGNIFICANT FIGURES

In multiplication and division, the calculated result should not be expressed to more significant figures than the factor in the calculation with the least number of significant figures. For example, if we perform the multiplication

$$8.3143 \times 298.2$$

on a hand calculator, the following result comes up on the calculator display:

$$2479.3243$$

▪ Your hand calculator usually will give many more digits than are significant.

Not all these figures are significant. The correct result is 2479 because the factor 298.2 has only four significant figures and thus the result cannot have more than four. The extra figures are not significant and should be discarded.

Example 1-9: Determine the result to the correct number of significant figures:

$$y = \frac{3.00 \times 0.08205 \times 298}{0.93}$$

Solution: Using a hand calculator, we obtain

$$y = 78.873871$$

The factor 0.93 has the least number of significant figures—only two. Thus the calculated result should not be expressed to more than two significant figures, and the correct result is $y = 79$, which is 78.873871 rounded off to two significant figures.

The number of figures after the decimal point in a number resulting from addition or subtraction can have no more figures after the decimal point than the least number of figures after the decimal point in any of the numbers that are being added or subtracted. Consider the sum

$$\begin{array}{r} 6.939 \\ +1.00797 \\ \hline 7.94697 \end{array} \quad \text{round off to } 7.947$$

The last two digits in 7.94697 are not significant because we know the value of the first number in the addition, 6.939, to only three digits beyond the decimal place. Thus the result cannot be accurate to more than three digits past the decimal. Therefore our result expressed to the correct number of significant figures is 7.947.

In rounding off insignificant figures, we use the following convention. If the figure following the last figure retained is a 5, 6, 7, 8, or 9, then the preceding figure should be increased by 1; otherwise (that is, for 0, 1, 2, 3, and 4), the preceding figure should be left unchanged.

Thus, rounding off the following numbers to three significant figures, we obtain

$$27.35 \rightarrow 27.4$$
$$27.34 \rightarrow 27.3$$

Note that in half of the cases (0, 1, 2, 3, 4) we discard the insignificant digit and in the other half (5, 6, 7, 8, 9) we increase the preceding digit by 1 when we discard the insignificant digits.

1-17. THE METRIC SYSTEM IS USED IN SCIENTIFIC WORK

When a number represents a measurement, the units of that measurement must always be indicated. For example, if we measure the thickness of a wire to be 1.35 millimeters (mm), we express the result as 1.35 mm. To say that the thickness of the wire is 1.35 would be meaningless.

The preferred system of units used in scientific work is the **metric system.** There are several sets of units in the metric system, and in recent years there has been a worldwide movement to express all measurements in terms of just one set of metric units called SI units (for *S*ystème *I*nternational). The metric system, and SI units in particular, are described in Appendix B. We discuss the individual units in detail when we encounter them in the text.

Only numbers that have the same units can be added or subtracted. If we add 2.12 centimeters (cm) and 4.73 cm, we obtain 6.85 cm. If we wish to add 76.4 cm to 1.19 meters (m), we must first convert centimeters to meters or meters to centimeters. We convert from one unit to another by using a **unit conversion factor.** Suppose we want to convert meters to centimeters. From Appendix B we find that

$$1 \text{ m} = 100 \text{ cm} \tag{1-1}$$

Equation (1-1) is a definition and is exact; there is no limit to the number of significant figures on either side of the equation. If we divide both sides of Equation (1-1) by 1 m, we get

$$1 = \frac{100 \text{ cm}}{1 \text{ m}} \tag{1-2}$$

Equation (1-2) is called a unit conversion factor because we can use it to convert meters to centimeters. A unit conversion factor, as expressed in Equation (1-2), is equal to unity, and thus we can multiply any quantity by a unit conversion factor without changing its intrinsic value. If we multiply 1.19 m by Equation (1-2), we obtain

$$(1.19 \cancel{\text{m}})\left(\frac{100 \text{ cm}}{1 \cancel{\text{m}}}\right) = 119 \text{ cm}$$

Notice that the units of meters cancel, giving the final result in centimeters. To convert centimeters to meters, we use the reciprocal of Equation (1-2):

$$(76.4 \cancel{\text{cm}})\left(\frac{1 \text{ m}}{100 \cancel{\text{cm}}}\right) = 0.764 \text{ m}$$

Notice in this case that the units of centimeters cancel, giving the final result in meters. Using these results, we see that the sum of 76.4 cm and 1.19 m is

$$76.4\text{ cm} + (1.19\,\cancel{\text{m}})\left(\frac{100\text{ cm}}{1\,\cancel{\text{m}}}\right) = 195\text{ cm}$$

or

$$(76.1\,\cancel{\text{cm}})\left(\frac{1\text{ m}}{100\,\cancel{\text{cm}}}\right) + 1.19\text{ m} = 1.95\text{ m}$$

Notice that both results are given to three significant figures.

As another example of converting from one set of units to another, let's convert 55 miles per hour (mph) to kilometers (km) per hour (h). From the inside back cover, we find that

$$1\text{ mile} = 1.61\text{ km} \tag{1-3}$$

Dividing both sides of Equation (1-3) by 1 mile yields the unit conversion factor:

$$1 = \frac{1.61\text{ km}}{1\text{ mile}} \tag{1-4}$$

Equation (1-4) is the unit conversion factor that can be used to convert a speed given in miles per hour to a speed in kilometers per hour. Thus

$$\left(\frac{55\,\cancel{\text{mile}}}{\text{h}}\right)\left(\frac{1.61\text{ km}}{1\,\cancel{\text{mile}}}\right) = \frac{89\text{ km}}{\text{h}}$$

Note that the use of the proper units for each quantity provides an internal check on the correctness of the calculation. We must *multiply* 55 mph by

$$\frac{1.61\text{ km}}{1\text{ mile}}$$

in order to obtain the result in the desired units of km/h. Note that if we had used the conversion

$$1\text{ km} = 0.62\text{ mile}$$

from the inside back cover, then the unit conversion would be

$$\left(\frac{55\,\cancel{\text{mile}}}{\text{h}}\right)\left(\frac{1\text{ km}}{0.62\,\cancel{\text{mile}}}\right) = 89\frac{\text{km}}{\text{h}}$$

so that the unit of mile cancels out.

Example 1-10: You are driving in a certain country whose monetary unit is called a peso (124 pesos = 1 U.S. dollar). If you pay 53.6 pesos per liter for gasoline, then what is the cost in dollars per gallon?

Solution: We wish to convert $\left(\frac{\text{peso}}{\text{liter}}\right)$ to $\left(\frac{\text{dollar}}{\text{gallon}}\right)$. We do this in a series of steps, using the facts that one quart (qt) of liquid is equivalent to 0.946 liter (L) and that 1 gallon (gal) contains exactly 4 qt. Thus, we write

$$\left(\frac{53.6\ \text{peso}}{1\ \text{L}}\right)\left(\frac{0.946\ \text{L}}{1\ \text{qt}}\right)\left(\frac{4\ \text{qt}}{1\ \text{gal}}\right)\left(\frac{1\ \text{dollar}}{124\ \text{peso}}\right) = 1.64\frac{\text{dollar}}{\text{gal}}$$

Note that we wrote the series of conversions in one line, working on each unit sequentially.

Many quantities are expressed in **compound units.** To see what we mean by a compound unit, consider the quantity density. The **density** of a substance is defined as the ratio of the mass of the substance to its volume:

$$\text{density} = \frac{\text{mass}}{\text{volume}} \tag{1-5}$$

Thus we say that the **dimensions** of density are mass per unit volume, which can be expressed in a variety of units. If we express the mass in grams and the volume in cubic centimeters, then the units of density are grams per cubic centimeter. For example, the density of ice is 0.92 g/cm^3, where the slash denotes "per." From algebra we know that

$$\frac{1}{a^n} = a^{-n}$$

where n is an exponent. Thus

▪ Exponents are reviewed in Appendix A.

$$\frac{1}{\text{cm}^3} = \text{cm}^{-3}$$

Therefore we also can express density as $g{\cdot}cm^{-3}$ instead of g/cm^3. In the expression $g{\cdot}cm^{-3}$, the dot serves to separate the g from the cm^{-3}. The use of dots in compound units is an SI convention (Appendix B) and is used to avoid ambiguities. For example, m·s denotes meter-second, whereas ms (without the dot) denotes millisecond.

We now consider an example of converting quantities in compound units. It has been estimated that all the gold that has ever been mined would occupy a cube 17 m on a side. Given that the density of gold is 18.9 $g{\cdot}cm^{-3}$, let's calculate the mass of all this gold. The volume of a cube 17 m on a side is

$$\text{volume} = (17\ \text{m})^3 = 4913\ \text{m}^3 \tag{1-6}$$

Although the volume calculated in Equation (1-6) is good to only two significant figures (because the value 17 m is good to only two significant figures), we shall carry extra significant figures through the calculation and then round off the final result to two significant figures. The mass of the gold is obtained by multiplying the density by the volume, but before doing this, we must convert cubic meters to cubic centimeters (because the density is given in $g{\cdot}cm^{-3}$). From Appendix B, we find that

$$1\ \text{m} = 100\ \text{cm}$$

By cubing both sides of this expression, we obtain

$$1\ \text{m}^3 = 10^6\ \text{cm}^3$$

and so the unit conversion factor is

$$1 = \frac{10^6 \text{ cm}^3}{1 \text{ m}^3}$$

Thus the volume in Equation (1-6) is

$$\text{volume} = (4913 \cancel{\text{m}^3})\frac{10^6 \text{ cm}^3}{1 \cancel{\text{m}^3}}$$
$$= 4.913 \times 10^9 \text{ cm}^3$$

If we multiply the volume of the gold by its density, we obtain the mass:

$$\text{mass} = \text{volume} \times \text{density}$$
$$= (4.913 \times 10^9 \text{ cm}^3)(18.9 \text{ g·cm}^{-3})$$
$$= 9.3 \times 10^{10} \text{ g}$$

The result is rounded off to two significant figures because, as we mentioned before, the side of the cube (17 m) is given to only two significant figures. In obtaining this result, we have used the fact that

$$(\text{cm}^3)(\text{cm}^{-3}) = 1$$

▪ The price of gold ("gold fix") varies from day to day.

Let's see what this mass of gold would be worth at $400 per troy ounce (oz). (Gold is sold by the troy ounce, which is about 10 percent heavier than the avoirdupois ounce, which is the unit used for foods.) There are 31.1 g in 1 troy oz and so the unit conversion factor is

$$1 = \frac{1 \text{ troy oz}}{31.1 \text{ g}}$$

The mass of gold in troy ounces is

$$\text{mass} = (9.3 \times 10^{10} \cancel{\text{g}})\left(\frac{1 \text{ troy oz}}{31.1 \cancel{\text{g}}}\right)$$
$$= 3.0 \times 10^9 \text{ troy oz}$$

At $400 per troy ounce, the value of all the gold ever mined is

$$\text{value} = (3.0 \times 10^9 \cancel{\text{troy oz}})\left(\frac{\$400}{1 \cancel{\text{troy oz}}}\right)$$
$$= \$1.2 \times 10^{12} = 1.2 \text{ trillion dollars}$$

If you carefully set up the necessary conversion factors and make certain that the appropriate units cancel to give the units you need for the answer, then you cannot go wrong in making unit conversions. With a little practice, the manipulation of units will become easy.

1-18. THE GUGGENHEIM NOTATION IS USED TO LABEL TABLE HEADINGS AND GRAPH AXES

In presenting tables of quantities with units, it is convenient to list the numerical values without their units and to use a column heading to specify the units. The least ambiguous way to do this is to write the name or symbol of the quantity followed by a slash and then followed by the symbol for the units. For example, the heading "distance/m" indicates that the units of the numerical entries are distances expressed in meters. This notation, is called the **Guggenheim notation** after E. A. Guggenheim, the British chemist who proposed its use.

Suppose that we wish to tabulate a number of masses, such as 1.604 g, 2.763 g, and 3.006 g. We use the Guggenheim notation in the heading—"Mass/g"—and list the masses as numbers without units, as shown in Table 1-10(a) (margin). Now suppose that later we wish to retrieve the values with their units. The heading indicates that the numbers in the column are masses divided by grams, so we write, for example,

Table 1-10. Tabulated data with headings using the Guggenheim notation

(a) Mass/g	(b) Mass/10^{-4} g
1.604	1.604
2.763	2.763
3.006	3.006

$$\text{mass/g} = \frac{\text{mass}}{\text{g}} = 1.604$$

We can multiply both sides of the equation by g to obtain

$$\cancel{\text{g}} \times \frac{\text{mass}}{\cancel{\text{g}}} = 1.604 \times \text{g}$$

In the resulting expression we can easily recognize each component of the data:

$$\underset{\text{property}}{\text{mass}} = \underset{\text{value}}{1.604}\ \underset{\text{unit of measure}}{\text{g}}$$

Note that the heading is treated as an algebraic quantity, and that we retrieve the data through an algebraic process.

Use of the Guggenheim notation is particularly convenient when the values to be tabulated are expressed in scientific notation. Suppose that we wish to tabulate the masses 1.604×10^{-4} g, 2.763×10^{-4} g, and 3.006×10^{-4} g. In this case we can use the heading "Mass/10^{-4} g" to simplify the tabulated data, as shown in Table 1-10(b). To retrieve the data from the table, we write, for example,

$$\frac{\text{mass}}{10^{-4}\ \text{g}} = 1.604$$

from which we get mass = 1.604×10^{-4} g.

Notice that although the unitless numbers listed in Table 1-10(a) and (b) are the same, the actual data are different, as indicated by the heading in each case.

Example 1-11: Consider the following tabulated data:

Time/10^{-5} s	Speed/10^5 m·s^{-1}
1.00	3.061
1.50	4.153
2.00	6.302
2.50	8.999

Retrieve the actual data—the values and their units for time and speed, respectively—as four pairs of data.

Solution: To find the actual times, we use, for example,

$$\frac{\text{time}}{10^{-5}\ \text{s}} = 1.00$$

from which we obtain time = 1.00×10^{-5} s. The corresponding speed is given by

$$\frac{\text{speed}}{10^{5}\ \text{m}\cdot\text{s}^{-1}} = 3.061$$

or speed = $3.061 \times 10^{5}\ \text{m}\cdot\text{s}^{-1}$. The other pairs of data are (1.50×10^{-5} s, $4.153 \times 10^{5}\ \text{m}\cdot\text{s}^{-1}$), ($2.00 \times 10^{-5}$ s, $6.302 \times 10^{5}\ \text{m}\cdot\text{s}^{-1}$), and ($2.50 \times 10^{-5}$ s, $8.999 \times 10^{5}\ \text{m}\cdot\text{s}^{-1}$).

Example 1-12: The SI unit for the quantity of electrical charge is a **coulomb** (C). Tabulate the following data: 7.05×10^{-15} C, 3.24×10^{-15} C, and 9.86×10^{-16} C.

Solution: In this case we use the heading "Charge/10^{-15} C" and tabulate the data as shown in the margin.

▪ **Charge/10^{-15} C**
7.05
3.24
0.986

If you look through this book, then you will see that the axes of graphs in figures are labeled like the column headings of tabulated data and that the numbers on the axes are unitless. For example, the vertical axis in Figure 4-6 is labeled V/L (volume divided by liters), and the numbers on the axis are 1.0, 2.0, and 3.0. For a point half way between 1.0 and 2.0, we would have

$$\frac{V}{\text{L}} = 1.5$$

or $V = 1.5$ L.

The Guggenheim notation has been adopted by the International Union of Pure and Applied Chemistry (IUPAC).* Although older and less convenient alternative conventions still are in use, the Guggenheim notation is being used increasingly in the scientific literature.

▪ For a thorough discussion of the advantages of the Guggenheim notation see "Notations in Physics and Chemistry" by E. A. Guggenheim, in *Journal of Chemical Education* [35, 606 (1958)]

*See the IUPAC *Manual of Symbols and Terminology for Physicochemical Quantities and Units,* 1979 Edition.

SUMMARY

Chemistry is an experimental science based on the scientific method. Scientific questions are answered by carrying out appropriate experiments. Scientific laws are concise summaries of large numbers of experimental observations. Scientific theories are designed to provide explanations for natural laws and observations.

The beginning of modern chemistry occurred in the late eighteenth century when Lavoisier, considered to be the founder of modern chemistry, introduced quantitative measurements into chemical research. Lavoisier's wor led directly to the discovery of the law of constant composition and then to Dalton's atomic theory. Dalton was able to use the atomic theory to determine the relative masses of atoms and molecules and to use these values in interpreting the results of chemical analyses. According to the atomic theory, the atoms in reactant molecules are separated and rearranged into product molecules in a chemical reaction. Because atoms are neither created nor destroyed in chemical reactions, chemical reactions obey the law of conservation of mass.

Elements are substances that consist of only one kind of atom. There are 108 known elements, about three quarters of which are metals. Elements combine to form compounds, whose constituent particles are called molecules, which are groups of atoms joined together. Chemists represent elements by chemical symbols and compounds by chemical formulas. The system of naming compounds is called chemical nomenclature.

Protons and neutrons form the nucleus of the atom, the small center containing all the positive charge and essentially all the mass of the atom. The number of protons in an atom is the atomic number (Z) of that atom. The total number of protons and neutrons in an atom is the mass number (A) of that atom.

Each element can be characterized by its atomic number. Nuclei with the same number of protons but different numbers of neutrons are called isotopes. Most elements occur naturally as mixtures of isotopes, and atomic masses are weighted averages of the isotopic masses. Isotopic abundances and masses can be determined by using a mass spectrometer. Because isotopes of an element have the same atomic number, they are chemically identical and they undergo the same reactions.

In a neutral atom, the number of electrons is equal to the number of protons. When an atom loses or gains electrons, the resulting species is called an ion. Positive ions (cations) have a deficiency of electrons, and negative ions (anions) have extra electrons.

To carry out scientific calculations, it is important to understand significant figures, units of measurement, and unit conversions. Significant figures represent the precision of a measurement. Units must always be included with numbers that represent measurements or else the numbers are meaningless. When different units are used in a measurement, one unit must be converted to the other before they can be used together in calculations. Unit conversions are carried out using unit conversion factors.

TERMS YOU SHOULD KNOW*

scientific method 2
scientific law 3
natural law 3
theory 3
hypothesis 3
definition 3
element 4
compound 4
metal 5
nonmetal 5
chemical symbol 5
atom 6
molecule 6
diatomic molecule 6
quantitative measurement 6
qualitative observation 7
law of conservation of mass 7
law of constant composition 8
mass 8
atomic theory 9
atom 6
molecule 6
atomic mass 10
atomic mass unit (amu) 10
chemical nomenclature 12
binary compound 12
chemical formula 12
molecular mass 14
subatomic particle 15
electron 15
radioactivity 15
radioactive 15
α-particle 16
β-particle 16
γ-ray 16
malleable 16
nucleus 16
proton 17
neutron 17
atomic number, Z 17
mass number, A 17
isotope 18
deuterium 18
heavy water 18
natural abundance 19
mass spectrometer 20
ion 21
cation 22
anion 22
ionic compound 22
isoelectronic 22
significant figure 23
metric system 25
unit conversion factor 25
compound unit 27
density 27
dimensions 27
Guggenheim notation 28
Coulomb 30

*These terms are listed in the order in which they appear in the text. Page numbers refer to the pages on which the terms are introduced. A complete glossary of these terms can be found in the *Study Guide/Solutions Manual* accompanying this text.

PROBLEMS*

CHEMICAL SYMBOLS

1-1. Give the chemical symbols for the following elements:

(a) selenium (b) indium (c) manganese
(d) thulium (e) mercury (f) krypton
(g) palladuim (h) thallium (i) uranium
(j) tungsten

1-2. Give the chemical symbols for the following elements:

(a) tin (b) gold (c) zirconium
(d) bismuth (e) ruthenium (f) rubidium
(g) bromine (h) neon (i) antimony
(j) arsenic

*Problems grouped by subject area are arranged in matched pairs (1-1 and 1-2, for example) such that both usually involve the same principle or operation. Answers to odd-numbered problems are given in Appendix F. Detailed solutions to odd-numbered problems can be found in the *Study Guide/Solutions Manual* accompanying this text. Problems given under the heading "Additional Problems" are not identified by subject area and are not paired.

1-3. Name the elements with the following chemical symbols:

(a) Ge (b) Sc
(c) Ir (d) Cs
(e) Sr (f) Am
(g) Mo (h) In
(i) Pu (j) Xe

1-4. Name the elements with the following chemical symbols:

(a) Pt (b) Te
(c) Pb (d) Ta
(e) Ba (f) Ti
(g) Re (h) La
(i) Eu (j) Pr

MASS PERCENTAGES IN COMPOUNDS

1-5. A 1.659-g sample of a compound of sodium and oxygen contains 0.978 g of sodium and 0.681 g of oxygen. Calculate the mass percentages of sodium and oxygen in the compound.

1-6. The compound lanthanum oxide is used in the production of optical glass and the fluorescent phosphors used to coat television screens. An 8.29-g sample is found to contain 7.08 g of lanthanum and 1.21 g of oxygen. Calculate the mass percentages of lanthanum and oxygen in lanthanum oxide.

1-7. A 1.28-g sample of copper is heated with sulfur to produce 1.60 g of a copper sulfide compound. Calculate the mass percentages of copper and sulfur in the compound.

1-8. Stannous fluoride, an active ingredient in some toothpastes that helps to prevent cavities, contains tin and fluorine. A 1.793-g sample was found to contain 1.358 g of tin. Calculate the mass percentages of tin and fluorine in stannous fluoride.

1-9. Potassium cyanide is used in extracting gold and silver from their ores. A 12.63-mg sample is found to contain 7.58 mg of potassium, 2.33 mg of carbon, and 2.72 mg of nitrogen. Calculate the mass percentages of potassium, carbon, and nitrogen in potassium cyanide.

1-10. Ethyl alcohol, the alcohol in alcoholic beverages, is a compound of carbon, hydrogen, and oxygen. A 3.70-g sample of ethyl alcohol contains 1.93 g of carbon and 0.49 g of hydrogen. Calculate the mass percentages of carbon, hydrogen, and oxygen in ethyl alcohol.

NOMENCLATURE

1-11. Name the following binary compounds:

(a) Li_2S (b) BaO
(c) Mg_3P_2 (d) CsBr

1-12. Name the following binary compounds:

(a) BaF_2 (b) Mg_3N_2
(c) CsCl (d) CaS

1-13. Name the following binary compounds:

(a) CaC_2 (b) GaP
(c) Al_2O_3 (d) $BeCl_2$

1-14. Name the following binary compounds:

(a) MgF_2 (b) AlN
(c) MgSe (d) Li_3P

1-15. Name the following pairs of compounds:

(a) ClF_3 and ClF_5 (b) SF_4 and SF_6
(c) KrF_2 and KrF_4 (d) BrO and BrO_2

1-16. Name the following pairs of compounds:

(a) $SbCl_3$ and $SbCl_5$ (b) ICl_3 and ICl_5
(c) SeO_2 and SeO_3 (d) CS and CS_2

MOLECULAR MASSES

1-17. Compute the molecular mass for each of the following oxides:

(a) TiO_2 (white pigment)
(b) Fe_2O_3 (rust)
(c) V_2O_5 (catalyst)
(d) P_4O_{10} (dehydrating agent)

1-18. Calculate the molecular mass for each of the following ores:

(a) $CaWO_4$ (scheelite, an ore of tungsten)
(b) Fe_3O_4 (magnetite)
(c) Na_3AlF_6 (cryolite)
(d) $Be_3Al_2Si_6O_{18}$ (beryl)
(e) Zn_2SiO_4 (willemite)

1-19. Compute the molecular mass for each of the following halogen compounds:

(a) BrN_3 (explosive)
(b) $NaIO_3$ (antiseptic)
(c) CCl_2F_2 (refrigerant)
(d) $C_{14}H_9Cl_6$ (DDT)

1-20. Calculate the molecular mass for each of the following vitamins:

(a) $C_{20}H_{30}O$ (vitamin A)
(b) $C_{12}H_{17}ClN_4OS$ (vitamin B_1, thiamine)
(c) $C_{17}H_{20}N_4O_6$ (vitamin B_2, riboflavin)
(d) $C_{56}H_{88}O_2$ (vitamin D_1)
(e) $C_6H_8O_6$ (vitamin C, ascorbic acid)

MASS PERCENTAGES FROM ATOMIC MASSES

1-21. Use the atomic masses given on the inside front cover of the text to calculate the mass percentages of bromine and fluorine in bromine pentafluoride.

1-22. Use the atomic masses given on the inside front cover of the text to calculate the mass percentages of nitrogen and oxygen in dinitrogen oxide.

1-23. Ordinary table sugar, whose common chemical name is sucrose, has the chemical formula $C_{12}H_{22}O_{11}$. Calculate the mass percentages of carbon, hydrogen, and oxygen in sucrose.

1-24. A key compound in the production of aluminum metal is Na_3AlF_6. Calculate the mass percentages of sodium, aluminum, and fluorine in this compound.

1-25. Calculate the number of grams of xenon in 2.000 g of the compound xenon tetrafluoride.

1-26. Calculate the number of grams of sulfur in 5.585 g of the compound sulfur trioxide.

PROTONS, NEUTRONS, AND ELECTRONS

1-27. The following isotopes are used widely in medicine or industry:

(a) iodine-131 (b) cobalt-60
(c) potassium-43 (d) indium-113

How many protons, neutrons, and electrons are there in each of these isotopes?

1-28. The following isotopes do not occur naturally but are produced in nuclear reactors:

(a) phosphorus-30 (b) technetium-97
(c) iron-55 (d) americium-240

How many protons, neutrons, and electrons are there in each of these isotopes?

1-29. Fill in the blanks in the following table:

Symbol	Atomic number	Number of neutrons	Mass number
$^{14}_{6}C$	——	——	——
$^{?}_{?}Am$	——	——	241
——	53	——	123
——	——	10	18

1-30. Fill in the blanks in the following table:

Symbol	Atomic number	Number of neutrons	Mass number
$^{?}_{?}Ca$	——	——	48
——	40	——	90
——	——	78	131
$^{?}_{?}Mo$	——	57	——

1-31. Fill in the blanks in the following table:

Symbol	Atomic number	Number of neutrons	Mass number
——	31	36	——
——	——	8	15
——	27	——	58
$^{?}_{?}Xe$	——	——	133

1-32. Fill in the blanks in the following table:

Symbol	Atomic number	Number of neutrons	Mass number
$^{39}_{19}K$	——	——	——
$^{?}_{?}Fe$	——	——	56
——	36	——	84
——	——	70	120

ISOTOPIC COMPOSITION

1-33. Naturally occurring hydrogen consists of three isotopes with the atomic masses and abundances given in Table 1-9. Calculate the atomic mass of hydrogen.

1-34. Naturally occurring magnesium consists of three isotopes with the atomic masses and abundances given in Table 1-9. Calculate the atomic mass of magnesium.

1-35. Naturally occurring neon is a mixture of three isotopes with the following atomic masses and abundances:

Mass number	Atomic mass	Abundance/%
20	19.99	90.51
21	20.99	0.27
22	21.99	9.22

Calculate the atomic mass of naturally occurring neon.

1-36. Naturally occurring silicon consists of three isotopes with the following atomic masses and abundances:

Mass number	Atomic mass	Abundance/%
28	27.977	92.23
29	28.977	4.67
30	29.974	3.10

Calculate the atomic mass of naturally occurring silicon.

1-37. Naturally occurring bromine consists of two isotopes, ^{79}Br and ^{81}Br, whose atomic masses are 78.9183 and 80.9163, respectively. Given that the observed atomic mass of bromine is 79.904, calculate the percentages of ^{79}Br and ^{81}Br in naturally occurring bromine.

1-38. Naturally occurring boron consists of two isotopes with the atomic masses 10.013 and 11.009. The observed atomic mass of boron is 10.811. Calculate the abundance of each isotope.

1-39. Nitrogen has two naturally occurring isotopes, ^{14}N and ^{15}N, whose atomic masses are 14.0031 and 15.0001, respectively. The atomic mass of nitrogen is 14.0067. Use these data to compute the percentage of ^{15}N in naturally occurring nitrogen.

1-40. Naturally occurring europium consists of two isotopes, ^{151}Eu and ^{153}Eu, whose atomic masses are 150.9199 and 152.9212, respectively. Given that the atomic mass of europium is 151.96, calculate its isotopic percentage composition.

IONS

1-41. How many electrons are there in the following ions?

(a) Cs^+ (b) I^-
(c) Se^{2-} (d) N^{3-}

1-42. How many electrons are there in the following ions?

(a) Br^- (b) P^{3-}
(c) Ag^+ (d) Pb^{4+}

1-43. Determine the number of electrons in the following ions:

(a) Ba^{2+} (b) Tl^{3+}
(c) Fe^{2+} (d) Ti^{4+}

1-44. Determine the number of electrons in the following ions:

(a) Te^{2-} (b) La^{3+}
(c) Au^+ (d) Ir^{3+}

1-45. Give three ions that are isoelectronic with each of the following:

(a) K^+ (b) Kr
(c) N^{3-} (d) I^-

1-46. Give three ions that are isoelectronic with each of the following:

(a) F^- (b) Se^{2-}
(c) Ba^{2+} (d) La^{3+}

SIGNIFICANT FIGURES

1-47. Determine the number of significant figures in each of the following:

(a) 0.0390 (b) 6.022×10^{23}
(c) 3.652×10^{-5} (d) the 1980 U.S. population of about 226,000,000
(e) the number 16

1-48. Determine the number of significant figures in each of the following:

(a) 578 (b) 0.000578
(c) There are 1000 m in 1 km. (d) The distance from the earth to the sun is 93,000,000 miles.

1-49. Use the atomic masses given in Appendix E to compute molecular masses of the following compounds to the greatest number of significant figures justified by the data:

(a) H_2O (b) $MgCl_2$
(c) AlI_3 (d) $^{98}TcBr_2$

1-50. Use the atomic masses given in Appendix E to compute molecular masses to the greatest number of significant figures justified by the data:

(a) CH_4 (b) CaF_2
(c) $TiCl_4$ (d) $^{243}AmCl_3$

1-51. Calculate the following to the correct number of significant figures:

(a) $656.29 - 654 =$
(b) $(27.5)^3 =$
(c) $\dfrac{51}{18.02} =$
(d) $(6.022 \times 10^{23})(5.6 \times 10^{-2}) =$

1-52. Calculate the following to the correct number of significant figures:

(a) $213.3642 + 17.54 + 32978 =$
(b) $373.26 - 119 =$
(c) $\dfrac{(6.626196 \times 10^{-34})(2.997925 \times 10^{8})}{(1.38062 \times 10^{-23})} =$
(d) $(9.109558 \times 10^{-31} + 1.67252 \times 10^{-27} - 1.67482 \times 10^{-27})(2.997925 \times 10^{8})^2 =$

UNIT CONVERSIONS

1-53. Use the information from the inside back cover to make the following conversions, expressing your results to the correct number of significant figures:

(a) 1.00 liter to quarts
(b) 186,000 miles per second to meters per second
(c) $8.314\ J\cdot K^{-1}\cdot mol^{-1}$ to $cal\cdot K^{-1}\cdot mol^{-1}$

1-54. Use the information from the inside back cover to make the following conversions, expressing your results to the correct number of significant figures:

(a) 325 feet to meters
(b) 1.54 angstroms (Å) to picometers and to nanometers
(c) 175 pounds to kilograms

1-55. A lightyear is the distance light travels in 1 year. The speed of light is 3.00×10^8 m·s^{-1}. Compute the distance in meters and in miles that light travels in 1 year.

1-56. In older U.S. cars, total cylinder volume is expressed in cubic inches. Compute the total cylinder volume in liters of a 454-cubic-inch engine.

1-57. You are shopping for groceries and see that one bottle of soda has a volume of 2.00 liters and costs $1.49, whereas a six-pack of 16-oz bottles costs $3.50. Which is the better buy?

1-58. Compute the speed in meters per second of a 90-mph fastball. The pitcher's mound on a regulation baseball diamond is 60 ft, 6 in. from home plate. Compute the time in seconds that it takes a 90-mph fastball to travel from the pitcher's mound to home plate.

ADDITIONAL PROBLEMS

1-59. Assume that the diameter of an atom is about 10^5 times larger than the diameter of the nucleus. If the nucleus were magnified to the size of a ping-pong ball (approximately 3 cm diameter), estimate the diameter of the atom. Construct an analogy in terms of everyday objects.

1-60. Naturally occurring lithium consists of two isotopes, ^6Li and ^7Li, whose atomic masses and natural abundances are 6.0151 (7.42 percent) and 7.0160 (92.58 percent), respectively. Lithium-6 is extracted from natural lithium for use in the manufacture of nuclear weapons. (a) Compute the atomic mass of naturally occurring lithium. (b) Compute the percentage of ^6Li in a lithium sample with an atomic mass of 7.000.

1-61. Which of the following compounds has the highest mass percentage of nitrogen?

(a) N_2O_3
(b) HNO_3
(c) NH_3
(d) NH_4Cl
(e) PbN_6

1-62. What volume of pure (absolute) ethyl alcohol would you add to 50.0 mL of water to obtain a 50-percent-by-mass solution of ethyl alcohol? Take the densities of water and ethyl alcohol to be 1.00 g·cm^{-3} and 0.816 g·cm^{-3}, respectively.

1-63. Magnesium metal is obtained from the mineral dolomite, which is a mixture of magnesium carbonate, $MgCO_3$, and calcium carbonate, $CaCO_3$. Calculate the number of grams of magnesium metal that can be obtained from 1000 kg of dolomite that contains 11.2 percent $MgCO_3$.

1-64. Sulfuric acid sold for laboratory use consists of 96.7 percent sulfuric acid, H_2SO_4, by mass. The density of the solution is 1.845 g·cm^{-3}. Compute the number of kilograms and pounds of H_2SO_4 in a 2.20-L bottle of laboratory sulfuric acid.

1-65. You have only a 100-mL graduated cylinder available. What volume of benzene should you use to have 55 g of benzene? The density of benzene is 0.879 g·cm^{-3}.

1-66. An approximate, theoretical formula for the radius of a nucleus is

$$r = (1.2 \times 10^{-15}\ \text{m})A^{1/3}$$

where A is the mass number of the nucleus. Given the following atomic radii, calculate the ratio of the atomic diameter to the nuclear diameter:

Atom	Radius/pm*
carbon-12	77
argon-40	94
silver-108	144
radium-226	220

*The "pm" denotes picometers (1 pm = 10^{-12} m).

1-67. Naturally occurring silicon consists of three isotopes, ^{28}Si, ^{29}Si, and ^{30}Si, whose atomic masses are 27.9769, 28.9765, and 29.9738, respectively. The most abundant isotope is ^{28}Si, which accounts for 92.23 percent of naturally occurring silicon. Given that the observed atomic mass of silicon is 28.0855, calculate the percentages of ^{29}Si and ^{30}Si in nature.

1-68. Lithium—in the form of lithium carbonate (Li_2CO_3), lithium acetate ($LiC_2H_3O_2$), lithium citrate ($Li_3C_6H_5O_7$), or lithium sulfate (Li_2SO_4)—is used in the treatment of manic-depressive disorders. All these compounds are equally effective, but lithium carbonate is used most often because it contains a greater mass percentage of lithium. Calculate and compare the mass percentage of lithium in each of these four compounds.

1-69. What volume of acetone has the same mass as 10.0 mL of mercury? Take the densities of acetone and mercury to be 0.792 g·cm^{-3} and 13.59 g·cm^{-3}, respectively.

1-70. Magnesium chloride often occurs as an impurity in table salt, NaCl, causing the salt to "cake." If a 0.4500-g sample of table salt is found to be 61.11 percent chlorine by mass, what is the percentage of $MgCl_2$ in the sample?

Separation of Mixtures

Potassium compounds are obtained commercially by evaporation of brines. This photo shows a solar evaporation pond. The blue color is due to a dye that is mixed with the solution to absorb heat faster and hence speed up evaporation.

When determining the physical and chemical characteristics of an element or a compound, chemists must be certain that the substance is pure. Most substances in nature are **mixtures,** in which the component substances exist together without combining chemically, and it is often necessary for chemists to be able to separate a mixture into its various pure components. In this interchapter, we discuss some of the techniques that chemists use to separate mixtures and to purify substances.

A-1. SOME MIXTURES OF SOLIDS CAN BE SEPARATED BY DISSOLVING ONE OR MORE COMPONENTS

Let's consider the problem of separating a mixture composed of sugar, sand, iron filings, and gold dust into its four pure components (Figure

(a) The components of the mixture cannot be determined by casual inspection.

(b) The pure, separated components of the mixture.

(c) A microscopic view of the mixture. Note that the mixture is heterogeneous, that is, not uniform from point to point, and that each of the four components is clearly distinguishable.

(d) A magnet can be used to separate iron filings from the mixture. The iron filings are attracted by the magnet, but the other three components are not.

Figure A-1 A mixture of sugar, sand, iron filings, and gold dust.

A-1). The first thing to recognize about the mixture is that it is **heterogeneous;** that is, it is not uniform from point to point. The heterogeneity of the mixture can be clearly seen with the aid of a microscope (Figure A-1c). The mixture could be separated into its four components by using a tweezers, a microscope, and a lot of time and patience; however, a much more rapid separation can be achieved with other methods. The iron filings can be separated from the mixture by using a magnet (Figure A-1d), which attracts the magnetic iron particles but has no effect on the other three components. The same technique is used on a much larger scale in waste recycling to separate ferrous metals (iron, steel, nickel) from nonferrous refuse, such as aluminum, glass, paper, and plastics.

After the iron has been removed from our mixture, the sugar can be separated by adding water. Only the sugar dissolves in the water, leaving the sand and gold particles at the bottom of the container. The solution-plus-solid mixture can then be separated by **filtration** (Figure A-2). The sugar-water solution passes readily through the small pores in the filter paper, but the solid particles are too large to pass through and are trapped on the paper. The sugar can be recovered from the sugar-water solution by evaporating the water, a process that leaves the recrystallized solid sugar in the container.

The sand and gold dust can be separated by panning or by sluice-box techniques, which rely on the differences in density of the two solids to achieve a separation. In simple panning, water is added to the mixture of sand and gold and the slurry is swirled in a shallow, saucer-shaped metal pan. The dense (18.9 $g\cdot cm^{-3}$) gold particles collect near the center of the pan, whereas the less dense sand particles (2 to 3 $g\cdot cm^{-3}$) are swirled out of the pan. In the sluice-box technique, running water is passed over an agitated sand-gold mixture and the less dense sand particles rise higher in the water than the gold and are swept away in the stream of water.

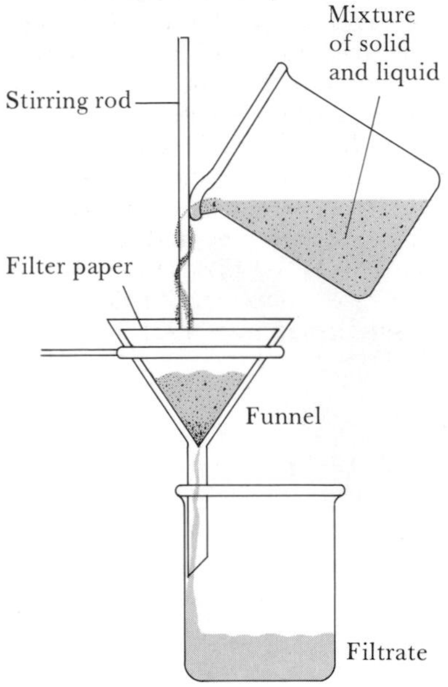

Mixture of solid and liquid
Stirring rod
Filter paper
Funnel
Filtrate

Figure A-2 Filtration can be used to separate a liquid from a solid. The liquid passes through filter paper, but the solid particles are too large to do so. Filter paper is available in a wide range of pore sizes, down to pores small enough (2.5×10^{-8}m) to remove bacteria (the smallest bacteria are about 1×10^{-7} m in diameter).

When fine gold particles are firmly attached to sand particles, the gold can be separated by shaking the mixture with liquid mercury, in which the gold dissolves. The sand, which floats on the mercury, is removed. The solution of gold in mercury is then separated by **distillation,** a process in which the mercury is boiled away and the solid gold is left behind. The mercury vapors are condensed (cooled and thereby converted back to liquid) in a **condenser.** A simple but typical distillation apparatus in shown in Figure A-3. Mercury distillation is usually carried out in an iron flask, and the mercury is collected and reused to extract more gold. Another example of distillation is the extraction of fresh water from seawater; the dissolved salts remain behind in the distillation flask after the water is boiled away.

The simple distillation apparatus shown in Figure A-3 is suitable for the separation of a liquid from a solution when a solid is dissolved in the liquid. The liquid is the only component that vaporizes or, in other words, is the only **volatile** component. The idea is that the liquid is boiled away, leaving the solid behind. If a solution contains two or more volatile components, however, such as alcohol and water, then the components can be separated by taking advantage of differences in boiling point. The separation of a solution with two or more volatile components is achieved by a process called fractional distillation (Chapter 11).

A-2. LIQUID OR GAS MIXTURES CAN BE SEPARATED BY CHROMATOGRAPHY

One of the most versatile and powerful separation techniques used by chemists is **chromatography,** of which there are several types. **Gas-liquid chromatography** (GLC) is used to separate mixtures of gases or volatile liquids. The mixture to be separated is vaporized and passed through a long, narrow column that is packed with fine solid particles thinly coated with some nonvolatile liquid. The gaseous mixture is swept through the column by an unreactive gas, such as helium, called the **carrier gas.** The different vapors move along the column at different rates because of their different degrees of interaction with the liq-

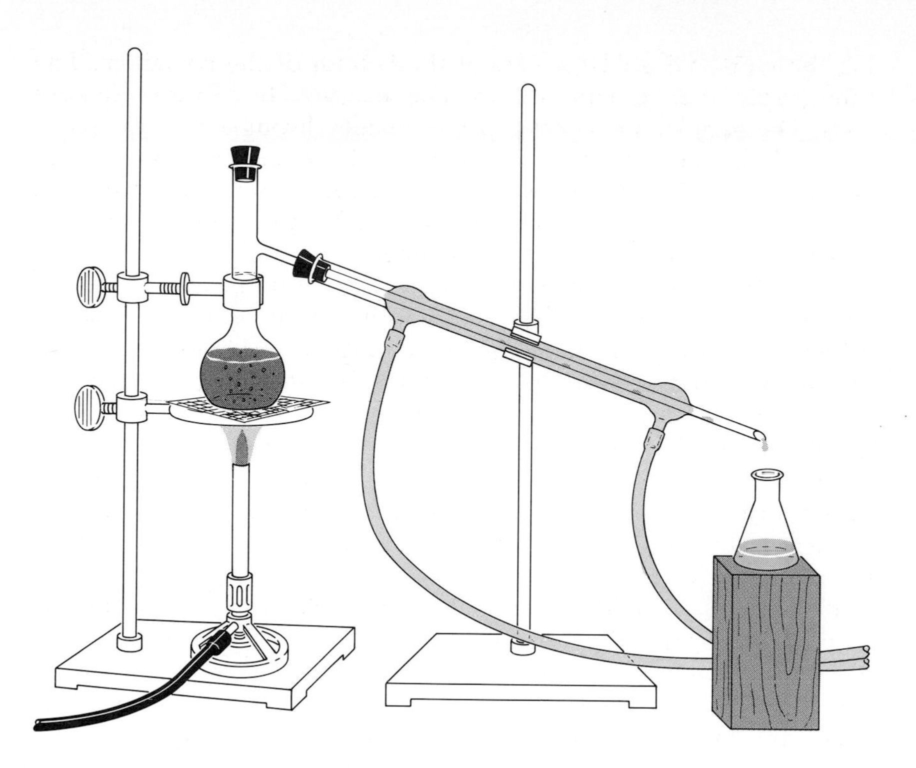

Figure A-3 A simple distillation apparatus can be used to separate a solid from a liquid in which it is dissolved. The solution in the distillation flask is heated, and the liquid is vaporized. The vapors rise in the distillation flask and pass into the condenser (the long, horizontal tube with the two hoses connected to it). The condenser is surrounded by a water jacket through which cooling water circulates. The vapor is cooled and condenses as it flows down the condenser tube and is collected in the flask at the right. The solid component of the solution remains behind in the distillation flask.

uid coating on the packed solid particles. Those vapors that interact least with the liquid coating travel through the column most rapidly and exit first. The vapors that interact most strongly with the liquid coating lag behind and exit last. This process is shown schematically in Figure A-4. As each component emerges, it passes through a detector and its presence is recorded on a moving piece of chart paper. The result is called a **chromatogram.** Gas chromatography is a powerful tool for separating complex mixtures in which the components are only very slightly different from each other (Figure A-5). Gasoline mixtures containing over 100 different compounds can be separated and characterized by gas chromatography.

Other types of chromatography are **paper chromatography** and **thin-layer chromatography** (TLC). Both methods are used to separate compounds in solution. They are based upon the fact that if a piece of porous paper is hung with one end dipping in a liquid, then the liquid will wick up along the paper. A drop of solution containing the substances to be separated is placed on one edge of the paper, and the liquid is allowed to pass over the spot. Those components that interact least strongly with the paper and most strongly with the liquid are drawn farthest up the paper with the moving liquid. This technique is used often in biochemistry to separate the products of a reaction. Figure A-6 shows a mixture of amino acids separated by thin-layer chromatography.

The name chromatography is derived from the original chromatographic separation method used in 1906 by Mikhail Tswett, a Russian botanist. Tswett separated a chlorophyll solution into various colored bands (chroma means "color") by passing the solution through a column packed with pulverized calcium carbonate. Because this method

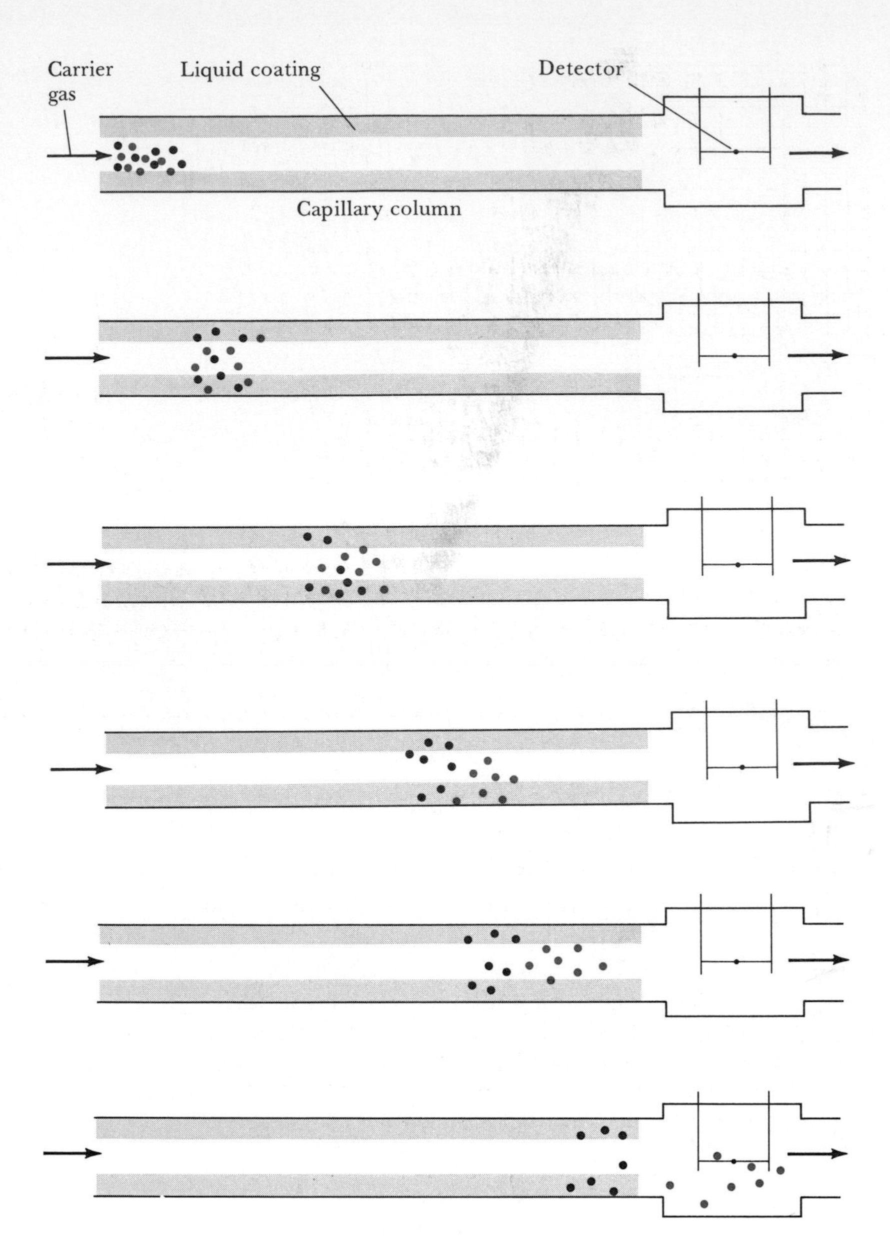

Figure A-4 Chromatographic separation takes place when a sample mixture (black and red circles) is driven by an unreactive carrier gas through a narrow column packed with a solid that is coated with a nonvolatile liquid. The sample mixture is separated because the liquid coating dissolves (and absorbs) various components of the sample to different degrees. After separation, each of the components passes through a detector and its presence is recorded on a chart, as in Figure A-5.

involves a liquid moving past a stationary solid, it is called **liquid-solid chromatography.** The different chlorophyll components move through the column at different rates because of their different degrees of interaction with the calcium carbonate. As additional solvent is passed through the column, the chlorophyll components that are more strongly adsorbed on the surface of the calcium carbonate crystals move more slowly down the column and the components are separated.

These are just a few of the methods that chemists use to separate mixtures and purify substances. It is possible to achieve extremely high degrees of purity. The purity required in the production of semiconductors is 99.99999 percent, and such purities are achieved routinely by semiconductor manufacturers.

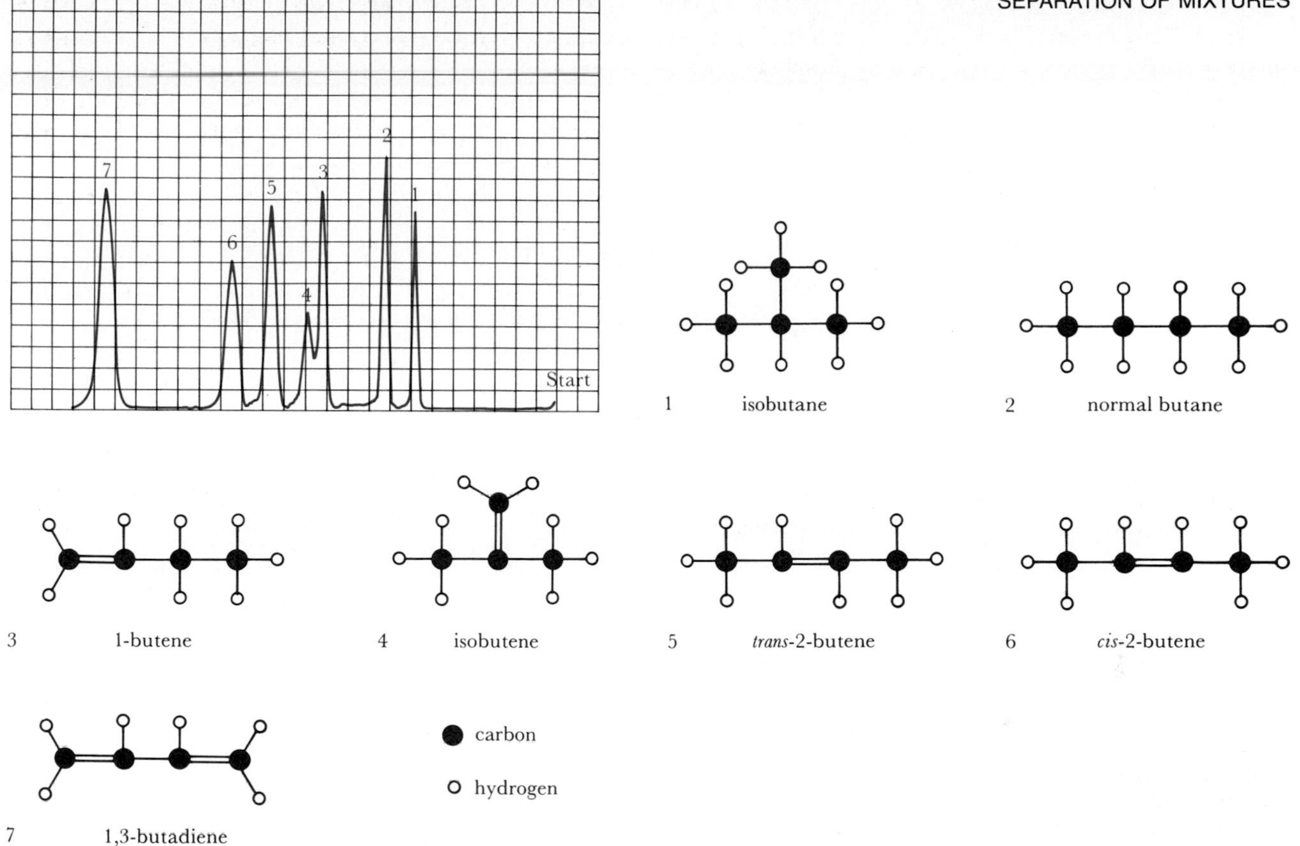

Figure A-5 A chromatogram of a mixture of seven hydrocarbons, all containing four carbon atoms. The names and molecular structures of the hydrocarbons appear to the right of the chromatogram. It is not important for you to understand these names and structures at this point. They just give an idea of relative sizes and shapes. The chromatogram shows how nicely these compounds are separated by a chromatography column. The separation of hydrocarbons is a common problem in the oil and synthetic rubber industries.

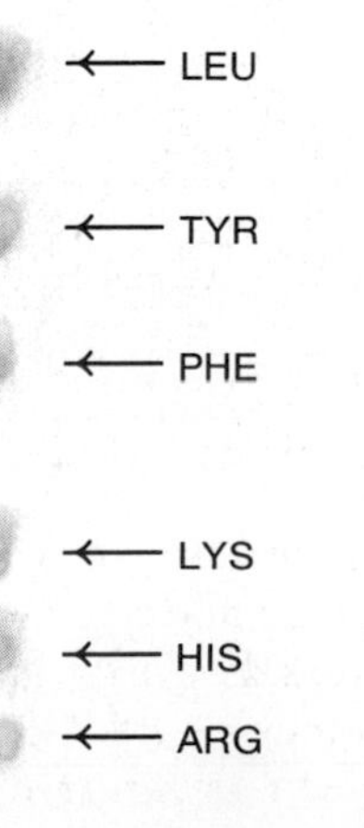

Figure A-6 Thin-layer chromatographic separation of an amino acid mixture. The identities of the particular amino acids are established by using known samples of the various amino acids in a comparison experiment.

TERMS YOU SHOULD KNOW

mixture 36
heterogeneous 37
filtration 38
distillation 38
condenser 38
volatile 38
chromatography 38
gas-liquid chromatography 38
carrier gas 38
chromatogram 39
paper chromatography 39
thin-layer chromatography 39
liquid-solid chromatography 40

QUESTIONS

A-1. Explain how you could separate iron filings from sand.

A-2. Explain how you could separate sugar from sand.

A-3. Describe filtration.

A-4. What are the contrasting physical properties of gold and sand on which panning of gold depends?

A-5. What is the role of the condenser in distillation?

A-6. What is meant when a liquid is said to be volatile?

A-7. What is distillation used for?

A-8. What is meant by a heterogeneous mixture?

A-9. What is the role of the carrier gas in gas-liquid chromatography?

A-10. Describe how two gases or vapors can be separated by gas-liquid chromatography.

A-11. Explain how paper chromatography works.

A-12. What is the origin of the term "chromatography"?

CHAPTER ▪ 2

CHEMICAL REACTIONS AND THE PERIODIC TABLE

When an aqueous solution of lead nitrate, $Pb(NO_3)_2$, is added to an aqueous solution of potassium iodide, KI, it produces lead iodide, PbI_2, a yellow precipitate.

There are 108 known chemical elements, and each undergoes a variety of characteristic chemical reactions. The study of these chemical reactions is simplified because there are patterns in the chemical properties of the elements. Instead of having to learn the chemical reactions of each individual element, we shall divide the elements into groups and study the reactions of each group. The arrangement of the elements into groups can be displayed in a form called the periodic table of the elements. The periodic table is the most useful guide to the study of chemistry and is the basis for understanding chemical reactions.

Throughout your study of chemistry you will encounter many chemical reactions. One classification scheme for chemical reactions consists of four categories: (1) combination reactions, (2) decomposition reactions, (3) single-replacement reactions, and (4) double-replacement reactions. Although not all chemical reactions fall into these four categories, a large number do, and it is helpful in first learning about chemical reactions to try to classify them into types rather than to view each one separately. In this chapter, we describe each of these four classes of reactions by way of examples. Our discussion of reaction types leads naturally to chemical nomenclature, to relative reactivities of metals, and to an introduction to the chemistry of acids and bases.

2-1. NEW SUBSTANCES ARE FORMED IN CHEMICAL REACTIONS

There are many different types of chemical reactions. Some of the simplest are those in which a metal and a nonmetal react directly with each other. For example, consider the reaction between sodium and chlorine. Sodium is a very reactive metal. It reacts spontaneously with the oxygen and water vapor in the air and must be stored under kerosene, an unreactive, oily liquid. Chlorine, a very reactive nonmetal, is a greenish yellow, irritating, toxic gas that attacks many metals. When sodium metal is dropped into a container of chlorine gas, there is a vigorous, spontaneous reaction. The product of the reaction is a white, crystalline solid, sodium chloride, which is ordinary table salt. We can represent the reaction of sodium with chlorine as

$$\underset{\substack{\text{very}\\\text{reactive}\\\text{metal}}}{\text{sodium metal}} + \underset{\substack{\text{very}\\\text{reactive}\\\text{nonmetal}}}{\text{chlorine gas}} \rightarrow \underset{\substack{\text{ordinary}\\\text{table}\\\text{salt}}}{\text{sodium chloride}} \qquad (2\text{-}1)$$

Reaction (2-1) illustrates the fact that the chemical properties of a product of a chemical reaction need bear no resemblance to the chemical properties of the reactants (Figure 2-1). *Entirely new substances are formed in chemical reactions.* Another chemical reaction that illustrates this fact is the reaction between hydrogen and oxygen to form water. Both hydrogen and oxygen are colorless, odorless gases. They form an explosive mixture that is set off easily by a spark or a flame. The reaction between hydrogen and oxygen can be written

$$\underset{\text{colorless gas}}{\text{hydrogen gas}} + \underset{\text{colorless gas}}{\text{oxygen gas}} \rightarrow \underset{\text{colorless liquid}}{\text{water}} \qquad (2\text{-}2)$$

The properties of water are radically different from those of hydrogen or oxygen.

Because hydrogen is the least dense gas, being about 15 times less dense than air, balloons and dirigibles used to be filled with hydrogen. This practice was discontinued after 1937, however, when the hydrogen-filled dirigible *Hindenburg* exploded.

It is cumbersome to write out the full names of the reactants and products in chemical reactions as we have been doing. Consequently, chemists have devised a shorthand way of describing the chemical changes that occur in chemical reactions. A chemical reaction is expressed in terms of the chemical formulas of the reactants and the

Figure 2-1 When sodium metal (a very reactive metal) reacts with chlorine gas (a very reactive nonmetal), the product is sodium chloride (ordinary table salt), illustrating the fact that entirely new substances are formed in chemical reactions.

products. The **reactants** are the substances that react with each other, and the **products** are the substances formed in the reaction. An arrow is used to indicate that the reactants are converted to products. For example, the reaction between sodium metal and chlorine gas to form sodium chloride is expressed by

$$Na(s) + Cl_2(g) \rightarrow NaCl(s) \qquad \text{(not balanced)} \qquad (2\text{-}3)$$

The symbol (*s*) after a chemical formula tells us that the substance is a solid, and the symbol (*g*) tells us that the substance is a gas.

The representation of a chemical reaction by writing the chemical formulas of the reactants and products separated by an arrow as we have done above is called a **chemical equation.** As noted, Equation (2-3) is not balanced because the number of chlorine atoms is not the same on the left (reactant) side as on the right (product) side.

2-2. A CHEMICAL EQUATION MUST BE BALANCED

An essential feature of chemical reactions is the conservation of each type of atom. Although new substances are formed in chemical reactions as a result of new arrangements of the atoms, *the individual atoms of various types are neither created nor destroyed in a chemical reaction.* The number of atoms of each element remains the same in a chemical reaction.

A chemical equation must always be balanced; that is, it must have the same number of each type of atom on both sides. Note that Equation (2-3) contains two atoms of chlorine on the left but only one atom of chlorine on the right. The law of conservation of matter requires that all the atoms that enter into a chemical reaction appear in the products. We can balance Equation (2-3) with respect to the chlorine atoms by placing a "2" in front of the NaCl(*s*) on the right-hand side of the equation:

$$Na(s) + Cl_2(g) \rightarrow 2NaCl(s) \qquad \text{(not balanced)}$$

Now there are two sodium atoms on the right but only one on the left. If we place a 2 in front of the Na(*s*) on the left, we obtain a **balanced chemical equation** for the reaction of sodium with chlorine:

$$2Na(s) + Cl_2(g) \rightarrow 2NaCl(s) \qquad \text{(balanced)} \qquad (2\text{-}4)$$

Both sides of Equation (2-4) contain the same number of each kind of atom.

We balance chemical equations by placing the appropriate numbers, called **balancing coefficients,** in front of the chemical formulas. The chemical formulas of the reactants and products themselves are fixed and should not be altered by changing the subscripts. Equation (2-3) should not be balanced by changing NaCl to $NaCl_2$. The chemical formula of sodium chloride is NaCl, not $NaCl_2$. In fact, there is no such compound as $NaCl_2$.

Let's consider now the reaction of hydrogen with oxygen. Recall that hydrogen and oxygen exist as diatomic molecules. We first write Reaction (2-2) as an unbalanced chemical equation:

$$H_2(g) + O_2(g) \rightarrow H_2O(l) \qquad \text{(not balanced)}$$

where the (l) after the formula tells us that the substance is a *l*iquid. In this equation, there are two oxygen atoms on the left (in O_2) and one oxygen atom on the right (in H_2O). If we place a 2 in front of the $H_2O(l)$, then the equation is balanced with respect to oxygen atoms:

$$H_2(g) + O_2(g) \rightarrow 2H_2O(l) \qquad \text{(not balanced)}$$

Now there are four hydrogen atoms on the right (2 × 2) and only two on the left. We balance the hydrogen atoms by placing a 2 in front of the $H_2(g)$:

$$2H_2(g) + O_2(g) \rightarrow 2H_2O(l) \qquad \text{(balanced)}$$

to obtain the balanced equation for the reaction of hydrogen with oxygen. Once again, note that a chemical equation is balanced by placing coefficients *in front of* the formulas of the reactants and products; the formulas themselves are not altered. Figure 2-2 is a schematic representation of the reaction between hydrogen and oxygen.

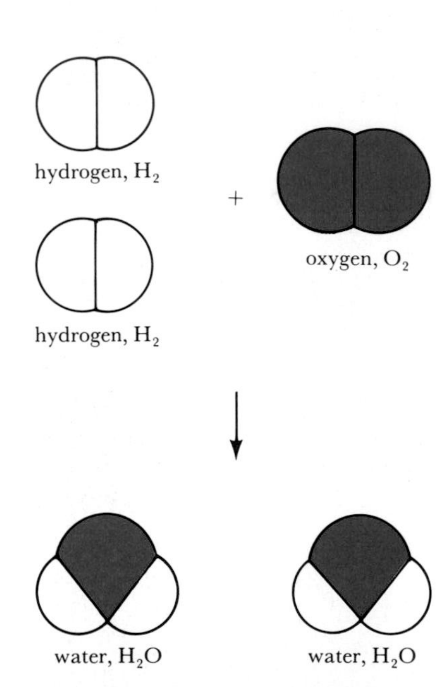

Figure 2-2 The chemical reaction between hydrogen and oxygen. The figure illustrates that a chemical reaction is a rearrangement of the atoms in reactant molecules into product molecules. The numbers of each kind of atom are the same before and after the reaction. The spatial arrangement of atoms in molecules is discussed in Chapter 9.

Example 2-1: Sodium metal reacts vigorously with water. The reactants and products of the reaction are

$Na(s)$	$+ H_2O(l) \rightarrow$	$NaOH(s)$	+	$H_2(g)$
sodium metal	water	sodium hydroxide		hydrogen gas

Balance this chemical equation.

Solution: Let's balance the equation with respect to hydrogen atoms first. If we place a 2 in front of $NaOH(s)$ and a 2 in front of $H_2O(l)$, there will be four hydrogen atoms on each side of the equation:

$$Na(s) + 2H_2O(l) \rightarrow 2NaOH(s) + H_2(g) \qquad \text{(not balanced)}$$

Now we need to balance the sodium atoms. If we place a 2 in front of $Na(s)$, we have a balanced equation:

$$2Na(s) + 2H_2O(l) \rightarrow 2NaOH(s) + H_2(g) \qquad \text{(balanced)}$$

Sodium hydroxide is a white, translucent solid used in the manufacture of paper and soaps and in petroleum refining. It is extremely corrosive to the skin and other tissues and is sometimes called caustic soda or lye.

This method of balancing chemical equations is called **balancing by inspection.** With a little practice, you can become proficient in balancing certain types of chemical equations this way. (Problems 2-1 through 2-8 should be worked for practice in balancing equations.)

2-3. ELEMENTS CAN BE GROUPED ACCORDING TO THEIR CHEMICAL PROPERTIES

By the 1860s, more than 60 elements had been discovered and many chemists had begun to notice some pattern in the chemical properties of certain elements. For example, consider the three metals lithium (Li), sodium (Na), and potassium (K). All three of these metals are less dense than water, are soft enough to be cut with a knife, have fairly low melting points (below 200°C), and are very reactive. They all react spontaneously with oxygen and water. Just as sodium reacts vigorously with chlorine, so do lithium and potassium:

$$2\text{Li}(s) + \text{Cl}_2(g) \rightarrow 2\text{LiCl}(s)$$

$$2\text{Na}(s) + \text{Cl}_2(g) \rightarrow 2\text{NaCl}(s)$$

$$2\text{K}(s) + \text{Cl}_2(g) \rightarrow 2\text{KCl}(s)$$

The product in all three cases is a white, unreactive, crystalline solid that dissolves readily in water.

Example 2-1 shows that sodium reacts with water to produce sodium hydroxide (NaOH) and hydrogen. Lithium and potassium (Figure 2-3a) undergo a similar reaction:

$$2\text{Li}(s) + 2\text{H}_2\text{O}(l) \rightarrow 2\text{LiOH}(s) + \text{H}_2(g)$$

$$2\text{K}(s) + 2\text{H}_2\text{O}(l) \rightarrow 2\text{KOH}(s) + \text{H}_2(g)$$

Just like sodium hydroxide, lithium hydroxide and potassium hydroxide are corrosive, white, translucent solids. Lithium, sodium, and potassium have similar chemical properties and can be considered as a group. Because the hydroxides of these metals are **alkaline,** a property discussed in Section 2-11, they are called **alkali metals.**

There are other groups of elements that share similar chemical properties. For example, magnesium (Mg), calcium (Ca), strontium (Sr), and barium (Ba) have many chemical properties in common. As a group, these metals are called **alkaline earth metals.** They all burn brightly when heated in oxygen to form white oxides:

$$2\text{Mg}(s) + \text{O}_2(g) \rightarrow 2\text{MgO}(s)$$

$$2\text{Ca}(s) + \text{O}_2(g) \rightarrow 2\text{CaO}(s)$$

$$2\text{Sr}(s) + \text{O}_2(g) \rightarrow 2\text{SrO}(s)$$

$$2\text{Ba}(s) + \text{O}_2(g) \rightarrow 2\text{BaO}(s)$$

Calcium (see Figure 2-3b), strontium, and barium react slowly with water to yield a metal hydroxide and hydrogen gas:

$$\text{Ca}(s) + 2\text{H}_2\text{O}(l) \rightarrow \text{Ca(OH)}_2(s) + \text{H}_2(g)$$

$$\text{Sr}(s) + 2\text{H}_2\text{O}(l) \rightarrow \text{Sr(OH)}_2(s) + \text{H}_2(g)$$

$$\text{Ba}(s) + 2\text{H}_2\text{O}(l) \rightarrow \text{Ba(OH)}_2(s) + \text{H}_2(g)$$

Magnesium undergoes a similar reaction at high temperatures. We see, then, that these four metals have similar chemical properties and can be placed into a group, just as lithium, sodium, and potassium can.

(a)

(b)

Figure 2-3 Reactions of the alkali metal potassium and the alkaline earth metal calcium with water. (a) Potassium reacts violently with water to produce potassium hydroxide and $\text{H}_2(g)$. (b) Calcium reacts slowly with cold water to produce calcium hydroxide and $\text{H}_2(g)$.

Example 2-2: Given that magnesium reacts with sulfur to form magnesium sulfide, MgS(s), write a balanced chemical equation for the reaction between calcium and sulfur.

Solution: The chemical equation for the reaction of Mg(s) with S(s) is

$$\text{Mg}(s) + \text{S}(s) \rightarrow \text{MgS}(s)$$

Therefore, by analogy, we predict

$$\text{Ca}(s) + \text{S}(s) \rightarrow \underset{\text{calcium sulfide}}{\text{CaS}(s)}$$

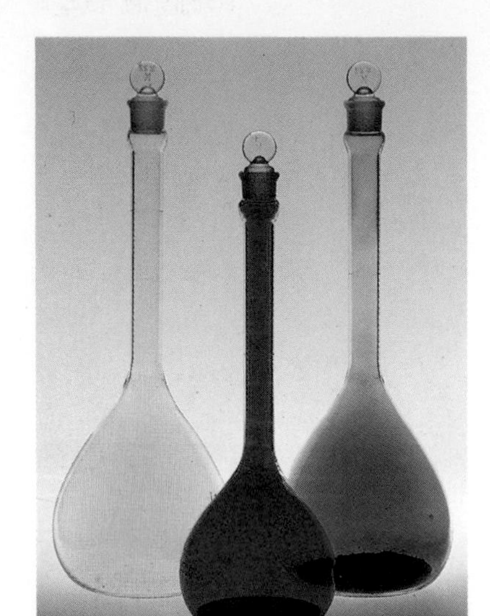

Figure 2-4 *Left to right:* Chlorine, bromine, and iodine, members of the halogen family.

Another group of elements having similar chemical properties consists of the nonmetals fluorine (F_2), chlorine (Cl_2), bromine (Br_2), and iodine (I_2) (Figure 2-4). These elements are all very reactive and react with most metals and nonmetals. As a group, they are called **halogens,** meaning "salt formers." The halogens react with the alkali metals to give white, crystalline solids called **halides.** The specific names of the halides are fluoride, chloride, bromide, and iodide. For example, in the case of sodium we have

$$2Na(s) + F_2(g) \rightarrow 2NaF(s) \qquad \text{sodium fluoride}$$

$$2Na(s) + Cl_2(g) \rightarrow 2NaCl(s) \qquad \text{sodium chloride}$$

$$2Na(s) + Br_2(l) \rightarrow 2NaBr(s) \qquad \text{sodium bromide}$$

$$2Na(s) + I_2(s) \rightarrow 2NaI(s) \qquad \text{sodium iodide}$$

The halogens react with the alkaline earth metals to give $MCl_2(s)$, $MBr_2(s)$, and $MI_2(s)$, where M is any alkaline earth metal.

Table 2-1 The chemical properties of some elements, listed in order of increasing atomic mass

Atomic mass	Element	Symbol	Properties	Formula of halogen compound*
6.9	lithium	Li	very reactive metal	LiX
9.0	beryllium	Be	reactive metal	BeX_2
10.8	boron	B	semimetal†	BX_3
12.0	carbon	C	nonmetallic solid	CX_4
14.0	nitrogen	N	nonmetallic diatomic gas	NX_3
16.0	oxygen	O	nonmetallic, moderately reactive diatomic gas	OX_2
19.0	fluorine	F	very reactive diatomic gas	FX
20.2	neon	Ne	very unreactive monatomic gas	none
	⋮			
23.0	sodium	Na	very reactive metal	NaX
24.3	magnesium	Mg	reactive metal	MgX_2
27.0	aluminum	Al	metal	AlX_3
28.1	silicon	Si	semimetal†	SiX_4
31.0	phosphorus	P	nonmetallic solid	PX_3
32.1	sulfur	S	nonmetallic solid	SX_2
35.5	chlorine	Cl	very reactive diatomic gas	ClX
39.9	argon	Ar	very unreactive monatomic gas	none

*X stands for F, Cl, Br, or I.

†Boron and silicon are called semimetals because they have properties that are intermediate between those of the metals and those of the nonmetals.

2-4. THE ELEMENTS SHOW A PERIODIC PATTERN WHEN LISTED IN ORDER OF INCREASING ATOMIC NUMBER

After Dalton proposed the atomic theory, the concept of atomic mass and the experimental determination of atomic masses took on increasing importance. In 1869, the Russian chemist Dmitri Mendeleev arranged the elements in order of increasing atomic mass and was able to show that the chemical properties of the elements exhibit periodic behavior. To illustrate the idea of Mendeleev's observation, let's start with the element lithium and arrange the succeeding elements in order of increasing atomic mass, as shown in Table 2-1. If we look at Table 2-1 carefully, we see that the chemical properties of the elements show a remarkably repetitive, or periodic, pattern. The variations in properties as we go from lithium to neon are repeated as we go from sodium to argon. The repeating pattern (or periodicity) is seen more clearly if we arrange the elements horizontally:

Li	Be	B	C	N	O	F	Ne
lithium	beryllium	boron	carbon	nitrogen	oxygen	fluorine	neon
Na	Mg	Al	Si	P	S	Cl	Ar
sodium	magnesium	aluminum	silicon	phosphorus	sulfur	chlorine	argon

Notice that elements with similar chemical properties, for example, lithium and sodium or fluorine and chlorine, are placed in the same column.

Figure 2-5 presents a modern version of Mendeleev's **periodic table of the elements.** The modern version is more complicated than the abbreviated version shown above, primarily because it contains many more elements than just the 16 listed in Table 2-1. In the modern periodic table in Figure 2-5, the elements are arranged in order of

Figure 2-5 A modern version of the periodic table of the elements. The elements are ordered according to increasing atomic number, and their chemical properties show a periodic pattern. Elements that appear in the same column have similar chemical properties.

Row	1	2																									3	4	5	6	7	8	Row
1	1 H																															2 He	1
2	3 Li	4 Be																									5 B	6 C	7 N	8 O	9 F	10 Ne	2
3	11 Na	12 Mg																									13 Al	14 Si	15 P	16 S	17 Cl	18 Ar	3
4	19 K	20 Ca															21 Sc	22 Ti	23 V	24 Cr	25 Mn	26 Fe	27 Co	28 Ni	29 Cu	30 Zn	31 Ga	32 Ge	33 As	34 Se	35 Br	36 Kr	4
5	37 Rb	38 Sr															39 Y	40 Zr	41 Nb	42 Mo	43 Tc	44 Ru	45 Rh	46 Pd	47 Ag	48 Cd	49 In	50 Sn	51 Sb	52 Te	53 I	54 Xe	5
6	55 Cs	56 Ba	57 La	58 Ce	59 Pr	60 Nd	61 Pm	62 Sm	63 Eu	64 Gd	65 Tb	66 Dy	67 Ho	68 Er	69 Tm	70 Yb	71 Lu	72 Hf	73 Ta	74 W	75 Re	76 Os	77 Ir	78 Pt	79 Au	80 Hg	81 Tl	82 Pb	83 Bi	84 Po	85 At	86 Rn	6
7	87 Fr	88 Ra	89 Ac	90 Th	91 Pa	92 U	93 Np	94 Pu	95 Am	96 Cm	97 Bk	98 Cf	99 Es	100 Fm	101 Md	102 No	103 Lr																7

increasing atomic number instead of increasing atomic mass. With a few exceptions, the order is the same in both cases. The idea of atomic number was not developed until the early 1900s, about 40 years after Mendeleev's first periodic table.

2-5. ELEMENTS IN THE SAME COLUMN IN THE PERIODIC TABLE HAVE SIMILAR CHEMICAL PROPERTIES

Notice in Figure 2-5 that lithium, sodium, and potassium occur in the far left column of the periodic table. All the elements in that column have similar chemical properties. Although we have not discussed rubidium (Rb), cesium (Cs), or francium (Fr), the fact that these elements occur in the same column as lithium, sodium, and potassium suggests that they undergo similar chemical reactions. Francium is a radioactive element not found in nature, but rubidium and cesium are light, soft, very reactive metals. Rubidium and cesium react vigorously with the halogens, water, hydrogen, oxygen, and many other substances. By analogy with the reactions of sodium that we discussed earlier, we can predict that, for example,

$$2Rb(s) + Cl_2(g) \rightarrow 2RbCl(s)$$

$$2Cs(s) + 2H_2O(l) \rightarrow 2CsOH(s) + H_2(g)$$

and that other reactions of rubidium and cesium are similar to those of, say, sodium.

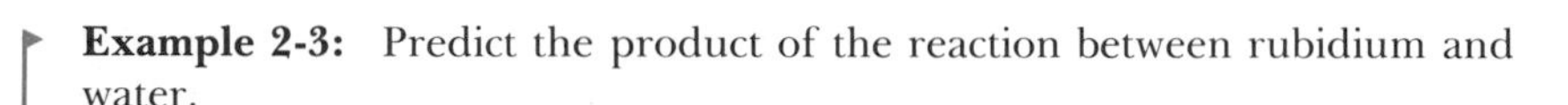

Example 2-3: Predict the product of the reaction between rubidium and water.

Solution: Because rubidium, like sodium, is an alkali metal, we predict that rubidium reacts with water to form rubidium hydroxide, $RbOH(s)$, and hydrogen gas. The balanced equation for the reaction is

$$2Rb(s) + 2H_2O(l) \rightarrow 2RbOH(s) + H_2(g)$$

Figure 2-6 When an electric discharge is passed through a noble gas, light of a characteristic color is emitted. The tubes in the photo contain helium (*left*), neon (*center*), and argon (*right*).

The far left column in the periodic table is labeled 1, and so the elements in that column are referred to as the **Group 1 metals.** As noted previously, the Group 1 metals are also called the alkali metals. The metals in the column labeled 2 are called the **Group 2 metals,** or the alkaline earth metals. All the elements in Group 2 are reactive metals and undergo similar chemical reactions.

Elements in the same column in the rest of the periodic table also share similar chemical properties. This similarity is particularly strong in the columns headed by numbers. The halogens, which we have seen behave similarly, occur in Group 7. The extreme right column of the periodic table (Group 8) contains the **noble gases,** which are characterized primarily by their lack of chemical reactivity. Prior to 1962, they were called the **inert gases** because no compounds of these gases were known (Figure 2-6). In 1962, however, xenon was shown to form compounds with fluorine and oxygen, the most reactive nonmetals. Krypton fluorides are also known, but no stable compounds of helium, neon, or argon have been made yet, and the noble gas elements are generally very unreactive.

Example 2-4: Phosphorus is a nonmetallic solid that occurs in white, red, and black forms. White phosphorus spontaneously bursts into flame in the presence of oxygen to produce the oxide P_4O_6. Use the periodic table to predict the reaction between arsenic (atomic number 33) and oxygen.

Solution: Arsenic occurs in the same group (Group 5) as phosphorus, and so we predict, by analogy with phosphorus, that arsenic reacts with oxygen to produce As_4O_6. The balanced chemical equation for the reaction is

$$4As(s) + 3O_2(g) \rightarrow As_4O_6(s)$$

Arsenic compounds are well known to be poisonous, the lethal dose of As_4O_6 being about 0.1 g for an average adult male. Small amounts of arsenic compounds, however, promote the growth of red blood cells in bone marrow, and the human body normally contains about 5 mg of arsenic.

The periodic table shown in Figure 2-5 differs from the original table proposed by Mendeleev. A number of elements had not yet been discovered in 1869, and Mendeleev had the genius to leave gaps in his periodic table to accommodate undiscovered elements. For example, in 1869 the element following zinc in atomic mass was arsenic. Yet Mendeleev felt that arsenic belonged in Group 5 rather than in Group 3. He boldly proposed that there were two as yet undiscovered elements between zinc and arsenic and predicted many of the properties of these two elements prior to their discovery. Table 2-2 compares Mendeleev's 1869 predictions with the actual properties of the element gallium (atomic number 31), which was not discovered until 1875. Gallium has an unusually low melting point for a metal (30°C) and melts when held in the hand (Figure 2-7).

The periodic table is the most important and useful concept in chemistry. Almost every general chemistry classroom and laboratory has a periodic table hanging on the wall. You may have noticed that the periodic table hanging in your classroom or laboratory is different from the one given in Figure 2-5. A more common version of the periodic table is shown in Figure 2-8. The only difference between Figures 2-5 and 2-8 is that elements 57 through 70, collectively called the **lanthanide series,** and elements 89 through 102, called the **acti-**

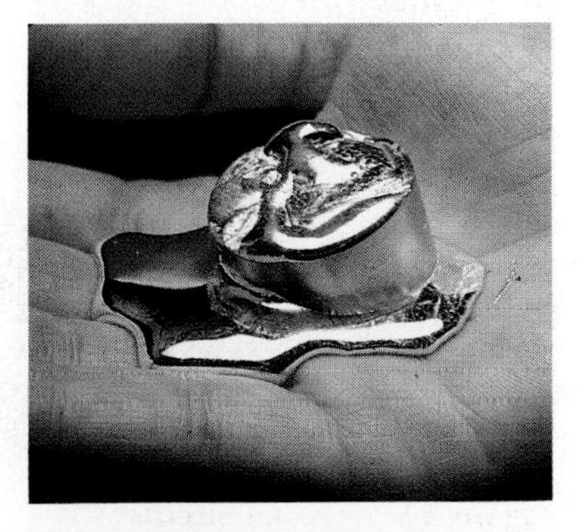

Figure 2-7 Gallium metal has a melting point of 30°C, and so a piece of gallium melts when held in the hand (human body temperature is 37°C).

Table 2-2 Comparison of Mendeleev's predictions and actual experimental values for the properties of gallium

Property	Predicted	Observed
atomic mass	69	69.7
density/$g \cdot cm^{-3}$	6.0	5.9
melting point	low	30°C
boiling point	high	2400°C
formula of oxide	M_2O_3	Ga_2O_3

▪ Most periodic tables have the elements cerium (58) through lutetium (71) and thorium (90) through lawrencium (103) placed below. See, however, the paper by William B. Jensen in the *Journal of Chemical Education* [*59,* 634 (1982)].

nide series, have been cut out and placed at the bottom of the table. Some versions of the periodic table place lanthanum (La) and actinium (Ac) in the table where we have lutetium (Lu) and lawrencium (Lr) and place elements 58 through 71 (cerium through lutetium) and elements 90 through 103 (thorium through lawrencium) below. In addition, many versions have headings for every column. For example, the periodic table in your lecture room or laboratory may have the heading 3A or 3B for the third column. The differences between the table shown in Figure 2-8 and any others are minor and of no real consequence.

The periodic table contains all the known chemical elements and shows the periodic relationships among them. Elements in the same column are said to belong to the same **group** or **family.** The horizontal rows in the periodic table often are called **periods.** Your progress in

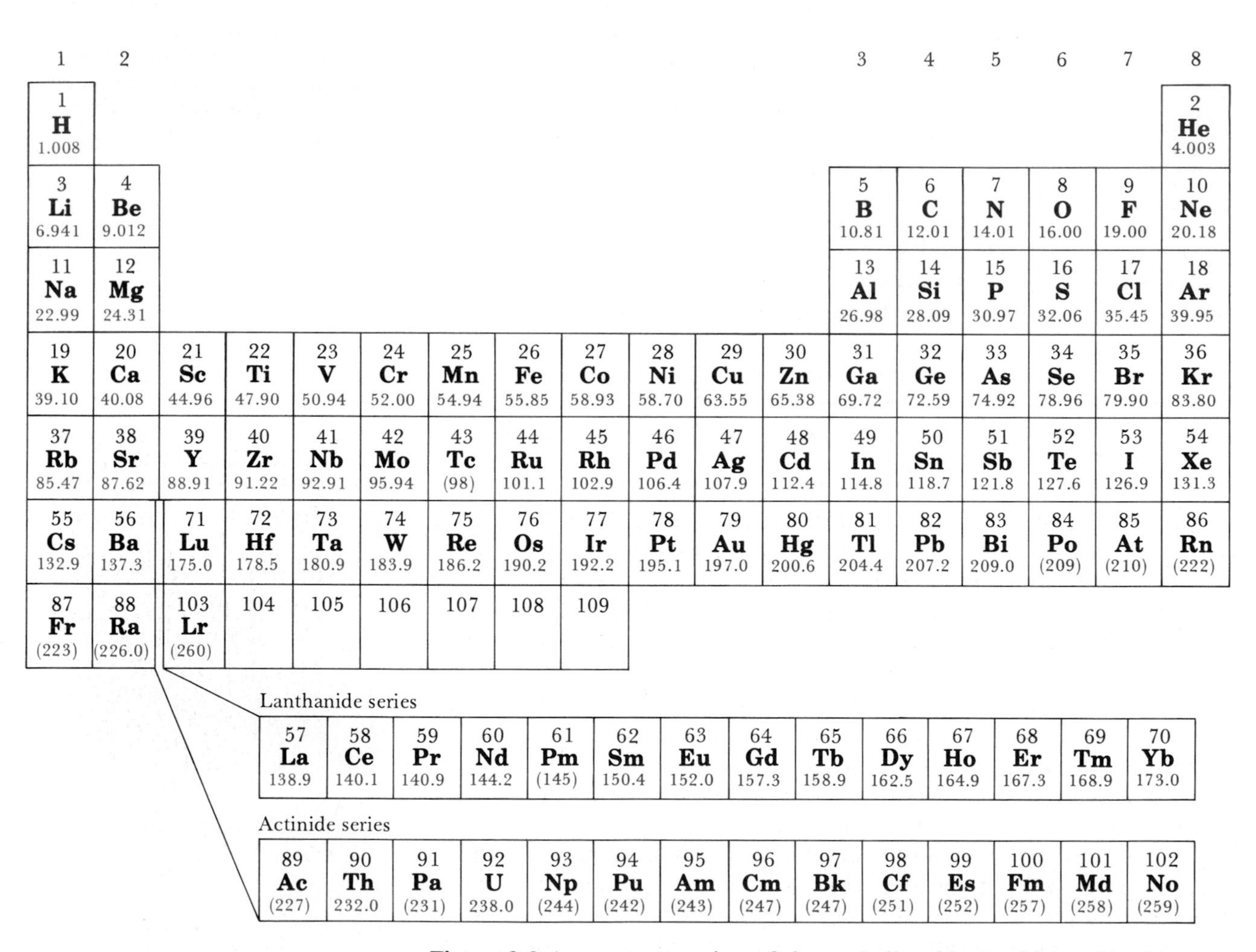

1	2											3	4	5	6	7	8
1 **H** 1.008																	2 **He** 4.003
3 **Li** 6.941	4 **Be** 9.012											5 **B** 10.81	6 **C** 12.01	7 **N** 14.01	8 **O** 16.00	9 **F** 19.00	10 **Ne** 20.18
11 **Na** 22.99	12 **Mg** 24.31											13 **Al** 26.98	14 **Si** 28.09	15 **P** 30.97	16 **S** 32.06	17 **Cl** 35.45	18 **Ar** 39.95
19 **K** 39.10	20 **Ca** 40.08	21 **Sc** 44.96	22 **Ti** 47.90	23 **V** 50.94	24 **Cr** 52.00	25 **Mn** 54.94	26 **Fe** 55.85	27 **Co** 58.93	28 **Ni** 58.70	29 **Cu** 63.55	30 **Zn** 65.38	31 **Ga** 69.72	32 **Ge** 72.59	33 **As** 74.92	34 **Se** 78.96	35 **Br** 79.90	36 **Kr** 83.80
37 **Rb** 85.47	38 **Sr** 87.62	39 **Y** 88.91	40 **Zr** 91.22	41 **Nb** 92.91	42 **Mo** 95.94	43 **Tc** (98)	44 **Ru** 101.1	45 **Rh** 102.9	46 **Pd** 106.4	47 **Ag** 107.9	48 **Cd** 112.4	49 **In** 114.8	50 **Sn** 118.7	51 **Sb** 121.8	52 **Te** 127.6	53 **I** 126.9	54 **Xe** 131.3
55 **Cs** 132.9	56 **Ba** 137.3	71 **Lu** 175.0	72 **Hf** 178.5	73 **Ta** 180.9	74 **W** 183.9	75 **Re** 186.2	76 **Os** 190.2	77 **Ir** 192.2	78 **Pt** 195.1	79 **Au** 197.0	80 **Hg** 200.6	81 **Tl** 204.4	82 **Pb** 207.2	83 **Bi** 209.0	84 **Po** (209)	85 **At** (210)	86 **Rn** (222)
87 **Fr** (223)	88 **Ra** (226.0)	103 **Lr** (260)	104	105	106	107	108	109									

Lanthanide series

57 **La** 138.9	58 **Ce** 140.1	59 **Pr** 140.9	60 **Nd** 144.2	61 **Pm** (145)	62 **Sm** 150.4	63 **Eu** 152.0	64 **Gd** 157.3	65 **Tb** 158.9	66 **Dy** 162.5	67 **Ho** 164.9	68 **Er** 167.3	69 **Tm** 168.9	70 **Yb** 173.0

Actinide series

89 **Ac** (227)	90 **Th** 232.0	91 **Pa** (231)	92 **U** 238.0	93 **Np** (244)	94 **Pu** (242)	95 **Am** (243)	96 **Cm** (247)	97 **Bk** (247)	98 **Cf** (251)	99 **Es** (252)	100 **Fm** (257)	101 **Md** (258)	102 **No** (259)

Figure 2-8 A common version of the periodic table. In this version the lanthanide series (elements 57 through 70) and the actinide series (elements 89 through 102) have been placed at the bottom of the table, for two reasons. First, the separation leads to a more compact table. Second, and more important, all the elements in each series have exceptionally similar chemical properties, and in effect, all can be assigned to just one position in the periodic table. The number under the symbol for each element is the atomic mass.

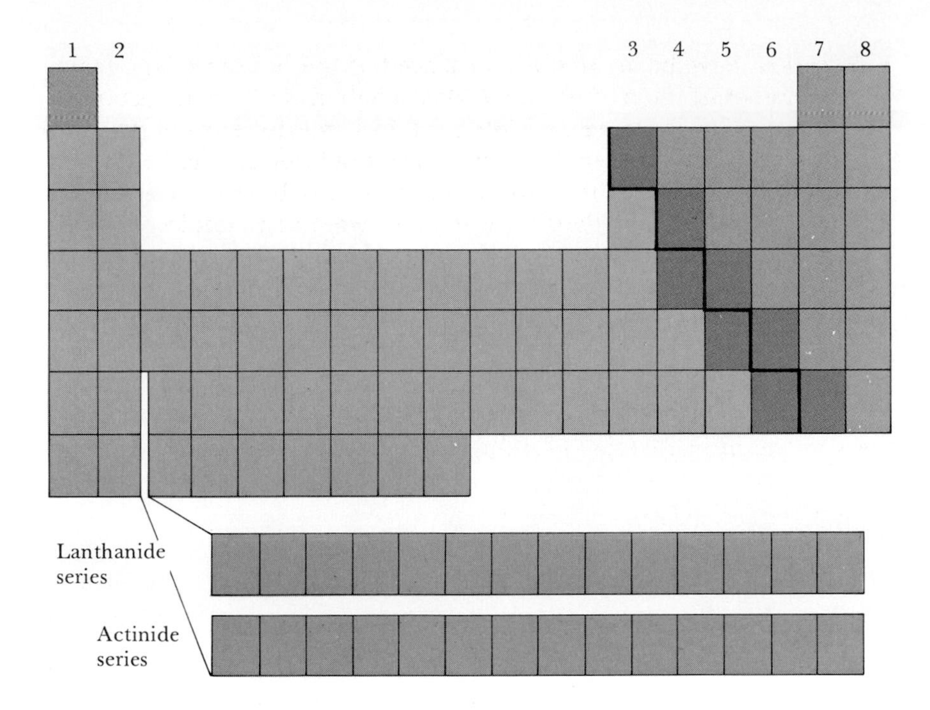

Figure 2-9 The position of metals (*light blue*) and nonmetals (*light red*) in the periodic table. The nonmetals appear only at the far right of the table, and the metals appear at the left. The elements along the steplike border between metals and nonmetals are the semimetals (*light purple*).

learning basic chemistry will be greatly aided by an understanding and appreciation of the periodic table. In the remainder of this chapter we study some of the general features and uses of the periodic table.

2-6. ELEMENTS ARE ARRANGED AS MAIN-GROUP ELEMENTS, TRANSITION METALS, AND INNER TRANSITION METALS

Metals and nonmetals occur in separate regions of the periodic table, as shown in Figure 2-9. The nonmetals are on the right-hand side and are separated from the metals by a zigzag line. As might be expected, the elements on the border between metals and nonmetals (light purple elements in Figure 2-9) have properties that are intermediate between those of metals and those of nonmetals. Such elements are called **semimetals** and are brittle, dull solids (Figure 2-10). Because semimetals conduct electricity and heat less well than metals but better than nonmetals, they are called **semiconductors.** The semimetals silicon and germanium are widely used in the manufacture of semiconducting devices and transistors (see Interchapter H).

For the elements in groups headed by a number, there is a continuous progression from metallic to nonmetallic properties as we move from left to right across the periodic table. The most reactive metals are the Group 1 metals, and the most reactive nonmetals are the halogens (Group 7). The noble gases, which are unusually nonreactive, occupy the far right column (Group 8). The elements in groups headed by numbers in Figure 2-8 are called the **main-group elements.**

The elements in the groups not headed by numbers are called **transition metals.** Within any row of transition metals, the chemical properties are more similar across the row than is the case for the main-group

Figure 2-10 The semimetals are brittle solids. Two semimetals, boron (*top*) and silicon (*bottom*), are shown here.

Figure 2-11 The $3d$ transition series metals. *Top row (left to right)*: Ti, Zn, Cu, Ni, and Co. *Bottom row:* Sc, V, Cr, Mn, and Fe. (The elements are not arranged in order of the periodic table.)

elements. Many of the transition metals are probably already familiar to you (Figure 2-11). Iron, nickel, chromium, copper, tungsten, and titanium are widely used in alloys for structural materials and play a key role in the world's technology. The precious metals—gold, platinum, and silver—are used as hard currency, in jewelry, and in high-quality electronic circuits. The transition metals vary greatly in abundance. Iron and titanium are plentiful, whereas rhenium (Re) and hafnium (Hf) are rare.

The characteristics of the transition metals vary from group to group; yet they all are characterized by high densities and high melting points. In addition, unlike the compounds of the Group 1 and Group 2 metals, many compounds of the transition metals are colored. The metals with the greatest densities, iridium (Ir), 22.65 $g{\cdot}cm^{-3}$, and osmium (Os), 22.61 $g{\cdot}cm^{-3}$, and the highest melting point, tungsten (W), 3410°C, are transition metals.

The elements in the lanthanide series and the actinide series—the two series that begin with lanthanum ($Z = 57$) and actinium ($Z = 89$) in Figure 2-8—together are called the **inner transition metals.** The elements in each of these two series have remarkably similar chemical properties. The lanthanides are also called the **rare-earth elements** because they were once thought to occur only in very small quantities. The actinides are radioactive elements, most of which do not occur in nature but are produced in nuclear reactions. Figure 2-12 indicates the position in the periodic table of the three main classes of elements.

Example 2-5: By referring to the periodic table, classify each of the following as a main-group element, a transition metal, or an inner transition metal. If it is a main-group element, indicate its group and whether it is a metal, a nonmetal, or a semimetal.

Fr Am Ge

Solution:

Symbol	Name	Atomic number	Classification
Fr	francium	87	main-group element (Group 1 metal)
Am	americium	95	inner transition metal (actinide)
Ge	germanium	32	main-group element (Group 4 semimetal)

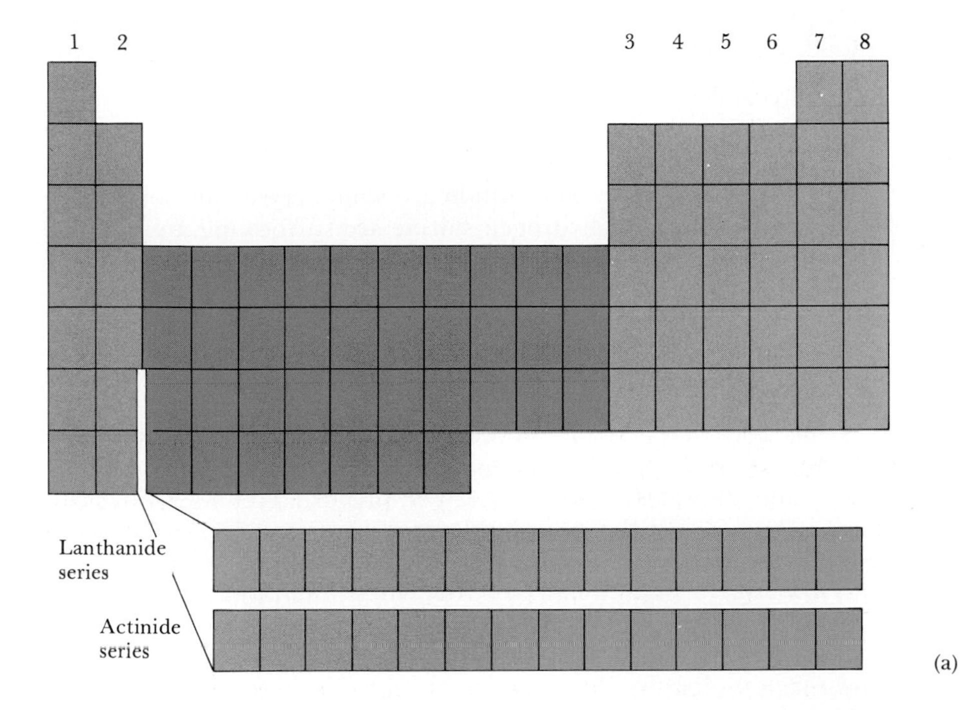

(a)

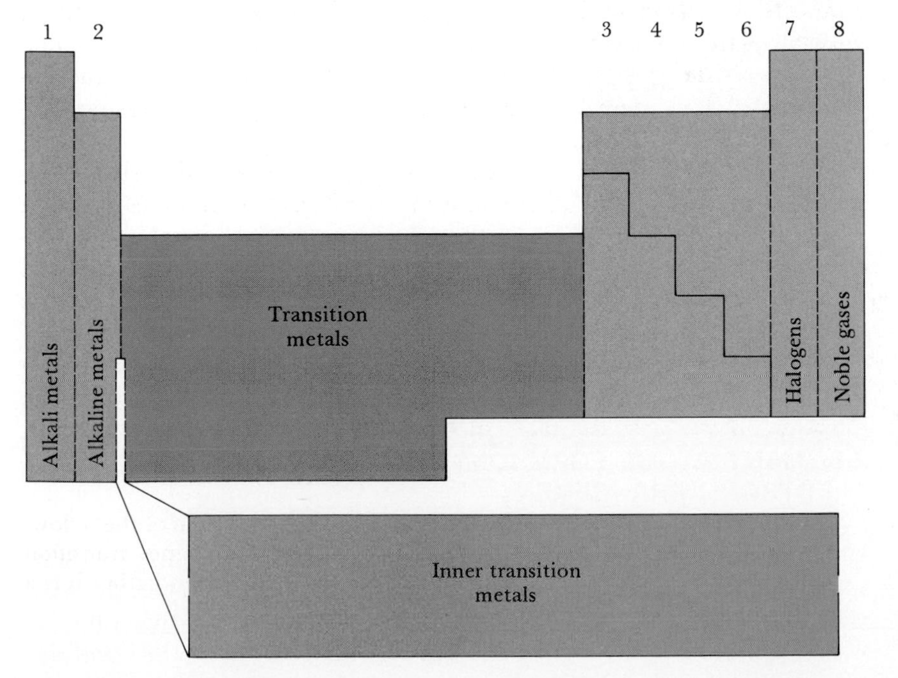

(b)

Figure 2-12 (a) The usual form of the periodic table, showing the main-group elements (*red*), the transition metals (*green*), and the inner transition metals (*yellow*). The main-group elements occur in columns headed by numbers, the transition metals occur in the other columns, and the inner transition metals are located at the bottom of the table. (b) A schematic of the periodic table indicating the major groups (families) of elements.

2-7. THE PERIODIC TABLE CONTAINS SOME IRREGULARITIES

Even though the periodic table is our most important guide to chemistry, it would be overly optimistic to expect that the great diversity of the chemical reactions of 108 elements could be summarized or condensed into a single diagram. To begin with, hydrogen is unusual because it does not fit nicely into any group. It is placed sometimes in Group 1 with the alkali metals and sometimes in Group 7 with the halogens. Some versions of the periodic table place hydrogen in both groups. Hydrogen is not a metal like the Group 1 metals; yet it forms many compounds whose formulas are similar to those of the Group 1 metal compounds. For example, we have

HCl	hydrogen chloride	H_2S	hydrogen sulfide
$NaCl$	sodium chloride	Na_2S	sodium sulfide

Sodium chloride and sodium sulfide are white, crystalline solids, and hydrogen chloride and hydrogen sulfide are suffocating, toxic gases.

Hydrogen is a diatomic gas like the halogens and forms many compounds similar to the halogen compounds:

NaH	sodium hydride	NH_3	ammonia
$NaCl$	sodium chloride	NCl_3	nitrogen trichloride

This analogy is superficial, however, because hydrogen and the halogens undergo many different chemical reactions.

Although the members of a given group usually undergo similar chemical reactions, the first member is somewhat nontypical. For example, lithium, the first member of the alkali metals, reacts directly with nitrogen at room temperature to form lithium nitride:

$$6Li(s) + N_2(g) \rightarrow 2Li_3N(s)$$

whereas the rest of the alkali metals do not react readily with nitrogen. The Group 5 elements phosphorus, arsenic, antimony, and bismuth react directly with chlorine (Figure 2-13):

$$2P(s) + 3Cl_2(g) \rightarrow 2PCl_3(l)$$

$$2As(s) + 3Cl_2(g) \rightarrow 2AsCl_3(l)$$

$$2Sb(s) + 3Cl_2(g) \rightarrow 2SbCl_3(s)$$

$$2Bi(s) + 3Cl_2(g) \rightarrow 2BiCl_3(s)$$

but nitrogen does not react directly with Cl_2. Five of the Group 2 metals react vigorously with dilute acids, but the first one, beryllium, reacts very slowly.

Figure 2-13 When antimony powder is placed in an atmosphere of chlorine, a vigorous reaction takes place.

2-8. MANY ATOMS FORM IONS THAT HAVE A NOBLE-GAS ELECTRON ARRANGEMENT

When Mendeleev formulated his version of the periodic table in 1869, some scientists pictured atoms as little solid spheres. However, the very existence of atoms was not accepted by all scientists at that time. It wasn't until the discovery of the electron by Thomson in 1897 and the proposal of the nuclear model of the atom by Rutherford in 1911 that the idea of the atom was generally accepted. As the structure of the atom became better understood, a great deal of research was directed toward an explanation of the periodic properties of the elements in terms of atomic structure.

The nuclear model of the atom pictures it as a very small, central nucleus surrounded by electrons (see Figure 1-8). When two atoms collide and undergo a chemical reaction, the outermost electrons of each atom interact and are primarily responsible for the reaction that occurs. In later chapters, we learn that the electrons are arranged in an atom in an orderly fashion that is characteristic of that atom. The electron arrangement in an atom is important because it determines the chemical properties of that element. We study the connection between the chemical properties and the electron arrangement of an element in Chapters 6 through 10, but even now we can deduce some important features of electron arrangement by using the periodic table as a guide.

The noble gases are unusual in that they are very nonreactive. Only a few noble-gas compounds exist. This lack of chemical reactivity suggests that the noble-gas atoms have an exceptionally stable arrangement of electrons about their nuclei. If we accept the stability of the noble gases as a working hypothesis, then we can account for a number of chemical properties of the elements we have discussed so far. Let's consider sodium, which we know is a very reactive metal. The atomic number of sodium is 11, and the atomic number of neon is 10. Thus, sodium follows the noble gas neon in the periodic table and has one more electron than a neon atom. If a sodium atom loses one electron, then it becomes a sodium ion, Na^+. A sodium ion has the same number of electrons around its nucleus as does a neon atom, and so we assume that a sodium ion, like a neon atom, is exceptionally stable. Thus there is a tendency, or a driving force, for a sodium atom to lose an electron. We can depict the loss of an electron by a sodium atom as

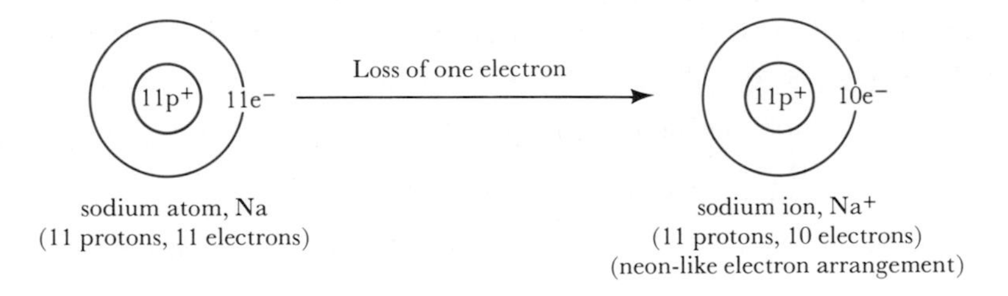

where p^+ denotes a proton and e^- denotes an electron. In a similar manner, the other elements in Group 1 can lose one electron to become +1 ions and achieve the electron arrangement of the preceding noble gas.

Example 2-6: Predict the charge on a calcium ion.

Solution: Calcium belongs to Group 2. A calcium atom has two more electrons than an argon atom. If it loses two electrons, then the Ca^{2+} ion achieves the relatively stable argon electron arrangement. When a calcium atom loses two electrons, the charge of the resulting calcium ion is +2. We can also conclude that the other Group 2 elements form stable +2 ions.

The atoms of an element that directly precedes a noble gas in the periodic table can gain electrons to achieve that noble-gas electron arrangement. For example, fluorine directly precedes neon and so we expect

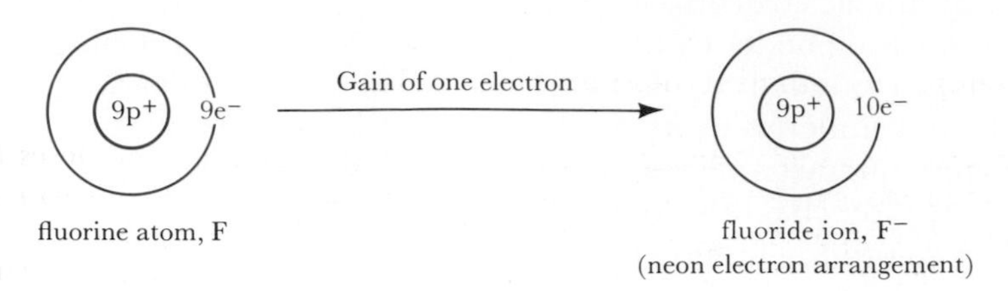

Note that each of the halogens comes right before a noble gas in the periodic table, and so we predict that all the halogen atoms can gain one electron to form halide ions with a charge of −1.

Example 2-7: Predict the charge on an oxide ion.

Solution: Oxygen occurs two positions before neon in the periodic table, and so an oxygen atom has two electrons fewer than a neon atom. An oxygen atom can achieve a neon electron arrangement by gaining two electrons to become an oxide ion, written O^{2-}.

The other elements in Group 6 form −2 ions, such as a sulfide ion, S^{2-}, and a selenide ion, Se^{2-}.

2-9. IONIC CHARGES CAN BE USED TO WRITE CHEMICAL FORMULAS

You may have noticed that metal atoms lose electrons to become cations and that nonmetal atoms gain electrons to become anions. This observation accounts for the chemical formulas of many of the binary compounds that form between the reactive metals and reactive nonmetals.

The chemical formula of potassium bromide is KBr; it is *not* KBr_2, K_2Br, or anything other than KBr. If we realize that a potassium atom readily gives up one electron and that a bromine atom readily takes on one electron, then we see that the chemical combination KBr is a natural result. We may depict the reaction by

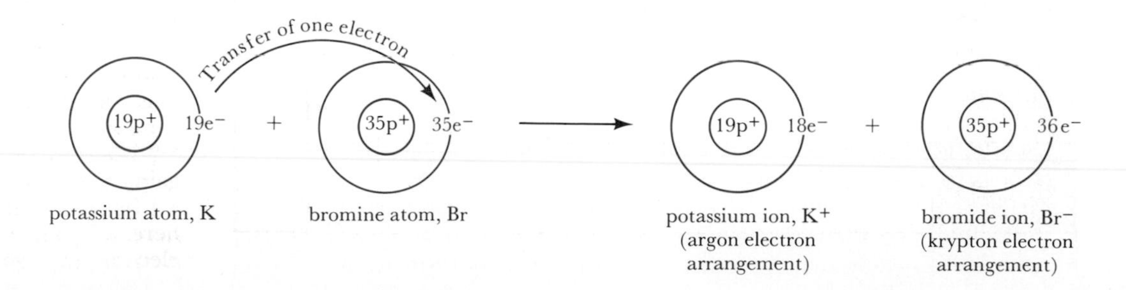

The compound KBr is called an **ionic compound** because it consists of potassium ions and bromide ions, K^+Br^-. Note that the **formula unit,** K^+Br^-, has no net charge. It is conventional to write the formula of potassium bromide as simply KBr instead of K^+Br^-.

Example 2-8: Predict the chemical formula of calcium chloride.

Solution: Calcium belongs to Group 2, and so a calcium atom loses two electrons to form a calcium ion, Ca^{2+}, which has an argon electron arrangement. Chlorine belongs to Group 7, and so a chlorine atom gains one electron to form a chloride ion, Cl^-. A chloride ion also has an argon electron arrangement and so has no tendency to gain more electrons. Thus *two* chlorine atoms are required to take on the two electrons that one calcium atom loses. The formula unit of calcium chloride consists of one calcium ion and two chloride ions, and the chemical formula is $CaCl_2$. Note that the total positive and negative charges in the formula unit are equal and $CaCl_2$ has no net charge.

Using this method, we could deduce the formulas of many compounds, but it is simpler to give a set of rules for writing correct chemical formulas directly. We assign a positive or negative number, called the **ionic charge,** to an element. The ionic charges of some common elements are shown in Figure 2-14. Note that the ionic charges in Figure 2-14 are simply the charges on the ions that have a noble-gas electron arrangement. Also note the correspondence between the ionic charge of an element and its position in the periodic table.

A correct chemical formula is obtained by combining elements in Figure 2-14 such that the total positive and negative charges are equal. For example, the correct formula of strontium fluoride is SrF_2 because a strontium ion has an ionic charge of +2 and a fluoride ion has an

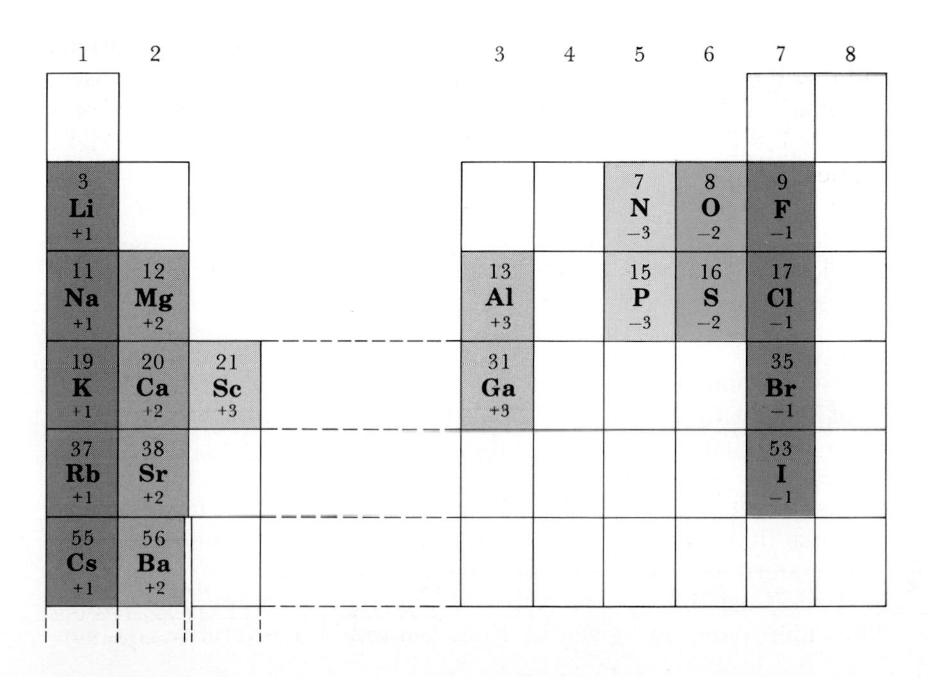

Figure 2-14 The ionic charges of some metals and nonmetals. The ionic charges of Group 1 elements are all +1, Group 2 elements are all +2, and Group 3 elements are all +3. Group 6 and 7 elements gain electrons to become negative ions, with Group 6 elements gaining two electrons to become −2 ions and Group 7 elements gaining one electron to become −1 ions. In every case depicted here, the ions have a noble-gas electron arrangement.

ionic charge of −1. It requires two fluoride ions to balance the +2 ionic charge of one strontium ion. Some other chemical formulas of ionic compounds are

LiI	lithium iodide	Ga_2S_3	gallium sulfide
AlF_3	aluminum fluoride	Na_2O	sodium oxide

Example 2-9: Use Figure 2-14 to write the chemical formula of (a) sodium sulfide and (b) aluminum oxide.

Solution: (a) According to Figure 2-14, the ionic charge of sodium is +1 and the ionic charge of sulfur is −2. We must combine sodium and sulfur into sodium sulfide in such a way that the total positive charge is equal to the total negative charge in the formula unit. Thus we combine two sodium ions and one sulfur ion, and the formula of sodium sulfide is Na_2S.

(b) The ionic charge of aluminum is +3, and that of oxygen is −2. We must combine two aluminum ions (total charge on two Al^{3+} ions is $2 \times (+3) = +6$) with three oxygen ions (total charge on three O^{2-} ions is $3 \times (-2) = -6$) in order to obtain a neutral formula unit for aluminum oxide. Thus Al_2O_3 is the formula of aluminum oxide. Notice that the subscript "2" on the Al in Al_2O_3 is equal to the ionic charge (without the minus sign) on oxygen and the subscript "3" on the O is equal to the ionic charge on aluminum. This can be illustrated schematically as

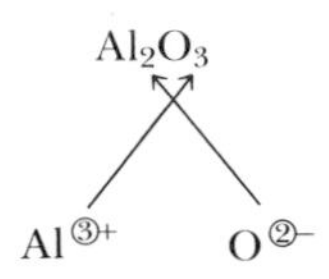

The use of ionic charges to write chemical formulas is limited, and we shall learn a more general method for writing chemical formulas later. Nevertheless, the ionic charges listed in Figure 2-14 can be used to write correct chemical formulas for many binary compounds.

Most transition metals can form ions having more than one possible charge, and so it is necessary to indicate the charge of the metal when naming transition-metal compounds. The ionic charges of some transition-metal ions are given in Table 2-3. For those metals with more than one possible charge, the charge is indicated by a roman numeral after the name of the metal. For example, according to Table 2-3, iron has two possible charges, +2 and +3. The two chlorides of iron (Figure 2-15) are

$FeCl_2$	iron(II) chloride	$FeCl_3$	iron(III) chloride

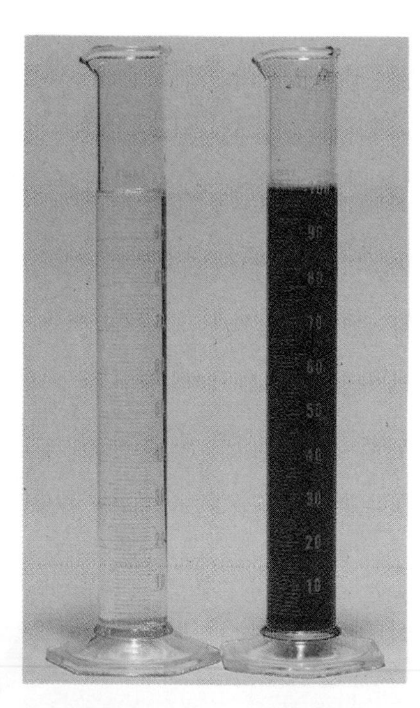

Figure 2-15 Aqueous solutions of iron(II) chloride (*left*) and iron(III) chloride.

An older nomenclature, which is still in use, is also given in Table 2-3. Note that in the older nomenclature, the *-ous* ending indicates the lower ionic charge and the *-ic* ending indicates the higher ionic charge. For example, iron(III) bromide is called ferric bromide in the older nomenclature. A compound that became famous as an ingredient in the first fluoride-containing toothpastes is stannous fluoride, SnF_2. The modern name for SnF_2 is tin(II) fluoride. In this book, we shall in most cases use the modern (systematic) names of compounds or ions. The older names are given in Table 2-3 only for reference.

Table 2-3 Some common metals whose ions have more than one possible ionic charge

Metal	Ionic charges	Symbols	Modern (systematic) names	Old names
chromium	+2 and +3	Cr^{2+} and Cr^{3+}	chromium(II) and chromium(III)	chromous and chromic
cobalt	+2 and +3	Co^{2+} and Co^{3+}	cobalt(II) and cobalt(III)	cobaltous and cobaltic
copper	+1 and +2	Cu^{+} and Cu^{2+}	copper(I) and copper(II)	cuprous and cupric
gold	+1 and +3	Au^{+} and Au^{3+}	gold(I) and gold(III)	aurous and auric
iron	+2 and +3	Fe^{2+} and Fe^{3+}	iron(II) and iron(III)	ferrous and ferric
mercury	+1 and +2	Hg_2^{2+} and Hg^{2+}	mercury(I) and mercury(II)	mercurous and mercuric
tin	+2 and +4	Sn^{2+} and Sn^{4+}	tin(II) and tin(IV)	stannous and stannic

Example 2-10: Give the systematic names of (a) $AuCl_3$ and (b) Fe_2O_3.

Solution: (a) The charge on a chloride ion is −1, and so the total charge on the three chloride ions is −3. Thus the charge on the gold ion is +3. The systematic name of $AuCl_3$ is gold(III) chloride.

(b) Taking the charge on an oxide ion as −2, we determine that the total charge on the three oxide ions is −6. This charge must be balanced by the total charge on the two iron ions. Thus the charge on each iron ion is +3. The systematic name of Fe_2O_3 is iron(III) oxide.

When ionic compounds dissolve in water, the species (that is, the chemical entities) in solution usually are the individual ions surrounded by water molecules. We can represent the process of dissolution by a chemical equation such as

$$NaCl(s) \xrightarrow[H_2O(l)]{} Na^+(aq) + Cl^-(aq)$$

The placement of $H_2O(l)$ below the reaction arrow indicates that water is acting as a solvent, the liquid in which the sodium chloride dissolves. The symbols $Na^+(aq)$ and $Cl^-(aq)$ indicate that the ions occur in **aqueous solution.**

2-10. A COMBINATION REACTION IS THE REACTION OF TWO SUBSTANCES TO FORM A SINGLE PRODUCT

It is helpful in first learning about chemical reactions to try to classify them into types rather than to view each one separately. One classification scheme for chemical reactions consists of four categories: (1) combination reactions, (2) decomposition reactions, (3) single-replacement reactions, and (4) double-replacement reactions. Although not all chemical reactions fall into these four categories, a large number do.

The simplest example of a **combination reaction** is that of a metal with a nonmetal. For example,

$$2Na(s) + Cl_2(g) \rightarrow 2NaCl(s)$$

$$2Ca(s) + O_2(g) \rightarrow 2CaO(s)$$

In each of these reactions, electrons are transferred from the metal atoms to the nonmetal atoms, resulting in ionic compounds in which the ions have a noble-gas arrangement. The transfer of electrons necessarily involves a change in the charge of the species. For example, consider the reaction between sodium and sulfur to give sodium sulfide, Na_2S:

$$2Na(s) + S(s) \rightarrow Na_2S(s)$$

A neutral sodium atom has no charge (a charge of zero). The sodium ion in Na_2S has a charge of +1, and so the charge on sodium changes from 0 to +1 in the reaction. Similarly, the charge on sulfur changes from 0 to −2.

When the charge on an atom increases, the atom is said to be **oxidized.** The term **oxidation** denotes a loss of electrons. When the charge on an atom decreases, the atom is said to be **reduced.** The term **reduction** denotes a gain of electrons. In the above reaction, sodium is oxidized and sulfur is reduced. The reaction itself is an example of an **oxidation-reduction reaction.** Notice that each sodium atom loses one electron and each sulfur atom gains two electrons; thus it requires two sodium atoms to react with each sulfur atom. The total number of electrons lost by the element that is oxidized must equal the total number of electrons gained by the element that is reduced (conservation of electrons).

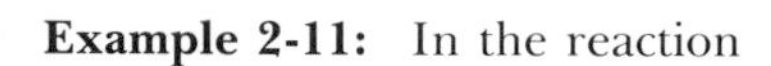

Example 2-11: In the reaction

$$2Mg(s) + O_2(g) \rightarrow 2MgO(s)$$

which atom is oxidized and which is reduced? How many electrons are transferred per formula unit in the reaction as written?

Solution: Magnesium atoms are neutral, and so have no charge. An oxygen molecule is neutral, and each oxygen atom in O_2 has no charge. In MgO, the ionic charges on magnesium and oxygen are +2 and −2, respectively. The charge on magnesium changes from 0 to +2, and the charge on each oxygen changes from 0 to −2. Thus, magnesium is oxidized and oxygen is reduced in this reaction. Two magnesium atoms each lose two electrons for a total of four electrons lost; each of the two oxygen atoms gains two electrons for a total of four electrons gained. The transfer of four electrons yields two MgO formula units and thus two electrons are transferred per formula unit.

Figure 2-16 Sulfur burns in oxygen with a blue flame. The production of SO_2 by burning sulfur is a key reaction in the manufacture of sulfuric acid. Sulfur dioxide is also produced when sulfur-containing coal or petroleum is burned.

Many combination reactions occur between two nonmetals. For example, carbon burns in oxygen to give carbon dioxide:

$$C(s) + O_2(g) \rightarrow CO_2(g)$$

Sulfur burns in oxygen with a blue flame (Figure 2-16) to form sulfur dioxide, a colorless, foul-smelling gas:

$$S(s) + O_2(g) \rightarrow SO_2(g)$$

A reaction in which a substance is burned in oxygen is called a **combustion reaction,** and these combination reactions of carbon and sulfur with oxygen are two examples of combustion reactions.

Carbon dioxide and sulfur dioxide are **molecular compounds** (as opposed to ionic compounds). Molecular compounds are composed of neutral molecules, rather than ions, and are generally much more volatile than ionic compounds; that is, they are more readily vaporized than ionic solids. Thus the molecular compounds CO_2 and SO_2 are gases at 25°C, whereas the ionic compounds NaCl and CaO are solids even at high temperatures.

Although the products in the above two reactions are molecular compounds, we shall see in Chapter 18 that even these reactions involve a transfer of electrons and thus are oxidation-reduction reactions.

The combination reactions presented so far are combinations of two elements. There are also combination reactions between compounds. For example, the reaction between sodium oxide and carbon dioxide,

$$Na_2O(s) + CO_2(g) \rightarrow \underset{\text{sodium carbonate}}{Na_2CO_3(s)}$$

is a combination reaction. This reaction can be used to remove CO_2 from air. Sulfur trioxide reacts with magnesium oxide to form a solid compound called magnesium sulfate:

$$MgO(s) + SO_3(g) \rightarrow \underset{\text{magnesium sulfate}}{MgSO_4(s)}$$

Solid ammonium chloride forms when the gases ammonia and hydrogen chloride are mixed (Figure 2-17):

$$NH_3(g) + HCl(g) \rightarrow \underset{\text{ammonium chloride}}{NH_4Cl(s)}$$

None of the products in these three reactions are binary compounds; nevertheless, when, for example, $MgSO_4$ is dissolved in water, only two ions per $MgSO_4$ formula unit result:

$$\underset{\text{magnesium sulfate}}{MgSO_4(s)} \xrightarrow[H_2O(l)]{} \underset{\text{magnesium ion}}{Mg^{2+}(aq)} + \underset{\text{sulfate ion}}{SO_4^{2-}(aq)}$$

The SO_4^{2-} ion remains intact as a unit in the solution. Magnesium sulfate, $MgSO_4$, consists of Mg^{2+} ions and SO_4^{2-} ions. A **polyatomic ion** is an ion that contains more than one atom. There are a number of groups of atoms, such as SO_4^{2-}, that form stable polyatomic ions. The SO_4^{2-} ion is called a sulfate ion, and thus the name for $MgSO_4$ is magnesium sulfate. Table 2-4 lists a number of common polyatomic ions and their charges. Compounds containing the ions in Table 2-4 are named according to the rules for binary compounds, with the polyatomic ions treated as simple monatomic ions. For example, NaOH is called sodium hydroxide, KCN is called potassium cyanide, and NH_4Cl is called ammonium chloride.

Figure 2-17 Ammonia and hydrogen chloride are colorless gases. They react in a combination reaction to produce the solid white compound ammonium chloride. The white cloud in the picture consists of small ammonium chloride particles formed where the gaseous NH_3 and HCl come into contact with each other. The $NH_3(g)$ comes from one bottle and the $HCl(g)$ comes from the other.

Example 2-12: Name the following compounds:

$RbMnO_4$ $\quad$ $Co(NO_2)_2$ $\quad$ $CrPO_4$

Solution: From Table 2-4 we see that MnO_4^- is called the permanganate ion, and so $RbMnO_4$ is called rubidium permanganate. The NO_2^- ion is called the nitrite ion, and so $Co(NO_2)_2$ is called cobalt(II) nitrite (see Table

2-3). The PO_4^{3-} ion is called the phosphate ion, and $CrPO_4$ is called chromium(III) phosphate (see Table 2-3).

Table 2-4 Some common polyatomic ions

Positive ions	
ammonium	NH_4^+
mercury(I)*	Hg_2^{2+}
Negative ions	
acetate	$C_2H_3O_2^-$
carbonate	CO_3^{2-}
chlorate	ClO_3^-
chromate	CrO_4^{2-}
cyanide	CN^-
dichromate	$Cr_2O_7^{2-}$
hydrogen carbonate ("bicarbonate")	HCO_3^-
hydrogen sulfate ("bisulfate")	HSO_4^-
hydroxide	OH^-
hypochlorite	ClO^-
nitrate	NO_3^-
nitrite	NO_2^-
perchlorate	ClO_4^-
permanganate	MnO_4^-
phosphate	PO_4^{3-}
sulfate	SO_4^{2-}
sulfite	SO_3^{2-}
thiosulfate	$S_2O_3^{2-}$

*Note that Hg_2^{2+} contains two Hg^+ ions joined together.

Example 2-13: Write the formula for each of the following compounds:

sodium thiosulfate   copper(II) perchlorate   calcium hydroxide

Solution: The formula for sodium thiosulfate is written by combining Na^+ and $S_2O_3^{2-}$. Because $S_2O_3^{2-}$ has an ionic charge of -2, it requires two Na^+ for each $S_2O_3^{2-}$, and so $Na_2S_2O_3$ is the formula for sodium thiosulfate.

Copper(II) perchlorate requires two ClO_4^- for each Cu^{2+}, and so the formula is $Cu(ClO_4)_2$. Note that the entire perchlorate ion is enclosed in parentheses and the subscript 2 lies outside the parentheses. One formula unit of $Cu(ClO_4)_2$ contains one copper atom, two chlorine atoms, and eight oxygen atoms.

Calcium hydroxide involves Ca^{2+} and OH^- ions, and thus has the formula $Ca(OH)_2$; once again, note the use of parentheses.

2-11. SOLUBLE METAL OXIDES YIELD BASES AND SOLUBLE NONMETAL OXIDES YIELD ACIDS WHEN DISSOLVED IN WATER

Because of the central role that water plays in chemistry, the combination reactions of metal oxides and nonmetal oxides with water are particularly important. If we dissolve solid sodium oxide in water, the water acts not only as the solvent but also as a reactant:

$$Na_2O(s) + H_2O(l) \xrightarrow[H_2O(l)]{} 2NaOH(aq)$$

Similarly

$$BaO(s) + H_2O(l) \xrightarrow[H_2O(l)]{} Ba(OH)_2(aq)$$

In aqueous solution, $NaOH(aq)$ exists as $Na^+(aq)$ and $OH^-(aq)$, and $Ba(OH)_2(aq)$ exists as $Ba^{2+}(aq)$ and $2OH^-(aq)$. Thus we can write

$$Na_2O(s) + H_2O(l) \xrightarrow[H_2O(l)]{} 2Na^+(aq) + 2OH^-(aq)$$

and

$$BaO(s) + H_2O(l) \xrightarrow[H_2O(l)]{} Ba^{2+}(aq) + 2OH^-(aq)$$

Compounds (such as sodium hydroxide and barium hydroxide) that yield hydroxide ions when dissolved in water are called **bases.** Oxides (such as sodium oxide and barium oxide) that yield bases when dissolved in water are called **basic anhydrides** (the word anhydride means without water).

Many metal oxides are not soluble in water, and so there is no reaction when the oxide is placed in contact with water. For example, aluminum oxide, Al_2O_3, which gives the surface of aluminum doors and window frames their dull appearance, does not dissolve in water, and so we write

$$Al_2O_3(s) + H_2O(l) \rightarrow \text{no reaction}$$

The only metal oxides that are soluble in water are the Group 1 metal oxides and some of the Group 2 metal oxides (Figure 2-18).

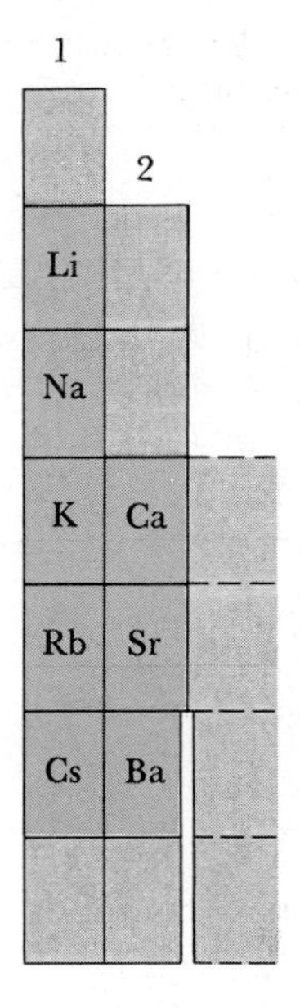

Figure 2-18 The metals whose oxides are water soluble.

Example 2-14: Write the chemical equation for the reaction for the dissolution of potassium oxide, K_2O, in water.

Solution: Potassium is a Group 1 metal, and so its oxide reacts with water much as sodium oxide does. Using the above reaction of Na_2O with water as a guide, we write

$$K_2O(s) + H_2O(l) \xrightarrow[H_2O(l)]{} 2K^+(aq) + 2OH^-(aq)$$

Because hydroxide ions are produced when K_2O is dissolved in water, K_2O is a basic anhydride.

Many water-soluble nonmetal oxides yield acids when dissolved in water. An **acid** is a compound that yields **hydrogen ions** when dissolved in water. Experimental evidence suggests that H^+ in aqueous solution exists in several forms, such as H_3O^+ and $H_9O_4^+$. The species H_3O^+, called the **hydronium ion,** appears to be the dominant species, however. Note that a hydronium ion is a combination of a hydrogen ion, H^+, and a water molecule. For simplicity, we often denote a hydrogen ion in water by $H^+(aq)$ instead of by $H_3O^+(aq)$. The (aq) notation emphasizes that the hydrogen ion is in contact with one or more water molecules.

Oxides that yield acids when dissolved in water are called **acidic anhydrides.** The acidic anhydride of nitric acid is dinitrogen pentoxide:

$$\underset{\text{dinitrogen pentoxide}}{N_2O_5(s)} + H_2O(l) \rightarrow \underset{\text{nitric acid}}{2HNO_3(l)}$$

Nitric acid is an acid because it yields $H^+(aq)$ when dissolved in water:

$$HNO_3(l) \xrightarrow[H_2O(l)]{} H^+(aq) + NO_3^-(aq)$$

It is not possible, in most cases, to tell if a compound is an acid or not simply from its chemical formula. The fact that a compound contains hydrogen atoms does not guarantee that it will yield $H^+(aq)$ ions in water. For example, H_2 is not an acid. The hydrogen atoms, or actually the protons, that result in $H^+(aq)$ ions when an acid is dissolved in water are called **acidic hydrogen atoms,** or **acidic protons.** Thus, nitric acid, HNO_3, is said to have one acidic proton, whereas sulfuric acid, H_2SO_4, is said to have two acidic protons:

$$H_2SO_4(l) \xrightarrow[H_2O(l)]{} 2H^+(aq) + SO_4^{2-}(aq)$$

The names of some common polyatomic acids and their corresponding anions are given in Table 2-5.

Example 2-15: Write a chemical equation showing the dissolution of perchloric acid in water.

Solution: According to Table 2-5, perchloric acid is $HClO_4$ and the corresponding anion is ClO_4^-. Consequently, we write

$$HClO_4(l) \xrightarrow[H_2O(l)]{} H^+(aq) + ClO_4^-(aq)$$

Table 2-5 Some common polyatomic acids and their anions

Acid	Formula	Anion	Formula
acetic	$HC_2H_3O_2$	acetate	$C_2H_3O_2^-$
carbonic	H_2CO_3	carbonate	CO_3^{2-}
nitric	HNO_3	nitrate	NO_3^-
perchloric	$HClO_4$	perchlorate	ClO_4^-
phosphoric	H_3PO_4	phosphate	PO_4^{3-}
sulfuric	H_2SO_4	sulfate	SO_4^{2-}

The acids listed in Table 2-5 are called **oxyacids** because they contain oxygen atoms. There is another group of acids called **binary acids.** Binary acids consist of two elements, one of which must be hydrogen. Binary acids are produced by dissolving certain binary hydrogen compounds (those containing only hydrogen and one other element) in water. The most important binary acid is hydrochloric acid, which is made by dissolving hydrogen chloride gas in water:

$$HCl(g) \xrightarrow[H_2O(l)]{} \underbrace{H^+(aq) + Cl^-(aq)}_{\text{hydrochloric acid}}$$

Some common binary acids are given in Table 2-6.

Table 2-6 Some common binary acids

Acid	Anion	Corresponding gas
hydrobromic	$Br^-(aq)$	hydrogen bromide, HBr
hydrochloric	$Cl^-(aq)$	hydrogen chloride, HCl
hydroiodic	$I^-(aq)$	hydrogen iodide, HI
hydrosulfuric	$S^{2-}(aq)$	hydrogen sulfide, H_2S

Figure 2-19 When mercury(II) oxide is heated, it decomposes into elemental mercury and oxygen gas. The red compound shown here is mercury(II) oxide. The elemental liquid mercury has condensed on the walls of the test tube.

2-12. IN A DECOMPOSITION REACTION, A SUBSTANCE IS BROKEN DOWN INTO TWO OR MORE SIMPLER SUBSTANCES

Decomposition reactions are the opposite of combination reactions because they involve the breaking down of a substance into simpler substances. They are easy to recognize because there is usually only one reactant and more than one product.

When heated, many metal oxides decompose by giving off oxygen gas. An example is (Figure 2-19):

$$2HgO(s) \xrightarrow{\text{high T}} 2Hg(l) + O_2(g)$$

where "high T" (T for temperature) over the arrow indicates that the reaction must be run at a high temperature. (In some cases we specify the actual temperature, as in $\xrightarrow{500°C}$.)

This is the reaction that was used by the English chemist, Joseph Priestley, in 1774 in the first preparation and identification of oxygen as an element.

Many metal carbonates decompose to the metal oxide and CO_2 gas upon heating. For example,

$$CaCO_3(s) \rightarrow CaO(s) + CO_2(g)$$

Calcium carbonate occurs in nature as the principal constituent of limestone and seashells (Figure 2-20). Many metal sulfites undergo a similar reaction:

$$CaSO_3(s) \rightarrow CaO(s) + SO_2(g)$$

Example 2-16: When potassium chlorate is heated, oxygen gas is evolved and solid potassium chloride formed. This reaction is used to generate small quantities of oxygen in the laboratory. Write a balanced chemical equation for the process.

Solution: The chemical formulas of the reactant and products are

$$KClO_3(s) \rightarrow KCl(s) + O_2(g) \qquad \text{(not balanced)}$$

Note that the equation as written is not balanced with respect to oxygen atoms; there are three oxygen atoms on the left and two on the right. We can balance the equation with respect to oxygen by placing a 2 in front of $KClO_3(s)$ ($2 \times 3 = 6$ oxygen atoms) and a 3 in front of $O_2(g)$ ($3 \times 2 = 6$ oxygen atoms):

$$2KClO_3(s) \rightarrow KCl(s) + 3O_2(g) \qquad \text{(not balanced)}$$

We now note that there are two K atoms and two Cl atoms on the left and only one K atom and one Cl atom on the right; thus we place a 2 in front of $KCl(s)$ to obtain the balanced equation:

$$2KClO_3(s) \rightarrow 2KCl(s) + 3O_2(g)$$

Note that the reaction is a decomposition reaction. It is called a **thermal decomposition** reaction because the decomposition is brought about by heating.

(a)

(b)

Figure 2-20 Calcium carbonate is the principal constituent of egg shells. (a) A magnification of an egg shell showing the shell and the underlyng membrane. (300×) (b) Further enlargement of the surface of the shell itself showing the crystalline structure. (12,000×)

2-13. IN A SINGLE-REPLACEMENT REACTION, ONE ELEMENT IN A COMPOUND IS REPLACED BY ANOTHER

Titanium is used to make lightweight, high-strength alloys for airplanes and missiles. Titanium metal is prepared by reacting titanium tetrachloride with molten magnesium:

$$2Mg(l) + TiCl_4(g) \rightarrow 2MgCl_2(s) + Ti(s)$$

Note that the magnesium takes the place of the titanium in the chloride. A reaction in which an element in a compound is replaced by another element is called a **single-replacement reaction** or a **substitution reaction.**

An important type of single-replacement reaction involves the reaction between a reactive metal, such as iron, and a dilute solution of an acid, such as sulfuric acid. As the reaction takes place, bubbles appear at the surface of the iron (Figure 2-21) and the iron slowly dissolves.

The equation for the reaction that occurs is

$$Fe(s) + 2H^+(aq) + SO_4^{2-}(aq) \rightarrow Fe^{2+}(aq) + SO_4^{2-}(aq) + H_2(g)$$

Although the sulfuric acid and iron(II) sulfate exist as ions in aqueous solution, the equation often is written in the following simplified manner:

$$Fe(s) + H_2SO_4(aq) \rightarrow FeSO_4(aq) + H_2(g)$$

Hydrogen gas is only slightly soluble in water and so appears as bubbles that escape from the solution. The iron metal is consumed because it reacts with the H_2SO_4 to produce the soluble compound $FeSO_4$. Note that the iron replaces the hydrogen atoms in H_2SO_4 to yield $FeSO_4$. One of the properties of acids is that dilute solutions of acids attack reactive metals to produce hydrogen gas.

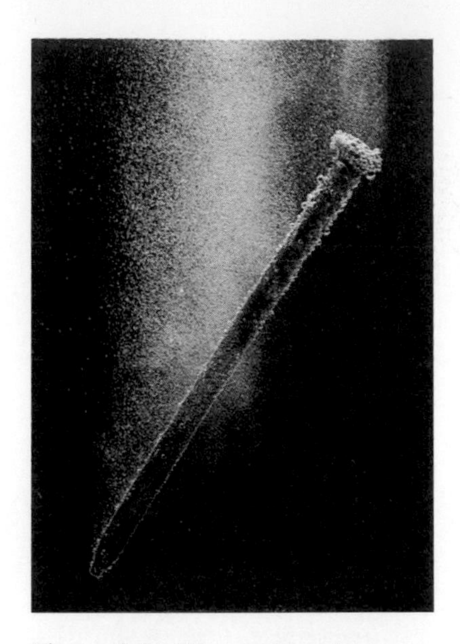

Figure 2-21 The reaction between iron metal and a dilute aqueous solution of sulfuric acid. The bubbles generated at the surface of the iron nail are hydrogen gas. The iron replaces the hydrogen in H_2SO_4 and thereby enters the solution as $FeSO_4(aq)$.

Example 2-17: Barium is a reactive Group 2 metal. Write the chemical equation that describes the reaction of barium with hydrobromic acid, $HBr(aq)$.

Solution: Barium reacts with hydrobromic acid to produce hydrogen. The chemical equation for the single-replacement reaction is

$$Ba(s) + 2HBr(aq) \rightarrow BaBr_2(aq) + H_2(g)$$

All reactions between metals and dilute acids that evolve hydrogen gas may be pictured as the replacement of the hydrogen atoms of the acid by the metal.

2-14. METALS CAN BE ORDERED IN TERMS OF RELATIVE REACTIVITY

Silver nitrate dissolves in water to form a colorless, clear solution. If we place a copper wire in the $AgNO_3(aq)$ solution, then the solution becomes blue (Figure 2-22). In addition, crystals of silver form at the copper wire. The equation for the reaction is

$$\underset{\text{}}{Cu(s)} + \underset{\text{colorless}}{2AgNO_3(aq)} \rightarrow \underset{\text{blue}}{Cu(NO_3)_2(aq)} + 2Ag(s)$$

Copper(II) nitrate dissolved in water forms a blue solution. We can see from the chemical equation that the copper replaces the silver in the nitrate compound. On the basis of this reaction, we conclude that copper metal is more reactive than silver metal because copper metal displaces silver from its nitrate salt.

If we carry out a similar reaction with zinc metal and $Cu(NO_3)_2(aq)$ solution, then we find that zinc metal displaces copper(II) from its nitrate salt:

$$Zn(s) + Cu(NO_3)_2(aq) \rightarrow Cu(s) + Zn(NO_3)_2(aq)$$

We can conclude from this reaction that zinc is more reactive than copper. Thus, on the basis of these reactions we can arrange zinc, copper, and silver in order of their relative reactivities:

Figure 2-22 A solution of silver nitrate, $AgNO_3$, in which a copper wire has been placed turns blue. (a) The original silver nitrate solution is clear and colorless. (b) When a copper wire is added, the solution slowly turns blue as a result of the formation of $Cu(NO_3)_2(aq)$. Note the silver metal crystals at the bottom of the beaker.

zinc is more reactive than copper

copper is more reactive than silver

Thus we have for the order of reactivities of Zn, Cu, and Ag

	increasing reactivity (↑)
Zn	
Cu	
Ag	

By performing additional experiments similar to those just described, we can determine the position of other metals in the **reactivity series** of the metals (Table 2-7). In general, a metal will displace from a compound any metal that lies below it in the reactivity series.

Table 2-7 Reactivity series for some common metals

increasing reactivity (↑)	
Na, Ca	react directly with cold water and vigorously with dilute acids
Al, Zn, Fe, Pb	do not react with cold water, but do react with dilute acids
Cu, Hg, Ag	do not react with water or dilute acids

When a zinc rod is placed in a copper nitrate solution, the zinc replaces the copper and elemental copper forms.

Example 2-18: Predict whether or not a reaction occurs in the following cases and complete and balance the chemical equation if a reaction does occur:

$$Zn(s) + HgCl_2(aq) \rightarrow$$

$$Zn(s) + Ca(ClO_4)_2(aq) \rightarrow$$

Solution: Zinc lies above mercury in the reactivity series given in Table 2-7, and so zinc replaces mercury from its chloride compound:

$$Zn(s) + HgCl_2(aq) \rightarrow ZnCl_2(aq) + Hg(l)$$

Zinc lies below calcium in the reactivity series and so will not replace calcium from the perchlorate compound:

$$Zn(s) + Ca(ClO_4)_2(aq) \rightarrow \text{no reaction}$$

Although the single-replacement reactions we have discussed so far involve the replacement of one metal by another, there are many other types of single-replacement reactions. One particularly important type is the reaction between a metal oxide and carbon:

$$3C(s) + 2Fe_2O_3(s) \rightarrow 4Fe(s) + 3CO_2(g)$$

$$C(s) + 2ZnO(s) \rightarrow 2Zn(s) + CO_2(g)$$

Reactions such as these are used in the large-scale production of metals from their ores, which are often either metal oxides or compounds readily convertible to oxides.

Figure 2-23 When $Cl_2(g)$ is bubbled into an aqueous NaBr solution, the chlorine replaces the bromine in NaBr, forming $NaCl(aq)$ and reddish $Br_2(aq)$.

2-15. THE REACTIVITY ORDER OF THE HALOGENS IS $F_2 > Cl_2 > Br_2 > I_2$

The reactions of the nonmetals are too diversified to allow for a single table similar to Table 2-7. However, the relative reactivities of the halogens can be established by means of single-replacement reactions. For example, if bromine is added to an aqueous solution of NaI, free iodine is produced:

$$Br_2(l) + 2NaI(aq) \rightarrow 2NaBr(aq) + I_2(s)$$

implying that bromine is more reactive than iodine.

Similarly, if chlorine gas is bubbled into an aqueous solution of NaBr, as shown in Figure 2.23, then free bromine is produced:

$$Cl_2(g) + 2NaBr(aq) \rightarrow 2NaCl(aq) + Br_2(aq)$$

implying that chlorine is more reactive than bromine. Finally, fluorine not only is the most reactive halogen but also is the most reactive element. Fluorine readily displaces chlorine from chlorides; for example,

$$2KCl(s) + F_2(g) \rightarrow 2KF(s) + Cl_2(g)$$

Thus the reactivity order of the halogens is

$$F_2 > Cl_2 > Br_2 > I_2$$

Generally, the reactivity of nonmetals (except for the noble gases) increases as we go *up* a group in the periodic table. This is in sharp contrast to the reactivity of metals, which increases as we go *down* a group.

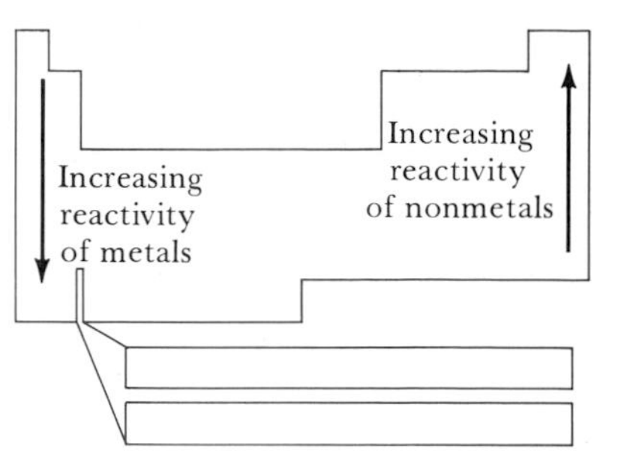

Reactivity trends of the main-group elements.

Example 2-19: Predict whether or not a reaction occurs in the following cases, and complete and balance the chemical equation if a reaction does occur (assume that any salt that is formed is water soluble):

$$\text{(a) } Br_2(l) + CaI_2(aq) \rightarrow$$

$$\text{(b) } Br_2(l) + KF(aq) \rightarrow$$

Solution: (a) Bromine is more reactive than I_2 and so will replace the iodide in CaI_2:

$$Br_2(l) + CaI_2(aq) \rightarrow CaBr_2(aq) + I_2(s)$$

(b) Fluorine is the most reactive halogen, and so none of the other halogens will liberate fluorine from fluoride salts. Thus we predict that

$$Br_2(l) + KF(aq) \rightarrow \text{no reaction}$$

which is correct.

2-16. IN A DOUBLE-REPLACEMENT REACTION, CATIONS AND ANIONS EXCHANGE TO FORM NEW COMPOUNDS

▪ Double-replacement reactions are also called metathesis reactions.

A simple and impressive **double-replacement reaction** is the reaction between an aqueous solution of NaCl and an aqueous solution of $AgNO_3$. Both solutions are clear; yet when they are mixed, a white **precipitate** forms immediately (Figure 2-24). A **precipitate** is an insoluble product of a reaction that occurs in solution. In the chemical reaction between sodium chloride and silver nitrate,

$$NaCl(aq) + AgNO_3(aq) \rightarrow NaNO_3(aq) + AgCl(s)$$

the white precipitate that forms is silver chloride. The reaction is a double-replacement reaction because the two cations, $Na^+(aq)$ and $Ag^+(aq)$, can be visualized as exchanging anions.

A double-replacement reaction involving the formation of a precipitate usually is called a **precipitation reaction.** It is instructive to analyze this precipitation reaction in terms of the ions involved. Sodium chloride and silver nitrate are soluble ionic compounds, and so, in aqueous solution, they consist of $Na^+(aq)$ and $Cl^-(aq)$ ions and $Ag^+(aq)$ and $NO_3^-(aq)$ ions, respectively. At the very instant the NaCl and $AgNO_3$ solutions are mixed, these four kinds of ions exist in one solution. As we shall learn in later chapters, the ions in a solution constantly move around and collide with water molecules and with each other. If a $Na^+(aq)$ ion collides with a $Cl^-(aq)$ ion, the ions simply drift apart because NaCl is soluble in water. Similarly, a collision between $Ag^+(aq)$ and $NO_3^-(aq)$ is of no consequence because $AgNO_3$ is also soluble in water. When a $Ag^+(aq)$ ion collides with a $Cl^-(aq)$ ion, however, AgCl is

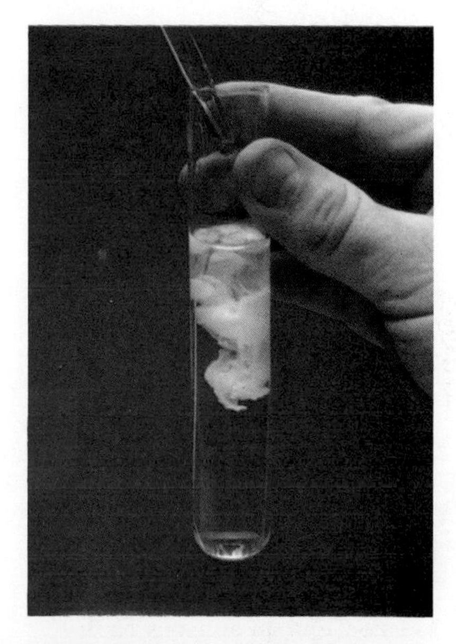

Figure 2-24 The reaction of $NaCl(aq)$ with $AgNO_3(aq)$ yields the white precipitate $AgCl(s)$.

formed, and it is not soluble in water. Consequently, the AgCl precipitates out of the solution. The $Ag^+(aq)$ and $Cl^-(aq)$ are depleted from the solution and appear as a white precipitate. The $Na^+(aq)$ and $NO_3^-(aq)$ remain in solution because $NaNO_3$ is soluble in water. We say that the driving force of the chemical reaction between NaCl and $AgNO_3$ is the formation of the AgCl precipitate.

▪ The formation of a precipitate drives a precipitation reaction.

It is convenient to use **net ionic equations** to describe double-replacement reactions that occur in solution. In the reaction between NaCl and $AgNO_3$, the $Na^+(aq)$ and $NO_3^-(aq)$ ions do not participate in the reaction, which is the formation of solid AgCl from $Ag^+(aq)$ and $Cl^-(aq)$ ions. We say that the $Na^+(aq)$ and $NO_3^-(aq)$ ions are **spectator ions;** that is, they are not involved directly in the formation of the AgCl precipitate. Thus, if we write the chemical equation in terms of ions, that is, as an **ionic equation,**

▪ Spectator ions do not appear in a net ionic equation.

$$Na^+(aq) + Cl^-(aq) + Ag^+(aq) + NO_3^-(aq) \rightarrow Na^+(aq) + NO_3^-(aq) + AgCl(s)$$

then we see that the spectator ions, $Na^+(aq)$ and $NO_3^-(aq)$, appear on both sides of the equation. Because they appear on both sides of the ionic equation, the spectator ions can be canceled:

$$\cancel{Na^+(aq)} + Cl^-(aq) + Ag^+(aq) + \cancel{NO_3^-(aq)} \rightarrow \cancel{Na^+(aq)} + \cancel{NO_3^-(aq)} + AgCl(s)$$

to give the net ionic equation:

$$Ag^+(aq) + Cl^-(aq) \rightarrow AgCl(s)$$

The net ionic equation describes the essence of the reaction, namely, the formation of solid AgCl from $Ag^+(aq)$ and $Cl^-(aq)$ ions.

Note that the net ionic equation corresponding to the double-replacement reaction between $KCl(aq)$ and $AgClO_4(aq)$ is the same as that for the reaction between $NaCl(aq)$ and $AgNO_3(aq)$:

ionic equation:

$$\cancel{K^+(aq)} + Cl^-(aq) + Ag^+(aq) + \cancel{ClO_4^-(aq)} \rightarrow \cancel{K^+(aq)} + \cancel{ClO_4^-(aq)} + AgCl(s)$$

net ionic equation:

$$Ag^+(aq) + Cl^-(aq) \rightarrow AgCl(s)$$

The main advantage to writing net ionic equations is that they focus on the reactant ions that are involved directly in the reaction.

In order to predict if a precipitate forms in a reaction occurring in aqueous solution, it is necessary to know whether a possible reaction product is soluble in water or not. We shall see in Chapter 17 that each compound has a definite solubility (for example, so many grams per liter of water), but in this chapter we shall simply indicate whether a compound is insoluble or not.

Example 2-20: Write the net ionic equation for the reaction between $Cd(NO_3)_2(aq)$ and $Na_2S(aq)$ given that cadmium sulfide, $CdS(s)$, is the only insoluble product.

Solution: The complete ionic equation is

$$Cd^{2+}(aq) + 2NO_3^-(aq) + 2Na^+(aq) + S^{2-}(aq) \rightarrow 2Na^+(aq) + 2NO_3^-(aq) + CdS(s)$$

We get the net ionic equation by canceling the $Na^+(aq)$ and $NO_3^-(aq)$ ions on both sides:

$$Cd^{2+}(aq) + S^{2-}(aq) \rightarrow \underset{\text{orange}}{CdS(s)}$$

This reaction is shown in the Frontispiece to this chapter.

We have seen that the driving force for double-replacement reactions can be the formation of a precipitate. It can also be the formation of an un-ionized (molecular) compound from ionic reactants. The most important example of such a reaction is the reaction between an acid and a base. For example,

$$HCl(aq) + NaOH(aq) \rightarrow NaCl(aq) + H_2O(l)$$

The complete ionic equation is

$$H^+(aq) + Cl^-(aq) + Na^+(aq) + OH^-(aq) \rightarrow Na^+(aq) + Cl^-(aq) + H_2O(l)$$

and the net ionic equation is

$$H^+(aq) + OH^-(aq) \rightarrow H_2O(l)$$

Because H_2O exists almost exclusively as neutral molecules, the $H^+(aq)$ and $OH^-(aq)$ are depleted from the reaction mixture by the formation of H_2O. Note the comparison between the two driving forces for double-replacement reactions: the formation of a precipitate and the formation of a molecular compound. In each case reactant ions are depleted from the reaction mixture.

The reaction between HCl and NaOH is amazing. We probably all know that acids like hydrochloric acid are corrosive. They react with many metals, and their concentrated solutions damage flesh, causing painful burns and blisters. Less familiar, perhaps, are the chemical properties of bases. Bases like sodium hydroxide are also very corrosive, causing painful burns and blisters on the skin, just as hydrochloric acid does. Sodium hydroxide is the principal ingredient in oven and drain cleaners. Thus the reaction between $HCl(aq)$ and $NaOH(aq)$ is one between two reactive and hazardous substances, but the products are sodium chloride and water, a harmless solution of table salt. We say that the acid (HCl) and the base (NaOH) have **neutralized** each other. The chemical reaction between an acid and a base is called a **neutralization reaction.** The ionic compound that is formed along with the water in a neutralization reaction is called a **salt.** For example,

▪ Sodium hydroxide is a key ingredient in Drāno and other cleaners.

$$\underset{\text{acid}}{H_2SO_4(aq)} + \underset{\text{base}}{2KOH(aq)} \rightarrow \underset{\text{salt}}{K_2SO_4(aq)} + \underset{\text{water}}{2H_2O(l)}$$

Most salts are crystalline solids.

Because acids and bases are so important in chemistry, we have summarized some of their properties in Table 2-8. As the table shows, **acidic solutions** taste sour. Vinegar tastes sour because it is a dilute solution of acetic acid, lemons taste sour because they contain citric

Table 2-8 Some properties of acids and bases

Acids	Bases
solutions taste sour (don't taste any, but recall the taste of vinegar)	solutions taste bitter and feel slippery to the touch (don't taste any)
produce hydrogen ions when dissolved in water	produce hydroxide ions when dissolved in water
neutralize bases to produce salts and water	neutralize acids to produce salts and water
solutions turn litmus paper red	solutions turn litmus paper blue
react with many metals to produce hydrogen gas	

acid, and rhubarb tastes sour because it contains oxalic acid. **Basic solutions** taste bitter and feel slippery; an example of a basic solution is soapy water. An interesting and important property of acidic and basic solutions is their effect on certain dyes and vegetable matter. **Litmus,** which is a vegetable substance obtained from lichens, is red in acidic solutions and blue in basic solutions. Paper impregnated with litmus, called **litmus paper,** serves as a quick test to see whether a solution is acidic or basic (Figure 2-25).

In summary, then, a chemical characteristic of an acid is that it produces $H^+(aq)$ in aqueous solutions and a chemical characteristic of a base is that it produces $OH^-(aq)$ in aqueous solutions. When an acid and a base neutralize each other, the $H^+(aq)$ and $OH^-(aq)$ react to produce $H_2O(l)$ and in doing so nullify the acidic and basic character of the separate solutions:

$$\underset{\text{acid}}{H^+(aq)} + \underset{\text{base}}{OH^-(aq)} \rightarrow \underset{\text{water}}{H_2O(l)}$$

Example 2-21: Complete and balance the equations for the following neutralization reactions and name the salt formed in each case:

$$H_2SO_4(aq) + LiOH(aq) \rightarrow$$

$$HNO_3(aq) + Ba(OH)_2(aq) \rightarrow$$

Solution: Each of these reactions is between an acid and a base, and so a salt and water are produced in each case. The first reaction is between sulfuric acid and lithium hydroxide, and so the salt formed is lithium sulfate:

$$\underset{\substack{\text{sulfuric acid}\\ \text{(an acid)}}}{H_2SO_4(aq)} + \underset{\substack{\text{lithium hydroxide}\\ \text{(a base)}}}{2LiOH(aq)} \rightarrow \underset{\substack{\text{lithium sulfate}\\ \text{(a salt)}}}{Li_2SO_4(aq)} + 2H_2O(l)$$

The second reaction is between nitric acid and barium hydroxide, and so the salt formed is barium nitrate (Table 2-4):

$$\underset{\substack{\text{nitric acid}\\ \text{(an acid)}}}{2HNO_3(aq)} + \underset{\substack{\text{barium hydroxide}\\ \text{(a base)}}}{Ba(OH)_2(aq)} \rightarrow \underset{\substack{\text{barium nitrate}\\ \text{(a salt)}}}{Ba(NO_3)_2(aq)} + 2H_2O(l)$$

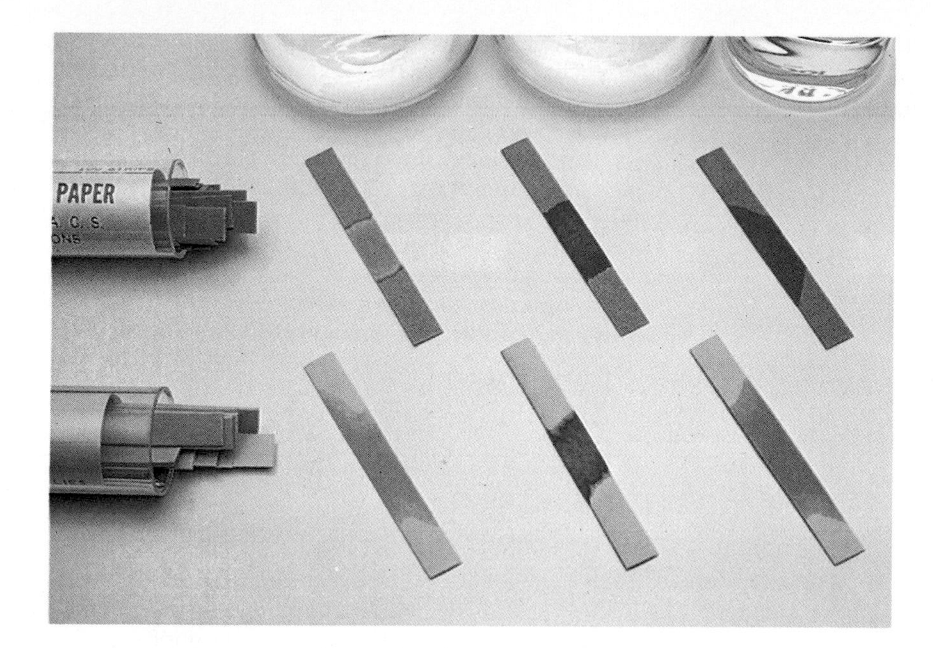

Figure 2-25 Litmus paper is used as a simple test for acidity or basicity. Blue litmus paper turns red when placed in an acidic solution. Red litmus paper turns blue when placed in a basic solution. In the upper row of strips, blue litmus paper is treated first with acid, then with base, and then with pure water. In the lower row of strips, red litmus paper is treated first with acid, then with base, and then with pure water.

One of the most difficult questions that a beginning student of chemistry faces is to predict the products of a reaction when only the reactants are given. This is often a difficult question even for a chemist. The classification of reaction types presented in this chapter is helpful in this regard, but it still requires additional chemical experience to develop confidence. As you see more chemical reactions throughout this book, think about each one and try to classify it according to the scheme developed in this chapter.

SUMMARY

Chemical reactions are written in terms of chemical formulas. The numbers of each kind of atom are the same on both sides of a balanced chemical equation (law of conservation of mass), although atoms are rearranged so that new substances are formed in chemical reactions.

When the elements are ordered according to atomic number, there is a repetitive, or periodic, pattern of chemical properties. These periodic patterns of chemical behavior are displayed in the periodic table of the elements, which contains all the known chemical elements. Elements are denoted as main-group elements, transition metals, or inner transition metals. Most elements are metals which appear on the left side of the periodic table. The nonmetals appear on the right side of the periodic table, and the semimetals appear on the border between metals and nonmetals.

It is possible to infer the charges of many stable ions from the periodic table. These ionic charges can be used to write formulas of many binary ionic compounds by combining the positive and negative ions such that the net charge on the formula unit is zero.

Many chemical reactions involve changes in ionic charge. When its charge increases, an atom is said to be oxidized. When its charge decreases, an atom is reduced. Such reactions are known as oxidation-reduction reactions.

Many chemical reactions can be classified as being one of four types:

1. combination: a reaction of two substances to form a single product.
2. decomposition: a reaction in which a substance breaks down into two or more simpler substances.
3. single-replacement: a reaction involving the substitution of one element in a compound by another (also called a substitution reaction).
4. double-replacement: a reaction in which the cations of two ionic compounds exchange anionic partners. The driving force can be the formation of a precipi-

tate or the formation of a molecular product such as water.

Net ionic equations are written by canceling the spectator ions from the two sides of the chemical equation.

Single-replacement reactions can be used to order the metals in terms of relative reactivity. The resulting reactivity series of the metals can be used to predict whether or not a single-replacement reaction occurs. The nonmetals are too varied to allow a correspondingly simple activity series, but the relative reactivities of the halogens are $F_2 > Cl_2 > Br_2 > I_2$.

Acids are substances that produce the species $H^+(aq)$ when dissolved in water; bases are substances that produce hydroxide ions, $OH^-(aq)$, when dissolved in water. Two classes of acids are binary acids and oxyacids. Solutions of oxyacids yield relatively stable polyatomic anions, which remain intact in many chemical reactions. Acids and bases react with each other to produce a salt and water; such reactions are called neutralization reactions. The names of salts are based upon the names of the acids that produce them (Table 2-4). Acids and bases cause certain dyes to change color. For example, litmus turns red in acidic solutions and blue in basic solutions.

TERMS YOU SHOULD KNOW

reactant 45
product 45
chemical equation 45
balanced chemical equation 45
balancing coefficient 45
balancing by inspection 46
alkaline 47
alkali metal 47
alkaline earth metal 47
halogen 48
halide 48
periodic table of the elements 49
Group 1 metal 50
Group 2 metal 50
noble gas (inert gas) 50
lanthanide series 51
actinide series 51
group (family) 52
period 52
semimetal 53
semiconductor 53
main-group element 53
transition metal 53
inner transition metal 54
rare-earth element 54
ionic compound 59
formula unit 59
ionic charge 59
(*aq*), aqueous solution 61
combination reaction 61
oxidized 62
oxidation 62
reduced 62
reduction 62
oxidation-reduction reaction 62
combustion reaction 62
molecular compound 63
polyatomic ion 63
base 64
basic anhydride 64
acid 65
hydrogen ion, $H^+(aq)$ 65
hydronium ion, $H_3O^+(aq)$ 65
acidic anhydride 65
acidic hydrogen atom (acidic proton) 65
oxyacid 66
binary acid 66
decomposition reaction 66
thermal decomposition 67
single-replacement reaction (substitution reaction) 67
reactivity series 69
double-replacement reaction (metathesis reaction) 71
precipitate 71
precipitation reaction 71
net ionic equation 72
spectator ion 72
ionic equation 72
neutralize 73
neutralization reaction 73
salt 73
acidic solution 73
basic solution 74
litmus 74
litmus paper 74

PROBLEMS

BALANCING EQUATIONS

2-1. Balance the following chemical equations:

(a) $P(s) + Br_2(l) \rightarrow PBr_3(l)$
(b) $H_2O_2(l) \rightarrow H_2O(l) + O_2(g)$
(c) $CoO(s) + O_2(g) \rightarrow Co_2O_3(s)$
(d) $PCl_5(s) + H_2O(l) \rightarrow H_3PO_4(l) + HCl(g)$

2-2. Balance the following chemical equations:

(a) $KHF_2(s) \rightarrow KF(s) + H_2(g) + F_2(g)$
(b) $C_3H_8(g) + O_2(g) \rightarrow CO_2(g) + H_2O(g)$
(c) $P_4O_{10}(s) + H_2O(l) \rightarrow H_3PO_4(l)$
(d) $N_2H_4(g) \rightarrow NH_3(g) + N_2(g)$

2-3. Balance the following chemical equations:

(a) $CaH_2(s) + H_2O(l) \rightarrow Ca(OH)_2(aq) + H_2(g)$
(b) $CaCO_3(s) + HCl(aq) \rightarrow CaCl_2(aq) + CO_2(g) + H_2O(l)$
(c) $C_6H_{12}O_2(aq) + O_2(g) \rightarrow CO_2(g) + H_2O(l)$
(d) $Li(s) + CO_2(g) + H_2O(g) \rightarrow LiHCO_3(s) + H_2(g)$

2-4. Balance the following chemical equations:

(a) $H_2SO_4(aq) + KOH(aq) \rightarrow K_2SO_4(aq) + H_2O(l)$
(b) $Li_3N(s) + H_2O(l) \rightarrow LiOH(aq) + NH_3(g)$

(c) $Al_4C_3(s) + HCl(aq) \rightarrow AlCl_3(aq) + CH_4(g)$
(d) $ZnS(s) + HBr(aq) \rightarrow ZnBr_2(aq) + H_2S(g)$

2-5. Balance the following chemical equations and name the reactants and products in each case:

(a) $NaH(s) + H_2O(l) \rightarrow NaOH(aq) + H_2(g)$
(b) $SO_2(g) + O_2(g) \rightarrow SO_3(g)$
(c) $H_2S(g) + LiOH(aq) \rightarrow Li_2S(aq) + H_2O(l)$
(d) $ZnO(s) + CO(g) \rightarrow Zn(s) + CO_2(g)$

2-6. Balance the following chemical equations and name the reactants and products in each case:

(a) $PCl_3(g) + Cl_2(g) \rightarrow PCl_5(s)$
(b) $Sb(s) + Cl_2(g) \rightarrow SbCl_3(s)$
(c) $GaBr_3(s) + Cl_2(g) \rightarrow GaCl_3(s) + Br_2(g)$
(d) $Mg_3N_2(s) + HCl(aq) \rightarrow MgCl_2(aq) + NH_3(g)$

2-7. Complete and balance the following chemical equations and name the product formed in each case:

(a) $Na(s) + I_2(s) \rightarrow$
(b) $Sr(s) + H_2(g) \rightarrow$
(c) $Ca(s) + N_2(g) \rightarrow$
(d) $Mg(s) + O_2(g) \rightarrow$

2-8. Complete and balance the following chemical equations and name the product(s) formed in each case:

(a) $Sr(s) + S(s) \rightarrow$
(b) $K(s) + H_2O(l) \rightarrow$
(c) $Ca(s) + H_2O(l) \rightarrow$
(d) $Al(s) + Cl_2(g) \rightarrow$

PERIODIC TABLE

2-9. Astatine is a radioactive halogen that concentrates in the thyroid gland. Predict the following properties of astatine:

(a) physical state at 25°C (solid, liquid, or gas)
(b) formula of sodium salt
(c) color of sodium salt
(d) formula of gaseous astatine
(e) color of solid astatine

2-10. Radon is a radioactive noble gas that has been used as a tracer in detecting gas leaks. Predict the following properties of radon:

(a) color
(b) odor
(c) formula of gaseous radon
(d) reaction with water

2-11. By referring to the periodic table, classify each of the following as a main-group element, a transition metal, or an inner transition metal. If a main-group element, indicate which group and whether the element is a metal or a nonmetal:

Tl Eu Xe Hf Ru Am B

2-12. By referring to the periodic table, classify each of the following as a main-group element, a transition metal, or an inner transition metal. If a main-group element, indicate which group and whether the element is a metal or a nonmetal:

Se As Mo Rn Ta Bi In

2-13. Radium is a brilliant white radioactive metal that was discovered by Pierre and Marie Curie. It was isolated from the mineral pitchblende from North Bohemia, and it took 7 tons of pitchblende to recover only one g of radium. Predict the reaction of radium with

(a) oxygen
(b) chlorine
(c) hydrogen chloride
(d) hydrogen
(e) sulfur

2-14. Francium is a radioactive metal that does not occur in nature. Predict the reaction of francium with

(a) bromine
(b) hydrogen
(c) water
(d) sulfur
(e) hydrogen chloride

NOMENCLATURE AND CHEMICAL FORMULAS

2-15. Indicate which of the following ions have a noble-gas electron arrangement and, for such ions, identify the corresponding noble gas:

(a) Cs^+
(b) Ga^{3+}
(c) S^{2-}
(d) P^{3-}
(e) Al^{3+}
(f) O^-

2-16. Indicate which of the following ions have a noble-gas electron arrangement and, for such ions, identify the corresponding noble gas:

(a) Ba^{2+}
(b) Ca^+
(c) Br^-
(d) N^{3-}
(e) Se^{2-}
(f) Fe^{2+}

2-17. Determine the charge on each atom in the following compounds, and name the compound.

(a) MgS
(b) AlP
(c) BaF_2
(d) Ga_2O_3

2-18. Determine the charge on each atom in the following compounds, and name the compound.

(a) Li_2O
(b) CaS
(c) Mg_3N_2
(d) Al_2S_3

2-19. Write the chemical formula for

(a) gallium selenide
(b) aluminum phosphide
(c) potassium iodide
(d) strontium fluoride

2-20. Write the chemical formula for

(a) aluminum sulfide
(b) sodium oxide
(c) barium fluoride
(d) potassium iodide

2-21. Write the chemical formulas for the following compounds:

(a) lithium nitride (b) gallium telluride
(c) barium nitride (d) magnesium bromide

2-22. Write the chemical formulas for the following compounds:

(a) cesium oxide (b) sodium selenide
(c) lithium sulfide (d) calcium iodide

2-23. Write the formula of the binary compound formed from each of the following pairs of ions:

(a) Fe^{3+} and O^{2-} (b) Cd^{2+} and S^{2-}
(c) Ru^{3+} and F^- (d) Tl^+ and S^{2-}

2-24. Write the formula of the binary compound formed from each of the following pairs of ions:

(a) Ga^{3+} and S^{2-} (b) Fe^{3+} and Se^{2-}
(c) Ba^{2+} and At^- (d) Zn^{2+} and N^{3-}

2-25. Name the following compounds:

(a) $Ca(CN)_2$ (b) $AgClO_4$
(c) $KMnO_4$ (d) $SrCrO_4$

2-26. Name the following compounds:

(a) $NaC_2H_3O_2$ (b) $Ca(ClO_3)_2$
(c) $(NH_4)_2CO_3$ (d) $Ba(NO_3)_2$

2-27. Name the following compounds, which are used as fertilizers:

(a) $(NH_4)_2SO_4$ (b) $(NH_4)_3PO_4$
(c) $Ca_3(PO_4)_2$ (d) K_3PO_4

2-28. Name the following compounds, which are used in photography:

(a) $(NH_4)_2S_2O_3$ (fixer) (b) Na_2SO_3 (preservative)
(c) K_2CO_3 (activator) (d) $Na_2S_2O_3$ (fixer)

2-29. Write the chemical formula for each of the following:

(a) sodium thiosulfate
(b) potassium hydrogen carbonate
(c) sodium hypochlorite
(d) calcium sulfite

2-30. Write the chemical formula for each of the following:

(a) acetic acid (b) chloric acid
(c) carbonic acid (d) perchloric acid

2-31. Give the chemical formula for each of the following:

(a) sodium sulfite (b) potassium phosphate
(c) silver sulfate (d) ammonium nitrate

2-32. Give the chemical formula for each of the following:

(a) sodium perchlorate (b) potassium permanganate
(c) calcium sulfite (d) lithium cyanide

2-33. Give the systematic name for each of the following:

(a) Hg_2Cl_2 (b) $Cr(NO_3)_3$
(c) $CoBr_2$ (d) $CuCO_3$

2-34. Give the systematic name for each of the following:

(a) $CrSO_4$ (b) $Co(CN)_2$
(c) $Sn(NO_3)_2$ (d) Cu_2CO_3

2-35. Write the chemical formula for each of the following:

(a) chromium(III) oxide (b) tin(II) hydroxide
(c) copper(II) acetate (d) cobalt(III) sulfate

2-36. Write the chemical formula for each of the following:

(a) mercury(I) acetate (b) mercury(II) cyanide
(c) iron(II) perchlorate (d) chromium(II) sulfite

CLASSIFICATION AND PREDICTION OF REACTIONS

2-37. Classify each of the following reactions as combination, decomposition, single-replacement, or double-replacement:

(a) $CaCO_3(s) \rightarrow CaO(s) + CO_2(g)$
(b) $NH_3(g) + HCl(g) \rightarrow NH_4Cl(s)$
(c) $2AgBr(s) + Cl_2(g) \rightarrow 2AgCl(s) + Br_2(l)$
(d) $Ag_2SO_4(s) + 2NaI(aq) \rightarrow 2AgI(s) + Na_2SO_4(aq)$

2-38. Classify each of the following reactions as combination, decomposition, single-replacement, or double-replacement:

(a) $Pb(NO_3)_2(aq) + 2NaI(aq) \rightarrow PbI_2(s) + 2NaNO_3(aq)$
(b) $2KClO_3(s) \rightarrow 2KCl(s) + 3O_2(g)$
(c) $2NaCl(s) + H_2SO_4(l) \rightarrow 2HCl(g) + Na_2SO_4(s)$
(d) $Fe(s) + 2HBr(aq) \rightarrow FeBr_2(aq) + H_2(g)$

2-39. Classify each of the following as a combination, a decomposition, a single-replacement, or a double-replacement reaction. If the equation is not balanced, then balance the equation.

(a) $BaCO_3(s) \rightarrow BaO(s) + CO_2(g)$
(b) $Fe(s) + O_2(g) \rightarrow Fe_2O_3(s)$
(c) $Al(s) + Mn_2O_3(s) \rightarrow Mn(s) + Al_2O_3(s)$
(d) $AgNO_3(aq) + H_2SO_4(aq) \rightarrow Ag_2SO_4(s) + HNO_3(aq)$
(e) $Ca(OH)_2(aq) + HBr(aq) \rightarrow CaBr_2(aq) + H_2O(l)$
(f) $Cd(s) + HCl(aq) \rightarrow CdCl_2(aq) + H_2(g)$

2-40. Classify each of the following as a combination, a decomposition, a single-replacement, or a double-replacement reaction. If the equation is not balanced, then balance the equation.

(a) $NaClO_3(s) \rightarrow NaCl(s) + O_2(g)$
(b) $CaO(s) + SO_3(g) \rightarrow CaSO_4(s)$
(c) $Pb(s) + HCl(aq) \rightarrow PbCl_2(s) + H_2(g)$
(d) $Pb(s) + CdSO_4(aq) \rightarrow Cd(s) + PbSO_4(s)$
(e) $H_2(g) + AgCl(s) \xrightarrow[H_2O(l)]{} Ag(s) + HCl(aq)$
(f) $Hg_2(NO_3)_2(aq) + HC_2H_3O_2(aq) \rightarrow Hg_2(C_2H_3O_2)_2(s) + HNO_3(aq)$

2-41. Complete and balance the equations for the following combination reactions:

(a) $Mg(s) + N_2(g) \rightarrow$
(b) $H_2(g) + S(s) \rightarrow$
(c) $K(s) + Br_2(l) \rightarrow$
(d) $Al(s) + O_2(g) \rightarrow$
(e) $MgO(s) + SO_2(g) \rightarrow$

2-42. Complete and balance the equations for the following combination reactions:

(a) $Na(s) + Cl_2(g) \rightarrow$
(b) $Li(s) + O_2(g) \rightarrow$
(c) $MgO(s) + CO_2(g) \rightarrow$
(d) $H_2(g) + O_2(g) \rightarrow$
(e) $N_2(g) + H_2(g) \rightarrow$

2-43. Complete and balance the equations for the following single-replacement reactions (assume that any salts formed are soluble in water):

(a) $Zn(s) + HBr(aq) \rightarrow$
(b) $Al(s) + Fe_2O_3(s) \rightarrow$
(c) $Pb(s) + Cu(NO_3)_2(aq) \rightarrow$
(d) $Br_2(l) + NaI(aq) \rightarrow$

2-44. Complete and balance the equations for the following single-replacement reactions (assume that any salts formed are soluble in water):

(a) $Ba(s) + H_2O(g) \rightarrow$
(b) $Fe(s) + H_2SO_4(aq) \rightarrow$
(c) $Ca(s) + HBr(aq) \rightarrow$
(d) $Pb(s) + HCl(aq) \rightarrow$

2-45. Indicate which element is oxidized and which is reduced in the following reactions:

(a) $Ca(s) + Cl_2(g) \rightarrow CaCl_2(s)$
(b) $4Al(s) + 3O_2(g) \rightarrow 2Al_2O_3(s)$
(c) $2Rb(s) + Br_2(l) \rightarrow 2RbBr(s)$
(d) $2Na(s) + S(s) \rightarrow Na_2S(s)$

2-46. Indicate which element is oxidized and which is reduced in the following reactions:

(a) $2Li(s) + Se(s) \rightarrow Li_2Se(s)$
(b) $2Sc(s) + 3I_2(s) \rightarrow 2ScI_3(s)$
(c) $Ga(s) + P(s) \rightarrow GaP(s)$
(d) $2K(s) + F_2(g) \rightarrow 2KF(s)$

2-47. In the oxidation-reduction reactions in Problem 2-45, indicate how many electrons are transferred in the formation of one formula unit of product.

2-48. In the oxidation-reduction reactions in Problem 2-46, indicate how many electrons are transferred in the formation of one formula unit of product.

NET IONIC EQUATIONS

2-49. Write the net ionic equation corresponding to each of the following equations:

(a) $Na_2S(aq) + 2HCl(aq) \rightarrow 2NaCl(aq) + H_2S(g)$
(b) $PbCl_2(aq) + Na_2S(aq) \rightarrow 2NaCl(aq) + PbS(s)$
(c) $H_2SO_4(aq) + 2KOH(aq) \rightarrow K_2SO_4(aq) + 2H_2O(l)$
(d) $Na_2O(s) + 2HCl(aq) \rightarrow 2NaCl(aq) + H_2O(l)$
(e) $NH_3(g) + HCl(aq) \rightarrow NH_4Cl(aq)$

2-50. Write the net ionic equation corresponding to each of the following equations:

(a) $HClO_3(aq) + KOH(aq) \rightarrow KClO_3(aq) + H_2O(l)$
(b) $Pb(NO_3)_2(aq) + Na_2CO_3(aq) \rightarrow 2NaNO_3(aq) + PbCO_3(s)$
(c) $2AgClO_4(aq) + (NH_4)_2SO_4(aq) \rightarrow 2NH_4ClO_4(aq) + Ag_2SO_4(s)$
(d) $K_2S(aq) + Zn(NO_3)_2(aq) \rightarrow 2KNO_3(aq) + ZnS(s)$
(e) $Hg_2(ClO_3)_2(aq) + SrCl_2(aq) \rightarrow Sr(ClO_3)_2(aq) + Hg_2Cl_2(s)$

2-51. In each case balance the equation and write out the corresponding net ionic equation.

(a) $Fe(NO_3)_3(aq) + NaOH(aq) \rightarrow Fe(OH)_3(s) + NaNO_3(aq)$
(b) $Zn(ClO_4)_2(aq) + K_2S(aq) \rightarrow ZnS(s) + KClO_4(aq)$
(c) $Pb(NO_3)_2(aq) + KOH(aq) \rightarrow Pb(OH)_2(s) + KNO_3(aq)$
(d) $Zn(NO_3)_2(aq) + Na_2CO_3(aq) \rightarrow ZnCO_3(s) + NaNO_3(aq)$
(e) $Cu(ClO_4)_2(aq) + Na_2CO_3(aq) \rightarrow CuCO_3(s) + NaClO_4(aq)$

2-52. In each case balance the equation and write out the corresponding net ionic equation:

(a) $AgNO_3(aq) + Na_2S(aq) \rightarrow Ag_2S(s) + NaNO_3(aq)$
(b) $H_2SO_4(aq) + Pb(NO_3)_2(aq) \rightarrow PbSO_4(s) + HNO_3(aq)$
(c) $Hg(NO_3)_2(aq) + NaI(aq) \rightarrow HgI_2(s) + NaNO_3(aq)$
(d) $CdCl_2(aq) + AgClO_4(aq) \rightarrow AgCl(s) + Cd(ClO_4)_2(aq)$
(e) $LiBr(aq) + Pb(ClO_4)_2(aq) \rightarrow PbBr_2(s) + LiClO_4(aq)$

ACIDS AND BASES

2-53. Decide, based on your personal experience and on Table 2-8, which of the following solutions are acidic and which are basic:

(a) carbonated soft drinks
(b) apple cider
(c) milk of magnesia (suspension)
(d) tomatoes
(e) soap

2-54. Decide, based on your personal experience and on Table 2-8, which of the following solutions are acidic and which are basic:

(a) laundry detergent in water
(b) orange juice
(c) jam
(d) bicarbonate of soda in water
(e) household ammonia

2-55. Complete and balance the equation for each of the following acid-base reactions and name the salt produced in each case:

(a) $HClO_3(aq) + Ba(OH)_2(aq) \rightarrow$
(b) $HC_2H_3O_2(aq) + KOH(aq) \rightarrow$
(c) $HI(aq) + Mg(OH)_2(s) \rightarrow$
(d) $H_2SO_4(aq) + RbOH(aq) \rightarrow$

2-56. (a) Red ants contain an appreciable amount of formic acid, $\underline{H}CHO_2$, where only the underlined proton is acidic. It is observed that ants sprayed with window cleaner containing ammonia die quickly. Write the neutralization reaction between formic acid and ammonia. (b) Vinegar, which is an aqueous solution of acetic acid, is used to remove deposits of calcium carbonate from automatic coffee makers. Write the reaction between acetic acid and calcium carbonate.

ADDITIONAL PROBLEMS

2-57. Write the balanced chemical equation for each of the following reactions:

(a) An aqueous solution of hydrochloric acid is added to an aqueous solution of potassium cyanide with the emission of toxic hydrogen cyanide gas. This reaction is extremely dangerous and must be done in a well-ventilated hood.
(b) Potassium metal reacts with water with the evolution of hydrogen gas.
(c) Hydrogen peroxide decomposes to produce oxygen gas and a colorless liquid.
(d) Excess hydrogen gas is injected into a flask containing a small amount of liquid bromine. Upon heating, the bromine liquid disappears and a colorless gas is formed.

2-58. Without using any references, list as many elements as you can from memory and classify each as a metal or a nonmetal. Check your results and score yourself as follows:

more than 95	hall-of-famer
80 to 95	major leaguer
60 to 79	triple A player
40 to 59	semipro player
fewer than 40	little leaguer

2-59. The metal the ancient Greeks produced from lead-containing ores was an alloy of lead and silver. They were able to separate the two metals by melting the alloy and blowing air over the molten metal. Write the equation for the reaction that takes place.

2-60. Arrange the following metals in order of decreasing reactivity:

Pt Ag K Zn

2-61. For each of the following combination reactions, write the chemical formulas and balance the equation:

(a) sodium + sulfur → sodium sulfide
(b) calcium + bromine → calcium bromide
(c) barium + oxygen → barium oxide
(d) sulfur dioxide + oxygen → sulfur trioxide
(e) magnesium + nitrogen → magnesium nitride

2-62. For each of the following reactions, write the chemical formulas and balance the equation:

(a) potassium + water → potassium hydroxide + hydrogen
(b) potassium hydride + water → potassium hydroxide + hydrogen
(c) silicon dioxide + carbon → silicon carbide + carbon monoxide
(d) silicon dioxide + hydrogen fluoride → silicon tetrafluoride + water
(e) phosphorus + chlorine → phosphorus trichloride

2-63. Mercury has been known since early times, although we do not know when it was first discovered. Probably because it is a liquid metal, mercury has been thought to have mystical properties throughout history. It occurs in the ore cinnabar as the sulfide, HgS. The metal was prepared by heating the ore and condensing the mercury vapor. Write the equation for this reaction.

2-64. For each of the following reactions, write the chemical formulas and balance the equations:

(a) carbon monoxide + oxygen → carbon dioxide
(b) cesium + bromine → cesium bromide
(c) nitrogen monoxide + oxygen → nitrogen dioxide
(d) ammonia + oxygen → nitrogen monoxide + water
(e) gallium + arsenic → gallium arsenide

2-65. Arrange the following metals in order of decreasing reactivity:

Na Au Fe Sn

2-66. Lead production became important in Roman times because of its use in making pipes to carry water to the famous Roman baths. The manufacture of lead had been developed by the Greeks as a by-product of the silver mines outside Athens. Lead occurs in silver ores as the sulfide, galena, PbS. The silver occurs as the oxide. The ore is first roasted, that is, heated in air to convert the lead to an oxide. The ore is then smelted, that is, heated with charcoal. Write the chemical equations for the reactions that occur in the roasting and smelting of the lead-silver ore.

2-67. For each of the following reactions, write the chemical formulas and balance the equations:

(a) sodium + hydrogen → sodium hydride
(b) aluminum + sulfur → aluminum sulfide
(c) steam + carbon → carbon monoxide + hydrogen
(d) carbon + hydrogen → methane
(e) phosphorus trichloride + chlorine → phosphorus pentachloride

2-68. Write the balanced chemical equation for each of the following reactions:

(a) Zinc sulfide dissolves in aqueous hydrochloric acid with the emission of foul-smelling and highly toxic hydrogen sulfide gas.
(b) A solution containing calcium chloride is added to dilute aqueous phosphoric acid, and solid white calcium phosphate forms.
(c) Hydrochloric acid is spilled on a laboratory bench and solid sodium carbonate is thrown over the spill to neutralize the acid. Some of the solid dissolves, and carbon dioxide is given off.
(d) An aqueous solution of zinc chloride is added to an aqueous solution of sodium sulfide to produce a white precipitate of zinc sulfide.

2-69. The first metal to be prepared from its ore was copper, perhaps as early as 6000 BC in the Middle East. The ore, which contained copper(II) oxide, was heated with charcoal, which was prepared by the incomplete burning of wood and is mainly elemental carbon. Later, iron and tin were prepared in the same way. Bronze was made by mixing copper-containing ore with tin ore and heating the mixture in the presence of charcoal. Write the chemical equations for the preparation of these metals. Assume the ores to be CuO, SnO_2, and Fe_2O_3.

2-70. The crystals of many salts have a fixed number of water molecules associated with each formula unit. For example, copper(II) sulfate crystallizes from aqueous solution as $CuSO_4 \cdot 5H_2O$. Such compounds are called hydrates, and the name of $CuSO_4 \cdot 5H_2O$ is copper(II) sulfate pentahydrate. Name the following hydrates:

(a) $NiSO_4 \cdot 6H_2O(s)$
(b) $PbCl_2 \cdot 2H_2O(s)$
(c) $LiOH \cdot H_2O(s)$
(d) $IrBr_3 \cdot 4H_2O(s)$
(e) $Li_2CrO_4 \cdot 2H_2O(s)$

The Alkali Metals

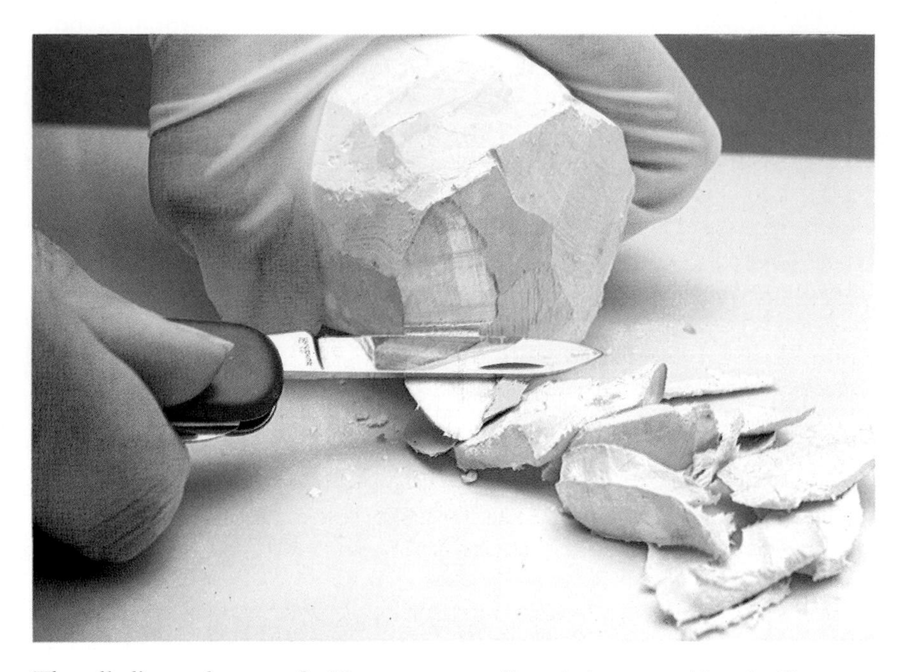

The alkali metals are soft. Here we see sodium being cut with a knife.

The alkali metals are lithium, sodium, potassium, rubidium, cesium, and francium. They occur in Group 1 of the periodic table and so have an ionic charge of +1 in their compounds. All the alkali metals are very reactive. None occur as the free metal in nature. They must be stored under an inert substance, such as kerosene, because they react spontaneously and rapidly with the oxygen and water vapor in the air.

B-1. THE HYDROXIDES OF THE ALKALI METALS ARE STRONG BASES

The alkali metals are all fairly soft and can be cut with a sharp knife (Frontispiece). When freshly cut they are bright and shiny, but they soon take on a dull finish because of their reaction with air. The alkali metals are so called because their hydroxides, MOH, are all soluble bases in water (alkaline means basic). The physical properties of the alkali metals are given in Table B-1. There are no stable isotopes of francium; all of them are radioactive.

The alkali metals can be obtained by **electrolysis,** which is a decomposition reaction driven by passing an electric current through a solution (Chapter 20). For example, sodium metal is obtained by electrolysis of molten mixtures of sodium chloride and calcium chloride.

▪ Sodium is the least expensive metal per unit volume.

$$2NaCl\ [\text{in } CaCl_2(l)] \xrightarrow[\text{electrolysis}]{600°C} 2Na(l) + Cl_2(g)$$

Chlorine gas is a useful by-product of the electrolysis. The $CaCl_2$ is added to the NaCl to lower the temperature necessary for the operation of the electrolysis cell. Pure NaCl melts at 800°C.

B-2. THE ALKALI METALS ARE VERY REACTIVE

The alkali metals react directly with all the nonmetals except the noble gases. The increasing reactivity of the alkali metals with increasing atomic number is demonstrated in a spectacular manner by their reaction with water. When metallic lithium reacts with water, hydrogen gas is evolved slowly, whereas sodium reacts vigorously with water. The reaction of potassium with water produces a fire because the heat generated by the reaction is sufficient to ignite the hydrogen gas evolved. Rubidium and cesium react with water with explosive violence.

The alkali metals react directly with oxygen. Molten lithium ignites in oxygen to form $Li_2O(s)$; the reaction is accompanied by a bright red flame. The reactions of the other alkali metals do not yield the oxides M_2O. With sodium the **peroxide** Na_2O_2 is formed, and with potassium, rubidium, and cesium the **superoxides** KO_2, RbO_2, and CsO_2 are formed.

Both potassium superoxide and sodium peroxide are used in self-contained breathing apparatus. The relevant reactions for KO_2 are

Figure B-1 Lithium floating on oil, which in turn is floating on water. Lithium has the lowest density of any element that is a solid or a liquid at 20°C.

Table B-1 The physical properties of the alkali metals

Property	Lithium	Sodium	Potassium	Rubidium	Cesium
chemical symbol	Li	Na	K	Rb	Cs
atomic number	3	11	19	37	55
atomic mass	6.941	22.98977	39.0983	85.4678	132.9054
melting point/°C	181	98	64	39	29
boiling point/°C	1347	892	774	696	670
density at 20°C/g·cm^{-3}	0.53	0.97	0.87	1.53	1.88
ionic radius/pm*	60	95	133	148	169

*pm = picometer = 10^{-12} m

$$4KO_2(s) + \underset{\text{exhaled air}}{2H_2O(g)} \rightarrow 4KOH(s) + 3O_2(g)$$

$$KOH(s) + \underset{\text{exhaled air}}{CO_2(g)} \rightarrow KHCO_3(s)$$

Figure B-2 Sylvite is a potassium chloride mineral that is found in extensive deposits in ancient lake and sea beds. It is used as a major source of potassium and its compounds.

The reactions for Na_2O_2 are

$$2Na_2O_2(s) + \underset{\text{exhaled air}}{2H_2O(g)} \rightarrow 4NaOH(s) + O_2(g)$$

$$NaOH(s) + \underset{\text{exhaled air}}{CO_2(g)} \rightarrow NaHCO_3(s)$$

Note that $O_2(g)$ is generated and that $H_2O(g)$ and $CO_2(g)$ are absorbed in each case.

The alkali metals react directly with hydrogen at high temperatures to form hydrides. For example,

$$2Na(l) + H_2(g) \xrightarrow{500°C} 2NaH(s)$$

The alkali metal hydrides are ionic compounds that contain the hydride ion, H^-. The hydrides react with water to liberate hydrogen,

$$NaH(s) + H_2O(l) \rightarrow NaOH(aq) + H_2(g)$$

and are used to remove traces of water from organic solvents. In such cases, the metal hydroxide precipitates from the solution.

Lithium is the only element that reacts directly with nitrogen at room temperature (Figure B-3):

$$6Li(s) + N_2(g) \xrightarrow{\text{room T}} \underset{\text{lithium nitride}}{2Li_3N(s)}$$

The reddish black lithium nitride reacts directly with water to form ammonia:

$$Li_3N(s) + 3H_2O(l) \rightarrow 3LiOH(aq) + NH_3(g)$$

This reaction can be used to prepare deuterated ammonia, ND_3:

$$Li_3N(s) + 3D_2O(l) \rightarrow 3LiOD(aq) + ND_3(g)$$

Some of the more common reactions of the alkali metals are summarized in Figure B-4.

Compounds of the alkali metals are for the most part white, high-melting ionic solids. With very few exceptions, alkali metal salts are soluble in water and the resulting solutions conduct an electric current, as a result of the dissociation of the salt into its constituent ions.

Not all the properties of lithium are analogous to those of the other members of the alkali metal family. For example, in contrast to the analogous salts of the other alkali metals, LiF and Li_2CO_3 are insoluble in water and LiCl is soluble in alcohols and ethers. The anomalous behavior of lithium is ascribed to the much smaller size (Table B-1) of the Li^+ ion.

The alkali metals have the unusual property of dissolving in liquid ammonia to yield blue solutions that conduct an electric current (Figure B-5). The properties of such a solution are interpreted in terms of **solvated electrons** and alkali metal ions:

Figure B-3 Lithium nitride, Li_3N.

$$M(s) \xrightarrow[NH_3(l)]{} M^+(amm) + e^-(amm)$$

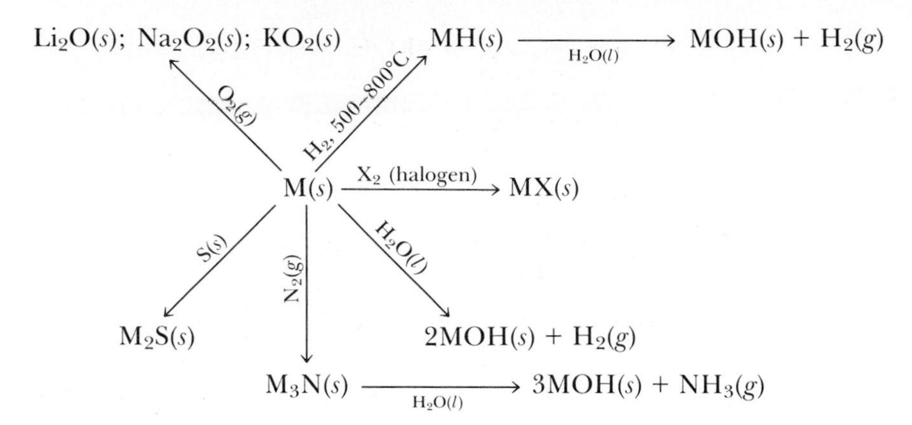

Figure B-4 Representative reactions of the Group 1 metals.

The solvated electrons are electrons surrounded by ammonia molecules. When the blue solutions are concentrated by evaporation, they become bronze in color and behave like liquid metals.

B-3. MANY ALKALI METAL COMPOUNDS ARE IMPORTANT COMMERCIALLY

Sodium hydroxide is the seventh ranked industrial chemical. Over 22 billion pounds of it is produced annually in the United States. Sodium hydroxide, sometimes called **caustic soda,** is prepared by the electrolysis of concentrated aqueous sodium chloride solutions:

$$2NaCl(aq) + 2H_2O(l) \xrightarrow{\text{electrolysis}} 2NaOH(aq) + H_2(g) + Cl_2(g)$$

or by the reaction between calcium hydroxide (called slaked lime) and sodium carbonate:

$$Na_2CO_3(aq) + Ca(OH)_2(aq) \rightarrow 2NaOH(aq) + CaCO_3(s)$$

The formation of the insoluble $CaCO_3$ is a driving force for this second reaction. The alkali metal hydroxides are white, translucent, corrosive solids that are extremely soluble in water; at 20°C the solubility of NaOH is 65 g per 100 mL of H_2O and that of KOH is 100 g per 100 mL of H_2O.

Sodium carbonate, which is called **soda ash,** is the tenth-ranked industrial chemical. The annual U.S. production of sodium carbonate exceeds 17 billion pounds. About 90 percent of the soda ash produced in the United States is obtained from natural deposits of the mineral **trona,** which has the composition $Na_2CO_3 \cdot NaHCO_3 \cdot 2H_2O(s)$. When trona is heated it yields sodium carbonate:

$$2Na_2CO_3 \cdot NaHCO_3 \cdot 2H_2O(s) \xrightarrow{\text{high T}} 3Na_2CO_3(s) + CO_2(g) + 3H_2O(g)$$

The carbon dioxide is recovered as a by-product.

Soda ash also is prepared from sodium chloride by the **Solvay process,** which was devised by the Belgian brothers Ernest and Edward Solvay in 1861. The Solvay process is used extensively in countries that lack large trona deposits. In this process, carbon dioxide is bubbled through a cooled solution of sodium chloride and ammonia. The reactions are

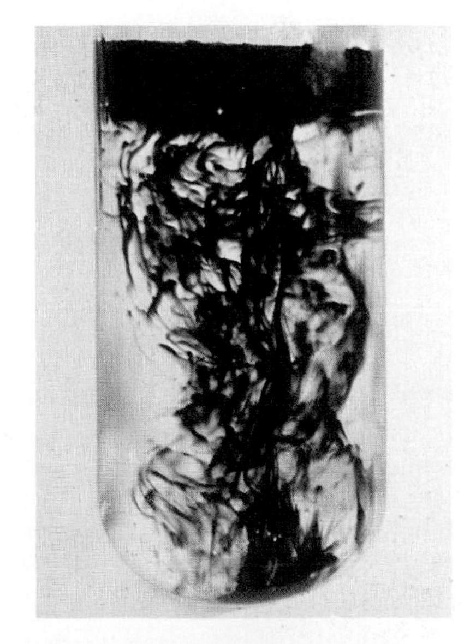

Figure B-5 Sodium dissolving in liquid ammonia. Solutions of sodium in liquid ammonia are deep blue in color and are able to conduct an electric current.

Figure B-6 Underground deposits of trona being mined at Tenneco's Green River, Wyoming, mine.

$$NH_3(aq) + CO_2(aq) + H_2O(l) \rightarrow NH_4^+(aq) + HCO_3^-(aq)$$

$$NaCl(aq) + NH_4^+(aq) + HCO_3^-(aq) \xrightarrow{15^\circ C} NaHCO_3(s) + NH_4Cl(aq)$$

At 15°C the sodium hydrogen carbonate precipitates from the solution. Part of the sodium hydrogen carbonate is converted to sodium carbonate by heating:

$$2NaHCO_3(s) \xrightarrow{80^\circ C} Na_2CO_3(s) + H_2O(l) + CO_2(g)$$

The carbon dioxide produced in this reaction is used again in the first reaction.

Table B-2 Some commercially important alkali metal compounds and their uses

Compound	Uses
lithium aluminum hydride, $LiAlH_4$	production of many pharmaceuticals, perfumes, and organic chemicals
sodium hydrogen carbonate (sodium bicarbonate), $NaHCO_3$	manufacture of effervescent salts and beverages, baking powder, gold plating
sodium carbonate, Na_2CO_3	manufacture of glass, pulp and paper, soaps and detergents, textiles
sodium hydroxide, NaOH	production of rayon, cellulose, paper, soaps, detergents, textiles, oven cleaner
sodium sulfate decahydrate (Glauber's salt), $Na_2SO_4 \cdot 10H_2O$	solar heating storage, air conditioning
potassium carbonate (potash), K_2CO_3	manufacture of special glass for optical instruments and electronic devices, soft soaps
potassium nitrate, KNO_3	pyrotechnics, explosives, matches

The commercial success of the Solvay process requires the recovery of the ammonia, which is relatively expensive. The ammonia is recovered from the NH_4Cl by the reaction

$$2NH_4Cl(aq) + Ca(OH)_2(s) \rightarrow 2NH_3(g) + CaCl_2(aq) + 2H_2O(l)$$

The calcium hydroxide and the carbon dioxide used in the process are obtained by heating limestone (primarily $CaCO_3$).

The raw materials of the Solvay process are sodium chloride, limestone, and water, all of which are inexpensive. The principal use of sodium carbonate is in the manufacture of glass (Interchapter H).

Some other important alkali metal compounds and their uses are given in Table B-2.

TERMS YOU SHOULD KNOW

electrolysis 83
peroxide 83
superoxide 83
solvated electron 84
caustic soda 85
soda ash 85
trona 85
Solvay process 85

QUESTIONS

B-1. Why must the alkali metals be stored under an inert liquid like kerosene or an inert gas like argon?

B-2. How are the alkali metals produced commercially?

B-3. Why do you think sodium metal is the least expensive metal per unit volume?

B-4. What is the only element that reacts directly with nitrogen at room temperature?

B-5. Outline, by means of balanced chemical equations, the Solvay process. What are the raw materials in the Solvay process?

B-6. Outline, by means of balanced chemical equations, the operation of a self-contained breathing apparatus charged with sodium peroxide.

B-7. Explain why the alkali metals cannot be stored in water.

B-8. Arrange the alkali metals in order of increasing reactivity.

B-9. Superoxide ion, peroxide ion, and oxide ion can be thought of as arising from oxygen by the transfer of the appropriate number of electrons to O_2: $O_2 + 1e^- \rightarrow O_2^-$; $O_2 + 2e^- \rightarrow O_2^{2-}$; and $O_2 + 4e^- \rightarrow 2O^{2-}$. What are the chemical formulas for potassium oxide, potassium peroxide, and potassium superoxide?

B-10. What is the major commercial source of sodium carbonate?

PROBLEMS

B-11. Complete and balance the following equations:

(a) $Na(s) + O_2(g) \rightarrow$
(b) $Na(s) + H_2O(l) \rightarrow$
(c) $Li(s) + N_2(g) \rightarrow$
(d) $NaH(s) + H_2O(l) \rightarrow$
(e) $Li_3N(s) + H_2O(l) \rightarrow$

B-12. Complete and balance the following equations:

(a) $Na_2CO_3(aq) + Ca(OH)_2(s) \rightarrow$
(b) $NaHCO_3(s) \xrightarrow{80°C}$
(c) $NH_3(aq) + CO_2(aq) + H_2O(l) \rightarrow$

B-13. Write chemical formulas for the following compounds:

(a) sodium hydrogen carbonate
(b) potassium hydroxide
(c) calcium hydroxide
(d) cesium hydride

B-14. Name the following alkali metal compounds (the major commercial uses are given in parentheses):

(a) $Na_3PO_4(s)$ (detergents)
(b) K_2SO_4 (glass manufacture, cathartic)
(c) $Na_2S_2O_3 \cdot 5H_2O$ ("hypo" solution used in photography)
(d) KNO_3 (gunpowder)

CHAPTER • 3

CHEMICAL CALCULATIONS

Starting at the upper left and going in a clockwise direction, we have molar quantities of acetylsalicylic acid (aspirin) (180 g), water (18.0 g), sucrose (table sugar) (342 g), mercury (201 g), iron (55.9 g), sodium chloride (58.5 g), and iodine (254 g). One mole of a substance is that quantity containing the number of grams numerically equal to its formula mass.

It is important to be able to calculate how much of a reaction product can be obtained from a given amount of reactant or how much reactant should be used to obtain a desired amount of product. In this chapter, we develop a systematic method for carrying out a variety of chemical calculations. For example, we shall determine simplest chemical formulas using mass percentages obtained from chemical analysis. We'll combine chemical analysis data with molecular mass information to determine molecular formulas. After discussing the determination of molecular formulas, we'll treat calculations involving chemical reactions. For example, we'll calculate how much of a compound can be prepared starting with a given amount of reactants. We will also learn how to carry out such calculations for reactions that occur in solution.

3-1. THE QUANTITY OF A SUBSTANCE THAT IS EQUAL TO ITS FORMULA MASS IN GRAMS IS CALLED A MOLE

In Chapter 1 we learned that the atomic mass of an element is a relative quantity; it is the mass of one atom of the element relative to the mass of one atom of carbon-12, which has a mass of exactly 12 amu. Consider the following table of four elements:

Element	Atomic mass
helium, He	4
carbon, C	12
titanium, Ti	48
molybdenum, Mo	96

This table shows that one carbon atom has a mass three times that of one helium atom, that one titanium atom has a mass four times that of one carbon atom, and that one molybdenum atom has a mass twice that of one titanium atom, eight times that of one carbon atom, and 24 times that of one helium atom. It is important to realize that we have not deduced the mass of any one atom; at this point we can determine only relative masses.

Consider 12 g of carbon and 48 g of titanium. One titanium atom has a mass four times that of one carbon atom, and therefore 48 g of titanium atoms must contain the same number of atoms as 12 g of carbon. Similarly, one molybdenum atom has twice the mass of one titanium atom, and so 96 g of molybdenum must contain the same number of atoms as 48 g of titanium. We conclude that 12 g of carbon, 48 g of titanium, and 96 g of molybdenum all contain the same number of atoms. If we continue this line of reasoning, then we find that the quantity of any element that is numerically equal to its atomic mass in grams contains the same number of atoms as the corresponding quantity of any other element. Thus, 10.8 g of boron, 23.0 g of sodium, 63.6 g of copper, and 200.6 g of mercury all contain the same number of atoms.

Samples of 12 g of carbon, 48 g of titanium *(single long rod)*, and 96 g of molybdenum *(two smaller rods)*. Each sample contains the same number of atoms.

All the substances we have considered so far are **atomic substances,** that is, substances whose constituent particles are atoms. Now consider the following **molecular substances:**

Substance	Molecular mass
methane, CH_4	$12.0 + (4 \times 1.0) = 16.0$
oxygen, O_2	$2 \times 16.0 = 32.0$
ozone, O_3	$3 \times 16.0 = 48.0$

Recall from Chapter 1 that the molecular mass of a substance is the sum of the atomic masses in the chemical formula. Like atomic masses, molecular masses are relative masses. A molecule of oxygen, O_2, has a mass (32.0 amu) twice that of a molecule of methane (16.0 amu). A molecule of ozone has a mass (48.0 amu) three times that of a molecule of methane. Using the same reasoning we used for atomic substances, we conclude that 16.0 g of methane, 32.0 g of oxygen, and 48.0 g of ozone must all contain the same number of *molecules*. In addition, because the atomic mass of titanium is equal to the molecular mass of ozone, the number of *atoms* in 48.0 g of titanium must equal the number of *molecules* in 48.0 g of ozone.

▪ A formula unit is the collection of atoms or ions indicated by the chemical formula.

By using the term **formula mass** instead of atomic mass or molecular mass, we can combine all the preceding statements. Given that the **formula unit** is the collection of atoms or ions indicated by the chemical formula, we say that *one formula mass in grams of any substance contains the same number of formula units as the corresponding quantity of any other substance.* Thus, 4.0 g of helium, 12.0 g of carbon, 16.0 g of methane, and 32.0 g of oxygen all contain the same number of formula units. The formula units are atoms in the cases of helium and carbon and molecules in the cases of methane and oxygen. We can now state our preliminary definition of a **mole:** *the quantity of a substance that is equal to its formula mass in grams is called a mole* (abbreviated mol). The **molar mass** of a compound is numerically equal to its formula mass or molecular mass but has the units grams per mole. Thus CH_4 has a formula mass of 16.04 and a molar mass of 16.04 $g{\cdot}mol^{-1}$. The Frontispiece to this chapter shows one mole of a number of common substances.

▪ The mole is one of the most important concepts in chemistry.

As you perform chemical calculations, you will often need to determine the number of moles in a given mass of a substance. For example, we might need to calculate the number of moles of methane in 50.0 g of methane. The formula mass of CH_4 is 16.04, and so there are 16.04 g per mole of methane. To determine how many moles there are in 50.0 g, we must convert grams of methane to moles of methane. We carry out this conversion using the unit conversion factor

$$1 = \frac{1 \text{ mol } CH_4}{16.04 \text{ g } CH_4}$$

Thus

$$\text{mol } CH_4 = (50.0 \text{ g } CH_4)\left(\frac{1 \text{ mol } CH_4}{16.04 \text{ g } CH_4}\right) = 3.12 \text{ mol } CH_4$$

Notice that we express our result to three significant figures and that the answer is expressed in moles of CH_4, as required.

You should also be able to calculate the mass of a certain number of moles of a substance. For example, let's calculate the mass of 2.16 mol of sodium chloride. The formula mass of NaCl is 58.44. The mass of NaCl in 2.16 mol is

$$\text{g of NaCl} = (2.16 \text{ mol NaCl})\left(\frac{58.44 \text{ g NaCl}}{1 \text{ mol NaCl}}\right) = 126 \text{ g NaCl}$$

Notice that we express the final result to three significant figures and that the units yield grams of NaCl.

Example 3-1: Calculate the number of moles in (a) 28.0 g of water (about 1 ounce) and (b) 325 mg of aspirin, $C_9H_8O_4$ (325 mg is the mass of aspirin in one 5-grain aspirin tablet).

Solution: (a) The formula mass of H_2O is 18.02. Consequently the number of moles of water in 28.0 g is

$$\text{mol } H_2O = (28.0 \text{ g } H_2O)\left(\frac{1 \text{ mol } H_2O}{18.02 \text{ g } H_2O}\right) = 1.55 \text{ mol } H_2O$$

(b) The chemical formula of aspirin is $C_9H_8O_4$, and so its formula mass is

$$(9 \times 12.01) + (8 \times 1.008) + (4 \times 16.00) = 180.15$$

The number of moles of aspirin in 325 mg is

$$\text{mol } C_9H_8O_4 = (325 \text{ mg } C_9H_8O_4)\left(\frac{1 \text{ g}}{1000 \text{ mg}}\right)\left(\frac{1 \text{ mol } C_9H_8O_4}{180.15 \text{ g } C_9H_8O_4}\right)$$
$$= 1.80 \times 10^{-3} \text{ mol } C_9H_8O_4$$

Notice that we have to convert milligrams to grams before dividing by 180.15 g.

Example 3-1 illustrates an important point. *In order to calculate the number of moles in a given mass of a chemical compound, it is necessary to know the chemical formula of the compound. A mole of any compound is defined only in terms of its chemical formula.*

3-2. ONE MOLE OF ANY SUBSTANCE CONTAINS AVOGADRO'S NUMBER OF FORMULA UNITS

It has been determined experimentally that one mol of any substance contains 6.022×10^{23} formula units. This number is called **Avogadro's number** after Amedeo Avogadro, an Italian scientist who was one of the earliest proponents of the atomic theory. Not only do we say that one mol of any substance contains Avogadro's number of formula units, but another widely used definition of a mole is that one mol is that mass of a substance that contains Avogadro's number of formula units or "elementary entities." For example, the molecular mass of water is 18.02, and thus 18.02 g of water contains 6.022×10^{23} molecules. A mole is, in essence, a certain number of "things" (that is, formula units), like a dozen. Instead of 12, as in a dozen, the number is 6.022×10^{23}. A molecular mass or formula mass is thus the mass in grams of 6.022×10^{23} molecules or formula units.

▪ Avogadro's number is 6.022×10^{23} formula units/mole.

Avogadro's number is an enormous number, much larger than any number you may have encountered. In order to appreciate the magnitude of Avogadro's number, let's compute how many years it would take to spend Avogadro's number of dollars at the rate of one million dollars per second. There are 3.15×10^7 seconds in 1 year; thus the number of years required is

$$(6.022 \times 10^{23} \text{ dollars})\left(\frac{1 \text{ s}}{10^6 \text{ dollars}}\right)\left(\frac{1 \text{ year}}{3.15 \times 10^7 \text{ s}}\right)$$
$$= 1.91 \times 10^{10} \text{ years}$$

or 19.1 billion years (1 billion = 10^9). This is over four times longer than the estimated age of the earth (4.6 billion years) and roughly equal to the conjectured age of the universe (13 to 20 billion years). This illustrates how large Avogadro's number is and, consequently, how small atoms and molecules are. Look again at the Frontispiece to this chapter. Each of those samples contains 6.022×10^{23} formula units of the indicated substance.

We can use Avogadro's number to calculate the mass of one atom or molecule.

Example 3-2: Using Avogadro's number, calculate the mass of one nitrogen molecule.

Solution: Recall that nitrogen occurs as a diatomic molecule. The formula of molecular nitrogen is N_2, and so its formula mass, or molecular mass, is 28.02. Thus there are 6.022×10^{23} molecules of nitrogen in 28.02 g of nitrogen. The mass of one nitrogen molecule is

$$\begin{aligned}\left(\begin{matrix}\text{mass of one}\\ \text{nitrogen molecule}\end{matrix}\right) &= \frac{\begin{matrix}\text{mass of Avogadro's number}\\ \text{of nitrogen molecules}\end{matrix}}{\text{Avogadro's number}}\\ &= \frac{\text{mass corresponding to 1 mol } N_2}{\text{Avogadro's number}}\\ &= \frac{28.02\text{ g } N_2}{6.022 \times 10^{23}\ N_2\text{ molecules}}\\ &= 4.653 \times 10^{-23}\text{ g}\cdot\text{molecule}^{-1}\end{aligned}$$

Avogadro's number can also be used to calculate the number of atoms or molecules in a given mass of a substance. The next example illustrates this type of calculation.

Example 3-3: Calculate how many methane molecules and how many hydrogen and carbon atoms there are in 25.0 g of methane.

Solution: The formula mass of CH_4 is $12.0 + (4 \times 1.0) = 16.0$, and so 25.0 g of CH_4 corresponds to

$$\text{mol } CH_4 = (25.0\text{ g } CH_4)\left(\frac{1\text{ mol } CH_4}{16.0\text{ g } CH_4}\right) = 1.56\text{ mol } CH_4$$

Since 1 mol of CH_4 contains 6.022×10^{23} molecules, 1.56 mol must contain

$$\begin{aligned}(\text{number of } CH_4\text{ molecules}) &= (1.56\text{ mol } CH_4)\left(\frac{6.022 \times 10^{23}\text{ molecules}}{1\text{ mol}}\right)\\ &= 9.39 \times 10^{23}\ CH_4\text{ molecules}\end{aligned}$$

Each molecule of methane contains one carbon atom and four hydrogen atoms, and so

$$\begin{aligned}\text{number of C atoms} &= (9.39 \times 10^{23}\ CH_4\text{ molecules})\left(\frac{1\text{ C atom}}{1\ CH_4\text{ molecule}}\right)\\ &= 9.39 \times 10^{23}\text{ C atoms}\end{aligned}$$

$$\text{number of H atoms} = (9.39 \times 10^{23}\ \text{CH}_4\ \text{molecules})\left(\frac{4\ \text{H atoms}}{1\ \text{CH}_4\ \text{molecule}}\right)$$

$$= 3.76 \times 10^{24}\ \text{H atoms}$$

Table 3-1 summarizes some relationships between molar quantities. In reading the table, recall that sodium chloride is an ionic compound, with one formula unit consisting of one sodium ion, Na^+, and one chloride ion, Cl^-. Similarly, the formula unit of barium fluoride consists of one barium ion, Ba^{2+}, and *two* fluoride ions, F^-.

We conclude this section with the official SI definition of mole. "The mole is the amount of substance of a system which contains as many elementary entities as there are atoms in exactly 0.012 kilogram of carbon-12. When the mole is used, the elementary entities must be specified; they may be atoms, molecules, ions, electrons, other particles, or specified groups of such particles." Note that because the atomic mass of ^{12}C is exactly 12 by definition, a mole of ^{12}C contains exactly 12 g (= 0.012 kg) of carbon. The SI definition of mole is equivalent to the other definitions given in this section.

3-3. SIMPLEST FORMULAS CAN BE DETERMINED BY CHEMICAL ANALYSIS

■ The word stoichiometry is derived from the Greek words *stoicheia*, meaning simplest components or parts, and *metrein*, meaning to measure.

Stoichiometry (stoi'kē om'i trē) is the calculation of the quantities of elements or compounds involved in chemical reactions. The concept of

Table 3-1 Some relationships between molar quantities

Substance	Formula	Formula mass	Molar mass/g·mol^{-1}	Number of particles in 1 mol	Moles
atomic chlorine	Cl	35.45	35.45	6.022×10^{23} chlorine atoms	1 mol Cl atoms
chlorine gas	Cl_2	70.90	70.90	6.022×10^{23} chlorine molecules	1 mol Cl_2 molecules
				12.044×10^{23} chlorine atoms	2 mol Cl atoms
water	H_2O	18.02	18.02	6.022×10^{23} water molecules	1 mol H_2O molecules
				12.044×10^{23} hydrogen atoms	2 mol H atoms
				6.022×10^{23} oxygen atoms	1 mol O atoms
sodium chloride	NaCl	58.44	58.44	6.022×10^{23} NaCl formula units	1 mol NaCl formula units
				6.022×10^{23} sodium ions	1 mol Na^+ ions
				6.022×10^{23} chloride ions	1 mol Cl^- ions
barium fluoride	BaF_2	175.3	175.3	6.022×10^{23} BaF_2 formula units	1 mol BaF_2 formula units
				6.022×10^{23} barium ions	1 mol Ba^{2+} ions
				12.044×10^{23} fluoride ions	2 mol F^- ions
nitrate ion	NO_3^-	62.01	62.01	6.022×10^{23} nitrate ions	1 mol NO_3^- ions
				6.022×10^{23} nitrogen atoms	1 mol N atoms
				18.066×10^{23} oxygen atoms	3 mol O atoms

a mole is central to carrying out stoichiometric calculations. For example, we can use the concept of a mole to determine the simplest chemical formula of a substance. Zinc oxide is found by chemical analysis to be 80.3 percent (by mass) zinc and 19.7 percent (by mass) oxygen. In working with mass percentages in chemical calculations, it is convenient to consider a 100-g sample so that the mass percentages can be easily converted to grams. For example, a 100-g sample of zinc oxide contains 80.3 g of zinc and 19.7 g of oxygen. We can write this schematically as

$$80.3 \text{ g Zn} \Leftrightarrow 19.7 \text{ g O}$$

where the symbol $\Leftrightarrow$ means "is stoichiometrically equivalent to" or, in this case, "combines with." If we divide 80.3 g by the atomic mass of zinc (65.38) and 19.7 g by the atomic mass of oxygen (16.00), then we find that

$$\text{mol Zn} = (80.3 \text{ g Zn})\left(\frac{1 \text{ mol Zn}}{65.38 \text{ g Zn}}\right) = 1.23 \text{ mol Zn}$$

and

$$\text{mol O} = (19.7 \text{ g O})\left(\frac{1 \text{ mol O}}{16.00 \text{ g O}}\right) = 1.23 \text{ mol O}$$

Thus, we have

$$1.23 \text{ mol Zn} \Leftrightarrow 1.23 \text{ mol O}$$

or, correspondingly,

$$1.00 \text{ mol Zn} \Leftrightarrow 1.00 \text{ mol O}$$

By dividing both sides by Avogadro's number, we get

$$1.00 \text{ atom of Zn} \Leftrightarrow 1.00 \text{ atom of O}$$

This says that one atom of zinc combines with one atom of oxygen and that the **simplest chemical formula** of zinc oxide is therefore ZnO. We call ZnO the simplest chemical formula of zinc oxide because chemical analysis provides us with only the *ratios* of atoms in a compound. Mass percentages alone cannot be used to distinguish among ZnO, Zn_2O_2, Zn_3O_3, or any other multiple of ZnO. Simplest formulas are often called **empirical formulas.** The following Example illustrates another calculation of a simplest, or empirical, formula.

▪ Rubbing alcohol is a topical antiseptic.

Example 3-4: The chemical name for rubbing alcohol is isopropyl alcohol. Chemical analysis shows that pure isopropyl alcohol is 60.0 percent carbon, 13.4 percent hydrogen, and 26.6 percent oxygen by mass. Determine the empirical formula of isopropyl alcohol.

Solution: As usual, we take a 100-g sample and write

$$60.0 \text{ g C} \Leftrightarrow 13.4 \text{ g H} \Leftrightarrow 26.6 \text{ g O}$$

We divide each value by the corresponding atomic mass and get

$$(60.0 \text{ g C})\left(\frac{1 \text{ mol C}}{12.01 \text{ g C}}\right) = 5.00 \text{ mol C} \Leftrightarrow (13.4 \text{ g H})\left(\frac{1 \text{ mol H}}{1.008 \text{ g H}}\right)$$

$$\approx (13.4 \text{ g H})\left(\frac{1 \text{ mol H}}{1.008 \text{ g H}}\right) = 13.3 \text{ mol H}$$

$$\approx (26.6 \text{ g O})\left(\frac{1 \text{ mol O}}{16.00 \text{ g O}}\right) = 1.66 \text{ mol O}$$

or

$$5.00 \text{ mol C} \approx 13.3 \text{ mol H} \approx 1.66 \text{ mol O}$$

To find a simple, whole-number relationship for these values, we divide through by the smallest value (1.66) and get

$$3.01 \text{ mol C} \approx 8.01 \text{ mol H} \approx 1.00 \text{ mol O}$$

Realizing that 3.01 and 8.01 can be rounded off to 3 and 8 within probable experimental error, we find that an isopropyl alcohol molecule consists of three carbon atoms, eight hydrogen atoms, and one oxygen atom. Thus the empirical formula is C_3H_8O.

The next example illustrates an experimental procedure for determining empirical formulas.

Example 3-5: A 0.450-g sample of magnesium metal is reacted completely in a nitrogen atmosphere to produce 0.623 g of magnesium nitride. Use these data to determine the empirical formula of magnesium nitride.

Solution: The 0.623 g of magnesium nitride contains 0.450 g of magnesium, and so the mass of nitrogen in the product is

$$\text{mass of N in product} = 0.623 \text{ g} - 0.450 \text{ g} = 0.173 \text{ g N}$$

We can convert 0.450 g of Mg to moles of Mg by dividing by its atomic mass (24.31):

$$\text{mol Mg} = (0.450 \text{ g Mg})\left(\frac{1 \text{ mol Mg}}{24.31 \text{ g Mg}}\right) = 0.0185 \text{ mol Mg}$$

Similarly, by dividing the mass of nitrogen by its atomic mass (14.01), we obtain

$$\text{mol N} = (0.173 \text{ g N})\left(\frac{1 \text{ mol N}}{14.01 \text{ g N}}\right) = 0.0123 \text{ mol N}$$

Thus we have the relation

$$0.0185 \text{ mol Mg} \approx 0.0123 \text{ mol N}$$

If we divide both quantities by the smaller number (0.0123), then we obtain

$$1.50 \text{ mol Mg} \approx 1.00 \text{ mol N}$$

Multiplying both sides by 2 to get whole numbers, we get

$$3.00 \text{ mol Mg} \approx 2.00 \text{ mol N}$$

Thus, we see that 3.00 mol of magnesium combines with 2.00 mol of nitrogen. Thus the empirical formula for magnesium nitride is Mg_3N_2.

Figure 3-1 Magnesium nitride, Mg_3N_2.

If we know the chemical formula of a compound, then it is a simple matter to compute the percentage by mass of any element in that compound. As an example, let's compute the mass percentage of nitrogen (as N) in the fertilizer ammonium nitrate, $NH_4NO_3(s)$. The formula mass of NH_4NO_3 is

$$(2 \times 14.01) + (4 \times 1.008) + (3 \times 16.00) = 80.05$$

Nitrogen (as N) makes a contribution of 2×14.01 to the formula mass of NH_4NO_3. The mass percentage of N in NH_4NO_3 is

$$\begin{pmatrix}\text{mass percentage}\\ \text{of nitrogen}\end{pmatrix} = \left(\frac{\text{mass due to N}}{\text{mass of NH}_4\text{NO}_3}\right) \times 100$$

$$= \left(\frac{2 \times 14.01}{80.05}\right)(100) = 35.00\%$$

Example 3-6: Compute the mass percentage of iron in iron(III) oxide, $Fe_2O_3(s)$.

Solution: The formula mass of $Fe_2O_3(s)$ is

$$(2 \times 55.85) + (3 \times 16.00) = 159.7$$

Iron makes a contribution of 2×55.85 to the formula mass of Fe_2O_3. Thus the mass percentage of iron in Fe_2O_3 is

$$\begin{pmatrix}\text{mass percentage}\\ \text{of iron}\end{pmatrix} = \left(\frac{\text{mass due to Fe}}{\text{mass of Fe}_2\text{O}_3}\right) \times 100$$

$$= \left(\frac{2 \times 55.85}{159.7}\right)(100) = 69.94\%$$

3-4. EMPIRICAL FORMULAS CAN BE USED TO DETERMINE AN UNKNOWN ATOMIC MASS

If we know the empirical formula of a compound, then we can determine the atomic mass of one of the elements in the compound if the atomic masses of the others are known. This is a standard experiment in many general chemistry laboratory courses.

Example 3-7: The empirical formula of magnesium oxide is MgO. A weighed quantity of magnesium, 0.490 g, is burned in oxygen, and the MgO produced is found to have a mass of 0.813 g. Given that the atomic mass of oxygen is 16.00, determine the atomic mass of magnesium.

Solution: The mass of oxygen in the magnesium oxide is

$$\text{mass of O in sample} = 0.813\text{ g MgO} - 0.490\text{ g Mg} = 0.323\text{ g O}$$

The number of moles of O is

$$\text{mol O} = (0.323\text{ g O})\left(\frac{1\text{ mol O}}{16.00\text{ g O}}\right) = 0.0202\text{ mol O}$$

The empirical formula MgO tells us that 0.0202 mol of Mg is combined with 0.0202 mol of O, and so we have

$$0.490\text{ g Mg} \approx 0.0202\text{ mol Mg}$$

The atomic mass of magnesium can be obtained if we determine how many grams of magnesium correspond to 1.00 mol. To determine this, we divide the stoichiometric correspondence by 0.0202 to get

$$24.3 \text{ g Mg} \approx 1.00 \text{ mol Mg}$$

or that the atomic mass of magnesium is 24.3, in excellent agreement with the accepted value.

You can see from these examples that it is possible to determine the empirical formula of a compound if the atomic masses are known and that it is possible to determine the atomic mass of an element if the empirical formula of one of its compounds and the atomic masses of the other elements that make up the compound are known. We are faced with a dilemma here. Atomic masses can be determined if empirical formulas are known (Example 3-7), but atomic masses must be known to determine empirical formulas (Examples 3-4 and 3-5). This was a serious problem in the early 1800s, shortly after Dalton formulated his atomic theory, because it was necessary to guess the empirical formulas of compounds—an incorrect guess led to an incorrect atomic mass. We shall see in Chapter 4 that it was the quantitative study of gases and of reactions between gases that was the key to resolving the difficulty in determining reliable values of atomic masses.

3-5. AN EMPIRICAL FORMULA ALONG WITH THE MOLECULAR MASS DETERMINES THE MOLECULAR FORMULA

Suppose that the chemical analysis of a compound gives 85.7 percent carbon and 14.3 percent hydrogen by mass. We then have

$$85.7 \text{ g C} \approx 14.3 \text{ g H}$$

$$7.14 \text{ mol C} \approx 14.2 \text{ mol H}$$

$$1 \text{ mol C} \approx 2 \text{ mol H}$$

and conclude that the empirical formula is CH_2. However, the actual formula might be C_2H_4, C_3H_6, or, generally, C_nH_{2n}. The chemical analysis gives us only ratios of numbers of atoms. If we know the molecular mass from another experiment, however, then we can determine the **molecular formula** unambiguously. For example, suppose we know that the molecular mass of our compound is 42. By listing the various possible formulas,

Formula	Formula mass
CH_2	14
C_2H_4	28
C_3H_6	42
C_4H_8	56

we see that the molecular formula of the compound is C_3H_6. This is why the formula deduced from chemical analysis is called the empirical formula or the simplest formula. It must be supplemented by molecular mass data to determine the molecular formula.

Example 3-8: There are many compounds, called **hydrocarbons,** that consist of only carbon and hydrogen. Gasoline is a mixture of over 100 different hydrocarbons. Chemical analysis of one of the constituents of gasoline yields 92.30 percent carbon and 7.70 percent hydrogen by mass. (a) Determine the simplest formula of this compound. (b) Given that the molecular mass is 78, determine its molecular formula.

Solution: (a) The determination of the simplest formula can be summarized by

$$92.30 \text{ g C} \approx 7.70 \text{ g H}$$

$$7.69 \text{ mol C} \approx 7.64 \text{ mol H}$$

$$1 \text{ mol C} \approx 1 \text{ mol H}$$

The simplest formula is CH, which has a formula mass of 13. (b) The molecular mass must be a multiple of 13, and indeed, 78 is 6 times 13. Thus, the molecular formula is C_6H_6.

3-6. THE COEFFICIENTS IN CHEMICAL EQUATIONS CAN BE INTERPRETED AS NUMBERS OF MOLES

A subject of great practical importance in chemistry is the determination of what quantity of product can be obtained from a given quantity of reactants. For example, the reaction between hydrogen and nitrogen to produce ammonia, NH_3, is described by the equation

$$3H_2(g) + N_2(g) \rightarrow 2NH_3(g)$$

▪ The production of ammonia from hydrogen and nitrogen requires high pressure and temperature and a catalyst to speed up the reaction (Interchapter I).

where the coefficients in the equation are called balancing coefficients or **stoichiometric coefficients.** We might wish to know how much NH_3 is produced when 100 g of H_2 reacts with excess N_2. To answer questions like this, we interpret the equation in terms of moles rather than molecules. The molecular interpretation of the hydrogen-nitrogen reaction is

$$3 \text{ molecules of hydrogen} + 1 \text{ molecule of nitrogen} \rightarrow 2 \text{ molecules of ammonia}$$

If we multiply both sides of this equation by Avogadro's number, then we obtain

$$3(6.022 \times 10^{23}) \text{ molecules of hydrogen} + 6.022 \times 10^{23} \text{ molecules of nitrogen} \rightarrow 2(6.022 \times 10^{23}) \text{ molecules of ammonia}$$

If we use the fact that 6.022×10^{23} molecules corresponds to 1 mol, then we also have

$$3 \text{ mol of } H_2 + 1 \text{ mol of } N_2 \rightarrow 2 \text{ mol of } NH_3$$

This is an important result. It tells us that the stoichiometric or balancing coefficients are the relative numbers of moles of each substance in a balanced chemical equation.

We can also interpret the hydrogen-nitrogen reaction in terms of masses. If we convert moles to masses by multiplying by the appropriate molar masses, then we get

$$6.05 \text{ g } H_2 + 28.02 \text{ g } N_2 \rightarrow 34.07 \text{ g } NH_3$$

Note that the total mass is the same on the two sides of the equation, in accord with the law of conservation of mass. Table 3-2 summarizes the various interpretations of the hydrogen-nitrogen reaction as well as those of the sodium-chlorine reaction.

We are now ready to calculate how much ammonia is produced when a given quantity of nitrogen or hydrogen is used.

▪ Calculations like the ones in Example 3-9 require that you understand the mole concept.

Example 3-9: (a) How many moles of NH_3 can be produced from 10.0 mol of N_2? (b) How many grams of NH_3 can be produced from 280 g of N_2?

Solution: (a) According to the balanced chemical equation,

$$3H_2(g) + N_2(g) \rightarrow 2NH_3(g)$$

2 mol of NH_3 are produced from each mole of N_2; thus, the stoichiometric unit conversion factor is

$$1 = \frac{2 \text{ mol } NH_3}{1 \text{ mol } N_2}$$

Therefore 10.0 mol of N_2 yields

$$(10.0 \text{ mol } N_2)\left(\frac{2 \text{ mol } NH_3}{1 \text{ mol } N_2}\right) = 20.0 \text{ mol } NH_3$$

(b) The molecular mass of N_2 is 28.0, and so 280 g of N_2 corresponds to

$$(280 \text{ g } N_2)\left(\frac{1 \text{ mol } N_2}{28.0 \text{ g } N_2}\right) = 10.0 \text{ mol } N_2$$

We have just seen that 10.0 mol of N_2 yields 20.0 mol of NH_3. The molecular mass of NH_3 is 17.0, and so 20.0 mol corresponds to

$$\text{g } NH_3 = (20.0 \text{ mol } NH_3)\left(\frac{17.0 \text{ g } NH_3}{1 \text{ mol } NH_3}\right) = 340 \text{ g } NH_3$$

Thus, 280 g of N_2 yields 340 g of NH_3.

Table 3-2 The various interpretations of two chemical equations

Interpretation	**$3H_2$**	**+ N_2**	**→ $2NH_3$**
molecular	3 molecules	+ 1 molecule	→ 2 molecules
molar	3 mol	+ 1 mol	→ 2 mol
mass	6.05 g	+ 28.02 g	→ 34.07 g
Interpretation	**2Na**	**+ Cl_2**	**→ 2NaCl**
molecular	2 atoms	+ 1 molecule	→ 2 ion pairs or 2 formula units
molar	2 mol	+ 1 mol	→ 2 mol
mass	45.98 g	+ 70.90 g	→ 116.88 g

Example 3-10: Propane, C_3H_8, a common fuel, burns in oxygen according to the equation

$$C_3H_8(g) + 5O_2(g) \rightarrow 3CO_2(g) + 4H_2O(l)$$

(a) How many grams of O_2 are required to burn 75.0 g of C_3H_8? (b) How many grams of H_2O and CO_2 are produced?

Solution: (a) The chemical equation states that 5 mol of O_2 is required to burn 1 mol of C_3H_8. The molecular mass of C_3H_8 is 44.09, and so 75.0 g of C_3H_8 corresponds to

$$\text{mol } C_3H_8 = (75.0 \text{ g } C_3H_8)\left(\frac{1 \text{ mol } C_3H_8}{44.09 \text{ g } C_3H_8}\right) = 1.70 \text{ mol}$$

The number of moles of O_2 required is

$$\text{mol } O_2 = (1.70 \text{ mol } C_3H_8)\left(\frac{5 \text{ mol } O_2}{1 \text{ mol } C_3H_8}\right) = 8.50 \text{ mol } O_2$$

To find out how many grams of O_2 this is, we multiply the number of moles by the molar mass:

$$\text{g } O_2 = (8.50 \text{ mol } O_2)\left(\frac{32.00 \text{ g } O_2}{1 \text{ mol } O_2}\right) = 272 \text{ g } O_2$$

Thus, we see that 272 g of O_2 is required to burn 75.0 g of C_3H_8.
(b) According to the chemical equation, 3 mol of CO_2 and 4 mol of H_2O are produced for each mole of C_3H_8 burned. Therefore the number of moles of CO_2 produced is

$$\text{mol } CO_2 = (1.70 \text{ mol } C_3H_8)\left(\frac{3 \text{ mol } CO_2}{1 \text{ mol } C_3H_8}\right) = 5.10 \text{ mol } CO_2$$

The number of moles of H_2O produced is

$$\text{mol } H_2O = (1.70 \text{ mol } C_3H_8)\left(\frac{4 \text{ mol } H_2O}{1 \text{ mol } C_3H_8}\right) = 6.80 \text{ mol } H_2O$$

The moles of CO_2 and H_2O are converted to grams by multiplying by the respective molar masses:

$$\text{g } CO_2 = (5.10 \text{ mol } CO_2)\left(\frac{44.01 \text{ g } CO_2}{1 \text{ mol } CO_2}\right) = 224 \text{ g } CO_2$$

$$\text{g } H_2O = (6.80 \text{ mol } H_2O)\left(\frac{18.02 \text{ g } H_2O}{1 \text{ mol } H_2O}\right) = 123 \text{ g } H_2O$$

These are the quantities of CO_2 and H_2O produced when 75.0 g of propane are burned.

We can summarize the results of this example by

$$\begin{array}{llll} C_3H_8(g) & + 5O_2(g) \rightarrow & 3CO_2(g) & + 4H_2O(l) \\ 75 \text{ g} & + 272 \text{ g} \rightarrow & 224 \text{ g} & + 123 \text{ g} \end{array}$$

Notice that the total masses on the two sides of the chemical reaction are the same, as they must be according to the principle of conservation of mass.

3-7. THE PERCENTAGE COMPOSITION OF MANY COMPOUNDS CAN BE DETERMINED BY COMBUSTION ANALYSIS

Example 3-10 shows that when propane, C_3H_8, is burned in oxygen the reaction products are CO_2 and H_2O. When compounds containing only carbon and hydrogen or carbon, hydrogen, and oxygen are burned in an excess of O_2, all the carbon in the original sample ends up in CO_2 and all the hydrogen ends up in H_2O. These facts are the basis of the determination of the percentage composition of such compounds by **combustion analysis.** In combustion analysis the sample is burned in excess oxygen and the resulting gaseous water and CO_2 are passed through chambers containing different substances, as shown in Figure 3-2. The water is absorbed in the magnesium perchlorate chamber by the reaction

$$\underset{\text{anhydrous}}{Mg(ClO_4)_2(s)} + 6H_2O(g) \rightarrow \underset{\text{hydrated}}{Mg(ClO_4)_2{\cdot}6H_2O(s)}$$

■ Many salts such as magnesium perchlorate form definite **hydrates** in which a fixed number of water molecules are associated with each formula unit in the crystal, as in $Mg(ClO_4)_2{\cdot}6H_2O(s)$. The corresponding anhydrous salt lacks the water, as in $Mg(ClO_4)_2(s)$.

After passing through the magnesium perchlorate chamber, the carbon dioxide reacts with the sodium hydroxide in the next chamber according to

$$NaOH(s) + CO_2(g) \rightarrow NaHCO_3(s)$$

The masses of water and carbon dioxide formed in the combustion reaction are determined by measuring mass increases in the magnesium perchlorate and sodium hydroxide chambers. The masses of CO_2 and H_2O produced in the combustion reaction are converted readily to the stoichiometrically equivalent masses of carbon and hydrogen:

$$\begin{aligned}\text{mass of C} &= \\ &(\text{mass of } CO_2 \text{ formed})(\text{fraction of the mass of } CO_2 \text{ due to C}) \\ &= (\text{mass of } CO_2 \text{ formed})\left(\frac{\text{atomic mass of C}}{\text{formula mass of } CO_2}\right)\end{aligned}$$

and

$$\begin{aligned}\text{mass of H} &= \\ &(\text{mass of } H_2O \text{ formed})(\text{fraction of the mass of } H_2O \text{ due to H}) \\ &= (\text{mass of } H_2O \text{ formed})\left(\frac{2 \times \text{atomic mass of H}}{\text{formula mass of } H_2O}\right)\end{aligned}$$

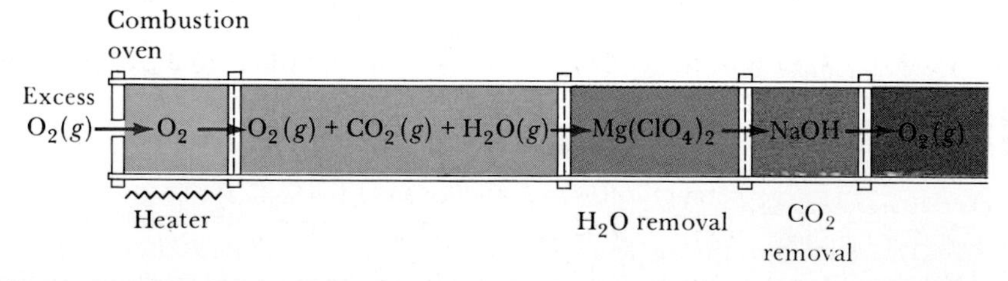

Figure 3-2 A schematic illustration of the removal of H_2O and CO_2 from combustion gases. Anhydrous magnesium perchlorate removes the water vapor:

$$Mg(ClO_4)_2(s) + 6H_2O(g) \rightarrow Mg(ClO_4)_2{\cdot}6H_2O(s)$$

and then sodium hydroxide removes the CO_2:

$$NaOH(s) + CO_2(g) \rightarrow NaHCO_3(s)$$

Excess O_2 gas is used as a carrier gas to "sweep" all the H_2O and CO_2 from the combustion zone (*color*).

The mass percentages of carbon and hydrogen in the original sample are then computed as follows:

$$\text{mass percentage of C} = \left(\frac{\text{mass of C}}{\text{mass of sample}}\right) \times 100$$

$$\text{mass percentage of H} = \left(\frac{\text{mass of H}}{\text{mass of sample}}\right) \times 100$$

If the original sample contains only carbon and hydrogen, then the mass percentages of C and H should sum to 100 percent, within experimental error. If the sample also contains oxygen, then the mass percentage of oxygen is determined by difference:

$$\text{mass percentage of O} = 100 - \text{mass percentage of C} - \text{mass percentage of H}$$

Example 3-11 outlines the workup of data from a combustion analysis.

Example 3-11: A 1.250-g sample of the compound responsible for the odor of cloves, which is known to contain only the elements C, H, and O, is burned in a combustion analysis apparatus. The mass of CO_2 produced is 3.350 g, and the mass of water produced is 0.8232 g. (a) Determine the percentage composition of the sample. (b) Determine the molecular formula of the compound, given that its molecular mass is 164.

Solution: (a) The masses of carbon and hydrogen in the original sample are

$$\text{mass of C} = (3.350\text{ g } CO_2)\left(\frac{12.01\text{ g C}}{44.01\text{ g } CO_2}\right) = 0.9142\text{ g C}$$

$$\text{mass of H} = (0.8232\text{ g } H_2O)\left(\frac{2 \times 1.008\text{ g H}}{18.02\text{ g } H_2O}\right) = 0.09210\text{ g H}$$

The mass percentages of C and H are

$$\text{mass percentage of C} = \left(\frac{0.9142\text{ g}}{1.250\text{ g}}\right) \times 100 = 73.14\%$$

$$\text{mass percentage of H} = \left(\frac{0.09210\text{ g}}{1.250\text{ g}}\right) \times 100 = 7.37\%$$

The mass percentage of oxygen is obtained by difference:

$$\text{mass percentage of O} = 100.00 - 73.14 - 7.37 = 19.49\%$$

(b) To determine the molecular formula, first we calculate the empirical formula from the mass percentages, as described in Section 3-3. As usual, we consider a 100-g sample and write

$$73.14\text{ g C} \Leftrightarrow 7.37\text{ g H} \Leftrightarrow 19.49\text{ g O}$$

$$6.090\text{ mol C} \Leftrightarrow 7.312\text{ mol H} \Leftrightarrow 1.218\text{ mol O}$$

$$5.00\text{ mol C} \Leftrightarrow 6.00\text{ mol H} \Leftrightarrow 1.00\text{ mol O}$$

Thus the empirical formula is C_5H_6O. The molecular mass is known to be 164, and so the molecular formula is $C_{10}H_{12}O_2$.

3-8. CALCULATIONS INVOLVING CHEMICAL REACTIONS ARE CARRIED OUT IN TERMS OF MOLES

For calculations involving chemical reactions, the procedure is to first convert mass to moles, then convert moles of one substance to moles of another by using the balancing coefficients in the chemical equation, and then convert moles into mass. An understanding of the flow chart in Figure 3-3 will allow you to do any calculations involving chemical equations. The following calculation illustrates the use of Figure 3-3.

Sulfuric acid, H_2SO_4, usually in the form of an aqueous solution, is the most widely used and important industrial chemical. About 40 million tons of sulfuric acid is produced annually in the United States, and most sulfuric acid is made by the **contact process.** First sulfur is burned in oxygen to produce sulfur dioxide:

$$S(s) + O_2(g) \rightarrow SO_2(g) \qquad (3\text{-}1)$$

Then the $SO_2(g)$ is mixed with more $O_2(g)$ and passed over vanadium pentoxide, a substance that increases the rate of the reaction:

$$2SO_2(g) + O_2(g) \xrightarrow{V_2O_5,\ 500^\circ C} 2SO_3(g) \qquad (3\text{-}2)$$

Vanadium pentoxide is a **catalyst** for this reaction. We shall learn more about catalysts in later chapters. At this point it is sufficient to realize that a catalyst is a substance that increases the rate of a reaction but is not consumed in the reaction. We denote a catalyst by placing the formula for it over the arrow in the chemical equation. The $SO_3(g)$ produced in Equation (3-2) is dissolved in sulfuric acid, which is then reacted with water. The overall reaction is

$$SO_3(g) + H_2O(l) \rightarrow H_2SO_4(l) \qquad (3\text{-}3)$$

Let's calculate how much $H_2SO_4(l)$ can be produced from 1 metric ton of sulfur. A **metric ton** is equal to 1000 kg, or 2200 pounds. From Equation (3-1), we see that 1 mol of $S(s)$ yields 1 mol of $SO_2(g)$; Equation (3-2) tells us that each mole of $SO_2(g)$ that reacts yields 1 mol of $SO_3(g)$; Equation (3-3) shows that 1 mol of $SO_3(g)$ produces 1 mol of $H_2SO_4(l)$. We can summarize these statements as follows:

$$1 \text{ mol S} \approx 1 \text{ mol } SO_2 \approx 1 \text{ mol } SO_3 \approx 1 \text{ mol } H_2SO_4$$

Thus we see that 1 mol of $H_2SO_4(l)$ is produced from 1 mol of $S(s)$:

$$1 \text{ mol S} \approx 1 \text{ mol } H_2SO_4$$

This result should not be surprising because one molecule of H_2SO_4 contains one atom of sulfur, and all the sulfur ends up in sulfuric acid. One metric ton of $S(s)$ corresponds to

$$\text{mol S} = (1.00 \times 10^6 \text{ g S})\left(\frac{1 \text{ mol S}}{32.06 \text{ g S}}\right) = 3.12 \times 10^4 \text{ mol S}$$

The chemical equations show that

$$3.12 \times 10^4 \text{ mol S} \approx 3.12 \times 10^4 \text{ mol } H_2SO_4$$

The molecular mass of H_2SO_4 is 98.08, and so the quantity of H_2SO_4 produced is

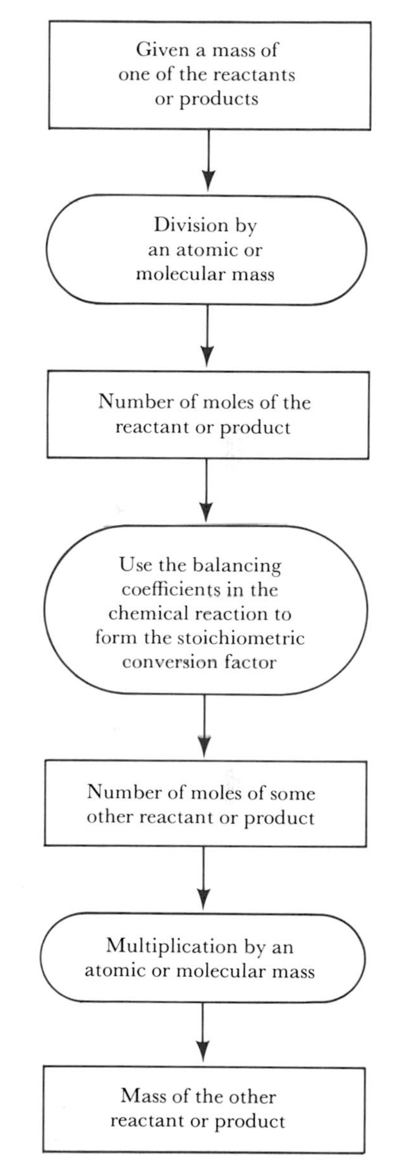

Figure 3-3 This flow diagram describes the procedure for calculations involving chemical equations. The essence of the method is to realize that we convert from moles of one substance to moles of another substance in a chemical equation by using the stoichiometric coefficients.

$$\text{g } H_2SO_4 = (3.12 \times 10^4 \text{ mol } H_2SO_4)\left(\frac{98.08 \text{ g } H_2SO_4}{1 \text{ mol } H_2SO_4}\right)$$
$$= 3.06 \times 10^6 \text{ g } H_2SO_4$$

$$\text{kg } H_2SO_4 = (3.06 \times 10^6 \text{ g } H_2SO_4)\left(\frac{1 \text{ kg}}{10^3 \text{ g}}\right)$$
$$= 3.06 \times 10^3 \text{ kg } H_2SO_4$$

$$\text{metric tons } H_2SO_4 = (3.06 \times 10^3 \text{ kg } H_2SO_4)\left(\frac{1 \text{ metric ton}}{10^3 \text{ kg}}\right)$$
$$= 3.06 \text{ metric tons } H_2SO_4$$

3-9. WHEN TWO OR MORE SUBSTANCES REACT, THE QUANTITY OF PRODUCT IS DETERMINED BY THE LIMITING REACTANT

If you look back over the examples in this chapter, you will notice that in no case did we start out with stated quantities of two reactants. Let's consider an example in which we do. Cadmium sulfide, which is used in light meters, solar cells, and other light-sensitive devices, can be made by the direct combination of the two elements:

$$Cd(s) + S(s) \rightarrow CdS(s) \qquad (3\text{-}4)$$

How much CdS is produced if we start out with 2.00 g of cadmium and 2.00 g of sulfur? The number of moles of each element is

$$\text{mol Cd} = (2.00 \text{ g Cd})\left(\frac{1 \text{ mol Cd}}{112.4 \text{ g Cd}}\right) = 0.0178 \text{ mol Cd}$$

$$\text{mol S} = (2.00 \text{ g S})\left(\frac{1 \text{ mol S}}{32.06 \text{ g S}}\right) = 0.0624 \text{ mol S}$$

According to the balanced chemical equation, 1 mol of cadmium requires 1 mol of sulfur, and so the 0.0178 mol of cadmium requires 0.0178 mol of sulfur. Thus, there is excess sulfur; only 0.0178 mol of sulfur reacts, and (0.0624 − 0.0178) mol = 0.0446 mol of sulfur remains. The cadmium reacts completely, and the moles of cadmium consumed determines how much CdS is produced. The reactant that is consumed completely and thereby limits the amount of product formed is called the **limiting reactant** (Cd in the example), and any other reactants are called **excess reactants** (S in the example). The initial quantity of limiting reactant must be used to calculate how much product is formed.

In Equation (3-4), 0.0178 mol of cadmium reacts with 0.0178 mol of sulfur to produce 0.0178 mol of CdS. The mass of CdS produced is

$$\text{g CdS} = (0.0178 \text{ mol CdS})\left(\frac{144.5 \text{ g CdS}}{1 \text{ mol CdS}}\right) = 2.57 \text{ g CdS}$$

The unused sulfur (0.0446 mol) has a mass of

$$\text{g unused S} = (0.0446 \text{ mol S})\left(\frac{32.06 \text{ g S}}{1 \text{ mol S}}\right) = 1.43 \text{ g S}$$

Note that before the reaction there is 2.00 g of cadmium and 2.00 g of sulfur, or 4.00 g of reactants. After the reaction, there is 2.57 g of cadmium sulfide and 1.43 g of sulfur, or a total of 4.00 g, as required by the law of conservation of mass.

When the masses of two or more reactants are given in a problem, the limiting reactant must be determined and is the only one to be used in the calculation of the quantity of product obtained.

Example 3-12: A mixture is prepared from 25.0 g of aluminum and 85.0 g of Fe_2O_3. The reaction that occurs is described by the equation

$$Fe_2O_3(s) + 2Al(s) \rightarrow Al_2O_3(s) + 2Fe(l)$$

How much iron is produced in the reaction?

Solution: Because the masses of both reactants are given, we must check to see which, if either, is a limiting reactant. The number of moles of Al and Fe_2O_3 available is

$$\text{mol Al} = (25.0 \text{ g Al})\left(\frac{1 \text{ mol Al}}{26.98 \text{ g Al}}\right) = 0.927 \text{ mol Al}$$

$$\text{mol } Fe_2O_3 = (85.0 \text{ g } Fe_2O_3)\left(\frac{1 \text{ mol } Fe_2O_3}{159.7 \text{ g } Fe_2O_3}\right) = 0.532 \text{ mol } Fe_2O_3$$

From the balanced chemical equation, we see that 0.927 mol of aluminum consumes only

$$(0.927 \text{ mol Al})\left(\frac{1 \text{ mol } Fe_2O_3}{2 \text{ mol Al}}\right) = 0.464 \text{ mol } Fe_2O_3$$

Thus, we see that the Fe_2O_3 is in excess. The amount of Fe_2O_3 in excess is

$$\text{mol excess } Fe_2O_3 = (0.532 - 0.464) \text{ mol } Fe_2O_3 = 0.068 \text{ mol } Fe_2O_3$$

The aluminum is the limiting reactant and the one to use in calculating how much iron is produced. From the balanced equation,

$$2 \text{ mol Al} \approx 2 \text{ mol Fe}$$

or

$$0.927 \text{ mol Al} \approx 0.927 \text{ mol Fe}$$

The mass of iron corresponding to 0.927 mol is

$$\text{g Fe} = (0.927 \text{ mol Fe})\left(\frac{55.85 \text{ g Fe}}{1 \text{ mol Fe}}\right) = 51.8 \text{ g Fe}$$

The reaction of aluminum metal with a metal oxide is called a **thermite reaction** (Figure 3-4) and has numerous applications. Thermite reactions were once used to weld railroad rails and are used in thermite grenades, which are employed by the military to destroy heavy equipment. In a thermite reaction the reaction temperature can exceed 3500°C.

There are many instances in which it is important to add reactants in stoichiometric proportions so as not to have any reactants left over. The propulsion of rockets and space vehicles serves as a good example. The Lunar Lander rocket engines were powered by a reaction similar to

Figure 3-4 A thermite reaction. A spectacular example of a single-replacement reaction is the reaction between powdered aluminum metal and iron(III) oxide

$$2Al(s) + Fe_2O_3(s) \rightarrow 2Fe(l) + Al_2O_3(s)$$

Once this reaction is initiated by a heat source such as a burning magnesium ribbon, it proceeds vigorously, producing so much heat that the iron is formed as a liquid.

▪ An easy way to see which reactant is the limiting reactant is to compare the number of moles given for each reactant divided by the corresponding stoichiometric coefficient; the smallest resulting number of moles identifies the limiting reactant. Thus, in Example 3-12 we have

$$\frac{\text{mol Al}}{2} = \frac{0.927 \text{ mol}}{2} = 0.464 \text{ mol}$$

and

$$\frac{\text{mol } Fe_2O_3}{1} = 0.532 \text{ mol}$$

This result tells us that the Al(*s*) is the limiting reactant. This procedure is equivalent to that used in Example 3-12.

$$\underset{\text{dinitrogen tetroxide}}{N_2O_4(l)} + \underset{\text{hydrazine}}{2N_2H_4(l)} \rightarrow 3N_2(g) + 4H_2O(g)$$

Dinitrogen tetroxide and hydrazine react explosively when brought into contact. These two reactants were kept in separate tanks and pumped through pipes into the rocket engines, where they reacted. The gases produced (H_2O is a gas at the exhaust temperatures of the rocket engines) exited through the exhaust chamber of the engine and propelled the rocket forward. The cost of carrying materials into space is enormous, and the two fuels must be combined in the correct proportions. It would be wasteful to carry any excess reactant.

Example 3-13: Suppose that a rocket is powered by the reaction between liquid dinitrogen tetroxide and hydrazine. If the tanks are designed to hold 50.0 metric tons of dinitrogen tetroxide, how much hydrazine should be carried?

Solution: Recall that a metric ton is 1000 kg. The molecular mass of N_2O_4 is 92.0, and the number of moles of N_2O_4 carried is

$$\text{mol } N_2O_4 = \left(\begin{matrix}50.0 \text{ metric tons} \\ \text{of } N_2O_4\end{matrix}\right)\left(\frac{10^3 \text{ kg}}{1 \text{ metric ton}}\right)\left(\frac{10^3 \text{ g}}{1 \text{ kg}}\right)\left(\frac{1 \text{ mol } N_2O_4}{92.0 \text{ g } N_2O_4}\right)$$

$$= 5.43 \times 10^5 \text{ mol } N_2O_4$$

According to the balanced chemical equation, it requires 2 mol of hydrazine for every mole of dinitrogen tetroxide. Therefore

$$\text{mol } N_2H_4 \text{ required} = \left(\frac{2 \text{ mol } N_2H_4}{1 \text{ mol } N_2O_4}\right)(5.43 \times 10^5 \text{ mol } N_2O_4)$$

$$= 1.09 \times 10^6 \text{ mol } N_2H_4$$

The mass of hydrazine required is

$$\text{mass } N_2H_4 \text{ required} = (1.09 \times 10^6 \text{ mol } N_2H_4)\left(\frac{32.05 \text{ g } N_2H_4}{1 \text{ mol } N_2H_4}\right)$$

$$= 3.49 \times 10^7 \text{ g } N_2H_4$$

$$= 34.9 \text{ metric tons } N_2H_4$$

3-10. FOR MANY CHEMICAL REACTIONS THE AMOUNT OF THE DESIRED PRODUCT OBTAINED IS LESS THAN THE THEORETICAL AMOUNT

The mass of a particular product that is calculated to result from the limiting reactant in a chemical reaction is called the **theoretical yield** of the reaction. The mass of the product that actually is obtained is called the **actual yield** of the reaction. The **percentage yield** of the reaction is defined as

$$\% \text{ yield} = \left(\frac{\text{actual yield}}{\text{theoretical yield}}\right) \times 100 \qquad (3\text{-}5)$$

The percentage yield cannot exceed 100 percent, but it is often less than 100 percent because (1) the reaction may not go to completion;

(2) there may be side reactions that give rise to undesired product; or (3) some of the desired product may not be readily recoverable or may be lost in the purification process. The industrial production of methyl alcohol, $CH_3OH(l)$, from the high-pressure reaction

$$CO(g) + 2H_2(g) \rightarrow CH_3OH(l)$$

serves to illustrate the difference between the theoretical yield and the actual yield of a reaction. For a variety of reasons, this reaction does not give a 100 percent yield. Suppose that 5.12 metric tons of $CH_3OH(l)$ is obtained from 1.00 metric ton of $H_2(g)$ reacting with an excess of $CO(g)$. Let's calculate the percentage yield of $CH_3OH(l)$. The theoretical yield is

$$\text{theoretical yield} = (1 \text{ metric ton } H_2)\left(\frac{10^6 \text{ g}}{1 \text{ metric ton}}\right)\left(\frac{1 \text{ mol } H_2}{2.016 \text{ g } H_2}\right) \times \left(\frac{1 \text{ mol } CH_3OH}{2 \text{ mol } H_2}\right)\left(\frac{32.03 \text{ g } CH_3OH}{1 \text{ mol } CH_3OH}\right)$$

$$= 7.94 \times 10^6 \text{ g } CH_3OH = 7.94 \text{ metric tons } CH_3OH$$

The percentage yield is given by Equation (3-5):

$$\%\text{ yield} = \left(\frac{\text{actual yield}}{\text{theoretical yield}}\right) \times 100 = \left(\frac{5.12 \text{ metric tons}}{7.94 \text{ metric tons}}\right) \times 100$$

$$= 64.5\%$$

Example 3-14: A 0.473-g sample of phosphorus is reacted with an excess of chlorine, and 2.12 g of phosphorus pentachloride, PCl_5, is collected. The equation for the reaction is

$$2P(s) + 5Cl_2(g) \rightarrow 2PCl_5(s)$$

What is the percentage yield of PCl_5?

Solution: The theoretical yield of PCl_5 is

$$\text{theoretical yield} = (0.473 \text{ g P})\left(\frac{1 \text{ mol P}}{30.97 \text{ g P}}\right)\left(\frac{1 \text{ mol } PCl_5}{1 \text{ mol P}}\right)\left(\frac{208.2 \text{ g } PCl_5}{1 \text{ mol } PCl_5}\right)$$

$$= 3.18 \text{ g } PCl_5$$

The percentage yield is

$$\%\text{ yield} = \left(\frac{\text{actual yield}}{\text{theoretical yield}}\right) \times 100 = \left(\frac{2.12 \text{ g}}{3.18 \text{ g}}\right) \times 100 = 66.7\%$$

We shall always assume that the percentage yield of a reaction is 100 percent unless we know otherwise.

3-11. MANY REACTIONS TAKE PLACE IN SOLUTION

Many important chemical and essentially all biological processes take place in solution, particularly in aqueous solution. A **solution** is a mixture of two or more substances that is uniform at the molecular level. A solution must be **homogeneous,** meaning that it has the same properties from one region to another. The most common examples of solutions involve a solid, such as NaCl, dissolved in water. Sodium chloride is an ionic compound, and the sodium ions (Na^+) and chloride ions (Cl^-) are uniformly dispersed throughout the water. The resulting solution is clear and homogeneous.

The solid that is dissolved is called the **solute,** and the liquid in which it is dissolved is called the **solvent.** In the case of NaCl dissolved in water, NaCl is the solute and water is the solvent. We designate a species in aqueous solution by writing (*aq*) after the species. For example, $Na^+(aq)$ and $Cl^-(aq)$ represent sodium ions and chloride ions in water. The process of dissolving NaCl(*s*) in water is represented by the equation

$$\text{NaCl}(s) \xrightarrow[\text{H}_2\text{O}(l)]{} \text{Na}^+(aq) + \text{Cl}^-(aq)$$

where $H_2O(l)$ *under* the arrow tells us that water is the solvent.

The **concentration** of solute in a solution describes the quantity of solute dissolved in a given quantity of solvent. The most common method of expressing the concentration of a solute is **molarity,** which is represented by the symbol M. Molarity is defined as the number of moles of solute per liter of solution:

$$\text{molarity} = \frac{\text{moles of solute}}{\text{liters of solution}} \tag{3-6}$$

In terms of symbols, Equation (3-6) is

$$M = \frac{n}{V} \tag{3-7}$$

where M is the molarity of the solution, n is the number of moles of solute dissolved in the solution, and V is the volume of the solution in liters. To see how to use Equation (3-7), let's calculate the molarity of a solution prepared by dissolving 62.3 g of sucrose, $C_{12}H_{22}O_{11}$, in enough water to form 0.500 L of solution. The formula mass of sucrose is 342, and so 62.3 g corresponds to

$$(62.3\text{ g sucrose})\left(\frac{1\text{ mol sucrose}}{342\text{ g sucrose}}\right) = 0.182\text{ mol sucrose}$$

This is the number of moles of sucrose dissolved in 0.500 L of solution. The molarity of the solution is

$$M = \frac{n}{V} = \frac{0.182\text{ mol}}{0.500\text{ L}} = 0.364\text{ mol·L}^{-1} = 0.364\text{ M}$$

We say that the concentration of the solution is 0.364 **molar.**

The definition of molarity involves the volume of the *solution, not* the volume of the solvent. Suppose we wish to prepare one liter of a 2.50 M aqueous solution of NaCl. We would prepare the solution by weighing out 2.50 mol (146 g) of NaCl, dissolving it in *less* than one liter of water, say, about 500 mL, and then adding water until the final volume of the *solution* is exactly one liter (Figure 3-5). Realize that it would be incorrect to add 2.50 mol of NaCl to one liter of water; the final volume of

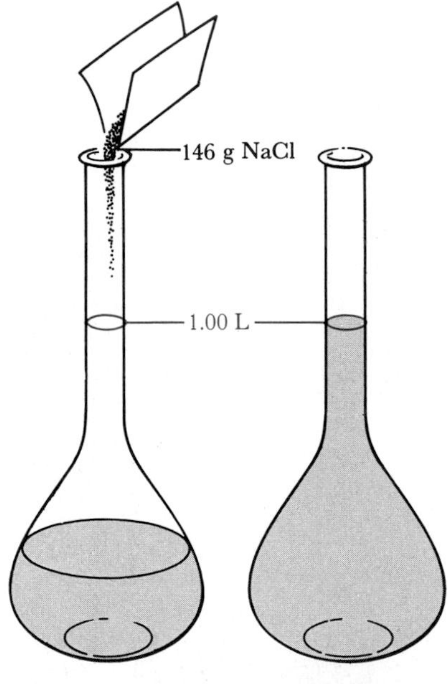

Figure 3-5 The procedure used to prepare one liter of a solution of a certain molarity, such as 2.50 M NaCl. The 2.50 mol of NaCl (146 g) is weighed out and placed in a 1-L volumetric flask that is only partially filled with water. The NaCl is dissolved, and then more water is added to bring the final volume up to the mark on the flask.

such a solution is not exactly one liter because the added NaCl changes the volume from 1.00 L to 1.04 L.

Solutions of known molarity must always be made by the procedure described in Figure 3-5. For such measurements we use a **volumetric flask,** which is a precision piece of glassware used to prepare precise volumes. The following example illustrates the procedure for making up a solution of a specified molarity.

Example 3-15: Potassium bromide, KBr, is used as a sedative and as an anticonvulsive agent. Explain how you would prepare 250 mL of a 0.600 M aqueous KBr solution.

Solution: From Equation (3-7) and the specified volume and concentration, we can calculate the number of moles of KBr required. Equation (3-7) can be written as

$$n = MV$$

and so

$$\begin{aligned}\text{mol KBr} &= (0.600\ \text{M})(0.250\ \text{L}) \\ &= (0.600\ \text{mol}\cdot\text{L}^{-1})(0.250\ \text{L}) \\ &= 0.150\ \text{mol}\end{aligned}$$

We can convert moles to grams by multiplying by the formula mass of KBr (119.0):

$$\begin{aligned}\text{g KBr} &= (0.150\ \text{mol KBr})\left(\frac{119.0\ \text{g KBr}}{1\ \text{mol KBr}}\right) \\ &= 17.9\ \text{g KBr}\end{aligned}$$

To prepare the solution, we dissolve 17.9 g of KBr in a 250-mL volumetric flask that is partially filled with distilled water, shake the flask until the salt is dissolved, and then dilute the solution to the 250-mL mark on the flask and shake again to assure uniformity. *Do not* add the KBr to 250 mL of water because the volume of the resulting solution will not necessarily be 250 mL.

The next example illustrates a calculation involving a reaction that takes place in solution.

Example 3-16: Zinc reacts with hydrochloric acid, HCl(*aq*) (Figure 3-6), according to the equation

$$\text{Zn}(s) + 2\text{HCl}(aq) \rightarrow \text{ZnCl}_2(aq) + \text{H}_2(g)$$

Calculate how many grams of zinc react with 50.0 mL of 6.00 M HCl(*aq*).

Solution: According to the equation for the reaction, 1 mol of Zn reacts with 2 mol of HCl(*aq*). We can use Equation (3-7) to calculate how many moles of HCl there are in 50.0 mL of a 6.00 M HCl solution:

$$\begin{aligned}\text{mol HCl} = MV &= (6.00\ \text{M})(50.0\ \text{mL})\left(\frac{1\ \text{L}}{1000\ \text{mL}}\right) \\ &= (6.00\ \text{mol}\cdot\text{L}^{-1})(0.0500\ \text{L}) \\ &= 0.300\ \text{mol}\end{aligned}$$

Figure 3-6 The reaction of zinc metal with an aqueous solution of hydrochloric acid. The bubbles are hydrogen gas escaping from the water.

By referring to the equation, we see that 0.150 mol of Zn reacts with 0.300 mol of HCl. The mass that corresponds to 0.150 mol of Zn (atomic mass 65.38) is

$$\text{g Zn} = (0.150 \text{ mol Zn})\left(\frac{65.38 \text{ g Zn}}{1 \text{ mol Zn}}\right)$$

$$= 9.81 \text{ g Zn}$$

It is often necessary in laboratory work to prepare a more dilute solution from a more concentrated stock solution. In such cases, a known volume of a solution of known molarity is diluted with a volume of pure solvent that produces the final solution with the desired molarity. The key point in carrying out such **dilution** calculations is that the number of moles of solute does not change on dilution with solvent. Thus from Equation (3-7) we have

$$\text{moles of solute} = n = M_1V_1 = M_2V_2$$

or

$$M_1V_1 = M_2V_2 \qquad \text{(dilution)} \tag{3-8}$$

where M_1 and V_1 are the initial molarity and initial volume, and M_2 and V_2 are the final molarity and final volume of the solution. The following example illustrates a dilution calculation.

Example 3-17: Compute the volume of 6.00 M $H_2SO_4(aq)$ required to produce 500 mL of 0.30 M $H_2SO_4(aq)$.

Solution: From Equation (3-8) we have

$$M_1V_1 = M_2V_2$$

$$(6.00 \text{ mol}\cdot\text{L}^{-1})V_1 = (0.30 \text{ mol}\cdot\text{L}^{-1})(0.500 \text{ L})$$

Thus

$$V_1 = \frac{0.30 \text{ mol}\cdot\text{L}^{-1} \times 0.500 \text{ L}}{6.00 \text{ mol}\cdot\text{L}^{-1}} = 0.025 \text{ L}$$

or $V_1 = 25$ mL. Thus we add 25 mL of 6.00 M $H_2SO_4(aq)$ to a 500-mL volumetric flask that is about half filled with water, swirl the solution, and dilute with water to the 500-mL mark on the flask. Finally, we swirl again to make the solution homogeneous.

3-12. THE CONCENTRATION OF AN ACID OR A BASE CAN BE DETERMINED BY TITRATION

We can utilize neutralization reactions to determine the concentration of solutions of acids or bases. Suppose we have a basic solution whose concentration is not known. We measure out a certain volume of the basic solution and then slowly add an acidic solution of known concentration until the base is completely neutralized. Such a process is called a **titration** and can be carried out with the apparatus shown in Figure 3-7. Knowing the volume and concentration of the acidic solution required to neutralize the base is sufficient to determine the concentration of the basic solution. As an example, suppose we find that it requires 27.25 mL of 0.150 M HCl solution to neutralize 30.00 mL of a NaOH solution. The number of moles of HCl required to neutralize the NaOH(*aq*) is given by Equation (3-7):

$$\text{number of moles} = (\text{molarity})(\text{volume of solution in liters})$$

$$n = MV$$

Thus

$$\text{mol HCl} = (0.150\ \text{mol}\cdot\text{L}^{-1})(27.25 \times 10^{-3}\ \text{L})$$
$$= 4.09 \times 10^{-3}\ \text{mol}$$

The reaction between HCl(*aq*) and NaOH(*aq*) is

$$HCl(aq) + NaOH(aq) \rightarrow NaCl(aq) + H_2O(l)$$

which indicates that 1 mol of HCl is required to neutralize 1 mol of NaOH. Therefore

$$\text{mol NaOH} = \text{mol HCl} = 4.09 \times 10^{-3}\ \text{mol}$$

There is 4.09×10^{-3} mol of NaOH in the 30.00 mL of the NaOH solution, and so the concentration of the solution is

$$M = \frac{n}{V} = \frac{4.09 \times 10^{-3}\ \text{mol}}{30.00 \times 10^{-3}\ \text{L}} = 0.136\ \text{M}$$

The following Example illustrates the calculation of the concentration of a solution from titration data.

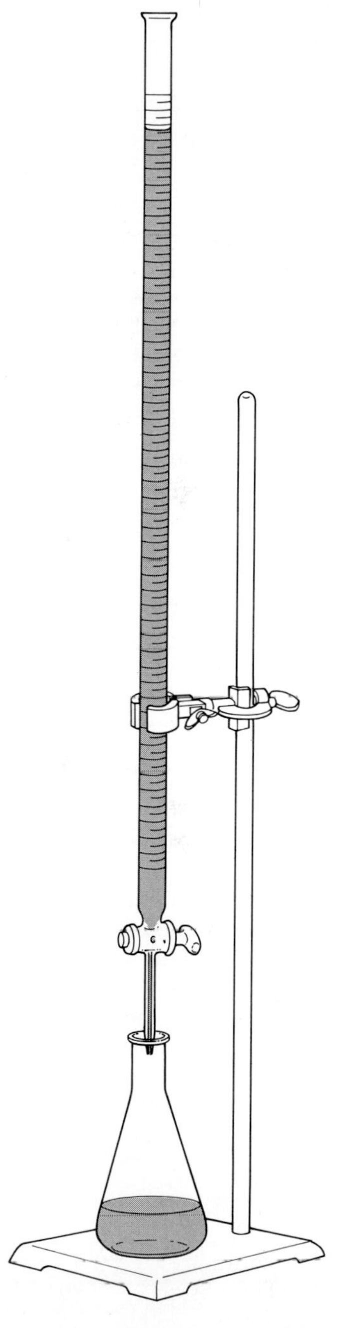

Figure 3-7 A **buret** is a precision-made piece of glassware used to measure accurately the volume of a solution. Burets are particularly useful for delivering accurate volumes of one solution to another.

Example 3-18: By titration, it is found that 37.60 mL of 0.210 M NaOH is required to neutralize 25.05 mL of $H_2SO_4(aq)$. Calculate the concentration of the H_2SO_4 solution.

Solution: The number of moles of NaOH required to neutralize the H_2SO_4 is

$$\text{mol NaOH} = MV = (0.210\ \text{mol}\cdot\text{L}^{-1})(37.60 \times 10^{-3}\ \text{L})$$
$$= 7.90 \times 10^{-3}\ \text{mol}$$

According to the equation for the reaction

$$H_2SO_4(aq) + 2NaOH(aq) \rightarrow Na_2SO_4(aq) + 2H_2O(l)$$

it requires 2 mol of NaOH to neutralize 1 mol of H_2SO_4. Consequently, we have

$$\text{mol } H_2SO_4 = (7.90 \times 10^{-3}\ \text{mol NaOH})\left(\frac{1\ \text{mol } H_2SO_4}{2\ \text{mol NaOH}}\right)$$
$$= 3.95 \times 10^{-3}\ \text{mol}$$

The concentration of the H_2SO_4 solution is

$$M = \frac{n}{V} = \frac{3.95 \times 10^{-3}\ \text{mol}}{25.05 \times 10^{-3}\ \text{L}} = 0.158\ \text{M}$$

The following Example illustrates the determination of the formula mass of an acid from titration data.

Example 3-19: A 1.50-g sample of an unknown acid is dissolved to make 100 mL of solution and neutralized with 0.200 M NaOH(*aq*). The volume of NaOH solution required to neutralize the acid is 75.0 mL. Assume that the acid has only one acidic proton per molecule and compute its formula mass.

Solution: The number of moles of base required to neutralize the acid is

$$\text{mol NaOH} = MV = (0.200\ \text{mol}\cdot\text{L}^{-1})(0.0750\ \text{L}) = 1.50 \times 10^{-2}\ \text{mol}$$

Therefore, the number of moles of acid present in the original 100 mL of solution is 1.50×10^{-2} mol. Thus we see that

$$1.50\ \text{g acid} \Bumpeq 1.50 \times 10^{-2}\ \text{mol acid}$$

■ Recall that the symbol $\Bumpeq$ means "combines with" or "is stoichiometrically equivalent to."

Upon dividing by 1.50×10^{-2}, we obtain

$$100\ \text{g acid} \Bumpeq 1.00\ \text{mol acid}$$

Thus the formula mass of the unknown acid is 100.

Although we have done a number of different types of calculations in this chapter, they are all unified by the concept of a mole, the chapter's central theme. The mole is one of the most important and useful concepts in chemistry.

SUMMARY

Stoichiometric calculations are based on the concept of a mole. The quantity of a substance that is numerically equal to its formula mass in grams is called a mole of that substance. Thus in order to calculate the number of moles in a given mass of a substance, it is necessary to know its chemical formula. Another definition of a mole is that mass of a substance that contains Avogadro's number (6.022×10^{23}) of formula units or elementary entities. By using the concept of a mole and Avogadro's number, it is possible to calculate the mass of individual atoms and molecules, to calculate how many atoms and molecules there are in a given mass, to determine chemical formulas from chemical analysis, and to calculate quantities of substances involved in chemical reactions. Concentrations of species in solution can be described in terms of the molarity, M, of a species. The molarity is the number of moles of solute per liter of solution. Addition of pure solvent to a solution decreases the molarity of a solute but does not change the number of moles of the solute. The molarity of a solution of an acid or a base can be determined by titration.

TERMS YOU SHOULD KNOW

atomic substance 89
molecular substance 89
formula mass 90
formula unit 90
mole (mol) 90
molar mass 90
Avogadro's number (6.022×10^{23}) 91
stoichiometry 93
$\Bumpeq$ (stoichiometrically equivalent to) 94
simplest formula 94
empirical formula 94
molecular formula 97
hydrocarbon 98
stoichiometric coefficient 98
combustion analysis 101
hydrated salt (hydrate) 101
contact process 103
catalyst 103
metric ton 103
limiting reactant 104
excess reactant 104
thermite reaction 105
theoretical yield 106
actual yield 106
percentage yield 106

solution 107
homogeneous 108
solute 108
solvent 108
concentration 108
molarity, M 108
molar 108
volumetric flask 109
dilution 110
titration 111
buret 111

EQUATIONS YOU SHOULD KNOW HOW TO USE

$$\% \text{ yield} = \left(\frac{\text{actual yield}}{\text{theoretical yield}}\right) \times 100$$ (3-5) (calculation of percent yield)

$$M = \frac{n}{V}$$ (3-7) (definition of molarity)

$$M_1V_1 = M_2V_2$$ (3-8) (dilution calculations)

PROBLEMS

NUMBER OF MOLES

3-1. Calculate the number of moles in

(a) 28.0 g of H_2O (1 ounce)
(b) 200 mg of diamond (C) (1 carat)
(c) 454 g of NaCl (1 pound)
(d) 1000 kg of CaO (1 metric ton)

3-2. Compute the number of moles in the recommended daily allowance of the following substances:

(a) 15 mg of zinc
(b) 60 mg of vitamin C, $C_6H_8O_6$
(c) 1.5 mg of vitamin A, $C_{20}H_{30}O$
(d) 6.0 μg of vitamin B_{12}, $C_{63}H_{88}CoN_{14}O_{14}P$

3-3. Calculate the number of moles in

(a) 1.00 kg of malathion, $C_{10}H_{19}O_6PS_2$
(b) 75.0 g of aluminum sulfate, $Al_2(SO_4)_3$
(c) 50.0 mg of oil of peppermint, $C_{10}H_{20}O$
(d) 2.756 g of potassium dichromate, $K_2Cr_2O_7$

3-4. Calculate the number of moles in

(a) 2.00 kg of parathion, $C_{10}H_{14}NO_5PS$
(b) 250.0 g of ammonium hydrogen phosphate, $(NH_4)_2HPO_4$
(c) 75.0 mg of oil of cinnamon, $C_{18}H_{14}O_3$
(d) 150.0 g of Epsom salt, $MgSO_4 \cdot 7H_2O$

AVOGADRO'S NUMBER

3-5. A baseball has a mass of 142 g. Calculate the mass of Avogadro's number of baseballs (1 mol) and compare your result with the mass of the earth, 6.0×10^{24} kg.

3-6. The U.S. population is about 230 million. If Avogadro's number of dollars were distributed equally among the population, how many dollars would each person receive?

3-7. Compute the mass of

(a) one CO_2 molecule
(b) one $C_6H_{12}O_6$ (glucose) molecule
(c) one $CaCl_2$ formula unit

3-8. Compute the mass of

(a) one O_2 molecule
(b) one $FeSO_4$ formula unit
(c) one $C_{12}H_{22}O_{11}$ (sucrose) molecule

3-9. Calculate the mass of

(a) 200 iron atoms
(b) 1.0×10^{16} water molecules
(c) 1.0×10^6 oxygen atoms
(d) 1.0×10^6 oxygen molecules (O_2)

3-10. Calculate the mass of

(a) 100 molecules of nitroglycerin, $C_3H_5N_3O_9$
(b) 5000 molecules of TNT, $C_7H_5N_3O_6$
(c) 10^{10} molecules of octane, C_8H_{18}
(d) 10 molecules of ozone, O_3

3-11. A 50.0-g sample of H_2O contains ____ moles of H_2O, ____ molecules of H_2O, and a total of ____ atoms.

3-12. A 450-g sample of CH_3OH contains ____ moles of CH_3OH, ____ molecules of CH_3OH, and a total of ____ atoms.

3-13. Calculate the number of molecules in

(a) 100 g of ammonia, NH_3
(b) 200 g of sucrose, $C_{12}H_{22}O_{11}$
(c) 400 g of sulfuric acid, H_2SO_4
(d) 100 mg of chlorophyll, $C_{55}H_{72}MgN_4O_5$

3-14. Calculate the number of carbon atoms in

(a) 25.0 g of sodium carbonate, Na_2CO_3
(b) 50 mg of cholesterol, $C_{27}H_{46}O$
(c) 2.50 g of tetraethyl lead, $(C_2H_5)_4Pb$
(d) 2.50 kg of paraquat, $C_{14}H_{20}N_2O_8S_2$

EMPIRICAL FORMULA

3-15. Calcium carbide produces acetylene when water is added to it. The acetylene evolved is burned to provide the light source on spelunkers' helmets. Chemical analysis shows that calcium carbide is 62.5 percent (by mass) calcium and 37.5 percent (by mass) carbon. What is the empirical formula of calcium carbide?

3-16. Rust occurs when iron metal reacts with the oxygen in the air. Chemical analysis shows that dry rust is 69.9 percent iron and 30.1 percent oxygen by mass. Determine the empirical formula of rust.

3-17. A 2.46-g sample of copper metal is reacted completely with chlorine gas to produce 5.22 g of copper chloride. Determine the empirical formula for this chloride.

3-18. A 3.78-g sample of iron metal is reacted with sulfur to produce 5.95 g of iron sulfide. Determine the empirical formula of this compound.

3-19. A 28.1-g sample of cobalt metal was reacted completely with excess chlorine gas. The mass of the compound formed was 61.9 g. Determine its empirical formula.

3-20. A 5.00-g sample of aluminum metal is burned in an oxygen atmosphere to provide 9.45 g of aluminum oxide. Use these data to determine the empirical formula of aluminum oxide.

3-21. Freons are gases used as refrigerants. Chemical analysis shows that a certain Freon is 9.9 percent carbon, 58.7 percent chlorine, and 31.4 percent fluorine by mass. Determine its empirical formula.

3-22. Lead sulfate is one of the components in lead storage batteries. Chemical analysis shows that it is 68.3 percent lead, 10.6 percent sulfur, and 21.1 percent oxygen by mass. What is its empirical formula?

3-23. Given the following mass percentages of the elements in certain compounds, determine the empirical formulas in each case.

(a) 46.45% Li   53.55% O
(b) 59.78% Li   40.22% N
(c) 14.17% Li   85.83% N
(d) 36.11% Ca   63.89% Cl

3-24. Given the following mass percentages of the elements in certain compounds, determine the empirical formula in each case.

(a) 71.89% Tl   28.11% Br
(b) 74.51% Pb   25.49% Cl
(c) 82.24% N   17.76% H
(d) 72.24% Mg   27.76% N

DETERMINATION OF ATOMIC MASS

3-25. A 1.443-g sample of metal is reacted with excess oxygen to yield 1.683 g of the oxide M_2O_3. Compute the atomic mass of the element M.

3-26. An element forms a chloride whose formula is XCl_4, which is known to consist of 75.0 percent chlorine by mass. Find the atomic mass of X and identify it.

3-27. A sample of a compound with the formula $MCl_2{\cdot}2H_2O$ has a mass of 0.642 g. When the compound is heated to remove the water of hydration (represented by $\cdot 2H_2O$ in the formula), 0.0949 g of water is collected. What element is M?

3-28. The formula of an acid is only partially known as HXO_3. The mass of 0.0133 mol of this acid is 1.123 g. Find the atomic mass of X and identify the element represented by X.

COMBUSTION ANALYSIS

3-29. Combustion analysis of a 1.000-g sample of a compound known to contain only carbon, hydrogen, and oxygen gave 1.500 g of CO_2 and 0.409 g of H_2O. Determine the empirical formula of the compound.

3-30. Combustion analysis of a 1.000-g sample of a compound known to contain only carbon, hydrogen, and iron gave 2.367 g of CO_2 and 0.4835 g of H_2O. Determine the empirical formula of the compound.

3-31. Diethyl ether, often called simply ether, is a common solvent that contains carbon, hydrogen, and oxygen. A 1.23-g sample was burned under controlled conditions to produce 2.92 g of CO_2 and 1.49 g of H_2O. Determine the empirical formula of diethyl ether.

3-32. Butylated hydroxytoluene, BHT, a food preservative, contains carbon, hydrogen, and oxygen. A 15.42-mg sample of BHT was burned in a stream of oxygen and yielded 46.20 mg CO_2 and 15.13 mg H_2O. Calculate the empirical formula of BHT.

3-33. Pyridine is recovered from coke-oven gases and is used extensively in the chemical industry, in particular, in the synthesis of vitamins and drugs. Pyridine contains carbon, hydrogen, and nitrogen. A 0.546-g sample was burned to produce 1.518 g of CO_2 and 0.311 g of H_2O. Determine the empirical formula of pyridine.

3-34. One of the additives to gasoline to prevent knocking was found to contain lead, carbon, and hydrogen. A 5.83-g sample was burned in an apparatus like that in

Figure 3-2, and 6.34 g of CO_2 and 3.26 g of H_2O were produced. Determine the empirical formula of this additive.

MOLECULAR FORMULAS

3-35. Acetone is an important chemical solvent; a familiar home use is as a nail polish remover. Chemical analysis shows that acetone is 62.0 percent carbon, 10.4 percent hydrogen, and 27.5 percent oxygen by mass. Determine the empirical formula of acetone. In a separate experiment, the molecular mass is found to be 58.1. What is the molecular formula of acetone?

3-36. Glucose, one of the main sources of energy used by living organisms, has a molecular mass of 180.2. Chemical analysis shows that glucose is 40.0 percent carbon, 6.71 percent hydrogen, and 53.3 percent oxygen by mass. Determine its molecular formula.

3-37. A class of compounds called sodium metaphosphates were used as additives to detergents to improve cleaning ability. One of them has a molecular mass of 612. Chemical analysis shows that this sodium metaphosphate consists of 22.5 percent sodium, 30.4 percent phosphorus, and 47.1 percent oxygen by mass. Determine the molecular formula of this compound.

3-38. A hemoglobin sample was found to be 0.373 percent (by mass) iron. Given that there are four iron atoms per hemoglobin molecule, determine the molecular mass of hemoglobin.

CALCULATIONS INVOLVING CHEMICAL REACTIONS

3-39. The combustion of propane occurs via the reaction

$$C_3H_8(g) + 5O_2(g) \rightarrow 3CO_2(g) + 4H_2O(g)$$

How many grams of oxygen are required to burn completely 10.0 g of propane?

3-40. Iodine is prepared both in the laboratory and commercially by adding $Cl_2(g)$ to an aqueous solution containing sodium iodide according to

$$2NaI(aq) + Cl_2(g) \rightarrow I_2(s) + 2NaCl(aq)$$

How many grams of sodium iodide must be used to produce 50.0 grams of iodine?

3-41. Small quantities of chlorine can be prepared in the laboratory by the reaction

$$MnO_2(s) + 4HCl(aq) \rightarrow MnCl_2(aq) + Cl_2(g) + 2H_2O(l)$$

How many grams of chlorine can be prepared from 100 grams of manganese dioxide?

3-42. Small quantities of oxygen can be prepared in the laboratory by heating potassium chlorate, $KClO_3$. The equation for the reaction is

$$2KClO_3(s) \rightarrow 2KCl(s) + 3O_2(g)$$

Calculate how many grams of O_2 can be produced from heating 10.0 grams of $KClO_3$.

3-43. Lithium nitride reacts with water to produce ammonia and lithium hydroxide:

$$Li_3N(s) + 3H_2O(l) \rightarrow NH_3(g) + 3LiOH(aq)$$

Heavy water is water with the isotope deuterium in place of ordinary hydrogen, and its formula is D_2O. The above reaction can be used to produce heavy ammonia, ND_3.

$$Li_3N(s) + 3D_2O(l) \rightarrow ND_3(g) + 3LiOD(aq)$$

Calculate how many grams of heavy water are required to produce 200 milligrams of ND_3. The atomic mass of deuterium is 2.014.

3-44. A common natural source of phosphorus is phosphate rock, an ore found in extensive deposits in areas that were originally ocean floor. The formula of one type of phosphate rock is $Ca_{10}(OH)_2(PO_4)_6$. Phosphate rock is converted to phosphoric acid by the reaction

$$Ca_{10}(OH)_2(PO_4)_6(s) + 10H_2SO_4(l) \rightarrow 6H_3PO_4(l) + 10CaSO_4(s) + 2H_2O(l)$$

Calculate how many metric tons of phosphoric acid can be produced from 100 metric tons of phosphate rock (1 metric ton = 1000 kg).

3-45. The most common ore of arsenic is mispickel, FeSAs. Upon heating this ore, free arsenic is obtained:

$$FeSAs(s) \rightarrow FeS(s) + As(s)$$

How many grams of FeSAs are required to produce 10.0 g of arsenic?

3-46. Glucose is used as an energy source by the human body. The overall reaction in the body is

$$C_6H_{12}O_6(aq) + 6O_2(g) \rightarrow 6CO_2(g) + 6H_2O(l)$$

Calculate the number of grams of oxygen required to convert 28 g of glucose to CO_2 and H_2O. Also compute the number of grams of CO_2 produced.

3-47. Zinc is produced from its principal ore, sphalerite (ZnS), by the two-step process

$$2ZnS(s) + 3O_2(g) \rightarrow 2ZnO(s) + 2SO_2(g)$$

$$ZnO(s) + C(s) \rightarrow Zn(s) + CO(g)$$

How many kilograms of zinc can be produced from 2.00×10^5 kg of ZnS?

3-48. Titanium is produced from its principal ore, rutile (TiO_2), by the two-step process

$$TiO_2(s) + 2Cl_2(g) + 2C(s) \rightarrow TiCl_4(g) + 2CO(g)$$

$$TiCl_4(g) + 2Mg(s) \rightarrow Ti(s) + 2MgCl_2(s)$$

How many kilograms of titanium can be produced from 4.10×10^3 kg of TiO_2?

3-49. Nitric acid, HNO_3, is made commercially from ammonia by the Ostwald process, which was developed by the German chemist Wilhelm Ostwald. The process consists of three steps:

$$4NH_3(g) + 5O_2(g) \rightarrow 4NO(g) + 6H_2O(g)$$

$$2NO(g) + O_2(g) \rightarrow 2NO_2(g)$$

$$3NO_2(g) + H_2O(l) \rightarrow 2HNO_3(aq) + NO(g)$$

How many kilograms of nitric acid can be produced from 6.40×10^4 kg of ammonia?

3-50. Antimony is usually found in nature as the mineral stibnite, Sb_2S_3. Pure antimony can be obtained by first converting the sulfide to an oxide and then heating the oxide with coke (carbon). The reactions are

$$2Sb_2S_3(s) + 9O_2(g) \rightarrow Sb_4O_6(s) + 6SO_2(g)$$

$$Sb_4O_6(s) + 6C(s) \rightarrow 4Sb(s) + 6CO(g)$$

How many grams of antimony are formed from 500 g of stibnite?

LIMITING REACTANT

3-51. Potassium nitrate is widely used as a fertilizer because it provides two essential elements, potassium and nitrogen. It is made by mixing potassium chloride and nitric acid in the presence of oxygen according to the equation

$$4KCl(aq) + 4HNO_3(aq) + O_2(g) \rightarrow 4KNO_3(aq) + 2Cl_2(g) + 2H_2O(l)$$

How many kilograms of potassium nitrate will be produced from 50.0 kg of potassium chloride and 50.0 kg of nitric acid? An important by-product is chlorine. How many kilograms of chlorine will be produced?

3-52. Phosphorus forms a compound similar to ammonia. The compound has the chemical formula PH_3 and is called phosphine. It can be prepared by the reaction

$$P_4(s) + 3NaOH(aq) + 3H_2O(l) \rightarrow PH_3(g) + 3NaH_2PO_2(aq)$$

If 20.0 g of phosphorus and 50.0 g of NaOH are reacted with $H_2O(l)$ in excess, how many grams of phosphine will be obtained?

3-53. Sodium hydroxide reacts with sulfuric acid according to the equation

$$2NaOH(aq) + H_2SO_4(aq) \rightarrow Na_2SO_4(aq) + 2H_2O(l)$$

Suppose that 60.0 g of sodium hydroxide is added to 20.0 g of sulfuric acid. How many grams of Na_2SO_4 will be produced?

3-54. Bromine can be prepared by adding $Cl_2(g)$ to an aqueous solution of sodium bromide. The reaction is

$$2NaBr(aq) + Cl_2(g) \rightarrow Br_2(l) + 2NaCl(aq)$$

How many grams of bromine are formed if 25.0 g of NaBr and 25.0 g of Cl_2 are reacted?

PERCENTAGE YIELD

3-55. Titanium dioxide is converted to titanium tetrachloride by reaction with chlorine gas and carbon:

$$TiO_2(s) + 2Cl_2(g) + 2C(s) \rightarrow TiCl_4(g) + 2CO(g)$$

Suppose 50.0 g of $TiO_2(s)$ is reacted with excess $Cl_2(g)$ and $C(s)$, and 55.0 g of $TiCl_4$ is obtained. Compute the percentage yield of $TiCl_4(g)$.

3-56. Antimony is produced by the reaction of antimony(III) oxide, Sb_4O_6, with carbon:

$$Sb_4O_6(s) + 6C(s) \rightarrow 4Sb(s) + 6CO(g)$$

Given that 600 g of Sb_4O_6 is reacted with excess $C(s)$, and 490 g of Sb is obtained, compute the percentage yield of antimony.

3-57. Ethyl propionate is obtained from the reaction between ethanol, C_2H_5OH, and propionic acid, $C_2H_5CO_2H$, using sulfuric acid as a catalyst:

$$C_2H_5OH(aq) + C_2H_5CO_2H(aq) \xrightarrow{H_2SO_4} C_2H_5O_2CC_2H_5(aq) + H_2O(l)$$

Ethyl propionate has a pineapple-like odor and is used as a flavoring agent in fruit syrup. In an experiment 349 g of ethyl propionate was obtained from 250 g of ethanol, with propionic acid in excess. Calculate the percentage yield of this reaction.

3-58. Ethyl alcohol is produced commercially from the reaction of water with ethylene. Ethylene is obtained from petroleum and is the sixth-ranked industrial chemical in the United States. It is the basis for the synthesis of a variety of important chemicals and polymers. The reaction for the synthesis of ethyl alcohol is

$$C_2H_4(g) + H_2O(l) \xrightarrow{H_2SO_4} C_2H_5OH(l)$$

Given that 13.5 kg of ethyl alcohol was produced from 10.0 kg of ethylene, calculate the percentage yield in this synthesis.

PREPARATION OF SOLUTIONS

3-59. A saturated solution of calcium hydroxide, $Ca(OH)_2$, contains 0.185 g per 100 mL of solution. Calculate the molarity of a saturated calcium hydroxide solution.

3-60. Calculate the molarity of a saturated solution of sodium hydrogen carbonate, $NaHCO_3$ (baking soda), which contains 69.0 g in 1.00 L of solution.

3-61. Sodium hydroxide is extremely soluble in water. A saturated solution contains 572 g of NaOH per liter of solution. Calculate the molarity of a saturated NaOH solution.

3-62. A cup of coffee may contain as much as 300 mg of caffeine, $C_8H_{10}N_4O_2$. Compute the molarity of caffeine in one cup of coffee (4 cups = 0.946 L).

3-63. Explain how you would prepare 500 mL of a 0.250 M aqueous solution of sucrose, $C_{12}H_{22}O_{11}$. This solution is used frequently in biological experiments.

3-64. How would you prepare 50.0 mL of 0.200 M $CuSO_4$ solution, starting with solid $CuSO_4 \cdot 5H_2O$?

3-65. Compute the number of moles of the solute in each of the following solutions:

(a) 25.46 mL of 0.1255 M $K_2Cr_2O_7(aq)$
(b) 50 μL of 0.020 M $C_6H_{12}O_6(aq)$

3-66. Compute the number of moles of the solute in each of the following solutions:

(a) 50.0 μL of a 0.200 M NaCl(*aq*) solution
(b) 2.00 mL of a 2.00 mM $H_2SO_4(aq)$ solution

3-67. Describe how you would prepare 500 mL of 0.050 M sodium dihydrogen phosphate, NaH_2PO_4, starting with a 1.0 M solution.

3-68. How many milliliters of 6.00 M $HNO_3(aq)$ are required to prepare 50.0 mL of 0.50 M $HNO_3(aq)$?

3-69. How many milliliters of 18.0 M $H_2SO_4(aq)$ are required to prepare 500 mL of 0.30 M $H_2SO_4(aq)$?

3-70. How many milliliters of 12.0 M HCl(*aq*) are required to prepare 250 mL of 1.0 M HCl(*aq*)?

CALCULATIONS INVOLVING SOLUTIONS

3-71. Zinc reacts with hydrochloric acid according to

$$Zn(s) + 2HCl(aq) \rightarrow ZnCl_2(aq) + H_2(g)$$

How many milliliters of 2.00 M HCl are required to react with 2.55 g of Zn?

3-72. Bromine is obtained commercially from natural brines from wells in Michigan and Arkansas by the reaction

$$Cl_2(g) + 2NaBr(aq) \rightarrow 2NaCl(aq) + Br_2(l)$$

If the concentration of NaBr is 4.00×10^{-3} M, how many grams of bromine can be obtained per cubic meter of brine? How many grams of chlorine are required?

3-73. Sodium hypochlorite, NaClO, is used as a bleaching agent in many commercial bleaches. Sodium hypochlorite can be prepared by the reaction

$$Cl_2(g) + 2NaOH(aq) \rightarrow NaClO(aq) + NaCl(aq) + H_2O(l)$$

How many grams of Cl_2 are required to react with 5.00 L of 6.00 M NaOH?

3-74. Silver chloride can be dissolved in an aqueous solution of ammonia according to

$$AgCl(s) + 2NH_3(aq) \rightarrow Ag(NH_3)_2^+(aq) + Cl^-(aq)$$

How many liters of 0.100 M NH_3 solution would be required to dissolve 0.231 g of AgCl?

NEUTRALIZATION AND TITRATION

3-75. By titration, it is found that 27.5 mL of 0.155 M NaOH(*aq*) is required to neutralize 25.0 mL of HCl(*aq*). Calculate the concentration of the hydrochloric acid solution.

3-76. By titration, it is found that 24.6 mL of 0.300 M $H_2SO_4(aq)$ is required to neutralize 20.0 mL of NaOH(*aq*). Calculate the concentration of the NaOH solution.

3-77. (a) What volume of 0.108 M HNO_3 solution is required to neutralize 15.0 μL of 0.010 M $Ca(OH)_2$? (b) What volume of 0.300 M H_2SO_4 solution is required to neutralize 25.0 mL of 0.200 M NaOH?

3-78. Commercial antacid tablets contain a base, often an insoluble metal hydroxide, that reacts with stomach acid. Two bases used for this are $Mg(OH)_2(s)$ and $Al(OH)_3(s)$. Given that stomach acid is about 0.10 M HCl(*aq*), compute the number of milliliters of stomach acid that can be neutralized by 500 mg of each of these bases.

3-79. A 40.0-g sample of KOH(*s*) is dissolved in water to a final volume of 0.200 L, and the resulting solution is added to 2.00 L of 0.125 M HCl(*aq*). Compute the molarity of KCl(*aq*) in the resulting solution.

3-80. A 500-mL sample of 0.200 M NaOH(*aq*) is added to 200 mL of 0.100 M HBr(*aq*). Compute the molarities of the various species after the reaction occurs.

3-81. A 0.365-g sample of a mixture of NaOH and NaCl requires 31.7 mL of 0.150 M HCl to react with all the NaOH. What is the mass percentage of NaOH in the mixture?

3-82. In order to test the purity of NaOH after its manufacture, 0.400 g is dissolved in enough water to make 100 mL of a 0.100 M solution. The solution is titrated with 0.100 M HCl to determine the actual concentration of NaOH. It is found that 25.0 mL of NaOH is neutralized by 23.2 mL of HCl. Calculate the purity of the solid NaOH. What assumption do you have to make?

3-83. A 1.00-g sample of an unknown acid is dissolved to make 100.0 mL of solution and neutralized with 0.250 M NaOH(*aq*). The volume of NaOH(*aq*) required to neutralize the acid was 66.6 mL. Assume that the acid has only one acidic proton per molecule and compute the formula mass of the acid.

3-84. A 1.00-g sample of an unknown acid is dissolved to make 100.0 mL of solution and neutralized with 0.250 M NaOH(*aq*). The volume of NaOH(*aq*) required to neutralize the acid was 86.9 mL. Assume that the acid has two acidic protons per molecule and compute the formula mass of the acid.

ADDITIONAL PROBLEMS

3-85. The following is one of the side reactions in the manufacture of rayon from wood pulp:

$$3CS_2(g) + 6NaOH(aq) \rightarrow 2Na_2CS_3(aq) + Na_2CO_3(aq) + 3H_2O(l)$$

How many grams of each product are formed when 1.00 kg of each reactant is used?

3-86. Chlorine is produced industrially by the electrolysis of brine, which is a solution of naturally occurring salt and consists mainly of sodium chloride:

$$2NaCl(aq) + 2H_2O(l) \xrightarrow{\text{elec}} 2NaOH(aq) + Cl_2(g) + H_2(g)$$

The other products, sodium hydroxide and hydrogen, are also valuable commercial compounds. How many kilograms of each product can be obtained from the electrolysis of 1.00 kg of salt that is 95 percent sodium chloride by mass?

3-87. For the reaction

$$2KMnO_4(aq) + 5H_2O_2(aq) + 6HCl(aq) \rightarrow 2MnCl_2(aq) + 5O_2(g) + 8H_2O(l) + 2KCl(aq)$$

(a) How many grams of $MnCl_2$ can be produced from reaction of 20.0 g $KMnO_4$, 10.0 g H_2O_2, and 1.00×10^2 g HCl?
(b) If the actual yield is 9.82 g $MnCl_2$, what is the percentage yield?

3-88. Lithium is the only Group 1 metal that yields the normal oxide, Li_2O, when it is burned in excess oxygen. The other alkali metals react with excess oxygen according to

(a) $2Na(s) + O_2(g) \rightarrow Na_2O_2(s)$ sodium peroxide

(b) $K(s) + O_2(g) \rightarrow KO_2(s)$ potassium superoxide

(c) $Rb(s) + O_2(g) \rightarrow RbO_2(s)$ rubidium superoxide

(d) $Cs(s) + O_2(g) \rightarrow CsO_2(s)$ cesium superoxide

Calculate how much product is formed when 0.600 g of each alkali metal is burned in oxygen.

3-89. The arsenic in an ore sample was converted into water soluble sodium arsenate, Na_3AsO_4. From the solution of Na_3AsO_4, insoluble silver arsenate, Ag_3AsO_4, was precipitated and weighed. A 5.00-g ore sample gave 3.09 g of silver arsenate. Calculate the mass percentage of arsenic in the ore sample.

3-90. A hydrated form of $CuSO_4$ is heated to drive off all the water. If we started with 10.0 g of hydrated salt and have 5.25 g of anhydrous $CuSO_4$ after heating, find the number of water molecules associated with each $CuSO_4$.

3-91. Table salt (NaCl) and sugar ($C_{12}H_{22}O_{11}$) are accidentally mixed. A 5.00-g sample is burned, and 2.20 g of CO_2 is produced. What is the mass percentage of salt in the mixture?

3-92. Ethyl alcohol is produced by the action of certain yeasts on sugars such as glucose:

$$\underset{\text{glucose}}{C_6H_{12}O_6(aq)} \rightarrow 2C_2H_5OH(aq) + 2CO_2(g)$$

Wine is made by adding yeast to grape juice. What concentration ($g \cdot L^{-1}$) of glucose must the grape juice contain to produce wine that is 11 percent ethyl alcohol by volume? (Take the density of ethyl alcohol to be $0.79\ g \cdot mL^{-1}$.)

3-93. A saturated hydrochloric acid solution has a density of $1.20\ g \cdot mL^{-1}$ and is 40 percent by mass HCl. What is the molarity of a saturated HCl solution?

3-94. A 0.450-g sample of impure $CaCO_3$ is dissolved in 50.0 mL of 0.150 M HCl. The reaction is

$$CaCO_3(s) + 2HCl(aq) \rightarrow CaCl_2(aq) + H_2O(l) + CO_2(g)$$

The excess HCl is titrated by 8.75 mL of 0.125 M NaOH. Calculate the mass percentage of $CaCO_3$ in the sample.

3-95. The reaction used in some breath analyzer devices is

$$3C_2H_5OH(aq) + 2Cr_2O_7^{2-}(aq) + 16H^+(aq) \rightarrow 3HC_2H_3O_2(aq) + 4Cr^{3+}(aq) + 11H_2O(l)$$

How many grams of $HC_2H_3O_2$ are produced by reaction of 55.0 mL of 0.560 M $Cr_2O_7^{2-}$ with 100.0 mL of 0.963 M C_2H_5OH?

3-96. An ore is analyzed for its lead content as follows. A sample is dissolved in water; then sodium sulfate is added to precipitate the lead as lead sulfate ($PbSO_4$). The reaction can be written

$$Pb^{2+}(aq) + SO_4^{2-}(aq) \rightarrow PbSO_4(s)$$

It was found that 13.73 g of lead sulfate was precipitated from a sample of ore having a mass of 53.92 g. How many grams of lead are there in the sample? What is the mass percentage of lead in the ore?

The Main-Group Metals

Magnesium metal burns vigorously in oxygen with a bright, white flame.

The metals that occur in Groups 1 through 5 in the periodic table are called the main-group metals. The chemical reactivities of these metals vary greatly, from the very reactive alkali metals, which combine spontaneously with the oxygen and water vapor in the air, to the relatively unreactive tin, which is used to make tinned cans. Most of the compounds of these elements are ionic. We discussed the alkali metals in Interchapter B, and here we discuss the other main-group metals.

C-1. THE ALKALINE EARTH METALS FORM IONIC COMPOUNDS CONSISTING OF M^{2+} IONS

The alkaline earth metals—beryllium, magnesium, calcium, strontium, barium, and radium—occur in Group 2 in the periodic table. Beryllium is a relatively rare element but occurs as localized surface deposits in the mineral **beryl** (Figure C-1). Essentially unlimited quantities of magnesium are readily available in seawater, where $Mg^{2+}(aq)$ occurs at a concentration of 0.054 M. Calcium, strontium, and barium rank 5th, 18th, and 19th in abundance in the earth's crust, occurring primarily as carbonates and sulfates. All isotopes of radium are radioactive. Some of the physical properties of the alkaline earth metals are given in Table C-1.

Figure C-1 The mineral beryl, $Be_3Al_2Si_6O_{18}$ is the chief source of beryllium and is used as a gem.

The chemistry of the Group 2 elements involves primarily the metals and the +2 ions. With few exceptions the reactivity of the Group 2 elements increases from beryllium to barium. As in all families of the main-group elements, the first member of the family differs in several respects from the other members of the family. The anomalous properties of beryllium are attributed to the small ionic radius of Be^{2+}.

The alkaline earth (Group 2) metals are too reactive to occur as the free metals in nature. The Group 2 metals are prepared by high-temperature electrolysis of the appropriate molten chloride; for example,

$$CaCl_2(l) \xrightarrow[\text{high T}]{\text{electrolysis}} Ca(l) + Cl_2(g)$$

The metals Mg, Ca, Sr, and Ba are silvery-white when freshly cut, but they tarnish rapidly in air to form the metal oxide, MO(*s*) (Figure C-2). The alkaline earth metals react rapidly with water, but the rates of these reactions are much slower than those for the alkali metals. Beryllium and magnesium react slowly with water at ordinary temperatures, although hot magnesium reacts violently with water.

The alkaline earth metals burn in oxygen to form the MO oxides, which are ionic solids. Magnesium is used as an incendiary in warfare because of its vigorous reaction with oxygen. It burns even more rapidly when sprayed with water and reacts with carbon dioxide at elevated temperatures via the reaction

$$2Mg(s) + CO_2(g) \rightarrow 2MgO(s) + C(s)$$

Covering burning magnesium with sand slows the combustion, but the molten magnesium reacts with the silicon dioxide (the principal component of sand) to form magnesium oxide:

$$2Mg(l) + SiO_2(s) \rightarrow 2MgO(s) + Si(s)$$

Table C-1 Physical properties of the alkaline earth metals

	Beryllium	Magnesium	Calcium	Strontium	Barium
chemical symbol	Be	Mg	Ca	Sr	Ba
atomic number	4	12	20	38	56
atomic mass	9.0218	24.305	40.08	87.62	137.33
melting point/°C	1278	651	845	769	725
boiling point/°C	2970	1107	1487	1384	1740
density at 20°C/$g \cdot cm^{-3}$	1.85	1.74	1.55	2.54	3.51
ionic radius of M^{2+} ion/pm	31	65	99	113	135

Magnesium ribbon is used in flashbulbs. The brilliant flash is produced by the light emitted in the reaction of magnesium with oxygen. Barium reacts with excess oxygen to form barium peroxide:

$$Ba(s) + O_2(g) \rightarrow BaO_2(s)$$

When barium is in excess, the product is primarily $BaO(s)$.

Except for beryllium, the alkaline earth metals react vigorously with dilute acids:

$$Mg(s) + 2HCl(aq) \rightarrow MgCl_2(aq) + H_2(g)$$

The alkaline earth metals react with most of the nonmetals to form ionic binary compounds. The reactions of the alkaline earth metals are summarized in Figure C-3.

▪ Beryllium is somewhat anomalous when compared to the other Group 2 metals in that its binary compounds have less ionic character. Also, Be does not react readily with water at room temperature

Many alkaline earth metal compounds are important commercially. Magnesium hydroxide is only slightly soluble in water, and suspensions of it are sold as the antacid Milk of Magnesia. Magnesium sulfate heptahydrate, $MgSO_4 \cdot 7H_2O$, known as Epsom salt, is used as a cathartic, or purgative. The name Epsom comes from the place where the compound was first discovered in 1695, in a natural spring in Epsom, England.

Calcium oxide, or quicklime, is made by heating limestone:

$$CaCO_3(s) \rightarrow CaO(s) + CO_2(g)$$

Calcium oxide is the fourth-ranked industrial chemical in the United States; over 32 billion pounds are produced annually. It is mixed with water to form calcium hydroxide, which is also called slaked lime:

$$CaO(s) + H_2O(l) \rightarrow Ca(OH)_2(aq)$$

Slaked lime is used to make cement, mortar, and plaster. Calcium, as the Ca^{2+} ion, is an essential constituent of bones and teeth, limestone, plants, and the shells of marine organisms. The Ca^{2+} ion plays a major role in muscle contraction, vision, and nerve excitation.

(a)

(b)

Figure C-2 Calcium is a very reactive metal and reacts with the oxygen and water vapor in the air. Consequently, fresh calcium turnings, shown in (a), corrode rapidly when exposed to air, as seen in (b).

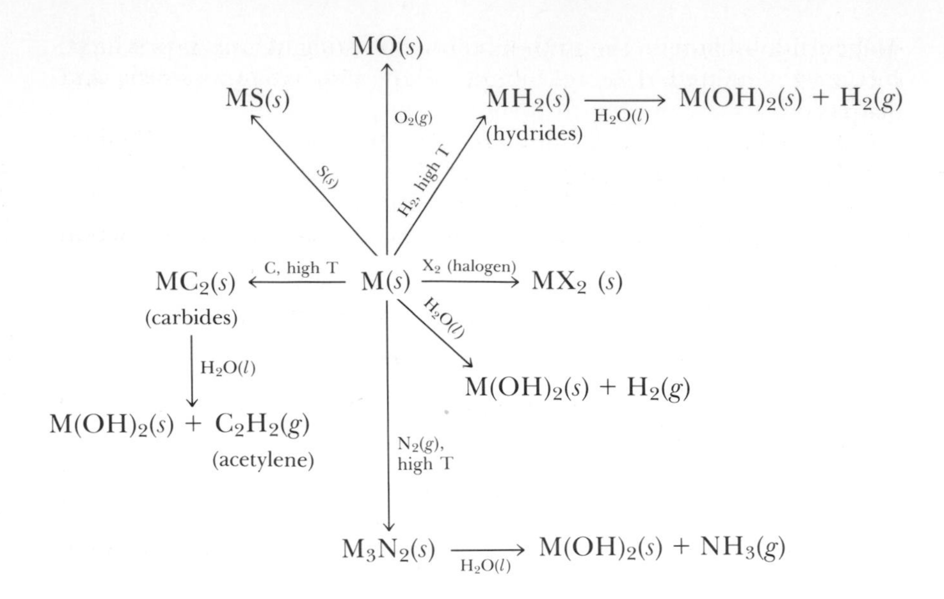

Figure C-3 Representative reactions of Group 2 metals.

Figure C-4 A red signal flare. The red color arises from light emitted by electronically excited strontium atoms.

Strontium salts are used in signal flares and fireworks (Figure C-4). The radioactive isotope strontium-90, which is produced in atomic bomb explosions, is a major health hazard because it behaves like calcium and becomes incorporated in bone marrow, causing various cancers.

C-2. ALUMINUM, GALLIUM, INDIUM, AND THALLIUM ARE GROUP 3 METALS

Some of the physical properties of Group 3 metals are presented in Table C-2.

Aluminum is the most abundant metallic element in the earth's crust, where it occurs in various silicates and in vast deposits of bauxite, AlO(OH) (Figure C-5), from which it is obtained by electrolysis in the Hall process (Chapter 20). Although aluminum reacts with oxygen to form aluminum oxide, aluminum metal resists corrosion by forming a protective, adherent layer of aluminum oxide. Freshly polished aluminum has a bright, silvery appearance, but weathered aluminum has a

Table C-2 Physical properties of the Group 3 metals

	Aluminum	Gallium	Indium	Thallium
chemical symbol	Al	Ga	In	Tl
atomic number	13	31	49	81
atomic mass	26.98154	69.72	114.82	204.37
melting point/°C	660	30	157	304
boiling point/°C	2467	2400	2070	1457
density at 20°C/g·cm^{-3}	2.70	5.90	7.30	11.85
ionic radius of M^{3+} ion/pm	50	62	81	95 (Tl^{3+}) 144 (Tl^{+})

dull tarnish because of the aluminum oxide coating. Aluminum is light, soft, and widely used in lightweight alloys with silicon, copper, and magnesium.

Gallium, indium, and thallium are soft, silvery-white metals. Gallium has the greatest liquid range (over 2000 °C) of any known substance (Figure 2-7). Its melting point is 30°C and its boiling point is 2400°C.

Aluminum and gallium dissolve in *both* acids and bases. The reaction of aluminum with an acid is

$$2Al(s) + 6H^+(aq) \rightarrow 2Al^{3+}(aq) + 3H_2(g)$$

and the reaction of aluminum with a base is

$$2Al(s) + 6H_2O(l) + 2OH^-(aq) \rightarrow \underset{\text{aluminate ion}}{2Al(OH)_4^-(aq)} + 3H_2(g)$$

Substances that dissolve in both acids and bases are called **amphoteric.**

The dissolution of aluminum in strong base is used as the basis of some drain cleaners, such as Drāno, to unplug drains by a combination of the grease-dissolving action of the strong base (NaOH) and the agitation produced by the bubbles of evolved hydrogen gas. Note that explosive mixtures of hydrogen and oxygen gas may form in the sink and all sparks and flames should be avoided.

In contrast to the other Group 3 metals, thallium exhibits both +3 and +1 ionic charges in aqueous media. Thallium(I) compounds are very poisonous, and even trace amounts can cause complete loss of body hair. The aqueous-solution chemistry of $Tl^+(aq)$ is similar to that of $Ag^+(aq)$; for example, as with Ag^+, the Cl^-, Br^-, I^-, and S^{2-} salts of Tl^+ are all insoluble and the halide salts darken on exposure to sunlight.

Figure C-5 Aluminum sources; bauxite (the principal ore of aluminum) and pellets of aluminum oxide (the principal constituent of bauxite).

C-3. TIN AND LEAD ARE GROUP 4 METALS

The principal compounds of both tin and lead involve M(II) and M(IV), although Pb(IV) is known only in the solid state, for example, in $PbO_2(s)$. Tin is found primarily in the mineral **cassiterite,** SnO_2, which occurs in rare but large deposits in Malaysia, China, the U.S.S.R., and the United States. Major deposits of tin sulfide ores are found in Bolivia. Total U.S. natural reserves of tin are very small. It is used in plating (tin-plated food cans) and in various alloys, including solders, type metal, pewter and bronze. Tin is also used to line distilled water tanks. Tin objects are subject to a condition called **tin disease,** which is the conversion of one crystalline form of tin, called white tin, to another crystalline form, called gray tin. This conversion occurs slowly below 13°C and results gray tin, which is a brittle, crumbly metal.

Lead is obtained primarily from the ore **galena,** PbS, by roasting with carbon:

$$PbS(s) + C(s) + 2O_2(g) \rightarrow Pb(l) + CO_2(g) + SO_2(g)$$

Other important lead ores are anglesite ($PbSO_4$) and cerussite ($PbCO_3$). Lead ranks fifth (behind iron, copper, aluminum, and zinc) among the metals in the amount used in manufactured goods (Figure C-6). Lead is resistant to corrosion, but because of its softness it is almost always used in alloy form. The most common lead-alloying material is antimony. Lead storage batteries, which utilize lead-antimony and lead-calcium

Figure C-6 Lead metal and the lead oxide Pb_3O_4. Lead metal is malleable and can be hammered and rolled into sheets.

alloys, constitute the major use of lead. The metal is also used in cable coverings, ammunition, and the synthesis of tetraethyl lead, $(CH_3CH_2)_4Pb$, which is used in leaded gasolines. Lead was once used in paints—$PbCrO_4$ is yellow and Pb_3O_4 is red—but lead salts constitute a health hazard, as they are cumulative poisons, and their use in paints has been discontinued. The Romans used lead vessels to store wine and other consumables and to conduct water in lead-lined aqueducts; thus lead poisoning may have been a factor in the collapse of the Roman Empire. The use of lead-containing glazes on pottery for food use is now prohibited in the United States.

▪ The chemistry of Pb(II) is similar to that of the Group 2 metals.

C-4. BISMUTH IS THE ONLY GROUP 5 METAL

Bismuth is a pink-white metal that occurs rarely as the free metal. The most common source of bismuth is the sulfide ore bismuthinite, Bi_2S_3. The principal compounds of bismuth contain either Bi(III) or Bi(V).

Bismuth metal (Figure C-7) is obtained from the ore by roasting the ore with carbon in air. Bismuth is also obtained as a by-product in lead and copper smelting. It burns in air with a bright blue flame, forming the yellow oxide Bi_2O_3 (Figure C-7) and is used in a variety of alloys, including pewter and low-melting alloys that are used in fire-extinguisher sprinkler-head plugs, electrical fuses, and relief valves for compressed-gas cylinders. Bismuth alloys contract on heating and thus find use in alloys that might otherwise crack because of thermal expansion when subjected to high temperatures.

The oxide Bi_2O_3 is soluble in strongly acidic aqueous solutions. The bismuthyl ion, $BiO^+(aq)$, and the bismuthate ion, $BiO_3^-(aq)$, are important in the aqueous-solution chemistry of bismuth. The bismuthyl ion forms insoluble compounds such as BiOCl and BiO(OH), whereas BiO_3^- is a powerful oxidizing agent. Bismuth pentafluoride, BiF_5, is a potent fluorinating agent that transfers fluorine to various compounds and is converted to the trifluoride, BiF_3.

Figure C-7 Bismuth metal and bismuth(III) oxide, Bi_2O_3.

TERMS YOU SHOULD KNOW

beryl 120
amphoteric 123
cassiterite 123
tin disease 123
galena 123
bismuthinite 124

QUESTIONS

C-1. How are the Group 2 metals obtained?

C-2. Why would you not throw water on a magnesium fire?

C-3. Write the balanced equation for the reaction that occurs when a flashbulb flashes.

C-4. Give the chemical formula for each of the following substances:

(a) Milk of Magnesia
(b) Epsom salt
(c) quicklime
(d) limestone
(e) slaked lime
(f) bauxite

C-5. What substance has the greatest liquid range?

C-6. Write balanced chemical equations for the dissolution of gallium metal in HCl(*aq*) and NaOH(*aq*).

C-7. List the metals of Groups 3, 4, and 5.

C-8. What is tin disease?

C-9. How is lead produced from galena?

C-10. Explain why the reaction of aluminum metal with concentrated NaOH(*aq*) poses an explosion hazard.

PROBLEMS

C-11. How many grams of calcium metal will react with 45.0 g of $O_2(g)$? How many grams of CaO(*s*) will be formed?

C-12. How many grams of In(*s*) will react with 50.0 mL of 6.00 M HBr(*aq*)?

C-13. Calculate the number of kilograms of lead that can be obtained from 1.00 kg of each of the following lead ores:

(a) PbS (galena)
(b) $PbSO_4$ (anglesite)
(c) $PbCO_3$ (cerussite)

C-14. Outline, using balanced chemical equations, a method for obtaining tin metal from the tin sulfide ore, SnS(*s*).

C-15. How many milliliters of 0.10 M HCl(*aq*) are required to neutralize 20 g of CaO(*s*)?

CHAPTER • 4

THE PROPERTIES OF GASES

Chemists can manipulate gases at various pressures using an apparatus called a vacuum rack. This photo shows a vacuum rack in a research laboratory at the Scripps Institute in San Diego.

In this chapter, we study the properties of gases. Many chemical reactions involve gases as reactants or products or both, and so we must learn how the properties of gases depend upon conditions such as temperature, pressure, volume, and number of moles. We shall see how gases respond to changes in pressure and temperature and then discuss how the pressure, temperature, and volume of a gas are related to each other. After presenting a number of experimental observations concerning gases, we shall discuss the kinetic theory of gases, which gives a nice insight into the molecular nature of gases.

4-1. MOST OF THE VOLUME OF A GAS IS EMPTY SPACE

To understand the nature of gases, we must first discuss the three physical states of matter: solid, liquid, and gas. A **solid** has a fixed

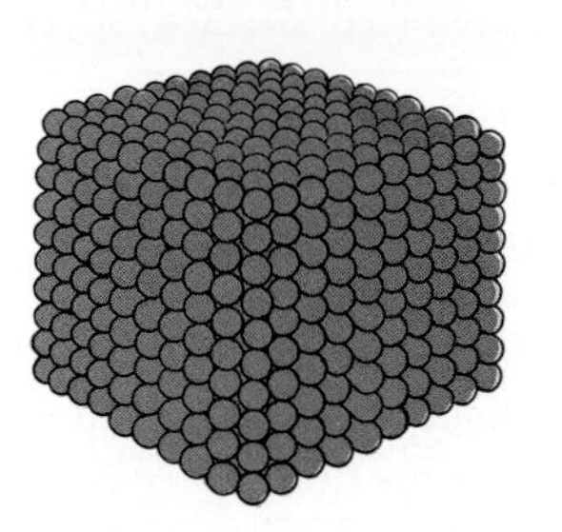

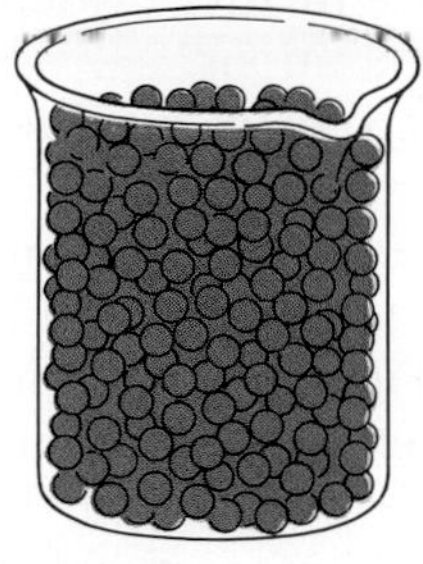

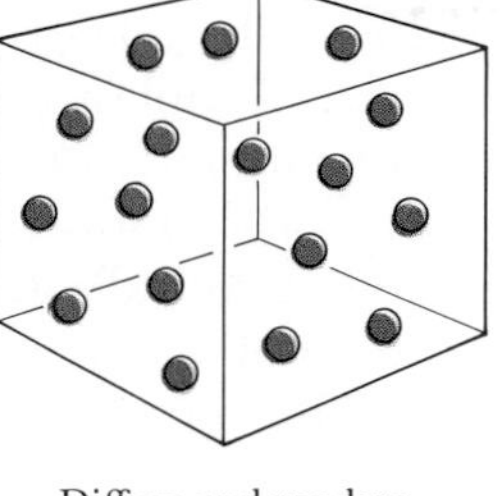
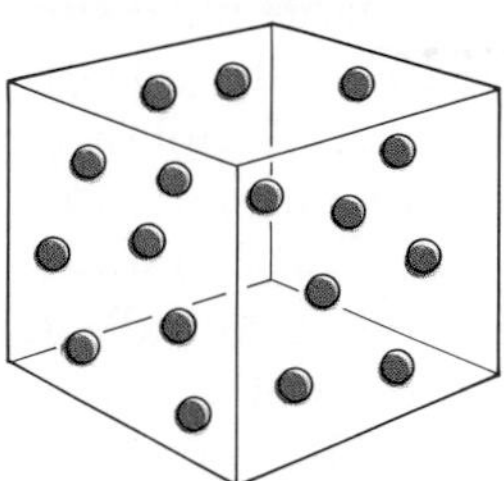

Figure 4-1 Molecular views of (a) a solid, (b) a liquid, and (c) a gas.

volume and shape. A **liquid,** on the other hand, has a fixed volume but assumes the shape of the container into which it is poured. A **gas** has neither a fixed volume nor a fixed shape; it always expands to occupy the entire volume of any closed container into which it is placed.

The molecular picture of a solid is that of a lattice (that is, an ordered array) of particles (atoms, molecules, or ions), as shown in Figure 4-1a. The individual particles vibrate about fixed lattice positions but are not free to move about (Figure 4-2a). The fixed lattice positions of the particles of a solid are reflected by the fixed volume and shape that characterize a solid.

A molecular view of a liquid (Figure 4-1b) is that the particles are in continuous contact with each other but are free to move about throughout the liquid. There is no orderly, fixed arrangement of particles in a liquid as there is in a solid. When a solid melts and becomes a liquid, the lattice array breaks down and the constituent particles are no longer held in fixed positions (Figure 4-2b). The fact that the densities of the solid phase and the liquid phase of any substance do not differ greatly from each other indicates that the amount of separation between the particles is similar in the two phases. Furthermore, the solid and liquid phases of a substance have similar, small **compressibilities,** meaning that their volume does not change appreciably with

Figure 4-2 Computer-calculated paths of particles appear as bright lines on the face of a cathode-ray tube coupled to a computer. (a) Motion of atoms in an atomic crystal (note that the atoms move only about fixed positions). (b) A crystal in the process of melting (note the breakdown of the ordered array). (c) A liquid and its vapor (the dark area represents a gas bubble surrounded by particles whose motions characterize a liquid).

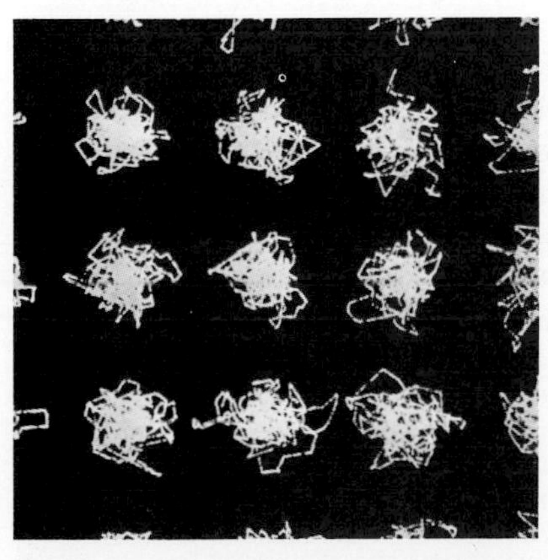

(a)

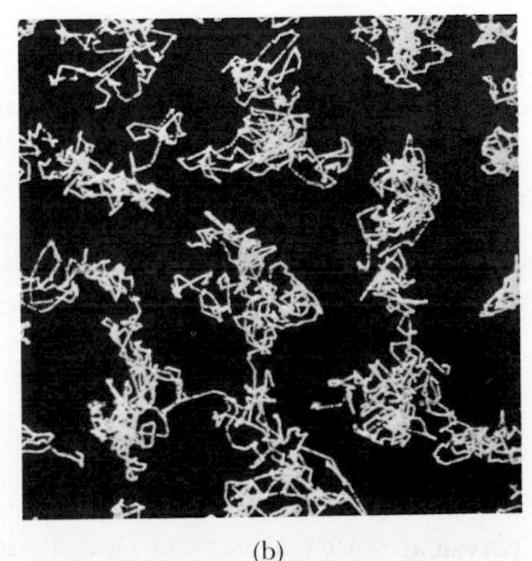

(b)

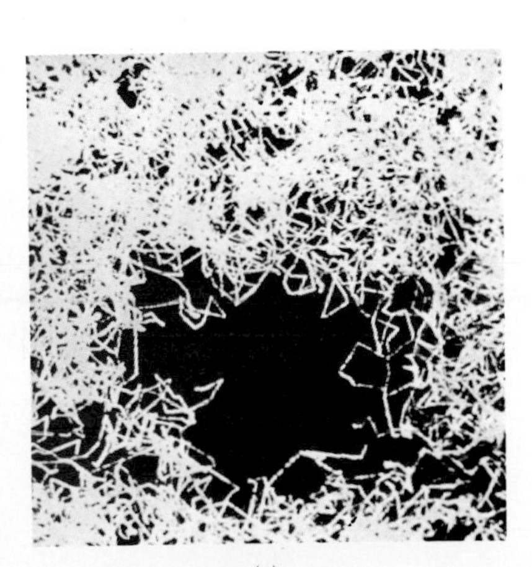

(c)

▪ Compressibility is a measure of how much the volume of a substance changes when the pressure is increased.

increasing pressure. The similar compressibilities of the solid and liquid phases of a substance are further evidence that the particles in the two phases have similar separations.

When a liquid is vaporized, there is a huge increase in volume. For example, 1 mol of liquid water occupies 17.3 mL at 100°C, whereas 1 mol of water vapor occupies over 30,000 mL under the same conditions. Upon vaporization, the molecules of a substance become widely separated, as indicated in Figure 4-1(c). The picture of a gas as being made up of widely separated particles accounts nicely for the relative ease with which gases can be compressed. The particles take up only a small fraction of the total space occupied by a gas; most of the volume of a gas is empty space. As we see later in this chapter, the volume of a gas decreases markedly with increasing pressure.

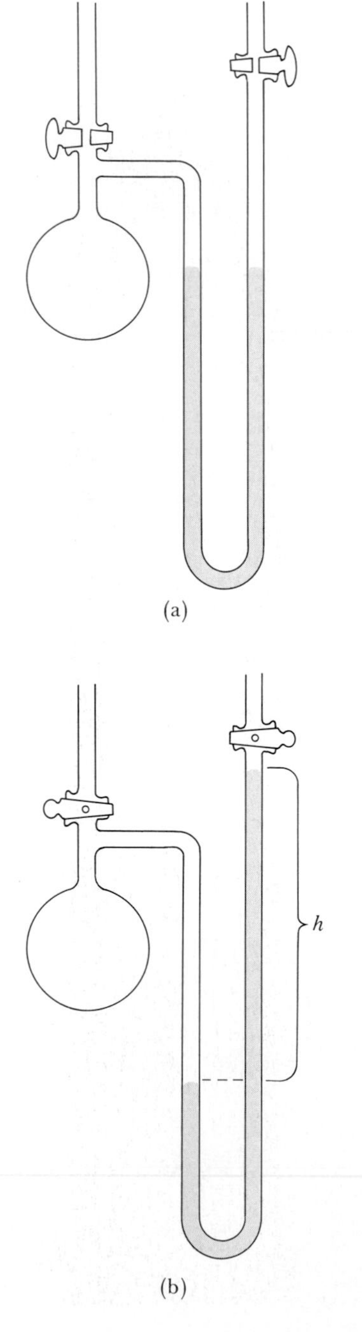

4-2. A MANOMETER IS USED TO MEASURE THE PRESSURE OF A GAS

As we have seen, a gas is mostly empty space, with the molecules widely separated from each other. The molecules are in constant motion, traveling about at high speeds and colliding with each other and with the walls of the container. It is the force of these incessant, numerous collisions with the walls of the container that is responsible for the **pressure** exerted by a gas.

A common laboratory setup used to measure the pressure exerted by a gas is a **manometer,** which is a glass U-shaped tube partially filled with a liquid (Figure 4-3). Mercury is commonly used as the liquid because it has a high density and is fairly unreactive. Figure 4-3 illustrates the measurement of gas pressure with a manometer. The height h of the column of mercury that is supported by the gas in the flask is directly proportional to the pressure of the gas. Because of this direct proportionality, it is convenient to express pressure in terms of the height of a column of mercury that the gas will support. This height is usually expressed in millimeters, and so pressure is expressed in terms of millimeters of mercury (mm Hg). The pressure unit mm Hg is called a **torr,** after the Italian scientist Evangelista Torricelli, who invented the **barometer,** which is a device similar to that shown in Figure 4-3 that is used in meteorology to measure barometric (atmospheric) pressure (Figure 4-4). Thus we say, for example, that the pressure of a gas is 600 torr.

Although mercury is most often used as the liquid in a manometer, other liquids can be used. The height of the column of liquid that can be supported by a gas is inversely proportional to the density of the liquid; that is, the less dense the liquid, the taller will be the column.

Figure 4-3 A mercury manometer. (a) Both stopcocks are open to the atmosphere, and so both columns are exposed to atmospheric pressure. Both columns are at the same height because the pressure is the same on both surfaces. (b) The two stopcocks are closed, and the air in the right-hand column has been evacuated so that there is essentially no pressure on the top of the right-hand column of mercury. As a result, the heights of the columns are no longer the same. The difference in heights is a direct measure of the pressure of the gas in the flask.

Example 4-1: Suppose the pressure of a gas is 760 torr. What is the height of a column of water that will be supported by the gas? Take the density of mercury to be 13.6 g·mL^{-1}, and that of water to be 1.00 g·mL^{-1}.

Solution: Mercury is 13.6 times as dense as water, and so the column of water supported will be 13.6 times higher than the column of mercury. The column of mercury supported is 760 mm, and so

$$\begin{pmatrix}\text{height of}\\ \text{column}\\ \text{of water}\end{pmatrix} = (760\ \text{mm})\left(\frac{13.6\ \text{g·mL}^{-1}}{1.00\ \text{g·mL}^{-1}}\right)\left(\frac{1\ \text{m}}{1000\ \text{mm}}\right) = 10.3\ \text{m}\ (= 33.8\ \text{ft})$$

4-3. A STANDARD ATMOSPHERE IS 760 TORR

The atmosphere surrounding the earth is a gas that exerts a pressure. The manometer pictured in Figure 4-3 can be used to demonstrate this pressure. If the flask is open to the atmosphere and the air in the right-hand side is evacuated, a column of mercury will be supported by the atmospheric (barometric) pressure. The height of the mercury column depends upon elevation above sea level, temperature, and climatic conditions, but at sea level on a clear day it is about 760 mm.

Several units can be used to express pressure. A pressure of 760 torr is defined as one **standard atmosphere** (atm). It is common to express pressure in terms of standard atmospheres or, more simply, atmospheres.

Example 4-2: Given that the measured barometric pressure on a given day at Boulder, Colorado, is 680 torr, express this pressure in atmospheres.

Solution: The conversion between torr and atmospheres is

$$1\ \text{atm} = 760\ \text{torr}$$

Therefore,

$$P = (680\ \text{torr})\left(\frac{1\ \text{atm}}{760\ \text{torr}}\right) = 0.895\ \text{atm}$$

The barometric pressure decreases with increasing altitude.

Strictly speaking, torr and atmosphere are not units of pressure because pressure is defined as a *force per unit area*. The SI unit of pressure is the **pascal** (Pa). The precise definition of a pascal is given in Appendix B; however, a simple, operational definition is that a pascal is the

Figure 4-4 The pressure exerted by the atmosphere can support a column of mercury which is about 760 mm high, as can be seen in the central tube of the barometer shown here. This barometer is at the National Maritime Museum in Greenwich, England.

pressure exerted on a 1-m^2 surface by a mass of 102 g. More important for us is the relation between torr, atmosphere, and pascal:

$$760 \text{ torr} = 1 \text{ atm} = 1.013 \times 10^5 \text{ Pa}$$

▪ The precise defintion of a pascal requires a knowledge of elementary physics.

Several common units for expressing pressure are summarized in Table 4-1. The units torr and atmosphere are so widely used by chemists that their replacement by the pascal will be extremely slow and painful. Consequently, in most cases, we shall use torr or atmosphere in this text, but a section of problems using SI pressure units is included at the end of this chapter.

Example 4-3: Convert 2280 torr to standard atmospheres and to kilopascals.

Solution: To convert from torr to standard atmospheres, we use the fact that 1 atm = 760 torr. Therefore,

$$(2280 \text{ torr})\left(\frac{1 \text{ atm}}{760 \text{ torr}}\right) = 3.00 \text{ atm}$$

To convert from torr to kilopascals, we use the conversion factor 760 torr = 101.3 kPa (Table 4-1):

$$(2280 \text{ torr})\left(\frac{101.3 \text{ kPa}}{760 \text{ torr}}\right) = 304 \text{ kPa}$$

Note that, because there are about 100 kPa in 1 atm, 1 kPa is almost equal to 0.01 atm. Thus 304 kPa is approximately 3.0 atm.

▪ It is useful to remember that one kilopascal is approximately equal to 0.01 atmosphere.

4-4. THE VOLUME OF A GAS IS INVERSELY PROPORTIONAL TO ITS PRESSURE AND DIRECTLY PROPORTIONAL TO ITS KELVIN TEMPERATURE

The first systematic study of the behavior of gases under different applied pressures was carried out in the 1660s by the Irish scientist

Table 4-1 Various units for expressing pressure

SI unit
  1 pascal (Pa) = pressure (force per unit area) exerted by a mass of 102 g on a 1-m^2 surface
  (see Appendix B for precise definition)

"convenience" unit
  height of a column of mercury supported by the pressure; commonly expressed as torr (1 torr = 1 mm Hg)

defined unit
  1 standard atmosphere = 1.013×10^5 Pa
  = 101.3 kPa
  = 760 torr
  = 14.7 lb·in.$^{-2}$

meterological unit
  1 bar = 10^5 Pa
  1 atm = 1.013 bar = 1013 mbar (the bar is derived from *bar*ometer)

Robert Boyle. Boyle was able to show that, at constant temperature, the volume of a given sample of gas is inversely proportional to the pressure:

$$V \propto \frac{1}{P}$$

In terms of an equation, we have

$$V = \frac{c}{P} \qquad \text{(constant temperature)} \qquad (4\text{-}1)$$

where c denotes a proportionality constant whose value in each case is determined by the amount and the temperature of the gas. The relationship between pressure and volume expressed in Equation (4-1) is known as **Boyle's law.** Equation (4-1) is plotted in Figure 4-5. Note that the greater the pressure on a gas, the smaller is the volume at constant temperature. If we double the pressure on a gas, then its volume decreases by a factor of two.

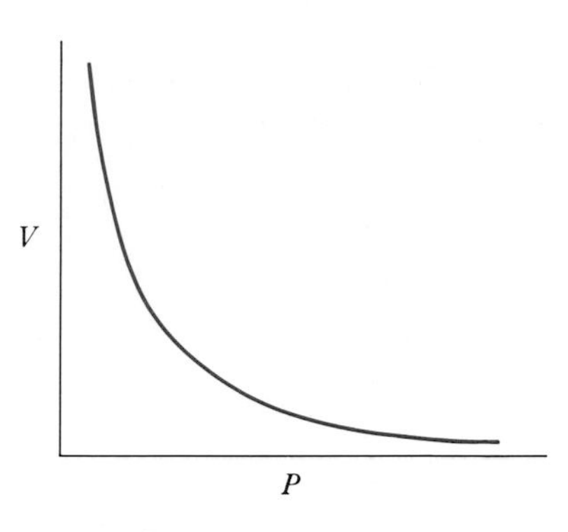

Figure 4-5 Boyle's law states that the volume of a gas is inversely proportional to its pressure.

Jacques Charles, the French scientist and adventurer, showed that there is a linear relationship between the volume of a gas and its temperature. Typical experimental data are plotted as volume versus temperature in Figure 4-6.

Not only does Figure 4-6 show that there is a linear relationship between volume and temperature, but, more important, it also suggests that we can define a new temperature scale that reflects this basic, or absolute, relationship by adding 273° to the Celsius scale. By doing so, all the curves in Figure 4-6 can be represented by the simple equation

$$V = kT \qquad \text{(constant pressure)} \qquad (4\text{-}2)$$

where T, called the **absolute temperature,** is related to t by

$$T = t + 273$$

and where k denotes a proportionality constant whose value in each case is determined by the pressure and quantity of the gas. The relation

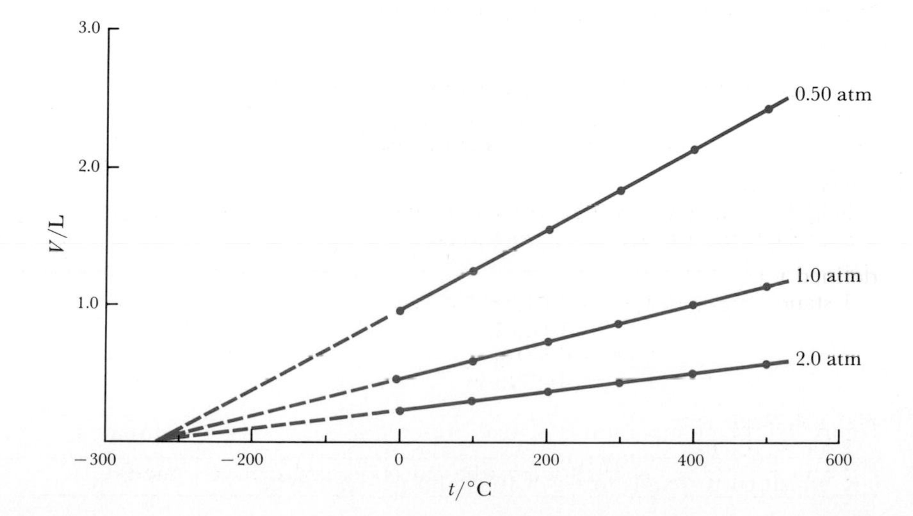

Figure 4-6 The volume (V) of 0.580 g of air plotted as a function of temperature (t) at three different pressures. Note that all three curves extrapolate to $V = 0$ at −273°C. These plots suggest that we can define a temperature scale that reflects the basic relationship of volume and temperature by adding 273 to the Celsius scale.

between volume and absolute temperature given by Equation (4-2) is called **Charles's law:** the volume of a fixed mass of gas at a fixed pressure is directly proportional to its absolute temperature.

The temperature scale introduced here is called the **absolute temperature scale** or the **Kelvin temperature scale** (after Lord Kelvin, the British scientist who proposed it); the unit for this scale is the **kelvin** (K). The precise relation between the two scales is

$$T(\text{in K}) = t(\text{in °C}) + 273.15 \tag{4-3}$$

The Kelvin scale is the fundamental temperature scale. Figure 4-7 shows the temperatures of a number of fixed points on three temperature scales. The lowest possible temperature on the Kelvin scale is 0 K, which corresponds to −273.15°C.

Example 4-4: A simple **gas thermometer** can be made by trapping a small sample of gas with a drop of mercury in a glass capillary tube that is sealed at one end and open at the other (Figure 4-8). Suppose that in such a thermometer, the gas occupies a volume of 0.180 mL at 0°C. The thermometer is then immersed in a liquid, and the final volume of the gas is 0.232 mL. What is the temperature of the liquid?

Solution: From Charles's law, we have $V = kT$, or

$$\frac{V}{T} = k$$

The sample of gas trapped in the tube is a fixed quantity, and so as long as the pressure on the gas remains the same, the value of k does not change. Consequently, we can write

$$\frac{V_i}{T_i} = k \qquad \text{and} \qquad \frac{V_f}{T_f} = k$$

where the subscripts i and f stand for *i*nitial and *f*inal, respectively. By equating these two expressions we get

$$\frac{V_f}{T_f} = \frac{V_i}{T_i} \tag{4-4}$$

We must always remember to use absolute temperatures in Charles's law:

$$T_i = 0 + 273 = 273 \text{ K}$$

We are seeking T_f, and so we solve Equation (4-4) for T_f to get

$$T_f = T_i\frac{V_f}{V_i} = (273 \text{ K})\left(\frac{0.232 \text{ mL}}{0.180 \text{ mL}}\right) = 352 \text{ K}$$

The corresponding temperature in degrees Celsius is

$$t_f = 352 - 273 = 79\text{°C}$$

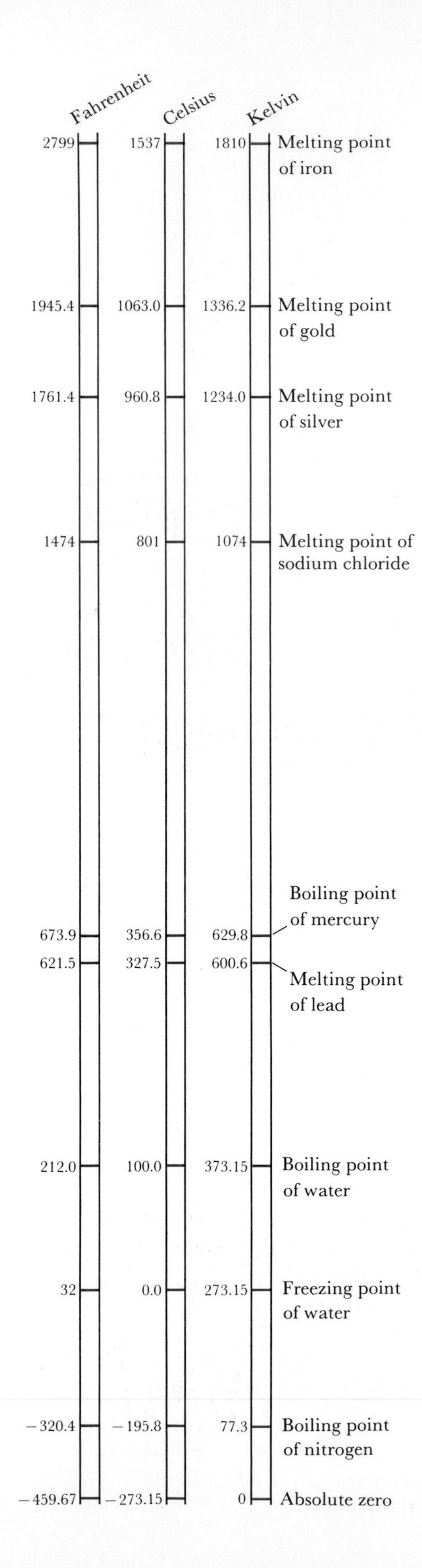

Figure 4-7 The temperatures of fixed reference points on the Fahrenheit, Celsius, and Kelvin temperature scales. The lowest possible temperature is 0 K, which corresponds to −273.15°C and to −459.67°F.

4-5. EQUAL VOLUMES OF GASES AT THE SAME PRESSURE AND TEMPERATURE CONTAIN EQUAL NUMBERS OF MOLECULES

Early experiments with gaseous reactions showed a remarkable property. It was observed by Gay-Lussac in 1809 that if all volumes are measured at the same pressure and temperature, then the volumes in which gases combine in chemical reactions are related to each other by simple whole numbers. For example,

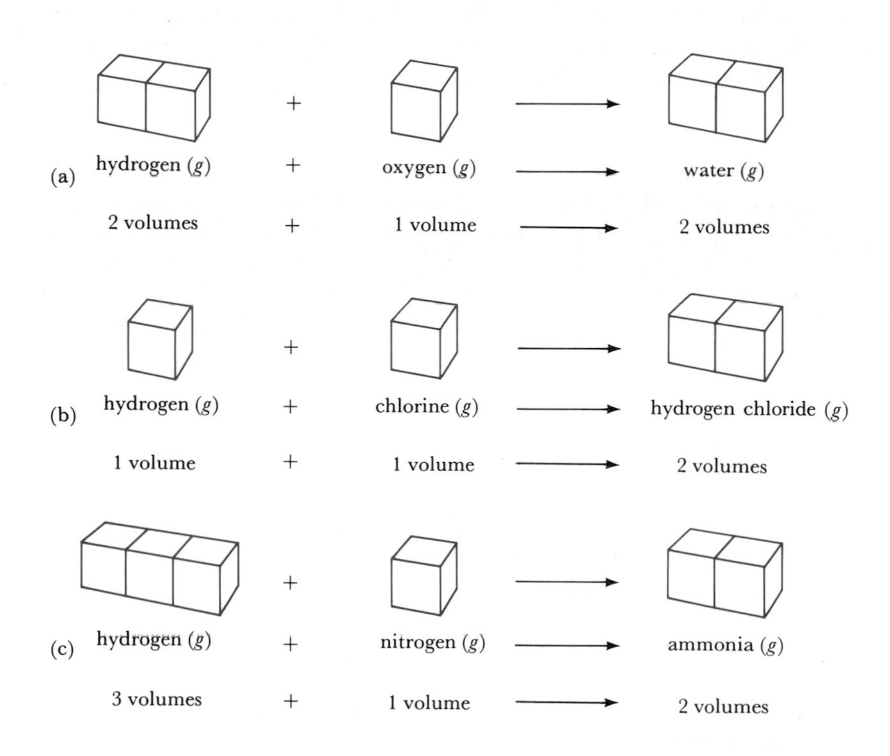

Note that in each of these cases the relative volumes of reactants and products are in the proportion of simple whole numbers. This observation is known as **Gay-Lussac's law of combining volumes** and was one of the earliest indications of the existence of atoms and molecules. The interpretation of Gay-Lussac's law by Avogadro in 1811 led to the realization that many of the common gaseous elements, such as hydrogen, oxygen, nitrogen, and chlorine, occur naturally as diatomic molecules (H_2, O_2, N_2, and Cl_2) rather than as single atoms. Let's review Avogadro's line of reasoning and see why this is so.

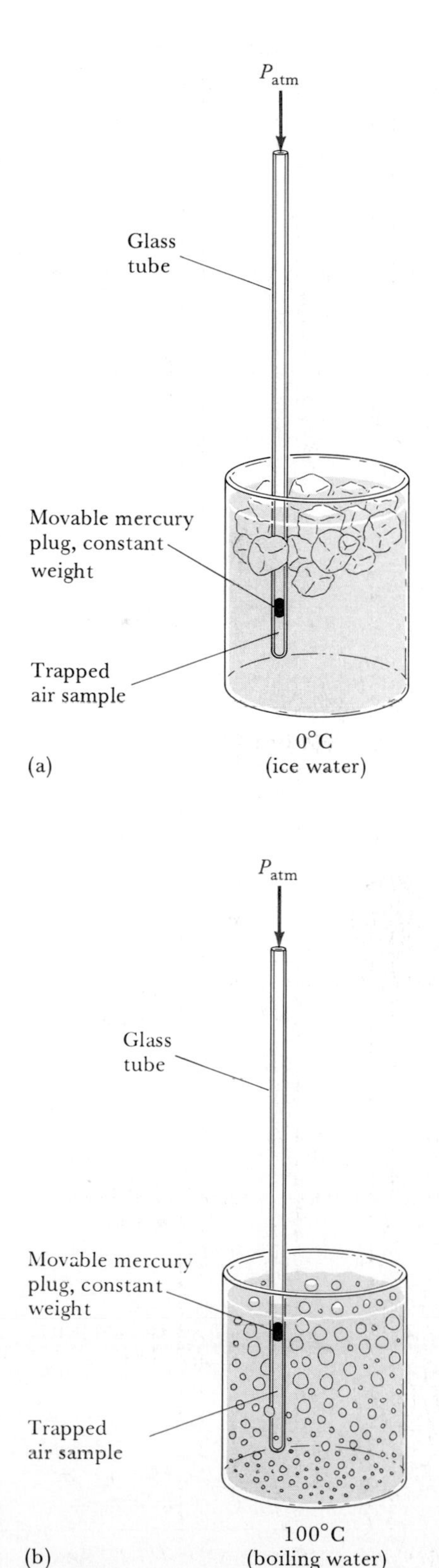

Figure 4-8 A gas thermometer. A sample of air is trapped by a drop of mercury in a capillary tube that is sealed at the bottom. According to Charles's law, the volume of the air is directly proportional to the Kelvin temperature. The atmosphere maintains a constant pressure on the air trapped below the mercury, which moves up or down to a position at which the pressure of the trapped air equals the atmospheric pressure. As the temperature increases, as in (b), the drop of mercury rises because the gas expands.

Following Gay-Lussac's observations, Avogadro postulated that equal volumes of gases at the same pressure and temperature contain equal numbers of molecules. This statement was known as **Avogadro's hypothesis** at the time, but now it is accepted as a law. Avogadro was the first to point out the distinction between atoms and molecules. Consider the reaction between hydrogen and chlorine to form hydrogen chloride. Recall that *one volume* of hydrogen reacts with *one volume* of chlorine to produce *two volumes* of hydrogen chloride. According to Avogadro's reasoning, this means that *one molecule* of hydrogen reacts with *one molecule* of chlorine to produce *two molecules* of hydrogen chloride (Figure 4-9). If one molecule of hydrogen can form two molecules of hydrogen chloride, then a hydrogen molecule must consist of two atoms (at least) of hydrogen. Avogadro pictured both hydrogen and chlorine as diatomic gases and was able to represent the reaction between them as

$$H_2(g) + Cl_2(g) \rightarrow 2HCl(g)$$

$$1 \text{ molecule} + 1 \text{ molecule} \rightarrow 2 \text{ molecules}$$

$$1 \text{ volume} + 1 \text{ volume} \rightarrow 2 \text{ volumes}$$

Prior to Avogadro's explanation, it was difficult to see how one volume of hydrogen could produce two volumes of hydrogen chloride. If hydrogen occurred simply as atoms, there would be no way to explain Gay-Lussac's law of combining volumes. Another example of Avogadro's law applied to a chemical reaction is

$$3H_2(g) + N_2(g) \rightarrow 2NH_3(g)$$

$$3 \text{ molecules} + 1 \text{ molecule} \rightarrow 2 \text{ molecules}$$

$$3 \text{ volumes} + 1 \text{ volume} \rightarrow 2 \text{ volumes}$$

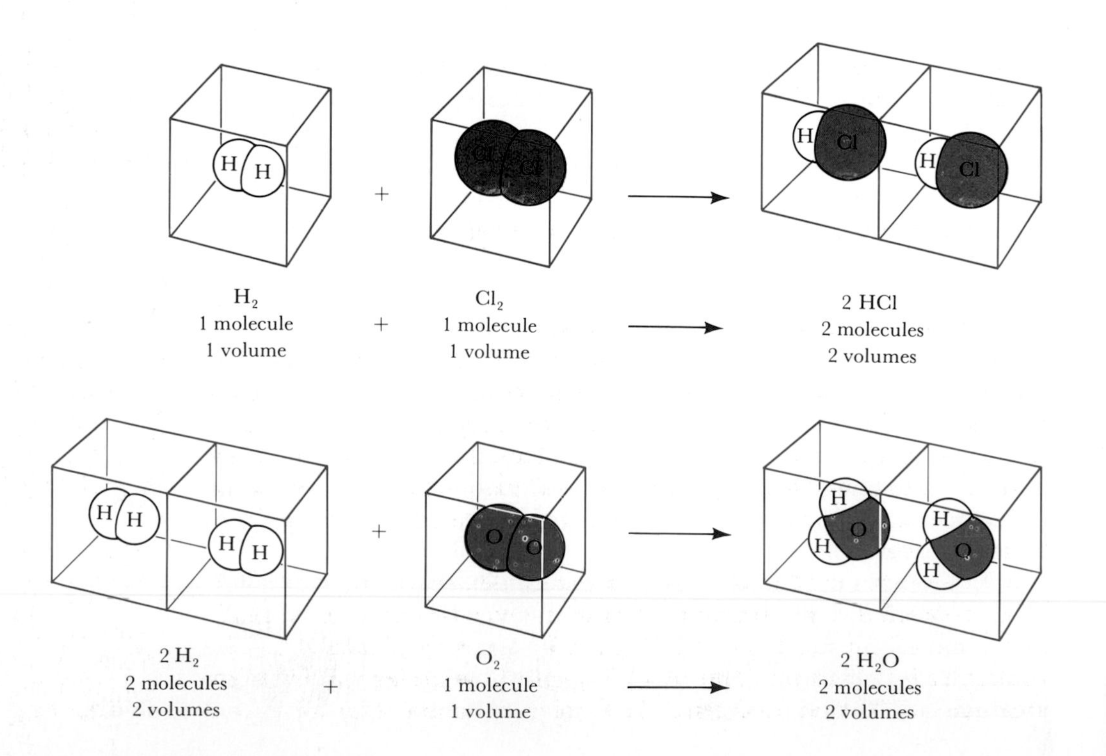

Figure 4-9 An illustration of Avogadro's explanation of Gay-Lussac's law of combining volumes.

It is interesting to note that, in spite of the beautiful simplicity of Avogadro's explanation of these reactions, his work was largely ignored, and chemists continued to confuse atoms and molecules and to use many incorrect chemical formulas. It wasn't until the mid 1800s that Avogadro's hypothesis was finally appreciated and generally accepted.

4-6. THE IDEAL GAS EQUATION IS A COMBINATION OF BOYLE'S, CHARLES'S, AND AVOGADRO'S LAWS

Avogadro postulated that equal volumes of gases at the same pressure and temperature contain the same number of molecules. This implies that equal volumes of gases at the same pressure and temperature contain equal numbers of moles, n. Thus, we can write Avogadro's law as

$$V \propto n \qquad (\text{fixed } P \text{ and } T)$$

Boyle's law and Charles's law are, respectively,

$$V \propto \frac{1}{P} \qquad (\text{fixed } T \text{ and } n)$$

$$V \propto T \qquad (\text{fixed } P \text{ and } n)$$

We can combine these three proportionality statements for V into one by writing

$$V \propto \frac{nT}{P}$$

Note how the three individual statements for V are all included in the combined statement. For example, if P and n are fixed, then only T can vary, and we see that $V \propto T$, which is Charles's law. If P and T are fixed, then only n can vary and we have Avogadro's law, $V \propto n$. Last, if T and n are fixed, we have Boyle's law, $V \propto 1/P$.

We can convert the combined proportionality statement for V to an equation by introducing a proportionality constant R:

$$V = \frac{RnT}{P}$$

This equation is equivalently, and more commonly, written as

$$PV = nRT \qquad (4\text{-}5)$$

and is called the **ideal-gas law** or **ideal-gas equation.** It is based upon Boyle's law, Charles's law, and Avogadro's law. Boyle's law and Charles's law are valid only at low pressures (less than a few atmospheres, say), and so Equation (4-5) is valid only at low pressures. It turns out, however, that most gases obey Equation (4-5) within a few percent up to tens of atmospheres, and so Equation (4-5) is very useful. Gases that satisfy the ideal-gas equation are said to behave ideally, or to be **ideal gases.**

Before we can use Equation (4-5), we must determine the value of R, which is called the **gas constant.** It has been determined experimentally that 1 mol of an ideal gas at 0°C and 1.00 atm occupies 22.4 L. The volume 22.4 L, shown in Figure 4-10, is called the **molar volume** of an ideal gas at 0°C and 1.00 atm. If we solve Equation (4-5) for R and

Figure 4-10 The volume of one mole of an ideal gas at 0°C and 1 atm. A volume of 22.4L can be represented by a cube whose edges measure 28.2 cm. A basketball in its carton is shown here for comparison.

substitute this information into the resulting equation, then we find that

$$R = \frac{PV}{nT} = \frac{(1.00 \text{ atm})(22.4 \text{ L})}{(1.00 \text{ mol})(273 \text{ K})}$$

$$= 0.0821 \text{ L·atm·mol}^{-1}\text{·K}^{-1} \quad (4\text{-}6)$$

▪ $R = 0.0821 \text{ L·atm·mol}^{-1}\text{·K}^{-1}$

▪ Notice that we say that T is in kelvins, not in degrees kelvin.

You should pay careful attention to the units of R. When the value $R = 0.0821 \text{ L·atm·mol}^{-1}\text{·K}^{-1}$ is used in Equation (4-5), P must be expressed in atmospheres, V in liters, n in moles, and T in kelvins.

Now that we have determined the value of the gas constant, we can use Equation (4-5) in many ways.

▪ Normal human body temperature is 37.0°C (98.6°F).

Example 4-5: The pressure of oxygen in inhaled air is about 160 torr. The total volume of the average adult lungs when expanded is about 6.0 L, and body temperature is 37°C. Calculate the mass of O_2 required to occupy a volume of 6.0 L at a pressure of 160 torr and a temperature of 37°C.

Solution: Note that we are given three quantities—V, P, and T—and that we wish to calculate the fourth—n. We must express P and T in the correct units:

$$P = (160 \text{ torr})\left(\frac{1 \text{ atm}}{760 \text{ torr}}\right) = 0.21 \text{ atm}$$

$$T = 37 + 273 = 310 \text{ K}$$

Solving Equation (4-5) for n, we get

$$n = \frac{PV}{RT} = \frac{(0.21 \text{ atm})(6.0 \text{ L})}{(0.0821 \text{ L·atm·mol}^{-1}\text{·K}^{-1})(310 \text{ K})}$$

$$= 0.050 \text{ mol}$$

The number of grams of O_2 is

$$\text{grams of } O_2 = (0.050 \text{ mol } O_2)\left(\frac{32.0 \text{ g } O_2}{1 \text{ mol } O_2}\right)$$

$$= 1.6 \text{ g}$$

Example 4-6: If all the nitrogen generated from the decomposition of 0.512 g of ammonium nitrite by the reaction

$$NH_4NO_2(s) \rightarrow 2H_2O(l) + N_2(g)$$

occupies a 100-mL container at 15°C, what will be the pressure of the nitrogen gas in the container?

Solution: The number of moles of $N_2(g)$ generated is

$$\text{mol } N_2 = (0.512 \text{ g } NH_4NO_2)\left(\frac{1 \text{ mol } NH_4NO_2}{64.05 \text{ g } NH_4NO_2}\right)\left(\frac{1 \text{ mol } N_2}{1 \text{ mol } NH_4NO_2}\right)$$

$$= 7.99 \times 10^{-3} \text{ mol}$$

The pressure is calculated from Equation (4-5):

$$P = \frac{nRT}{V} = \frac{(7.99 \times 10^{-3} \text{ mol})(0.0821 \text{ L·atm·mol}^{-1}\text{·K}^{-1})(288 \text{ K})}{(0.100 \text{ L})}$$

$$= 1.89 \text{ atm}$$

In Examples 4-5 and 4-6 we were given three quantities and had to calculate a fourth. Another type of application of the ideal-gas equation involves changes from one set of conditions to another.

Example 4-7: One mole of O_2 gas occupies 22.4 L at 0°C and 1.00 atm. What volume does it occupy at 100°C and 4.00 atm?

Solution: Note that in this problem we are given V at one set of conditions (T and P) and asked to calculate V under another set of conditions (that is, a different T and P). Because R is a constant, and because n is a constant in this problem, we can write the ideal-gas equation as

$$\frac{PV}{T} = nR = \text{constant}$$

This equation says that the ratio PV/T remains constant, and so we can write

$$\frac{P_iV_i}{T_i} = \frac{P_fV_f}{T_f}$$

where the subscripts i and f denote *i*nitial and *f*inal, respectively. Recall that we wish to calculate a final volume, and so we solve this equation for V_f:

$$V_f = V_i\left(\frac{P_i}{P_f}\right)\left(\frac{T_f}{T_i}\right) \tag{4-7}$$

If we substitute the given quantities into this equation, then we obtain

$$V_f = (22.4\text{ L})\left(\frac{1.00\text{ atm}}{4.00\text{ atm}}\right)\left(\frac{373\text{ K}}{273\text{ K}}\right)$$

$$= 7.65\text{ L}$$

Note that the increase in pressure (from 1.00 atm to 4.00 atm) decreases the gas volume (the gas is compressed), whereas the increase in temperature increases it. The pressure increases by a factor of 4.00, whereas the temperature increases only by a factor of 373/273 = 1.37; thus the *net* effect is a decrease in the volume of the gas.

Note that in Example 4-7 we multiplied the initial volume (22.4 L) by a pressure ratio and a temperature ratio. The pressure increased from 1.00 atm to 4.00 atm, and the pressure ratio used was 1.00/4.00, resulting in a smaller volume, as you would expect. Similarly, the temperature increased from 273 K to 373 K, and the temperature ratio used was 373/273, resulting in an increased volume. A "common sense" method of solving this problem is to write

▪ An increase in pressure decreases the volume of a gas. An increase in temperature increases the volume of a gas.

$$V_f = V_i \times \text{pressure ratio} \times \text{temperature ratio}$$

and to decide by simple reasoning whether each ratio to be used is greater or less than unity.

4-7. THE IDEAL-GAS EQUATION CAN BE USED TO CALCULATE THE MOLAR MASSES OF GASES

Some of the most important applications of the ideal-gas equation involve the calculation of densities and molar masses. As noted in Chap-

■ Molar mass is numerically equal to molecular mass but has units of $g{\cdot}mol^{-1}$.

ter 3, the **molar mass,** M, of a substance is simply the mass in grams of one mole of that substance. Molar masses are numerically equal to formula masses, but have units of $g{\cdot}mol^{-1}$. For example, the formula mass of H_2O is 18.0, and its molar mass is 18.0 $g{\cdot}mol^{-1}$.

If Equation (4-5) is solved for n/V, then we get

$$\frac{n}{V} = \frac{P}{RT}$$

The ratio n/V is equal to gas density in the units moles per liter. We can convert from moles per liter to grams per liter by multiplying both sides of the equation by the molar mass. If we denote the density in grams per liter by the symbol ρ (the Greek letter rho), then we can write

$$\rho = \frac{Mn}{V} = \frac{MP}{RT} \tag{4-8}$$

Note that gas density increases as pressure increases and as temperature decreases.

Example 4-8: Calculate the density of nitrogen dioxide gas at 0°C and 1.00 atm.

Solution: The molar mass of NO_2 is 46.01 $g{\cdot}mol^{-1}$. Using Equation (4-8) with appropriate units gives

$$\rho = \frac{MP}{RT} = \frac{(46.01\ g{\cdot}mol^{-1})(1.00\ atm)}{(0.0821\ L{\cdot}atm{\cdot}mol^{-1}{\cdot}K^{-1})(273\ K)}$$
$$= 2.05\ g{\cdot}L^{-1}$$

Figure 4-11 Like liquids, gases have flow properties and can be poured from one container to another if they are denser than air. This photo shows $NO_2(g)$ being poured.

Another important application of the ideal-gas equation involves the calculation of molar masses.

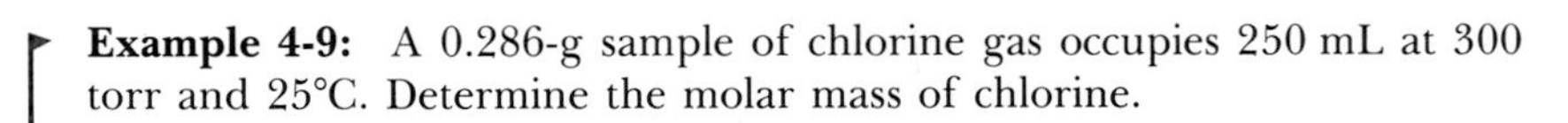

Example 4-9: A 0.286-g sample of chlorine gas occupies 250 mL at 300 torr and 25°C. Determine the molar mass of chlorine.

Solution: We are given V, P, and T, and so we can use Equation (4-5) to calculate n:

$$n = \frac{PV}{RT} = \frac{(300\ torr)\left(\frac{1\ atm}{760\ torr}\right)(0.250\ L)}{(0.0821\ L{\cdot}atm{\cdot}mol^{-1}{\cdot}K^{-1})(298\ K)}$$
$$= 4.03 \times 10^{-3}\ mol$$

Thus 0.286 g of chlorine gas corresponds to 4.03×10^{-3} mol:

$$0.286\ g \Leftrightarrow 4.03 \times 10^{-3}\ mol$$

We would like to have this stoichiometric correspondence read

$$\text{a certain number of grams} \Leftrightarrow 1.00\ mol$$

We can achieve this by dividing both sides by 4.03×10^{-3}:

$$\frac{0.286\ g}{4.03 \times 10^{-3}} \Leftrightarrow \frac{4.03 \times 10^{-3}\ mol}{4.03 \times 10^{-3}}$$

$$71.0\ g \Leftrightarrow 1.00\ mol$$

Thus, we find that the molar mass of chlorine is 71.0 $g \cdot mol^{-1}$, or that the molecular mass of chlorine is 71.0. This implies that chlorine is a diatomic gas (Cl_2) because the atomic mass of chlorine is 35.45.
An alternate solution is to use the relation

$$\text{number of moles} = \frac{\text{mass in grams}}{\text{molar mass}}$$

$$n = \frac{m}{M}$$

Solving for M we obtain

$$M = \frac{m}{n} = \frac{0.286 \text{ g}}{4.03 \times 10^{-3} \text{ mol}} = 71.0 \text{ g} \cdot \text{mol}^{-1}$$

We can combine a problem like Example 4-9 with a determination of the empirical (simplest) formula of a compound from chemical analysis to determine the molecular formula.

Example 4-10: Chemical analysis shows that the gas acetylene is 92.3 percent carbon and 7.70 percent hydrogen by mass. It has a density of 0.711 $g \cdot L^{-1}$ at 20°C and 500 torr. Use these data to determine the molecular formula of acetylene.

Solution: The determination of the empirical formula from chemical analysis is explained in Section 3-3. Following the procedure given there, we write

$$92.3 \text{ g C} \approx 7.70 \text{ g H}$$

Dividing the left side by 12.01 g C/mol C and the right side by 1.008 g H/mol H gives

$$7.69 \text{ mol C} \approx 7.64 \text{ mol H}$$

Dividing both sides by 7.64 and rounding off yields

$$1 \text{ mol C} \approx 1 \text{ mol H}$$

and so the empirical formula of acetylene is CH.

We now use the density data to determine the molar mass of acetylene. Solving Equation (4-8) for M gives

$$M = \frac{\rho RT}{P} = \frac{(0.711 \text{ g} \cdot \text{L}^{-1})(0.0821 \text{ L} \cdot \text{atm} \cdot \text{mol}^{-1} \cdot \text{K}^{-1})(293 \text{ K})}{(500 \text{ torr})\left(\frac{1 \text{ atm}}{760 \text{ torr}}\right)}$$

$$= 26.0 \text{ g} \cdot \text{mol}^{-1}$$

The formula mass of acetylene is 26.0 and its empirical formula is CH ($M = 13.0 \text{ g} \cdot \text{mol}^{-1}$); therefore, the molecular formula of acetylene is C_2H_2 ($M = 26.0 \text{ g} \cdot \text{mol}^{-1}$).

4-8. THE TOTAL PRESSURE OF A MIXTURE OF IDEAL GASES IS THE SUM OF THE PARTIAL PRESSURES OF ALL THE GASES IN THE MIXTURE

Up to this point we have not considered explicitly mixtures of gases, and yet mixtures of gases are of great importance. For example, air is a mixture of nitrogen (78 percent), oxygen (20 percent), and argon (1 percent) with lesser amounts of other gases, such as carbon dioxide. Many industrial processes involve gaseous mixtures. For example, the commercial production of ammonia involves the reaction

$$3H_2(g) + N_2(g) \xrightarrow[300\ \text{atm}]{500°\text{C}} 2NH_3(g)$$

and thus the reaction vessel contains a mixture of N_2, H_2, and NH_3.

In a mixture of ideal gases, each gas exerts a pressure as if it were present alone in the container. For a mixture of two ideal gases, we have

$$P_{\text{total}} = P_1 + P_2 \tag{4-9}$$

The pressure exerted by each gas is called its **partial pressure,** and Equation (4-9) is known as **Dalton's law of partial pressures.** Each of the gases obeys the ideal-gas equation, and so

$$P_1 = \frac{n_1RT}{V} \qquad P_2 = \frac{n_2RT}{V} \tag{4-10}$$

Notice that the volume occupied by each gas is V, because each gas in a mixture occupies the entire container. If the partial pressures P_1 and P_2 are substituted into Equation (4-9), then we get for our two-gas mixture,

$$\begin{aligned} P_{\text{total}} &= \frac{n_1RT}{V} + \frac{n_2RT}{V} \\ &= (n_1 + n_2)\frac{RT}{V} \\ &= n_{\text{total}}\frac{RT}{V} \end{aligned} \tag{4-11}$$

The total pressure exerted by a mixture of gases is determined by the total number of moles of gas in the mixture.

If we divide either of the Equations (4-10) by Equation (4-11), then we obtain

$$\frac{P_i}{P_{\text{total}}} = \frac{n_i}{n_{\text{total}}} = \frac{n_i}{n_1 + n_2} \tag{4-12}$$

▪ A mole fraction may also be thought of as a molecular fraction. It represents the fraction of molecules of a given species in a mixture or a solution.

where the subscript i equals 1 or 2. The ratio on the right-hand side of Equation (4-12) is called the **mole fraction** of species i and is denoted by X_i:

$$X_i = \frac{n_i}{n_1 + n_2} \qquad i = 1, 2 \tag{4-13}$$

Notice that a mole fraction is unitless (it is a fraction) and that

$$X_1 + X_2 = 1$$

Equation (4-12) can be written as

$$P_i = X_iP_{\text{total}} \qquad i = 1, 2 \tag{4-14}$$

which expresses the partial pressure of the ith species in terms of its mole fraction and the total pressure.

Example 4-11: Suppose we have a mixture containing 2.00 mol of helium, 1.00 mol of nitrogen, and 1.50 mol of methane at a total pressure of 650 torr. Assuming ideal behavior, calculate the partial pressure of each gas in the mixture.

Solution: The total number of moles of gases is 4.50 mol. Thus, the partial pressure of each gas is

$$P_{He} = X_{He}P_{total} = \left(\frac{2.00\text{ mol}}{4.50\text{ mol}}\right)(650\text{ torr}) = 289\text{ torr}$$

$$P_{N_2} = X_{N_2}P_{total} = \left(\frac{1.00\text{ mol}}{4.50\text{ mol}}\right)(650\text{ torr}) = 144\text{ torr}$$

$$P_{CH_4} = X_{CH_4}P_{total} = \left(\frac{1.50\text{ mol}}{4.50\text{ mol}}\right)(650\text{ torr}) = 217\text{ torr}$$

The sum of the partial pressures of each gas in the mixture is equal to the total pressure.

Practical applications of Dalton's law of partial pressures arise often in the laboratory. A standard method for determining the quantity of a water-insoluble gas evolved in a chemical reaction is diagramed in Figure 4-12. The gas displaces the water from an inverted beaker that is initially filled with water. When the reaction is completed, the beaker is raised or lowered until the water levels inside and outside are the same. When the two levels are the same, the pressure inside the beaker is equal to the atmospheric pressure. The pressure inside the beaker, however, is not due just to the gas collected; there is also water vapor present. Thus, the pressure inside the beaker is

$$P_{total} = P_{gas} + P_{H_2O} = P_{atmospheric}$$

The vapor pressure of H_2O depends only upon the temperature; Table 11-2 gives the vapor pressure of H_2O at various temperatures.

▪ We shall study the pressure due to water vapor more fully in Chapter 11.

Example 4-12: The reaction

$$2KClO_3(s) \rightarrow 2KCl(s) + 3O_2(g)$$

represents a common laboratory procedure for producing small quantities

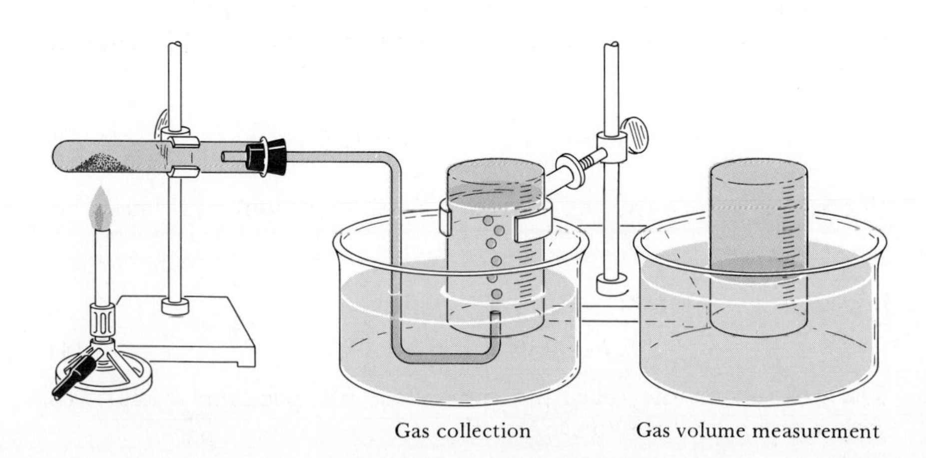

Figure 4-12 The collection of a gas over water. When the water levels inside and outside the container are equal, the pressure inside the container and the atmospheric pressure must be equal.

of pure oxygen. A 0.250-L flask is filled with oxygen that has been collected over water at an atmospheric pressure of 729 torr (see Figure 4-12). The gas temperature is 14°C. Compute the molar volume of dry oxygen at 0°C and 760 torr. The vapor pressure of H_2O at 14°C is 12.0 torr.

Solution: To calculate the volume of dry oxygen, we must first determine its partial pressure. The atmospheric pressure is 729 torr, and the vapor pressure of H_2O at 14°C is 12.0 torr; therefore

$$P_{O_2} = P_{\text{total}} - P_{H_2O} = (729 - 12.0)\text{ torr} = 717\text{ torr}$$

We now calculate the number of moles of O_2 produced using the ideal-gas equation:

$$n = \frac{PV}{RT} = \frac{(717\text{ torr})\left(\dfrac{1\text{ atm}}{760\text{ torr}}\right)(0.250\text{ L})}{(0.0821\text{ L·atm·mol}^{-1}\text{·K}^{-1})(287\text{ K})}$$
$$= 0.0100\text{ mol}$$

We wish to calculate the molar volume of O_2 at 0°C and 760 torr, and so we use Equation (4-7):

$$V_f = V_i\left(\frac{P_i}{P_f}\right)\left(\frac{T_f}{T_i}\right) = (0.250\text{ L})\left(\frac{717\text{ torr}}{760\text{ torr}}\right)\left(\frac{273\text{ K}}{287\text{ K}}\right)$$
$$= 0.224\text{ L}$$

The molar volume is the volume occupied by one mole of gas. Thus

$$\text{molar volume} = \frac{0.224\text{ L}}{0.0100\text{ mol}} = 22.4\text{ L·mol}^{-1}$$

4-9. THE MOLECULES OF A GAS HAVE A DISTRIBUTION OF SPEEDS

The fact that Boyle's, Charles's, and Dalton's laws and the ideal-gas equation are valid for all gases suggests that these laws reflect the fundamental nature of gases. As we have seen, the space occupied by a gas is mostly empty, with the molecules being widely separated from each other and in constant motion.

By applying the laws of physics to the motion of the molecules, it is possible to calculate the pressure exerted by gas molecules on the walls of a container and to show that the pressure is given by the ideal-gas equation. Because this theory focuses on the motion of the molecules, it

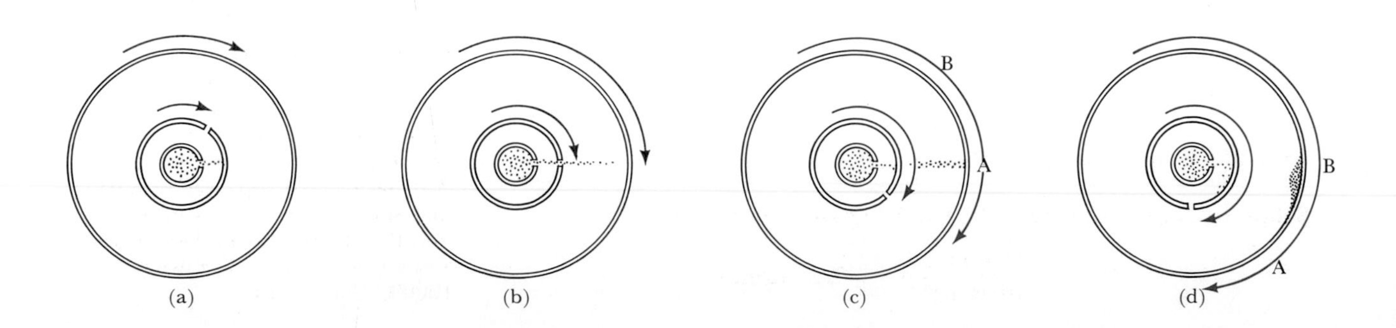

Figure 4-13 An experimental setup that was used not only to demonstrate that there is a distribution of molecular speeds in a gas but also to measure that distribution. (a) The apparatus consists of three concentric evacuated cylindrical drums. The two outermost drums rotate together at the same angular speed and the innermost drum is stationary. The innermost drum contains a gas or vapor, say, silver vapor. The two innermost drums have small slits, and when these momentarily line up, as shown in (b), a beam of silver atoms is directed to the inner surface of the outermost drum. The silver atoms with the greatest speed reach the outermost drum first, at point *A* in (c). By the time the slower silver atoms reach the outer drum, it will have rotated some and so the deposit of silver atoms will be spread out, as seen in (d). The thickness of the silver deposit is proportional to the number of silver atoms with a certain speed, and so the variation in thickness represents the actual distribution of speeds of the silver atoms in the vapor.

is called the **kinetic theory of gases.** The kinetic theory predicts that, as long as the volume taken up by the molecules is much smaller than the volume of the container and as long as the molecules do not repel or attract each other, then the gas behaves ideally.

A body in motion has an energy by virtue of the fact that it is in motion. The energy associated with the motion of a body is called the **kinetic energy** E and is given by the formula

$$E = \tfrac{1}{2}mv^2 \tag{4-15}$$

▪ The word kinetic is derived from the Greek word *kinetikos,* meaning motion or moving.

where m is the mass of the body and v is its speed. If m is expressed in kilograms and v in meters per second, then E has the units $\text{kg}\cdot\text{m}^2\cdot\text{s}^{-2}$. This combination of units is called a **joule** (J), which is the SI unit of energy: $1\ \text{J} = 1\ \text{kg}\cdot\text{m}^2\cdot\text{s}^{-2}$. (See Appendix B.)

The molecules in a gas do not all have the same speed. As the caption to Figure 4-13 explains, there is a distribution of molecular speeds in a gas. Figure 4-14 shows the distribution of molecular speeds for N_2 gas at two different temperatures. Notice that, as the temperature increases, more molecules travel at higher speeds.

A fundamental result of the kinetic theory of gases is that the **average kinetic energy** per mole of a gas E_{av} is given by

$$E_{av} = \tfrac{3}{2}RT \tag{4-16}$$

The R in this equation is the same R (the gas constant) that appears in the ideal-gas equation. Because E_{av} is expressed in $\text{J}\cdot\text{mol}^{-1}$, we must express R in $\text{J}\cdot\text{mol}^{-1}\cdot\text{K}^{-1}$ when we use Equation (4-16). It is shown in Appendix B that $R = 8.314\ \text{J}\cdot\text{mol}^{-1}\cdot\text{K}^{-1}$.

We can define an average speed, v_{av}, by the relation

$$E_{av} = \tfrac{1}{2}M_{kg}v_{av}^2 \tag{4-17}$$

▪ Several different definitions of average speeds occur in the kinetic theory of gases. The quantity v_{av} defined by Equation (4-17) is the only average speed that we shall use in this book.

Because E_{av} is the average kinetic energy *in joules per mole,* the mass on the right-hand side, M_{kg}, is the mass in *kilograms* of a mole of molecules, which is equal to the molar mass of the gas divided by $1000\ \text{g}\cdot\text{kg}^{-1}$. If we substitute Equation (4-16) into Equation (4-17) and solve for v_{av}, then we get

$$v_{av}^2 = \frac{2E_{av}}{M_{kg}} = \frac{2(\frac{3}{2}RT)}{M_{kg}} = \frac{3RT}{M_{kg}}$$

Taking the square root of both sides yields

$$v_{av} = \left(\frac{3RT}{M_{kg}}\right)^{1/2} \tag{4-18}$$

We can use Equation (4-18) to calculate v_{av}, which is an **average speed** of the molecules in a gas.

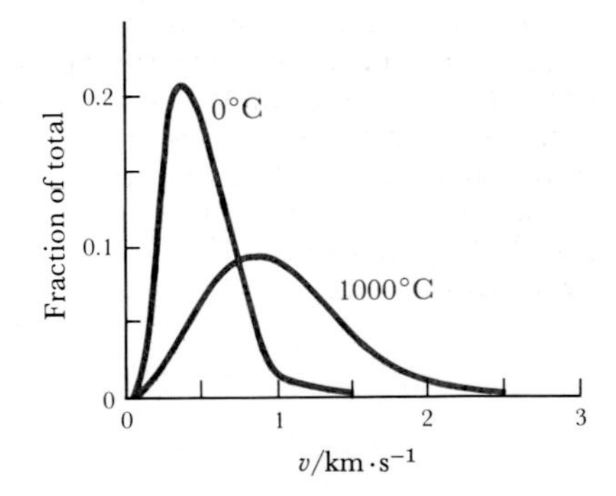

Figure 4-14 The distribution of speeds for nitrogen gas molecules at 0°C and 1000°C. The distribution is represented by the fraction of nitrogen molecules that have speed v plotted versus that speed. Note, for example, that the fraction of molecules with a speed of $1\text{km}\cdot\text{s}^{-1}$ is greater at 1000°C than at 0°C.

Example 4-13: Calculate v_{av} for N_2 at 20°C.

Solution: The molar mass of N_2 in kilograms per mole is

$$M_{kg} = \frac{28.0\ \text{g}\cdot\text{mol}^{-1}}{1000\ \text{g}\cdot\text{kg}^{-1}} = 0.0280\ \text{kg}\cdot\text{mol}^{-1}$$

▪ 1 J = 1 $kg{\cdot}m^2{\cdot}s^{-2}$

Thus, using Equation (4-18) we compute

$$v_{av} = \left(\frac{3RT}{M_{kg}}\right)^{1/2}$$

$$= \left[\frac{(3)(8.314\ \text{J}{\cdot}\text{mol}^{-1}{\cdot}\text{K}^{-1})(293\ \text{K})}{0.028\ \text{kg}{\cdot}\text{mol}^{-1}}\right]^{1/2}$$

$$= (2.61 \times 10^5\ \text{J}{\cdot}\text{kg}^{-1})^{1/2}$$

$$= (2.61 \times 10^5\ \text{m}^2{\cdot}\text{s}^{-2})^{1/2} = 511\ \text{m}{\cdot}\text{s}^{-1}$$

A speed of 511 $m{\cdot}s^{-1}$ is equivalent to about 1100 mph.

Values of v_{av} for several gases are given in Table 4-2. Note that v_{av} decreases with increasing molecular mass at constant temperature, as is required by Equation (4-18).

A sound wave is a pressure wave that travels through a substance. The speed with which a sound wave travels through a gas depends upon the speeds of the molecules in the gas. It can be shown from the kinetic theory of gases that the speed of sound through a gas is about $0.7v_{av}$. The speed of sound in air at 20°C and 1 atm is about 760 mph, or 340 $m{\cdot}s^{-1}$.

We can use Equation (4-18) to derive a formula for the relative rates at which gases leak from a container through a small hole, a process called **effusion.** For two gases at the same pressure and temperature, the rate of effusion is directly proportional to the average speed of the molecules. We let $v_{av,A}$ and $v_{av,B}$ be the average speeds of two gases A and B, and we use Equation (4-18) to write

$$v_{av,A} = \left(\frac{3RT}{M_{kg,A}}\right)^{1/2} \quad \text{and} \quad v_{av,B} = \left(\frac{3RT}{M_{kg,B}}\right)^{1/2}$$

The temperature does not have a subscript because both gases are at the same temperature. If we divide $v_{av,A}$ by $v_{av,B}$, we obtain

$$\frac{v_{av,A}}{v_{av,B}} = \left(\frac{M_{kg,B}}{M_{kg,A}}\right)^{1/2} = \left(\frac{M_B}{M_A}\right)^{1/2}$$

where M_A and M_B are the molar masses of A and B. The rate of effusion is directly proportional to v_{av}, and so

$$\frac{\text{rate}_A}{\text{rate}_B} = \left(\frac{M_B}{M_A}\right)^{1/2} \tag{4-19}$$

Table 4-2 Values of v_{av} for some gases at 20°C and 1000°C

		$v_{av}/m{\cdot}s^{-1}$	
Molecule	**Formula mass**	**t = 20°C**	**t = 1000°C**
H_2	2.0	1900	4000
N_2	28.0	510	1060
O_2	32.0	480	1000
CO_2	44.0	410	850

▪ The molecules of a gas move at average speeds of hundreds of meters per second.

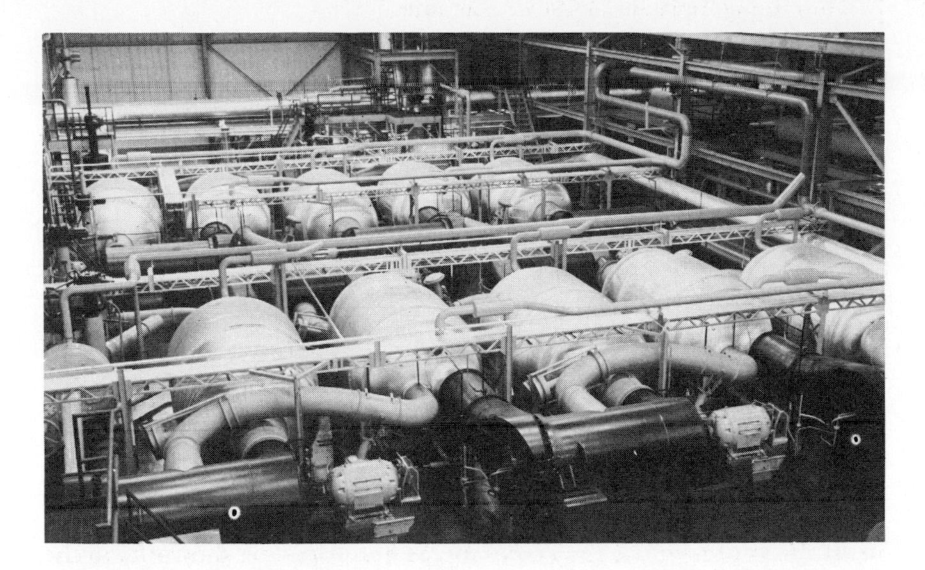

Figure 4-15 Isotopes of uranium can be separated by making the gaseous compound UF_6 and then using Graham's law of effusion, which says that the lighter isotopic compound will effuse more quickly than the heavier one. This photo shows the large process equipment in which effusion is carried out in stages to achieve isotopic enrichment.

This relation was observed experimentally by Graham in the 1840s and is called **Graham's law of effusion.** The process of effusion was used to separate isotopes of uranium in the production of the atomic bomb in World War II (Figure 4-15).

4-10. THE AVERAGE DISTANCE A MOLECULE TRAVELS BETWEEN COLLISIONS IS CALLED THE MEAN FREE PATH

Although the molecules in a gas at 1 atm and 20°C travel with speeds of hundreds of meters per second, they do not travel any appreciable distances that rapidly. We all have observed that it may take several minutes for an odor to spread through a room. The explanation for this lies in the fact that the molecules in a gas undergo many collisions, and so their actual path is a chaotic, zigzag path like that shown in Figure 4-16. Between collisions, gas molecules travel with speeds of hundreds of meters per second, but their net progress is quite slow. The average distance traveled between collisions is called the **mean free path** (l). According to the kinetic theory, the mean free path is given by

$$l = (3.1 \times 10^{7}\ \text{pm}^{3}\cdot\text{atm}\cdot\text{K}^{-1})\frac{T}{\sigma^{2}P} \tag{4-20}$$

where T is the temperature in kelvins, σ (the Greek letter sigma) is the diameter of a molecule in picometers, and P is the pressure in atmospheres. Table 4-3 lists **molecular diameters** for various gas molecules.

We can use Equation (4-20) to calculate the mean free path in N_2 at 1.00 atm and 20°C. For N_2, $\sigma = 370$ pm (Table 4-3), and so we compute

$$l = (3.1 \times 10^{7}\ \text{pm}^{3}\cdot\text{atm}\cdot\text{K}^{-1})\left(\frac{293\ \text{K}}{(370\ \text{pm})^{2}(1.00\ \text{atm})}\right)$$

$$= 6.6 \times 10^{4}\ \text{pm}$$

$$= 6.6 \times 10^{-8}\ \text{m}$$

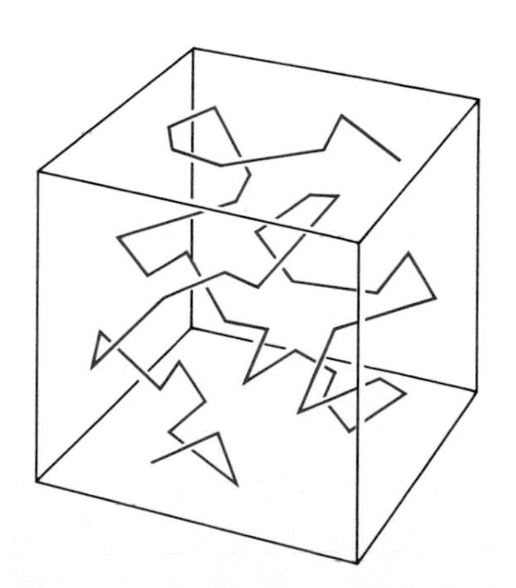

Figure 4-16 A typical path followed by a gas molecule. The molecule travels in a straight line until it collides with another molecule, at which point its direction is changed in an almost random manner. At 0°C and 1 atm, a molecule undergoes about 10^{10} collisions per second.

where we have used the fact that 1 pm = 10^{-12} m. A distance of 66,000 pm is over 175 times the diameter of a nitrogen molecule. Thus, we see that at 1.00 atm and 20°C a nitrogen molecule travels an average distance of over 175 molecular diameters between collisions.

If we divide v_{av} by the mean free path, we get an estimate of the number of collisions (z) that one molecule undergoes per second. This quantity is called the **collision frequency** and is given by

$$z = \text{number of collisions per second per molecule}$$

$$\approx \frac{\text{distance traveled per second}}{\text{distance traveled per collision}} = \frac{\text{collisions}}{\text{second}} = \frac{v_{av}}{l} \quad (4\text{-}21)$$

Thus, for N_2 at 20°C and 1.00 atm we compute

$$z = \frac{v_{av}}{l} = \frac{511 \text{ m}\cdot\text{s}^{-1}}{6.6 \times 10^{-8} \text{ m}\cdot\text{collision}^{-1}}$$

$$= 7.7 \times 10^{9} \text{ collisions}\cdot\text{s}^{-1}$$

Thus we see that one nitrogen molecule undergoes about 8 billion collisions per second at 20°C and 1 atm.

The kinetic theory of gases gives us a detailed picture of the molecular nature of gases. From it we can calculate the properties of gases in terms of molecular quantities. The kinetic theory has numerous applications throughout chemistry, and we shall refer to it frequently in later chapters.

Table 4-3 Atomic or molecular diameters for some common gases

Gas	Diameter/pm
He	210
Ne	250
Ar	360
Kr	410
Xe	470
H_2	280
N_2	370
O_2	370

4-11. THE VAN DER WAALS EQUATION ACCOUNTS FOR DEVIATIONS FROM GAS IDEALITY

The ideal-gas equation is valid for all gases at sufficiently low densities and sufficiently high temperatures. As the pressure on a given quantity of gas is increased or the temperature is decreased, however, deviations from the ideal-gas equation appear. These deviations can be displayed graphically by plotting PV/RT as a function of pressure, as shown in Figure 4-17. For 1 mol of an ideal gas, PV/RT is equal to unity for any value of P, and so deviations from ideal-gas behavior occur as deviations of the ratio PV/RT from unity. The extent of **deviation from ideality** at a given pressure depends upon the temperature and upon the nature of the gas. The closer the gas is to the point at which it liquefies, the larger will be the deviation from ideal behavior. The kinetic theory of gases assumes that the molecules of a gas are simply point masses and that they have no attraction for each other. The deviations from ideality shown in Figure 4-17 are due to the inaccuracy of these two assumptions.

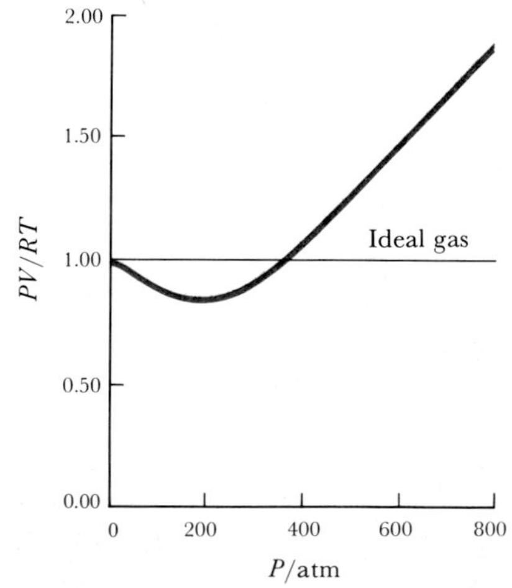

The behavior shown in Figure 4-17 for methane is typical for all gases. As the pressure is increased, the deviation from ideal-gas behavior first lies below the ideal-gas prediction. This negative deviation from ideality can be explained by recognizing that there is an intrinsic attraction between molecules. This attraction is responsible for the condensation of a gas to a liquid as the temperature is lowered. We discuss the attraction between molecules in more detail in Chapter 11, but here we need to realize only that molecules do attract each other. When the

Figure 4-17 A plot of (PV/RT) versus P for 1 mol of methane at 300 K. This figure shows that the ideal-gas equation is not valid at high pressures.

molecules in a gas collide with each other, their mutual attraction causes them to stay together somewhat longer than if they did not attract each other. Consequently, the number of molecules, or moles, is effectively reduced. There are fewer collisions with the walls of the container than there would be if the gas were ideal (that is, if there were no attraction between the molecules), and so the pressure and the product PV are less than predicted by the ideal-gas equation.

▪ The fact that gases condense to liquids suggests that molecules attract each other.

Figure 4-17 also shows that at higher pressures the deviation from ideal-gas behavior lies above the ideal-gas prediction. This positive deviation from ideality can be understood by recognizing that molecules have a finite size. At high pressures, the volume of the gas molecules is not negligible relative to the volume of the container (Figure 4-18). Consequently, the volume that is *available* to any given molecule is less than the total volume of the container. If we let b be the volume of the molecules, then $V - b$ is the volume available to any molecule. This free volume, $V - b$, instead of simply V should be used in calculations. If we replace V by $V - b$ in the ideal-gas equation, then we obtain

$$P(V - b) = RT$$

If we divide by RT and then by $V - b$, we obtain

$$\frac{P}{RT} = \frac{1}{V - b}$$

We now multiply by V to get the following expression for PV/RT:

$$\frac{PV}{RT} = \frac{V}{V - b} > 1$$

▪ Do you see that the ratio $V/(V - b)$ is always greater than unity because b is positive?

The ratio $V/(V - b)$ is greater than unity, and so PV/RT is greater than unity, as in Figure 4-17. Thus, the positive deviations from ideal-gas behavior at high pressures are due to the finite size of the gas molecules.

There are many equations that modify the ideal-gas equation to account for the attraction between molecules and for their finite size. The best-known is called the **van der Waals equation:**

▪ An equation that connects P, V, T, and n is called an equation of state. The van der Waals equation is one of many equations of state.

$$\left(P + \frac{n^2 a}{V^2}\right)(V - nb) = nRT \tag{4-22}$$

where a and b are constants, called **van der Waals constants,** whose values depend upon the particular gas (Table 4-4). The quantity b is proportional to the volume of a gas molecule. The quantity a is related to the attraction between the molecules; the more strongly the molecules attract each other, the larger a is.

The form of the van der Waals equation can be rationalized as follows. We start with an equation of the form of the ideal-gas equation:

$$P_{id}V_{id} = nRT \tag{4-23}$$

where P_{id} and V_{id} are interpreted as an ideal pressure and an ideal volume, respectively. We use the expression for the free volume of a gas that we deduced earlier for the ideal volume:

$$V_{id} = V - nb \tag{4-24}$$

where V is the actual volume occupied by the gas. Figure 4-19 can be

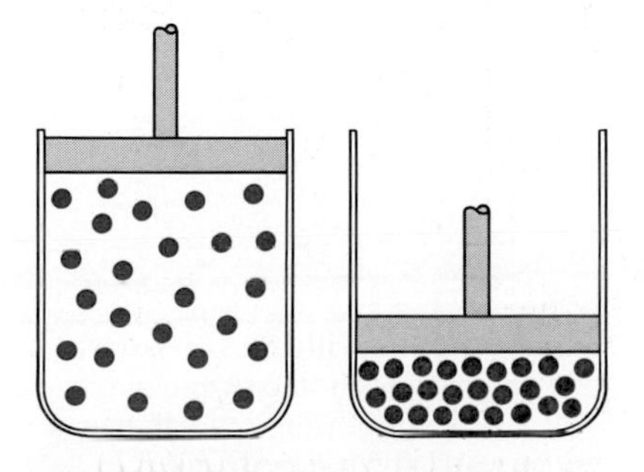

Figure 4-18 At high pressures, the volume of the molecules of a gas is no longer negligible relative to the volume of the container.

Table 4-4 The van der Waals constants of some gases

Name	Formula	a/L^2·atm·mol^{-2}	b/L·mol^{-1}
ammonia	NH_3	4.170	0.0371
carbon dioxide	CO_2	3.592	0.0427
methane	CH_4	2.253	0.0428
neon	Ne	0.211	0.0171
nitrogen	N_2	1.390	0.0391
oxygen	O_2	1.360	0.0380
propane	C_3H_8	8.664	0.0844

used to argue a relationship between the ideal pressure, P_{id}, and the actual pressure, P. Figure 4-19 shows a gas molecule near the wall of the container. As this molecule approaches the wall, the attraction due to the other gas molecules slows it down, and consequently it strikes the wall with less force than if there were no attractions (as in an ideal gas). Thus the actual pressure is less than if the gas were ideal, or $P < P_{id}$. The difference between these pressures is proportional to the density of molecules striking the wall *and* to the density of molecules in the interior of the gas, that is, proportional to the density squared. We let the proportionality constant be a and write

$$P_{id} = P + a\left(\frac{n}{V}\right)^2 \tag{4-25}$$

If we substitute Equations (4-24) and (4-25) into Equation (4-23), then we obtain the van der Waals equation (Equation (4-22)).

Let's use the van der Waals equation to calculate the pressure exerted at 300 K by 1.00 mol of methane occupying a 250-mL container. From Table 4-4 we find that $a = 2.253$ L^2·atm·mol^{-2} and $b =$ 0.0428 L·mol^{-1}. If we divide Equation (4-22) by $V - nb$ and solve for P, then we obtain

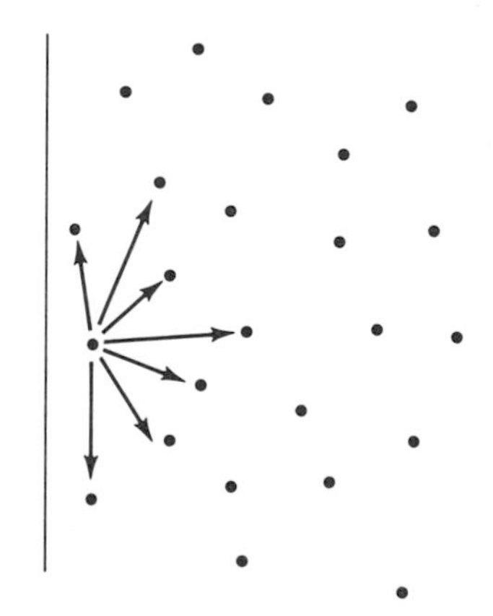

Figure 4-19 An illustration of the fact that a gas molecule near a wall of its container experiences a net inward drag due to its attraction by the other molecules. Consequently, it strikes the wall with less force than if the gas were ideal.

$$P = \frac{nRT}{V - nb} - \frac{n^2a}{V^2} \tag{4-26}$$

Substituting $n = 1.00$ mol, $R = 0.0821$ L·atm·mol^{-1}·K^{-1}, $T = 300$ K, $V = 0.0250$ L, and the values of a and b into Equation (4-26), we obtain

$$P = \frac{(1.00\text{ mol})(0.0821\text{ L·atm·mol}^{-1}\text{·K}^{-1})(300\text{ K})}{0.250\text{ L} - (1.00\text{ mol})(0.0428\text{ L·mol}^{-1})} - \frac{(1.00\text{ mol})^2(2.253\text{ L}^2\text{·atm·mol}^{-2})}{(0.250\text{ L})^2} = 82.8\text{ atm}$$

By comparison, the ideal-gas equation predicts that

$$P = \frac{nRT}{V} = \frac{(1.00\text{ mol})(0.0821\text{ L·atm·mol}^{-1}\text{·K}^{-1})(300\text{ K})}{0.250\text{ L}} = 98.5\text{ atm}$$

The prediction of the van der Waals equation is in good agreement with the experimental value.

SUMMARY

In a gas, the particles are widely separated and travel throughout the entire volume of their container in a chaotic manner, colliding with each other and with the walls of the container. The pressure exerted by a gas is due to the incessant collisions of the gas molecules on the walls of the container. At constant temperature, the volume and the pressure of a gas are related by Boyle's law, which says that volume and pressure are inversely related. The relation between the volume of a gas and its temperature is given by Charles's law, which serves also to define the absolute or Kelvin temperature scale.

The experimental study of the combining volumes of reacting gases led to Gay-Lussac's law of combining volumes. Gay-Lussac's law leads to Avogadro's law, which states that equal volumes of gases at the same pressure and temperature contain equal numbers of molecules. Boyle's, Charles's, and Avogadro's laws can be combined into one law, the ideal-gas law, which is the relation among the pressure, volume, temperature, and number of moles of a gas.

All the experimental gas laws can be explained by the kinetic theory of gases. One of the central results of the kinetic theory of gases is that the average kinetic energy of a gas is directly proportional to its absolute temperature. The kinetic theory provides equations that can be used to calculate the average speed of a molecule, the mean free path, the rate of effusion, and other molecular quantities.

The ideal-gas equation is valid for all gases at sufficiently low densities and sufficiently high temperatures. As the pressure on a given quantity of gas is increased, however, deviations from the ideal-gas equation are observed. The van der Waals equation, which takes into account that the molecules of a gas attract each other and have a finite size, is valid at higher densities and lower temperatures than is the ideal-gas equation.

TERMS YOU SHOULD KNOW

solid 126
liquid 127
gas 127
compressibility 127
pressure 128
manometer 128
torr (mm Hg) 128
barometer 128
standard atmosphere (atm) 129
pascal (Pa) 129
Boyle's law 131
absolute temperature, T 131
Charles's law 132
absolute (Kelvin) temperature scale 132
kelvin 132
gas thermometer 132
Gay-Lussac's law of combining volumes 133
Avogadro's law 134
ideal-gas equation (ideal-gas law) 135
ideal gas 135
gas constant, R 135
molar volume 135
molar mass 138
partial pressure 140
Dalton's law of partial pressures 140
mole fraction, X_i 140
kinetic theory of gases 143
kinetic energy, E 143
joule, J 143
average kinetic energy, E_{av} 143
average speed, v_{av} 143
effusion 144
Graham's law of effusion 145
mean free path, l 145
molecular diameter, σ 145
collision frequency, z 146
deviation from ideality 146
van der Waals equation 147
van der Waals constants 147

EQUATIONS YOU SHOULD KNOW HOW TO USE

$PV = nRT$ (4-5) (ideal-gas equation)

$\rho = \frac{MP}{RT}$ (4-8) (density of an ideal gas)

$P_{total} = P_1 + P_2 + P_3 + \cdots$ (4-9) (Dalton's law of partial pressures)

$X_i = \frac{n_i}{n_{total}}$ (4-13) (mole fraction)

$P_i = X_i P_{total}$ (4-14) (partial pressure)

$$E_{av} = \frac{3}{2}RT \qquad (4\text{-}16) \quad \text{(average kinetic energy per mole of a gas)}$$

$$v_{av} = \left(\frac{3RT}{M_{kg}}\right)^{1/2} \qquad (4\text{-}18) \quad \text{(average speed of gas molecules)}$$

$$\frac{\text{rate}_A}{\text{rate}_B} = \frac{t_B}{t_A} = \left(\frac{M_B}{M_A}\right)^{1/2} \qquad (4\text{-}19) \quad \text{(Graham's law of effusion)}$$

$$l = (3.1 \times 10^7 \text{ pm}^3\cdot\text{atm}\cdot\text{K}^{-1})\frac{T}{\sigma^2 P} \qquad (4\text{-}20) \quad \text{(mean free path of gas molecules)}$$

$$z = \frac{v_{av}}{l} \qquad (4\text{-}21) \quad \text{(collision frequency of gas molecules)}$$

$$\left(P + \frac{n^2 a}{V^2}\right)(V - nb) = nRT \qquad (4\text{-}22) \quad \text{(van der Waals equation)}$$

PROBLEMS

PRESSURE AND BOYLE'S LAW

4-1. Carry out the following unit conversions:

(a) The atmospheric pressure at the surface of Venus is about 75 atm. Convert 75 atm to torr and to bars.
(b) The atmospheric pressure in Mexico City is about 580 torr. Convert this pressure to atmospheres and to millibars.
(c) The pressure of carbon dioxide in a gas cylinder is 5.2 atm. Convert this pressure to pascals and kilopascals.
(d) The pressure of a sample of nitrogen gas is 920 torr. Convert this pressure to pascals and to atmospheres.

4-2. Carry out the following unit conversions:

(a) The partial pressure of water vapor in air saturated with water at 25°C is 24 torr. Convert 24 torr to atmospheres and to pascals.
(b) The atmospheric pressure due to a low-pressure front is 990 mbar. Convert this pressure to atmospheres and to torr.
(c) The pressure of a hydrogen gas sample is 3.75 atm. Convert this pressure to pascals and to kilopascals.
(d) The pressure of an automobile tire is 39 psi. Convert this pressure to atmospheres and to kilopascals.

4-3. A gas bubble has a volume of 0.650 mL at the bottom of a lake, where the pressure is 3.46 atm. What is the volume of the bubble at the surface of the lake, where the pressure is 1.00 atm? What are the diameters of the bubble at the two depths? Assume that the temperature is constant and that the bubble is spherical.

4-4. Suppose we wish to inflate a weather balloon with helium. The balloon has a volume of 100 m^3, and we wish to inflate it to a pressure of 0.10 atm. If we use 50-L cylinders of compressed helium gas at a pressure of 100 atm, how many cylinders do we need? Assume that the temperature remains constant.

4-5. The volume of one cylinder in a particular automobile engine is 0.44 L. The cylinder is filled with a mixture of gasoline and air at 1.0 atm. The cylinder is compressed to 0.073 L prior to ignition of the combustible mixture. What pressure must be applied to produce this compression? Assume that the temperature remains constant.

4-6. A human adult breathes in approximately 0.50 L of air at 1.00 atm with each breath. If an air tank holds 50 L of air at 200 atm, how many breaths will the tank supply? Assume a temperature of 37°C.

TEMPERATURE AND CHARLES'S LAW

4-7. Convert the following temperatures to the Kelvin scale:

(a) body temperature, 37°C
(b) room temperature, 20°C
(c) the freezing point of hydrogen, −259°C
(d) the boiling point of ethylene glycol, 199°C

4-8. Convert the following temperatures to the Kelvin scale:

(a) −183°C (the melting point of oxygen)
(b) 6000°C (temperature at the surface of the sun)
(c) −269°C (the boiling point of helium)
(d) 800°C (the melting point of sodium chloride)

4-9. Suppose that in a gas thermometer the gas occupies 14.7 mL at 0°C. The thermometer is immersed in boiling water (100°C). What is the volume of the gas at 100°C?

4-10. Suppose that in a gas thermometer the gas occupies 12.6 mL at 20°C. The thermometer is immersed in a container of solid carbon dioxide chips ("Dry Ice"), and the gas then occupies 8.4 mL. What is the temperature of the Dry Ice?

4-11. Methane burns according to the equation

$$CH_4(g) + 2O_2(g) \rightarrow CO_2(g) + 2H_2O(g)$$

What volume of air, which is 20 percent oxygen by volume, is required to burn 5.0 L of methane when both are the same temperature and pressure?

4-12. Hydrogen and oxygen react violently with each other once the reaction is initiated. For example, a spark can set off the reaction and cause the mixture to explode. What volume of oxygen will react with 0.55 L of hydrogen if both are at 300°C and 1 atm? What volume of water will be produced at 300°C and 1 atm?

IDEAL-GAS LAW

4-13. Calculate the volume that 0.65 mol of ammonia gas occupies at 37°C and 600 torr.

4-14. Calculate the number of grams of propane, C_3H_8, in a 50-L container at a pressure of 7.5 atm and a temperature of 25°C.

4-15. Calculate the pressure exerted by 18 g of steam (H_2O) confined to a volume of 18 L at 100°C. What volume would the water occupy if the steam were condensed to liquid water at 25°C? The density of liquid water is 1.00 $g \cdot mL^{-1}$ at 25°C.

4-16. Calculate the volume in liters occupied by 0.55 kg of dimethyl ether, C_2H_6O, at 950 torr and 15°C.

4-17. Calculate the number of atoms of helium in 1.0 L at −200°C and 0.0010 atm. Compare this value with the number in 1.0 L at 0°C and 1.0 atm.

4-18. Calculate the number of molecules of SO_3 in 100 mL at 100°C and 1.00 atm.

4-19. The ozone molecules in the stratosphere absorb much of the ultraviolet radiation from the sun. The temperature of the stratosphere is −23°C, and the pressure due to the ozone is 1.4×10^{-7} atm. Calculate the number of ozone molecules present in 1.0 mL.

4-20. A low pressure of 1.0×10^{-3} torr is readily obtained in the laboratory by means of a vacuum pump. Calculate the number of molecules in 1.00 mL of gas at this pressure and 20°C.

4-21. A 0.500-L container is occupied by nitrogen at a pressure of 800 torr and a temperature of 0°C. If the highest pressure the container can withstand is 3.0 atm, what is the highest temperature the gas should be heated to?

4-22. A weather balloon is partially filled with helium at 20°C to a volume of 43.7 L and a pressure of 1.16 atm. The balloon rises to the stratosphere, where the temperature and pressure are −23.0°C and 6.00×10^{-3} atm. Calculate the volume of the balloon in the stratosphere.

4-23. Acetylene is prepared by the reaction of calcium carbide with water:

$$CaC_2(s) + 2H_2O(l) \rightarrow Ca(OH)_2(s) + C_2H_2(g)$$

What volume of acetylene can be obtained from 100 g of calcium carbide and 100 g of water at 0°C and 1.00 atm? What volume results when the temperature is 120°C and the pressure is 1.00 atm?

4-24. Lithium metal reacts with nitrogen at room temperature (20°C) according to the equation

$$6Li(s) + N_2(g) \rightarrow 2Li_3N(s)$$

A sample of lithium metal was placed under a nitrogen atmosphere in a sealed 1.00-L container at a pressure of 1.23 atm. One hour later the pressure dropped to 0.92 atm. Calculate the number of grams of nitrogen that reacted with the lithium metal. Assuming that all the lithium reacted, calculate the mass of lithium originally present.

4-25. Cellular respiration occurs according to the overall equation

$$\underset{\text{glucose}}{C_6H_{12}O_6(s)} + 6O_2(g) \rightarrow 6CO_2(g) + 6H_2O(l)$$

Calculate the volume of $CO_2(g)$ produced at 37°C (body temperature) and 1.00 atm when 1.00 g of glucose is metabolized.

4-26. Chlorine is produced by the electrolysis of an aqueous solution of sodium chloride:

$$2NaCl(aq) + 2H_2O(l) \xrightarrow{\text{electrolysis}} 2NaOH(aq) + H_2(g) + Cl_2(g)$$

The hydrogen gas and chlorine gas are collected separately at 10.0 atm and 25°C. What volume of each can be obtained from 2.50 kg of sodium chloride?

4-27. Chlorine gas can be prepared in the laboratory by the reaction of manganese dioxide with hydrochloric acid

$$MnO_2(s) + 4HCl(aq) \rightarrow MnCl_2(aq) + 2H_2O(l) + Cl_2(g)$$

How much MnO_2 should be added to excess HCl to obtain 500 mL of chlorine gas at 25°C and 750 torr?

4-28. About 50 percent of U.S. and most Canadian sulfur is produced by the Claus process, in which sulfur is obtained from the $H_2S(g)$ that occurs in natural gas deposits or is produced when sulfur is removed from petroleum. The reactions are described by the equations

$$2H_2S(g) + 3O_2(g) \rightarrow 2SO_2(g) + 2H_2O(g)$$

$$SO_2(g) + 2H_2S(g) \rightarrow 3S(l) + 2H_2O(g)$$

How many metric tons of sulfur can be produced from 2.00 million liters of $H_2S(g)$ at 6.00 atm and 200°C?

4-29. Calculate the density of water in the gas phase at 100°C and 1.00 atm. Compare this value with the density of liquid water at 100°C and 1.00 atm (0.958 $g \cdot mL^{-1}$).

4-30. Calculate the density of the gas CF_2Cl_2 at 0°C and 1.00 atm.

4-31. The pressure due to a 2.42-g sample of a Freon gas in a sealed 250-mL container is 1473 torr at 22°C. Determine the molar mass of the Freon gas.

4-32. A 2.22-g sample of dimethylamine, a widely used substance in the chemical industry, was vaporized in a sealed 500-mL container at 55°C. The pressure due to dimethylamine was 2.65 atm. Determine the molar mass of this compound.

4-33. A 0.271-g sample of an unknown vapor occupies 294 mL at 100°C and 765 torr. The simplest formula of the compound is CH_2. What is the molecular formula of the compound?

4-34. Upon chemical analysis, a gaseous hydrocarbon is found to contain 88.82 percent C and 11.18 percent H by mass. A 62.6-mg sample of the gas occupies 34.9 mL at 772 torr and 100°C. Determine the molecular formula of the hydrocarbon.

4-35. Ethylene is a gas produced in petroleum cracking and is used to synthesize a variety of important chemicals, such as polyethylene and polyvinylchloride. Chemical analysis shows that ethylene is 85.60 percent carbon and 14.40 percent hydrogen by mass. It has a density of 0.9588 $g \cdot L^{-1}$ at 25°C and 635 torr. Use these data to determine the molecular formula of ethylene.

4-36. Benzene is the fifteenth most widely used chemical in the United States, and its principal source is petroleum. Benzene has a wide range of uses, including its use as a solvent and in the synthesis of nylon and detergents. Chemical analysis shows that benzene is 92.24 percent carbon and 7.76 percent hydrogen by mass. A 2.334-g sample of benzene was vaporized in a sealed 500-mL container at 100°C, producing a pressure of 1.83 atm. Use these data to determine the molecular formula of benzene.

PARTIAL PRESSURES

4-37. A gaseous mixture consists of 0.513 g H_2 and 16.1 g N_2 and occupies 10.0 L at 20°C. Calculate the partial pressures of H_2 and N_2 in the mixture.

4-38. A 2.0-L sample of H_2 at 1.00 atm, an 8.0-L sample of N_2 at 3.00 atm, and a 4.0-L sample of Kr at 0.50 atm are transferred to a 10.0-L container. Calculate the partial pressure of each gas and the total final pressure. Assume constant temperature conditions.

4-39. A mixture of O_2 and N_2 is reacted with white phosphorus, which removes the oxygen. If the volume of the mixture decreases from 50.0 mL to 35.0 mL, calculate the partial pressures of O_2 and N_2 in the mixture. Assume that the total pressure remains constant at 740 torr.

4-40. A gaseous mixture of three volumes of carbon dioxide and one volume of water vapor at 200°C and 2.00 atm is cooled to 10°C, thereby condensing the water vapor. If the total volume remains the same, what is the pressure of the carbon dioxide at 10°C? Assume that there is no water vapor present after condensation.

4-41. Nitroglycerin decomposes according to the equation

$$4C_3H_5(NO_3)_3(s) \rightarrow 12CO_2(g) + 10H_2O(l) + 6N_2(g) + O_2(g)$$

What is the total volume of gases produced when collected at 1.0 atm and 25°C from 10 g of nitroglycerin? What pressure is produced if the reaction is confined to a volume of 0.50 L at 25°C? Assume that you can use the ideal-gas equation. Neglect any pressure due to water vapor.

4-42. Explosions occur when a substance decomposes very rapidly with the production of a large volume of gases. When detonated, TNT (trinitrotoluene) decomposes according to the equation

$$2C_7H_5(NO_2)_3(s) \rightarrow 12CO(g) + 2C(s) + 5H_2(g) + 3N_2(g)$$

What is the total volume of gases produced from 1.00 kg of TNT if collected at 0°C and 1.0 atm? What pressure is produced if the reaction is confined to a 50-L container at 500°C? Assume that you can use the ideal-gas equation.

MOLECULAR SPEEDS

4-43. Calculate the average speed, v_{av}, of a fluorine molecule at 25°C.

4-44. Calculate the average speed, v_{av} for N_2O at 20°C, 200°C, and 2000°C.

4-45. If the temperature of a gas is doubled, how much is the average speed of the molecules increased?

4-46. The speed of sound in air at sea level at 20°C is about 760 mph. Compare this value with the average speed of N_2 and O_2 gas molecules at 20°C (see Table 4-2).

4-47. Arrange the following gas molecules in order of increasing average speed at the same temperatures: O_2, N_2, H_2O, CO_2, NO_2, $^{235}UF_6$, and $^{238}UF_6$.

4-48. Consider a mixture of $H_2(g)$ and $I_2(g)$. Compute the ratio of the average speeds of H_2 and I_2 molecules in the reaction mixture.

MEAN FREE PATH

4-49. Calculate the mean free path and the collision frequency for one molecule of oxygen at 37°C and 0.20 atm.

4-50. Using the atomic diameters given in Table 4-3, calculate the mean free path for helium and krypton at 1.00 torr and at 760 torr when the temperature is 20°C.

4-51. Interstellar space has an average temperature of about 10 K and an average density of hydrogen atoms of about one hydrogen atom per cubic meter. Compute the mean free path of hydrogen atoms in interstellar space. Take $\sigma = 100$ pm.

4-52. Calculate the pressures at which the mean free path of a hydrogen molecule will be 1.00 μm, 1.00 mm, and 1.00 m at 20°C.

4-53. Calculate the number of collisions per second of one hydrogen molecule at 20°C and 1.0 atm.

4-54. Calculate the number of collisions per second that one molecule of nitrogen undergoes at 20°C and 1.0×10^{-3} torr.

GRAHAM'S LAW OF EFFUSION

4-55. Two identical balloons are filled, one with helium and one with nitrogen, at the same temperature and pressure. If the nitrogen leaks out from its balloon at the rate of 75 $\text{mL}\cdot\text{h}^{-1}$, then what will be the rate of leakage from the helium-filled balloon?

4-56. Two identical porous containers are filled, one with hydrogen and one with carbon dioxide, at the same temperature and pressure. After one day, 1.50 mL of carbon dioxide had leaked out of its container. How much hydrogen has leaked out in one day?

4-57. It takes 145 s for 1.00 mL of N_2 to effuse from a certain porous container. Given that it takes 230 s for 1.00 mL of an unknown gas to effuse under the same temperature and pressure, calculate the molecular mass of the unknown gas.

4-58. Suppose that it takes 175 s for 1.00 mL of N_2 to effuse from a porous container under a certain temperature and pressure and that it takes 200 s for 1.00 mL of a CO-CO_2 mixture to effuse under the same conditions. What is the volume percentage of CO in the mixture?

VAN DER WAALS EQUATION

4-59. Use the van der Waals equation to calculate the pressure exerted by 24.5 g of NH_3 confined to a 2.15-L container at 300 K. Compare your answer with the pressure calculated using the ideal-gas equation.

4-60. Use the van der Waals equation to calculate the pressure exerted by 45 g of propane, C_3H_8, confined to a 2.2-L container at 300°C. Compare your answer with the pressure calculated using the ideal gas equation.

IDEAL-GAS LAW USING PASCALS

4-61. Calculate the number of Cl_2 gas molecules in a volume of 5.00 mL at 40°C and 2.15×10^4 Pa.

4-62. Calculate the pressure in pascals that is exerted by 6.15 mg of CO_2 occupying 2.10 mL at 75°C.

4-63. A sample of radon occupies 7.12 μL at 22°C and 8.72×10^4 Pa. Calculate the volume at 0°C and 1.013×10^5 Pa. What is the mass of the radon?

4-64. Calculate the mass of $N_2O(g)$ that occupies 2.10 L at a pressure of 4.50×10^4 Pa and a temperature of 15°C.

4-65. Calculate the density of $ND_3(g)$ at 0°C and 2.00×10^3 Pa. Take $M = 20.06\ \text{g}\cdot\text{mol}^{-1}$.

4-66. Using the fact that 1 atm $= 1.013 \times 10^5$ Pa, calculate the value of the gas constant in SI units, $\text{J}\cdot\text{mol}^{-1}\cdot\text{K}^{-1}$.

ADDITIONAL PROBLEMS

4-67. A pressure of exactly one atmosphere is defined to be the pressure that supports a column of mercury 760.0 mm high. Pressure is force per unit area, and the force, F, that a mass, m, exerts on a surface is

$$F = mg$$

where $g = 9.806\ \text{m}\cdot\text{s}^{-2}$ is a constant called the gravitational constant. Given that the density of mercury is $13.59\ \text{g}\cdot\text{cm}^{-3}$, show that 1 atm $= 1.013 \times 10^5\ \text{N}\cdot\text{m}^{-2}$, or 1.013×10^5 Pa.

4-68. Gallium metal can be used as a manometer fluid at high temperatures because of its wide liquid range (30 to 2400°C). Compute the height of the liquid gallium column in a gallium manometer when the temperature is 850°C and the pressure is 1300 torr. Take the density of liquid gallium to be 6.0 $\text{g}\cdot\text{mL}^{-1}$.

4-69. Given below are pressure-volume data for a sample of 0.28 g of N_2 at 25°C. Verify Boyle's law for these data. Plot the data so that a straight line is obtained.

P/atm	V/L	P/atm	V/L
0.26	0.938	2.10	0.116
0.41	0.595	2.63	0.093
0.83	0.294	3.14	0.078
1.20	0.203		

4-70. A gas tank containing carbon dioxide has a volume of 50 L. Originally the carbon dioxide was at a pressure of 10.0 atm at 25°C. After the tank had been used for a month at 25°C, the pressure dropped to 4.7 atm. Calculate the number of grams of carbon dioxide used during this period of time.

4-71. Recent measurements have shown that the atmosphere of Venus is mostly carbon dioxide. At the surface, the temperature is about 800°C and the pressure is about 75 atm. In the unlikely chance that a resident of Venus defined a standard temperature and pressure and took those values as standard conditions, what value would he find for the volume of a mole of ideal gas at Venusian standard conditions?

4-72. The organic compound di-*n*-butylphthalate, $C_{16}H_{22}O_4$, is sometimes used as a low-density ($1.043\ g\cdot mL^{-1}$) manometer fluid. Compute the pressure in torr of a gas that supports a 500-mm column of di-*n*-butylphthalate.

4-73. Several television commercials state that it requires 10,000 gallons of air to burn 1 gal of gasoline. Using octane, C_8H_{18}, as the chemical formula of gasoline and using the fact that air is 20 percent oxygen by volume, calculate the volume of air at 0°C and 1.0 atm that is required to burn one gal of gasoline. The density of octane is $0.70\ g\cdot mL^{-1}$.

4-74. Lactic acid is produced by the muscles when insufficient oxygen is available and is responsible for muscle cramps during vigorous exercising. It also provides the acidity found in dairy products. Chemical analysis shows that lactic acid is 39.99 percent carbon, 6.73 percent hydrogen, and 53.28 percent oxygen by mass. A 0.3338-g sample of lactic acid was vaporized in a sealed 300-mL container at 150°C, producing a pressure of 326 torr. Use these data to determine the molecular formula of lactic acid.

4-75. Show that the van der Waals equation reduces to the ideal-gas equation at low densities, where the molecules are very far apart from each other.

4-76. The van der Waals constant b can be argued to be four times the volume of the molecules in the container. Using the data in Table 4-4, calculate the radius of a neon atom.

4-77. A 0.428-g sample of a mixture of KCl and $KClO_3$ is heated; 80.7 mL of O_2 is collected over water at 18°C and 756 torr. Calculate the mass percentage of $KClO_3$ in the mixture. ($P_{H_2O} = 15.5$ torr at 18°C.)

4-78. Nitrous oxide, N_2O, sometimes called laughing gas, is used as an anesthetic. It is prepared by the decomposition of ammonium nitrate with heat

$$NH_4NO_3(s) \xrightarrow{\text{high T}} N_2O(g) + 2H_2O(g)$$

What volume of nitrous oxide will be produced when 10.0 g of ammonium nitrate is heated to 200°C at 1.00 atm? The gases are then cooled to 0°C to condense $H_2O(g)$ to water. What volume will the nitrous oxide occupy at 0°C and 1.00 atm? Neglect any pressure due to water vapor.

4-79. Given that air is 20 percent oxygen and 80 percent nitrogen by volume, calculate the density of air at 20°C and 760 torr and the effective molecular mass of air.

4-80. Calculate the temperature at which a carbon dioxide molecule would have the same average speed as a neon atom at 100°C.

4-81. Calculate the volume of 0.200 M NaOH required to prepare 150 mL of $H_2(g)$ at 10°C and 750 torr from the reaction

$$2Al(s) + 2NaOH(aq) + 2H_2O(l) \rightarrow 2NaAlO_2(aq) + 3H_2(g)$$

4-82. In many everyday situations, atmospheric pressure is reported in "inches of mercury." Calculate the pressure of a standard atmosphere in inches of mercury.

4-83. Ammonia gas reacts with hydrogen chloride gas as described by the equation

$$NH_3(g) + HCl(g) \rightarrow NH_4Cl(s)$$

Suppose 5.0 g of NH_3 is reacted with 10.0 g of $HCl(g)$ in a 1.00-L vessel at 75°C. Compute the final pressure (in atm) of gas in the vessel.

4-84. Elemental phosphorus vapor, P_4, is produced according to the equation

$$4Ca_5F(PO_4)_3(s) + 18SiO_2(s) + 15C(s) \rightarrow 18CaSiO_3(s) + 2CaF_2(s) + 15CO_2(g) + 3P_4(g)$$

How many liters of $P_4(g)$ measured at 600°C and 1.00 atm can be produced if 5.00 kg of each reactant is used?

4-85. Compare the mass of oxygen that you utilize to the mass of solid food that you consume each day. Assume that you breathe in 0.5 L of air with each breath, that your respiratory rate is 14 times per minute, that you utilize about 25 percent of the oxygen inhaled, and that you eat 2 pounds of solid food daily.

4-86. Using the data given in Table 4-4, predict which compound of those listed shows the largest deviation from ideal-gas behavior.

4-87. What must be the temperature in order that CO_2 molecules have an average speed of $1000\ m\cdot s^{-1}$?

4-88. Compare the values of the average speed, mean free path, and collision frequency for helium and krypton atoms.

4-89. The speed of sound in a diatomic gas is given by

$$v_{sound} = \left(\frac{7RT}{5M_{kg}}\right)^{1/2}$$

Calculate the speed of sound in $N_2(g)$ at 0°C. Compare your result with the value of v_{av} in $N_2(g)$ at 0°C. Can you give a physical argument for why the speed of sound in a gas is comparable to v_{av} for that gas?

Hydrogen and Oxygen

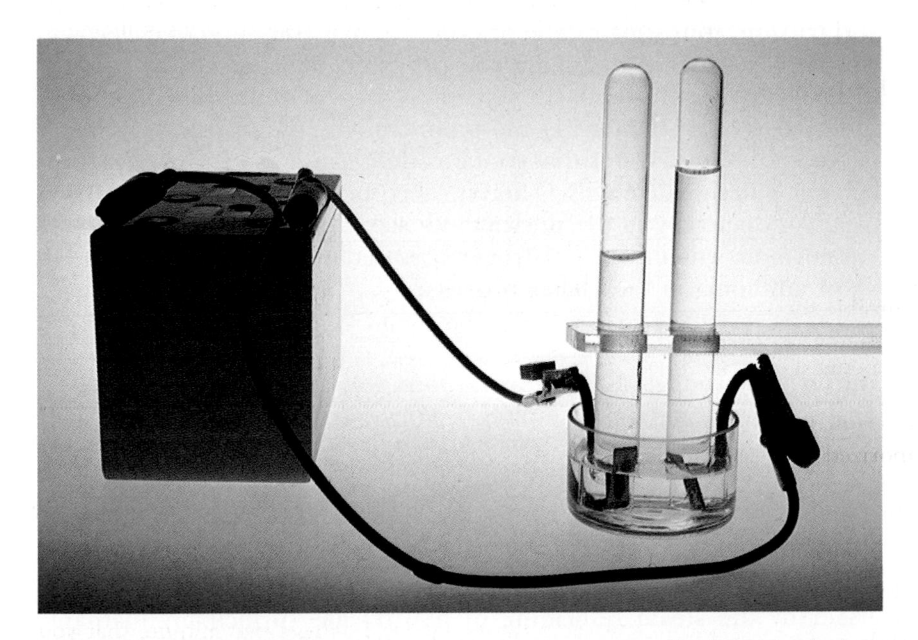

Preparation of hydrogen by electrolysis of an aqueous sulfuric acid solution. Hydrogen gas is liberated at the cathode, and oxygen gas is liberated at the anode. Note, as is required by the reaction stoichiometry, that the volume of $H_2(g)$ liberated is twice as great as that of $O_2(g)$.

Now that we have studied gases in Chapter 4, we shall discuss two important gaseous elements, hydrogen and oxygen. Hydrogen (atomic number 1, atomic mass 1.0079) is the most abundant element in the universe and is the ninth most abundant element in the earth's crust. Hydrogen constitutes over 10 percent of water by mass and occurs in petroleum and all organic matter.

Oxygen (atomic number 8, atomic mass 15.9994) is the most abundant element on earth and the third most abundant element in the universe, ranking behind only hydrogen and helium. Most rocks contain a large amount of oxygen. For example, sand is predominantly silicon dioxide (SiO_2) and consists of more than 50 percent oxygen by mass. Almost 90 percent of the mass of the oceans and two thirds of the mass of the human body are oxygen. Air is 21 percent oxygen by volume. We can live weeks without food, days without water, but only minutes without oxygen.

D-1. MOST HYDROGEN IS USED TO PRODUCE AMMONIA AND METHYL ALCOHOL

Free hydrogen occurs in nature as a diatomic molecule, H_2. It is a colorless, odorless, tasteless gas with a boiling point of $-253°C$ (20 K). Only helium has a lower boiling point (4.3 K). Hydrogen is the least dense substance ($0.09\ g{\cdot}L^{-1}$ at 0°C and 1 atm), being twice as light as helium under the same conditions. There are three isotopes of hydrogen: ordinary hydrogen, consisting of one proton and one electron; **deuterium,** consisting of one proton, one neutron, and one electron; and **tritium,** consisting of one proton, two neutrons, and one electron. Naturally occurring hydrogen consists of 99.985 percent by mass of ordinary hydrogen and 0.015 percent by mass of deuterium. Tritium is radioactive (see Chapter 21) and occurs only in the upper atmosphere in trace amounts. Water that contains deuterium as its hydrogen constituent is designated by D_2O and is called **heavy water.** Heavy water is used commercially in the nuclear industry.

Enormous quantities of hydrogen are used industrially in the synthesis of ammonia in the **Haber process:**

$$3H_2(g) + N_2(g) \xrightarrow{500°C,\ 300\ atm} 2NH_3(g)$$

and in the production of hydrocarbons and alcohols by the **Fischer-Tropsch process**—for example,

$$CO(g) + 2H_2(g) \xrightarrow[\text{catalyst}]{500°C,\ 200\ atm} \underset{\text{methyl alcohol}}{CH_3OH(l)}$$

The explosion of the German dirigible *Hindenburg* during landing in Lakehurst, New Jersey. Today helium, a nonflammable lighter-than-air gas, is used in place of hydrogen in lighter-than-air craft in order to eliminate the possibility of a similar disaster.

The hydrogen used in these large-scale processes is produced on site either by the **steam reforming of natural gas** (principally methane, CH_4):

$$CH_4(g) + H_2O(g) \xrightarrow[30\ atm]{1000°C} CO(g) + 3H_2(g)$$

or by the **water-gas reaction:**

$$\underset{\text{coal}}{C(s)} + H_2O(g) \xrightarrow{800°C} CO(g) + H_2(g)$$

Hydrogen is also used industrially in the hydrogenation of fats and vegetable oils (Chapter 23), in the manufacture of electronic components, in the preparation of HCl and HBr, and in the reduction of metals from their oxides at high temperature by reactions such as

$$Cr_2O_3(s) + 3H_2(g) \rightarrow 2Cr(s) + 3H_2O(g)$$

D-2. HYDROGEN IS REACTIVE AT HIGH TEMPERATURES

Small quantities of hydrogen can be prepared for laboratory use by the reaction of many metals with acids:

$$Zn(s) + 2HCl(aq) \rightarrow ZnCl_2(aq) + H_2(g)$$

or by the reaction of reactive metals such as sodium or calcium with water:

$$Ca(s) + 2H_2O(l) \rightarrow Ca(OH)_2(aq) + H_2(g)$$

or by the reaction of many metallic hydrides with water.

$$NaH(s) + H_2O(l) \rightarrow NaOH(aq) + H_2(g)$$

or by passing an electric current through water (electrolysis) using an apparatus like the one shown in the Frontispiece. The chemical equation for the electrolytic decomposition of water is

$$2H_2O(l) \xrightarrow{\text{electrolysis}} 2H_2(g) + O_2(g)$$

The energy necessary to decompose water into hydrogen and oxygen by electrolysis is supplied by a battery or other external power source. Electrolysis is described in more detail in Chapter 20.

Hydrogen is not particularly reactive under ordinary conditions, but at high temperatures and pressures it is quite reactive and forms compounds with many elements, from the alkali metals:

$$2Na(s) + H_2(g) \xrightarrow{400^\circ C} 2NaH(s)$$

to the halogens:

$$H_2(g) + Cl_2(g) \rightarrow 2HCl(g)$$

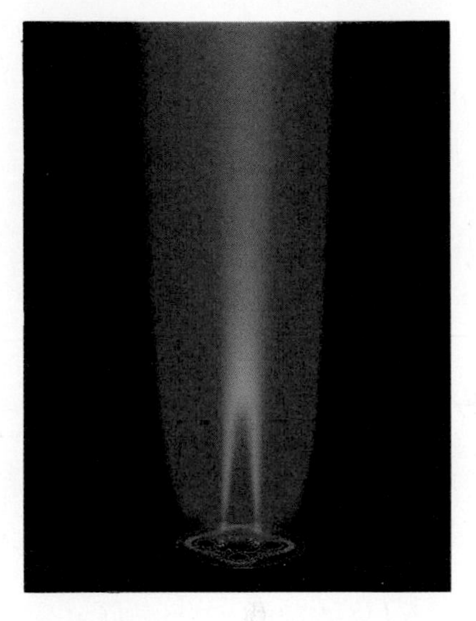

Figure D-1 An oxy-hydrogen torch. The flame temperature is about 2500°C.

The reaction of $H_2(g)$ and $Cl_2(g)$ does not occur in the dark but occurs explosively if the reaction mixture is exposed to sunlight or is sparked. Hydrogen and oxygen also form very dangerous mixtures; they do not react unless sparked or heated to a high temperature, but then they do so explosively. Hydrogen also burns in air, and the flame of an oxyhydrogen blow torch (Figure D-1) has a temperature of about 2500°C, which is high enough to melt platinum.

D-3. OVER THIRTY BILLION POUNDS OF OXYGEN ARE SOLD ANNUALLY IN THE UNITED STATES

Oxygen in air exists primarily as the diatomic molecule O_2. It is a colorless, odorless, tasteless gas with a boiling point of −183°C and a freezing point of −218°C. Although oxygen is colorless as a gas, both liquid and solid oxygen are pale blue (Figure D-2).

Industrially, oxygen is produced by the fractional distillation of liquid air (Interchapter A and Chapter 11), a method that exploits the difference in the boiling points of nitrogen and oxygen, the principal components of air. The nitrogen can be separated from the oxygen because nitrogen boils at −196°C, whereas oxygen boils at −183°C. The pure oxygen thus obtained is compressed in steel cylinders to a pressure of about 150 atm. Approximately 30 billion pounds of oxygen are sold annually in the United States, making it the sixth highest ranked industrial chemical. The major commercial use of oxygen is in the blast furnaces used to manufacture steel. (To save transportation costs, the oxygen is produced on site.) Oxygen is also used in hospitals, in oxyhydrogen and oxyacetylene torches for welding metals, at ath-

Figure D-2 Although gaseous oxygen is colorless, liquid oxygen is pale blue.

letic events, and to facilitate breathing at high altitudes and under water. Tremendous quantities of oxygen are used directly from air as a reactant in the combustion of hydrocarbon fuels, which supply 93 percent of the energy consumed in the United States. In terms of total usage (pure oxygen and oxygen used directly from air), oxygen is the number two chemical, ranking behind water.

Most of the oxygen in the atmosphere is the result of **photosynthesis,** the process by which green plants combine $CO_2(g)$ and $H_2O(l)$ into carbohydrates and $O_2(g)$ under the influence of visible light. The carbohydrates appear in the plants as starch, cellulose, and sugars. The net reaction is

▪ Carbohydrates are so named because their empirical formulas are CH_2O (*carbo-* from carbon and *-hydrate* from water.)

$$CO_2(g) + H_2O(g) \xrightarrow{\text{visible light}} \text{carbohydrate} + O_2(g)$$

▪ Photosynthesis is an active area of chemical research.

Although many details of photosynthesis have yet to be worked out, it is known that the light is first absorbed by the chlorophyll molecules of the plants. In one year, more than 10^{10} metric tons of carbon is incorporated worldwide into carbohydrates by photosynthesis. Photosynthesis is the source of most of the oxygen in the earth's atmosphere.

D-4. OXYGEN REACTS DIRECTLY WITH MOST ELEMENTS

The most frequently used method for preparing oxygen in the laboratory involves the thermal decomposition of potassium chlorate, $KClO_3$ (Figure D-3). The chemical equation for the reaction is

$$2KClO_3(s) \xrightarrow{\text{high T}} 2KCl(s) + 3O_2(g)$$

This reaction requires a fairly high temperature (400°C), but if a small

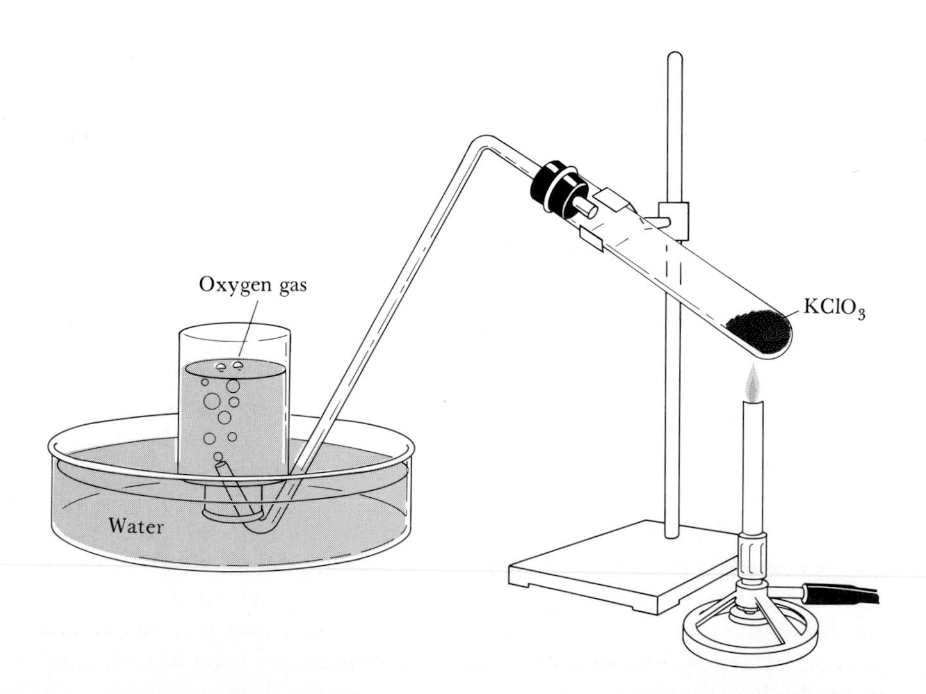

Figure D-3 A typical experimental setup for the production of small amounts of oxygen. Because it is only slightly soluble in water, the oxygen is collected by the displacement of water from an inverted bottle. You probably will collect gases by this method in your own laboratory work.

amount of manganese dioxide, MnO_2, is added, the reaction occurs rapidly at a lower temperature (250°C). The manganese dioxide speeds up the reaction, yet is not a reactant itself. We shall meet examples of this behavior often. A substance that facilitates a reaction but is not consumed in the reaction is called a **catalyst.**

▪ Catalysts are discussed in Chapter 13.

An alternate method for the laboratory preparation of oxygen is to add sodium peroxide, Na_2O_2, to water (Interchapter B):

$$2Na_2O_2(s) + 2H_2O(l) \rightarrow 4NaOH(aq) + O_2(g)$$

This rapid and convenient reaction does not require heat. Oxygen also can be prepared by the electrolytic decomposition of water (see Frontispiece).

Oxygen is a very reactive element. It reacts directly with all the other elements except the halogens, the noble gases, and some of the less reactive metals to form a wide variety of compounds. Only fluorine reacts with more elements than oxygen. Compounds containing oxygen constitute 30 of the top 50 industrial chemicals. The ranking of the first few of these compounds, in terms of quantities sold, is shown in Table D-1.

Oxygen forms oxides with many elements. Most metals react rather slowly with oxygen at ordinary temperatures but react more rapidly as the temperature is increased. For example, iron, in the form of steel wool, burns vigorously in pure oxygen but does not burn in air (Figure D-4).

Fuels are compounds that burn in oxygen and release large quantities of heat. Such a reaction is called a **combustion reaction.** Most fuels are **hydrocarbons,** which are compounds that contain only carbon and hydrogen. Methane, the main constituent of natural gas, burns in oxygen according to the equation

Table D-1. Major oxygen-containing compounds sold in the United States

Rank	Compound	Name	Annual U.S. production/ billions of pounds (1984)
1	H_2SO_4	sulfuric acid	79
4	CaO	calcium oxide (lime)	32
6	O_2	oxygen	31
7	$NaOH$	sodium hydroxide	22
8	H_3PO_4	phosphoric acid	22
10	Na_2CO_3	sodium carbonate	17
11	HNO_3	nitric acid	16
14	NH_4NO_3	ammonium nitrate	14
13	H_2NCONH_2	urea	14
18	CH_3OH	methanol	8.3
19	CO_2	carbon dioxide	7.8

Source: Chemical and Engineering News, June 10, 1985, p. 25.

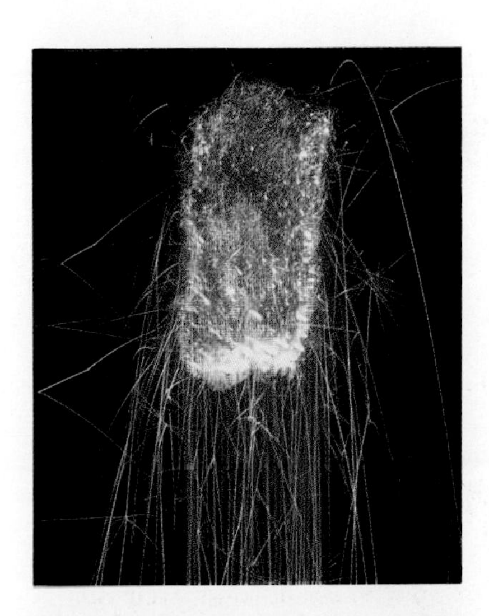

Figure D-4 Although steel wool does not burn in air, it burns vigorously in pure oxygen.

$$\underset{\text{methane}}{CH_4(g)} + 2O_2(g) \rightarrow CO_2(g) + 2H_2O(g)$$

All hydrocarbons burn in oxygen to give carbon dioxide and water. For example, gasoline is a mixture of hydrocarbons. Using octane, C_8H_{18}, as a typical hydrocarbon in gasoline, we write the combustion of gasoline as

$$\underset{\text{octane}}{2C_8H_{18}(l)} + 25O_2(g) \rightarrow 16CO_2(g) + 18H_2O(g)$$

The energy released in hydrocarbon combustion reactions is used to power machinery and to produce electricity (Chapter 5 and Interchapter E).

A mixture of acetylene and oxygen is burned in the oxyacetylene torch. The chemical equation for the combustion of acetylene is

$$\underset{\text{acetylene}}{2C_2H_2(g)} + 5O_2(g) \rightarrow 4CO_2(g) + 2H_2O(g)$$

▪ Different flames have different temperatures. The temperature of a hot region of a candle flame is about 1200°C, that of a Bunsen burner flame is about 1800°C, and that of an oxyacetylene torch flame is about 2400°C.

The flame temperature of an oxyacetylene welding torch is about 2400°C, which is sufficient to melt iron and steel. A combustion reaction with which we are all familiar is the burning of a candle. The wax in a candle is composed of high-molar-mass hydrocarbons, such as $C_{20}H_{42}$. The molten wax rises up the wick to the combustion zone the way ink rises in a piece of blotting paper.

Although most metals yield oxides when they react with oxygen, some of the very reactive metals, such as sodium and barium, yield **peroxides.** With potassium and cesium, **superoxides** are obtained (Interchapter B). One of the most important peroxides is hydrogen peroxide, H_2O_2, a colorless, syrupy liquid. Hydrogen peroxide can be prepared in the laboratory by adding cold dilute sulfuric acid to barium peroxide:

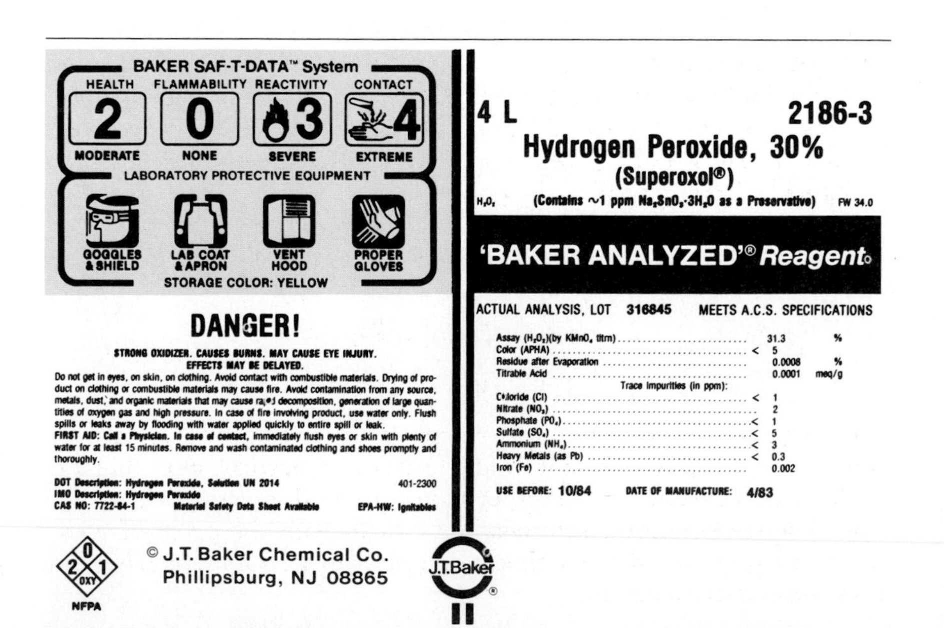

Figure D-5 The labels on reagent bottles carry information concerning the handling of the reagents. Here we see that 30% hydrogen peroxide is very reactive and corrosive.

$$BaO_2(s) + H_2SO_4(aq) \rightarrow BaSO_4(s) + H_2O_2(l)$$

A 3% aqueous solution is sold in drugstores and used as a mild antiseptic and as a bleach. It is sold in brown bottles because hydrogen peroxide decomposes in light according to the equation

$$2H_2O_2(aq) \xrightarrow{\text{light}} 2H_2O(l) + O_2(g)$$

More concentrated solutions (30%) of hydrogen peroxide (Figure D-5) are used industrially. Some of the industrial applications of hydrogen peroxide are as a bleaching agent for feathers, hair, flour, bone, and textile fibers; in renovating old paintings and engravings; in the artificial aging of wines and liquor; in refining oils and fats; and in photography as a fixative eliminant. Concentrated solutions of hydrogen peroxide are extremely corrosive and explosive and must be handled with great care.

D-5. OZONE IS A TRIATOMIC OXYGEN MOLECULE

When a spark is passed through oxygen, some of the oxygen is converted to ozone, O_3:

$$3O_2(g) \rightarrow 2O_3(g)$$

Ozone is a light blue gas at room temperature. It has a sharp, characteristic odor, which often is noticed after electrical storms or near high-voltage generators. Liquid ozone (boiling point $-112°C$) is a deep blue, explosive liquid (Figure D-6). Ozone is so reactive that it cannot be transported, but must be generated as needed. Relatively unreactive metals, such as silver and mercury, which do not react with oxygen, react with ozone to form oxides. Ozone is used as a bleaching agent and is being considered as a replacement for chlorine in water treatment because of the environmental problem involving chlorinated hydrocarbons.

Oxygen and ozone are called **allotropes.** Allotropes are two different forms of an element that have a different number or arrangement of the atoms in the molecules. Many other elements have allotropic forms. For example, graphite and diamond are allotropes. They both consist of carbon atoms, but the atoms are arranged differently in the two substances (Section 11-9).

Ozone plays a vital role in the earth's atmosphere. The action of sunlight on oxygen in the upper atmosphere leads to the production of ozone:

$$O_2(g) \xrightarrow{\text{sunlight}} 2O(g)$$

$$O_2(g) + O(g) \rightarrow O_3(g)$$

The ozone produced in the upper atmosphere absorbs the ultraviolet radiation from sunlight that would otherwise destroy most life on earth. Without ozone in the upper atmosphere, there could be no life as we know it on earth. Interchapter F deals with the earth's atmosphere and describes the role that ozone plays in screening us from the sun's ultraviolet radiation.

Figure D-6 Solid ozone is dark blue.

TERMS YOU SHOULD KNOW

deuterium 156
tritium 156
Haber process 156
Fischer-Tropsch process 156
steam reforming of natural gas 156
water-gas reaction 156
photosynthesis 158
catalyst 159
combustion reaction 159
hydrocarbon 159
peroxide 160
superoxide 160
allotrope 161

QUESTIONS

D-1. What is the most abundant element in the universe?

D-2. Name the naturally occurring isotopes of hydrogen.

D-3. What is heavy water?

D-4. What is the Haber process?

D-5. Write the chemical equation that represents the steam reforming of natural gas.

D-6. Complete and balance the following equations:

(a) $Mg(s) + HCl(aq) \rightarrow$
(b) $Na(s) + H_2O(l) \rightarrow$
(c) $CaH_2(s) + H_2O(l) \rightarrow$

D-7. What is electrolysis?

D-8. Sand is predominantly what oxide?

D-9. How is oxygen produced commercially?

D-10. Describe two methods used to produce small quantities of oxygen in the laboratory.

D-11. What is a catalyst?

D-12. What is combustion?

D-13. Write the chemical equation for the combustion of the principal component of natural gas.

D-14. Write the chemical equation for the heat-producing reaction of an oxyacetylene torch.

D-15. Describe the chemical process involved in the burning of a candle.

D-16. What is an allotrope?

D-17. Name two allotropes of oxygen.

PROBLEMS

D-18. What volumes of $H_2(g)$ and $O_2(g)$ at 700 torr and 25°C can be obtained from the electrolysis of one gram of water?

D-19. What volume of $H_2(g)$ at 760 torr and 0°C will be produced by the reaction of 0.450 g $NaH(s)$ with excess water?

D-20. What volume of $H_2(g)$ at 760 torr and 0°C will be produced by the reaction of 1.62 g $Zn(s)$ with 35.0 mL of 0.600 M $HCl(aq)$?

D-21. What volume of $O_2(g)$ at 655 torr and 15°C will be produced from the thermal decomposition of 3.06 g $KClO_3(s)$?

THERMOCHEMISTRY

Blastoff of a space shuttle. The energy released by the three main engines is equivalent to the output of 23 Hoover Dams. The two side solid rocket boosters consume 11,000 pounds of fuel per second and generate a combined thrust of 44 million horse power, which is equivalent to the horse power of 15,000 six-axle diesel locomotives or 400,000 subcompact cars.

In this chapter we begin our discussion of thermodynamics, which is that part of chemistry that deals with energy changes. Thermochemistry is the branch of thermodynamics that deals with the evolution or absorption of energy as heat in chemical processes. Most chemical reactions are accompanied by a change in energy. These changes may occur in the form of the absorption or the evolution of heat. Energy evolved as heat in the burning of petroleum, natural gas, and coal supplies over 90 percent of the energy used annually in the United States. All living organisms store energy in the form of certain chemicals. Rockets, missiles, and explosives all derive their power from the energy of chemical reactions. Our primary objective in this chapter is to develop an understanding of the energy changes involved in chemical reactions.

5-1. ENERGY IS CONSERVED

When chemical reactions take place, they almost always involve a transfer of energy between the reaction system and its surroundings. **Thermodynamics** (thermo = heat, dynamics = changes) is the study of these energy transfers. One of the fundamental principles of thermodynamics is the **first law of thermodynamics,** which states that energy is neither created nor destroyed but is simply converted from one form to another. The first law of thermodynamics is the law of **conservation of energy.**

There are basically two ways in which energy can be transferred between a reaction system and its surroundings: either as work or as heat. The transfer of energy as **work** involves the action of a force that causes a displacement within the system, that is, a change in a measurable dimension, such as distance or volume. For any such dimension X, the displacement is denoted by ΔX. The Greek letter delta, Δ, indicates a change in a quantity. The value of ΔX is determined by subtracting the initial value of the quantity X from the final value: thus, if X denotes a distance, then a displacement is given by $\Delta X = X_{\text{final}} - X_{\text{initial}}$.

For simplicity we can visualize a system as the contents of a cylinder equipped with a piston, as shown in Figure 5-1. Suppose that the application of a constant force, F, to the piston causes a displacement of the system by an amount ΔX. We know from physics that the work done on a system is given by

$$\text{work} = \text{force} \times \text{displacement}$$

Thus we can express the work done on the system, w, by

$$w = -F\Delta X \tag{5-1}$$

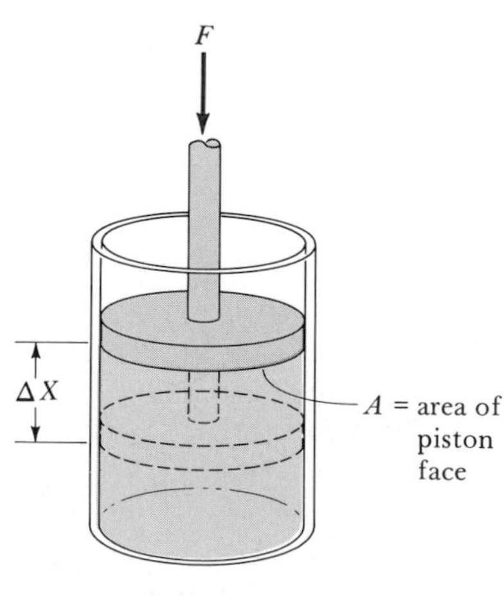

Figure 5-1 A system enclosed in a cylinder equipped with a piston. A force F is applied to a piston causing a displacement $\Delta X = X_{\text{final}} - X_{\text{initial}}$ of the piston. The area of the piston face is A.

The minus sign occurs in Equation (5-1) because our sign convention for work is that work done on a system is a positive quantity. Note that when the piston in Figure 5-1 is pushed inward, $\Delta X = X_{\text{final}} - X_{\text{initial}}$ has a negative value, because $X_{\text{final}} < X_{\text{initial}}$ (compression). The minus sign in Equation (5-1) makes the right-hand side of the equation a positive value, as required for work done upon compressing a system. Conversely, the work done on a system upon expansion is a negative quantity.

Some algebraic manipulations of Equation (5-1) give an interesting and useful result. First we can simultaneously multiply and divide the right-hand side by A, the area of the piston face, to get

$$w = -\left(\frac{F}{A}\right)A\Delta X \tag{5-2}$$

We know from Chapter 4 that pressure is defined as force per unit area, or $P = F/A$, so we can substitute P for the F/A term that appears in Equation (5-2). We also note that the volume of a cylinder is the area of the base times the height, so if $V = AX$, then the change in volume of the system when the piston is displaced by ΔX is $\Delta V = A\Delta X$. Thus Equation (5-2) can be rewritten as

$$w = -P\Delta V = -P(V_f - V_i) \quad (5\text{-}3)$$

where V_f is the final volume and V_i is the initial volume. Equation (5-3) gives the work done on a system that is compressed or expanded at a constant pressure P. Because $V_f < V_i$ upon compression, the work done on a system due to compression is a positive quantity. Conversely, because $V_f > V_i$ upon expansion, the work done on a system due to expansion is a negative quantity. Equation (5-3) is useful in thermochemistry because many reactions take place in vessels open to the atmosphere, which exerts a constant pressure during the course of the reaction (Figure 5-2).

The transfer of energy as **heat** does not require the application of a force. Energy transfer as heat occurs when there is a temperature difference between the system and the surroundings. Energy flows spontaneously as heat from regions of higher temperature to regions of lower temperature. We denote the transfer of energy as heat by q. It should be recognized that a system does not contain heat or work. Rather, heat and work are ways in which energy is transferred.

In thermodynamics energy is denoted by the symbol U. We denote the **energy change of a reaction** by ΔU_{rxn}, where

▪ The subscript *rxn* means reaction.

$$\Delta U_{rxn} = U_f - U_i$$

The energy of a system can change as a result of a transfer of energy as heat or as work. Application of the first law of thermodynamics to a chemical reaction yields

$$\Delta U_{rxn} = q + w \quad (5\text{-}4)$$

where q is the energy transferred as heat and w is the energy transferred as work. For a reaction that occurs at constant pressure, we can substitute Equation (5-3) into Equation (5-4) to obtain

$$\Delta U_{rxn} = q - P\Delta V \quad (5\text{-}5)$$

If a reaction occurs in a rigid, closed container, then there is no change in volume. Therefore, no energy is transferred as work, and $\Delta V = 0$ in Equation (5-5). In this case we have

$$\Delta U_{rxn} = q_V = \begin{pmatrix}\text{heat evolved or absorbed} \\ \text{at constant volume}\end{pmatrix} \quad (5\text{-}6)$$

The V subscript on q emphasizes that q_V is the heat transferred when the volume of the reaction system is constant.

For reactions that occur at constant pressure, it is convenient to introduce a thermodynamic function H, called the **enthalpy,** which is defined by the equation

$$H = U + PV \quad (5\text{-}7)$$

where U is the energy, P is the pressure, and V is the volume. Application of Equation (5-7) to a chemical reaction yields

$$\Delta H_{rxn} = H_f - H_i = U_f - U_i + P_fV_f - P_iV_i \quad (5\text{-}8)$$

where ΔH_{rxn} is the **enthalpy change** for a chemical reaction. For a reaction that occurs at a constant pressure, $P_f = P_i$, and so Equation (5-8) becomes

$$\Delta H_{rxn} = \Delta U_{rxn} + P\Delta V \quad (5\text{-}9)$$

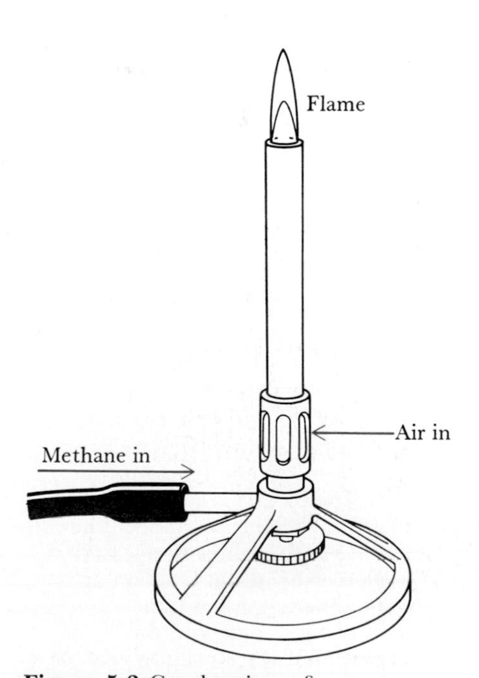

Figure 5-2 Combustion of methane gas in air (in a Bunsen burner). Because the reaction is open to the atmosphere, it is a constant-pressure reaction.

Substitution of Equation (5-5) into (5-9) yields

$$\Delta H_{rxn} = q_P - P\Delta V + P\Delta V = q_P$$

The P subscript on q_P emphasizes that q_P is the heat transferred when the reaction takes place at constant pressure. Thus we see that the enthalpy has the useful property that

$$\Delta H_{rxn} = q_P = \begin{pmatrix}\text{heat evolved or absorbed}\\ \text{at constant pressure}\end{pmatrix} \tag{5-10}$$

Note that Equation (5-10) is the constant-pressure analog of Equation (5-6). Because most reactions occur at constant pressure, ΔH_{rxn} is often called the **heat of reaction.** Equation (5-10) is a key equation in **thermochemistry.**

5-2. CHEMICAL REACTIONS EVOLVE OR ABSORB ENERGY AS HEAT

There are many types of combustible **fuels,** which are reactants that burn in oxygen to provide heat. For example, the equation for the combustion of methane is

$$CH_4(g) + 2O_2(g) \rightarrow CO_2(g) + 2H_2O(l)$$

This reaction releases energy as heat and thus is called an **exothermic reaction** (exo = out). All combustion reactions are highly exothermic.

Reactions that absorb energy as heat are called **endothermic reactions** (endo = in). An exothermic and an endothermic reaction are illustrated schematically in Figure 5-3. In Figure 5-3(a), the exothermic reaction, the enthalpy of the reactants is greater than the enthalpy of the products and so heat is evolved as the reaction proceeds. In Figure 5-3(b), the endothermic reaction, the enthalpy of the reactants is less than the enthalpy of the products and so heat must be supplied in order for the reaction to proceed. In a sense, the heat must be supplied to push the reaction "uphill" with respect to enthalpy.

The enthalpy change for a chemical reaction is the total enthalpy of the products minus the total enthalpy of the reactants:

$$\Delta H_{rxn} = H_{prod} - H_{react} \tag{5-11}$$

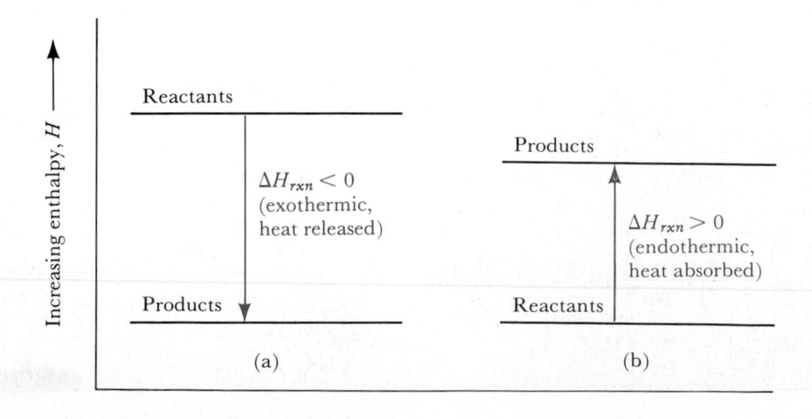

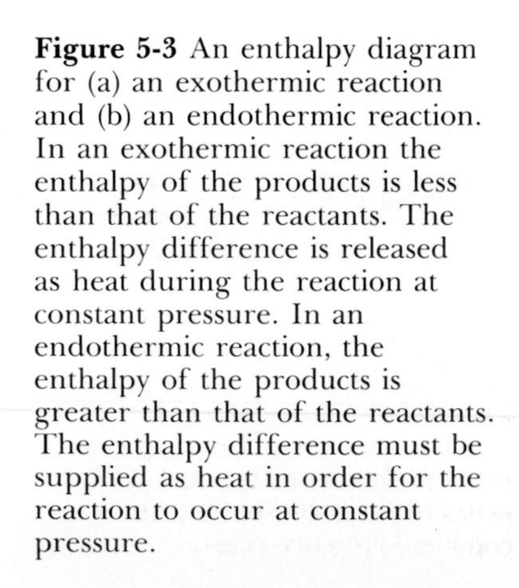
Figure 5-3 An enthalpy diagram for (a) an exothermic reaction and (b) an endothermic reaction. In an exothermic reaction the enthalpy of the products is less than that of the reactants. The enthalpy difference is released as heat during the reaction at constant pressure. In an endothermic reaction, the enthalpy of the products is greater than that of the reactants. The enthalpy difference must be supplied as heat in order for the reaction to occur at constant pressure.

For an exothermic reaction, H_{prod} is less than H_{react} and so $\Delta H_{rxn} < 0$. For an endothermic reaction, H_{prod} is greater than H_{react} and so $\Delta H_{rxn} > 0$.

When specifying a value for the enthalpy change of a particular chemical reaction that takes place at 1 atm, we use the notation ΔH°_{rxn}. The superscript degree sign tells us that the enthalpy change is the **standard enthalpy change** for the reaction. The standard enthalpy change refers to pure reactants and products at 1 atm pressure and solutions at 1 M concentration. For the combustion of methane, the value of ΔH°_{rxn} is -890 kJ at 25°C. The negative value of ΔH°_{rxn} tells us that the reaction gives off heat and is therefore exothermic.

An example of an endothermic reaction is the **water-gas reaction,** in which steam is passed over hot carbon to produce a mixture of carbon monoxide and hydrogen (Figure 5-4).

$$C(s) + H_2O(g) \rightarrow CO(g) + H_2(g)$$

For this reaction, $\Delta H^\circ_{rxn} = +131$ kJ at 25°C and so heat must be supplied. Both CO and H_2 are combustible gases, and thus the water-gas reaction can be used to convert coal to a combustible, gaseous mixture that can be transported through pipelines.

You may have noticed that we specified the temperature (25°C) when we gave the values of ΔH°_{rxn} for the combustion of methane and for the water-gas reaction. We did this because the value of ΔH°_{rxn} depends upon the temperature at which a reaction occurs. The variation in ΔH°_{rxn} with temperature usually is not great, but strictly speaking, we should specify the reaction temperature. In this book we do not study the variation in ΔH°_{rxn} with temperature; instead we shall neglect the dependence of ΔH°_{rxn} on temperature, which is a satisfactory approximation for our purposes.

Figure 5-4 Tenneco's Great Plains coal gasification plant in North Dakota, where coal is converted into a combustible gaseous mixture, as in the water-gas reaction. This plant is able to produce over 400 million liters of combustible gas per day.

5-3. ENTHALPY CHANGES FOR CHEMICAL EQUATIONS ARE ADDITIVE

One of the most useful properties of ΔH°_{rxn} values is that we can add the ΔH°_{rxn} values for two or more chemical equations to obtain the ΔH°_{rxn} value for another chemical equation. This property is best illustrated by example. Consider the two chemical equations

▪ Chemical equations can be added and subtracted like algebraic equations.

$$(1)\quad C(s) + \tfrac{1}{2}O_2(g) \rightarrow CO(g) \qquad \Delta H^\circ_{rxn}(1) = -110.5 \text{ kJ}$$

$$(2)\quad CO(g) + \tfrac{1}{2}O_2(g) \rightarrow CO_2(g) \qquad \Delta H^\circ_{rxn}(2) = -283.0 \text{ kJ}$$

If we add these two chemical equations as if they were algebraic equations, then we get

$$(3)\quad C(s) + CO(g) + \tfrac{1}{2}O_2(g) + \tfrac{1}{2}O_2(g) \rightarrow CO(g) + CO_2(g)$$

If we cancel $CO(g)$ from both sides, then we get

$$(3)\qquad C(s) + O_2(g) \rightarrow CO_2(g)$$

The additive property of ΔH°_{rxn} says that ΔH°_{rxn} for Equation (3) is simply

$$\Delta H^\circ_{rxn}(3) = \Delta H^\circ_{rxn}(1) + \Delta H^\circ_{rxn}(2)$$
$$= -110.5 \text{ kJ} + (-283.0 \text{ kJ}) = -393.5 \text{ kJ} \qquad (5\text{-}12)$$

The additivity property of ΔH°_{rxn} values is known as **Hess's law:** if two or more chemical equations are added together, then the value of ΔH°_{rxn} for the resulting equation is equal to the sum of the ΔH°_{rxn} values for the separate equations.

Example 5-1: Given the following ΔH°_{rxn} values

$$(1)\qquad SO_2(g) \rightarrow S(s) + O_2(g) \qquad \Delta H^\circ_{rxn}(1) = +296.8 \text{ kJ}$$

$$(2)\qquad 2S(s) + 3O_2(g) \rightarrow 2SO_3(g) \qquad \Delta H^\circ_{rxn}(2) = -791.4 \text{ kJ}$$

compute ΔH°_{rxn} for the equation

$$(3)\qquad 2SO_2(g) + O_2(g) \rightarrow 2SO_3(g)$$

Solution: To obtain Equation (3) from Equations (1) and (2), it is first necessary to multiply Equation (1) through by 2 because Equation (3) involves 2 mol of $SO_2(g)$ as a reactant. If we multiply an equation through by 2, then its ΔH°_{rxn} must also be multiplied by 2 because twice as many moles of reactants are consumed and twice as many moles of products are produced. Thus the amount of heat evolved or absorbed is doubled. For the equation

$$(4)\qquad 2SO_2(g) \rightarrow 2S(s) + 2O_2(g)$$

we have

$$\Delta H^\circ_{rxn} = 2\Delta H^\circ_{rxn}(1) = 2 \times 296.8 \text{ kJ} = +593.6 \text{ kJ}$$

Addition of Equations (2) and (4) yields

$$2S(s) + 3O_2(g) + 2SO_2(g) \rightarrow 2S(s) + 2O_2(g) + 2SO_3(g)$$

If we cancel $2S(s)$ and $2O_2(g)$ from both sides, then we get Equation (3):

$$2SO_2(g) + O_2(g) \rightarrow 2SO_3(g)$$

The corresponding value of ΔH°_{rxn} is

$$\Delta H^\circ_{rxn}(3) = 2\Delta H^\circ_{rxn}(1) + \Delta H^\circ_{rxn}(2)$$
$$= +593.6 \text{ kJ} - 791.4 \text{ kJ}$$
$$= -197.8 \text{ kJ}$$

Note that the conversion of sulfur dioxide to sulfur trioxide is an exothermic reaction. This is the reaction that takes place in the catalytic converters of automobiles that run on unleaded gasoline and in the manufacture of sulfuric acid (Interchapter L).

Another algebraic property of ΔH°_{rxn} values, which is useful in applying Hess's law, follows from Equation (5-11). If we reverse a chemical equation, then the reactants become the products and the products become the reactants. Thus

$$\Delta H^\circ_{rxn}(\text{reverse rxn}) = -\Delta H^\circ_{rxn}(\text{forward rxn}) \qquad (5\text{-}13)$$

The value of ΔH°_{rxn} for the reaction given by

(1) $$CO_2(g) \rightarrow C(s) + O_2(g)$$

which is the reverse of the reaction given by

(2) $$C(s) + O_2(g) \rightarrow CO_2(g) \qquad \Delta H_{rxn} = -393.5 \text{ kJ}$$

is

$$\Delta H^\circ_{rxn}(1) = -\Delta H^\circ_{rxn}(2)$$
$$= -(-393.5 \text{ kJ}) = +393.5 \text{ kJ}$$

Hess's law enables us to calculate a ΔH°_{rxn} value for a chemical equation from ΔH°_{rxn} values for related equations. For example, suppose we are given that

(1) $$2P(s) + 3Cl_2(g) \rightarrow 2PCl_3(l) \qquad \Delta H^\circ_{rxn}(1) = -640 \text{ kJ}$$

(2) $$2P(s) + 5Cl_2(g) \rightarrow 2PCl_5(s) \qquad \Delta H^\circ_{rxn}(2) = -886 \text{ kJ}$$

and we wish to calculate ΔH°_{rxn} for the equation

$$PCl_3(l) + Cl_2(g) \rightarrow PCl_5(s)$$

In this case we combine Equation (2) and the reverse of Equation (1) to obtain

(3) $$2PCl_3(l) + 2Cl_2(g) \rightarrow 2PCl_5(s)$$

$$\Delta H^\circ_{rxn}(3) = \Delta H^\circ_{rxn}(2) - \Delta H^\circ_{rxn}(1)$$
$$= -886 \text{ kJ} + 640 \text{ kJ} = -246 \text{ kJ}$$

We now divide Equation (3) through by 2 to obtain our final result:

(4) $$PCl_3(l) + Cl_2(g) \rightarrow PCl_5(s)$$

$$\Delta H^\circ_{rxn}(4) = (\tfrac{1}{2})\Delta H^\circ_{rxn}(3)$$
$$= -\frac{246 \text{ kJ}}{2} = -123 \text{ kJ}$$

5-4. HEATS OF REACTION CAN BE CALCULATED FROM TABULATED HEATS OF FORMATION

The value of ΔH°_{rxn} for the reaction of carbon with oxygen is

$$C(s) + O_2(g) \rightarrow CO_2(g) \qquad \Delta H^\circ_{rxn} = -393.5 \text{ kJ}$$

Because 1 mol of $CO_2(g)$ is formed directly from the elements carbon and oxygen, the value of ΔH°_{rxn} for this equation is *defined* to be the **standard molar enthalpy of formation** of $CO_2(g)$ from its elements. We denote the standard molar enthalpy of formation, often called simply the **heat of formation,** of one mole of a substance from its constituent elements in their most stable form at 25°C and 1 atm by ΔH°_f. For the formation of carbon dioxide from its constituent elements,

$$\begin{aligned}\Delta H^\circ_{rxn} &= \Delta H^\circ_f[CO_2(g)] = -393.5 \text{ kJ}\cdot\text{mol}^{-1}\\ &= H^\circ[CO_2(g)] - H^\circ[C(s)] - H^\circ[O_2(g)]\end{aligned}$$

The subscript f on ΔH°_f stands for *f*ormation from the elements and also indicates that the value is for 1 mol. A $\Delta H^\circ_f[CO_2(g)]$ value of -393.5 kJ tells us that 1 mol of $CO_2(g)$ lies 393.5 kJ "downhill" on the enthalpy scale relative to its constituent elements (Figure 5-5a).

The molar enthalpies of formation of water and acetylene from their elements are equal to the ΔH°_{rxn} values for the reactions in which one mole of each compound is formed from its elements (Figure 5-5b and c):

▪ Note that we have used $\frac{1}{2}$ as a balancing coefficient in the equation $H_2(g) + \frac{1}{2}O_2(g) \rightarrow H_2O(l)$. We do this because we want the equation for the formation of *one mole* of $H_2O(l)$ from its constituent elements.

$$H_2(g) + \tfrac{1}{2}O_2(g) \rightarrow H_2O(l) \qquad \Delta H^\circ_{rxn} = \Delta H^\circ_f[H_2O(l)] = -285.8 \text{ kJ}\cdot\text{mol}^{-1}$$

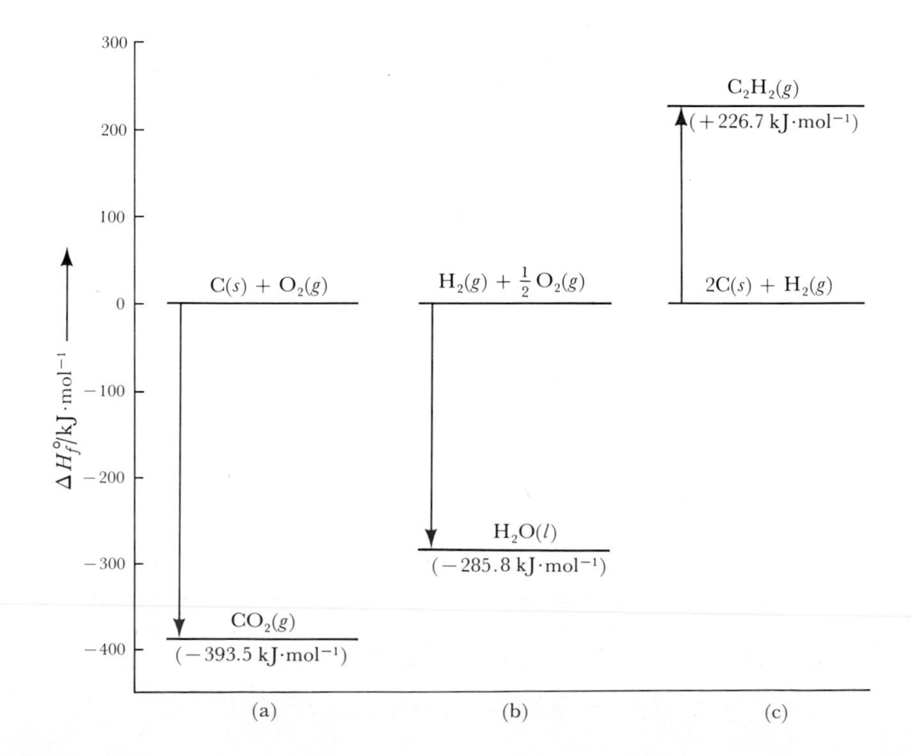

Figure 5-5 Enthalpy changes involved in the formation of $CO_2(g)$, $H_2O(l)$, and $C_2H_2(g)$ from their elements. Note that (a) $CO_2(g)$ lies 393.5 kJ·mol^{-1} (on the enthalpy scale) below the elements; (b) $H_2O(l)$ lies 285.8 kJ·mol^{-1} below the elements; and (c) $C_2H_2(g)$ lies 226.7 kJ·mol^{-1} above the elements.

$$2C(s) + H_2(g) \rightarrow C_2H_2(g) \qquad \Delta H^\circ_{rxn} = \Delta H^\circ_f[C_2H_2(g)]$$
$$= +226.7 \text{ kJ}\cdot\text{mol}^{-1}$$

As suggested by Figure 5-5, we can set up a table of ΔH°_f values for compounds by setting the ΔH°_f values for the elements equal to zero. That is, for each element in its normal physical state at 25°C and 1 atm, we set ΔH°_f equal to zero. Thus, for example,

▪ The form of an element for which we take $\Delta H^\circ_f = 0$ is the most stable form of the element at that temperature.

$$\Delta H^\circ_f[Fe(s)] = 0 \qquad \Delta H^\circ_f[S(s)] = 0$$
$$\Delta H^\circ_f[Hg(l)] = 0 \qquad \Delta H^\circ_f[Cl_2(g)] = 0$$
$$\Delta H^\circ_f[O_2(g)] = 0 \qquad \Delta H^\circ_f[C(s), \text{graphite}] = 0$$

We can use Hess's law to understand how heats of formation are used to calculate enthalpy changes. Consider the general chemical equation

$$aA + bB \rightarrow yY + zZ$$

where a, b, y, and z are the numbers of moles of the respective compounds. We can calculate ΔH°_{rxn} in two steps, as shown in the following diagram:

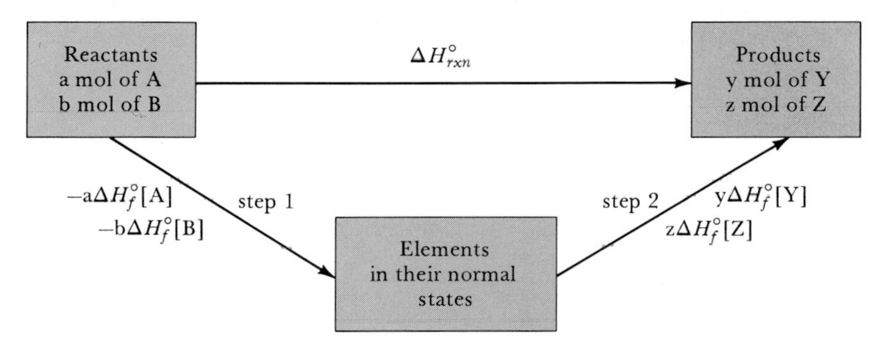

First we decompose compounds A and B into their constituent elements (step 1) and then combine the elements to form the compounds Y and Z (step 2). In the first step, we have

$$\Delta H^\circ_{rxn}(1) = -a\Delta H^\circ_f[A] - b\Delta H^\circ_f[B]$$

The minus signs occur here because the reaction involved is the reverse of the formation of the compounds from their elements; we are forming the elements from the compounds. In the second step, we have

$$\Delta H^\circ_{rxn}(2) = y\Delta H^\circ_f[Y] + z\Delta H^\circ_f[Z]$$

The sum of $\Delta H^\circ_{rxn}(1)$ and $\Delta H^\circ_{rxn}(2)$ gives ΔH°_{rxn} for the general equation:

$$\Delta H^\circ_{rxn} = y\Delta H^\circ_f[Y] + z\Delta H^\circ_f[Z] - a\Delta H^\circ_f[A] - b\Delta H^\circ_f[B] \qquad (5\text{-}14)$$

Note that the right-hand side of Equation (5-14) is the total enthalpy of the products minus the total enthalpy of the reactants (see Equation 5-11).

In using Equation (5-14), it is necessary to specify whether each substance is a gas, liquid, or solid because the value of ΔH°_f depends on the physical state of the substance. Using Equation (5-14), we determine the value of ΔH°_{rxn} at 25°C for the reaction

$$2C_2H_2(g) + 5O_2(g) \rightarrow 4CO_2(g) + 2H_2O(l)$$

to be

$$\Delta H^\circ_{rxn} = 4\Delta H^\circ_f[CO_2(g)] + 2\Delta H^\circ_f[H_2O(l)] - 2\Delta H^\circ_f[C_2H_2(g)] - 5\Delta H^\circ_f[O_2(g)]$$

Using data in Figure 5-5, we obtain

$$\begin{aligned}\Delta H^\circ_{rxn} &= (4\text{ mol})(-393.5\text{ kJ}\cdot\text{mol}^{-1}) + (2\text{ mol})(-285.8\text{ kJ}\cdot\text{mol}^{-1}) \\ &\quad - (2\text{ mol})(+226.7\text{ kJ}\cdot\text{mol}^{-1}) - (5\text{ mol})(0\text{ kJ}\cdot\text{mol}^{-1}) \\ &= -2599.0\text{ kJ}\end{aligned}$$

Note that $\Delta H^\circ_f[O_2(g)] = 0$ because the ΔH°_f value of any element in its normal state at 25°C is zero. Table 5-1 lists ΔH°_f values for a variety of substances at 25°C.

Example 5-2: Use the ΔH°_f data in Table 5-1 to compute ΔH°_{rxn} for the combustion of ethyl alcohol:

$$C_2H_5OH(l) + 3O_2(g) \rightarrow 2CO_2(g) + 3H_2O(l)$$

Solution: Referring to Table 5-1, we find

$\Delta H^\circ_f[CO_2(g)] = -393.5\text{ kJ}\cdot\text{mol}^{-1}$ $\quad$ $\Delta H^\circ_f[H_2O(l)] = -285.8\text{ kJ}\cdot\text{mol}^{-1}$

$\Delta H^\circ_f[O_2(g)] = 0$ $\quad$ $\Delta H^\circ_f[C_2H_5OH(l)] = -277.7\text{ kJ}\cdot\text{mol}^{-1}$

and thus

$$\begin{aligned}\Delta H^\circ_{rxn} &= 2\Delta H^\circ_f[CO_2(g)] + 3\Delta H^\circ_f[H_2O(l)] \\ &\quad - \Delta H^\circ_f[C_2H_5OH(l)] - 3\Delta H^\circ_f[O_2(g)] \\ &= (2\text{ mol})(-393.5\text{ kJ}\cdot\text{mol}^{-1}) + (3\text{ mol})(-285.8\text{ kJ}\cdot\text{mol}^{-1}) \\ &\quad - (1\text{ mol})(-277.7\text{ kJ}\cdot\text{mol}^{-1}) - (3\text{ mol})(0) \\ &= -1366.7\text{ kJ}\end{aligned}$$

The reaction is highly exothermic.

5-5. THE VALUE OF ΔH°_{rxn} IS DETERMINED PRIMARILY BY THE DIFFERENCE IN THE BOND ENTHALPIES OF THE REACTANT AND PRODUCT MOLECULES

In this section we learn what molecular properties of reactants and products give rise to the observed value of ΔH°_{rxn}. The enthalpy change for the equation

$$H_2O(g) \rightarrow O(g) + 2H(g)$$

is $\Delta H^\circ_{rxn} = +925$ kJ. Although we have not studied bonding yet, we shall learn in Chapters 8 and 9 that a water molecule has the structure

```
   O
  / \
 H   H
```

There are two oxygen-hydrogen bonds, represented by the lines joining the H atoms to the O atom, in a water molecule. At constant pressure an input of 925 kJ of energy as heat is required to break the two oxygen-hydrogen bonds in 1 mol of water molecules. Let's denote the

Table 5-1 Standard molar enthalpies of formation, ΔH_f°, for various substances at 25°C

Substance	Formula	ΔH_f°/kJ·mol^{-1}
acetylene	$C_2H_2(g)$	+226.7
ammonia	$NH_3(g)$	−46.19
benzene	$C_6H_6(l)$	+49.03
bromine vapor	$Br_2(g)$	+30.91
carbon dioxide	$CO_2(g)$	−393.5
carbon monoxide	$CO(g)$	−110.5
carbon tetrachloride	$CCl_4(l)$	−135.4
	$CCl_4(g)$	−103.0
diamond	$C(s)$	+1.897
ethane	$C_2H_6(g)$	−84.68
ethanol (ethyl alcohol)	$C_2H_5OH(l)$	−277.7
ethene (ethylene)	$C_2H_4(g)$	+52.28
glucose	$C_6H_{12}O_6(s)$	−1260
graphite	$C(s)$	0
hydrazine	$N_2H_4(l)$	+50.6
hydrogen bromide	$HBr(g)$	−36.4
hydrogen chloride	$HCl(g)$	−92.31
hydrogen fluoride	$HF(g)$	−271.1
hydrogen iodide	$HI(g)$	+26.1
hydrogen peroxide	$H_2O_2(l)$	−187.8
iodine vapor	$I_2(g)$	+62.4
magnesium carbonate	$MgCO_3(s)$	−1096
magnesium oxide	$MgO(s)$	−601.7
methane	$CH_4(g)$	−74.86
methanol (methyl alcohol)	$CH_3OH(l)$	−238.7
	$CH_3OH(g)$	−200.7
nitrogen oxide	$NO(g)$	+90.37
nitrogen dioxide	$NO_2(g)$	+33.85
dinitrogen tetroxide	$N_2O_4(g)$	+9.66
	$N_2O_4(l)$	−19.5
propane	$C_3H_8(g)$	−103.8
sodium carbonate	$Na_2CO_3(s)$	−1131
sodium oxide	$Na_2O(s)$	−418.0
sucrose	$C_{12}H_{22}O_{11}(s)$	−2220
sulfur dioxide	$SO_2(g)$	−296.8
sulfur trioxide	$SO_3(g)$	−395.7
water	$H_2O(l)$	−285.8
	$H_2O(g)$	−241.8

▪ The normal form of bromine at 25°C is $Br_2(l)$, and this is the form of bromine at 25°C for which we take $\Delta H_f^\circ[Br_2(l)] = 0$. Heat must be applied to convert $Br_2(l)$ to $Br_2(g)$, and thus $\Delta H_f^\circ[Br_2(g)]$ is equal to ΔH_{rxn}° for the reaction $Br_2(l) \rightarrow Br_2(g)$.

Elemental forms for which we take $\Delta H_f^\circ = 0$ at 25°C

$H_2(g)$	$Na(s)$
$O_2(g)$	$Br_2(l)$
$N_2(g)$	$Cl_2(g)$
$C(s, \textit{graphite})$*	$F_2(g)$
$S(s, \textit{rhombic})$	$I_2(s)$
$Mg(s)$	

*Elemental carbon occurs both as diamond and as graphite at 25°C. We take $\Delta H_f^\circ = 0$ for graphite because it is the more stable form at 25°C and 1 atm. Note that energy is required to convert graphite to diamond.

▪ The formation of a chemical bond releases energy as heat.

average oxygen-hydrogen bond enthalpy per mole of oxygen-hydrogen bonds in water by H(O—H). The average **molar bond enthalpy** H(oxygen-hydrogen) is equal to one half the total energy input required to break the two oxygen-hydrogen bonds:

$$H(\text{O—H}) = \frac{925\ \text{kJ}}{2} = 463\ \text{kJ}$$

The values of the oxygen-hydrogen molar bond enthalpy in a variety of compounds are approximately the same as the value for water. For example, the enthalpy change for the reaction

$$\begin{array}{ccc} \text{H} & & \text{H} \\ | & & | \\ \text{H—C—O—H} & \rightarrow & \text{H—C—O} + \text{H} \\ | & & | \\ \text{H} & & \text{H} \end{array}$$

in which a single oxygen-hydrogen bond is broken, is $\Delta H^\circ_{rxn} = 464$ kJ. Molar bond enthalpies are often referred to simply as **bond energies.**

In Figure 5-6 we picture a chemical reaction as taking place in two steps. In step 1 the reactant molecules are broken down into their constituent atoms, and in step 2 the atoms are rejoined to form the product molecules. Step 1 requires an input of energy to break all the bonds in the reactant molecules, whereas step 2 evolves energy as the bonds in the product molecules are formed. The total enthalpy change for the equation is

$$\Delta H^\circ_{rxn} \approx \begin{pmatrix}\text{heat energy input to}\\ \text{break all bonds in}\\ \text{reactants}\end{pmatrix} - \begin{pmatrix}\text{energy evolved as heat}\\ \text{on formation of all}\\ \text{bonds in products}\end{pmatrix}$$

$$\approx H(\text{bond})_R - H(\text{bond})_P \qquad (5\text{-}15)$$

where $H(\text{bond})_R$ represents the sum of the molar bond enthalpies for all the reactant bonds and $H(\text{bond})_P$ represents the sum of the molar bond enthalpies for all the product bonds in the chemical equation for the reaction. (The reason for the approximately equals sign in Equation (5-15) is discussed later in this section.)

If more energy is released on formation of the product bonds than is required to break the reactant bonds, then the value of ΔH°_{rxn} is negative (energy released). If less energy is released on the formation of the product bonds than is required to break the reactant bonds, then the value of ΔH°_{rxn} is positive (energy consumed). The value of ΔH°_{rxn} is determined primarily by the difference in the bond enthalpies of the reactants and products. Values of average bond enthalpies for a variety of chemical bonds are given in Table 5-2.

Table 5-2 Average molar bond enthalpies ("bond energies")

Bond	Molar bond enthalpy, H(bond)/kJ·mol^{-1}
O—H	464
O—O	142
C—O	351
C═O	730
C—C	347
C═C	615
C≡C	811
C—H	414
C—F	439
C—Cl	331
C—Br	276
C—N	293
C═N	615
C≡N	890
N—H	390
N—N	159
N═N	418
N≡N	945
F—F	155
Cl—Cl	243
Br—Br	192
H—H	435
H—F	565
H—Cl	431
H—Br	368
H—S	364

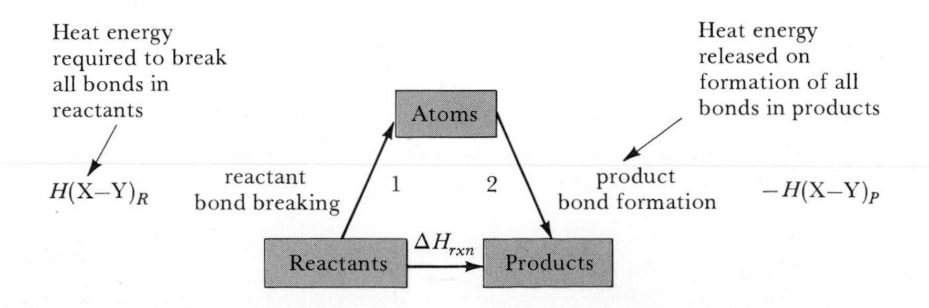

Figure 5-6 The enthalpy change of a reaction is given approximately by the difference between the heat energy required to break all the chemical bonds in the reactants and the energy released as heat on the formation of all the chemical bonds in the products. X and Y are any elements.

Example 5-3: Chemical reactions can be used to produce flames for heating. The nonnuclear reaction with the highest attainable flame temperature (approximately 6000°C, about the surface temperature of the sun) is that between hydrogen and fluorine:

$$\text{H—H}(g) + \text{F—F}(g) \rightarrow 2\text{H—F}(g)$$

Use the molar bond enthalpies in Table 5-2 to estimate ΔH°_{rxn} for this equation.

Solution: The reaction involves the rupture of one hydrogen-hydrogen and one fluorine-fluorine bond and the formation of two hydrogen-fluorine bonds; thus:

$$\Delta H^\circ_{rxn} \approx H(\text{H—H}) + H(\text{F—F}) - 2H(\text{H—F})$$

and

$$\begin{aligned}\Delta H^\circ_{rxn} &\approx (1\ \text{mol})(435\ \text{kJ}\cdot\text{mol}^{-1}) + (1\ \text{mol})(155\ \text{kJ}\cdot\text{mol}^{-1}) \\ &\quad - (2\ \text{mol})(565\ \text{kJ}\cdot\text{mol}^{-1}) \\ &\approx -540\ \text{kJ}\end{aligned}$$

The value of ΔH°_{rxn} we would obtain using data from Table 5-1 is -542.2 kJ, which is in nice agreement with what we have found here using bond energies.

Although the value of ΔH°_{rxn} is determined primarily by the difference in bond energies of the reactants and the products, interactions (attractive forces) *between* chemical species in the liquid and in solid phases (Chapter 11) can also make significant contributions to the value of ΔH°_{rxn}. As an example, consider the vaporization of liquid water at 25°C:

$$H_2O(l) \rightarrow H_2O(g) \qquad \Delta H^\circ_{rxn} = +44.0\ \text{kJ}$$

Note that no internal oxygen-hydrogen bonds are broken in this process. In the vaporization of water, it is the attractive forces *between* the water molecules that must be overcome. This is one of the reasons that ΔH°_{rxn} values obtained from bond enthalpies, rather than from ΔH°_f values, are only approximate. Another reason is that the bond energies given in Table 5-2 are *average* values.

5-6. HEAT CAPACITY MEASURES THE ABILITY OF A SUBSTANCE TO TAKE UP ENERGY AS HEAT

It is possible to measure heats of reactions experimentally. We shall see how this is done in Section 5-7, but first we must introduce a quantity called heat capacity.

The **heat capacity** of a substance is defined as the heat required to raise the temperature of the substance by one degree Celsius or by one kelvin. If the substance is heated at constant pressure, then the heat capacity is denoted by c_P, where the subscript P denotes constant pressure. In terms of an equation, we can write

▪ The heat capacity of a substance is always positive.

$$c_P = \frac{q_P}{\Delta T} \qquad (5\text{-}16)$$

where q_P is the energy added as heat and ΔT is the increase in temperature of the substance arising from the heat input. All that is necessary to determine the heat capacity of a substance is to add a known quantity of energy as heat and then measure the resulting increase in temperature. The following example illustrates the use of Equation (5-16).

Example 5-4: When 421.2 J of heat is added to 36.0 g of liquid water, the temperature of the water increases from 10.000°C to 12.800°C. Compute the heat capacity of the 36.0 g of $H_2O(l)$.

Solution: We start with Equation (5-16):

$$c_P = \frac{q_P}{\Delta T}$$

ΔT is equal to 12.800°C − 10.000°C = 2.800°C. Because a kelvin and a Celsius degree are the same size, then we have

$$c_P = \frac{421.2\ \text{J}}{2.800\ \text{K}} = 150.4\ \text{J·K}^{-1}$$

for the heat capacity of 36.0 g of $H_2O(l)$.

▪ The specific heat of a substance is defined as the heat capacity per gram of that substance.

We denote the heat capacity per mole, or the **molar heat capacity,** of any substance by C_P, where the capital C denotes a molar heat capacity. The heat capacity per mole of water can be computed from the c_P value of 150.4 J·K^{-1} calculated in Example 5-4 for 36.0 g of water. Since 1.00 mol of water has a mass of 18.0 g, 36.0 g contains 2.00 mol of water. Thus the value of C_P for liquid water is

$$C_P = \frac{150.4\ \text{J·K}^{-1}}{2.00\ \text{mol}} = 75.2\ \text{J·K}^{-1}\text{·mol}^{-1}$$

Table 5-3 Constant-pressure molar heat capacities

Formula	$C_P/\text{J·K}^{-1}\text{·mol}^{-1}$
$NH_3(g)$	35.1
$CO_2(g)$	37.1
$Cu(s)$	24.5
$C_2H_4(g)$	42.0
$He(g)$	20.8
$H_2(g)$	28.8
$H_2O(s)$	37.7
$H_2O(l)$	75.2
$Li(l)$	40.0
$Hg(l)$	28.0
$N_2(g)$	29.2
$O_2(g)$	29.4
$Na(l)$	30.8

Example 5-5: Compute the energy as heat required to raise the temperature of 151 kg of water (40 gal, the volume of a typical home water heater) from 18.0°C to 60.0°C, assuming no loss of energy to the surroundings.

Solution: The heat energy required can be computed from Equation (5-16):

$$\begin{aligned} q_P &= c_P\Delta T \\ &= (75.2\ \text{J·K}^{-1}\text{·mol}^{-1})(151\ \text{kg})\left(\frac{1000\ \text{g}}{1\ \text{kg}}\right)\left(\frac{1\ \text{mol}}{18.0\ \text{g}}\right)(333 - 291)\ \text{K} \\ &= 2.65 \times 10^7\ \text{J} = 26.5\ \text{MJ} \end{aligned}$$

where 1 MJ = 1 megajoule = 1×10^6 J. If natural gas provides the heat energy at a cost of about 80 cents per 100 MJ, it costs about 20 cents to heat 40 gal of water from 18°C to 60°C, assuming that all the heat goes into the water. In actual practice, only about half the heat is absorbed by the water; the remainder is lost to the surroundings, and so the actual cost is about 40 cents.

The molar heat capacities for a variety of substances are given in Table 5-3.

5-7. A CALORIMETER IS A DEVICE USED TO MEASURE THE AMOUNT OF HEAT EVOLVED OR ABSORBED IN A REACTION

The value of ΔH°_{rxn} for a chemical reaction can be measured in a device called a **calorimeter.** A simple calorimeter consisting of a **Dewar flask** ("thermos bottle") equipped with a high-precision thermometer is shown in Figure 5-7. A calorimeter works on the principle of energy conservation, which requires that the total energy is always conserved. Recall that the principle of energy conservation (first law of thermodynamics) states that energy cannot be created or destroyed but can only be transferred from one system to another.

For a chemical reaction occurring at fixed pressure, the value of ΔH_{rxn} is equal to the heat evolved or absorbed in the process (Equation 5-10):

$$\Delta H_{rxn} = q_P$$

Consider an exothermic reaction run in a Dewar flask. The heat evolved by the reaction cannot escape from the flask and thus is absorbed by the calorimeter contents (reaction mixture, thermometer, stirrer, and so on). The heat absorption leads to an increase in the temperature of the calorimeter contents. Because all the heat evolved by the reaction is absorbed by the calorimeter contents, we can write

$$\Delta H_{rxn} = -\Delta H_{\text{calorimeter}} \tag{5-17}$$

From the definition of heat capacity, Equation (5-16), we have

$$q_P = \Delta H_{\text{calorimeter}} = c_{P,\ \text{calorimeter}}\ \Delta T \tag{5-18}$$

where ΔT is the observed temperature change. Substitution of Equation (5-18) into Equation (5-17) yields

$$\Delta H_{rxn} = -c_{P,\ \text{calorimeter}}\ \Delta T \tag{5-19}$$

Equation (5-19) tells us that if we run a chemical reaction in a calorimeter with a known heat capacity ($c_{P,\ \text{calorimeter}}$) and if we determine the temperature change, then the value of ΔH_{rxn} can be computed. The value of $c_{P,\ \text{calorimeter}}$ can be determined by electrical resistance heating when the reaction has gone to completion. The value of ΔH_{rxn} for reactions involving solutions depends on concentration, but this effect is usually not large. We shall ignore the effect of concentration on ΔH_{rxn} values and assume that $\Delta H_{rxn} = \Delta H^\circ_{rxn}$.

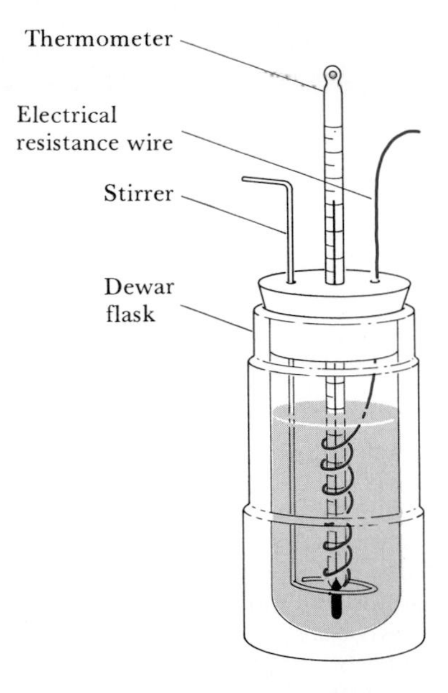

Figure 5-7 A simple calorimeter, consisting of a Dewar flask with cover, which prevents a significant loss or gain of heat from the surroundings; a high-precision thermometer, which gives the temperature to within ± 0.001 K; a simple ring-type stirrer; and an electrical resistance heater. One reactant is placed in the Dewar flask, and then the other reactant, at the same temperature, is added. The reaction mixture is stirred and the change in temperature is measured.

Example 5-6: A 0.500-L sample of 0.200 M NaCl(*aq*) is added to 0.500 L of 0.200 M $AgNO_3(aq)$ in a calorimeter with $c_P = 4.60 \times 10^3\ \text{J}\cdot\text{K}^{-1}$. The observed ΔT is +1.423 K. Compute the value of ΔH°_{rxn} for the equation

$$AgNO_3(aq) + NaCl(aq) \rightarrow AgCl(s) + NaNO_3(aq)$$

Solution: The observed increase in temperature arises from the formation of the precipitate AgCl(*s*). The number of moles of AgCl(*s*) formed in the reaction is given by:

$$\text{mol AgCl}(s)\ \text{formed} = (0.500\ \text{L})(0.200\ \text{mol}\cdot\text{L}^{-1}) = 0.100\ \text{mol}$$

The value of ΔH°_{rxn} for the formation of 0.100 mol of AgCl(*s*) is

▪ As you may know from physics, when an electric current, *I*, flows through a resistance, *R*, for a time, *t*, the amount of heat produced is given by I^2Rt.

$$\Delta H^\circ_{rxn} = -c_{P,\ \text{calorimeter}}\ \Delta T$$
$$= -(4.60 \times 10^3\ \text{J·K}^{-1})(1.423\ \text{K}) = -6550\ \text{J}$$

For the formation of 1.00 mol of AgCl(s), the value of ΔH°_{rxn} is 10 times the value for 0.100 mol; thus

$$\Delta H^\circ_{rxn} = \left(-\frac{6550\ \text{kJ}}{0.100\ \text{mol}}\right)(1.00\ \text{mol}) = -65500\ \text{J} = -65.5\ \text{kJ}$$

5-8. COMBUSTION REACTIONS ARE USED AS ENERGY SOURCES

The heat of combustion of a substance, which is a special case of the heat of reaction, can be determined in a **bomb calorimeter** like that shown in Figure 5-8. A known mass of the substance whose heat of combustion is to be determined is loaded into the bomb calorimeter along with an ignition wire. The calorimeter is then pressurized with excess oxygen gas at about 30 atm. The combustion reaction is then started by passing a short burst of high-voltage current through the ignition wire. The heat of combustion is determined by measuring the temperature increase of the calorimeter and of the water in which it is immersed. From the known heat capacity of the calorimeter assembly and the observed ΔT, the heat of combustion can be computed in a manner analogous to that described for the Dewar-flask calorimeter.

A bomb calorimeter like that shown in Figure 5-8 has a fixed total volume, and thus the reaction takes place at constant volume. When the volume is constant, no work is done, and Equation (5-6) applies:

$$\Delta U_{rxn} = q_V = \begin{pmatrix}\text{heat evolved or absorbed}\\ \text{at constant volume}\end{pmatrix}$$

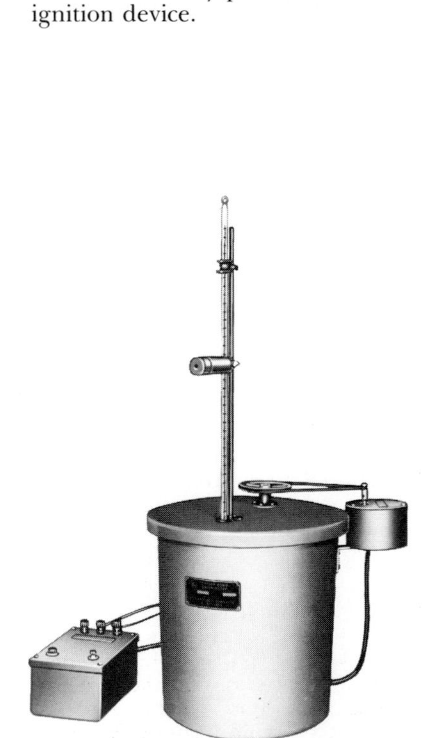

(a)

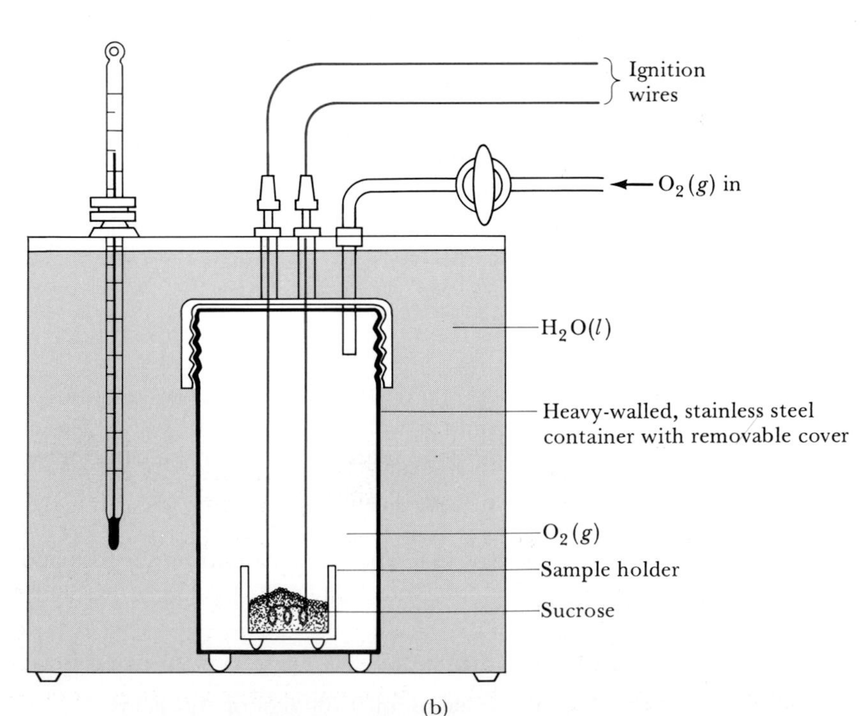

(b)

Figure 5-8 (a) A bomb calorimeter. (b) Cross section of a bomb calorimeter. The container ("bomb") in which the combustion reaction occurs is placed in water within the large container, as shown here. The thermometer, which is equipped with a small telescopic eyepiece for more precise reading, sticks through the top of the outer container and into the water surrounding the bomb. The device attached to the drive belt is the stirrer motor. The stirrer blades are immersed in the water. The small box to the left is the battery-powered ignition device.

The energy evolved or absorbed as heat in a reaction run in a sealed, rigid container is equal to ΔU_{rxn}, the energy change for the reaction.

The difference between a constant-pressure process and a constant-volume process can be seen by considering a reaction in which the volume of the reaction mixture increases during the course of the reaction. In this case, the mixture has to expend some energy to push back the surrounding atmosphere as it expands. Consequently, the energy evolved or absorbed as heat in a constant-pressure reaction is not the same as that for the same reaction run at constant volume. The difference between ΔU_{rxn} and ΔH_{rxn} is just the energy that is required for the system to expand against a constant atmospheric pressure. It turns out that the numerical difference between ΔH_{rxn} and ΔU_{rxn} is usually small, and we shall assume that $\Delta H_{rxn} \approx \Delta U_{rxn}$ and that $\Delta H^\circ_{rxn} \approx \Delta U^\circ_{rxn}$.

Example 5-7: A 1.00-g sample of octane, $C_8H_{18}(l)$, is burned in a calorimeter like that shown in Figure 5-8, and the observed temperature increase is 1.837 K. The heat capacity of the calorimeter assembly is $c_{calorimeter} = 28.46\ kJ\cdot K^{-1}$. Compute the heat of combustion per gram and per mole of $C_8H_{18}(l)$. Assume that $\Delta H^\circ_{rxn} \approx \Delta U^\circ_{rxn}$.

Solution: The equation for the combustion reaction is

$$C_8H_{18}(l) + \tfrac{25}{2}O_2(g) \rightarrow 8CO_2(g) + 9H_2O(l)$$

The heat of combustion is

$$\Delta H^\circ_{rxn} \approx \Delta U^\circ_{rxn} = -c_{calorimeter}\ \Delta T$$

Using this equation, we find that the value of ΔH°_{rxn} for 1.00 g of $C_8H_{18}(l)$ is

$$\Delta H^\circ_{rxn}[\text{per g } C_8H_{18}(l)] = -(28.46\ kJ\cdot K^{-1})(1.837\ K\cdot g^{-1})$$
$$= -52.28\ kJ\cdot g^{-1}$$

because the observed ΔT is for the combustion of 1.00 g of octane. The value of ΔH°_{rxn} per mole of octane (molecular mass = 114.2) is

$$\Delta H^\circ_{rxn}[\text{per mol } C_8H_{18}(l)] = -(52.28\ kJ\cdot g^{-1})(114.2\ g\cdot mol^{-1})$$
$$= -5970\ kJ\cdot mol^{-1}$$

Combustible fuels are used extensively as thermal energy sources (Interchapter E).

5-9. FOOD IS FUEL

The food we eat constitutes the fuel necessary to maintain body temperature and accomplish other physiological functions and to provide the energy we need to move about. To maintain body weight, a normally active, healthy adult must take in about 130 kJ of food energy per kilogram of body weight per day. The common unit for the energy content of food is the **calorie,** a term you undoubtedly have heard of. The calorie used to be the unit of energy for scientific work, but it is slowly being replaced by the joule. Nevertheless, nutritionists, physicians, and the popular press still use the term calorie, and so it is neces-

▪ A calorie is defined as the amount of heat required to raise the temperature of one gram of water from 14.5°C to 15.5°C (one degree Celsius).

sary to be able to convert from one unit to another. There are 4.184 J in 1 cal, and so the unit conversion factor is

$$\frac{4.184\ \mathrm{J}}{1\ \mathrm{cal}} = 1$$

In this section we express all energy values in kilojoules and kilocalories. Incidentally, the popular term calorie is actually a kilocalorie and is sometimes written Calorie. The total daily energy intakes required to maintain various body weights are

Body weight	**Required daily energy input**	
110 lb (50 kg)	6600 kJ	1600 kcal
175 lb (80 kg)	10,600 kJ	2500 kcal
250 lb (114 kg)	15,000 kJ	3600 kcal

About 100 kJ per kilogram of body weight per day is required to keep the body functioning at a minimal level. If we consume more food than we require for our normal activity level, then the excess that is not eliminated is stored in the body as fat. One gram of body fat contains about 39 kJ of stored energy. The per-gram chemical energy values for fats, proteins, and carbohydrates are

Food	**Energy value**	
fats	$39\ \mathrm{kJ \cdot g^{-1}}$	$9.3\ \mathrm{kcal \cdot g^{-1}}$
proteins	$17\ \mathrm{kJ \cdot g^{-1}}$	$4.1\ \mathrm{kcal \cdot g^{-1}}$
carbohydrates	$16\ \mathrm{kJ \cdot g^{-1}}$	$3.8\ \mathrm{kcal \cdot g^{-1}}$

Note that carbohydrates and proteins have less than one half the energy content of fats.

Exercise is great for improving muscle tone and thereby firming up sagging tissue, but exercise is not an effective way to lose weight. A one-hour brisk walk over average terrain consumes only about 700 kJ of stored energy, which corresponds to about one fourth of the energy content of a quarter-pound hamburger on a roll. The energy values of some common foods are given in Table 5-4. The message is that calories (or better, kilojoules) count.

SUMMARY

A chemical reaction involves a transfer of energy between the system of interest and its surroundings. Commonly, the energy is either absorbed or released as heat. Reactions that give off energy as heat are called exothermic reactions, and reactions that take in energy as heat are called endothermic reactions. The total energy involved is always conserved; that is, energy is neither created nor destroyed but is simply transferred from one system to another. This principle of energy conservation is known as the first law of thermodynamics. The study

Table 5-4 Approximate energy values of some common foods

Food	kJ/100 g	kcal/4 oz†
green vegetables*	115	31
beer	200	54
fruits*	250	67
milk	300	80
seafood (steamed)*	400	107
cottage cheese, low-fat yogurt	450	120
chicken (broiled, meat only)	600	160
yogurt	1000	270
liquor (80 proof)	1000	270
ice cream (bulk)	1100	295
bread, cheese*	1200	320
steak (broiled, no fat)	1400	350
hamburger, hot dogs, popcorn, sugar	1600	430
potato chips, nuts (roasted)	2400	640
butter, cream, margarine, mayonnaise	3000	800
fat	3900	1045

*Average values for the more common varieties; variations within categories are as much as 10 to 15 percent.
†1 kcal = 4.184 kJ; 1 oz = 28 g.

of energy changes is called thermodynamics, and thermochemistry is the branch of thermodynamics that deals with the heat evolved or absorbed in chemical reactions.

For a constant-pressure process, the heat evolved or absorbed, q_P, is equal to the enthalpy change; thus $\Delta H_{rxn} = q_P$ for a reaction run at constant pressure. For a reaction run at constant volume, the heat evolved or absorbed, q_V, is equal to the change in energy for the reaction; thus $\Delta U_{rxn} = q_V$.

The standard molar enthalpy of formation of a compound from its elements, ΔH_f°, is determined by assigning a value of zero to the standard enthalpies of formation of the normal state of each element at 25°C and 1 atm. A table of ΔH_f° values can be used to compute ΔH_{rxn}° values. The value of ΔH_{rxn}° for a particular reaction is determined primarily by the difference in the bond energies of the reactant and the product molecules. The breaking of chemical bonds in the reactant molecules consumes energy, and the formation of chemical bonds in the product molecules releases energy. The difference between these values is ΔH_{rxn}°.

The heat capacity of a substance is a measure of its capacity to take up energy as heat. The higher the heat capacity of a substance, the smaller the resulting temperature increase for a given amount of heat energy added. A reactant used to provide heat energy is called a fuel. Combustion reactions are those in which a fuel burns in oxygen. The amount of energy absorbed or evolved as heat by a chemical reaction can be measured in a calorimeter.

TERMS YOU SHOULD KNOW

thermodynamics 164
first law of thermodynamics 164
conservation of energy 164
work, w 164
heat, q 165
energy change, ΔU_{rxn} 165
enthalpy 165
enthalpy change, ΔH_{rxn} 165
heat of reaction, ΔH_{rxn} 166
thermochemistry 166
fuel 166
exothermic reaction 166
endothermic reaction 166
standard enthalpy change, ΔH_{rxn}° 167
water-gas reaction 167
Hess's law 168

standard molar enthalpy of formation, ΔH_f° 170
heat of formation 170
molar bond enthalpy, H(bond) 174
bond energy 174
heat capacity, c_P 175
molar heat capacity, C_P 176
specific heat, c_{sp} 176
calorimeter 177
Dewar flask 177
bomb calorimeter 178
calorie 179

EQUATIONS YOU SHOULD KNOW HOW TO USE

$w = -P\Delta V$	(constant pressure)	(5-3)	(expansion or compression work)
$\Delta U_{rxn} = q + w$		(5-4)	(first law of thermodynamics)
$\Delta U_{rxn} = q_V$	(constant volume)	(5-6)	(heat transferred at constant volume)
$\Delta H_{rxn} = \Delta U_{rxn} + P\Delta V$	(constant pressure)	(5-9)	(relation between ΔH_{rxn} and ΔU_{rxn})
$\Delta H_{rxn} = q_P$	(constant pressure)	(5-10)	(heat transferred at constant pressure)
$\Delta H^\circ_{rxn}(3) = \Delta H^\circ_{rxn}(1) + \Delta H^\circ_{rxn}(2)$		(5-12)	(Hess's law)
$\Delta H^\circ_{rxn}(\text{reverse rxn}) = -\Delta H^\circ_{rxn}(\text{forward rxn})$		(5-13)	(a form of Hess's law)

For the reaction $a\text{A} + b\text{B} \rightarrow y\text{Y} + z\text{Z}$,

$\Delta H^\circ_{rxn} = y\Delta H_f^\circ[\text{Y}] + z\Delta H_f^\circ[\text{Z}] - a\Delta H_f^\circ[\text{A}] - b\Delta H_f^\circ[\text{B}]$	(5-14)	(ΔH°_{rxn} in terms of standard molar heats of formation)
$\Delta H^\circ_{rxn} \approx H(\text{bond})_R - H(\text{bond})_P$	(5-15)	(relation between ΔH°_{rxn} and bond enthalpies)
$c_P = \dfrac{q_P}{\Delta T}$	(5-16)	(definition of heat capacity)
$\Delta H_{rxn} = -c_{P,\ \text{calorimeter}}\,\Delta T$	(5-19)	(calorimetric determination of ΔH°_{rxn})

PROBLEMS

HEAT AND ENERGY

5-1. When 30.0 g of methane burns in oxygen, 1503 kJ of heat is evolved. Calculate the amount of heat (in kilojoules) evolved when 1.00 mol of methane burns.

5-2. When 2.46 g of barium reacts with chlorine, 15.4 kJ is evolved. Calculate the heat evolved (in kilojoules) when 1.00 mol of barium chloride is formed from barium and chlorine.

5-3. When 1.280 g of carbon reacts with sulfur to give carbon disulfide, 9.52 kJ is absorbed. Calculate the heat absorbed (in kilojoules) when 1.00 mol of carbon disulfide is formed from carbon and sulfur.

5-4. When 0.165 g of magnesium is burned in oxygen, 4.08 kJ is evolved. Calculate the heat evolved (in kilojoules) when 1.00 mol of magnesium oxide is formed from magnesium and oxygen.

HESS'S LAW

5-5. The ΔH°_{rxn} value for the equation

$$2\text{ZnS}(s) + \text{O}_2(g) \rightarrow 2\text{ZnO}(s) + 2\text{S}(s)$$

is -290.8 kJ. Compute the value of ΔH°_{rxn} for the equation

$$\text{ZnO}(s) + \text{S}(s) \rightarrow \text{ZnS}(s) + \tfrac{1}{2}\text{O}_2(g)$$

5-6. The ΔH°_{rxn} value for the equation

$$\text{CaO}(s) + \text{H}_2\text{O}(l) \rightarrow \text{Ca(OH)}_2(s)$$

is -56.27 kJ. Compute the amount of heat (in kilojoules) required to convert 1.00 g of $\text{Ca(OH)}_2(s)$ to $\text{CaO}(s)$.

5-7. Given that

$$\text{C}_2\text{H}_5\text{OH}(l) + 3\text{O}_2(g) \rightarrow 2\text{CO}_2(g) + 3\text{H}_2\text{O}(g) \qquad \Delta H^\circ_{rxn} = -1234.7\ \text{kJ}$$

$$\text{CH}_3\text{OCH}_3(l) + 3\text{O}_2(g) \rightarrow 2\text{CO}_2(g) + 3\text{H}_2\text{O}(g) \qquad \Delta H^\circ_{rxn} = -1328.3\ \text{kJ}$$

calculate ΔH°_{rxn} for the equation

$$\text{C}_2\text{H}_5\text{OH}(l) \rightarrow \text{CH}_3\text{OCH}_3(l)$$

5-8. Use the values of ΔH°_{rxn} given for these equations

$$\text{Cu}(s) + \text{Cl}_2(g) \rightarrow \text{CuCl}_2(s) \qquad \Delta H^\circ_{rxn} = -206\ \text{kJ}$$

$$2\text{Cu}(s) + \text{Cl}_2(g) \rightarrow 2\text{CuCl}(s) \qquad \Delta H^\circ_{rxn} = -136\ \text{kJ}$$

to calculate ΔH°_{rxn} for the equation

$$CuCl_2(s) + Cu(s) \rightarrow 2CuCl(s)$$

5-9. The ΔH°_{rxn} values for the following equations are

$$2Fe(s) + \tfrac{3}{2}O_2(g) \rightarrow Fe_2O_3(s) \quad \Delta H^\circ_{rxn} = -823.41 \text{ kJ}$$

$$3Fe(s) + 2O_2(g) \rightarrow Fe_3O_4(s) \quad \Delta H^\circ_{rxn} = -1120.48 \text{ kJ}$$

Use these data to compute ΔH°_{rxn} for the equation

$$3Fe_2O_3(s) \rightarrow 2Fe_3O_4(s) + \tfrac{1}{2}O_2(g)$$

5-10. Given that

$$H_2(g) + F_2(g) \rightarrow 2HF(g) \quad \Delta H^\circ_{rxn} = -542.2 \text{ kJ}$$

$$2H_2(g) + O_2(g) \rightarrow 2H_2O(l) \quad \Delta H^\circ_{rxn} = -571.6 \text{ kJ}$$

calculate ΔH°_{rxn} for

$$2F_2(g) + 2H_2O(l) \rightarrow 4HF(g) + O_2(g)$$

5-11. The standard molar heats of combustion at 25°C for sucrose, glucose, and fructose are

Compound	ΔH°_{rxn}/kJ·mol^{-1}
$C_{12}H_{22}O_{11}$, sucrose	−5646.7
$C_6H_{12}O_6$, glucose	−2815.8
$C_6H_{12}O_6$, fructose	−2826.7

Use Hess's law to compute ΔH°_{rxn} at 25°C for the equation

$$\underset{\text{sucrose}}{C_{12}H_{22}O_{11}(s)} + H_2O(l) \rightarrow \underset{\text{glucose}}{C_6H_{12}O_6(s)} + \underset{\text{fructose}}{C_6H_{12}O_6(s)}$$

5-12. The standard molar heats of combustion of the isomers *m*-xylene and *p*-xylene, both $(CH_3)_2C_6H_4$, are -4553.9 kJ·mol^{-1} and -4556.8 kJ·mol^{-1}, respectively. Use these data, together with Hess's law, to compute ΔH°_{rxn} for the reaction

$$m\text{-xylene} \rightarrow p\text{-xylene}$$

5-13. Given that

$$4NH_3(g) + 5O_2(g) \rightarrow 4NO(g) + 6H_2O(l) \quad \Delta H^\circ_{rxn} = -1170 \text{ kJ}$$

$$4NH_3(g) + 3O_2(g) \rightarrow 2N_2(g) + 6H_2O(l) \quad \Delta H^\circ_{rxn} = -1530 \text{ kJ}$$

calculate ΔH°_{rxn} for the equation

$$N_2(g) + O_2(g) \rightarrow 2NO(g)$$

5-14. Given that

$$Xe(g) + F_2(g) \rightarrow XeF_2(s) \quad \Delta H^\circ_{rxn} = -123 \text{ kJ}$$

$$Xe(g) + 2F_2(g) \rightarrow XeF_4(s) \quad \Delta H^\circ_{rxn} = -262 \text{ kJ}$$

calculate ΔH°_{rxn} for the equation

$$XeF_2(s) + F_2(g) \rightarrow XeF_4(s)$$

5-15. Given that $\Delta H^\circ_f = 142$ kJ·mol^{-1} for $O_3(g)$ and $\Delta H^\circ_f = 247.5$ kJ·mol^{-1} for $O(g)$, calculate ΔH°_{rxn} for the equation

$$O_2(g) + O(g) \rightarrow O_3(g)$$

This is one of the reactions that produce ozone in the atmosphere.

5-16. The ΔH°_f values for $Cu_2O(s)$ and $CuO(s)$ are -169.0 kJ·mol^{-1} and -157.3 kJ·mol^{-1}, respectively. Compute ΔH°_{rxn} for the equation

$$CuO(s) + Cu(s) \rightarrow Cu_2O(s)$$

5-17. Use the ΔH°_f data in Table 5-1 to compute ΔH°_{rxn} for the following equations:

(a) $N_2H_4(l) + O_2(g) \rightarrow N_2(g) + 2H_2O(g)$
(b) $C_2H_4(g) + H_2O(l) \rightarrow C_2H_5OH(l)$
(c) $CH_4(g) + 4Cl_2(g) \rightarrow CCl_4(l) + 4HCl(g)$

In each case state whether the reaction is endothermic or exothermic.

5-18. Use the ΔH°_f data in Table 5-1 to compute ΔH°_{rxn} for the following equations:

(a) $2H_2O_2(l) \rightarrow 2H_2O(l) + O_2(g)$
(b) $MgO(s) + CO_2(g) \rightarrow MgCO_3(s)$
(c) $4NH_3(g) + 5O_2(g) \rightarrow 4NO(g) + 6H_2O(g)$

In each case state whether the reaction is endothermic or exothermic.

5-19. Use the ΔH°_f data in Table 5-1 to compute ΔH°_{rxn} for the following combustion reactions:

(a) $C_2H_5OH(l) + 3O_2(g) \rightarrow 2CO_2(g) + 3H_2O(l)$
(b) $C_2H_6(g) + \tfrac{7}{2}O_2(g) \rightarrow 2CO_2(g) + 3H_2O(l)$

Compare the heat of combustion per gram of the fuels $C_2H_5OH(l)$ and $C_2H_6(g)$.

5-20. Use the ΔH°_f data in Table 5-1 to compute ΔH°_{rxn} for the following combustion reactions:

(a) $CH_3OH(l) + \tfrac{3}{2}O_2(g) \rightarrow CO_2(g) + 2H_2O(l)$
(b) $N_2H_4(l) + O_2(g) \rightarrow N_2(g) + 2H_2O(l)$

Compare the heat of combustion per gram of the fuels $CH_3OH(l)$ and $N_2H_4(l)$.

5-21. Given that $\Delta H^\circ_{rxn} = -2826.7$ kJ for the combustion of 1.00 mol of fructose:

$$C_6H_{12}O_6(s) + 6O_2(g) \rightarrow 6CO_2(g) + 6H_2O(l)$$

use the ΔH°_f data in Table 5-1 together with the given ΔH°_{rxn} value to compute the value of ΔH°_f for fructose.

5-22. Use the fact that $\Delta H^\circ_{rxn} = -5646.7$ kJ for the combustion of 1.00 mol of sucrose,

$$C_{12}H_{22}O_{11}(s) + 12O_2(g) \rightarrow 12CO_2(g) + 11H_2O(l)$$

plus the ΔH_f° data in Table 5-1 to compute ΔH_f° for sucrose.

5-23. Calculate ΔH_f° for the atomic species for each of the following:

(a) $N_2(g) \rightarrow 2N(g)$ $\qquad \Delta H_{rxn}^\circ = +945.2\ \text{kJ}$
(b) $F_2(g) \rightarrow 2F(g)$ $\qquad \Delta H_{rxn}^\circ = +158.0\ \text{kJ}$
(c) $H_2(g) \rightarrow 2H(g)$ $\qquad \Delta H_{rxn}^\circ = +436.0\ \text{kJ}$
(d) $Cl_2(g) \rightarrow 2Cl(g)$ $\qquad \Delta H_{rxn}^\circ = +243.4\ \text{kJ}$

Which of these diatomic molecules has the greatest bond strength?

5-24. Using Table 5-1, calculate ΔH_{rxn}° for

(a) $H_2(g) + F_2(g) \rightarrow 2HF(g)$
(b) $2CO(g) + O_2(g) \rightarrow 2CO_2(g)$
(c) $3H_2(g) + N_2(g) \rightarrow 2NH_3(g)$
(d) $2NO(g) + O_2(g) \rightarrow 2NO_2(g)$

State whether each reaction is endothermic or exothermic.

5-25. Using Table 5-1, calculate the heat required to vaporize 1.00 mol of $CCl_4(l)$ at 25°C.

5-26. Using Table 5-1, calculate the heat required to vaporize 1.00 mol of water at 25°C.

BOND ENTHALPIES

5-27. The enthalpy change for the equation

$$ClF_3(g) \rightarrow Cl(g) + 3F(g)$$

is 514 kJ. Calculate the average chlorine-fluorine bond energy in ClF_3.

The bonding in ClF_3 is

$$\begin{array}{c} \text{F—Cl—F} \\ | \\ \text{F} \end{array}$$

5-28. The enthalpy change for the equation

$$OF_2(g) \rightarrow O(g) + 2F(g)$$

is 368 kJ. Calculate the average oxygen-fluorine bond energy of OF_2.

The bonding in OF_2 is F—O—F.

5-29. Use the bond enthalpy data in Table 5-2 to estimate the ΔH_{rxn}° for the equation

$$CCl_4(g) + 2F_2(g) \rightarrow CF_4(g) + 2Cl_2(g)$$

The bonding in CF_4 and CCl_4 is

$$\begin{array}{c} \text{F} \\ | \\ \text{F—C—F} \\ | \\ \text{F} \end{array} \qquad \begin{array}{c} \text{Cl} \\ | \\ \text{Cl—C—Cl} \\ | \\ \text{Cl} \end{array}$$

5-30. Use the bond enthalpy data in Table 5-2 to estimate ΔH_{rxn}° for the reaction

$$CH_3OH(g) + F_2(g) \rightarrow FCH_2OH(g) + HF(g)$$

The bonding in CH_3OH and FCH_2OH is

$$\begin{array}{c} \text{O—H} \\ | \\ \text{H—C—H} \\ | \\ \text{H} \end{array} \qquad \begin{array}{c} \text{O—H} \\ | \\ \text{H—C—H} \\ | \\ \text{F} \end{array}$$

5-31. The formation of water from oxygen and hydrogen involves the reaction

$$2H_2(g) + O_2(g) \rightarrow 2H_2O(g)$$

Use the bond energies given in Table 5-2 and the ΔH_f° value for $H_2O(g)$ given in Table 5-1 to compute the oxygen-oxygen bond energy in $O_2(g)$.

The bonding in H_2O is H—O—H.

5-32. The formation of ammonia from hydrogen and nitrogen involves the reaction

$$N_2(g) + 3H_2(g) \rightarrow 2NH_3(g)$$

Use the bond energies given in Table 5-2 and the ΔH_f° value for $NH_3(g)$ given in Table 5-1 to compute the bond energy in $N_2(g)$.

The bonding in NH_3 is

$$\begin{array}{c} \text{H—N—H.} \\ | \\ \text{H} \end{array}$$

5-33. Given that

$$\Delta H_f^\circ[H(g)] = 218\ \text{kJ·mol}^{-1}$$
$$\Delta H_f^\circ[C(g)] = 709\ \text{kJ·mol}^{-1}$$
$$\Delta H_f^\circ[CH_4(g)] = -74.86\ \text{kJ·mol}^{-1}$$

calculate the average carbon-hydrogen bond energy in CH_4.

The bonding in CH_4 is

$$\begin{array}{c} \text{H} \\ | \\ \text{H—C—H} \\ | \\ \text{H} \end{array}$$

5-34. Given that

$$\Delta H_f^\circ[Cl(g)] = 128\ \text{kJ·mol}^{-1}$$
$$\Delta H_f^\circ[C(g)] = 709\ \text{kJ·mol}^{-1}$$
$$\Delta H_f^\circ[CCl_4(g)] = -103\ \text{kJ·mol}^{-1}$$

calculate the average carbon-chlorine bond energy in CCl_4.

The bonding in CCl_4 is

$$\begin{array}{c} \text{Cl} \\ | \\ \text{Cl—C—Cl} \\ | \\ \text{Cl} \end{array}$$

HEAT CAPACITY

5-35. When 1105 J of heat is added to 36.5 g of ethyl alcohol, C_2H_5OH, the temperature increases by 12.3°C. Compute the molar heat capacity of ethyl alcohol.

5-36. When 285 J of heat is added to 33.6 g of hexane, C_6H_{14}, a component of gasoline, the temperature rises from 25.00°C to 28.74°C. Calculate the molar heat capacity of hexane.

5-37. A 10.0-kg sample of liquid water is used to cool an engine. Calculate the heat removed (in joules) from the engine when the temperature of the water is raised from 20.0°C to 100.0°C. Take $C_P = 75.2\ \text{J·K}^{-1}\text{·mol}^{-1}$ for $H_2O(l)$.

5-38. Liquid sodium is being considered as an engine coolant. How many grams of sodium are needed to absorb 1.00 MJ of heat if the temperature of the sodium is not to increase by more than 10°C? Take $C_P = 30.8\ \text{J·K}^{-1}\text{·mol}^{-1}$ for $Na(l)$.

5-39. A 25.0-g sample of copper at 90.0°C is placed in 100.0 g of water at 20.0°C. The copper and water quickly come to the same temperature by the process of heat transfer from water to copper. Calculate the final temperature of the water in °C. The heat capacity of copper is $24.5\ \text{J·K}^{-1}\text{·mol}^{-1}$.

5-40. If a 50.0-g piece of copper is heated to 100°C and then put into a vessel containing 250 mL of water at 0°C, what will be the final temperature (in °C) of the water? The heat capacity of copper is $24.5\ \text{J·K}^{-1}\text{·mol}^{-1}$.

5-41. A 1.00-kg block of aluminum metal ($C_P = 24.2\ \text{J·K}^{-1}\text{·mol}^{-1}$) at 500°C is placed in contact with a 1.00-kg block of copper ($C_P = 24.5\ \text{J·K}^{-1}\text{·mol}^{-1}$) at 10°C. What will be the final temperature (in °C) of the two blocks? Assume that no heat is lost to the surroundings.

5-42. A 50.0-g sample of a metal alloy at 25.0°C is placed in 99.9 g of water at 55.0°C. The final temperature of the water and metal is 48.1°C. Calculate the heat capacity per gram of the alloy.

CALORIMETRY

5-43. A 100-mL sample of 0.200 M aqueous hydrochloric acid is added to 100 mL of 0.200 M aqueous ammonia in a calorimeter whose heat capacity is $480\ \text{J·K}^{-1}$. The temperature increase is 2.34 K. Calculate ΔH°_{rxn} (in kJ) for the reaction

$$HCl(aq) + NH_3(aq) \rightarrow NH_4Cl(aq)$$

which occurs when the two solutions are mixed.

5-44. A 0.0500-L sample of 0.500 M barium nitrate is added to 0.0500 L of 0.500 M magnesium sulfate in a calorimeter with a heat capacity of $455\ \text{J·K}^{-1}$. The observed increase in temperature is 1.43 K. Calculate ΔH°_{rxn} (in kJ) for the reaction

$$Ba(NO_3)_2(aq) + MgSO_4(aq) \rightarrow BaSO_4(s) + Mg(NO_3)_2(aq)$$

which occurs when the two solutions are mixed.

5-45. Under the right conditions, such as high temperature, ammonium nitrate is an explosive, decomposing according to the equation

$$2NH_4NO_3(s) \rightarrow 2N_2(g) + 4H_2O(g) + O_2(g)$$

A 1.00-g sample of NH_4NO_3 is detonated in a calorimeter whose heat capacity is $4.92\ \text{kJ·K}^{-1}$. The temperature increase is 0.300 K. Calculate the heat of reaction for the decomposition of 1.00 kg of ammonium nitrate.

5-46. Calcium hydroxide is prepared by adding calcium oxide (lime) to water. It is important to know how much heat is evolved in order to provide for adequate cooling. A 10.0-g sample of calcium oxide is added to 1.00 L of water in a calorimeter whose heat capacity is $4.37\ \text{kJ·K}^{-1}$. The observed increase in temperature is 2.70 K. Calculate ΔH°_{rxn} for the formation of 1.00 mol of $Ca(OH)_2$. The equation for the reaction is

$$CaO(s) + H_2O(l) \rightarrow Ca(OH)_2(s)$$

5-47. A 5.00-g sample of potassium chloride is dissolved in 1.00 L of water in a calorimeter whose heat capacity is $4.51\ \text{kJ·K}^{-1}$. The temperature decreases 0.256 K. Calculate the molar heat of solution of potassium chloride.

5-48. A 5.00-g sample of nitric acid is dissolved in 1.00 L of water in a calorimeter whose heat capacity is $5.16\ \text{kJ·K}^{-1}$. The temperature increases 0.511 K. Calculate the molar heat of solution of nitric acid in water.

COMBUSTION OF FUELS

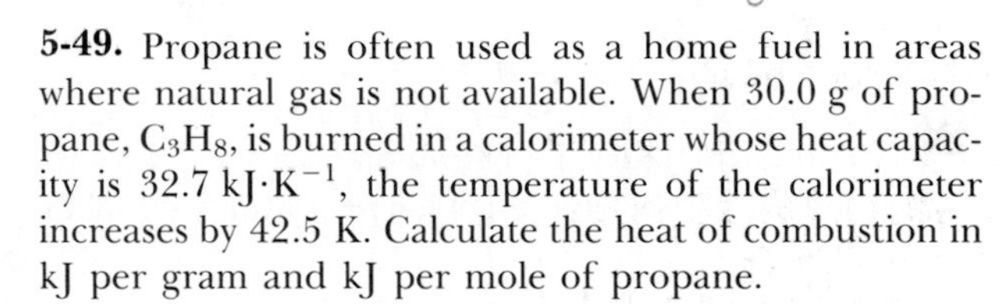

5-49. Propane is often used as a home fuel in areas where natural gas is not available. When 30.0 g of propane, C_3H_8, is burned in a calorimeter whose heat capacity is $32.7\ \text{kJ·K}^{-1}$, the temperature of the calorimeter increases by 42.5 K. Calculate the heat of combustion in kJ per gram and kJ per mole of propane.

5-50. Fructose, a sugar found in fruits, is a source of energy for the body. The combustion of fructose takes place according to the reaction

$$C_6H_{12}O_6(s) + 6O_2(g) \rightarrow 6CO_2(g) + 6H_2O(l)$$

When 5.00 g of fructose is burned in a calorimeter with a heat capacity of $29.7\ \text{kJ·K}^{-1}$, the temperature of the calorimeter increases by 2.635 K. Calculate the heat of combustion per gram and per mole of fructose. How much heat (in kJ) is released when 1.00 g of fructose is converted to $CO_2(s)$ and $H_2O(l)$ in the body?

5-51. When 2.50 g of oxalic acid, $H_2C_2O_4$, is burned in a calorimeter whose heat capacity is $8.75\ \text{kJ·K}^{-1}$, the temperature increases 0.780 K. Calculate the heat of combustion in kilojoules per mole of oxalic acid. Using the ΔH_f° values for $CO_2(g)$ and $H_2O(l)$ given in Table 5-1, calculate the heat of formation of oxalic acid at 25°C.

5-52. When 2.62 g of lactic acid, $C_3H_6O_3$, is burned in a calorimeter whose heat capacity is $21.7\ kJ \cdot K^{-1}$, the temperature of the calorimeter increases 1.800 K. Calculate the heat of combustion (in kJ) per mole of lactic acid. Using the ΔH_f° values for $CO_2(g)$ and $H_2O(l)$ given in Table 5-1, calculate ΔH_f° for lactic acid. Lactic acid is produced in muscle when there is a shortage of oxygen, such as during vigorous exercise. A buildup of lactic acid is responsible for muscle cramps.

ADDITIONAL PROBLEMS

5-53. One proposal for an effortless method of losing weight is to drink large amounts of cold water. Water has no food value. The body must provide heat in order to bring the temperature of the water to body temperature, 37°C. This heat is provided by the burning of stored carbohydrates or fat. How much heat must the body provide to warm 1.0 L of water at 0°C to 37°C? How many grams of body fat must be burned to provide this heat? Compute the amount of ice at 0°C that must be consumed to produce the same effect. Take the heat of fusion (melting) of ice to be $6.0\ kJ \cdot mol^{-1}$.

5-54. Glucose is used as fuel in the body according to the reaction

$$C_6H_{12}O_6(aq) + 6O_2(g) \rightarrow 6CO_2(g) + 6H_2O(l) \qquad \Delta H^\circ_{rxn} = -2820\ kJ \cdot mol^{-1}$$

How many grams of glucose must be burned to raise the temperature of the body from 34°C to 37°C for an 82-kg person? Assume that all the heat of combustion is used to heat the body. Assume also that the heat capacity per gram (specific heat) of the body is that of water.

5-55. The French chemists Pierre L. Dulong and Alexis T. Petit noted in 1819 that the molar heat capacity of many solids at ordinary temperatures is proportional to the number of atoms per formula unit of the solid. They quantified their observations in what is known as Dulong and Petit's rule, which says that the molar heat capacity of a solid can be expressed as

$$C_P \simeq N \times 25\ J \cdot K^{-1} \cdot mol^{-1}$$

where N is the number of atoms per formula unit. The observed heat capacity per gram of a compound containing thallium and chlorine is $0.208\ J \cdot K^{-1} \cdot g^{-1}$. Use Dulong and Petit's rule to determine the formula of the compound.

5-56. The heat capacity per gram of an oxide of rubidium is $0.64\ J \cdot K^{-1} \cdot g^{-1}$. Use Dulong and Petit's rule (Problem 5-55) to determine the formula of the compound.

5-57. The mineral stilleite contains zinc and selenium. The observed heat capacity per gram is $0.348\ J \cdot K^{-1} \cdot g^{-1}$. Use Dulong and Petit's rule (Problem 5-55) to determine the formula of stilleite.

5-58. The mineral, matlockite, contains lead, fluorine, and chlorine. The observed heat capacity per gram (specific heat) of matlockite is $0.290\ J \cdot K^{-1} \cdot g^{-1}$. Use Dulong and Petit's rule (Problem 5-55) to determine the formula of matlockite.

5-59. Bicycle riding at 13 mph (a moderate pace) consumes $2000\ kJ \cdot h^{-1}$ for a 150-lb person. How many miles must this person ride in order to lose 1 lb of body fat?

5-60. Calculate the heat of combustion per gram of $C_2H_2(g)$, $C_2H_4(g)$, and $C_2H_6(g)$ at 25°C.

5-61. For each of the following reactions,

Reaction	ΔH°_{rxn}/kJ	ΔV/L (at 1.00 atm)
(a) $2H_2(g) + O_2(g) \rightarrow 2H_2O(l)$	−572	−67.2
(b) $Sn(s) + 2Cl_2(g) \rightarrow SnCl_4(l)$	−545	−44.8
(c) $H_2(g) + Cl_2(g) \rightarrow 2HCl(g)$	−180	0

calculate ΔU°_{rxn} at 25°C and 1.00 atm and compare the result to ΔH°_{rxn}.

5-62. Compute ΔH°_{rxn} for the equation

$$3Cu(s) + 8HNO_3(aq) \rightarrow 3Cu(NO_3)_2(aq) + 2NO(g) + 4H_2O(l)$$

(Note: $\Delta H_f^\circ[H^+(aq)] = 0$, $\Delta H_f^\circ[NO_3^-(aq)] = -207.35\ kJ \cdot mol^{-1}$ and $\Delta H_f^\circ[Cu^{2+}(aq)] = 64.39\ kJ \cdot mol^{-1}$. Additional data can be found in Table 5-1).

5-63. The Apollo Lunar Module was powered by a reaction similar to the following:

$$\underset{\text{hydrazine}}{2N_2H_4(l)} + \underset{\text{dinitrogen tetroxide}}{N_2O_4(l)} \rightarrow 3N_2(g) + 4H_2O(g)$$

Using the values of ΔH_f° given in Table 5-1, calculate ΔH°_{rxn} for this reaction.

5-64. A 1.53-g sample of octane, $C_8H_{18}(l)$, is burned in a bomb calorimeter and the observed temperature increase is 1.86 K. The heat capacity of the calorimeter assembly is $42.70\ kJ \cdot K^{-1}$. Calculate the heat of combustion per mole and per gram of octane.

5-65. Ethylamine undergoes an endothermic dissociation producing ethylene and ammonia:

```
    H  H                   H       H
    |  |                    \     /
 H—C—C—N—H(g)  ⟶          C=C     (g) + H—N—H(g)
    |  |  |                 /     \          |
    H  H  H                H       H         H
```

Using the data in Table 5-2, calculate ΔH°_{rxn}.

Energy Utilization

A satellite photo of North America. The heavily populated areas are readily identified by the numerous lights.

Ninety-three percent of the energy utilized in the United States and 70 percent of that in Canada is generated from the combustion of oil, coal, and natural gas (fossil fuels). The remainder is supplied primarily by hydroelectric and nuclear power. At present, other, so-called alternative energy sources, such as solar, biomass, geothermal, and wind, make relatively minor contributions to total energy utilization, except in some localized areas where conditions are especially favorable.

E-1. THE UNITED STATES UTILIZES ABOUT 84 QUADRILLION KILOJOULES OF ENERGY PER YEAR

In 1984 the United States utilized 84×10^{15} kJ of energy, which corresponds to a per-person annual consumption of

$$\frac{84 \times 10^{15}\ \mathrm{kJ \cdot year^{-1}}}{220 \times 10^{6}\ \mathrm{people}} = 3.8 \times 10^{8}\ \mathrm{kJ \cdot year^{-1} \cdot person^{-1}}$$

Table E-1 Energy utilization by type in the United States and Canada, 1980

Energy source	Amount of energy utilized/10^{15} kJ		Percent of total	
	United States	Canada	United States	Canada
Petroleum liquids	40	3.9	48	42
Natural gas	21	2.2	25	23
Coal	17	0.82	20	8.7
Nuclear power	2.9	0.30	3.5	3.3
Hydroelectric power	2.9	2.2	3.5	23
	84	9.4	100	100

or, because there are 3.15×10^7 seconds in a year, the rate of U.S. energy consumption is 12 $kJ \cdot s^{-1} \cdot person^{-1}$. The corresponding figure for Canada is about the same.

The rate of production or utilization of energy is a quantity called **power.** The SI unit of power is a **watt** (W), which is equal to exactly one joule per second. A kilowatt (kW) is equal to one kilojoule per second. Thus the rate of energy consumption in the United States is 12 kW per person. A power usage of 12 kW per person is equivalent to one-hundred twenty 100-W light bulbs burning continuously for each person. An annual energy use of 12 kW per person is also equivalent to 62 barrels of crude oil per person per year.

At the present time, the United States produces about 80 percent of its total annual energy requirements: the 20 percent shortfall is made up for the most part by imported oil. Oil furnishes about 48 percent of the total U.S. energy supply (Table E-1), and about 43 percent of the oil used in the United States is imported. The major ways in which energy is utilized in the United States are industry (35 percent), transportation (22 percent), and space heating (18 percent). Canada imports about 20 percent of the oil it uses, and exports natural gas in quantities equal to about half of its own internal use.

Although U.S. reserves of oil and gas are sufficient to last about only 25 to 30 years at present rates of consumption, the United States has over 240 billion tons of recoverable coal reserves. The known U.S. coal reserves are sufficient to supply total U.S. energy needs at present rates of consumption for over 300 years.

The Kemmerer Coal Mine in Wyoming. This is the largest open-pit surface coal mine in the United States. The large drill, which is dwarfed by the coal seams, bores holes in which explosive charges are placed.

E-2. ENERGY IS OBTAINED FROM FOSSIL FUELS BY COMBUSTION REACTIONS

Fossil fuels are essentially mixtures of hydrocarbons. Thermal energy is obtained from such fuels by burning them in air. The heats of combustion in kilojoules per gram for a variety of fuels are given in Table E-2. An essential criterion for the quality of a substance that is used as a fuel in a combustion reaction is the quantity of heat evolved per gram of fuel when the fuel is burned. The more energy evolved as heat per gram of fuel, the greater the quality of the fuel as an energy source.

Other important criteria for fuels are cost, ease of transport, and utilization hazards.

Natural gas is essentially methane (CH_4), with traces of other hydrocarbons, and is second only to hydrogen in the energy released per gram of fuel burned. Propane (C_3H_8), which is used as a fuel in areas not serviced by natural gas mains, is stored on site as a liquid in tanks. The vapor over the liquid propane has a sufficiently high pressure to flow out of the tank to the combustion region when a valve is opened. Conventional liquid fuels such as gasoline, jet fuel, diesel fuel, and heating oils are complex mixtures of hydrocarbons. For example, gasolines consist of mixtures of over 100 different hydrocarbon compounds in variable proportions. The hydrocarbons have from 4 to 14 carbon atoms per formula unit. Various blends are produced, depending on the environmental conditions of use and the quality of the gasoline. Diesel fuel and heating oils contain various hydrocarbons that have from 10 to 20 carbon atoms per formula unit.

Gasohol is a mixture of gasoline and ethyl alcohol (C_2H_5OH). The most common blend is 90 percent conventional gasoline plus 10 percent ethyl alcohol. Ethyl alcohol is produced by fermentation of sugars from plants using yeasts as the fermentation agent; for example,

$$\underset{\substack{\text{glucose} \\ \text{(a sugar)}}}{C_6H_{12}O_6(aq)} \xrightarrow{\text{yeast}} \underset{\text{ethyl alcohol}}{2C_2H_5OH(aq)} + 2CO_2(g)$$

The heat of combustion per gram for ethyl alcohol is significantly less than for a hydrocarbon because ethyl alcohol is already partially oxidized relative to a hydrocarbon:

$$\underset{\text{ethane}}{C_2H_6(g)} + \tfrac{1}{2}O_2(g) \rightarrow C_2H_5OH(l) \qquad \Delta H^\circ_{rxn} = -193 \text{ kJ}$$

As a consequence an automobile running on pure ethyl alcohol fuel uses about 75 percent more fuel per mile than does an automobile running on gasoline.

Table E-2 Heats of combustion of fuels

Fuel	ΔH°_{rxn}/kJ·g^{-1}
hydrogen, $H_2(g)$	−131
methane, $CH_4(g)$	−50
propane, $C_3H_8(l)$	−48
carbon, $C(s)$	−33
ethyl alcohol, $C_2H_5OH(l)$	−27
refined heating oil	−44
gasoline, kerosene, or diesel fuel	−48
coal	−28
dry seasoned wood	−25

*Enthalpy changes refer to the reactions where gaseous products, $CO_2(g)$ and $H_2O(g)$, are formed.

E-3. COAL CAN BE CONVERTED TO A GASEOUS FUEL

Coal is a complex substance that contains highly variable amounts of many different elements. Sulfur is a major impurity in many coals, and the SO_2 produced when the coal is burned in a power plant must be removed before the combustion gases exit the plant. The need to remove a wide range of environmental pollutants from coal combustion gases has increased greatly the costs of building coal-fired power plants.

Coal is not as convenient or versatile an energy source as petroleum liquids because it is a solid. Coal is much more expensive to transport than oil, which is moved through pipelines, and is most economically used on a large scale close to where it is mined. Because petroleum liquids are in short supply relative to coal and also because petroleum liquids are much more valuable as transportation fuels, many oil-fired power plants have been converted to coal-fired plants in the last decade.

Coal can be converted to a gaseous fuel by various methods of **coal gasification,** which involves the reaction of coal with steam and further reactions involving intermediate products to generate combustible gases. For simplicity, we can represent the carbon in coal as C(*s*). In the first step of a coal-gasification process, the carbon reacts with steam at about 1000°C to produce carbon monoxide and hydrogen via the **water-gas reaction:**

$$C(s) + H_2O(g) \rightarrow CO(g) + H_2(g) \qquad \Delta H^\circ_{rxn} = +131\ \text{kJ} \qquad \text{(E-1)}$$

Reaction (E-1) is very endothermic, and the energy as heat necessary to drive the reaction is supplied by burning some of the available coal, which further reduces the extent of conversion of coal. The product mixture in Reaction (E-1) is combustible (to CO_2 and H_2O) and is known as **synthesis gas.** Synthesis gas can be used directly as a fuel, but because of its toxicity (CO) and highly explosive nature (H_2), and the fact that it cannot be burned in appliances designed to use methane, further conversion is necessary.

In the second step, additional steam is added to convert some of the CO to CO_2 and H_2 via the **water-gas shift reaction:**

$$CO(g) + H_2O(g) \rightarrow CO_2(g) + H_2(g) \qquad \Delta H^\circ_{rxn} = -41\ \text{kJ} \qquad \text{(E-2)}$$

Reaction (E-2) is controlled by adjusting the ratio of H_2O to CO such that the gas mixture is converted from the 1-to-1 H_2-CO ratio in synthesis gas to a 3-to-1 ratio. The 3-to-1 H_2-CO ratio is the correct ratio for the next step in the process:

$$CO(g) + 3H_2(g) \xrightarrow{\text{catalyst}} H_2O(g) + CH_4(g) \qquad \Delta H^\circ_{rxn} = -207\ \text{kJ} \qquad \text{(E-3)}$$

Removal of water from the H_2O-CH_4 mixture yields a product consisting primarily of methane. This product is directly substitutable for natural gas. Synthesis gas can be converted to hydrocarbons and alcohols in the **Fischer-Tropsch synthesis**; for example,

$$CO(g) + 2H_2(g) \xrightarrow{\text{catalyst}} CH_3OH(l)$$

$$2CO(g) + 4H_2(g) \xrightarrow{\text{catalyst}} C_2H_4(g) + 2H_2O(g)$$

This reaction is discussed in Chapter 19. The Fischer-Tropsch synthesis was used in Germany during World War II to manufacture gasolines from coal.

E-4. ROCKETS UTILIZE HIGHLY EXOTHERMIC REACTIONS WITH GASEOUS PRODUCTS

The heat-of-combustion data in Table E-2 show that the most energy-rich fuel on a mass basis is hydrogen, which has an energy content per gram of well over twice that of the next-best fuel. Because of its unusually high energy content per gram, liquid hydrogen was used in the first stage of the Apollo series spaceships that traveled to the moon (the fuel must be lifted as part of the space vehicle). The main disadvantages of liquid hydrogen as a fuel is that H_2 can be maintained as a liquid only at very low temperatures (about 20 K at 1 atm) and that hydrogen forms potentially explosive mixtures with air. The second and third stages of the Apollo spaceships were powered by the reaction between kerosene and **liquid oxygen (LOX),** both of which were stored on the spaceship.

The Apollo Lunar Lander spaceships (Figure E-1) were powered to and from the surface of the moon by the energy released on the reaction of *N,N*-dimethylhydrazine, $H_2NN(CH_3)_2(l)$, with nitrogen tetroxide, $N_2O_4(l)$:

$$H_2NN(CH_3)_2(l) + 2N_2O_4(l) \rightarrow 3N_2(g) + 2CO_2(g) + 4H_2O(g)$$

$$\Delta H^\circ_{rxn} = -29 \text{ kJ per gram of fuel}$$

This reaction is especially suitable for a lunar escape vehicle because the reaction starts spontaneously on mixing. No battery or spark plugs, with associated electrical circuitry, are required.

The principal solid fuel used in the Minuteman ICBMs (*i*nter*c*ontinental *b*allistic *m*issiles) and the Polaris submarine missiles (Figure E-2) and the airplane-to-airplane Sidewinder missiles typically consists of a mixture of 70 percent ammonium perchlorate, 18 percent aluminum metal powder, and 12 percent binder. Ammonium perchlorate is a self-contained solid fuel—the fuel NH_4^+ and the oxidizer ClO_4^- are together in the solid. The overall equation for the NH_4ClO_4 decomposition reaction is

$$2NH_4ClO_4(s) \rightarrow N_2(g) + 2HCl(g) + 3H_2O(g) + \tfrac{5}{2}O_2(g)$$

Note that the NH_4ClO_4 is oxygen rich; that is, O_2 is a reaction product. To utilize this available oxygen and thereby to provide more rocket thrust, aluminum powder and a binder are added. The aluminum is oxidized to aluminum oxide:

$$4Al(s) + 3O_2(g) \rightarrow 2Al_2O_3(s)$$

and the binder is oxidized to CO_2 and H_2O. The aluminum powder also promotes a more rapid and even decomposition of the NH_4ClO_4.

Figure E-1 An Apollo Lunar Lander rocket.

E-5. SUNLIGHT IS A MAJOR ENERGY SOURCE

A major energy source that remains largely untapped, at least by humans, is sunlight. The sunlight that reaches the surface of the United States on a clear day has a power level of about 1 $kW \cdot m^{-2}$. The energy requirements of a typical U.S. home are about 30 kilowatt-hours per day. A **kilowatt-hour** ($kW \cdot h$) is the energy produced by a 1-kW power source operating for 1 h. If 30 percent of the incident solar energy could be collected and utilized, then for 8 h of sunlight in a day, all the household energy requirements could be satisfied by about 12 m^2 (say, 3 m × 4 m) of collector surface (Problem E-17).

One of the major drawbacks of solar energy is the fact that the sunshine is inherently intermittent in nature. An energy storage device is necessary so that the energy can be used at night or when the day is overcast. The energy collected can be stored in basically two ways: (1) by heating up a large mass of substance, such as water or rocks, and (2) by storing the energy in chemicals that, on reaction, will release the stored energy. In effect, a suitable chemical reaction is driven uphill energetically by the solar energy input.

Salt hydrates are the simplest types of chemicals that can be used to store energy in a chemical process. An example is **Glauber's salt,**

Figure E-2 Underwater firing of a Polaris A-3 missile from a nuclear submarine.

$Na_2SO_4 \cdot 10H_2O(s)$, which at 32.3°C (90°F) dissolves in its own waters of hydration to form a solution:

$$Na_2SO_4 \cdot 10H_2O(s) \rightarrow (Na_2SO_4 + 10H_2O)_{soln}$$

When the temperature of the salt solution drops below 32.3°C, the reverse reaction occurs, and the $Na_2SO_4 \cdot 10H_2O(s)$ crystallizes out of the solution with the evolution of about 354 kilojoules per liter of salt solution. As the salt crystallizes, the heat can be drawn off and used for space or water heating.

Another approach to solar energy conversion is to use mirrors to concentrate the sunlight on a reactor vessel and use the high temperatures produced to drive an endothermic chemical reaction. An example is the reaction

$$CO_2(g) + CH_4(g) \xrightarrow{750°C} 2CO(g) + 2H_2(g) \qquad \Delta H_{rxn} = 247 \text{ kJ}$$

The 247 kJ of stored energy per mole of CO_2 or CH_4 reacted can be released by running the reaction in reverse. The energy evolved can be used to provide steam for power generation. Focused sunlight can also be used to convert liquid water to steam that is then used to drive a turbine that produces electricity.

Electricity obtained from solar cells, which convert sunlight to an electric current, can be used to decompose compounds by electrolysis; for example,

$$2HBr(aq) \xrightarrow{\text{electrolysis}} H_2(g) + Br_2(l)$$

The reaction can then be run in reverse to release energy when needed.

Solar energy holds great promise as a major energy source of the future (Figure E-3). Development is slow at present because petroleum, coal, and natural gas are less expensive energy sources readily utilized in existing large-scale technologies.

Figure E-3 A solar power tower. A large receiver structure is located on top of a high tower at the focal point of a large array of sun-tracking mirrors called heliostats. The heliostat array covers 6000 square meters for each megawatt of thermal energy produced using steam generated in the receiver. "Solar One," shown here, is located near Barstow, California, in the Mojave Desert. The power tower generates 10 megawatts peak power from 1818 heliostats, each of which is 430 square feet in area. The tower is 300 feet high.

power 188
watt, W ($J \cdot s^{-1}$) 188
kilowatt, kW ($kJ \cdot s^{-1}$) 188
fossil fuel 188
gasohol 189
coal gasification 189
water-gas reaction 190
synthesis gas 190
water-gas shift reaction 190
Fischer-Tropsch synthesis 190
liquid oxygen (LOX) 190
kilowatt-hour (kW·h) 191
Glauber's salt 191

QUESTIONS

E-1. How does diesel fuel differ from gasoline?

E-2. What is gasohol?

E-3. Economic and environmental considerations aside, explain why petroleum liquids are more desirable as fuels than coal.

E-4. Describe briefly the Fischer-Tropsch process.

E-5. What is the basic idea in coal-gasification schemes?

E-6. List the major advantages and disadvantages of solar energy as a home energy source.

E-7. In the United States, air conditioning accounts for 3 percent of the total energy use, whereas space heating accounts for 18 percent of the total. Suggest an explanation for this difference.

E-8. What is synthesis gas?

E-9. What is the major disadvantage of hydrogen as a fuel?

E-10. Describe two ways in which solar energy can be collected and stored by means of chemical reactions.

PROBLEMS

E-11. Using the data on page 188, estimate the number of barrels of oil imported per day by the United States.

E-12. Natural gas use in the United States is billed in therms. A therm is defined as 100,000 Btu. Given that 1 Btu = 1.05 kJ, estimate the cost per mole of CH_4 where natural gas costs 60 cents per therm.

E-13. Calculate the number of kilojoules that can be stored at 32.3°C in 100 kilograms of $Na_2SO_4 \cdot 10H_2O(l)$. Take the density of $Na_2SO_4 \cdot 10H_2O(l)$ as $1.5\ g \cdot mL^{-1}$.

E-14. Using the data in Table 5-1, calculate the heat of combustion per gram of ethyl alcohol.

E-15. As stated in Section E-2, ethane can be partially oxidized to ethyl alcohol. Given that the heat of combustion of ethane is $-1427\ kJ \cdot mol^{-1}$, calculate the heat of combustion per mole of ethyl alcohol.

E-16. Convert 30 kW·h to kJ.

E-17. Using the data given in Section E-5, show that the energy requirements of a typical U.S. home could be satisfied if 30 percent of the incident solar energy in an 8-h period per day were collected by 12 m^2 of collector surface.

E-18. Consider a swimming pool 25 feet long by 12 feet wide with an average depth of 6 feet. Take the specific heat of water as $4.2\ J \cdot K^{-1} \cdot g^{-1}$ and the density of water as $1.0\ g \cdot mL^{-1}$, and compute the number of kilojoules of heat required to raise the temperature of the pool water by 10°C, assuming no heat loss to the surroundings. Given that the cost of natural gas is 60 cents per therm (1 therm = 10^5 Btu; 1 Btu = 1.05 kJ) and assuming that 25 percent of the heat is lost to the surroundings during the heating cycle, compute the cost of raising the pool temperature by 10°C.

THE QUANTUM THEORY AND ATOMIC STRUCTURE

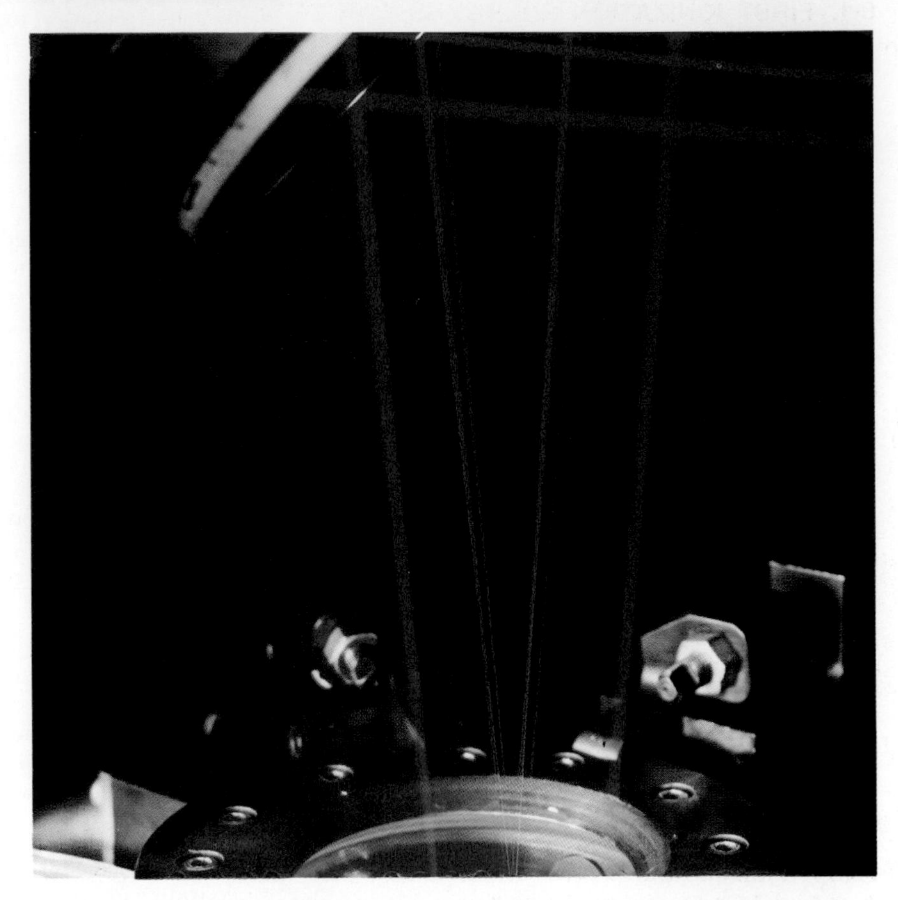

Argon ion laser light being used to study combustion processes occurring in a reaction cylinder (lower portion of photo) equipped with a window. The wavelengths of the laser light are 514.5 nm (green) and 488.0 nm (blue). The laser is located to the right and a mirror (top center) is used to direct the beams into the reaction cylinder.

We learned in Chapter 1 that atoms consist of protons, neutrons, and electrons, with the protons and neutrons making up the dense, central nucleus and the electrons arranged in some unspecified manner around the nucleus. Our model pictures an atom as mostly empty space, with the diameter of the nucleus being roughly 10^{-5} times that of the whole atom. We shall learn in this chapter that a description of the arrangement of electrons around a nucleus requires a new and unexpected way of looking at nature. This new perspective is given by what is called the quantum theory. One of the principal results of the quantum theory is that the electronic energies of atoms are quantized, meaning that they can take on only certain discrete values. In this chapter we trace the development of the quantum theory and apply it to the electronic structure of atoms. We introduce the Pauli exclusion principle and use it to correlate electronic structures with the chemical properties of the elements and their periodic properties.

6-1. FIRST IONIZATION ENERGY IS ONE OF MANY PERIODIC PROPERTIES OF THE ELEMENTS

The periodic table offers a great deal of insight into the **electronic structure** of atoms. For example, elements in the same column in the periodic table are similar chemically, and so we might expect that their electronic arrangements, particularly those of the outermost and hence most chemically important electrons, are similar.

A direct indication of the arrangement of electrons about a nucleus is given by the **ionization energies** of the atom or ion. The ionization energy of an atom or an ion is the minimum energy required to remove an electron completely from the gaseous atom or ion; this energy can be determined experimentally. The **first ionization energy** of an atom is the minimum energy required to remove an electron from the neutral atom, A, to produce a positively charged ion, A^+:

$$A(g) \rightarrow A^+(g) + e^-(g)$$

The **second ionization energy** is the minimum energy required to remove an electron from the A^+ ion to produce an A^{2+} ion:

$$A^+(g) \rightarrow A^{2+}(g) + e^-(g)$$

We can go on to define and measure third, fourth, and successive ionization energies. We denote the first ionization energy by I_1, the second ionization energy by I_2, and so forth. We expect the second and higher ionization energies to be greater than the first ionization energy because, in removing successive electrons from an atom, we must overcome an increasingly greater electrical attraction between the positively charged ion and the electron that is being removed. Thus we find that $I_1 < I_2 < I_3 < I_4$, and so forth, for any given atom.

If we plot the first ionization energies of the elements against atomic number (Figure 6-1), then we find that there is a periodic pattern in these data. Note that the noble gases have relatively large first ionization energies. This means that it is relatively difficult to remove electrons from noble-gas atoms and hence suggests that the electronic structures of these atoms are more stable than those of the elements that precede and follow them in the periodic table. Furthermore, the alkali metals have relatively low ionization energies (Figure 6-1), in accord with their extremely reactive nature. Thus we see that ionization energies as well as chemical properties display a periodic character because both depend upon electronic structure. Figure 6-2 illustrates the trend in first ionization energies in the periodic table.

6-2. THE VALUES OF SUCCESSIVE IONIZATION ENERGIES OF ATOMS SUGGEST A SHELL STRUCTURE

We can gain more insight into electronic structure by listing not just the first ionization energies but successive ionization energies, as in Table

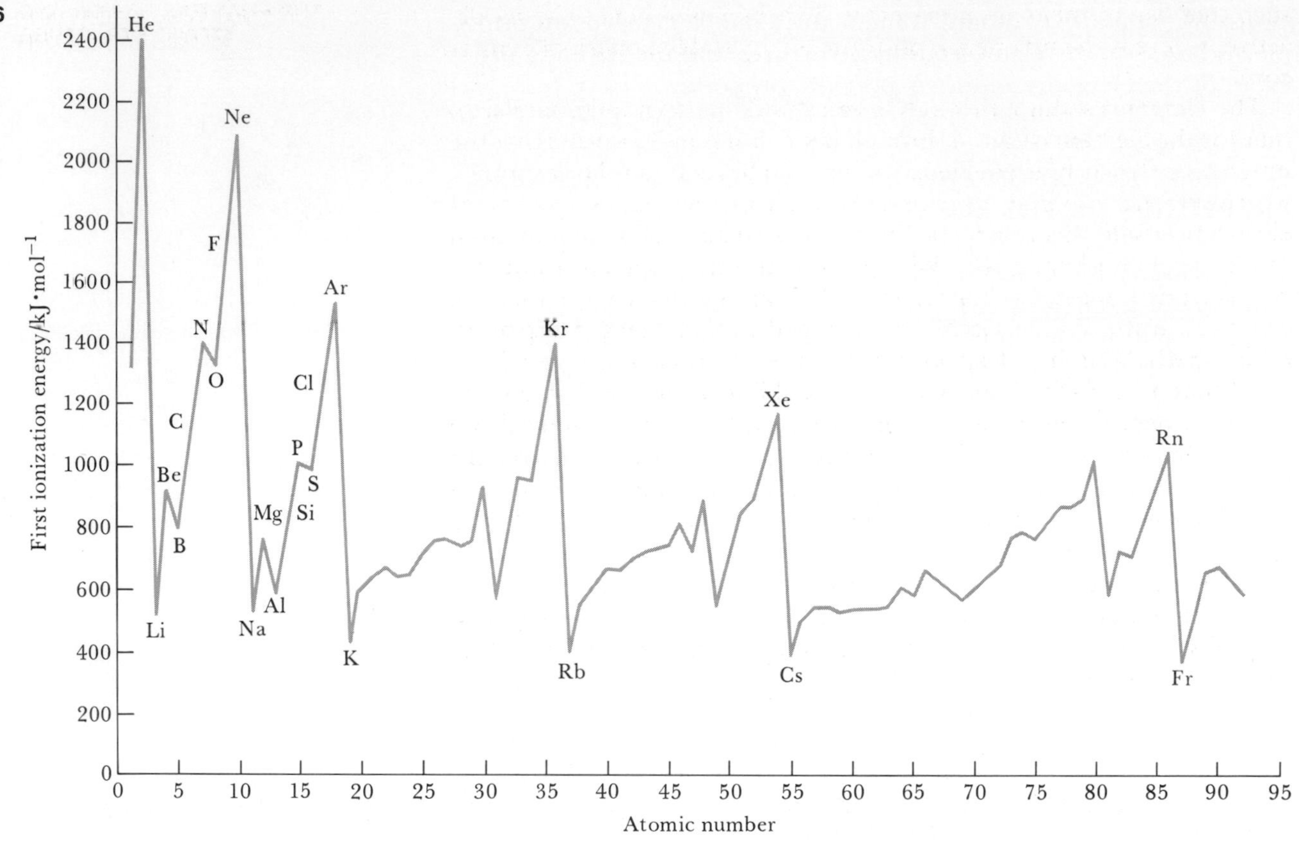

Figure 6-1 A graph of first ionization energy versus atomic number shows clearly the periodic nature of the properties of the elements.

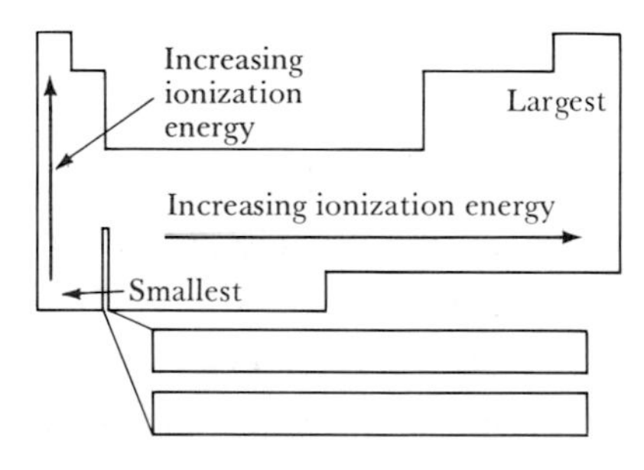

Figure 6-2 The trend in first ionization energy in the periodic table. Ionization energy increases as we go from left to right across the rows and as we go up the columns.

6-1, where I_1 through I_{10} are tabulated for the elements hydrogen through argon.

Let's look at helium first. The first ionization energy is much greater than that of hydrogen or lithium, once again indicating the extraordinary stability of the helium atom. The second ionization energy is even higher, being more than twice as large as the first. Realize, however, that here we are removing an electron from a positively charged He^+ ion, and so we should expect I_2 to be greater than I_1 because of the attraction between the positively charged ion and the negatively charged electron we are removing.

The case of lithium is more interesting than that of helium. The first ionization energy is 0.52 $MJ\cdot mol^{-1}$, and the second is 7.30 $MJ\cdot mol^{-1}$. The value for I_2 is far greater than can be accounted for by the electrical attraction between an electron and the resultant Li^+ ion and indicates that the species Li^+ has an extraordinary stability toward ionization. Once Li^+ is ionized to Li^{2+}, the next ionization energy is regular, as with helium. This pattern of ionization energies suggests that the lithium atom has one chemically important electron; when it loses that electron, the result is a Li^+ ion with two electrons and helium-like stability. The ion Li^+ usually takes part in chemical reactions as a spectator ion.

Table 6-1 shows that for beryllium there is a large jump from I_2 to I_3. This jump suggests that the four electrons in beryllium are arranged such that two of them are more easily detached, and hence chemically active, whereas the other two constitute a very stable helium-like inner core.

The elements sodium through argon show a pattern quite similar to that for the elements lithium through neon, but now it appears that the inner-core structure is like neon rather than like helium. For example, we can picture a sodium atom as a neon-like core with a loosely bound electron outside this core. The electronic structure of a sodium atom can be seen by plotting successive ionization energies against number of electrons removed, as in Figure 6-3. This figure suggests that the electrons in the sodium atom are arranged in three separate groups, called **shells.** The first electron is relatively easily removed to give Na^+, which has a neon-like stable core. This stability is indicated by a large jump between I_1 and I_2. After nine electrons have been removed, we have only two electrons left, which are arranged in a tightly bound helium-like core. There is a large jump in energy in going from I_9 to I_{10}, which suggests that the last two electrons to be removed from the sodium atom constitute a third shell.

Table 6-1 suggests that the electrons in atoms are arranged in shells consisting of noble-gas-like cores and chemically active **outer electrons.** We can summarize these ideas by presenting the atoms as in Table 6-2. In the second column we indicate the noble-gas-like core by using the

Table 6-1 Successive ionization energies of the elements hydrogen through argon. The lines separate regions of relatively low and relatively high ionization energies.

		Ionization energy/MJ·mol^{-1}									
Z	Element	I_1	I_2	I_3	I_4	I_5	I_6	I_7	I_8	I_9	I_{10}
1	H	1.31									
2	He	2.37	5.25								
3	Li	0.52	7.30	11.81							
4	Be	0.90	1.76	14.85	21.01						
5	B	0.80	2.42	3.66	25.02	32.82					
6	C	1.09	2.35	4.62	6.22	37.83	47.28				
7	N	1.40	2.86	4.58	7.48	9.44	53.27	64.36			
8	O	1.31	3.39	5.30	7.47	10.98	13.33	71.33	84.08		
9	F	1.68	3.37	6.05	8.41	11.02	15.16	17.87	92.04	106.43	
10	Ne	2.08	3.95	6.12	9.37	12.18	15.24	20.00	23.07	115.38	131.43
11	Na	0.50	4.56	6.91	9.54	13.35	16.61	20.11	25.49	28.93	141.37
12	Mg	0.74	1.45	7.73	10.54	13.62	17.99	21.70	25.66	31.64	35.46
13	Al	0.58	1.82	2.74	11.58	14.83	18.38	23.30	27.46	31.86	38.46
14	Si	0.79	1.58	3.23	4.36	16.09	19.78	23.79	29.25	33.87	38.73
15	P	1.06	1.90	2.91	4.96	6.27	21.27	25.40	29.85	35.87	40.96
16	S	1.00	2.25	3.36	4.56	7.01	8.49	27.11	31.67	36.58	43.14
17	Cl	1.26	2.30	3.82	5.16	6.54	9.36	11.02	33.60	38.60	43.96
18	Ar	1.52	2.67	3.93	5.77	7.24	8.78	11.99	13.84	40.76	46.19

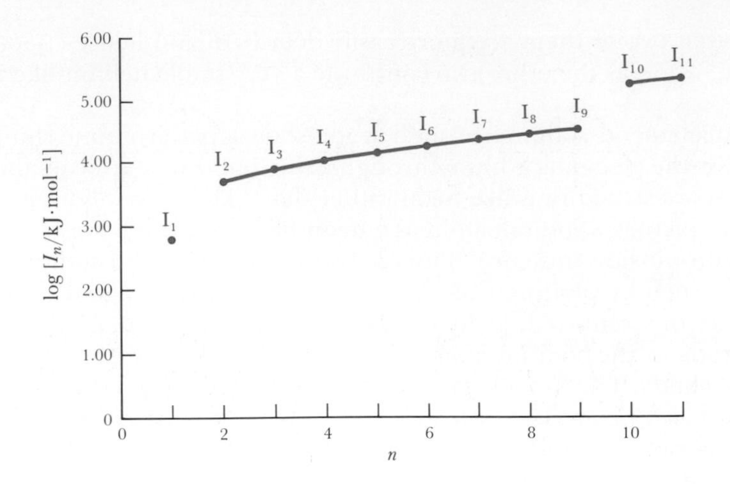

Figure 6-3 The logarithms of the 11 ionization energies of the sodium atom versus the number of electrons removed (n). This graph suggests that the electrons in a sodium atom are arranged in three shells. (Logarithms of the ionization energies are plotted simply in order to compress the vertical scale.)

Table 6-2 A simple representation of the first 18 elements, indicating their noble-gas-like inner core and their outer electrons

Symbol	Inner-core representation	Lewis electron-dot formula
H	H·	H·
He	[He]	·He·
Li	[He]·	Li·
Be	·[He]·	·Be·
B	·[He]·	·B·
C	·[He]·	·C·
N	·[He]:	·N·
O	·[He]·	·O·
F	:[He]:	:F:
Ne	[Ne]	:Ne:
Na	[Ne]·	Na·
Mg	·[Ne]·	·Mg·
Al	·[Ne]·	·Al·
Si	·[Ne]·	·Si·
P	·[Ne]·	·P·
S	·[Ne]·	·S·
Cl	:[Ne]:	:Cl:
Ar	[Ar]	:Ar:

symbol for the gas enclosed in brackets; the outer-shell electrons are indicated by the dots. Outer electrons are also called **valence electrons,** and the entries in Table 6-2 highlight the valence electrons. Thus, for example, we represent beryllium by a helium-like inner core with two electrons outside this core. The placement of the dots is arbitrary at this point.

The third column in Table 6-2 is an abbreviated version of the second column. Only the valence electrons are indicated, and the appropriate noble-gas core is understood. Note that the number of valence electrons increases from one to eight as we go across a row in the periodic table from alkali metal to noble gas. This pattern repeats itself from row to row. It is the valence electrons that play the key role in the chemical properties of the elements; thus, fluorine and chlorine, for example, are chemically similar because they both have seven valence electrons. This representation, called a **Lewis electron-dot formula,** was introduced in 1916 by G. N. Lewis, one of the greatest American chemists. Lewis formulas show only the valence electrons, which are the chemically important electrons.

6-3. THE REGIONS OF THE ELECTROMAGNETIC SPECTRUM ARE CHARACTERIZED BY RADIATION OF DIFFERENT WAVELENGTHS

In order to picture more clearly the electronic structure of atoms, we must first briefly consider electromagnetic radiation. Visible light, ultraviolet and infrared light, radio waves, and X-rays are all forms of electromagnetic radiation. For many years scientists disagreed over whether electromagnetic radiation exists as beams of particles or as waves. Many experiments supported one viewpoint or the other, but toward the end of the nineteenth century most evidence favored a wave picture.

Figure 6-4 depicts typical waves, whose motion can be visualized as moving across the page. The distance between successive crests or troughs is called the **wavelength** and is denoted by the Greek letter

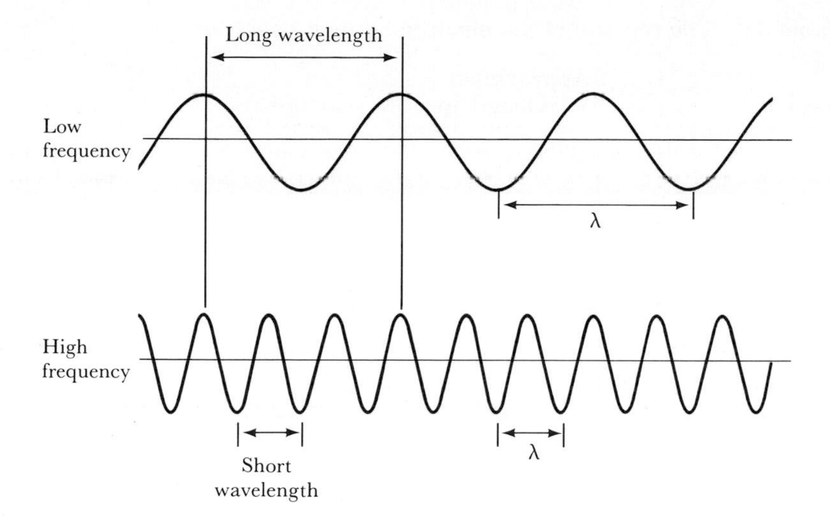

Figure 6-4 Two wave forms, one with a wavelength (λ) three times as large as the other. If both waves moved across the page with the same speed, three crests of the bottom wave would pass by a given point for every one crest of the top wave. Thus the frequency of the bottom wave is three times greater than the frequency of the top wave.

lambda, λ. If we picture each wave as moving across the page, the number of crests that pass a given point per second is called the **frequency** and is denoted by the Greek letter nu, ν. Although the units of wavelength are meters per cycle and the units of frequency are cycles per second, the term cycle is understood and so is omitted from both units. Thus the units of wavelength are meters and the units of frequency are reciprocal seconds. The product of the wavelength and frequency, $\lambda\nu$, is the speed at which the wave travels. All forms of electromagnetic radiation travel at a speed of $3.00 \times 10^8\ \text{m·s}^{-1}$, which is usually called simply the **speed of light.** If we denote the speed of light by c, we can write

▪ The unit of reciprocal second is 1/second, or s^{-1}.

▪ You may know the speed of light as 186,000 miles per second.

$$\lambda\nu = c \tag{6-1}$$

The various forms of electromagnetic radiation differ only in their frequency or their wavelength. For example, a look at a radio dial shows that 100 MHz (megahertz) is in the middle of the FM dial. The unit **hertz** is the same as the unit reciprocal second, and so 100 MHz is equal to $100 \times 10^6\ s^{-1}$. The wavelength of a 100-MHz signal is

▪ The prefix mega- means million.

▪ $1\ \text{Hz} = 1\text{s}^{-1}$

$$\lambda = \frac{c}{\nu} = \frac{3.00 \times 10^8\ \text{m·s}^{-1}}{100 \times 10^6\ \text{s}^{-1}} = 3.00\ \text{m}$$

Example 6-1: A CO_2 laser produces an intense beam of radiation whose wavelength is 10.6 μm (micrometers; 1 $\mu\text{m} = 10^{-6}$ m). Calculate the frequency of this radiation.

Solution: Because $\lambda = 10.6 \times 10^{-6}$ m, we have, using Equation 6-1

$$\nu = \frac{c}{\lambda} = \frac{3.00 \times 10^8\ \text{m·s}^{-1}}{10.6 \times 10^{-6}\ \text{m}}$$

$$= 2.83 \times 10^{13}\ \text{s}^{-1} = 2.83 \times 10^{13}\ \text{Hz}$$

We see from these calculations that the range of wavelengths and frequencies of electromagnetic radiation, called the **electromagnetic**

Table 6-3 The regions of the electromagnetic spectrum

Region	Approximate wavelength/m	Comments
radio waves	1000 – 1	AM band 190–560 m; shortwave band 14– 75 m; FM band 2.8– 3.4 m; television bands 5.6–4.2 m (channels 2, 3, 4), 4.0–3.4 m (channels 5, 6), 1.7–1.4 m (channels 7–13)
microwaves	1 – 10^{-4}	includes radar; used to probe rotational motion of molecules and in microwave ovens
infrared light	10^{-4} – 7×10^{-7}	can be felt as heat; used to probe vibrational motions of molecules
visible light	7×10^{-7} – 4×10^{-7}	consists of the colors red, orange, yellow, green, blue, indigo, and violet
ultraviolet light	4×10^{-7} – 10^{-8}	causes sunburn; kills bacteria
X-rays	10^{-8} – 10^{-11}	penetrate human tissue and other matter; have important medical applications
gamma rays	10^{-11} – 10^{-13}	emitted by energetic nuclei; very penetrating radiation; used to kill cancer cells
cosmic rays	10^{-13} – 10^{-14}	very high-energy, penetrating radiation of cosmic origin

spectrum, is truly enormous. Table 6-3 gives the wavelengths and frequencies associated with certain regions in the electromagnetic spectrum.

6-4. THE SPECTRA EMITTED BY ATOMS ARE LINE SPECTRA

When white light is passed through a prism, we see that the light is made up of many colors. The same effect can be seen in a rainbow, where white light from the sun passes through water droplets in the atmosphere and is separated into its component colors. The wavelengths of the radiation in white light range from about 400 nm (nanometers) to 700 nm (1 nm = 10^{-9} m). This is the region of the electromagnetic spectrum to which the human eye is sensitive and is called the **visible region.** The short-wavelength end of the visible region (400 nm) is violet and the long-wavelength end (700 nm) is red.

The spectrum of white light has no gaps in it and is called a **continuous spectrum.** Yet, if we examine radiation emitted from a glass tube containing a gas through which an electric spark is passed, we find that the resultant spectrum is not continuous but consists of several separate lines (Figure 6-5). This type of spectrum is called a **line spectrum** and is characteristic of the particular gas used in the discharge tube. If the gas consists of atoms, then the emitted spectrum is called an **atomic spectrum.** The simplest atomic spectrum is that of the hydrogen atom. Part of this spectrum is shown in Figure 6-5.

The line spectrum associated with an element serves as a fingerprint for that element. The study and analysis of the spectral lines emitted by atoms, or, more generally, the study of the interaction of electromagnetic radiation and atoms, is called **atomic spectroscopy.** We can identify each type of atom present in a spectroscopic sample by comparing the observed spectrum with those in a handbook of atomic spectra. Atomic spectroscopy is a standard technique of analytical chemistry, which is that part of chemistry principally involved with chemical analysis. Spectroscopy is used also in agriculture, archaeology, art, criminal investigation, and many other fields. For example, we have probably all seen a movie or television show in which the crime lab analyzes soil taken from a suspect's shoes and compares it with soil collected from the scene of a crime.

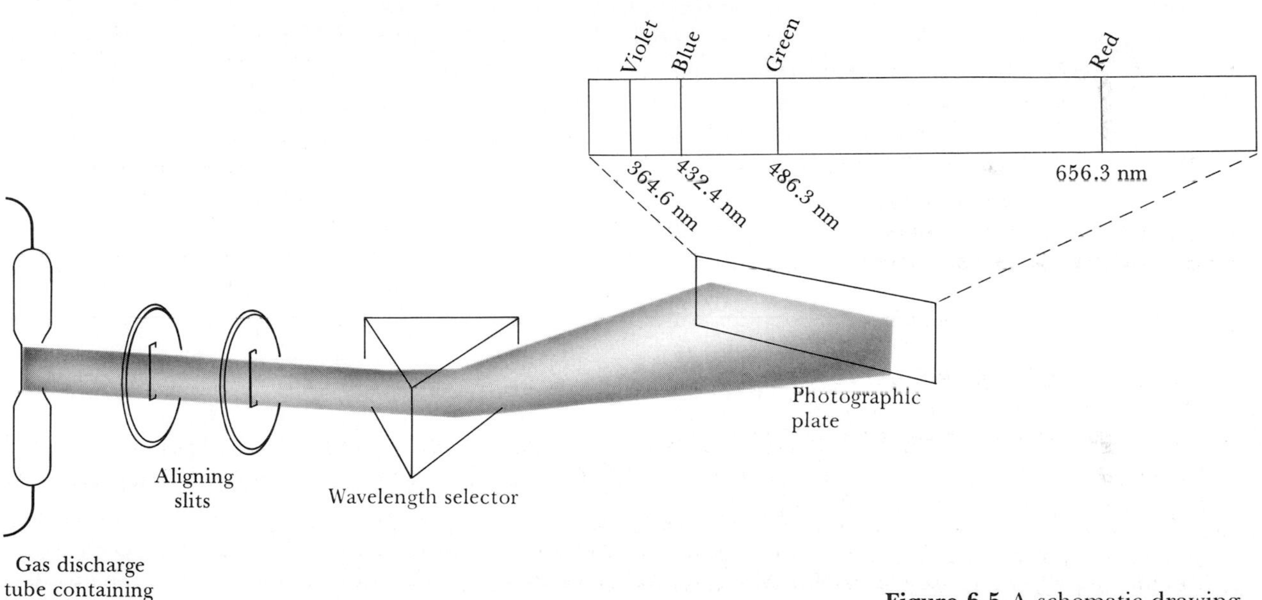

Figure 6-5 A schematic drawing of an atomic spectrometer. When a spark is passed through the discharge tube, hydrogen molecules are dissociated into hydrogen atoms, which emit radiation of only certain frequencies. This radiation is collimated by the slits, passed through a wavelength selector such as a prism, and projected onto a photographic plate. The result is a line spectrum. The wavelengths observed in a line spectrum depend upon the particular gas in the discharge tube.

6-5. ELECTROMAGNETIC RADIATION CAN BE VIEWED AS A BEAM OF PHOTONS

The end of the 1800s saw a number of experiments whose results were impossible to interpret within the framework of the physics of that time, which is now called **classical physics.** The theoretical analysis of these experiments stimulated some unconventional ideas about the nature of matter and energy that directly contradicted accepted theory. As we shall see later in the chapter, these new perspectives led eventually to a general theory of atomic structure that is in excellent agreement with our picture of the electronic structures of atoms and with the periodic properties of the elements.

The first person to break with the ideas of classical physics was the German physicist Max Planck (Figure 6-6). In 1900, Planck was trying to analyze experiments that dealt with blackbody radiation, which is the radiation emitted by solid bodies when they are heated to high temperatures. The problem was that the classical physics approach failed to reproduce the observed experimental values. Prior attempts at analysis assumed that the radiation is emitted in a continuous manner. Planck, however, had the great revolutionary insight to break away from this thinking: he assumed that the radiation is emitted only in little packets, called **quanta.** Furthermore, he proposed that the energy associated with these quanta is proportional to the frequency of the radiation. In terms of an equation, Planck proposed that

Figure 6-6 Max Planck (1858–1947), the German physicist who is considered to be the father of the quantum theory. Planck's personal life was clouded by tragedy. His two daughters died in childbirth, one son died in World War I, and another son was executed in World War II for his part in the attempt to assassinate Hitler in 1944.

$$E = h\nu \qquad (6\text{-}2)$$

where E is the energy of the radiation, h is a proportionality constant now called **Planck's constant,** and ν is the frequency of the radiation. Planck was able to reproduce all the data on blackbody radiation by choosing h to be 6.626×10^{-34} J·s. Planck's constant is one of the most famous and fundamental constants of science.

Planck's theory of blackbody radiation, with its unconventional assumption that a heated body can emit radiation only in small packets, was not widely accepted in spite of its success in interpreting the experimental data. Most scientists considered Planck's theory to be somewhat of a curiosity and believed that in time a more satisfactory, which is to say, classical, theory of blackbody radiation would emerge.

In the 1880s it was discovered that electrons are ejected from the surface of a metal when the surface is exposed to ultraviolet radiation. This phenomenon is now known as the **photoelectric effect.** (Typical experimental data are shown in Figure 6-7.) It was also found that each metal requires exposure to ultraviolet radiation at a certain, characteristic minimum frequency in order for electrons to be ejected. This frequency is called the **threshold frequency** and is denoted by ν_0. The value of ν_0 depends upon the metal used in the experiment. Below the threshold frequency no electrons are ejected, while above this frequency electrons are readily ejected. In addition, it was observed (Figure 6-7) that the graph of the kinetic energy of the ejected electrons versus the frequency of the radiation is a straight line when the frequency exceeds the threshold frequency ν_0. Like the data on blackbody radiation, these data defied theoretical explanation for many years, and it wasn't until 1905 when Einstein, using the ideas behind Equation (6-2), was able to give an interpretation of the experimental results. Following Planck's thinking, Einstein proposed that the incident radiation consists of little packets of energy $E = h\nu$, which he called **photons.** It is often convenient to picture electromagnetic radiation as a beam of photons with an energy per photon of $E = h\nu$.

Example 6-2: Calculate the frequency and the wavelength of a photon that has an energy equal to the ionization energy of one hydrogen atom.

Solution: According to Table 6-1, the ionization energy of one mole of hydrogen atoms is 1.31 $MJ{\cdot}mol^{-1}$. The ionization energy of one hydrogen atom is

$$E = \frac{1.31 \times 10^6 \text{ J}\cdot\text{mol}^{-1}}{6.022 \times 10^{23} \text{ atom}\cdot\text{mol}^{-1}} = 2.18 \times 10^{-18} \text{ J}\cdot\text{atom}^{-1}$$

We use Equation (6-2) to calculate the frequency:

$$\nu = \frac{E}{h} = \frac{2.18 \times 10^{-18} \text{ J}}{6.626 \times 10^{-34} \text{ J}\cdot\text{s}} = 3.29 \times 10^{15} \text{ s}^{-1}$$

The wavelength is calculated by using Equation (6-1):

$$\lambda = \frac{c}{\nu} = \frac{3.00 \times 10^8 \text{ m}\cdot\text{s}^{-1}}{3.29 \times 10^{15} \text{ s}^{-1}} = 9.12 \times 10^{-8} \text{ m} = 91.2 \text{ nm}$$

6-6. EINSTEIN APPLIED CONSERVATION OF ENERGY TO THE PHOTOELECTRIC EFFECT

Energy is required to remove an electron from the surface of a metal just as energy is required to remove an electron from an atom. The minimum energy required to remove an electron from the surface of a metal is called the **work function** of the metal and is denoted by the Greek letter phi, Φ. Because Φ is an energy, it has units of joules.

In order to eject an electron, we must supply at least an energy Φ. This is the minimum energy required, but since energy and frequency are related through the equation $E = h\nu$, we see that there is a minimum frequency, the threshold frequency, ν_0, required as well. Thus we write

$$\Phi = h\nu_0 \tag{6-3}$$

If we supply more energy than Φ by using radiation with a higher frequency, the excess energy goes into the kinetic energy of the ejected electrons. If the incident radiation has a frequency ν, then the incident energy in excess of Φ is $h\nu - \Phi$, or $h\nu - h\nu_0$. Because the excess energy goes into the kinetic energy of the ejected electrons, we can write the equation

$$\text{K.E.} = h\nu - \Phi = h\nu - h\nu_0 \tag{6-4}$$

Einstein developed Equation (6-4) by applying the principle of conservation of energy. The energy of the incident photon ($h\nu$) is equal to the energy just required to remove the electron from the surface (Φ) plus the energy that becomes the kinetic energy of the electron. Equation (6-4) gives a straight line like that in Figure 6-7 when K.E. is plotted versus ν.

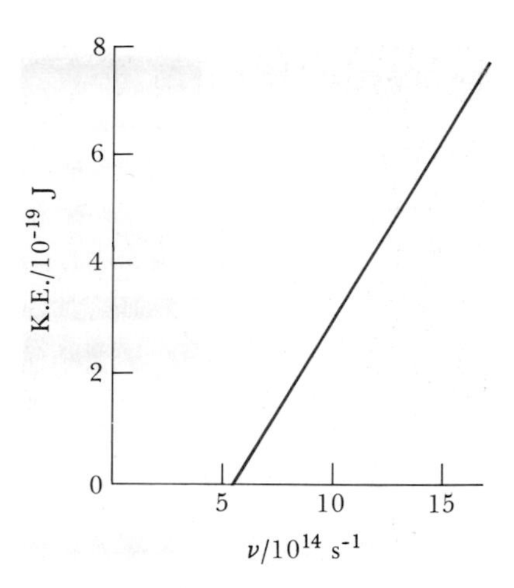

Figure 6-7 When a metallic surface is exposed to ultraviolet radiation of frequency ν, electrons are ejected from the surface. Here we plot the kinetic energy, K.E., of these electrons versus ν for sodium. The frequency below which no electrons are ejected is called the threshold frequency and is equal to 5.51×10^{14} Hz for sodium metal.

Example 6-3: Given that the work function of sodium is 3.65×10^{-19} J, determine the threshold frequency for sodium.

Solution: Using Equation (6-3), we get

$$\nu_0 = \frac{\Phi}{h} = \frac{3.65 \times 10^{-19} \text{ J}}{6.626 \times 10^{-34} \text{ J}\cdot\text{s}}$$
$$= 5.51 \times 10^{14} \text{ s}^{-1}$$

Figure 6-8 Albert Einstein is most famous for his theory of relativity, but his theory of the photoelectric effect also is recognized as a key step in the development of the quantum theory.

Example 6-4: The threshold frequency of sodium is $5.51 \times 10^{14}\ s^{-1}$. Calculate the kinetic energy of the electrons ejected when the surface of sodium is exposed to ultraviolet radiation of wavelength 180 nm.

Solution: In order to use Equation (6-4), we must first calculate the frequency of the incident radiation. This is given by Equation (6-1) as

$$\nu = \frac{c}{\lambda} = \frac{3.00 \times 10^{8}\ \text{m}\cdot\text{s}^{-1}}{180 \times 10^{-9}\ \text{m}} = 1.67 \times 10^{15}\ \text{s}^{-1}$$

The kinetic energy of the ejected electrons is

$$\begin{aligned} \text{K.E.} &= h\nu - \Phi \\ &= h\nu - h\nu_0 \\ &= (6.626 \times 10^{-34}\ \text{J}\cdot\text{s})[1.67 \times 10^{15}\ \text{s}^{-1} - 5.51 \times 10^{14}\ \text{s}^{-1}] \\ &= 7.41 \times 10^{-19}\ \text{J} \end{aligned}$$

By repeating this calculation for a number of frequencies, we can plot the results and obtain excellent agreement with the experimental data in Figure 6-7.

As simple as Einstein's theory of the photoelectric effect may appear today, it was an exceedingly profound proposal in 1905. The idea of energy existing in discrete little packets, that is, the **quantization of energy,** was contradictory to classical physics and not at all well accepted by the scientific community of the time. Nevertheless, in the two very different sets of experimental data for blackbody radiation and the photoelectric effect, which had previously defied explanation, the very same quantization constant, h, arose. These early theories were the first steps in the development of the **quantum theory,** where energies are allowed to take on only certain discrete values. Planck, who first recognized this in 1900, is called the father of the quantum theory. We shall see that all phenomena that occur on molecular and submolecular levels must be described by the quantum theory.

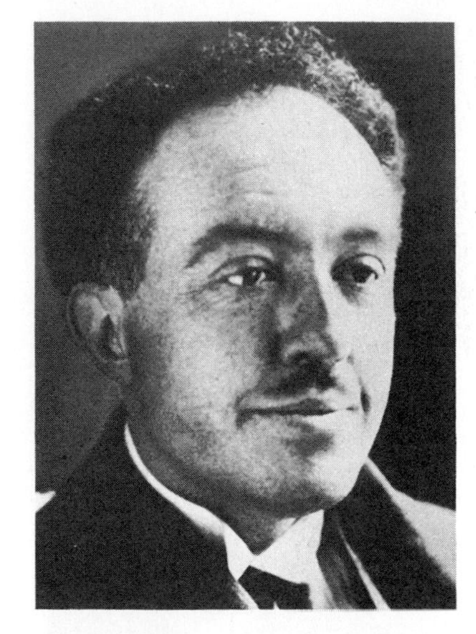

Figure 6-9 Born in 1892, Louis de Broglie studied history as an undergraduate. His interest turned to science as a result of his assignment to radio communications in World War I. De Broglie received a doctoral degree in physics from the University of Paris in 1924 and was awarded the Nobel Prize in physics in 1929.

6-7. DE BROGLIE WAS THE FIRST TO PROPOSE THAT MATTER HAS WAVELIKE PROPERTIES

Scientists have always had difficulty describing the nature of light. In many experiments light exhibits a definite wavelike character, but in many others it seems to behave as a stream of little particles. Because light appears sometimes to be wavelike and sometimes to be particle-like, we talk of the **wave-particle duality** of light. In 1924, Louis de Broglie, a young French physicist, proposed for his doctoral thesis that, if light can display this wave-particle duality, then matter, which certainly appears to be particle-like, might also display wavelike properties under certain conditions. This is a rather strange proposal at first sight, but it does suggest a nice symmetry in nature. Certainly if light can appear to be particle-like at times, why shouldn't matter appear to be wavelike at times?

De Broglie put this idea into a quantitative scheme by proposing that both light *and* matter obey the equation

$$\lambda = \frac{h}{mv} \tag{6-5}$$

where h is Planck's constant, m is the mass of a particle, and v is its speed. Equation (6-5) predicts that a particle having mass m and moving with a speed v has a **de Broglie wavelength** $\lambda = h/mv$.

Example 6-5: Calculate the de Broglie wavelength of an electron traveling at 1.00 percent of the speed of light.

Solution: The mass of an electron is 9.11×10^{-31} kg (inside back cover). Its speed is

$$v = (0.0100)(3.00 \times 10^{8}\ \mathrm{m{\cdot}s^{-1}}) = 3.00 \times 10^{6}\ \mathrm{m{\cdot}s^{-1}}$$

and so

$$mv = (9.11 \times 10^{-31}\ \mathrm{kg})(3.00 \times 10^{6}\ \mathrm{m{\cdot}s^{-1}}) = 2.73 \times 10^{-24}\ \mathrm{kg{\cdot}m{\cdot}s^{-1}}$$

The de Broglie wavelength of this electron is

$$\lambda = \frac{h}{mv} = \frac{6.626 \times 10^{-34}\ \mathrm{J{\cdot}s}}{2.73 \times 10^{-24}\ \mathrm{kg{\cdot}m{\cdot}s^{-1}}} = 2.43 \times 10^{-10}\ \mathrm{m} = 0.243\ \mathrm{nm}$$

We have used the fact that $1\ \mathrm{J} = 1\ \mathrm{kg{\cdot}m^2{\cdot}s^{-2}}$ (Chapter 4). By referring to Table 6-3, we see that the wavelength of the electron in this example corresponds to the wavelength of X-rays.

Problem 6-20 asks you to show that the wavelength for a golf ball traveling at 120 mph is 2.6×10^{-34} m, which is an extremely small wavelength. Thus, although Equation (6-5) is of trivial consequence for a macroscopic object like a baseball, it predicts that electrons can act like X-rays.

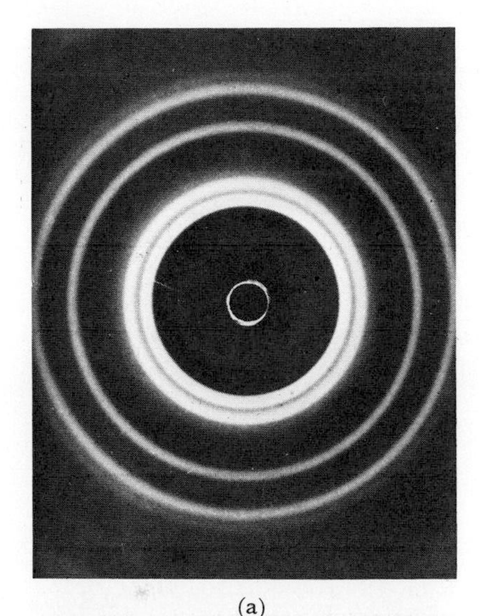

(a)

(b)

Figure 6-10 Bombarding aluminum foil with X-rays and with electrons produces characteristic diffraction patterns. (a) The X-ray diffraction pattern. (b) The electron diffraction pattern. The similarity of the patterns shows that electrons can behave like X-rays and display wavelike properties.

6-8. THE ELECTRON MICROSCOPE UTILIZES THE WAVELIKE PROPERTIES OF ELECTRONS

When a beam of X-rays is directed at a thin foil of a crystalline substance, the beam is scattered in a definite manner characteristic of the atomic structure of the crystalline substance. This phenomenon, called **X-ray diffraction,** happens because the length of the interatomic spacings in the crystal is close to the wavelength of the X-rays. A similar pattern, called an **electron diffraction** pattern, occurs when a beam of electrons is used (Figure 6-10). The similarity of the two patterns demonstrates that both X-rays and electrons do indeed behave similarly in these experiments.

The wavelike property of electrons is used in **electron microscopes.** Electron wavelengths can be controlled by an applied voltage; the small deBroglie wavelengths attainable provide a more precise probe than that possible with an ordinary light microscope. In addition, in contrast to electromagnetic radiation of similar wavelengths (X-rays and ultraviolet radiation), electron beams can be readily focused by using electric and magnetic fields. Electron microscopes are now routine tools in chemical and biological investigations of molecular structures.

An interesting fact in the concept of the wave-particle duality of matter is that it was J. J. Thomson who first showed, in 1895, that the electron is a subatomic particle and G. P. Thomson who was one of the first to show experimentally, in 1926, that the electron could act as a wave. These two Thomsons were father and son. The father received a Nobel Prize in 1906 for showing that the electron is a particle, and the son received a Nobel Prize in 1937 for showing that it is a wave.

6-9. THE ENERGY OF THE ELECTRON IN A HYDROGEN ATOM IS QUANTIZED

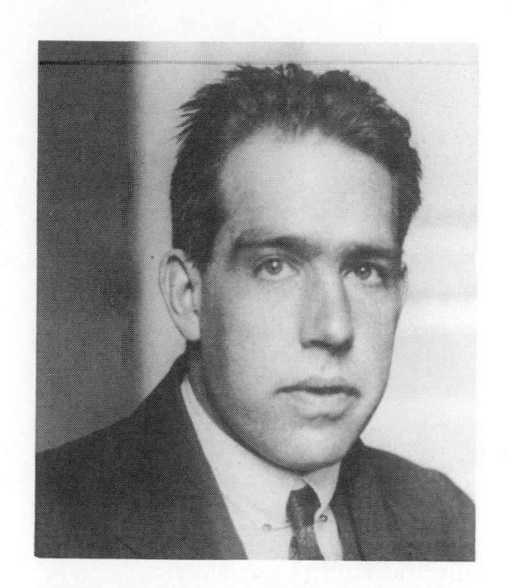

Figure 6-11 The Danish physicist, Niels Bohr (1885–1962), was a key figure in the development of the quantum theory. He was director of the Institute of Theoretical Physics in Copenhagen, a center of quantum theoretical research in the 1920s and 1930s. The Institute is still famous and is supported by the Carlsberg brewery.

In 1913, a young Danish physicist named Niels Bohr (Figure 6-11) formulated a description of the hydrogen atom that very successfully explained the observed atomic spectrum of hydrogen. A key postulate of the Bohr theory is that the electron in a hydrogen atom is restricted to only certain circular orbits about the nucleus. Although the Bohr theory preceded de Broglie's work by more than ten years, this postulate is equivalent to assuming that a stable orbit results only when the de Broglie wave associated with the electron matches, or is in phase, as the electron makes a complete revolution. Otherwise there would be some cancellation of amplitude upon each revolution, and the wave would progressively disappear, as shown schematically in Figure 6-12. For the wave pattern around an orbit to be stable, an integral number of complete wavelengths must fit around the circumference of the orbit. For an orbit of radius r, the circumference is $2\pi r$, and so we have the **quantum condition**

$$2\pi r = n\lambda \qquad n = 1, 2, 3, \ldots \tag{6-6}$$

Using what is essentially this condition, Bohr showed that the energies of the electron in these orbits is given by the equation

$$E_n = -\frac{2.18 \times 10^{-18}\ \text{J}}{n^2} \qquad n = 1, 2, 3, \ldots \tag{6-7}$$

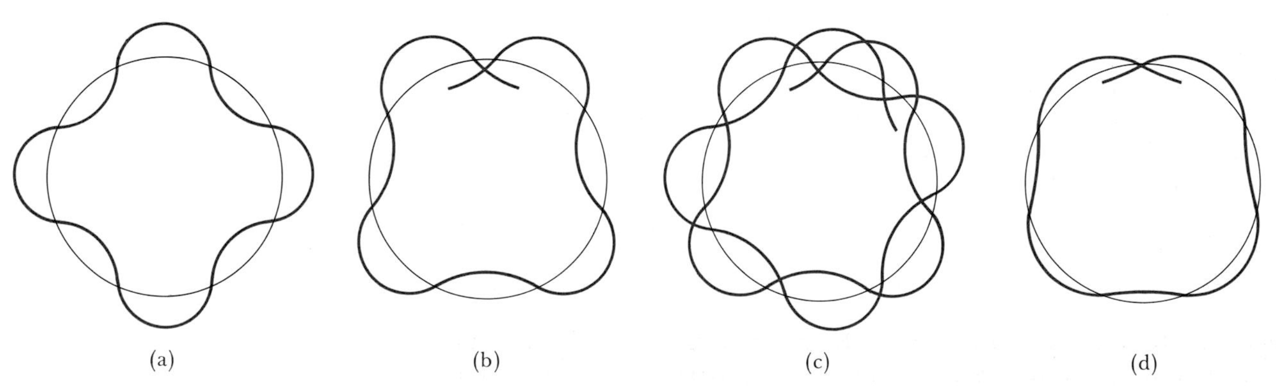

Figure 6-12 An illustration of matching and mismatching de Broglie waves traveling in Bohr orbits. If the wavelengths of the de Broglie waves are such that an integral number of them fit around the circle, then they match after a complete revolution, as shown in (a). If a wave does not match after a complete revolution, cancellation will result and the wave will progressively disappear, as seen in (b) through (d).

Note that the value of n is restricted to integer values. When $n = 1$, $E_1 = -2.18 \times 10^{-18}$ J; when $n = 2$, $E_2 = -0.545 \times 10^{-18}$ J; and so on. We say that the energy is **quantized** because E_n can take on only discrete values. These allowed **energy states** are presented schematically in Figure 6-13.

Because of the minus sign, all the energies given by Equation (6-7) are negative. The lowest energy is E_1, and $E_1 < E_2 < E_3$, and so on. The state of zero energy occurs when $n = \infty$. In the state of zero energy, the proton and electron are so far apart that they do not attract each other at all, and so we take their interaction energy to be zero. At closer distances, the proton and electron attract each other because they have opposite charges. A negative energy state is more stable than a state of zero energy.

Allowed energy states are called **stationary states** in the quantum theory. The stationary state of lowest energy is called the **ground state,** and the states of higher energies are called **excited states.** The state with $n = 1$ is the ground state; the state with $n = 2$ is the **first excited state;** the state with $n = 3$ is the **second excited state;** and so on.

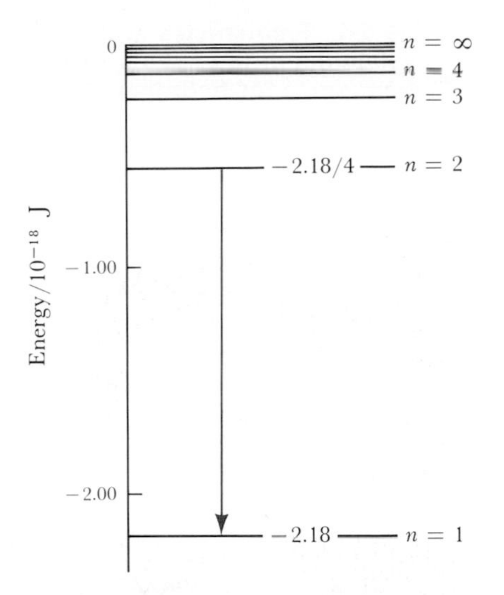

Figure 6-13 The energy states of the electron in the hydrogen atom according to the quantum theory. The electron can have only the energies given by $E_n = (-2.18 \times 10^{-18} \text{J})/n^2$, $n = 1, 2, 3, \ldots$, and no others. The red vertical line shows the electron in a hydrogen atom undergoing a transition from the $n = 2$ state to the $n = 1$ state. The energy of the $n = 1$ state is less than that of the $n = 2$ state, and so this transition is accompanied by the emission of electromagnetic radiation.

6-10. ATOMS EMIT OR ABSORB ELECTROMAGNETIC RADIATION WHEN THEY UNDERGO TRANSITIONS FROM ONE STATIONARY STATE TO ANOTHER

Bohr assumed that when an atom is in a stationary state, it does not absorb or emit electromagnetic radiation. When it undergoes a transition from one stationary state to another, however, it emits or absorbs electromagnetic radiation of only certain frequencies, producing a line spectrum. Consider a hydrogen atom that undergoes a transition from the $n = 2$ state to the $n = 1$ state (Figure 6-13). In this case the electron goes from a higher energy state to a lower energy state. The energy released is given by $E_2 - E_1$ and is emitted as electromagnetic radiation. The frequency of the emitted radiation satisfies the equation

$$\Delta E = E_2 - E_1 = h\nu_{2\rightarrow 1} \tag{6-8}$$

Equation (6-8) is identical to Planck's original quantum hypothesis (Equation 6-2).

Equation (6-8) can be written in a form that emphasizes the law of conservation of energy. Initially, the hydrogen atom is in state 2, with energy E_2. It then makes a transition to state 1, with energy E_1, and emits a photon of energy $h\nu_{2\rightarrow 1}$. The total energy after the transition is $E_1 + h\nu_{2\rightarrow 1}$, and the total energy before the transition is E_2. According to the law of conservation of energy,

$$E_2 = E_1 + h\nu_{2\rightarrow 1} \tag{6-9}$$

Equation (6-9) is exactly the same as Equation (6-8). We can obtain the frequency of the electromagnetic radiation emitted by solving Equation (6-8) or (6-9) for $\nu_{2\rightarrow 1}$:

$$\nu_{2\rightarrow 1} = \frac{E_2 - E_1}{h}$$

If we examine the transition from an arbitrary excited state ($n =$

2, 3, 4, . . .) to the ground state ($n = 1$), then we find a series of lines whose frequencies are

$$\nu_{n\to1} = \frac{E_n - E_1}{h} \qquad n = 2, 3, 4, \ldots \qquad (6\text{-}10)$$

Table 6-4 Frequencies and wavelengths of the lines in the Lyman series for hydrogen

n	$\nu_{n\to1}/10^{15}\ s^{-1}$	$\lambda_{n\to1}/nm$
2	2.47	122
3	2.92	103
4	3.08	97.3
5	3.16	95.0
6	3.20	93.7
⋮	⋮	⋮
∞	3.29	91.2

If we substitute Equation (6-7) into Equation (6-10), then we get

$$\nu_{n\to1} = \left(\frac{2.18 \times 10^{-18}\ J}{6.626 \times 10^{-34}\ J\cdot s}\right)\left(\frac{1}{1^2} - \frac{1}{n^2}\right) \qquad n = 2, 3, 4, \ldots$$

$$= (3.29 \times 10^{15}\ s^{-1})\left(\frac{1}{1^2} - \frac{1}{n^2}\right) \qquad n = 2, 3, 4, \ldots \qquad (6\text{-}11)$$

Equation (6-11) predicts that there is a series of lines in the hydrogen atom emission spectrum that correspond to transitions from state n to state 1. The frequencies of these lines are given by Equation (6-11) with $n = 2$, 3, 4, and so on. The values of these frequencies are given in Table 6-4. This series of lines occurs in the ultraviolet region and is called the **Lyman series.** The agreement between the frequencies or wavelengths calculated from Equation (6-11) and those observed experimentally is excellent.

Equation (6-11) indicates that the lines of the Lyman series emission spectrum are caused by transitions from excited states ($n > 1$) to the ground state ($n = 1$). Figure 6-14 shows that there is also a series due to transitions from higher excited states to the $n = 2$ state. In this case Equation (6-10) becomes

$$\nu_{n\to2} = \frac{E_n - E_2}{h} \qquad n = 3, 4, 5, \ldots$$

and substitution of Equation (6-7) into this equation gives

$$\nu_{n\to2} = (3.29 \times 10^{15}\ s^{-1})\left(\frac{1}{2^2} - \frac{1}{n^2}\right) \qquad n = 3, 4, 5, \ldots \ (6\text{-}12)$$

Equation (6-12) predicts the series of lines due to transitions from state n to state 2. This series of lines is called the **Balmer series.**

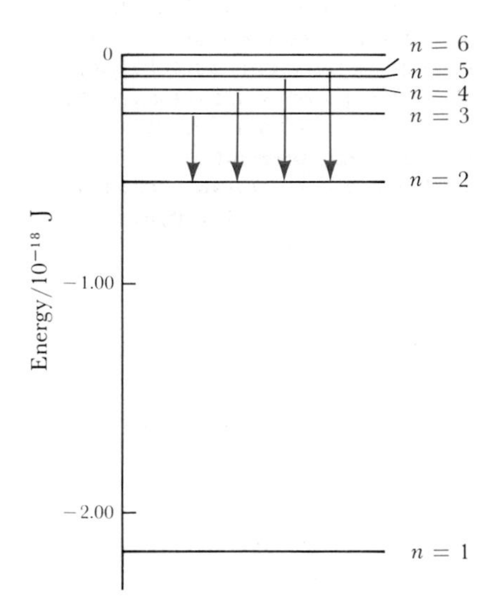

Figure 6-14 Transitions from higher energy states to the $n = 2$ state for a hydrogen atom. Each transition is accompanied by the emission of a photon. This series of lines is called the Balmer series.

Example 6-6: Calculate the frequency and wavelength of the 3→2 transition in the Balmer series.

Solution: We use Equation (6-12) with $n = 3$:

$$\nu_{3\to2} = (3.29 \times 10^{15}\ s^{-1})\left(\frac{1}{4} - \frac{1}{9}\right) = 4.57 \times 10^{14}\ s^{-1}$$

To calculate the corresponding wavelength, we use Equation (6-1):

$$\lambda_{3\to2} = \frac{c}{\nu_{3\to2}} = \frac{3.00 \times 10^8\ m\cdot s^{-1}}{4.57 \times 10^{14}\ s^{-1}} = 6.56 \times 10^{-7}\ m = 656\ nm$$

The agreement of this prediction with the experimental observation is excellent (Figure 6-5). The frequencies and wavelengths of the other $n\to2$ transitions are calculated in the same way.

As we said before, line spectra are obtained when a gas is placed in a discharge tube and a spark is discharged through the gas. The discharge is a pulse of energy that promotes the atoms into excited states. As the atoms return to their ground state, the electrons fall down through the allowed energy states and produce the observed spectrum, which is called an **emission spectrum.**

An **absorption spectrum** is observed experimentally when an atomic gas is irradiated with electromagnetic radiation containing all frequencies (Figure 6-15). The radiation promotes electrons to excited energy states. Only radiation of certain frequencies is absorbed by the gas.

Let's consider transitions from the ground state ($n = 1$) of a hydrogen atom to some excited state ($n > 1$). Before the transition, we have a hydrogen atom in its ground state with energy E_1 and a photon of energy $h\nu_{1\rightarrow n}$. After the transition, we have a hydrogen atom in state n with energy E_n. Conservation of energy requires that the total energy before the transition ($E_1 + h\nu_{1\rightarrow n}$) be equal to the total energy after the transition (E_n):

$$h\nu_{1\rightarrow n} + E_1 = E_n \qquad n = 2, 3, 4, \ldots \qquad (6\text{-}13)$$

If we solve Equation (6-13) for $\nu_{1\rightarrow n}$, then we find that

$$\nu_{1\rightarrow n} = \frac{E_n - E_1}{h} \qquad n = 2, 3, 4, \ldots \qquad (6\text{-}14)$$

Note that the frequency of absorption is the same as the frequency of emission given by Equation (6-10). Equation (6-14) represents the Lyman series in the absorption spectrum of a hydrogen atom. Figure 6-16 shows the transitions that correspond to the Lyman series in an absorption spectrum.

Example 6-7: Use Equation (6-7) to calculate the ionization energy of the hydrogen atom.

Solution: The ionization energy is the energy required to completely remove the electron from the ground state of the atom. Therefore, the ionization must correspond to the transition from $n = 1$ to $n = \infty$. The energy associated with this transition is $E_\infty - E_1$. If we use Equation (6-7) for E_n, then we obtain for the ionization energy

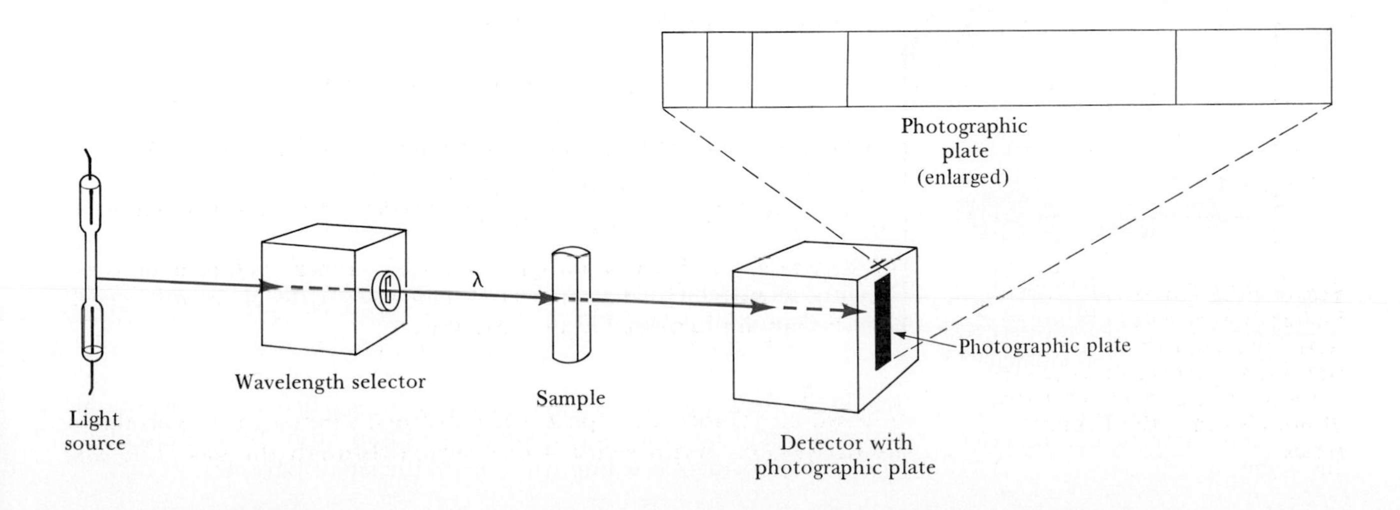

Figure 6-15 A schematic diagram of an atomic absorption experiment. Electromagnetic radiation is passed through a gas sample and then through a wavelength selector (a prism). The transmitted beam is then directed onto a detector, such as a photographic plate. Only certain wavelengths are absorbed by the sample gas, and the photographic plate is exposed at all wavelengths except those absorbed by the sample. The wavelengths absorbed by the sample appear as unexposed lines on the plate.

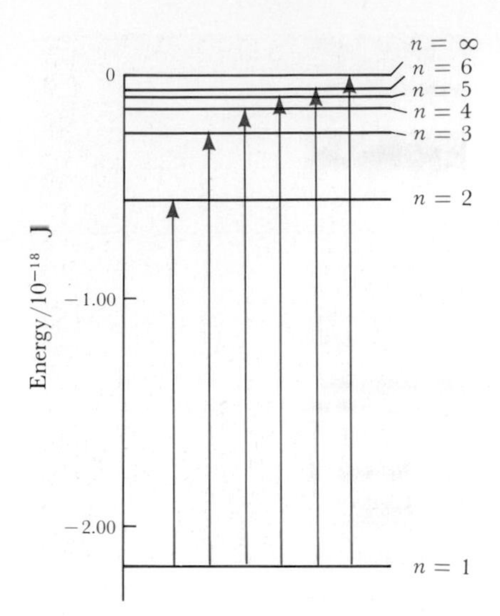

Figure 6-16 A hydrogen atom in its ground electronic state ($n = 1$) can absorb electromagnetic radiation. When this occurs, the electron is promoted to an excited state.

$$E_\infty - E_1 = 0 - \left(\frac{-2.18 \times 10^{-18}\,\text{J}}{1^2}\right) = I = 2.18 \times 10^{-18}\,\text{J}$$

This is the ionization energy per atom. If we multiply this result by Avogadro's number, then we obtain the ionization energy per mole:

$$I = \left(2.18 \times 10^{-18}\frac{\text{J}}{\text{atom}}\right)\left(\frac{6.022 \times 10^{23}\text{ atom}}{1\text{ mol}}\right) = 1310\text{ kJ}\cdot\text{mol}^{-1}$$

which is in excellent agreement with the value given in Table 6-1.

6-11. THE BOHR THEORY IS NOT CONSISTENT WITH THE HEISENBERG UNCERTAINTY PRINCIPLE

As successful as the Bohr theory was in explaining the atomic spectrum of hydrogen, it could not be extended to atoms other than hydrogen. Furthermore, later work showed that it was inconsistent with a fundamental principle of nature called the **Heisenberg uncertainty principle.** In the mid 1920s, a young German physicist, Werner Heisenberg (Figure 6-17), showed that it is not possible to measure accurately *both* the position *and* the momentum (mv, mass times velocity) of a particle simultaneously. This uncertainty is *not* due to poor measurement or poor experimental technique, but is a fundamental property of the act of measurement itself. Heisenberg was able to show that if Δx is the distance within which we locate a particle and if Δp is the range of possible values of its momentum, then Δx and Δp, the uncertainties in the position and momentum of the particle, are related through the relation

$$(\Delta x)(\Delta p) \simeq h \qquad (6\text{-}15)$$

This relation expresses the Heisenberg uncertainty principle, which imposes a fundamental limitation on the accuracy of determining simultaneously the position and the momentum of a particle.

Figure 6-17 Werner Heisenberg (1901–1976) was one of Germany's greatest scientists. He was a leader in the development of the quantum theory in the 1920s and formulated the uncertainty principle that bears his name.

Example 6-8: Suppose that we wish to locate an electron to within 5.0×10^{-11} meter, which is a few percent of the size of an atom. Estimate the corresponding uncertainty in the velocity of the electron according to the Heisenberg uncertainty principle.

Solution: We use Equation (6-15) to estimate Δp:

$$\Delta p \simeq \frac{h}{\Delta x} \simeq \frac{6.626 \times 10^{-34}\,\text{J}\cdot\text{s}}{5.0 \times 10^{-11}\,\text{m}} \simeq 1.3 \times 10^{-23}\,\text{kg}\cdot\text{m}\cdot\text{s}^{-1}$$

where we have used the fact that $1\,\text{J} = 1\,\text{kg}\cdot\text{m}^2\cdot\text{s}^{-2}$. Because the mass of an electron is 9.11×10^{-31} kg, the uncertainty in its velocity is given by

$$\Delta v = \frac{\Delta p}{m} = \frac{1.3 \times 10^{-23}\,\text{kg}\cdot\text{m}\cdot\text{s}^{-1}}{9.11 \times 10^{-31}\,\text{kg}} = 1.4 \times 10^7\,\text{m}\cdot\text{s}^{-1}$$

This is an extremely large uncertainty in the velocity, being about 5 percent of the speed of light.

Just as the de Broglie wavelength of a moving particle is important only for very small particles such as electrons and atoms, the Heisenberg uncertainty principle is important only for small particles.

In assuming that the electron in the hydrogen atom is restricted to discrete, sharp orbits, the Bohr theory attempts to provide a too-detailed picture of the motion of the electron. In 1926 Erwin Schrödinger (Figure 6-18) first presented what has become one of the most important equations in science: the **Schrödinger equation,** the central equation of the quantum theory. This equation is consistent with both the wave nature of particles and the Heisenberg uncertainty principle. Furthermore, unlike the Bohr theory, it correctly predicts the properties of multielectron atoms and molecules. The Schrödinger equation is too complicated to present here, but we must discuss some of its consequences. When we solve the Schrödinger equation for a hydrogen atom, we find that the energy of the electron is restricted to a discrete set of values that is the same as that predicted by the Bohr theory. In other words, the energy of the electron is quantized and is restricted to the values given by Equation (6-7). This is why the Bohr theory gave such excellent agreement with the hydrogen atomic spectrum.

Figure 6-18 Erwin Schrödinger, a Viennese physicist, formulated the modern quantum theory of atoms and molecules in 1925. His theory is summarized by what is called the Schrödinger equation, which describes the motion of extremely small particles, such as electrons, atoms, and molecules.

The Bohr theory and the Schrödinger equation differ completely, however, in their descriptions of the location of the electron about the nucleus. Instead of restricting the electron to certain, sharp orbits, the Schrödinger equation provides one or more functions, called **wave functions** or **orbitals,** associated with each allowed energy. Wave functions are customarily denoted by the Greek letter psi, ψ, and are functions of the position of the electron. We emphasize this dependence by writing $\psi = \psi(x, y, z)$, where x, y, and z are coordinates that are used to denote the position of the electron. The square of a wave function, ψ^2, has a direct physical interpretation. The value of the square of a wave function, $\psi^2(x, y, z)$, is a **probability density** in the sense that $\psi^2 \Delta V$ is the probability that the electron will be found in a small volume ΔV surrounding the point (x, y, z). This is a profound statement because it says that we cannot locate the electron precisely; we only can assign a *probability* that the electron is in a certain region.

The integer n that specifies the energy of the electron in a hydrogen atom is called a **quantum number.** Although only one quantum number is needed to specify the energy of the electron, three quantum numbers are required to specify the wave functions. These three quantum numbers are denoted by n, ℓ, and m_ℓ. Let's look at the significance of each of these quantum numbers in turn.

The quantum number n is called the **principal quantum number.** It alone determines the energy of the electron in a hydrogen atom, and we have already seen that n can take on the values $n = 1, 2, 3,$ and so on.

When $n = 1$, the energy is the lowest allowed value. This is the ground state of the hydrogen atom. The wave function that describes the ground state of the hydrogen atom depends upon the distance of the electron from the proton and can be written $\psi(r)$, where r is the distance of the electron from the nucleus. For reasons that we'll soon know, the ground-state wave function of the hydrogen atom is denoted by ψ_{1s} rather than by just ψ_1. The probability density, $\psi^2_{1s}(r)$, is plotted in Figure 6-19. Note that probability density falls off rapidly with distance. Because ψ^2_{1s} depends upon only the magnitude of r and not upon the direction of r in space, ψ^2_{1s} is said to be **spherically symmetric.**

Several other representations of wave functions, or orbitals, are more lucid than simply plotting the square of the wave function. For

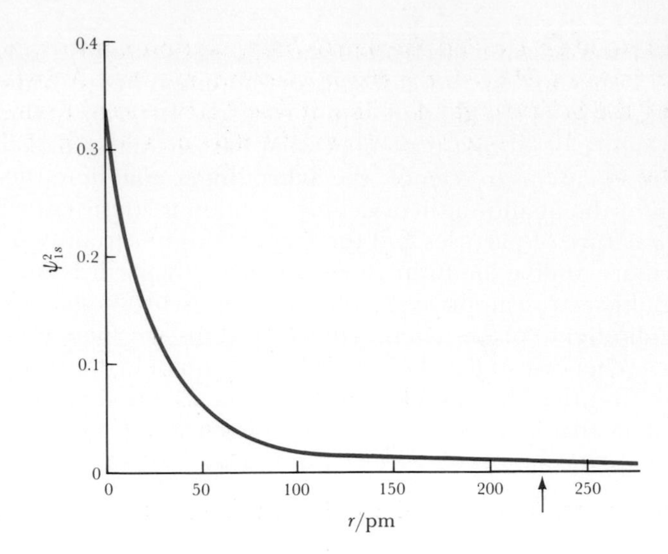

Figure 6-19 A graph of ψ_{1s}^2 versus r. Even though the electron is most likely to be found near the nucleus, the curve never quite falls to zero as r increases. Thus there is a nonzero probability, however small, of finding the electron at *any* distance from the nucleus. The arrow indicates the distance beyond which there is only a 1 percent chance of finding the electron.

example, we can represent the 1s orbital by the stippled diagram in Figure 6-20(a). The number of dots in a volume ΔV is proportional to the probability of finding the electron in that volume. The relation between a plot of ψ_{1s}^2 versus r and the stippled representation of Figure 6-20 is shown in Figure 6-21.

Another representation of the 1s orbital shows the volume within which the electron has a certain chance of being found. The sphere in Figure 6-20(b) represents the volume within which there is a 99 percent chance of finding the electron. The representation in Figure 6-20(b) has the advantage of portraying clearly the three-dimensional shape of the orbital.

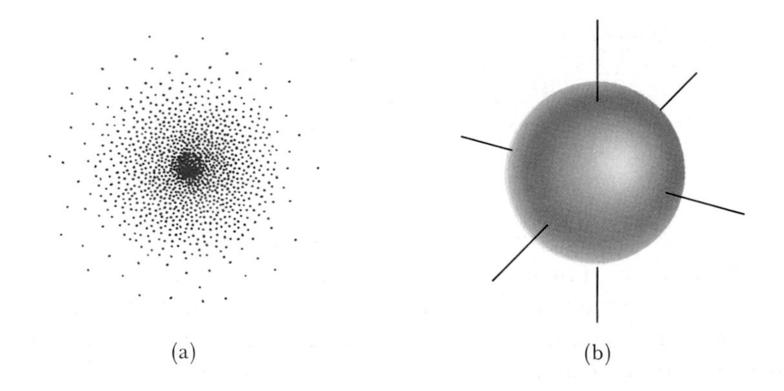

Figure 6-20 Two different representations of a hydrogen 1s orbital, or wave function. (a) The density of the stippled dots in any small region is proportional to the probability of finding the electron in that region. (b) The sphere encloses a volume in which there is a 99 percent probability of finding the electron. Realize that the 1s orbital is spherically symmetric and that (a) represents a cross section through a sphere.

6-12. THE SHAPE OF AN ORBITAL DEPENDS UPON THE VALUE OF THE AZIMUTHAL QUANTUM NUMBER

The principal quantum number, n, specifies the size, or the extent, of an orbital. The quantum number ℓ specifies the shape of an orbital. Orbitals with different values of ℓ have different shapes. This second quantum number is called the **azimuthal quantum number,** although we could just as well call it the shape quantum number. A direct result of solving the Schrödinger equation is that ℓ is restricted to the values, 0, 1, . . . , $n - 1$. The allowed values of ℓ depend upon the value of n according to

n	ℓ
1	0
2	0, 1
3	0, 1, 2
4	0, 1, 2, 3
⋮	⋮

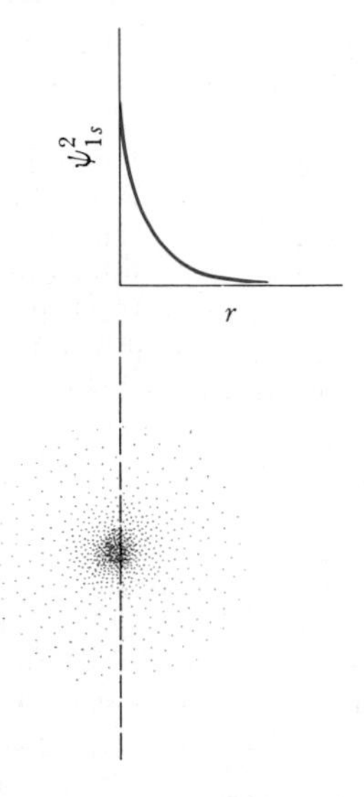

Figure 6-21 The relation between a plot of ψ^2_{1s} versus r and the stippled representation of a 1s orbital. Both show that the probability of finding the electron around some point decays rapidly with distance from the nucleus.

For historical reasons, the values of ℓ are designated by letters:

ℓ	0	1	2	3 . . .
Designation	*s*	*p*	*d*	*f* . . .

The letters *s, p, d,* and *f* stand for *s*harp, *p*rincipal, *d*iffuse, and *f*undamental, which are the designations of the series in the atomic emission spectra of the alkali metals. For $\ell = 4$ and greater, the letters follow alphabetical order after *f*.

Orbitals are denoted by first writing the numerical value of n (1, 2, 3, . . .) and then following this by the letter designation for the value of ℓ (*s, p, d, f,* . . .). For example, an orbital for which $n = 1$ and $\ell = 0$ is called a 1*s* orbital. An orbital for which $n = 3$ and $\ell = 2$ is called a 3*d* orbital. Table 6-5 lists the orbitals for $n = 1$ through 4.

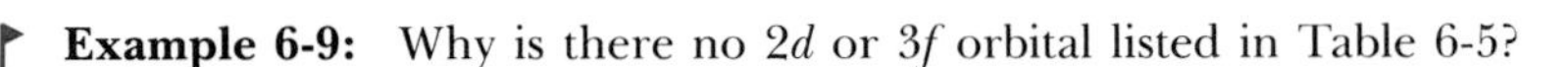

Example 6-9: Why is there no 2*d* or 3*f* orbital listed in Table 6-5?

Solution: When $n = 2$, ℓ can have only the values 0 and 1 because $\ell = 0, 1, 2, \ldots, n - 1$ and $n - 1 = 1$ when $n = 2$. Thus there is no such orbital as a 2*d* orbital. Similarly, when $n = 3$, ℓ can have only the values 0, 1, and 2. A 3*f* orbital would require that $n = 3$ and $\ell = 3$, and so there is no 3*f* orbital.

An electron described by a 1*s* orbital has an energy that is obtained from Equation (6-7) by setting $n = 1$. When $n = 2$, ℓ can be 0 or 1, and so we have two possibilities, a 2*s* and a 2*p* orbital. Both of these orbitals have a principal quantum number $n = 2$, and so an electron described by either of these orbitals has an energy E_2 in Equation (6-7). These two orbitals have different shapes, however, because they are associated with different values of ℓ. All *s* orbitals are spherically symmetric. In Figure 6-22, ψ^2_{2s} is plotted versus r. By comparing this graph with Figure 6-21, we see that a 2*s* orbital extends farther from the nucleus than does a 1*s* orbital. The radius of a sphere that encloses a 99 percent probability of finding the electron in a 2*s* orbital is about 500 pm; the corresponding radius for a 1*s* orbital is about 200 pm. Figure 6-22 shows that the probability density is zero over a spherical surface whose radius is 106 pm. It is common for orbitals to have surfaces on which

Table 6-5 The designation of orbitals by letters

n	ℓ	Designation
1	0	1*s*
2	0	2*s*
	1	2*p*
3	0	3*s*
	1	3*p*
	2	3*d*
4	0	4*s*
	1	4*p*
	2	4*d*
	3	4*f*

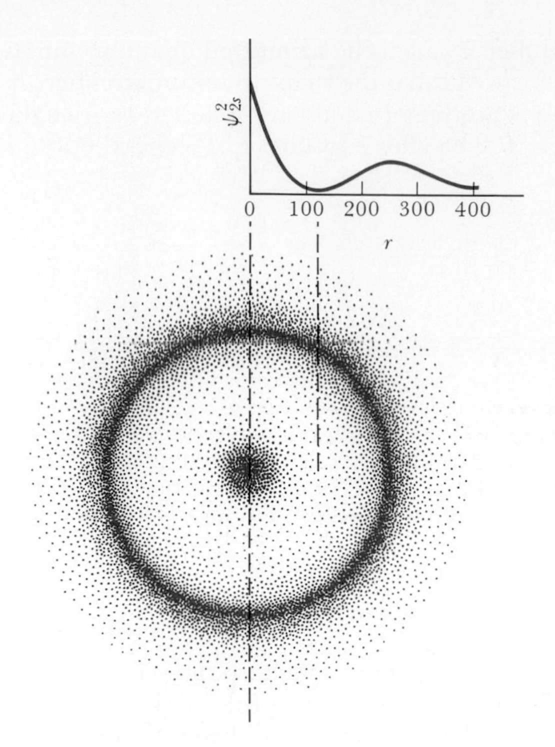

Figure 6-22 The relationship between a plot of ψ_{2s}^2 versus r and a stippled representation of a $2s$ orbital. The distance at which the plot of ψ_{2s}^2 touches zero indicates a spherical nodal surface in the orbital. Remember that s orbitals are spherically symmetric and that this is a cross section of a three-dimensional diagram.

the probability density is zero. Such surfaces are called **nodal surfaces.** Figure 6-22 illustrates the relation between a plot of ψ_{2s}^2 versus r and a stippled diagram representing the $2s$ orbital probability density. The representations shown in Figure 6-20(b) for a $1s$ orbital would look the same, only larger, for a $2s$ orbital.

The surface of 99 percent probability for a $3s$ orbital looks the same as that of a $2s$ orbital, simply larger. Figure 6-23 shows the 99 percent contour surfaces for $1s$, $2s$, and $3s$ orbitals. We see from this figure that n determines the size, or spatial extent, of an orbital.

We also have a $2p$ orbital to consider when $n = 2$. Figure 6-24(a) shows a stippled diagram of a $2p$ orbital. The most obvious feature of a

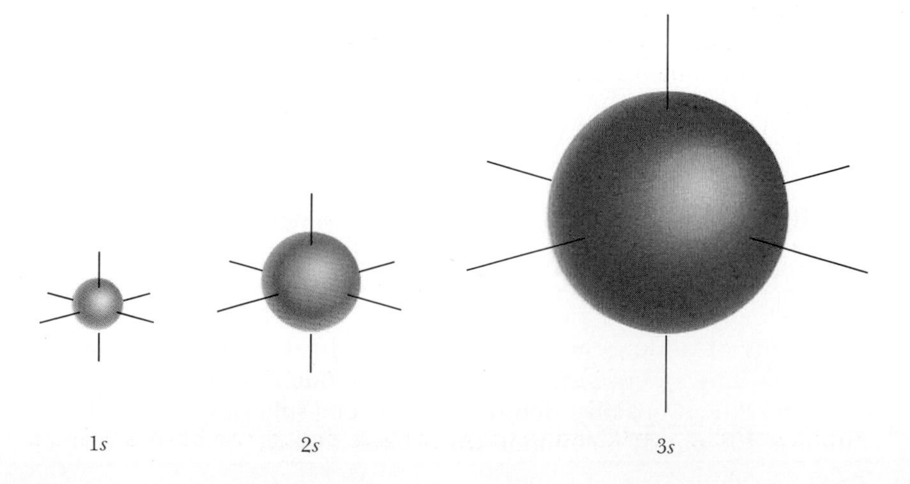

Figure 6-23 The surfaces that enclose a 99 percent probability of finding the electron in a $1s$, $2s$, and $3s$ orbital. Because s orbitals are spherically symmetric, these surfaces are spherical. The radii of the spheres depicted here are in a ratio of about 1:2:5.

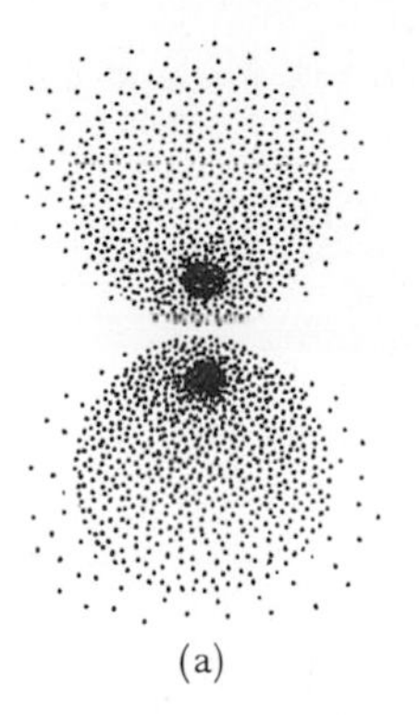
(a)

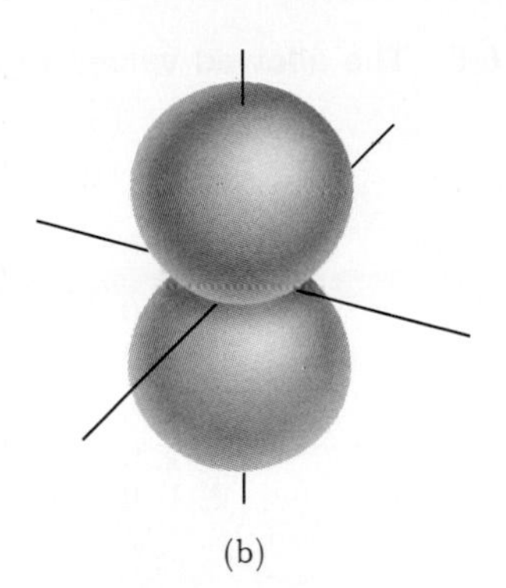
(b)

Figure 6-24 Two representations of a 2*p* orbital: (a) a stippled representation and (b) the surfaces that enclose a 99 percent probability of finding the 2*p* electron within them. The representation in (b) clearly depicts the shape of a 2*p* orbital, which is *not* spherically symmetric. The first representation is a cross section, and the three-dimensional representation is obtained by rotating the figure around the *z* axis. The resulting orbital is said to be cylindrically symmetric.

2*p* orbital is that it is *not* spherically symmetric. The representation shown in Figure 6-24(b) shows the three-dimensional shape of a 2*p* orbital. When viewed along the *z* axis, the 2*p* orbital appears to be circular. We say that the 2*p* orbital is **cylindrically symmetric** about its long axis (the *z* axis in Figure 6-24). Note that the *xy* plane that bisects the 2*p* orbital is a nodal surface; the 2*p* orbital vanishes everywhere on that surface. Just as all *s* orbitals are spherically symmetric, all *p* orbitals are cylindrically symmetric about their long axis. A 3*p* orbital differs from a 2*p* orbital in being larger (because *n* is larger) and in having more nodal surfaces. The most important property of *p* orbitals for our purposes is that they are directed along an axis, as shown in Figure 6-24.

6-13. THE SPATIAL ORIENTATION OF AN ORBITAL DEPENDS UPON THE VALUE OF THE MAGNETIC QUANTUM NUMBER

The third quantum number, m_ℓ, called the **magnetic quantum number,** determines the spatial orientation of an orbital. It turns out that the magnetic quantum number can assume only the values $\ell, \ell - 1, \ell - 2, \ldots, 0, -1, -2, \ldots, -\ell$. The allowed values of m_ℓ depend upon the value of ℓ according to

ℓ	m_ℓ
0	0
1	1, 0, −1
2	2, 1, 0, −1, −2
3	3, 2, 1, 0, −1, −2, −3

For an *s* orbital, $\ell = 0$, and so the only value that m_ℓ can have is 0. For a

Table 6-6 The allowed values of ℓ and m_ℓ for n = 1 through 4

n	ℓ	m_ℓ	Orbital	Number of orbitals
1	0	0	1*s*	1
2	0	0	2*s*	1
	1	1, 0, −1	2*p*	3
3	0	0	3*s*	1
	1	1, 0, −1	3*p*	3
	2	2, 1, 0, −1, −2	3*d*	5
4	0	0	4*s*	1
	1	1, 0, −1	4*p*	3
	2	2, 1, 0, −1, −2	4*d*	5
	3	3, 2, 1, 0, −1, −2, −3	4*f*	7

p orbital, $\ell = 1$ and so m_ℓ can have the values +1, 0, and −1. Table 6-6 summarizes the allowed values of ℓ and m_ℓ for $n = 1$ through $n = 4$.

Example 6-10: Without referring to Table 6-6, list all the values of ℓ and m_ℓ that are allowed for $n = 3$.

Solution: When $n = 3$, ℓ can have the values 0, 1, and 2. Thus we have a 3*s* orbital ($\ell = 0$), a 3*p* orbital ($\ell = 1$), and a 3*d* orbital ($\ell = 2$). For the 3*s* orbital, $\ell = 0$ and so m_ℓ must also equal 0. For the 3*p* orbital, $\ell = 1$ and so m_ℓ can be +1, 0, or −1. For the 3*d* orbital, $\ell = 2$ and so m_ℓ can be +2, +1, 0, −1, or −2.

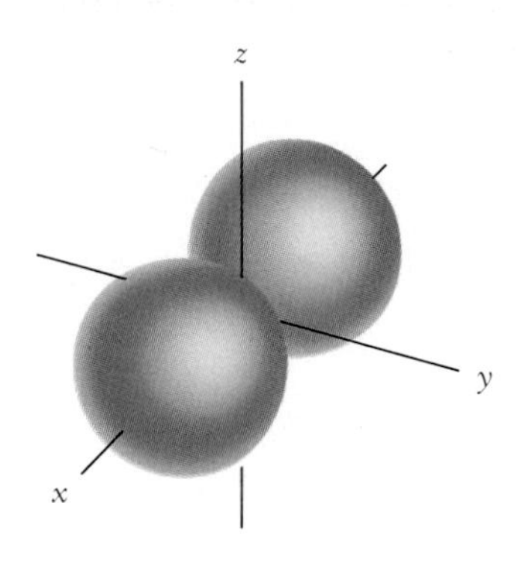

Table 6-6 shows that there is only one *s* orbital for each value of *n*, three *p* orbitals for $n \geq 2$, five *d* orbitals for $n \geq 3$, and seven *f* orbitals for $n \geq 4$. Let *np* denote a 2*p* orbital, a 3*p* orbital, and so on. The three *np* orbitals differ by the value of the magnetic quantum number. All *np* orbitals have the same shape because they all have the same value of ℓ (= 1), but they have different orientations because they all have different values of m_ℓ. The three 2*p* orbitals are shown in Figure 6-25. One 2*p* orbital is directed along the *z* axis, as in Figure 6-24. The other two have the same shape as the one directed along the *z* axis but are directed along the *x* axis and the *y* axis. The *p* orbitals are designated by p_x, p_y, and p_z, with the subscripts indicating the axis along which the orbital is directed.

We could go on to consider *d* orbitals and *f* orbitals, but for most of the topics we shall discuss, *s* orbitals and *p* orbitals will be sufficient (although *d* orbitals are used in Chapters 10 and 22).

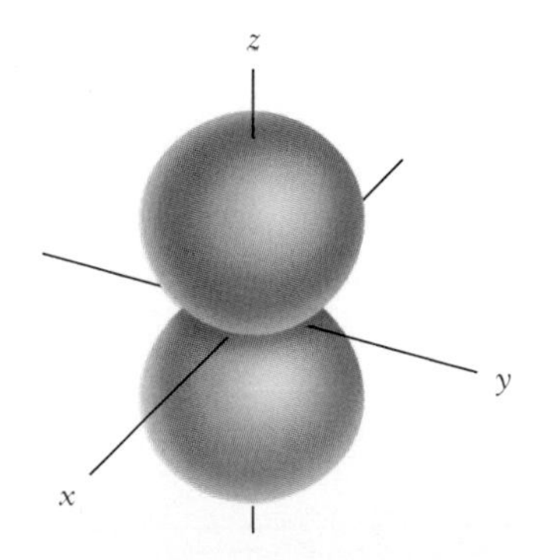

Figure 6-25 The three 2*p* orbitals. Each has the same shape, but their spatial orientations differ. This is because they have the same azimuthal quantum number ($\ell = 1$) but different magnetic quantum numbers. Recall that the shape of an orbital depends upon the value of ℓ and that its orientation depends upon the value of m_ℓ. The three *p* orbitals are directed along the *x*, *y*, and *z* axes and are designated by p_x, p_y, and p_z.

6-14. AN ELECTRON HAS AN INTRINSIC SPIN

As we know, the Schrödinger equation yields three quantum numbers, n, ℓ, and m_ℓ. When first presented, this equation explained a great deal of experimental data, but some scattered observations still did not fit into the picture. For example, close examination of some atomic spectral lines shows that they actually consist of two closely spaced lines. Even though a fine detail, the splitting of spectral lines was perplexing. In 1926, Wolfgang Pauli (Figure 6-26), a young German physicist, argued that this splitting could be explained if the electron exists in two different states. Shortly after this, two Dutch scientists, George Uhlenbeck and Samuel Goudsmit, identified these two different states with a property called the **intrinsic electron spin.** Simply interpreted, an electron spins like a top in one of two directions about its axis. The intrinsic spin of an electron introduces a fourth quantum number, called the **spin quantum number,** denoted by m_s. It designates the spin state of the electron and takes on one of two possible values: $+\frac{1}{2}$ or $-\frac{1}{2}$.

Figure 6-26 Wolfgang Pauli (1900–1958) received his Ph.D. from the University of Munich at the age of 21. Pauli mastered Einstein's papers on relativity while in high school and wrote a highly acclaimed monograph on the theory of relativity when he was only 20 years old. He was famous for his sharp and critical judgment.

Example 6-11: Deduce the possible sets of the four quantum numbers (n, ℓ, m_ℓ, and m_s) when $n = 2$.

Solution: When $n = 2$, ℓ can be 0 or 1. Let's consider the case $\ell = 0$ first. If $\ell = 0$, then $m_\ell = 0$. Thus, we have so far that $n = 2$, $\ell = 0$, and $m_\ell = 0$. The spin quantum number can have the value $+\frac{1}{2}$ or $-\frac{1}{2}$, regardless of the values of the other three quantum numbers. We thus have two possible sets of quantum numbers;

n	ℓ	m_ℓ	m_s
2	0	0	$+\frac{1}{2}$
2	0	0	$-\frac{1}{2}$

Now consider the case $n = 2$ and $\ell = 1$. When $\ell = 1$, m_ℓ can be $+1$, 0, or -1. Thus we have

n	ℓ	m_ℓ
2	1	1
2	1	0
2	1	-1

Each of these sets of three quantum numbers can have $m_s = +\frac{1}{2}$ or $-\frac{1}{2}$, and so

n	ℓ	m_ℓ	m_s
2	1	1	$+\frac{1}{2}$
2	1	1	$-\frac{1}{2}$
2	1	0	$+\frac{1}{2}$
2	1	0	$-\frac{1}{2}$
2	1	-1	$+\frac{1}{2}$
2	1	-1	$-\frac{1}{2}$

There are eight possible sets of the four quantum numbers when $n = 2$.

The introduction of the spin quantum number implies that it takes four quantum numbers to specify the state of the electron in a hydrogen atom. These quantum numbers are

$$n = 1, 2, 3, \ldots$$

$$\ell = 0, 1, 2, \ldots, n - 1$$

$$m_\ell = \ell, \ell - 1, \ldots, 0, -1, \ldots, -\ell$$

$$m_s = +\tfrac{1}{2} \text{ or } -\tfrac{1}{2}$$

Table 6-7 summarizes the allowed combinations of the four quantum numbers for $n = 1$ through $n = 3$.

Table 6-7 The allowed combinations of the four quantum numbers for $n = 1$ through 3

n	ℓ	m_ℓ	m_s
1	0	0	$+\frac{1}{2}$ or $-\frac{1}{2}$
2	0	0	$+\frac{1}{2}$ or $-\frac{1}{2}$
	1	1	$+\frac{1}{2}$ or $-\frac{1}{2}$
		0	$+\frac{1}{2}$ or $-\frac{1}{2}$
		−1	$+\frac{1}{2}$ or $-\frac{1}{2}$
3	0	0	$+\frac{1}{2}$ or $-\frac{1}{2}$
	1	1	$+\frac{1}{2}$ or $-\frac{1}{2}$
		0	$+\frac{1}{2}$ or $-\frac{1}{2}$
		−1	$+\frac{1}{2}$ or $-\frac{1}{2}$
	2	2	$+\frac{1}{2}$ or $-\frac{1}{2}$
		1	$+\frac{1}{2}$ or $-\frac{1}{2}$
		0	$+\frac{1}{2}$ or $-\frac{1}{2}$
		−1	$+\frac{1}{2}$ or $-\frac{1}{2}$
		−2	$+\frac{1}{2}$ or $-\frac{1}{2}$

6-15. THE ENERGY STATES OF ATOMS WITH TWO OR MORE ELECTRONS DEPEND UPON THE VALUES OF BOTH n AND ℓ

As Equation (6-7) indicates, the energy of an electron in a hydrogen atom depends upon only the principal quantum number n, and not upon the other quantum numbers ℓ, m_ℓ, or m_s. Consequently, orbitals having the same value of n, such as the 3*s*, 3*p*, and 3*d* orbitals, have the same energy, as shown in Figure 6-27(a). This is *not* the case, however, for atoms with more than one electron. In multielectron atoms there are not only electron-nucleus interactions but also electron-electron interactions. Because of these electron-electron interactions, the relationship between energy and quantum numbers is more complicated for multielectron atoms than that given by Equation (6-7). The electronic energies of multielectron atoms depend in a complicated way on the azimuthal quantum number ℓ as well as on the principal quantum number n (see Problem 6-86). Thus, for example, the 2*s* and 2*p* orbitals for atoms other than hydrogen have different energies. The ordering of the orbital energies, shown in Figure 6-27(b), is $1s < 2s < 2p < 3s < 3p < 4s < 3d < \ldots$. Note that, as n increases, the dependence of the energy upon ℓ becomes so pronounced that the energy of the 4*s* orbital is less than that of the 3*d* orbital. As in the case of the hydrogen atom, the orbital energies bunch together as n increases, and so this type of "reversal" becomes even more pronounced at higher energies. Fortunately, there is a simple mnemonic for remembering the order of the orbitals, which is shown in Figure 6-28.

6-16. THE PAULI EXCLUSION PRINCIPLE STATES THAT NO TWO ELECTRONS IN THE SAME ATOM CAN HAVE THE SAME SET OF FOUR QUANTUM NUMBERS

Before we can correlate electronic structure with the periodic table, we must learn how to assign the electrons to the various orbitals. It was Wolfgang Pauli, in 1926, who first determined how to make this assignment. In what is now called the **Pauli exclusion principle,** he proposed that no two electrons in the same atom can have the same set of four quantum numbers. The Pauli exclusion principle is a fundamental principle of physics and can be used to understand the periodic table and the other periodic properties of the elements.

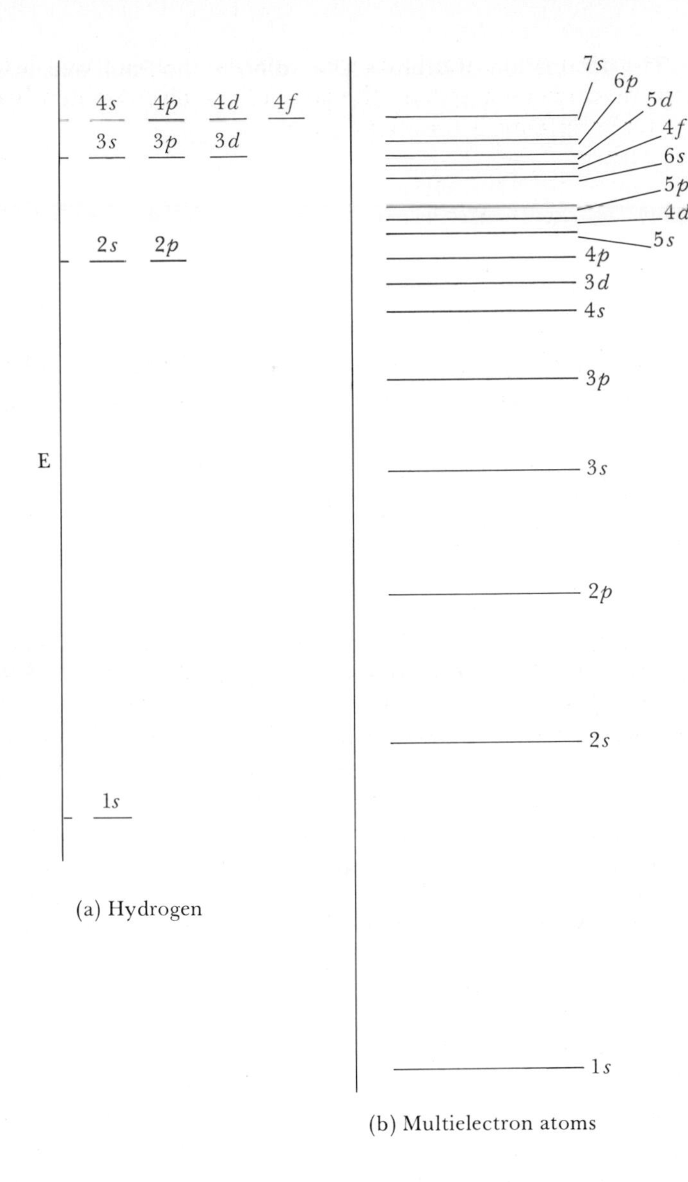

Figure 6-27 The relative energies of atomic orbitals. (a) For hydrogen, the energy depends upon only the principal quantum number; thus orbitals with the same value of n have the same energy. (b) For atoms containing more than one electron, the orbital energies depend upon both the principal quantum number n and the azimuthal quantum number ℓ. Thus orbitals with the same value of n but different values of ℓ have different energies.

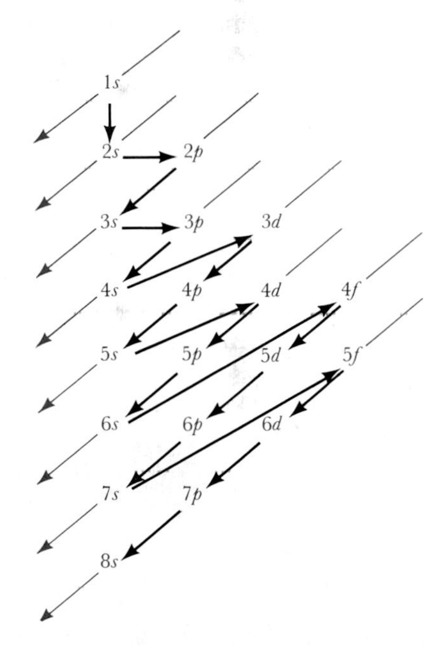

Figure 6-28 A mnemonic for the order of the orbital energies of atoms containing more than one electron. The orbitals are arranged as shown, and then diagonal lines are drawn through them. The correct order of the orbital energies of atoms beyond hydrogen is obtained by going down a line as far as possible and then jumping to the top of the next line.

Table 6-8 lists the allowed sets of four quantum numbers (n, ℓ, m_ℓ, m_s). Note that there are only two allowed combinations for $n = 1$: (1, 0, 0, $+\frac{1}{2}$) and (1, 0, 0, $-\frac{1}{2}$). Both combinations have $n = 1$ and $\ell = 0$ and so correspond to two electrons in a 1s orbital. The two electrons differ only in their spin quantum numbers. We can represent this pictorially by a circle enclosing two vertical arrows:

The circle represents the orbital, and the two arrows represent the two electrons with different spin quantum numbers. The arrow pointing upward represents $m_s = +\frac{1}{2}$, and the arrow pointing downward represents $m_s = -\frac{1}{2}$. This pictorial representation is so ingrained that chemists often use the terms **spin up** and **spin down** to refer to electrons with $m_s = +\frac{1}{2}$ and $m_s = -\frac{1}{2}$, respectively. The Pauli exclusion principle

Table 6-8 The occupation of orbitals according to the Pauli exclusion principle

n	ℓ	m_ℓ	m_s
1 (K shell) (2 electrons)	0 (s subshell) (2 electrons)	0	$+\frac{1}{2}$ or $-\frac{1}{2}$
2 (L shell) (8 electrons)	0 (s subshell) (2 electrons)	0	$+\frac{1}{2}$ or $-\frac{1}{2}$
	1 (p subshell) (6 electrons)	+1	$+\frac{1}{2}$ or $-\frac{1}{2}$
		0	$+\frac{1}{2}$ or $-\frac{1}{2}$
		−1	$+\frac{1}{2}$ or $-\frac{1}{2}$
3 (M shell) (18 electrons)	0 (s subshell) (2 electrons)	0	$+\frac{1}{2}$ or $-\frac{1}{2}$
	1 (p subshell) (6 electrons)	+1	$+\frac{1}{2}$ or $-\frac{1}{2}$
		0	$+\frac{1}{2}$ or $-\frac{1}{2}$
		−1	$+\frac{1}{2}$ or $-\frac{1}{2}$
	2 (d subshell) (10 electrons)	+2	$+\frac{1}{2}$ or $-\frac{1}{2}$
		+1	$+\frac{1}{2}$ or $-\frac{1}{2}$
		0	$+\frac{1}{2}$ or $-\frac{1}{2}$
		−1	$+\frac{1}{2}$ or $-\frac{1}{2}$
		−2	$+\frac{1}{2}$ or $-\frac{1}{2}$
4 (N shell) (32 electrons)	0 (s subshell) (2 electrons)	0	$+\frac{1}{2}$ or $-\frac{1}{2}$
	1 (p subshell) (6 electrons)	+1	$+\frac{1}{2}$ or $-\frac{1}{2}$
		0	$+\frac{1}{2}$ or $-\frac{1}{2}$
		−1	$+\frac{1}{2}$ or $-\frac{1}{2}$
	2 (d subshell) (10 electrons)	+2	$+\frac{1}{2}$ or $-\frac{1}{2}$
		+1	$+\frac{1}{2}$ or $-\frac{1}{2}$
		0	$+\frac{1}{2}$ or $-\frac{1}{2}$
		−1	$+\frac{1}{2}$ or $-\frac{1}{2}$
		−2	$+\frac{1}{2}$ or $-\frac{1}{2}$
	3 (f subshell) (14 electrons)	+3	$+\frac{1}{2}$ or $-\frac{1}{2}$
		+2	$+\frac{1}{2}$ or $-\frac{1}{2}$
		+1	$+\frac{1}{2}$ or $-\frac{1}{2}$
		0	$+\frac{1}{2}$ or $-\frac{1}{2}$
		−1	$+\frac{1}{2}$ or $-\frac{1}{2}$
		−2	$+\frac{1}{2}$ or $-\frac{1}{2}$
		−3	$+\frac{1}{2}$ or $-\frac{1}{2}$

states that it is not possible for the spin quantum numbers of the electrons in a given orbital to be the same; if they were, the electrons would have the same set of four quantum numbers. Thus the representations (↑↑) and (↓↓) are not allowed; that is, they are forbidden.

▪ No orbital can be occupied by more than two electrons.

The $n = 1$ level is complete with two electrons because there are only two possible sets of four quantum numbers with $n = 1$. When $n = 2$, there are two possible values of ℓ, namely, 0 and 1. The $\ell = 0$ value corresponds to a 2s orbital, which can hold two electrons of opposite spins. The $\ell = 1$ value corresponds to the three 2p orbitals ($m_\ell = 1$, $m_\ell = 0$, $m_\ell = -1$), each of which can hold two electrons of opposite spins, giving in all six electrons in the three p orbitals. The $n = 2$ level, then, can hold eight electrons (two in the 2s orbital and six in the 2p orbitals):

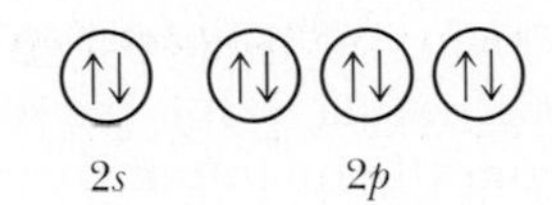

For historical reasons, the levels designated by n are called shells. The $n = 1$ shell is called the K shell, the $n = 2$ shell is called the L shell, the $n = 3$ shell is called the M shell, and so forth. The groups of orbitals designated by ℓ values within these shells are called **subshells.** For $n = 2$, there are two subshells: the *s* subshell, which can contain a maximum of two electrons, and the *p* subshell, which can contain a maximum of six electrons.

For $n = 3$, we have the 3*s*, 3*p*, and 3*d* orbitals. The only new feature here is the *d* subshell. Because there are five *d* orbitals and each one can contain 2 electrons with opposite spins, the *d* subshell can contain up to 10 electrons. Thus, as Table 6-8 shows, the $n = 3$ level, or M shell, can contain 18 electrons. The only new feature for $n = 4$ is the *f* subshell. Because there are seven *f* orbitals and each one can contain 2 electrons with opposite spins, the *f* subshell can contain up to 14 electrons, giving a total capacity of 32 (2 + 6 + 10 + 14) electrons for the $n = 4$ level.

n	shell
1	K
2	L
3	M
4	N
⋮	⋮

6-17. ELECTRONIC CONFIGURATIONS DESIGNATE THE OCCUPANCY OF ELECTRONS IN ATOMIC ORBITALS

We are now ready to use Table 6-8 to understand some of the principal features of the periodic table in terms of electronic structure. We first consider the helium atom. The lowest energy state of the helium atom is achieved by placing both electrons in the 1*s* orbital because this is the orbital with the lowest energy. Thus we can represent the **ground electronic state** in helium by $(\uparrow\downarrow)$ or by $1s^2$. The latter notation is standard. The 1*s* means that we are considering a 1*s* orbital, and the superscript denotes that there are two electrons in the orbital. It is understood that the electrons have different spin quantum numbers, or opposite spins. If we are depicting five electrons in the 3*p* orbitals, then we write $3p^5$. The arrangement of electrons in the orbitals is called the **electron configuration** of the atom. We say that the electron configuration of the ground state of helium is $1s^2$.

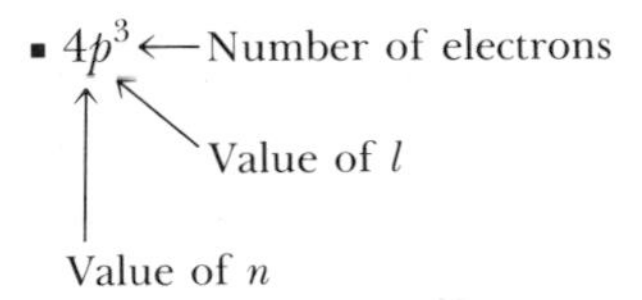

Let's go on now and consider the case of lithium with its three electrons. It is not possible for three electrons in a 1*s* orbital to have different sets of the four quantum numbers. The 1*s* orbital is completely filled by two of the electrons, and so the third electron must be assigned to the next available orbital, the 2*s* orbital. The electron in the 2*s* orbital can have $m_s = +\frac{1}{2}$ or $-\frac{1}{2}$, and so we can represent the lithium atom by

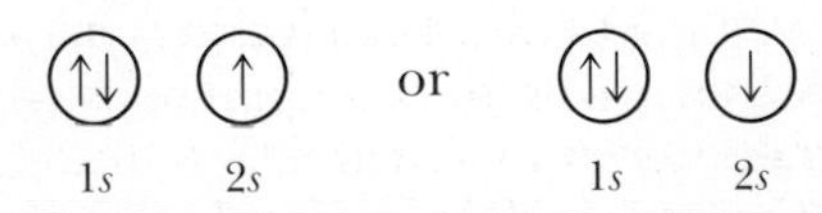

The direction of the arrow in the 2*s* orbital is not important here, and it is customary to use the spin up picture. The more standard notation is $1s^2 2s^1$.

We used the experimental values of the ionization energies for lithium in Table 6-2 to argue that lithium can be represented as a helium core with one outer electron. In Table 6-2, we represent the lithium atom by the electron-dot formula ·[He]. We see that this same conclusion follows naturally from the quantum theory.

The ground state of beryllium ($Z = 4$) is obtained by placing the fourth electron in the 2*s* orbital such that the two electrons there have opposite spins. Pictorially we have for the ground state of the beryllium atom

beryllium (↑↓) (↑↓)
1*s* 2*s*

and the standard notation for this ground-state electron configuration is $1s^2 2s^2$.

In boron ($Z = 5$) both the 1*s* and 2*s* orbitals are filled, and so we must use the 2*p* orbitals. Thus we have for boron

boron (↑↓) (↑↓) (↑)( )( )
1*s* 2*s* 2*p*

The three *p* orbitals for the hydrogen atom have the same energy in the absence of any external electric or magnetic fields, and this is also true for multielectron atoms. Thus, it does not matter into which of the three *p* orbitals we place the electron. In addition, recall that the direction of the arrow in the 2*p* orbital is not important and that it is customary to draw such unpaired electrons as spin up. The ground-state electron configuration of boron is written $1s^2 2s^2 2p^1$.

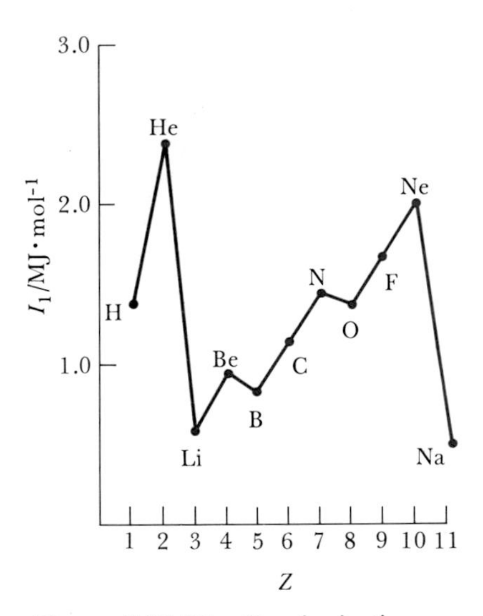

Figure 6-29 The first ionization energy I_1 of the elements H through Na. Notice that I_1 is relatively high for the noble gases (He and Ne), indicating that a completed shell is extraordinarily stable.

Example 6-12: The ground-state electron configuration of ions can be described by the same notation that we have discussed for atoms. What is the ground-state electron configuration of B^+?

Solution: B^+ has four electrons ($Z = 5$ for B, and $Z - 1 = 4$ electrons for B^+). The ground electronic state is obtained by placing two of these electrons in the 1*s* orbital and two in the 2*s* orbital:

B^+ (↑↓) (↑↓)

or simply $1s^2 2s^2$.

We might guess that the unpaired electron in the boron atom is relatively easier to detach than is a paired 2*s* electron in beryllium. Figure 6-29 is a plot of the first ionization energies of the elements hydrogen through sodium. Note that these data confirm this guess.

6-18. HUND'S RULE IS USED TO PREDICT GROUND-STATE ELECTRON CONFIGURATIONS

For a carbon atom ($Z = 6$) we have three distinct choices for the placement of the two 2*p* electrons. The three configurations that obey the

Pauli exclusion principle are shown in the margin. There are, however, small differences in the energies of these three configurations. In configuration I, both electrons are in the same p orbital and hence are restricted, on the average, to the same region in space. In the other two cases, the two electrons are in different p orbitals and so are, on the average, in different regions. Because electrons have the same charge and so repel each other, the placement of the two electrons into different p orbitals and hence different regions allows repulsion between electrons to be minimized. Thus we conclude that configurations II and III have lower energies and so are favored over configuration I. It has been determined experimentally that the configuration in which the two p electrons are placed in different p orbitals with **parallel spins** leads to the lowest-energy, or ground-state, configuration. Therefore, the ground-state configuration of the carbon atom is

I (↑↓) (↑↓) (↑↓)( )( )
1s 2s 2p

II (↑↓) (↑↓) (↑)(↑)( )
1s 2s 2p

III (↑↓) (↑↓) (↑)(↓)( )
1s 2s 2p

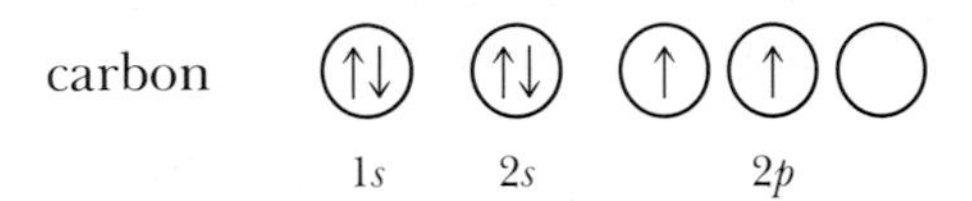

These arguments can be generalized to give what is called **Hund's rule,** which states that, for any set of orbitals of the same energy, that is, for any subshell, the ground-state electron configuration is obtained by placing the electrons in different orbitals of this set with parallel spins. No orbital in the subshell contains two electrons until each one contains one electron. Using Hund's rule, we write for nitrogen ($Z = 7$)

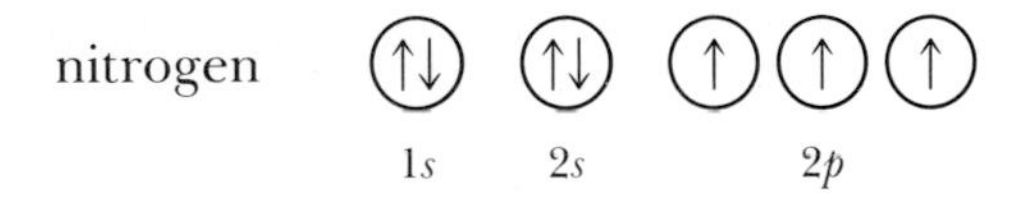

The more standard notation is $1s^22s^22p_x^12p_y^12p_z^1$. This often is condensed to $1s^22s^22p^3$. In both cases the reader is assumed to know that the three $2p$ electrons have parallel spins.

For an oxygen atom ($Z = 8$) we begin to pair up the p electrons and have

oxygen (↑↓) (↑↓) (↑↓)(↑)(↑)
1s 2s 2p

or $1s^22s^22p_x^22p_y^12p_z^1$, or simply $1s^22s^22p^4$. Realize that it does not matter into which p orbital we place the paired electrons. The electron configurations $1s^22s^22p_x^12p_y^22p_z^1$ and $1s^22s^22p_x^12p_y^12p_z^2$ are equivalent to each other and to $1s^22s^22p_x^22p_y^12p_z^1$.

Example 6-13: What is the ground-state electron configuration of O^+?

Solution: The O^+ ion has seven electrons (for O, $Z = 8$; for O^+ we have $8 - 1 = 7$ electrons). Four of the electrons are in the 1*s* and 2*s* orbitals. The other three are in the 2*p* orbitals. According to Hund's rule, the three 2*p* electrons are in different 2*p* orbitals and all have the same spin. The electron configuration is $1s^22s^22p_x^12p_y^12p_z^1$, or simply $1s^22s^22p^3$.

The ground-state electron configurations of the first 10 elements are shown in Table 6-9 (margin). Note that helium has a filled $n = 1$ shell and that neon has a filled $n = 2$ shell. These electron configurations are for the ground electronic state of the atom, that is, the state of lowest energy. The ground-state electron configuration is obtained by filling up the atomic orbitals of lowest energy in accord with the Pauli exclusion principle and Hund's rule.

Table 6-9 Ground-state electron configurations for the first ten elements

Element	Electron configuration
H	$1s^1$
He	$1s^2$
Li	$1s^2 2s^1$
Be	$1s^2 2s^2$
B	$1s^2 2s^2 2p^1$
C	$1s^2 2s^2 2p^2$
N	$1s^2 2s^2 2p^3$
O	$1s^2 2s^2 2p^4$
F	$1s^2 2s^2 2p^5$
Ne	$1s^2 2s^2 2p^6$

We saw in Section 6-10 that an atom can absorb electromagnetic radiation. In this process an electron is promoted to an orbital of higher energy, and the atom is said to be in an **excited state.** For example, a lithium atom absorbs electromagnetic radiation of wavelength 671 nm according to

$$\mathrm{Li}(1s^2 2s^1) + h\nu \rightarrow \mathrm{Li}(1s^2 2p^1)$$

We see that the electron in the $2s$ orbital is promoted to a $2p$ orbital in the process. The resulting lithium atom is in an excited state, and its electron configuration is $1s^2 2p^1$. We are interested primarily in ground electronic states, but we should realize that the ground state is just the lowest of a set of allowed atomic energy states.

Example 6-14: What is the electron configuration of the first excited state of neon?

Solution: The first excited state is obtained by promoting the electron of highest energy in the ground state to the next available orbital. The ground-state electron configuration of neon is $1s^2 2s^2 2p^6$. The electron of highest energy is any one of the $2p$ electrons. The next available orbital is the $3s$ orbital, and so

$$\mathrm{Ne}^* \text{ (first excited state)} \qquad 1s^2 2s^2 2p^5 3s^1$$

The asterisk indicates an excited state.

6-19. ELEMENTS IN THE SAME COLUMN OF THE PERIODIC TABLE HAVE SIMILAR VALENCE-ELECTRON CONFIGURATIONS

Figure 6-30 is a periodic table indicating the orbitals that are used in building up the electron configuration of each element. Note that after neon we must go to the $3s$ and $3p$ orbitals to obtain the next row of the periodic table:

Element	Ground-state configuration	Abbreviated form of ground-state configuration
sodium	$1s^2 2s^2 2p^6 3s^1$	$[\mathrm{Ne}]3s^1$
magnesium	$1s^2 2s^2 2p^6 3s^2$	$[\mathrm{Ne}]3s^2$
aluminum	$1s^2 2s^2 2p^6 3s^2 3p^1$	$[\mathrm{Ne}]3s^2 3p^1$
silicon	$1s^2 2s^2 2p^6 3s^2 3p^2$	$[\mathrm{Ne}]3s^2 3p^2$
phosphorus	$1s^2 2s^2 2p^6 3s^2 3p^3$	$[\mathrm{Ne}]3s^2 3p^3$
sulfur	$1s^2 2s^2 2p^6 3s^2 3p^4$	$[\mathrm{Ne}]3s^2 3p^4$
chlorine	$1s^2 2s^2 2p^6 3s^2 3p^5$	$[\mathrm{Ne}]3s^2 3p^5$
argon	$1s^2 2s^2 2p^6 3s^2 3p^6$	$[\mathrm{Ne}]3s^2 3p^6$ or [Ar]

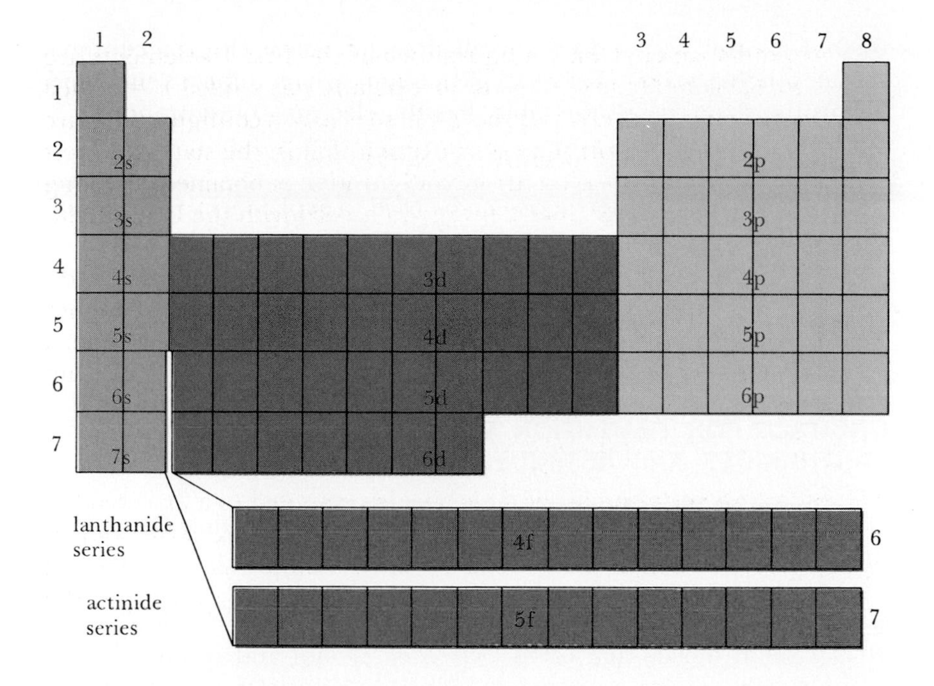

Figure 6-30 A periodic table indicating which orbitals are occupied by the valence electrons of each element. Blocks of elements with the same color have the same valence electron subshells.

In this series of elements we are filling up the 3s and 3p orbitals outside a neon inner-shell structure. It is therefore common practice to use the abbreviated form of the electron configurations shown in the right-hand column. If we compare the Na through Ar electron configurations with the Li through Ne electron configurations (Table 6-9), then we see why these two series of elements have a periodic correlation in chemical properties. Their valence (outer-electron) electron configurations range from ns^1 to ns^2np^6 ($n = 2$ and $n = 3$) in the same manner. Reference to Figure 6-30 shows that the number at the top of each column in the periodic table is equal to the number of valence electrons.

Figure 6-27(b) shows that, after argon, the next available orbital is the 4s orbital. Thus the electron configurations of the next two elements after argon, namely, potassium and calcium, are

potassium	[Ar]$4s^1$
calcium	[Ar]$4s^2$

where [Ar] denotes the ground-state electron configuration of an argon atom. If we consider the ground-state electron configurations of lithium, sodium, and potassium, then we can see why they fall naturally into the same column of the periodic table. Each has an ns^1 configuration outside a noble-gas configuration, that is,

lithium	[He]$2s^1$
sodium	[Ne]$3s^1$
potassium	[Ar]$4s^1$

Also note that the principal quantum number of the outer s orbital coincides with the number of the row of the periodic table (Figure

6-30). Each row starts off with an alkali metal, whose electron configuration is [noble gas]ns^1. For example, cesium, which follows xenon and begins the sixth row of the table, has the electron configuration

$$\text{cesium} \qquad [\text{Xe}]6s^1$$

▪ beryllium	[He]$2s^2$
magnesium	[Ne]$3s^2$
calcium	[Ar]$4s^2$
strontium	[Kr]$5s^2$
barium	[Xe]$6s^2$
radium	[Rn]$7s^2$

The same type of observation can be used to explain why the alkaline earths all occur in the second column in the periodic table. The electron configuration of an alkaline earth metal is [noble gas]ns^2 (see margin).

6-20. THE OCCUPIED ORBITALS OF HIGHEST ENERGY ARE *d* ORBITALS FOR TRANSITION METALS AND *f* ORBITALS FOR LANTHANIDES AND ACTINIDES

Once we reach calcium ($Z = 20$), the $4s$ orbital is completely filled, Figure 6-27(b) shows that the next available orbitals are the five $3d$ orbitals. Each of these can be occupied by 2 electrons of opposite spins, giving a total of 10 electrons in all. Note that this corresponds perfectly with the 10 transition metals that occur between calcium and gallium in the periodic table. Thus we see that in the first set of transition metals there is the sequential filling of the five $3d$ orbitals. Because of this, the first set of transition metals is called the **3*d* transition-metal series.**

▪ scandium	[Ar]$4s^2 3d^1$
titanium	[Ar]$4s^2 3d^2$
vanadium	[Ar]$4s^2 3d^3$
chromium	[Ar]$4s^1 3d^5$
manganese	[Ar]$4s^2 3d^5$
iron	[Ar]$4s^2 3d^6$
cobalt	[Ar]$4s^2 3d^7$
nickel	[Ar]$4s^2 3d^8$
copper	[Ar]$4s^1 3d^{10}$
zinc	[Ar]$4s^2 3d^{10}$

You may think that the ground-state electron configurations of these 10 elements go smoothly from [Ar]$4s^2 3d^1$ to [Ar]$4s^2 3d^{10}$, but this is not so. The ground-state electron configurations of the $3d$ transition metals are given in the margin.

We see that chromium and copper have only one $4s$ electron. Note that in each case an electron has been taken from the $4s$ orbital in order to either half-fill or completely fill all of the $3d$ orbitals. This happens because an extra stability is realized by the electron configurations

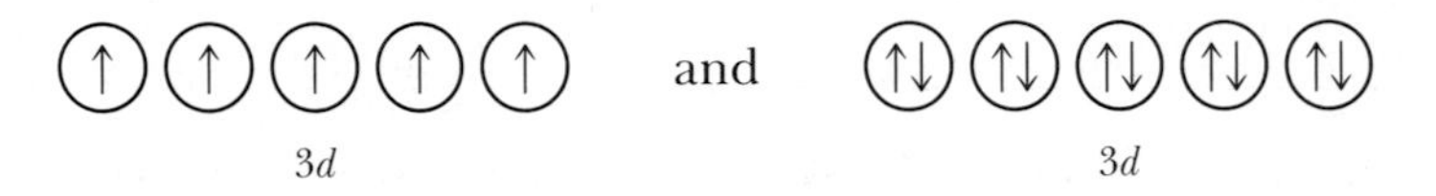

relative to the *incorrect* $4s^2 3d^4$ and $4s^2 3d^9$ ground-state configurations for these elements. It so happens that the energies of the $4s$ and $3d$ electrons are rather close to each other (Figure 6-27b), and the exchange of an electron between these two types of orbitals occurs easily. This is one reason that the transition metals exhibit ions with different charges, such as Fe^{2+} and Fe^{3+}.

After the $3d$ orbitals are filled, the next available orbitals are the $4p$ orbitals, which fill up as follows:

gallium	[Ar]$4s^2 3d^{10} 4p^1$
germanium	[Ar]$4s^2 3d^{10} 4p^2$
arsenic	[Ar]$4s^2 3d^{10} 4p^3$
selenium	[Ar]$4s^2 3d^{10} 4p^4$
bromine	[Ar]$4s^2 3d^{10} 4p^5$
krypton	[Ar]$4s^2 3d^{10} 4p^6$

For these six elements the $4p$ orbitals are sequentially filled, and these elements fall naturally into the fourth row of the periodic table under the sequence of elements B through Ne and Al through Ar, which fill the $2p$ and $3p$ orbitals, respectively (Figure 6-30).

Krypton, like all the noble gases, has a completely filled set of p orbitals whose principal quantum number corresponds to the row in the periodic table. Figure 6-28 shows that the $5s$ orbital follows the $4p$ orbital, and so we are back to the left-hand column of the periodic table with the alkali metal rubidium and the alkaline earth metal strontium. These two metals have the ground-state electron configurations $[Kr]5s^1$ and $[Kr]5s^2$, respectively. The next available orbitals are the $4d$ orbitals, which lead to the second transition-metal series, or the **$4d$ transition-metal series,** Y through Cd. Figure 6-31 gives the ground-state electron configurations of the outer electrons of these 10 metals and shows irregularities like those found in the $3d$ transition-metal series. After cadmium, $[Kr]5s^24d^{10}$, the $5p$ orbitals are filled to give the six elements indium through the noble gas xenon, which has the ground-state electron configuration $[Kr]5s^24d^{10}5p^6$. As before, the completion of a set of p orbitals leads to a noble gas located in the right-hand column of the periodic table. The two reactive metals cesium and barium follow

Figure 6-31 A periodic table showing the ground-state electron configurations of the valence (outer) electrons of the elements. The general valence-electron configurations of the various groups are given above each group. Thus, the alkali metals have the outer electron configuration ns^1, the alkaline earths ns^2, and so on.

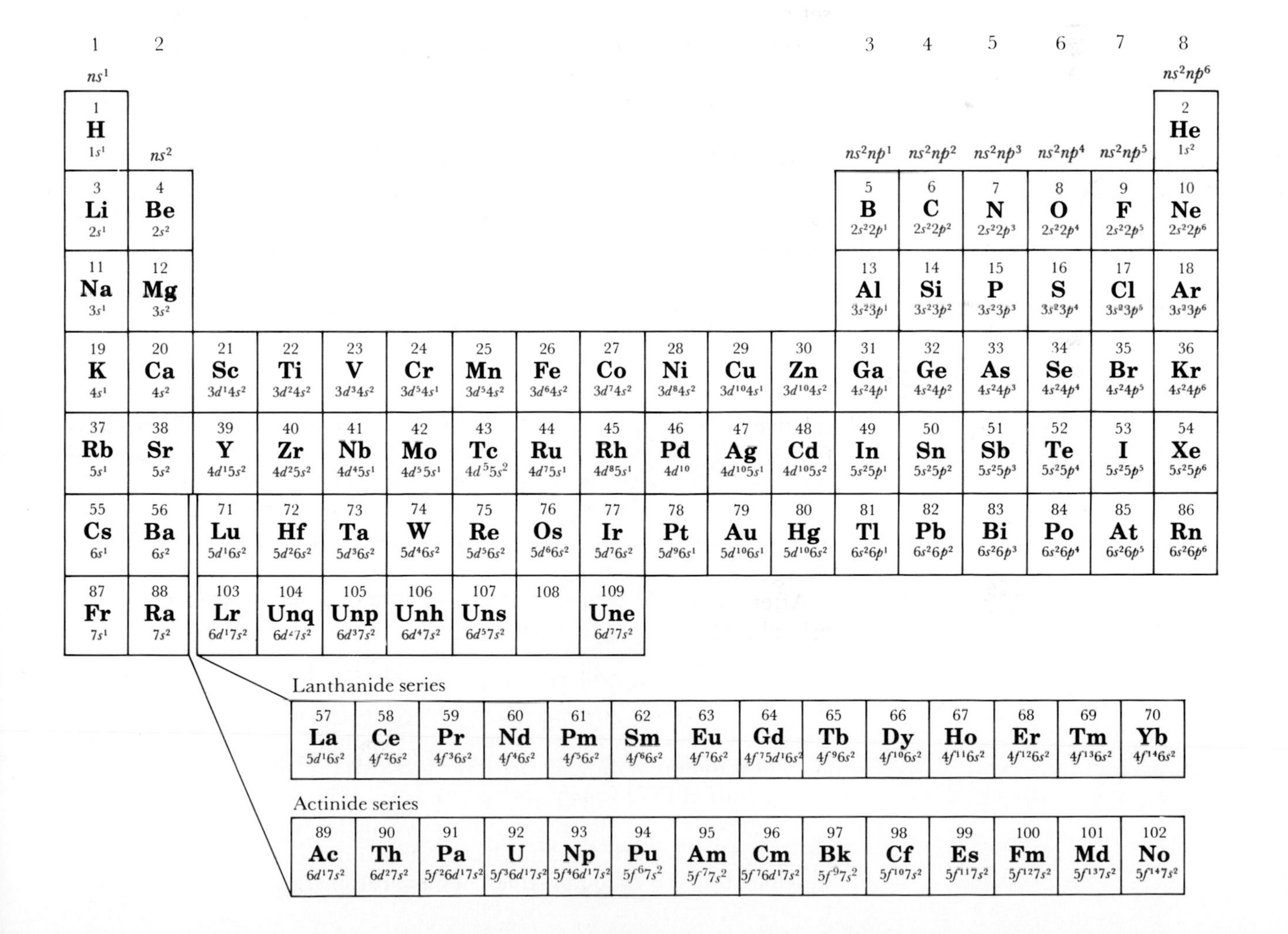

xenon by filling the $6s$ orbital to give the ground-state electron configurations $[Xe]6s^1$ and $[Xe]6s^2$, respectively.

After filling the $6s$ orbital, we use the seven $4f$ orbitals. Because each of these 7 orbitals can hold 2 electrons of opposite spin, we expect that the next 14 elements should involve the filling of the $4f$ orbitals. The elements lanthanum ($Z = 57$) through ytterbium ($Z = 70$) constitute what is called the **lanthanide series** because it begins with the element lanthanum in the periodic table. The chemistry of these elements is so similar that for many years it proved very difficult to separate them from the naturally occuring mixtures. However, separations are possible using modern chromatographic methods (Interchapter A).

Example 6-15: By referring to Figure 6-30, predict the ground-state electron configuration of a neodymium atom ($Z = 60$).

Solution: Neodymium occurs in the sixth row of the periodic table. The noble gas preceding this row is xenon. The ground-state electron configuration of barium, the element that precedes the lanthanides, is $[Xe]6s^2$. Neodymium is the fourth member of the lanthanides, and so we predict that it has four $4f$ electrons. The predicted ground-state electron configuration is

$$\text{Nd} \qquad [\text{Xe}]6s^2 4f^4$$

Figure 6-31 shows that this result is correct. Notice, however, that there are several irregularities in the electron configurations of the lanthanides.

▪ On the average, $4f$ electrons are closer to the nucleus than are $6s$ or $5p$ electrons.

If we consider that the lanthanides differ only in the number of electrons in the $4f$ subshells, with the $6s$ and $5p$ subshells already filled, the reason for their chemical similarity becomes clear. According to the quantum theory, the average distance of an electron from a nucleus depends upon both the principal quantum number n and the azimuthal quantum number ℓ. Although the average distance from the nucleus increases with n, it decreases as ℓ increases, and so electrons with large values of ℓ are on the average closer to the nucleus than are electrons with smaller values of ℓ. Therefore, we can conclude that the average distance of $4f$ electrons from the nucleus is smaller than that of $6s$ or $5p$ electrons. For the $4f$ electrons, not only is n smaller but ℓ is larger than for $6s$ or $5p$ electrons. The $4f$ electrons, then, tend to lie deeper in the interior of the atom and so have little effect on the chemical activity of the atom, which is dominated by the outer (valence) electrons. For this reason, the lanthanides are also called **inner transition metals.** The outer electron configuration, which plays a principal role in determining chemical activity, is the same for all the lanthanides ($5p^6 6s^2$) and accounts for their similar chemical properties.

Following the lanthanides is a third transition metal series (the $5d$ transition-metal series) consisting of the elements lutetium ($Z = 71$) through mercury ($Z = 80$). This series, in which the $5d$ orbitals are filled, is followed by the six elements thallium ($Z = 81$) through radon ($Z = 86$). Radon, a radioactive noble gas with the ground-state electron configuration $[Xe]6s^2 4f^{14} 5d^{10} 6p^6$, finishes the sixth row of the table.

The next two elements, the radioactive metals francium, $[Rn]7s^1$, and radium, $[Rn]7s^2$, are followed by another inner transition-metal series in which the $5f$ orbitals are filled. This series begins with actinium ($Z =$

89) and ends with nobelium ($Z = 102$) and is called the **actinide series.** All the elements in this series occur only in radioactive form, and in fact, with the exception of trace quantities of plutonium, the elements beyond uranium ($Z = 92$) have not been found in nature. They are synthesized in nuclear reactors and are called the **transuranium elements** (Chapter 21).

Figure 6-31 shows the ground-state outer electron configurations of all the elements. It is a good exercise to go through the periodic table and predict the electron configuration of each element. A few irregularities occur as n increases, but the general features should be apparent.

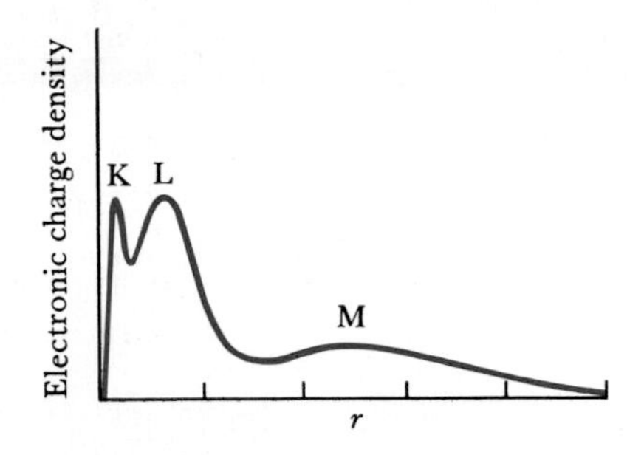

Figure 6-32 The distribution of electronic charge density versus distance from the nucleus for an argon atom can be obtained by solving the Schrödinger equation with a computer. Note that there appear to be three shells. Two of these are well defined and close to the nucleus (the K and L shells). The third, outermost shell (the M shell) is more diffuse.

6-21. ATOMIC RADIUS IS ANOTHER PERIODIC PROPERTY

Because electrons distribute themselves about a nucleus in a diffuse, cloudlike manner, there is no sharp boundary at the "edge" of an atom. Although the Schrödinger equation is complicated for multielectron atoms, it can be solved with a computer. The results of such a calculation for argon are sketched in Figure 6-32. We can discern clearly three shells: The inner two shells, the K shell and the L shell, are well defined. The third, outermost shell, the M shell, is more diffuse.

Even though atoms do not have well-defined radii, we can propose operational definitions for **atomic radii** that are based on models. For example, the atoms in a crystal of an element are arranged in ordered arrays. A simple version of such an ordered array is shown in Figure 6-33. The atoms are arranged in a simple cubic array, and we can propose that one half of the distance between adjacent nuclei be used as an effective atomic radius. Real crystals usually exist in more complicated geometric patterns than simple cubic, but effective atomic radii can still be deduced. Atomic radii obtained in this manner are called **crystallographic radii.** The crystallographic radii of the elements are plotted versus atomic number in Figure 6-34, indicating the periodic dependence of the radii on atomic number.

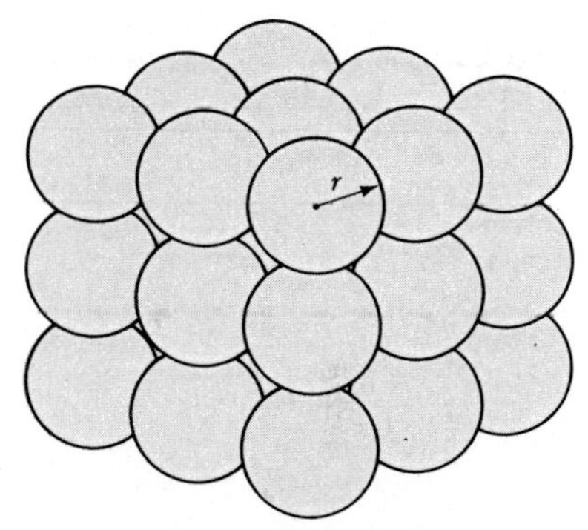

Figure 6-33 A simple cubic arrangement of atoms in a crystal.

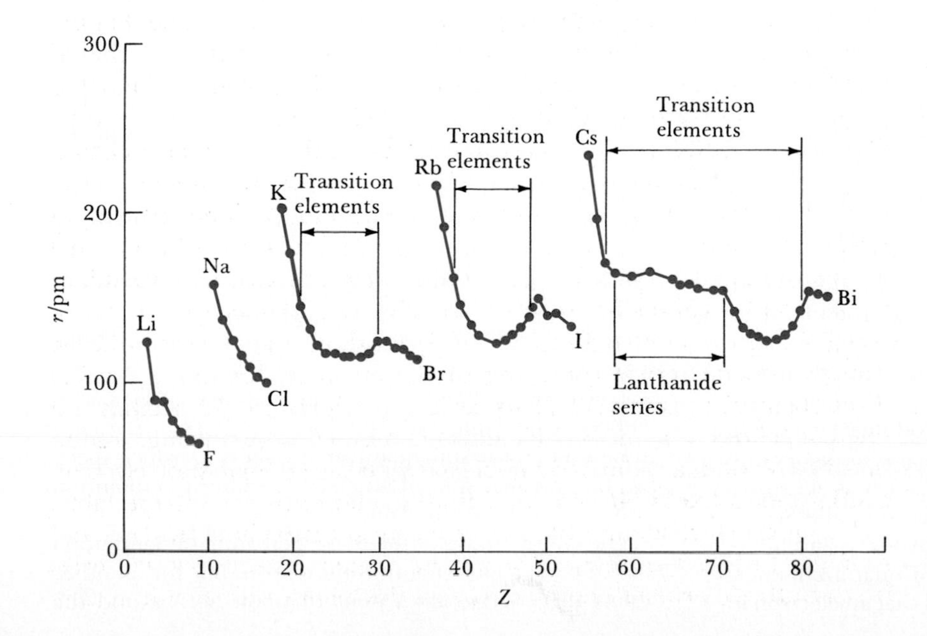

Figure 6-34 Crystallographic radii of the elements versus atomic number. Note that atomic radius is a periodic property.

The crystallographic radii of the elements lithium through fluorine decrease uniformly from left to right across this row of the periodic table because the nuclear charge increases and attracts the electrons more strongly. Both the K shell and L shell contract, giving a smaller effective radius as the atomic number increases. This same trend is seen in Figure 6-34 for the other rows of the periodic table. Atomic radii of the main group elements usually decrease as we go from left to right in a row across the periodic table.

The crystallographic radii of the alkali metal group increase as we go down the periodic table from lithium to cesium. Although the nuclear charge increases, the outermost electrons begin new shells, and this effect outweighs the increased nuclear attraction. Similar behavior is found for other groups in the periodic table. Atomic radii usually increase as we go down the periodic table within a group, as shown in Figure 6-35.

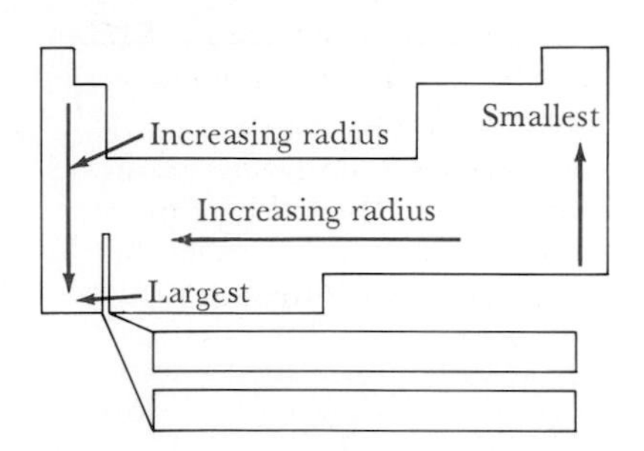

Figure 6-35 The trend of atomic radii in the periodic table.

The reasoning we have just used to explain the variation of atomic radii in the periodic table can be used to explain variations in first ionization energies (Figure 6-1). The steady decrease with increasing atomic number within a group is due to the increase in atomic radius as we go down the periodic table. The farther the electron is from the nucleus, the less the nuclear attraction, and so the electron is more easily removed. Therefore, ionization energies decrease as we go down the periodic table within a group.

SUMMARY

First ionization energy is a periodic property. The values of successive ionization energies suggest that electrons in atoms are arranged in a shell structure (Figure 6-3). The shell structures deduced from a study of successive ionization energies are depicted in Lewis electron-dot formulas given in Table 6-2.

The quantum theory was initiated by Planck in 1900, who assumed that electromagnetic radiation could be emitted from heated bodies only in little packets of energy, given by $E = h\nu$. This idea was used five years later by Einstein to describe the photoelectric effect.

In 1911 Niels Bohr developed a model of the hydrogen atom that was able to account for the atomic spectrum of hydrogen. It was later shown, however, that the Bohr theory was inconsistent with the Heisenberg uncertainty principle. In 1925 Erwin Schrödinger proposed what is now called the Schrödinger equation, which is the central equation of the quantum theory and governs the motion of small particles, such as electrons, atoms, and molecules. One consequence of the Schrödinger equation is that the electrons in atoms and molecules can have only certain discrete, or quantized, energies. In addition, Schrödinger showed that an electron in an atom or molecule must be described by a wave function, or orbital, which is obtained by solving the Schrödinger equation. The square of a wave function gives the probability density associated with finding the electron in some region of space. The hydrogen atom wave functions serve as the prototype for all other atoms. The wave functions are specified by three quantum numbers: n, the principal quantum number; ℓ, the azimuthal quantum number; and m_ℓ, the magnetic quantum number. Orbitals with $\ell = 0$ are called s orbitals, orbitals with $\ell = 1$ are called p orbitals, orbitals with $\ell = 2$ are called d orbitals, and orbitals with $\ell = 3$ are called f orbitals. For a given value of n, there is one s orbital and there may be three p orbitals, five d orbitals, and seven f orbitals (Table 6-8).

To explain certain fine details in atomic spectra, Uhlenbeck, Goudsmit, and Pauli introduced a fourth quantum number, the spin quantum number, m_s, which specifies the intrinsic spin of an electron. The spin quantum number can have the value $+\frac{1}{2}$ or $-\frac{1}{2}$.

The energy states of the hydrogen atom depend upon only the principal quantum number n, but for all other atoms the energy states depend upon both n and the azimuthal quantum number ℓ. According to the Pauli exclusion principle, no two electrons in an atom can have the same set of four quantum numbers (n, ℓ, m_ℓ, m_s). Using this principle and the order of the energy states given in Figures 6-27(b) and 6-28, we are able to write ground-state electron configurations and to correlate these with the periodic table. Electron configurations enable us to discuss atomic radii and the trend of atomic radii within the periodic table.

TERMS YOU SHOULD KNOW

electronic structure 195
ionization energy 195
first ionization energy 195
second ionization energy 195
shell 197
outer electron 197
valence electron 198
Lewis electron-dot formula 198
wavelength, λ 198
frequency, ν 199
speed of light, c 199
hertz, Hz 199
electromagnetic spectrum 199, 200
visible region 200
continuous spectrum 200
line spectrum 201
atomic spectrum 201
atomic spectroscopy 201
classical physics 201
quanta 202
Planck's constant, h 202
photoelectric effect 202
threshold frequency, ν_0 202
photon 202
work function, Φ 203
quantization of energy 204
quantum theory 204
wave-particle duality 204
de Broglie wavelength 205
X-ray diffraction 205
electron diffraction 205
electron microscope 205
quantum condition 206
quantized 207
energy state 207
stationary state 207
ground state 207
excited state 207
first excited state 207
second excited state 207
Lyman series 208
Balmer series 208
emission spectrum 209
absorption spectrum 209
Heisenberg uncertainty principle 210
Schrödinger equation 211
wave function, ψ 211
orbital 211
probability density 211
quantum number 211
principal quantum number, n 211
spherically symmetric 211
azimuthal quantum number, ℓ 213
s orbital 213
p orbital 213
d orbital 213
f orbital 213
nodal surface 214
cylindrically symmetric 215
magnetic quantum number, m_ℓ 215
intrinsic electron spin 217
spin quantum number, m_s 217
Pauli exclusion principle 218
spin up 219
spin down 219
opposite spins 220
shell 221
subshell 221
ground electronic state 221
electron configuration 221
parallel spins 223
Hund's rule 223
excited state 224
3d transition-metal series 226
4d transition-metal series 227
lanthanide series 228
inner transition metal 228
actinide series 229
transuranium elements 229
atomic radius 229
crystallographic radius 229

EQUATIONS YOU SHOULD KNOW HOW TO USE

$\lambda\nu = c$ (6-1) (relation between wavelength and frequency)

$E = h\nu$ (6-2) (energy of a photon)

$\Phi = h\nu_0$ (6-3) (threshold frequency, work function)

$\text{K.E.} = h\nu - \Phi$ (6-4) (photoelectric effect)

$\lambda = \dfrac{h}{mv}$ (6-5) (de Broglie wavelength)

$E_n = -\dfrac{2.18 \times 10^{-18}\ \text{J}}{n^2} \quad n = 1, 2, 3, \ldots$ (6-7) (energies of the electron in a hydrogen atom)

$\nu_{n\to 1} = (3.29 \times 10^{15}\ \text{s}^{-1})\left(\dfrac{1}{1^2} - \dfrac{1}{n^2}\right) \quad n = 2, 3, 4, \ldots$ (6-11) (Lyman series frequencies)

$$\nu_{n\to 2} = (3.29 \times 10^{15}\ \text{s}^{-1})\left(\frac{1}{2^2} - \frac{1}{n^2}\right) \qquad n = 3, 4, 5, \ldots \qquad \text{(6-12) (Balmer series frequencies)}$$

$$(\Delta p)(\Delta x) \simeq h \qquad \text{(6-15) (Heisenberg uncertainty principle)}$$

$$\left.\begin{aligned} &n = 1, 2, \ldots \\ &\ell = 0, 1, 2, \ldots, n-1 \\ &m_\ell = \ell, \ell-1, \ell-2, \ldots, 0, -1, -2, \ldots, -\ell \\ &m_s = +\tfrac{1}{2} \text{ or } -\tfrac{1}{2} \end{aligned}\right\} \quad \text{(quantum numbers)}$$

PROBLEMS

IONIZATION ENERGIES

6-1. Arrange the following species in order of increasing first ionization energy:

He $\quad Be^+ \quad$ Kr $\quad$ Ne

6-2. Arrange the following species in order of increasing first ionization energy:

Ca $\quad$ Mg $\quad$ Ba $\quad$ Sr

6-3. Use the data in Table 6-1 to plot the logarithms of the ionization energies of the boron atom versus the number of electrons removed. What does the plot suggest about the electronic structure of boron?

6-4. Use the data in Table 6-1 to plot the logarithms of the ionization energies of beryllium versus the number of electrons removed. Compare your plot to Figure 6-3.

LEWIS ELECTRON-DOT FORMULAS

6-5. Write Lewis electron-dot formulas for all the alkali metal atoms and for all the halogen atoms. What is the similarity in all the alkali metal atom formulas? in all the halogen formulas?

6-6. Write the Lewis electron-dot formulas for the Group 6 elements.

6-7. Write the Lewis electron-dot formula for each of the following species:

Ar $\quad$ S $\quad S^{2-} \quad Al^{3+} \quad Cl^-$

6-8. Write the Lewis electron-dot formula for each of the following species:

$B^+ \quad N^{3-} \quad F^- \quad O^{2-} \quad Na^+$

ELECTROMAGNETIC RADIATION

6-9. A helium-neon laser produces light of wavelength 633 nm. What is the frequency of this light?

6-10. The radiation given off by a sodium lamp, which is used in streetlights, has a wavelength of 589.2 nm. What is the frequency of this radiation?

6-11. The first ionization energy of potassium is 419 kJ·mol^{-1}. What is the wavelength of light that is just sufficient to ionize one potassium atom?

6-12. The first ionization energy of argon is 1.52 MJ·mol^{-1}. Do X-rays with a wavelength of 80 nm have sufficient energy to ionize argon?

6-13. The human eye can detect as little as 2.35×10^{-18} J of green light of wavelength 510 nm. Calculate the minimum number of photons that can be detected by the human eye.

6-14. Calculate the energy of 1.00 mol of X-ray photons of wavelength 210 pm. The energy of one mol of photons is called an einstein; its value depends upon the energy of the photons.

PHOTOELECTRIC EFFECT

6-15. The work function of gold metal is 7.7×10^{-19} J. Will ultraviolet radiation of wavelength 200 nm eject electrons from the surface of metallic gold?

6-16. Photocells that are used in "electric eye" door openers are applications of the photoelectric effect. A beam of light strikes a metal surface, from which electrons are emitted, producing an electric current. When the beam of light is blocked by a person walking through the beam, the electric circuit is broken, thereby opening the door. If the source of light is a sodium vapor lamp that emits light at a wavelength of 589 nm, would copper be a satisfactory metal to use in the photocell? The work function of copper is 6.69×10^{-19} J.

6-17. Given that the work function of cesium metal is 2.90×10^{-19} J, calculate the kinetic energy of an electron ejected from the surface of cesium metal when it is irradiated with light of wavelength 400 nm.

6-18. The work function of a metal can be determined from measurements of the speed of the ejected electrons. Electrons were ejected from a metal with a speed of 5.00×10^5 m·s^{-1} when irradiated by light having a wavelength of 390 nm. Find the work function of this metal and the threshold frequency.

DE BROGLIE WAVELENGTH

6-19. Calculate the de Broglie wavelength of a proton traveling at a speed of 1.00×10^5 m·s^{-1}. The mass of a proton is 1.67×10^{-27} kg.

6-20. A golf ball has a mass of 1.68 oz. Calculate the de Broglie wavelength of a golf ball traveling at 120 mph.

6-21. Calculate the de Broglie wavelength of a hydrogen molecule traveling with a speed of 2000 m·s^{-1}.

6-22. The de Broglie wavelength of electrons used in an experiment utilizing an electron microscope is 96.0 pm. What is the speed of one of these electrons?

HYDROGEN ATOMIC SPECTRUM

6-23. How much energy is required for an electron in a hydrogen atom to make a transition from the $n = 2$ state to the $n = 3$ state? What is the wavelength of a photon having this energy?

6-24. A line in the Lyman series of hydrogen has a wavelength of 1.03×10^{-7} m. Find the original energy level of the electron.

6-25. A ground-state hydrogen atom absorbs a photon of light having a wavelength of 97.2 nm. It then gives off a photon having a wavelength of 486 nm. What is the final state of the hydrogen atom?

6-26. Use Equation (6-7) to compute the ionization energy of a hydrogen atom in which the electron is in the first excited state.

6-27. The energy levels of one-electron ions, such as He^+ and Li^{2+}, are given by the equation

$$E_n = -\frac{(2.18 \times 10^{-18}\ \text{J})Z^2}{n^2}$$

where Z is the atomic number. Compare the measured ionization energies (Table 6-1) for He^+, Li^{2+}, and Be^{3+} ions with the values calculated from this equation.

6-28. A helium ion is called hydrogen-like because it consists of one electron and one nucleus. The Schrödinger equation can be applied to He^+, and the result that corresponds to Equation (6-7) is

$$E_n = -\frac{8.72 \times 10^{-18}\ \text{J}}{n^2}$$

Show that the spectrum of He^+ consists of a number of separate series, just as the spectrum of atomic hydrogen does.

QUANTUM NUMBERS AND ORBITALS

6-29. Indicate which of the following atomic orbital designations are impossible:

(a) 7*s* (b) 1*p* (c) 5*d* (d) 2*d* (e) 4*f*

6-30. Give all the possible sets of four quantum numbers for an electron in a 5*d* orbital.

6-31. Give the corresponding atomic orbital designations (that is, 1*s*, 3*p*, and so on) for electrons with the following sets of quantum numbers:

	n	ℓ	m_ℓ	m_s
(a)	4	1	0	$-\frac{1}{2}$
(b)	3	2	0	$+\frac{1}{2}$
(c)	4	2	-1	$-\frac{1}{2}$
(d)	2	0	0	$-\frac{1}{2}$

6-32. Give the corresponding atomic orbital designations for electrons with the following sets of quantum numbers:

	n	ℓ	m_ℓ	m_s
(a)	3	1	-1	$+\frac{1}{2}$
(b)	5	0	0	$+\frac{1}{2}$
(c)	2	1	0	$+\frac{1}{2}$
(d)	4	3	-2	$+\frac{1}{2}$

6-33. If $\ell = 2$, what can you deduce about n? If $m_\ell = 3$, what can you say about ℓ?

6-34. Indicate which of the following sets of quantum numbers are allowed (that is, possible) for an electron in an atom:

	n	ℓ	m_ℓ	m_s
(a)	2	1	0	$+\frac{1}{2}$
(b)	3	0	$+1$	$-\frac{1}{2}$
(c)	3	2	-2	$-\frac{1}{2}$
(d)	1	1	0	$+\frac{1}{2}$
(e)	2	1	0	0

ORBITALS AND ELECTRONS

6-35. Give all the possible sets of four quantum numbers for an electron in a 3*d* orbital.

6-36. Give all the possible sets of four quantum numbers for an electron in a 4*f* orbital.

6-37. Without referring to the text, deduce the maximum number of electrons that can occupy an *s* orbital, a subshell of *p* orbitals, a subshell of *d* orbitals, and a subshell of *f* orbitals.

6-38. Without referring to the text, deduce the maximum number of electrons that can occupy a K shell, an L shell, an M shell, and an N shell.

6-39. Explain why there are 10 members of each d transition series.

6-40. Explain why there are 14 members of each f transition series.

ELECTRON CONFIGURATIONS OF ATOMS

6-41. Indicate which of the following electron configurations are ruled out by the Pauli exclusion principle:

(a) $1s^22s^22p^7$ (b) $1s^22s^22p^63s^3$
(c) $1s^22s^22p^63s^23p^64s^23d^{12}$ (d) $1s^22s^22p^63s^23p^6$

6-42. Explain why the following ground-state electron configurations are not possible:

(a) $1s^22s^32p^3$ (b) $1s^22s^22p^33s^6$
(c) $1s^22s^22p^73s^23p^8$ (d) $1s^22s^22p^63s^23p^14s^23d^{14}$

6-43. Write the corresponding electron configuration for each of the following pictorial representations. Name the element, assuming that the configuration describes a neutral atom:

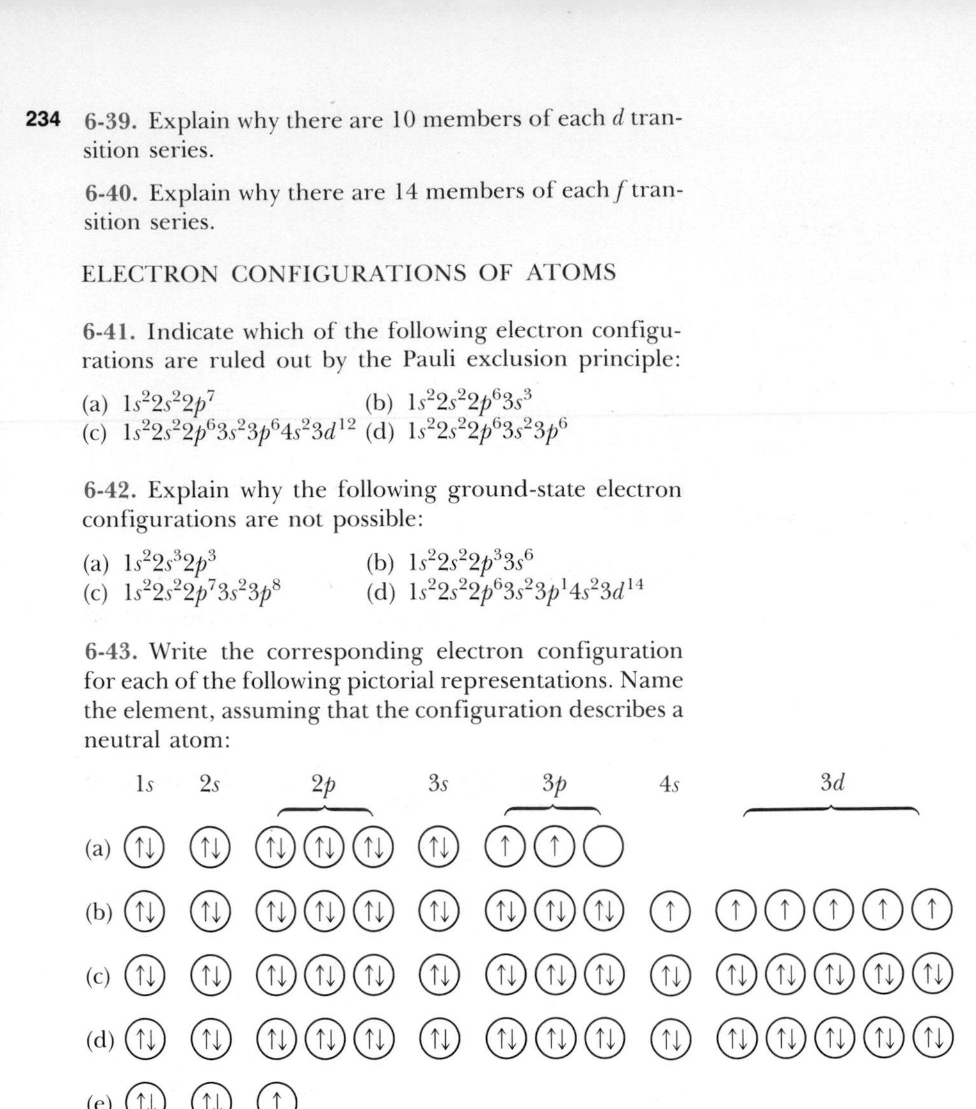

6-44. Write the corresponding electron configuration for each of the following pictorial representations. Name the element that each represents, assuming neutral atoms:

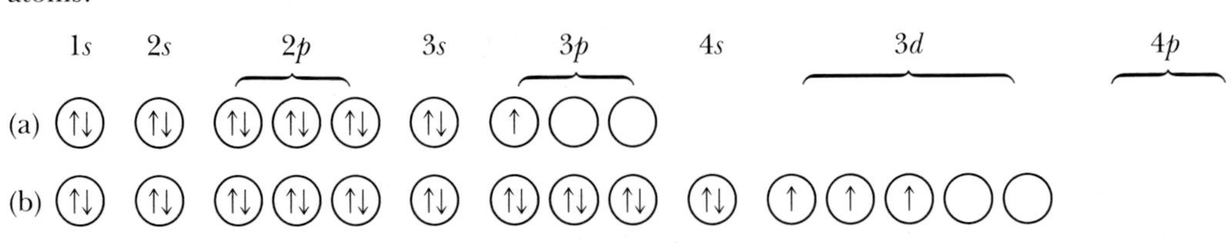

6-45. Write the ground-state electron configurations for the following elements:

(a) Ti (b) K
(c) Fe (d) As

6-46. Write the ground-state electron configurations for the following neutral atoms:

(a) Si (b) Ni
(c) Se (d) Cd

ELECTRON CONFIGURATIONS AND THE PERIODIC TABLE

6-47. Referring only to the periodic table in Figure 2-8, write the ground-state electron configuration of

(a) Ca (b) Br
(c) Ag (d) Zn

6-48. Referring only to the periodic table in Figure 2-8, give two examples of

(a) an atom with a half-filled subshell
(b) an atom with a completed outer shell
(c) an atom with its outer electrons occupying a half-filled subshell and a filled subshell

6-49. Referring only to the periodic table in Figure 2-8, indicate which elements have an outer

(a) *s* electron configuration
(b) *p* electron configuration
(c) *d* electron configuration
(d) *f* electron configuration

Some chemists call these various elements the *s*-block, *p*-block, *d*-block, and *f*-block elements.

6-50. Referring only to the periodic table in Figure 2-8, determine the element of lowest atomic number whose ground state contains

(a) an *f* electron (b) three *d* electrons
(c) a complete *d* subshell (d) ten *p* electrons

6-51. How many unpaired electrons are there in the ground state of each of the following atoms?

(a) Ge (b) Se
(c) V (d) Fe

6-52. How many unpaired electrons are there in the ground state of each of the following ions?

(a) Cl^- (b) O^+
(c) Al^{3+} (d) Xe^+

6-53. Nonmetals add electrons under certain conditions in order to attain a noble-gas electron configuration. How many electrons must be gained in this process by the following elements? Write the Lewis electron-dot formula for each ion that is formed. What noble-gas electron configuration is attained in each case?

(a) H (b) O
(c) C (d) S

6-54. Metals lose electrons to attain a noble-gas electron configuration. How many electrons are lost by the following elements when they attain such a configuration? Write the Lewis electron-dot formula for the ion thus formed. What is the corresponding noble-gas-like inner core in each case?

(a) Ca (b) Li
(c) Na (d) Mg

GROUND-STATE ELECTRON CONFIGURATIONS OF IONS

6-55. Write the ground-state electron configuration for the following ions:

(a) P^{3-} (b) Br^-
(c) Se^{2-} (d) Ba^{2+}

What do these electron configurations have in common?

6-56. Use Hund's rule to write ground-state electron configurations for

(a) O^+ (b) C^-
(c) F^+ (d) O^{2+}

6-57. Determine the number of unpaired electrons in the ground state of the following species:

(a) F^+ (b) Sn^{2+} (c) Bi^{3+} (d) Ar^+

6-58. Arrange the following species into groups of isoelectronic species (see Example 1-7):

F^-	Sc^{3+}	Be^{2+}	Rb^+	O^{2-}	Na^+
Ti^{4+}	Ar	B^{3+}	He	Se^{2-}	Y^{3+}

6-59. Describe the following processes in terms of the electronic configurations of the species involved:

(a) $O(g) + 2e^- \rightarrow O^{2-}(g)$
(b) $Ca(g) + Sr^{2+}(g) \rightarrow Sr(g) + Ca^{2+}(g)$

6-60. Describe the following processes in terms of the electronic configurations of the species involved:

(a) $I(g) + e^- \rightarrow I^-(g)$
(b) $K(g) + F(g) \rightarrow K^+(g) + F^-(g)$

EXCITED-STATE ELECTRON CONFIGURATIONS

6-61. Write the electron configuration for the first excited state of each of the following ions:

(a) Be^{2+} (b) He^+
(c) F^- (d) O^{2-}

6-62. Which of the following electron configurations of neutral atoms represent excited states:

(a) $1s^2 2s^2 2p^5 3s^1$ (b) $1s^2 2s^2 2p^5 3s^2$
(c) $1s^1 2s^1$ (d) $1s^2 2s^2 2p^6 3s^2 3p^6 3d^1$

6-63. Determine the member of each of the following pairs of atoms that has the larger radius (do not use any references except the periodic table):

(a) N and P (b) P and S
(c) S and Ar (d) Ar and Kr

6-64. Arrange the following sets of atoms in order of increasing atomic radius:

(a) Kr, He, Ar, Ne
(b) K, Na, Rb, Li
(c) Be, Ne, F, N, B

6-65. Without using any references except the periodic table, arrange the members of the following groups in order of increasing size:

(a) Li, Na, Cs, Rb
(b) Al, Na, Mg, P
(c) Ca, Ba, Sr

6-66. The radii of lithium and its ions are

Species	Radius/pm
Li	135
Li^+	60
Li^{2+}	18

Explain why the radii decrease from Li to Li^{2+}.

IONIZATION ENERGIES

6-67. The first ionization energies of the alkaline earth metals are

Metal	Ionization energy/$MJ \cdot mol^{-1}$
Be	0.899
Mg	0.738
Ca	0.590
Sr	0.549
Ba	0.503

Explain why the ionization energies decrease from beryllium to barium.

6-68. Arrange the following sets of atoms in order of increasing ionization energy:

(a) B, O, Ne, F (b) Te, I, Sb, Xe
(c) K, Ca, Rb, Cs (d) Ar, Na, S, Al

ADDITIONAL PROBLEMS

6-69. The heat capacity of water is $4.18\ J \cdot K^{-1} \cdot g^{-1}$. Estimate the number of infrared photons with a wavelength of 900 nm that is required to raise the temperature of 1.00 L of water by 1.00°C.

6-70. A watt is a unit of energy per unit time, and one watt (W) is equal to one joule per second ($J \cdot s^{-1}$). A 100-W light bulb produces about 10 percent of its energy as visible light. Assuming that the light has an average wavelength of 510 nm, calculate how many such photons are emitted per second by a 100-W light bulb.

6-71. Below are some data for the photoelectric effect of silver:

Frequency of incident radiation/$10^{15}\ s^{-1}$	Kinetic energy of ejected electrons/10^{-19} J
2.00	5.90
2.50	9.21
3.00	12.52
3.50	15.84
4.00	19.15

Plot the kinetic energy of the ejected electrons versus the frequency of the incident radiation and determine the value of Planck's constant and the threshold frequency.

6-72. Sunlight reaches the earth's surface at Madison, Wisconsin, with an average power of about $1.0\ kJ \cdot s^{-1} \cdot m^{-2}$. If the sunlight consists of photons with an average wavelength of 510 nm, how many photons strike a 1-cm^2 area per second?

6-73. A carbon dioxide laser produces radiation of wavelength 10.6 μm. Calculate the energy of one photon produced by this laser. If the laser produces about 1 J of energy per pulse, how many photons are produced per pulse?

6-74. Neutrons that are at equilibrium at a temperature T are called thermal neutrons. Calculate the de Broglie wavelength of a thermal neutron at 1000 K. The mass of a neutron is 1.67×10^{-27} kg.

6-75. In order to resolve an object in the electron microscope, the wavelength of the electrons must be close to the diameter of the object. What kinetic energy must the electrons have in order to resolve a DNA molecule, which is 2.00 nm in diameter? The mass of an electron is 9.11×10^{-31} kg.

6-76. For a particle moving in a circular orbit, the quantity mvr (mass × velocity × radius of the orbit) is a fundamental quantity called the angular momentum of the particle. Show that Equation (6-6) is equivalent to the condition that the angular momentum of the electron in a hydrogen atom must be an integral multiple of $h/2\pi$.

6-77. Make a graph of frequency versus $1/n^2$ for the lines in the Lyman series of atomic hydrogen.

6-78. Compute the energy necessary to completely remove an electron from the $n = 2$ level of an He^+ ion (see Problem 6-27).

6-79. Estimate the value of ΔH_{rxn} for the following reactions using the data given in Table 6-1.

(a) $Li(g) + Na^+(g) \rightarrow Li^+(g) + Na(g)$
(b) $Mg^{2+}(g) + Mg(g) \rightarrow 2Mg^+(g)$
(c) $Al^{3+}(g) + 3e^- \rightarrow Al(g)$

6-80. Without counting the total number of electrons, determine the neutral atom whose ground-state electron configuration is

(a) $1s^22s^22p^63s^23p^64s^23d^8$
(b) $1s^22s^22p^63s^23p^64s^23d^{10}4p^65s^14d^{10}$
(c) $1s^22s^22p^63s^23p^4$
(d) $1s^22s^22p^63s^23p^64s^23d^{10}4p^65s^24d^{10}5p^66s^24f^{14}5d^{10}6p^2$

6-81. Without looking at a periodic table, deduce the atomic numbers of the other elements that are in the same family as the element with atomic number (a) 16 and (b) 11.

6-82. Name each of the atoms with the following ground-state electron configuration for its valence electrons:

(a) $3s^23p^1$ (b) $2s^22p^4$
(c) $4s^23d^{10}$ (d) $4s^24p^6$

6-83. For elements of atomic number (a) 15, (b) 26, and (c) 32 in their ground states, answer the following questions without reference to the text or to a periodic table:

How many d electrons?
How many electrons having quantum number $\ell = 1$?
How many unpaired electrons?

6-84. Draw what the periodic table would look like if the order of the energies of atomic orbitals were regular; that is, if the order were $1s < 2s < 2p < 3s < 3p < 3d < 4s$, and so on.

6-85. How would the ground-state electron configurations of the elements in the second row of the periodic table differ if the $2s$ and $2p$ orbitals had the same energy, as they do for a hydrogen atom?

6-86. The order of the orbitals given in Figure 6-27(b) can be deduced by the following argument. The energy of an orbital increases with the sum $n + \ell$. For orbitals with the same value of $n + \ell$, those with the smaller value of n have lower energies. This observation is also known as Hund's rule (there are several Hund's rules). Show that this rule is consistent with the order given in Figure 6-27(b).

The Chemistry of the Atmosphere

The earth from space. The cloud cover is clearly visible and gives the earth a bright appearance.

The atmosphere is the sea of gas that envelops the earth. We live at the bottom of this gaseous sea and our existence depends on its chemical properties. The major emphasis of this interchapter is on the chemical composition and reactions of the atmosphere and their influence on life on earth.

F-1. THE EARTH'S ATMOSPHERE CAN BE DIVIDED INTO FOUR DISTINCT REGIONS

Because of the effect of the earth's gravitational field, the pressure of the atmosphere decreases with increasing altitude, from a maximum of about 760 torr (1 atm) at sea level to effectively zero at several hundred kilometers. Most properties of the atmosphere change gradually with altitude, but it is nonetheless convenient to divide the atmosphere into four separate regions, which can be defined by considering the temperature as a function of altitude (Figure F-1).

The lowest region, in which the temperature decreases steadily with increasing altitude, is the **troposphere.** At the top of the troposphere (about 10 km), the temperature is about 218 K (−55°C). The troposphere accounts for more than 80 percent of the mass and virtually all the water vapor, clouds, and precipitation in the earth's atmosphere. It is characterized by strong vertical mixing. For example, in clear air, a molecule can traverse the entire depth of the troposphere in a few days; during severe thunderstorms, the traversal may occur in minutes. All our weather takes place in the troposphere.

Above the troposphere lies the **stratosphere.** In this region, the temperature increases with altitude until it reaches about −10°C at roughly 50 km. The troposphere and stratosphere together account for 99.9 percent of the mass of the atmosphere. Compared with the troposphere, the stratosphere is relatively calm. It is characterized by very little vertical mixing. Debris from nuclear explosions and dust from volcanic eruptions remain in the stratosphere for years before becoming mixed with the troposphere.

Beyond the stratosphere is the region called the **mesosphere** (literally, the middle sphere). Like the troposphere, it is a region in which temperature decreases with altitude.

The fourth region is called the **ionosphere.** As the name implies, the ionosphere contains ions and electrons, which are produced by high-energy solar radiation.

Geologists believe that much of the earth's atmosphere was formed from gases that were discharged from the earth's interior through volcanic activity. Its present composition is fairly uniform up to about 100 km (Table F-1) and is roughly 78 percent nitrogen and 21 percent oxygen, whereas the gaseous emissions from volcanos are a mixture of about 85 percent water vapor, 10 percent carbon dioxide, and a few percent nitrogen (as N_2) and sulfur compounds. Elemental oxygen is notably absent from volcanic emissions; most of the oxygen in the atmosphere is a result of photosynthesis. Nitrogen is a fairly unreactive

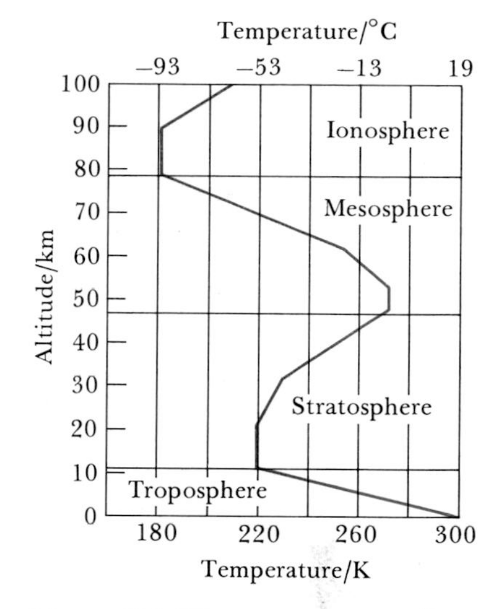

Figure F-1 The temperature of the earth's atmosphere as a function of height above sea level. The average temperature over the surface of the earth is about 15°C. The temperature drops at the rate of 7°C/km for the first 10 km or so, and then stays constant for about 10 kilometers. It increases with altitude from 20 km up to about 50 km, after which it decreases until 80 km and then increases again. This temperature profile delineates four separate regions of the atmosphere.

Table F-1 The composition of the earth's atmosphere below 100 km

Constituent	Content in fraction of total molecules or ppm*	Constituent	Content in fraction of total molecules or ppm*
nitrogen (N_2)	0.7808	helium	5 ppm
oxygen (O_2)	0.2095	methane	2 ppm
argon	0.0093	krypton	1 ppm
water vapor	0–0.04	hydrogen (H_2)	0.5 ppm
carbon dioxide	325 ppm	dinitrogen oxide	0.5 ppm
neon	18 ppm	xenon	0.1 ppm

*Parts per million; for example, 325 ppm of carbon dioxide means that of each 1 million molecules 325 are CO_2.

molecule, and so most of the nitrogen released by volcanic activity remains in the atmosphere. Consequently, nitrogen has become the dominant constituent of the earth's atmosphere.

The atmosphere was capable of holding only a small fraction of the water vapor that resulted from volcanic eruptions. Eventually the accumulated water vapor gave rise to clouds and rain and subsequently to the bodies of water on the earth's surface. The main source of atmospheric water vapor today is evaporation from the earth's surface. The evaporated water is incorporated into clouds and then returned by precipitation. The average time a water molecule spends in the atmosphere is about one week. The concentration of water vapor is highest near the ground and drops to very low values above 10 km.

F-2. THE CONCENTRATION OF CARBON DIOXIDE IN THE EARTH'S ATMOSPHERE IS SLOWLY INCREASING

Carbon dioxide is produced not only by volcanic eruptions but also by respiration, the decay of organic matter, and the combustion of fossil fuels. It is removed from the atmosphere by photosynthesis, dissolution in the oceans, and the formation of shales and carbonate rocks (primarily as $CaCO_3$ and $MgCO_3$). There is evidence that the rate of removal of carbon dioxide from the earth's atmosphere is not high enough to keep pace with the ever-increasing rate at which it is added to the atmosphere as a result of the combustion of fossil fuels. The concentration of carbon dioxide in the atmosphere has increased by almost 8 percent since the beginning of this century. Between 1958 and 1975 the increase was 5 percent (Figure F-2). The present worldwide rate of increase of carbon dioxide concentration is about half the rate at which it is produced by the combustion of fossil fuels.

This steady increase in the concentration of carbon dioxide in the atmosphere could have serious consequences. The solar radiation that penetrates the earth's atmosphere is mostly visible radiation, with a little ultraviolet and infrared. This radiation is absorbed by the ground and the oceans. The earth's surface is thus warmed and as a result emits heat (infrared) radiation. Carbon dioxide, as well as water vapor, absorbs infrared radiation strongly. Hence, instead of passing through

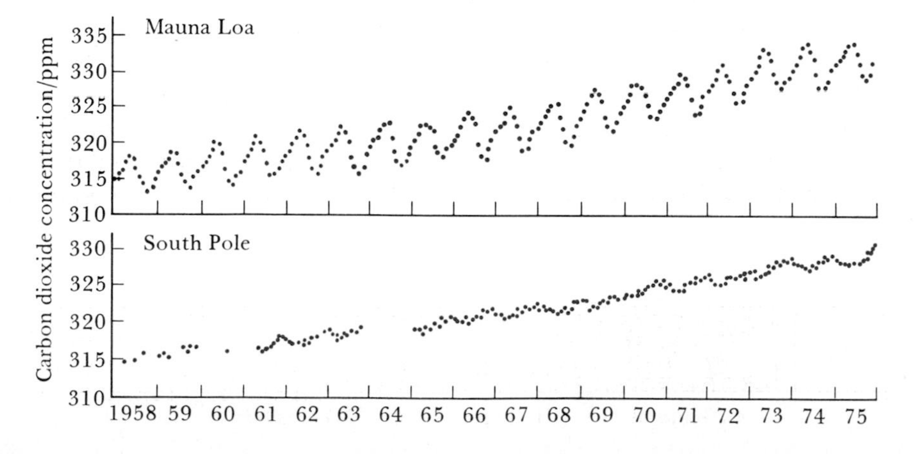

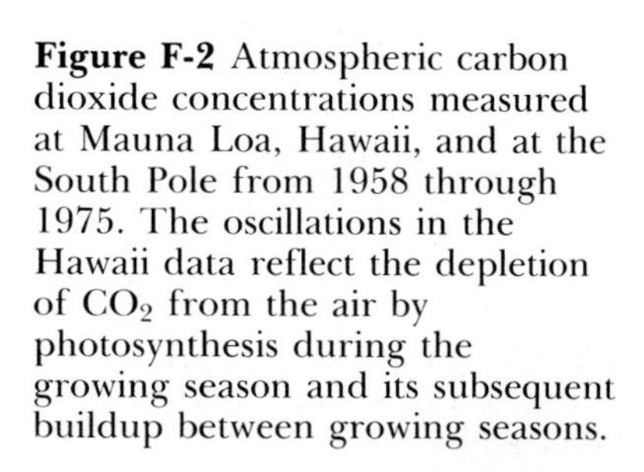
Figure F-2 Atmospheric carbon dioxide concentrations measured at Mauna Loa, Hawaii, and at the South Pole from 1958 through 1975. The oscillations in the Hawaii data reflect the depletion of CO_2 from the air by photosynthesis during the growing season and its subsequent buildup between growing seasons.

the atmosphere, the radiated energy is absorbed by the atmosphere, and the temperature of the atmosphere is increased. This phenomenon is called the **greenhouse effect** because it is similar to the trapping of the sun's heat in a greenhouse. It has been estimated that the greenhouse effect is responsible for an average increase of about 45 C° in the temperature at the earth's surface. Without this effect, the average surface temperature of the earth would be −30°C instead of 15°C. The planet Venus, which has a very dense atmosphere of carbon dioxide, has a surface temperature of about 500°C.

Calculations indicate that a doubling of the concentration of carbon dioxide in the atmosphere would cause an increase in the average earth temperature of several degrees Celsius. (Recall that the concentration of carbon dioxide has increased by almost 8 percent in this century). An increase in the average earth temperature by a few degrees is not trivial; such an increase could cause a large part of the polar ice caps to melt, raising the level of the oceans and flooding coastal cities. The calculations that lead to this dire forecast, however, are controversial.

The greenhouse effect due to water vapor in the atmosphere is a major factor in the weather of deserts and other regions. Although desert days may be very hot, there is little water vapor in the atmosphere to absorb the reradiated infrared radiation at night and so the nights are very cool.

F-3. THE NOBLE GASES WERE NOT DISCOVERED UNTIL 1893

In 1893, the English physicist Lord Rayleigh noticed a small discrepancy between the density of nitrogen obtained by the removal of oxygen, water vapor, and carbon dioxide from air and the density of nitrogen prepared by chemical reaction, such as the thermal decomposition of ammonium nitrite:

$$NH_4NO_2(s) \rightarrow N_2(g) + 2H_2O(l)$$

One liter of nitrogen at 0°C and 1 atm obtained by the removal of all the other known gases from air (Figure F-3) has a mass of 1.2565 g, whereas one liter of dry nitrogen obtained from ammonium nitrite has a mass of 1.2498 g under the same conditions. This slight difference led Lord Rayleigh to suspect that some other gas was present in the sample of nitrogen obtained from air.

In the late 1890s, the English chemist William Ramsay found that if hot calcium metal is placed in a sample of nitrogen obtained from air, about one percent of the gas fails to react to form calcium nitride. Pure nitrogen would react completely. Because of the inertness of the residual gas, Ramsay gave it the name argon (Greek, idle). He then liquefied the residual gas and, upon measuring its boiling point, discovered that it consisted of five components, each with its own characteristic boiling

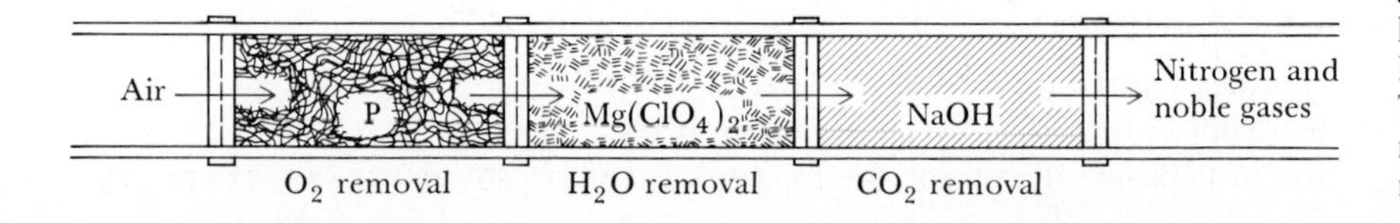

Figure F-3 A schematic illustration of the removal of O_2, H_2O, and CO_2 from air. First the oxygen is removed by allowing the air to pass over phosphorus: $P_4(s) + 5O_2(g) \rightarrow P_4O_{10}(s)$. The residual air is passed through anhydrous magnesium perchlorate to remove the water vapor: $Mg(ClO_4)_2(s) + 6H_2O(g) \rightarrow Mg(ClO_4)_2{\cdot}6H_2O(s)$ and then through sodium hydroxide to remove the CO_2: $NaOH(s) + CO_2(g) \rightarrow NaHCO_3(s)$. The gas that remains is primarily nitrogen with about 1 percent noble gases.

Table F-2 Properties of the noble gases

Gas	ppm in the air	Density*/$g \cdot L^{-1}$	Melting point/°C	Boiling point/°C	Cost†
helium	5.2	0.179	−272.2	−268.9	$115
neon	18.2	0.900	−248.6	−245.9	$100
argon	9340	1.78	−189.3	−185.8	$75
krypton	1.1	3.75	−157.1	−152.9	$400
xenon	0.08	5.90	−112.0	−108.1	$1400

*At 0°C and 1 atm.
†For 100 L at about 40 atm, research grade (1983).

point (Table F-2). The component present in the greatest amount retained the name argon. The others were named helium (sun), neon (new), krypton (hidden), and xenon (stranger). Helium was named after the Greek word for sun (helios) because its presence in the sun had been detected earlier by spectroscopic methods.

The noble gases in the atmosphere are thought to have arisen as by-products of the decay of radioactive elements in the earth's crust (Chapter 21). For their work in discovering and characterizing an entire new family of elements, Rayleigh received the 1904 Nobel Prize in physics and Ramsay received the 1904 Nobel Prize in chemistry.

▪ Helium has 93 percent of the lifting power of hydrogen.

All the noble gases are colorless, odorless, and relatively inert. Helium is used in lighter-than-air craft, despite the fact that it is denser and hence has less lifting power than hydrogen, because it is nonflammable. Helium is also used in welding to provide an inert atmosphere around the welding flame and thus reduce corrosion of the heated metal. Neon is used in neon signs, which are essentially discharge tubes (Section 6-10) filled with neon or a neon-argon mixture. When placed in a discharge tube, neon emits an orange-yellow glow that penetrates fog very well. Argon, the most plentiful and least expensive noble gas, often is used in place of nitrogen in incandescent light bulbs of high candlepower because it does not react with the hot filament. Krypton and xenon are too scarce and costly to find much application, although they are used in lasers, timing lights, and flashtubes for high-speed photography.

F-4. OXIDES OF SULFUR ARE MAJOR POLLUTANTS OF THE ATMOSPHERE

Except for the carbon dioxide produced from the combustion of fossil fuels, all the atmospheric constituents discussed up to this point arise from natural sources, meaning sources not due to human activities. Many machines and industries introduce harmful and irritating substances into the atmosphere, which collectively are referred to as pollutants. Some common pollutants are the oxides of sulfur (denoted by SO_x) that result from the combustion of sulfur-containing coal and oil, oxides of nitrogen (NO_x) and carbon monoxide (CO) produced in the internal combustion engine, and hydrocarbons, which arise from the incomplete combustion of gasoline and other hydrocarbon fuels.

Two oxides of sulfur, SO_2 and SO_3, are major pollutants in industrial and urban areas. Most coal and petroleum contain some sulfur, which becomes SO_2 when burned. Concentrations of SO_2 as low as 0.1 to 0.2 ppm can be incapacitating to persons suffering from respiratory conditions such as emphysema and asthma. Although SO_2 is not easily oxidized to SO_3, the presence of dust particles and other particulate matter or ultraviolet radiation facilities the conversion. The SO_3 then reacts with water vapor to form a very fine sulfuric acid mist. Such a mist is also produced in automobile catalytic converters. Both sulfuric acid and sulfurous acid, H_2SO_3, which arises from the reaction

$$SO_2(g) + H_2O(g) \rightarrow H_2SO_3(\text{mist})$$

produce **acid rain,** which is rain that is up to 1000 times more acidic than normal rain. Acid rain occurs commonly in northern Europe and in the northeastern United States. Many lakes in these regions are so acidic that the fish life and plant life are disappearing.

Acid rain has a devastating effect on limestone and marble, both of which contain $CaCO_3$. One of the reactions that occurs is

$$CaCO_3(s) + H_2SO_4(aq) \rightarrow CaSO_4(s) + H_2O(l) + CO_2(g)$$

The formation of powdered calcium sulfate breaks down the limestone or marble structure. The decomposition of carbonates by acid rain is a major cause of the deterioration of the ancient buildings and monuments of Europe (Figure F-4).

There have been three major disasters attributed to air polluted with oxides of sulfur. In 1952, a gray fog highly polluted with oxides of sulfur settled over London for several days and was reportedly responsible for 4000 deaths. Such a **London fog,** as it is now called, also caused hundreds of deaths in Donora, Pennsylvania, in 1948 and along the Meuse Valley in Belgium in 1930.

Several methods can be used to control the amount of SO_2 introduced into the atmosphere. One obvious way is to burn low-sulfur coal and petroleum. Nigerian oil and some Middle East oil is low in sulfur, whereas Venezuelan oil is high in sulfur. In general, coal from east of the Mississippi River is higher in sulfur than western coal. One method for removing SO_2 from fossil fuel combustion products involves passing the effluent gases through a device called a **scrubber,** where the gases are sprayed with an aqueous suspension of calcium oxide (lime). The scrubbing eliminates most of the SO_2 but produces large amounts of $CaSO_3$ and $CaSO_4$ that must be disposed of.

(a)

(b)

Figure F-4 Many of the limestone and marble monuments of Europe, such as this one at the gateway of the castle in Herten, Westphalia, are being damaged by acid rain. The gateway was constructed in 1702; (a) and (b) are photographs taken in 1908 and 1969, respectively.

F-5. HYDROCARBONS AND OXIDES OF NITROGEN ARE THE PRIMARY INGREDIENTS OF PHOTOCHEMICAL SMOG

Under ordinary conditions, nitrogen and oxygen do not react with each other. When combined at high pressure and temperature, however, as in the cylinders of an automobile engine, they react to form nitrogen oxide, NO, which then reacts with O_2 to produce nitrogen dioxide. Ordinarily this reaction occurs too slowly at the low concentrations of NO in the atmosphere to account for any significant concentration of $NO_2(g)$, but the reaction occurs rapidly in sunny, urban atmospheres, as a result of various **photochemical** (light-induced) processes.

Nitrogen dioxide is a red-brown noxious gas that is partially responsible for the yellow-brown color of smog, first made famous in Los Angeles, but now common in many urban areas.

The problem of NO_2 is not so much its primary toxicity but the fact that it is dissociated by radiation to produce atomic oxygen:

$$NO_2(g) \xrightarrow{\text{392-nm light}} NO(g) + O(g)$$

Because the dissociation of the NO_2 is caused by radiation (light), it is called **photodissociation.** The atomic oxygen then reacts with molecular oxygen to produce ozone. These two reactions account for the fact that ozone levels are higher on sunny days than on cloudy days. Ozone in the atmosphere makes up about 90 percent of the general category of pollutants called oxidants, which are now measured continually in many cities.

The atomic oxygen produced by the photodissociation of NO_2 also attacks the hydrocarbons introduced into the atmosphere by the incomplete combustion of gasoline and diesel fuel. The reaction of atomic oxygen with hydrocarbons initiates a complicated sequence of chemical reactions. The end products of these reactions are a number of substances that attack living tissue and lead to great discomfort, if not serious disorders. These substances make up what is called **photochemical smog,** so called because the entire process is initiated by the photodissociation of NO_2. A particularly irritating product is PAN (peroxyacetylnitrate), which causes eyes to tear and smart, something that people who live in smoggy cities experience often.

The control of photochemical smog requires controlling the emission of its two principal ingredients, NO and hydrocarbons from automobile exhausts. The Congressional Clean Air Act of 1967, with its amendments in 1970 and 1977, imposed limitations on exhaust emissions. Although there are indications that smog has lessened in some cities, in many others smog and other types of pollution problems are still increasing.

F-6. OZONE IS A LIFE-SAVING CONSTITUENT OF THE STRATOSPHERE

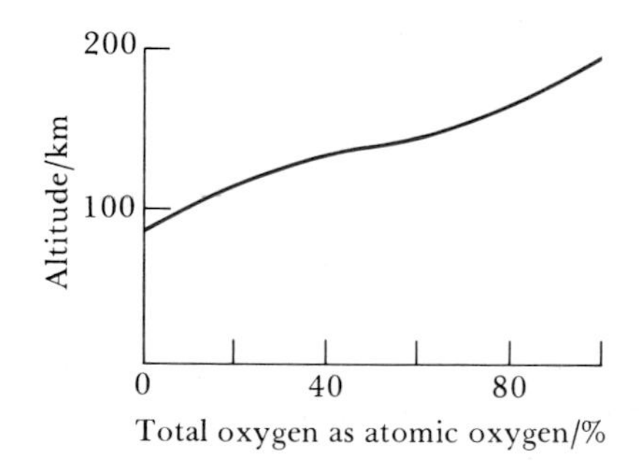

Figure F-5 The concentration of atomic oxygen as a function of altitude. Atomic oxygen is produced by the photodissociation of molecular oxygen.

Oxygen molecules in the stratosphere and higher regions undergo photodissociation by absorption of solar radiation of wavelengths less than 240 nm:

$$O_2(g) \xrightarrow{\lambda < 240\text{ nm}} 2O(g)$$

Virtually all the solar radiation in the region between 100 and 200 nm is absorbed by O_2. The atomic oxygen produced by this reaction is a major atmospheric constituent above 100 km; the total fraction of oxygen that exists as atomic oxygen is plotted versus altitude in Figure F-5.

Small mass spectrometers have been part of the experimental apparatus sent up in satellites and have provided data like those in Figure F-5. Because the gas density is so low above 100 km, oxygen atoms produced at greater altitudes can exist for a long time. At lower altitudes, where the total gas density is higher, the oxygen atoms are able to react not only with each other but also, and more important, with oxygen molecules to produce ozone. Below 70 km or so, atomic oxygen produced by photodissociation of O_2 reacts almost immediately with O_2 to produce O_3. Figure F-6 shows the concentration of ozone in the atmosphere versus altitude. Note that most of the ozone is found between 15 and 30 km; this region is called the **ozone layer.**

It should be emphasized that the concentration of atmospheric ozone is indeed small, never exceeding 10 ppm. If the entire ozone layer were compressed to a pressure of 1 atm at 0°C, its thickness would be only 3 mm.

Ultraviolet radiation with wavelengths longer than 240 nm is not absorbed until it encounters the ozone layer, where it is then absorbed by photodissociation of O_3:

$$O_3(g) \xrightarrow{240\ \text{nm} < \lambda < 310\ \text{nm}} O_2(g) + O(g)$$

The oxygen atom produced in this reaction quickly recombines with another oxygen molecule:

$$O(g) + O_2(g) \rightarrow O_3(g)$$

By adding these two equations, we see that there is no net chemical change but there is an absorption of radiation. Through this pair of reactions repeated many times, a single ozone molecule ultimately leads to the absorption of many photons of radiation. Trace amounts of ozone in the stratosphere can absorb virtually all the solar ultraviolet radiation in the region 240 nm $< \lambda <$ 310 nm. When the solar spectrum is measured at the far reaches of the earth's atmosphere and at the earth's surface, it is seen that the ozone in the stratosphere and oxygen in the ionosphere absorb nearly all the sun's ultraviolet radiation. Were it not for the formation of ozone in the upper atmosphere, life as it currently exists on earth would not be possible because of the deleterious effect of ultraviolet radiation on most living organisms.

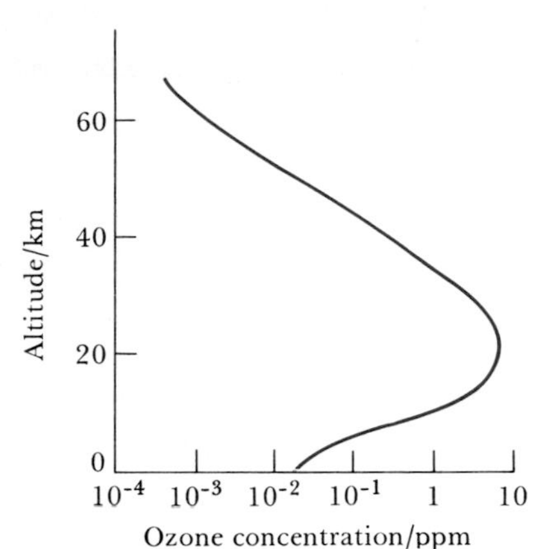

Figure F-6 The concentration of ozone in the earth's atmosphere as a function of altitude. Note that even in the ozone layer the concentration of ozone does not exceed 10 ppm.

TERMS YOU SHOULD KNOW

troposphere 239
stratosphere 239
mesosphere 239
ionosphere 239
greenhouse effect 241
acid rain 243
London fog 243
scrubber 243
photochemical 243
photodissociation 244
photochemical smog 244
ozone layer 245

QUESTIONS

F-1. Name the four regions into which the earth's atmosphere can be divided.

F-2. In which region of the atmosphere does all our weather take place?

F-3. What is the origin of the nitrogen in our atmosphere?

F-4. What is the potential problem concerning the continuous increase of the concentration of carbon dioxide in the atmosphere?

F-5. Why are clear nights colder than cloudy nights?

F-6. Describe how the noble gases were discovered.

F-7. What is photochemical smog? Why is it so called?

F-8. Write a chemical equation for the reaction of the deterioration of limestone and marble structures by acid rain.

F-9. Why do we say that ozone is a life-saving constituent of the stratosphere?

F-10. What is meant by the ozone layer?

F-11. What is photodissociation? Discuss the life-saving role of photodissociation in the ionosphere.

F-12. What is acid rain?

PROBLEMS

F-13. Oxygen molecules undergo photodissociation by absorption of solar radiation of wavelengths less than 240 nm. Calculate the energy of this radiation in $J \cdot mol^{-1}$.

F-14. According to Table F-1, the composition of dry air in units of fraction of total molecules is 0.7808 $N_2(g)$, 0.2095 $O_2(g)$, and 0.0093 $Ar(g)$. Calculate the partial pressure of each of these gases in a sample of dry air at 1.000 atm.

F-15. Calculate the mass of $CO_2(g)$ produced by the burning of a 16-gallon tank of gasoline. Assume that gasoline consists entirely of octane (C_8H_{18}) and that the density of octane is 0.80 $g \cdot mL^{-1}$.

F-16. The density at 0°C and 1 atm of $N_2(g)$ obtained by the removal of all the other known gases from air is 1.2565 $g \cdot L^{-1}$, whereas that obtained from the thermal decomposition of $NH_4NO_2(s)$ is 1.2498 $g \cdot L^{-1}$. Assume that all the residual gas in the "nitrogen" obtained from air is argon. Calculate the mass percentage of argon in the "nitrogen." To do this problem, you need to use the more accurate atomic masses 14.0067 for N and 39.948 for Ar and take 22.414 L to be the molar volume of a gas at 0°C and 1 atm.

F-17. Sulfur dioxide can be removed from the effluents of smoke stacks by passing it through a slurry of lime, $CaO(s)$. How much lime is required to remove one metric ton of $SO_2(g)$?

CHAPTER • 7

IONIC BONDS AND COMPOUNDS

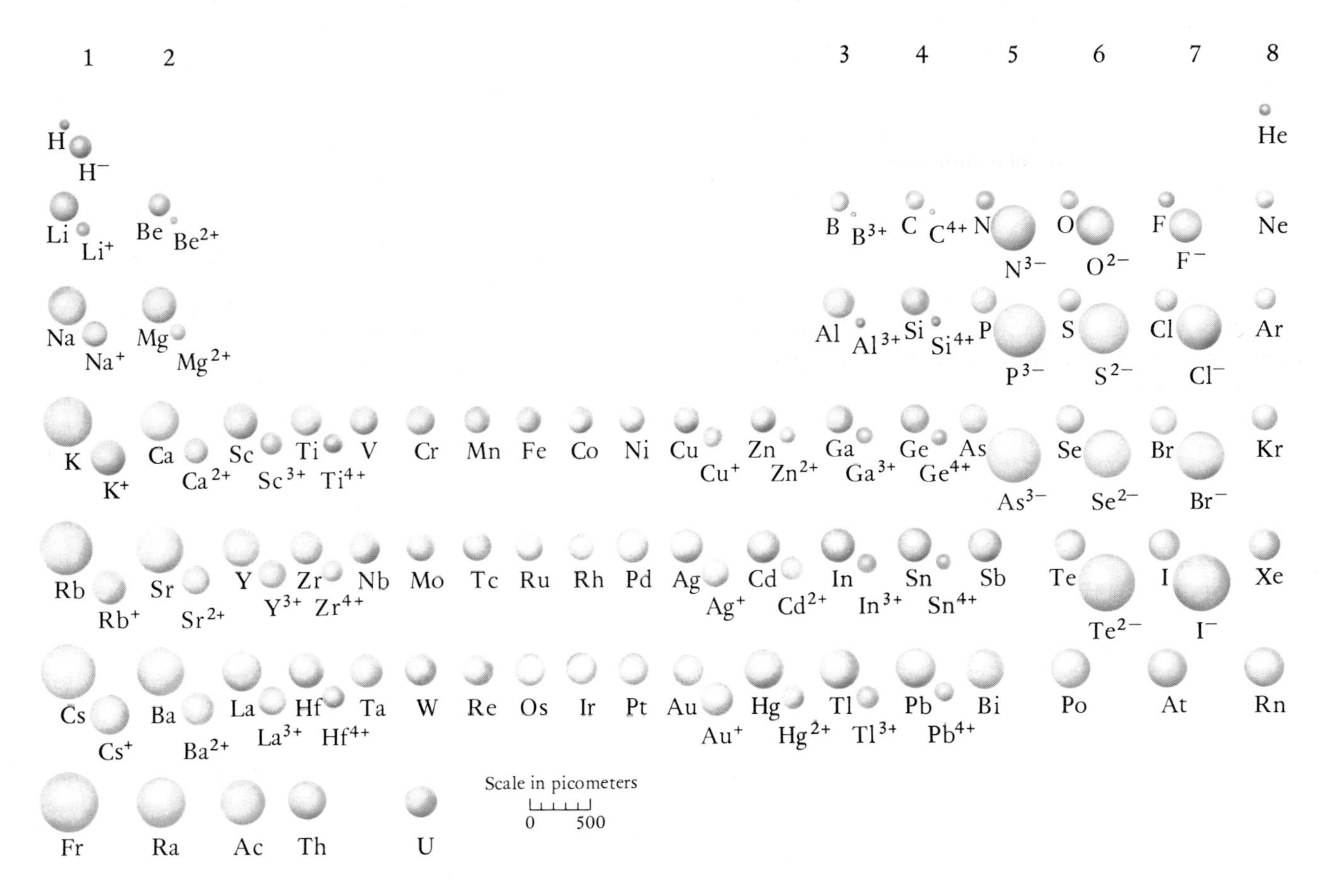

The relative sizes of atoms and ions.

In Chapter 6 we learned about the relationship of the electronic structure of atoms and the periodic table. It seems reasonable to suppose that an understanding of atomic structure should lead to an understanding of the chemical bonding that occurs between atoms. For example, it is possible to use the electron configurations of the sodium atom and the chlorine atom to understand why one atom of sodium combines with just one atom of chlorine to form sodium chloride. Why is the result NaCl, instead of $NaCl_2$ or Na_2Cl? Why is NaCl an ionic compound, capable of conducting an electric current when it is dissolved in water or melted? Why do carbon and hydrogen combine to form the stable molecule methane, whose formula is CH_4 instead of

CH or CH_2? Why is nitrogen a diatomic gas, N_2, at room temperature, whereas sodium chloride is a solid with a melting point of 800°C? Why doesn't a solution of sucrose dissolved in water conduct an electric current? These are the kind of questions that we answer in this and the next chapter.

7-1. SOLUTIONS THAT CONTAIN IONS CONDUCT AN ELECTRIC CURRENT

Before we discuss ionic bonds and ionic bonding, we should discuss an important experimental property of aqueous solutions of ionic compounds. When most ionic compounds dissolve in water, they yield free, mobile ions. For example, an aqueous solution of NaCl consists of $Na^+(aq)$ and $Cl^-(aq)$ ions that move throughout the water. If electrodes connected to the poles of a battery are dipped into a solution containing ions (Figure 7-1), then the positive ions are attracted to the negative electrode and the negative ions are attracted to the positive electrode. The motion of the ions through the solution constitutes an electric current.

Compounds that yield neutral molecules when they dissolve in water are not able to conduct an electric current because there are no charge carriers present, as there are in solutions of ionic compounds. For example, an aqueous solution of sucrose (table sugar, $C_{12}H_{22}O_{11}$) contains neutral sucrose molecules and so does not conduct an electric current. Substances like NaCl or $CaCl_2$, whose aqueous solutions conduct an electric current, are called **electrolytes,** whereas substances such as sucrose, whose aqueous solutions do not conduct an electric current, are called **nonelectrolytes.**

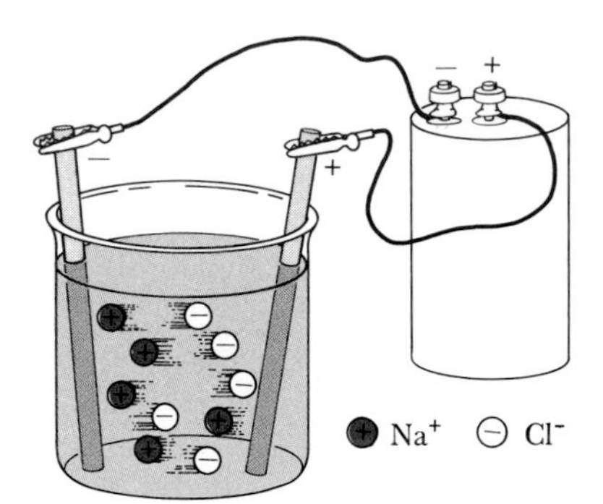

Figure 7-1 The conduction of an electric current by an aqueous solution of NaCl. An electric voltage is applied by dipping metal strips (electrodes) attached to the poles of a battery into the solution. Like the poles of a battery, one of the electrodes is positive and the other is negative. The positively charged sodium ions are attracted to the negative electrode, and the negatively charged chloride ions are attracted to the positive electrode. Thus the Na^+ ions migrate to the left in the figure and the Cl^- ions migrate to the right. The migration of the ions constitutes an electric current through the solution.

Not all solutions of electrolytes conduct an electric current to the same extent. For example, a 0.10 M $HgCl_2(aq)$ solution is a much poorer conductor of electricity than is a 0.10 M $CaCl_2$ solution. We call $CaCl_2$ a **strong electrolyte** and $HgCl_2$ a **weak electrolyte.** When a strong electrolyte such as $CaCl_2$ dissolves in water, essentially all the $CaCl_2$ formula units dissociate into free ions in solution and are available to conduct an electric current. However, when a weak electrolyte such as $HgCl_2$ dissolves in water, only a small fraction of the $HgCl_2$ formula units dissociate into ions; most exist as molecular $HgCl_2$ units. Thus, there are far fewer ions than in a corresponding $CaCl_2(aq)$ solution to conduct a current, and so $HgCl_2(aq)$ is a poorer conductor than $CaCl_2(aq)$ at the same concentration.

Some helpful guidelines for determining whether a substance is a strong electrolyte, a weak electrolyte, or a nonelectrolyte are

1. The acids HCl, HBr, HI, HNO_3, H_2SO_4, and $HClO_4$ are strong electrolytes. Most other acids are weak electrolytes.
2. The soluble hydroxides (hydroxides of the Group 1 and 2 metals except beryllium) are strong electrolytes. Most other bases, and particularly ammonia, are weak electrolytes.
3. Most soluble salts are strong electrolytes.
4. The halides and cyanides of metals with high atomic numbers (for example, mercury and lead) are often weak electrolytes.

5. Most organic compounds, that is, compounds that consist of carbon, hydrogen and possibly other atoms, are nonelectrolytes. Notable exceptions are organic acids and bases, which are usually weak electrolytes.

Example 7-1: Classify each of the following compounds as a strong electrolyte, a weak electrolyte, or a nonelectrolyte: $KClO_3$, C_6H_6 (benzene), $Al(OH)_3$, H_2SO_3.

Solution: Since $KClO_3$ is a soluble salt, it is a strong electrolyte. Benzene, C_6H_6, is an organic compound and so is a nonelectrolyte. The compound $Al(OH)_3$ is a hydroxide but not one of a Group 1 or 2 metal, and so it is a weak electrolyte. Since H_2SO_3 is not among the acids classified as strong electrolytes, it is a weak electrolyte.

7-2. THE ELECTROSTATIC ATTRACTION THAT BINDS OPPOSITELY CHARGED IONS TOGETHER IS CALLED AN IONIC BOND

To understand ionic bonds, let's first consider the reaction between a sodium atom and a chlorine atom. The electron configurations of the sodium and chlorine atoms are

$$\text{Na} \quad [\text{Ne}]3s^1 \qquad \text{Cl} \quad [\text{Ne}]3s^23p^5$$

Note that the electron configuration of a sodium atom consists of a neon-like inner core with a $3s$ electron outside the core. If the sodium atom loses the $3s$ electron, then the result is a sodium ion, with an electron configuration like that of the noble gas neon. We can describe the ionization process by the equation

$$\text{Na}([\text{Ne}]3s^1) \rightarrow \text{Na}^+([\text{Ne}]) + \text{e}^-$$

Once a sodium atom loses its $3s$ electron, the resultant sodium ion has a neon-like electron configuration and is relatively stable to further ionization.

If a chlorine atom accepts an electron, then the result is a chloride ion, with an electron configuration like that of the noble gas argon. We write this as

$$\text{Cl}([\text{Ne}]3s^23p^5) + \text{e}^- \rightarrow \text{Cl}^-([\text{Ar}])$$

Thus we see that both a sodium atom and a chlorine atom can achieve a noble-gas electron configuration through the transfer of an electron from the sodium atom to the chlorine atom. We can describe the electron transfer by the equation

▪ Noble-gas electron configurations are relatively stable.

$$\text{Na}([\text{Ne}]3s^1) + \text{Cl}([\text{Ne}]3s^23p^5) \rightarrow \text{Na}^+([\text{Ne}]) + \text{Cl}^-([\text{Ar}])$$

or, in terms of Lewis electron-dot formulas,

$$\text{Na}\cdot + \cdot\ddot{\underset{\cdot\cdot}{\text{Cl}}}\!: \longrightarrow \underbrace{\text{Na}^+ + :\ddot{\underset{\cdot\cdot}{\text{Cl}}}\!:^-}_{\text{Na}^+\text{Cl}^-}$$

The sodium ion and the chloride ion have opposite charges and so attract each other. The electrostatic attraction binds the ions together and is called an **ionic bond.**

We have seen that noble-gas electron configurations are relatively stable to the gain or loss of additional electrons. Because both Na^+ and Cl^- have achieved a noble-gas electron configuration, the above reaction will occur easily and there is no tendency for additional electron transfer. We already can answer several of the questions posed in the introduction to this chapter. We have seen that a sodium atom has one and only one electron that it loses readily, and that a chlorine atom readily gains one and only one electron. Therefore, when a sodium atom reacts with a chlorine atom, the transfer of an electron from the sodium atom to the chlorine atom occurs readily and the result is one sodium ion and one chloride ion. The chemical formula of sodium chloride, then, is NaCl and not $NaCl_2$, Na_2Cl, or anything other than NaCl. Furthermore, NaCl is an **ionic compound,** that is, a compound composed of ions.

The reaction between sodium and chlorine is an example of a reaction between a reactive metal and a reactive nonmetal. These atoms can lose or gain electrons relatively easily to achieve a noble-gas electron configuration. Figure 7-2 shows some of the common atoms that lose or gain electrons to achieve a noble-gas electron configuration. All the ions in Figure 7-2 have an outer electron configuration of ns^2np^6. We shall see that eight electrons in an outer shell is a particularly stable arrangement and that there is a strong tendency for it to occur. This is especially true for the elements in the first two rows of the periodic table. Note that metallic elements lose electrons to become positively charged ions (called **cations**) and nonmetallic elements gain electrons to become negatively charged ions (called **anions**), and that the charges on these ions correspond exactly to the ionic charges discussed in Chapter 2. In fact, the rules developed there for writing correct formu-

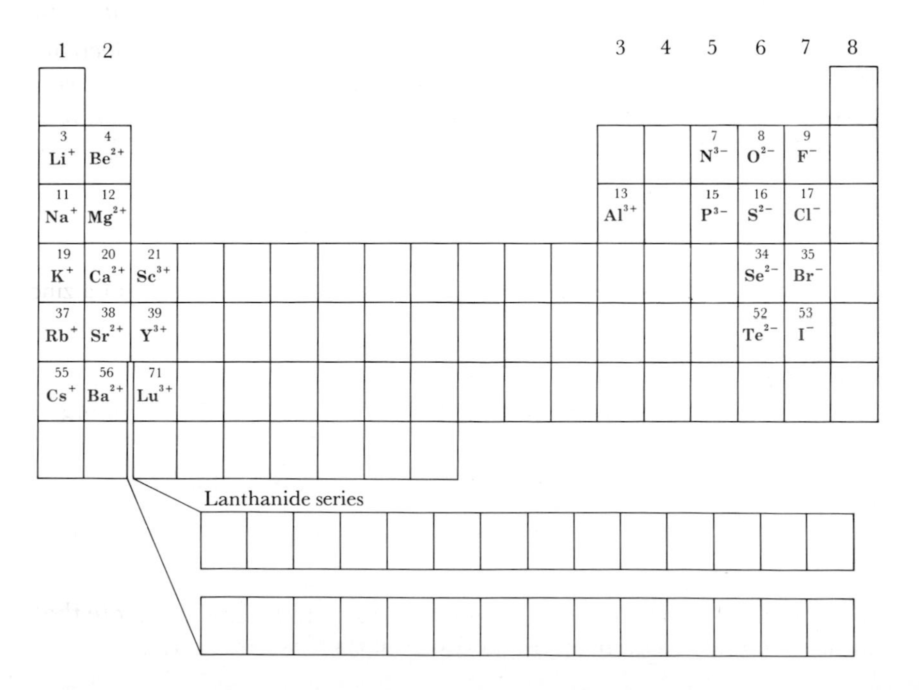

Figure 7-2 Some common ions with a noble-gas outer electron configuration ns^2np^6.

las for simple chemical compounds reflect the fact that the group of ions indicated by the chemical formula must have no net electrical charge.

Example 7-2: Write the ground-state electron configurations of Ca^{2+} and Se^{2-}, and predict the formula for calcium selenide.

Solution: The ground-state electron configurations of Ca and Se are

$$\text{Ca:} \quad 1s^2 2s^2 2p^6 3s^2 3p^6 4s^2$$

$$\text{Se:} \quad 1s^2 2s^2 2p^6 3s^2 3p^6 4s^2 3d^{10} 4p^4$$

To form Ca^{2+}, a calcium atom loses its two 4*s* electrons, and to form Se^{2-}, a selenium atom gains two 4*p* electrons.

$$Ca^{2+}: \quad 1s^2 2s^2 2p^6 3s^2 3p^6 \quad \text{or} \quad [\text{Ar}]$$

$$Se^{2-}: \quad 1s^2 2s^2 2p^6 3s^2 3p^6 4s^2 3d^{10} 4p^6 \quad \text{or} \quad [\text{Kr}]$$

In each case, the resulting ion has a noble-gas electron configuration. The formula for calcium selenide is CaSe.

7-3. THE ORBITALS OF TRANSITION-METAL IONS ARE FILLED IN A REGULAR ORDER

There are many metals not listed in Figure 7-2. For example, let's consider silver ($Z = 47$), which has the electron configuration $[\text{Kr}]5s^1 4d^{10}$. The silver atom would have to lose 11 electrons or gain 7 to achieve a noble-gas electron configuration. Table 6-1 shows that the first of these alternatives would require an enormous amount of energy. The energy required for the addition of seven electrons is also prohibitively large. Each successive electron would have to overcome a larger and larger repulsion as the negative charge on the ion is increased. Consequently, atomic ions with charge greater than three are uncommon.

Although a silver atom cannot achieve a noble-gas configuration, its outer electron configuration will be $4s^2 4p^6 4d^{10}$ if it loses its 5*s* electron. This configuration, with 18 electrons in the outer shell, is relatively stable and is sometimes called an **18-electron outer electron configuration.** Thus, silver forms a unipositive ion according to the equation

$$\text{Ag}([\text{Kr}]5s^1 4d^{10}) \rightarrow \text{Ag}^+([\text{Kr}]4d^{10}) + e^-$$

▪ The 18-electron outer electron configuration is $ns^2 np^6 nd^{10}$.

Example 7-3: Predict the electron configuration and the charge of a zinc ion.

Solution: A zinc atom has a $1s^2 2s^2 2p^6 3s^2 3p^6 4s^2 3d^{10}$ configuration. The zinc atom can achieve the 18-electron outer configuration $3s^2 3p^6 3d^{10}$ by losing its two 4*s* electrons. Thus the electron configuration of the zinc ion is

$$Zn^{2+}(1s^2 2s^2 2p^6 3s^2 3p^6 3d^{10}) \quad \text{or} \quad Zn^{2+}([\text{Ar}]3d^{10})$$

and its charge is +2. Note that a Zn^{2+} ion has a completely filled M shell.

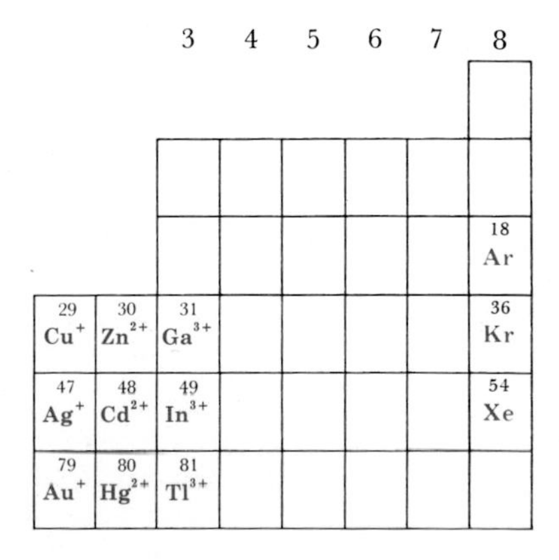

Figure 7-3 Some ions with an 18-electron outer configuration $ns^2 np^6 nd^{10}$.

Other elements that behave like zinc are shown in Figure 7-3. Note that these elements occur near the end of the *d* transition series.

Another outer electron configuration that is often found in ions can be illustrated by the element indium ($Z = 49$). The electron configuration of In is $[Kr]5s^24d^{10}5p^1$. Loss of the $5p$ electron yields the electron configuration $[Kr]5s^24d^{10}$ for the In^+ ion. Although In^+ does not have a noble-gas electron configuration or an 18-electron outer configuration, it does have all its subshells completely filled, and this is also a relatively stable electron configuration. Other elements that behave like indium are shown in Figure 7-4. Note that these elements are in Groups 3, 4, and 5. The unipositive ions in Figure 7-4 can lose two more electrons to achieve an 18-electron outer configuration. For example, if In^+ loses its two $5s$ electrons, then the resulting configuration for In^{3+} is $[Kr]4d^{10}$, which appears in Figure 7-3. Thus we see that indium and thallium have two possible ionic charges, +1 and +3, whereas gallium has only one possible ionic charge, +3.

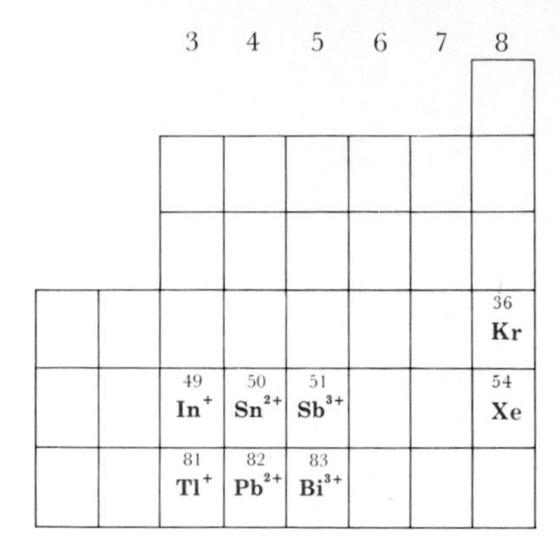

Figure 7-4 Some ions with the outer electron configuration [noble gas]$(n + 1)s^2nd^{10}$.

The ground-state electron configurations of transition-metal ions are relatively easy to deduce. We learned in Chapter 6 that the $3d$ orbitals are filled after the $4s$ orbital. When the transition-metal atoms lose electrons, we might expect that the $3d$ electrons are lost first, but this is not the case. For example, the ground-state electron configuration of Ni^{2+} is $[Ar]3d^8$ and *not* $[Ar]4s^23d^6$. The reason for this is that the energy levels shown in Figures 6-27(b) and 6-28 are for the *neutral* atoms. The charges on ions alter the order of the orbital energies so that, in Ni^{2+} and the other transition-metal ions, the energy of the $4s$ orbital is greater than that of the $3d$ orbitals.

Example 7-4: Predict the ground-state electron configuration of Ti^{3+}.

Solution: The ground-state electron configuration of the titanium atom is

$$[Ar]4s^23d^2$$

Although the $4s$ orbital is filled before the $3d$ orbitals in neutral atoms, this is not true for ions. In Ti^{3+} the $3d$ orbitals are used first. Therefore, the ground-state electron configuration of Ti^{3+}, which has three fewer electrons than neutral Ti, is

$$[Ar]3d^1$$

Note that the electron configuration of Ti^{3+} is regular in the sense that it is

$$1s^22s^22p^63s^23p^63d^1$$

The orbitals of the transition-metal ions are filled in the regular order.

▪ The ground-state electron configurations of transition-metal ions follows the regular sequence $1s^22s^22p^63s^23p^63d^{10}4s^24p^64d^{10}4f^{14}5s^2 \ldots$

Example 7-5: Referring only to the periodic table, predict the electron configurations of the ions Fe^{2+} and Cu^{2+}.

Solution: According to Example 7-4, the electron configurations of the transition-metal ions are regular. The atomic number of iron is 26, and so Fe^{2+} has 24 electrons. The electron configuration of Fe^{2+} is

$$1s^22s^22p^63s^23p^63d^6$$

or simply $[Ar]3d^6$.

The atomic number of copper is 29, and so Cu^{2+} has 27 electrons and its electron configuration is

$$1s^2 2s^2 2p^6 3s^2 3p^6 3d^9$$

or simply $[Ar]3d^9$. Do you see a relation between the atomic number of a $3d$ transition metal and the number of $3d$ electrons in the M^{2+} ion?

7-4. CATIONS ARE SMALLER AND ANIONS ARE LARGER THAN THEIR CORRESPONDING NEUTRAL ATOMS

Because atoms and ions are different species, we should expect atomic radii and **ionic radii** to have different values. The $3s$ electron in a sodium atom is, on the average, farther away from the nucleus than the $1s$, $2s$, and $2p$ electrons because the $3s$ electron is in the M shell. When a sodium atom loses its $3s$ electron, only the K and L shells remain, and so the resulting sodium ion is smaller than a sodium atom. In addition, the excess positive charge draws the remaining electrons toward the nucleus, and the K and L shells contract. The relative sizes of the alkali metal atoms and ions are shown in the first column of the Frontispiece. The Group 2 metals lose two outer s electrons in becoming ions. The excess positive charge of +2 contracts the remaining shells even more than in the case of the Group 1 metals, as you can see in the second column of the Frontispiece.

The atoms of nonmetals gain electrons in becoming ions. The addition of an extra electron increases the electron-electron repulsion and causes the electron cloud to expand. Negative ions are always larger than their corresponding atoms for this reason. The relative atomic and ionic sizes of the halogen atoms are shown in the column headed by 7 in the Frontispiece. Numerical values of ionic radii are given in Table 7-1.

Example 7-6: Predict which of the isoelectronic species is the larger ion, Na^+ or F^-.

Table 7-1 Some ionic radii

Ion	*Radius/ pm*	*Ion*	*Radius/ pm*	*Ion*	*Radius/ pm*	*Ion*	*Radius/ pm*	*Ion*	*Radius/ pm*	*Ion*	*Radius/ pm*	*Ion*	*Radius/ pm*
Cations								**Anions**					
Ag^+	126	Ba^{2+}	135	Al^{3+}	50	Ce^{4+}	101	Br^-	195	O^{2-}	140	N^{3-}	171
Cs^+	169	Ca^{2+}	99	B^{3+}	20	Ti^{4+}	68	Cl^-	181	S^{2-}	184	P^{3-}	212
Cu^+	96	Cd^{2+}	97	Cr^{3+}	65	U^{4+}	97	F^-	136	Se^{2-}	196		
K^+	133	Co^{2+}	82	Fe^{3+}	67	Zr^{4+}	80	H^-	154	Te^{2-}	221		
Li^+	60	Cu^{2+}	70	Ga^{3+}	62			I^-	216				
Na^+	95	Fe^{2+}	78	In^{3+}	81								
NH_4^+	148	Hg^{2+}	110	La^{3+}	115								
Rb^+	148	Ni^{2+}	69	Tl^{3+}	95								
Tl^+	144	Sr^{2+}	113	Y^{3+}	93								
		Zn^{2+}	74										

Solution: The electron configuration of both Na^+ and F^- is $1s^22s^22p^6$. The excess positive charge of Na^+ contracts the K and L shells, and the excess negative charge of F^- leads to an enlargement of the shells, and so we predict that F^- is larger than Na^+. Note also that both ions have the same number of electrons (10) but sodium has a nuclear charge of +11 and fluorine has a nuclear charge of only +9. The radius of Na^+ is 95 pm and that of F^- is 136 pm.

▪ Isoelectronic species are species that have the same number of electrons.

7-5. COULOMB'S LAW IS USED TO CALCULATE THE ENERGY OF AN ION PAIR

Up to now our discussion of ionic bonds has been qualitative. We now show by calculation that, when an ionic bond is formed, the energy of the ionic products is lower than that of the atomic reactants. Let's consider the reaction

$$\mathrm{Na}(g) + \mathrm{Cl}(g) \rightarrow \mathrm{Na}^+\mathrm{Cl}^-(g)$$

The net energy change for this reaction can be calculated by breaking the reaction down into three separate steps and applying Hess's law:

1. The electron is removed from the sodium atom (ionization). The energy required to ionize sodium atoms is 496 $\mathrm{kJ \cdot mol^{-1}}$.
2. The electron is added to the chlorine atom. Energy is released in the process, and this energy is called the **electron affinity** of chlorine. We discuss the idea of electron affinity below. The electron affinity of atomic chlorine is $-348\ \mathrm{kJ \cdot mol^{-1}}$.
3. The sodium ion and chloride ion are brought together as shown in Figure 7-5. From Table 7-1 we see that the radius of Na^+ is 95 pm and that of Cl^- is 181 pm. Thus the centers of the two ions are 181 + 95 = 276 pm apart when the two ions are just touching.

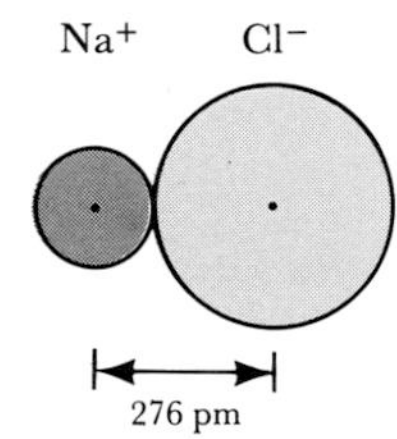

Figure 7-5 A solid-sphere representation of the ion-pair NaCl. According to Table 7-1, the sodium ion can be represented as a sphere of radius 95 pm and the chloride ion can be represented as a sphere of radius 181 pm. Because the two ions have opposite charges, they draw together until they touch, at a distance between their centers of 95 + 181 = 276 pm. They are bound together at this distance in an ionic bond.

Step 1 involves ionization energy, which we discussed in Chapter 6. We now need to discuss the energy associated with steps 2 and 3. Consider the process of adding an electron to an isolated atom:

$$\mathrm{A}(g) + \mathrm{e}^- \rightarrow \mathrm{A}^-(g)$$

The energy associated with this process is called the **first electron affinity,** EA_1, of the atom A(g). For example, in the case of chlorine we have

$$\mathrm{Cl}(g) + \mathrm{e}^- \rightarrow \mathrm{Cl}^-(g) \qquad EA_1 = -348\ \mathrm{kJ \cdot mol^{-1}}$$

The value of EA_1 in this case is negative because energy is released in the process. Notice that the electron affinity of an atom is the negative of the first ionization energy of the ion, $\mathrm{A}^-(g)$. In an equation, we have

$$\mathrm{A}^-(g) \rightarrow \mathrm{A}(g) + \mathrm{e}^- \qquad I_1 = -EA_1$$

Thus, we see that the first ionization energy of an isolated chloride ion, $\mathrm{Cl}^-(g)$, is $+348\ \mathrm{kJ \cdot mol^{-1}}$.

Just as we define successive ionization energies, we define successive electron affinities. For example, the first two electron affinities for an oxygen atom are defined by

$$O(g) + e^- \rightarrow O^-(g) \qquad EA_1 = -136\ \text{kJ}\cdot\text{mol}^{-1}$$

$$O^-(g) + e^- \rightarrow O^{2-}(g) \qquad EA_2 = +780\ \text{kJ}\cdot\text{mol}^{-1}$$

Notice that the value of the **second electron affinity,** EA_2, is positive; it requires energy to overcome the repulsion of the negatively charged $O^-(g)$ ion and the electron. The most important electron affinities for our purposes are those of the reactive nonmetals (Table 7-2).

Table 7-2 Electron affinities of the atoms of some reactive nonmetals

Atom	$EA/\text{kJ}\cdot\text{mol}^{-1}$
H	−72
F	−333
Cl	−348
Br	−324
I	−295
O	−136
	+780 (EA_2)
S	−200
	+590 (EA_2)
Se	−210
	+420 (EA_2)
N	−58
	+800 (EA_2)
	+1300 (EA_3)

So far then, we can write for steps 1 and 2

$$\text{step 1:} \quad Na(g) \rightarrow Na^+(g) + e^- \qquad I_1 = 496\ \text{kJ}\cdot\text{mol}^{-1}$$

$$\text{step 2:} \quad Cl(g) + e^- \rightarrow Cl^-(g) \qquad EA_1 = -348\ \text{kJ}\cdot\text{mol}^{-1}$$

If we add these two equations, then we find that

$$\text{steps 1 and 2:} \quad Na(g) + Cl(g) \rightarrow Na^+(g) + Cl^-(g)$$

$$\Delta H^\circ_{rxn} = I_1 + EA_1 = +148\ \text{kJ}\cdot\text{mol}^{-1}$$

Notice that an energy input of 148 $\text{kJ}\cdot\text{mol}^{-1}$ is required for this reaction.

Each of the species in this process is in the gaseous phase, which means in particular that Na^+ and Cl^- are so far apart from each other that they are effectively isolated entities. We now must calculate the energy required to bring these two ions to their equilibrium separation of 276 pm (Figure 7-5). We use **Coulomb's law** to calculate this energy. Coulomb's law states that the energy of two ions whose centers are separated by a distance d is

$$E = (2.31 \times 10^{-16}\ \text{J}\cdot\text{pm})\frac{Z_1 Z_2}{d} \tag{7-1}$$

where E is in joules, d is in picometers, and Z_1 and Z_2 are the charges of the two ions (Figure 7-6). If the charges of the ions have the same sign, then E in Equation (7-1) is positive. If the ions are oppositely charged, E is negative. In our case, $Z_1 = +1\ (Na^+)$, $Z_2 = -1\ (Cl^-)$, and $d =$ 276 pm, so that

$$E = \frac{(2.31 \times 10^{-16}\ \text{J}\cdot\text{pm})(+1)(-1)}{276\ \text{pm}}$$

$$= -8.37 \times 10^{-19}\ \text{J}$$

The minus sign means that the ions attract each other, so that the energy at 276 pm is less than it is when the ions are isolated from each other; therefore, energy is released when Na^+ and Cl^- ions are brought together. The quantity -8.37×10^{-19} J is for one ion-pair. For 1 mol of NaCl ion-pairs, the energy released is

$$E = \left(\frac{-8.37 \times 10^{-19}\ \text{J}}{\text{ion-pair}}\right)\left(6.02 \times 10^{23}\frac{\text{ion-pairs}}{\text{mol}}\right)$$

$$= -504\ \text{kJ}\cdot\text{mol}^{-1}$$

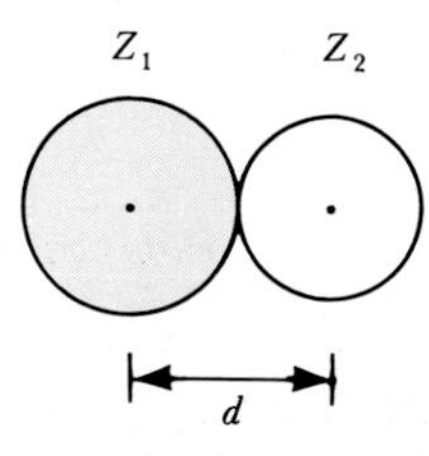

$$E = (2.31 \times 10^{-16}\ \text{J}\cdot\text{pm})\frac{Z_1 Z_2}{d}$$

Figure 7-6 Two ions separated by a distance d. The charges of the ions are Z_1 and Z_2. The energy of interaction of two ions is given by Coulomb's law, shown in Equation (7-1).

In terms of a chemical equation, we write

$$\text{step 3:} \quad Na^+(g) + Cl^-(g) \rightarrow Na^+Cl^-(g) \qquad d = 276\ \text{pm}$$

$$\Delta H^\circ_3 = -504\ \text{kJ}\cdot\text{mol}^{-1}$$

If we add this equation to the sum of steps 1 and 2, we obtain

$$Na(g) + Cl(g) \rightarrow Na^+Cl^-(g) \qquad \Delta H^\circ_{rxn} = -356\ kJ{\cdot}mol^{-1}$$
$$d = 276\ pm$$

Thus we calculate that 356 kJ is released in the formation of 1 mol of NaCl ion-pairs from 1 mol of sodium atoms and 1 mol of chlorine atoms. The fact that energy is released in the process means that the energy of the ion-pair is lower than that of the two separated atoms. Because the ion-pair has a lower energy, it is stable with respect to the atoms. The overall process is illustrated in Figure 7-7.

Realize that this calculation is for gaseous NaCl, which exists as widely separated ion-pairs. The energy released in the formation of crystalline NaCl is greater than for gaseous NaCl because each ion in crystalline NaCl is surrounded by six ions of opposite charge, thus giving an additional stability. We shall discuss the energy released upon crystal formation in the next section.

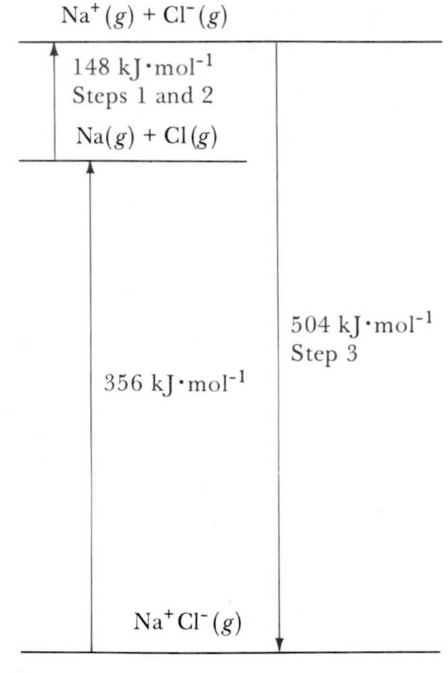

Figure 7-7 A summary of the steps used to calculate the energy evolved in the process $Na(g) + Cl(g) \rightarrow Na^+Cl^-(g)$. The process is broken down into two steps. First the atoms are converted to ions (steps 1 and 2), and then the two ions are brought together to a distance equal to the sum of their ionic radii (step 3). This first step uses the ionization energy of sodium and the electron affinity of chlorine. The second step uses Coulomb's law to calculate the energy involved in bringing the two isolated ions together.

Example 7-7: Calculate the energy released in the reaction

$$Cs(g) + Cl(g) \rightarrow Cs^+Cl^-(g)$$

given that the first ionization energy of cesium is 376 kJ·mol⁻¹.

Solution: We must break this reaction down into three steps (1) the ionization of Cs, (2) the addition of an electron to Cl, and (3) the bringing together of Cs^+ and Cl^- to their ion-pair separation.

1. From the first ionization energy of Cs, we write

$$Cs(g) \rightarrow Cs^+(g) + e^- \qquad I_1 = +376\ kJ{\cdot}mol^{-1}$$

2. From Table 7-2, we see that the first electron affinity of Cl is $-348\ kJ{\cdot}mol^{-1}$, and so

$$Cl(g) + e^- \rightarrow Cl^-(g) \qquad EA_1 = -348\ kJ{\cdot}mol^{-1}$$

If we add the results of steps 1 and 2, we get

$$Cs(g) + Cl(g) \rightarrow Cs^+(g) + Cl^-(g) \qquad \Delta H^\circ_{rxn} = +28\ kJ{\cdot}mol^{-1}$$

3. We now must calculate the energy involved in bringing Cs^+ and Cl^- to their separation as an ion-pair. According to Table 7-1, the radius of Cs^+ is 169 pm and that of Cl^- is 181 pm. Their separation as an ion-pair, then, is 350 pm. We now use Equation (7-1):

$$E = (2.31 \times 10^{-16}\ J{\cdot}pm)\frac{Z_1 Z_2}{d}$$
$$= \frac{(2.31 \times 10^{-16}\ J{\cdot}pm)(+1)(-1)}{350\ pm}$$
$$= -6.60 \times 10^{-19}\ J$$

This is the energy released by the formation of one ion-pair. For 1 mol of CsCl, we multiply this result by Avogadro's number:

$$E = \left(\frac{-6.60 \times 10^{-19}\ J}{\text{ion-pair}}\right)\left(6.02 \times 10^{23}\frac{\text{ion-pairs}}{\text{mol}}\right)$$
$$= -397\ kJ{\cdot}mol^{-1}$$

The minus sign indicates that energy is released in the process. We can express this result in the form of an equation:

$$Cs^+(g) + Cl^-(g) \rightarrow Cs^+Cl^-(g) \qquad \Delta H^\circ_{rxn} = -397\ kJ\cdot mol^{-1}$$
$$(d = 350\ pm)$$

If we combine this with the net result of steps 1 and 2, that is, with

$$Cs(g) + Cl(g) \rightarrow Cs^+(g) + Cl^-(g) \qquad \Delta H^\circ_{rxn} = +28\ kJ\cdot mol^{-1}$$

we get

$$Cs(g) + Cl(g) \rightarrow Cs^+Cl^-(g) \qquad \Delta H^\circ_{rxn} = -369\ kJ\cdot mol^{-1}$$
$$(d = 350\ pm)$$

Thus 369 kJ is evolved in the formation of 1 mol of CsCl ion-pairs from 1 mol of cesium atoms and 1 mol of chlorine atoms.

7-6. THE FORMATION OF IONIC SOLIDS FROM THE ELEMENTS IS AN EXOTHERMIC PROCESS

Many of the reactions that we have considered so far in this chapter have been simplified in the sense that we have discussed only reactions between atoms. Chlorine, for example, exists as a diatomic molecule at room temperature. In addition, sodium is a solid at room temperature and so the reaction we should consider is

$$Na(s) + \tfrac{1}{2}Cl_2(g) \rightarrow NaCl(s)$$

instead of

$$Na(g) + Cl(g) \rightarrow NaCl(g)$$

We can break the first of these reactions down into five steps:

1. Vaporize 1 mol of sodium metal so that the sodium atoms are far apart and effectively isolated from each other. The energy required for this step is the energy of vaporization of sodium, which is 93 $kJ\cdot mol^{-1}$ at room temperature. We can write this process as

$$Na(s) \rightarrow Na(g) \qquad \Delta H^\circ_{vap} = +93\ kJ\cdot mol^{-1}$$

2. Dissociate 0.5 mol of $Cl_2(g)$ into 1 mol of chlorine atoms. The energy required for this process is 122 $kJ\cdot mol^{-1}$ (one half the value in Table 5-2), and so we write

$$\tfrac{1}{2}Cl_2(g) \rightarrow Cl(g) \qquad \Delta H^\circ_{diss} = +122\ kJ\cdot mol^{-1}$$

3. Ionize the mole of Na(*g*). The energy required is the first ionization energy of sodium, which is 496 $kJ\cdot mol^{-1}$ (Table 6-1). Therefore, we write

$$Na(g) \rightarrow Na^+(g) + e^- \qquad I_1 = +496\ kJ\cdot mol^{-1}$$

4. Add the mole of electrons generated in step 3 to 1 mol of chlorine atoms. This is step 2 in Example 7-7 and is

$$Cl(g) + e^- \rightarrow Cl^-(g) \qquad EA_1 = -348\ kJ\cdot mol^{-1}$$

5. Bring the mole of isolated sodium ions and the mole of isolated chloride ions together to form the NaCl crystal shown in Figure 7-8. Energy is released in this step. This energy, called the **lattice energy,** is known to be $-780\ kJ\cdot mol^{-1}$ for NaCl(*s*) (Table 7-3). The equation describing this process is

$$Na^+(g) + Cl^-(g) \rightarrow NaCl(s) \qquad \Delta H^\circ_{LE} = -780\ kJ\cdot mol^{-1}$$

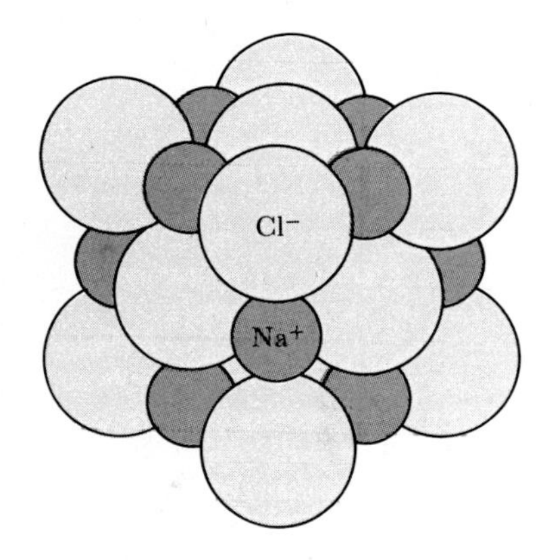

Figure 7-8 The crystalline structure of NaCl. Each Na^+ ion is surrounded by six Cl^- ions, and each Cl^- ion is surrounded by six Na^+ ions. This drawing shows the ions as spheres drawn to scale and illustrates the packing that occurs in the crystal.

Table 7-3 The lattice energies of some alkali halide crystals

Salt	Lattice energy/$kJ\cdot mol^{-1}$
NaF	−919
NaCl	−780
NaBr	−740
NaI	−704
KF	−813
KCl	−709
KBr	−688
RbCl	−686
RbBr	−670
RbI	−622

Notice that the lattice energy of NaCl is greater than the energy released upon gaseous ion-pair formation. As we said above, each Na^+ and Cl^- in the crystal is surrounded by six oppositely charged ions, and so the crystal has a lower energy than the gas of ion-pairs. We can summarize the five steps by the equations

1. $Na(s) \rightarrow Na(g)$ $\qquad \Delta H^\circ_{vap} = +93\ kJ \cdot mol^{-1}$
2. $\frac{1}{2}Cl_2(g) \rightarrow Cl(g)$ $\qquad \Delta H^\circ_{diss} = +122\ kJ \cdot mol^{-1}$
3. $Na(g) \rightarrow Na^+(g) + e^-$ $\qquad I_1 = +496\ kJ \cdot mol^{-1}$
4. $Cl(g) + e^- \rightarrow Cl^-(g)$ $\qquad EA_1 = -348\ kJ \cdot mol^{-1}$
5. $Na^+(g) + Cl^-(g) \rightarrow NaCl(s)$ $\qquad \Delta H^\circ_{LE} = -780\ kJ \cdot mol^{-1}$

If we add these five equations, then the net result is

$$Na(s) + \tfrac{1}{2}Cl_2(g) \rightarrow NaCl(s) \qquad \Delta H^\circ_{rxn} = -\ 417\ kJ \cdot mol^{-1}$$

Thus we see that the formation of $NaCl(s)$ from $Na(s)$ and $Cl_2(g)$ is accompanied by the evolution of a large amount of energy. Notice that it is the large value of the lattice energy (step 5) that is primarily responsible for the strongly exothermic character of the overall reaction. Generally, the magnitude of the lattice energies of ionic crystals is large enough to produce exothermic reactions for the formation of ionic solids from their elements. The various steps in the process are displayed in Figure 7-9. The complete cycle of steps shown in Figure 7-9 is called a **Born-Haber cycle.**

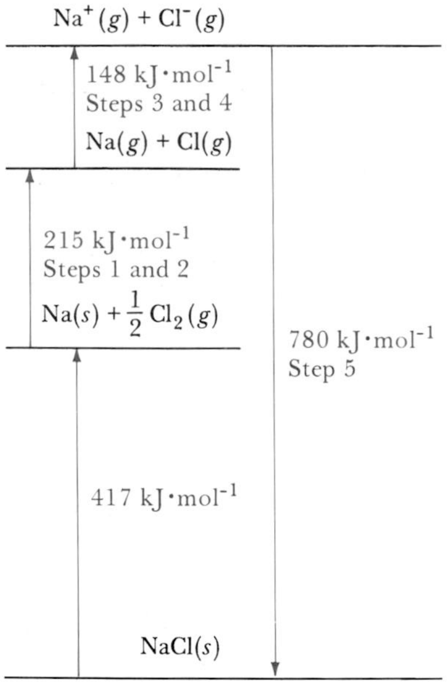

$Na^+(g) + Cl^-(g)$
148 kJ·mol^{-1}
Steps 3 and 4
$Na(g) + Cl(g)$
215 kJ·mol^{-1}
Steps 1 and 2
$Na(s) + \frac{1}{2}Cl_2(g)$
780 kJ·mol^{-1}
Step 5
417 kJ·mol^{-1}
$NaCl(s)$

Figure 7-9 A Born-Haber cycle: A summary of the steps used to calculate the energy evolved in the process $Na(s) + \frac{1}{2}Cl_2(g) \rightarrow NaCl(s)$. The process is broken down into three stages. First the sodium is vaporized and the chlorine is dissociated to produce $Na(g)$ and $Cl(g)$ (steps 1 and 2). Then these atoms are converted to ions (steps 3 and 4). Last, the $Na^+(g)$ and $Cl^-(g)$ are brought together to form the NaCl crystal lattice (step 5) shown in Figure 7-8. The energy involved in this last step is called the lattice energy of sodium chloride.

SUMMARY

An ionic bond is the electrostatic attraction that binds oppositely charged ions together. An ionic compound is formed between two reactants if one of them has a relatively low ionization energy and the other has a relatively high electron affinity. This is the case when a reactive metal reacts with a reactive nonmetal. When this occurs, one or more electrons are transferred completely from the metal atom to the nonmetal atom, resulting in an ionic bond.

Many ions are stable because they achieve a noble-gas electron configuration, although there are stable ions that have other types of electron configurations as well, such as the 18-electron outer configurations. Ions are completely different species from atoms, and so ionic radii are not the same as atomic radii. The relative sizes of ions and their trend in the periodic table can be understood in terms of electron configurations. The chapter concludes with a quantitative discussion of the energetics of the formation of ionic compounds from their respective elements. These calculations use the concepts of ionization energy and electron affinity and are summarized in Figures 7-7 and 7-9.

TERMS YOU SHOULD KNOW

electrolyte 248
nonelectrolyte 248
strong electrolyte 248
weak electrolyte 248
ionic bond 250
ionic compound 250
cation 250
anion 250
18-electron outer configuration 251
ionic radius 253
electron affinity 254
first electron affinity, EA_1 254
second electron affinity, EA_2 255
Coulomb's law 255
lattice energy 257
Born-Haber cycle 258

AN EQUATION YOU SHOULD KNOW HOW TO USE

$$E = (2.31 \times 10^{-16}\ \mathrm{J \cdot pm})\frac{Z_1 Z_2}{d} \quad \text{(7-1) (Coulomb's law)}$$

PROBLEMS

ELECTRON CONFIGURATIONS AND CHEMICAL REACTIONS

7-1. Describe the following equations in terms of the electron configurations of the various species involved:

(a) $Ca(g) + 2F(g) \rightarrow CaF_2(g)$
(b) $Sr(g) + 2Br(g) \rightarrow SrBr_2(g)$
(c) $2Al(g) + 3O(g) \rightarrow Al_2O_3(g)$

7-2. Use electron configurations to describe the formation of the following ionic compounds from the atoms:

(a) $GaF_3(g)$ (b) $AgCl(g)$ (c) $Li_3N(g)$

7-3. Describe the following equations in terms of Lewis electron-dot formulas for the various species involved:

(a) $3Li(g) + N(g) \rightarrow Li_3N(g)$
(b) $Na(g) + H(g) \rightarrow NaH(g)$
(c) $Al(g) + 3I(g) \rightarrow AlI_3(g)$

7-4. Predict the products of the following reactions from a consideration of the Lewis electron-dot formulas of the reactants and the achievement of noble-gas electron configurations in the product ions:

(a) calcium and nitrogen (as N)
(b) aluminum and chlorine (as Cl)
(c) lithium and oxygen (as O)

ELECTRON CONFIGURATIONS OF IONS

7-5. Predict the ground-state electron configuration of

(a) Cr^{2+} (b) Cu^{2+}
(c) Co^{3+} (d) Mn^{2+}

7-6. Predict the ground-state electron configuration of

(a) Ru^{2+} (b) W^{3+}
(c) Pd^{2+} (d) Ag^{2+}

7-7. Which *d* transition-metal ions with a +2 charge have

(a) six *d* electrons (b) ten *d* electrons
(c) one *d* electron (d) five *d* electrons

Do you see a connection between the number of *d* electrons in the +2 ion and the position of the ion in its transition-metal series?

7-8. How many *d* electrons are there in

(a) Fe^{2+} (b) Zn^{2+}
(c) V^{2+} (d) Ni^{2+}

Can you see a pattern between the number of *d* electrons and the position of these ions in the first transition-metal series?

7-9. Using only a periodic table, predict the ground-state 18-electron outer configuration and the charge of the following ions:

(a) cadmium ion (b) indium(III) ion
(c) thallium(III) ion (d) zinc ion

7-10. Predict the ground-state 18-electron outer configuration and the charge of the following ions:

(a) copper(I) ion (b) gallium ion
(c) mercury(II) ion (d) gold(I) ion

7-11. Determine which of the following salts are composed of isoelectronic cations and anions:

(a) LiF (b) NaF
(c) KBr (d) KCl
(e) BaI_2 (f) AlF_3

7-12. Determine which of the following salts are composed of isoelectronic cations and anions:

(a) NaCl (b) RbBr
(c) $SrCl_2$ (d) $SrBr_2$
(e) MgF_2 (f) KI

CHEMICAL FORMULAS OF IONIC COMPOUNDS

7-13. Write the chemical formula for each of the following ionic compounds:

(a) yttrium sulfide (b) lanthanum bromide
(c) magnesium telluride (d) rubidium nitride
(e) aluminum selenide (f) calcium oxide

7-14. Write the chemical formula for each of the following compounds:

(a) thallium chloride (b) cadmium sulfide
(c) zinc nitride (d) copper(I) bromide
(e) gallium oxide (f) tin(II) fluoride

IONIC RADII

7-15. The following pairs of ions are isoelectronic. Predict which is the larger ion in each pair:

(a) K^+ and Cl^- (b) Ag^+ and Cd^{2+}
(c) Cu^+ and Zn^{2+} (d) F^- and O^{2-}

7-16. List the following ions in order of increasing radius:

$$Na^+ \quad O^{2-} \quad Mg^{2+} \quad F^- \quad Al^{3+}$$

IONIZATION ENERGIES AND ELECTRON AFFINITIES

7-17. List the following atoms in order of the ease with which they gain an electron to form an anion:

Br I H Cl

7-18. List the following atoms in order of the ease with which they lose electron(s) to form cations:

Ca K Na Al Li

7-19. Use the electron affinity data for Cl, Br, and I in Table 7-2, together with your knowledge of atomic periodicity trends, to estimate the electron affinity of astatine.

7-20. Explain why the magnitude of the first electron affinity of a chlorine atom is greater than that of a sulfur atom.

7-21. Using Tables 6-1 and 7-2, calculate ΔH°_{rxn} for the following equations:

(a) $Li(g) + Br(g) \rightarrow Li^+(g) + Br^-(g)$
(b) $I^-(g) + Cl(g) \rightarrow I(g) + Cl^-(g)$
(c) $Na(g) + H(g) \rightarrow Na^+(g) + H^-(g)$

7-22. Using Tables 6-1 and 7-2, calculate ΔH°_{rxn} for the following equations:

(a) $2Na(g) + S(g) \rightarrow 2Na^+(g) + S^{2-}(g)$
(b) $Mg(g) + O(g) \rightarrow Mg^{2+}(g) + O^{2-}(g)$
(c) $Mg(g) + 2Br(g) \rightarrow Mg^{2+}(g) + 2Br^-(g)$

CALCULATIONS INVOLVING COULOMB'S LAW

7-23. Use Coulomb's law to calculate the energy of a zinc ion and an oxide ion that are just touching.

7-24. Use Coulomb's law to calculate the energy of a sodium ion and a fluoride ion that are just touching.

7-25. Calculate the energy released (in $kJ\cdot mol^{-1}$) in the reaction

$$K(g) + Br(g) \rightarrow K^+Br^-(g)$$

The first ionization energy of potassium is 419 $kJ\cdot mol^{-1}$.

7-26. Calculate the energy released (in $kJ\cdot mol^{-1}$) in the reaction

$$Mg(g) + O(g) \rightarrow Mg^{2+}O^{2-}(g)$$

7-27. Calculate the energy released (in $kJ\cdot mol^{-1}$) in the reaction

$$Na(g) + H(g) \rightarrow Na^+H^-(g)$$

7-28. Calculate ΔH°_{rxn} for the following equation:

$$Zn(g) + S(g) \rightarrow Zn^{2+}S^{2-}(g)$$

The ionization energy for the process

$$Zn(g) \rightarrow Zn^{2+}(g) + 2e^-$$

is 2.640 $MJ\cdot mol^{-1}$.

7-29. Construct a diagram like that shown in Figure 7-7 for $Li^+F^-(g)$ (see Tables 6-1, 7-1, and 7-2 for the necessary data).

7-30. Construct a diagram like that shown in Figure 7-7 for $Na^+H^-(g)$ (see Tables 6-1, 7-1, and 7-2 for the necessary data).

LATTICE ENERGIES

7-31. Calculate the energy released (in $kJ\cdot mol^{-1}$) in the reaction

$$Na(s) + \tfrac{1}{2}F_2(g) \rightarrow NaF(s)$$

The heat of vaporization of Na is 93 $kJ\cdot mol^{-1}$, the dissociation energy of one mole of F_2 is 155 $kJ\cdot mol^{-1}$, and the lattice energy of NaF is -919 $kJ\cdot mol^{-1}$.

7-32. Calculate the energy released (in $kJ\cdot mol^{-1}$) in the reaction

$$K(s) + \tfrac{1}{2}Br_2(l) \rightarrow KBr(s)$$

The heat of vaporization of potassium is 89 $kJ\cdot mol^{-1}$, the first ionization energy of potassium is 419 $kJ\cdot mol^{-1}$, the sum of the dissociation and vaporization energies for $Br_2(l)$ is 223 $kJ\cdot mol^{-1}$, and the lattice energy of KBr is -688 $kJ\cdot mol^{-1}$.

7-33. Calculate the energy released (in $kJ\cdot mol^{-1}$) when NaI(*s*) is formed in the reaction

$$Na(s) + \tfrac{1}{2}I_2(s) \rightarrow NaI(s)$$

The energy of vaporization of Na(*s*) is 93 kJ·mol^{-1}. The sum of the heats of dissociation and vaporization of $I_2(s)$ is 214 kJ·mol^{-1}, and the lattice energy of NaI is −704 kJ·mol^{-1}.

7-34. Calculate the energy released (in kJ·mol^{-1}) when LiH(*s*) is formed in the reaction

$$Li(s) + \frac{1}{2}H_2(g) \rightarrow LiH(s)$$

The heat of vaporization of lithium is 161 kJ·mol^{-1}, the dissociation energy of H_2 is 436 kJ·mol^{-1}, and the lattice energy of LiH is −917 kJ·mol^{-1}.

7-35. Calculate the energy released in the reaction

$$Ca(s) + Cl_2(g) \rightarrow CaCl_2(s)$$

The heat of vaporization of Ca(*s*) is 193 kJ·mol^{-1}, the dissociation energy of Cl_2 is 244 kJ·mol^{-1}, the lattice energy of $CaCl_2(s)$ is −2266 kJ·mol^{-1}, the first ionization energy of Ca(*g*) is 590 kJ·mol^{-1} and the second ionization energy of Ca(*g*) is 1140 kJ·mol^{-1}.

7-36. Given the following data

$K(s) + \frac{1}{2}Br_2(l) \rightarrow KBr(s)$ $\qquad \Delta H^{\circ}_{rxn} = -392$ kJ·mol^{-1}

$K(g) + Br(g) \rightarrow KBr(g)$ $\qquad \Delta H^{\circ}_{rxn} = -329$ kJ·mol^{-1}

calculate the heat of vaporization for KBr(*s*) (see Problem 7-32 for additional data).

ADDITIONAL PROBLEMS

7-37. List three ions that are isoelectronic with F^-.

7-38. Predict the charge on (a) a lutetium ion; (b) a lawrencium ion.

7-39. The ionic radius of K^+ is 133 pm, while the ionic radius of Cu^+ is 96 pm. Explain why the radius of Cu^+ is smaller than that of K^+.

7-40. Calculate the lattice energy of LiCl(*s*) given the following data:

chlorine-chlorine bond energy: 244 kJ·mol^{-1}

first ionization energy of Li(*g*): 519 kJ·mol^{-1}

the heat of vaporization of Li(*s*): 161 kJ·mol^{-1}

the electron affinity of Cl: −348 kJ·mol^{-1}

heat of formation of LiCl(*s*): −408 kJ·mol^{-1}.

7-41. Calculate the electron affinity for chlorine given the following data:

chlorine-chlorine bond energy: 244 kJ·mol^{-1}

first and second ionization energies of Ca: 590 kJ·mol^{-1} and 1140 kJ·mol^{-1}

heat of vaporization of Ca(*s*): 193 kJ·mol^{-1}

heat of formation of $CaCl_2(s)$: −795 kJ·mol^{-1}

lattice energy of $CaCl_2(s)$: −2266 kJ·mol^{-1}.

7-42. A transition-metal ion with *x* outer-shell *d* electrons is said to be a d^x ion. Which +3 transition metal ions are d^4 ions?

7-43. How many unpaired electrons are there in a Fe^{2+} ion? in a Zn^{2+} ion?

7-44. Explain why the electron affinity of S is negative, but that of S^- is positive.

7-45. Explain why the energy evolved in the process

$$Na(g) + X(g) \rightarrow Na^+X^-(g)$$

decreases in the order X = Cl < X = Br < X = I.

7-46. Which +3 transition-metal ions are d^6 ions?

7-47. Which +1 transition-metal ions are d^3 ions?

7-48. Give three examples of each of the following:

(a) d^{10} +2 transition-metal ions.
(b) d^0 +4 transition-metal ions.

7-49. Classify each of the following compounds as a strong electrolyte, a weak electrolyte or a nonelectrolyte in aqueous solution:

(a) KNO_3 (b) $Pb(CN)_2$
(c) C_2H_5OH (ethyl alcohol) (d) CH_3OCH_3 (acetone)

7-50. Classify each of the following compounds as a strong electrolyte, a weak electrolyte or a nonelectrolyte in aqueous solution:

(a) HClO (b) Na_2SO_4
(c) $Hg(CN)_2$ (d) C_3H_7OH (rubbing alcohol)

CHAPTER ▪ 8

LEWIS FORMULAS

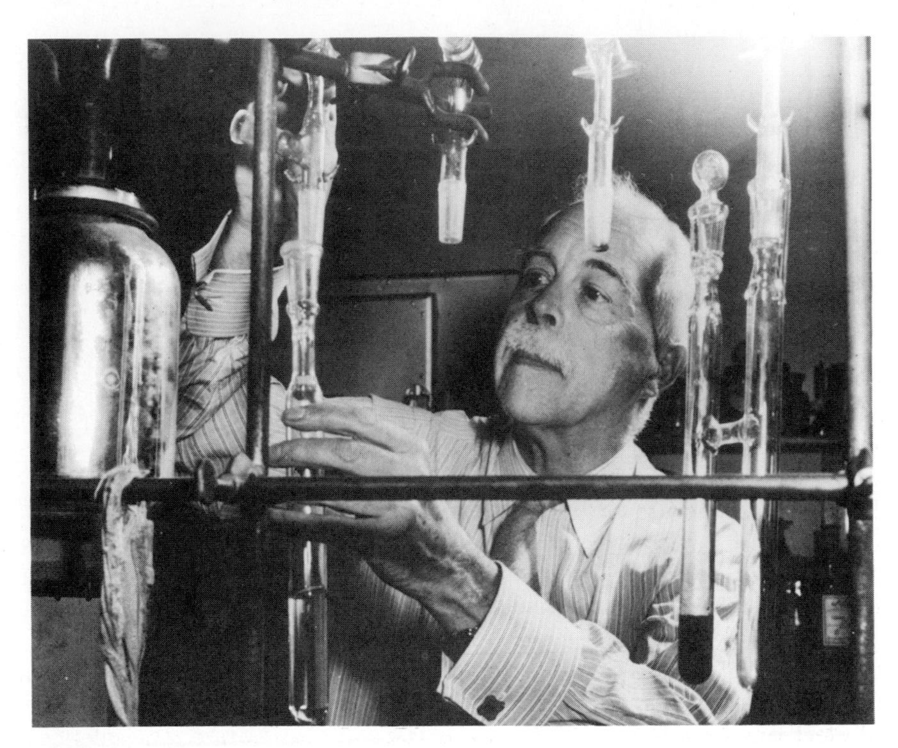

Gilbert N. Lewis was one of America's greatest chemists. He not only introduced the Lewis method of writing molecular formulas that is the subject of this chapter but also was one of the pioneers of chemical thermodynamics. He spent most of his career at the University of California at Berkeley, where he built up one of the strongest chemistry departments in the world.

In Chapter 7 we saw that, when metals react with nonmetals, valence electrons (that is, outer-shell electrons) of the metallic atoms are transferred completely to the nonmetallic atoms, resulting in ionic compounds. There is an enormous class of compounds that are not ionic. Many of these compounds are gases or liquids at room temperature, and they are poor conductors of an electric current. The bonding in these compounds differs from that in ionic compounds and is called covalent bonding. Although an understanding of covalent bonding requires a quantum theoretical description, it is possible to gain a qualitative and intuitive insight into covalent bonding in molecules by studying a method of writing molecular formulas that was introduced by the

American chemist G. N. Lewis. A full decade before the quantum theory was formulated by Schrödinger in 1925, Lewis postulated that a covalent bond can be described as a pair of electrons that is shared between two atoms. The electron-pair bond idea was later given a firm theoretical basis by the quantum theory of chemical bonding, which we discuss in Chapter 10. The molecular formulas introduced by Lewis are called Lewis formulas and are one of the most useful and important concepts of chemistry.

8-1. A COVALENT BOND CAN BE DESCRIBED AS A PAIR OF ELECTRONS SHARED BY TWO ATOMS

Consider the molecule Cl_2. The Lewis electron-dot formula for a chlorine atom is

$$:\ddot{\underset{..}{Cl}}\cdot$$

Recall that a Lewis electron-dot formula for an atom shows only the **valence electrons** (the outer-shell electrons). The chlorine atom is one electron short of having eight electrons in its outer shell and achieving an argon-like electron configuration. A chlorine atom in Cl_2 could get this electron from the other chlorine atom, but certainly that chlorine atom does not wish to give up an electron. In a sense there is a stalemate with respect to electron transfer because both atoms have the same driving force to gain an electron. From a more quantitative point of view, although a chlorine atom has a large electron affinity (-348 $kJ \cdot mol^{-1}$), it has a much higher ionization energy (1260 $kJ \cdot mol^{-1}$) and does not lose an electron very easily. Ionic bonds result only when one atomic reactant is a metal (relatively low ionization energy) and the other is a nonmetal (relatively high electron affinity).

Although we have ruled out the formation of an ionic bond in Cl_2, there is a way for the two chlorine atoms to achieve an argon-like electron configuration *simultaneously*. If the two chlorine atoms *share* a pair of electrons between them, then the resulting distribution of valence electrons can be pictured as

$$:\ddot{\underset{..}{Cl}}\cdot + \cdot\ddot{\underset{..}{Cl}}: \rightarrow :\ddot{\underset{..}{Cl}}:\ddot{\underset{..}{Cl}}:$$

Notice that *each* chlorine atom has eight electrons in its outer shell:

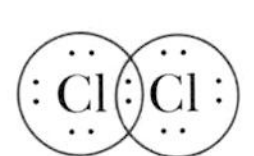

Thus by sharing a pair of electrons, each chlorine atom is able to achieve the stable argon-like outer electron configuration of eight electrons. According to Lewis's picture, the shared electron pair is responsible for holding the two chlorine atoms together as a chlorine molecule. The bond formed between two atoms by a shared electron pair is called a **covalent bond.**

The electron-dot formula depicted for Cl_2 is called a **Lewis formula.** It is conventional to indicate the electron-pair bond as a line joining the two atoms and the other electrons as pairs of dots surrounding the atoms:

$$:\ddot{\underset{..}{Cl}}-\ddot{\underset{..}{Cl}}:$$

▪ The halogens have the Lewis formula $:\ddot{\underset{..}{X}}-\ddot{\underset{..}{X}}:$, where X is F, Cl, Br, or I.

The pairs of electrons that are not shared between the chlorine atoms are called **lone electron pairs,** or simply **lone pairs.** A Lewis formula correctly depicts a covalent bond as a pair of electrons shared between two atoms.

Figure 8-1 The regular arrangement of the chlorine molecules in crystalline Cl_2. This pattern is repeated throughout the crystal. The chlorine molecules are neutral and so do not attract each other as strongly as neighboring ions in an ionic lattice. Consequently, molecular crystals like Cl_2 usually have lower melting points than ionic crystals. The melting point of chlorine is −101°C. For comparison, the melting point of NaCl(*s*) is 800°C.

When Cl_2 is solidified (its freezing point is −101°C), it forms a **molecular crystal** (Figure 8-1). In contrast to an ionic crystal, the constituent particles of a molecular crystal are molecules, in this case Cl_2 molecules. The low melting point of chlorine indicates that the attraction between the molecules is weak relative to the attraction between ions in a crystal. The Cl_2 molecules are neutral, and thus there is no net electrostatic attraction between them in the crystal. The interactions of (neutral) molecules are discussed fully in Section 11-3.

Figure 8-2, which shows molecular models of the halogen molecules, can be used to help define **bond length.** The drawings suggest that the nuclei of the two halogen atoms are held a fixed distance apart. In a real molecule the atoms vibrate about these positions, but they vibrate about a well-defined average bond distance. Table 8-1 shows the average bond lengths of the halogens. Note that the bond lengths in the diatomic halogen molecules increase as we go down the periodic table.

Table 8-1 The bond lengths of the halogen molecules

Molecule	Bond length/pm
F_2	128
Cl_2	198
Br_2	228
I_2	266

8-2. WE ALWAYS TRY TO SATISFY THE OCTET RULE WHEN WRITING LEWIS FORMULAS

When F_2 is bubbled through an aqueous solution of NaOH, the pale yellow gas oxygen difluoride, OF_2, is formed. Oxygen difluoride reacts explosively with the other halogens. We can deduce the Lewis formula for OF_2 by first writing the Lewis electron-dot formulas for the atoms:

$$:\ddot{\underset{..}{F}}\cdot \qquad \cdot\ddot{\underset{..}{O}}\cdot \qquad \cdot\ddot{\underset{..}{F}}:$$

We wish to join these three atoms such that each has eight electrons in its outer shell. Pictorially, we wish to join these atoms such that each one can be written with eight valence electrons surrounding the nucleus. By bringing the fluorine atoms in toward the oxygen atom, we see that the electron-dot formula

$$:\ddot{\underset{..}{F}}:\ddot{\underset{..}{O}}:\ddot{\underset{..}{F}}:$$

allows all three atoms to be surrounded simultaneously by eight electrons. Thus we conclude that a satisfactory Lewis formula for OF_2 is

$$:\ddot{\underset{..}{F}}—\ddot{\underset{..}{O}}—\ddot{\underset{..}{F}}:$$

As a final check of this formula, note that there are 20 valence electrons indicated in the Lewis formula for the molecule and that there is a total of 20 valence electrons [(2 × 7) + 6] in the Lewis electron-dot formulas for the individual atoms.

The fact that the Lewis formula for OF_2 depicts the two fluorine atoms attached to a central oxygen atom suggests that this is the actual bonding in the molecule. One great utility of Lewis formulas is that they suggest which atoms are bonded to which in a molecule.

When writing Lewis formulas, we always try to satisfy the **octet rule.** The octet rule states that many elements form covalent bonds so as to end up with eight electrons in their outer shells. We shall see that, although there are exceptions to the octet rule, it is still useful because of the large number of compounds that do obey it. In general, we do not violate the octet rule in writing a Lewis formula unless it is impossible to avoid doing so. The octet rule has its origin in the special stability of the noble-gas electron configuration. Thus, for example, carbon, nitrogen, oxygen, and fluorine achieve a neon-like electron configuration when they are surrounded by eight valence electrons.

We can write Lewis formulas in a systematic manner by using the following four-step procedure:

1. Arrange the symbols of the atoms that are bonded together in the molecule next to one another. For OF_2 we would write

   F O F

   Although this may seem like a difficult step to you at this stage, you will become more confident with experience. Often, if there is only one of a particular atom, it is a good first try to assume that this atom is the central atom (as in OF_2) and that the other atoms are bonded to it. Sometimes the correct arrangement is found by trial and error.

2. Compute the total number of valence electrons in a molecule by adding the numbers of valence electrons for all the atoms in the molecule. If the species is an ion rather than a molecule, then you must take the charge of the ion into account by adding electrons if it is a negative ion or subtracting electrons if it is a positive ion. For example,

Ion	Total number of valence electrons
NO_3^-	$(1 \times 5) + (3 \times 6) + 1 = 24$
SO_4^{2-}	$(1 \times 6) + (4 \times 6) + 2 = 32$
NH_4^+	$(1 \times 5) + (4 \times 1) - 1 = 8$

3. Represent a two-electron covalent bond by placing a line between the atoms that are assumed to be bonded to each other. For OF_2, we have

   F—O—F

4. Now arrange the remaining valence electrons as lone pairs about each atom so that the octet rule is satisfied about each atom

   $:\ddot{\underset{..}{F}}—\ddot{\underset{..}{O}}—\ddot{\underset{..}{F}}:$

   The use of this procedure is illustrated in the following Examples.

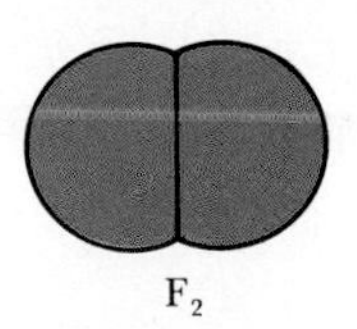

F_2

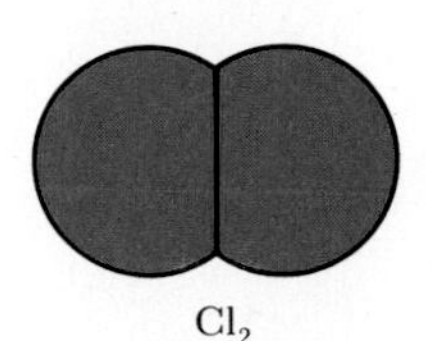

Cl_2

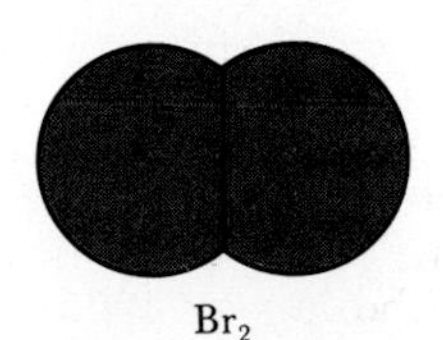

Br_2

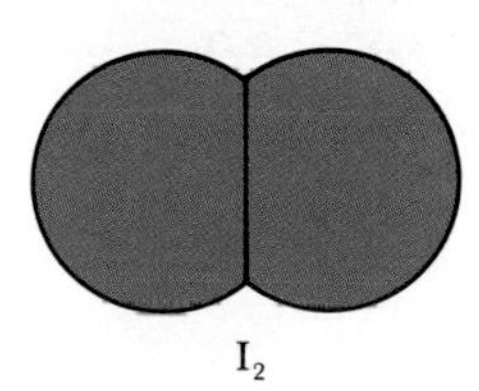

I_2

Figure 8-2 Molecular models of the halogen molecules, drawn to scale to indicate the relative sizes of the atoms. Note that the halogens become larger as we go down the group in the periodic table.

Example 8-1: Write the Lewis formula for carbon tetrachloride, CCl_4.

Solution: Because carbon is the unique atom in this molecule, we shall assume that it is central and that each chlorine atom is attached to it:

$$\begin{matrix} & Cl & \\ Cl & C & Cl \\ & Cl & \end{matrix}$$

The total number of valence electrons is $(1 \times 4) + (4 \times 7) = 32$. We use eight of these electrons to form carbon-chlorine bonds, which satisfies the octet rule about the carbon atom. We place the remaining 24 valence electrons as lone pairs on the chlorine atoms to satisfy the octet rule on the chlorine atoms. The Lewis formula is

```
         ..
        :Cl:
    ..   |   ..
   :Cl—C—Cl:
    ..   |   ..
        :Cl:
         ..
```

Example 8-2: Write the Lewis formula for nitrogen trifluoride, NF_3.

Solution: Because nitrogen is the unique atom in this molecule, we shall assume that it is the central atom and that each fluorine atom is attached to it:

```
F  N  F

   F
```

The total number of valence electrons is $(1 \times 5) + (3 \times 7) = 26$. We use six of these valence electrons to form nitrogen-fluorine bonds. We now place valence electrons as lone pairs on each fluorine atom (accounting for 18 of the 20 valence electrons) and the remaining two valence electrons as a lone pair on the nitrogen atom. The completed Lewis formula is

```
 ..  ..  ..
:F—N—F:
 ..  |   ..
    :F:
     ..
```

Notice that all four atoms satisfy the octet rule.

8-3. HYDROGEN ATOMS ARE ALWAYS TERMINAL ATOMS IN LEWIS FORMULAS

We have said that there are exceptions to the octet rule. One important exception is the hydrogen atom. The noble gas closest to hydrogen in the periodic table is helium. We might expect, then, that hydrogen needs only two electrons in order to attain a noble-gas electron configuration. For example, let's consider H_2 itself. The electron-dot formula for hydrogen is ·H. Each hydrogen atom can be surrounded by two electrons if the two hydrogen atoms share electrons:

$$H\cdot + \cdot H \rightarrow H:H \qquad \text{or} \qquad H—H$$

In this way each hydrogen atom achieves a helium-like electron configuration.

The Lewis formulas for the hydrogen halides are obtained directly from the electron-dot formulas of the individual atoms. If we let X be F, Cl, Br, or I, we can write

```
        ..           ..
H· + ·X:  →  H—X:
        ..           ..
```

The hydrogen atom has two electrons surrounding it, and the halogen atom has eight. Molecular models of the hydrogen halides are shown in Figure 8-3. The bond lengths of the hydrogen halides are given in Table 8-2.

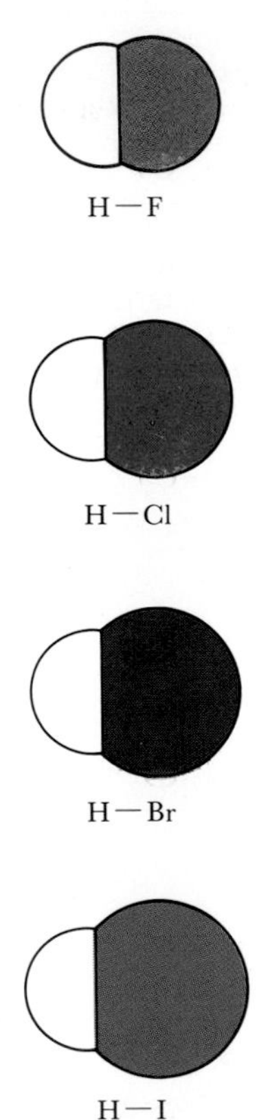

Figure 8-3 Molecular models of the hydrogen halides.

Example 8-3: Write the Lewis formula for NH_4^+.

Solution: We first arrange the atoms as

```
   H
H  N  H
   H
```

There is a total of $(4 \times 1) + 5 - 1 = 8$ valence electrons. We use all eight of these to form the hydrogen-nitrogen bonds:

```
    H
    |
H — N — H
    |
    H
```

We indicate that this species has a charge of +1 by writing

```
⌈    H     ⌉+
|    |     |
| H — N — H |
|    |     |
⌊    H     ⌋
```

Note that the nitrogen atom has eight electrons around it and each hydrogen atom has two.

Because a hydrogen atom completes its valence shell with a total of two electrons, hydrogen atoms almost always form a covalent bond to only one other atom and so are always *terminal* atoms in Lewis formulas.

Table 8-2 The bond lengths of the hydrogen halides

Compound	Bond length/pm
HF	92
HCl	127
HBr	141
HI	161

Example 8-4: Write the Lewis formula for chloroform, $HCCl_3$.

▪ At one time chloroform was used extensively as an inhalation anesthetic and analgesic.

Solution: The chloroform molecule has three different types of atoms, and thus the first step is to decide how to arrange them in the Lewis formula. Hydrogen is always a terminal atom in a Lewis formula. Of the remaining four atoms (CCl_3), the carbon atom is unique, and therefore we guess that it is the central atom. Thus we have the postulated arrangement

```
    H
Cl  C  Cl
    Cl
```

There is a total of $4 + (3 \times 7) + 1 = 26$ valence electrons to be accommodated in the Lewis formula. If we use 8 of these electrons to form 4 bonds

```
     H
     |
Cl — C — Cl
     |
     Cl
```

then the remaining 18 (26 − 8) can be accommodated as 9 lone pairs (3 lone pairs on each of the 3 Cl atoms). Thus the Lewis formula of $HCCl_3$ is

```
        H
  ..    |    ..
 :Cl —  C  — Cl:
  ..    |    ..
       :Cl:
        ..
```

Note that the octet rule is satisfied for the carbon atom and for each chlorine atom.

Up to this juncture we have considered only molecules for which the arrangement of atoms in the Lewis formula is based on placing the unique atom (other than hydrogen, which is always a terminal atom) in the central position. However, consider the problem of writing a Lewis formula for hydrazine, N_2H_4. For this compound, there is no unique atom to place in the central position, and so we assume that the two nitrogen atoms must be bonded to each other with the four hydrogen atoms in terminal positions. Thus we write

▪ Hydrazine is a colorless, oily liquid that fumes in air. It has a penetrating odor resembling that of ammonia and is very poisonous.

```
H—N—N—H
  |  |
  H  H
```

There is a total of $(4 \times 1) + (2 \times 5) = 14$ valence electrons. The 5 bonds require a total of 10 valence electrons, and the remaining 4 valence electrons are placed as lone pairs on the nitrogens in order to complete the octets on the nitrogen atoms and simultaneously accommodate the 14 valence electrons.

```
   ..  ..
H—N—N—H
  |  |
  H  H
```

▪ Methyl alcohol is also called wood alcohol because it can be obtained by heating wood in the absence of air.

Example 8-5: Write the Lewis formula for methyl alcohol, CH_3OH.

Solution: Because the hydrogen atoms must be terminal, the carbon and oxygen atoms must be bonded to each other:

C—O

We must now position the four hydrogen atoms. An oxygen atom has six valence electrons and thus usually completes its octet by forming two bonds, whereas a carbon atom has only four valence electrons and usually completes its octet by forming four bonds. Thus we write

```
   H
   |
H—C—O—H
   |
   H
```

The CH_3OH molecule contains $(4 \times 1) + 4 + 6 = 14$ valence electrons. The 5 bonds require a total of 10 valence electrons. The remaining four valence electrons are accommodated in two lone pairs on the oxygen atom, which are also necessary to complete the octet on the oxygen atom. Thus the Lewis formula for methyl alcohol is

```
   H
   |  ..
H—C—O—H
   |  ..
   H
```

8-4. FORMAL CHARGES CAN BE ASSIGNED TO ATOMS IN LEWIS FORMULAS

It is often convenient to assign a charge to each atom in a molecule or ion. Such charges are called **formal charges** because they are assigned by a set of arbitrary rules and do not necessarily represent the actual charges on the atoms. To assign formal charges, we *assume* that each pair of shared electrons is shared *equally* between the two atoms and assign one of these electrons to each atom. Lone-pair electrons are assigned to the atom on which they are located. The formal charge is the assigned net charge on the atom. The formal charge on an atom in a Lewis formula is calculated by using the equation

$$\begin{pmatrix}\text{formal charge}\\ \text{on an atom in}\\ \text{a Lewis formula}\end{pmatrix} = \begin{pmatrix}\text{total number of}\\ \text{valence electrons}\\ \text{in the free atom}\end{pmatrix} - \begin{pmatrix}\text{total number}\\ \text{of lone-pair}\\ \text{electrons}\end{pmatrix} - \frac{1}{2}\cdot\begin{pmatrix}\text{total number}\\ \text{of shared}\\ \text{electrons}\end{pmatrix} \qquad (8\text{-}1)$$

Consider NH_4^+. A hydrogen atom has one valence electron, and there are no lone-pair electrons in NH_4^+. Each hydrogen atom shares two electrons, and so the formal charge assigned to each hydrogen atom is, from Equation (8-1),

$$\text{formal charge on H in } NH_4^+ = 1 - 0 - \frac{1}{2}(2) = 0$$

A nitrogen atom has five valence electrons. The nitrogen atom in NH_4^+ shares eight electrons, and so

$$\text{formal charge on N in } NH_4^+ = 5 - 0 - \frac{1}{2}(8) = +1$$

Thus NH_4^+ is written

```
     H
     | ⊕
  H—N—H
     |
     H
```

where the ⊕ on N denotes a formal charge of +1 on the nitrogen atom. Note that the sum of the formal charges on the various atoms is equal to the net charge on the molecular ion.

Example 8-6: Assign formal charges to the Lewis formula for ClO_3^-.

Solution: The Lewis formula for ClO_3^- is

```
 ⎡     ‥       ⎤ −
 ⎢    :O:      ⎥
 ⎢ ‥   |   ‥   ⎥
 ⎢:O—Cl—O:     ⎥
 ⎣ ‥   ‥   ‥   ⎦
```

The formal charges on chlorine and oxygen in ClO_3^- are calculated by using Equation (8-1):

$$\text{formal charge on Cl} = 7 - 2 - \frac{1}{2}(6) = +2$$

$$\text{formal charge on O} = 6 - 6 - \frac{1}{2}(2) = -1$$

Thus the Lewis formula for ClO_3^- with the formal charges indicated is

```
        ‥
       :O: ⊖
    ‥   |   ‥
 ⊖:O—Cl—O: ⊖
    ‥   ‥   ‥
       (+2)
```

Once again, notice that the sum of the formal charges on the various atoms is equal to the net charge on the species, which is −1 for ClO_3^-.

We should emphasize that formal charges are *not* real charges, but they can be used to predict and correlate various chemical properties.

We discussed the molecule oxygen difluoride, OF_2, in Section 8-2 and wrote its Lewis formula as

(1) $$:\ddot{\underset{\cdot\cdot}{F}}-\ddot{\underset{\cdot\cdot}{O}}-\ddot{\underset{\cdot\cdot}{F}}:$$

The OF_2 molecule was one of the examples we used to illustrate that it is a good first try to place the unique atom in the center. Note that the formal charge on each atom in formula (1) is zero. There is another Lewis formula for OF_2, however, that satisfies the octet rule:

(2) $$:\ddot{\underset{\cdot\cdot}{F}}-\overset{\oplus}{\ddot{\underset{\cdot\cdot}{F}}}-\overset{\ominus}{\ddot{\underset{\cdot\cdot}{O}}}:$$

These two Lewis formulas predict entirely different bonding in OF_2. The first predicts that the oxygen atom is in the center of the molecule and that there are two oxygen-fluorine bonds. The second predicts that one of the fluorine atoms is in the center and that there is one fluorine-fluorine bond and one oxygen-fluorine bond.

We can use formal charges to select one of these Lewis formulas for OF_2 over the other. Although formal charges do not represent the actual charges on the atoms in a molecule, it is sometimes convenient to consider them as if they were real. For example, consider formula (2). Although it satisfies the octet rule, the central fluorine atom is assigned a formal charge of +1. Fluorine is the most reactive nonmetal and so gains electrons instead of losing them. Consequently, the formal charge of +1 on the fluorine atom is not chemically reasonable, and so we reject formula (2). We predict correctly, then, that the actual structure of OF_2 is represented by formula (1), with the oxygen atom being central. Usually, the Lewis formula with the lower formal charges represents the preferred (lowest-energy) Lewis formula.

8-5. IT IS NOT ALWAYS POSSIBLE TO SATISFY THE OCTET RULE BY USING ONLY SINGLE BONDS

In all the molecules that we have discussed so far, there were exactly the correct number of valence electrons remaining after step 3 (page 265) to be used in step 4. In this section we shall consider the case in which there are not enough electrons to satisfy the octet about each atom by using only single bonds. The molecule ethylene, C_2H_4, serves as a good example. Using the same reasoning that we did for hydrazine, we arrange the atoms as

$$\begin{array}{cccc} H & C & C & H \\ & H & H & \end{array}$$

There is a total of $(4 \times 1) + (2 \times 4) = 12$ valence electrons. We use ten of them to join the atoms:

H—C—C—H
   | |
   H H

If we only use single bonds, it is not possible to satisfy the octet rule about each carbon atom with only the two remaining valence electrons. We are short two electrons. When this situation occurs, we add one more bond for each two electrons that we are short. In the case of ethylene, we add another bond between the carbon atoms to get

H—C=C—H
   | |
   H H

Notice that now the octet rule is satisfied for each carbon atom. When two atoms are joined by two pairs of electrons, we say that there is a **double bond** between the atoms. A double bond between two atoms is shorter and stronger than a single bond between the same two atoms. The carbon-carbon bond in C_2H_4 is indeed shorter and stronger than, for example, the carbon-carbon single bond in ethane, C_2H_6. Table 8-3 gives typical bond lengths and bond energies for some single and double bonds.

It is also possible to have a **triple bond,** as we now show for N_2. There are 10 valence electrons in N_2. If we add one bond and then try to satisfy the octet rule about each nitrogen atom, we find that we are four electrons short. For example,

N—N (violates the octet rule)

In this case using only a single bond leaves us short four electrons to satisfy the octet rule about each nitrogen atom. Thus, we add two more bonds to obtain

N≡N

The four remaining valence electrons are now added according to step 4 to obtain

:N≡N:

Table 8-3 Average bond lengths and bond energies of single, double, and triple bonds

Bond	Average bond length/pm	Average bond energy/ $kJ \cdot mol^{-1}$
C—O	143	351
C=O	120	730
C—C	154	350
C=C	134	615
C≡C	120	810
N—N	145	160
N=N	125	420
N≡N	110	950

Example 8-7: Write a Lewis formula for CO_2.

Solution: We arrange the atoms with the carbon atom in the center:

O C O

There is a total of $(1 \times 4) + (2 \times 6) = 16$ valence electrons. If we add one bond between each oxygen atom and carbon atom, and then try to satisfy the octet rule about each atom, we find that we are four electrons short. For example,

:O—C—O: (violates the octet rule)

Thus, we go back to step 3 and add two more bonds:

O=C=O

Now we use step 4 and arrange the remaining 8 valence electrons as lone pairs to satisfy the octet rule about each atom:

:O=C=O:

The Lewis formula of CO_2 shows two carbon-oxygen double bonds.

Incidentally, we could also have made one single bond and one triple bond instead of two double bonds. Usually we choose a more symmetric formula and the one with the lowest formal charges, if possible.

Example 8-8: Write the Lewis formula for the gas hydrogen cyanide, HCN.

Solution: Either the carbon atom or the nitrogen atom might be the central atom in this case. When in doubt, arrange the atoms as the formula is written:

H C N

Use four of the 10 valence electrons to write

H—C—N

We are four electrons short of satisfying the octet rule on both the carbon atom and the nitrogen atom, so we add two more bonds. The hydrogen atom already has two electrons around it, and so in this case we must form a triple bond between the carbon atom and nitrogen atom:

H—C≡N

The remaining two valence electrons are placed on the nitrogen atom as a lone pair, so that both the carbon atom and the nitrogen atom satisfy the octet rule. The Lewis formula is

H—C≡N:

8-6. A RESONANCE HYBRID IS A SUPERPOSITION OF LEWIS FORMULAS

There are a number of molecules and ions for which it is possible to write two or more satisfactory Lewis formulas. For example, let's consider the nitrite ion, NO_2^-. One Lewis formula for NO_2^- is

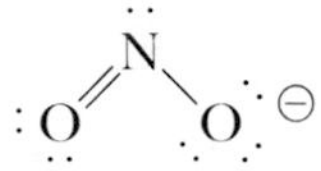

One of the oxygen atoms, the right-hand one as written, has a formal charge of -1, having three lone pairs and one bond. Another equally acceptable Lewis formula for NO_2^- is

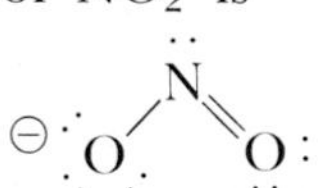

In this case the negative formal charge is on the other oxygen atom. Both of these Lewis formulas satisfy the octet rule. When it is possible to write two or more satisfactory Lewis formulas *without altering the positions of the nuclei,* the actual formula is viewed as an average or as a superposition of the individual formulas. Each of the individual Lewis formulas is said to be a **resonance form,** and the use of multiple Lewis formulas is called **resonance.** We indicate resonance forms by means of a two-headed arrow, as in

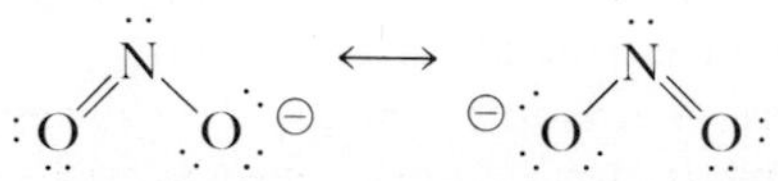

⟷

Neither of the individual Lewis formulas taken separately accurately reflects the actual formula. Two separate Lewis formulas are necessary to describe the bonding in NO_2^-.

There is no generally accepted way to represent resonance pictorially, but one way is to write the formula as

$$\left[O \overset{..}{\underset{}{=\!\!=}} N \overset{}{=\!\!=} O \right]^-$$

where the two dashed lines taken together represent a pair of bonding electrons. Such a superimposed formula is called a **resonance hybrid** because it is a hybrid of the various resonance forms. Each of the nitrogen-oxygen bonds in NO_2^- can be thought of as an *average* of a single bond and a double bond. The superimposed Lewis formulas for NO_2^- suggest that the two nitrogen-oxygen bonds in NO_2^- are equivalent, and this is in accord with experimental observation: the two bonds do have exactly the same length, 113 pm. Notice that the individual Lewis formulas suggest that the two nitrogen-oxygen bonds are not equivalent, one being a single bond and the other being a double bond.

Another example of resonance occurs in the nitrate ion, NO_3^-. Three equally satisfactory Lewis formulas for NO_3^- are

$$\ominus\!:\!\ddot{O}\!-\!\overset{\overset{\ddot{O}}{\|}}{N^{\oplus}}\!-\!\ddot{O}\!:\!\ominus \;\longleftrightarrow\; :\!\ddot{O}\!=\!\overset{\overset{:\ddot{O}:\ominus}{|}}{N^{\oplus}}\!-\!\ddot{O}\!:\!\ominus \;\longleftrightarrow\; \ominus\!:\!\ddot{O}\!-\!\overset{\overset{:\ddot{O}:\ominus}{|}}{N^{\oplus}}\!=\!\ddot{O}\!:$$

Because each of these Lewis formulas is equally satisfactory, the actual structure is viewed as a superposition or an average of the three formulas and can be represented pictorially as the resonance hybrid.

$$\left[O \overset{\overset{O}{\|}}{=\!\!=\!N\!=\!\!=} O \right]^-$$

where the three dashed lines taken together represent a pair of bonding electrons. In this case each of the nitrogen-oxygen bonds is an average of a double bond and two single bonds. As the superimposed representation suggests, the three nitrogen-oxygen bonds are equivalent, which is in agreement with experimental observations (each nitrogen-oxygen bond is 122 pm in length). Furthermore, there are no known chemical reactions that can be used to distinguish one oxygen atom from another in a nitrate ion.

▪ A nitrate ion is planar, with each NO bond pointing to a vertex of an equilateral triangle.

The need for resonance forms arises from the fact that Lewis formulas involve electron-pair bonds. If the species involves a bond intermediate between a single and a double bond, then we need to write two or more Lewis formulas to describe the bonding in the molecule.

The following Example is particularly important to understand because it involves several of the important concepts that we have discussed in this chapter.

Example 8-9: Draw Lewis formulas for the two resonance forms of SO_2. Indicate formal charges and discuss the bonding in this molecule.

Solution: We arrange the atoms as

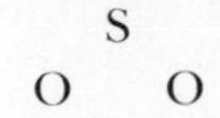

S
O O

There is a total of 18 valence electrons in the molecule. The two resonance forms are

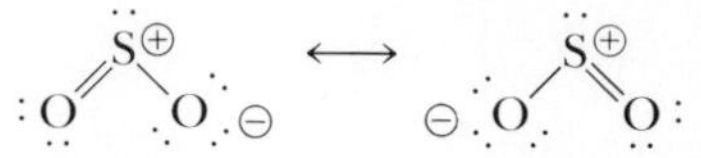

$$:\ddot{O}=\ddot{S}^{\oplus}-\ddot{O}:^{\ominus} \longleftrightarrow {}^{\ominus}:\ddot{O}-\ddot{S}^{\oplus}=\ddot{O}:$$

The formal charges indicated are calculated according to Equation (8-1):

$$\text{formal charge on S in } SO_2 = 6 - 2 - \frac{1}{2}(6) = +1$$

$$\text{formal charge on singly bonded O in } SO_2 = 6 - 6 - \frac{1}{2}(2) = -1$$

$$\text{formal charge on doubly bonded O in } SO_2 = 6 - 4 - \frac{1}{2}(4) = 0$$

These two Lewis formulas constitute resonance forms, and so the actual formula is an average of the two, which can be represented by the resonance hybrid

$$O\overset{..}{S}O$$

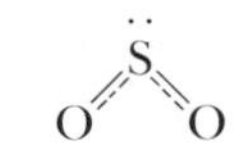

This formula suggests that the two sulfur-oxygen bonds in SO_2 are equivalent, or that the two sulfur-oxygen bond lengths in SO_2 are equal. This is in agreement with experiment.

An important example of the idea of resonance and its consequences is provided by benzene, C_6H_6, a clear, colorless, highly flammable liquid (its boiling point is 80.1°C) with a characteristic odor. Benzene is obtained from petroleum and coal tar and has many chemical uses. The benzene molecule has two major resonance forms:

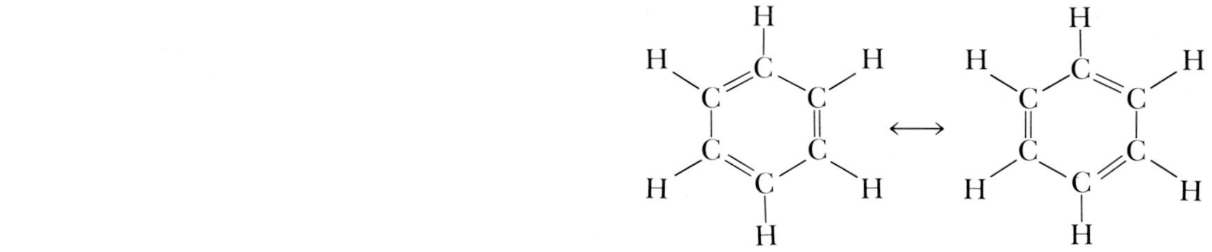

[Two Kekulé structures of benzene, C_6H_6, connected by ⟷]

We depict the superposition or average of these two Lewis formulas by the resonance hybrid

[Resonance hybrid of benzene with dashed circle inside ring; C–C–C angle 120°]

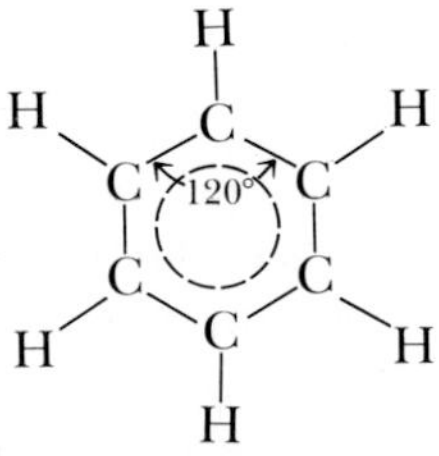

All the carbon-carbon bonds in benzene are equivalent, and each is the average of a single bond and a double bond. The carbon-carbon bond

distance in benzene is 140 pm, which is intermediate between the usual carbon-carbon single-bond (154 pm) and double-bond (134 pm) distances. Benzene is a **planar molecule** (all the atoms lie in the same plane), with the ring of carbon atoms constituting a perfect hexagon with 120° interior carbon-carbon bond angles. A commonly used representation of the benzene molecule is simply

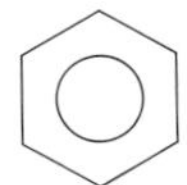

where each vertex represents a carbon atom attached to a hydrogen atom. The benzene ring is part of the chemical formulas of a great many organic compounds. We shall learn in Chapter 23 that benzene behaves chemically as a substance with no double bonds and is a relatively unreactive molecule. The unusual stability of a benzene molecule is ascribed to what chemists call **resonance stabilization**: the energy of the actual molecule, represented by a superposition of Lewis formulas, is lower than the energy of any of its (hypothetical) individual Lewis formulas.

8-7. A SPECIES WITH ONE OR MORE UNPAIRED ELECTRONS IS CALLED A FREE RADICAL

As useful as the octet rule is, there are some cases in which the rule cannot be satisfied. For example, consider the molecule nitrogen oxide, NO. The electron-dot formulas of nitrogen and oxygen are

$$\cdot\ddot{\mathrm{N}}\cdot \quad\text{and}\quad \cdot\ddot{\underset{\cdot\cdot}{\mathrm{O}}}\cdot$$

If we try to write a Lewis formula for NO, we find that it is not possible to satisfy the octet rule. The best that we can do is

$$\ddot{\mathrm{N}}\!=\!\ddot{\underset{\cdot\cdot}{\mathrm{O}}} \quad\text{or}\quad \overset{\ominus}{\ddot{\underset{\cdot\cdot}{\mathrm{N}}}}\!=\!\overset{\oplus}{\ddot{\underset{\cdot}{\mathrm{O}}}}$$

The difficulty here is that the total number of valence electrons is an odd number (11), and so it is impossible to pair up all the electrons as we have been doing. A species that has one or more unpaired electrons is called a **free radical.** Because of the unpaired electrons, free radicals are usually very reactive.

Another example of a free radical is chlorine dioxide, ClO_2. The chlorine atom has seven valence electrons, and each oxygen atom has six valence electrons. Therefore, ClO_2 has an odd number (19) of valence electrons. Two resonance forms for ClO_2 are

▪ Chlorine dioxide is a yellow to reddish yellow gas with an unpleasant odor similar to that of chlorine. It reacts explosively with many substances.

$$\overset{\ominus}{}:\ddot{\underset{\cdot\cdot}{\mathrm{O}}}\!-\!\overset{\oplus}{\ddot{\underset{\cdot\cdot}{\mathrm{Cl}}}}\!-\!\ddot{\underset{\cdot\cdot}{\mathrm{O}}}\cdot \quad\text{and}\quad \cdot\ddot{\underset{\cdot\cdot}{\mathrm{O}}}\!-\!\overset{\oplus}{\ddot{\underset{\cdot\cdot}{\mathrm{Cl}}}}\!-\!\ddot{\underset{\cdot\cdot}{\mathrm{O}}}:\overset{\ominus}{}$$

The ClO_2 free radical is viewed as a hybrid of the resonance forms, and the two chlorine-oxygen bonds in ClO_2 are equivalent.

The molecules NO and ClO_2 are free radicals. They have an odd number of electrons and so cannot satisfy the octet rule. There is also a class of compounds that have an even number of outer electrons but do not have enough electrons to form octets about each atom. Compounds of beryllium and boron serve as particularly important examples of **electron-deficient compounds.** Consider the molecule BeH_2. The electron-dot formulas for Be and H are

$$\mathrm{H}\cdot \quad \text{and} \quad \cdot\mathrm{Be}\cdot$$

A Lewis formula for BeH_2 is

$$\mathrm{H{-}Be{-}H}$$

The beryllium atom is four electrons short of satisfying the octet rule.

Example 8-10: Suggest a Lewis formula for BF_3 that satisfies the octet rule. Give a reason why the electron-deficient formula is preferred.

Solution: Each fluorine atom has 7 valence electrons and the boron atom has 3, for a total of 24. A Lewis formula for BF_3 using 12 electron pairs is

```
 ..  ⊖  ..
:F—B—F:
 ..  ‖  ..
    .F.: ⊕
```

The formal charge of +1 on the very reactive nonmetallic fluorine atom can be used to decide that this Lewis formula is less favorable than the formula

```
 ..     ..
:F—B—F:
 ..  |  ..
    :F:
     ..
```

for which the formal charges are all zero.

Electron-deficient compounds are usually highly reactive species. For example, the electron-deficient compound BF_3 readily reacts with NH_3 to form H_3NBF_3:

```
                ..                  ..
      H        :F:            H   :F:
      |         |  ..        ⊕|    |⊖ ..
   H—N:  +    B—F:   →   H—N—B—F:
      |         |  ..         |    |  ..
      H        :F:            H   :F:
                ..                  ..
```

The lone electron pair in NH_3 can be shared between the nitrogen atom and the boron atom so that the octet rule is satisfied. A covalent bond that is formed when one atom contributes both electrons is called a **coordinate-covalent bond.** The product of the foregoing reaction, H_3NBF_3, is called a **donor-acceptor complex.**

8-8. ATOMS OF ELEMENTS BELOW CARBON THROUGH NEON IN THE PERIODIC TABLE CAN EXPAND THEIR VALENCE SHELLS

In using step 4 (page 265) we have not yet considered the case in which there are more valence electrons than are needed to satisfy the octet rule on each atom. When this situation occurs, we assign the "extra" electrons as lone pairs to atoms of elements whose valence electrons have $n > 3$ as the principal quantum number. This will usually be the central atom. Such elements generally lie below the second-row elements carbon, nitrogen, oxygen, and fluorine in the periodic table.

As an example, let's write the Lewis formula for sulfur tetrafluoride, SF_4. First we arrange the atoms as

$$\begin{array}{ccc} & F & \\ F & S & F \\ & F & \end{array}$$

Of the $6 + (4 \times 7) = 34$ valence electrons, we use 8 electrons to form four sulfur-fluorine bonds. We can satisfy the octet rule about each atom by using only 24 of the remaining valence electrons:

$$\begin{array}{ccc} & :\ddot{F}: & \\ & | & \\ :\ddot{F}- & S & -\ddot{F}: \\ & | & \\ & :\underset{\cdot\cdot}{F}: & \end{array} \qquad \text{(two valence electrons missing)}$$

There still are two valence electrons to be accounted for. Sulfur lies in the third row of the periodic table, so we add these as a lone pair to the sulfur atom. Thus the Lewis formula is

$$\begin{array}{ccc} & :\ddot{F}: & \\ & | & \\ :\ddot{F}- & \underset{\cdot\cdot}{S} & -\ddot{F}: \\ & | & \\ & :\underset{\cdot\cdot}{F}: & \end{array}$$

The exact position of the lone pair on the sulfur atom is not important; for example, we could just as well have placed the pair at the upper right of the sulfur atom.

In a Lewis formula such as that for SF_4 we say that the sulfur atom has an **expanded valence shell.** Sulfur expands its valence shell by using its *d* orbitals. It is not possible for the atoms of elements in the second row of the periodic table to expand their valence shells beyond eight electrons because second-row elements complete the L shell when they satisfy the octet rule. Second-row elements would have to use orbitals in the M shell to accommodate more electrons, and the energies of the orbitals in the M shell are much higher than those of the orbitals in the L shell. Thus, although SF_4 has been synthesized, OF_4 has never been observed.

Example 8-11: Xenon difluoride, XeF_2, was one of the first xenon compounds to be prepared. Write the Lewis formula for the XeF_2 molecule.

Solution: We arrange the atoms as

$$F \quad Xe \quad F$$

Of the $8 + (2 \times 7) = 22$ valence electrons, 4 are used to form two xenon-fluorine bonds. We can use 12 of the remaining 18 valence electrons to satisfy the octet rule on each fluorine atom:

$$:\underset{\cdot\cdot}{\ddot{F}}-Xe-\underset{\cdot\cdot}{\ddot{F}}: \qquad \text{(missing 6 valence electrons)}$$

The "extra" six valence electrons are placed on the xenon atom as three lone pairs. The Lewis formula is

$$:\ddot{\underset{..}{F}}-\ddot{\underset{..}{Xe}}-\ddot{\underset{..}{F}}:$$

Some other examples in which using step 4 results in more than eight electrons around the central atom are

xenon tetrafluoride   bromine trifluoride

triiodide ion   iodine dioxide difluoride ion

Because the atoms of elements below the second row of the periodic table can accommodate more than eight electrons in the valence shells, they are able to bond to more than four atoms. Some examples are

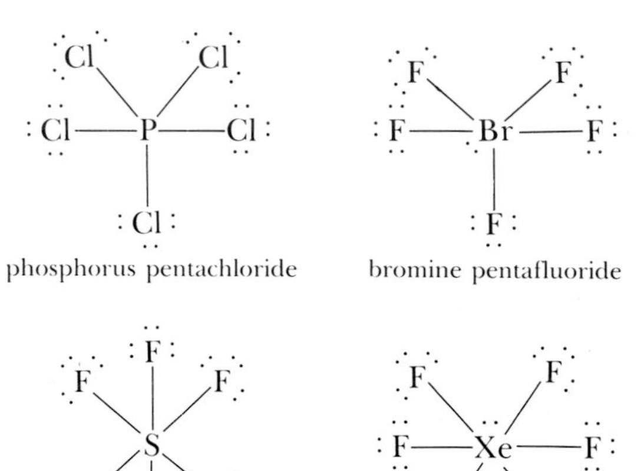

phosphorus pentachloride   bromine pentafluoride

sulfur hexafluoride   xenon hexafluoride

▪ Sulfur hexafluoride is one of the most dense gases, having a density about five times as great as air. Tennis balls are filled with a mixture of about 50% sulfur hexafluoride and 50% air.

The fact that atoms of elements below the second row of the periodic table can expand their valence shells leads to additional resonance formulas for many of the compounds involving those atoms. For example, consider sulfuryl chloride, SO_2Cl_2. According to the rules that we have presented, the Lewis formula for SO_2Cl_2 is

However, because the sulfur atom can expand its valence shell, we can write the additional Lewis formulas

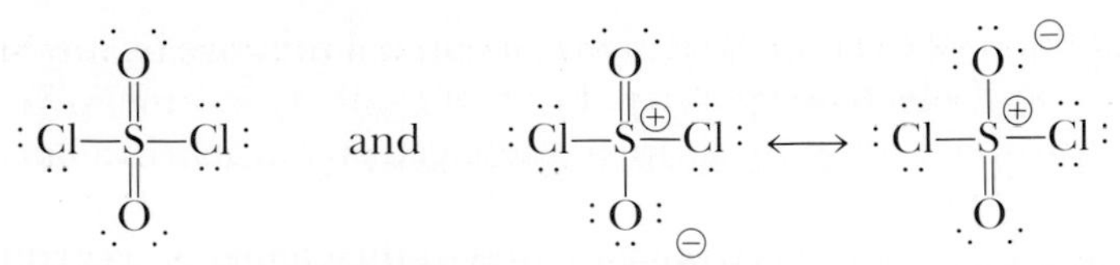

: Cl—S—Cl : and : Cl—S—Cl : ⟷ : Cl—S—Cl :

All four of these Lewis formulas are resonance forms of SO_2Cl_2, and so we have

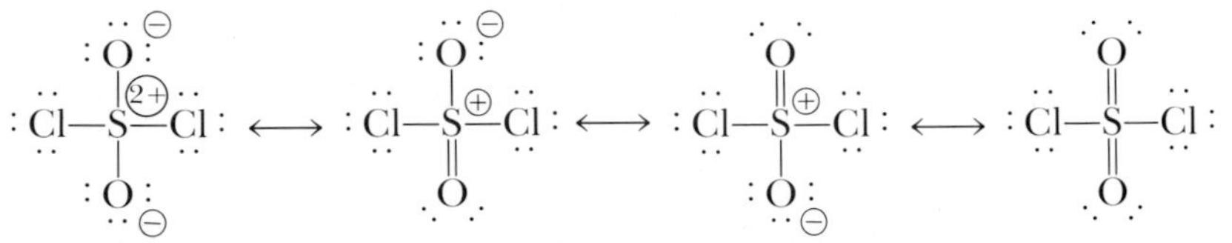

: Cl—S—Cl : ⟷ : Cl—S—Cl : ⟷ : Cl—S—Cl : ⟷ : Cl—S—Cl :

The resonance hybrid of these four resonance forms suggests that the two sulfur-oxygen bonds are equivalent, with a bond length intermediate between that of a single bond and that of a double bond. Notice that we would expect these bonds to be single bonds if we did not recognize that the sulfur atom can expand its valence shell.

As another example, consider chloric acid, $HClO_3$. The Lewis formula of $HClO_3$ in which the octet rule is satisfied is

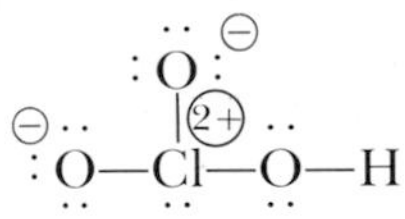

: O—Cl—O—H

If we recognize that the chlorine atom can expand its valence shell, then we can write the additional Lewis formulas

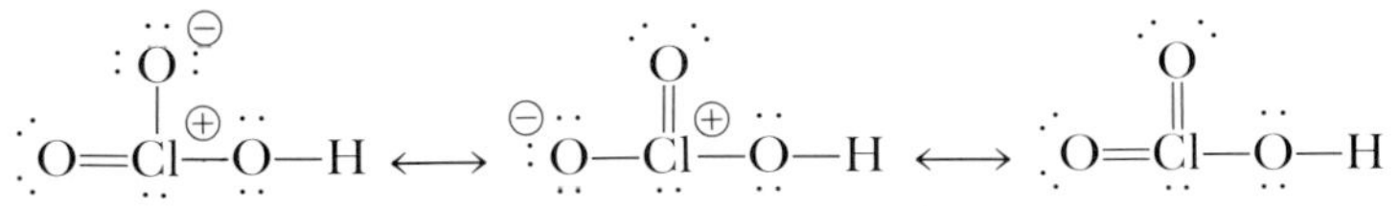

O=Cl—O—H ⟷ : O—Cl—O—H ⟷ O=Cl—O—H

All four of these Lewis formulas are acceptable resonance forms of $HClO_3$. The resonance hybrid suggests that two of the chlorine-oxygen bonds in $HClO_3$ are equivalent, and that the bond length is intermediate between that of a single bond and that of a double bond.

8-9. MOST CHEMICAL BONDS ARE INTERMEDIATE BETWEEN PURELY IONIC AND PURELY COVALENT

Although we have discussed ionic bonds and covalent bonds in this and the previous chapter as distinct cases, most chemical bonds are neither purely ionic nor purely covalent, but are intermediate between the two. The bond in HCl serves as a good example to discuss this point.

When we introduced the concept of formal charge, we *arbitrarily* assigned one of the electrons in the covalent bond to each atom. The formal charges of H and Cl in HCl are zero. We emphasize here that this is a formal and arbitrary, but useful, procedure. We are tacitly assuming that the electrons in the covalent bond are shared equally by the hydrogen atom and the chlorine atom. We know, however, that different isolated atoms have different ionization energies and different electron affinities. It seems reasonable that different atoms will attract electrons differently even if they are bonded to each other by a covalent bond. We can make this discussion quantitative by introducing the concept of **electronegativity.** Electronegativity is a measure of the force with which an atom attracts the electrons in its covalent bonds with other atoms.

▪ Electronegativity as defined by Pauling is given by

$$|X_A - X_B| = 0.208[H_{AB} - (H_{AA}H_{BB})^{1/2}]$$

where X is an electronegativity and H is a bond energy. The vertical bars mean "take the difference $X_A - X_B$ as a positive quantity."

Figure 8-4 gives the electronegativities of the elements according to a procedure devised by Linus Pauling. The larger the electronegativity of an atom, the greater is the attraction of the atom for the electrons in its covalent bonds. Note that electronegativities increase from left to right across the short (second and third) rows of the periodic table. This left-to-right increase in electronegativities reflects the increasingly nonmetallic nature of the elements toward the right-hand side of the table. Note also that electronegativities increase as we go up a column. The reason for the bottom-to-top increase in electronegativities is that the nuclear attraction of the outer electrons increases as the size of the atom decreases. The periodic trend of electronegativity is sketched in Figure 8-5. Note that fluorine is the most electronegative atom and that cesium and francium are the least electronegative. The order for the atoms with the greatest electronegativities is

$$F > O > N \approx Cl > C \approx S > H \approx P$$

It is the *difference* in electronegativities of the two atoms in a covalent bond that determines how the electrons in the bond are shared. If the electronegativities are the same, then the electrons in the bond are shared equally, and the bond is called a **pure covalent bond,** or a **non-polar bond.** Equal sharing of bonding electrons occurs in homonuclear diatomic molecules (that is, molecules consisting of two identical atoms). If the electronegativities of the two atoms differ, then electrons in the bond are not shared equally and the bond is said to be a **polar bond.** The extreme case of a polar bond occurs when the difference in

Figure 8-4 Electronegativities of the elements as calculated by Linus Pauling. Note that the electronegativities of the elements in the second and third rows increase from left to right, and from bottom to top in a given column.

1 H 2.1																	2 He -
3 Li 1.0	4 Be 1.5											5 B 1.9	6 C 2.5	7 N 3.0	8 O 3.5	9 F 4.0	10 Ne -
11 Na 0.9	12 Mg 1.2											13 Al 1.5	14 Si 1.8	15 P 2.1	16 S 2.5	17 Cl 3.0	18 Ar -
19 K 0.8	20 Ca 1.0	21 Sc 1.3	22 Ti 1.5	23 V 1.6	24 Cr 1.6	25 Mn 1.5	26 Fe 1.8	27 Co 1.8	28 Ni 1.8	29 Cu 1.9	30 Zn 1.5	31 Ga 1.6	32 Ge 1.8	33 As 2.0	34 Se 2.4	35 Br 2.8	36 Kr -
37 Rb 0.8	38 Sr 1.0	39 Y 1.2	40 Zr 1.4	41 Nb 1.6	42 Mo 1.8	43 Tc 1.9	44 Ru 2.2	45 Rh 2.2	46 Pd 2.2	47 Ag 1.7	48 Cd 1.4	49 In 1.7	50 Sn 1.8	51 Sb 1.9	52 Te 2.1	53 I 2.5	54 Xe -
55 Cs 0.7	56 Ba 0.9	57-71 1.1-1.2	72 Hf 1.3	73 Ta 1.5	74 W 1.7	75 Re 1.9	76 Os 2.2	77 Ir 2.2	78 Pt 2.2	79 Au 2.4	80 Hg 1.9	81 Tl 1.8	82 Pb 1.8	83 Bi 1.8	84 Po 2.0	85 At 2.2	86 Rn -
87 Fr 0.7	88 Ra 0.9	89 Ac 1.1	90 Th 1.3	91 Pa 1.5	92 U 1.7	93-103 Np-Lr 1.3											

electronegativities is large, say greater than about 1.7. For such a case, the electron pair ends up completely on the more electronegative atom, giving a **pure ionic bond.**

A polar bond can be illustrated by HCl. The electronegativity of H is 2.1 and that of Cl is 3.0, and the difference between them is 0.9. Thus the electrons in the bond are not shared equally. Because the electronegativity of Cl is greater than that of H, the chlorine atom attracts the electron pair more strongly than does the hydrogen atom. The bonding electrons are shifted a little toward the chlorine atom and so it acquires a *partial* negative charge, which leaves the hydrogen with a partial positive charge; that is, the bond is polar. We indicate partial charges by the lowercase Greek letter delta, δ, and we write

$$\overset{\delta+}{\text{H}}\text{—}\overset{\delta-}{\text{Cl}}$$

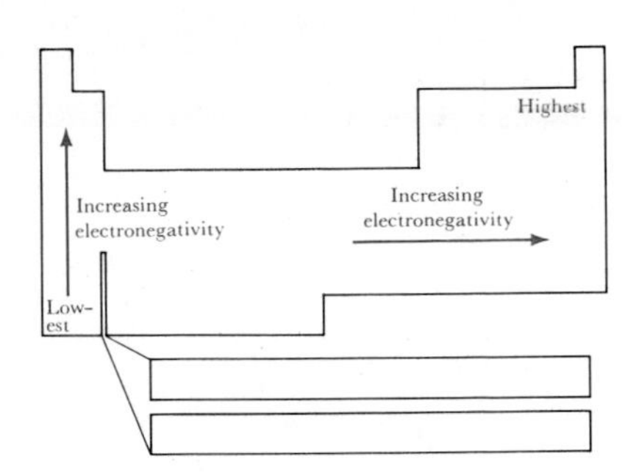

Figure 8-5 The trends of electronegativities in the periodic table.

It is important to understand that δ+ or δ− represents only the *fraction* of an electronic charge that results from the unequal sharing of the electrons in the covalent bond. The numerical value of δ is of no importance to us at this stage. It denotes only that the hydrogen atom is slightly positively charged and that the chlorine atom is slightly negatively charged. From a quantum theoretical point of view, δ represents the fact that the two electrons in the covalent bonds are more likely to be found near the chlorine atom than near the hydrogen atom.

Example 8-12: Describe the charge distribution in the interhalogen compound ClF.

Solution: According to Figure 8-4, the electronegativity of F is 4.0 and that of Cl is 3.0, and so the chlorine-fluorine bond is a polar bond. The electron pair is somewhat more likely to be found near the fluorine atom than near the chlorine atom. Thus the fluorine atom has a slightly negative charge, δ−, and the chlorine atom has a slightly positive charge, δ+. We can represent this polar bond by writing

$$\overset{\delta+}{\text{Cl}}\text{—}\overset{\delta-}{\text{F}}$$

8-10. POLYATOMIC MOLECULES WITH POLAR BONDS MAY BE NONPOLAR

A quantity that is a measure of the polarity of a molecule is its **dipole moment.** The dipole moment of a diatomic molecule customarily is represented as an arrow (⟼) pointing along the bond from δ+ to δ−. The cross on the tail of the arrow indicates the positive charge. For HCl and ClF we have

$$\overset{+\!\!\longrightarrow}{\text{H—Cl}} \quad \text{and} \quad \overset{+\!\!\longrightarrow}{\text{Cl—F}}$$

This notation indicates the direction of the dipole moment. The magnitude of the dipole moment is the product of the length of the bond and the amount of net electric charge that is separated. Dipole moments can be measured experimentally, and the dipole moments of the

Table 8-4 The dipole moments of the gaseous hydrogen halides

Molecule	Electronegativity difference	Dipole moment*/10^{-30} C·m
HF	1.9	6.36
HCl	0.9	3.43
HBr	0.7	2.63
HI	0.4	1.27

*The units of dipole moment are charge × distance, or coulombs × meters (C·m) in SI units.

hydrogen halides are given in Table 8-4. Note the dependence of the dipole moment on the electronegativity difference. The larger the difference in electronegativity, the larger the dipole moment.

▪ Dipole moments are vector quantities.

In a polyatomic molecule (a molecule with more than two atoms), the polarity of each bond, like the dipole moment of a diatomic molecule, can be represented by an arrow and is a quantity that has both magnitude *and* direction. A quantity that has magnitude and direction is called a **vector.** A familiar example of a vector is a force. A force must be described by both the direction in which it is applied *and* its magnitude. Vectors have the important property that they cancel if applied with the same magnitude in opposite directions. In a stalemated tug of war, both teams are pulling with the same magnitude of force but in opposing directions. The net result is an effective cancellation. This can be illustrated pictorially by

←——•——→ = no net force

If the forces are not applied in opposing directions, then there is a resultant net force:

It is not necessary for us to be able to calculate the magnitude of the net force. We only need to see pictorially that the direction of the net force is as shown above.

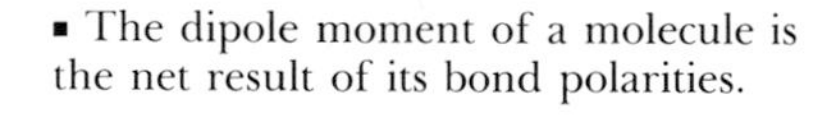

▪ The dipole moment of a molecule is the net result of its bond polarities.

Bond polarities have the same properties as forces. In CO_2,

$$:\ddot{O}=C=\ddot{O}:$$

each carbon-oxygen bond is polar because oxygen is more electronegative than carbon, but the bond polarities point in opposite directions:

$$\overset{\longleftarrow\!+\quad+\!\longrightarrow}{:\ddot{O}=C=\ddot{O}:}$$

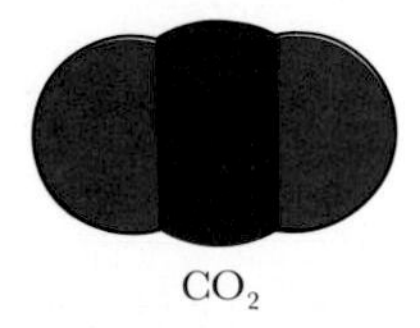

CO_2

Figure 8-6 A molecular model of carbon dioxide, a linear molecule. The O—C—O bond angle is 180°.

The bond polarities cancel exactly, so the CO_2 molecule has no dipole moment. A molecule with no dipole moment is a **nonpolar molecule.** Because CO_2 has a zero dipole moment, the molecule must be a **linear molecule**—that is, all the atoms lie on a straight line (Figure 8-6).

Let's consider H_2O as another example. The oxygen-hydrogen bonds are polar because oxygen is more electronegative than hydrogen, and we can display this by

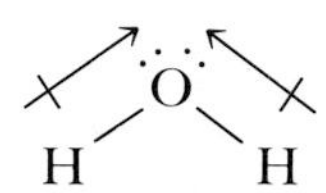

This diagram implies that water is a nonpolar molecule. In fact, H_2O has a large dipole moment and so is a **polar molecule**. The discrepancy is due to our assumption that H_2O is a linear molecule. We see in the next chapter that the water molecule is bent. The H—O—H bond angle is not 180°; rather it is about 104° (Figure 8-7). We can illustrate the bent structure by

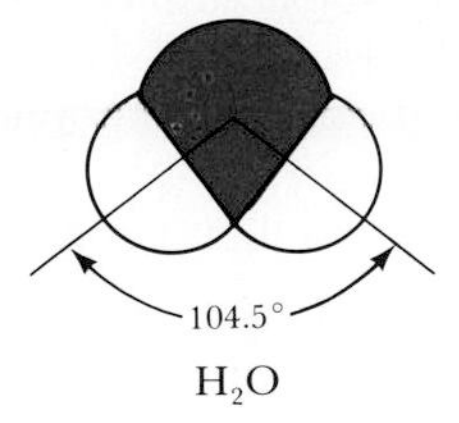

Figure 8-7 A molecular model of water, a nonlinear molecule. The H—O—H bond angle is 104.5°.

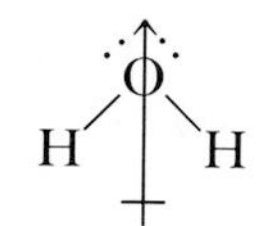

Thus the net dipole moment in H_2O has the orientation

There is an important lesson for us here. The Lewis formulas that we have learned to write in this chapter suggest which atoms are bonded to which in the molecule. Although the formulas are very useful in this regard, they are *not* meant to indicate or suggest the three-dimensional arrangement of the atoms in a molecule. In the next chapter we learn some simple, useful rules for predicting the shapes of molecules using Lewis formulas.

SUMMARY

In a Lewis formula, a covalent bond is represented as a pair of electrons shared between two atoms. A Lewis formula shows the arrangement of the valence electrons in a molecule and suggests which atoms are bonded to which. The octet rule is a useful rule for writing Lewis formulas, especially for compounds involving the elements carbon, nitrogen, oxygen, and fluorine.

Although there are a number of exceptions to the octet rule, it is still a very useful guide for writing Lewis formulas for compounds. Another aid for writing Lewis formulas is the idea of formal charge. Formal charges can be assigned to the various atoms in a Lewis formula and used to select preferred Lewis formulas. Sometimes it is possible to write two or more Lewis formulas for a molecule without altering the position of the atoms. When this is possible, each individual formula is said to be a resonance form, and the accepted Lewis formula, called a resonance hybrid, is an average or a superposition of the individual formulas.

Most chemical bonds are neither purely ionic nor purely covalent. The degree of ionic character in a bond can be estimated by means of the electronegativity. When the electronegativities of the two atoms joined by a covalent bond are different, the bond is said to be polar.

Lewis formulas indicate not the shape of a molecule but only the bonding within a molecule.

TERMS YOU SHOULD KNOW

valence electron 263
covalent bond 263
Lewis formula 263
lone electron pair 264
lone pair 264
molecular crystal 264
bond length 264
octet rule 264
formal charge 268
double bond 271
triple bond 271
resonance form 272

AN EQUATION YOU SHOULD KNOW HOW TO USE

$$\begin{pmatrix}\text{formal charge}\\ \text{on an atom in}\\ \text{a Lewis formula}\end{pmatrix} = \begin{pmatrix}\text{total number of}\\ \text{valence electrons}\\ \text{in the free atom}\end{pmatrix} - \begin{pmatrix}\text{total number}\\ \text{of lone-}\\ \text{pair electrons}\end{pmatrix} - \frac{1}{2}\begin{pmatrix}\text{total number}\\ \text{of shared}\\ \text{electrons}\end{pmatrix} \quad \text{(8-1) (calculations of formal charges)}$$

PROBLEMS

COMPOUNDS INVOLVING SINGLE BONDS

8-1. Write the Lewis formula for

(a) SCl_2 (b) $GeCl_4$
(c) $AsBr_3$ (d) PH_3

8-2. Write the Lewis formula for

(a) PBr_3 (b) SiF_4
(c) NI_3 (d) H_2Se

8-3. Pure hydrogen peroxide, H_2O_2, is a colorless liquid that is caustic to the skin, but a 3% aqueous solution is a mild bleaching agent. Write the Lewis formula for H_2O_2.

8-4. Tetrafluorohydrazine, N_2F_4, is a colorless liquid that is used as rocket fuel. Write the Lewis formula for tetrafluorohydrazine.

8-5. Write the Lewis formula for

(a) methane, CH_4
(b) fluoromethane, CH_3F
(c) aminomethane, CH_3NH_2

8-6. Write the Lewis formula for

(a) methyl mercaptan, CH_3SH
(b) dimethyl ether, CH_3OCH_3
(c) trimethyl amine, $N(CH_3)_3$

8-7. Hydrocarbons are compounds that contain only hydrogen and carbon. In one type of hydrocarbon, the carbon atoms are connected to each other in a chain. Write the Lewis formulas for the following "straight-chain" hydrocarbons:

(a) propane, C_3H_8
(b) butane, C_4H_{10}
(c) octane, C_8H_{18}

8-8. In branched-chain hydrocarbons, the carbon atoms are not connected to each other in a straight chain (see Problem 8-7). Write the Lewis formula for the following branched-chain hydrocarbons:

(a) isobutane, $(CH_3)_3CH$
(b) neopentane, $(CH_3)_4C$
(c) isopentane, $C_2H_5CH(CH_3)_2$

8-9. An alcohol is an organic compound containing an OH group. Write the Lewis formula for the following alcohols:

(a) ethyl alcohol, CH_3CH_2OH
(b) *n*-propyl alcohol, $CH_3CH_2CH_2OH$
(c) isopropyl alcohol, $(CH_3)_2CHOH$

8-10. Write all possible Lewis formulas for $C_2H_4Cl_2$.

8-11. A hydroxide ion, OH^-, results when a proton is removed from H_2O. The ion that results when a proton is removed from NH_3 is called an amide ion, NH_2^-. Write the Lewis formula for an amide ion. Name the following ionic compounds: $NaNH_2$ and $Ba(NH_2)_2$.

8-12. A methoxide ion, CH_3O^-, results when a proton is removed from methanol, CH_3OH. Write the Lewis formula for CH_3O^-. Name the compounds $KOCH_3$ and $Al(OCH_3)_3$.

MULTIPLE BONDS

8-13. Write the Lewis formula for

(a) acetylene, C_2H_2
(b) diazine, N_2H_2
(c) phosgene, $COCl_2$

8-14. Write the Lewis formula for

(a) nitrous acid, HNO_2
(b) silicon dioxide, SiO_2
(c) propylene, CH_3CHCH_2

8-15. Formic acid is a colorless liquid with a penetrating odor. It is the irritating ingredient in the bite of ants. Its chemical formula is HCOOH. Write the Lewis formula for formic acid.

8-16. Hydrazoic acid, HN_3, is a dangerously explosive, colorless liquid. Azides of heavy metals explode when struck sharply and are used in detonation caps. Write the Lewis formula for hydrazoic acid (HN_3) and for the azide ion (N_3^-).

8-17. Vinyl chloride is an important industrial chemical used in the manufacture of polyvinyl chloride. Its chemical formula is C_2H_3Cl. Write the Lewis formula for vinyl chloride.

8-18. Acetone is an organic compound widely used in the chemical industry as a solvent, for example, in paints and varnishes. You may be familiar with its sweet odor because it is used as a fingernail-polish remover. Its chemical formula is CH_3COCH_3. Write the Lewis formula for acetone.

FORMAL CHARGE

8-19. Use formal charge considerations to rule out the Lewis formula for NF_3 in which the nitrogen atom and the three fluorine atoms are connected in a row.

8-20. Use formal charge considerations to predict the arrangement of the atoms in NOCl.

8-21. Laughing gas, an anesthetic and a propellant in whipped-cream-dispensing cans, has the composition N_2O. Use Lewis formulas and formal charge considerations to predict which structure, NNO or NON, is the more likely.

8-22. Use formal charge considerations to rule out the Lewis formula for NO_2^- in which the arrangement of the atoms is O-O-N.

RESONANCE

8-23. Write Lewis formulas for the resonance forms of the formate ion, $HCOO^-$. Indicate formal charges and discuss the bonding of this ion.

8-24. Write Lewis formulas for the resonance forms of the acetate ion, CH_3COO^-. Indicate formal charges and discuss the bonding in this ion.

8-25. Write Lewis formulas for the resonance forms of the carbonate ion, CO_3^{2-}. Indicate formal charges and discuss the bonding in this ion.

8-26. Write Lewis formulas for the resonance forms of ozone, O_3. Indicate formal charges and discuss the bonding in this molecule.

8-27. Write the Lewis formula or formulas for each of the following species. Give resonance forms where appropriate, and indicate formal charges on each.

(a) CS_3^{2-} (b) $C_2O_4^{2-}$
(c) NCS^- (d) O_2^-

8-28. Write the Lewis formula for each of the following nitrogen oxides. Indicate formal charge and resonance forms.

Formula	**Name**	**Form at 25°C, 1 atm**
N_2O	dinitrogen oxide	colorless gas
NO	nitrogen oxide	colorless gas
N_2O_3	dinitrogen trioxide	dark blue gas
NO_2	nitrogen dioxide	brown gas
N_2O_4	dinitrogen tetroxide	colorless gas
N_2O_5	dinitrogen pentoxide	white solid

8-29. Write the Lewis formula for the following benzene derivatives:

(a) chlorobenzene, C_6H_5Cl
(b) aminobenzene, $C_6H_5NH_2$
(c) benzoic acid, C_6H_5COOH
(d) phenol, C_6H_5OH

8-30. Naphthalene, which has the characteristic odor of mothballs, has the formula $C_{10}H_8$. Given that its structure is two benzene rings fused together along one carbon-carbon bond, write its Lewis formula.

OCTET RULE VIOLATIONS

8-31. Which of the following species contain an odd number of electrons?

(a) NO_2 (b) CO
(c) O_3^- (d) O_2^-

Write a Lewis formula for each of these species.

8-32. Which of the following species contain an odd number of electrons?

(a) BrO_3 (b) SO_3
(c) HNO (d) HO_2

Write a Lewis formula for each of these species.

8-33. Nitrosamines are carcinogens that are found in tobacco smoke. They can also be formed in the body from the nitrites and nitrates used to preserve processed meats, especially bacon and sausage. The simplest nitrosamine is methylnitrosamine, H_3CNNO. Write the Lewis formula for this molecule. Is it a free radical?

8-34. Many free radicals combine to form molecules that do not contain any unpaired electrons. The driving force for the radical-radical combination reaction is the formation of a new electron-pair bond. Write Lewis formulas for the reactant and product species in the following reactions:

(a) $CH_3(g) + CH_3(g) \rightarrow H_3CCH_3(g)$
(b) $N(g) + NO(g) \rightarrow NNO(g)$
(c) $2OH(g) \rightarrow H_2O_2(g)$

8-35. Write the Lewis formula for

(a) PCl_6^- (b) I_3^- (c) SiF_6^{2-}

8-36. Write a Lewis formula for each of the following compounds of xenon:

Compound	Form at 25°C	Melting point/°C
XeF_2	colorless crystals	129
XeF_4	colorless crystals	117
XeF_6	colorless crystals	50
$XeOF_4$	colorless liquid	−46
XeO_2F_2	colorless crystals	31

DIPOLE MOMENTS

8-37. Write the Lewis formula for bromine chloride and indicate its dipole moment.

8-38. Arrange the following groups of molecules in order of increasing dipole moment:

(a) HCl HF HI HBr
(b) PH_3 NH_3 AsH_3 (tripod-shaped molecules)
(c) Cl_2O F_2O H_2O (bent molecules)
(d) ClF_3 BrF_3 IF_3 (T-shaped molecules)
(e) H_2O H_2S H_2Te H_2Se (bent molecules)

8-39. Describe the charge distribution in

(a) nitrogen trifluoride, NF_3
(b) oxygen difluoride, OF_2
(c) oxygen dibromide, OBr_2

8-40. Describe the charge distribution in

(a) hydrogen fluoride, HF
(b) phosphine, PH_3
(c) hydrogen sulfide, H_2S

ADDITIONAL PROBLEMS

8-41. Write the Lewis formula for

(a) the sulfate ion, SO_4^{2-}
(b) the phosphate ion, PO_4^{3-}
(c) the acetate ion, $CH_3CO_2^-$

Show the formal charges and discuss the bonding in each ion.

8-42. The halogens form a number of interhalogen compounds. For example, chlorine pentafluoride, ClF_5, can be prepared by the reaction

$$KCl(s) + 3F_2(g) \rightarrow \underset{\text{colorless}}{ClF_5(g)} + KF(s)$$

The halogen fluorides are very reactive, combining explosively with water, for example. Write the Lewis formula for each of the following halogen fluoride species:

(a) ClF_5 (b) IF_3
(c) IF_7 (d) IF_4^+

8-43. Write the Lewis formula for

(a) tetrafluoroammonium ion, NF_4^+
(b) tetrafluorochlorinium ion, ClF_4^+
(c) phosphonium ion, PH_4^+
(d) hexafluoroarsenate ion, AsF_6^-
(e) tetrafluorobromate ion, BrF_4^-

8-44. Write the Lewis formula for each of the following oxychlorine species:

(a) perchlorate ion, ClO_4^-
(b) chlorine oxide, ClO
(c) chlorate ion, ClO_3^-
(d) chlorine dioxide, ClO_2
(e) hypochlorite ion, ClO^-

8-45. Write the Lewis formula for each of the following acids:

(a) $HClO_3$ (b) HNO_2
(c) HIO_4 (d) $HBrO_2$

8-46. Write the Lewis formula for each of the following oxyacids of sulfur:

(a) sulfuric acid, H_2SO_4
(b) thiosulfuric acid, $H_2S_2O_3$
(c) disulfuric acid, $H_2S_2O_7$
(d) dithionic acid, $H_2S_2O_6$ (has an S—S bond)
(e) peroxydisulfuric acid, $H_2S_2O_8$ (has an O—O bond)

8-47. The Group 6 elements form a number of halides. Some of them are

(a) SCl_2 (b) SCl_4
(c) SeF_6 (d) S_2F_4

Write a Lewis formula for each of these halides.

8-48. Phosphorus forms a number of oxohalides, X_3PO,

in which X may be F, Cl or Br. The commonest, phosphoryl chloride, is obtained by the reaction

$$2PCl_3(g) + O_2(g) \rightarrow 2Cl_3PO(g)$$

Write the Lewis formula for the phosphoryl halides.

8-49. The dichromate ion, $Cr_2O_7^{2-}$, has no metal-metal and no oxygen-oxygen bonds. Write a Lewis formula for $Cr_2O_7^{2-}$ (consider Cr to have six valence electrons).

8-50. In the P_4O_6 molecule each phosphorus atom is bonded to three oxygen atoms and each oxygen atom is bonded to two phosphorus atoms. Write a Lewis formula for P_4O_6.

8-51. Indicate whether or not the following species have a triple bond.

(a) CH_2CHCN (b) HOOCCOOH
(c) C_2^{2-} (d) CH_3CHCH_2

8-52. Write all the possible Lewis formulas for $C_3H_5Cl_3$.

8-53. Solid sulfur consists of eight-membered rings of sulfur atoms. Write the Lewis formula for S_8.

8-54. Write all the possible Lewis formulas for butadiene, C_4H_6, which contains two double bonds.

8-55. Some transition metals form covalent compounds. Write the Lewis formula for the following covalent species:

(a) MnO_4^-, permanganate ion
(b) CrO_3, chromium trioxide or chromium(VI) oxide
(c) $TiCl_4$, titanium tetrachloride
(d) VO_2^+, dioxovanadium(V) ion

8-56. Some transition metals form covalent compounds. Write the Lewis formula for the following covalent species:

(a) MoO_3 (b) $VOCl_3$
(c) $TiOCl_4^{2-}$ (d) CrO_4^{2-}

8-57. Use Lewis formulas and formal charge considerations to suggest that the structure of N_2F_2 is FNNF rather than FFNN.

8-58. Use Lewis formulas and formal charge considerations to suggest that the structure of hydrocyanic acid is HCN rather than HNC.

8-59. Use Lewis formulas and formal charge considerations to suggest that the structure of formaldehyde is H_2CO rather than HCOH or COH_2.

VSEPR THEORY

J. H. van't Hoff, the Dutch chemist who first proposed the tetrahedral geometry of methane and related compounds, built cardboard models to illustrate molecular shapes. Van't Hoff received the first Nobel Prize for chemistry ever awarded, in 1901.

Molecules come in a variety of shapes. In this chapter we devise a set of simple, systematic rules that allow us to predict the shapes of thousands of molecules. These rules are based upon the Lewis formulas that we developed in Chapter 8 and are collectively called the valence-shell electron-pair repulsion (VSEPR) theory. In spite of its rather imposing name, VSEPR theory is easy to understand, easy to apply, and remarkably reliable.

9-1. LEWIS FORMULAS DO NOT REPRESENT THE SHAPES OF MOLECULES

Lewis formulas indicate which atoms are bonded to which in a molecule but do not indicate the molecule's actual shape. Consider the molecule dichloromethane, CH_2Cl_2. One Lewis formula for dichloromethane is

$$\text{(1)} \qquad :\ddot{\underset{..}{Cl}}-\underset{\underset{H}{|}}{\overset{\overset{H}{|}}{C}}-\ddot{\underset{..}{Cl}}:$$

If we infer from this Lewis formula that dichloromethane is flat, or **planar,** then we must conclude that the Lewis formula

$$\text{(2)} \qquad H-\underset{\underset{H}{|}}{\overset{\overset{:\ddot{Cl}:}{|}}{C}}-\ddot{\underset{..}{Cl}}:$$

represents a different geometry for dichloromethane. In formula (1), the two chlorine atoms lie 180° apart, whereas in formula (2) they lie 90° apart. Molecules that have the same chemical formula, CH_2Cl_2 in this case, but different geometric arrangements, are called **geometric isomers.** Geometric isomers are different molecular species and so have different chemical and physical properties. For example, their boiling points differ, so they can be separated by distillation.

Two isomers of dichloromethane have never been observed; therefore, our assumption that dichloromethane is a planar molecule is incorrect. The fact that there is only one kind of dichloromethane molecule suggests that the four bonds around the central atom are oriented such that there is only one distinct way of bonding two hydrogen atoms and two chlorine atoms to a central carbon atom. A geometric arrangement that explains why there are no geometric isomers of dichloromethane, as well as many other similar observations, was proposed by the Dutch chemist Jacobus van't Hoff and the French chemist Joseph Le Bel independently in 1874. They proposed that the four bonds about a central carbon atom in a molecule such as methane, CH_4, are directed toward the corners of a regular **tetrahedron** (Figure 9-1). A regular tetrahedron is a four-sided figure that has four equivalent vertices and four identical faces, each of which is an equilateral triangle (Figure 9-2). Appendix D gives instructions for building a regular tetrahedron out of cardboard.

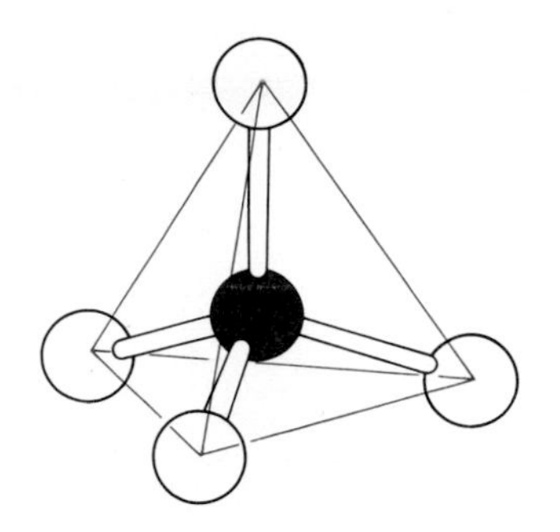

Figure 9-1 Each carbon-hydrogen bond in a methane molecule points toward the vertex of a regular tetrahedron. The positions of all four hydrogen atoms in CH_4 are equivalent by symmetry. All the H—C—H bond angles are the same, 109.5°.

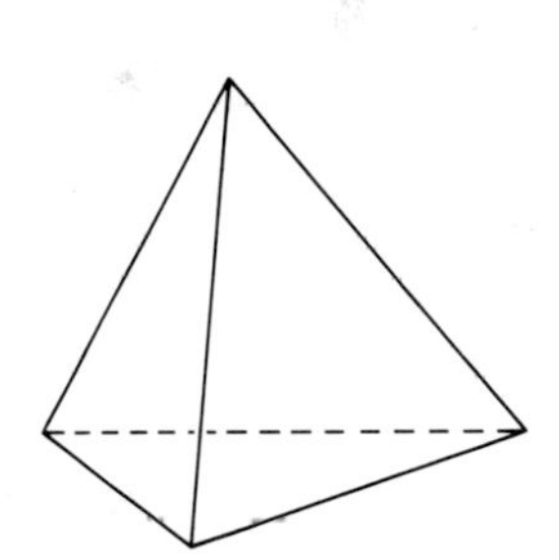

Figure 9-2 A regular tetrahedron is a symmetric body consisting of four equivalent vertices and four equivalent faces. Each face is an equilateral triangle. Note that a tetrahedron differs from the more familiar square pyramid, which has a square base.

9-2. ALL FOUR VERTICES OF A REGULAR TETRAHEDRON ARE EQUIVALENT

You can see from a model or from Figures 9-1 and 9-2 that the four vertices of a tetrahedron are equivalent. Because of this there is only one way of bonding two hydrogen atoms and two chlorine atoms directly to a central carbon atom, in accord with the experimental fact that dichloromethane has no isomers.

The model in Figure 9-3 is called a **space-filling molecular model** and gives a fairly accurate representation of the angles between bonds and of the relative sizes of the atoms in a molecule. A less realistic model, but perhaps one in which the geometry is easier to see, is the **ball-and-stick molecular model** shown in Figure 9-1.

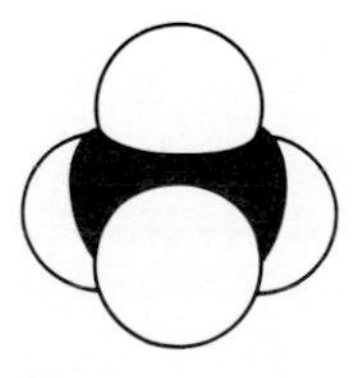

Figure 9-3 A space-filling molecular model of methane.

In a tetrahedral molecule like methane, all the H—C—H bond angles are equal to 109.5°, which is called the **tetrahedral bond angle.** A carbon atom that is bonded to four other atoms is called a **tetravalent** carbon atom. The hypothesis of van't Hoff and Le Bel that the bonds of a tetravalent carbon atom are tetrahedrally oriented was the beginning of what is called **structural chemistry,** the area of chemistry in which the shapes and sizes of molecules are studied. Many experimental methods have been developed to determine molecular geometries. Most of the methods involve the interaction of electromagnetic radiation or electrons with molecules (Interchapter G). Using such methods, it is possible to measure bond lengths and bond angles in molecules. It turns out that there is a great variety of molecular shapes. We saw in Chapter 8 that CO_2 is a linear molecule and that H_2O is bent. Methane is an example of a tetrahedral molecule. Some examples of other molecular geometries are shown below.

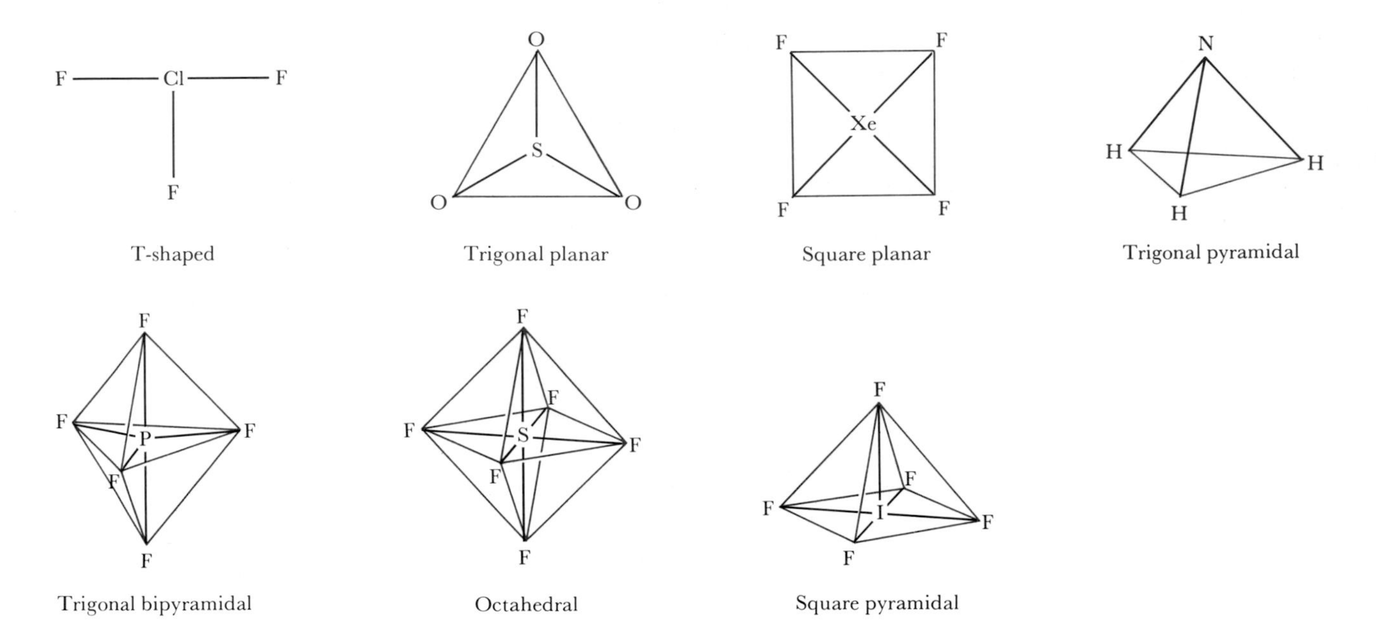

The experimentally observed shapes of various molecules.

9-3. VALENCE-SHELL ELECTRON-PAIR REPULSION THEORY IS USED TO PREDICT THE SHAPES OF MOLECULES

A set of rules has been devised that enables us to predict the shapes of the molecules shown above and the shapes of many other molecules as well. The method is based upon the total number of bonds and lone electron pairs in the valence shell of the central atom in a molecule. The postulate is that the shape of a molecule is determined by the mutual repulsion of the electron pairs in the valence shell of the central atom, and so the method that we are about to describe is called the **valence-shell electron-pair repulsion theory,** or the **VSEPR theory.**

Consider the electron-deficient compound beryllium chloride, $BeCl_2$. The Lewis formula for the $BeCl_2$ molecule is

$$:\ddot{\underset{..}{Cl}}—Be—\ddot{\underset{..}{Cl}}:$$

Although the central beryllium atom has no lone electron pairs, it does have two covalent bonds and thus has two electron pairs in its valence shell. These valence-shell electron pairs repel each other and can minimize their mutual repulsion by being as far apart as possible. If we visualize the central beryllium atom as a sphere and the two valence-shell electron pairs (the two covalent bonds) as being on the surface of the sphere, then the two bonds minimize their mutual repulsion by being at opposite poles of the sphere. Thus the two bonds are on opposite sides of the central beryllium atom and the Cl—Be—Cl bond angle is 180°. The shape of a molecule is determined by the positions of the atomic nuclei in the molecule, and so we say that $BeCl_2$ is a linear molecule. This is in accord with experimental studies of $BeCl_2$ in the gas phase. The positioning of the two valence-shell electron pairs on opposite sides of the central atom is shown in Figure 9-4(a).

We should point out that beryllium chloride is a solid at room temperature (20°C), but the vapor consists of individual beryllium chloride molecules with the Lewis formula $:\ddot{\underset{..}{Cl}}—Be—\ddot{\underset{..}{Cl}}:$. VSEPR theory applies only to individual molecules, as in the gas phase. Throughout this chapter we always refer to individual molecules in the gas phase even though the substance may be a solid at room temperature.

Consider now a molecule with three electron pairs in the valence shell of the central atom. An example is the electron-deficient compound boron trifluoride, whose Lewis formula is

▪ Boron trifluoride is a colorless gas with a pungent odor. It acts as a catalyst for a number of reactions.

```
     ..
    :F:
     |
     B
   /   \
 :F:   :F:
  ..    ..
```

The three valence-shell electron pairs (the three covalent bonds) surrounding the boron atom can minimize their mutual repulsion by max-

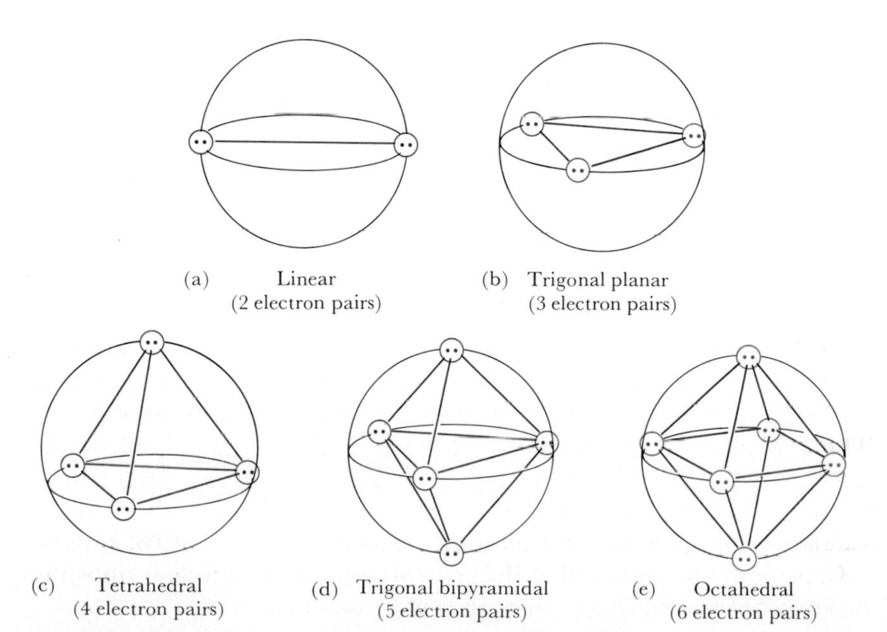

Figure 9-4 Sets of electron pairs (• •) arranged on the surfaces of spheres so as to minimize the mutual repulsion between them. (a) Two electron pairs lie at opposite poles of a sphere. (b) Three electron pairs lie on an equator at the verticies of an equilateral triangle. (c) Four electron pairs at the verticies of a regular tetrahedron. (d) Five electron pairs are arranged such that two lie at poles and the other three lie on the equator at the verticies of an equilateral triangle. (e) Six electron pairs lie at the vertices of a regular octahedron.

imizing their mutual separation. This results in a planar triangular, or **trigonal planar,** structure (Figure 9-4b). We predict, then, that BF_3 is a planar, symmetric molecule with F—B—F bond angles equal to 120°. This is in complete agreement with experiment.

9-4. THE NUMBER OF VALENCE-SHELL ELECTRON PAIRS DETERMINES THE SHAPE OF A MOLECULE

An example of a molecule with four covalent bonds surrounding a central atom is methane, CH_4. These four electron pairs minimize their mutual repulsion by pointing toward the vertices of a regular tetrahedron (Figure 9-4c). We see, then, that the tetrahedral geometry of methane is a result of the mutual repulsion of the four electron pairs making up its four covalent bonds. All the H—C—H bond angles in methane are equal to the tetrahedral bond angle, 109.5°.

Example 9-1: Silicon lies below carbon in the periodic table and is also tetravalent. Predict the geometry of silane, SiH_4.

▪ Silane is a gas that is used in the preparation of extremely pure silicon for semiconductors (Interchapter H).

Solution: The Lewis formula for silane is

```
     H
     |
  H—Si—H
     |
     H
```

There are four valence-shell electron pairs (four covalent bonds) about the central silicon atom, and so we predict correctly that silane is tetrahedral and that the H—Si—H bond angles are 109.5°.

The following Example shows that VSEPR theory can be applied to molecular ions.

Example 9-2: Predict the geometry of the ammonium ion, NH_4^+.

Solution: The Lewis formula for NH_4^+ is

```
 ┌    H    ┐+
 |    |    |
 | H—N—H   |
 |    |    |
 └    H    ┘
```

Because the valence shell of the nitrogen atom contains a total of four electron pairs (four covalent bonds), we predict that the ammonium ion is tetrahedral, which is the observed structure of NH_4^+.

Many molecules have five electron pairs in the form of five covalent bonds in the valence shell of the central atom. An example is phosphorus pentachloride, PCl_5, whose Lewis formula is

:Cl:
Cl
:Cl—P
Cl
:Cl:

The five electron pairs in the valence shell of the phosphorus atom are positioned to minimize their mutual repulsion, outlining a **trigonal bipyramid** (Figures 9-4d and 9-5). Notice that the three vertices on the equator of the sphere in Figure 9-4(d) form an equilateral triangle and that the vertices lying at the poles lie above and below the center of the equilateral triangle. The five vertices of a trigonal bipyramid are *not* equivalent. The three vertices lying on the equator in Figure 9-4(d) are equivalent and are called **equatorial vertices;** the two vertices lying at the poles are equivalent and are called **axial vertices.** Some other examples of trigonal bipyramidal molecules are antimony pentachloride, $SbCl_5$, and arsenic pentafluoride, AsF_5.

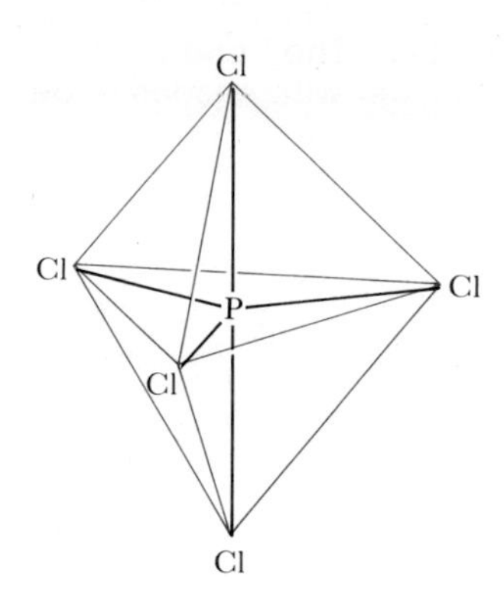

Figure 9-5 The shape of the phosphorus pentachloride molecule.

Consider the molecule sulfur hexafluoride, SF_6, whose Lewis formula is

:F:
F F
S
F F
:F:

This Lewis formula shows that there are six electron pairs in the valence shell of the central sulfur atom. These six electron pairs mutually repel each other. The repulsion is minimized if the six electron pairs, or covalent bonds in this case, point toward the corners of a regular **octahedron.** An octahedron (Figure 9-6) has six vertices and eight faces. All eight faces are identical equilateral triangles. An important property of a regular octahedron is that all six vertices are equivalent. This can be seen easily by building an octahedron using the directions given in Appendix D. We see, then, that SF_6 is octahedral and that the six fluorine atoms are equivalent. There is no way, by either chemical or physical methods, to distinguish among the six sulfur-fluorine bonds in SF_6.

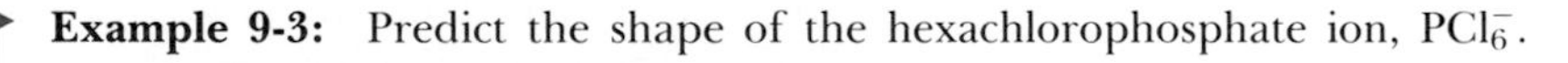

Example 9-3: Predict the shape of the hexachlorophosphate ion, PCl_6^-.

Solution: The Lewis formula for PCl_6^- is

:Cl:
Cl Cl
P
Cl Cl
:Cl:
(bracketed, charge −)

The six covalent bonds are directed toward the vertices of a regular octahedron, and we predict correctly that the PCl_6^- ion is octahedral.

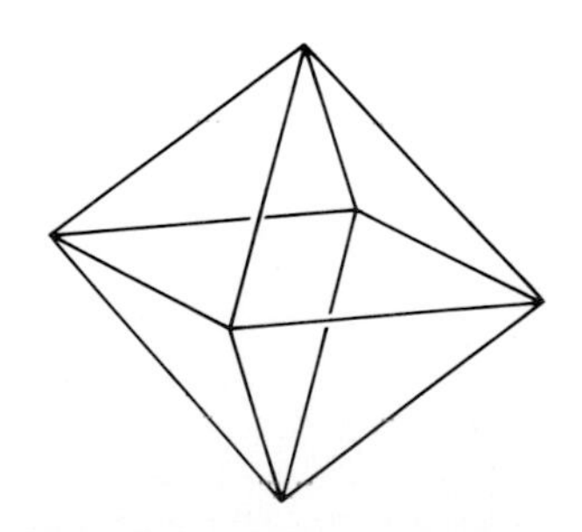

Figure 9-6 A regular octahedron is a symmetric body consisting of six equivalent vertices and eight identical faces that are equilateral triangles.

Table 9-1 shows the bond angles associated with the molecular shapes we have discussed thus far.

Table 9-1 The bond angles associated with shapes shown in Figure 9-4

Shape	Structure
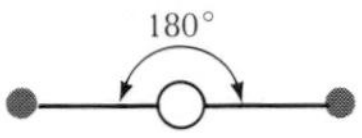 180°	linear
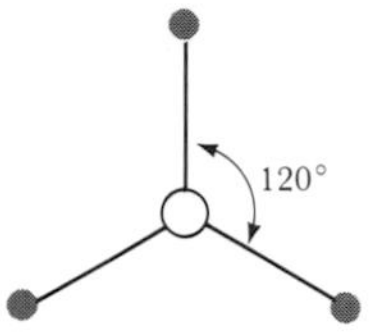 120°	trigonal planar
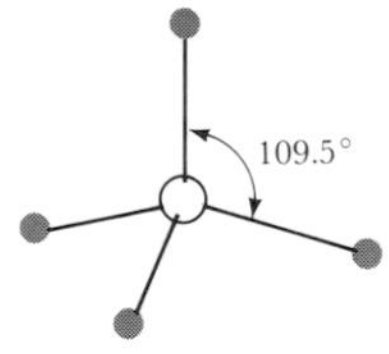 109.5°	tetrahedral
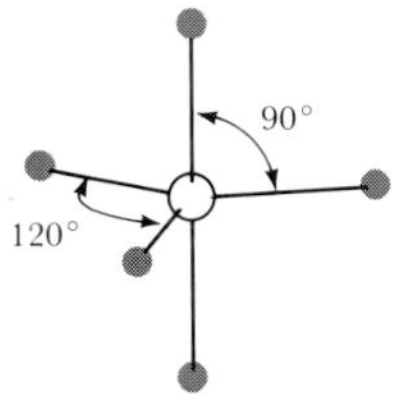 90°, 120°	trigonal bipyramidal
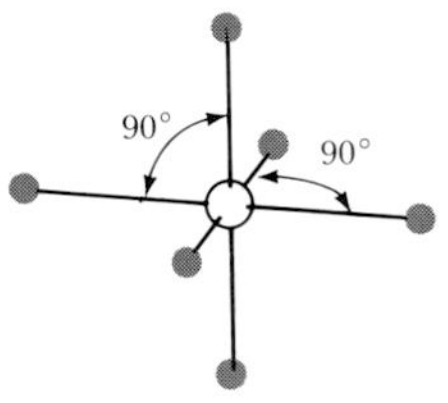 90°, 90°	octahedral

9-5. LONE ELECTRON PAIRS IN THE VALENCE SHELL AFFECT THE SHAPES OF MOLECULES

In each case we have discussed so far, all the electron pairs in the valence shell of the central atom have been in covalent bonds. Let's now consider cases in which there are lone pairs of electrons as well as covalent bonds in the valence shell of the central atom. As an example, consider the ammonia molecule. The Lewis formula for NH_3 is

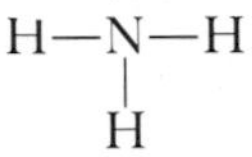

$$\begin{array}{c} \cdot\cdot \\ \text{H—N—H} \\ | \\ \text{H} \end{array}$$

There are four electron pairs in the valence shell of the nitrogen atom. Three of them are in covalent bonds, and one is a lone pair. These four valence-shell electron pairs mutually repel each other and so are directed toward the corners of a tetrahedron (Figure 9-7b). The three hydrogen atoms form an equilateral triangle, and the nitrogen atom sits in the center of and above the plane of the triangle. Such a structure is called a triangular pyramid or **trigonal pyramid.** The ammonia molecule is shaped like a tripod, with the three N—H bonds forming the legs of the tripod. A space-filling molecular model of NH_3 is shown in Figure 9-8. It is important to keep in mind that the shape of a molecule is given by the positions of the nuclei in the molecule. The electrons are diffuse and cloudlike and are spread over the molecule. The much more massive nuclei are relatively fixed, and it is the positions of the nuclei that define what we mean by the shape or geometry of a molecule.

If the four electron pairs in NH_3 pointed to the corners of a regular tetrahedron, then the H—N—H bond angles would be 109.5°. The four electron pairs in this case are not all of the same type, however. Three of them occur in covalent bonds, and the fourth is a lone pair. Thus we might expect some distortion from purely regular tetrahedral geometry. Electron pairs in covalent bonds are shared between two atoms and are localized between them. A lone pair of electrons, on the other hand, is associated with only the central atom and so is not as localized as the pair of electrons in a covalent bond. Thus a lone pair of electrons is more spread out and is bulkier; that is, it takes up more room around a central atom than does a covalent bond. This means that the repulsion between a lone pair of electrons and the electron pair in a covalent bond is greater than the repulsion between the electron pairs in two covalent bonds. This causes the H—N—H bond angles in NH_3 to decrease slightly from the 109.5° regular tetrahedral angle to 107.3°. Although VSEPR theory is not able to give actual numerical values of bond angles, it can be used to make qualitative statements. We cannot predict that the H—N—H bond angles in NH_3 are 107.3°, but we can predict that they are slightly less than the ideal tetrahedral angle of 109.5°. We say that VSEPR is a qualitative rather than a quantitative theory. There is no simple, reliable quantitative theory of chemical bonding and molecular structure.

The example of NH_3 shows that it is the *total* number of electron

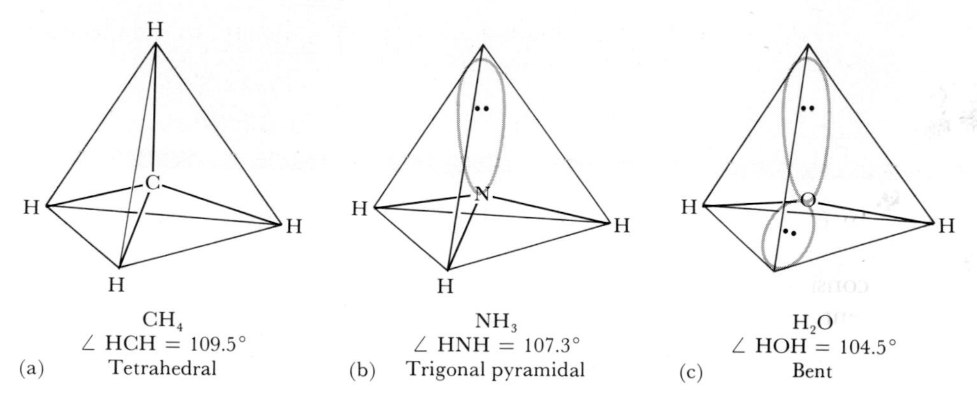

Figure 9-7 The role of bonding and nonbonding electron pairs in determining molecular geometry.

pairs in the valence shell of the central atom that determines the shape of a molecule. For example, the Lewis formula for H_2O is

$$\ddot{\underset{H\quad H}{\overset{..}{O}}}$$

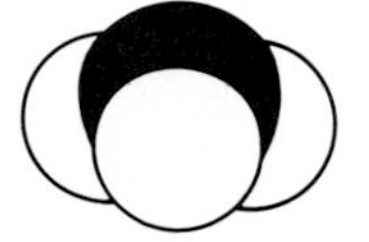

Figure 9-8 A space-filling molecular model of ammonia.

The four valence-shell electron pairs are directed toward the corners of a tetrahedron (Figure 9-7c) and so we see that H_2O is a bent or V-shaped molecule. The lone pairs take up more room around the oxygen atom than the two pairs in the covalent bonds, and so we expect the repulsion between the lone electron pairs to be greater than either that between a lone pair and a covalent bond or that between two covalent bonds. We can summarize this by writing

lone-pair–lone-pair repulsion >
lone-pair–bond-pair repulsion >
bond-pair–bond-pair repulsion (9-1)

We predict, then, that the H—O—H bond angle in H_2O is less than the regular tetrahedral angle of 109.5° and that it is even smaller than the H—N—H bond angle (107.3°) in NH_3. The experimentally measured bond angle in H_2O is 104.5°.

Each of the molecules CH_4, NH_3, and H_2O has four electron pairs in the valence shell of the central atom. The four electron pairs are directed toward the corners of a tetrahedron, as shown in Figure 9-7. The shape of a molecule is described by giving the positions of the nuclei. Thus, CH_4 is tetrahedral, NH_3 is trigonal pyramidal, and H_2O is bent.

We can classify molecules by introducing the following terminology. Let A represent a central atom, X an atom bonded to the central atom, and E a lone pair of electrons. We call an atom bonded to a central atom a **ligand.** The molecules that we are discussing in this chapter can be classified as $\mathbf{AX}_m\mathbf{E}_n$, where m is the number of ligands and n is the number of lone electron pairs in the valence shell of the central atom A. Therefore, the methane molecule belongs to the class AX_4, the ammonia molecule belongs to the class AX_3E, and the water molecule belongs to the class AX_2E_2. The classes of molecules that we discuss in this chapter are given in Figure 9-9.

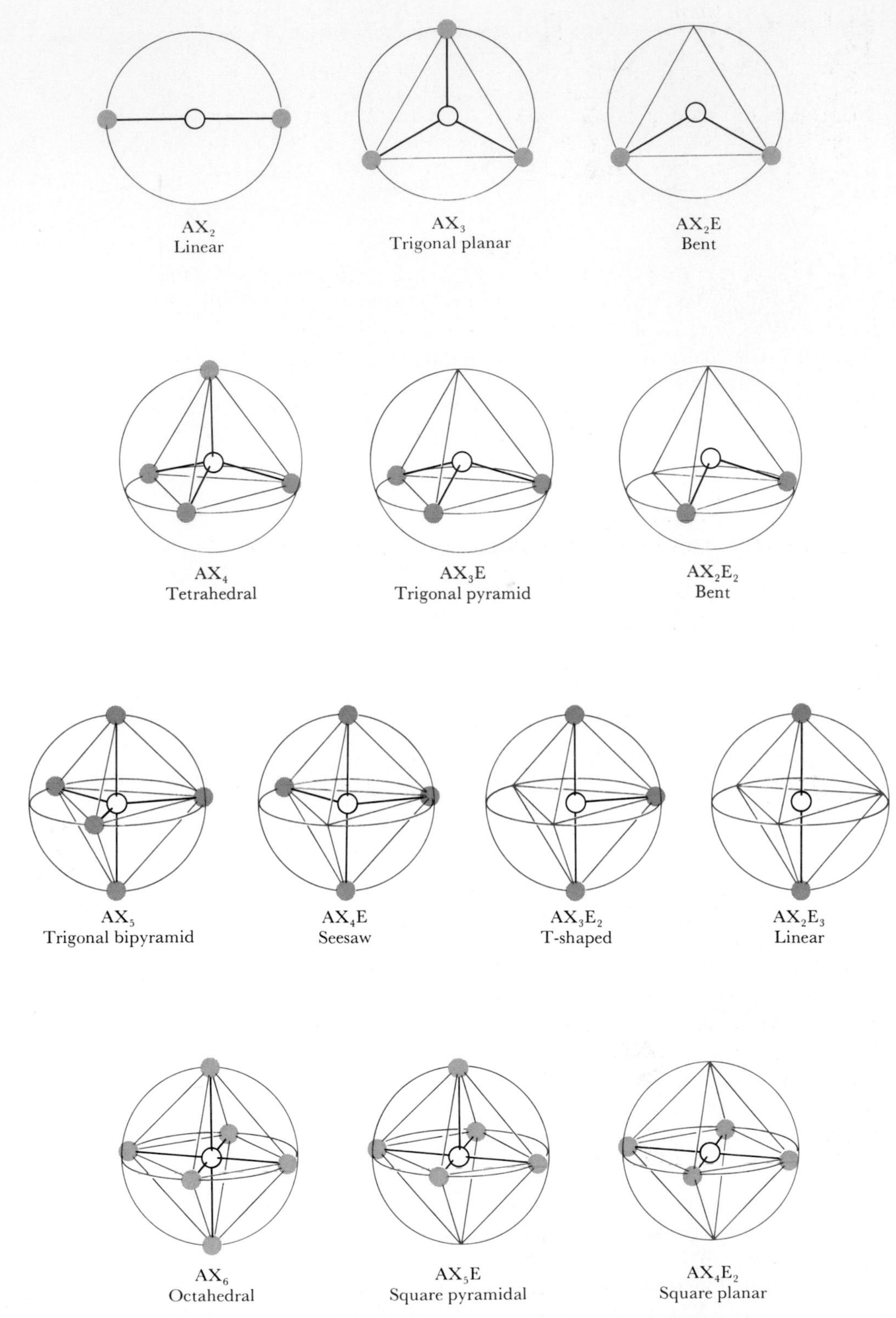

Figure 9-9 A summary of the various molecular shapes that result when m ligands (X) and n lone electron pairs (E) surround a central atom (A) to form an AX_mE_n molecule. The open circle at the center of each sphere represents the central atom, and the gray dots on the surfaces of the spheres represent the attached ligands. The lone electron pairs are at the vertices without dots.

9-6. VSEPR THEORY IS APPLICABLE TO MOLECULES THAT CONTAIN MULTIPLE BONDS

In predicting the shapes of molecules by VSEPR theory, a double or triple bond is counted simply as one group of electrons connecting the ligand X to the central atom A. For example, the Lewis formula for CO_2 is

$$\ddot{\underset{\cdot\cdot}{O}}=C=\ddot{\underset{\cdot\cdot}{O}}$$

Thus CO_2 is classified as an AX_2 molecule, and we predict that it is linear. Another example of an AX_2 molecule that contains a multiple bond is HCN, the Lewis formula for which is

$$H-C\equiv N:$$

There are two groups of electrons around the central carbon atom, and so we predict that HCN is a linear molecule.

Example 9-4: Predict the shape of a formaldehyde molecule, H_2CO.

Solution: The Lewis formula for H_2CO is

$$\begin{matrix} H \\ & \diagdown \\ & & C=\ddot{\underset{\cdot\cdot}{O}} \\ & \diagup \\ H \end{matrix}$$

We treat the double bond as any one group of electrons and classify H_2CO as an AX_3 molecule. We predict correctly that H_2CO is a trigonal planar molecule, meaning that all four atoms lie in a single plane. Figure 9-10 shows a model of the formaldehyde molecule.

Although a double or triple bond is treated as a single group of electrons when a molecule is placed in an AX_mE_n class, we must recognize that multiple bonds contain a higher electronic charge density than that in single covalent bonds. Consequently, multiple bonds repel single bonds more strongly than single bonds repel other single bonds. In this regard, multiple bonds act like lone electron pairs. Because of this, we predict that the H—C—H bond angle in H_2CO (Example 9-4) is slightly less than 120° and that the H—C—O bond angles are equal but are slightly larger than 120°. The actual experimental values are 116° and 122°, respectively:

$$\begin{matrix} H & 122^\circ \\ 116^\circ & C=\ddot{\underset{\cdot\cdot}{O}} \\ H & 122^\circ \end{matrix}$$

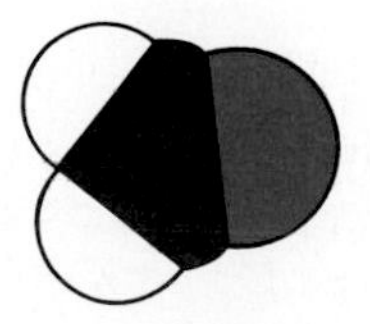

Figure 9-10 A space-filling molecular model of formaldehyde.

Example 9-5: Compare the shapes of the molecules $COCl_2$, phosgene, and $SOCl_2$, thionyl chloride.

Solution: The Lewis formulas for these two molecules are

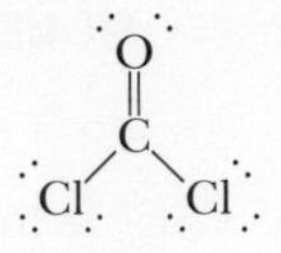

phosgene and thionyl chloride

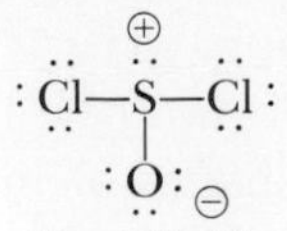

Phosgene belongs to the class AX_3 and so is planar, like formaldehyde. The lone pair on the sulfur atom in thionyl chloride puts this molecule in the AX_3E class. Thus, thionyl chloride is trigonal pyramidal, with the two chlorine atoms and the oxygen atom lying in a plane and the sulfur atom lying above the plane.

Notice that even though the chemical formulas for phosgene and thionyl chloride are similar, the shapes of the two molecules are different, because of the lone pair on the sulfur atom.

▪ Phosgene is a colorless, highly toxic gas. When diluted with air, it has an odor resembling newly mown hay.

▪ The different shapes of $COCl_2$ and $SOCl_2$ result from the presence of a lone pair on the sulfur atom.

VSEPR theory applies to molecules that are described by a resonance hybrid as well as it does to molecules that can be represented by just one Lewis formula, as illustrated in the following Example.

Example 9-6: Predict the shape of the carbonate ion, CO_3^{2-}.

Solution: There are three resonance forms of CO_3^{2-}:

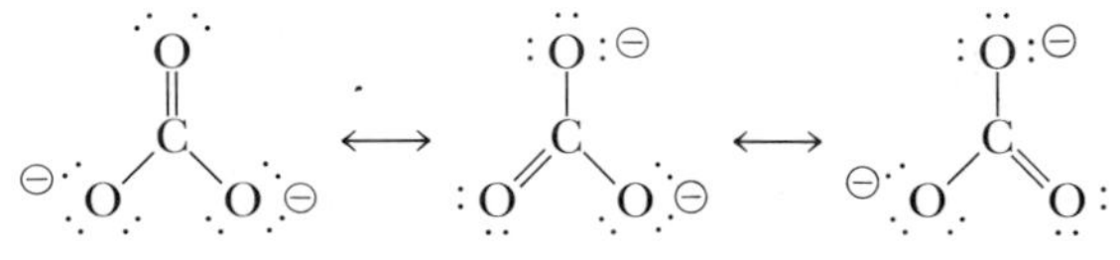

The resonance hybrid is

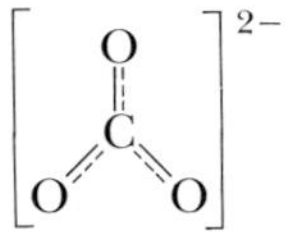

This shows that CO_3^{2-} is of the class AX_3, and so we predict that the carbonate ion is trigonal planar and that the three O—C—O bond angles are equal to 120°. This prediction is in agreement with the experimental result.

9-7. LONE-PAIR ELECTRONS OCCUPY THE EQUATORIAL VERTICES OF A TRIGONAL BIPYRAMID

Recall (Figure 9-4d) that the five vertices of a trigonal bipyramid are not equivalent to each other. They form a set of three equivalent equatorial vertices and two equivalent axial vertices. Consequently, in considering the class of molecules designated by AX_4E, for example, we have two nonequivalent choices for the position of the lone pair. We can place it at an equatorial vertex or at an axial vertex.

Figure 9-11(a) shows that an equatorial pair has only two nearest neighbors at 90°, and Figure 9-11(b) shows that an axial electron pair has three nearest neighbors at 90°. The two other neighbors in Figure 9-11(a) lie at 120° from the position labeled E and thus are far enough away that their interaction with E is much less than those at 90°. Consequently, the repulsion due to a lone electron pair is minimized by placing it at an equatorial vertex rather than at an axial vertex. Thus, mole-

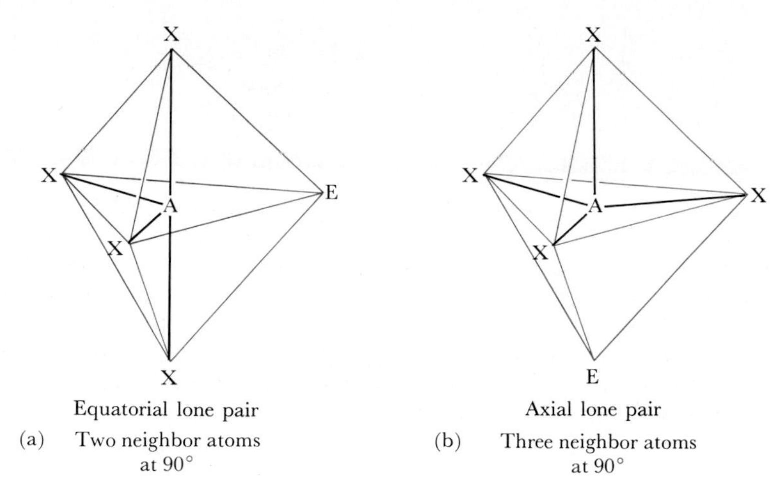

Figure 9-11 Because the five vertices of a trigonal bipyramid fall into two distinct classes, there are two nonequivalent positions available for the lone electron pair in a molecule of the class AX_4E. (a) A lone pair at an equatorial position has only two nearest neighbors at 90°, whereas (b) a lone pair at an axial position has three nearest neighbors at 90°. Consequently, the repulsion due to a lone electron pair is minimized by placing it at an equatorial vertex, and molecules with the general formula AX_4E are shaped like a seesaw, as shown in (a).

cules that belong to the classes AX_4E, AX_3E_2, and AX_2E_3 have the lone electron pairs at the equatorial positions, as shown in Figure 9-12. Because the shape of a molecule is determined by the positions of only the atomic nuclei, we see from Figure 9-12 that an AX_4E molecule is shaped like a seesaw, an AX_3E_2 molecule is T-shaped, and an AX_2E_3 molecule is linear.

The Lewis formula for sulfur tetrafluoride

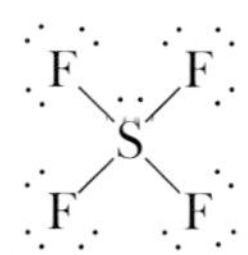

shows that this molecule belongs to the class AX_4E. The lone pair is placed at one of the equatorial positions of a trigonal bipyramid, and so the ideal shape of SF_4 is as shown in Figure 9-13(a). Note that the shape

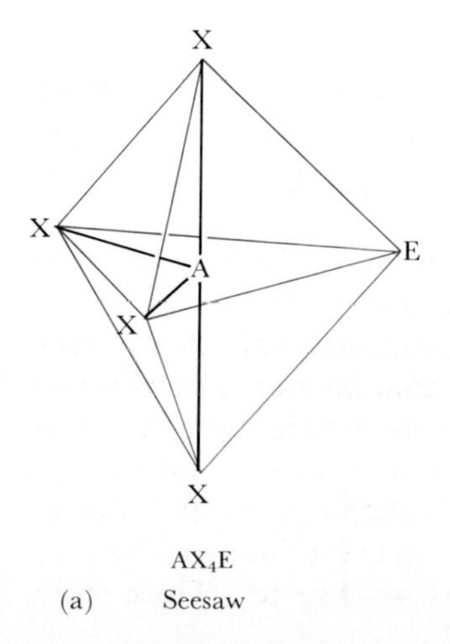

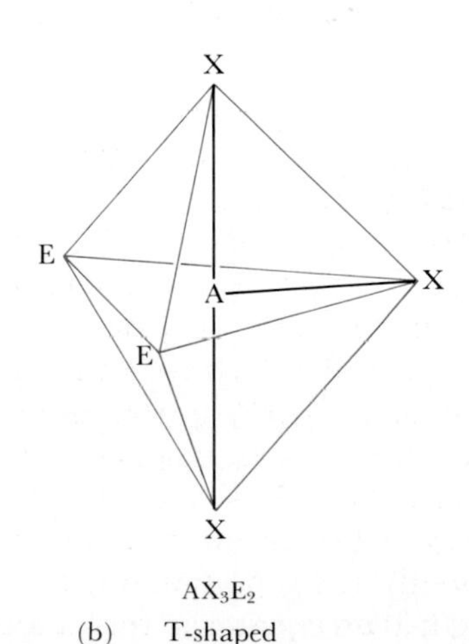

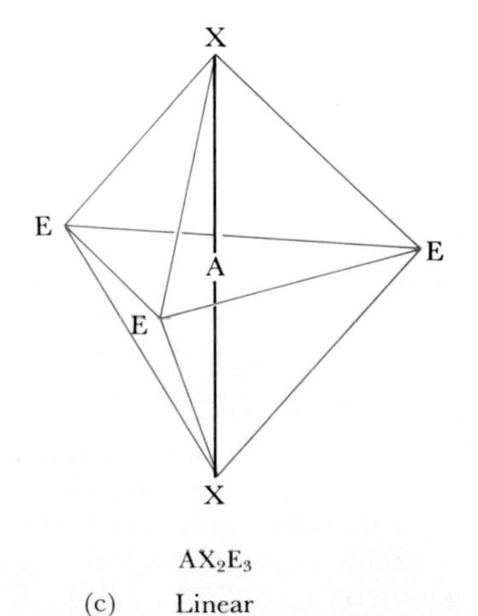

Figure 9-12 The shapes of molecules that belong to the classes (a) AX_4E, (b) AX_3E_2, and (c) AX_2E_3. The lone pairs occupy the equatorial positions in each case.

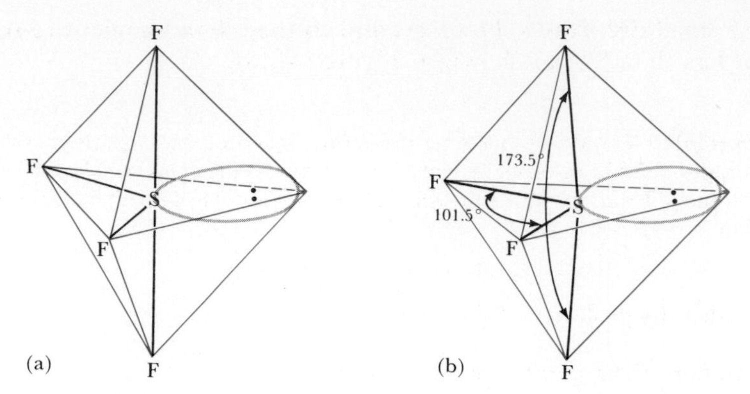

Figure 9-13 The geometry of a sulfur tetraflouride molecule, which belongs to the class AX_4E. (a) The ideal shape of the molecule. (b) The lone electron pair at the equatorial position repels the four covalent sulfur-fluorine bonds and distorts the molecule away from ideal geometry.

of SF_4 is like that of a seesaw. This ideal shape predicts that the F(axial)—S—F(axial) bond angle is 180° and that the F(equatorial)—S—F(equatorial) bond angle is 120°. The lone pair at the equatorial position causes a small distortion from ideal behavior, however, and the actual shape of SF_4 is that shown in Figure 9-13(b). The experimentally observed bond angles in SF_4, which are indicated in Figure 9-13(b), are in accord with the rule that says lone pairs take up more space than single covalent bonds.

Table 9-2 Some interhalogen compounds

AX	AX_3E_2	AX_5E
IF	IF_3	IF_5
BrF	BrF_3	BrF_5
ClF	ClF_3	ClF_5
ICl		
BrCl		
IBr		

There are a number of compounds formed between halogen atoms in which a less electronegative central halogen atom is bonded to more electronegative halogen atoms. Some of the **interhalogen compounds** are listed in Table 9-2. All known interhalogen molecules obey the predictions of VSEPR theory. Consider the molecule chlorine trifluoride, ClF_3, whose Lewis formula is

$$:\ddot{F}-\ddot{\underset{|}{Cl}}-\ddot{F}:$$
$$:\underset{..}{F}:$$

This Lewis formula shows that ClF_3 belongs to the class AX_3E_2. The ideal shape of ClF_3 is shown in Figure 9-14(a). Notice that the ClF_3 molecule is T-shaped. The ideal shape predicts that the F(axial)—Cl—F(equatorial) bond angles are 90°. The two lone pairs

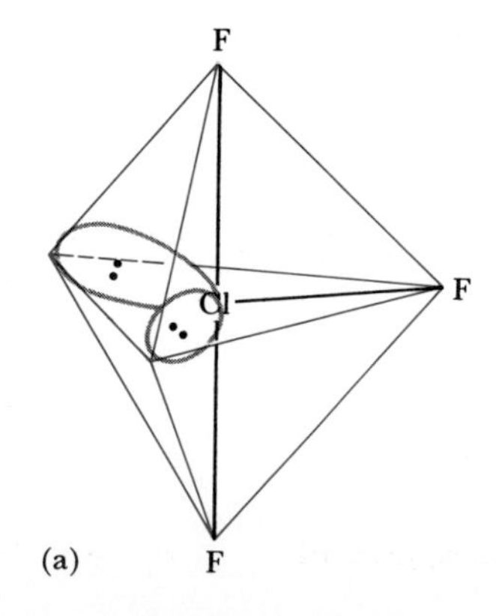

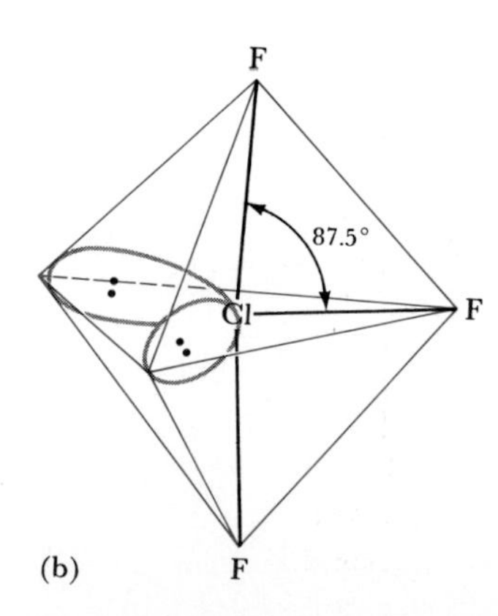

Figure 9-14 The geometry of a chlorine trifluoride molecule, which belongs to the class AX_3E_2. (a) The ideal shape of the molecule. (b) The two lone electron pairs at the equatorial positions repel the chlorine-fluorine bonds and distort the molecule away from ideal geometry.

cause small distortions, however, and so these bond angles are somewhat less than 90°, as shown in Figure 9-14(b).

Example 9-7: Although elemental iodine, I_2, is not very soluble in water, it is quite soluble in aqueous solutions of potassium iodide. The increased solubility is due to the formation of the triiodide ion, I_3^-, through the reaction

$$I_2(aq) + I^-(aq) \rightarrow I_3^-(aq)$$

Predict the geometry of the triiodide ion.

Solution: The Lewis formula for I_3^- is

$$\ominus$$
$$:\ddot{I}-\ddot{\underset{..}{I}}-\ddot{I}:$$

This Lewis formula shows that I_3^- belongs to the class AX_2E_3. Figure 9-12(c) indicates that the three lone pairs occupy equatorial positions in a trigonal bipyramid, and thus I_3^- is a linear ion.

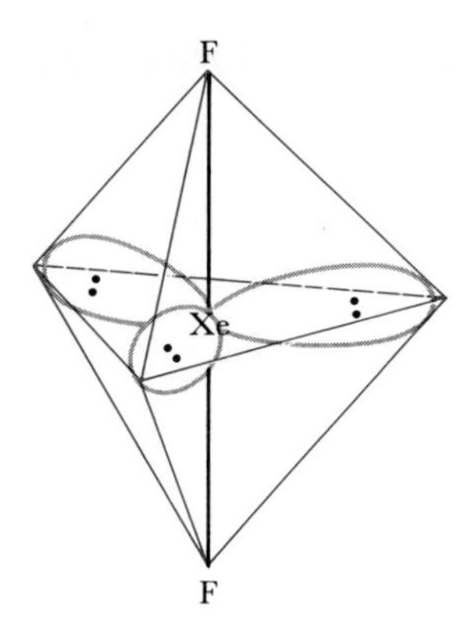

Figure 9-15 The molecule XeF_2 belongs to the class AX_2E_3. The three lone pairs occupy the equatorial vertices of a trigonal bipyramid. The two fluorine atoms occupy the axial positions, and so XeF_2 is a linear molecule.

One of the impressive successes of VSEPR theory is its correct prediction of the structures of noble-gas compounds. Some noble-gas compounds that have been synthesized are the xenon fluorides (XeF_2, XeF_4, and XeF_6), xenon oxyfluorides ($XeOF_4$, XeO_2F_2), xenon oxides (XeO_3, XeO_4), and krypton fluorides (KrF_2, KrF_4).

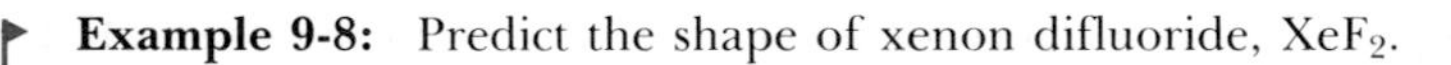

Example 9-8: Predict the shape of xenon difluoride, XeF_2.

Solution: The Lewis formula for XeF_2 is

$$:\ddot{\underset{..}{F}}-\ddot{\underset{..}{Xe}}-\ddot{\underset{..}{F}}:$$

This shows that XeF_2 belongs to the class AX_2E_3. Figure 9-15 shows the linear shape of XeF_2.

9-8. TWO LONE ELECTRON PAIRS OCCUPY OPPOSITE VERTICES OF AN OCTAHEDRON

The octahedral classes AX_6, AX_5E, AX_4E_2, and AX_3E_3 are shown in Figure 9-16. Because all six vertices of a regular octahedron are equiva-

Figure 9-16 The ideal shapes associated with the classes (a) AX_6, (b) AX_5E, (c) AX_4E_2, and (d) AX_3E_3. In (c) and (d), two of the lone electron pairs occupy opposite vertices because this placement minimizes the relatively strong lone-pair–lone-pair repulsion.

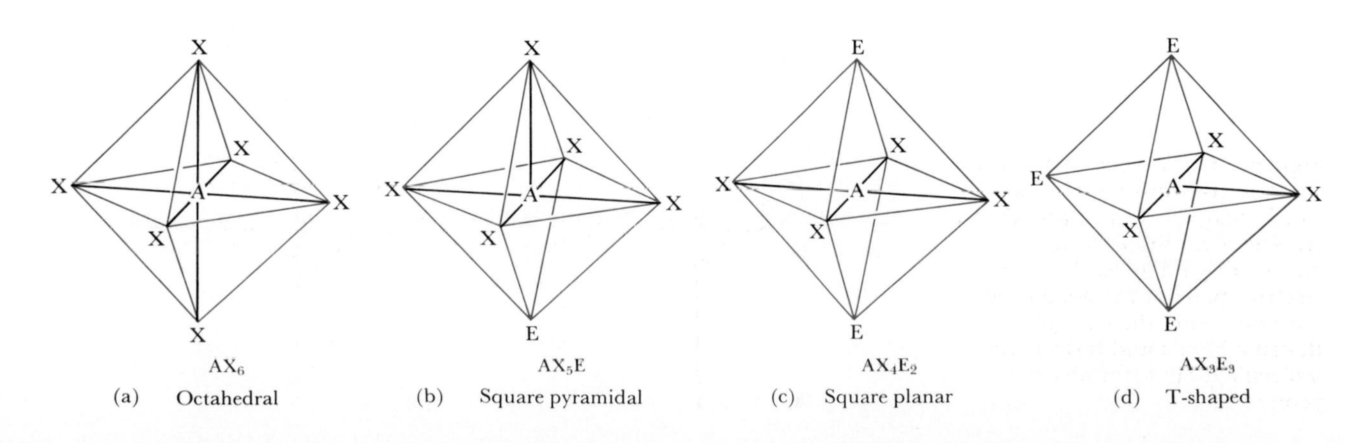

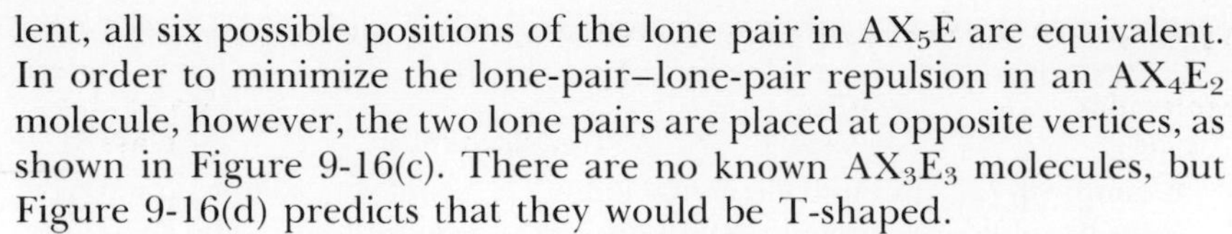
lent, all six possible positions of the lone pair in AX_5E are equivalent. In order to minimize the lone-pair–lone-pair repulsion in an AX_4E_2 molecule, however, the two lone pairs are placed at opposite vertices, as shown in Figure 9-16(c). There are no known AX_3E_3 molecules, but Figure 9-16(d) predicts that they would be T-shaped.

An example of an AX_6 molecule is SF_6, which has the predicted octahedral shape [Figure 9-16(a)]. The Lewis formula for the interhalogen compound BrF_5 is

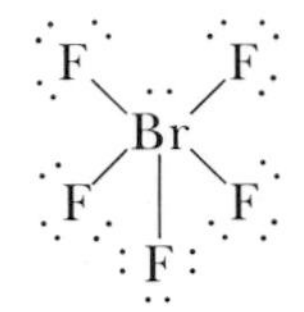

This shows that BrF_5 belongs to the class AX_5E. According to Figure 9-16(b), then, we predict that BrF_5 has a **square pyramidal** shape. The shape of BrF_5 is shown in Figure 9-17. The F—Br—F bond angles are slightly less than the ideal 90° because of the lone pair sitting at one vertex. The following Example shows that XeF_4 is an AX_4E_2 molecule and so is **square planar.**

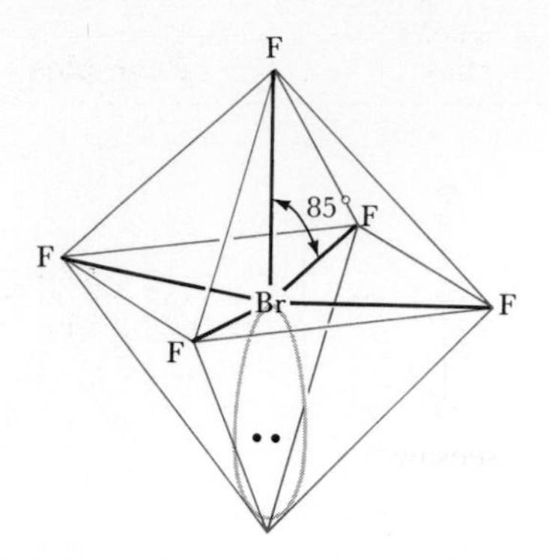

Figure 9-17 The shape of the interhalogen molecule bromine pentafluoride, BrF_5. The lone electron pair repels the bromine-fluorine bonds, causing the bromine atom to lie below the plane formed by four of the fluorine atoms. The BrF_5 molecule has a shape somewhat like an opened umbrella with its handle pointing upward.

Example 9-9: Xenon tetrafluoride is prepared by heating Xe and F_2 at 400°C at a pressure of 6 atm in a nickel container. The reaction is

$$Xe(g) + 2F_2(g) \rightarrow XeF_4(s)$$

Predict the shape of XeF_4.

Solution: The Lewis formula for XeF_4 is

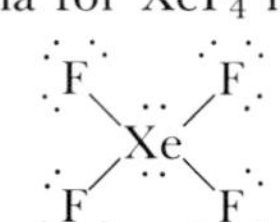

This molecule belongs to the class AX_4E_2, and so according to Figure 9-16(c), we predict that XeF_4 is a square planar molecule. This is indeed the observed structure of xenon tetrafluoride, as shown in Figure 9-18.

Table 9-3 summarizes all the cases that we have discussed in this chapter.

9-9. SYMMETRY DETERMINES WHETHER OR NOT A MOLECULE HAS A NET DIPOLE MOMENT

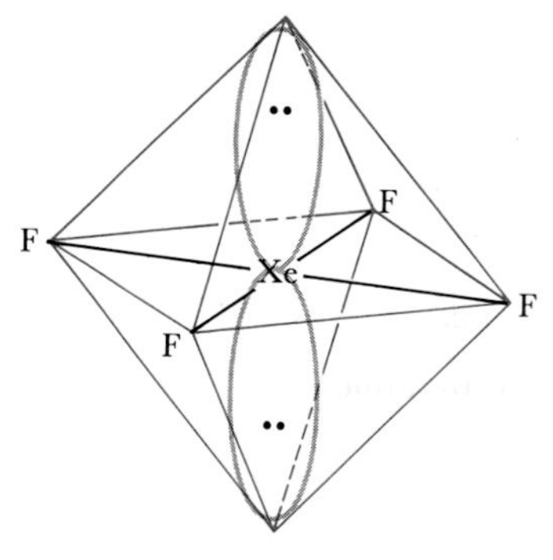

Figure 9-18 The geometry of xenon tetrafluoride, XeF_4, which belongs to the class AX_4E_2. The two lone electron pairs occupy opposite vertices of the octahedron, and so XeF_4 is a planar molecule.

We saw at the end of Chapter 8 that although both bonds in CO_2 are polar, the molecule itself does not have a dipole moment. Because CO_2 is a linear molecule, the two polar bonds, which can be treated as vector quantities, oppose each other exactly. Similar reasoning applies to a square planar molecule such as XeF_4. Although the four bonds in XeF_4 are polar, the square planar **symmetry** of the molecule

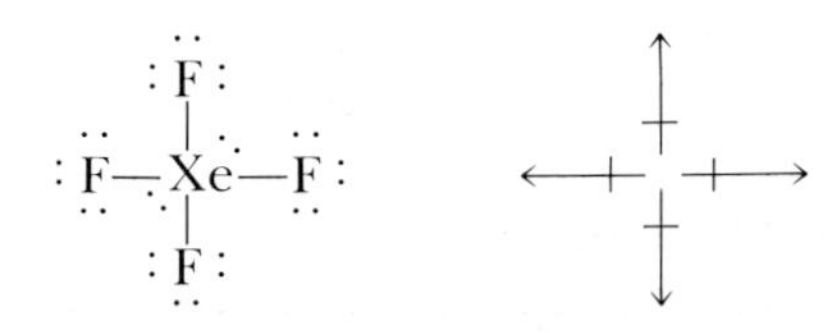

Table 9-3 A summary of molecular shapes

Molecular class	Ideal shape	Examples
AX_2	linear	CO_2, HCN, $BeCl_2$
AX_3	trigonal planar	SO_3, BF_3, NO_3^-, CO_3^{2-}
AX_2E	bent	SO_2, O_3, PbX_2, SnX_2 (where X is a halogen)
AX_4	tetrahedral	SiH_4, CH_4, SO_4^{2-}, ClO_4^-, PO_4^{3-}, XeO_4
AX_3E	trigonal pyramidal	NH_3, PF_3, $AsCl_3$, ClO_3^-, H_3O^+, XeO_3
AX_2E_2	bent	H_2O, OF_2, SF_2
AX_5	trigonal bipyramidal	PCl_5, AsF_5, SOF_4
AX_4E	seesaw	SF_4, XeO_2F_2, IF_4^+, $IO_2F_2^-$
AX_3E_2	T-shaped	ClF_3, BrF_3
AX_2E_3	linear	XeF_2, I_3^-, IF_2^-
AX_6	octahedral	SF_6, IOF_5,
AX_5E	square pyramidal	IF_5, TeF_5^-, $XeOF_4$
AX_4E_2	square planar	XeF_4, ICl_4^-

results in XeF_4 having no net dipole moment. These two examples show that even though a molecule has polar bonds, the symmetry of the molecule may cause it to have no net dipole moment. Such molecules are said to be **nonpolar molecules.** Some other examples of nonpolar molecules are

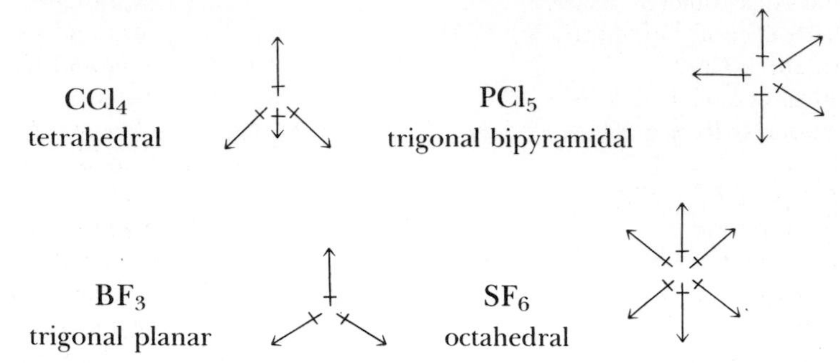

Some of these examples may not be immediately evident, but a study of the structures (Table 9-3) shows that in each case the symmetry of the directions of the polar bonds produces a zero net dipole moment.

Some examples of **polar molecules** (those possessing a net dipole moment) are

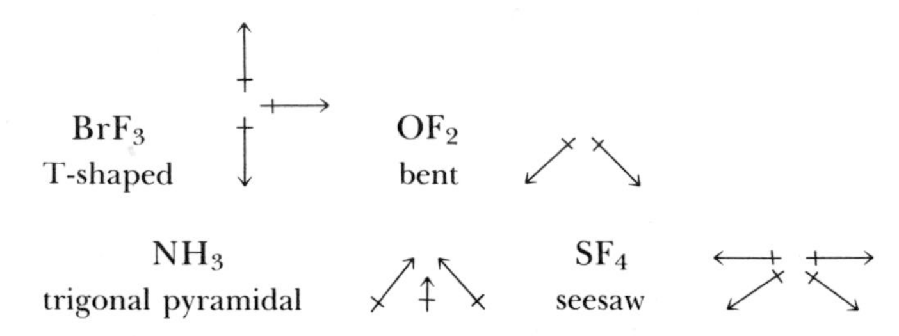

The VSEPR theory is simple and useful, and the shapes of thousands of molecules and ions can be predicted from it. As we mentioned before, VSEPR is a qualitative theory, in that it does not, in all cases, predict precise numerical values of bond angles and it cannot be used to predict bond lengths. Note, however, that although the theory cannot predict that the H—N—H bond angles in NH_3 are 107.3°, it can predict that they are slightly less than the ideal tetrahedral angle of 109.5°.

VSEPR theory does not give us much insight into the nature of chemical bonds. For example, it does not tell us why two atoms bond together in the first place. In Chapter 10 we shall study chemical bonding from a more fundamental point of view.

SUMMARY

The valence-shell electron-pair repulsion (VSEPR) theory is used primarily to predict the shapes of molecules in which there is a central atom bonded to ligands. The theory is based upon the premise that the valence-shell electron pairs around a central atom are arranged so as to minimize their mutual repulsion.

The procedure for using VSEPR theory to predict molecular shapes can be summarized as shown below:

1. Use the Lewis formula to determine the class AX_mE_n to which the molecule or ion belongs.
2. Given the class AX_mE_n to which the molecule or ion belongs, use Table 9-3 (page 303) to predict its shape.

TERMS YOU SHOULD KNOW

planar 289
geometric isomer 289
tetrahedron 289
space-filling molecular model 289
ball-and-stick molecular model 289
tetrahedral bond angle (109.5°) 290
tetravalent 290
structural chemistry 290
valence-shell electron-pair repulsion (VSEPR) theory 290
trigonal planar 292
trigonal bipyramid 293
equatorial vertex 293
axial vertex 293
octahedron 293
trigonal pyramid 294
ligand 295
AX_mE_n 295
linear 296
trigonal planar 296
bent 296
tetrahedral 296
trigonal pyramidal (tripod) 296
trigonal bipyramidal 296
seesaw 296
T-shaped 296
octahedral 296
square pyramidal 296
square planar 296
interhalogen compound 300
dipole moment 302
symmetry 302
nonpolar molecule 304
polar molecule 304

AN EQUATION YOU SHOULD KNOW HOW TO USE

lone-pair–lone-pair repulsion > lone-pair–bond-pair repulsion > bond-pair–bond-pair repulsion (9-1)

PROBLEMS

MOLECULES AND IONS INVOLVING ONLY SINGLE BONDS

9-1. Which of the following molecules have bond angles of 90°?

(a) TeF_6 (b) $AsBr_5$ (c) GaI_3 (d) XeF_4

9-2. Which of the following species have bond angles of 90°?

(a) NH_4^+ (b) PF_5 (c) AlF_6^{3-} (d) $SiCl_4$

9-3. Which of the following species have 120° bond angles?

(a) ClF_3 (b) $SbBr_6^-$ (c) $SbCl_5$ (d) $InCl_3$

9-4. Which of the following species have 180° bond angles?

(a) SeF_6 (b) BrF_2^- (c) SCl_2 (d) $SiCl_4$

9-5. Which of the following triatomic molecules are linear? Which are bent?

(a) TeF_2 (b) $SnBr_2$ (c) KrF_2 (d) OF_2

9-6. Which of the following triatomic ions are linear? Which are bent?

(a) NH_2^- (b) PF_2^+ (c) IF_2^+ (d) Br_3^-

9-7. Which of the following molecules are tetrahedral?

(a) XeF_4 (b) XeO_2F_2 (c) NF_3O (d) SeF_4

9-8. Which of the following molecules are trigonal pyramidal?

(a) NH_2Cl (b) ClF_3 (c) PF_3 (d) BF_3

9-9. Which of the following species are trigonal bipyramidal?

(a) BrF_5 (b) $SbCl_5$ (c) GeF_5^- (d) SF_5^-

9-10. Name the geometry (tetrahedral, seesaw, or square planar) that describes the shape of each of the following ions:

(a) IBr_4^- (b) PCl_4^+ (c) BF_4^- (d) IF_4^+

9-11. Indicate which, if any, of the listed bond angles occur in the following species:

(a) TeF_6 (1) 90°
(b) $SbCl_5$ (2) 109.5°
(c) ICl_4^- (3) 120°
(d) $InBr_3$

9-12. Indicate which, if any, of the listed bond angles occur in the following ions:

(a) BF_4^- (1) 90°
(b) SiF_6^{2-} (2) 109.5°
(c) SiF_3^+ (3) 120°
(d) $SnCl_6^{2-}$

9-13. Indicate which, if any, of the listed bond angles occur in the following ions:

(a) AlH_4^-	(1) 90°
(b) SbF_6^-	(2) 109.5°
(c) BrF_4^-	(3) 120°
(d) $AsCl_4^+$	

9-14. From the accompanying list, select the appropriate description(s) of the bond angles that occur in each of the following molecules:

(a) $GeCl_4$	(1) exactly 90°
(b) $SbCl_3$	(2) slightly less than 90°
(c) TeF_6	(3) exactly 109.5°
(d) SF_4	(4) slightly less than 109.5°
	(5) exactly 120°
	(6) slightly less than 120°
	(7) slightly greater than 120°
	(8) exactly 180°
	(9) slightly less than 180°

9-15. VSEPR theory has been successful in predicting the geometry of interhalogen molecules and ions. Predict the shapes of the interhalogen molecules given in Table 9-2.

9-16. Predict the shapes of the following iodofluorine ions:

(a) IF_2^+ (b) IF_6^+ (c) IF_4^+ (d) IF_4^-

MOLECULES OR IONS THAT MAY INVOLVE MULTIPLE BONDS

9-17. Write a Lewis formula for each of the following molecules and predict their shapes:

(a) $SeOCl_2$ (b) SO_2Cl_2 (c) SOF_4 (d) ClO_3F

9-18. Write a Lewis formula for each of the following species and predict their shapes:

(a) $XeOF_4$ (b) IOF_5 (c) $PO_2F_2^-$ (d) PO_3F^{2-}

9-19. Write a Lewis formula for each of the following species and predict their shapes:

(a) CCl_2O (b) NSF_3 (c) N_3^- (d) SbOCl

9-20. Write a Lewis formula for each of the following species and predict their shapes:

(a) $IO_2F_2^-$ (b) ClO_2^- (c) NOCl (d) NO_2Cl

9-21. Predict the shapes of the following ions:

(a) BrO_2^- (b) TeF_5^- (c) SO_3Cl^- (d) SF_3^+

9-22. Predict the shapes of the following molecules:

(a) NF_3O (b) GeO_2 (c) $AsOCl_3$ (d) XeO_2

9-23. Write Lewis formulas for the following molecules and predict their shapes:

(a) XeO_2F_4 (b) IO_2F_3 (c) IO_2F (d) IO_3F

9-24. Write Lewis formulas for the following ions and predict their shapes:

(a) TlF_4^- (b) IO_2^- (c) CS_3^{2-} (d) BrO_3^-

9-25. The species NO_2^+ and NO_2^- have O—N—O bond angles of 180° and 115°, respectively. Use VSEPR theory to explain the difference in bond angles.

9-26. Compare the shapes of the oxynitrogen ions

(a) NO_2^- (b) NO_3^- (c) NO_2^+ (d) NO_4^{3-}

MOLECULAR SHAPES AND DIPOLE MOMENTS

9-27. For each of the following molecules, write a Lewis formula and predict the shape. Indicate which ones have a dipole moment.

(a) XeF_2 (b) AsF_5 (c) $TeCl_4$ (d) Cl_2O

9-28. For each of the following molecules, write a Lewis formula and predict the shape. Indicate which ones have a dipole moment.

(a) $GeCl_4$ (b) SCl_2 (c) PoF_6 (d) BrF_3

9-29. For each of the following molecules, write a Lewis formula and predict the shape. Indicate which ones have a dipole moment.

(a) $GaCl_3$ (b) $TeCl_2$ (c) TeF_4 (d) $SbCl_5$

9-30. For each of the following molecules, write a Lewis formula and predict the shape. Indicate which ones have a dipole moment.

(a) TeF_6 (b) ClF_5 (c) $SiCl_4$ (d) $SeCl_2$

9-31. Predict which of these molecules are polar:

(a) CF_4 (b) AsF_3 (c) XeF_4 (d) SeF_4

9-32. Predict which of the following molecules are polar:

(a) $TeCl_4$ (b) BCl_3 (c) SF_6 (d) PCl_5

9-33. Describe the bond polarities in the following molecules

(a) nitrogen trifluoride, NF_3
(b) oxygen difluoride, OF_2
(c) oxygen dibromide, OBr_2

9-34. Describe the bond polarities in the following molecules.

(a) CCl_4 (b) PCl_3 (c) ClF_3

PREDICTING THE NUMBER OF GEOMETRIC ISOMERS

9-35. Describe the possible geometric isomers of

(a) a tetrahedral molecule AX_3Y
(b) a tetrahedral molecule AX_2YZ
(c) a square planar molecule AX_3Y
(d) a square planar molecule AX_2Y_2

where X and Y are different ligands.

9-36. The molecule $Pt(NH_3)_2Cl_2$ is square planar, with the platinum atom in the center of the square. How many isomers of $Pt(NH_3)_2Cl_2$ exist?

9-37. Describe the possible geometric isomers of an octahedral molecule whose formula is

(a) AX_5Y (b) AX_4Y_2 (c) AX_3Y_3

9-38. The species $Co(NH_3)_4Cl_2^+$ and $Co(NH_3)_3Cl_3$ are both octahedral, with the cobalt in the center. (a) How many isomers of $Co(NH_3)_4Cl_2^+$ are there? (b) How many isomers of $Co(NH_3)_3Cl_3$ are there?

9-39. The ions in the following series are octahedral, with the platinum in the center. How many isomers of each are there?

(a) $Pt(NH_3)_6^{4+}$
(b) $Pt(NH_3)_5Cl^{3+}$
(c) $Pt(NH_3)_4Cl_2^{2+}$
(d) $Pt(NH_3)_3Cl_3^+$

9-40. Describe the possible geometric isomers of a trigonal bipyramidal molecule whose formula is

(a) AX_4Y (b) AX_3Y_2 (c) AX_2Y_3

ADDITIONAL PROBLEMS

9-41. Name the geometry (tetrahedral, seesaw, or square planar) that describes the shape of each of the following halides:

(a) SF_4 (b) KrF_4 (c) CF_4 (d) $GeCl_4$

9-42. From the accompanying list, select the appropriate description(s) of the bond angles that occur in each of the following fluorides:

(a) SeF_6
(b) GeF_4
(c) BrF_3
(d) IF_5

(1) exactly 90°
(2) slightly less than 90°
(3) exactly 109.5°
(4) slightly less than 109.5°
(5) exactly 120°
(6) slightly less than 120°
(7) slightly greater than 120°
(8) exactly 180°
(9) slightly less than 180°

9-43. Give one example of each of the following:

(a) bent molecule
(b) bent ion
(c) tetrahedral ion
(d) octahedral molecule

9-44. Give one example of each of the following:

(a) trigonal planar molecule
(b) trigonal pyramidal molecule
(c) T-shaped molecule
(d) octahedral ion

9-45. VSEPR theory has been successful in predicting the molecular geometry of noble-gas compounds. Predict the molecular geometry of each of the following xenon species:

(a) XeO_3 (b) XeO_4 (c) XeO_2F_2 (d) XeO_6^{4-}

9-46. Predict the geometry of each of the following phosphorus-containing species:

(a) POF_3 (b) $POCl$ (c) PH_2^- (d) PCl_4^+

9-47. Compare the shapes of the following oxysulfur ions:

(a) sulfoxylate ion, SO_2^{2-}
(b) sulfite ion, SO_3^{2-}
(c) sulfate ion, SO_4^{2-}

9-48. Compare the shapes of the following oxychloro ions:

(a) chlorite, ClO_2^-
(b) chlorate, ClO_3^-
(c) perchlorate, ClO_4^-

9-49. Arrange the following groups of molecules in order of increasing dipole moment:

(a) HCl HF HI HBr (b) PH_3 NH_3 AsH_3
(c) ClF_3 BrF_3 IF_3 (d) H_2O H_2S H_2Te H_2Se

9-50. Which of the following molecules is trigonal pyramidal?

(a) SOF_2 (b) ClF_3 (c) NO_2Cl (d) BF_3

9-51. Which of the following fluorides has 90° bond angles?

(a) XeF_4 (b) CF_4 (c) SF_2 (d) XeF_2

9-52. Which of the following molecules is linear?

(a) $CdCl_2$ (b) O_3 (c) OCl_2 (d) NOF

9-53. Predict the shapes of the following covalent species involving transition metals:

(a) TiF_6^{2-} (b) VO_2^+
(c) $VOCl_3$ (d) CrO_4^{2-}

9-54. Predict the shapes of the following covalent species involving mercury:

(a) $HgCl_2$ (b) $HgCl_4^{2-}$ (c) $HgCl_3^-$

9-55. Predict the shapes of the following covalent species involving transition metals:

(a) $TiBr_5^-$ (b) MoF_6 (c) VF_6^- (d) MnO_4^-

9-56. Predict the shapes of the following bromofluoride ions:

(a) BrF_2^- (b) BrF_4^- (c) BrF_2^+ (d) BrF_4^+

9-57. Predict the shapes of the following oxyfluoro compounds of sulfur:

(a) SOF_4 (b) SOF_2 (c) SO_2F_2

Spectroscopy

Nuclear magnetic resonance spectrometers, like the one shown here, are commonly used to determine the structures of molecules. All chemistry departments with graduate research programs and chemical research laboratories have at least one NMR spectrometer.

In the previous chapter we learned how to predict the shapes of simple molecules. Chemists use a variety of techniques for determining the structures of molecules. Many of these techniques involve spectroscopy, which is based on the interaction of electromagnetic radiation with molecules to produce a characteristic spectrum. Spectroscopic methods are also routinely used in the identification of unknown substances, a problem that chemists face frequently. Such problems range from very simple to extremely complex. Fortunately, there are numerous powerful analytical methods available for determining the chemical composition of unknown substances. Many of these analytical methods establish the identity of chemical compounds by showing that certain characteristic properties of the unknown compound are identical to those of a known compound. The analytical methods we discuss in this interchapter are infrared spectroscopy, nuclear magnetic resonance spectroscopy, and mass spectrometry.

▪ The branch of chemistry that deals with the identification of unknown substances and the determination of the amounts of substances in samples is called **analytical chemistry.**

G-1. THE INFRARED SPECTRUM OF A COMPOUND GIVES CONSIDERABLE INFORMATION ABOUT ITS STRUCTURE

Many methods that are used to identify compounds involve the interaction of electromagnetic radiation with matter. In this section we discuss the interaction of infrared radiation with matter.

The infrared region of the electromagnetic spectrum lies between the visible and radio wave regions, that is, in the wavelength region between 10^{-6} and 10^{-4} m (Table 6-3). When a molecule is irradiated with infrared radiation of various wavelengths, some of the radiation is absorbed by the molecule. The energy of the absorbed radiation is stored in the molecule as **molecular vibrations,** that is, as vibrations of the various nuclei in the molecule. A plot of the amount of radiation absorbed versus the frequency or wavelength of the radiation is called a **spectrum**. If the radiation is in the infrared region, then the spectrum is called an **infrared spectrum.**

The infrared spectrum of an organic molecule gives a great deal of information about the structure of the molecule. Figure G-1 shows the infrared spectra of carbon tetrachloride and chloroform. The Lewis formulas for these two compounds are

```
      ..                    ..
     :Cl:                  :Cl:
 ..   |   ..                |   ..
:Cl — C — Cl:    and   H — C — Cl:
 ..   |   ..                |   ..
     :Cl:                  :Cl:
      ..                    ..
carbon tetrachloride     chloroform
```

Note from Figure G-1 that the substitution of one chlorine atom in CCl_4 by a hydrogen atom causes a dramatic change in the infrared spectrum. The absorption maximum (transmission minimum) that occurs around 13 μm in both spectra is due to the back-and-forth stretching motion of a carbon-chlorine bond. The absorption around 3.3 μm in the chloroform spectrum is due to the back-and-forth stretching motion of the carbon-hydrogen bond. The various vibrational motions

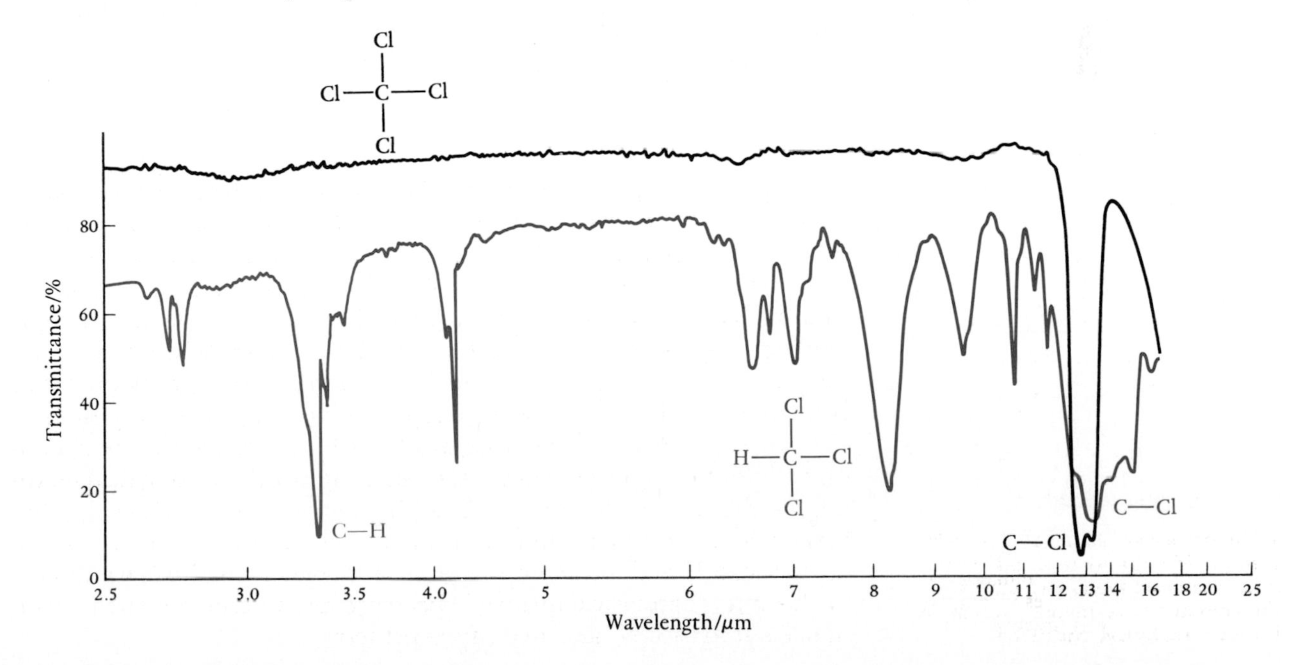

Figure G-1 The infrared spectra of carbon tetrachloride (upper curve) and chloroform (lower curve). The absorption of the infrared radiation at a given wavelength appears as a downward-pointing peak at that wavelength. The absorption is due to vibrational motion of the atoms in the molecule. The substitution of one chlorine atom in carbon tetrachloride by a hydrogen atom leads to a completely different infrared spectrum.

of groups of atoms in molecules lead to characteristic absorption of infrared radiation and can be used to verify the presence of certain bonds. For example, molecules that contain one or more carbon-hydrogen bonds absorb at around 3.3 μm.

The infrared spectra of three different compounds with the same chemical formula (C_4H_8O) are shown in Figure G-2; note that the spectra are very different. Each compound has a characteristic infrared spectrum that is determined by the types and arrangements of the bonds in the molecules. In effect, the infrared spectrum of a compound constitutes a "fingerprint" of the molecule. If the infrared spectrum of an unknown compound is known, then the compound can be identified by comparing its spectrum with the spectra of various known compounds. The infrared spectrum provides a variety of important clues to the structure of the molecule. Analytical methods that involve spectra are called **spectroscopic methods.**

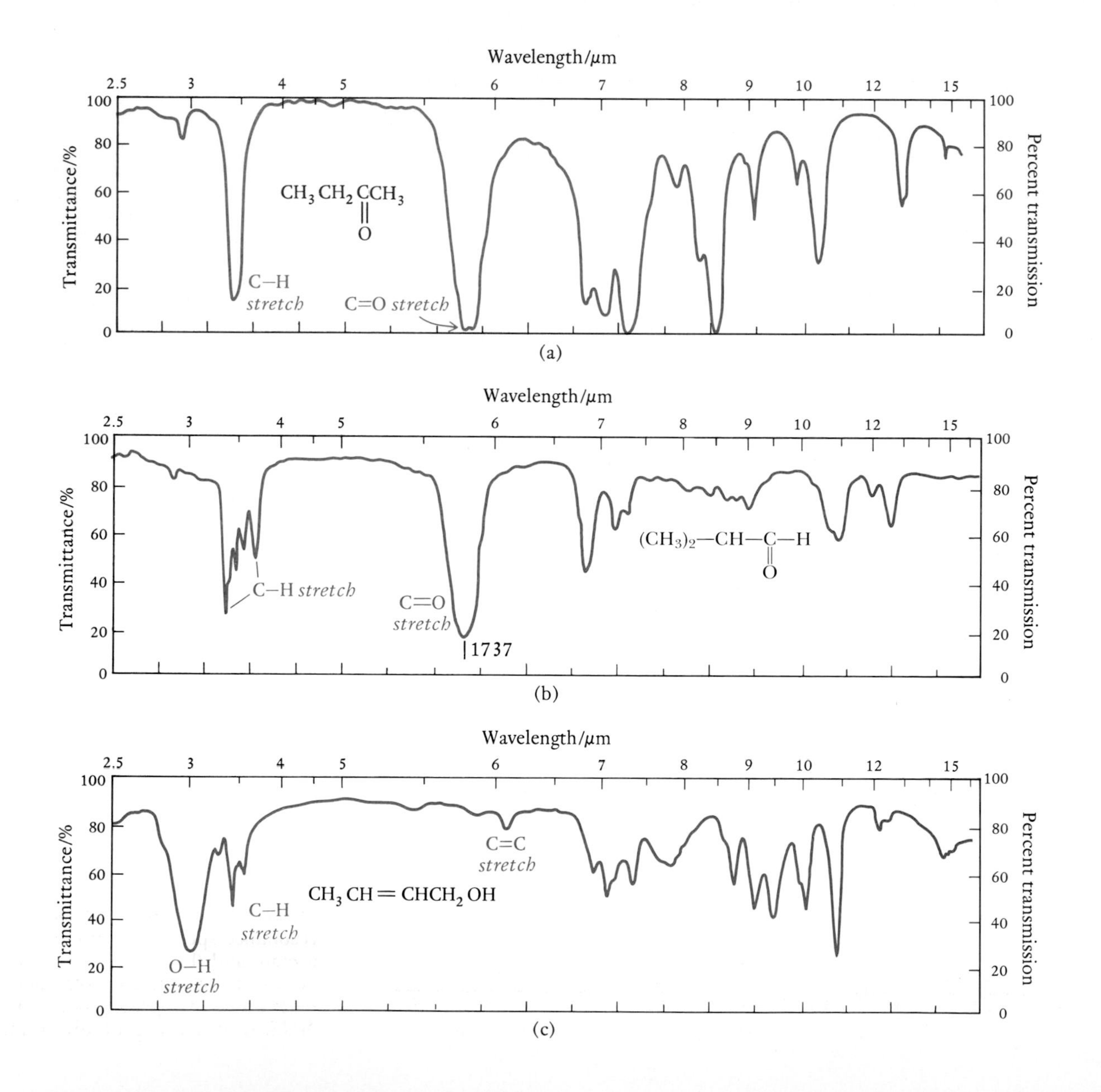

Figure G-2 The infrared spectra of three different compounds with the same chemical formula, C_4H_8O. Although the three compounds have the same chemical formula, their infrared spectra are easily distinguished, reflecting the different arrangements of the atoms in the molecules.

G-2. NUCLEAR MAGNETIC RESONANCE SPECTRA INDICATE THE POSITIONS OF HYDROGEN ATOMS IN A MOLECULE

When a sample of a compound containing hydrogen atoms is placed in a magnetic field and irradiated with electromagnetic energy in the radio wave region, the hydrogen nuclei absorb part of the radiation. This phenomenon is called **nuclear magnetic resonance** (NMR), and the resulting spectrum is called an **NMR spectrum.** The wavelength at which a hydrogen atom absorbs is strongly dependent on its neighboring atoms. A hydrogen atom in $CHCl_3$, for example, absorbs at a different wavelength than a hydrogen atom in $CHBr_3$. The absorption wavelength can be correlated with the position of a hydrogen atom in a molecule, so that an observed absorption can be identified with a particular hydrogen atom. Figure G-3 shows the NMR spectra of two molecules with the same chemical formula, $C_3H_6O_2$. The Lewis formulas and names of these two compounds are

▪ In practice, the wavelength of the electromagnetic radiation is fixed, and the strength of the magnetic field is varied in an NMR spectrometer.

```
   H  :O:     H                  :O:    H  H
   |  ||  ..  |                   ||  ..  |  |
H—C—C—O—C—H        and        H—C—O—C—C—H
   |      ..  |                       ..  |  |
   H          H                         H  H
   methyl acetate                 ethyl formate
```

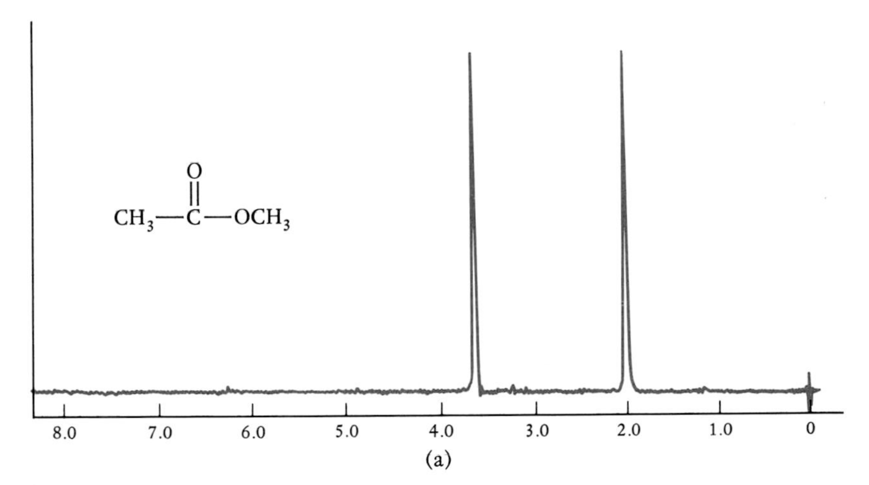

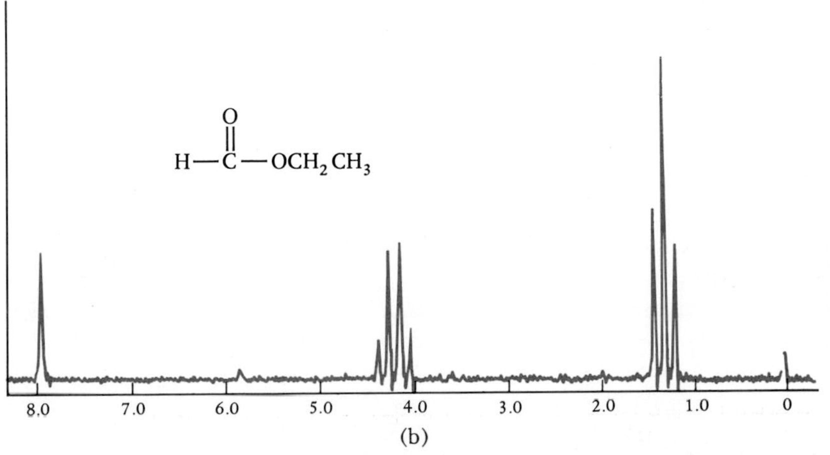

Figure G-3 The nuclear magnetic resonance spectra of (a) methyl acetate and (b) ethyl formate. The empirical formula for both these compounds is $C_3H_6O_2$, but their NMR spectra are quite different.

Note that the two spectra, and hence the two compounds, are easily distinguished from each other.

Figure G-4 shows the NMR spectra of two isomeric compounds, whose Lewis formulas are

```
   :Cl: H                    H  H
    |   |                ..  |  |  ..
H — C — C — H    and    :Cl — C — C — Cl:
    |   |                ¨¨  |  |  ¨¨
   :Cl: H                    H  H
```

These two molecules have the same chemical formula ($C_2H_4Cl_2$) and differ only in whether the two chlorine atoms are attached to the same carbon atom or to different carbon atoms. Yet their NMR spectra are easily distinguished from each other. With just a little experience, it is possible to interpret NMR spectra and deduce the positions of hydrogen atoms (and even other atoms) in molecules, and so learn a great

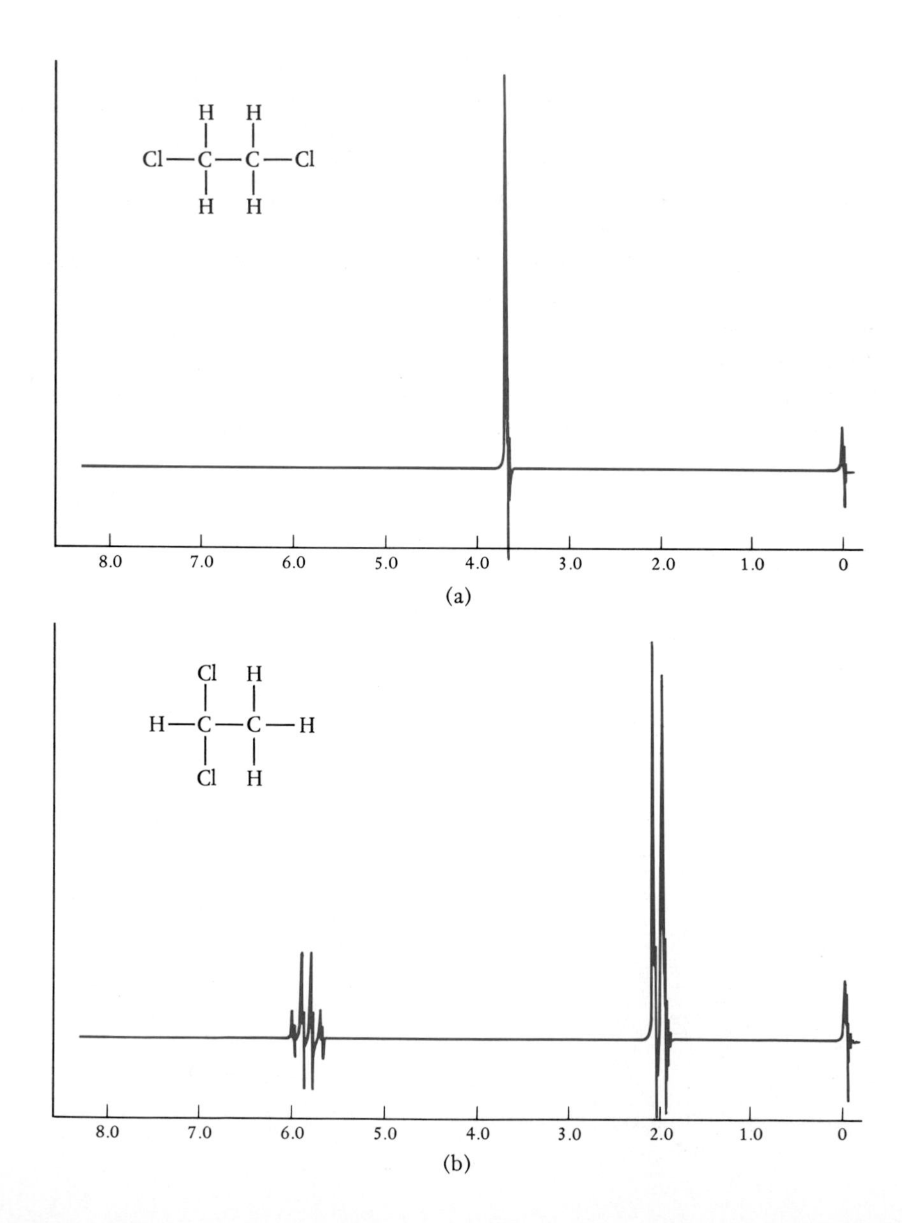

Figure G-4 The nuclear magnetic resonance spectra of two isomers, each with the formula $C_2H_4Cl_2$. These two very different spectra suggest that NMR is a sensitive probe of the structure of a molecule.

deal about molecular structure. NMR spectroscopy is the most widely used analytical technique in organic chemistry.

G-3. COMPOUNDS CAN BE IDENTIFIED BY THEIR CHARACTERISTIC MASS FRAGMENTATION PATTERNS

If electrons of sufficiently high energy are directed at molecules in the gas phase, then the molecules are ionized to positive ions. For example,

$$\underset{C_5H_{12}}{CH_3-\overset{\displaystyle CH_3}{\underset{\displaystyle CH_3}{\overset{|}{\underset{|}{C}}}}-CH_3} + e^- \text{ (high-energy)} \rightarrow \left[\underset{C_5H_{12}^{+}}{CH_3-\overset{\displaystyle CH_3}{\underset{\displaystyle CH_3}{\overset{|}{\underset{|}{C}}}}-CH_3}\right]^{\oplus} + 2e^-$$

The molecular ion $C_5H_{12}^{+}$ is unstable and fragments into several other ions:

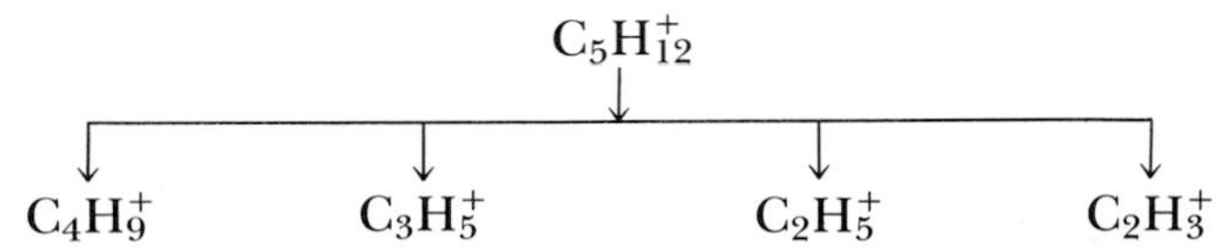

The various molecular ions produced in the electron bombardment are accelerated and colimated by an electric field (positive ions move toward the negative plate) and then passed into a magnetic field (Figure G-5). The ion beam, which constitutes an electric current, is deflected by the magnetic field. The greater the mass of an ion, the less its deflection in the magnetic field. By varying the magnitude of the electric or the magnetic field, the relative proportions of ions of a particular mass number are measured as an electric current by a detector.

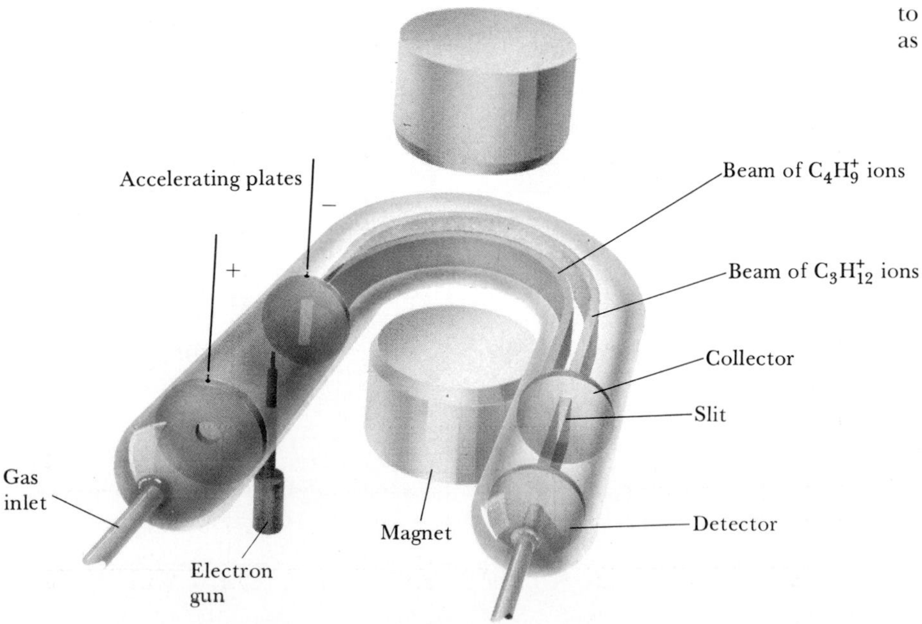

Figure G-5 A schematic drawing of a mass spectrometer. Gas molecules are ionized by electron bombardment, and the ions are accelerated and colimated by an electric field. The ion beam is passed through a magnetic field, where it is resolved into component beams of ions of equal mass. Light ions are deflected more strongly than heavy ions by the magnetic field. In a beam containing $C_5H_{12}^{+}$ and $C_4H_9^{+}$ ions, the lighter $C_4H_9^{+}$ ions are deflected more than the heavier $C_5H_{12}^{+}$ ions. The mass spectrometer depicted here is adjusted to detect the $C_5H_{12}^{+}$ ions. By changing the magnitude of the magnetic or electric field, the beam of $C_4H_9^{+}$ can be moved to strike the collector at the slit, where it would then pass through to the detector and be measured as a current.

A mass spectrometer fragmentation pattern, that is, a **mass spectrum,** is characteristic of a particular compound. An unknown compound often can be identified by comparison of its mass spectrum with tabulated mass spectra of known compounds. If two mass spectra match peak for peak (mass numbers and relative peak heights), then it is essentially certain that the two compounds are identical. Matching mass spectra of compounds is another example of molecular identification by a "fingerprint" method.

Figure G-6 shows the mass spectra of two compounds with the chemical formula C_8H_{18}, whose structures are

```
  H  H  H  H  H  H  H  H                       H        H
  |  |  |  |  |  |  |  |                       |        |
H—C—C—C—C—C—C—C—C—H                         H—C—H    H—C—H
  |  |  |  |  |  |  |  |                 H     |    H   |     H
  H  H  H  H  H  H  H  H       and       |     |    |   |     |
                                       H—C—————C————C———C—————C—H
                                         |     |    |   |     |
                                         H     |    H   H     H
                                             H—C—H
                                               |
                                               H
```

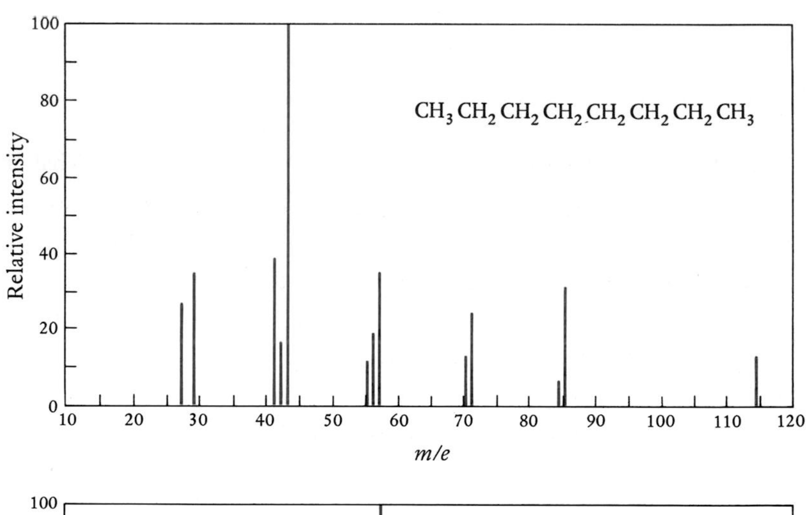

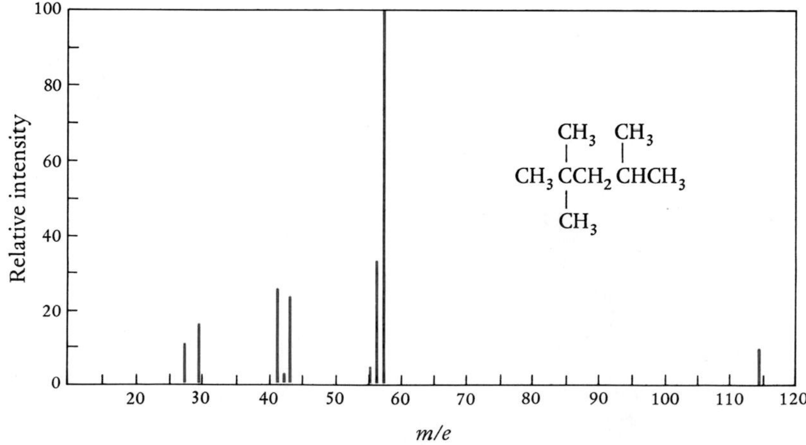

Figure G-6 Mass spectra of two molecules that have the same empirical formula, C_8H_{18}. The horizontal axis is the mass-to-charge ratio of a fragment.

Note how different the two spectra are even though the two compounds have the same total mass. The different molecular structures fragment in different ways.

If a compound is unknown, then its mass spectrum is useful for obtaining the molecular mass (which would be the largest observed mass value) and details about the molecular structure. This structural information is obtained by comparing the fragmentation pattern with accumulated knowledge about how various types of molecules tend to fragment.

Mass spectrometry has many practical applications. For example, it is used to identify compounds in the blood of people found unconscious as a result of ingestion of toxic substances.

TERMS YOU SHOULD KNOW

analytical chemistry 308
molecular vibrations 309
spectrum 309
infrared spectrum 309
spectroscopic method 310
nuclear magnetic resonance (NMR) 311
NMR spectrum 311
mass spectrum 314
mass spectrometry 315

QUESTIONS

G-1. What is an infrared spectrum?

G-2. Describe how infrared spectroscopy can be used to distinguish two substances from each other.

G-3. What type of molecular motion leads to the absorption of infrared radiation?

G-4. What do the initials NMR stand for?

G-5. Why do you think that nuclear magnetic resonance is sometimes called proton magnetic resonance?

G-6. What is a mass spectrum?

G-7. How can mass spectrometry be used to identify an unknown compound?

G-8. Describe how a mass spectrometer works (see Figure G-5).

COVALENT BONDING IN MOLECULES

Linus Pauling, one of America's greatest chemists, was a pioneer in the theory and understanding of chemical bonding. His book, *The Nature of the Chemical Bond,* first published in 1939, is one of the most influential chemistry texts of the twentieth century. During the 1950s, Pauling was in the forefront of the fight against nuclear bomb testing. He was awarded the Nobel Prize for chemistry in 1954 and the Nobel Peace Prize in 1963.

We learned in Chapter 6 that the electrons in atoms are described in terms of atomic orbitals. In this chapter we shall learn that the electrons in molecules are described in terms of molecular orbitals. Polyatomic molecules can be pictured as a group of atoms held together by covalent bonds. When covalent bonds are localized between pairs of atoms, they are called localized covalent bonds. We shall learn how to describe localized covalent bonds in terms of localized bond orbitals, which are formed by combining atomic orbitals. This description of the bonding in molecules is a simplified version of valence-bond theory, which was developed in the 1930s by Linus Pauling. Valence-bond theory makes it possible to translate Lewis formulas into the mathematical formulas of the quantum theory. In a polyatomic molecule, the localized bond orbitals are occupied by valence electrons and account for the bonding in the molecule. The atomic orbitals we use to construct bond orbitals allow us to describe the geometry of molecules in terms of their bond orbitals. Chemical bonding and molecular geometry are closely related.

In the final sections of the chapter we present a theory of bonding that serves as a basis for localized bond orbitals. We apply this theory, called molecular orbital theory, to the homonuclear diatomic molecules H_2 through Ne_2, learning how to write electron configurations for these molecules and how to predict relative bond lengths and bond energies.

10-1. A MOLECULAR ORBITAL IS A COMBINATION OF ATOMIC ORBITALS ON DIFFERENT ATOMS

The simplest neutral molecule is H_2, which has only two electrons. The Schrödinger equation that applies to H_2 can be solved with a computer. The results for H_2 are interesting because they are similar to the results for more complicated molecules. As a first step in setting up the Schrödinger equation for H_2, the two nuclei are fixed at some given separation. Then the two electrons are included, and the equation is solved to give the wave functions and energies that describe the two electrons. The wave function that corresponds to the lowest energy is the **ground-state wave function** and can be used to compute contour diagrams that show the distribution of the electron density around the two nuclei.

Figure 10-1 shows ground-state electron density contour diagrams as a function of the internuclear separation of two hydrogen atoms. Note that at large separations the two atoms hardly interact with each other and so the electron density is just that of two electrons, each in a 1*s* orbital about each of the hydrogen atoms. As the separation decreases, however, the two 1*s* orbitals combine into one orbital that is distributed around both nuclei. Such an orbital is called a **molecular orbital** because it extends over both nuclei in the molecule. The buildup of electron density between the nuclei results in a covalent bond that attracts

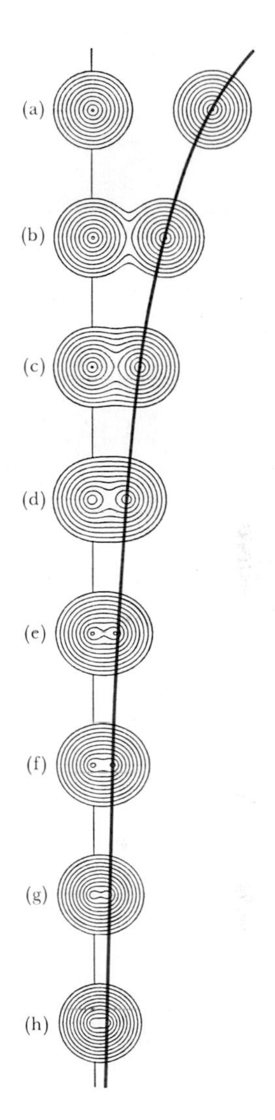

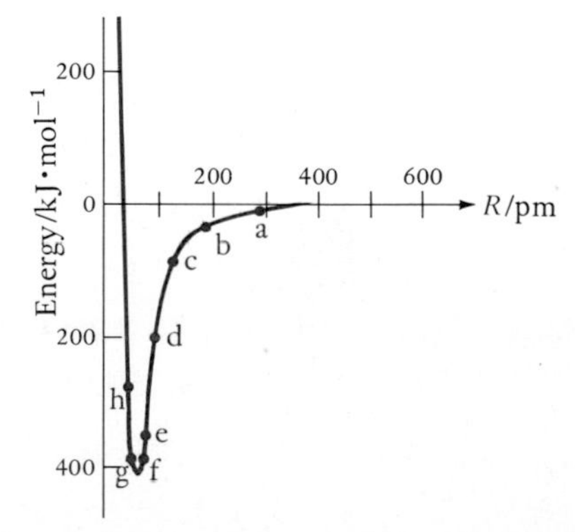

Figure 10-1 Electron density contour diagrams of two hydrogen atoms as a function of their separation (upper part). At large separations, as in (a), the two orbitals appear simply as those of two separate atoms. As the atoms come together, the two separate atomic orbitals combine into one molecular orbital encompassing both nuclei, as in (b) through (h). The lower part shows the energy of two hydrogen atoms as a function of their separation *R*. The labels (a) through (h) correspond to those in the upper part of the figure. At large distances (a), the two hydrogen atoms do not interact and so their interaction energy is zero. As the two come together, they attract each other and so their interaction energy becomes negative. When they come less than 74 pm apart, the interaction energy increases and they repel each other. The bond length of H_2 is the distance at which the energy is a minimum, that is, 74 pm. The energy at this distance is $-436\ kJ\cdot mol^{-1}$, which is the energy required to dissociate H_2 into two separate hydrogen atoms.

the two nuclei together. Note how the detailed quantum theoretical results shown in Figure 10-1 correspond to our notion that a covalent bond results when an electron pair is shared between two nuclei.

The lower part of Figure 10-1 shows the energies that correspond to the electron densities of the upper part. Notice that interaction energies have negative values for any separations at which the atoms attract each other. These negative values mean that energy is released when the H_2 bond is formed. The graph shows that, for H_2, the interaction energy versus separation R has a minimum at $R = 74$ pm. The internuclear separation at the minimum energy is the predicted length of a bond. The experimentally measured value of the bond length in H_2 is 74 pm, in excellent agreement with the calculated value.

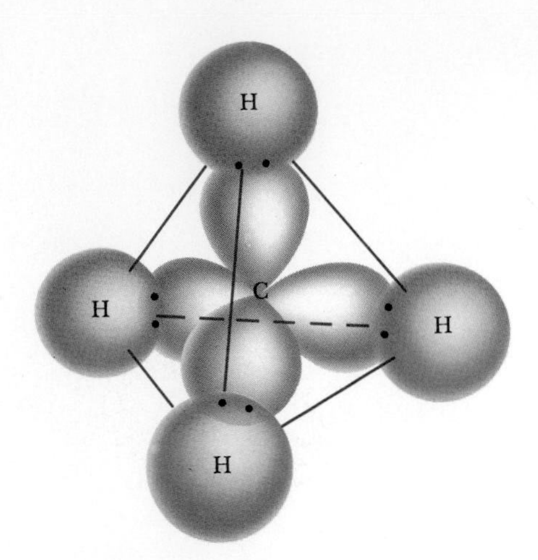

Figure 10-2 The bond orbitals in a methane molecule can be pictured as four carbon-hydrogen bond orbitals, directed toward the vertices of a tetrahedron. A localized bond orbital that is occupied by two electrons with opposite spins constitutes a covalent bond localized between two atoms.

10-2. THE BONDING IN POLYATOMIC MOLECULES CAN BE DESCRIBED IN TERMS OF BOND ORBITALS

Most molecules with more than two atoms, called **polyatomic molecules,** can be viewed as a group of atoms that are held together by covalent bonds. As a first approximation, the bonding in many polyatomic molecules can be analyzed in terms of orbitals that are localized between pairs of bonded atoms.

For example, consider the methane molecule, CH_4. Each hydrogen atom is joined to the central carbon atom by a covalent bond formed from an atomic orbital on carbon and the 1*s* orbital on hydrogen. As Figure 10-2 suggests, the bonding electrons, and hence the orbitals that describe them, are localized along the carbon-hydrogen bonds. The orbitals that describe the electrons in localized covalent bonds, such as those in methane, are called **localized bond orbitals** and are concentrated primarily in the region between the two atoms that are joined by the bond. The two electrons that occupy a localized bond orbital constitute a localized covalent bond. Note the similarity between the bonding picture in CH_4 and the four bonding electron pairs in the Lewis formula for CH_4:

```
    H
    |
H—C—H
    |
    H
```

The localized bond orbital approach that we are describing is a simplified version of the **valence-bond theory** of molecular bonding, which was developed by Linus Pauling in the 1930s. The valence-bond theory makes it possible to translate Lewis formulas into the mathematical formulas of the quantum theory. We shall often use Lewis formulas as a guide to constructing bond orbitals.

10-3. HYBRID ORBITALS ARE COMBINATIONS OF ATOMIC ORBITALS ON THE SAME ATOM

The simplest neutral polyatomic molecule is BeH_2. Beryllium hydride is an example of an electron-deficient compound, which means that its

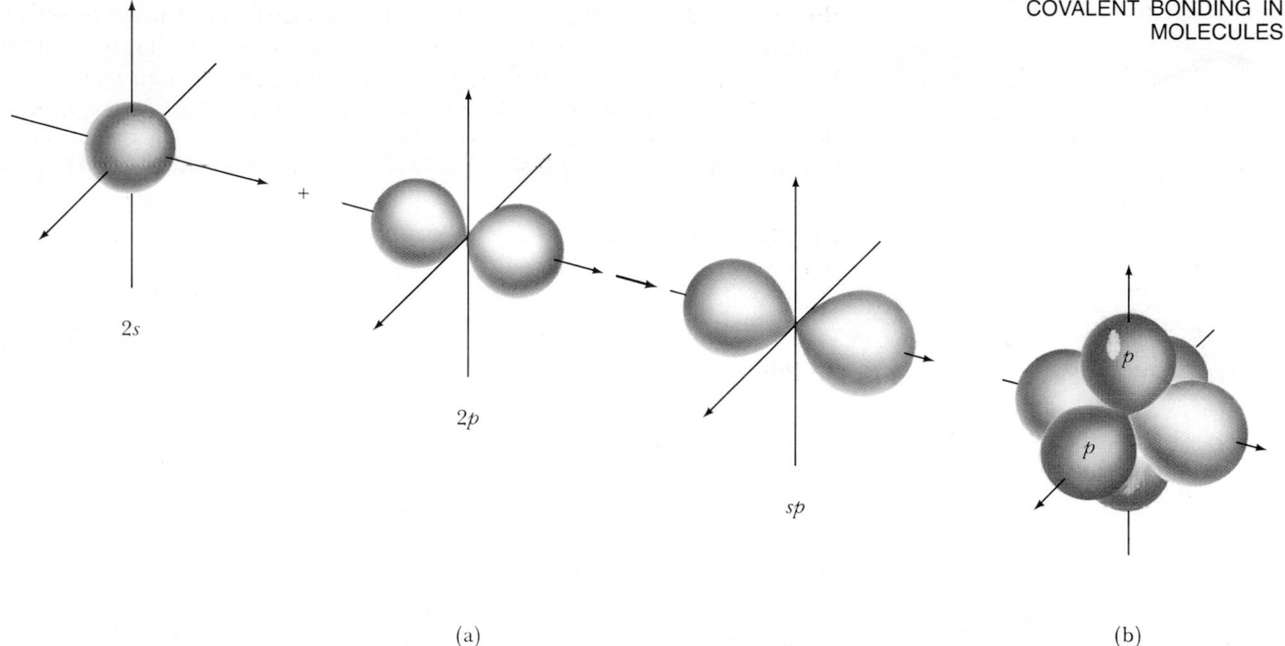

Lewis formula, H—Be—H, does not satisfy the octet rule. Beryllium hydride is a symmetric linear molecule; the two Be—H bonds are directed 180° from each other and are equivalent. To describe the bonding in BeH_2, we must form two equivalent localized bond orbitals.

Each of the localized bond orbitals in BeH_2 can be constructed by combining the valence orbitals on the beryllium atom with the 1*s* orbital of one of the hydrogen atoms. The valence orbitals in a beryllium atom are the 2*s*, $2p_x$, $2p_y$, and $2p_z$ orbitals. We must determine how we can use the valence orbitals on the beryllium atom to describe the two equivalent localized bond orbitals in BeH_2. The solution to this problem is to use the combination of the valence orbitals of the beryllium atom that produces the greatest overlap with each of the 1*s* hydrogen orbitals. This idea was proposed by Linus Pauling in 1931 and is motivated by the results in Figure 10-1, which suggest that a covalent bond orbital in H_2 is formed because the two 1*s* hydrogen orbitals combine, or **overlap.** Pauling was first to use the **principle of maximum overlap** of orbitals to explain the bonding in molecules. The idea of maximizing the overlap of orbitals is convenient because it is pictorial, and chemists often view chemical bonds in terms of overlapping orbitals.

In forming BeH_2, what combination of the valence orbitals of the beryllium atom provides the maximum overlap with the hydrogen 1*s* orbitals? There is a mathematical procedure for solving this problem, and the answer is to take only the 2*s* orbital and one of the 2*p* orbitals. This produces two equivalent orbitals on the beryllium atom, which are shown in Figure 10-3. These orbitals have two important features: (1) each one provides a large region to overlap with a hydrogen 1*s* orbital, and (2) they are directed 180° from each other. The two equivalent orbitals on the beryllium atom are called ***sp* orbitals** because they are formed from the 2*s* orbital and one 2*p* orbital. Orbitals that are

Figure 10-3 The formation of *sp* hybrid orbitals results from combining the 2*s* orbital and one 2*p* orbital on an atom. The two *sp* orbitals are equivalent and are directed 180° from each other. In (a), for simplicity, only the 2*p* orbital that is combined with the 2*s* orbital is shown. In (b), all three 2*p* orbitals are shown. The two 2*p* orbitals that are not combined with the 1*s* orbital are perpendicular to each other and to the line formed by the *sp* orbitals.

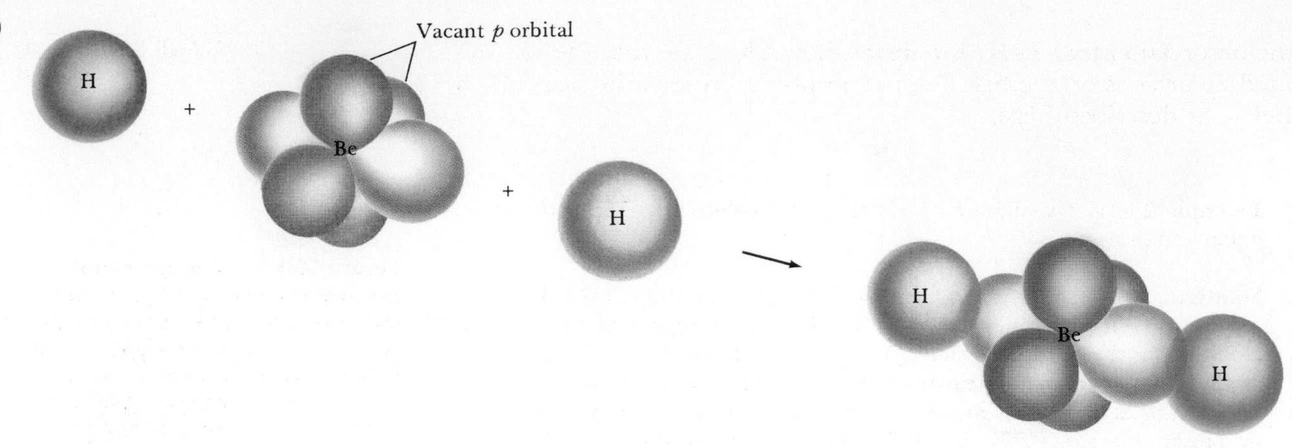

Figure 10-4 The formation of the two equivalent localized σ bond orbitals in BeH_2. Each bond orbital is formed by the overlap of a beryllium *sp* orbital and a hydrogen 1*s* orbital. There are four valence electrons in BeH_2: two from the beryllium atom and one from each of the two hydrogen atoms. The four valence electrons occupy the two localized bond orbitals, forming the two localized beryllium-hydrogen bonds in BeH_2.

made up of two different types of atomic orbitals are often called **hybrid atomic orbitals.** The two 2*p* orbitals that are not used to form the hybrid *sp* orbitals are perpendicular to each other and to the line formed by the *sp* orbitals (Figure 10-3b).

Each beryllium *sp* orbital is combined with (overlaps) a hydrogen 1*s* orbital to form two equivalent localized bond orbitals, as shown in Figure 10-4. Each beryllium-hydrogen bond orbital is cylindrically symmetric; that is, the bond orbital has a circular cross section when viewed along a line between the hydrogen and beryllium nuclei. Because atomic *s* orbitals have circular cross sections, bond orbitals or molecular orbitals that have circular cross sections when viewed along an internuclear axis are called ***σ* orbitals** (sigma orbitals; σ is the Greek letter corresponding to *s*). The beryllium-hydrogen bond orbitals in beryllium hydride are σ orbitals.

We can denote the two σ bond orbitals in BeH_2 by σ_1 and σ_2. There is a total of four valence electrons in BeH_2. Following the usual procedure of filling orbitals with electrons two at a time according to the Pauli exclusion principle, each bond orbital is occupied by two valence electrons of opposite spins. When a σ orbital is occupied by two electrons of opposite spins, the result is called a ***σ* bond.** We can write the valence-shell electron configuration of BeH_2 as $(\sigma_1)^2(\sigma_2)^2$, and because both σ_1 and σ_2 represent localized bond orbitals, the electron configuration $(\sigma_1)^2(\sigma_2)^2$ describes the two covalent bonds in BeH_2. The fact that two bond orbitals turn out to be 180° from each other agrees with

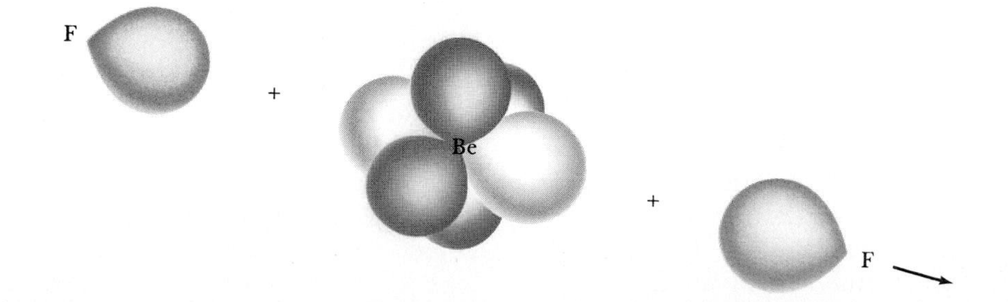

Figure 10-5 An illustration of the bonding in BeF_2. Each of the beryllium-fluorine σ bond orbitals is formed by the overlap of a beryllium *sp* orbital and a fluorine 2*p* orbital. The two localized σ bond orbitals are occupied by four of the valence electrons, in accord with the Pauli exclusion principle, to form two σ bonds.

the observation that BeH_2 is a linear molecule. Note that the two unused $2p$ orbitals on the beryllium atom play no role in the bonding in BeH_2, as described here.

Example 10-1: Describe the localized bond orbitals in BeF_2, which is a linear molecule.

Solution: Because BeF_2 is a linear molecule, we can use *sp* hybrid orbitals on the beryllium atom to form localized bond orbitals with the fluorine atoms. Thus the central beryllium atom in BeF_2 can be depicted as in Figure 10-4. The ground-state electron configuration of a fluorine atom is $1s^22s^22p_x^22p_y^22p_z^1$. The only orbital on the fluorine atom that can be occupied by another electron is the $2p_z$ orbital. The $2p_z$ valence-shell orbital on each fluorine atom can overlap with an *sp* orbital on the beryllium atom to form two localized σ bond orbitals, as shown in Figure 10-5. The beryllium *sp* orbitals and the fluorine $2p_z$ orbitals join colinearly in order to maximize their overlap. Each bond orbital is occupied by two electrons of opposite spin to form two σ bonds.

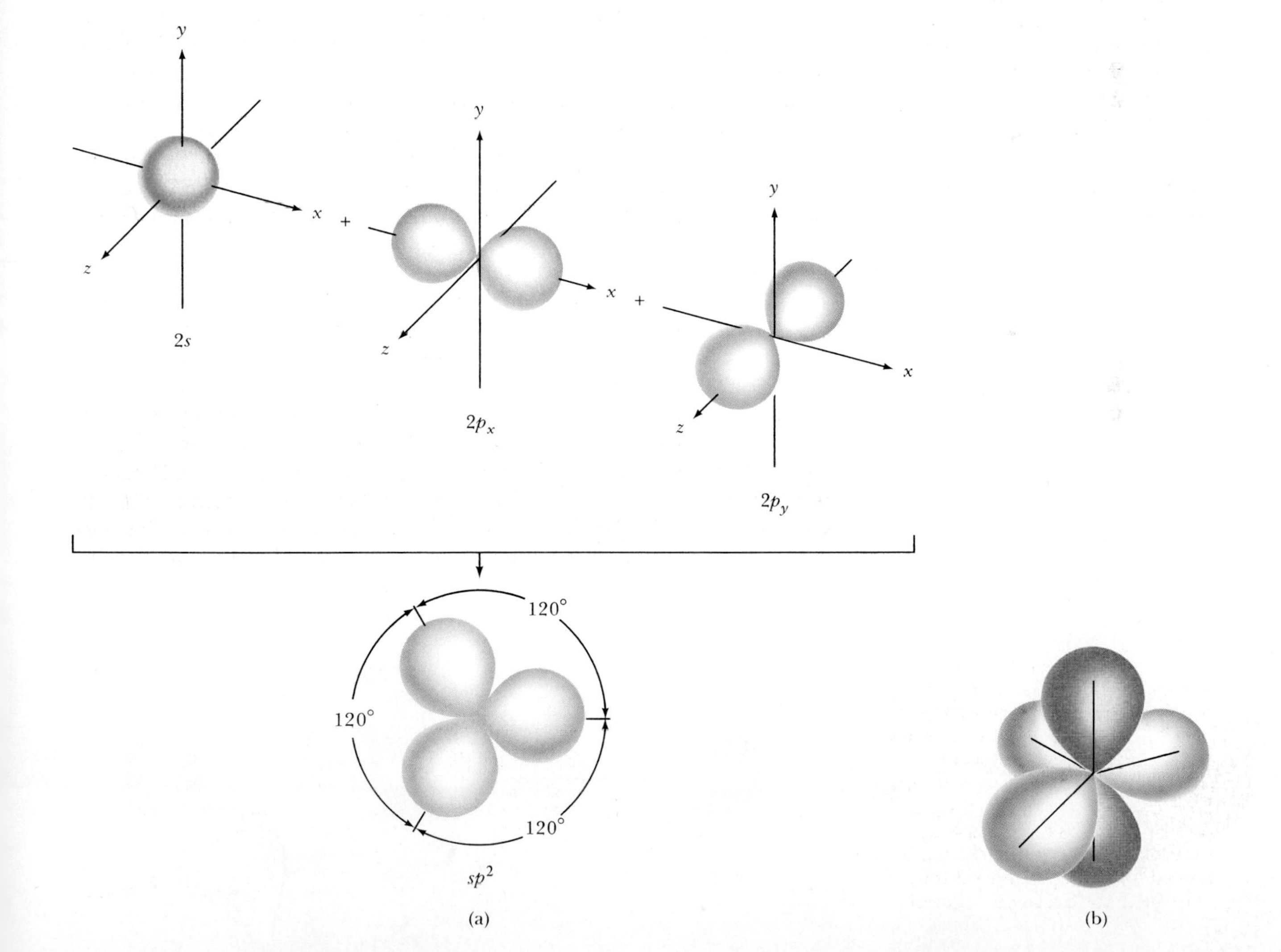

Figure 10-6 (a) The formation of sp^2 hybrid orbitals by combining the $2s$ orbital and two $2p$ orbitals on an atom. The three sp^2 orbitals formed are equivalent, lie in a plane, and are directed 120° from each other. For simplicity, only the two $2p$ orbitals that are combined with the $2s$ orbital are shown. (b) The $2p$ orbital that is not combined with the $2s$ orbital is perpendicular to the plane formed by the three sp^2 orbitals.

10-4. sp^2 HYBRID ORBITALS HAVE TRIGONAL PLANAR SYMMETRY

An example of a molecule with three localized covalent bonds is BF_3. Boron trifluoride is an electron-deficient compound whose Lewis formula is

$$\begin{array}{ccc} :\ddot{F}: & & :\ddot{F}: \\ & \diagdown \; \diagup & \\ & B & \\ & | & \\ & :\underset{..}{F}: & \end{array}$$

We have written the Lewis formula of BF_3 as we have because, according to Chapter 9, BF_3 is trigonal planar, with each F—B—F bond angle being 120°. The three B—F bonds are equivalent, and so we must construct three equivalent bond orbitals with trigonal planar symmetry. The valence-shell orbitals on a boron atom are $2s$, $2p_x$, $2p_y$, and $2p_z$. It turns out that if we combine the $2s$ orbital with two of the $2p$ orbitals, then three equivalent hybrid orbitals result that lie in a plane at 120° angles (Figure 10-6). Because the hybrid orbitals are formed from the $2s$ and two of the $2p$ orbitals, they are called ***sp*² orbitals.**

We now use the three sp^2 hybrid orbitals on the boron atom to form three equivalent localized bond orbitals with the three fluorine atoms by overlapping each sp^2 orbital on the boron atom with a $2p$ orbital on a fluorine atom. These resulting bond orbitals are cylindrically symmetric and so are σ orbitals. Each localized σ bond orbital is occupied by two electrons of opposite spin to form a σ bond. The bonding in BF_3 is shown in Figure 10-7. The remaining $2p$ orbital on the boron atom is not used in forming the boron-fluorine σ bonds.

10-5. sp^3 HYBRID ORBITALS POINT TOWARD THE VERTICES OF A TETRAHEDRON

As predicted by VSEPR theory, methane, CH_4, is a tetrahedral molecule, and its four carbon-hydrogen bonds are equivalent. Thus we must construct four equivalent bond orbitals to describe the bonding in methane. If we combine the $2s$ orbital and all three $2p$ orbitals on the carbon atom, we get four equivalent hybrid orbitals that point to the vertices of a tetrahedron (Figure 10-8). Because the four equivalent hybrid orbitals result from combining the $2s$ and all three $2p$ orbitals on the carbon atom, they are called ***sp*³ orbitals.**

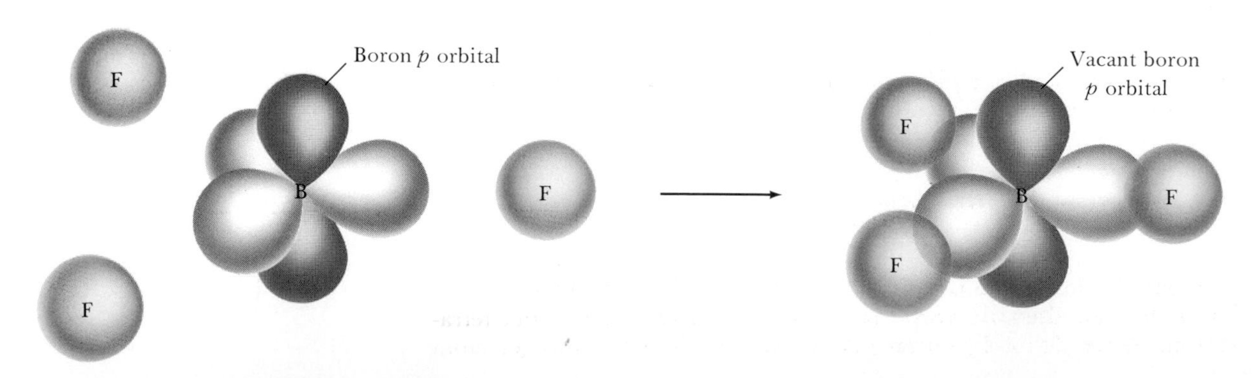

Figure 10-7 A schematic illustration of the bonding in BF_3. Each of the three boron-fluorine bond orbitals is formed by the overlap of a boron sp^2 orbital and a fluorine $2p$ orbital. The three localized boron-fluorine bond orbitals are occupied by six of the valence electrons and constitute the three covalent boron-fluorine bonds in BF_3.

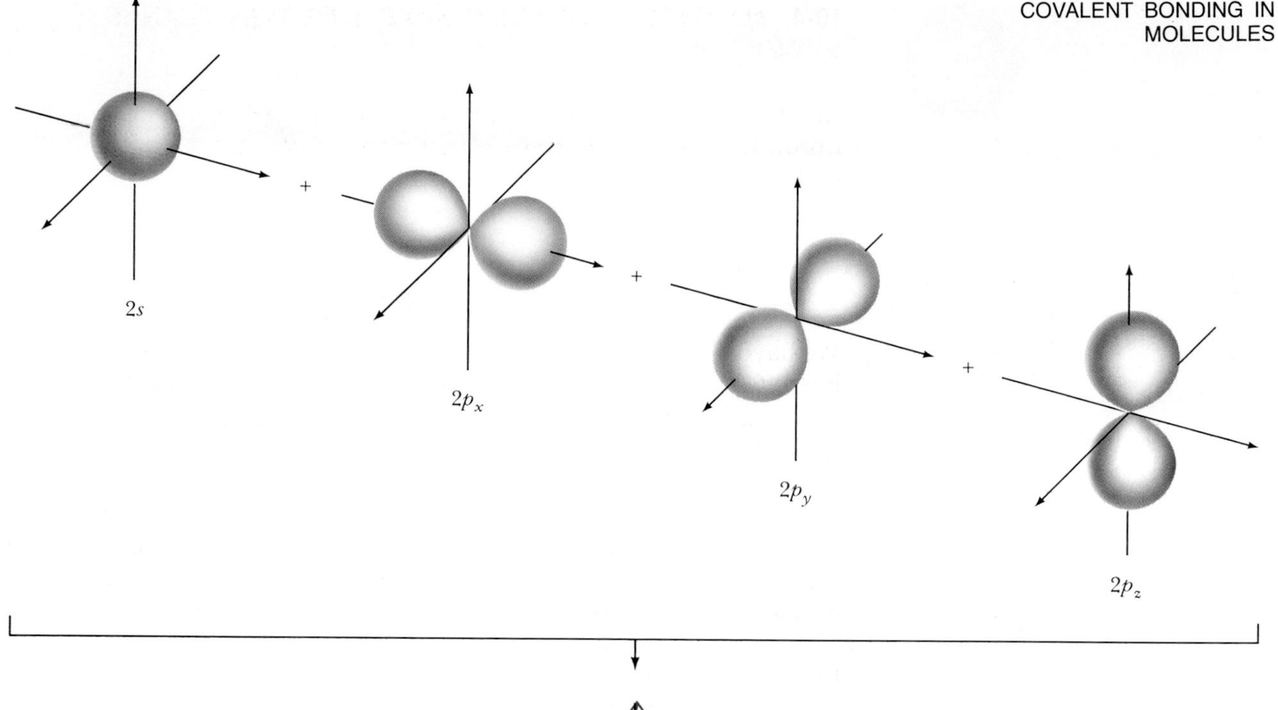

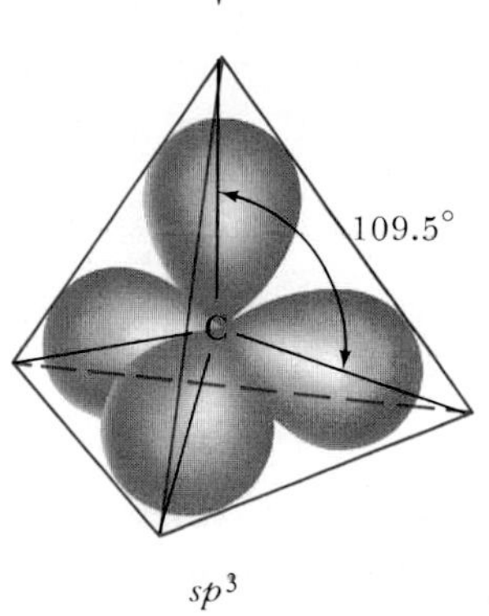

Figure 10-8 The 2*s* and three 2*p* orbitals on an atom can be combined to give four sp^3 hybrid orbitals, which are all equivalent and point toward the vertices of a tetrahedron. The angle between sp^3 orbitals is the tetrahedral bond angle, 109.5°.

Four equivalent localized σ bond orbitals in CH_4 are formed by overlapping each sp^3 orbital with a hydrogen 1*s* orbital (Figure 10-9). There are $4 + (4 \times 1) = 8$ valence electrons in CH_4. The eight valence electrons occupy the four equivalent localized bond orbitals to form the four equivalent covalent bonds in CH_4.

The tetrahedral arrangement of the four sp^3 hybrid orbitals on the carbon atom is consistent with the tetrahedral shape of a methane molecule.

Example 10-2: Describe the bonding in the ammonium ion, NH_4^+, whose Lewis formula is

$$\left[\begin{array}{c} \text{H} \\ | \\ \text{H—N—H} \\ | \\ \text{H} \end{array}\right]^+$$

Solution: We learned in Chapter 9 that NH_4^+ is tetrahedral. Thus we wish to form four localized bond orbitals that point toward the vertices of a tetrahedron. We can do this by forming sp^3 hybrid orbitals on the nitrogen atom

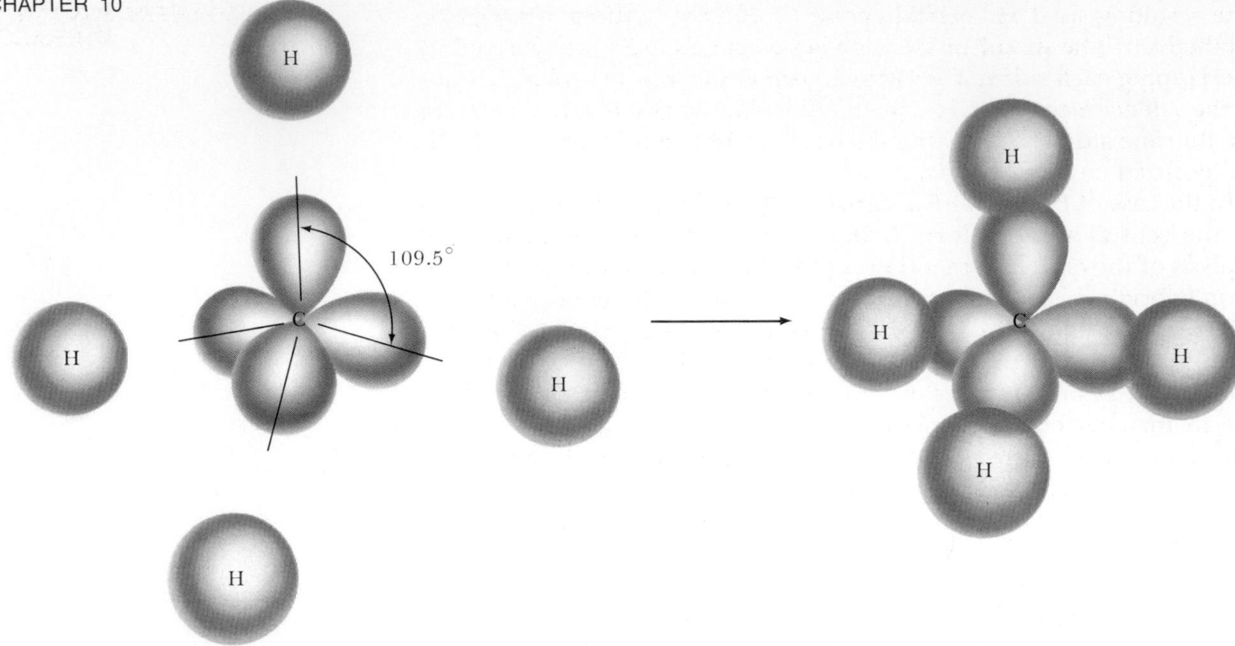

Figure 10-9 Four equivalent localized bond orbitals in CH_4 are formed by combining each of the four carbon sp^3 orbitals with a hydrogen $1s$ orbital. There are eight valence electrons in CH_4 (four from the carbon atom and one from each hydrogen atom). Each of the four localized bond orbitals is occupied by a pair of electrons of opposite spin, accounting for the four localized carbon-hydrogen bonds in CH_4.

by combining the nitrogen $2s$ orbital and all three of the nitrogen $2p$ orbitals. The sp^3 hybrid orbitals on a nitrogen atom are similar to the carbon atom sp^3 orbitals shown in Figure 10-8. We can form four equivalent localized bond orbitals by combining each sp^3 orbital on the nitrogen atom with a hydrogen $1s$ orbital. There are eight valence electrons in NH_4^+. Two valence electrons of opposite spins occupy each of the four bond orbitals, accounting for the four covalent bonds in NH_4^+. The bonding and the shape of the ammonium ion are similar to that shown for methane in Figure 10-9.

10-6. HYBRID ATOMIC ORBITALS CAN INVOLVE *d* ORBITALS

In Chapter 9 we learned about molecules that are trigonal bipyramidal (e.g., PCl_5) and octahedral (e.g., SF_6). The central atom in each of these molecules has an expanded valence shell, and so we must include *d* orbitals in the construction of hybrid orbitals in order to describe the bonding in these molecules. The combination of a $3s$ orbital, three $3p$ orbitals, and one $3d$ orbital gives five hybrid atomic orbitals that have trigonal bipyramidal symmetry. These five ***dsp*³ orbitals** have the interesting property that they are not equivalent to each other. In fact, they form two sets of equivalent orbitals: a set of three equivalent equatorial orbitals and a set of two equivalent axial orbitals, in accord with the experimental fact that the five chlorine atoms in PCl_5 are not equivalent (Section 9-10). The five phosphorus-chlorine bond orbitals are formed by overlapping each phosphorus dsp^3 hybrid orbital with a chlorine $3p$ orbital. Ten of the valence electrons (five from the phosphorus atom and one from each of the chlorine atoms) occupy the five localized σ bond orbitals (two electrons in each orbital) to form the five localized covalent bonds.

To describe the bonding in the octahedral SF_6 molecule, we combine the $3s$ orbital, three $3p$ orbitals, and two $3d$ orbitals on the sulfur atom.

The resulting six d^2sp^3 **orbitals** point toward the vertices of a regular octahedron. The six sulfur-fluorine bond orbitals in SF_6 are formed by overlapping each sulfur d^2sp^3 orbital with a fluorine $2p$ orbital. Twelve of the valence electrons (six from the sulfur atom and one from each of the fluorine atoms) occupy the six localized σ bond orbitals to form the six localized covalent bonds.

In the case of PCl_5 and SF_6, we use $3s$, $3p$, and some of the $3d$ orbitals on the central atom to form hybrid atomic orbitals. A mathematical analysis of the quantum theoretical procedure of combining orbitals to form hybrid orbitals shows that only orbitals of similar energy combine effectively. This means that we can combine $3s$, $3p$, and $3d$ orbitals because they have similar energies, but the combination of $3d$ orbitals with $2s$ and $2p$ orbitals does not produce hybrid orbitals that are effective in forming bonds. Consequently, for atoms of elements such as phosphorus and sulfur, whose valence electrons occupy $3s$ and $3p$ orbitals, we can use their $3d$ orbitals to expand their valence shells, but for atoms of second-row elements such as carbon and nitrogen, whose valence electrons occupy $2s$ and $2p$ orbitals, we do not use $3d$ orbitals to form hybrid orbitals.

Table 10-1 summarizes the hybrid atomic orbitals that we have introduced. Note that in each case the number of resulting hybrid orbitals is equal to the number of atomic orbitals used to construct them. These are examples of the principle of the **conservation of orbitals:** if we combine orbitals to make new orbitals, then the total number of new orbitals is equal to the number of orbitals used to make them. Even when we combine atomic orbitals on two different atoms to form a localized bond orbital, we actually end up with two localized bond orbitals, one of lower energy and one of higher energy. Because these orbitals are occupied by only two electrons when we make the corresponding covalent bond, we need be concerned only with the orbital of lowest energy, which are the bonding orbitals pictured in each of Figures 10-4, 10-5, 10-7, and 10-9.

Table 10-1 Properties of hybrid orbitals

Hybrid	Number	Molecular geometry	Bond angle	Examples
sp	2	linear	180°	BeH_2, BeF_2
sp^2	3	trigonal planar	120°	BF_3
sp^3	4	tetrahedral	109.5°	CH_4, CCl_4, NH_4^+
dsp^3	5	trigonal bipyramidal	90°, 120°	PCl_5
d^2sp^3	6	octahedral	90°	SF_6

10-7. WE USE sp^3 ORBITALS TO DESCRIBE THE BONDING IN MOLECULES THAT HAVE FOUR ELECTRON PAIRS ABOUT THE CENTRAL ATOM

Consider H_2O,

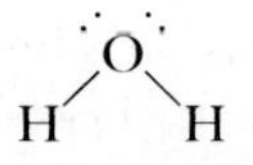

The two bond orbitals in a water molecule can be pictured as a combination of each hydrogen 1*s* orbital with one of the sp^3 orbitals on the oxygen atom to produce two oxygen-hydrogen σ bond orbitals (Figure 10-10). There are eight valence electrons in H_2O, six from the oxygen atom and one from each hydrogen atom. Four of the valence electrons occupy the two oxygen-hydrogen σ bond orbitals. The other four occupy the two nonbonded sp^3 orbitals and constitute the two lone electron pairs on the oxygen atom. We would predict on the basis of this bonding that the H—O—H bond angle is 109.5°, whereas the experimental value is 104.5°. The 5° difference can be attributed to the fact that the four orbitals about the oxygen atom are not used in the same way. Two are used to form bonds with the hydrogen atoms, and two are used for the lone-pair electrons. Recall from our discussion of VSEPR theory that we predicted the H—O—H bond angle in H_2O to be somewhat less than the tetrahedral value of 109.5° because the lone electron pairs repel the two hydrogen-oxygen bonds.

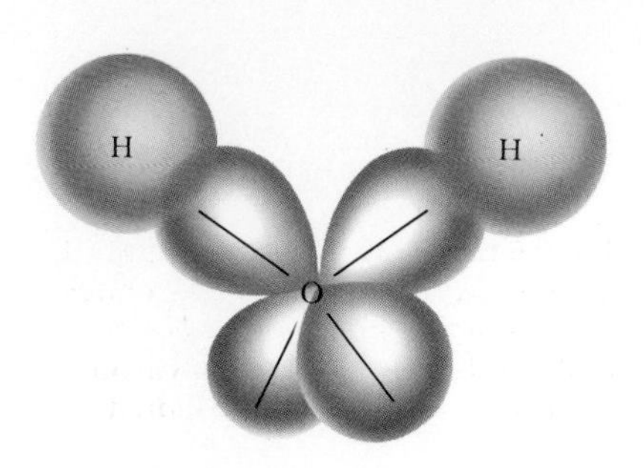

Figure 10-10 Bonding in the water molecule. The two hydrogen-oxygen bond orbitals are formed by the combination of two sp^3 orbitals on the oxygen atom with the hydrogen 1*s* orbitals. Of the eight valence electrons in the molecule, four occupy the two bond orbitals and four occupy the two nonbonded sp^3 orbitals on the oxygen atom. The latter are lone electron pairs.

Example 10-3: The ammonia molecule is trigonal pyramidal,

(Structure: N with lone pair bonded to three H atoms; angle 107°)

with H—N—H bond angles of 107°. Use hybrid orbitals to describe the bonding in NH_3.

Solution: Ammonia has three covalent bonds and one lone pair of electrons. We know from VSEPR theory that the four electron pairs in the valence shell of the nitrogen atom point toward the vertices of a tetrahedron. We can describe the bonding in NH_3 by assuming that the 2*s* and three 2*p* orbitals of the nitrogen atom form four sp^3 orbitals. Three of these sp^3 orbitals form localized bond orbitals by overlapping with the hydrogen 1*s* orbitals. Thus we can describe the bonding in an ammonia molecule in terms of three localized bond orbitals and a lone-pair (nonbonded) sp^3 orbital on the nitrogen atom.

There are eight valence electrons in NH_3. Six of them occupy the three localized bond orbitals and two occupy the nonbonded sp^3 orbital. The three fully occupied bond orbitals describe the three covalent bonds in NH_3; the fully occupied nonbonded sp^3 orbital on the nitrogen atom describes the lone pair of electrons in NH_3 (Figure 10-11). The use of sp^3 orbitals predicts that the H—N—H bond angles are 109.5°. The four valence orbitals in NH_3 are not used equivalently (one describes a lone pair), however, and we should expect to find small deviations from a regular tetrahedral shape. The observed H—N—H bond angles in NH_3 are 107°.

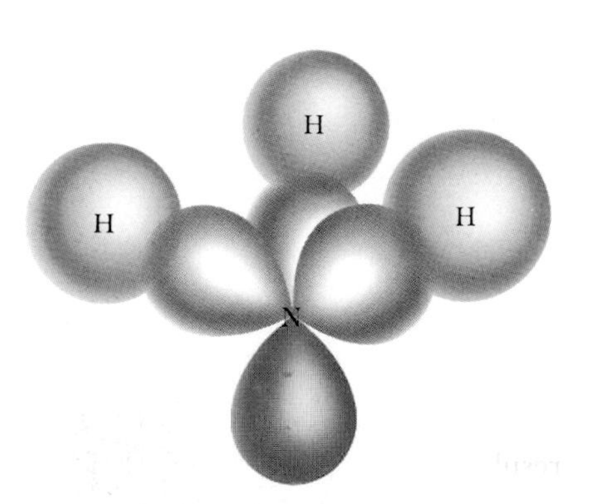

Figure 10-11 The use of sp^3 hybrid orbitals on nitrogen to describe the bonding in NH_3. Three of the nitrogen sp^3 orbitals are combined with hydrogen 1*s* orbitals to form three equivalent localized bond orbitals. The fourth nitrogen sp^3 orbital is a nonbonded orbital and is occupied by the lone pair of electrons in ammonia.

We can use sp^3 orbitals to describe the bonding in molecules that have no single central atom. For example, consider the hydrocarbon ethane, C_2H_6, whose Lewis formula is

```
   H  H
   |  |
H—C—C—H
   |  |
   H  H
```

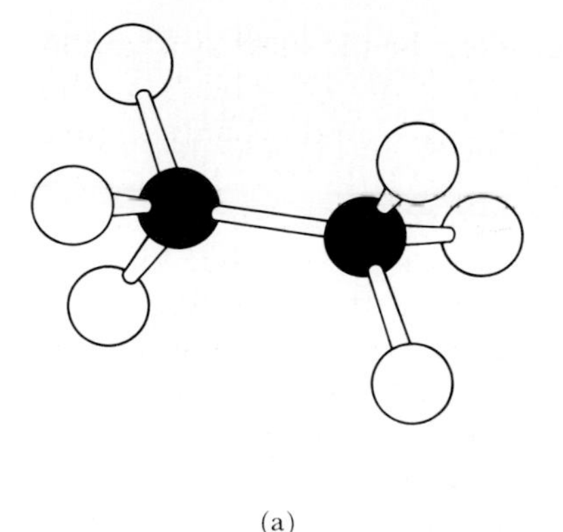

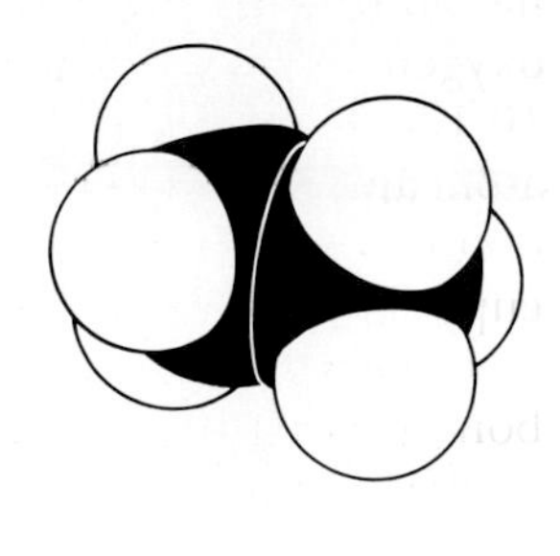

(a) (b)

Figure 10-12 Molecular models of ethane. (a) A ball-and-stick model. (b) A space-filling model. Note that the bonds about each carbon atom are tetrahedrally oriented.

Ball-and-stick and space-filling models of ethane are shown in Figure 10-12. The shape of an ethane molecule can be described in terms of sp^3 hybrid orbitals on the carbon atoms. The carbon-carbon bond orbital in ethane is formed by the overlap of two sp^3 orbitals, one from each carbon atom; the six carbon-hydrogen bond orbitals in ethane result from the overlap of the three remaining sp^3 orbitals on each carbon atom with the hydrogen 1*s* atomic orbitals. Note that there are seven bond orbitals in ethane. Each carbon atom has 4 valence electrons and each hydrogen atom has 1, giving a total of 14 valence electrons in ethane. The 14 valence electrons occupy the 7 bond orbitals in ethane such that each bond orbital has 2 electrons of opposite spins. The resulting bonding in ethane is shown in Figure 10-13.

Example 10-4: Describe the bonding in a hydrogen peroxide molecule, H_2O_2. The Lewis formula for H_2O_2 is

$$\mathrm{H{-}\ddot{\underset{..}{O}}{-}\ddot{\underset{..}{O}}{-}H}$$

Solution: Each oxygen atom has four sp^3 orbitals available for bonding. The oxygen-oxygen bond orbital is formed by the overlap of an sp^3 orbital on each oxygen atom. Each hydrogen-oxygen bond orbital is formed by the overlap of an oxygen sp^3 orbital and a hydrogen 1*s* orbital.

There is a total of 14 valence electrons in H_2O_2; each oxygen atom has six valence electrons and each hydrogen atom has one. Six of the valence elec-

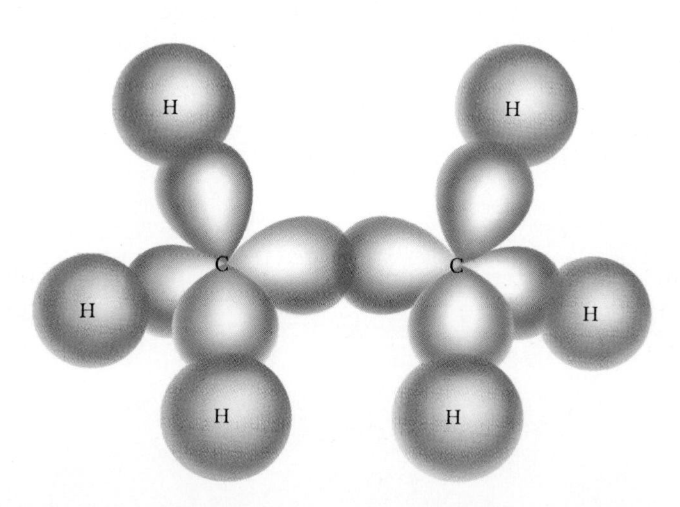

Figure 10-13 The six carbon-hydrogen bond orbitals in ethane result from the combination of sp^3 orbitals on the carbon atoms and 1*s* orbitals on the hydrogen atoms. The carbon-carbon bond orbital results from the combination of two sp^3 orbitals, one from each carbon atom. There are 14 valence electrons in ethane. Each of the seven bond orbitals is occupied by two valence electrons of opposite spins, accounting for the seven bonds in ethane.

trons occupy the three bonding orbitals, giving the three bonds in the Lewis formula for H_2O_2. The other eight valence electrons occupy the four sp^3 orbitals, two on each oxygen atom, to give the two lone electron pairs shown on each oxygen atom in the Lewis formula. Figure 10-14 shows the electron distribution in H_2O_2.

We can use sp^3 orbitals on oxygen to describe the bonding in alcohols, which are organic compounds involving an —OH group bonded to a carbon atom. The two simplest alcohols are

```
     H                   H  H
     |   ..              |  |   ..
  H—C—O—H          H—C—C—O—H
     |   ..              |  |   ..
     H                   H  H
```

methyl alcohol, CH_3OH ethyl alcohol, CH_3CH_2OH

The bonding in methyl alcohol is shown in Figure 10-15. The bonding in a methyl alcohol molecule can be described in terms of sp^3 orbitals on both the carbon atom and the oxygen atom. The carbon-oxygen bond orbital results from the overlap of sp^3 orbitals from each atom. The carbon-hydrogen and oxygen-hydrogen bonds result from the overlap of sp^3 and hydrogen 1*s* orbitals. Note that there is a total of five bond orbitals in CH_3OH. Ten of the 14 valence electrons in CH_3OH occupy the five bond orbitals. The other four valence electrons occupy the two remaining sp^3 orbitals as lone pairs on the oxygen atom.

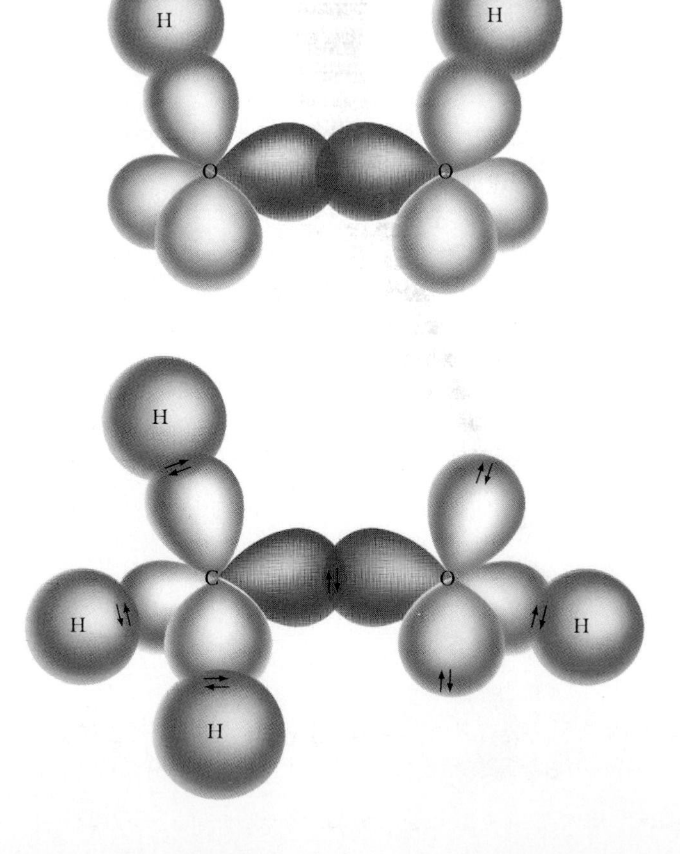

Figure 10-14 A schematic representation of the bonding orbitals in hydrogen peroxide, H_2O_2. The oxygen-oxygen bond orbital is formed by the overlap of an sp^3 orbital from each oxygen atom; each hydrogen-oxygen bond orbital is formed by the overlap of an oxygen sp^3 orbital and a hydrogen 1*s* orbital.

Figure 10-15 A schematic representation of the bonding orbitals in methyl alcohol. We use sp^3 orbitals on both the carbon atom and the oxygen atom.

10-8. A DOUBLE BOND CAN BE REPRESENTED BY A σ BOND AND A π BOND

All the molecules that we have discussed so far in this chapter have only single bonds. One of the simplest molecules in which there is a double bond is ethene, C_2H_4, which is more commonly known as ethylene. Its Lewis formula is

■ Ethylene is a gas at room temperature. It has the unusual property of causing fruit to ripen.

```
H       H
 \     /
  C=C
 /     \
H       H
```

The geometry of an ethylene molecule is quite different from that of ethane or the other alkanes. All six atoms in an ethylene molecule lie in one plane, and each carbon atom is bonded to three other atoms. In Section 10-4 we saw that such trigonal planar geometry can be described in terms of sp^2 orbitals. We shall describe the σ bonding in ethylene using sp^2 orbitals on each carbon atom.

The first step in describing the bonding in ethylene is to join the two carbon atoms by combining an sp^2 orbital from each, as shown in Figure 10-16. The resulting carbon-carbon bond is a σ bond. The four hydrogen atoms are bonded, two to each carbon atom, by combining the hydrogen 1*s* orbitals with the four remaining sp^2 orbitals on the carbon atoms, as shown in Figure 10-17. All five bonds formed so far are σ bonds, and Figure 10-17 shows the **σ-bond framework** in ethylene.

Recall that there is an unused 2*p* orbital on each carbon atom perpendicular to each H—C—H plane (Figure 10-17). The two ends of the molecule can now rotate about the σ carbon-carbon bond so as to maxi-

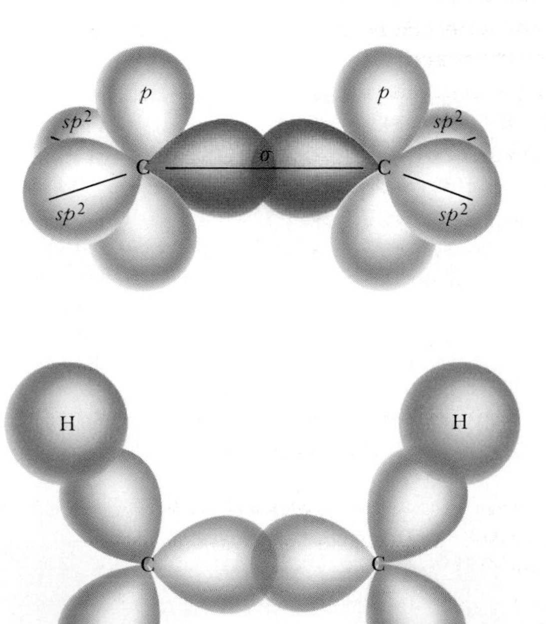

Figure 10-16 Two carbon atoms joined by the combination of an sp^2 orbital from each. The resulting bond orbital is cylindrically symmetric around the carbon-carbon axis and so is a σ bond orbital. The carbon-carbon σ bond orbital constitutes part of the double bond in ethylene.

Figure 10-17 The σ-bond framework in ethylene. The carbon-carbon bond orbital results from the combination of two sp^2 orbitals, one from each carbon atom. The four carbon-hydrogen bond orbitals result from the combination of carbon sp^2 orbitals and hydrogen 1*s* orbitals. The remaining *p* orbitals on the carbon atoms are not shown but are perpendicular to the page.

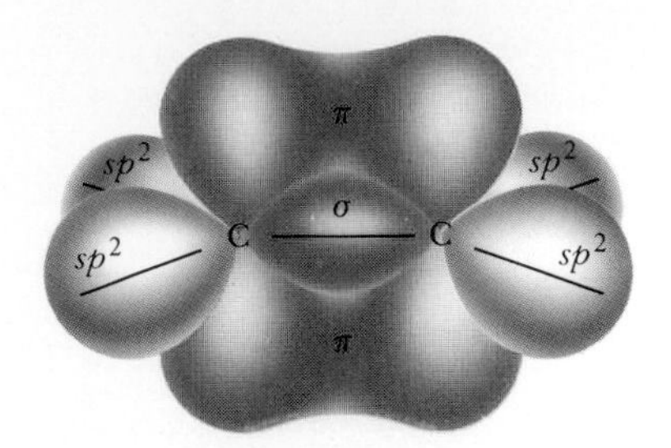

Figure 10-18 A double bond consists of a σ bond and a π bond. The σ bond results from the combination of two sp^2 orbitals, one from each atom. The π bond results from the combination of two p orbitals, one from each atom. The π orbital maintains the σ-bond framework in a planar shape and prevents rotation about the double bond.

mize the overlap of the $2p$ orbitals from each carbon atom. This occurs when the $2p$ orbitals are parallel to each other and results in the orbital denoted by the symbol π in Figure 10-18. The orbital that we denote by π in Figure 10-18 is not cylindrically symmetric along the bond axis. It does not have a circular cross section. Instead, its cross section is similar to that of an atomic p orbital, and so it is called a **π orbital** (pi orbital; π is the Greek letter that corresponds to p). The double bond in ethylene is described by the σ orbital *and* the π orbital in Figure 10-18. When two electrons occupy a π orbital, the result is a **π bond.** A σ bond and a π bond do not have the same energy, and so a double bond, although much stronger than a single bond, is not twice as strong as a single bond. Single carbon-carbon bond energies are about 350 $kJ{\cdot}mol^{-1}$, whereas double carbon-carbon bond energies are about 600 $kJ{\cdot}mol^{-1}$.

Figures 10-17 and 10-18 show the six bond orbitals in an ethylene molecule. Five are σ bond orbitals (Figure 10-17) and one is a π bond orbital (Figure 10-18). There are 12 valence electrons in ethylene (4 from each carbon atom and 1 from each hydrogen atom), and they occupy the six bond orbitals in pairs to give four single bonds and one double bond.

Example 10-5: Describe the bonding in the formaldehyde molecule, which has the Lewis formula

```
H
 \
  C=Ö:
 /
H
```

Solution: From VSEPR theory we conclude that the formaldehyde molecule is planar with a trigonal geometry around the carbon atom. Because the bond angles are about 120°, we use sp^2 hybrid orbitals on the carbon atom. Further, because there are also three groups of electrons around the oxygen atom (the double bond and two lone pairs), we also use sp^2 hybrid orbitals on the oxygen atom. An sp^2 orbital on the carbon atom is combined with an sp^2 orbital on the oxygen atom to form a carbon-oxygen σ bond. The remaining two sp^2 orbitals on the carbon atom are combined with the $1s$ orbitals on the hydrogen atom to form the two carbon-hydrogen σ bonds. The remaining two sp^2 orbitals on the oxygen atom are used to house the two lone pairs on the oxygen atom. The remaining p orbital on carbon and the remaining p orbital on the oxygen atom, both of which are perpendicular to the plane of the molecule, are combined to form a carbon-oxygen π bond. Thus, the carbon-oxygen double bond is composed of a σ bond and a π bond. The bonding in formaldehyde is shown in Figure 10-19.

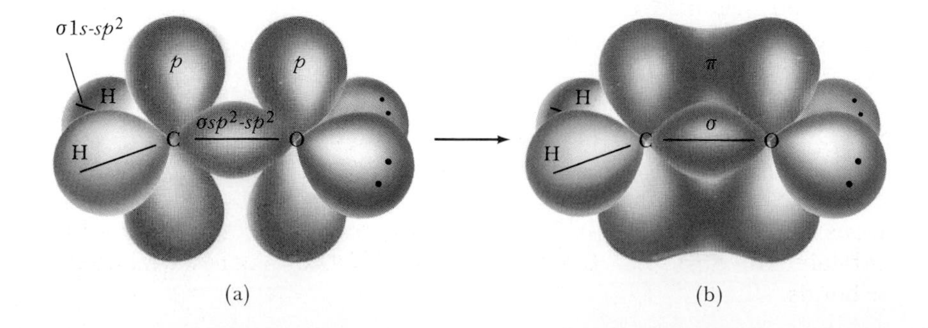

Figure 10-19 The bonding in formaldehyde, H_2CO. (a) The σ-bond framework, showing the unused $2p$ orbitals that are perpendicular to the plane formed by the four atoms. These two $2p$ orbitals combine to form a π bond. (b) The carbon-oxygen double bond consists of one σ bond and one π bond.

The double bond in ethylene consists of a σ bond and a π bond, each containing two electrons. The π bond locks the molecule into a planar shape, and so ethylene is a flat, or planar, molecule. There is restricted rotation about carbon-carbon double bonds. Consider the molecule 1,2-dichloroethene, ClCH=CHCl. Because there is no rotation about the carbon-carbon double bond, there are two distinct forms, or isomers, of 1,2-dichloroethene:

```
:Cl:        H                 :Cl:       :Cl:
    \      /                      \      /
     C=C          and              C=C
    /      \                      /      \
  H        :Cl:                  H        H
  trans isomer                    cis isomer
```

The first of these is called *trans*-1,2-dichloroethene because the chlorine atoms lie across (trans means across) the double bond from each other. The other is called *cis*-1,2-dichloroethene because the chlorine atoms lie on the same side (cis means on the same side) of the double bond. The 1,2- notation tells us that the chlorine atoms are attached to different carbon atoms. Molecules with the same atom-to-atom bonding but different spatial arrangements are called **stereoisomers.** The particular type of stereoisomerism that is displayed by 1,2-dichloroethene is called **cis-trans isomerism.** Stereoisomers, and cis-trans isomers in particular, have different physical properties. For example, the boiling point of the trans isomer of 1,2-dichloroethene is 48°C and that of the cis isomer is 60°C.

10-9. A TRIPLE BOND CAN BE REPRESENTED BY ONE σ BOND AND TWO π BONDS

Let us next consider a molecule that contains a triple bond. A good example is ethyne, C_2H_2, which is more commonly called acetylene. The Lewis formula for the acetylene molecule is

■ Acetylene is a colorless gas with a penetrating odor. One of its most important uses is as a fuel in oxyacetylene torches.

$$\mathrm{H{-}C{\equiv}C{-}H}$$

The acetylene molecule is linear, with each carbon atom bonded to only two other atoms. We have seen in Section 10-3 that *sp* hybrid orbitals can be used to describe the bonding of an atom that forms two bonds directed 180° from each other (Figure 10-4). We shall describe the bonding in the acetylene molecule using *sp* orbitals on each carbon atom.

We can build a σ-bond framework for the acetylene molecule in two steps. We first form a carbon-carbon bond orbital by combining two *sp* orbitals, one from each carbon atom. Then we form the carbon-hydrogen bond orbitals by overlapping a hydrogen 1*s* orbital with the remaining *sp* orbital on each carbon atom. The σ-bond framework of acetylene is shown in Figure 10-20.

The remaining carbon 2*p* orbitals are perpendicular to the H—C—C—H axis, as shown in Figure 10-21. These orbitals can combine to produce two π bond orbitals. There are five bond orbitals in acetylene, three σ bonds and two π bonds. There are 10 valence electrons in acetylene $[(2 \times 4) + (2 \times 1)]$, and they occupy the five bonding orbitals. The carbon-carbon triple bond consists of one σ bond and two π bonds.

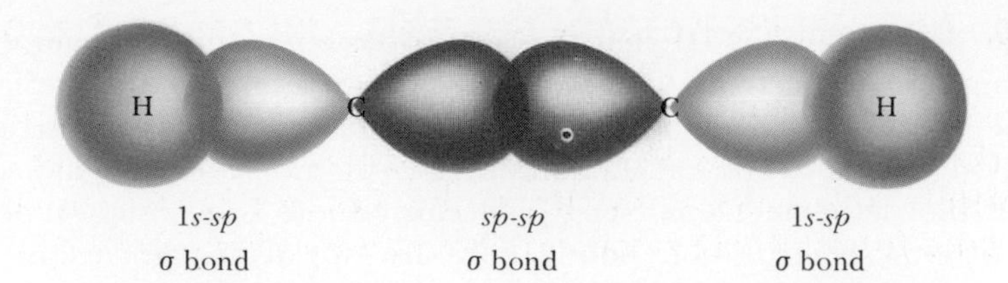

Figure 10-20 The σ-bond framework of acetylene. The carbon-carbon σ bond orbital results from combining two *sp* orbitals, one from each carbon atom. Each of the two carbon-hydrogen bond orbitals results from combining a carbon *sp* orbital and a hydrogen 1*s* orbital.

Example 10-6: Compare the bonding in acetylene with that in hydrogen cyanide, HCN.

Solution: The Lewis formula for HCN is

$$\mathrm{H{-}C{\equiv}N:}$$

Because a triple bond consists of one σ bond and two π bonds, we must use *sp* orbitals on both the carbon atom and the nitrogen atom in HCN. The σ-bond framework of HCN can be written as

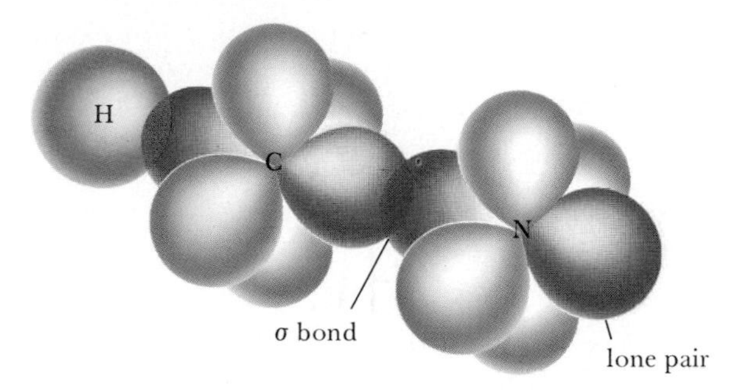

We see that HCN is a linear molecule and that its σ-bond framework is similar to that of C_2H_2 (Figures 10-20 and 10-21). The unused 2*p* orbitals of the carbon and nitrogen atoms combine to form the two π bond orbitals. There are four bond orbitals in HCN; two are σ bond orbitals and two are π bond orbitals. There are 10 valence electrons in HCN: eight occupy the four bond orbitals, and two occupy the nitrogen *sp* orbital and constitute a lone electron pair on the nitrogen atom.

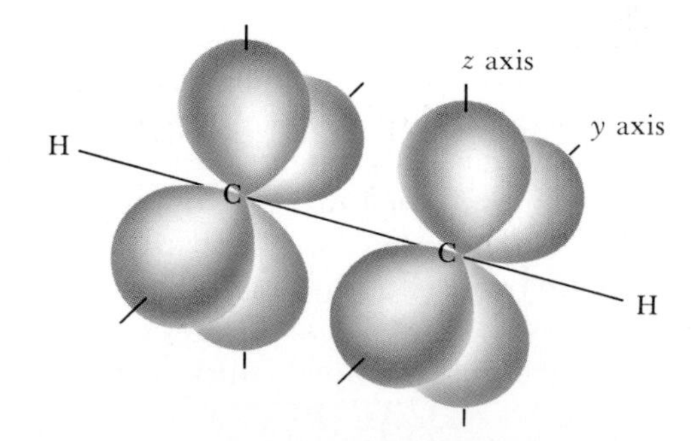

Figure 10-21 The 2*p* orbitals on the carbon atoms in acetylene. The 2*p* orbitals that are directed along the *z* axis combine to form one π bond orbital, and the 2*p* orbitals directed along the *y* axis overlap to form another π bond orbital.

10-10. THE π ELECTRONS IN BENZENE ARE DELOCALIZED

There are many molecules that have π orbitals that extend over more than two adjacent atoms. One of the most important examples of such a molecule is benzene, C_6H_6. Recall that benzene has two principal resonance forms, whose resonance hybrid is (Section 8-6)

We can describe the bonding in benzene in terms of σ and π bonds. The bonding about each carbon atom in benzene is trigonal, and so we assign sp^2 orbitals to each. This leads directly to the σ-bond framework shown in Figure 10-22. The angles in a regular hexagon are 120°, and this fits nicely with the 120° bond angles associated with sp^2 orbitals if the σ-bond framework in benzene is planar and if the carbon atoms form a regular hexagon (Figure 10-22). Note that there are 12 σ bond

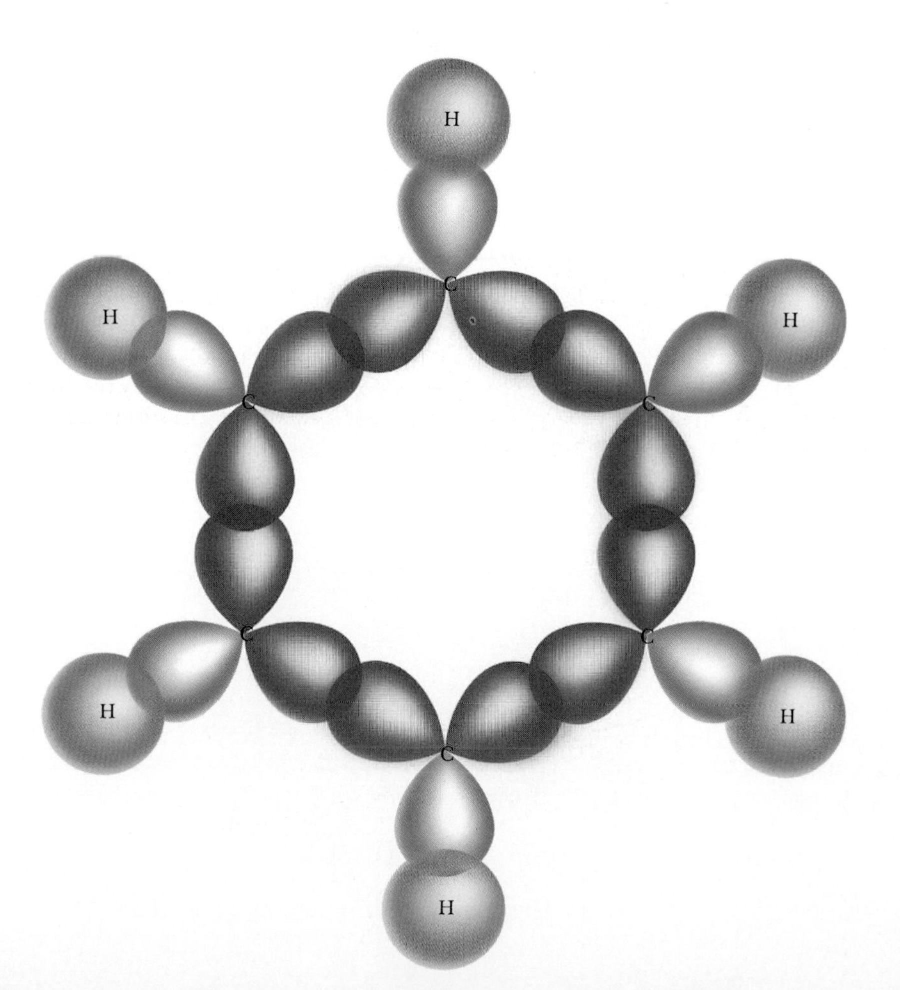

Figure 10-22 The σ-bond framework in a benzene molecule. Each carbon-carbon bond orbital results from the combination of sp^2 orbitals, and each carbon-hydrogen bond orbital results from the combination of a carbon sp^2 orbital and a hydrogen $1s$ orbital. All 12 atoms lie in a single plane, so benzene is a planar molecule. The six carbon atoms form a regular hexagon.

orbitals in benzene. Each carbon atom has a $2p$ orbital that is perpendicular to the hexagonal plane (Figure 10-23). These six p orbitals, which are perpendicular to the plane of the benzene ring, combine to give a total of six π orbitals (conservation of orbitals), each of which is spread over the ring. The π orbitals in benzene are not associated with any particular pair of carbon atoms, so they are said to be **delocalized orbitals.** There are $(6 \times 4) + (6 \times 1) = 30$ valence electrons in benzene; the 12 σ bonds depicted in Figure 10-22 are occupied by 24 of the valence electrons, and the 6 remaining valence electrons occupy three of the π orbitals. The resulting total π electron charge density is depicted in Figure 10-23(b). The delocalization of the π electrons around the benzene ring is an example of **charge delocalization.** Charge delocalization in our quantum theoretical description corresponds to resonance in writing Lewis formulas, and confers an extra degree of stability on a molecule relative to the hypothetical molecule with localized double bonds.

Example 10-7: Describe the bonding in a nitrate ion, NO_3^-, whose resonance hybrid is

Solution: As VSEPR theory correctly predicts, the nitrate ion is trigonal planar, with 120° bond angles. We shall use sp^2 hybrid orbitals on all four atoms, and build a σ-bond framework by overlapping an sp^2 orbital from each oxygen atom with an sp^2 orbital on the nitrogen atom. Each atom has an unused p orbital that is perpendicular to the NO_3^- plane, and these p orbitals combine to form four delocalized π orbitals that are spread over the ion. There are $(1 \times 5) + (3 \times 6) + 1 = 24$ valence electrons in a nitrate ion. Six of these occupy the three σ bond orbitals; four occupy the two unused sp^2 orbitals on each oxygen atom and constitute two lone pairs on each oxygen atom; and the remaining six valence electrons occupy three of the delo-

Figure 10-23 Each carbon atom in a benzene ring has a $2p$ orbital perpendicular to the ring. These six $2p$ orbitals combine to form six π orbitals that are spread uniformly over the entire ring.

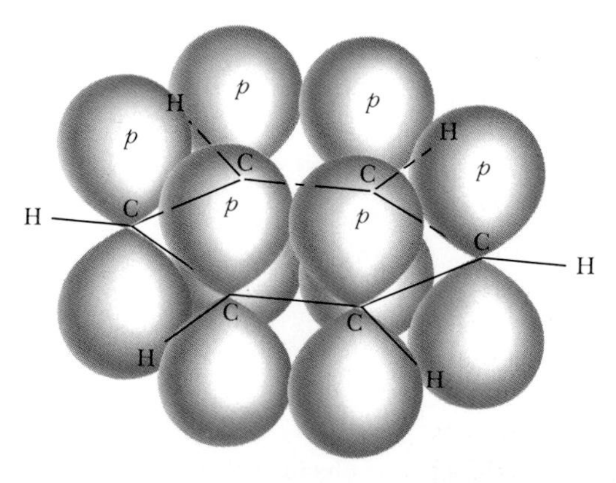

(a) Individual p orbitals in benzene

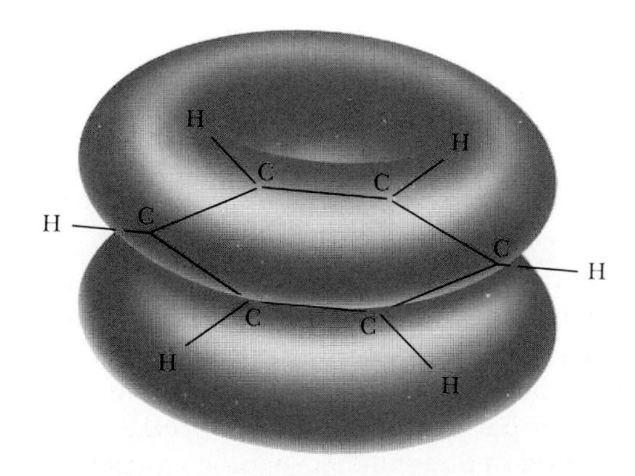

(b) The total π electron density in benzene

calized π orbitals. This resulting charge delocalization corresponds to the fact that the Lewis formula of NO_3^- is represented by a resonance hybrid.

10-11. THE HYDROGEN MOLECULAR ION IS THE SIMPLEST DIATOMIC SPECIES

Although we have been able to describe the bonding in a variety of molecules, we still have no insight into such questions as why, for example, two hydrogen atoms join to form a stable molecule whereas two helium atoms do not. In these sections we discuss a theory of bonding, called **molecular orbital theory,** that will answer such questions for us. Like the valence-bond theory, molecular orbital theory was developed in the 1930s. Although molecular orbital theory can be applied to all molecules, for simplicity, we shall consider only **homonuclear diatomic molecules,** that is, molecules consisting of two identical atoms.

Recall that we describe the electronic structure of atoms in terms of atomic orbitals, which are based on the set of orbitals that were given for a hydrogen atom. Because a hydrogen atom has only one electron, its atomic orbitals are relatively simple and serve as approximate orbitals for more complicated atoms. A one-electron system that applies to homonuclear diatomic molecules is the **hydrogen molecular ion,** H_2^+, which consists of two protons and one electron. The H_2^+ ion is stable relative to a separated H and H^+; its bond length is 106 pm and its dissociation energy is 255 $kJ \cdot mol^{-1}$.

▪ The H_2^+ molecular orbitals form the basis of diatomic molecular orbitals, just as hydrogen atomic orbitals form the basis for the atomic orbitals of other atoms.

When we solve the Schrödinger equation for H_2^+, we obtain a set of wave functions, or orbitals, and a corresponding set of energies. Because these orbitals extend over both nuclei in H_2^+, they are molecular orbitals. Just as we discussed the shapes of the various hydrogen atomic orbitals and then used them to build up more complicated atoms, we discuss the various H_2^+ molecular orbitals and then use them to build up more complicated diatomic molecules. Figure 10-24 shows the shapes of the first few molecular orbitals of H_2^+. Each shape represents the three-dimensional surface that encloses a certain probability of finding the electron within the volume enclosed by that surface.

10-12. MOLECULAR ORBITALS ARE BONDING OR ANTIBONDING

The first two H_2^+ molecular orbitals shown in Figure 10-24 are σ orbitals and so have circular cross sections when viewed along the internuclear axis. Figure 10-24 shows, however, that these two orbitals are different when viewed perpendicular to the internuclear axis. The first molecular orbital, the one that corresponds to the lowest energy in H_2^+, is concentrated between the two nuclei. Electrons in this molecular orbital are likely to be found in the region between the two nuclei and so draw the two nuclei toward each other. This effective attraction results in a covalent bond and provides a molecular orbital interpretation of the electron-pair bond in Lewis formulas. A molecular orbital that is concentrated in a region between two nuclei is called a **bonding orbital** because electrons in such an orbital act to bond the two nuclei together.

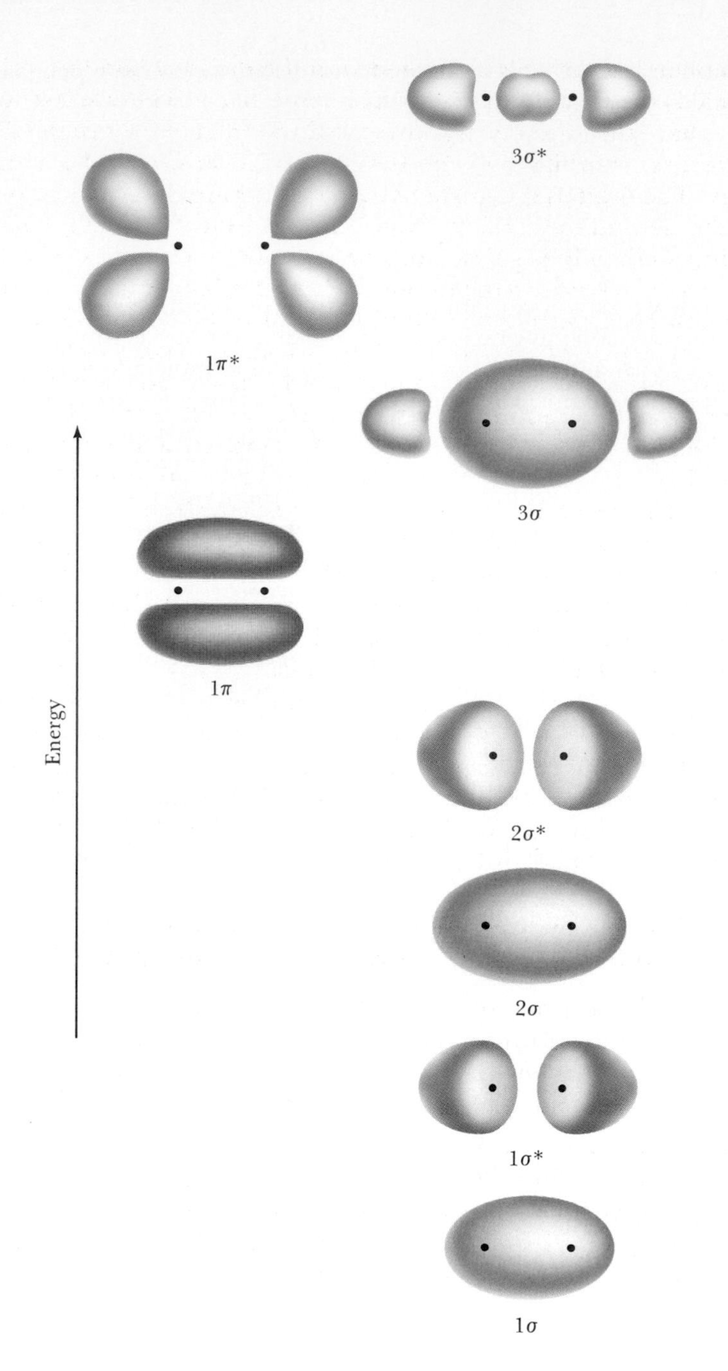

Figure 10-24 The three-dimensional surfaces that depict the shapes but not the relative sizes of the first few H_2^+ molecular orbitals. The orbitals are listed in order of increasing energy. Note that some molecular orbitals have zero values in the region between the two nuclei, which are shown as heavy dots. Also note that the two molecular orbitals designated by 1π have the same energy and that the two designated by $1\pi^*$ have the same energy.

The H_2^+ molecular orbital that corresponds to the second lowest energy, that is, the second lowest one in Figure 10-24, is zero in the region between the nuclei. It is concentrated more on the far sides of the two nuclei. Electrons in this orbital are likely to be found not in the region between the two nuclei but on the far sides of the nuclei. Thus, electrons in this orbital tend to draw the nuclei apart. A molecular orbital that is zero in the region between two nuclei and concentrated on the far sides of the nuclei is called an **antibonding orbital.** A bonding orbital is designated simply by its Greek letter, and an antibonding orbital is designated by its Greek letter followed by an asterisk. For example the antibonding orbitals associated with the σ and π bonding orbitals are called **σ* orbitals** and **π* orbitals,** respectively. The σ and σ^* orbitals that have the lowest energy are called the 1σ and $1\sigma^*$ orbitals. They are the first σ orbital and the first σ^* orbital in order of increasing energy. The orbitals designated by 2σ and $2\sigma^*$ are the second σ orbital and the second σ^* orbital. Note that molecular orbitals come in bonding-antibonding pairs. Thus we have both a 1π (first π orbital) and a $1\pi^*$ (first π^* orbital) orbital, a 3σ and a $3\sigma^*$ orbital, and so on.

▪ A σ * orbital is called a "sigma-star orbital" and a π * orbital is called a "pi-star orbital."

We now construct electron configurations of homonuclear diatomic molecules by placing electrons in these orbitals in accord with the Pauli exclusion principle. The hydrogen molecule, H_2, has two electrons. According to the Pauli exclusion principle, we place two electrons of opposite spins in the 1σ orbital and write the electron configuration of H_2 as $(1\sigma)^2$. The ground-state electron configuration of H_2 is illustrated in Figure 10-25, where, for simplicity, only the first two energy levels are shown. The two electrons with opposite spins occupy the 1σ orbital. The two electrons in the bonding orbital constitute a bonding pair of electrons and account for the single-bond character of H_2.

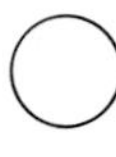

Figure 10-25 The electron configuration of H_2. The two electrons occupy the molecular orbital corresponding to the lowest energy and have opposite spins in accord with the Pauli exclusion principle.

Now let's consider the species He_2, which has four electrons. Two of the electrons of He_2 occupy the 1σ orbital, and two occupy the $1\sigma^*$ orbital (Figure 10-26). Thus, there are two electrons in a bonding orbital and two in an antibonding orbital. Electrons in a bonding orbital tend to draw the nuclei together, whereas electrons in an antibonding orbital tend to draw the nuclei apart. The result is that the effect of the antibonding electrons cancels the effect of the bonding electrons and there is no net bonding. This is in accord with the fact that the species He_2 has never been observed experimentally.

Figure 10-26 The electron configuration of the hypothetical molecule He_2. There are two electrons in a bonding orbital and two in an antibonding orbital, and so He_2 has no net bonding. The molecule He_2 has never been detected experimentally.

Table 10-2 summarizes the properties of the molecular species, H_2^+, H_2, He_2^+ and He_2. The property **bond order** in Table 10-2 is defined as

$$\text{bond order} = \frac{\begin{pmatrix}\text{number of}\\ \text{electrons in}\\ \text{bonding orbitals}\end{pmatrix} - \begin{pmatrix}\text{number of}\\ \text{electrons in}\\ \text{antibonding orbitals}\end{pmatrix}}{2} \qquad (10\text{-}1)$$

A bond order of 1/2 indicates a one-electron bond, a bond order of 1 indicates a single bond, a bond order of 2 indicates a double bond, and so on. The bond order of 0 for He_2 indicates that there is no helium-helium covalent bond. Note from Table 10-2 that bond lengths decrease and bond energies increase with increasing bond order.

Table 10-2 Molecular properties of H_2^+, H_2, He_2^+, and He_2

Species	Number of electrons	Ground-state electron configuration	Bond order	Bond length/pm	Bond energy/kJ·mol^{-1}
H_2^+	1	$(1\sigma)^1$	$\frac{1}{2}$	106	255
H_2	2	$(1\sigma)^2$	1	74	436
He_2^+	3	$(1\sigma)^2(1\sigma^*)^1$	$\frac{1}{2}$	108	251
He_2	4	$(1\sigma)^2(1\sigma^*)^2$	0	not observed	not observed

10-13. MOLECULAR ORBITAL THEORY PREDICTS MOLECULAR ELECTRON CONFIGURATIONS

Figure 10-27 shows an energy-level diagram for the molecular orbitals 1σ to $3\sigma^*$. We can use Figure 10-27 to write ground-state electron configurations for the homonuclear diatomic molecules Li_2 through Ne_2.

Lithium vapor contains diatomic lithium molecules, Li_2. A lithium atom has three electrons, and so Li_2 has a total of six electrons. In the ground state of Li_2, the six electrons occupy the lowest three molecular orbitals in Figure 10-27 in accord with the Pauli exclusion principle. The ground-state electron configuration of Li_2 is $(1\sigma)^2(1\sigma^*)^2(2\sigma)^2$. There is a net of two bonding electrons, and so the bond order is 1. Thus we predict that Li_2 is stable relative to two separated lithium atoms. Table 10-3 shows that Li_2 has a bond length of 267 pm and a bond energy of 101 $kJ \cdot mol^{-1}$. The process

$$Li_2(g) \rightarrow 2Li(g) \qquad \Delta H^\circ_{rxn} = 101 \text{ kJ}$$

is endothermic.

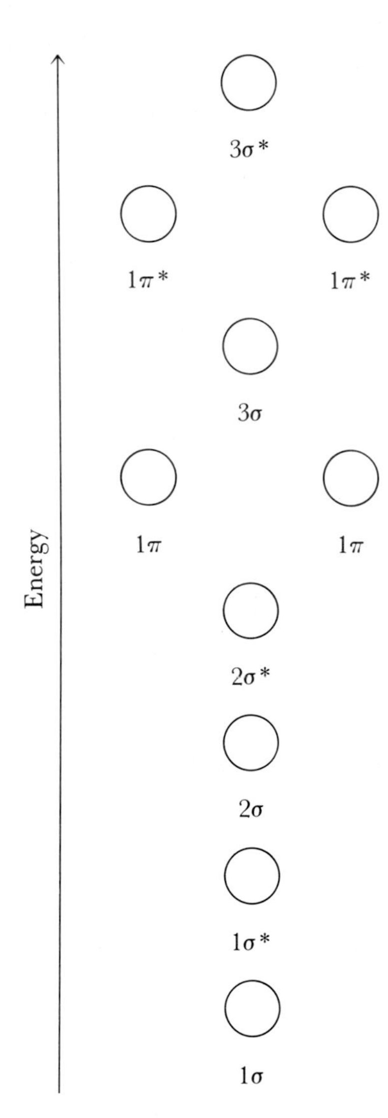

Figure 10-27 An energy-level diagram to be used for the homonuclear diatomic molecules H_2 through Ne_2. The orbitals are listed in order of increasing energy, $1\sigma < 1\sigma^* < 2\sigma < 2\sigma^* < 1\pi < 3\sigma < 1\pi^* < 3\sigma^*$. Electrons occupy these orbitals in accord with the Pauli exclusion principle.

Example 10-8: Use Figure 10-27 to write the ground-state electron configuration of N_2. Calculate the bond order of N_2 and compare your result with the Lewis formula for N_2.

Solution: There are 14 electrons in N_2, and, using Figure 10-27, we see that its ground-state electron configuration is $(1\sigma)^2(1\sigma^*)^2(2\sigma)^2(2\sigma^*)^2(1\pi)^4(3\sigma)^2$. According to Equation (10-1), the bond order in N_2 is

$$\text{bond order} = \frac{10 - 4}{2} = 3$$

The Lewis formula for N_2, :N≡N:, is thus in agreement with the more fundamental molecular orbital theory result. The triple bond in N_2 accounts for its short bond length (110 pm) and its unusually large bond energy (941 $kJ \cdot mol^{-1}$). The bond in N_2 is one of the strongest known bonds.

Example 10-9: Use molecular orbital theory to explain why neon does not form a stable diatomic molecule.

Solution: Neon has 10 electrons, and so Ne_2 would have 20 electrons. According to Figure 10-27, its ground-state electron configuration would be $(1\sigma)^2(1\sigma^*)^2(2\sigma)^2(2\sigma^*)^2(1\pi)^4(3\sigma)^2(1\pi^*)^4(3\sigma^*)^2$. The bond order associated with this electron configuration is

$$\text{bond order} = \frac{10 - 10}{2} = 0$$

indicating that there is no neon-neon bond and that neon does not form a diatomic molecule.

The prediction of the distribution of the electrons in an oxygen molecule is one of the most impressive successes of molecular orbital theory.

Oxygen molecules are **paramagnetic.** This means that oxygen is a magnetic substance and so is attracted to a region between the poles of a magnet.

Each oxygen atom has 8 electrons; thus, O_2 has a total of 16 electrons. When the 16 electrons are placed according to the molecular orbital diagram given in Figure 10-27, the last two electrons go into the $1\pi^*$ orbitals. As in the atomic case, we apply **Hund's rule** (Section 6-18) and place one electron in each $1\pi^*$ orbital such that the two electrons have unpaired spins. This is shown in Figure 10-28. The ground-state electron configuration of O_2 is $(1\sigma)^2(1\sigma^*)^2(2\sigma)^2(2\sigma^*)^2(1\pi)^4(3\sigma)^2(1\pi^*)^2$. Because each $1\pi^*$ orbital is occupied by one electron such that the spins are unpaired, an oxygen molecule has a net electron spin and so acts as a tiny magnet. These tiny magnets cause O_2 to be drawn between the poles of a magnet.

The amount of oxygen in air can be monitored by measuring the paramagnetism of a sample of air. Because oxygen is the only major component in air that is paramagnetic, the measured paramagnetism of air is directly proportional to the amount of oxygen present. Linus Pauling developed this method, which was used to monitor oxygen levels in submarines and airplanes in World War II. It is still used by physicians to monitor the oxygen content in blood during anesthesia.

The Lewis formula of O_2 does not account for the unpaired electrons and hence does not account for the paramagnetism of O_2. According to the octet rule, we should write $\ddot{O}{=}\ddot{O}$ for the Lewis formula of O_2, but this implies that all the electrons are paired, which is in disagreement with the paramagnetism of O_2. The oxygen molecule is an exception to the utility of Lewis formulas, whereas the more fundamental molecular orbital theory is able to account successfully for the distribution of the electrons in O_2.

The ground-state electron configurations of the homonuclear diatomic molecules Li_2 through Ne_2 are given in Table 10-3. Molecular orbital theory can also be applied to **heteronuclear diatomic molecules** (that is, diatomic molecules in which the two atoms are different) and to polyatomic molecules; however, we do not discuss those extensions here.

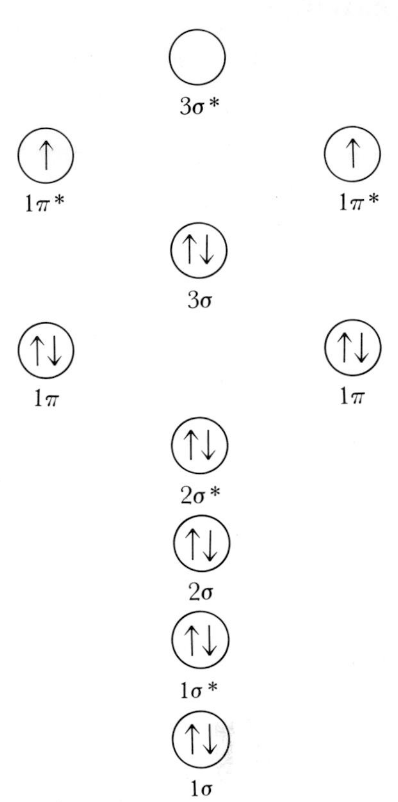

Figure 10-28 The ground-state electron configuration of O_2. There are 16 electrons in O_2, and they occupy the molecular orbitals as shown. Note that two of the electrons occupy the $1\pi^*$ orbitals in accord with Hund's rule, being placed in separate orbitals with unpaired spins. The molecule itself has a net electron spin and so acts as a tiny magnet.

Table 10-3 Properties of homonuclear diatomic molecules of elements in the second row of the periodic table

Species	Ground-state electron configuration	Bond order	Bond length/pm	Bond energy/kJ·mol^{-1}
Li_2	$(1\sigma)^2(1\sigma^*)^2(2\sigma)^2$	1	267	101
Be_2	$(1\sigma)^2(1\sigma^*)^2(2\sigma)^2(2\sigma^*)^2$	0	not observed	not observed
B_2	$(1\sigma)^2(1\sigma^*)^2(2\sigma)^2(2\sigma^*)^2(1\pi)^2$	1	159	289
C_2	$(1\sigma)^2(1\sigma^*)^2(2\sigma)^2(2\sigma^*)^2(1\pi)^4$	2	124	599
N_2	$(1\sigma)^2(1\sigma^*)^2(2\sigma)^2(2\sigma^*)^2(1\pi)^4(3\sigma)^2$	3	110	941
O_2	$(1\sigma)^2(1\sigma^*)^2(2\sigma)^2(2\sigma^*)^2(1\pi)^4(3\sigma)^2(1\pi^*)^2$	2	121	494
F_2	$(1\sigma)^2(1\sigma^*)^2(2\sigma)^2(2\sigma^*)^2(1\pi)^4(3\sigma)^2(1\pi^*)^4$	1	142	154
Ne_2	$(1\sigma)^2(1\sigma^*)^2(2\sigma)^2(2\sigma^*)^2(1\pi)^4(3\sigma)^2(1\pi^*)^4(3\sigma^*)^2$	0	not observed	not observed

10-14. PHOTOELECTRON SPECTRA ARE CONSISTENT WITH MOLECULAR ORBITAL THEORY

To many students beginning their study of chemistry, the idea of atomic orbitals and molecular orbitals is rather abstract and sometimes appears to be far removed from "reality." It so happens, however, that the electron configurations of molecules can be verified experimentally. If high-energy electromagnetic radiation, such as X-radiation, is directed into a gas, then electrons are ejected from the molecules of the gas. The energy required to eject an electron from a molecule in the gas is a direct measure of how strongly the electron is bound in the molecule. The energy with which an electron is bound in a molecule is called the **binding energy** of that electron. The binding energy of an electron in a molecule depends upon which molecular orbital the electron occupies. The lower the energy of the molecular orbital that the electron occupies, the more energy it takes to remove the electron from the molecule.

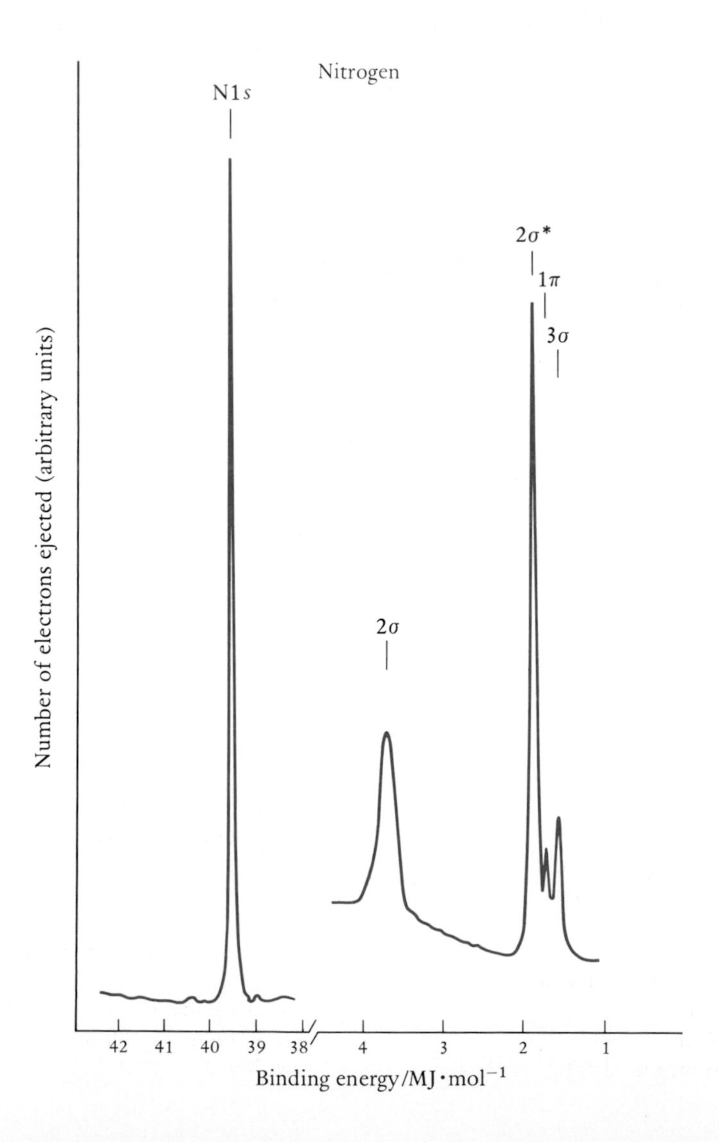

Figure 10-29 The photoelectron spectrum of N_2. When gas-phase molecules are irradiated with high-energy radiation, electrons are ejected from the molecules. By measuring the energies of the ejected electrons, we can deduce the energies of the electrons in the molecules. The peaks in this plot are interpreted as arising from electrons that are ejected from various molecular orbitals.

The measurement of the energies of the electrons ejected by radiation incident to gaseous molecules is called **photoelectron spectroscopy.** A **photoelectron spectrum** of N_2 is shown in Figure 10-29. The peaks in this figure correspond to the energies of the electrons in N_2. Note how they correspond to the indicated molecular orbital energies. Also note that the 1*s* electrons in N_2 are much more tightly bound than the other electrons. Photoelectron spectra provide striking experimental support for the molecular orbital picture developed in this chapter.

▪ Photoelectron spectroscopy can be viewed as a molecular analog of the photoelectric effect.

SUMMARY

Electrons in molecules are described by molecular orbitals, which are orbitals that are spread over two or more atoms. The bonding in polyatomic molecules often can be described in terms of localized bond orbitals, which are formed by combining atomic orbitals from atoms that are bonded together. Bonding in terms of localized bond orbitals is closely related to Lewis formulas.

The bonding in all the polyatomic molecules discussed in this chapter can be described in terms of hybrid orbitals. Different types of hybrid orbitals are used to describe different molecular geometries. Table 10-1 summarizes the associated geometries of various hybrid atomic orbitals.

Bond orbitals for single bonds occur from the overlap of (hybrid) orbitals from each of the atoms that are bonded together. These bond orbitals are cylindrically symmetric about the bond axis, and the resulting bonds are called σ bonds.

Double bonds can be represented by one σ bond and one π bond. The π bond is formed by the combination of *p* orbitals on adjacent atoms. Double bonds have a significant effect on the shape of a molecule. The atoms bonded directly to double-bonded atoms all lie in one plane. In addition, there is no rotation about double bonds, which leads to the possibility of cis and trans isomers. Triple bonds can be represented by one σ bond and two π bonds.

In some molecules, such as benzene, the π bond orbitals are spread uniformly over many atoms and are said to be delocalized. The electrons that occupy delocalized orbitals are also delocalized, giving what is called charge delocalization. Charge delocalization confers an extra degree of stability on a molecule or ion and accounts for the relative stability of benzene.

A set of diatomic molecular orbitals can be obtained for the hydrogen molecular ion, H_2^+, and these can be used to write electron configurations of diatomic molecules. This is in analogy with the atomic case, where we used hydrogen atomic orbitals to write electron configurations of other atoms. There are two types of molecular orbitals, bonding and antibonding. The bonding properties of molecules depend upon the number of electrons in bonding and antibonding orbitals. The scheme used in this chapter is called molecular orbital theory. Molecular orbital theory correctly predicts that the diatomic molecules He_2 and Ne_2 do not exist and that O_2 is paramagnetic.

TERMS YOU SHOULD KNOW

ground-state wave function 317
molecular orbital 317
polyatomic molecule 318
localized bond orbital 318
valence-bond theory 318
overlap 319
principle of maximum overlap 319
sp orbital 319
hybrid atomic orbital 320
σ orbital 320
σ bond 320
sp^2 orbital 322
sp^3 orbital 323
dsp^3 orbital 324
d^2sp^3 orbital 325
conservation of orbitals 325
σ-bond framework 329
π orbital 330
π bond 330
stereoisomer 331
cis-trans isomerism 331
delocalized orbital 334
charge delocalization 334
molecular orbital theory 335
homonuclear diatomic molecule 335
hydrogen molecular ion, H_2^+ 335
bonding orbital 335

antibonding orbital 336
σ^* orbital 337
π^* orbital 337
bond order 337
paramagnetic 339
Hund's rule 339
heteronuclear diatomic molecule 339
binding energy 340
photoelectron spectroscopy 341
photoelectron spectrum 341

AN EQUATION YOU SHOULD KNOW HOW TO USE

$$\text{bond order} = \frac{\begin{pmatrix}\text{number of}\\ \text{electrons in}\\ \text{bonding orbitals}\end{pmatrix} - \begin{pmatrix}\text{number of electrons}\\ \text{in antibonding}\\ \text{orbitals}\end{pmatrix}}{2} \qquad (10\text{-}1) \quad \text{(definition of bond order)}$$

PROBLEMS

BOND FORMATION

10-1. How many bonds are there in propane? The Lewis formula is

```
    H  H  H
    |  |  |
 H—C—C—C—H
    |  |  |
    H  H  H
```

How many valence electrons are there in three carbon atoms and eight hydrogen atoms?

10-2. How many bonds are there in butane? The Lewis formula is

```
    H  H  H  H
    |  |  |  |
 H—C—C—C—C—H
    |  |  |  |
    H  H  H  H
```

How many valence electrons are there in 4 carbon atoms and 10 hydrogen atoms?

10-3. How many valence electrons, localized bonds and lone pairs are there in each of the following molecules?

(a) C_2H_6 (b) PH_3 (c) N_2H_4 (d) SCl_2

10-4. How many valence electrons, localized bonds and lone pairs are there in each of the following species?

(a) SF_4 (b) H_2O_2 (c) NH_4^+ (d) SF_6

MOLECULES OF THE FORMULA AX_m

10-5. How many valence electrons are there in BH_3? Describe the bonding in BH_3 in terms of hybrid orbitals.

10-6. How many valence electrons are there in $HgCl_2$? Describe the covalent bonding in $HgCl_2$ in terms of hybrid orbitals.

10-7. How many valence electrons are there in CF_4? Use hybrid orbitals to describe the bonding in CF_4.

10-8. How many valence electrons are there in PCl_6^-? Use hybrid orbitals to describe the bonding in PCl_6^-.

10-9. Discuss the bonding in aluminum chloride, $AlCl_3$. What type of hybrid orbital best describes the bonding in this molecule?

10-10. How many valence electrons are there in chloroform, $HCCl_3$? Describe the bonding in the chloroform molecule.

MOLECULES OF THE FORMULA AX_mE_n

10-11. The hydronium ion, H_3O^+, is trigonal pyramidal with H—O—H bond angles of 110°. Describe the bonding in H_3O^+.

10-12. Describe the bonding in OF_2. Which atomic orbitals on the fluorine atoms are used to form localized bond orbitals?

10-13. Discuss the bonding in NF_3. Which atomic orbitals are used to form the localized bond orbitals in this molecule?

10-14. How many valence electrons are there in SF_4? Use hybrid orbitals to describe the bonding in SF_4.

10-15. How many valence electrons are there in PCl_3? Use hybrid orbitals to describe the bonding in PCl_3.

10-16. How many valence electrons are there in $SnCl_2$? Use hybrid orbitals to describe the bonding in $SnCl_2$.

MOLECULES WITH NO UNIQUE CENTRAL ATOM

10-17. Describe the bonding in hydrazine, N_2H_4.

10-18. A class of organic compounds called alcohols may be viewed as derived from HOH by replacing one of the hydrogen atoms by an alkyl group, which are hydrocarbon groups, such as $—CH_3$ (methyl) and

—CH_2CH_3 (ethyl). A simple alcohol is ethyl alcohol, whose Lewis formula is

```
    H  H
    |  |  ..
 H—C—C—O—H
    |  |  ..
    H  H
```

Dicuss the bonding and shape of ethyl alcohol.

10-19. A class of organic compounds called amines may be viewed as derived from NH_3 with one or more hydrogen atoms replaced by alkyl groups (see Problem 10-18). Examples of amines are

CH_3NH_2	$(CH_3)_2NH$	$(CH_3)_3N$
methylamine	dimethylamine	trimethylamine

Discuss the bonding and the shape of methylamine.

10-20. Discuss the bonding and shape of dimethylamine (Problem 10-19). How many σ bonds are there? How many lone pairs of electrons? How many valence shell electrons are there in the constituent atoms?

10-21. If both hydrogen atoms in HOH are replaced by alkyl groups (see Problem 10-18) the result is an ether, ROR′, where R and R′ are alkyl groups that may or may not be different. The simplest ether is dimethyl ether

```
    H       H
    |  ..   |
 H—C—O—C—H
    |  ..   |
    H       H
```

dimethyl ether

Discuss the bonding in dimethyl ether.

10-22. Discuss the bonding in ethyl methyl ether, whose Lewis formula is

```
    H  H      H
    |  |  ..  |
 H—C—C—O—C—H
    |  |  ..  |
    H  H      H
```

MULTIPLE BONDS

10-23. How many σ bonds and π bonds are there in each of the following molecules?

(a) $Cl_2C{=}CH_2$
(b) $H_2C{=}CHCH{=}CH_2$
(c) CH_3COOH
(d)

```
      CH2
    /     \
 HC         CH
  ||        ||
 HC         CH
    \     /
      CH2
```

10-24. How many σ bonds and π bonds are there in each of the following molecules?

(a) $F_2C{=}CF_2$
(b) HOOC—COOH
(c) $H_2C{=}C{=}CCl_2$
(d)

```
 H       H
   \   /
    C=C
    | |
    C=C
   /   \
 H       H
```

10-25. How many σ bonds are there in ethylacetylene, $CH_3CH_2C{\equiv}CH$? How many π bonds? How many valence electrons?

10-26. How many σ bonds and π bonds are there in methyl cyanide, CH_3CN? How many valence electrons?

10-27. Describe the bonding in carbon monoxide, CO.

10-28. Describe the bonding in the acetylide ion, C_2^{2-}.

DELOCALIZED BONDS

10-29. Phenol (carbolic acid), a derivative of benzene in which one hydrogen atom is replaced by —OH, has the formula C_6H_5OH and is used as an antiseptic and disinfectant. Write a Lewis formula for phenol and discuss its bonding.

10-30. Aniline, $C_6H_5NH_2$, a derivative of benzene in which one hydrogen atom is replaced by —NH_2, is used in the manufacture of numerous dyes. Write a Lewis formula for aniline and discuss its bonding.

10-31. Write the complete Lewis formula for and discuss the bonding in naphthalene, $C_{10}H_8$,

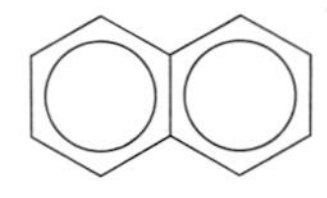

a white, crystalline solid with an odor characteristic of mothballs.

10-32. Write the complete Lewis formula for and discuss the bonding in anthracene, $C_{14}H_{10}$,

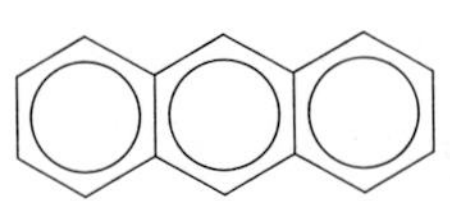

a yellow, crystalline solid found in coal tar.

10-33. Describe the bonding in a carbonate ion, CO_3^{2-}. How many delocalized orbitals are there in CO_3^{2-}? How many electrons are there in the delocalized orbitals?

10-34. Describe the bonding in a formate ion, CHO_2^-. How many delocalized orbitals are there in CHO_2^-? How many electrons are there in the delocalized orbitals?

10-35. Use molecular orbital theory to explain why diatomic beryllium does not exist.

10-36. Use molecular orbital theory to predict whether or not diatomic boron is paramagnetic.

10-37. Use molecular orbital theory to explain why the bond energy of N_2 is greater than that of N_2^+, but the bond energy of O_2 is less than that of O_2^+.

10-38. Use molecular orbital theory to predict the relative bond energies and bond lengths of F_2 and F_2^+.

10-39. Use molecular orbital theory to predict the relative bond energies and bond lengths of diatomic carbon, C_2, and the acetylide ion, C_2^{2-}.

10-40. Molecular orbital theory can be applied to heteronuclear diatomic molecules. The energy-level scheme in Figure 10-27 can be used if the atomic numbers of the two atoms in the heteronuclear diatomic molecule differ by only one or two atomic numbers. Use Figure 10-27 to determine the ground-state electron configurations and bond orders of NF, NF^+, and NF^-. Which of these species do you predict to be paramagnetic?

10-41. Molecular orbital theory can be applied to heteronuclear diatomic molecules. The energy-level scheme in Figure 10-27 can be used if the atomic numbers of the two atoms in the heteronuclear diatomic molecule differ by only one or two atomic numbers. Use Figure 10-27 to determine the bond order of carbon monoxide and compare your result with the Lewis formula for CO.

10-42. Molecular orbital theory can be applied to heteronuclear diatomic molecules. The energy-level scheme in Figure 10-27 can be used if the atomic numbers of the two atoms in the heteronuclear diatomic molecule differ by only one or two atomic numbers. Use Figure 10-27 to determine the bond order of the cyanide ion, CN^-, and compare your result with the Lewis formula for this ion.

10-43. Write the ground-state electron configurations and determine the bond orders of the following ions:

(a) O_2^{2-} (b) C_2^+ (c) Be_2^+ (d) Ne_2^+

10-44. For each of the following molecular electron configurations, decide whether it describes a ground electronic state or an excited electronic state.

(a) $(1\sigma)^2(1\sigma^*)^2(2\sigma^*)^1$
(b) $(1\sigma)^2(1\sigma^*)^2(2\sigma)^2(2\sigma^*)^1$
(c) $(1\sigma)^2(1\sigma^*)^2(2\sigma)^2(2\sigma^*)^2(1\pi)^3(3\sigma)^1$
(d) $(1\sigma)^2(1\sigma^*)^2(2\sigma)^2(2\sigma^*)^2(1\pi)^4(3\sigma)^2$

10-45. In some cases the removal of an electron from a species can result in a stronger net bonding (e.g., O_2^+ versus O_2). Give an example in which the addition of an electron to a species produces a stronger net bonding.

10-46. One of the excited states of C_2 has the electron configuration $(1\sigma)^2(1\sigma^*)^2(2\sigma)^2(2\sigma^*)^1(1\pi)^4(3\sigma)^1$. Would you expect the bond length in this excited state to be longer or shorter than that in the ground state?

10-47. Which of the following species are paramagnetic?

(a) C_2 (b) B_2 (c) C_2^{2+} (d) F_2^{2+}

ADDITIONAL PROBLEMS

10-48. The H—As—H bond angles in AsH_3 are about 90°. What atomic orbitals would you use to form the localized bond orbitals in this molecule?

10-49. Aldehydes are organic compounds that have the general Lewis formula

$$\begin{matrix} R \\ & \diagdown \\ & & C=\ddot{\underset{\cdot\cdot}{O}} \\ & \diagup \\ H \end{matrix}$$

where R is either a hydrogen atom (giving formaldehyde) or an alkyl group such as $-CH_3$ (methyl) or $-CH_2CH_3$ (ethyl). Discuss the bonding in acetaldehyde, CH_3CHO.

10-50. Ketones are organic compounds with the general Lewis formula

$$\begin{matrix} R' \\ & \diagdown \\ & & C=\ddot{\underset{\cdot\cdot}{O}} \\ & \diagup \\ R \end{matrix}$$

where R and R′ are alkyl groups that may or may not be different. The simplest ketone is acetone, $(CH_3)_2CO$, one of the most important solvents. Discuss the bonding and shape of acetone.

10-51. Discuss the bonding and shape of methylacetylene, $CH_3C{\equiv}CH$. How many σ and π bonds are there? How many valence electrons?

10-52. Discuss the bonding and shape of ethyl cyanide, CH_3CH_2CN. How many σ and π bonds are there? How many valence electrons?

10-53. The bond angle in H_2Te is about 90°. What atomic orbitals would you use to form the localized bond orbitals in this molecule?

10-54. Use hybrid orbitals to describe the bonding in CO_2.

10-55. Explain why there is no rotation around a double bond, but there is about a single bond.

10-56. Use hybrid orbitals to describe the bonding in O_3. How many delocalized π orbitals are there in O_3?

10-57. Formamide, H_2NCHO, is known to be planar. Discuss the bonding in formamide.

LIQUIDS AND SOLIDS

Solid iodine in equilibrium with gaseous iodine at 40°C (lighter-colored vapor) and at 90°C (darker-colored vapor). The equilibrium vapor pressure increases with increasing temperature.

The molecules in a gas are widely separated from each other and travel distances equal to many molecular diameters between collisions. The molecules in solids and liquids are in contact with each other, and so the properties of solids and liquids depend upon how strongly the molecules interact with, or attract, each other. The attractions between molecules hold solids and liquids in the condensed state. If molecules did not attract each other, gases would not condense to form liquids. In this chapter we study molecular interactions and how they affect the properties of liquids and solids. Some of the properties we shall study are the enthalpies of fusion and vaporization and vapor pressure. We also shall study various types of crystals, which are characterized by a repetitive, ordered arrangement of atoms or molecules. We learn that the underlying orderly arrangement of the atoms or molecules in a crystal can be determined by the patterns that result when X-rays are passed through the crystal. We also study phase diagrams, which are diagrams that give the regions of stability of the solid, liquid, and gas phases of a substance on one graph.

11-1. THE PROCESSES OF MELTING AND BOILING APPEAR AS HORIZONTAL LINES ON A HEATING CURVE

Let's consider an experiment in which a pure substance is converted from a solid to a liquid to a gas by the application of heat at a constant rate. Figure 11-1 shows how the temperature of the substance—water in this case—varies with time when it is heated at a constant rate. Such a plot is called a **heating curve.** Initially the water is at a temperature of −10°C and so is in the form of ice. As heat is added, the temperature of the ice increases until 0°C, the melting point of ice, is reached. At 0°C, the temperature remains constant for 60 min even though the ice is being heated at a constant rate of 100 $J \cdot min^{-1}$. The heat being added at 0°C melts the ice to liquid water. The temperature of the ice-water mixture remains at 0°C until all the ice is melted. The energy absorbed as heat that is required to melt one mole of any substance is called the **molar enthalpy of fusion** and is denoted by ΔH_{fus}. The experimental data plotted in Figure 11-1 show that it requires 60 min to melt 1 mol of

▪ The enthalpy of fusion is often called the heat of fusion.

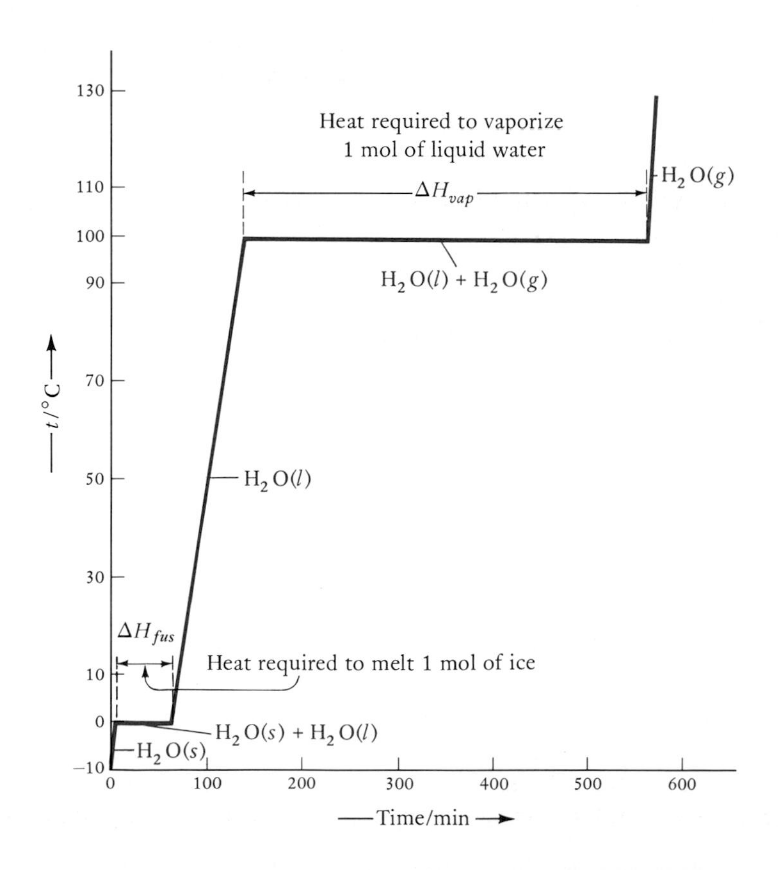

Figure 11-1 The heating curve for one mole of water starting with ice at −10°C. The energy is added as heat at a constant rate of 100 $J \cdot min^{-1}$. The most noteworthy features of the heating curve are the horizontal portions, which represent the heat of fusion, ΔH_{fus}, and the heat of vaporization, ΔH_{vap}. Note that the heat of vaporization is much larger than the heat of fusion.

ice when heat is added at the rate of 100 $J \cdot min^{-1}$, and so the enthalpy of fusion for ice is given by

$$\Delta H_{fus} = (60\ min \cdot mol^{-1})(100\ J \cdot min^{-1})$$
$$= 6000\ J \cdot mol^{-1}$$
$$= 6.0\ kJ \cdot mol^{-1}$$

After all the ice is melted, and not until then, the temperature increases from 0°C. The temperature continues to increase until 100°C, the boiling point of water, is reached. At 100°C, the temperature remains constant for 407 min even though heat is being added at a rate of 100 $J \cdot min^{-1}$. The heat being absorbed at 100°C is vaporizing the liquid water to water vapor. The temperature of the liquid-vapor mixture remains at 100°C until all the water is vaporized. The energy absorbed as heat that is required to vaporize one mole of any substance is called the **molar enthalpy of vaporization** and is denoted by ΔH_{vap}. According to Figure 11-1, ΔH_{vap} for water is

▪ The enthalpy of vaporization is often called the heat of vaporization.

$$\Delta H_{vap} = (407\ min \cdot mol^{-1})(100\ J \cdot min^{-1})$$
$$= 40{,}700\ J \cdot mol^{-1}$$
$$= 40.7\ kJ \cdot mol^{-1}$$

Once all the water is vaporized, the temperature increases from 100°C, as shown in the figure.

The horizontal portions of the heating curve of a pure substance represent the heat of fusion and the heat of vaporization. The other regions represent pure phases being heated at a constant rate. We learned in Chapter 5 that it requires energy to raise the temperature of a substance. The heat absorbed in raising the temperature of a substance from T_1 to T_2 without a change in phase is given by (see Equation 5-16)

$$q_P = nC_P(T_2 - T_1) \qquad (11\text{-}1)$$

where C_P is the molar heat capacity at constant pressure and n is the number of moles. The heat capacity is the measure of the ability of a substance to take up energy as heat. The different rates of temperature increase for $H_2O(s)$, $H_2O(l)$, and $H_2O(g)$ that appear in Figure 11-1 (steep segments of the heating curve) are a result of the fact that

▪ Heat capacity is discussed in Section 5-6.

$$C_P[H_2O(l)] > C_P[H_2O(s)] > C_P[H_2O(g)]$$

11-2. IT REQUIRES ENERGY TO MELT A SOLID AND TO VAPORIZE A LIQUID

The melting, or fusion, of ice can be represented by

$$H_2O(s) \rightarrow H_2O(l) \qquad \Delta H_{fus} = 6.0\ kJ$$

or, in general, we write

$$X(s) \rightarrow X(l) \qquad \Delta H_{fus}$$

▪ $\Delta H_{fus} > 0$

where X stands for any element or compound. The enthalpy of fusion is necessarily positive because it requires energy to break up the crystal lattice. Recall that a positive value of ΔH means that heat is absorbed in the process. The vaporization of water can be represented by

$$H_2O(l) \rightarrow H_2O(g) \qquad \Delta H_{vap} = 40.7 \text{ kJ}$$

▪ $\Delta H_{vap} > 0$

or, in general,

$$X(l) \rightarrow X(g) \qquad \Delta H_{vap}$$

The value of ΔH_{vap} is always positive because it requires energy to separate the molecules in a liquid from each other. Gas-phase molecules are so far apart from each other that they interact only very weakly relative to liquid-phase molecules. Essentially all the energy put in as heat of vaporization is required to separate the molecules of the liquid from one another. Because the temperature does not change during vaporization, the average kinetic energy of the molecules does not change. The molar enthalpies of fusion and vaporization of several compounds are given in Table 11-1.

The consumption of heat in the vaporization of a liquid is one of the mechanisms used by the human body to regulate its temperature at about 310 K. When the body becomes overheated, sweating begins; the evaporation of water from the surface of the body consumes 2.3 kJ per gram of sweat that is evaporated. The rate of evaporation can be en-

Table 11-1 Melting points, boiling points, and molar enthalpies of vaporization and fusion for several substances

Compound	Chemical formula	Melting point/K	Boiling point/K	ΔH_{fus}/kJ·mol^{-1}	ΔH_{vap}/kJ·mol^{-1}
ammonia	NH_3	195	240	5.65	23.4
argon	Ar	84	87	1.17	6.52
bromine	Br_2	266	332	10.6	29.5
carbon dioxide	CO_2	217	(195) sublimes	8.33	(25.2) sublimes
chlorine	Cl_2	172	239	6.41	20.4
formaldehyde	H_2CO	181	252	—	24.5
helium	He	—	4.2	0.014	0.081
hydrogen	H_2	14	20	0.12	0.90
iodine	I_2	387	458	15.5	41.9
krypton	Kr	116	121	1.63	9.03
lithium bromide	LiBr	823	1583	17.6	148.1
mercury	Hg	234	630	2.30	59.1
methane	CH_4	91	112	0.94	8.17
neon	Ne	24	27	0.33	1.76
nitrogen	N_2	63	77	0.72	5.58
oxygen	O_2	54	90	0.44	6.82
water	H_2O	273	373	6.01	40.7
xenon	Xe	160	166	2.30	12.63

hanced greatly by continuously sweeping away the water-saturated air near the skin surface, as occurs in a breeze.

It is possible for a solid to be converted directly to a gas without passing through the liquid phase. This process is called **sublimation.** The **molar enthalpy of sublimation,** ΔH_{sub}, is the energy absorbed as heat when one mol of a solid is sublimed at constant pressure. Essentially all the energy put in as heat in the sublimation process is used to separate the molecules in the solid from one another. The larger the value of ΔH_{sub}, the stronger are the intermolecular attractions in the solid. The best-known example of sublimation is the conversion of Dry Ice, $CO_2(s)$, to carbon dioxide gas:

▪ The enthalpy of sublimation is often called the heat of sublimation.

$$CO_2(s) \rightarrow CO_2(g) \qquad \Delta H_{sub} = 25.2\ \text{kJ}\cdot\text{mol}^{-1}$$

▪ $\Delta H_{sub} > 0$

The name Dry Ice is used because the CO_2 does not become a liquid at 1 atm pressure. Dry Ice at 1 atm has a temperature of $-78°C$ and is widely used as a one-time low-temperature refrigerant. The sublimation of 44 g (one mole) of Dry Ice requires 25.2 kJ of heat.

▪ Like carbon dioxide, iodine and moth balls pass directly from the solid to the vapor phase at 1 atm.

Ice sublimes at temperatures below its melting point (0°C). For the sublimation of ice,

$$H_2O(s) \rightarrow H_2O(g) \qquad \Delta H_{sub} = 46.7\ \text{kJ}\cdot\text{mol}^{-1}$$

Snow often sublimes, and so does the ice in the freezer compartment of your refrigerator. This is why ice cubes left in the freezer get smaller as time passes.

Example 11-1: Compute the energy released as heat when 28 g of liquid water at 18°C is converted to ice at 0°C. (An ice cube contains about 1 oz of water, and 1 oz is equivalent to 28 g.) The molar heat capacity of $H_2O(l)$ is $C_P = 75.3\ \text{J}\cdot\text{K}^{-1}\cdot\text{mol}^{-1}$, and $\Delta H_{fus} = 6.0\ \text{kJ}\cdot\text{mol}^{-1}$ for ice.

▪ 454 g = 1 lb
16 oz = 1 lb
1 oz = 28 g

Solution: The overall process must be broken down into two steps. We must first bring the $H_2O(l)$ from 18°C to 0°C (the freezing point of water) and then consider the process $H_2O(l) \rightarrow H_2O(s)$ at 0°C:

$$\underset{\text{at } 18°C}{28\text{ g } H_2O(l)} \xrightarrow{\text{step 1}} \underset{\text{at } 0°C}{28\text{ g } H_2O(l)} \xrightarrow{\text{step 2}} \underset{\text{at } 0°C}{28\text{ g } H_2O(s)}$$

For step 1 we have, from Equation (11-1), where n is the number of moles of water,

$$\begin{aligned} q_P &= nC_P(T_2 - T_1) \\ &= (28\text{ g})\left(\frac{1\text{ mol } H_2O}{18\text{ g } H_2O}\right)(75.3\ \text{J}\cdot\text{K}^{-1}\cdot\text{mol}^{-1})\left(\frac{1\text{ kJ}}{1000\text{ J}}\right)(-18\text{ K}) \\ &= -2.1\text{ kJ} \end{aligned}$$

The negative sign for q_P reflects the fact that energy must be removed to lower the temperature of the water. For step 2, where n is the number of moles of H_2O,

$$q_P = n(-\Delta H_{fus}) = (28\text{ g})\left(\frac{1\text{ mol } H_2O}{18\text{ g } H_2O}\right)(-6.0\ \text{kJ}\cdot\text{mol}^{-1}) = -9.3\text{ kJ}$$

where the minus sign in front of ΔH_{fus} arises because freezing is the reverse of fusion, that is, $\Delta H_{freezing} = -\Delta H_{fus}$. The total amount of energy that must be *removed* as heat from the 28 g of water is 2.1 kJ + 9.3 kJ = 11.4 kJ.

11-3. VAN DER WAALS FORCES ARE ATTRACTIVE FORCES BETWEEN MOLECULES

In the process of vaporization or sublimation, the molecules of the liquid or solid, which are in contact with each other, become separated from each other and widely dispersed. The value of ΔH_{vap} or ΔH_{sub} reflects how strongly the molecules attract each other in the liquid or solid. The more strongly the molecules attract each other, the greater the value of ΔH_{vap} or ΔH_{sub}.

The simplest force at the atomic or molecular level is the force between ions. We studied this force in Chapter 7 when we discussed ionic compounds. Two ions with opposite charges attract each other; ions with like charges repel each other. The force between ions is a relatively strong force; it requires a relatively large amount of energy to separate ions of opposite charge. The enthalpies of vaporization of ionic compounds are much larger than those of nonionic compounds. Enthalpies of vaporization of ionic compounds are typically at least $100\ kJ \cdot mol^{-1}$. In addition, the boiling points of ionic compounds are higher than those of nonionic compounds.

▪ Ionic compounds have relatively high boiling points and large molar enthalpies of vaporization.

Most of the compounds listed in Table 11-1 are molecular compounds. The fact that all substances can be liquefied means that even neutral molecules attract each other. In Chapter 9 we learned that some molecules have dipole moments, or are polar. An example of a polar molecule is formaldehyde, whose Lewis formula is

$$H_2C{=}\ddot{O}\!:$$

Because the electronegativity of an oxygen atom is greater than that of a carbon atom, the oxygen atom has a small negative charge and the carbon atom has a small positive charge. The electronegativities of a carbon atom and a hydrogen atom are almost equal, and so carbon-hydrogen bonds are not polar. A formaldehyde molecule has partial charges on the carbon and oxygen atoms, and we write

$$H_2\overset{\delta+}{C}{=}\overset{\delta-}{\ddot{O}}\!:$$

Even though a formaldehyde molecule is electrically neutral overall, it has a positively charged end and a negatively charged end and thus is a polar molecule.

Polar molecules attract each other (Figure 11-2), and the attraction between them is called a **dipole-dipole attraction.** Charges in polar molecules are considerably smaller than the full electronic charges on ions, and so dipole-dipole forces are smaller than ion-ion forces. Enthalpies of vaporization for many polar compounds are around $20\ kJ \cdot mol^{-1}$.

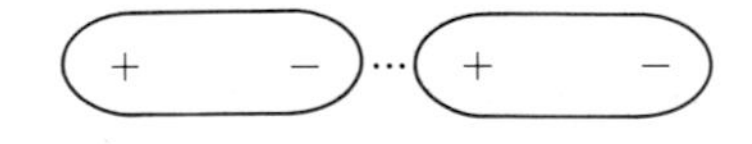

Figure 11-2 Even though polar molecules are electrically neutral overall, they attract each other by a dipole-dipole force. The molecules orient themselves as shown because the positive end of one attracts the negative end of another. The dipoles are said to be oriented head to tail.

A particularly important dipole-dipole attraction occurs when one or more hydrogen atoms are bonded to an electronegative atom, as in the case of water and ammonia. Let's consider the molar enthalpies of vaporization of water, ammonia, and methane:

H_2O	NH_3	CH_4
40.7 $kJ \cdot mol^{-1}$	23.4 $kJ \cdot mol^{-1}$	8.2 $kJ \cdot mol^{-1}$

These three compounds have approximately the same molecular mass (18, 17, 16, respectively), but the amounts of energy required to separate the molecules of the liquids are very different.

We can represent a water molecule by

$$\overset{2\delta-}{\mathrm{O}} \quad {}^{\delta+}\mathrm{H}{-}\mathrm{O}{-}\mathrm{H}^{\delta+}$$

Water molecules in liquid water attract each other through the electrostatic interaction between a hydrogen atom and the oxygen atom on a different molecule:

$$\mathrm{H}_{\delta+}{-}\overset{2\delta-}{\mathrm{O}}{-}\mathrm{H}\cdots\overset{2\delta-}{\mathrm{O}}(\mathrm{H}^{\delta+})(\mathrm{H}_{\delta+})$$

hydrogen bond

The electrostatic attraction that occurs between molecules in which a hydrogen atom is covalently bonded to a highly electronegative atom, such as O, N, or F, is called **hydrogen bonding.** Because a hydrogen atom is so small, the charge on it is highly concentrated, and so it strongly attracts electronegative atoms in neighboring molecules.

Hydrogen bonds are a particularly strong form of dipole-dipole attraction. The pattern shown in Figure 11-3 extends throughout liquid water and gives water its large value of ΔH_{vap} (40.7 $kJ \cdot mol^{-1}$) and its high boiling point. We shall see that hydrogen bonding gives water many special properties.

Hydrogen bonding greatly affects the structure of ice (Figure 11-4), which is described as an *open structure* because of the significant fraction of the space that is unoccupied. The open structure of ice is a direct consequence of the fact that each hydrogen atom is hydrogen-bonded to an oxygen atom of an adjacent molecule. Note the tetrahedral arrangement of the oxygen atoms in ice. Every oxygen atom sits in the center of a tetrahedron formed by four other oxygen atoms.

The structure of liquid water is less open than the structure of ice because when ice melts the total number of hydrogen bonds decreases. Unlike most other substances, water *increases* in density on going from solid to liquid because of a partial breakdown of the hydrogen-bonded structure. The extent of hydrogen bonding in liquid water is only about 80 percent, whereas in ice 100 percent of the oxygen atoms are hydrogen-bonded. The extent of hydrogen bonding in water decreases as the temperature increases.

Hydrogen bonding occurs also in liquid ammonia, but the individual hydrogen bonds are weaker than in water because nitrogen is less electronegative than oxygen and the fractional charges on the nitrogen and hydrogen atoms in NH_3 are less than those on the oxygen and hydrogen atoms in H_2O. Furthermore, there are fewer hydrogen

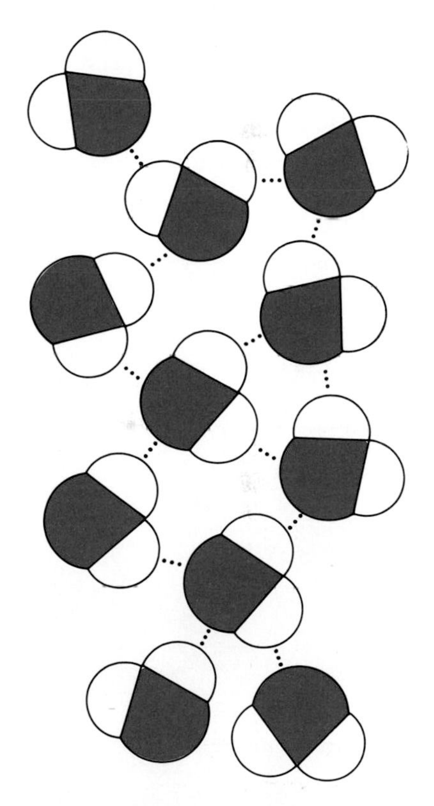

Figure 11-3 There are many hydrogen bonds in liquid water; each oxygen atom can form two hydrogen bonds because each oxygen atom has two lone pairs of electrons. Thus each water molecule has the ability to form four hydrogen bonds. At 25°C about 80 percent of the hydrogen atoms in water are hydrogen-bonded.

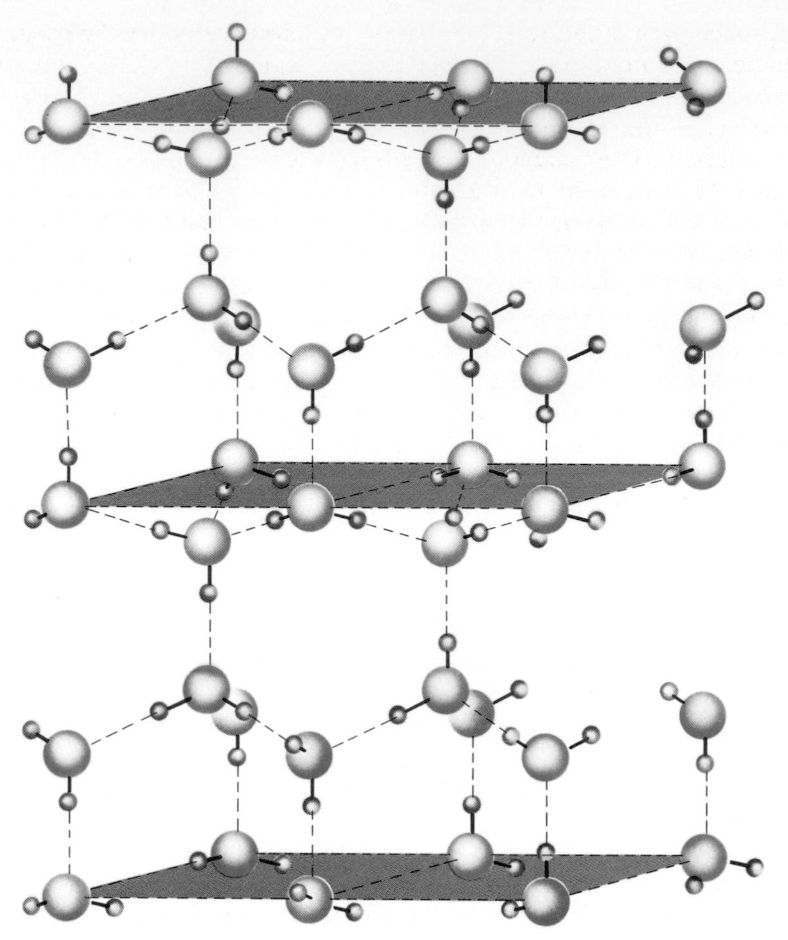

Figure 11-4 The crystalline structure of ice. Each water molecule can form four hydrogen bonds. Note the hydrogen bonds and the tetrahedral arrangement of the oxygen atoms. Each oxygen atom is located in the center of a tetrahedron formed by four other oxygen atoms. The entire structure is held together by hydrogen bonds.

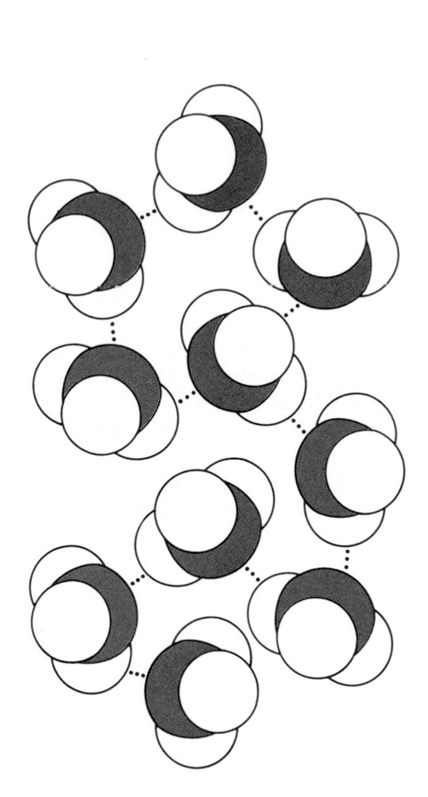

Figure 11-5 There are fewer hydrogen bonds in liquid NH_3 than in H_2O because each nitrogen atom can form only one hydrogen bond. Because NH_3 has only one lone pair, there is only one vertex at which to form a hydrogen bond.

bonds in NH_3 because each nitrogen atom can form only one hydrogen bond (Figure 11-5). Consequently, ΔH_{vap} for ammonia is only about half that for water.

Methane is a nonpolar molecule because it is tetrahedral. Furthermore, because the electronegativities of a carbon atom and a hydrogen atom are almost the same, there is no hydrogen bonding or dipole-dipole attraction in methane. Thus, of H_2O, NH_3, and CH_4, CH_4 has the lowest value of ΔH_{vap}. There is one other attractive force that we must consider in this case, however. Methane liquefies at 91 K and has an enthalpy of vaporization of 8.2 $kJ\cdot mol^{-1}$; therefore, methane molecules must attract each other. Even the noble gases, which consist of single, spherical atoms, can be liquefied.

How neutral, nonpolar molecules attract each other was not understood until the quantum theory was developed. Let's consider two argon atoms. As electrons move about in each atom, an instantaneous dipole moment is set up in each atom (Figure 11-6). When the atoms approach each other, the electrons on one atom influence the electrons on the other atom in such a way that the instantaneous dipole moments on each atom are head to tail (Figure 11-7). The electronic distribution then has a lower energy than if the instantaneous dipole moments were not aligned. The instantaneous dipole-dipole attraction depicted in Figure 11-7 accounts for the attractive force between nonpolar molecules as well as atoms. This force was first explained by the German physicist Fritz London in 1930 and is now called a **London force.**

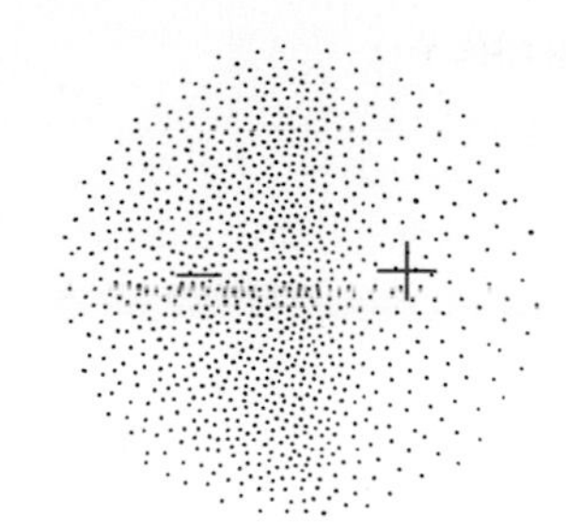

Figure 11-6 If it were possible to take an instantaneous view of an atom, it might look like this drawing. The instantaneous position of the electrons leads to an instantaneous dipole moment. The negative charge is due to a greater-than-average electronic-charge density, and the positive charge is due to a less-than-average electronic-charge density. As the electrons move around, the dipole moment points in all directions equally and averages out to zero.

Because London forces are due to the motion of electrons, their strength depends upon the number of electrons. The more electrons there are in the two interacting molecules, the stronger their attraction for each other. Therefore, we expect the value of ΔH_{vap} to increase with the number of electrons, or even with the size of the molecules. The following data support this prediction:

Substance	$\Delta H_{vap}/\mathrm{kJ \cdot mol^{-1}}$	Substance	$\Delta H_{vap}/\mathrm{kJ \cdot mol^{-1}}$
He	0.08	F_2	6.5
Ne	1.76	Cl_2	20.4
Ar	6.52	Br_2	29.5
Kr	9.03	I_2	41.9
Xe	12.63		

Note that, within each group, the value of ΔH_{vap} increases with the number of electrons.

Figure 11-8 shows plots of the melting points and boiling points of the noble gases and the hydrides of the nonmetallic elements. The hydrogen-bonded compounds (H_2O, NH_3, and HF) have unusually high melting points and boiling points. Except for the hydrogen-bonded compounds, there is a general increase of melting point and boiling point with increasing molecular mass. This increase is due to the increase in London forces with an increase in the number of electrons in a molecule. The attractive forces between molecules, be they dipole-dipole forces or London forces, are collectively called **van der Waals forces.**

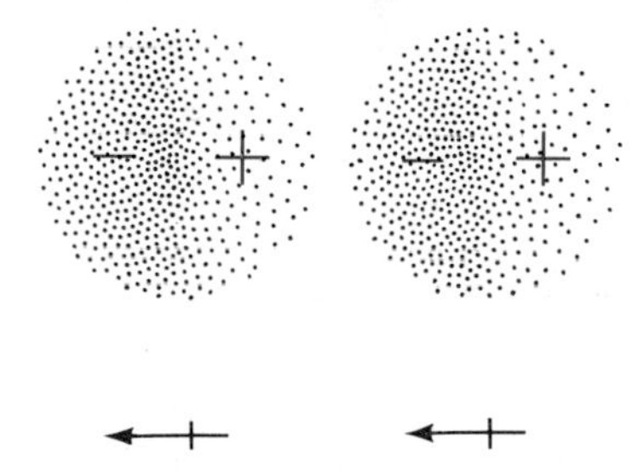

Figure 11-7 When two atoms are near each other, the motion of the electrons in the two atoms affect each other so that the instantaneous dipole moments are head to tail. This leads to an instantaneous dipole-dipole attraction between the two atoms.

Example 11-2: Without referring to any sources, rank the following substances in order of increasing enthalpies of vaporization and boiling points:

$$NaCl \qquad C_2H_4 \qquad CH_3OH$$

Solution: The only ionic compound listed is NaCl. Thus we predict that NaCl has the largest value of ΔH_{vap} and the highest boiling point. The Lewis formula for CH_3OH, shown in the margin, indicates that CH_3OH is hydrogen-bonded. The Lewis formula for C_2H_4 (margin) indicates that C_2H_4 is a nonpolar molecule, and so we predict that

$$\Delta H_{vap}[\mathrm{NaCl}] > \Delta H_{vap}[\mathrm{CH_3OH}] > \Delta H_{vap}[\mathrm{C_2H_4}]$$

The boiling points are in the same order. The actual values are 170 kJ·mol^{-1}, 35.3 kJ·mol^{-1}, and 13.5 kJ·mol^{-1}, respectively. The boiling points are 1690 K for NaCl, 337 K for CH_3OH, and 170 K for C_2H_4.

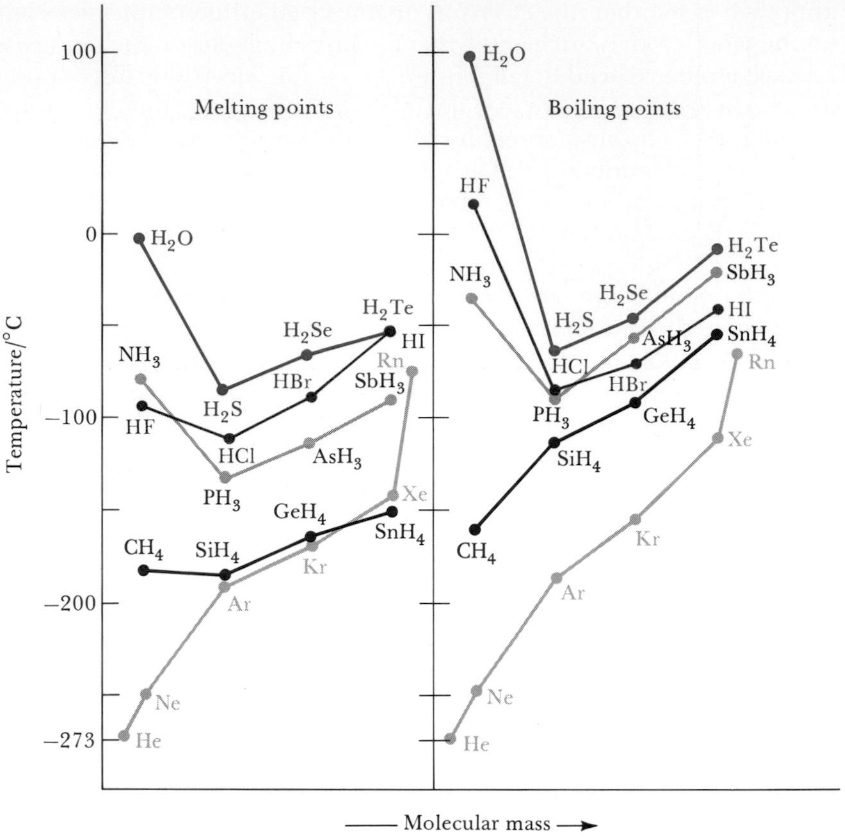

Figure 11-8 Melting and boiling points of the noble gases and hydrides of the nonmetallic elements. Note the abnormally high values for hydrogen fluoride, water, and ammonia, which are the result of hydrogen bonds in the liquid and solid phases.

11-4. A LIQUID HAS A UNIQUE EQUILIBRIUM VAPOR PRESSURE AT EACH TEMPERATURE

Figure 11-9 When a liquid is placed in a closed container that has been evacuated, eventually the pressure of the vapor above the liquid reaches a constant value that depends upon the particular liquid and the temperature.

Suppose that a liquid is placed in an evacuated container with a vapor space and that the container is closed (Figure 11-9). It is observed that the pressure of the vapor over the liquid increases rapidly at first and then progressively more slowly until a constant pressure is reached. Initially, molecules in the liquid phase escape from the surface of the liquid and go into the vapor phase. The number of molecules that leave the surface of the liquid is proportional to the surface area of the liquid. Because the surface area is constant, the rate of evaporation is constant (Figure 11-10). There are no molecules in the vapor phase initially, and so there is no condensation from the vapor phase to the liquid phase. As the concentration of molecules in the vapor phase increases, the pressure of the vapor increases and the number of vapor-phase molecules that collide with the liquid surface increases. As a result, the rate of condensation of the vapor increases. Eventually a point is reached where the rate of evaporation from the liquid surface is equal to the rate of condensation from the vapor phase. The pressure of the vapor no longer increases but takes on a constant value. The evaporation-condensation process appears to have stopped, and we say that the system is at **equilibrium,** meaning that there is no apparent change taking place (Figure 11-11).

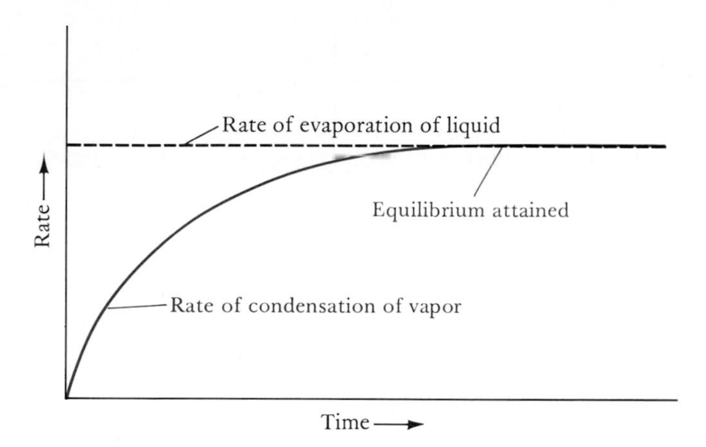

Figure 11-10 Equilibrium is attained when the rate of evaporation of the liquid equals the rate of condensation of the vapor. At equilibrium the vapor pressure of the liquid is constant and is called the equilibrium vapor pressure.

The equilibrium between the liquid and the vapor is a **dynamic equilibrium.** That is, the liquid continues to evaporate and the vapor continues to condense, but the rate of evaporation is exactly equal to the rate of condensation and thus there is no *net* change. In an equation, we have

▪ We shall see that the concept of dynamic equilibrium occurs frequently in chemistry.

$$\text{rate of evaporation} = \text{rate of condensation}$$

▪ A rate is a measure of how fast a quantity changes with time.

The pressure of the vapor at equilibrium is called the **equilibrium vapor pressure.** We shall see that the value of the equilibrium vapor pressure depends upon the particular liquid and the temperature.

Let's consider the approach to a dynamic liquid-vapor equilibrium at two different temperatures. The higher the temperature is, the more rapidly the molecules in the liquid phase are moving and so the higher the rate of evaporation. Figure 11-12 shows that, because the rate of evaporation at T_2 is greater than the rate of evaporation at T_1 (given that $T_2 > T_1$), the equilibrium vapor pressure at T_2 is greater than that at T_1. Thus we see that the value of the equilibrium vapor pressure of a liquid increases with increasing temperature. At each temperature, a

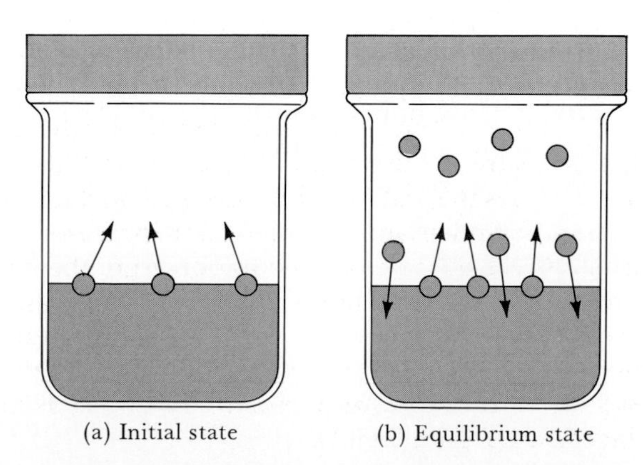

Figure 11-11 When a liquid is placed in a closed container, the rate at which molecules escape from the surface is constant, but the rate at which molecules enter the liquid from the vapor is proportional to the number of molecules in the vapor. When the number of molecules in the vapor is such that the rate of escape from the surface is equal to the rate of condensation from the vapor, the liquid and vapor are in equilibrium with each other.

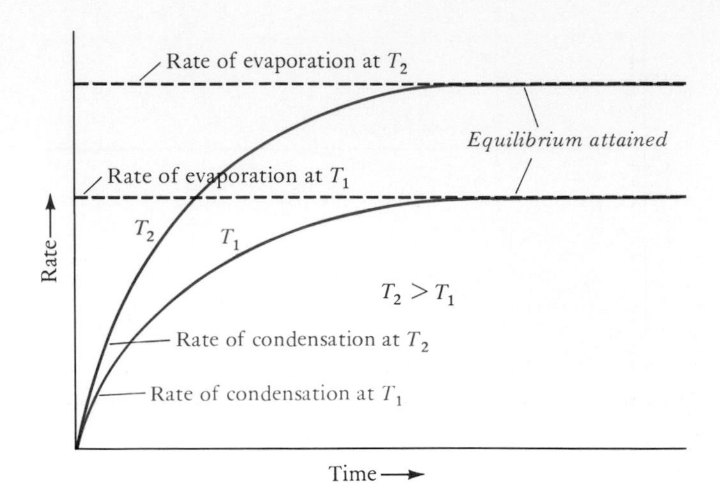

Figure 11-12 The change in vapor pressure with time for a liquid as it approaches equilibrium. Because the rate of evaporation increases with increasing temperature, the equilibrium vapor pressure increases with increasing temperature ($T_2 > T_1$).

Table 11-2 Equilibrium vapor pressure of water as a function of temperature

t/°C	P/atm	P/torr
0	0.0060	4.6
5	0.0086	6.5
10	0.0121	9.2
15	0.0168	12.8
20	0.0230	17.4
25	0.0313	23.8
30	0.0418	31.6
35	0.0555	42.2
40	0.0728	55.3
45	0.0946	71.9
50	0.122	92.5
55	0.155	118.0
60	0.197	149.4
65	0.247	187.5
70	0.308	233.7
75	0.380	289.1
80	0.467	355.1
85	0.571	433.6
90	0.692	525.8
95	0.834	633.9
100	1.000	760.0
105	1.192	906.1
110	1.414	1074.6
120	2.00	1520

liquid has a definite equilibrium vapor pressure. The equilibrium vapor pressure of water as a function of the absolute temperature, which is called its **vapor pressure curve,** is given in Figure 11-13 and in Table 11-2. Figure 11-13 also shows the equilibrium vapor pressure curve for ethanol, CH_3CH_2OH.

The boiling point of a liquid is the temperature at which its vapor pressure equals atmospheric pressure. The **normal boiling point** (that at exactly 1 atm) of water is 373 K (100°C). If the atmospheric pressure is less than 1.00 atm, then the temperature at which the vapor pressure of liquid water equals atmospheric pressure is less than 100°C. For example, the elevation at Vail, Colorado, is about 8000 feet. The atmospheric pressure at this elevation is about 0.75 atm, and so water boils at 92°C. In a pressure cooker, on the other hand, when the pressure is 2.0 atm, water boils at 120°C. Because atmospheric pressure decreases with increasing elevation and because the rate at which food cooks depends on the temperature, it requires a significantly longer time to cook food by boiling at high elevations than at sea level. An egg must be boiled for about 5 min at 9200 ft in order to be cooked to the same extent as one boiled for 3 min at sea level.

11-5. RELATIVE HUMIDITY IS BASED UPON THE VAPOR PRESSURE OF WATER

The amount of water vapor in the atmosphere is expressed in terms of **relative humidity.** Relative humidity is the ratio of the partial pressure of the water vapor in the atmosphere to the equilibrium vapor pressure of water at the same temperature times 100. In terms of an equation we have

$$\text{relative humidity} = \frac{P_{H_2O}}{P^{\circ}_{H_2O}} \times 100 \qquad (11\text{-}2)$$

where P_{H_2O} is the partial pressure of the water vapor in the air and $P^{\circ}_{H_2O}$ is the equilibrium vapor pressure of water at the same temperature. At 20°C, the equilibrium vapor pressure of water is 17.4 torr. If

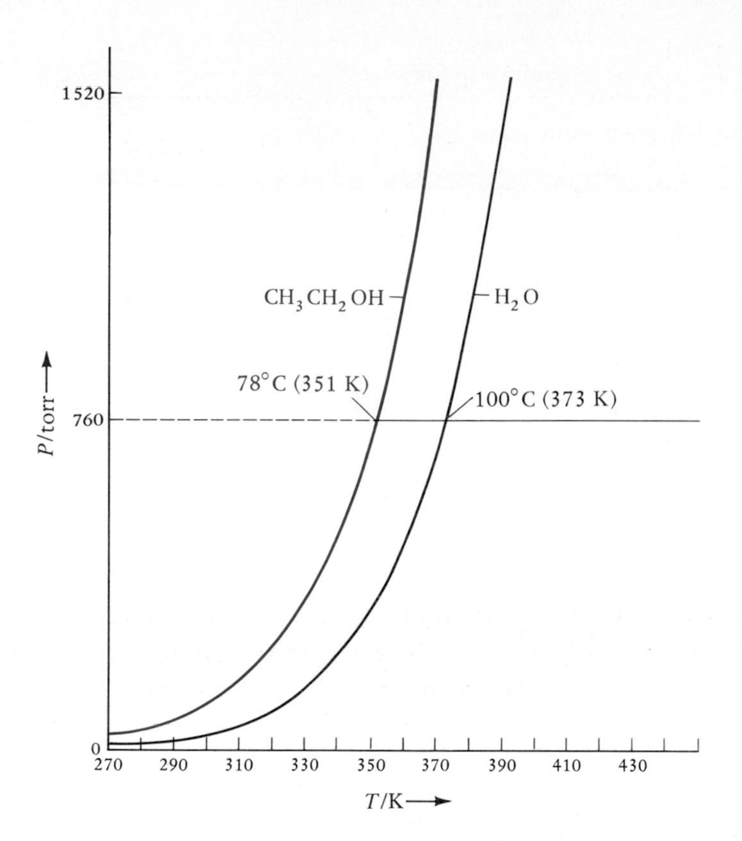

Figure 11-13 The equilibrium vapor pressure curves for water and ethanol over the temperature range 270 to 430 K (−3 to 157°C). Note the rapid increase in vapor pressure with increasing temperature. The equilibrium vapor pressure curve for ethanol lies above that for water because ethanol has a higher equilibrium vapor pressure than water at the same temperature.

the partial pressure of the water vapor in the air is 11.2 torr, then the relative humidity is

$$\text{relative humidity} = \frac{11.2 \text{ torr}}{17.4 \text{ torr}} \times 100 = 64.4\%$$

If the temperature of the air is lowered to 13°C, where the equilibrium vapor pressure of water is 11.2 torr, then the relative humidity is

$$\text{relative humidity} = \frac{11.2 \text{ torr}}{11.2 \text{ torr}} \times 100 = 100\%$$

At 13°C, air that contains water vapor at a partial pressure of 11.2 torr is saturated with water vapor. At this temperature, the water vapor begins to condense as dew or fog, which consists of small droplets of water. The air temperature at which the relative humidity reaches 100 percent is called the **dew point.** Most people begin to feel uncomfortable when the dew point rises above 20°C, and air with a dew point above 24°C is generally regarded as extremely humid or muggy.

▪ Dew forms when the night temperature drops below the dew point. Frost is frozen dew.

Example 11-3: Calculate the relative humidity and the dew point if the partial pressure of water vapor in the air is 22.2 torr and the temperature of the air is 30°C. The equilibrium vapor pressure of water at 30°C is 31.6 torr (Table 11-2).

Solution: The relative humidity, given by Equation (11-2), is

$$\text{relative humidity} = \frac{P_{H_2O}}{P^\circ_{H_2O}} \times 100 = \frac{22.2 \text{ torr}}{31.6 \text{ torr}} \times 100 = 70.3\%$$

The dew point is the temperature at which the equilibrium vapor pressure of water is equal to 22.2 torr. According to Table 11-2, this is about 24°C. Such a day would be considered very uncomfortable.

11-6. THE TEMPERATURE DEPENDENCE OF THE EQUILIBRIUM VAPOR PRESSURE IS GIVEN BY THE CLAPEYRON-CLAUSIUS EQUATION

The dependence of equilibrium vapor pressure on temperature, as shown in Figure 11-13 or Table 11-2 for water, is given by an equation called the **Clapeyron-Clausius equation,** which is

$$\log\left(\frac{P_2}{P_1}\right) = \frac{\Delta H_{vap}}{2.30R}\left(\frac{T_2 - T_1}{T_1 T_2}\right) \tag{11-3}$$

(The properties of logarithms are reviewed in Appendix A-2.) In this equation P_2 is the equilibrium vapor pressure at the kelvin temperature T_2, P_1 is the equilibrium vapor pressure at the kelvin temperature T_1, R is the molar gas constant, $8.314\ \text{J}\cdot\text{K}^{-1}\cdot\text{mol}^{-1}$ and ΔH_{vap} is the molar enthalpy of vaporization. Given the value of the heat of vaporization of a liquid, we can use Equation (11-3) to calculate the vapor pressure at one temperature if we know the vapor pressure at some other temperature. Let's calculate the vapor pressure of water at 110°C, given that the normal boiling point of water is 100°C. Thus we let the vapor pressure at 110°C (T_2 = 383 K) be P_2, and we have that P_1 = 1.000 atm at T_1 = 373 K. Given that $\Delta H_{vap} = 40.7\ \text{kJ}\cdot\text{mol}^{-1}$, we write

$$\log\left(\frac{P_2}{1.000 \text{ atm}}\right) = \frac{40.7 \times 10^3\ \text{J}\cdot\text{mol}^{-1}}{(2.30)(8.314\ \text{J}\cdot\text{K}^{-1}\cdot\text{mol}^{-1})}\left(\frac{383\ \text{K} - 373\ \text{K}}{(373\ \text{K})(383\ \text{K})}\right)$$

$$= 0.149$$

Therefore

$$\frac{P_2}{1.000 \text{ atm}} = 10^{0.149} = 1.41$$

or

$$P_2 = 1.41 \text{ atm}$$

in good agreement with the entry at 110°C in Table 11-2.

An assumption in deriving the Clapeyron-Clausius equation is that ΔH_{vap} does not vary with temperature. In fact, however, a slight variation is observed for water; for example, $\Delta H_{vap} = 40.7\ \text{kJ}\cdot\text{mol}^{-1}$ at 100°C, whereas $\Delta H_{vap} = 44.0\ \text{kJ}\cdot\text{mol}^{-1}$ at 25°C. Thus, the value of ΔH_{vap} to be used in Equation (11-3) should be an average over the temperature range involved. Otherwise, the values calculated from Equation (11-3) will disagree somewhat with experimental values.

Example 11-4: The Clapeyron-Clausius equation can be used to calculate equilibrium vapor pressures of a solid by using ΔH_{sub} instead of ΔH_{vap}. Given that $\Delta H_{sub} = 25.2\ \text{kJ}\cdot\text{mol}^{-1}$ for $CO_2(s)$ and that P = 1.000 atm at

$-78°C$, calculate the equilibrium vapor pressure of $CO_2(s)$ at $-100°C$. The experimental value is 0.138 atm.

Solution: We let $P_1 = 1.000$ atm, $T_1 = 195$ K, and $T_2 = 173$ K and write

$$\log\left(\frac{P_2}{1.000\text{ atm}}\right) = \frac{25.2 \times 10^3\text{ J}\cdot\text{mol}^{-1}}{(2.30)(8.314\text{ J}\cdot\text{K}^{-1}\cdot\text{mol}^{-1})}\left(\frac{173\text{ K} - 195\text{ K}}{(173\text{ K})(195\text{ K})}\right)$$

$$= -0.859$$

or

$$\frac{P_2}{1.000\text{ atm}} = 10^{-0.859} = 0.138$$

or

$$P_2 = 0.138\text{ atm}$$

which is in good agreement with the experimental result.

11-7. A PHASE DIAGRAM DISPLAYS THE REGIONS OF ALL THE PHASES OF A PURE SUBSTANCE SIMULTANEOUSLY

The vapor pressure curve, the sublimation pressure curve, and the melting point curve of a pure substance can be combined into a single diagram called a **phase diagram.** The phase diagram of water is shown in Figure 11-14.

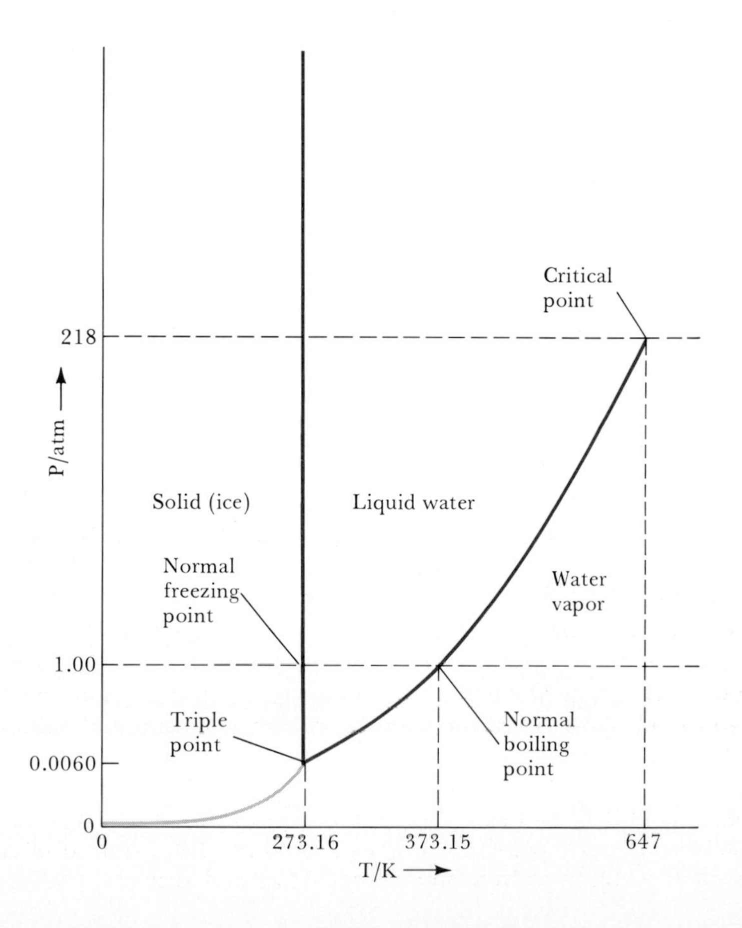

Figure 11-14 The phase diagram of water (not to scale), which displays simultaneously the sublimation pressure curve *(gray)*, the vapor pressure curve *(red)*, and the melting point curve *(blue)*. The triple point, the critical point, the normal boiling point, and the normal freezing point are indicated in the figure. The phase equilibrium lines are the boundaries between the regions of stability of the solid, liquid, and vapor phases. The regions of stability of the various phases are labeled solid (ice), liquid water, and water vapor. The scale of the figure has been distorted in order to show the various features more clearly.

Along the vapor pressure curve (the red curve in Figure 11-14), liquid and vapor exist together at equilibrium. To the left of this curve, at lower temperatures, the water exists as a liquid. To the right of the curve, at higher temperatures, the water exists as a vapor. The equilibrium vapor pressure of a liquid increases with temperature up to the **critical point** (Figure 11-14), where the vapor pressure curve terminates abruptly. A liquid cannot exist at a temperature greater than the **critical temperature** of the substance. A gas above its critical temperature cannot be liquefied no matter how high a pressure is applied. The critical point for water occurs at 218 atm and 647 K.

Along the **sublimation pressure curve** (gray curve in Figure 11-14), solid and vapor exist together at equilibrium. To the left of this curve, at lower temperatures, water exists as a solid (ice). To the right of this curve, at higher temperatures, water exists as a vapor.

Along the **melting point curve** (blue curve in Figure 11-14), solid and liquid exist together in equilibrium. To the left of this curve, water exists as ice and to the right of the curve water exists as liquid. Melting points are only weakly dependent on pressure, and so a melting point curve is an almost vertical line (Figure 11-14). For almost all substances, the melting point increases with increasing pressure at a rate of 0.01 to 0.03 $K{\cdot}atm^{-1}$. Water is anomalous in that its melting point decreases with increasing pressure. The melting point of ice decreases by 0.01 K per atmosphere of applied pressure. Consequently, unlike most other solids, ice can be melted by the application of pressure.

Notice that the three curves in Figure 11-14 separate regions in which water exists as a solid, a liquid, or a vapor. Let's use Figure 11-14 to follow the behavior of water as it is heated from −50°C to 200°C at a constant pressure of 1 atm. At −50°C (223 K) and 1 atm, water exists as ice. As we heat the ice at 1 atm, we move horizontally from left to right along the dashed line in Figure 11-14. At 0°C, we cross the melting point curve and pass from the solid region into the liquid region. At 100°C, we cross the equilibrium vapor pressure curve and pass from the liquid region into the vapor region.

The three curves in Figure 11-14 intersect at a point called the **triple point.** At the triple point, and only at the triple point, all three phases—solid, liquid, and gas—coexist in equilibrium. The triple point for water occurs at 4.58 torr and 273.16 K. Notice that if we heat ice at a constant pressure less than 4.58 torr (0.0060 atm), then the ice sublimes rather than melts.

Example 11-5: Use the phase diagram of water given in Figure 11-14 to predict the result of increasing the pressure of water vapor initially at 1 atm and 500 K, keeping the temperature constant.

Solution: At 1 atm and 500 K water exists as a vapor. As the pressure is increased, we cross the liquid-vapor curve below the critical point at a pressure of about 150 atm, and the vapor condenses to a liquid.

The phase diagram of carbon dioxide is shown in Figure 11-15. Although it looks similar to that of water, there are several important differences. The melting point curve of carbon dioxide goes up and to

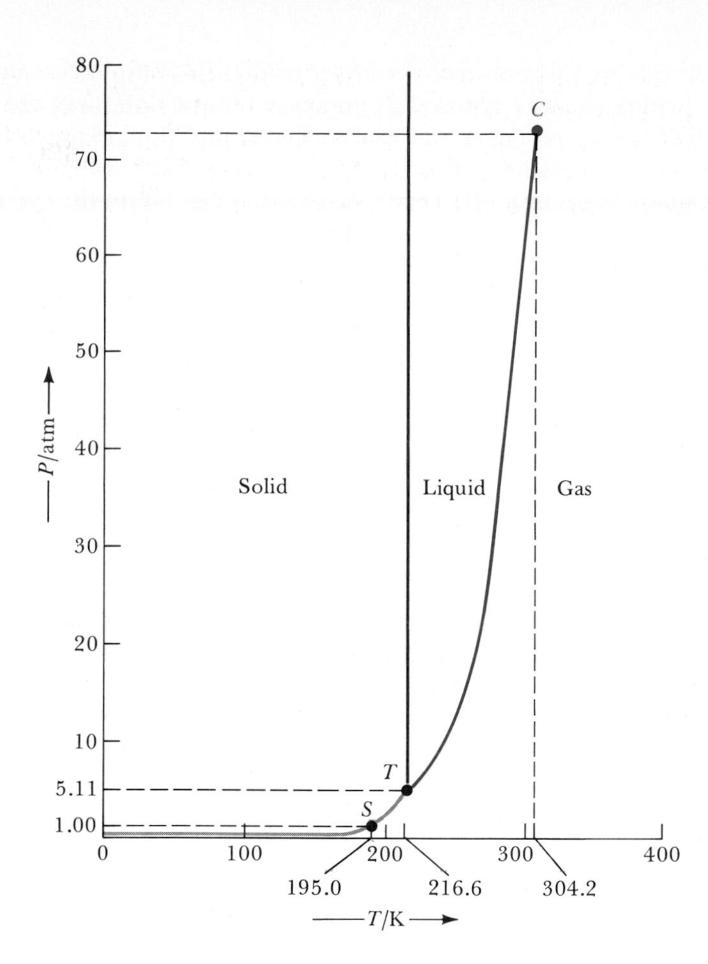

Figure 11-15 The phase diagram of carbon dioxide. The point *C* is the critical point, and the point *T* is the triple point. Note that the triple point lies above 1 atm and thus $CO_2(s)$ at 1 atm does not melt—it sublimes. The point *S* is the normal sublimation point of CO_2.

the right, indicating that the melting point of CO_2 increases with increasing pressure. Recall that the melting point curve of water points up and slightly to the left, indicating that the melting point of H_2O decreases with increasing pressure.

Another difference between Figures 11-14 and 11-15 is in the triple point. The triple point for CO_2 occurs at 5.11 atm and 216.6 K. Because the pressure at the triple point is greater than 1 atm, CO_2 does not melt when it is heated at 1 atm. Instead, it sublimes. The **normal sublimation point** of CO_2 is 195 K (−78°C). This is the temperature of solid CO_2 at 1 atm. Liquid CO_2 can be obtained by compressing $CO_2(g)$ at a temperature below its critical point (31°C). A pressure of about 60 atm is required to liquefy CO_2 at 25°C. A carbon dioxide–filled fire extinguisher at 25°C contains liquid CO_2 at a pressure of about 60 atm.

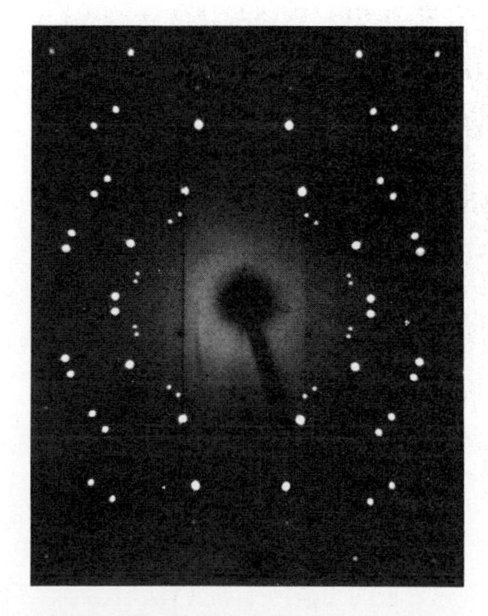

Figure 11-16 The X-ray diffraction pattern produced by a crystal of sodium chloride. The symmetry and spacing of the dots carry detailed information regarding the arrangement of atoms in the crystals.

11-8. X-RAY DIFFRACTION PATTERNS YIELD INFORMATION ABOUT THE STRUCTURES OF CRYSTALS

In contrast to gases and liquids, a distinguishing characteristic of crystals is the ordered nature of the crystal lattice. When X-rays with a wavelength comparable to the nearest-neighbor distance between atoms in a crystal pass through the crystal, an **X-ray diffraction pattern** results (Figure 11-16). The presence of a definite ordered array of atoms in the crystal produces a characteristic diffraction pattern of the X-rays, which can be recorded as an array of spots on an X-ray film. X-ray diffraction patterns are used to determine the arrangement of the atoms and molecules in a crystal.

The smallest subunit of a crystal lattice that can be used to generate the entire lattice is called a **unit cell.** A crystal lattice is a repeating pattern of unit cells (Figure 11-17). There are three types of cubic unit cells which are shown in Figure 11-18. The crystalline structures of many metals fall into one of these classes. Only one metal, polonium, has a **simple cubic** lattice. Some examples of metals that have **body-centered cubic** lattices are Ba, Cs, K, Li, Mo, Na, Ta, U, and V. **Face-centered cubic** lattices are found in Ag, Al, Au, Cu, Sr, Ni, Pb, Pt, and the noble gases.

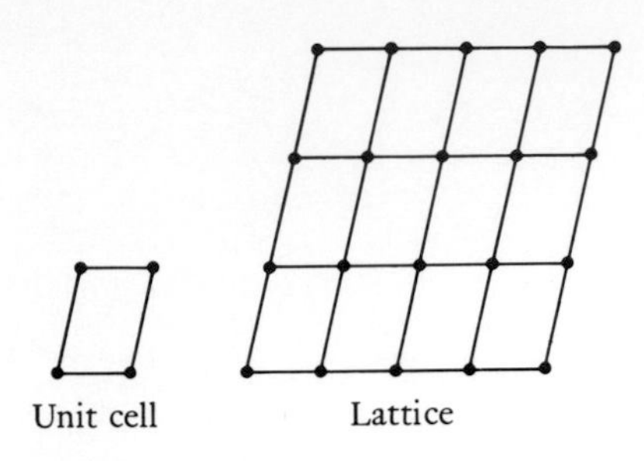

Figure 11-17 A two-dimensional illustration of the generation of a crystal lattice by a unit cell.

Example 11-6: Copper exists as a face-centered cubic lattice. How many copper atoms are there in a unit cell?

Solution: Reference to Figure 11-18 shows that each of the eight copper atoms at the corners of the unit cell are shared by eight unit cells, and so we assign one copper atom ($8 \times 1/8 = 1$) to each unit cell. Each of the six atoms at the faces is shared by two unit cells, and so we assign three more copper atoms ($6 \times 1/2 = 3$) to each unit cell, giving a total of four copper atoms in a unit cell. Figure 11-18 also illustrates the counting process for the simple cubic and body-centered cubic unit cells.

Example 11-7: Copper, which crystallizes as a face-centered cubic lattice, has a density of $8.930\ \text{g}\cdot\text{cm}^{-3}$ at 20°C. Calculate the radius of a copper atom.

Solution: The molar volume of copper is given by

$$V = \frac{\text{molar mass}}{\text{density}} = \frac{63.55\ \text{g}\cdot\text{mol}^{-1}}{8.930\ \text{g}\cdot\text{cm}^{-3}} = 7.116\ \text{cm}^3\cdot\text{mol}^{-1}$$

The volume due to one copper atom is thus

$$\frac{7.116\ \text{cm}^3\cdot\text{mol}^{-1}}{6.022 \times 10^{23}\ \text{atom}\cdot\text{mol}^{-1}} = 1.182 \times 10^{-23}\ \text{cm}^3\cdot\text{atom}^{-1}$$

In Example 11-6 we found that the unit cell of copper contains four copper atoms. The volume occupied by the unit cell, then, is given by

$$v = \left(\frac{4\ \text{atoms}}{\text{unit cell}}\right)\left(1.182 \times 10^{-23}\ \frac{\text{cm}^3}{\text{atom}}\right)$$
$$= 4.728 \times 10^{-23}\ \frac{\text{cm}^3}{\text{unit cell}}$$

Because the unit cell is cubic, the length of one edge is

$$l = (4.727 \times 10^{-23}\ \text{cm}^3)^{1/3} = 3.616 \times 10^{-8}\ \text{cm} = 361.6\ \text{pm}$$

The length of a diagonal of any face of the unit cell is

$$\text{diagonal} = \sqrt{2}(361.6\ \text{pm}) = 511.4\ \text{pm}$$

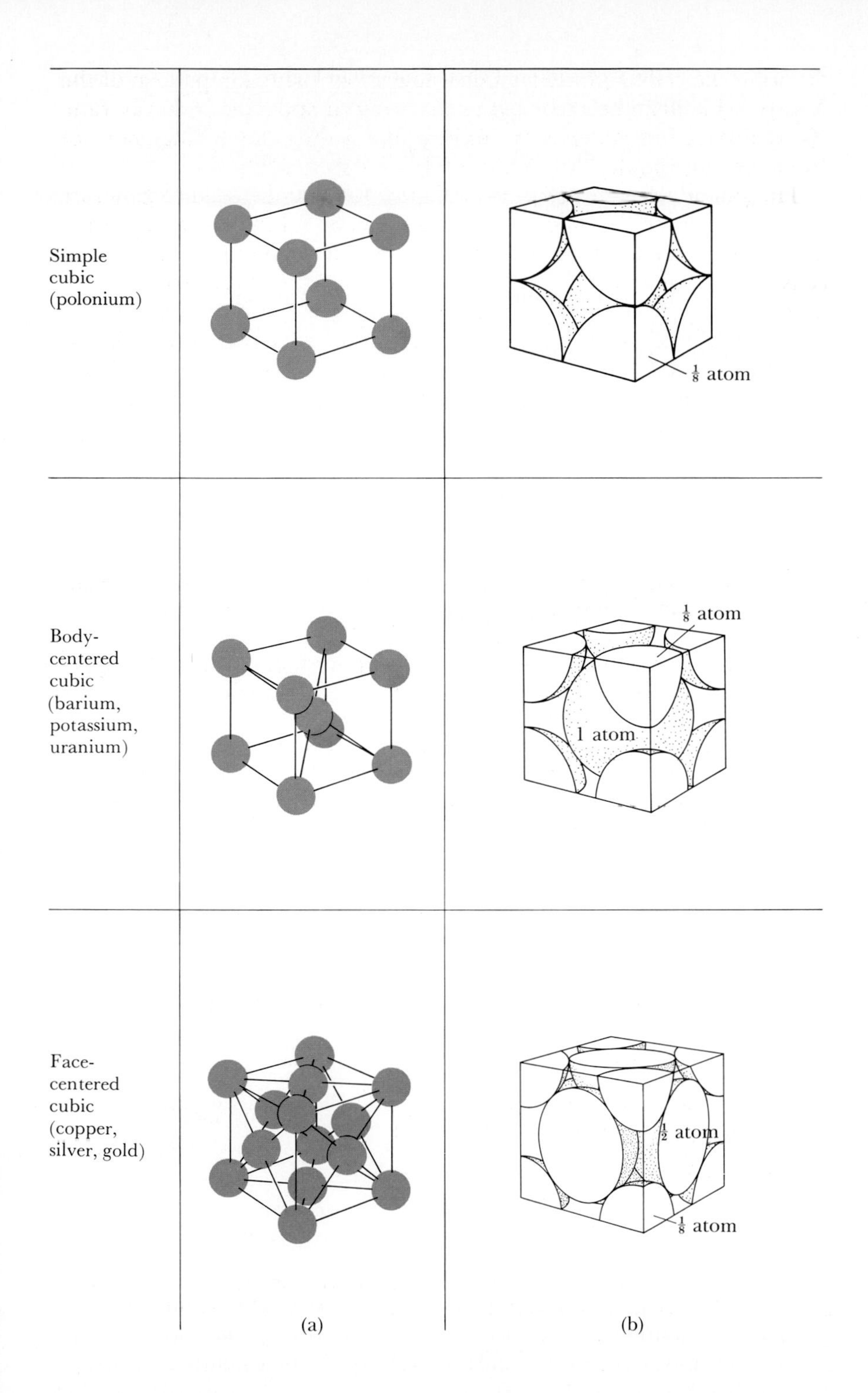

Figure 11-18 The three cubic unit cells: simple cubic, body-centered cubic, and face-centered cubic. (a) An open perspective of the unit cells. Note that the body-centered cubic unit cell has an atom at the center of the unit cell and that a face-centered unit cell has an atom at the center of each face. (b) The sharing of atoms by adjacent unit cells.

According to the face-centered cubic unit cell in Figure 11-18(b), the diameter of a copper atom is

$$\text{diameter} = \frac{511.4\ \text{pm}}{2} = 255.7\ \text{pm}$$

and thus the radius is 127.9 pm.

If the density and the dimensions of a unit cell are known from X-ray analysis, then we can determine the value of Avogadro's number, as shown in the following Example.

Example 11-8: Potassium crystallizes in a body-centered cubic lattice, and the length of a unit cell is 533.3 pm. Given that the density of potassium is 0.8560 $\text{g}\cdot\text{cm}^{-3}$, calculate Avogadro's number.

Solution: The molar volume of potassium is

$$V = \frac{\text{molar mass}}{\text{density}} = \frac{39.10\ \text{g}\cdot\text{mol}^{-1}}{0.8560\ \text{g}\cdot\text{cm}^{-3}} = 45.68\ \text{cm}^3\cdot\text{mol}^{-1}$$

The volume of a unit cell is

$$v = (533.3\ \text{pm})^3 = (5.333 \times 10^{-8}\ \text{cm})^3 = 1.517 \times 10^{-22}\ \text{cm}^3$$

The number of unit cells per mole is

$$\text{unit cells per mole} = \frac{V}{v} = \frac{45.68\ \text{cm}^3\cdot\text{mol}^{-1}}{1.517 \times 10^{-22}\ \text{cm}^3/\text{unit cell}}$$
$$= 3.011 \times 10^{23} \frac{\text{unit cells}}{\text{mol}}$$

There are two atoms per unit cell in this case (Figure 11-18b), however, and so the number of atoms per mole is

$$\left(3.011 \times 10^{23} \frac{\text{unit cells}}{\text{mol}}\right)\left(\frac{2\ \text{atoms}}{\text{unit cell}}\right) = 6.022 \times 10^{23}\ \text{atom}\cdot\text{mol}^{-1}$$

In 1976, scientists at the National Bureau of Standards used very precise X-ray measurements on ultra-pure silicon to obtain a value of $(6.022098 \pm 6) \times 10^{23}$ for Avogadro's number. The ± 6 indicates the uncertainty in the last digit.

11-9. CRYSTALS CAN BE CLASSIFIED ACCORDING TO THE FORCES BETWEEN THE CONSTITUENT PARTICLES

Crystal structures are determined by the size of the atoms, ions, or molecules making up the lattice, and by the nature of the forces that act between these particles. Crystals, or solids, can be classified according to the forces between the particles. The examples that we discussed in the previous section consisted of particles of the same size. Such crystals are called **atomic crystals,** a good example being crystals of the noble gases, which crystallize as face-centered cubic crystals. In this section we shall discuss ionic crystals, molecular crystals, covalent network crystals, and the conduction properties of metals.

Ionic crystals are held together by the electrostatic attraction between ions of opposite charge. The crystalline structure of ionic crystals

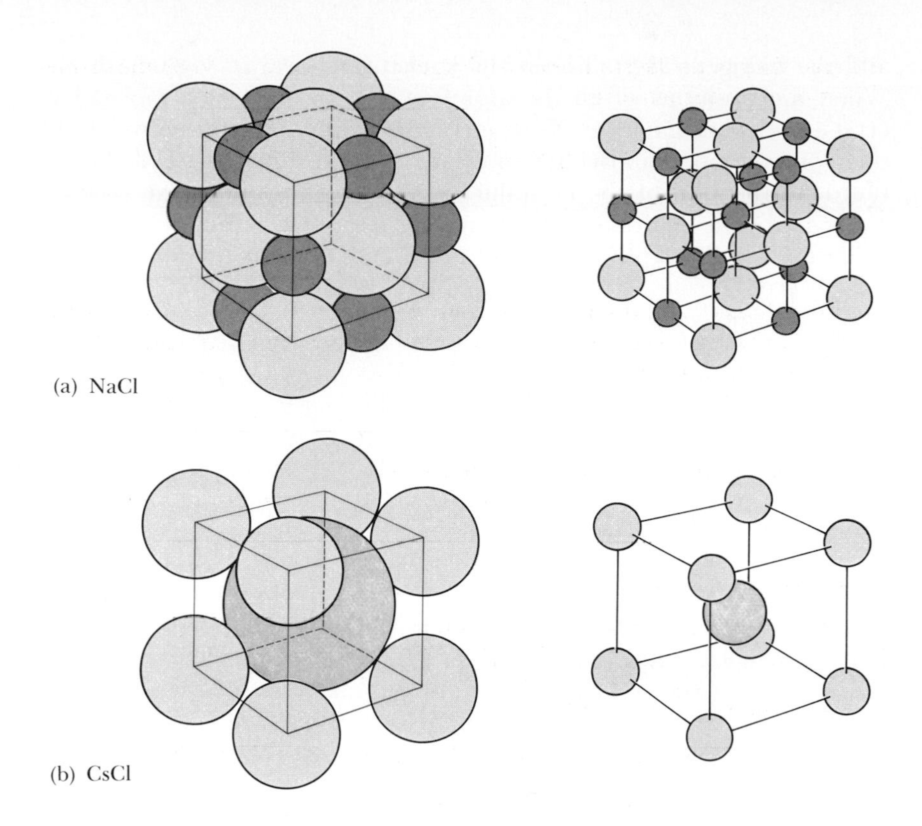

Figure 11-19 Different representations of the unit cells of (a) NaCl and (b) CsCl. The different crystalline structures in the two cases are a direct consequence of the relative sizes of the cations and the anions. Recall that cations are positively charged ions and that anions are negatively charged ions.

often depends upon how the cations and anions can be packed together to form a lattice. Consequently, the difference in the sizes of the cation and the anion plays a key role. For example, the unit cells of the salts NaCl and CsCl are shown in Figure 11-19. Notice that each sodium ion in NaCl is surrounded by an octahedral arrangement of chloride ions, and that each cesium ion in CsCl is surrounded by a cubic arrangement of chloride ions. The different packing arrangements for NaCl and CsCl are a direct consequence of the fact that cesium ions are larger than sodium ions.

Each ion in an ionic crystal is surrounded by ions of opposite charge. The total electrostatic interaction energy of the lattice accounts for much of the lattice energy, a quantity that we used in Section 7-7. Because electrostatic interactions are relatively strong, ionic solids usually have high melting points and low vapor pressures.

Crystals composed of neutral molecules are called **molecular crystals.** The forces that hold the molecules together in molecular crystals are the types of forces that we discussed in Section 11-3. These forces range in strength from the weak London attraction between small nonpolar molecules such as H_2 and CH_4 to the relatively strong dipole-dipole attraction between large, polar molecules. Consequently, the melting points of molecular crystals range from 14 K for H_2 to over 500 K for large polar molecules. Generally, molecular crystals have lower melting points and higher vapor pressures than those of ionic crystals.

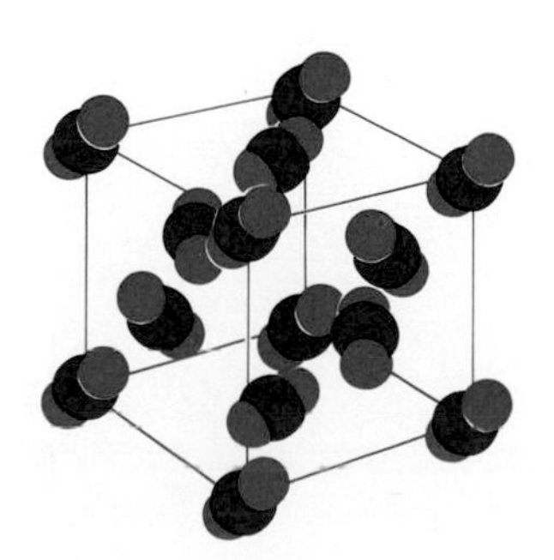

Figure 11-20 The unit cell of CO_2. The molecules have been reduced in size for clarity.

Molecular crystals come in a great variety of types. For example, methane crystallizes in a face-centered cubic structure like the noble gases, carbon dioxide crystals have the unit cell shown in Figure 11-20, and the halogens crystallize in the structure shown in Figure 11-21. When the positions of all the atoms in the unit cell of a molecular crystal are determined by **X-ray crystallography,** the positions of all the atoms within an individual molecule are determined as well. Thus, the determination of the crystalline structure of a molecular solid is equivalent to a determination of the structure of a single molecule.

X-ray crystallography was used to produce the electron density contour map of benzoic acid shown in Figure 11-22. The general outline of the planar molecule

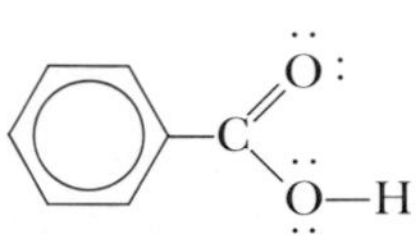

benzoic acid

is clearly discernible. The characteristic shape of a benzene ring (page 275) is also evident. Such information is typical of that obtainable with X-ray crystallographic techniques. Precise details of structure such as those shown in Figure 11-23 for the anthracene molecule are also readily determined. X-ray crystallography is one of the most powerful methods available for the determination of molecular structure and is used extensively by chemists.

A few substances form **covalent network crystals,** in which the constituent particles are held together by covalent bonds. Carbon, which in

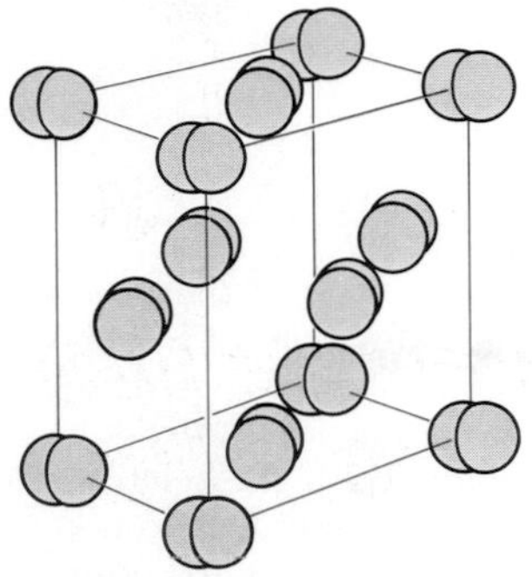

Figure 11-21 The unit cells of Cl_2, Br_2, and I_2. The molecules have been reduced in size for clarity.

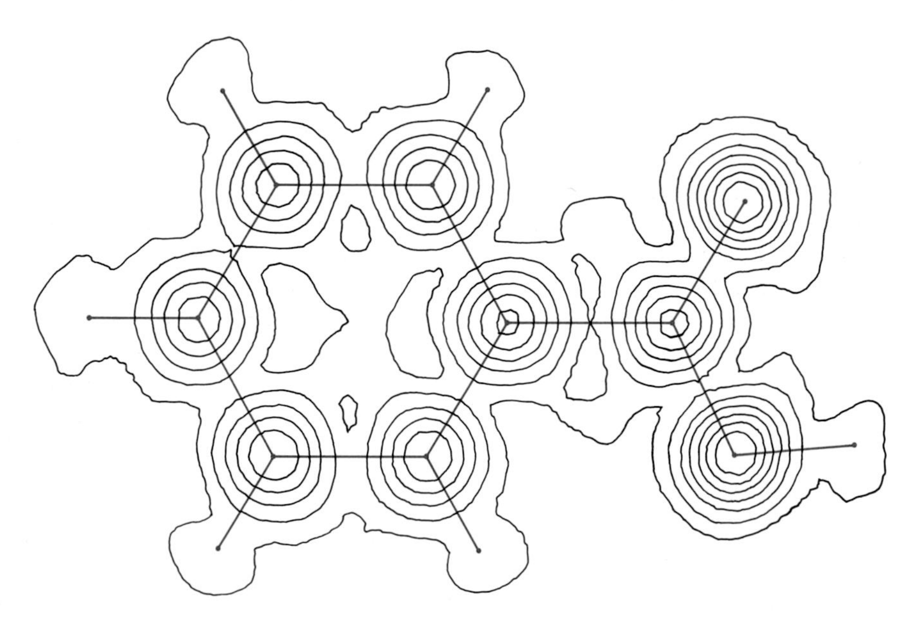

Figure 11-22 An electron density contour map for benzoic acid obtained from X-ray diffraction measurements. The positions of the nuclei in the molecule are readily deduced from this contour map. Notice the hexagonal array of carbon atoms. Also notice the —COOH group that is attached to the benzene ring. The overall molecule is planar. Figure courtesy of Professor H. Hope, University of California, Davis.

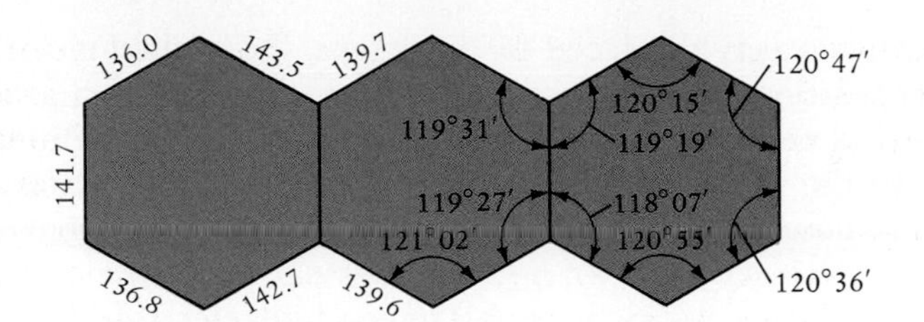

Figure 11-23 The dimensions of an individual anthracene molecule. This is the type of information that can be obtained from X-ray diffraction experiments. The distances given are in picometers.

pure form can exist as diamond or graphite, is a good example of such a substance. Diamond has an extended, covalently bonded tetrahedral structure. Each carbon atom lies at the center of a tetrahedron formed by four other carbon atoms (Figure 11-24). The carbon-carbon bond distance is 154 pm, which is the same as the carbon-carbon bond distance in ethane. The diamond crystal is, in effect, a gigantic molecule. The hardness of diamond is due to the fact that each carbon atom throughout the crystal is covalently bonded to four others and thus many strong covalent bonds must be broken in order to cleave a diamond. Graphite has the unusual layered structure shown in Figure 11-25. The carbon-carbon bond distance within a layer is 139 pm, which is close to the carbon-carbon bond distance in benzene. The distance between layers is about 340 pm. The bonding within a layer is covalent, but the interaction between layers is weak. Therefore, the layers easily slip past each other, producing the molecular basis for the lubricating action of graphite. The "lead" in lead pencils is actually graphite. Layers of the graphite slide from the pencil onto the paper.

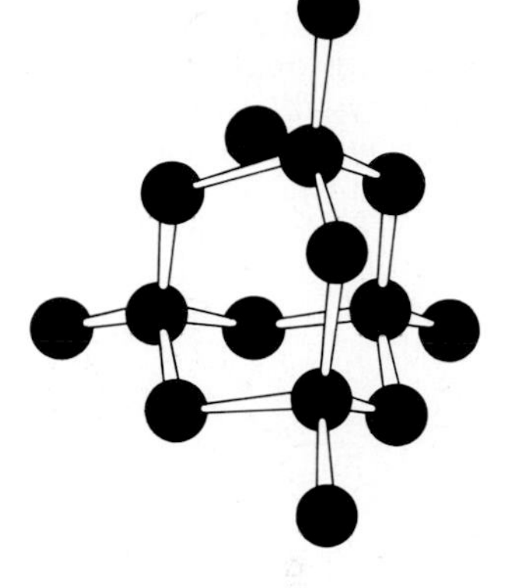

Figure 11-24 The crystalline structure of diamond. Each carbon atom is covalently bonded to four other carbon atoms, forming a tetrahedral network. A diamond crystal is essentially one gigantic molecule.

11-10. THE ELECTRONS IN METALS ARE DELOCALIZED THROUGHOUT THE CRYSTAL

A quantum theoretical treatment of a **metallic crystal** shows that the lattice sites are occupied by ions, formed by removing the valence-shell electrons from the atoms of the metal, and that the valence electrons are delocalized over the entire crystal lattice. Thus, we can picture a metal as an array of fixed ionic cores, consisting of the inner-core electrons and nuclei, immersed in a sea of (valence) electrons. Because the valence electrons are delocalized, and not bound to any particular nu-

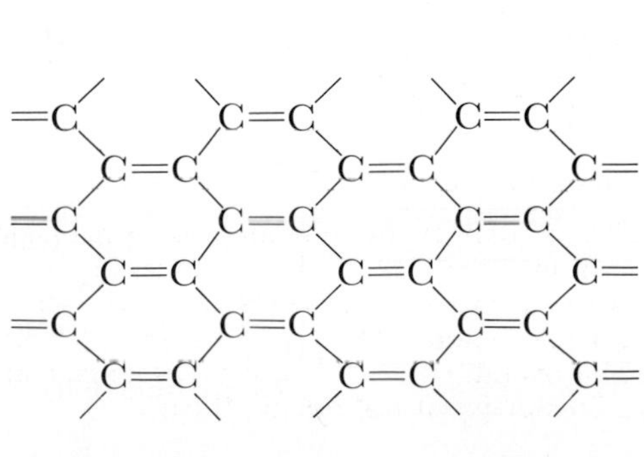

Figure 11-25 The layered structure of graphite; each layer resembles a network of benzene rings joined together. The bonding within a layer is covalent and strong. The interaction between layers, however, is due only to London forces and so is relatively weak. Consequently, the layers easily slip past each other, giving graphite its slippery feel.

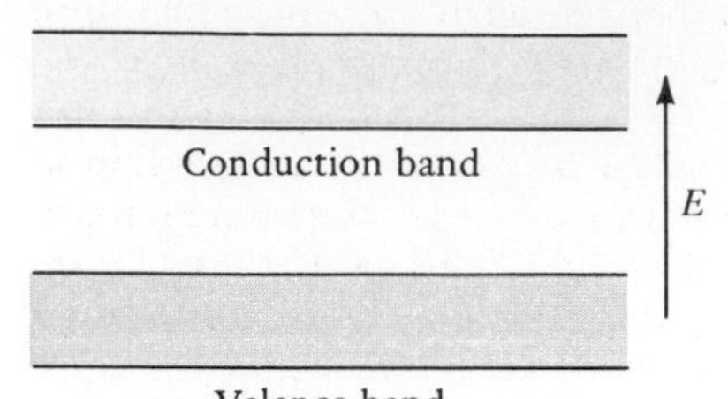

Figure 11-26 When the atoms of a crystal are brought together to form the crystal lattice, the valence orbitals of the atoms combine to form two sets of energy levels, called the valence band and the conduction band.

cleus, they are easily displaced by an externally applied electric field. This delocalization of electrons is what gives metals their characteristic high electrical conductivity. The crystalline structure of a metal is determined by the packing of its ionic cores, and most metals have one of the cubic structures shown in Figure 11-18.

The high electrical conductivity of metals is one of their most characteristic properties. To see why metals have a high electrical conductivity whereas insulators have a low electrical conductivity, we must discuss the electronic energy levels in crystals.

In a crystal, there are two sets of energy levels because of the combination of the valence orbitals of all the atoms. These two sets of energy levels are analogous to the two orbitals that result when orbitals from just two atoms are combined. The lower set of energy levels is called the **valence band** and is occupied by the valence electrons of the atoms. The higher set corresponds to antibonding orbitals and is called the **conduction band** (Figure 11-26). Electrons in the conduction band can move readily throughout the crystal.

An electric current is carried in a solid by the electrons in the conduction band, which are called the **conduction electrons.** In an insulator (such as a nonmetal), there are essentially no electrons in the conduction band because the energies there are much higher than the energies in the valence band. Metals are excellent electrical conductors because there is no energy gap between the conduction band and the valence band. The valence electrons in a metal are conduction electrons. In a semiconductor the energy separation between the conduction band and the valence band is comparable to thermal energies, and thus some of the valence electrons can be thermally excited into the conduction band. Thus a semiconductor has electrical properties intermediate between those of metals and insulators. Figure 11-27 illustrates the difference between a metal, an insulator, and a semiconductor.

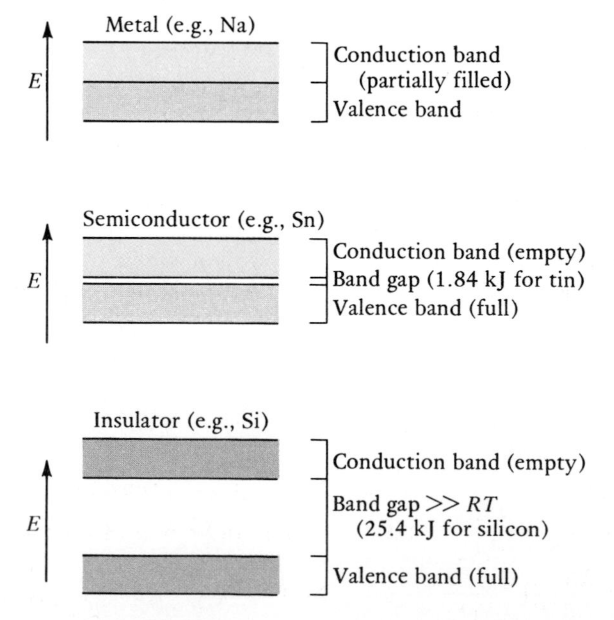

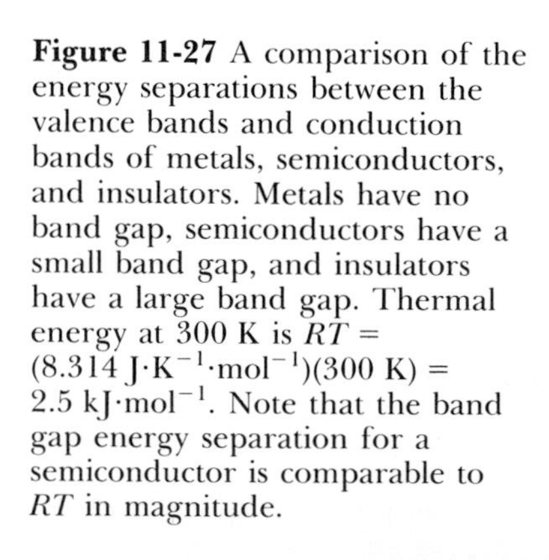

Figure 11-27 A comparison of the energy separations between the valence bands and conduction bands of metals, semiconductors, and insulators. Metals have no band gap, semiconductors have a small band gap, and insulators have a large band gap. Thermal energy at 300 K is $RT = (8.314\ \mathrm{J{\cdot}K^{-1}{\cdot}mol^{-1}})(300\ \mathrm{K}) = 2.5\ \mathrm{kJ{\cdot}mol^{-1}}$. Note that the band gap energy separation for a semiconductor is comparable to RT in magnitude.

SUMMARY

Molecular interactions in solids and liquids are much greater than those in gases. The values of ΔH_{vap} and ΔH_{sub} are quantitative measures of the strength of molecular interactions in liquids and solids, respectively. All molecules attract each other. If the molecules are polar, then they attract each other by dipole-dipole forces. The polar molecules orient themselves so that their dipole moments are head to tail; this is the orientation that minimizes the energy of the interacting molecules. Two nonpolar molecules attract each other because the electrons in the molecules redistribute such that the instantaneous dipole moments on each molecule are head to tail. The resulting attraction is called a London force. All the attractions between molecules are collectively called van der Waals forces.

Pure liquids have a unique equilibrium vapor pressure at each temperature; the equilibrium vapor pressure increases with increasing temperature. The Clapeyron-Clausius equation describes the dependence of the equilibrium vapor pressure of a liquid on the temperature. A liquid boils when its vapor pressure equals the pressure of the atmosphere. The normal boiling point of a liquid is the temperature at which the vapor pressure of the liquid is equal to 1 atm.

The regions of the various phases of a pure substance can be displayed simultaneously on a phase diagram. The normal melting point is the temperature at which the pure solid and liquid phases are in equilibrium at 1 atm. The normal boiling point is the temperature at which the pure liquid and gas phases are in equilibrium at 1 atm. The triple point is the only point at which the solid, liquid, and gas phases are simultaneously in equilibrium with one another. The critical point is the temperature above which a gas cannot be liquefied by the application of pressure.

A crystalline structure can be determined from the diffraction pattern of X-rays passed through the crystal. The diffraction pattern yields information regarding the positions of the atoms in the crystal. Crystals can be classified according to the type of forces between the constituent particles. Important types of crystals are atomic crystals, ionic crystals, molecular crystals, covalent network crystals, and metallic crystals. Metals are good conductors of electricity because the valence electrons are delocalized over the entire crystal.

TERMS YOU SHOULD KNOW

heating curve 346
molar enthalpy of fusion, ΔH_{fus} 346
molar enthalpy of vaporization, ΔH_{vap} 347
sublimation 349
molar enthalpy of sublimation, ΔH_{sub} 349
dipole-dipole attraction 350
hydrogen bonding 351
London force 353
van der Waals forces 353
equilibrium 354
dynamic equilibrium 355
equilibrium vapor pressure 355
vapor pressure curve 356
normal boiling point 356
relative humidity 356
dew point 357
Clapeyron-Clausius equation 358
phase diagram 359
critical point 360
critical temperature 360
sublimation pressure curve 360
melting point curve 360
triple point 360
normal sublimation point 361
X-ray diffraction pattern 361
unit cell 362
simple cubic 362
body-centered cubic 362
face-centered cubic 362
atomic crystal 364
ionic crystal 364
molecular crystal 365
X-ray crystallography 366
covalent network crystal 366
metallic crystal 367
valence band 368
conduction band 368
conduction electrons 368

EQUATIONS YOU SHOULD KNOW HOW TO USE

$q_P = nC_P(T_2 - T_1)$ (11-1) (heat required to change the temperature of a substance)

$\text{relative humidity} = \dfrac{P_{H_2O}}{P^\circ_{H_2O}} \times 100$ (11-2) (definition of relative humidity)

$\log\left(\dfrac{P_2}{P_1}\right) = \dfrac{\Delta H_{vap}}{2.30R}\left(\dfrac{T_2 - T_1}{T_1T_2}\right)$ (11-3) (Clapeyron-Clausius equation)

HEATS OF VAPORIZATION, FUSION, AND SUBLIMATION

11-1. Ammonia is used as a refrigerant in some industrial refrigeration units. The molar enthalpy of vaporization of liquid ammonia is 23.4 $kJ \cdot mol^{-1}$. Compute the amount of heat absorbed in the vaporization of 5.00 kg of $NH_3(l)$.

11-2. Given that 26.2 kJ of heat is required to completely vaporize 60.0 g of benzene, C_6H_6, at 80.1°C, calculate the molar heat of vaporization, ΔH_{vap}, of benzene.

11-3. Calculate the energy released as heat when 20.1 g of liquid mercury at 25°C is converted to solid mercury at its melting point. The heat capacity of Hg(l) is 28.0 $J \cdot K^{-1} \cdot mol^{-1}$.

11-4. The heat of vaporization of the refrigerant Freon-12, CCl_2F_2, is 155 $J \cdot g^{-1}$. Estimate the number of grams of Freon-12 that must be evaporated to freeze a tray of 16 one-ounce (1 oz = 28 g) ice cubes with the water initially at 18°C.

11-5. The metal gallium melts when held in the hand; its melting point is 29.0°C. How much heat is removed from the hand when 5.00 g of gallium initially at 20.0°C melts? The value of ΔH_{fus} is 5.59 $kJ \cdot mol^{-1}$ and the heat capacity per gram of gallium is 0.37 $J \cdot K^{-1} \cdot g^{-1}$. Take the final temperature to be 29.0°C.

11-6. The heat of vaporization of einsteinium was determined to be 128 $kJ \cdot mol^{-1}$ using only a 100-μg sample. How much heat is required to vaporize 100 μg of einsteinium?

11-7. Calculate the heat absorbed by the sublimation of 100.0 g of solid carbon dioxide.

11-8. Compute the number of moles of water at 0°C that can be frozen by 1.00 mol of solid carbon dioxide. See Table 11-1 for necessary data.

HEATING CURVES

11-9. Sketch a heating curve for 7.50 g of mercury from 200 to 800 K using a heat input rate of 100 $J \cdot min^{-1}$. Refer to Table 11-1 for some of the necessary data for mercury. The molar heat capacities of solid, liquid, and gaseous mercury are 27.2 $J \cdot K^{-1} \cdot mol^{-1}$, 28.0 $J \cdot K^{-1} \cdot mol^{-1}$, and 20.8 $J \cdot K^{-1} \cdot mol^{-1}$, respectively.

11-10. What would take longer, heating 10.0 g of water at 50.0°C to 100.0°C or vaporizing the 10.0 g if the rate of heating in both cases is 5 $J \cdot s^{-1}$?

11-11. Heat was added to 25.0 g of solid sodium chloride, NaCl, at the rate of 3.00 $kJ \cdot min^{-1}$. The temperature remained constant at 801°C, the melting point of NaCl, for 250 s. Calculate the molar enthalpy of fusion of NaCl.

11-12. Heat was added to a 45.0-g sample of liquid propane, $C_3H_8(l)$, at the rate of 500 $J \cdot min^{-1}$. The temperature remained constant at −42.1°C, the normal boiling point of propane, for 38.4 min. Calculate the molar enthalpy of vaporization of propane.

VAN DER WAALS FORCES

11-13. Which of the following molecules have polar interactions?

Cl_2 $\quad$ ClF $\quad$ NF_3 $\quad$ F_2

11-14. Which of the following molecules involve hydrogen bonding?

H_2 $\quad$ HF $\quad$ CH_4 $\quad$ CH_3OH

11-15. Arrange the following compounds in order of increasing boiling point:

KBr $\quad$ C_2H_5OH $\quad$ C_2H_6 $\quad$ He

11-16. Arrange the following compounds in order of increasing boiling point:

Ar $\quad$ NH_3 $\quad$ PH_3 $\quad$ NaCl

11-17. Arrange the following molecules in order of increasing molar enthalpy of vaporization:

CH_4 $\quad$ C_2H_6 $\quad$ CH_3OH $\quad$ C_2H_5OH

11-18. Arrange the following molecules in order of increasing molar enthalpy of vaporization:

CCl_4 $\quad$ $SiCl_4$ $\quad$ CH_4 $\quad$ $SiBr_4$

VAPOR PRESSURE

11-19. The temperature of the human body is about 37°C. Use data in Table 11-2 and Figure 11-13 to estimate the vapor pressure of water in exhaled air. Assume that air in the lungs is saturated with water vapor.

11-20. Calculate the concentration in $mol \cdot L^{-1}$ of water in air saturated with water vapor at 25°C.

11-21. A 0.75-g sample of ethyl alcohol is placed in a sealed 400-mL container. Is there any liquid present when the temperature is held at 60°C?

11-22. Mexico City lies at an elevation of 7400 ft (2300 m). Water boils at 93°C in Mexico City. What is the normal atmospheric pressure there?

11-23. A sample of ethyl alcohol vapor in a vessel of constant volume exerts a pressure of 300 torr at 75.0°C. Use the ideal gas law to plot pressure versus temperature of the vapor between 75.0°C and 40.0°C. Assume no condensation. Compare your result with the vapor pressure curve for ethyl alcohol shown in Figure 11-13. Estimate the temperature at which condensation occurs upon cooling from 75.0°C.

11-24. Atmospheric pressure decreases with altitude. Plot the following data to obtain the relationship between pressure and altitude:

Altitude/ft	Atmospheric pressure/atm
5000	0.83
10,000	0.70
15,000	0.58
20,000	0.47

Using your plot and the vapor pressure curve of water (Figure 11-13), estimate the boiling point of water at the following locations:

Location	Altitude/ft
Denver	5280
Mount Kilimanjaro	19,340
Mt. Washington	6290
The Matterhorn	14,690

11-25. Compare the dew points of two days with the same relative humidity of 70 percent but with temperatures of 20°C and 30°C, respectively.

11-26. The relative humidity in a greenhouse at 40°C is 92 percent. Calculate the vapor pressure of water vapor in the greenhouse.

11-27. The relative humidity is 65 percent on a certain day on which the temperature is 30°C. As the air cools during the night, what will be the dew point?

11-28. Moisture often forms on the outside of a glass containing a mixture of ice and water. Use the principles developed in this chapter to explain this phenomenon.

CLAPEYRON-CLAUSIUS EQUATION

11-29. Acetone, a widely used solvent (as nail-polish remover, for example), has a normal boiling point of 56.2°C and a molar enthalpy of vaporization of 31.97 $kJ \cdot mol^{-1}$. Calculate the equilibrium vapor pressure of acetone at 20.0°C.

11-30. Diethyl ether is a volatile liquid whose vapor is highly combustible. The equilibrium vapor pressure over ether at 20.0°C is 380 torr. Calculate the vapor pressure over ether when it is stored in the refrigerator at 4.0°C ($\Delta H_{vap} = 29.1\ kJ \cdot mol^{-1}$).

11-31. The heat of vaporization of benzene (C_6H_6) is 32.3 $kJ \cdot mol^{-1}$. Given that the vapor pressure of benzene is 387 torr at 60.0°C, calculate the normal boiling point of benzene.

11-32. Mercury is an ideal substance to use in manometers and in studying the effect of pressure on the volume of gases. Its surface is fairly inert and few gases are soluble in mercury. We now consider whether the partial pressure of mercury vapor contributes significantly to the pressure of the gas above mercury. Using the data in Table 11-1, calculate the vapor pressure of mercury at 25°C and at 100°C.

11-33. Carbon tetrachloride, CCl_4, has a vapor pressure of 92.68 torr at 23.50°C and 221.6 torr at 45.00°C. Calculate ΔH_{vap} for CCl_4.

11-34. The vapor pressure of bromine is 133 torr at 20.0°C and 48.10 torr at 0.00°C. Calculate ΔH_{vap} for bromine.

11-35. The molar heat of vaporization of lead is 178 $kJ \cdot mol^{-1}$. Calculate the ratio of the vapor pressure of lead at 1300°C to that at 500°C.

11-36. The molar heat of vaporization of NaCl is 180 $kJ \cdot mol^{-1}$. Calculate the ratio of the vapor pressure of NaCl at 1100°C to that at 900°C.

PHASE DIAGRAMS

11-37. Determine whether water is a solid, liquid, or gas at the following pressure and temperature combinations (use Figure 11-14):

(a) 373 K, 0.70 atm (b) −5°C, 0.006 atm
(c) 400 K, 200 atm (d) 0°C, 300 atm

11-38. Referring to Figure 11-15, state the phase of CO_2 under the following conditions:

(a) 127°C, 8 atm (b) −50°C, 40 atm
(c) 50°C, 1 atm (d) −80°C, 5 atm

11-39. Sketch the phase diagram for oxygen using the following data:

	Triple point	Critical point
temperature/K	54.3	154.6
pressure/torr	1.14	37,823

The normal melting point and normal boiling point of oxygen are −218.4°C and −182.9°C. Does oxygen melt under an applied pressure as water does?

11-40. Sketch the phase diagram for nitrogen given the following data:
triple point, 63.156 K and 139 torr

normal melting point, 63.29 K

normal boiling point, 77.395 K

critical point, 126.1 K and 33.49 atm

11-41. Potassium exists as a body-centered cubic lattice. How many potassium atoms are there per unit cell?

11-42. Crystalline potassium fluoride has the NaCl-type structure shown in Figure 11-19. How many potassium ions and fluoride ions are there per unit cell?

11-43. The density of silver is $10.50\ g \cdot cm^{-3}$ at 20°C. Given that the unit cell of silver is face-centered cubic, calculate the length of an edge of a unit cell.

11-44. The density of tantalum is $16.69\ g \cdot cm^{-3}$ at 20°C. Given that the unit cell of tantalum is body-centered cubic, calculate the length of an edge of a unit cell.

11-45. Polonium is the only metal that exists as a simple cubic lattice. Given that the length of a side of the unit cell of polonium is 334.7 pm at 25°C, calculate the density of polonium.

11-46. The unit cell of lithium is body-centered cubic, and the length of an edge of a unit cell is 351 pm at 20°C. Calculate the density of lithium at 20°C.

11-47. Copper crystallizes in a face-centered cubic lattice with a density of $8.93\ g \cdot cm^{-3}$. Given that the length of an edge of a unit cell is 361.6 pm, calculate Avogadro's number.

11-48. Chromium crystallizes in a body-centered cubic lattice with a density of $7.20\ g \cdot cm^{-3}$. Given that the length of an edge of a unit cell is 288.4 pm, calculate Avogadro's number.

11-49. Crystalline potassium fluoride has the NaCl-type structure shown in Figure 11-19. Given that the density of KF is $2.481\ g \cdot cm^{-3}$ at 20°C, calculate the unit cell length and the nearest-neighbor distance in KF. (The nearest-neighbor distance is the shortest distance between any two ions in the lattice.)

11-50. Crystalline cesium bromide has the CsCl-type structure shown in Figure 11-19. Given that the density of CsBr is $4.44\ g \cdot cm^{-3}$ at 20°C, calculate the unit cell length and the nearest-neighbor distance (see Problem 11-49) in CsBr.

ADDITIONAL PROBLEMS

11-51. Suppose you are stranded in a mountain cabin by a snowstorm. You have some food and fuel, but you wish to conserve them as long as possible. You remember reading that you should melt snow to get water to drink and not eat the snow directly because the body expends energy when it has to melt the snow. Explain why this is so and estimate how much energy is used per gram of snow.

11-52. Trouton's rule states that the molar enthalpy of vaporization of a liquid that does not involve strong molecular interactions such as hydrogen bonding or ion-ion attractions is given by

$$\Delta H_{vap} = (85\ J \cdot K^{-1} \cdot mol^{-1}) T_b$$

where T_b is the normal boiling point of the liquid in kelvins. Use Trouton's rule to estimate ΔH_{vap} for argon, given that T_b is 87 K.

11-53. Apply Trouton's rule, given in Problem 11-52, to water and suggest a molecular explanation for any discrepancy with the value of ΔH_{vap} given in Table 11-1.

11-54. What is the minimum amount of data necessary to make a rough sketch of the phase diagram of a substance exhibiting (a) a single solid phase? (b) two solid phases?

11-55. What is the relationship between ΔH_{fus}, ΔH_{vap}, and ΔH_{sub} in the vicinity of the triple point?

11-56. Why is H_2S a gas at −10°C whereas H_2O is a solid at this temperature?

11-57. Explain why silicon carbide, SiC, and boron nitride, BN, are about as hard as diamond.

11-58. Although the temperature may not exceed 0°C, the amount of ice on a sidewalk decreases owing to sublimation. A source of heat for the sublimation is solar radiation. The average daily solar radiation in February for Boston is $8.1\ MJ \cdot m^{-2}$. Calculate how much ice will disappear from a $1.0\text{-}m^2$ area in one day assuming that all the radiation is used to sublime the ice. The density of ice is $0.917\ g \cdot cm^{-3}$.

11-59. British surveyors were prevented from extending their survey of India into the Himalayas because entry into Tibet was banned by the Chinese emperor. In 1865, the Indian Nain Singh secretly entered Lhasa, the capital city of Tibet, and determined its correct location for map placement. Singh was not able to bring instruments for measuring altitude with him, but he did have a thermometer. He estimated that Lhasa was 3420 m above sea level. (Its true elevation is 3540 m.) Describe how he was able to estimate the altitude from a measurement of the boiling point of water.

11-60. A 0.677-g sample of zinc reacts completely with sulfuric acid:

$$Zn(s) + H_2SO_4(aq) \rightarrow ZnSO_4(aq) + H_2(g)$$

A volume of 263 mL of hydrogen is collected over water; the water level in the collecting vessel is the same as the outside level. Atmospheric pressure is 756 torr and the temperature is 25°C. Calculate the atomic mass of zinc.

11-61. A cylinder of chlorine contains 68 kg of Cl_2 at a pressure of 9.6 atm at 20°C. The volume of the cylinder is 56 L. The melting point of chlorine is −103°C; its normal boiling point is −35°C; and the critical point is 144°C and 76 atm. In what phase or phases does the chlorine exist in the cylinder?

11-62. The vapor pressures (in torr) of solid and liquid uranium hexafluoride are given by

$$\log P_s = 10.646 - \frac{2559.1\text{ K}}{T} \qquad \text{(solid)}$$

$$\log P_l = 7.538 - \frac{1511\text{ K}}{T} \qquad \text{(liquid)}$$

where T is the absolute temperature. Calculate the temperature and pressure at the triple point of UF_6.

11-63. The vapor pressures (in torr) of solid and liquid chlorine are given by

$$\log P_s = 10.560 - \frac{1640\text{ K}}{T} \qquad \text{(solid)}$$

$$\log P_l = 7.769 - \frac{1159\text{ K}}{T} \qquad \text{(liquid)}$$

where T is the absolute temperature. Calculate the temperature and pressure at the triple point of chlorine.

11-64. The following table gives the vapor pressure of solid argon from −208°C to −190°C (its normal melting point):

t/°C	*P*/torr
−208	20.4
−203	57.1
−196	192.5
−190	463.5

Calculate the heat of sublimation of argon.

11-65. The vapor pressure of potassium from 260°C to 760°C (its normal boiling point) can be represented by the empirical formula

$$\log P = -\frac{4021\text{ K}}{T} + 6.774$$

where P is the vapor pressure in torr and T is the kelvin temperature. Use this formula to determine the heat of vaporization of potassium.

11-66. The polar caps of Mars are thought to be composed of solid carbon dioxide. Spectroscopic measurements by space vehicles show that the Martian atmosphere has a CO_2 gas pressure of 4 torr. Use these data, together with the Clapeyron-Clausius equation and the sublimation data in Table 11-1, to estimate the temperature of the Martian polar caps.

11-67. The unit cell of calcium oxide is one of the three cubic crystalline structures. Given that the density of CaO is 3.25 $g \cdot cm^{-3}$ and that the length of an edge of a unit cell is 481 pm, determine how many formula units of CaO there are in a unit cell. Does the unit cell have a NaCl or a CsCl structure?

11-68. The unit cell of potassium bromide is one of the three cubic crystalline structures. Given that the density of KBr is 2.75 $g \cdot cm^{-3}$ and that the length of an edge of a unit cell is 654 pm, determine how many formula units of KBr there are in a unit cell. Does the unit cell have a NaCl or a CsCl structure?

11-69. Cesium chloride has the crystal structure shown in Figure 11-19. The length of a side of a unit cell is determined by X-ray diffraction to be 412.1 pm. What is the density of cesium chloride?

11-70. Sodium chloride has the crystal structure shown in Figure 11-19. By X-ray diffraction, it is determined that the distance between a sodium ion and a chloride ion is 282 pm. Using the fact that the density of sodium chloride is 2.163 $g \cdot cm^{-3}$, calculate Avogadro's number.

11-71. The equilibrium vapor pressure of benzene was measured as a function of temperture, with the following results:

t/°C	*P*/torr
20.6	77.3
31.0	124.7
39.1	175.9
49.1	261.7
60.8	402.4
74.0	627.9
80.9	779.3

Plot the data in the form $\log P$ versus $1/T$. From your plot, determine the normal boiling point and the molar heat of vaporization of benzene.

11-72. Commercial refrigeration units in the United States are rated in tons. A 1-ton unit is capable of removing, during 24 hours of operation, an amount of heat equal to that released when 1.00 ton of water at 0°C is converted to ice. Calculate the number of kilojoules of heat per hour that can be removed by a 4-ton home air conditioner.

Silicon: A Semimetal

The silicon computer chip revolutionized the electronics industry and is directly responsible for the microminiaturized electronic devices that are so common today. One small chip may consist of over a million transistors and perform the same function as rooms full of electronic equipment did 20 years ago.

Silicon (Group 4, atomic number 14, atomic mass 28.0855) constitutes 28 percent of the mass of the earth's mantle and is the second most abundant element in the mantle, being exceeded only by oxygen. Silicon does not occur as the free element in nature. It occurs primarily as the oxide and in various silicates. Elemental silicon has a gray, metallic luster. It is fairly inert and reacts at normal temperatures only with the halogens and with dilute aqueous alkalis. The major use of silicon is in the manufacture of glass and transistors. Pure silicon is an insulator, but it can be converted to a semiconductor by addition of trace amounts of certain elements.

H-1. SILICON IS USED TO MAKE TRANSISTORS

▪ Amorphous silicon is a solid form of silicon that does not have a crystalline structure. It is used in the fabrication of some solar-energy devices because of its relatively low cost.

Silicon is the most important industrial semimetal, its major use being in the manufacture of electronic components such as transistors. It has a density of 2.3 $g \cdot cm^{-3}$, which is about the same as that of graphite, and melts at 1420°C. Elemental silicon is made by the high-temperature reduction of silicon dioxide (the major constituent of numerous sands) with carbon:

$$SiO_2(l) + C(s) \xrightarrow{3000^\circ C} Si(l) + CO_2(g)$$

The 98-percent-pure silicon prepared by this reaction must be further purified before it can be used to make transistors. It is converted to the liquid silicon tetrachloride by reaction with chlorine:

$$Si(s) + 2Cl_2(g) \rightarrow SiCl_4(l)$$

The silicon tetrachloride is further purified by repeated distillation and then converted to silicon by reaction with magnesium:

$$SiCl_4(g) + 2Mg(s) \rightarrow 2MgCl_2(s) + Si(l)$$

The resulting silicon is purified still further by a special method of recrystallization called **zone refining.** In this process, solid silicon is packed in a tube that is mounted in a vertical position (Figure H-1) with an electric heating loop around the base of the tube. The solid near the heating loop is melted by passing a current through the loop, and the tube is then lowered very slowly through the loop. As the melted solid cools slowly in the region of the tube below the heating loop, pure crystals separate out, leaving most of the impurities behind in the moving molten zone. The process can be repeated as often as necessary to achieve the desired purity of the recrystallized solid. Purities up to 99.9999 percent are possible with zone refining.

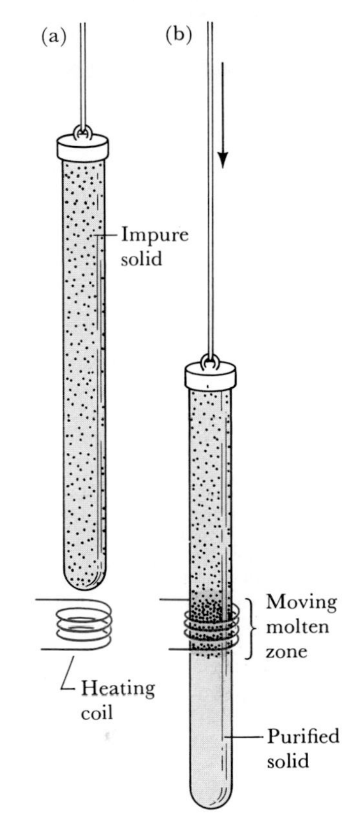

Figure H-1 Zone refining. An impure solid is packed tightly in a glass tube, and the tube is lowered slowly through a heating coil that melts the solid. Pure solid crystallizes out from the bottom of the melted zone, and the impurities concentrate in the moving molten zone.

In Chapter 11 we discussed the differences between metals, insulators, and semiconductors. Pure silicon is an insulator, but it can be converted to a semiconductor by addition of selected impurity atoms, a process called **doping.** For example, an ***n*-type** (*n* for negative) **semiconductor** is produced when trace amounts of elements whose atoms have five valence electrons, such as phosphorus or antimony, are added to silicon, which has four valence electrons (Figure H-2a). The excess valence electrons on the impurity atoms, which substitute for some of the silicon atoms in the crystal, become the conduction electrons in the crystal (Figure H-2b). A ***p*-type** (*p* for positive) **semiconductor** is produced when trace amounts of elements whose atoms have three valence electrons, such as boron or indium, are added to silicon. The deficiency of valence electrons on the impurity atoms functions to create "holes" by means of which electrons can "hop" through the silicon crystal (Figure H-2c). Because impurity atoms have a major effect on the electrical properties of semiconductors, it is necessary to use extremely pure ($\geq$99.9999 percent) silicon and to add precise amounts of impurities of carefully controlled composition to the crystal in order to obtain the desired electrical properties.

H-2. SILICATES OCCUR WIDELY IN NATURE

Silicates occur in numerous minerals and in asbestos, mica, and clays. Cement, bricks, tiles, porcelains, glass, and pottery are all made from silicates. All silicates involve silicon-oxygen single bonds, of which there

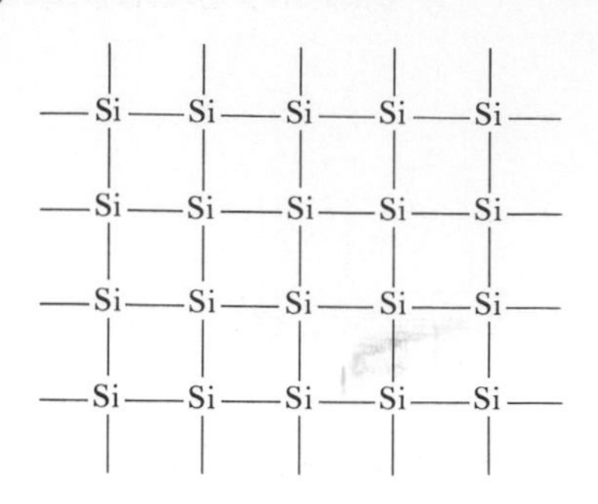

(a) Normal silicon (insulator)

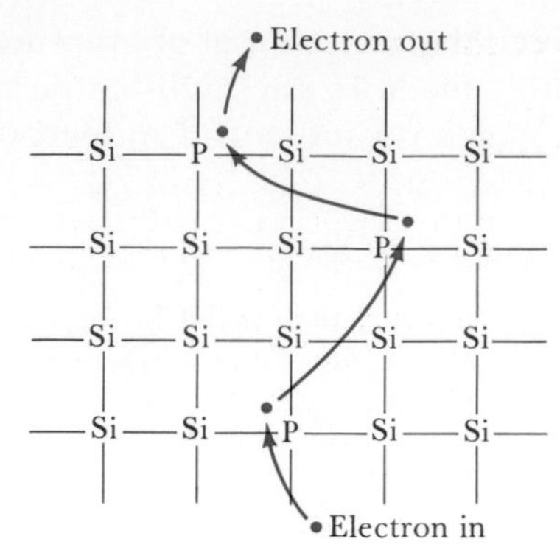

(b) Silicon with phosphorus impurity (n-type semiconductor)

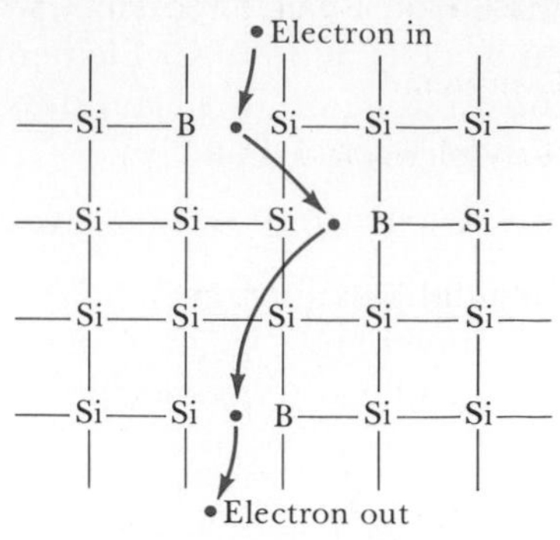

(c) Silicon with boron impurity (p-type semiconductor)

Figure H-2 Comparison of pure silicon and n-type, and p-type doped silicon. (a) Silicon has four valence electrons, and each silicon atom forms four 2-electron bonds to other silicon atoms. (b) Phosphorus has five valence electrons, and thus when a phosphorus atom substitutes for a silicon atom in a silicon crystal, there is an unused valence electron on each phosphorus atom that can become a conduction electron. (c) Boron has only three valence electrons, and thus when a boron atom substitutes for a silicon atom in a silicon crystal, there results an electron vacancy (a "hole"). Electrons from the silicon valence bond can move through the crystal by hopping from one vacancy site to another.

are two types. Terminal —Si—O bonds involve oxygen bonded to silicon and no other atoms, and bridging —Si—O—Si— bonds involve oxygen linking two silicon atoms.

The simplest silicate anion is the tetrahedral **orthosilicate ion,** SiO_4^{4-}. The SiO_4^{4-} ion is found in the minerals zircon, $ZrSiO_4$, and willemite, Zn_2SiO_4 (Figure H-3) and also in sodium silicate, which, when dissolved in water, is called **water glass,** $Na_4SiO_4(aq)$.

The minerals enstatite, $MgSiO_3$, and spodumene, $LiAl(SiO_3)_2$ (Figure H-3), are silicates that contain long, straight-chain silicate polyanions involving the SiO_3^{2-} chain unit (Figure H-4):

$$
\begin{array}{ccccccccccc}
 & O^{\ominus} & & O^{\ominus} & & O^{\ominus} & & O^{\ominus} & & O^{\ominus} & \\
 & | & & | & & | & & | & & | & \\
-O- & Si & -O- & Si & -O- & Si & -O- & Si & -O- & Si & - \\
 & | & & | & & | & & | & & | & \\
 & O_{\ominus} & & O_{\ominus} & & O_{\ominus} & & O_{\ominus} & & O_{\ominus} & \\
 & \underbrace{\quad}_{SiO_3^{2-}} & & \underbrace{\quad}_{SiO_3^{2-}} & & \underbrace{\quad}_{SiO_3^{2-}} & & \underbrace{\quad}_{SiO_3^{2-}} & & \underbrace{\quad}_{SiO_3^{2-}} &
\end{array}
$$

Figure H-3 The minerals enstatite (*right*), willemite (*rear*), spodumene (*front*), and zircon (*left*).

Table H-1 Some important compounds of silicon

Compound	Uses
fluorosilicic acid, $H_2SiF_6(s)$	water fluoridation, sterilizing agent in the brewing industry; hardener in cement and ceramics
sodium silicate, $Na_4SiO_4(s)$	soaps and detergents; silica gels; adhesives; water treatment; sizing of textiles and paper; waterproofing cement; flame retardant, preservative
silicon carbide (carborundum), $SiC(s)$	abrasive for cutting and grinding metals
silicon dioxide (silica), $SiO_2(s)$	glass manufacture, abrasives, refractory material, cement
silicones, $\left(-\underset{R'}{\overset{R}{Si}}-O-\right)_n$	lubricants, adhesives, protective coatings, coolant, waterproofing agent, cosmetics

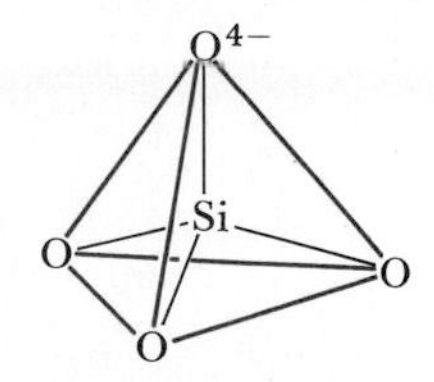

The orthosilicate ion.

Structures that result from joining many smaller units together are called **polymers.** The straight-chain silicate anions shown in the preceding structure are called **silicate polyanions** because they result from joining together many silicate anions. The mineral beryl, $Be_3Al_2Si_6O_{18}$, contains the cyclic polysilicate anion $Si_6O_{18}^{12-}$ (Figure H-5a). These cyclic polysilicate anions can themselves be joined together to form polymeric, cyclic polysilicate anions with the composition $(Si_4O_{11}^{6-})_n$ and the structure shown in Figure H-5(b). The best example of a mineral containing polymeric, cyclic polysilicate chains is **asbestos** (Figure H-6). The fibrous character of asbestos is a direct consequence of the molecular structure of the $(Si_4O_{11}^{6-})_n$ polymeric chains.

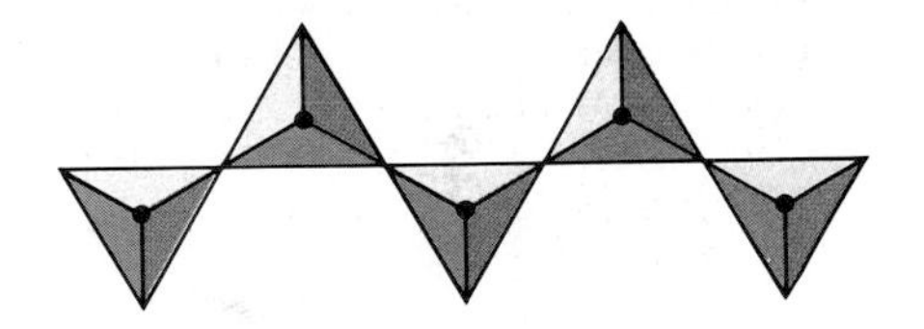

Figure H-4 Tetrahedral SiO_4^{4-} units are linked together through oxygen atoms that are shared by tetrahedra to form straight-chain silicate polyanions involving the SiO_3^{2-} chain unit.

The silicate minerals **mica** and **talc** contain two-dimensional, polymeric silicate sheets with the overall silicate composition $Si_2O_5^{2-}$. The structure of these sheets is illustrated in Figure H-5(c), and Figure H-7 shows how mica can easily be fractured into thin sheets. Talc has the composition $Mg_3(OH)_2(Si_2O_5)_2$, whereas micas have a variety of compositions, one example of which is lepidolite, $KLi_2Al(Si_2O_5)_2(OH)$. The ease with which mica can be separated into thin sheets and the slippery feel of talcum powder arise from the layered structure of the silicates in these minerals. Some commercially important compounds of silicon are given in Table H-1.

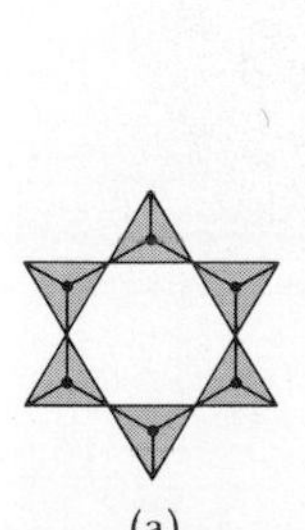

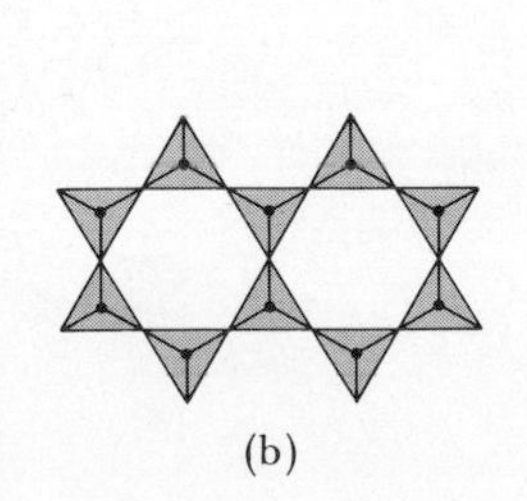

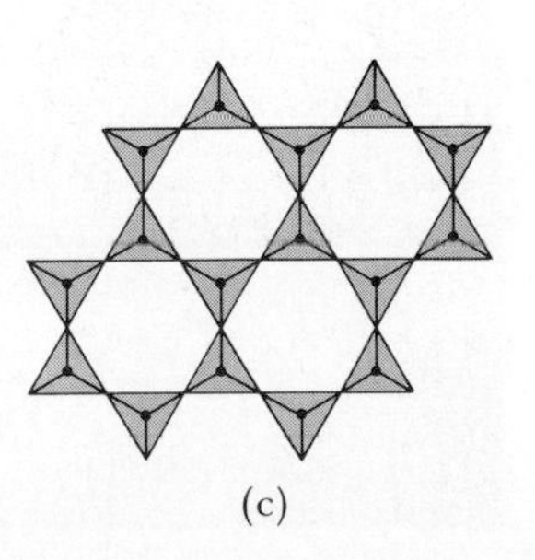

Figure H-5 (a) The cyclic polysilicate ion $Si_6O_{18}^{12-}$, which occurs in the mineral beryl. Six SiO_4^{4-} tetrahedral units are joined in a ring with the tetrahedra linked by shared oxygen atoms. (b) The cyclic polysilicate ion $Si_6O_{18}^{12-}$ can form a polymeric cyclic network like that shown here. The composition of the cyclic network is $(Si_4O_{11}^{6-})_n$. Asbestos has this structure. (c) Structure of polysilicate sheets composed of $(Si_2O_5^{2-})_n$ subunits. Mica has this structure.

H-3. MOST GLASSES ARE SILICATES

Figure H-6 Asbestos. The fibrous character of this mineral is a direct consequence of its $(Si_4O_{11}^{6-})_n$ polymeric chains.

Quartz is a crystalline material with the composition SiO_2 and the crystalline structure shown in Figure H-8(b). When crystalline quartz is melted and then cooled quickly to prevent the formation of crystals, there is formed a disordered three-dimensional array of polymeric chains, sheets, and other three-dimensional clusters. The resulting material is called quartz glass. Most glasses consist of a random array of these clusters.

Glass manufacturing is a 10-billion-dollar-per-year industry in the United States. The major component of most glasses is almost pure quartz sand. Among the other components of glass, soda (Na_2O) comes from soda ash (Na_2CO_3), lime (CaO) comes from limestone ($CaCO_3$), and aluminum oxide comes from feldspars, which have the general formula $M_2O \cdot Al_2O_3 \cdot 6SiO_2$, where M is K or Na. All the components of glass are fairly inexpensive chemicals.

A wide variety of glass properties can be produced by varying the glass composition. For example, partial replacement of CaO and Na_2O by B_2O_3 gives a glass that does not expand on heating or contract on cooling and is thus used in making glass utensils meant to be heated. Colored glass is made by adding a few percent of a colored transition-metal oxide, such as CoO to make blue "cobalt" glass and Cr_2O_3 to make orange glass. Lead glass, which contains PbO, has attractive optical properties and is used to make decorative, cut-glass articles.

The addition of K_2O increases the hardness of glass and makes it easier to grind to precise shapes. Optical glass contains about 12 percent K_2O. The **photochromic glass** used in eyeglasses that darken in sunlight contains a small amount of silver chloride dispersed throughout and trapped in the glass. When sunlight strikes photochromic glass, the tiny AgCl grains decompose into opaque clusters of silver atoms and chlorine atoms:

$$\underset{\text{clear}}{AgCl} \underset{\text{dark}}{\overset{\text{sunlight}}{\rightleftharpoons}} \underset{\text{opaque}}{Ag + Cl}$$

Figure H-7 The ease with which the mineral mica can be separated into thin sheets is a direct consequence of the existence of polymeric silicate sheets with the composition $(Si_2O_5^{2-})_n$.

(a)

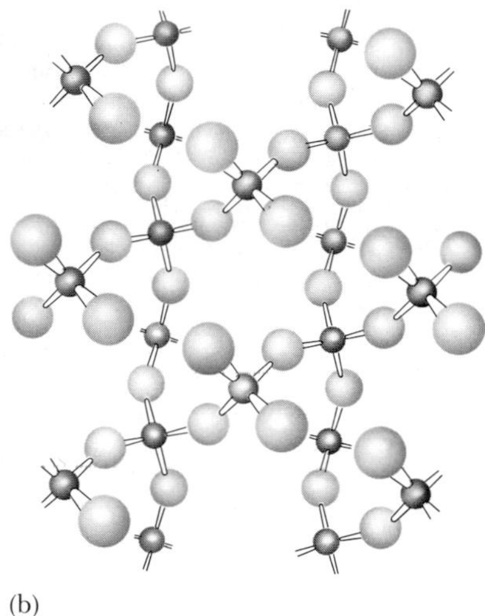

(b)

Figure H-8 Quartz. (a) Quartz often forms large, beautiful crystals. (b) The crystalline structure of quartz. Note that each silicon atom is surrounded by four oxygen atoms. The silicon atoms are linked by the oxygen atoms.

The atoms are trapped in the crystal lattice, and they recombine in the dark to form silver chloride, which causes the glass to become clear.

The etching of glass by hydrofluoric acid, HF(*aq*), is a result of the reaction

$$SiO_2(s) + 6HF(aq) \rightarrow H_2SiF_6(s) + 2H_2O(l)$$

and this reaction is used to "frost" the inside surface of lightbulbs.

Porcelain has a much higher percentage of Al_2O_3 than glass and as a result is a heterogeneous substance. Porcelain is stronger than glass because of this hetereogeneity and is also more chemically resistant than glass. Earthenware is similar in composition to porcelain but is more porous because it is fired at a lower temperature.

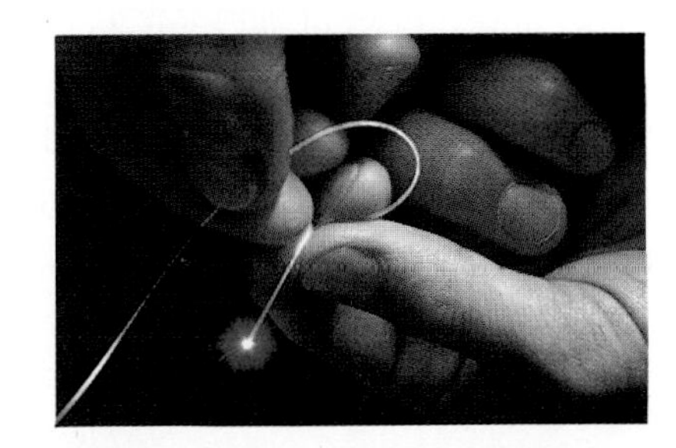

Figure H-9 Fiber optics. Light is transmitted through thin strands of silica glass that form an optical-waveguide fiber bundle.

TERMS YOU SHOULD KNOW

zone refining 375
doping 375
n-type semiconductor 375
p-type semiconductor 375
orthosilicate ion, SiO_4^{4-} 376
water glass 376
polymer 377
silicate polyanion (polysilicate anion) 377
asbestos 377
mica 377
talc 377
photochromic glass 378

H-1. Describe, using balanced chemical equations, how silicon is produced.

H-2. Describe the process of zone refining. What purity of silicon is obtainable by zone refining?

H-3. Discuss the difference between *p*-type and *n*-type semiconductors.

H-4. Use VSEPR theory to predict the shapes of SiO_4^{4-} and SiO_6^{2-}.

H-5. Describe how photochromic glass works.

H-6. Sketch a straight-chain silicate polyanion. How are the SiO_4^{4-} units linked to each other?

H-7. Sketch the cyclic polysilicate ion, $Si_6O_{18}^{12-}$. In which mineral does this structure occur?

H-8. Discuss the silicate polyanion structures of mica and asbestos.

H-9. What are the four principal components of glass?

H-10. Describe the reaction by which HF(*aq*) etches glass.

H-11. Describe the silicate anion structures in the following minerals:

(a) $ZrSiO_4(s)$, zircon
(b) $MgSiO_3(s)$, enstatite
(c) $Mg_3(OH)_2(Si_2O_5)_2(s)$, talc

H-12. Draw structures for the following ions:
(a) $Si_2O_7^{6-}$ (b) $Si_3O_{10}^{8-}$ (c) $Si_6O_{18}^{12-}$

H-13. Give two examples not mentioned in the text of elements that can be added to silicon to make *n*-type and *p*-type semiconductors, respectively.

H-14. Suggest an explanation for the observation that heterogeneity confers strength to materials, as in, for example, porcelain versus glass.

PROBLEMS

H-15. Silicon occurs as three principal isotopes, whose atomic masses are 27.976927, 28.976495 and 29.973770. Given that the abundance of silicon-28 is 92.23 percent, calculate the abundances of silicon-29 and silicon-30. The atomic mass of silicon is 28.0855.

H-16. How many grams of chlorine must be used to prepare one gram of silicon tetrachloride by the reaction

$$Si(s) + 2Cl_2(g) \rightarrow SiCl_4(l)$$

What volume does this quantity of chlorine occupy at 250°C and 2.00 atm?

H-17. Assuming a yield of 75 percent, how many metric tons of silicon can be obtained from 500 metric tons of silicon dioxide?

H-18. Given that $T_{mp} = 1420$ K, $\Delta H_{fus} = 43.4\ kJ\cdot mol^{-1}$, $C_p(s) = 25.8\ J\cdot K^{-1}\cdot mol^{-1}$, and $C_p(l) = 31.0\ J\cdot K^{-1}\cdot mol^{-1}$ for silicon, calculate the heat required to raise the temperature of one mole silicon from 1000 K to 2000 K.

PROPERTIES OF SOLUTIONS

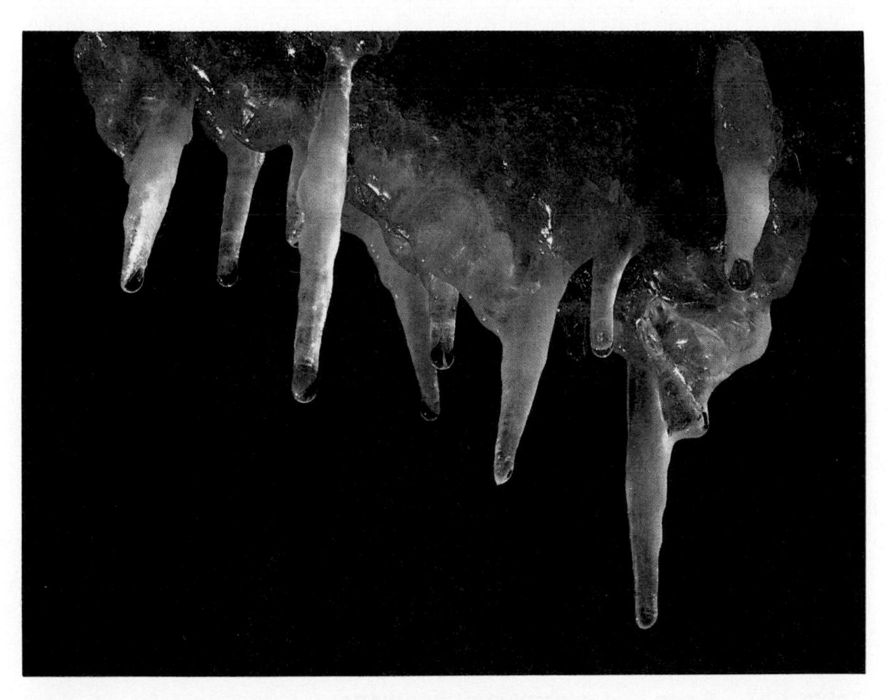

Stalactites are deposits of $CaCO_3(s)$ that form from water containing $Ca^{2+}(aq)$ and $HCO_3^-(aq)$ that evaporates slowly from the drops at the tips of the stalactites.

In this chapter we discuss some of the elementary properties of solutions from a molecular viewpoint. We introduce the concept of solubility, which tells us how much of one substance can be dissolved in another substance. The major emphasis of this chapter is on the colligative properties of solutions, which are properties that depend primarily on the ratio of the number of solute particles to the number of solvent particles in the solution. The colligative properties of solutions are vapor pressure lowering, boiling point elevation, freezing point depression, and osmotic pressure.

We learned in Chapter 11 that pure water has a unique equilibrium vapor pressure at each temperature. We learn in this chapter that the equilibrium vapor pressure of pure water always decreases when a substance is dissolved in it, which is the key to understanding the colligative properties of solutions.

12-1. A SOLUTION IS A HOMOGENEOUS MIXTURE OF TWO OR MORE SUBSTANCES

Most chemical and biological processes take place in solution, particularly in aqueous solution. A **solution** is a mixture of two or more substances that is homogeneous at the molecular level. A solution must be **homogeneous,** meaning that it must have the same properties from one region to another. The most common examples of solutions involve a solid, such as NaCl, dissolved in water. The resulting solution is clear and homogeneous. From a molecular point of view, the sodium ions and chloride ions are uniformly dispersed among the water molecules. The **components** of a solution are the pure substances that are mixed to form the solution. The components do not have to be a solid and a liquid. There are many other types of solutions (Table 12-1).

The component of a solution that has the same state (gas, liquid, or solid) as the resulting solution is called the **solvent.** If two or more components have the same state as the resulting solution, then the component present in excess is the solvent. The other components of a solution are called **solutes.** A solute is often considered to be a substance that is dissolved in a solvent. The terms solvent and solute are merely terms of convenience; all components of a solution are uniformly dispersed throughout the solution.

12-2. SOLUBILITY INVOLVES A DYNAMIC EQUILIBRIUM

Consider a solid, such as crystals of NaCl, at the bottom of a beaker of water. As we know from Chapter 11, the molecular picture of a solid is that of an ordered lattice, with the individual particles restricted to

Table 12-1 Types and examples of solutions

State of component 1	State of component 2	State of resulting solution	Examples
gas	gas	gas	air; vaporized gasoline-air mixture in the combustion chambers of a car
gas	liquid	liquid	oxygen in water; carbon dioxide in carbonated beverages
gas	solid	solid	hydrogen in palladium and platinum
liquid	liquid	liquid	water and alcohol
liquid	solid	solid	mercury in gold or silver
solid	liquid	liquid	sodium chloride in water
solid	solid	solid	metal alloys

lattice sites. The molecules of a liquid, on the other hand, move about continuously, colliding very frequently with each other. The water molecules in our example collide not only with each other but also with the surfaces of the sodium chloride crystals at the bottom of the beaker. In doing so, they jar sodium ions and chloride ions loose, allowing them to enter into solution (Figure 12-1). The sodium and chloride ions interact strongly with water molecules, and this interaction facilitates the solution process.

The ions that dissolve are free to move about throughout the solution, colliding with water molecules, other ions, the walls of the beaker, and the surfaces of the NaCl crystals remaining at the bottom of the beaker. Some of these ion-crystal collisions result in sodium and chloride ions sticking to the crystal surfaces and being incorporated back into the crystalline structure. We can represent the solution process by

$$\mathrm{NaCl}(s) \underset{\text{crystallization}}{\overset{\text{solution}}{\rightleftharpoons}} \mathrm{Na}^+(aq) + \mathrm{Cl}^-(aq)$$

Notice that the reaction proceeds both forward (solution) and backward (crystallization). According to our molecular picture, both processes are occurring constantly. As the number of ions in solution increases, the number of collisions with the crystal surfaces increases, and so more dissolved ions are incorporated back into the solid phase. Eventually a balance is reached, where at any instant the number of ions going into solution is equal to the number of ions being incorporated into the crystal.

Let's call the number of ions that leave the crystals and enter the solution per second the **rate of solution** and the number of ions that leave the solution and attach to the crystals per second the **rate of crystallization.** When the rate of solution is equal to the rate of crystallization, then the quantity of NaCl at the bottom of the beaker no longer changes and it appears to the naked eye that the solution process has ceased. From a molecular point of view, however, some ions are still entering the solution and others are being deposited onto the crystal surfaces. The rates of solution and crystallization are equal, and so no *net* change is observed. To emphasize that there is still a great deal of molecular activity even though there is no net progress, we say that the solution and crystallization processes are in **dynamic equilibrium** (Figure 12-2).

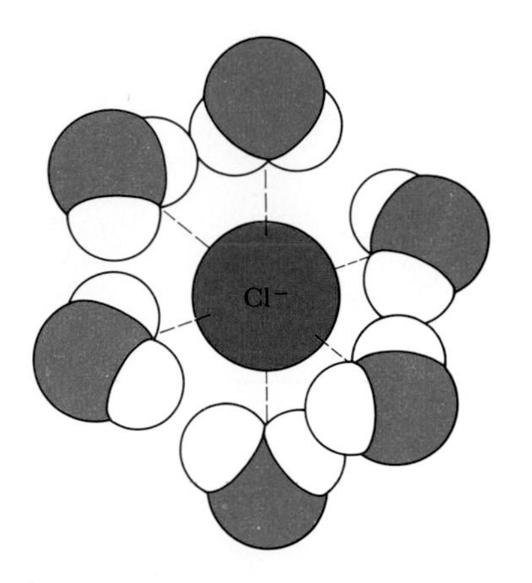

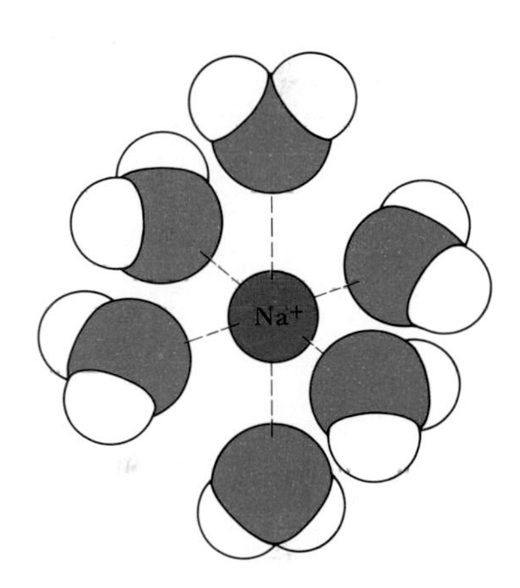

Figure 12-1 Ions in aqueous solutions are stabilized by the interaction with water molecules. Such ions are said to be solvated.

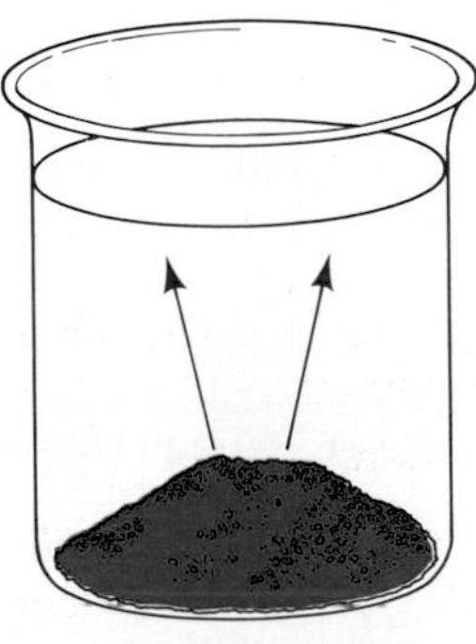

(a) Initial state

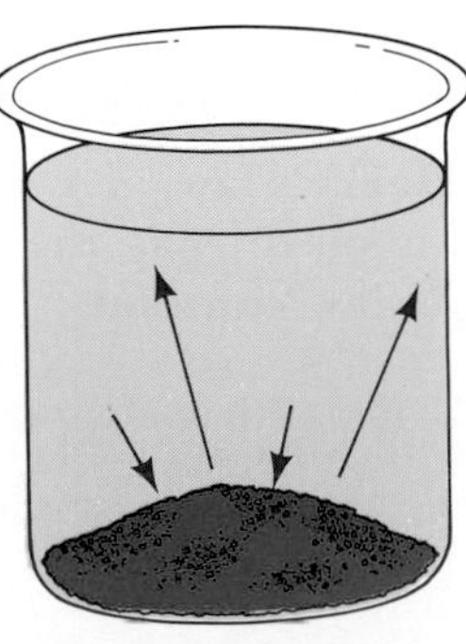

(b) Ions begin to go into solution

(c) Dynamic equilibrium

Figure 12-2 Dynamic equilibrium between ions going into solution and ions being incorporated back into the crystal lattice. When the rates of these two processes are the same, then the system is said to be in dynamic equilibrium.

Dynamic equilibrium occurs when the number of ions being deposited on the surface of the crystals is equal to the number of ions entering the solution at any instant. At this equilibrium point, no more ions can be accommodated by the solution; more ions in solution result only in more collisions with the crystal surfaces and thus more ions deposited. The solution is then said to be a **saturated solution,** and the quantity of solute dissolved is called the **solubility** of that solute. Solubility can be expressed in a variety of units, such as grams of solute per 100 g of solvent, or as molarity (moles of solute per liter of solution). For example, we can say that the solubility of KCl in water at 20°C is about 32 g per 100 g of H_2O, or that the solubility is 3.8 M.

It is important to realize that the solubility of a substance is the maximum quantity that can be dissolved in a given quantity of solvent at a particular temperature. The solubility of NaCl at 20°C is about 35 g per 100 g of H_2O. If we add 50 g of NaCl to 100 g of H_2O at 20°C, then 35 g dissolves and 15 g is left as undissolved NaCl(*s*). The solution is saturated. If we add 25 g of NaCl to 100 g of H_2O, then all the NaCl dissolves to form what is called an **unsaturated solution,** which is a solution in which we can dissolve more solute.

In most cases the solubility of a substance depends upon temperature. The effect of temperature on the solubility of several salts in water is shown in Figure 12-3. Almost all salts become more soluble in water as the temperature increases. For example, potassium nitrate is about five times more soluble in water at 40°C than in water at 0°C. At higher temperatures both the ions in the crystal and the water molecules in the solvent have higher average energies, and thus the breakdown of the crystal lattice that must occur in order for solution to take place is more easily accomplished.

▪ Salts can be purified by forming a saturated solution at a high temperature and then cooling the solution. The decreased solubility as the temperature decreases results in the precipitation of some of the salt. The soluble impurities tend to remain in solution.

12-3. THE SOLUBILITY OF A GAS IN A LIQUID IS DIRECTLY PROPORTIONAL TO THE PRESSURE OF THE GAS OVER THE LIQUID

The solubility of a gas in a liquid is directly proportional to the partial pressure of the gas in contact with the liquid. If we express the solubility as molarity, M_{gas}, and the partial pressure of the gas as P_{gas}, then we can write

$$M_{gas} \propto P_{gas}$$

This relationship is usually written the other way around as

$$P_{gas} = k_h M_{gas} \qquad (12\text{-}1)$$

Equation (12-1) is called **Henry's law,** and the proportionality constant k_h, whose value depends upon the gas, the solvent, and the temperature, is called the **Henry's law constant.**

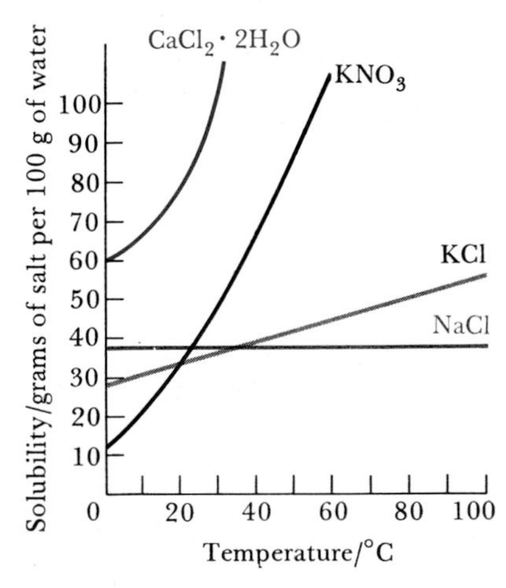

Figure 12-3 The solubility of several salts as a function of temperature. These curves show that solubility depends upon temperature and usually increases with increasing temperature.

Example 12-1: Compute the concentration of O_2 in water that is in equilibrium with air at 25°C. The Henry's law constant for O_2 in water at 25°C is 780 atm·M^{-1}.

Solution: The partial pressure of O_2 in the atmosphere is 0.20 atm, and thus, from Equation (12-1),

$$M_{O_2} = \frac{P_{O_2}}{k_h} = \frac{0.20 \text{ atm}}{780 \text{ atm} \cdot \text{M}^{-1}} = 2.6 \times 10^{-4} \text{ M}$$

Figure 12-4 shows the solubility of oxygen in water as a function of the pressure of the oxygen in contact with the water. The resulting straight line is in accord with Henry's law.

Henry's law constants for several common gases are given in Table 12-2. The smaller the value of this constant for a gas, the greater the solubility of the gas because $M_{gas} = P_{gas}/k_h$.

Carbonated beverages are pressurized with CO_2 gas at a pressure above 1 atm; champagne is pressurized at 4 to 5 atm. The CO_2 pressure is responsible for the rush of escaping gas that causes the "pop" when the carbonated drink container is opened. The loss of CO_2 from the solution begins because the average atmospheric partial pressure of CO_2 is only 3×10^{-4} atm. The bubbles that form in the liquid are mostly CO_2 plus some water vapor at about 1 atm total pressure.

The air that is breathed by a diver under water is significantly above atmospheric pressure because the diver must exhale the air into an environment that has a pressure greater than atmospheric pressure. For example, at a depth of 90 ft the pressure is about 3.7 atm, and so the diver must breathe air at 3.7 atm. At this pressure the solubilities of N_2 and O_2 in the blood are 3.7 times as great as at sea level. If a diver ascends too rapidly, then the sudden pressure drop causes the dissolved nitrogen to form numerous small gas bubbles in the blood. This phenomenon, which is extremely painful and can result in death, is called the bends because it causes the afflicted person to bend over in pain. Oxygen, which is readily metabolized, does not accumulate as bubbles in the blood. However, pure oxygen cannot be used for breathing because at high oxygen pressures the need to breathe is greatly reduced and so CO_2 accumulates in the bloodstream and leads to CO_2 asphyxiation. The solution to this problem was proposed by the chemist Joel Hildebrand and consists of substituting helium for nitrogen. Helium is only about half as soluble in blood as nitrogen, and thus the magnitude of the problem is cut in half. Divers' "air" tanks contain a mixture of He and O_2 adjusted so that the pressure of O_2 is about 0.20 atm at maximum dive depth.

The solubility of gases in liquids decreases with increasing temperature. This effect is depicted in Figure 12-5 for O_2 dissolved in water that is in equilibrium with air. Cold water in equilibrium with air has a higher concentration of dissolved oxygen than warm water. The decreased solubility of oxygen with increased temperature is the reason why most fish, especially the more active ones, prefer cooler water.

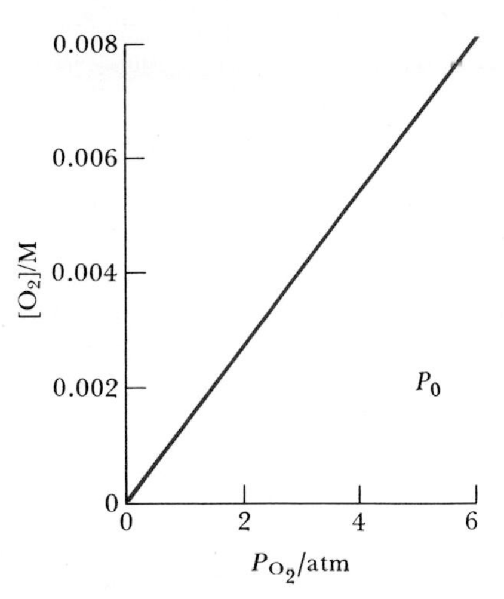

Figure 12-4 The solubility of oxygen in water at 25°C plotted against the pressure of the oxygen in contact with the water. The linear plot confirms that Henry's law holds for oxygen in water over the range of 0 to 6 atm $O_2(g)$.

▪ Joel Hildebrand was a professor of chemistry at the University of California for about 70 years. He was fully active in chemical research well into his 90s, over 20 years after his formal retirement.

Table 12-2 Henry's law constants for gases in water at 25°C

Gas	k_h/atm·M^{-1}
He	2.7×10^3
N_2	1.6×10^3
O_2	7.8×10^2
CO_2	29
H_2S	10

12-4. THE EQUILIBRIUM VAPOR PRESSURE OF A PURE LIQUID IS ALWAYS DECREASED WHEN A SUBSTANCE IS DISSOLVED IN THE LIQUID

Consider a solution of a nonvolatile solute such as sucrose dissolved in a volatile solvent such as water. As Figure 12-6 suggests, the vapor pressure of the solvent over the solution will be less than the vapor pressure of pure solvent at the same temperature.

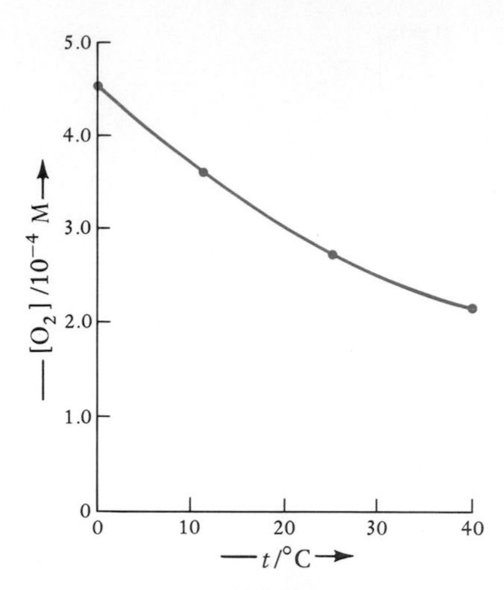

Figure 12-5 Solubility of oxygen in water as a function of temperature. The partial pressure of O_2 in each case is 0.20 atm, which is its partial pressure in air. Note that the solubility of O_2 *decreases* as the temperature of the solution increases. This result is generally observed for the solubility of a gas in a liquid. The decreasing solubility of O_2 in water with increasing temperature has important ramifications for plant and animal life.

The equilibrium vapor pressure results when the rate of evaporation of the solvent from the solution is equal to the rate of condensation of the solvent from the vapor. The rate of evaporation of the solvent from a solution is less than that of the pure solvent because the presence of solute molecules at or near the surface of the solution allows fewer solvent molecules per unit area of surface (Figure 12-6). The rate of condensation is directly proportional to the number of molecules in the vapor, which in turn is proportional to the pressure. The lower rate of evaporation, then, is balanced by a lower rate of condensation, or by a lower vapor pressure (Figure 12-7).

As Figure 12-6 implies, the vapor pressure of the solvent over a solution is directly proportional to the fraction of solvent particles at the surface. Recall from Section 4-8 that we defined a quantity called a mole fraction. In a solution containing n_1 moles of solvent and n_2 moles of solute, the mole fraction of the solvent, X_1, is

$$X_1 = \frac{n_1}{n_1 + n_2} \tag{12-2}$$

Because of the proportionality between the number of moles and the number of particles, X_1 can also be considered as a particle fraction. Therefore, we can write the relation between the equilibrium vapor pressure of the solvent P_1 and the mole fraction of solvent X_1 as

$$P_1 \propto X_1$$

By introducing a proportionality constant, k, we can write

$$P_1 = kX_1 \tag{12-3}$$

The value of the proportionality constant can be determined as follows. When the mole fraction of the solvent is unity, that is, when we have pure solvent, then P_1 is equal to the vapor pressure of the pure solvent, which we denote by P_1°. Thus when $X_1 = 1$ in Equation (12-3), we have $P_1 = P_1^\circ$. If we substitute $X_1 = 1$ and $P_1 = P_1^\circ$ into Equation (12-3), then we see that $k = P_1^\circ$. Thus we can rewrite Equation (12-3) as

$$P_1 = X_1 P_1^\circ \tag{12-4}$$

which is known as **Raoult's law.** Raoult's law relates the equilibrium vapor pressure of the solvent to the mole fraction of the solvent.

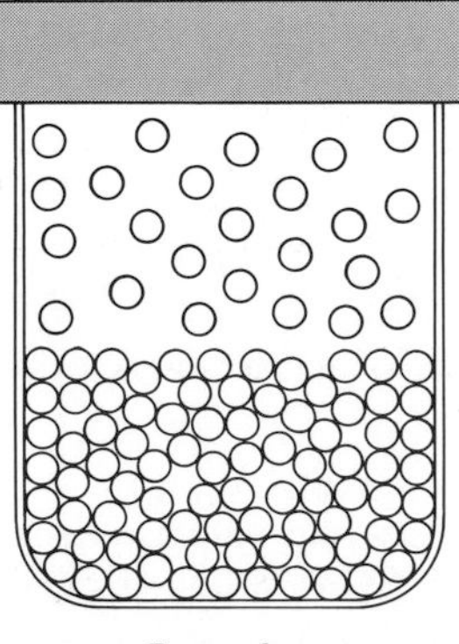

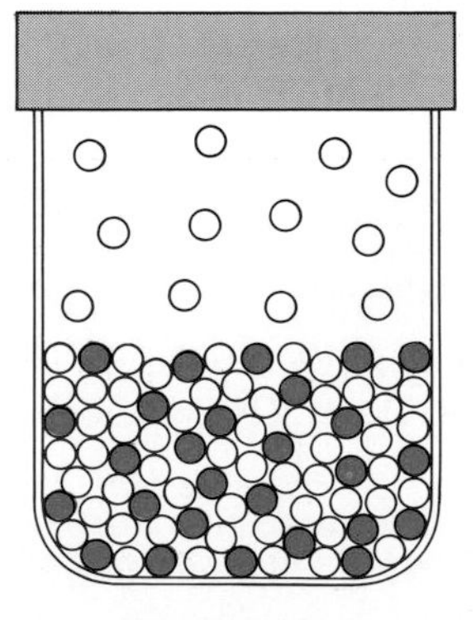

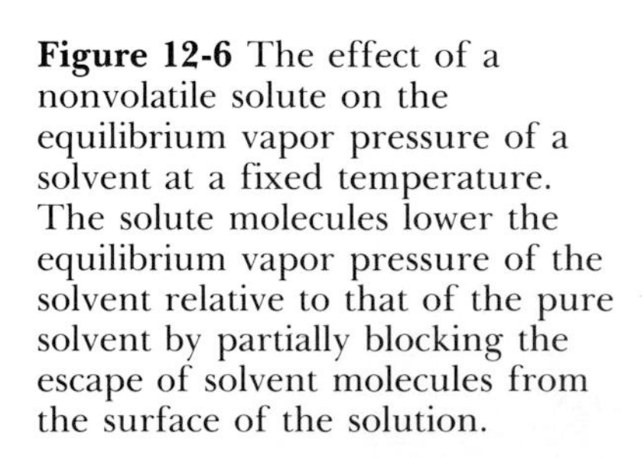

Figure 12-6 The effect of a nonvolatile solute on the equilibrium vapor pressure of a solvent at a fixed temperature. The solute molecules lower the equilibrium vapor pressure of the solvent relative to that of the pure solvent by partially blocking the escape of solvent molecules from the surface of the solution.

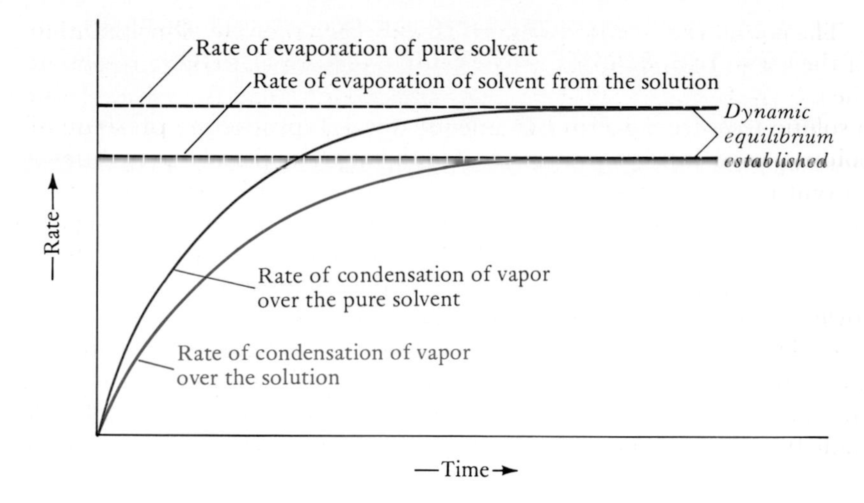

Figure 12-7 The effect of a nonvolatile solute on the solvent evaporation rate. The decreased solvent evaporation rate leads to a lower equilibrium vapor pressure for the solvent.

▪ Raoult's law tells us that the vapor pressure of a solvent is directly proportional to the mole fraction of the solvent in the solution.

The solvent always obeys Raoult's law if the solution is sufficiently dilute. Solutions for which all constituents obey Raoult's law for *all* concentrations are called **ideal solutions.** We shall discuss ideal solutions in Section 12-9, where we treat solutions consisting of two liquids.

The amount by which the vapor pressure of a solution is less than the vapor pressure of the pure solvent, that is, $P_1^\circ - P_1$, is called the **vapor pressure lowering.** Using Equation (12-4), we can write for the vapor pressure lowering, ΔP_1,

$$\Delta P_1 = P_1^\circ - P_1 = P_1^\circ - X_1 P_1^\circ = (1 - X_1)\, P_1^\circ = X_2 P_1^\circ \qquad (12\text{-}5)$$

where we have used the fact that $X_1 + X_2 = 1$.

Example 12-2: The vapor pressure of benzene at 30°C is 120 torr. Use Raoult's law to calculate the vapor pressure of a solution made by dissolving 8.71 g of the relatively nonvolatile solid naphthalene, $C_{10}H_8$, in 32.6 g of benzene, C_6H_6. Also calculate the vapor pressure lowering of benzene.

Solution: The mole fraction of benzene in the solution is

$$X_{benz} = \frac{n_{benz}}{n_{benz} + n_{naph}}$$

$$= \frac{(32.6\text{ g})\left(\dfrac{1\text{ mol benzene}}{78.11\text{ g benzene}}\right)}{(32.6\text{ g})\left(\dfrac{1\text{ mol benzene}}{78.11\text{ g benzene}}\right) + (8.71\text{ g})\left(\dfrac{1\text{ mol naphthalene}}{128.2\text{ g naphthalene}}\right)}$$

$$= 0.860$$

The equilibrium vapor pressure of benzene over the solution is given by Equation (12-4):

$$P_{benz} = X_{benz} P_{benz}^\circ = (0.860)(120\text{ torr}) = 103\text{ torr}$$

The vapor pressure lowering of benzene produced by the dissolved naphthalene is

$$\Delta P_{benz} = P_{benz}^\circ - P_{benz} = 120\text{ torr} - 103\text{ torr} = 17\text{ torr}$$

12-5. COLLIGATIVE PROPERTIES OF SOLUTIONS DEPEND ONLY ON THE SOLUTE PARTICLE CONCENTRATION

Vapor pressure lowering is one of several properties of solutions known as **colligative properties.** Other major colligative properties of solutions, which we discuss later in the chapter, are boiling point elevation, freezing-point depression, and osmotic pressure. Colligative properties depend primarily upon the fractions of solute particles and solvent particles and not upon the chemical nature of the solute. The mole fraction of the solvent in a dilute solution is usually very close to one, and so mole fraction is not a convenient concentration unit for dilute solutions. It is more convenient to use a concentration unit called **molality,** which is directly proportional to the mole fraction of solute in a dilute solution. We define the molality m of a solute as the number of moles of solute per 1000 g of solvent:

▪ Molality is proportional to the mole fraction of the solute in a dilute solution (see Problem 12-64).

$$\text{molality} = \frac{\text{moles of solute}}{1000 \text{ g of solvent}} = \frac{\text{moles of solute}}{\text{kilogram of solvent}}$$

For example, a solution prepared by dissolving 1.00 mol of sodium chloride (58.44 g) in 1.00 kg of water is 1.00 molal (1.00 m) in NaCl.

Example 12-3: Compute the molality of a solution prepared by dissolving 20.0 g of $CaCl_2(s)$ in 500 g of water.

Solution: The number of moles of $CaCl_2$ in 20.0 g is

$$(20.0 \text{ g})\left(\frac{1 \text{ mol } CaCl_2}{111.0 \text{ g } CaCl_2}\right) = 0.180 \text{ mol}$$

When 0.180 mol of $CaCl_2$ is dissolved in 0.500 kg of water, the molality of $CaCl_2$ in the resulting solution is

$$m = \frac{0.180 \text{ mol}}{0.500 \text{ kg}} = 0.360 \text{ mol} \cdot \text{kg}^{-1} = 0.360 \text{ m}$$

The molality concentration scale, which is denoted by m, is not the same as the molarity concentration scale, denoted by M (Section 3-11). To prepare 500 mL of a 0.360 M solution of $CaCl_2$ in water, we dissolve 0.180 mol of $CaCl_2$ in less than 500 mL of water and dilute the resulting solution with enough water to yield exactly 500 mL *of solution*. Compare this procedure with that described in Example 12-3 for the preparation of a 0.360 m solution. Furthermore, note that molality is independent of temperature, whereas molarity varies with temperature because the volume of the solution varies with temperature.

It is essential for the understanding of colligative properties to recognize that it is the fraction of solute particles or solvent particles that determines the magnitude of a colligative effect. A 0.10 m aqueous NaCl solution has twice as many solute particles per mole of water as a 0.10 m aqueous glucose solution because NaCl is a strong electrolyte and dissociates completely in water to $Na^+(aq)$ and $Cl^-(aq)$, whereas glucose exists in solution as intact $C_6H_{12}O_6$ molecules. We can express this result as an equation by writing

$$m_c = im \tag{12-6}$$

where i is the number of particles per formula unit that occur in solution. We thus distinguish between the molality, denoted by m, and the **colligative molality,** denoted by m_c, of a solute. The distinction is illustrated numerically in Table 12-3.

Table 12-3 Comparison of molality and colligative molality of aqueous solutions

Solute	Solute molality/m	Solute particles per formula unit	Colligative molality/m_c
$C_6H_{12}O_6$	0.10	1	0.10
NaCl	0.10	2	0.20
$CaCl_2$	0.10	3	0.30

Example 12-4: What is the colligative molality of the $CaCl_2$ solution in Example 12-3?

Solution: Because one formula unit of $CaCl_2$ produces one $Ca^{2+}(aq)$ and two $Cl^-(aq)$ in aqueous solution, the colligative molality is three times the molality calculated in Example 12-3; that is, $m_c = im = 3 \times 0.360\ \text{m} = 1.08\ \text{m}_c$.

Example 12-5: A 1.0-mol sample of each of the following substances is dissolved in 1000 g of water:

$$CH_3OH \qquad AgNO_3 \qquad K_2SO_4$$

Determine the colligative molality of each of the resulting solutions.

Solution: Molality is the number of moles of solute per kilogram of solvent, and so the molality of each of these substances is 1.0 m. Methanol is a nonelectrolyte, and thus the colligative molality is the same as the molality. Silver nitrate is a strong electrolyte in water, yielding $Ag^+(aq)$ and $NO_3^-(aq)$, and thus $m_c = 2\ m = 2.0\ \text{m}_c$. Potassium sulfate is a strong electrolyte in water, yielding $2K^+(aq) + SO_4^{2-}(aq)$, and thus $m_c = 3\ m = 3.0\ \text{m}_c$.

12-6. NONVOLATILE SOLUTES INCREASE THE BOILING POINT OF A LIQUID

Recall that the boiling point is the temperature at which the equilibrium vapor pressure equals the atmospheric pressure. We know from Section 12-4 that the equilibrium vapor pressure of the solvent over a solution containing a nonvolatile solute is less than that for the pure solvent at the same temperature. Therefore, the temperature at which the equilibrium vapor pressure reaches atmospheric pressure is higher for the solution than for the pure solvent (Figure 12-8). In other words, the boiling point of the solution, T_b, is higher than the boiling point of the pure solvent, T_b°. The **boiling point elevation,** $T_b - T_b^\circ$, is proportional to the magnitude of the vapor pressure lowering, $P_1^\circ - P_1$:

$$(T_b - T_b^\circ) \propto (P_1^\circ - P_1)$$

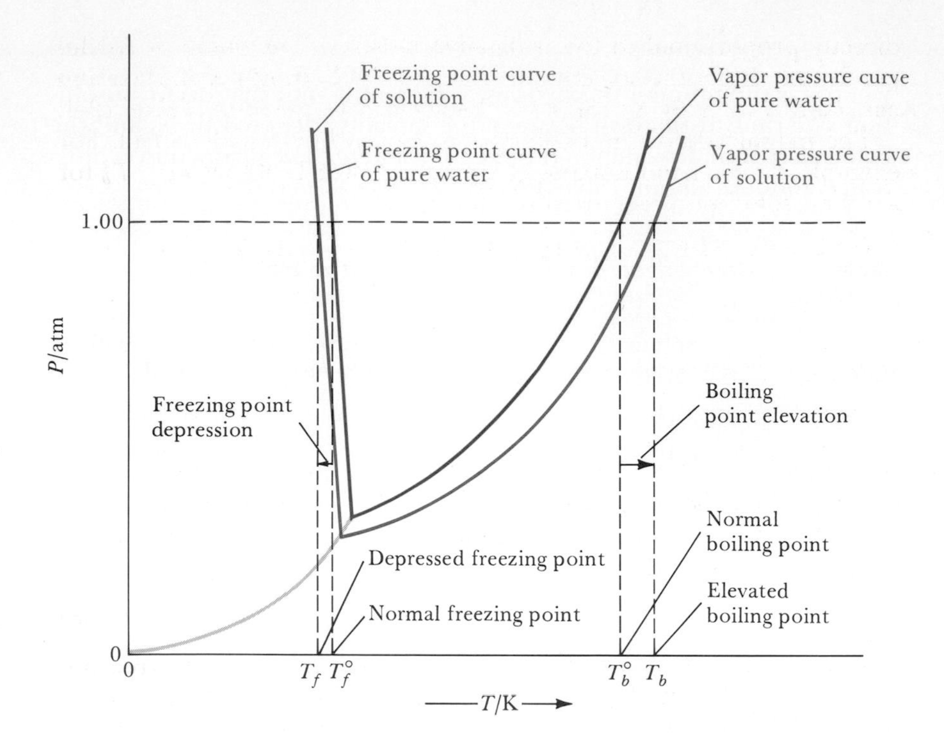

Figure 12-8 Phase diagrams for pure water (blue lines) and for water containing a nonvolatile solute (red lines). The presence of the solute lowers the vapor pressure of the solvent. The reduced vapor pressure of the solvent results in an *increase* in the boiling point of the solution relative to that of the pure solvent and a *decrease* in the freezing point of the solution relative to that of the pure solvent.

The vapor pressure lowering is proportional to the mole fraction of solute or to the colligative molality of the solute:

$$(P_1^\circ - P_1) \propto m_c$$

Because the boiling point elevation is proportional to the vapor pressure lowering, it is also proportional to the colligative molality. Thus we can write

$$T_b - T_b^\circ = K_b m_c \tag{12-7}$$

where K_b is a proportionality constant. Equation (12-7) tells us that the increase in the boiling point of a solution containing a nonvolatile solute is directly proportional to the solute (particle) concentration, that is,

Table 12-4 Boiling point elevation constants (K_b) and freezing point depression constants (K_f) for various solvents

Solvent	Boiling point/°C	K_b/K·m_c^{-1}	Freezing point/°C	K_f/K·m_c^{-1}
water	100.00	0.52	0.00	1.86
acetic acid	117.9	2.93	16.6	3.90
benzene	80.0	2.53	5.50	5.10
chloroform	61.2	3.63	−63.5	4.68
cyclohexane	80.7	2.79	6.5	20.2
nitrobenzene	210.8	5.24	5.7	6.87
camphor	(208.0)*	5.95	179.8	40.0

*Sublimation point

directly proportional to the colligative molality. The value of K_b depends only upon the solvent and is called the **boiling point elevation constant.** Values of K_b for several solvents are given in Table 12-4.

The magnitude of the boiling point elevation is usually small. For example, we see from Equation (12-7) and Table 12-4 that $T_b - T_b^\circ$ for a 1.0 m$_c$ solution of glucose in water is

$$T_b - T_b^\circ = [0.52\ \text{K}\cdot\text{m}_c^{-1}](1.0\ \text{m}_c) = 0.52\ \text{K}$$

▪ Note that the boiling point elevation is a small effect for a 1.0 m$_c$ solution.

12-7. SOLUTES DECREASE THE FREEZING POINT OF A LIQUID

By arguments analogous to those used to obtain the boiling point elevation equation, the magnitude of the **freezing point depression** produced by a solute with colligative molality m_c is found to be given by an equation similar to Equation (12-7):

$$T_f^\circ - T_f = K_f m_c \qquad (12\text{-}8)$$

where T_f° is the freezing point of the pure solvent, T_f is the freezing point of the solution ($T_f^\circ > T_f$), and K_f is the **freezing point depression constant.** The value of K_f depends only upon the solvent.

The value of the freezing point depression constant of water is 1.86 K·m$_c^{-1}$ (Table 12-4). Thus we predict that an aqueous solution with a colligative molality of 0.50 m$_c$ has a freezing point depression of

$$T_f^\circ - T_f = K_f m_c = (1.86\ \text{K}\cdot\text{m}_c^{-1})(0.50\ \text{m}_c) = 0.93\ \text{K}$$

and the freezing point is −0.93°C.

The freezing point depression due to a dissolved substance is the basis of the action of **antifreeze.** The most commonly used antifreeze is ethylene glycol,

```
     H  H
     |  |
HO—C—C—OH
     |  |
     H  H
```

whose boiling point is 197°C and whose freezing point is −17.4°C. The addition of ethylene glycol to water depresses the freezing point and elevates the boiling point of the water. An ethylene glycol–water solution in which ethylene glycol is 50 percent by volume has a freezing point of −36°C.

Example 12-6: Estimate the freezing point of a 5.0 m$_c$ solution of ethylene glycol in water.

Solution: The freezing point of the solution is computed from Equation (12-8):

$$T_f^\circ - T_f = K_f m_c$$
$$= (1.86\ \text{K}\cdot\text{m}_c^{-1})(5.0\ \text{m}_c) = 9.3\ \text{K}$$

The freezing point of the solution is 263.9 K, or −9.3°C.

The freezing point of aqueous ethylene glycol solutions as a function of the molality of ethylene glycol is given in Figure 12-9. The effectiveness of ethylene glycol as an antifreeze is a result of several factors, among which are its high boiling point and chemical stability and the tendency of the ice that freezes out of the solution to form a slushy mass rather than a solid block. In the absence of a sufficient amount of antifreeze, the 9 percent volume expansion of water on freezing generates a force of 30,000 lb/in^2 at -22°C, which is sufficient to rupture a radiator or even a metal engine block.

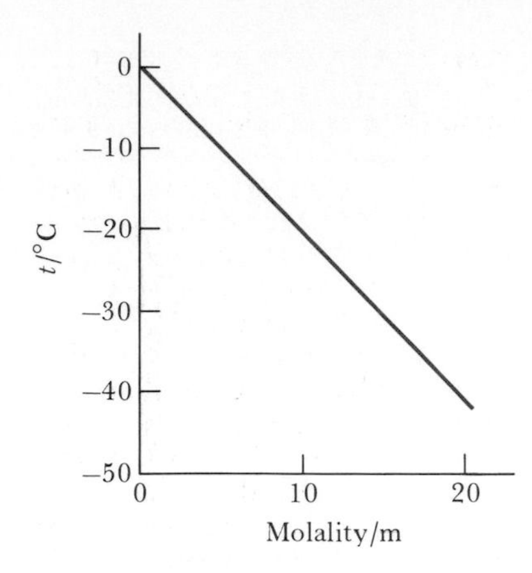

Figure 12-9 Freezing point versus molality for aqueous ethylene glycol (antifreeze) solutions. Note that $m_c = m$ for ethylene glycol.

12-8. OSMOTIC PRESSURE REQUIRES A SEMIPERMEABLE MEMBRANE

Suppose we place pure water in one beaker and an equal volume of seawater in another beaker and then place both beakers under a bell jar, as shown in Figure 12-10(a). We observe that as time passes the volume of pure water decreases and the volume of seawater increases (Figure 12-10b). The pure water has a higher equilibrium vapor pressure than the seawater, and thus the rate of condensation of the water into the seawater is greater than the rate of evaporation of the water from the seawater. The net effect is the transfer of water, via the vapor phase, from the beaker with pure water to the beaker with seawater. This transfer continues until no pure liquid water remains, and the seawater ends up diluted.

If pure water and seawater are separated by a membrane that is permeable to water but not to the ions in seawater, then the water passes directly through the membrane from the pure water side of the membrane to the seawater side (Figure 12-11). Such a membrane is called a **semipermeable membrane,** and the tendency of the water to pass through the membrane is called **escaping tendency.** The escaping tendency of water from pure water is greater than the escaping tendency of water from seawater because pure water has a higher vapor

▪ The higher the vapor pressure of a substance, the greater is its escaping tendency.

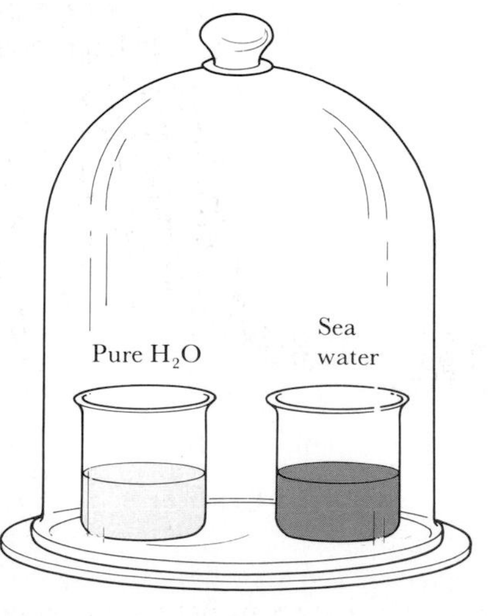

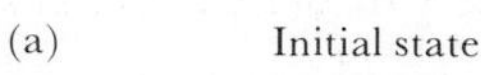

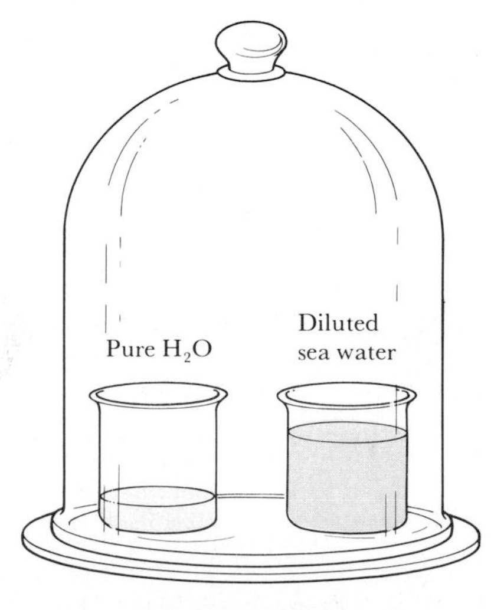

Figure 12-10 One beaker contains pure water, and the other contains seawater. (a) The equilibrium vapor pressure of the pure water is greater than that of the water over the seawater solution. (b) As time passes, pure water is transferred via the vapor phase from the beaker containing pure water to the beaker containing seawater, thereby diluting the seawater. If we wait long enough, all the pure liquid water will transfer to the seawater beaker.

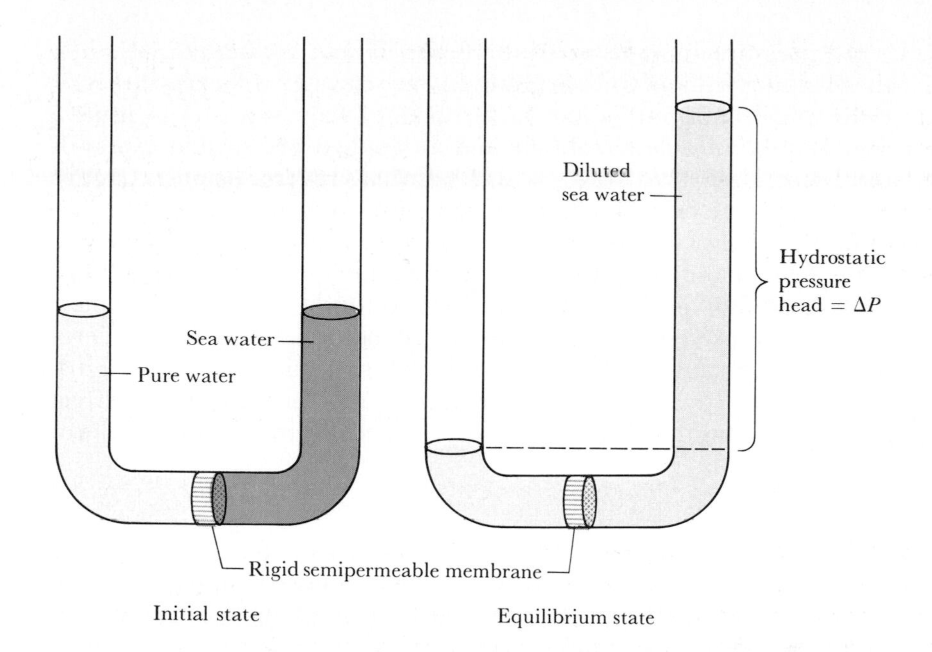

Figure 12-11 Passage of water through a rigid semipermeable membrane separating pure water from seawater. The water passes through the membrane until the escaping tendency of the water from the seawater equals the escaping tendency of the pure water. The escaping tendency of water from the seawater side of the membrane increases as the seawater is diluted and also as a result of the increased hydrostatic pressure head on the seawater, which results from the increase in the height of the seawater column.

pressure than seawater. As water passes through the membrane to the seawater side, the escaping tendency of the water in the seawater increases, not only because the seawater is being diluted but also because of the increased pressure on the seawater side of the membrane. This pressure increase arises from the hydrostatic pressure head of the solution. The column of seawater rises until the escaping tendency of the water in the seawater is equal to the escaping tendency of the pure water. At this point, equilibrium is reached and the column no longer rises. The hydrostatic pressure head produced in this process is called the **osmotic pressure** (Figure 12-11). The spontaneous passage of solvent through a semipermeable membrane from one solution to a more concentrated solution is called **osmosis.**

The osmotic pressure π of a solution is given by the equation

$$\pi = RTM_c \tag{12-9}$$

where R is the gas constant, 0.0821 $L{\cdot}atm{\cdot}K^{-1}{\cdot}mol^{-1}$; T the absolute temperature; and M_c the **colligative molarity,** that is, the molarity multiplied by the number of particles per formula unit.

Example 12-7: As a rough approximation, seawater can be regarded as a 0.55 M aqueous NaCl solution. Estimate the osmotic pressure of seawater at 15°C.

Solution: The colligative molarity of seawater is approximately 2 × 0.55 M because NaCl dissociates in water to yield two ions per formula unit. Thus the osmotic pressure of seawater is

$$\begin{aligned}\pi &= RTM_c \\ &= (0.0821\ L{\cdot}atm{\cdot}K^{-1}{\cdot}mol^{-1})(288\ K)(2 \times 0.55\ mol{\cdot}L^{-1}) \\ &= 26\ atm\end{aligned}$$

Note that the osmotic pressure of seawater is 26 times higher than atmospheric pressure.

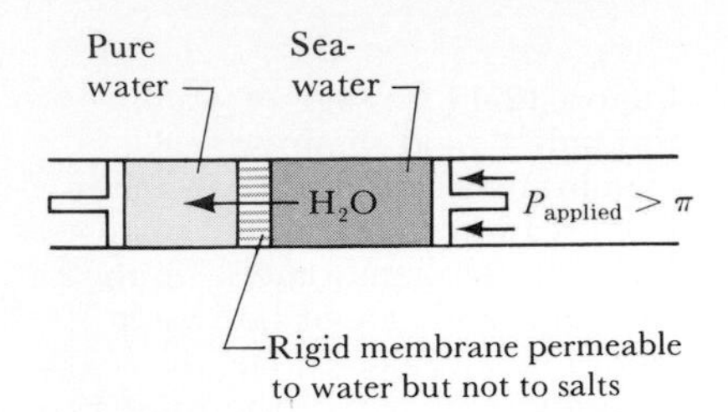

Figure 12-12 Reverse osmosis. A rigid semipermeable membrane separates pure water from seawater. A pressure in excess of the osmotic pressure of seawater (26 atm at 15°C) is applied to the seawater, and this increases the escaping tendency of water from the seawater to a value above that of pure water. Under these conditions, the net flow of water is from the seawater side through the semipermeable membrane to the pure water side. The net effect is the production of fresh water from seawater.

If a pressure in excess of 26 atm is applied to seawater at 15°C, then the escaping tendency of the water in the seawater will exceed that of pure water. Thus pure water can be obtained from seawater by using a rigid semipermeable membrane and an applied pressure in excess of the osmotic pressure. This process is known as **reverse osmosis** (Figure 12-12). Reverse osmosis units are commercially available and are used to obtain fresh water from brine.

Because of the large values of osmotic pressure, osmotic pressure measurement is an especially good method for determining the molecular mass of proteins. Proteins have large molecular masses and therefore yield a relatively small number of solute particles for a given dissolved mass. In fact, osmotic pressure is the only colligative effect sufficiently sensitive to provide useful molecular information on proteins.

Example 12-8: A 4.00-g sample of human hemoglobin was dissolved to make 0.100 L of solution, and the osmotic pressure of the solution at 7°C was found to be 10.0 torr, or 0.0132 atm. Estimate the molecular mass of the hemoglobin.

Solution: From Equation (12-9), the concentration of the hemoglobin in the aqueous solution is

$$M_c = \frac{\pi}{RT} = \frac{0.0132 \text{ atm}}{(0.0821 \text{ L·atm·K}^{-1}\text{·mol}^{-1})(280 \text{ K})}$$
$$= 5.74 \times 10^{-4} \text{ mol·L}^{-1}$$

The molecular mass can be computed from the concentration of the protein because the dissolved mass is known. We have that

$$5.74 \times 10^{-4} \text{ mol·L}^{-1} \approx \frac{4.00 \text{ g}}{0.100 \text{ L}} = 40.0 \text{ g·L}^{-1}$$

and therefore

$$5.74 \times 10^{-4} \text{ mol} \approx 40.0 \text{ g}$$

By dividing both sides of this stoichiometric correspondence by 5.74×10^{-4}, we find that

$$1 \text{ mol} \approx 69{,}700 \text{ g}$$

The molecular mass of the hemoglobin is 69,700. Protein molecular masses can be as large as 1,000,000.

Both plant cells and animal cells have membranes that are permeable to water but not, for example, to sucrose. The colligative concentration of the solution inside a typical biological cell is approximately 0.3 M_c. Most animal cells have about the same internal colligative molarity as the extracellular fluid in which the cells exist.

▪ The greater the colligative molarity is, the less the escaping tendency of the water from the solution.

Water passes spontaneously through a cell membrane from the side with the lower colligative molarity (higher water escaping tendency) to the side with the higher colligative molarity (lower water escaping tendency). The entry of water into an animal cell causes the cell to expand, and the exit of water from the cell causes the cell to contract. The cell assumes its normal volume when it is placed in a solution with a colligative molarity of 0.3 M_c. More concentrated solutions cause the cell to contract, and less concentrated solutions cause it to expand (Figure 12-13). When cells are placed in distilled water at 27°C, equilibrium is obtained at an internal cell pressure equal to the osmotic pressure of a 0.30 M_c solution, that is,

$$\pi = RTM_c = (0.0821\ \text{L·atm·K}^{-1}\text{·mol}^{-1})(300\ \text{K})(0.30\ \text{mol·L}^{-1})$$
$$= 7.4\ \text{atm}$$

Because a pressure of 7.4 atm cannot be maintained by animal cell membranes, the cells burst. Plant cell walls are rigid, and they can tolerate a pressure of 7.4 atm. Thus the entry of water into plant cells gives nonwoody plants the rigidity required to stand erect.

12-9. IDEAL SOLUTIONS CONSISTING OF TWO LIQUIDS OBEY RAOULT'S LAW

In the previous sections of this chapter we have discussed only dilute solutions of nonvolatile solutes in volatile solvents. In this section we shall discuss solutions consisting of two volatile liquids. It is convenient to define an ideal solution as follows. A solution of two components, A and B, is said to be ideal if the interactions between A and B molecules are the same as those between A molecules or B molecules. In an ideal solution the A and B molecules are randomly distributed throughout the solution, including the region near the surface (Figure 12-14). When the molecules of the two components are very similar, then the solution is essentially ideal. For example, benzene and toluene form essentially ideal solutions.

In an ideal solution of two liquids, the vapor pressure of each component is given by Raoult's law, so that

$$P_A = X_A P_A^\circ \qquad \text{and} \qquad P_B = X_B P_B^\circ \tag{12-10}$$

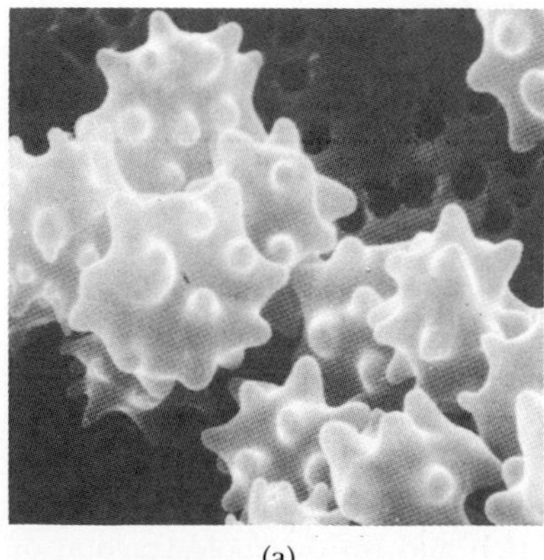

(a)

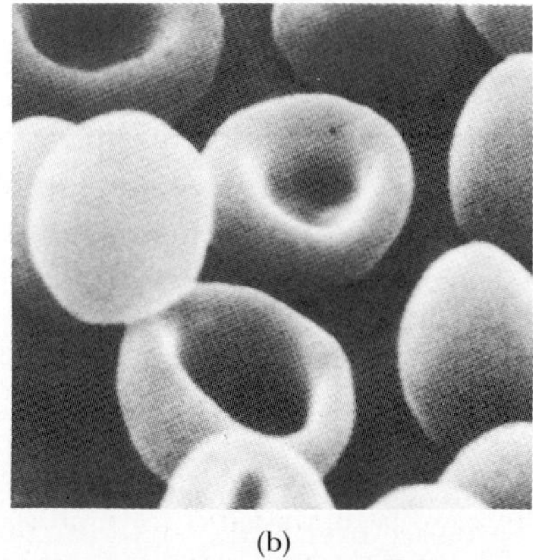

(b)

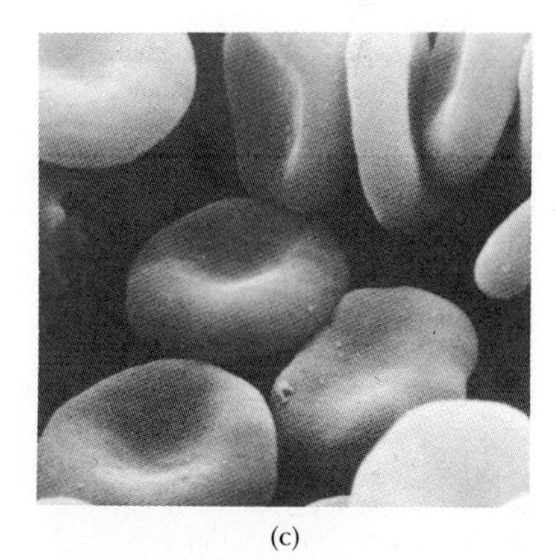

(c)

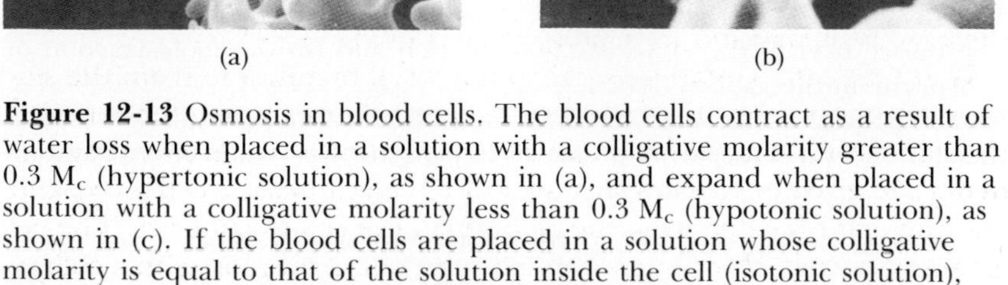

Figure 12-13 Osmosis in blood cells. The blood cells contract as a result of water loss when placed in a solution with a colligative molarity greater than 0.3 M_c (hypertonic solution), as shown in (a), and expand when placed in a solution with a colligative molarity less than 0.3 M_c (hypotonic solution), as shown in (c). If the blood cells are placed in a solution whose colligative molarity is equal to that of the solution inside the cell (isotonic solution), then they neither contract nor expand, as seen in (b).

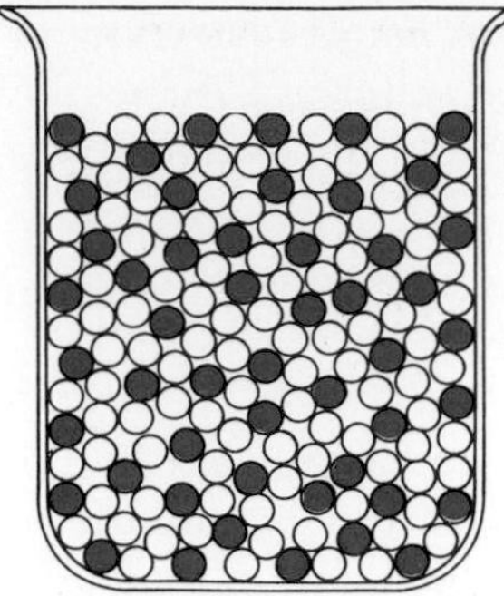

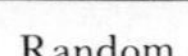

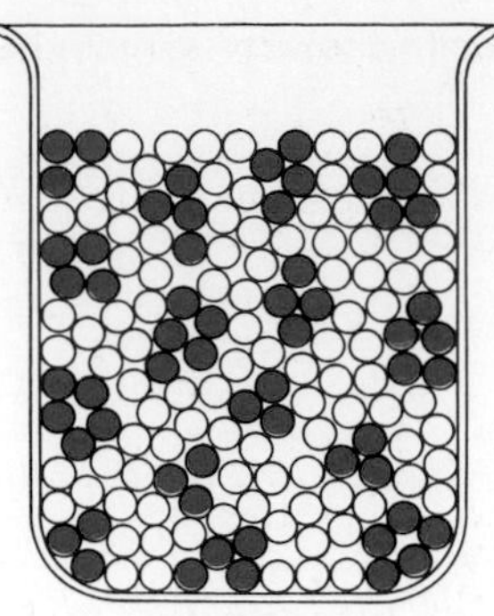

Figure 12-14 Random and nonrandom distribution of two types of molecules, A (open circles) and B (solid circles), in solution. Note in the nonrandom case that AAA . . . and BBB . . . clusters form in the solution. The random distribution represents an ideal solution.

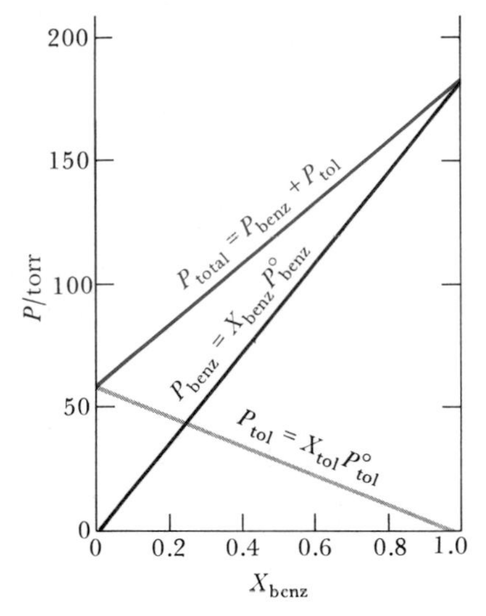

Figure 12-15 Equilibrium vapor pressure at 40°C versus the mole fraction of benzene in solutions of benzene and toluene. The solutions are essentially ideal, and the equilibrium vapor pressures of benzene and toluene are given by Raoult's law (Equation 12-4). The total pressure is a linear function of X_{benz} when Raoult's law holds for both components, that is, when the solution is ideal.

The vapor over such a solution consists of both components, and so the total vapor pressure is

$$P_{total} = P_A + P_B \qquad (12\text{-}11)$$

Substituting Equation (12-10) into Equation (12-11) gives

$$P_{total} = X_A P^{\circ}_A + X_B P^{\circ}_B \qquad (12\text{-}12)$$

Because $X_A + X_B = 1$, we can set $X_B = 1 - X_A$ and write Equation (12-12) as

$$P_{total} = P^{\circ}_B + X_A(P^{\circ}_A - P^{\circ}_B) \qquad (12\text{-}13)$$

Notice that when $X_A = 0$ (pure B), $P_{total} = P^{\circ}_B$, and that when $X_A = 1$ (pure A), $P_{total} = P^{\circ}_A$. Notice also that if one of the components, say B, is nonvolatile, then $P^{\circ}_B = 0$ and Equation (12-13) becomes the same as Equation (12-4).

If we plot P_{total} versus X_A for an ideal solution, then we obtain a straight line. You may have learned in algebra that the equation for a straight line is

$$y = mx + b$$

where m and b are constants. The quantity b is the **intercept** of the straight line with the y axis, and m is the **slope** of the line. In Equation (12-13), if we let $P_{total} = y$ and $X_A = x$, then the intercept is P°_B and the slope is $P^{\circ}_A - P^{\circ}_B$. The total equilibrium vapor pressure of a benzene-toluene solution is plotted against the mole fraction of benzene in Figure 12-15. These data confirm that a benzene-toluene solution is an ideal solution.

■ $y = mx + b$ (m ↑ slope, b ↑ intercept)
Plotting data in the form of a straight line is discussed in Appendix A and in the Study Guide/Solutions Manual.

Example 12-9: Given that the equilibrium vapor pressures of benzene and toluene are 183 torr and 59.2 torr, respectively, calculate (a) the total vapor pressure over a $X_{benz} = X_{tol} = 0.500$ solution and (b) the mole fraction of benzene in the vapor.

Solution: (a) The partial pressures of benzene and toluene over the solution are given by Raoult's law:

$$P_{benz} = X_{benz}P^{\circ}_{benz} = (0.500)(183 \text{ torr}) = 91.5 \text{ torr}$$

$$P_{tol} = X_{tol}P^{\circ}_{tol} = (0.500)(59.2 \text{ torr}) = 29.6 \text{ torr}$$

The total vapor pressure is the sum of the partial pressures:

$$P_{total} = P_{benz} + P_{tol} = 91.5 \text{ torr} + 29.6 \text{ torr} = 121.1 \text{ torr}$$

(b) The pressure of a gas is directly proportional to the number of moles of the gas, and so the mole fraction of benzene in the vapor is given by

$$Y_{benz} = \frac{n_{benz}}{n_{benz} + n_{tol}} = \frac{P_{benz}}{P_{benz} + P_{tol}} = \frac{91.5 \text{ torr}}{121.1 \text{ torr}} = 0.756$$

where we have used Y for the mole fraction in the vapor to distinguish it from the mole fraction in solution.

From the previous Example, we see that the vapor over a benzene-toluene solution is richer in benzene, the more volatile component, than is the solution. If this vapor is condensed and then re-evaporated, the resulting vapor will be even richer in benzene. If this condensation-evaporation process is repeated many times, a separation of the benzene and toluene is achieved. Such a process is called **fractional distillation** and is carried out automatically in a single fractional distillation column (Figure 12-16). A fractional distillation column differs from an ordinary distillation column in that the former is packed with glass beads, glass rings, or glass wool. The packing material provides a large surface area for the repeated condensation-evaporation process.

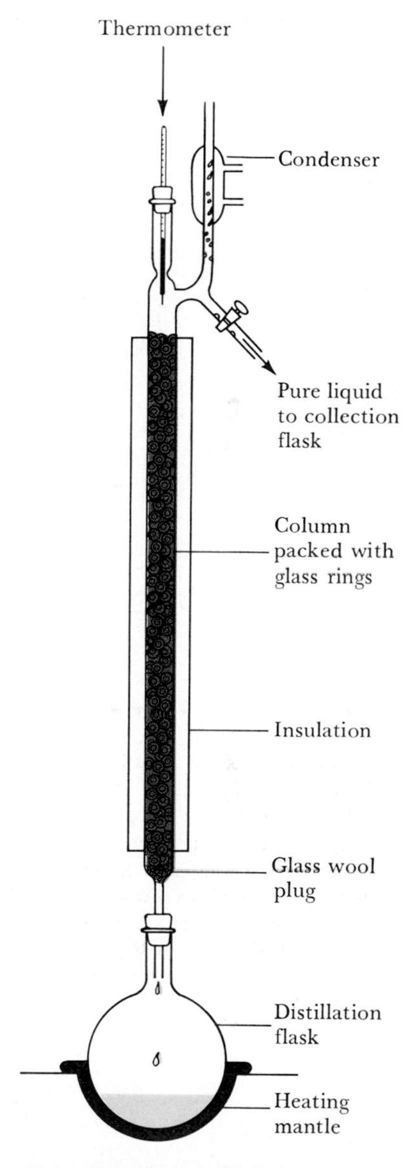

Figure 12-16 A simple fractional distillation column. Repeated condensation and re-evaporation occurs along the entire column. The vapor becomes progressively richer in the more volatile component as it moves up the column.

Remarkable separations can be achieved with elaborate fractional distillation units. Fractional distillation techniques are used to separate heavy water (D_2O) from regular water. The heavy water is used on a large scale in nuclear reactors and as a coolant in heavy-water nuclear power plants. Regular water has a normal boiling point of 100.00°C, whereas heavy water has a normal boiling point of 101.42°C. Only 0.015 percent of the hydrogen atoms in regular water are the deuterium isotope. Nonetheless, a modern heavy-water distillation plant produces almost pure D_2O from regular water at a total cost of around $400 per kilogram of D_2O. These distillation plants have over 300 successive distillation stages and require an input of over 1 metric ton of water per gram of D_2O produced. Canada, which uses D_2O extensively in its nuclear reactors, has two heavy-water plants with a combined D_2O output capability of 1600 tons per year.

If a two-component solution is not ideal, then two cases arise. If the attractions between the A and B molecules are greater than those between A molecules or B molecules, then the A and B molecules prefer to be in solution and so the vapor pressure will be less than that calculated from Equation (12-13). An example of such a solution is a carbon disulfide–chloromethane (CS_2–CH_3Cl) solution, whose total vapor pressure is plotted in Figure 12-17(a). The deviations from Raoult's law, shown as dashed lines in Figure 12-17(a), are called **negative deviations** from ideal behavior. If the attractions between A and B molecules are less than those between A molecules or B molecules, then the A and B molecules prefer not to be in solution and so the vapor pressure will be greater than that calculated from Equation (12-12). An example is an ethyl alcohol–water (C_2H_5OH–H_2O) solution, whose total vapor pressure is plotted in Figure 12-17(b). In this case, we see **positive deviations** from ideal behavior. Thus, plots such as those in Figure 12-17 can give us insight into molecular interactions.

Ethyl alcohol–water solutions constitute an especially interesting case of nonideal solution behavior. If we attempt to separate alcohol from water by distillation, an amazing result is obtained. No matter how effective a distillation column we use, the distillate issuing from the column has a maximum alcohol content of 95 percent; water is 5 percent of the distillate. A 95 percent alcohol-plus-water solution distills as if it were a pure liquid. A solution that distills without change in composition is called an **azeotrope.**

Alcohol-water solutions obtained by fermentation have a maximum alcohol content of about 12 percent. Yeasts that produce alcohol as a waste product cannot survive in solutions with more than 12 percent alcohol. In effect, the yeasts are poisoned by their own waste products. The **proof** of an alcohol-water solution is defined as two times the percent alcohol by volume in the solution. An 86 proof liquor is 43 percent alcohol. The proof can be increased by simple distillation up to 190 (95 percent). **Absolute alcohol** (100 percent ethanol) can be prepared from 95 percent alcohol by dehydration with calcium oxide, which removes the water from the solution by the reaction

$$CaO(s) + H_2O(soln) \rightarrow Ca(OH)_2(s)$$

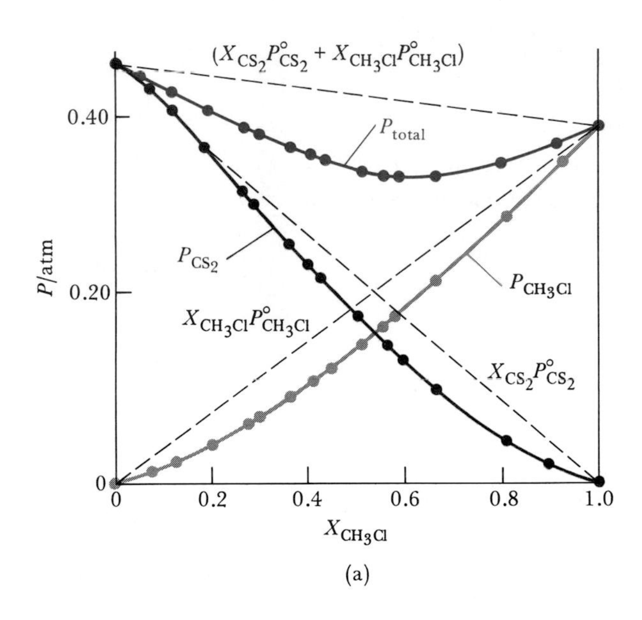

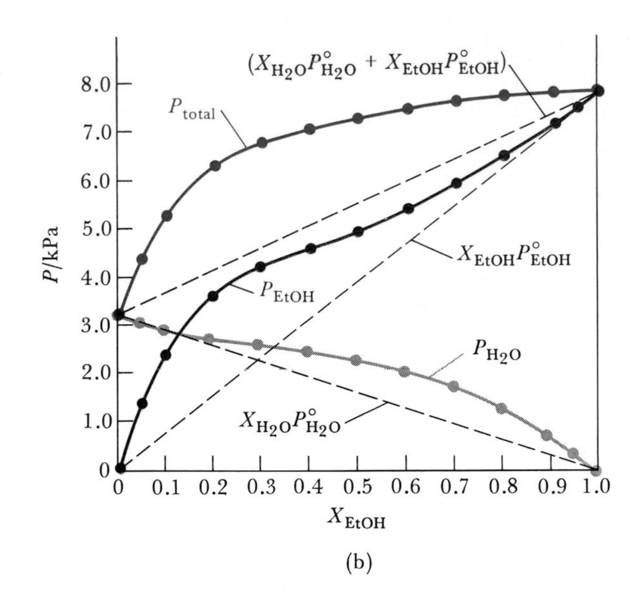

Figure 12-17 Nonideal solutions. (a) Equilibrium vapor pressures of carbon disulfide–methyl chloride solutions at 25°C. The dashed lines represent the behavior expected if Raoult's law holds. These curves illustrate negative deviations from ideal behavior. (b) Equilibrium vapor pressures of alcohol-water solutions at 25°C. The dashed lines represent the behavior expected if Raoult's law holds. These curves illustrate positive deviations from ideal behavior.

SUMMARY

A solution is a mixture of two or more substances that is homogeneous at the molecular level. Often the substance that is considered to be dissolved is called the solute and the substance doing the dissolving is called the solvent. The solubility of a solute in a solvent is the equilibrium quantity of solute that can be dissolved in the solvent under the given conditions. The solubilities of most solids in water increase with temperature. Henry's law states that the solubility of a gas in a liquid is directly proportional to the pressure of the gas over the solution. Gas solubility decreases with increasing temperature.

Vapor pressure lowering, boiling point elevation, freezing point depression, and osmotic pressure are colligative properties. The key point in understanding the colligative properties of solutions is that the equilibrium vapor pressure of a solvent is reduced when a solute is dissolved in it. If a solution is sufficiently dilute, then the equilibrium vapor pressure of any component is given by Raoult's law. Colligative properties of solutions depend only on the solute particle concentration and are independent of the chemical nature of the solute.

Osmotic pressure is the largest of the colligative effects and can be used to determine the molecular mass of proteins. Osmotic pressure effects are important in biological systems in that osmotic pressure keeps living cells inflated.

Raoult's law applies to ideal solutions consisting of two volatile liquids. Ideal solutions result when the interactions between unlike molecules in the solution are the same as the interactions between like molecules. If this is not the case, then the solution is not ideal and deviations from Raoult's law are observed.

TERMS YOU SHOULD KNOW

solution 382
homogeneous 382
components 382
solvent 382
solute 382
rate of solution 383
rate of crystallization 383
dynamic equilibrium 383
saturated solution 384
solubility 384
unsaturated solution 384
Henry's law 384
Henry's law constant, k_h 384
Raoult's law 386
ideal solution 387
vapor pressure lowering 387
colligative property 388
molality, m 388
colligative molality, m_c 389
boiling point elevation 389
boiling point elevation constant, K_b 391
freezing point depression 391
freezing point depression constant, K_f 391
antifreeze 391
semipermeable membrane 392
escaping tendency 392
osmotic pressure, π 393
osmosis 393
colligative molarity, M_c 393
reverse osmosis 394
intercept 396
slope 396
fractional distillation 397
positive deviation from Raoult's law 397
negative deviation from Raoult's law 397
azeotrope 398
proof (alcohol) 398
absolute alcohol 398

EQUATIONS YOU SHOULD KNOW HOW TO USE

$P_{gas} = k_h M_{gas}$	(12-1)	(Henry's law)
$X_1 = \dfrac{n_1}{n_1 + n_2}$	(12-2)	(mole fraction)
$P_1 = X_1 P_1^\circ$	(12-4)	(Raoult's law)
$m_c = im$	(12-6)	(colligative molality)
$T_b - T_b^\circ = K_b m_c$	(12-7)	(boiling point elevation)
$T_f^\circ - T_f = K_f m_c$	(12-8)	(freezing point depression)
$\pi = RTM_c$	(12-9)	(osmotic pressure)
$P_{total} = X_A P_A^\circ + X_B P_B^\circ$	(12-12)	
$P_{total} = P_B^\circ + X_A(P_A^\circ - P_B^\circ)$	(12-13)	(Raoult's law)

PROBLEMS

HENRY'S LAW

12-1. Calculate the concentration of nitrogen in water at a nitrogen gas pressure of 0.79 atm and a temperature of 25°C.

12-2. Of the gases N_2, O_2, and CO_2, which has the highest concentration in water at 25°C when each gas has a pressure of 1.0 atm?

12-3. The Henry's law constant for CO_2 in water at 25°C is 29 atm·M^{-1}. Estimate the concentration of dissolved CO_2 in a carbonated soft drink pressurized with 2.0 atm of CO_2 gas.

12-4. Calculate the masses of oxygen and nitrogen that are dissolved in 1.00 L of aqueous solution in equilibrium with air at 25°C and 760 torr. Assume that air is 20 percent oxygen and 79 percent nitrogen by volume.

12-5. Every 33 ft under water the pressure increases by 1 atm. How many feet would a diver have to descend before the blood concentration of oxygen reaches 1.28×10^{-3} M, assuming that compressed air is used (20 percent O_2 by volume)?

12-6. Compare the concentrations of helium and nitrogen in water when each gas is at a pressure of 4.0 atm at 25°C.

MOLE FRACTION

12-7. Calculate the mole fractions in a solution that is made up of 20.0 g of ethyl alcohol, C_2H_5OH, and 80.0 g of water.

12-8. A solution of 40 percent formaldehyde, H_2CO, 10 percent methyl alcohol, CH_3OH, and 50 percent water by mass is called formalin. Calculate the mole fractions of formaldehyde, methyl alcohol, and water in formalin. Formalin is used to disinfect dwellings, ships, storage houses, and so forth.

12-9. Describe how you would prepare 1.00 kg of an aqueous solution of acetone, CH_3COCH_3, in water in which the mole fraction of acetone is 0.19.

12-10. Describe how you would prepare 500 g of a solution of sucrose, $C_{12}H_{22}O_{11}$, in water in which the mole fraction of sucrose is 0.125.

12-11. Calculate the mole fraction of isopropyl alcohol, C_3H_7OH, in a solution that is 70.0 percent isopropyl alcohol and 30.0 percent water by volume. Take the density of water as $1.000\ g{\cdot}cm^{-3}$ and the density of isopropyl alcohol as $0.785\ g{\cdot}cm^{-3}$.

12-12. So-called "maintenance-free" car batteries use a lead-calcium alloy (solid solution). Given that the alloy is 95 percent lead and 5.0 percent calcium by mass, calculate the mole fraction of calcium in the solid solution.

RAOULT'S LAW

12-13. The vapor pressure of pure water at 37°C is 47.1 torr. Use Raoult's law to estimate the vapor pressure of an aqueous solution at 37°C containing 20.0 g of glucose, $C_6H_{12}O_6$, dissolved in 500.0 g of water. Also compute the vapor pressure lowering.

12-14. Water at 37°C has a vapor pressure of 47.1 torr. Calculate the vapor pressure of water if 50.0 g of glycerin, $C_3H_8O_3$, is added to 100.0 mL of water. The density of water at 37°C is $0.993\ g{\cdot}mL^{-1}$. Also calculate the vapor pressure lowering.

12-15. The vapor pressure of pure water at 25°C is 23.76 torr. Use Raoult's law to estimate the vapor pressure of an aqueous solution at 25°C containing 20.0 g of sucrose, $C_{12}H_{22}O_{11}$, dissolved in 195 g of water. Also calculate the vapor pressure lowering.

12-16. The vapor pressure of pure water at 100°C is 1.00 atm. Use Raoult's law to estimate the vapor pressure of water over an aqueous solution at 100°C containing 50.0 g of ethylene glycol, $C_2H_6O_2$, dissolved in 100.0 g of water. Also calculate the vapor pressure lowering for water.

12-17. Calculate the vapor pressure of an ethyl alcohol solution at 25°C containing 20.0 g of the nonvolatile solute urea, $(NH_2)_2CO$, dissolved in 100.0 g of ethyl alcohol, C_2H_5OH. The vapor pressure of ethyl alcohol at 25°C is 59.2 torr. Also calculate the vapor pressure lowering.

12-18. Estimate the vapor pressure of ethyl alcohol, C_2H_5OH, over 80 proof (40 percent by volume) vodka at 19°C. The vapor pressure of pure ethyl alcohol at 19°C is 40 torr. The density of ethyl alcohol is $0.79\ g{\cdot}mL^{-1}$ and the density of water is $1.00\ g{\cdot}mL^{-1}$.

MOLALITY

12-19. Describe how you would prepare a solution of formic acid in acetone, $(CH_3)_2CO$, that is 2.50 m in formic acid, $HCHO_2$.

12-20. Describe how you would prepare an aqueous solution that is 1.75 m in $Ba(NO_3)_2$.

12-21. The solubility of iodine in carbon tetrachloride is 2.603 g per 100.0 g of CCl_4 at 35°C. Calculate the molality of iodine in a saturated solution.

12-22. How many kilograms of water would have to be added to 18.0 g of oxalic acid, $H_2C_2O_4$, to prepare a 0.050 m solution?

12-23. A 1.0-mol sample of each of the following substances is dissolved in 1000 g of water. Determine the colligative molality of the substance in each case.

(a) $MgSO_4$ (b) $Cu(NO_3)_2$
(c) C_2H_5OH (d) $Al_2(SO_4)_3$

12-24. A 1.0-mol sample of each of the following substances is dissolved in 1000 g of water. Determine the colligative molality of the substance in each case.

(a) CH_3OH (b) $Al(NO_3)_3$
(c) $Fe(NO_3)_2$ (d) $(NH_4)_2Cr_2O_7$

12-25. Given that the vapor pressure of water is 17.54 torr at 20°C, calculate the vapor pressure lowering of aqueous solutions that are 0.25 m in

(a) NaCl (b) $CaCl_2$
(c) sucrose, $C_{12}H_{22}O_{11}$ (d) $Al(ClO_4)_3$

12-26. The colligative molality of the contents of a typical human cell is about 0.30 m_c. Compute the equilibrium vapor pressure of water at 37°C for the cell solution. Take $P^{\circ}_{H_2O} = 47.1$ torr at 37°C.

BOILING POINT ELEVATION

12-27. Estimate the boiling point of a 2.0 m aqueous solution of $Sc(ClO_4)_3$.

12-28. How much NaCl would have to be dissolved in 1000 g of water in order to raise the boiling point by 1.0°C?

12-29. Estimate the boiling point of a solution of 10.0 g of picric acid, $C_6H_2(OH)(NO_2)_3$, dissolved in 100.0 g of cyclohexane, C_6H_{12}. Assume that the colligative molality and the molality are the same for picric acid in cyclohexane.

12-30. The colligative molality of seawater is about 1.10 m_c. Estimate the boiling point of seawater at 1.00 atm and its vapor pressure at 15°C. The vapor pressure of pure water at 15°C is 12.79 torr.

12-31. Estimate the boiling point of a solution containing 25.0 g of urea, $(H_2N)_2CO$, dissolved in 1500 g of nitrobenzene, $C_6H_5NO_2$.

12-32. Estimate the boiling point of a solution of 25.0 g of urea, $(H_2N)_2CO$, plus 25.0 g of thiourea, $(H_2N)_2CS$, in 500 g of chloroform, $CHCl_3$.

FREEZING POINT DEPRESSION

12-33. Estimate the freezing point of an aqueous solution of 60.0 g of glucose, $C_6H_{12}O_6$, dissolved in 200.0 g of water.

12-34. Estimate the freezing point of a 0.15 m aqueous solution of NaCl.

12-35. Estimate the freezing point of a solution of 22.0 g of carbon tetrachloride dissolved in 800 g of benzene.

12-36. Estimate the freezing point of an aqueous solution of 20.0 g of $Ca(NO_3)_2$ dissolved in 500 g of water.

12-37. Calculate the freezing point of a solution of 5.00 g of diphenyl, $C_{12}H_{10}$, and 7.50 g of naphthalene, $C_{10}H_8$, dissolved in 200 g of benzene.

12-38. Quinine is a natural product extracted from the bark of the cinchona tree, which is native to South America. Quinine is used as an antimalarial agent. When 1.00 g of quinine is dissolved in 10.0 g of cyclohexane, the freezing point is lowered 6.23 K. Calculate the molecular mass of quinine.

12-39. Vitamin K is involved in normal blood clotting. When 0.500 g of vitamin K is dissolved in 10.0 g of camphor, the freezing point of the solution is lowered by 4.43 K. Calculate the molecular mass of vitamin K.

12-40. Polychlorinated biphenyls (PCBs) are highly resistant to decomposition and have been used as coolants in transformers; however, PCBs are carcinogenic and are being phased out of use. A 0.100-g sample of a PCB dissolved in 10.0 g of camphor depressed the freezing point 1.22 K. Calculate the molecular mass of the compound.

12-41. Don Juan Pond in the Wright Valley of Antarctica freezes at −57°C. The major solute in the pond is $CaCl_2$. Estimate the concentration of $CaCl_2$ in the pond water.

12-42. Menthol is a crystalline substance with a peppermint taste and odor. A solution of 6.54 g of menthol per 100.0 g of cyclohexane freezes at −1.95°C. Determine the formula mass of menthol.

12-43. A solution of mercury(II) chloride, $HgCl_2$, is a poor conductor of electricity. A 40.7-g sample of $HgCl_2$ is dissolved in 100.0 g of water, and the freezing point of the solution is found to be −2.83°C. Explain why $HgCl_2$ in solution is a poor conductor of electricity.

12-44. Mayer's reagent, K_2HgI_4, is used in analytical chemistry. In order to determine its extent of dissociation in water, its effect on the freezing point of water is investigated. A 0.25 m aqueous solution is prepared, and its freezing point is found to be −1.41°C. Suggest a possible dissociation reaction that takes place when K_2HgI_4 is dissolved in water.

OSMOTIC PRESSURE

12-45. Calculate the osmotic pressure of a 0.25 M aqueous solution of sucrose, $C_{12}H_{22}O_{11}$, at 37°C.

12-46. Calculate the osmotic pressure of seawater at 37°C. Take $M_c = 1.10\ \text{mol·L}^{-1}$ for seawater.

12-47. Insulin is a small protein hormone that regulates carbohydrate metabolism by decreasing blood glucose levels. A deficiency of insulin leads to diabetes. A 20.0-mg sample of insulin is dissolved in enough water to make 10.0 mL of solution, and the osmotic pressure of the solution at 25°C is found to be 6.48 torr. Estimate the molecular mass of insulin.

12-48. Pepsin is the principal digestive enzyme of gastric juice. A 3.00-mg sample of pepsin is dissolved in enough water to make 10.0 mL of solution, and the osmotic pressure of the solution at 25°C is found to be 2.20 mm H_2O. Estimate the molecular mass of pepsin.

12-49. In reverse osmosis, water flows out of a salt solution until the osmotic pressure of the solution equals the applied pressure. If a pressure of 100 atm is applied to seawater, what will be the final concentration of the seawater at 20°C when reverse osmosis stops? Given that seawater is a 1.1 M_c solution of NaCl(*aq*), calculate how many liters of seawater are required to produce 10 L of fresh water at 20°C with an applied pressure of 100 atm.

12-50. What is the minimum pressure that must be applied at 25°C to obtain pure water by reverse osmosis from water that is 0.15 M in NaCl and 0.015 M in $MgSO_4$?

RAOULT'S LAW FOR TWO COMPONENTS

12-51. Given that the equilibrium vapor pressures of benzene and toluene at 81°C are 768 torr and 293 torr, respectively, (a) compute the total vapor pressure at 81°C over a benzene-toluene solution with $X_{benz} = 0.250$, and (b) compute the mole fraction of benzene in the vapor phase over the solution.

12-52. Propyl alcohol, $CH_3CH_2CH_2OH$, and isopropyl alcohol, $CH_3CHOHCH_3$, form ideal solutions in all proportions. Calculate the partial pressure of each component in equilibrium at 25°C with a solution of compositions $X_{prop} = 0.25$, 0.50, and 0.75, given that $P^{\circ}_{prop} = 20.9$ torr and $P^{\circ}_{iso} = 45.2$ torr at 25°C. Calculate the composition of the vapor phase also.

ADDITIONAL PROBLEMS

12-53. Immunoglobulin G, formerly called gamma globulin, is the principal antibody in blood serum. A 0.50-g sample of immunoglobulin G is dissolved in enough water to make 0.100 L of solution, and the osmotic pressure of the solution at 25°C is found to be 8.42 mm H_2O. Estimate the molecular mass of immunoglobulin G.

12-54. Most wines are about 12% ethyl alcohol by volume, and many hard liquors are about 80 proof. Assuming that the only major nonaqueous constituent of wine and vodka is ethyl alcohol, estimate the freezing points of wine and vodka. Take the density of ethyl alcohol as $0.79\ g{\cdot}mL^{-1}$ and the density of water as $1.00\ g{\cdot}mL^{-1}$. The formula of ethyl alcohol is C_2H_5OH.

12-55. The boiling point of ethylene glycol is 197°C, whereas the boiling point of ethyl alcohol is 78°C. Ethylene glycol is called a "permanent" antifreeze and ethyl alcohol a "temporary" one. Explain the difference between "permanent" and "temporary" antifreezes.

12-56. What volume percent of oxygen should be used in a diver's air tanks to make the partial pressure of oxygen in the air supply 0.20 atm at a water depth of 90 ft? (See Problem 12-5.)

12-57. Estimate the molality of sucrose, $C_{12}H_{22}O_{11}$, in a solution prepared by dissolving 2.00 teaspoons of sucrose in 1.00 cup of water (3 teaspoons = 1 tablespoon = 0.50 ounce = 14 g and 4 cups = 1 quart = 0.946 L). Take the density of water as $1.00\ g{\cdot}mL^{-1}$.

12-58. Suppose that an aqueous solution is observed to begin freezing when the temperature is +2.0°C. What can you say about the composition of the solid that freezes out of the solution?

12-59. An aqueous solution of ethylene glycol in which ethylene glycol is 50 percent by volume has a freezing point of −36°C. What is the molality of ethylene glycol? Calculate the boiling point of this solution. Take the density of ethylene glycol to be $1.116\ g{\cdot}mL^{-1}$ and the density of water to be $1.00\ g{\cdot}mL^{-1}$.

12-60. Recently scientists have discovered that some insects produce an antifreeze in cold weather; the antifreeze is glycerol, $HOCH_2CH_2OH$. How much glycerol must an insect produce per gram of body fluid (taken to be water) to survive at −5.0°C?

12-61. Calculate the mole fractions and molalities of methyl alcohol, CH_3OH, and ethyl alcohol, CH_3CH_2OH, if 305 mg of CH_3OH and 275 mg of CH_3CH_2OH are dissolved in 10.0 g H_2O.

12-62. Let the mole fraction of component A in the vapor over a two-component ideal solution be Y_A. Show that Y_A is given by

$$Y_A = \frac{X_A P^{\circ}_A}{X_A(P^{\circ}_A - P^{\circ}_B) + P^{\circ}_B}$$

Show that $Y_A > X_A$ if $P^{\circ}_A > P^{\circ}_B$; in other words, show that the vapor phase is richer than the liquid phase in the more volatile component.

12-63. A semipermeable membrane separates two aqueous solutions at 20°C. For each of the following cases, name the solution into which a net flow of water (if any) will occur:

(a) 0.10 M NaCl(*aq*) and 0.10 M KBr(*aq*)
(b) 0.10 M $Al(NO_3)_3$(*aq*) and 0.20 M $NaNO_3$(*aq*)
(c) 0.10 M $CaCl_2$(*aq*) and 0.50 M $CaCl_2$(*aq*)

12-64. Show that the relation between the mole fraction of solute and the molality of the solution is given by

$$X_2 = \frac{M_1 m/1000}{1 + \dfrac{M_1 m}{1000}}$$

where M_1 is the molar mass of the solvent and m is the molality. Now argue that

$$X_2 \simeq \frac{M_1 m}{1000}$$

if the solution is dilute.

12-65. The density of a glycerol-water solution that is 40.0 percent glycerol by mass is $1.101\ g{\cdot}mL^{-1}$ at 20°C.

Calculate the molality and the molarity of glycerol in the solution at 20°C. What is the molality at 0°C? The formula of glycerol is $C_3H_8O_3$.

12-66. When 2.87 g of an organic compound, which is known to be 39.12 percent carbon, 8.76 percent hydrogen, and 52.12 percent oxygen, is dissolved in 65.3 g of camphor, the freezing point of the solution is 160.7°C. Determine the molecular formula of the compound.

12-67. Compute the molality, the colligative molality, the freezing point, and the boiling point for each of the following solutions:

(a) 5.00 g of K_2SO_4 in 250 g of water
(b) 5.00 g of ethyl alcohol, C_2H_5OH, in 250 g of water
(c) 1.00 g of aniline, $C_6H_5NH_2$, in 50.0 g of camphor

12-68. Estimate the vapor pressures of carbon tetrachloride, CCl_4, and ethyl acetate, $C_4H_8O_2$, in a solution at 50°C containing 25.0 g of carbon tetrachloride dissolved in 100.0 g of ethyl acetate. The vapor pressures of pure CCl_4 and $C_4H_8O_2$ at 50°C are 306 torr and 280 torr, respectively.

12-69. When 2.74 g of phosphorus is dissolved in 100 mL of carbon disulfide, the boiling point is 46.71°C. Given that the normal boiling point of pure carbon disulfide is 46.30°C, that its density is 1.261 $g{\cdot}mL^{-1}$, and that its boiling point elevation constant is $K_b = 2.34$ $K{\cdot}m_c^{-1}$, determine the molecular formula of phosphorus.

12-70. A 2.0-g sample of the polymer polyisobutylene, $[CH_2C(CH_3)_2]_x$, is dissolved in enough cyclohexane to make 10.0 mL of solution at 20°C and produces an osmotic pressure of 2.0×10^{-2} atm. Determine the formula mass and the number of units (x) in the polymer.

12-71. It is possible to convert from molality to molarity if the density of a solution is known. The density of a 2.00 m NaOH aqueous solution is 1.22 $g{\cdot}mL^{-1}$. Calculate the molarity of this solution.

12-72. In many fields outside chemistry, solution concentrations are expressed in mass percent. Calculate the molality of an aqueous solution that is 24.0 percent potassium chromate, K_2CrO_4, by mass. Given that the density of the solution is 1.21 $g{\cdot}mL^{-1}$, calculate the molarity.

12-73. It is claimed that radiator antifreeze also provides "antiboiling" protection for automobile cooling systems. Estimate the boiling point of a solution composed of 50.0 g of water and 50.0 g of ethylene glycol. Assume that the vapor pressure of ethylene glycol is negligible at 100°C. The formula of ethylene glycol is $C_2H_6O_2$.

12-74. Use Figure 12-9 to determine the molality of ethylene glycol, $C_2H_6O_2$, in water that is necessary to give antifreeze protection down to −40°C. Compare the result obtained from Figure 12-9 with that calculated from the Equation (12-8).

12-75. Concentrated sulfuric acid, H_2SO_4, is sold as a solution that is 98 percent sulfuric acid and 2.0 percent water by mass. Given that the density is 1.84 $g{\cdot}mL^{-1}$, calculate the molarity of concentrated H_2SO_4.

12-76. Concentrated phosphoric acid is sold as a solution of 85 percent phosphoric acid and 15 percent water by mass. Given that its molarity is 15 M, calculate the density of concentrated phosphoric acid.

RATES AND MECHANISMS OF CHEMICAL REACTIONS

The colors in a computer-enhanced photomicrograph show the various densities in a sample of Engelhard Corporation's OCTAFINING™ catalyst. OCTAFINING catalysts are used to make polymers such as polyesters and nylon (Chapter 24). The different colors represent different densities of the catalyst particles. The catalyst is a zeolite material, which consists of a three-dimensional framework containing cavities and channels. Reactant molecules of the right size and shape can enter the cavities and undergo reaction.

Chemists study the rates of chemical reactions to determine the conditions under which a particular reaction can be made to proceed at a favorable rate. The reaction rate law, which must be determined by experiment, tells us how the rate of a reaction depends on the concentration of the reactants and of other added substances, such as catalysts. The reaction rate law provides the most important clue to the reaction mechanism, which is the sequence of steps by which the reactants are converted to products. An understanding of reaction mechanisms enables us to adjust reaction conditions in order to produce a desired reaction rate and increases our understanding of how chemical reactions occur at the molecular level.

All chemical reactions proceed toward an equilibrium state in which the forward and reverse reaction rates are equal and the concentrations show no further change with time. Most chemical systems of interest are not at chemical equilibrium, and thus a study of chemical reaction rates (chemical kinetics) is an essential prerequisite to understanding the chemistry of such systems.

13-1. A RATE TELLS US HOW FAST A QUANTITY IS CHANGING WITH TIME

Let's consider the rate of a chemical reaction. Suppose that the reaction is

$$A \rightarrow P$$

where A represents the reactant and P represents the product. We define the rate at which the product P is formed as

$$\text{rate} = \frac{\Delta[P]}{\Delta t} = \frac{[P]_2 - [P]_1}{t_2 - t_1} \tag{13-1}$$

▪ The most common units of reaction rate are moles per liter per unit time, such as $mol \cdot L^{-1} \cdot s^{-1}$, or $M \cdot s^{-1}$.

where $[P]_1$ is the concentration of P at time t_1 and $[P]_2$ is the concentration of P at some later time t_2. Note that the value of $t_2 - t_1$ is positive because $t_2 > t_1$ and that the value of $[P]_2 - [P]_1$ is positive because the concentration P increases as the reaction takes place. Thus $\Delta[P]/\Delta t$ is a positive quantity, and so the rate of a reaction is a positive quantity. Note that the units of rate are moles per liter per unit time ($mol \cdot L^{-1} \cdot s^{-1}$, for example).

We also can express the rate in terms of the concentrations of reactants. The rate at which the reactant A is consumed in the reaction is defined as

$$\text{rate} = -\frac{\Delta[A]}{\Delta t} = -\left(\frac{[A]_2 - [A]_1}{t_2 - t_1}\right) \tag{13-2}$$

▪ The rate of a chemical reaction is always a positive quantity.

where $[A]_1$ is the concentration of A at time t_1 and $[A]_2$ is the concentration of A at some later time t_2. The value of $t_2 - t_1$ is positive because $t_2 > t_1$, but $[A]_2$ is less than $[A]_1$ because A is being consumed by the reaction. Thus $[A]_2 - [A]_1 < 0$. A minus sign is inserted in front of $\Delta[A]/\Delta t$, which is negative, to make $-\Delta[A]/\Delta t$ (= rate) a positive quantity. A **reaction rate** is the rate at which a reactant is consumed (moles per liter reacted per unit time) or the rate at which a product is formed (moles per liter produced per unit time). The relation between Equations (13-1) and (13-2) is

$$\text{rate} = -\frac{\Delta[A]}{\Delta t} = \frac{\Delta[P]}{\Delta t}$$

The rate of a chemical reaction can be studied in numerous ways. For example, suppose that the reaction is

$$2H_2O_2(aq) \rightarrow 2H_2O(l) + O_2(g)$$

Because a gaseous product, O_2, is formed, we can determine the reaction rate by measuring the increase in pressure over the reaction mixture as a function of time.

The rate at which O_2 is produced is equal to one-half the rate at which H_2O_2 is consumed, because two H_2O_2 molecules react for each O_2 molecule produced. Thus we have

$$\text{rate} = \frac{\Delta[O_2]}{\Delta t} = -\frac{1}{2}\frac{\Delta[H_2O_2]}{\Delta t}$$

Example 13-1: Let the rate of the reaction

$$N_2O_5(g) \rightarrow 2NO_2(g) + \tfrac{1}{2}O_2(g)$$

be expressed as $-\Delta[N_2O_5]/\Delta t$. Express the rate of the reaction in terms of the other species involved.

Solution: The rate of loss of N_2O_5 is twice as great as the rate of production of O_2 because two N_2O_5 molecules are consumed for each O_2 molecule produced. Thus, we have

$$\text{rate} = -\frac{\Delta[N_2O_5]}{\Delta t} = \frac{2\Delta[O_2]}{\Delta t}$$

The rate of loss of N_2O_5 is one half the rate of production of NO_2 because two NO_2 molecules are formed for each N_2O_5 molecule that decomposes. Thus

$$\text{rate} = -\frac{\Delta[N_2O_5]}{\Delta t} = \frac{1}{2}\frac{\Delta[NO_2]}{\Delta t}$$

The rate for the decomposition of $N_2O_5(g)$ in Example 13-1 can be expressed as

$$\text{rate} = -\frac{1}{1}\frac{\Delta[N_2O_5]}{\Delta t} = \frac{1}{2}\frac{\Delta[NO_2]}{\Delta t} = \frac{1}{1/2}\frac{\Delta[O_2]}{\Delta t}$$

Notice that the numerical factor in the denominator of each expression is the same as the balancing coefficient of the species in the chemical equation. In general, for the reaction

$$a\text{A} \rightarrow b\text{B}$$

we define the rate to be

$$\text{rate} = -\frac{1}{a}\frac{\Delta[\text{A}]}{\Delta t} = \frac{1}{b}\frac{\Delta[\text{B}]}{\Delta t} \tag{13-3}$$

Thus for the reaction

$$2O_3(g) \rightarrow 3O_2(g)$$

we define the rate to be

$$\text{rate} = -\frac{1}{2}\frac{\Delta[O_3]}{\Delta t} = \frac{1}{3}\frac{\Delta[O_2]}{\Delta t}$$

Consider the decomposition of gaseous nitrogen pentoxide to gaseous nitrogen dioxide and oxygen given in Example 13-1:

$$\underset{\text{colorless}}{N_2O_5(g)} \rightarrow \underset{\text{brown}}{2NO_2(g)} + \underset{\text{colorless}}{\tfrac{1}{2}O_2(g)} \tag{13-4}$$

Table 13-1 Concentration of $N_2O_5(g)$, $NO_2(g)$, and $O_2(g)$ as a function of time at 45°C for an initial concentration of $[N_2O_5]_0 = 1.24 \times 10^{-2}$ M in the reaction $N_2O_5(g) \rightarrow 2NO_2(g) + 1/2O_2(g)$

t/min	$[N_2O_5]$/M	$[NO_2]$/M	$[O_2]$/M
0	1.24×10^{-2}	0	0
10	0.92×10^{-2}	0.64×10^{-2}	0.16×10^{-2}
20	0.68×10^{-2}	1.12×10^{-2}	0.28×10^{-2}
30	0.50×10^{-2}	1.48×10^{-2}	0.37×10^{-2}
40	0.37×10^{-2}	1.74×10^{-2}	0.44×10^{-2}
50	0.28×10^{-2}	1.92×10^{-2}	0.48×10^{-2}
60	0.20×10^{-2}	2.08×10^{-2}	0.52×10^{-2}
70	0.15×10^{-2}	2.18×10^{-2}	0.55×10^{-2}
80	0.11×10^{-2}	2.26×10^{-2}	0.57×10^{-2}
90	0.08×10^{-2}	2.32×10^{-2}	0.58×10^{-2}
100	0.06×10^{-2}	2.36×10^{-2}	0.59×10^{-2}

The rate of this reaction can be determined by measuring the increase in the intensity of the brown color of NO_2 in the reaction mixture as a function of time. The concentrations of $N_2O_5(g)$, $NO_2(g)$, and $O_2(g)$ in a reaction mixture at 45°C with the initial concentrations $[N_2O_5]_0 = 1.24 \times 10^{-2}$ M, $[NO_2]_0 = 0$, and $[O_2]_0 = 0$ are given in Table 13-1. The subscript zero denotes an initial concentration. The data in Table 13-1 are plotted in Figure 13-1, which illustrates both the decrease with time of the reactant concentration, $[N_2O_5]$, and the increase with time of the product concentrations, $[NO_2]$ and $[O_2]$. Note that $[NO_2]$ increases twice as fast as $[N_2O_5]$ decreases because 2 mol of NO_2 are produced for each mole of N_2O_5 that decomposes. The value of $[O_2]$ increases only half as fast as $[N_2O_5]$ decreases because only 0.5 mol of O_2 is produced for each mole of N_2O_5 that decomposes.

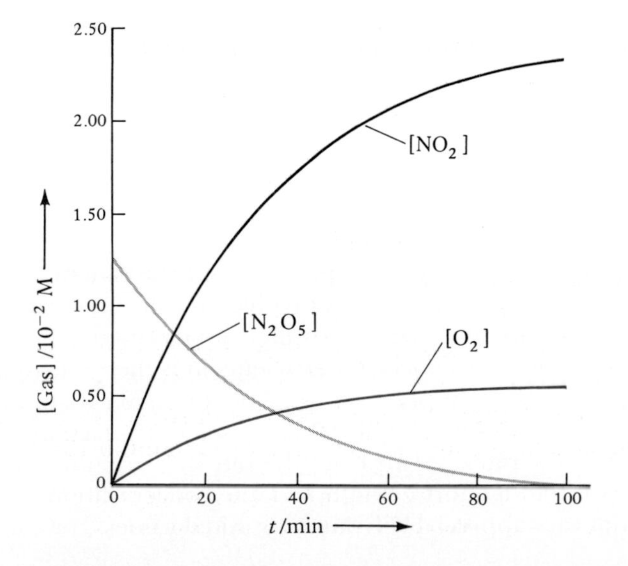

Figure 13-1 The change in concentration of $N_2O_5(g)$, $NO_2(g)$, and $O_2(g)$ as a function of time for the reaction $N_2O_5(g) \rightarrow 2NO_2(g) + \frac{1}{2}O_2(g)$ at 45°C. The initial concentrations are $[N_2O_5]_0 = 1.24 \times 10^{-2}$ M, $[NO_2]_0 = 0$, and $[O_2]_0 = 0$. Note that both $[NO_2]$ and $[O_2]$ increase with time because NO_2 and O_2 are reaction products.

An inspection of the data in Table 13-1 shows that the reaction rate decreases as $[N_2O_5]$ decreases. For example, the rate of decomposition of N_2O_5 over the first 10 min of the reaction is

$$\text{rate} = -\frac{\Delta[N_2O_5]}{\Delta t} = \frac{-(0.92 \times 10^{-2} - 1.24 \times 10^{-2})\ \text{M}}{(10 - 0)\ \text{min}}$$
$$= 3.2 \times 10^{-4}\ \text{mol·L}^{-1}\text{·min}^{-1}$$

whereas the rate over the period 10 min to 20 min is

$$\text{rate} = \frac{-(0.68 \times 10^{-2} - 0.92 \times 10^{-2})\ \text{M}}{(20 - 10)\ \text{min}}$$
$$= 2.4 \times 10^{-4}\ \text{mol·L}^{-1}\text{·min}^{-1}$$

Example 13-2: Use the data in Table 13-1 to calculate the rate of production of NO_2 over the first 10 min of the reaction

$$N_2O_5(g) \rightarrow 2NO_2(g) + \tfrac{1}{2}O_2(g)$$

Solution: Because NO_2 is a product species, the rate of production of NO_2 is given by $\Delta[NO_2]/\Delta t$. For the first 10 min of the reaction

$$\frac{\Delta[NO_2]}{\Delta t} = \frac{(0.64 \times 10^{-2} - 0)\ \text{M}}{(10 - 0)\ \text{min}} = 6.4 \times 10^{-4}\ \text{M·min}^{-1}$$

Note that the rate of production of NO_2 over the first 10 min is twice as great as the rate of consumption of N_2O_5 over the same time period.

13-2. THE RATE LAW OF A REACTION CAN BE DETERMINED BY THE METHOD OF INITIAL RATES

The rates of chemical reactions usually depend upon the concentrations of the reactants. Numerous experiments show that the rate of a reaction usually is proportional to the concentrations of the reactants raised to small integer powers. For the case of the thermal decomposition of N_2O_5, we assume that

$$\text{rate} \propto [N_2O_5]^x$$

We can write this proportionality as an equation by inserting a proportionality constant:

$$\text{rate} = k[N_2O_5]^x \qquad (13\text{-}5)$$

The value of x in Equation (13-5) must be determined experimentally; it is *not* necessarily the same as the balancing coefficient of N_2O_5 in the chemical equation.

The equation that gives the dependence of the rate of a reaction on the concentration(s) of the species involved is called the **rate law** of the reaction. Rate laws must be determined experimentally. There is no relation between the balancing coefficients in a chemical equation and the reaction rate law. In practice, rate laws often are determined from data involving reaction rates at the early stage of reactions. Using the **method of initial rates,** we measure the rate of a reaction over an initial time interval that is short enough that the concentrations of the reactants do not vary appreciably from their initial values. If we use zeros as

▪ The initial rate is the rate at the start of the reaction, that is, at $t = 0$.

subscripts to denote the initial rate and initial values, then for the general reaction described by the equation

$$aA + bB + cC \rightarrow \text{products}$$

we can write

$$(\text{rate})_0 = k[A]_0^x[B]_0^y[C]_0^z \quad (13\text{-}6)$$

We can then determine the values of x, y, and z by varying the initial concentration of each reactant in turn, while keeping the other reactant concentrations constant.

As a simple example, let's consider the N_2O_5 decomposition reaction, in which there is only one reactant:

$$N_2O_5(g) \rightarrow 2NO_2(g) + \tfrac{1}{2}O_2(g)$$

Suppose that we run this reaction using successively doubled concentrations of N_2O_5. We obtain the following initial-rate data:

Run	**$[N_2O_5]_0$/mol·L^{-1}**	**$(\text{rate})_0$/mol·L^{-1}·h^{-1}**
1	0.010	0.018
2	0.020	0.036
3	0.040	0.072

Notice that when the initial concentration, $[N_2O_5]_0$, is increased by a factor of 2, the initial rate, $(\text{rate})_0$, also increases by a factor of 2. These results show that $(\text{rate})_0$ is directly proportional to $[N_2O_5]_0$, and so we can write

$$(\text{rate})_0 = k[N_2O_5]_0 \quad (13\text{-}7)$$

showing that $x = 1$ in Equation (13-6). Extensive experimental studies show that in most cases the reaction rate law determined from initial-rate data also describes the reaction rate as the reaction proceeds. Thus, we can drop the subscript zero in Equation (13-7) to obtain the rate law:

$$\text{rate} = k[N_2O_5] \quad (13\text{-}8)$$

where k is the **rate constant** of the reaction. Because $[N_2O_5]$ is raised to the first power in Equation (13-8), the rate law is said to be a **first-order rate law,** and the reaction is said to be a **first-order reaction.** We say that the rate law is first order in $[N_2O_5]$.

▪ A rate law tells us how a reaction rate depends on concentration.

We can determine the numerical value of the rate constant k in Equation (13-8) by using the tabulated initial-rate data. If we substitute the fact that $(\text{rate})_0 = 0.072$ mol·L^{-1}·h^{-1} when $[N_2O_5]_0 = 0.040$ mol·L^{-1}, then we obtain

$$k = \frac{(\text{rate})_0}{[N_2O_5]_0} = \frac{0.072\ \text{mol·L}^{-1}\text{·h}^{-1}}{0.040\ \text{mol·L}^{-1}} = 1.8\ \text{h}^{-1}$$

Notice that the units of the rate constant for a first-order reaction are time^{-1}.

We have seen that if the initial concentration of a reactant is doubled and the initial reaction rate doubles, then the rate law is first order in the concentration of that reactant. This same idea can be applied to rate laws for reactions involving more than one reactant. For example, consider the following initial-rate data for the reaction

$$2NO_2(g) + F_2(g) \rightarrow 2NO_2F(g)$$

Run	Initial concentration		
	$[NO_2]_0$/M	$[F_2]_0$/M	$(\text{rate})/\text{M}\cdot\text{s}^{-1}$
1	1.00	1.00	1.00×10^{-4}
2	2.00	1.00	2.00×10^{-4}
3	1.00	2.00	2.00×10^{-4}
4	2.00	2.00	4.00×10^{-4}

A comparison of the data for runs 1 and 2 shows that a twofold increase in the initial concentration of NO_2 (with $[F_2]_0$ held fixed) increases the initial rate by a factor of 2, and thus the rate must be proportional to the first power of $[NO_2]_0$:

$$(\text{rate})_0 \propto [NO_2]_0$$

A comparison of the data for runs 1 and 3 shows that a twofold increase in the initial concentration of F_2 (with $[NO_2]_0$ held fixed) increases the initial rate by a factor of 2, and thus the rate must also be proportional to $[F_2]_0$, that is,

$$(\text{rate})_0 \propto [F_2]_0$$

We can combine these two results to get the reaction rate law:

$$(\text{rate})_0 = k[NO_2]_0[F_2]_0$$

Comparison of runs 1 and 4 shows that a simultaneous twofold increase in $[NO_2]_0$ and $[F_2]_0$ increases the initial rate by a factor of 4, which is consistent with the preceding rate law. The value of the reaction rate constant is (from run 1)

▪ In the method of initial rates we usually vary the concentration of only one species at a time.

$$k = \frac{(\text{rate})_0}{[NO_2]_0[F_2]_0} = \frac{1.00 \times 10^{-4}\ \text{M}\cdot\text{s}^{-1}}{[1.00\ \text{M}][1.00\ \text{M}]} = 1.00 \times 10^{-4}\ \text{M}^{-1}\cdot\text{s}^{-1}$$

Thus, assuming that the rate law does not change with time, we take the rate law for the reaction

$$2NO_2(g) + F_2(g) \rightarrow 2NO_2F(g)$$

to be

$$\text{rate} = (1.00 \times 10^{-4}\ \text{M}^{-1}\cdot\text{s}^{-1})[NO_2][F_2]$$

This rate law is first order in both $[NO_2]$ and $[F_2]$. The *overall* order of the rate law is determined by summing the exponents on the concentration terms. More generally, if a reaction rate law is of the form

$$\text{rate} = k[A]^x[B]^y[C]^z$$

then the rate law is x order in [A], y order in [B], and z order in [C] with an overall order of $x + y + z$. For the case we are considering, the sum of $x(= 1)$ and $y(= 1)$ is 2, and so the rate law is second order overall, and the reaction is a **second-order reaction.** Note that the units of k for a second-order rate law are $M^{-1}\cdot s^{-1}$.

Example 13-3: At 325°C, $NO_2(g)$ reacts with $CO(g)$ to yield $NO(g)$ and $CO_2(g)$:

$$NO_2(g) + CO(g) \rightarrow NO(g) + CO_2(g)$$

It is observed that the initial reaction rate does not depend upon the concentration of $CO(g)$, and so the reaction rate law does not involve $[CO]_0$. The rate does depend upon $[NO_2]_0$, however, and the following data are obtained:

▪ A reactant that does not appear in the rate law has no effect on the reaction rate.

Run	$[NO_2]_0/\text{mol}\cdot L^{-1}$	$(\text{rate})_0/\text{mol}\cdot L^{-1}\cdot s^{-1}$
1	0.15	0.011
2	0.30	0.045
3	0.60	0.18

Determine the rate law and the value of the rate constant.

Solution: The rate is *not* directly proportional to $[NO_2]_0$. As we double $[NO_2]_0$ in going from 0.15 M to 0.30 M, we quadruple the rate. In going from 0.15 M to 0.60 M, a fourfold increase in concentration, we increase the rate by a factor of 16. Thus, we see that the rate is proportional to the square of $[NO_2]_0$, and so we can write the rate law at $t = 0$ as

$$(\text{rate})_0 = k[NO_2]_0^2$$

Assuming that the rate law expression does not change with time, we take the rate law to be

$$\text{rate} = k[NO_2]^2$$

Because the exponent on the concentration term is 2, the rate law is a second-order rate law. We can evaluate k by using

$$k = \frac{(\text{rate})_0}{[NO_2]_0^2} = \frac{0.045\ \text{mol}\cdot L^{-1}\cdot s^{-1}}{(0.30\ \text{mol}\cdot L^{-1})^2} = 0.50\ \text{mol}^{-1}\cdot L\cdot s^{-1}$$

Notice, once again, that the units of k for a second-order rate law are $M^{-1}\cdot s^{-1}$.

Example 13-4: The following initial-rate data were obtained for the reaction

$$2NO(g) + Br_2(g) \rightarrow 2NOBr(g)$$

Run	$[NO]_0/M$	$[Br_2]_0/M$	$(\text{rate})_0/M\cdot \text{min}^{-1}$
1	1.0	1.0	1.30×10^{-3}
2	2.0	1.0	5.20×10^{-3}
3	4.0	2.0	4.16×10^{-2}

Determine the reaction rate law and the value of the rate constant.

Solution: From run 1 to run 2, $[NO]_0$ doubles and $[Br_2]_0$ remains the same.

The initial rate quadruples, and so the rate law is second order in $[NO]_0$. For the initial rate we have

$$(\text{rate})_0 = k[NO]_0^2[Br_2]_0^x$$

From run 2 to run 3, both $[NO]_0$ and $[Br_2]_0$ double and the initial rate increases eightfold. We already know that the rate law is second order in $[NO]_0$, and so the initial rate increases by a factor of 4 as a result of the increase in $[NO]_0$ alone. The rest of the increase in the initial rate, a factor of 2, is due to the increase in $[Br_2]_0$. Thus, assuming that the rate law does not change during the course of the reaction, we have

$$\text{rate} = k[NO]^2[Br_2]$$

The value of the rate constant is (run 2)

$$k = \frac{(\text{rate})_0}{[NO]_0^2[Br_2]_0} = \frac{5.20 \times 10^{-3}\ \text{M}\cdot\text{min}^{-1}}{(2.0\ \text{M})^2(1.0\ \text{M})} = 1.3 \times 10^{-3}\ \text{M}^{-2}\cdot\text{min}^{-1}$$

The rate law is third order overall.

13-3. THE HALF-LIFE FOR A FIRST-ORDER RATE LAW IS INDEPENDENT OF THE INITIAL CONCENTRATION

By using calculus, it is possible to show that if a rate law is first order in the concentration of a reactant A, then the dependence of the concentration of A on time is given by

$$\log\,[A] = \log\,[A]_0 - \frac{kt}{2.30} \qquad (13\text{-}9)$$

where $[A]_0$ is the concentration of A at $t = 0$.

It is convenient to plot Equation (13-9) as a straight line. Recall from Section 12-9 that the mathematical equation of a straight line is

$$y = mx + b$$

where m and b are constants. The quantity b is the **intercept** of the line with the y axis, and m is the **slope** of the line. The slope of a straight line is a measure of its steepness. Equation (13-9) may not look like it is of the form $y = mx + b$, but if we let $y = \log\,[A]$ and $x = t$, then Equation (13-9) is a straight line whose intercept is $\log\,[A]_0$ and whose slope is $-k/2.30$. Therefore, a plot of $\log\,[A]$ versus t is a straight line with a slope of $-k/2.30$. Furthermore, if the rate law is not first order in [A], then a plot of $\log\,[A]$ versus t will not be linear.

The $[N_2O_5]$ values given in Table 13-1 are converted to $\log\,[N_2O_5]$ data in Table 13-2. The data in Table 13-2 are plotted in the form $\log\,[N_2O_5]$ versus t in Figure 13-2. The fact that the plot of $\log\,[N_2O_5]$ versus t is linear confirms that the rate law is first order in $[N_2O_5]$.

Table 13-2 $[N_2O_5]$ **and** $\log\,[N_2O_5]$ **as a function of time for the reaction** $N_2O_5(g) \rightarrow 2NO_2(g) + \frac{1}{2}O_2(g)$

t/min	$[N_2O_5]$/M	$\log\,([N_2O_5]/\text{M})$
0	1.24×10^{-2}	-1.91
10	0.92×10^{-2}	-2.04
20	0.68×10^{-2}	-2.17
30	0.50×10^{-2}	-2.30
40	0.37×10^{-2}	-2.43
50	0.28×10^{-2}	-2.55
60	0.20×10^{-2}	-2.70
70	0.15×10^{-2}	-2.82
80	0.11×10^{-2}	-2.96
90	0.08×10^{-2}	-3.10
100	0.06×10^{-2}	-3.22

Example 13-5: The rate law for the decomposition of aqueous hydrogen peroxide at 70°C

$$2H_2O_2(aq) \rightarrow 2H_2O(l) + O_2(g)$$

is first order in $[H_2O_2]$ with a rate constant $k = 0.0347\ \text{min}^{-1}$. Given that the initial concentration is $[H_2O_2]_0 = 0.30$ M, compute the value of $[H_2O_2]$ 60 min after the solution is prepared.

Solution: We use Equation 13-9 rearranged to the form

$$\log \frac{[H_2O_2]}{[H_2O_2]_0} = -\frac{kt}{2.30}$$

(Recall that $\log a - \log b = \log (a/b)$), and so

$$\log \frac{[H_2O_2]}{0.30\ M} = \frac{-(0.0347\ min^{-1})(60\ min)}{2.30} = -0.905$$

or

$$\frac{[H_2O_2]}{0.30\ M} = 10^{-0.905} = 0.124$$

Thus

$$[H_2O_2] = (0.124)(0.30\ M) = 0.037\ M$$

Let's return to the data in Table 13-1 on the $N_2O_5(g)$ decomposition reaction. Figure 13-3 shows a plot of $[N_2O_5]$ versus time on a larger scale than that shown in Figure 13-1. The initial concentration of $N_2O_5(g)$ is $[N_2O_5]_0 = 1.24 \times 10^{-2}$ M. The time required for $[N_2O_5]$ to decrease from 1.24×10^{-2} M to $\frac{1}{2}(1.24 \times 10^{-2}\ M) = 0.62 \times 10^{-2}$ M can be read off the plot in Figure 13-3 as $t_{1/2} = 23$ min. Note also in Figure 13-3 that it requires another 23 min for $[N_2O_5]$ to decrease from 0.62×10^{-2} M to $0.31 \times 10^{-2}\ M = \frac{1}{2}(0.62 \times 10^{-2}\ M)$. Note that

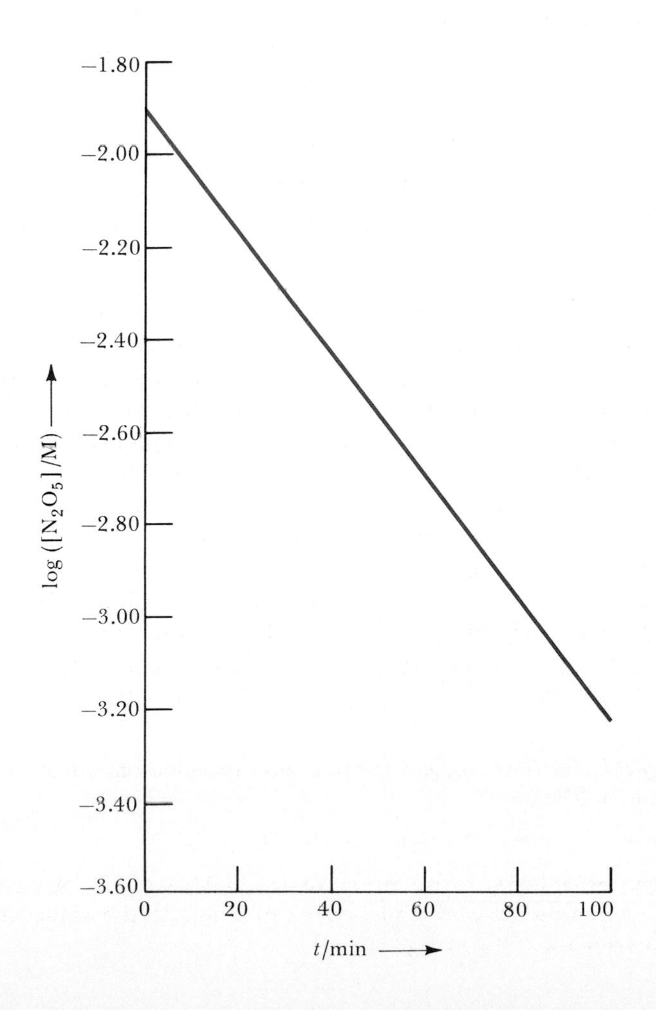

Figure 13-2 Plot of log $[N_2O_5]$ versus time (t) for the reaction $N_2O_5(g) \rightarrow 2NO_2(g) + \frac{1}{2}O_2(g)$. The fact that the plot is a straight line confirms that the reaction is first order.

Table 13-3 Half-life values for the N_2O_5 decomposition reaction

Twofold decrease in $[N_2O_5]$	$t_{1/2}$/min	Total time/min
1.24×10^{-2} M → 0.62×10^{-2} M	23	23
0.62×10^{-2} M → 0.31×10^{-2} M	23	46 (2 × 23)
0.31×10^{-2} M → 0.16×10^{-2} M	23	69 (3 × 23)
0.16×10^{-2} M → 0.08×10^{-2} M	23	92 (4 × 23)

the time required for the concentration of N_2O_5 to decrease by a factor of two for the first-order N_2O_5 decomposition reaction is *independent* of the concentration of unreacted $N_2O_5(g)$. This result is shown in Table 13-3.

The equation that describes the dependence of concentration on time for a first-order reaction rate, Equation (13-9), is consistent with the data in Figure 13-3 and Table 13-3. Equation (13-9) can be rearranged to the form

$$\log \frac{[A]}{[A]_0} = -\frac{kt}{2.30} \tag{13-10}$$

The **half-life,** $t_{1/2}$, of a reactant is defined as the time that it takes for the concentration of the reactant to decrease by a factor of 2. Thus

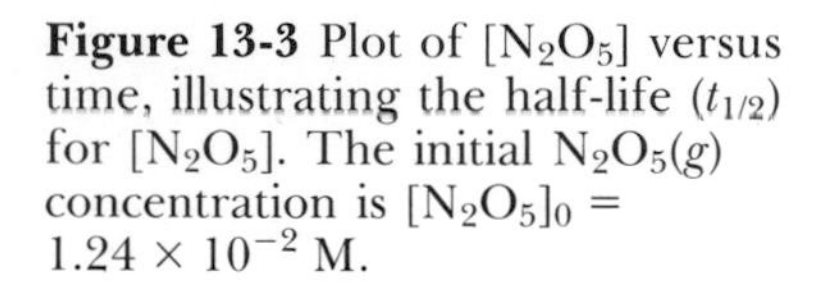

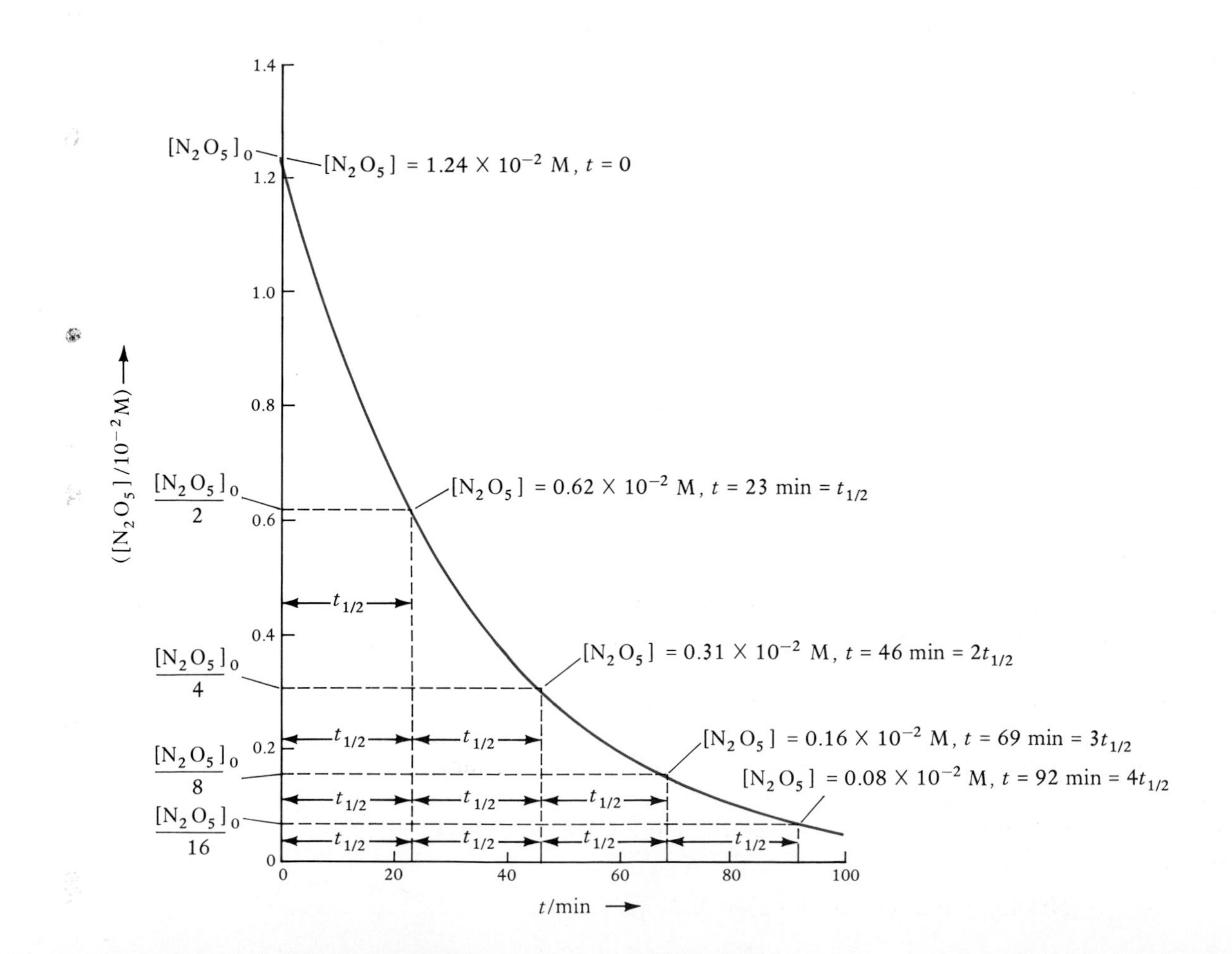

Figure 13-3 Plot of $[N_2O_5]$ versus time, illustrating the half-life ($t_{1/2}$) for $[N_2O_5]$. The initial $N_2O_5(g)$ concentration is $[N_2O_5]_0 = 1.24 \times 10^{-2}$ M.

when $t = t_{1/2}$, $[A] = [A]_0/2$ in Equation (13-10). Substituting $t = t_{1/2}$ and $[A] = [A]_0/2$ in Equation (13-10) gives

$$\log\left(\frac{1}{2}\right) = -0.301 = -\frac{kt_{1/2}}{2.30}$$

Solving this equation for $t_{1/2}$ yields

$$t_{1/2} = \frac{0.693}{k} \qquad (13\text{-}11)$$

Equation (13-11) says that the half-life for a first-order decomposition reaction is *independent of the initial concentration of the reactant.* The half-life is independent of the initial concentration of the reactant only for a first-order reaction. Therefore, if it is observed that $t_{1/2}$ for a reactant species A is independent of $[A]_0$, then the rate law must be first order.

Example 13-6: The rate law for the decomposition of aqueous hydrogen peroxide at 70°C,

$$2H_2O_2(aq) \rightarrow 2H_2O(l) + O_2(g)$$

is first order in $[H_2O_2]$. The half-life for the $H_2O_2(aq)$ decomposition at 70°C is $t_{1/2} = 20$ min. Given that the initial concentration is $[H_2O_2]_0 = 0.30$ M, compute the value of $[H_2O_2]$ 60 min after the 0.30 M $H_2O_2(aq)$ solution is prepared.

Solution: A reaction time of 60 min corresponds to three half-lives:

$$\frac{60 \text{ min}}{20 \text{ min/half-life}} = 3 \text{ half-lives}$$

The concentration of a reactant A that remains unreacted after n half-lives is

$$[A] = [A]_0\left(\frac{1}{2}\right)^n \qquad (13\text{-}12)$$

where $[A]_0$ is the initial reactant concentration. Using Equation (13-12), we compute

$$[H_2O_2] = [H_2O_2]_0\left(\frac{1}{2}\right)^n = (0.30\text{ M})\left(\frac{1}{2}\right)^3 = \frac{0.30\text{ M}}{8} = 0.038\text{ M}$$

Note that this is the same problem as is Example 13-5, but worked in a different way.

▪ If $[A]/[A_0]$ is known, then Equation (13-12) can be solved for n, the number of half-lives. Also, noninteger values of n can be used in Equation (13-12).

13-4. A RATE LAW CANNOT BE DEDUCED FROM THE REACTION STOICHIOMETRY

A balanced chemical equation gives us the relative numbers of moles of reactants consumed and moles of products formed. If a chemical reaction occurs in a single step, then the reaction is called an **elementary process.** Two examples of elementary processes are the reaction of gaseous nitrogen dioxide to produce nitrogen monoxide and nitrogen trioxide,

$$NO_2(g) + NO_2(g) \rightarrow NO(g) + NO_3(g)$$

for which

$$\text{rate} = k[NO_2]^2$$

and the reaction of $NO_3(g)$ with carbon monoxide to yield carbon dioxide and NO_2,

$$NO_3(g) + CO(g) \rightarrow CO_2(g) + NO_2(g)$$

for which

$$\text{rate} = k[NO_3][CO]$$

▪ An elementary process is a chemical reaction that occurs in a single step.

When we say that this second reaction is an elementary process, we mean that the conversion of NO_3 and CO to CO_2 and NO_2 can occur directly when an NO_3 molecule collides with a CO molecule; there are no other detectable chemical species involved in the reaction. The rate law for an elementary process can be written directly from the stoichiometry of the process; the rate is proportional to the product of the reactant concentrations (see the two examples just presented).

▪ Only for elementary processes can the rate law be deduced from the reaction stoichiometry.

Most chemical reactions, however, are not elementary processes. In most chemical reactions, the conversion of reactants to products requires more than one step. For example, the reaction

$$NO_2(g) + CO(g) \rightarrow CO_2(g) + NO(g)$$

is known to involve at least two steps; the first step is slow and the second step is fast:

$$NO_2(g) + NO_2(g) \rightarrow NO(g) + NO_3(g) \qquad \text{slow}$$
$$NO_3(g) + CO(g) \rightarrow CO_2(g) + NO_2(g) \qquad \text{fast}$$

A series of elementary processes that add up to give an overall reaction is called a **reaction mechanism.** A reaction mechanism is a detailed description of the pathway of a chemical reaction. One of the goals of chemical kinetics is to deduce the mechanisms of reactions. As we shall see, the idea is to construct a mechanism that is consistent with the observed rate law and the overall reaction stoichiometry. It is relatively easy to show that a certain rate law is consistent with a given mechanism. It takes more experience to construct a mechanism that agrees with a given rate law.

▪ A rate-determining step acts as a bottleneck that controls the reaction rate.

If one step in a reaction mechanism is much slower than any other steps, then the slow step controls the overall reaction rate and is called the **rate-determining step.** The rate of the slow step in the $NO_2 + CO$ reaction is

$$\text{rate} = k[NO_2]^2$$

Because the rate of this step controls the rate for the overall reaction, this is also the rate law for the overall reaction. In effect, the CO molecules have to wait around for NO_3 molecules to be formed in the slow step. The NO_3 molecules, once formed, are consumed very rapidly by reaction with CO.

▪ Intermediates are often present only at low concentrations and may be difficult to detect experimentally.

Note that the overall reaction stoichiometry does not disclose the involvement of the intermediate species NO_3. An **intermediate** is a species formed from the reactants that is involved in the conversion of reactants to products, but which does not appear as either reactant or product in the overall reaction. It is the possible involvement of intermediates in the reaction process that makes it impossible to deduce the reaction mechanism solely from the overall reaction stoichiometry.

Example 13-7: The reaction

$$2NO_2(g) + F_2(g) \rightarrow 2NO_2F(g)$$

is thought to proceed via the following two-step mechanism:

$$NO_2 + F_2 \xrightarrow{k_1} NO_2F + F \qquad \text{slow}$$

$$F + NO_2 \xrightarrow{k_2} NO_2F \qquad \text{fast}$$

where k_1 and k_2 are rate constants. Identify any species that are intermediates in the reaction mechanism and derive the rate law for this mechanism.

Solution: The fluorine atom is an intermediate in the reaction mechanism. Note that fluorine atoms are produced in the first step and consumed in the second. Also note that the sum of the two elementary processes yields the overall reaction stoichiometry.

The rate law is given by the slow elementary step, and so

$$\text{rate} = k_1[NO_2][F_2]$$

Once again, notice that this rate law could not be deduced directly from the overall reaction stoichiometry.

▪ A reaction mechanism is a postulated sequence of elementary processes by which the reactants are converted to products.

13-5. THE ACTIVATION ENERGY IS AN ENERGY BARRIER THAT THE REACTANTS MUST SURMOUNT IN ORDER TO REACT

A basic postulate of the **collision theory** of reaction rates is that a chemical reaction takes place only if the energy of the collision between two molecules is sufficient to break chemical bonds. In a mixture of the gases A and B with $P_A = P_B = 1.0$ atm, the collision frequency between A and B molecules is about 10^{33} collisions·s^{-1}·mL^{-1}. If every collision led to a reactive event, then the initial reaction rate would be about 10^{14} mol·L^{-1}·s^{-1}. Although there are a few reactions that occur at this very high rate, most occur at a much lower rate. The inescapable conclusion is that most collisions do not lead to reaction. Rather, the colliding molecules simply bounce off one another and remain unchanged. Only the more energetic collisions that occur with the necessary relative orientations lead to reaction.

▪ Many free-radical reactions have a reaction rate approximately equal to the collision rate; for example,

$$\cdot CH_3 + \cdot CH_3 \rightarrow H_3C{-}CH_3$$

In such reactions, unpaired electrons pair up to form a new bond.

In the collision theory of reaction rates, the reaction rate for the elementary process

$$A + B \rightarrow C + D$$

is given by

$$\text{rate} = \begin{pmatrix} \text{fraction of collisions} \\ \text{in which molecules have} \\ \text{the required relative} \\ \text{orientations} \end{pmatrix} \times \begin{pmatrix} \text{fraction of} \\ \text{collisions with} \\ \text{the required} \\ \text{energy} \end{pmatrix} \times \begin{pmatrix} \text{collision} \\ \text{frequency} \end{pmatrix}$$

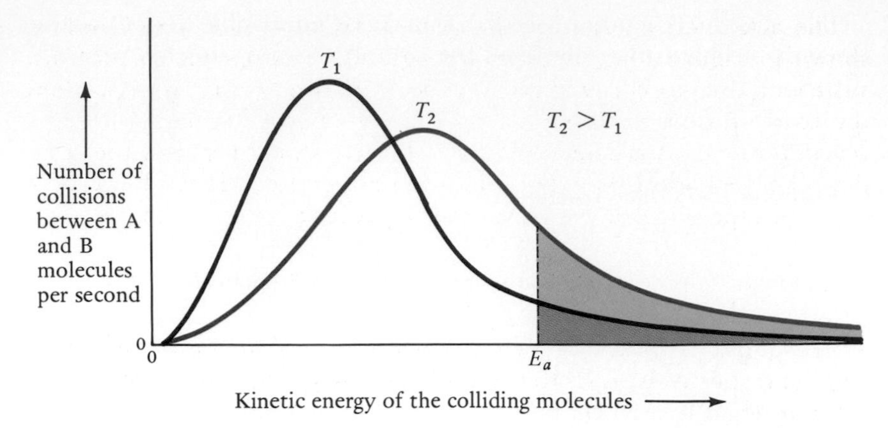

Figure 13-4 Plot of the number of collisions between A and B molecules per unit time versus the kinetic energy of the colliding molecules. The plot is made for two different temperatures, T_1 and T_2 ($T_2 > T_1$). Note that the hatched area under the curve beyond E_a is greater for the T_2 curve than for the T_1 curve. The number of collisions with an energy E_a or greater increases with increasing temperature.

▪ The activation energy, E_a, is a positive quantity.

A plot of the number of collisions per second of the A and B molecules versus the kinetic energy of the colliding molecules has the form shown in Figure 13-4. The **activation energy,** E_a, is the minimum energy necessary to cause a reaction between colliding molecules. Molecules that collide with a kinetic energy less than E_a simply bounce off one another. Molecules that collide with a kinetic energy greater than E_a can react, provided they collide with the required relative orientation (Figure 13-5). The fraction of the collisions that involve an energy equal to or greater than the activation energy increases with increasing temperature, because, as shown in Chapter 4, the average molecular speed increases with increasing temperature.

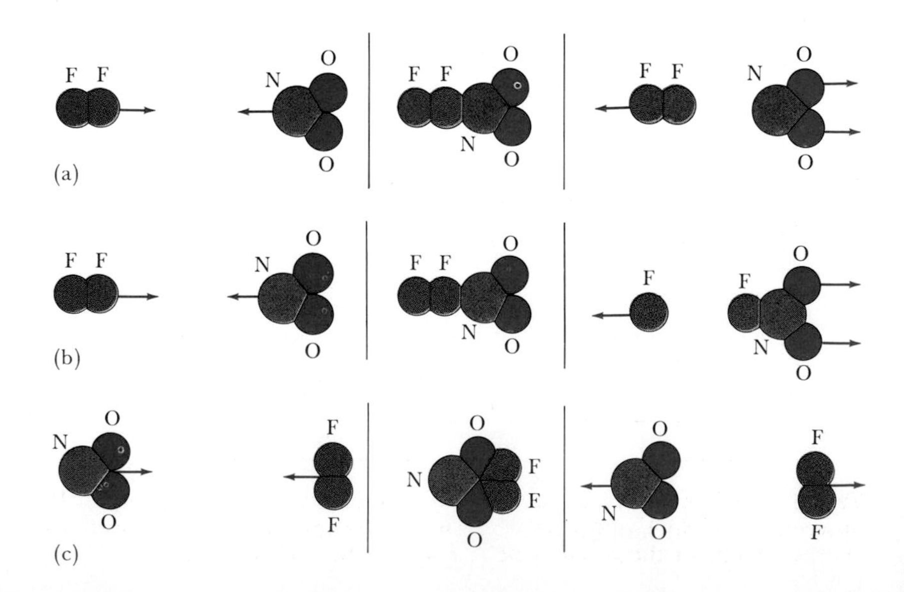

Figure 13-5 Nonreactive and reactive collisions of the molecules NO_2 and F_2 in the reaction $NO_2(g) + F_2(g) \rightarrow FNO_2(g) + F(g)$. (a) Nonreactive collision. Molecules bounce off one another without reacting because the kinetic energy of the colliding particles is less than the activation energy. (b) Reactive collision. Molecules collide with a kinetic energy greater than E_a and react. (c) Nonreactive collision. Molecules collide with a kinetic energy greater than E_a but do not react because they do not have the correct orientation for reaction.

The activation energy for the general reaction A + B → C + D is shown in Figure 13-6. If the colliding molecules do not collide with sufficient kinetic energy to go over the activation energy "hump," then the reaction does not take place. The temperature dependence of a reaction rate constant depends upon the activation energy of the reaction. The equation that describes the temperature dependence of a rate constant is called the **Arrhenius equation:**

▪ Svante Arrhenius, a native of Sweden, was awarded the Nobel Prize in chemistry in 1903.

$$\log\left(\frac{k_2}{k_1}\right) = \frac{E_a}{2.30R}\left(\frac{T_2 - T_1}{T_1T_2}\right) \qquad (13\text{-}13)$$

where k_1 and k_2 are the rate constants at the absolute temperatures T_1 and T_2, respectively; E_a is the activation energy; and R is the molar gas constant ($8.314\ \mathrm{J{\cdot}K^{-1}{\cdot}mol^{-1}}$).

Example 13-8: The activation energy for the reaction

$$2NO_2(g) + F_2(g) \rightarrow 2NO_2F(g)$$

is $E_a = 43.5\ \mathrm{kJ{\cdot}mol^{-1}}$. Estimate the increase in the rate constant of the reaction for an increase in temperature from 300 K to 310 K.

Solution: Using Equation (13-13) we have

▪ Logarithms are discussed in Appendix A2.

$$\log\left(\frac{k_2}{k_1}\right) = \frac{E_a}{2.30R}\left(\frac{T_2 - T_1}{T_1T_2}\right)$$

Inserting the quantities $E_a = 43.5\ \mathrm{kJ{\cdot}mol^{-1}}$, $T_1 = 300$ K, and $T_2 = 310$ K into this equation yields

$$\log\left(\frac{k_2}{k_1}\right) = \frac{43.5 \times 10^3\ \mathrm{J{\cdot}mol^{-1}}}{2.30 \times 8.314\ \mathrm{J{\cdot}K^{-1}{\cdot}mol^{-1}}}\left[\frac{310\ \mathrm{K} - 300\ \mathrm{K}}{(310\ \mathrm{K})(300\ \mathrm{K})}\right] = 0.245$$

Thus $$\frac{k_2}{k_1} = 10^{0.245} = 1.8$$

Thus the reaction rate constant increases by about a factor of 2 for the 10 K temperature increase.

▪ Rule of thumb: around room temperature a 10°C increase in temperature leads to a twofold increase in reaction rate.

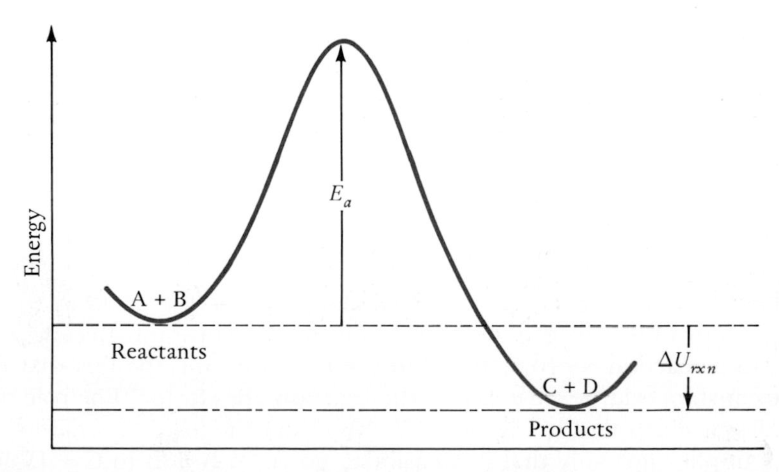

Figure 13-6 The energy of the molecules as the reaction A + B → C + D proceeds. The value of E_a has no relationship to the value of the overall energy change of the reaction $\Delta U_{rxn} = U(\mathrm{C + D}) - U(\mathrm{A + B})$.

Rate constants always increase with temperature because the activation energy is necessarily a positive quantity. The proof goes as follows. If $T_2 > T_1$ in Equation (13-13), then the right-hand side is necessarily positive. If the right-hand side is positive, then k_2 must be greater than k_1, in order that the left-hand side also be positive. Thus we see that $k_2 > k_1$ if $T_2 > T_1$ or, in other words, that the reaction rate constant increases with increasing temperature. The reaction rate also increases with increasing temperature because the rate is proportional to the rate constant.

13-6. A CATALYST IS A SUBSTANCE THAT INCREASES THE REACTION RATE BUT IS NOT CONSUMED AS A REACTANT

The rates of many reactions are increased by catalysts. A **catalyst** is a reaction facilitator that acts by providing a different and faster reaction mechanism than is possible in the absence of the catalyst. For example, the reaction rate law for the reaction

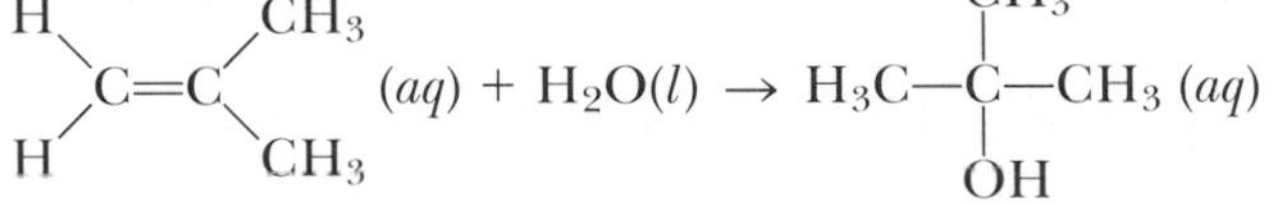

$$(H)_2C{=}C(CH_3)_2(aq) + H_2O(l) \rightarrow H_3C{-}C(CH_3)(OH){-}CH_3(aq)$$

an alkene — an alcohol

is

$$\text{rate} = k[\text{alkene}][H^+]$$

The solvated hydrogen ion, $H^+(aq)$, does not appear as a reactant in this reaction, but nevertheless the reaction rate is proportional to $[H^+]$. The $H^+(aq)$ ion presumably facilitates the reaction by attaching to the carbon atom that is bonded to the two hydrogen atoms. Thus, a plausible mechanism for this reaction is as follows:

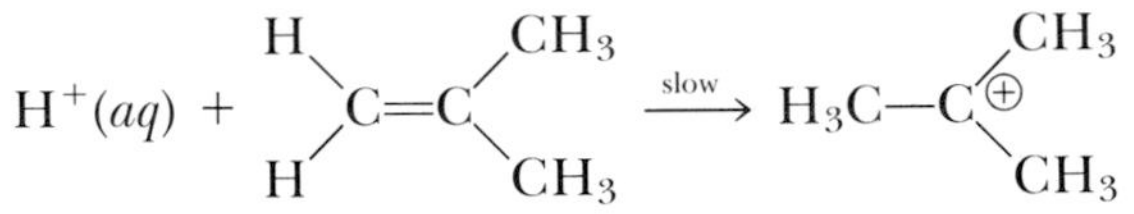

$$H^+(aq) + (H)_2C{=}C(CH_3)_2 \xrightarrow{\text{slow}} H_3C{-}C^{\oplus}(CH_3)_2$$

intermediate

The intermediate reacts rapidly with water to form the alcohol:

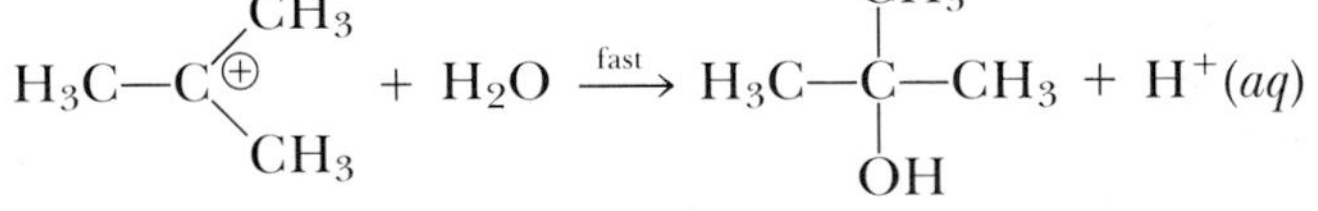

$$H_3C{-}C^{\oplus}(CH_3)_2 + H_2O \xrightarrow{\text{fast}} H_3C{-}C(CH_3)(OH){-}CH_3 + H^+(aq)$$

Note that $H^+(aq)$ is regenerated in the second step, and so is not consumed by the reaction.

The catalyst $H^+(aq)$ acts by providing a new reaction pathway with a lower activation energy, and thus a larger rate constant. Because the rate of a reaction is proportional to the rate constant, the fact that the rate constant is larger means that the reaction goes faster. The role of a catalyst is illustrated in Figure 13-7. Note that the lower activation energy implies not only that it is easier to go from A + B to C + D, but also that it is easier to go from C + D to A + B. We shall see in Section 13-7 that as the concentrations of products build up during a reaction, the reverse reaction becomes significant. Eventually the rate of the for-

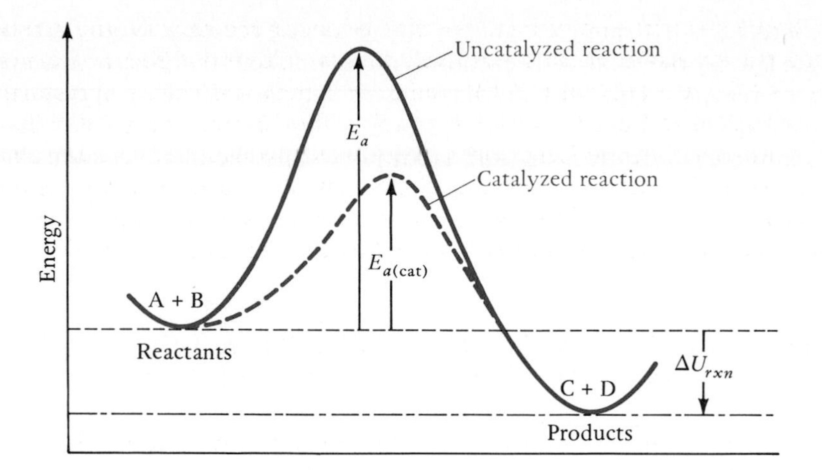

Figure 13-7 A comparison of the activation energies for the uncatalyzed, E_a, and catalyzed, E_a (cat), reaction A + B → C + D. The catalyst lowers the activation energy barrier to the reaction and thereby increases the reaction rate.

ward reaction and the rate of the reverse reaction become equal, at which point there is no *net* reaction, and a state of dynamic equilibrium exists, much as in the cases of the equilibrium vapor pressure of a liquid (Section 11-4) and of the dissolution of a solid in a liquid (Section 12-1). A catalyst increases the rates of both the forward and reverse reactions and hence increases the rate at which the reaction proceeds toward equilibrium. Nevertheless, a catalyst does not affect the relative amounts of reactants and products at equilibrium. In effect, a catalyst helps to get the job done faster, but the final result is the same.

Consider the reaction of aqueous cerium(IV) ions, $Ce^{4+}(aq)$, with aqueous thallium(I) ions, $Tl^{+}(aq)$:

$$2Ce^{4+}(aq) + Tl^{+}(aq) \rightarrow 2Ce^{3+}(aq) + Tl^{3+}(aq)$$

Although this reaction goes essentially to completion, the reaction rate is very slow. The rate law is

$$\text{rate} = k[Tl^{+}][Ce^{4+}]^2$$

The low reaction rate is thought to be a consequence of the requirement that the reactive event, that is, the simultaneous transfer of two electrons from a $Tl^{+}(aq)$ ion to two different $Ce^{4+}(aq)$ ions, requires that two $Ce^{4+}(aq)$ ions be present simultaneously near a $Tl^{+}(aq)$ ion. Three-body encounters are only about 10^{-5} times as likely to occur as two-body encounters, and thus the reaction rate is low. The $2Ce^{4+}(aq) + Tl^{+}(aq)$ reaction is catalyzed by $Mn^{2+}(aq)$. The catalytic action of Mn^{2+} has been attributed to the availability of the Mn^{3+} and Mn^{4+} oxidation states, which provide a new reaction pathway involving a three-step sequence of two-body elementary processes:

$$Ce^{4+} + Mn^{2+} \xrightarrow{\text{slow}} Mn^{3+} + Ce^{3+} \qquad \text{(rate-determining)}$$

$$Ce^{4+} + Mn^{3+} \xrightarrow{\text{fast}} Mn^{4+} + Ce^{3+}$$

$$Tl^{+} + Mn^{4+} \xrightarrow{\text{fast}} Mn^{2+} + Tl^{3+} \qquad \text{(two electrons transferred)}$$

The sum of these three equations corresponds to the overall reaction stoichiometry. The rate law for the $Mn^{2+}(aq)$-catalyzed reaction is determined by the slowest step in the mechanism, and thus the rate law is

$$\text{rate} = k_{cat}[Ce^{4+}][Mn^{2+}]$$

where k_{cat} is the rate constant for the catalyzed reaction. Note that the rate law for the catalyzed reaction is different from that for the uncatalyzed reaction. Different mechanisms usually (but not always) give rise to different rate laws.

Innumerable industrial- and laboratory-scale chemical reactions are carried out in the presence of catalysts. For example, platinum and palladium are used as surface catalysts for a variety of reactions, such as the hydrogenation of double bonds:

$$H_2C{=}CH_2(g) + H_2(g) \xrightarrow{Pt(s)} H{-}\underset{H}{\overset{H}{C}}{-}\underset{H}{\overset{H}{C}}{-}H(g)$$

The first step in this hydrogenation involves the adsorption of hydrogen onto the platinum surface, and this is followed by dissocation of the adsorbed H_2 into adsorbed hydrogen atoms.

$$H_2(\text{surface}) \rightarrow 2H(\text{surface})$$

The adsorbed hydrogen atoms can move around on the platinum surface and eventually react stepwise with adsorbed ethylene, H_2CCH_2, to form ethane, H_3CCH_3:

$$H_2C{=}CH_2(\text{surface}) + H(\text{surface}) \rightarrow H_2\dot{C}{-}CH_3(\text{surface})$$

$$H_2\dot{C}{-}CH_3(\text{surface}) + H(\text{surface}) \rightarrow H_3C{-}CH_3(g)$$

The ethane produced does not interact strongly with the platinum surface and thus leaves the surface immediately after it is formed.

Platinum metal also catalyzes a wide variety of oxygenation reactions, including the oxidation of SO_2 to SO_3 in the production of sulfuric acid (Chapter 2). The first step is the adsorption of O_2 onto the platinum surface. The adsorbed O_2 dissociates into O atoms to form a surface layer of reactive O atoms. The final step involves the rapid reaction of SO_2 with surface O atoms (Figure 13-8):

$$SO_2 + O(\text{surface}) \rightarrow SO_3$$

Enzymes are proteins that catalyze chemical reactions in living systems (proteins are discussed in Section 24-4). Without enzymes, most biochemical reactions are too slow to be of any consequence. Cells contain thousands of different enzymes. The absence or an insufficient

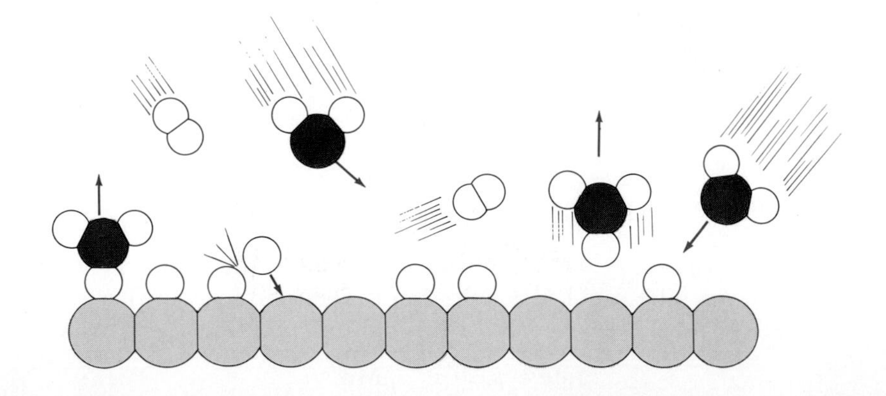

Figure 13-8 Heterogeneous (contact) catalysis by a platinum metal surface. The catalyzed reaction is $2SO_2(g) + O_2(g) \xrightarrow{Pt(s)} 2SO_3(g)$. The platinum surface catalyzes the reaction by causing the dissociation of adsorbed O_2 molecules into O atoms. The surface-bound O atoms then react with SO_2 to give SO_3.

quantity of an enzyme can lead to serious physiological disorders. Albinism, cystinosis, and phenylketonuria are three of many examples.

A remarkable property of enzymes is their extraordinary catalytic specificity. An enzyme usually catalyzes a single chemical reaction or a set of closely related reactions. The exact manner in which many enzymes function is still a topic of research, but one simple picture of enzyme activity is given by the **lock-and-key theory** (Figure 13-9). In this picture, the enzyme binds the **substrate,** which is the substance that is reacting, to its surface. The substrate molecule is bound in such a way that the substrate is susceptible to chemical attack. The binding site on the enzyme is such that it can bind only one substrate or a closely related one. The particular shape of and the nature and location of atoms at the binding site account for the extraordinary specificity of enzymes.

13-7. CHEMICAL REACTIONS REACH A STATE OF EQUILIBRIUM

Consider the dissociation of nitrosyl bromide, NOBr, the equation for which is

$$2NOBr(g) \rightarrow 2NO(g) + Br_2(g) \qquad (13\text{-}14)$$

The rate law for this reaction is

$$(\text{rate})_f = k_f[NOBr]^2 \qquad (13\text{-}15)$$

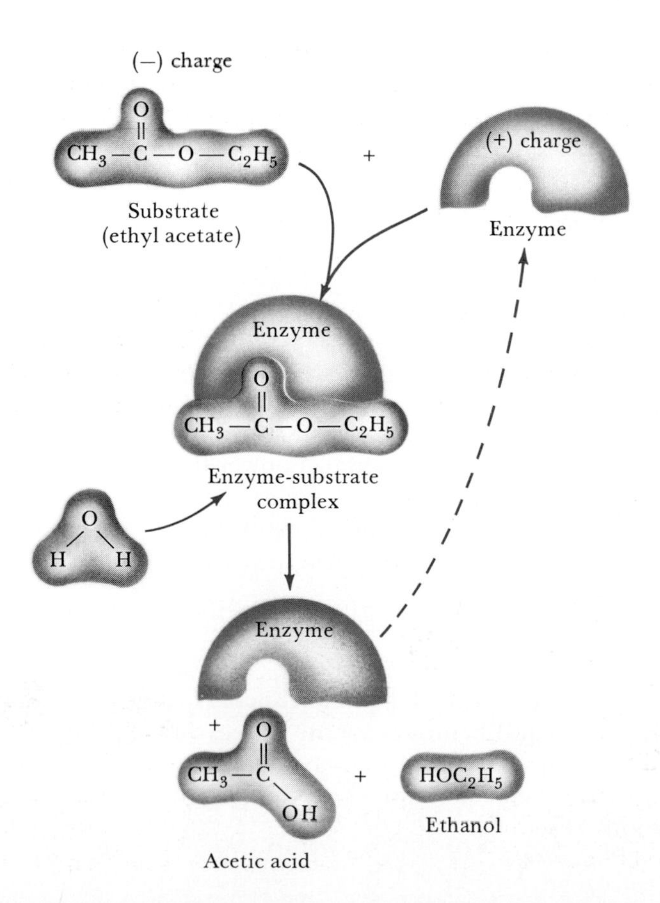

Figure 13-9 The lock-and-key theory of enzyme catalyzed reactions. The reaction in this case is the hydrolysis of ethyl acetate: $CH_3COOC_2H_5 + H_2O \rightarrow C_2H_5OH + HC_2H_3O_2$. The shape of and the charges at the binding site of the enzyme are perfectly suited to accommodate the substrate, ethyl acetate. Because of the specific way the enzyme binds the ethyl acetate, the ester bond is exposed and readily attacked by a water molecule. By holding the substrate in this way, the enzyme speeds up the reaction rate.

where the subscript f on (rate) denotes the **forward reaction rate,** and k_f is the rate constant for the forward reaction. If we start with pure $NOBr(g)$, then the initial concentrations of $NO(g)$ and $Br_2(g)$ are zero. As the reaction proceeds, the concentration of $NOBr(g)$ decreases and the concentrations of $NO(g)$ and $Br_2(g)$ increase. Because [NOBr] decreases with time, the forward reaction rate, which is proportional to $[NOBr]^2$, also decreases with time. As the reaction described by Equation (13-14) proceeds, the concentrations of $NO(g)$ and $Br_2(g)$ build up, and the reverse reaction, the equation for which is

$$2NO(g) + Br_2(g) \rightarrow 2NOBr(g) \tag{13-16}$$

becomes significant. The rate law for this reaction, which is the reverse of Equation (13-14), is

$$(\text{rate})_r = k_r[NO]^2[Br_2] \tag{13-17}$$

The subscript r on (rate) denotes the **reverse reaction rate,** and k_r is the rate constant for the reverse reaction. As the concentrations of $NO(g)$ and $Br_2(g)$ increase with time, the rate of this reverse reaction increases with time. Thus the rate of the forward reaction decreases with time while the rate of the reverse reaction increases with time. Eventually we will reach a point at which the two rates are equal (Figure 13-10), and we have

$$(\text{rate})_f = (\text{rate})_r \tag{13-18}$$

From this point on, there is no net change in the concentrations of $NOBr(g)$, $NO(g)$, and $Br_2(g)$. We say that the reaction has reached an **equilibrium state,** and the concentrations have the equilibrium values $[NOBr]_{eq}$, $[NO]_{eq}$, and $[Br_2]_{eq}$. If we equate the forward and reverse reaction rates at equilibrium, then we have

$$k_f[NOBr]^2_{eq} = k_r[NO]^2_{eq}[Br_2]_{eq}$$

We can rewrite this condition as

$$\frac{k_f}{k_r} = \frac{[NO]^2_{eq}[Br_2]_{eq}}{[NOBr]^2_{eq}} \tag{13-19}$$

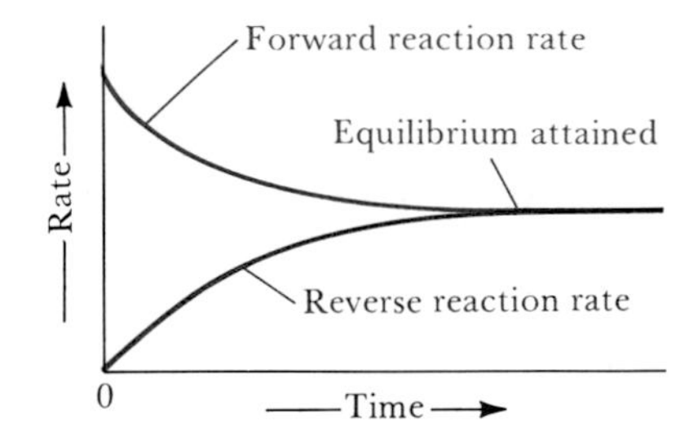

Figure 13-10 At equilibrium the forward and reverse reaction rates are equal. When the reaction starts, the forward rate is large compared with the reverse rate, but as the reaction proceeds, the forward rate decreases and the reverse rate increases until the two rates become equal at equilibrium. In the equilibrium state there is no further change in concentrations of the reactants and products.

The left-hand side of this equation is a constant at a given temperature. Notice that the right-hand side is a ratio of the concentrations of the products to the concentration of the reactant in the reaction described by Equation (13-14), each raised to a power that is equal to the stoichiometric coefficient in the chemical equation as written. Equation (13-19) says that this particular ratio of concentrations is a constant at a given temperature. We denote this constant by K and write

$$K = \frac{[NO]^2_{eq}[Br_2]_{eq}}{[NOBr]^2_{eq}} \tag{13-20}$$

The constant K is called the **equilibrium constant,** and Equation (13-20) is the **equilibrium-constant expression** for Equation (13-14). In the next chapter we shall study reactions at equilibrium and develop the implications and consequences of equilibrium-constant expressions for various reactions.

SUMMARY

A reaction rate tells us how fast a reactant is being consumed or how fast a product is being formed in a chemical reaction. The reaction rate law describes the dependence of the reaction rate on the concentrations of the various species involved in the reaction. A first-order rate law is proportional to the first power of the reactant concentration; a second-order rate law is proportional to the second power of the reactant concentration. The reaction rate law cannot be deduced solely from the reaction stoichiometry.

An elementary process is a chemical reaction that occurs in a single step. In this case, the rate law can be deduced from the stoichiometry. Most chemical reactions are not elementary processes, and so their rate laws cannot be deduced from the reaction stoichiometry. The reaction mechanism is the sequence of elementary processes by which reactants are converted to products. A reaction mechanism that involves more than one elementary process necessarily involves reaction intermediates. An intermediate is a species that is formed from the reactants and is involved in the conversion of reactants to products but does not appear in the overall reaction stoichiometry.

In order to react, molecules have to collide. However, only those collisions in which the molecules have the correct orientation and have a combined kinetic energy equal to or greater than the activation energy for the reaction lead to products. The temperature dependence of a rate constant is given by the Arrhenius equation.

A catalyst is a species that increases the rate of a reaction but is not consumed as a reactant. A catalyst provides a different and faster reaction pathway than is possible in its absence. A catalyst lowers the activation energy of a reaction but has no effect on the equilibrium concentration of a species.

As a reaction proceeds, the rate of the forward reaction decreases and the rate of the reverse reaction increases until equilibrium is attained, at which point the forward and reverse rates are equal. The equilibrium-constant expression governs the concentrations of reactants and products at equilibrium.

TERMS YOU SHOULD KNOW

reaction rate 405
rate law 408
method of initial rates 408
rate constant 409
first-order rate law 409
first-order reaction 409
second-order reaction 411
intercept 412
slope 412
half-life, $t_{1/2}$ 414
elementary process 415
reaction mechanism 416
rate-determining step 416
intermediate 416
collision theory 417
activation energy, E_a 418
Arrhenius equation 419
catalyst 420
enzyme 422
lock-and-key theory 423
substrate 423
forward reaction rate 424
reverse reaction rate 424
equilibrium state 424
equilibrium constant 424
equilibrium-constant expression 424

EQUATIONS YOU SHOULD KNOW HOW TO USE

$(\text{rate})_0 = k[\text{A}]_0^x[\text{B}]_0^y[\text{C}]_0^z$ (13-6) ($x + y + z$ = overall order of reaction)

$\text{rate} = k[\text{A}]$ (13-8) (first order rate law)

$\log\,[\text{A}] = \log\,[\text{A}]_0 - \dfrac{kt}{2.30}$ (13-9) (first-order rate law expression)

$t_{1/2} = \dfrac{0.693}{k}$ (13-11) (half life for a first-order reaction)

$[\text{A}] = [\text{A}]_0\left(\dfrac{1}{2}\right)^n$ (13-12) (an alternate first-order rate law expression)

$$\log\left(\frac{k_2}{k_1}\right) = \frac{E_a}{2.30R}\left(\frac{T_2 - T_1}{T_1 T_2}\right) \qquad (13\text{-}13) \quad \text{(Arrhenius equation)}$$

$$(\text{rate})_f = (\text{rate})_r \qquad (13\text{-}18) \quad \text{(condition of dynamic equilibrium)}$$

PROBLEMS

REACTION RATES

13-1. Suggest a method for measuring the rate of the following reactions:

(a) $I_2(aq) + 2S_2O_3^{2-}(aq) \rightarrow S_4O_6^{2-}(aq) + 2I^-(aq)$
(yellow) (colorless) (colorless) (colorless)

(b) $4PH_3(g) \rightarrow P_4(g) + 6H_2(g)$

13-2. Suggest a method for measuring the rate of the following reactions:

(a) $2H_2O_2(aq) \rightarrow 2H_2O(l) + O_2(g)$

(b) $2HBr(g) \rightarrow H_2(g) + Br_2(g)$

13-3. The rate law for the reaction

$$O_3(g) + NO(g) \rightarrow O_2(g) + NO_2(g)$$

at 310 K is

$$\text{rate} = (3.0 \times 10^6\ M^{-1}\cdot s^{-1})[O_3][NO]$$

Given that $[O_3] = 6.0 \times 10^{-4}$ M and $[NO] = 4.0 \times 10^{-5}$ M at $t = 0$, compute the rate of the reaction at $t = 0$.

13-4. The rate law for the reaction

$$2NO(g) + Br_2(g) \rightarrow 2NOBr(g)$$

is

$$\text{rate} = (1.3 \times 10^{-3}\ M^{-2}\cdot \text{min}^{-1})[NO]^2[Br_2]$$

Compute the initial reaction rate when $[NO]_0 = [Br_2]_0 = 3.0 \times 10^{-4}$ M.

13-5. For the reaction

$$2O_3(g) \rightarrow 3O_2(g)$$

$-\Delta P_{O_3}/\Delta t$ was found to be 5.0×10^{-4} atm·s^{-1}. Determine the value of $\Delta P_{O_2}/\Delta t$ in atm·s^{-1} during this period of time.

13-6. For the reaction

$$2C_5H_6(g) \rightarrow C_{10}H_{12}(g)$$

$-\Delta[C_5H_6]/\Delta t$ was found to be 2.3 torr·s^{-1}. Determine the value of $\Delta[C_{10}H_{12}]/\Delta t$ in torr·s^{-1} during this period of time.

13-7. Given the following data on the decomposition of N_2O_5,

$$2N_2O_5(soln) \rightarrow 4NO_2(soln) + O_2(g)$$

at 45°C in carbon tetrachloride, calculate the average reaction rate for each successive time interval.

t/s	$[N_2O_5]$/M
0	1.48
175	1.32
506	1.07
845	0.87

13-8. The value of $[H^+]$ in the reaction
$CH_3OH(aq) + H^+(aq) + Cl^-(aq) \rightarrow CH_3Cl(aq) + H_2O(l)$

was measured over a period of time:

t/min	$[H^+]$/M
0	2.12
31	1.90
61	1.78
121	1.61

Find the average rate of disappearance of $H^+(aq)$ for the time interval between each measurement. What is the average rate of disappearance of CH_3OH for the same time intervals? The average rate of appearance of CH_3Cl?

INITIAL RATES

13-9. Sulfuryl chloride decomposes according to the equation

$$SO_2Cl_2(g) \rightarrow SO_2(g) + Cl_2(g)$$

Using the following initial-rate data, determine the order of the reaction with respect to SO_2Cl_2:

$[SO_2Cl_2]_0$/mol·L^{-1}	(rate)$_0$/mol·L^{-1}·s^{-1}
0.10	2.2×10^{-6}
0.20	4.4×10^{-6}
0.30	6.6×10^{-6}
0.40	8.8×10^{-6}

13-10. Nitrosyl bromide decomposes according to the equation

$$2NOBr(g) \rightarrow 2NO(g) + Br_2(g)$$

Using the following initial-rate data, determine the order of the reaction with respect to NOBr(*g*):

$[NOBr]_0/mol \cdot L^{-1}$	$(rate)_0/mol \cdot L^{-1} \cdot s^{-1}$
0.20	0.80
0.40	3.20
0.60	7.20
0.80	12.80

13-11. The reaction

$$C_2H_5Cl(g) \rightarrow C_2H_4(g) + HCl(g)$$

was studied at 300 K, and the following data were collected:

Run	Initial concentration, $[C_2H_5Cl]_0/M$	Initial rate of formation of $C_2H_4/M \cdot s^{-1}$
1	0.33	2.40×10^{-30}
2	0.66	4.80×10^{-30}
3	1.32	9.60×10^{-30}

Determine the rate law, the order, and the rate constant for the reaction.

13-12. The reaction

$$2C_5H_6(g) \rightarrow C_{10}H_{12}(g)$$

was studied at 373 K, and the following data were collected:

Run	Initial pressure, $[P_{C_5H_6}]_0/torr$	Initial rate of formation of $C_{10}H_{12}/torr \cdot s^{-1}$
1	200	5.76
2	400	23.04
3	800	92.2

Determine the rate law in terms of pressure rather than concentration, and compute the rate constant for the reaction.

13-13. The reaction

$$2NOCl(g) \rightarrow 2NO(g) + Cl_2(g)$$

was studied at 400 K, and the following data were collected:

Run	Initial concentration, $[NOCl]_0/M$	Initial rate of formation of $NO/M \cdot s^{-1}$
1	0.25	1.75×10^{-6}
2	0.50	7.00×10^{-6}
3	0.75	1.57×10^{-5}

Determine the rate law, the order, and the rate constant for the reaction.

13-14. The following data were obtained for the decomposition of N_2O_3:

$$N_2O_3(g) \rightarrow NO(g) + NO_2(g)$$

Initial pressure, $P_{N_2O_3}/torr$	Initial rate of formation of $NO_2/torr \cdot s^{-1}$
0.91	5.5
1.4	8.4
2.1	13

Determine the rate law for the reaction, expressed in terms of $P_{N_2O_3}$ rather than $[N_2O_3]$. Calculate the rate constant for the reaction.

13-15. Consider the reaction

$$Cr(H_2O)_6^{3+}(aq) + SCN^-(aq) \rightarrow Cr(H_2O)_5NCS^{2+}(aq) + H_2O(l)$$

for which the following initial rate data were obtained at 25°C:

$[Cr(H_2O)_6^{3+}]_0/M$	$[SCN^-]_0/M$	$(rate)_0/M \cdot s^{-1}$
1.0×10^{-4}	0.10	2.0×10^{-11}
1.0×10^{-3}	0.10	2.0×10^{-10}
1.5×10^{-3}	0.20	6.0×10^{-10}
1.5×10^{-3}	0.50	1.5×10^{-9}

Determine the rate law and the rate constant for the reaction.

13-16. The reaction

$$CoBr(NH_3)_5^{2+}(aq) + OH^-(aq) \rightarrow Co(NH_3)_5OH^{2+}(aq) + Br^-(aq)$$

was studied at 25°C, and the following initial rate data were collected:

$[CoBr(NH_3)_5^{2+}]_0/M$	$[OH^-]_0/M$	$(rate)_0/M \cdot s^{-1}$
0.030	0.030	1.37×10^{-3}
0.060	0.030	2.74×10^{-3}
0.030	0.090	4.11×10^{-3}
0.090	0.090	1.23×10^{-2}

Determine the rate law, the overall order of the rate law, and the value of the rate constant for the reaction.

13-17. Given the following initial-rate data at 300 K for the reaction

$$2NO_2(g) + O_3(g) \rightarrow N_2O_5(g) + O_2(g)$$

determine the reaction rate law and the value of the rate constant:

$[NO_2]_0/M$	$[O_3]_0/M$	$(\text{rate})_0/M \cdot s^{-1}$
1.00	1.00	5.0×10^4
2.00	1.00	1.0×10^5
2.00	2.00	2.0×10^5

13-18. Given the following initial-rate data for the reaction

$$CH_3COCH_3(aq) + Br_2(aq) \xrightarrow{H^+(aq)} CH_3COCH_2Br(aq) + H^+(aq) + Br^-(aq)$$

determine the reaction rate law and the value of the rate constant:

$[CH_3COCH_3]_0/M$	$[Br_2]_0/M$	$[H^+]_0/M$	$(\text{rate})_0/M \cdot s^{-1}$
1.00	1.00	1.00	4.0×10^{-3}
2.00	1.00	1.00	8.0×10^{-3}
2.00	2.00	1.00	8.0×10^{-3}
1.00	2.00	2.00	8.0×10^{-3}

FIRST-ORDER REACTIONS

13-19. The reaction

$$SO_2Cl_2(g) \rightarrow SO_2(g) + Cl_2(g)$$

is first order with a rate constant of $2.2 \times 10^{-5}\ s^{-1}$ at 320°C. What fraction of a sample of SO_2Cl_2 will remain if it is heated for 5.0 h at 320°C?

13-20. The rate constant for the first-order reaction

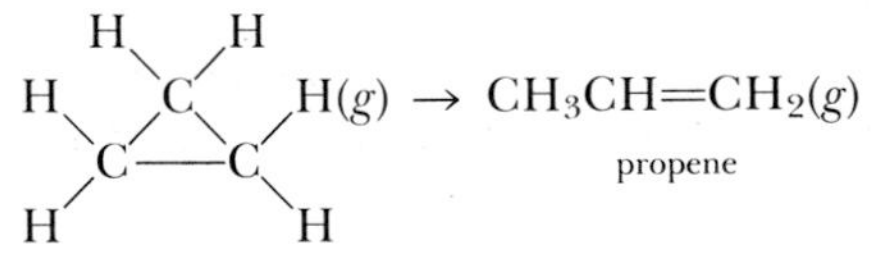

$$\text{cyclopropane}(g) \rightarrow CH_3CH{=}CH_2(g)$$

cyclopropane   propene

at 500°C is $5.5 \times 10^{-4}\ s^{-1}$. Compute the half-life of cyclopropane at 500°C. Given an initial cyclopropane concentration of 1.00×10^{-3} M at 500°C, compute the concentration of cyclopropane that remains after 2.0 h.

13-21. Azomethane, $CH_3N_2CH_3$, decomposes according to the equation

$$CH_3N_2CH_3(g) \rightarrow CH_3CH_3(g) + N_2(g)$$

Given that the decomposition is a first-order process with $k = 4.0 \times 10^{-4}\ s^{-1}$ at 300°C, calculate the fraction of azomethane that remains after one hour.

13-22. Methyl iodide, CH_3I, decomposes according to the equation

$$2CH_3I(g) \rightarrow C_2H_6(g) + I_2(g)$$

Given that the decomposition is a first-order process with $k = 1.5 \times 10^{-4}\ s^{-1}$ at 300°C, calculate the fraction of methyl iodide that remains after one millisecond.

13-23. Peroxydisulfate ion, $S_2O_8^{2-}$, decomposes in aqueous solution according to the equation

$$S_2O_8^{2-}(aq) + H_2O(l) \rightarrow 2SO_4^{2-}(aq) + \tfrac{1}{2}O_2(g) + 2H^+(aq)$$

Given the following data from an experiment with $[S_2O_8^{2-}]_0 = 0.100$ M in a solution with $[H^+]$ fixed at 0.100 M, determine the reaction rate law and calculate the rate constant:

t/min	$[S_2O_8^{2-}]/M$
0	0.100
17	0.050
34	0.025
51	0.012

13-24. At 400 K oxalic acid decomposes according to

$$H_2C_2O_4(g) \rightarrow CO_2(g) + HCHO_2(g)$$

The rate of this reaction can be studied by measurement of the total pressure. Determine the rate law and the rate constant of this reaction from the following measurements, which give the total pressure reached after 2.00×10^4 s from the given starting pressure of oxalic acid:

$P_{H_2C_2O_4}$/torr (at $t = 0$)	P_{total}/torr (at $t = 2.00 \times 10^4$ s)
5.0	7.2
7.0	10
8.4	12

REACTION MECHANISMS

13-25. Write the rate law for each of the following elementary reactions:

(a) $N_2O(g) + O(g) \rightarrow 2NO(g)$
(b) $O(g) + O_3(g) \rightarrow 2O_2(g)$
(c) $ClCO(g) + Cl_2(g) \rightarrow Cl_2CO(g) + Cl(g)$

13-26. Write the rate law for each of the following elementary reactions:

(a) $K(g) + HCl(g) \rightarrow KCl(g) + H(g)$

(b) $Cl(g) + ICl(g) \rightarrow Cl_2(g) + I(g)$
(c) $NO_3(g) + CO(g) \rightarrow NO_2(g) + CO_2(g)$

13-27. The reaction of CO_2 with hydroxide ion in aqueous solution is postulated to occur according to the mechanism

$$CO_2(aq) + OH^-(aq) \rightarrow HCO_3^- \quad \text{slow}$$
$$HCO_3^-(aq) + OH^-(aq) \rightarrow CO_3^{2-}(aq) + H_2O(l) \quad \text{fast}$$

The rate law for the disappearance of $CO_2(aq)$ was found experimentally to be

$$\text{rate} = k[CO_2][OH^-]$$

Is this mechanism consistent with the observed rate law? Explain your answer.

13-28. The reaction

$$2NO_2(g) \rightarrow 2NO(g) + O_2(g)$$

is postulated to occur via the mechanism

$$NO_2(g) + NO_2(g) \rightarrow NO(g) + NO_3(g) \quad \text{slow}$$
$$NO_3(g) \rightarrow NO(g) + O_2(g) \quad \text{fast}$$

The rate law for the reaction is

$$\text{rate} = k[NO_2]^2$$

Is this mechanism consistent with the observed rate law? Explain your answer.

ARRHENIUS EQUATION

13-29. The reaction

$$2N_2O_5(soln) \rightarrow 2N_2O_4(soln) + O_2(g)$$

takes place in carbon tetrachloride at room temperature. The rate constant is $2.35 \times 10^{-4}\ s^{-1}$ at 293 K and $9.15 \times 10^{-4}\ s^{-1}$ at 303 K. Calculate E_a for this reaction.

13-30. The rate constant for the reaction

$$H_2(g) + I_2(g) \rightarrow 2HI(g)$$

was determined to be $0.0234\ M^{-1}\cdot s^{-1}$ at 400°C and $0.750\ M^{-1}\cdot s^{-1}$ at 500°C. Calculate E_a for this reaction.

13-31. The activation energy for the reaction

$$C_4H_8(g) \rightarrow 2C_2H_4(g)$$

is $262\ kJ\cdot mol^{-1}$. At 600 K the rate constant is $6.07 \times 10^{-8}\ s^{-1}$. What is the rate constant at 800 K?

13-32. The activation energy for the reaction

$$2N_2O_5(g) \rightarrow 4NO_2(g) + O_2(g)$$

is $102\ kJ\cdot mol^{-1}$. At 45°C the rate constant is $5.0 \times 10^{-4}\ s^{-1}$. What is the rate constant at 65°C?

13-33. The denaturation of a certain virus is a first-order process with an activation energy of $586\ kJ\cdot mol^{-1}$. The half-life of the reaction at 29.6°C is 4.5 h. Compute the half-life at 37.0°C.

13-34. Cryosurgical procedures involve lowering the body temperature of the patient prior to surgery. Given that the activation energy for the beating of the heart muscle is about 30 kJ, estimate the pulse rate at 72.0°F. Assume the pulse rate at 98.6°F (310 K) to be 75 $beats\cdot min^{-1}$.

CATALYSIS

13-35. Can a catalyst affect equilibrium concentrations in a chemical reaction? Explain.

13-36. Explain why a catalyst for a forward reaction must also be a catalyst for the reverse reaction.

13-37. The aqueous decomposition of hydrogen peroxide in the presence of $Br^-(aq)$ and $H^+(aq)$,

$$2H_2O_2(aq) \rightarrow 2H_2O(l) + O_2(g)$$

has the rate law

$$\text{rate} = k[H_2O_2][H^+][Br^-]$$

(a) Identify the catalysts for the reaction.
(b) What is the overall order of the reaction?
(c) Suppose $[H_2O_2]_0 = 0.10$ M, $[H^+]_0 = 1.00 \times 10^{-3}$ M, and $[Br^-]_0 = 1.00 \times 10^{-3}$ M. Sketch the concentrations of these three species as a function of time given that $k = 1.0 \times 10^3\ M^{-2}\cdot s^{-1}$

13-38. Given the following initial-rate data for the decomposition of $H_2O_2(aq)$,

$$2H_2O_2(aq) \rightarrow 2H_2O(l) + O_2(g)$$

$[H_2O_2]_0/M$	$[I^-]_0/M$	$[H^+]_0/M$	Initial rate of formation of $O_2/M\cdot s^{-1}$
0.20	0.010	0.010	2.0×10^{-3}
0.40	0.010	0.010	4.0×10^{-3}
0.40	0.020	0.010	8.0×10^{-3}
0.20	0.020	0.020	1.6×10^{-2}

(a) determine the rate law.
(b) pick out the catalyst(s), if any.

13-39. It is suspected that a reaction is catalyzed by the wall of the reaction vessel. What experiments would you perform to check out this possibility?

13-40. It is suspected that a reaction rate is influenced by light. What experiments would you perform to check out this possibility?

13-41. Figure 13-8 outlines the platinum-catalyzed mechanism for the reaction

$$2SO_2(g) + O_2(g) \rightarrow 2SO_3(g)$$

It is observed that, except for very low pressures, the rate of the catalyzed reaction is independent of the pressures of $SO_2(g)$ and $O_2(g)$. That is, the rate law is zero order in both reactants:

$$\text{rate} = k$$

Explain how the mechanism outlined in Figure 13-8 leads to this rate law.

13-42. The rate of decomposition of gases on hot metal surfaces often is found to be independent of the concentration of the gas in the gas phase; that is, the rate law is zero order in the reactant gas:

$$\text{rate} = k$$

Such a situation is found for the catalytic decomposition of ammonia on tungsten,

$$2NH_3(g) \xrightarrow{W(s)} N_2(g) + 3H_2(g)$$

and for the catalytic decomposition of nitrous oxide on platinum,

$$2N_2O(g) \xrightarrow{Pt(s)} 2N_2(g) + O_2(g)$$

How do you explain these observations in mechanistic terms?

FORWARD AND REVERSE RATES

13-43. The reaction

$$C_2H_5Br(g) \underset{k_r}{\overset{k_f}{\rightleftharpoons}} C_2H_4(g) + HBr(g)$$

has a rate constant $k_f = 6.7 \times 10^{-7}\ s^{-1}$ and an equilibrium constant $K = 0.14$ M at 600 K. Evaluate k_r.

13-44. For the reaction

$$CO_2(aq) + OH^-(aq) \underset{k_r}{\overset{k_f}{\rightleftharpoons}} HCO_3^-(aq)$$

$K = 8.4 \times 10^7\ M^{-1}$ and $k_f = 8.4 \times 10^3\ M^{-1}{\cdot}s^{-1}$. Evaluate k_r.

13-45. Consider the reaction

$$2HI(g) \rightarrow H_2(g) + I_2(g)$$

Given that the rate law for the forward reaction is

$$(\text{rate})_f = k_f[HI]^2$$

deduce the rate law for the reverse reaction.

13-46. Consider the reaction

$$C_2H_5I(g) \rightleftarrows C_2H_4(g) + HI(g)$$

Given that the rate law for the forward reaction is

$$(\text{rate})_f = k_f[C_2H_5I]$$

deduce the rate law for the reverse reaction.

13-47. The reaction

$$2NO_2(g) + O_3(g) \rightleftarrows N_2O_5(g) + O_2(g)$$

has the forward reaction rate law

$$(\text{rate})_f = k_f[NO_2][O_3]$$

Derive the rate law for the reverse reaction.

13-48. The reaction

$$C_2H_5Br(aq) + OH^-(aq) \rightleftarrows C_2H_5OH(aq) + Br^-(aq)$$

has the forward reaction rate law

$$(\text{rate})_f = k_f[C_2H_5Br][OH^-]$$

Derive the rate law for the reverse reaction.

REACTION MECHANISMS INVOLVING EQUILIBRIUM STEPS

13-49. Write the overall equation for the following reaction mechanism:

$$2N_2O_5(g) \rightleftharpoons 2NO_2(g) + 2NO_3(g) \qquad \text{fast}$$

$$NO_2(g) + NO_3(g) \rightarrow NO(g) + O_2(g) + NO_2(g) \qquad \text{slow}$$

$$NO(g) + NO_3(g) \rightarrow 2NO_2(g) \qquad \text{fast}$$

Write the rate law for the disappearance of $N_2O_5(g)$.

13-50. The proposed mechanism for a reaction is given by the following sequence:

$$H_2O_2(aq) + I^-(aq) \rightarrow HOI(aq) + OH^-(aq) \qquad \text{slow}$$

$$HOI(aq) + I^-(aq) \rightleftharpoons I_2(aq) + OH^-(aq) \qquad \text{fast}$$

$$OH^-(aq) + H^+(aq) \rightarrow H_2O(l) \qquad \text{fast}$$

Determine the overall reaction and write the rate law for the disappearance of H_2O_2.

13-51. A proposed mechanism for the rate of formation of phosgene, Cl_2CO, is

$$Cl_2(g) \rightleftharpoons 2Cl(g) \qquad \text{fast}$$

$$Cl(g) + CO(g) \rightleftharpoons ClCO(g) \qquad \text{fast}$$

$$ClCO(g) + Cl_2(g) \rightarrow Cl_2CO(g) + Cl(g) \qquad \text{slow}$$

Show that a rate law that is consistent with this mechanism is

$$\text{rate} = k[Cl_2]^{3/2}[CO]$$

13-52. The rate law for the decomposition of phosgene

$$COCl_2(g) \rightarrow CO(g) + Cl_2(g)$$

is known to be

$$\text{rate} = k[COCl_2][Cl_2]^{1/2}$$

Show that the following mechanism is consistent with this rate law:

$$Cl_2(g) \rightleftharpoons 2Cl(g) \qquad \text{fast}$$

$$COCl_2(g) + Cl(g) \rightarrow COCl(g) + Cl_2(g) \qquad \text{slow}$$

$$COCl(g) \rightleftharpoons CO(g) + Cl(g) \qquad \text{fast}$$

13-53. The available kinetic data for the reaction

$$2NO(g) + O_2(g) \rightarrow 2NO_2(g)$$

are consistent with the following reaction mechanism:

$$NO(g) + O_2(g) \rightleftharpoons NO_3(g) \qquad \text{fast}$$
$$NO_3(g) + NO(g) \rightarrow 2NO_2(g) \qquad \text{slow}$$

Obtain the rate law for the reaction from the above mechanism. Express your rate law in terms of the reactant concentrations in the reaction.

13-54. The rate law for the reaction

$$Hg_2^{2+}(aq) + Tl^{3+}(aq) \rightarrow 2Hg^{2+}(aq) + Tl^{+}(aq)$$

is

$$\text{rate} = k\frac{[Hg_2^{2+}][Tl^{3+}]}{[Hg^{2+}]}$$

Show that the following mechanism is consistent with the observed rate law:

$$Hg_2^{2+}(aq) \rightleftharpoons Hg(aq) + Hg^{2+}(aq) \qquad \text{fast}$$
$$Hg(aq) + Tl^{3+}(aq) \rightarrow Tl^{+}(aq) + Hg^{2+}(aq) \qquad \text{slow}$$

ADDITIONAL PROBLEMS

13-55. The rate of the reaction

$$2CO(g) \rightarrow CO_2(g) + C(s)$$

was studied by injecting some CO(*g*) into a reaction vessel and measuring the total pressure while maintaining a constant reaction volume:

P_{total}/torr	*t*/s
250	0
238	398
224	1002
210	1801

Determine the reaction rate constant.

13-56. Given the following data for the reaction

$$ClO_3^-(aq) + 9I^-(aq) + 6H^+(aq) \rightarrow 3I_3^-(aq) + Cl^-(aq) + 3H_2O(l)$$

determine the reaction rate law:

$[I^-]_0$/M	$[ClO_3^-]_0$/M	$[H^+]_0$/M	$(\text{rate})_0/M\cdot s^{-1}$
0.10	0.10	0.10	*x*
0.10	0.20	0.10	2*x*
0.20	0.20	0.10	4*x*
0.20	0.20	0.20	16*x*

13-57. A reaction of importance in the formation of smog is that between ozone and nitrogen monoxide:

$$O_3(g) + NO(g) \rightarrow O_2(g) + NO_2(g)$$

The rate law for this reaction is

$$\text{rate} = k[O_3][NO]$$

Given that $k = 2.99 \times 10^6\ M^{-1}\cdot s^{-1}$ at 310 K, compute the initial reaction rate when $[O_3]$ and [NO] remain essentially constant at the values $[O_3]_0 = 2.0 \times 10^{-6}$ M and $[NO]_0 = 6.0 \times 10^{-5}$ M, owing to continuous production from separate sources. Compute the number of moles of $NO_2(g)$ produced per hour per liter of air.

13-58. Calculate the time required for the concentration to decrease to $^1/_{10}$ of its initial value for a first-order reaction with a rate constant $k = 10\ s^{-1}$.

13-59. Suppose that you place 100 bacteria into a flask containing nutrients for the bacteria and that you find the following data at 37°C:

t/min	Number of bacteria
0	100
15	200
30	400
45	800
60	1600

What is the order of the rate of production of the bacteria? How many bacteria do you predict there will be after 2 h? What is the rate constant for the process?

13-60. Show that for a first-order reaction the time required for 99.9 percent of the reaction to take place is about ten times that required for 50 percent of the reaction to take place.

13-61. Assuming that the loss of ability to recall learned material is a first-order process with a half-life of 70 days, compute the number of days required to forget 90 percent of the material that you have learned in preparation for an exam. (Assume constant temperature and no further reference to the learned material during the decay period.)

13-62. Identify in each of the following cases the order of the reaction rate law with respect to the reactant A, where A → products:

(a) The half-life of A is independent of the initial concentration of A.
(b) The rate of decrease of A is a constant.
(c) A twofold increase in the initial concentration of A leads to a 1.41-fold increase in the initial rate.
(d) A twofold increase in the initial concentration of A leads to a fourfold increase in the initial rate.
(e) The time required for $[A]_0$ to decrease to $[A]_0/2$ is equal to the time required for [A] to decrease from $[A]_0/2$ to $[A]_0/4$.

13-63. Many radical-radical recombination reactions, such as

$$2CH_3(g) \rightarrow C_2H_6(g)$$

 proceed at the diffusion-controlled limit; that is, essentially every collision leads to a reactive event. Why do you think this is so? (Explain in terms of chemical bonds.)

13-64. Given the following rate-constant data on the gas-phase decomposition of NO_2 [$2NO_2(g) \rightarrow 2NO(g) + O_2(g)$], plot log k versus $1/T$ and calculate E_a for the reaction.

T/K	k/M^{-1}·s^{-1}
600	0.70
625	1.83
650	4.46
700	21.8

Use your plot to estimate k at 500 K.

13-65. Consider the following reaction:

$$SO_2Cl_2(g) \rightarrow SO_2(g) + Cl_2(g)$$

A study of the rate of the reaction in the gas phase gives the following data:

t/s	$P_{SO_2Cl_2}$/torr
0	760
5000	680
10,000	610

Determine the value of the rate constant for the reaction.

13-66. Given the following data for the reaction

$$BrO_3^-(aq) + 9I^-(aq) + 6H^+(aq) \rightarrow 3I_3^-(aq) + Br^-(aq) + 3H_2O(l)$$

determine the reaction rate law:

$[I^-]_0$/M	$[BrO_3^-]_0$/M	$[H^+]_0$/M	(rate)$_0$/M·s^{-1}
0.10	0.10	0.10	x
0.20	0.20	0.10	$4x$
0.10	0.20	0.10	$2x$
0.20	0.20	0.20	$4x$

13-67. The rate law for the reaction

$$C_2H_4Br_2(aq) + 3I^-(aq) \rightarrow C_2H_4(g) + 2Br^-(aq) + I_3^-(aq)$$

at 300 K is

$$\text{rate} = (5.0 \times 10^{-3}\ M^{-1}\cdot s^{-1})[C_2H_4Br_2][I^-]$$

Fill in the missing entries in the following table:

Run	$[C_2H_4Br_2]_0$/M	$[I^-]_0$/M	Initial rate of formation of C_2H_4/M·s^{-1}
1	0.20	0.20	
2	0.20		4.0×10^{-4}
3		0.20	8.0×10^{-4}

13-68. Calculate the time required for the concentration to decrease by 10 percent of its initial value for a first-order reaction with $k = 10\ s^{-1}$.

13-69. The U.S. Public Health Service requires that milk fresh from a pasteurizer may contain no more than 20,000 bacteria per milliliter. It has been reported that bacteria in milk stored at 40°F may double in 39 h. If a milk sample had 20,000 bacteria per milliliter after pasteurization, what is the bacteria count per milliliter after 10 days?

13-70. Show that for a first-order reaction, the time required for 99.99 percent of the reaction to take place is twice as long as the time required for 99.0 percent of the reaction to take place.

13-71. A rough estimate for the number of nerve cells in the average human brain is 20 billion (2×10^{10}). This number reaches a maximum around 30 years of age and then decreases at a constant rate of 2×10^5 cells per day.

(a) Determine the rate law for the disappearance of nerve cells after 30 years of age.
(b) Estimate the age in years at which the number of cells has dropped to 80 percent of the maximum value.

13-72. Consider the following mechanism for the decomposition of $N_2O_5(g)$ to $NO_2(g)$ and $O_2(g)$:

$$N_2O_5(g) + N_2O_5(g) \rightleftharpoons N_2O_5^*(g) + N_2O_5(g) \qquad \text{fast}$$
$$N_2O_5^*(g) \rightarrow NO_2(g) + NO_3(g) \qquad \text{slow}$$
$$NO_3(g) + N_2O_5(g) \rightarrow 3NO_2(g) + O_2(g) \qquad \text{fast}$$

where $N_2O_5^*$ denotes a vibrationally excited molecule. Show that this mechanism gives rise to a rate law that is first order in N_2O_5 even though the mechanism involves collision of two N_2O_5 molecules.

13-73. The decomposition of dinitrogen pentoxide is a first-order process. The temperature dependence of the observed rate constant is given by

t/°C	k/10^{-5} s^{-1}
0	0.0787
25	3.46
45	49.8
65	487

Plot log k versus $1/T$ and calculate E_a for the reaction. Use your plot to estimate k at 50°C.

13-74. The kinetics of the decomposition of phosphine at 950 K was followed by measuring the total pressure in the system as a function of time:

$$4PH_3(g) \rightarrow P_4(g) + 6H_2(g)$$

The following data were obtained in a run where the reaction chamber contained only pure phosphine at the start of the reaction:

t/min	0	40	80	120
P_{total}/torr	100	150	167	172

Determine the reaction rate law and calculate the rate constant.

13-75. The first-order rate constant (in min^{-1}) for the decomposition of $CH_3CH_2Br(g)$ is given by:

$$\log k = 14.58 - \frac{99.96 \times 10^3 \ J \cdot mol^{-1}}{RT}$$

Calculate the fraction of a sample of $CH_3CH_2Br(g)$ that will remain if it is heated for 30.0 min at 700 K.

13-76. The following kinetic data were obtained on the reaction

$$2HI(g) \underset{k_r}{\overset{k_f}{\rightleftharpoons}} H_2(g) + I_2(g)$$

T/K	$k_f/10^{-4}\ M^{-1}\cdot s^{-1}$	$k_r/10^{-2}\ M^{-1}\cdot s^{-1}$
647	0.858	0.522
700	11.7	6.42
781	395.	133.5

Calculate the activation energies of the forward and reverse reactions. Evaluate the equilibrium constant at 700 K for the reaction.

13-77. The rate law for a second-order reaction has the form

$$\text{rate} = k[A]^2$$

In such a case, the dependence of the concentration of the reactant A is given by

$$\frac{1}{[A]} = \frac{1}{[A]_0} + kt$$

where $[A]_0$ is the initial concentration. Show that the half-life for a second-order process is given by

$$t_{1/2} = \frac{1}{k[A]_0}$$

13-78. The rate law for the reaction

$$2N_2O(g) \rightarrow 2N_2(g) + O_2(g)$$

is second order in $[N_2O]$. The reaction was carried out at 900 K with an initial concentration of N_2O of 2.0×10^{-2} M. It took 4500 s for N_2O to fall to half its initial concentration. What is the rate constant for this reaction? (See Problem 13-77).

13-79. The rate constant for the reaction

$$H^+(aq) + OH^-(aq) \rightarrow H_2O(l)$$

has been determined to be $1.3 \times 10^{11}\ M^{-1}\cdot s^{-1}$. Calculate the half-life of the neutralization of HCl by NaOH when both are originally at a concentration of 1.0×10^{-3} M. When $[H^+]_0 = [OH^-]_0$, the rate law can be written as

$$\text{rate} = k[H^+]^2 = k[OH^-]^2$$

(See Problem 13-77.)

CHEMICAL EQUILIBRIUM

Effect of temperature on the reaction equilibrium

$$\underset{\text{colorless}}{N_2O_4(g)} \rightleftharpoons \underset{\text{brown}}{2NO_2(g)}$$

An increase in temperature from 0°C (ice water) to 25°C converts some of the N_2O_4 to NO_2 and results in a darker color for the reaction mixture.

Thus far we have used the concept of dynamic equilibrium to describe the equilibrium vapor pressure of a liquid (Chapter 11), to describe the dissolution of a solid in a liquid (Chapter 12), and to describe the state achieved when the forward and reverse rates of a chemical reaction process are equal (Chapter 13). In this chapter we show how the concept of dynamic equilibrium can be extended to cover all types of chemical reactions. We learn how to describe a chemical system at equilibrium quantitatively. We shall see that a chemical reaction equilibrium is characterized by a quantity called an equilibrium constant. Two key questions arise in the analysis of reaction equilibria: (1) In which direction (left to right or right to left) does a reaction at equilibrium shift in response to a change in conditions that disturbs the equilibrium? (2) If we prepare a nonequilibrium mixture of reactants and products, then in what direction does the reaction proceed toward equilibrium and what are the equilibrium values of the concentrations? We learn in this chapter how to answer these questions and thereby how to predict the reaction conditions necessary to maximize the amount of a desired product of a chemical reaction.

14.1 A CHEMICAL EQUILIBRIUM IS DYNAMIC

We learned in Chapter 11 that a dynamic equilibrium between a liquid and its vapor is attained when the rate of evaporation from the liquid phase equals the rate of condensation from the vapor phase. If the liquid is water, then we can express the equilibrium process by the equation

$$H_2O(l) \rightleftharpoons H_2O(g)$$

We call the rate of evaporation from the liquid phase the **forward rate,** and the rate of condensation from the vapor phase the **reverse rate.** The double arrows denote a reaction equilibrium. In general, a state of **chemical equilibrium** is attained when the rates of the forward and reverse processes are equal. Furthermore, a true equilibrium can be attained from either direction. For example, the equilibrium vapor pressure of water at a particular temperature is exactly the same whether we start with pure liquid water that evaporates until the vapor pressure is constant or with pure water vapor that is initially at a pressure in excess of the equilibrium value. In the latter case, water vapor condenses until the equilibrium vapor pressure is attained and then the *net* condensation of water vapor ceases.

Chemical reactions also attain equilibrium states. For example, consider the chemical reaction described by the equation

▪ At equilibrium the forward reaction rate equals the reverse reaction rate.

$$\underset{\text{colorless}}{N_2O_4(g)} \rightleftharpoons \underset{\text{brown}}{2NO_2(g)} \qquad (14\text{-}1)$$

in which the colorless gas dinitrogen tetroxide, $N_2O_4(g)$, dissociates into the reddish-brown gas nitrogen dioxide, $NO_2(g)$. The reaction described by Equation (14-1), like all chemical reactions, is really two opposing ones. The forward reaction is the dissociation of N_2O_4 molecules into NO_2 molecules, and the reverse reaction is the association of NO_2 molecules into N_2O_4 molecules.

Suppose that we start with only $N_2O_4(g)$. Initially the reaction mixture is colorless. As N_2O_4 molecules dissociate into NO_2 molecules, the reaction mixture becomes reddish brown. As the concentration of NO_2 increases, more and more NO_2 molecules associate back into N_2O_4 molecules. Thus the reverse rate of the process described by Equation (14-1) increases with time. Eventually, the forward rate and the reverse rate become equal and a state of equilibrium exists (Figure 14-1). The concentrations of N_2O_4 and NO_2 no longer change with time. The equilibrium is a **dynamic equilibrium** because N_2O_4 molecules are still dissociating into NO_2 molecules and NO_2 molecules are still associating into N_2O_4 molecules. The rates of these two processes are exactly the

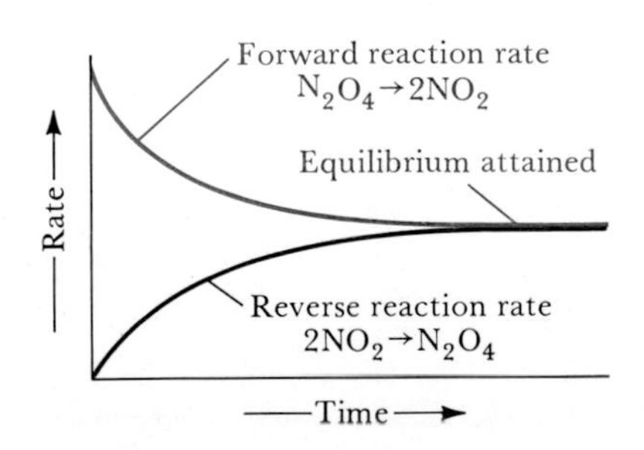

Figure 14-1 At equilibrium the forward reaction rate equals the reverse reaction rate and the concentrations of reactants and products do not change with time.

same, however, and so there is no net change in the concentrations of N_2O_4 and NO_2.

14-2. A CHEMICAL EQUILIBRIUM IS APPROACHABLE FROM EITHER DIRECTION

We can study the approach to equilibrium of the reaction described by Equation (14-1) quantitatively. If we determine the value of the NO_2 concentration, denoted as $[NO_2]$, in the reaction mixture and if we know the concentration of N_2O_4 that we started with, then it is a simple matter to determine the value of the N_2O_4 concentration, denoted as $[N_2O_4]$, in the reaction mixture. For example, suppose that we start with 1.00 $mol{\cdot}L^{-1}$ of N_2O_4 and no NO_2. We can denote these initial concentrations by

$$[N_2O_4]_0 = 1.00\ \text{M} \qquad \text{and} \qquad [NO_2]_0 = 0$$

where the subscript 0 indicates an initial concentration. Suppose also that at equilibrium we find that

$$[NO_2] = 0.40\ \text{M}$$

Then from the reaction stoichiometry

$$\left(\begin{array}{c}\text{moles per liter}\\ \text{of } N_2O_4 \text{ consumed}\end{array}\right) = \left(\begin{array}{c}\text{moles per liter}\\ \text{of } NO_2 \text{ produced}\end{array}\right)\left(\frac{1\ \text{mol}\ N_2O_4}{2\ \text{mol}\ NO_2}\right)$$

$$= \left(\frac{0.40\ \text{mol}\ NO_2}{\text{L}}\right)\left(\frac{1\ \text{mol}\ N_2O_4}{2\ \text{mol}\ NO_2}\right)$$

$$= 0.20\ \text{mol}{\cdot}\text{L}^{-1}$$

Therefore, the value of $[N_2O_4]$ when $[NO_2] = 0.40$ M is

$$[N_2O_4] = \underset{\substack{\text{initial}\\ \text{value}}}{1.00\ \text{M}} - \underset{\substack{\text{amount}\\ \text{consumed}}}{0.20\ \text{M}} = 0.80\ \text{M}$$

The following Example illustrates the calculation of **equilibrium concentrations** in reaction mixtures.

Example 14-1: Suppose that the methanol synthesis reaction

$$CO(g) + 2H_2(g) \rightleftharpoons CH_3OH(g)$$

is carried out with the following initial concentrations:

$$[CO]_0 = 2.00\ \text{M} \qquad [H_2]_0 = 0.50\ \text{M} \qquad [CH_3OH]_0 = 0$$

and that at equilibrium we find that $[CH_3OH] = 0.20$ M. Compute the equilibrium values of $[CO]$ and $[H_2]$.

Solution: From the reaction stoichiometry we have

$$\left(\begin{array}{c}\text{moles per liter}\\ \text{of CO consumed}\end{array}\right) = \left(\begin{array}{c}\text{moles per liter}\\ \text{of } CH_3OH \text{ produced}\end{array}\right)\left(\frac{1\ \text{mol CO}}{1\ \text{mol}\ CH_3OH}\right)$$

$$= \left(\frac{0.20\ \text{mol}\ CH_3OH}{\text{L}}\right)\left(\frac{1\ \text{mol CO}}{1\ \text{mol}\ CH_3OH}\right) = 0.20\ \text{M}$$

Therefore, the value of [CO] when $[CH_3OH] = 0.20$ M is

$$[CO] = \underset{\text{initial value}}{2.00\ M} - \underset{\text{amount consumed}}{0.20\ M} = 1.80\ M$$

Similarly, we compute for $[H_2]$ at equilibrium:

$$[H_2] = 0.50\ M - \left(\frac{0.20\ \text{mol}\ CH_3OH}{L}\right)\left(\frac{2\ \text{mol}\ H_2}{1\ \text{mol}\ CH_3OH}\right)$$

$$= \underset{\text{initial value}}{0.50\ M} - \underset{\text{amount consumed}}{0.40\ M} = 0.10\ M$$

Note that twice as much H_2 as CO is consumed in the reaction, as is required by the stoichiometry of the reaction.

Let's continue our discussion of the N_2O_4–NO_2 reaction. If we introduce 1.00 mol of N_2O_4 into a 1.00-L reaction vessel held at 100°C and then determine the concentration of $[NO_2]$, and thus that of $[N_2O_4]$, as a function of time, then we obtain the results shown in Figure 14-2. Note that the value of $[N_2O_4]$, which starts out at 1.00 M (blue curve) decreases to a constant value of 0.80 M. The value of $[NO_2]$, which starts out at zero (blue curve) increases to a constant value of 0.40 M. Now let's start with all $NO_2(g)$ and no $N_2O_4(g)$. The conditions that

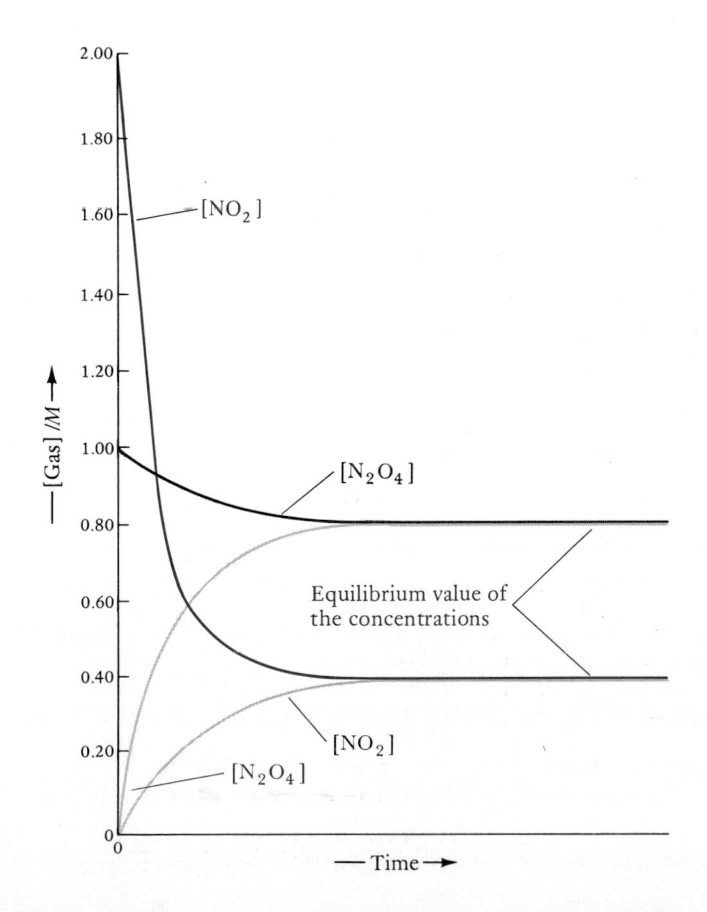

Figure 14-2 Plot of $[N_2O_4]$ and $[NO_2]$ as a function of time at 100°C for the reaction $N_2O_4(g) \rightleftharpoons 2NO_2(g)$. Starting from the reactant side (blue curves), we have at time zero $[N_2O_4]_0 = 1.00$ M and $[NO_2]_0 = 0$. As the reaction proceeds, $[N_2O_4]$ decreases from 1.00 M to 0.80 M and then remains constant, whereas $[NO_2]$ increases from zero to 0.40 M and then remains constant. Starting from the product side (red curves), we have at time zero $[NO_2]_0 = 2.00$ M and $[N_2O_4]_0 = 0$. As the reaction proceeds, $[N_2O_4]$ increases from zero to 0.80 M and then remains constant, whereas $[NO_2]$ decreases from 2.00 M to 0.40 M and then remains constant. The values $[N_2O_4] = 0.80$ M and $[NO_2] =$ 0.40 M are equilibrium concentrations. When the reaction mixture attains equilibrium $[N_2O_4]$ and $[NO_2]$ remain constant.

Table 14-1. Initial and equilibrium values at 100°C of $[N_2O_4]$ and $[NO_2]$ for the reaction $N_2O_4(g) \rightleftharpoons 2NO_2(g)$

Initial concentration		Equilibrium concentration		Value at equilibrium of the quantity
$[N_2O_4]_0/M$	$[NO_2]_0/M$	$[N_2O_4]/M$	$[NO_2]/M$	$\{[NO_2]^2/[N_2O_4]\}/M$
1.00	0	0.80	0.40	0.20
2.00	0	1.71	0.58	0.20
0	2.00	0.80	0.40	0.20
0	1.00	0.36	0.27	0.20

correspond stoichiometrically to $[N_2O_4]_0 = 1.00$ M and $[NO_2]_0 = 0$ are $[N_2O_4]_0 = 0$ and $[NO_2]_0 = 2.00$ M. In this case we find that the concentrations $[NO_2]$ and $[N_2O_4]$ change with time in the manner shown (red curves) in Figure 14-2. Note that the equilibrium values of $[N_2O_4]$ and $[NO_2]$ are the same in both cases.

▪ At equilibrium the reactant and product concentrations are constant.

The equilibrium values of $[N_2O_4]$ and $[NO_2]$ for several sets of initial conditions are given in Table 14-1. The most remarkable feature of the data in Table 14-1 is shown in the last column. The value of the ratio $[NO_2]^2/[N_2O_4]$ at equilibrium for the reaction

$$N_2O_4(g) \rightleftharpoons 2NO_2(g)$$

is equal to a constant. From the data in Table 14-1, we see that

$$\frac{[NO_2]^2}{[N_2O_4]} = 0.20 \text{ M (at equilibrium at 100°C)}$$

Thus, for example, if we know that $[NO_2] = 0.175$ M, at equilibrium, then $[N_2O_4]$ would have to be 0.150 M so that $[NO_2]^2/[N_2O_4]$ equals 0.20 M. Table 14-1 shows that the constant value of $[NO_2]^2/[N_2O_4]$ at equilibrium *is independent of the initial values,* $[NO_2]_0$ *and* $[N_2O_4]_0$. The reason behind this result is given in the next section.

The data in Figure 14-2 illustrate two essential facets of chemical equilibrium:

1. At equilibrium, the reactant and product concentrations show no further change with time.
2. The same equilibrium state is attained starting either from the reactant side or the product side of the equation.

14-3. THE EQUILIBRIUM-CONSTANT EXPRESSION FOR A CHEMICAL EQUATION IS EQUAL TO THE RATIO OF PRODUCT CONCENTRATION TERMS TO REACTANT CONCENTRATION TERMS

The general solution to the problem of formulating equilibrium constant expressions for chemical reactions was put forth by the Norwegian chemists Cato Guldberg and Peter Waage in 1864 and is referred to in this book as the **law of concentration action.**

Consider the balanced general chemical equation

$$aA(g) + bB(soln) + cC(s) \rightleftharpoons xX(g) + yY(soln) + zZ(l) \qquad (14\text{-}2)$$

for a reaction at equilibrium. Guldberg and Waage, on the basis of their experimental observations on a variety of chemical reactions, postulated that for Equation (14-2) the ratio on the right hand side of the expression

$$K_c = \frac{[X]^x[Y]^y}{[A]^a[B]^b} \qquad (14\text{-}3)$$

is a constant called the **equilibrium constant** (at a given temperature). In other words, the **equilibrium-constant expression** for a reaction is the ratio of product concentrations to reactant concentrations, with each concentration factor raised to a power equal to the stoichiometric coefficient of that species in the balanced equation. Note that pure liquids and solids, whose concentration cannot be varied, do not appear in the equilibrium constant expression. Because their concentrations cannot be varied, pure solids and liquids exert a constant effect on the reaction equilibrium and so are not included in Equation (14-3).

▪ The law of concentration action tells us how to write the equilibrium constant expression for a chemical equation.

The subscript c in Equation (14-3) emphasizes that the equilibrium constant is expressed in terms of concentrations. We shall soon see that we can also express equilibrium constants involving gaseous species in terms of pressures, in which case we write K_p instead of K_c. The crucial point here is that the value of K_c in Equation (14-3) is equal to a constant at a given temperature.

Application of the law of concentration action to the equilibrium equation

$$N_2O_4(g) \rightleftharpoons 2NO_2(g)$$

yields

$$K_c = \frac{[NO_2]^2}{[N_2O_4]} \qquad (14\text{-}4)$$

▪ In most books the Guldberg and Waage law of concentration action is referred to as the law of mass action or the mass action law. If we interpret Guldberg and Waage's "active masses" as concentrations, which was apparently their intent, then the designation law of concentration action is a more informative description.

From the results in the preceding section, we have for the $N_2O_4(g)$ dissociation reaction that $K_c = 0.20$ M at 100°C.

Some additional examples of the application of the law of concentration action (Equation 14-3) follow:

$$C(s) + H_2O(g) \rightleftharpoons CO(g) + H_2(g)$$

$$K_c = \frac{[CO][H_2]}{[H_2O]}$$

Note that C(s) does not appear in the K_c expression because it is a pure solid.

$$HNO_2(aq) \rightleftharpoons H^+(aq) + NO_2^-(aq)$$

$$K_c = \frac{[H^+][NO_2^-]}{[HNO_2]}$$

Example 14-2: Use the law of concentration action to write the equilibrium-constant expressions for the following chemical equations:

(a) $N_2(g) + 3H_2(g) \rightleftharpoons 2NH_3(g)$

(b) $C(s) + CO_2(g) \rightleftharpoons 2CO(g)$

(c) $PCl_3(l) + Cl_2(g) \rightleftharpoons PCl_5(s)$

Solution: In the reaction described by Equation (a), all the reactants and products are gases, and so we have

$$K_c = \frac{[NH_3]^2}{[N_2][H_2]^3}$$

whereas for Equation (b) we have

$$K_c = \frac{[CO]^2}{[CO_2]}$$

Note that C(*s*) does not appear in the K_c expression for Equation (b) because carbon is a pure solid. For Equation (c) we have

$$K_c = \frac{1}{[Cl_2]}$$

Note that, in effect, we enter unity in the K_c expression for a pure solid or pure liquid.

The following Example illustrates the calculation of K_c from equilibrium values of the concentrations.

▪ The reaction equilibrium in Example 14-3 can be studied by observing the intensity of the purple color, which is due to $I_2(g)$; $H_2(g)$ and HI(*g*) are colorless.

Example 14-3: Suppose that a sample of HI(*g*) is injected into a closed reaction vessel and the following reaction equilibrium is established:

$$2HI(g) \rightleftharpoons H_2(g) + I_2(g)$$

Analysis of the equilibrium reaction mixture yields the following results:

$$[HI] = 0.27\ M \qquad [H_2] = [I_2] = 0.86\ M$$

Use these results to calculate K_c for the reaction.

Solution: Application of the law of concentration action to the equation yields

$$K_c = \frac{[H_2][I_2]}{[HI]^2}$$

Substituting the values of the equilibrium concentrations into the K_c expression yields

$$K_c = \frac{(0.86\ M)(0.86\ M)}{(0.27\ M)^2} = 10$$

▪ From the ideal-gas law

$$P = \frac{nRT}{V}$$

the concentration of the gas is

$$[gas] = \frac{n}{V}$$

Thus

$$P = [\text{gas}]RT$$

The pressure of a gas is directly proportional to the concentration of the gas, and therefore we can express the equilibrium constant for a reaction involving gases in terms of gas partial pressures rather than gas concentrations. An equilibrium-constant expression written in terms of gas partial pressures is denoted by K_p. Thus the K_p expression for the equation

$$C(s) + CO_2(g) \rightleftharpoons 2CO(g)$$

is

$$K_p = \frac{P_{CO}^2}{P_{CO_2}}$$

In general, for a given equation, $K_p \neq K_c$ because the concentration of a gas is not equal to the pressure of a gas; rather, [gas] = P_{gas}/RT.

▪ The general relation between K_p and K_c is $K_p = K_c(RT)^{\Delta n}$, where Δn is the number of moles of gaseous products minus the number of moles of gaseous reactants (Problem 14-61).

14-4. EQUILIBRIUM CONSTANTS ARE USED IN A VARIETY OF CALCULATIONS

Equilibrium calculations are best illustrated by examples. The following three Examples illustrate several types of equilibrium calculations.

Example 14-4: Suppose that 2.00 mol of $N_2O_4(g)$ is injected into a 1.00-L reaction vessel held at 100°C and that equilibrium is attained:

$$N_2O_4(g) \rightleftharpoons 2NO_2(g) \qquad K_c = 0.20\ M$$

Compute the equilibrium values of $[N_2O_4]$ and $[NO_2]$.

Solution: From the law of concentration action we have

$$K_c = \frac{[NO_2]^2}{[N_2O_4]} = 0.20\ M$$

It is helpful in working equilibrium calculations to set the problem up in tabular form with the initial concentrations and equilibrium concentrations of each species placed in separate rows directly below the species in the chemical equation. Thus from the data given we have

	$N_2O_4(g)$ $\rightleftharpoons$	$2NO_2(g)$
initial concentration	2.00 M	0
equilibrium concentration		

We now use the reaction stoichiometry to obtain expressions for the equilibrium concentrations of N_2O_4 and NO_2. Let the number of moles per liter of $N_2O_4(g)$ that dissociates into $NO_2(g)$ be x; then the equilibrium concentration of N_2O_4 is

$$[N_2O_4] = 2.00\ M - x$$

Each mole of N_2O_4 that dissociates produces 2 mol of NO_2, and so the equilibrium concentration of $NO_2(g)$ expressed in terms of x is

$$[NO_2] = 2x$$

Thus our completed table for the calculation is

	$N_2O_4(g)$ $\rightleftharpoons$	$2NO_2(g)$
initial concentration	2.00 M	0
equilibrium concentration	2.00 M − x	$2x$

Substituting the equilibrium concentration data from this table into the K_c expression yields

$$\frac{(2x)^2}{(2.00\ \text{M} - x)} = 0.20\ \text{M}$$

or

$$4x^2 = (0.20\ \text{M})(2.00\ \text{M} - x)$$
$$= 0.40\ \text{M}^2 - (0.20\ \text{M})x$$

We can rearrange the terms in this equation to get the standard form of the quadratic equation:

$$4x^2 + (0.20\ \text{M})x - 0.40\ \text{M}^2 = 0$$

▪ The quadratic formula is discussed in detail in Appendix A and in the *Study Guide/Solutions Manual.*

The two roots of this equation are given by the **quadratic formula*** and thus

$$x = \frac{-0.20\ \text{M} \pm \sqrt{(0.20\ \text{M})^2 - (4)(4)(-0.40\text{M}^2)}}{(2)(4)} = 0.29\ \text{M}$$

We have rejected the negative value of x (-0.34 M) as physically unacceptable because only positive values of concentrations have physical meaning. From the value $x = 0.29$ M, we compute that at equilibrium

$$[NO_2] = 2x = (2)(0.29\ \text{M}) = 0.58\ \text{M}$$

$$[N_2O_4] = 2.00\ \text{M} - x = 2.00\ \text{M} - 0.29\ \text{M} = 1.71\ \text{M}$$

As a final check, we note that $[NO_2]^2/[N_2O_4]$ is equal to 0.20 M, which is the value of K_c.

The quantity K_c defined in Equation (14-3) is an equilibrium constant in terms of concentrations and generally has units involving concentration. For example, Example 14-4 shows that K_c has units of M^2/M, or M, for the reaction $N_2O_4(g) \rightleftharpoons 2NO_2(g)$. More advanced treatments of chemical equilibria use equilibrium constants that are defined in such a way that they are unitless. In recognition of this, some general chemistry textbooks do not include units for K_c, but this is inconsistent with the definition in Equation 14-3. In this text we always include the units of K_c. We do this not only because it is correct, but also because the units in calculations like those in Example 14-4 follow logically. A similar argument applies to K_p, which generally has units involving pressure.

Example 14-5: At a certain temperature, the equilibrium constant for the reaction

$$2ICl(g) \rightleftharpoons I_2(g) + Cl_2(g)$$

is $K_c = 0.11$. Calculate the equilibrium concentrations of ICl, I_2, and Cl_2 when 0.33 mol of I_2 and 0.33 mol of Cl_2 are added to a 1.5-L reaction vessel.

*The **quadratic equation** $ax^2 + bx + c = 0$ has the solutions

$$x = \frac{-b \pm \sqrt{b^2 - 4ac}}{2a}$$

Only positive values of pressure or concentration have physical significance.

Solution: From the law of concentration action we have

$$K_c = \frac{[Cl_2][I_2]}{[ICl]^2} = 0.11$$

We set up a table of initial concentrations and equilibrium concentrations. Let x be the number of moles per liter of I_2 or of Cl_2 that react. From the reaction stoichiometry the number of moles per liter of ICl produced is $2x$. Each mole of I_2 or Cl_2 that reacts produces two moles of ICI.

	$2ICl(g) \rightleftharpoons$	$I_2(g)$	+	$Cl_2(g)$
initial concentration	0	$\frac{0.33\text{ mol}}{1.5\text{ L}} = 0.22\text{ M}$		$\frac{0.33\text{ mol}}{1.5\text{ L}} = 0.22\text{ M}$
equilibrium concentration	$2x$	$0.22\text{ M} - x$		$0.22\text{ M} - x$

Substituting the equilibrium concentration expressions in the K_c expression yields

$$\frac{(0.22\text{ M} - x)^2}{(2x)^2} = 0.11$$

Because the left side of this equation is a perfect square, we can take the square root of both sides, rather than use the quadratic equation. The result is

$$\frac{0.22\text{ M} - x}{2x} = 0.33$$

or

$$0.22\text{ M} - x = (0.33)(2x) = 0.66x$$

Solving for x yields

$$x = \frac{0.22\text{ M}}{1.66} = 0.13\text{ M}$$

Thus, at equilibrium, $[Cl_2] = [I_2] = 0.22\text{ M} - 0.13\text{ M} = 0.09\text{ M}$ and $[ICl] = 0.26\text{ M}$.

Example 14-6: Given the equation for the reaction of $CO_2(g)$ with $C(s)$

$$C(s) + CO_2(g) \rightleftharpoons 2CO(g) \qquad K_p = 1.90\text{ atm}$$

and that at equilibrium the total pressure in the reaction vessel is 2.00 atm, compute P_{CO_2} and P_{CO} at equilibrium. Note that this problem involves an equilibrium constant expressed in pressure units rather than concentration units.

Solution: We are given only the total pressure, which is equal to the sum of the partial pressures of $CO(g)$ and $CO_2(g)$:

$$P_{total} = P_{CO} + P_{CO_2} = 2.00\text{ atm}$$

The reaction equilibrium fixes the ratio P_{CO}^2/P_{CO_2} at 1.90 atm, and so for the K_p expression we have

$$K_p = \frac{P_{CO}^2}{P_{CO_2}} = \frac{P_{CO}^2}{2.00\text{ atm} - P_{CO}} = 1.90\text{ atm}$$

We can rearrange the terms in this equation to get the standard form of a quadratic equation:

$$P_{CO}^2 + (1.90\ \text{atm})P_{CO} - 3.80\ \text{atm}^2 = 0$$

The positive root that we find is $P_{CO} = 1.22$ atm. The pressure of $CO_2(g)$ at equilibrium is thus

$$P_{CO_2} = 2.00\ \text{atm} - P_{CO} = 2.00\ \text{atm} - 1.22\ \text{atm} = 0.78\ \text{atm}$$

14-5. EQUILIBRIUM CONSTANTS FOR CHEMICAL EQUATIONS CAN BE COMBINED TO OBTAIN EQUILIBRIUM CONSTANTS FOR OTHER EQUATIONS

It is sometimes necessary to compute the equilibrium constant for a chemical equation from the equilibrium constants for other, algebraically related chemical equations. As an example, consider the pair of equations

(1) $$CO(g) + 2H_2(g) \rightleftharpoons CH_3OH(g) \qquad K_f = \frac{[CH_3OH]}{[CO][H_2]^2}$$

(2) $$CH_3OH(g) \rightleftharpoons CO(g) + 2H_2(g) \qquad K_r = \frac{[CO][H_2]^2}{[CH_3OH]}$$

The reaction described by Equation (2) is simply the reverse of the reaction described by Equation (1). Comparison of the equilibrium-constant expressions for the equations for the *forward* reaction, K_f, and the *reverse* reaction, K_r, leads to the conclusion that the equilibrium constant of the equation for the reverse reaction is equal to the reciprocal of the equilibrium constant of the equation for the forward reaction:

$$K_r = \frac{1}{K_f} \qquad (14\text{-}5)$$

Equation (14-5) is a general result that is easily verified for any particular case.

Example 14-7: The value of K_c at 500°C for the equation

(1) $$ZnO(s) + CO(g) \rightleftharpoons Zn(l) + CO_2(g)$$

is $K_c = 3.0 \times 10^3$. Compute the value of K_c at 500°C for the equation

(2) $$Zn(l) + CO_2(g) \rightleftharpoons ZnO(s) + CO(g)$$

Solution: The reaction described by Equation (2) is the reverse of the reaction described by Equation (1), and thus, using Equation (14-5), we have for Equation (2)

$$K_r = \frac{1}{K_f} = \frac{1}{3.0 \times 10^3} = 3.3 \times 10^{-4}$$

Thus $K_c = 3.3 \times 10^{-4}$ for Equation (2).

Consider the two reactions described by the equations

(1) $$C(s) + H_2O(g) \rightleftharpoons CO(g) + H_2(g)$$

(2) $$CO_2(g) + 2H_2(g) \rightleftharpoons 2H_2O(g) + C(s)$$

If we add Equation (2) to Equation (1), then we obtain

$$C(s) + H_2O(g) + CO_2(g) + 2H_2(g) \rightleftharpoons CO(g) + H_2(g) + 2H_2O(g) + C(s)$$

Cancellation of like terms on the two sides of this equation yields

(3) $$CO_2(g) + H_2(g) \rightleftharpoons CO(g) + H_2O(g)$$

Multiplication of the equilibrium-constant expressions for Equations (1) and (2) yields

$$K_1K_2 = \frac{[CO][H_2]}{[H_2O]} \times \frac{[H_2O]^2}{[CO_2][H_2]^2} = \frac{[CO][H_2O]}{[CO_2][H_2]}$$

The equilibrium-constant expression for Equation (3) is

$$K_3 = \frac{[CO][H_2O]}{[CO_2][H_2]}$$

Comparison of the K_3 expression with the expression for K_1K_2 shows

$$K_3 = K_1K_2 \qquad (14\text{-}6)$$

Equation 14-6 is a general result. If we add two equations to obtain a third, then the equilibrium constant of the third equation is equal to the product of the equilibrium constants of the two equations that are added together.

Example 14-8: Given the following equations and their equilibrium constants

(1) $H^+(aq) + NO_2^-(aq) \rightleftharpoons HNO_2(aq)$ $\quad K_1 = 2.22 \times 10^3\ M^{-1}$

(2) $H_2O(l) \rightleftharpoons H^+(aq) + OH^-(aq)$ $\quad K_2 = 1.00 \times 10^{-14}\ M^2$

compute the value of K for the equation

(3) $$NO_2^-(aq) + H_2O(l) \rightleftharpoons HNO_2(aq) + OH^-(aq)$$

Solution: Equation (3) is obtained by adding together Equations (1) and (2) and canceling like terms. Therefore, the equilibrium constant for Equation (3) is equal to the product K_1K_2 (Equation 14-6):

$$K_3 = K_1K_2 = (2.22 \times 10^3 M^{-1})(1.00 \times 10^{-14}\ M^2) = 2.22 \times 10^{-11}\ M$$

where

$$K_3 = \frac{[HNO_2][OH^-]}{[NO_2^-]}$$

14-6. LE CHATELIER'S PRINCIPLE IS USED TO PREDICT THE DIRECTION OF SHIFT IN A CHEMICAL REACTION DISPLACED FROM EQUILIBRIUM

▪ Henri Le Châtelier (pronounced luh shat´elyay), a French physical chemist, is most famous for postulating the principle that now bears his name. He also investigated gas combustion and platinum alloys.

▪ In the application of Le Châtelier's principle, we consider the chemical reaction to be initially at equilibrium and then subject the reaction system to a change in conditions that displaces the reaction from equilibrium.

Consider a chemical reaction initially at equilibrium. When the reaction system is subjected to a change in conditions that displaces the reaction from equilibrium, the reaction shifts toward one direction or the other (left to right or right to left in the written equation) as it proceeds to a new equilibrium state. The direction of this shift can be predicted by using **Le Châtelier's principle.** We do not need to know the numerical value of the equilibrium constant K in order to apply Le Châtelier's principle, which can be stated as follows:

If a chemical reaction at equilibrium is subjected to a change in conditions that displaces it from equilibrium, then the reaction proceeds toward a new equilibrium state in the direction that—at least partially—offsets the change in conditions.

The conditions that can affect a reaction equilibrium are

1. concentration of a reactant or product
2. reaction volume or applied pressure
3. temperature

Although Le Châtelier's principle sounds imposing, it is actually simple to apply. A key point to recognize is that the equilibrium constant depends only on the temperature; it does not change when the reactant or product concentrations, the reaction volume, or the applied pressure is changed.

Consider the reaction equilibrium described by the equation

$$C(s) + CO_2(g) \rightleftharpoons 2CO(g)$$

If we disturb the equilibrium by injecting some additional $CO_2(g)$ into the reaction vessel, then the concentration of $CO_2(g)$ is increased. In response to the change in conditions, the reaction equilibrium shifts from left to right because this is the direction in which $CO_2(g)$ is consumed and this leads to a partial decrease of the increase in the CO_2 concentration. In the new equilibrium state, the concentration of CO_2 and the concentration of CO are both greater than in the original equilibrium state, but the concentration of CO_2 in the new equilibrium state is less than it was immediately after the additional CO_2 was injected.

If we disturb the reaction equilibrium by injecting some additional $CO(g)$ into the reaction vessel, then the concentration of CO is increased and the reaction equilibrium shifts from right to left because this is the direction that decreases the CO concentration.

If we inject some additional $C(s)$ into the reaction vessel, then there is *no* shift in the reaction equilibrium because the *concentration* of a solid is independent of the amount present. In other words, the injection or removal of some $C(s)$ does not displace the reaction from equilibrium. Generally, the further addition or partial removal of any solid reactant or product does not shift the reaction equilibrium.

Henri Le Châtelier (1850-1936), a French physical chemist who formulated the principle of equilibrium systems that is now known by his name.

Example 14-9: Consider the chemical equilibrium described by the equation

$$C(s) + CO_2(g) \rightleftharpoons 2CO(g)$$

Use Le Châtelier's principle to predict the effect on the equilibrium concentration of $CO_2(g)$ produced by a decrease in the concentration of $CO(g)$.

Solution: If we remove some $CO(g)$ from the equilibrium reaction mixture, then the reaction equilibrium shifts from left to right to produce more $CO(g)$ because this is the direction that partially offsets the change in conditions. The concentration of $CO_2(g)$ in the new equilibrium state is less than that in the original equilibrium state.

In general, a decrease in volume shifts a reaction equilibrium toward the side with the smaller number of moles of gas. If the number of moles of gaseous products is greater than the number of moles of gaseous reactants, as in $N_2O_4(g) \rightleftharpoons 2NO_2(g)$, then a decrease in volume shifts the reaction equilibrium from right to left. If the number of moles of gaseous products is less than the number of moles of gaseous reactants, as in $N_2(g) + 3H_2(g) \rightleftharpoons 2NH_3(g)$, then a decrease in volume shifts the reaction equilibrium from left to right. In both cases the observed shift partially offsets the increased pressure that results when the volume is decreased, because a shift to the side of the reaction with the smaller number of moles of gas decreases the total number of molecules in the reaction system. If the number of moles of gas is the same on both sides, as in the equation $H_2(g) + I_2(g) \rightleftharpoons 2HI(g)$, then a change in volume has no effect on the reaction equilibrium, because there is no direction in which the reaction can shift to change the total number of gas molecules per unit volume.

Example 14-10: Should the total volume of the equilibrium reaction mixture described by the equation

$$C(s) + CO_2(g) \rightleftharpoons 2CO(g)$$

be increased or decreased in order to increase the extent of conversion of carbon and carbon dioxide to carbon monoxide?

Solution: An increase in total volume favors the side of the equation with the greater number of moles of gas. Thus, an increase in volume would shift this equilibrium from left to right, thereby increasing the production of $CO(g)$ from $C(s)$ and $CO_2(g)$.

Note that, in assessing the effect of a volume change on a reaction equilibrium, we do not have to consider pure liquid and solid phases because the change in total gas volume does not affect the concentration of species in solid and liquid phases.

From Boyle's law we know that the volume of an ideal gas is inversely proportional to the applied pressure. Thus, the results that we have just deduced for changes in volume are the opposite of what we would

Figure 14-3 Two allotropes of carbon are graphite and diamond.

predict for changes in applied pressure. For example, in Example 14-10, Le Châtelier's principle says that we should decrease the pressure on the reaction system in order to produce more $CO(g)$. A decrease in the applied pressure is equivalent to an increase in the reaction volume.

The interconversion of diamond and graphite, the two allotropic forms of carbon (Figure 14-3) discussed in Section 11-9, illustrates the application of Le Châtelier's principle to an equilibrium involving only condensed phases. Graphite is the stable form at ordinary pressures, whereas diamond is the stable form at high pressures, such as thousands of atmospheres. The density of graphite is 2.2 $g{\cdot}cm^{-3}$, and that of diamond is 3.5 $g{\cdot}cm^{-3}$. Consequently, the molar volume of graphite is 5.5 $cm^3{\cdot}mol^{-1}$ and the molar volume of diamond is 3.4 $cm^3{\cdot}mol^{-1}$. When a very high pressure (about 15,000 atm) is applied to graphite, it can relieve some of the effect of the pressure by converting to diamond, which has a smaller volume. Any substance that can exist in two different forms with different densities exists as the higher-density, or smaller-volume, form under high pressure. This is why ice melts under pressure; the liquid water has a higher density, or a smaller volume, than ice. Recall that ice floats on water.

14-7. AN INCREASE IN TEMPERATURE SHIFTS A REACTION EQUILIBRIUM IN THE DIRECTION IN WHICH HEAT IS ABSORBED

The value of ΔH°_{rxn} for the chemical equation

$$CaCO_3(s) \rightleftharpoons CaO(s) + CO_2(g)$$

is +158 kJ. In other words, the conversion of 1.00 mol of $CaCO_3(s)$ to 1.00 mol of $CaO(s)$ plus 1.00 mol of $CO_2(g)$ requires an input of energy of 158 kJ as heat. Thus we have

$$CaCO_3(s) + 158 \text{ kJ} \rightleftharpoons CaO(s) + CO_2(g)$$

or simply

$$CaCO_3(s) + \text{heat} \rightleftharpoons CaO(s) + CO_2(g)$$

An increase in temperature increases the availability of energy as heat to the reaction, and thus Le Châtelier's principle tells us that the equilibrium shifts in the direction that counteracts this change, that is, from left to right. The left-to-right shift produces an increase in the equilibrium concentration of $CO_2(g)$ in the system. When the temperature increases, the system absorbs energy as heat in an attempt to reduce its temperature to the original value.

For an endothermic reaction ($\Delta H^\circ_{rxn} > 0$), a reaction equilibrium shifts to the right when the temperature is increased and to the left when the temperature is decreased. For an exothermic reaction ($\Delta H^\circ_{rxn} < 0$), the equilibrium shifts to the left when the temperature is increased and to the right when the temperature is decreased

Example 14-11: For the reaction described by the equation

$$N_2(g) + 3H_2(g) \rightleftharpoons 2NH_3(g)$$

$\Delta H^\circ_{rxn} = -92$ kJ. Will an increase in the temperature increase or decrease the extent of conversion of N_2 and H_2 to NH_3?

Solution: The value of ΔH°_{rxn} is negative, and thus the reaction evolves energy as heat:

$$N_2(g) + 3H_2(g) \rightleftharpoons 2NH_3(g) + 92 \text{ kJ}$$

An increase in temperature favors the absorption of energy as heat and thus shifts the equilibrium to the left, which decreases the yield of ammonia.

Unlike the effect of a change in the concentration of any of the reactants or products, or a change in the reaction volume or applied pressure, a change in temperature changes the value of the equilibrium constant. The equation that governs how the value of an equilibrium constant varies with temperature is called the **van't Hoff equation.** If K_2 is the value of the equilibrium constant at a temperature T_2, and K_1 is the value at T_1, then we have

$$\log\left(\frac{K_2}{K_1}\right) = \frac{\Delta H^\circ_{rxn}}{2.30R}\left(\frac{T_2 - T_1}{T_1 T_2}\right) \qquad (14\text{-}7)$$

where R is the molar gas constant, $8.314 \text{ J}\cdot\text{K}^{-1}\cdot\text{mol}^{-1}$. We shall not derive this equation until Section 19-11, but we shall show how to use it here.

Example 14-12: For the chemical equation

$$H_2(g) + CO_2(g) \rightleftharpoons CO(g) + H_2O(g)$$

the value of ΔH°_{rxn} is $34.60 \text{ kJ}\cdot\text{mol}^{-1}$. Given that the equilibrium constant $K = 0.64$ at 700°C, calculate the value of K at 1000°C.

Solution: We use Equation (14-7) with $K_1 = 0.64$, $T_1 = 973$ K, and $T_2 = 1273$ K:

$$\log\left(\frac{K_2}{0.64}\right) = \frac{(34.60 \times 10^3\ \text{J}\cdot\text{mol}^{-1})}{(2.30)(8.314\ \text{J}\cdot\text{K}^{-1}\cdot\text{mol}^{-1})}\left[\frac{1273\ \text{K} - 973\ \text{K}}{(973\ \text{K})(1273\ \text{K})}\right]$$

$$= 0.438$$

Thus

$$\frac{K_2}{0.64} = 10^{0.438} = 2.74$$

or

$$K = 1.75$$

Note that the value of K increases with increasing temperature in this case.

Let's consider Equation (14-7) for the two cases $\Delta H^\circ_{rxn} > 0$ and $\Delta H^\circ_{rxn} < 0$. If $T_2 > T_1$ and $\Delta H^\circ_{rxn} > 0$, then

$$\log\left(\frac{K_2}{K_1}\right) > 0 \qquad \text{or} \qquad K_2 > K_1$$

Thus the value of K increases with increasing temperature for an endothermic reaction (see Example 14-12). If $T_2 > T_1$ and $\Delta H^\circ_{rxn} < 0$, then

$$\log\left(\frac{K_2}{K_1}\right) < 0 \qquad \text{or} \qquad K_2 < K_1$$

Thus we see that the value of K decreases with increasing temperature for an exothermic reaction. The higher the temperature, the larger is the value of K for a reaction that absorbs energy as heat, and the smaller is the value of K for a reaction that evolves energy as heat.

Ammonia is produced commercially by the **Haber process,** which is described in Interchapter I. Because of the tremendous scale of NH_3 production, it is important that the reaction be run under the most favorable conditions, that is, conditions that maximize the yield of product on a commercial time scale. The ammonia production reaction is

$$N_2(g) + 3H_2(g) \rightleftharpoons 2NH_3(g) \qquad \Delta H^\circ_{rxn} = -92\ \text{kJ}$$

Because there are more moles of gaseous species on the left of this equation, a decrease in the total reaction volume or, equivalently, an increase in the total reaction pressure favors the conversion of reactants to products. Thus, the percent conversion of N_2 and H_2 to NH_3 (the yield) increases as the total pressure increases. The value of ΔH°_{rxn} for the reaction is negative, and so a decrease in temperature favors the production of ammonia.

The application of Le Châtelier's principle to the ammonia synthesis reaction leads to the prediction that *at equilibrium* the yield of ammonia is greater the higher the total pressure and the lower the temperature. However, the *rate* of the reaction at 25°C is negligibly slow. A high yield is of no commercial value if it takes forever to achieve the conversion. The rates of most reactions increases with increasing temperature. The ammonia production reaction is run at an elevated temperature (500°C), even though the equilibrium yield is not as favorable as at lower temperatures, in order to make the reaction proceed at an economically feasible rate. The low value of K at 500°C is offset by using a very high pressure (300 atm). The Haber process is thus based on a compromise between equilibrium (yield) and rate (speed of reaction) considerations. The commercial-scale Haber process involves the use of an iron-molybdenum catalyst to increase the reaction rate. In the absence of the catalyst, the reaction rate is too low even at 500°C to make the process economically feasible.

14-8. CHEMICAL REACTIONS ALWAYS PROCEED TOWARD EQUILIBRIUM

At 100°C the equilibrium constant for the chemical reaction described by the equation

$$N_2O_4(g) \rightleftharpoons 2NO_2(g)$$

is

$$K_c = \frac{[NO_2]^2}{[N_2O_4]} = 0.20 \text{ M} \qquad (14\text{-}8)$$

We now define the **reaction quotient** Q_c as the quantity that has exactly the same algebraic form as the equilibrium-constant expression for a reaction, but has *arbitrary* concentrations for the substances involved. Thus, for the N_2O_4–NO_2 reaction, we have the Q_c expression

$$Q_c = \frac{[NO_2]_0^2}{[N_2O_4]_0} \qquad (14\text{-}9)$$

where the zero subscripts denote either arbitrary or initial concentrations. The difference between Equation (14-8) and Equation (14-9) is that only equilibrium values of the concentrations can be used in Equation (14-8), because Equation (14-8) applies only at equilibrium. However, arbitrary values of the concentrations can be used in Equation (14-9). The reaction quotient Q_c is not a constant; instead, it takes on whatever value results from the way we prepare the reaction system. For example, suppose we mix 2.00 mol of $N_2O_4(g)$ with 2.00 mol of $NO_2(g)$ in a 1.00-L reaction vessel at 100°C. The value of Q_c is then equal to

$$Q_c = \frac{[NO_2]_0^2}{[N_2O_4]_0} = \frac{(2.00 \text{ M})^2}{(2.00 \text{ M})} = 2.00 \text{ M}$$

The value of K_c at 100°C for this reaction is 0.20 M; therefore, $Q_c \neq K_c$ and the reaction mixture is not at equilibrium. At equilibrium the ratio of $[NO_2]^2$ to $[N_2O_4]$ must equal 0.20 M:

$$K_c = \frac{[NO_2]^2}{[N_2O_4]} = 0.20 \text{ M} \qquad \text{(at equilibrium at 100°C)}$$

In order for the reaction system with $Q_c = 2.00$ M to attain equilibrium, the value of Q_c must decrease from 2.00 M to 0.20 M. Because $Q_c = [NO_2]_0^2/[N_2O_4]_0$, we see that Q_c is decreased if $[NO_2]_0$ decreases and $[N_2O_4]_0$ increases. Consequently, the reaction in which $[N_2O_4]_0 =$ 2.00 M and $[NO_2]_0 = 2.00$ M proceeds from right to left toward equilibrium because this is the direction in which the value of $[NO_2]$ decreases and the value of $[N_2O_4]$ increases. When a reaction system reaches equilibrium, the value of Q_c is equal to the value of K_c; likewise, at equilibrium $Q_p = K_p$ for pressure units. That is, for any reaction at equilibrium,

$$\frac{Q}{K} = 1 \quad \text{or} \quad Q = K \qquad \text{at equilibrium}$$

The **direction of reaction spontaneity** is always toward equilibrium. The numerical value of the ratio Q/K tells us the direction (left to right or right to left) in which a reaction system not at equilibrium spontaneously proceeds toward equilibrium. The various possibilities are as follows:

Value of (Q/K)	Direction the reaction proceeds toward equilibrium
$Q/K < 1$ or $Q < K$	$\longrightarrow$
$Q/K > 1$ or $Q > K$	$\longleftarrow$
$Q/K = 1$ or $Q = K$	no net change (equilibrium state)

In other words, a system that is not in equilibrium proceeds toward equilibrium in the direction in which Q approaches K in magnitude. If Q is greater than K, then the value of Q decreases toward K and the reaction proceeds from right to left as the reaction moves toward equilibrium. If Q is smaller than K, then the value of Q increases toward K and the reaction proceeds from left to right as the reaction moves toward equilibrium. If $Q = K$, then the reaction is at equilibrium and no further net change occurs.

Both Q and K must be expressed in the same units. Thus if we express K in concentration units (K_c), then Q also must be expressed in the same concentration units (Q_c).

Example 14-13: Suppose that $CO_2(g)$ and $CO(g)$ are brought into contact with $C(s)$ at 1000 K at $P_{CO_2} = 2.00$ atm and $P_{CO} = 0.50$ atm. Is the reaction described by the equation

$$C(s) + CO_2(g) \rightleftharpoons 2CO(g) \qquad K_p = 1.90 \text{ atm}$$

at equilibrium? If not, in what direction will the reaction proceed toward equilibrium?

Solution: The value of Q_p for the reaction system as prepared is

$$Q_p = \frac{P_{CO}^2}{P_{CO_2}} = \frac{(0.50 \text{ atm})^2}{(2.00 \text{ atm})} = 0.13 \text{ atm}$$

The value of K_p is given as 1.90 atm, and so

$$\frac{Q_p}{K_p} = \frac{0.13}{1.90} < 1$$

Therefore, the reaction system is not at equilibrium. Because $Q_p/K_p < 1$, the reaction proceeds toward equilibrium from left to right, with P_{CO} increasing and P_{CO_2} decreasing until $P_{CO}^2/P_{CO_2} = 1.90$ atm, that is, until equilibrium is attained.

The use of Q/K values to predict the direction in which a system proceeds toward equilibrium is related closely to, but is not, in general, the same as the use of Le Châtelier's principle. Le Châtelier's principle applies to a system initially at equilibrium that is displaced from equilibrium by a change in conditions. In contrast, the use of Q/K values does not presume an initial equilibrium state, and Q/K values can be applied to systems that are not at equilibrium in the initial state as well as to systems that are displaced from an initial equilibrium state. In effect, the Q/K formulation is a more general statement of Le Châtelier's original principle.

SUMMARY

A chemical reaction equilibrium is dynamic. At equilibrium the rates of the forward and reverse reactions are balanced and there is no net change in the system. A chemical reaction equilibrium is characterized quantitatively by the equilibrium constant expression for the reaction. The law of concentration action tells us how to formulate the equilibrium-constant expression for a chemical reaction.

The equilibrium constant of the equation for the reverse reaction is equal to the reciprocal of the equilibrium constant of the equation for the forward reaction. The equilibrium constant for a chemical equation obtained by algebraically adding two chemical equations is equal to the product of the equilibrium constants for the two equations that are added together.

The direction in which an established dynamic chemical reaction equilibrium shifts in response to a change in conditions, such as a change in reactant or product concentration, in temperature, or in volume, is predicted by Le Châtelier's principle. The temperature dependence of an equilibrium constant is given by the van't Hoff equation. The direction in which a reaction mixture not at equilibrium proceeds toward equilibrium is predicted from the ratio of the reaction quotient, Q, to the reaction equilibrium constant, K.

In selecting a set of reaction conditions to maximize the yield of a desired product, both equilibrium factors and reaction rate factors must be considered.

TERMS YOU SHOULD KNOW

forward rate 435
reverse rate 435
chemical equilibrium 435
dynamic equilibrium 435
equilibrium concentration 436
law of concentration action 438
equilibrium constants, K_c and K_p 439
equilibrium-constant expression 439
quadratic formula 442
quadratic equation 442
Le Châtelier's principle 446
van't Hoff equation 449
Haber process 450
reaction quotients, Q_c and Q_p 451
Q/K 452
direction of reaction spontaneity 452

EQUATIONS YOU SHOULD KNOW HOW TO USE

$$K_c = \frac{[X]^x[Y]^y}{[A]^a[B]^b}$$ (14-3) (definition of equilibrium constant)

$$K_r = \frac{1}{K_f}$$ (14-5) (equilibrium constants of forward and reverse reactions)

$$K_3 = K_1K_2$$ (14-6) equilibrium constant of the sum of two chemical equations)

$$\log\left(\frac{K_2}{K_1}\right) = \frac{\Delta H^\circ_{rxn}}{2.30\,R}\left(\frac{T_2 - T_1}{T_1 T_2}\right) \qquad (14\text{-}7) \quad \text{(van't Hoff equation)}$$

Value of $\frac{Q}{K}$ compared with unity (criterion of reaction spontaneity)

PROBLEMS

EQUILIBRIUM-CONSTANT EXPRESSION

14-1. Use the law of concentration action to write the equilibrium-constant expression (K_c) for the following equations:

(a) $ZnO(s) + CO(g) \rightleftharpoons Zn(l) + CO_2(g)$
(b) $2O_3(g) \rightleftharpoons 3O_2(g)$
(c) $2C_5H_6(g) \rightleftharpoons C_{10}H_{12}(g)$
(d) $2N_2O_5(soln) \rightleftharpoons 4NO_2(soln) + O_2(g)$

14-2. Use the law of concentration action to write the equilibrium-constant expression (K_c) for the following equations:

(a) $CO(g) + 2H_2(g) \rightleftharpoons CH_3OH(g)$
(b) $2NaHCO_3(s) \rightleftharpoons Na_2CO_3(s) + CO_2(g) + H_2O(g)$
(c) $N_2(g) + O_2(g) \rightleftharpoons 2NO(g)$

14-3. Write the equilibrium-constant expression (K_c) for each of the following equations:

(a) $SO_2Cl_2(g) \rightleftharpoons SO_2(g) + Cl_2(g)$
(b) $2H_2O_2(g) \rightleftharpoons 2H_2O(l) + O_2(g)$
(c) $(CaSO_4)_2{\cdot}H_2O(s) + 3H_2O(g) \rightleftharpoons 2CaSO_4{\cdot}2H_2O(s)$

14-4. Write the equilibrium-constant expression (K_c) for each of the following equations:

(a) $NH_2COONH_4(s) \rightleftharpoons 2NH_3(g) + CO_2(g)$
(b) $2HgO(s) \rightleftharpoons 2Hg(l) + O_2(g)$
(c) $N_2(g) + 2O_2(g) \rightleftharpoons N_2O_4(g)$

14-5. Write K_p expressions for the chemical equations in Problem 14-3.

14-6. Write K_p expressions for the chemical equations in Problem 14-4.

CALCULATION OF EQUILIBRIUM CONSTANTS

14-7. Phosgene, a toxic gas used in the synthesis of a variety of organic compounds, decomposes according to the equation

$$COCl_2(g) \rightleftharpoons CO(g) + Cl_2(g)$$

A sample of $COCl_2$ at an initial concentration of 0.500 M is heated at 527°C in a reaction vessel. At equilibrium, the concentration of CO was found to be [CO] = 0.046 M. Calculate the equilibrium constant for the reaction at 527°C.

14-8. The decomposition of phosphorus pentachloride is described by the equation

$$PCl_5(g) \rightleftharpoons PCl_3(g) + Cl_2(g)$$

A sample of PCl_5 at an initial concentration of 1.10 M is placed in a reaction vessel held at 250°C. When equilibrium is attained, the concentration of PCl_5 is 0.33 M. Calculate K_c for the reaction.

14-9. Consider the chemical equation

$$CuSO_4{\cdot}4NH_3(s) \rightleftharpoons CuSO_4{\cdot}2NH_3(s) + 2NH_3(g)$$

At 20°C, the equilibrium pressure of NH_3 is 62 torr. Compute K_p (in atm units) for this equation.

14-10. At 1000°C, methane and water react according to

$$CH_4(g) + H_2O(g) \rightleftharpoons CO(g) + 3H_2(g)$$

At equilibrium, it was found that $P_{CH_4} = 0.31$ atm, $P_{H_2O} = 0.83$ atm, $P_{CO} = 0.57$ atm, and $P_{H_2} = 2.26$ atm. Calculate K_p for this equation.

14-11. A mixture of 1.00 mol of $H_2(g)$ and 1.00 mol of $I_2(g)$ is placed in a 2.00-L container held at a constant temperature. After equilibrium is attained, 1.56 mol of HI(g) is found. Calculate K_c for the chemical equation

$$H_2(g) + I_2(g) \rightleftharpoons 2HI(g)$$

14-12. Nitrogen dioxide decomposes at high temperatures according to

$$2NO_2(g) \rightleftharpoons 2NO(g) + O_2(g)$$

Suppose initially we have pure $NO_2(g)$ at 1000 K and 0.500 atm. If the total pressure is 0.732 atm when equilibrium is reached, what is the value of K_p?

EQUILIBRIUM CALCULATIONS

14-13. Given that $[Ni(CO)_4] = 0.85$ M at equilibrium for the equation

$$Ni(s) + 4CO(g) \rightleftharpoons Ni(CO)_4(g) \qquad K_c = 5.0 \times 10^4\ M^{-3}$$

calculate the concentration of CO(g) at equilibrium.

14-14. The equilibrium constant for the chemical equation

$$C(s) + CO_2(g) \rightleftharpoons 2CO(g)$$

at 1000 K is 1.90 atm. If the equilibrium pressure of CO is 1.50 atm, what is the equilibrium pressure of CO_2?

14-15. Phosphorus pentachloride decomposes according to

$$PCl_5(g) \rightleftharpoons PCl_3(g) + Cl_2(g) \qquad K_c = 1.8\ \mathrm{M}\ \text{at}\ 250°\mathrm{C}$$

A 0.50-mol sample of PCl_5 is injected into a 2.0-L reaction vessel held at 250°C. Calculate the concentrations of PCl_5 and PCl_3 at equilibrium.

14-16. Carbon disulfide is prepared by heating sulfur and charcoal. The chemical equation is

$$S_2(g) + C(s) \rightleftharpoons CS_2(g) \qquad K_c = 9.40\ \text{at}\ 900\ \mathrm{K}$$

How much CS_2 can be prepared by heating 10.0 mol of sulfur (S_2) with excess carbon in a 5.00-L reaction vessel held at 900 K until equilibrium is attained?

14-17. At 1200°C, $K_c = 2.5 \times 10^4$ for the equation

$$H_2(g) + Cl_2(g) \rightleftharpoons 2HCl(g)$$

If 0.50 mol of H_2 and 0.50 mol of Cl_2 are introduced initially into a reaction vessel, how many moles of HCl are there at equilibrium?

14-18. At 1000°C, $K_p = 0.263\ \mathrm{atm}^{-1}$ for the equation

$$C(s) + 2H_2(g) \rightleftharpoons CH_4(g)$$

Calculate the equilibrium pressure of $CH_4(g)$ if 0.250 mol of CH_4 is placed in a 4.00-L container at 1000°C.

14-19. Ammonium hydrogen sulfide decomposes according to

$$NH_4HS(s) \rightleftharpoons NH_3(g) + H_2S(g)$$

The equilibrium constant, K_c, is $1.81 \times 10^{-4}\ \mathrm{M}^2$ at 25°C. If $NH_4HS(s)$ is placed in an evacuated reaction vessel at 25°C, what is the total gas pressure in the vessel when equilibrium is attained?

14-20. Sodium hydrogen carbonate, commonly called sodium bicarbonate, is used in baking soda and in fire extinguishers as a source of CO_2. It decomposes according to

$$2NaHCO_3(s) \rightleftharpoons Na_2CO_3(s) + CO_2(g) + H_2O(g)$$

Given that $K_p = 0.25\ \mathrm{atm}^2$ at 125°C, calculate the partial pressures of CO_2 and H_2O at equilibrium when $NaHCO_3$ is heated to 125°C in a closed vessel.

14-21. The equilibrium constant for the equation

$$2ICl(g) \rightleftharpoons I_2(g) + Cl_2(g)$$

is $K_c = 0.11$. Calculate the equilibrium concentrations of ICl, I_2, and Cl_2 when 0.65 mol of I_2 and 0.33 mol of Cl_2 are added to a 1.5-L reaction vessel.

14-22. Suppose that 5.00-mol of $CO(g)$ is mixed with 2.50 mol of $Cl_2(g)$ in a 10.0-L reaction vessel and the following reaction attains equilibrium:

$$CO(g) + Cl_2(g) \rightleftharpoons COCl_2(g)$$

Given that $K_c = 4.0\ \mathrm{M}^{-1}$, compute the equilibrium values of [CO], [Cl_2], and [$COCl_2$].

14-23. Suppose that N_2O_4 and NO_2 are mixed together in a reaction vessel and that the total pressure at equilibrium is 1.45 atm. Calculate $P_{N_2O_4}$ and P_{NO_2} at equilibrium when the value of K_p is 4.90 atm for the equation

$$N_2O_4(g) \rightleftharpoons 2NO_2(g)$$

14-24. Given that $H_2(g)$ reacts with $I_2(s)$ according to

$$H_2(g) + I_2(s) \rightleftharpoons 2HI(g) \qquad K_p = 8.6\ \mathrm{atm}$$

and that at equilibrium the total pressure in the reaction vessel is 4.5 atm, calculate P_{HI} and P_{H_2} at equilibrium. (Neglect the vapor pressure of $I_2(s)$.)

14-25. Zinc metal is produced by the reaction of its oxide with carbon monoxide at high temperature. The chemical equation is

$$ZnO(s) + CO(g) \rightleftharpoons Zn(s) + CO_2(g) \qquad K_p = 600$$

At equilibrium the total pressure in the reaction vessel is 1.80 atm. Calculate P_{CO_2} and P_{CO} at equilibrium.

14-26. At equilibrium, the total pressure in the reaction vessel for the reaction between carbon and hydrogen

$$C(s) + 2H_2(g) \rightleftharpoons CH_4(g)$$

is 2.11 atm. Given that $K_p = 0.263\ \mathrm{atm}^{-1}$ at 1000°C, compute P_{H_2} and P_{CH_4}.

LE CHATELIER'S PRINCIPLE

14-27. Consider the chemical equilibrium

$$H_2(g) + CO_2(g) \rightleftharpoons H_2O(g) + CO(g)$$

Use Le Châtelier's principle to predict the effect on the equilibrium pressure of CO_2 and of CO produced by

(a) an increase in the pressure of $H_2O(g)$
(b) an increase in the reaction volume

14-28. Consider the chemical equilibrium

$$2NO(g) + Br_2(g) \rightleftharpoons 2NOBr(g)$$

Use Le Châtelier's principle to predict the effect on the equilibrium concentration of NOBr and of NO produced by

(a) an increase in the concentration of $Br_2(g)$
(b) a two-fold decrease in the reaction volume

14-29. Consider the chemical equilibrium

$$C(s) + 2H_2(g) \rightleftharpoons CH_4(g) \qquad \Delta H^\circ_{rxn} = -75\ \mathrm{kJ}$$

Predict the way in which the equilibrium will shift in response to each of the following changes in conditions (if the equilibrium is unaffected by the change, then write *no change*):

(a) decrease in temperature
(b) decrease in reaction volume
(c) decrease in P_{H_2}
(d) increase in P_{CH_4}
(e) addition of $C(s)$

14-30. For the chemical equilibrium

$$Ni(s) + 4CO(g) \rightleftharpoons Ni(CO)_4(g) \qquad \Delta H^\circ_{rxn} < 0$$

predict the way in which the equilibrium will shift in response to each of the following changes in conditions (if the equilibrium is unaffected by the change, then write *no change*):

(a) increase in temperature
(b) increase in reaction volume
(c) removal of $Ni(CO)_4(g)$
(d) addition of $Ni(s)$

14-31. For the chemical equilibrium

$$2SO_2(g) + O_2(g) \rightleftharpoons 2SO_3(g) \qquad \Delta H^\circ_{rxn} = -198 \text{ kJ}$$

predict the direction in which the equilibrium will shift in response to each of the following changes in conditions:

(a) increase in temperature
(b) increase in reaction volume
(c) decrease in $[O_2]$
(d) increase in $[SO_2]$

14-32. For the chemical equilibrium

$$N_2(aq) \rightleftharpoons N_2(g) \qquad \Delta H^\circ_{rxn} > 0$$

in which direction will the equilibrium shift in response to the following changes in conditions?

(a) increase in temperature
(b) increase in volume over the solution
(c) addition of $H_2O(l)$
(d) addition of $N_2(g)$

14-33. Several key reactions in coal gasification are

1. the synthesis gas reaction:

$$C(s) + H_2O(g) \rightleftharpoons CO(g) + H_2(g) \qquad \Delta H^\circ_{rxn} = +131 \text{ kJ}$$

2. the water-gas-shift reaction:

$$CO(g) + H_2O(g) \rightleftharpoons CO_2(g) + H_2(g) \qquad \Delta H^\circ_{rxn} = -41 \text{ kJ}$$

3. the catalytic methanation reaction:

$$CO(g) + 3H_2(g) \rightleftharpoons H_2O(g) + CH_4(g) \qquad \Delta H^\circ_{rxn} = -206 \text{ kJ}$$

(a) Write the equilibrium constant expressions in terms of concentrations, K_c, for each of these equations.
(b) Predict the direction in which each equilibrium shifts in response to (i) an increase in temperature and (ii) a decrease in reaction volume.

14-34. An important modern chemical problem is the liquefication of coal because it is still relatively abundant whereas oil is a dwindling resource. The first step is heating the coal with steam to produce synthesis gas:

$$C(s) + H_2O(g) \rightleftharpoons CO(g) + H_2(g) \qquad \Delta H^\circ_{rxn} = 131 \text{ kJ}$$

Carbon monoxide can be hydrogenated to form the important chemical, methyl alcohol:

$$CO(g) + 2H_2(g) \rightleftharpoons CH_3OH(g) \qquad \Delta H^\circ_{rxn} = -128 \text{ kJ}$$

Use Le Châtelier's principle to suggest conditions that maximize the yield of CH_3OH from $CO(g)$ and $H_2(g)$.

REACTION QUOTIENT CALCULATIONS

14-35. At 900 K the equilibrium constant for the equation

$$2SO_2(g) + O_2(g) \rightleftharpoons 2SO_3(g)$$

is 13 M^{-1}. If we mix the following concentrations of the three gases, predict in which direction the reaction will proceed toward equilibrium:

	$[SO_2]$/M	$[O_2]$/M	$[SO_3]$/M
(a)	0.40	0.20	0.10
(b)	0.05	0.10	0.30

14-36. Suppose that $H_2(g)$ and $CH_4(g)$ are brought into contact with $C(s)$ at 500°C with $P_{H_2} = 0.20$ atm and $P_{CH_4} = 3.0$ atm. Is the reaction described by the equation

$$C(s) + 2H_2(g) \rightleftharpoons CH_4(g) \qquad K_p = 2.69 \times 10^3 \text{ atm}^{-1}$$

at equilibrium? If not, in what direction will the reaction proceed toward equilibrium?

14-37. Suppose we have a mixture of the gases H_2, CO_2, CO, and H_2O at 1260 K, with $P_{H_2} = 0.55$ atm, $P_{CO_2} = 0.20$ atm, $P_{CO} = 1.25$ atm, and $P_{H_2O} = 0.10$ atm. Is the reaction described by the equation

$$H_2(g + CO_2(g) \rightleftharpoons CO(g) + H_2O(g) \qquad K_p = 1.59$$

at equilibrium? If not, in what direction will the reaction proceed toward equilibrium?

14-38. Suppose $S_2(g)$ and $CS_2(g)$ are brought into contact with solid carbon at 900 K with $P_{S_2} = 1.78$ atm and $P_{CS_2} = 0.794$ atm. Is the reaction described by the equation

$$S_2(g) + C(s) \rightleftharpoons CS_2(g) \qquad K_p = 9.40$$

at equilibrium? If not, in what direction will the reaction proceed toward equilibrium?

14-39. The equilibrium constant for the chemical equation

$$2SO_2(g) + O_2(g) \rightleftharpoons 2SO_3(g)$$

is $K_p = 0.14\ atm^{-1}$ at 900 K. Suppose the reaction system is prepared at 900 K with the initial pressures $P_{O_2} = 0.50$ atm, $P_{SO_2} = 0.30$ atm, and $P_{SO_3} = 0.20$ atm.

(a) Compute the value of Q for the reaction with these pressures.
(b) Indicate the direction in which the reaction proceeds toward equilibrium.

14-40. Given that $K_p = 2.25 \times 10^4 atm^{-2}$ at 25°C for the equation

$$2H_2(g) + CO(g) \rightleftharpoons CH_3OH(g)$$

predict the direction in which a reaction mixture for which $P_{CH_3OH} = 10.0$ atm, $P_{H_2} = 0.010$ atm, and $P_{CO} = 0.0050$ atm proceeds toward equilibrium.

EQUILIBRIUM CONSTANTS OF EQUATIONS FOR COMBINATIONS OF REACTIONS

14-41. Given that

$$CO(g) + H_2O(g) \rightleftharpoons CO_2(g) + H_2(g) \qquad K_p = 1.44$$

$$CH_4(g) + H_2O(g) \rightleftharpoons CO(g) + 3H_2(g) \qquad K_p = 25.6\ atm^2$$

calculate K_p for the equation

$$CH_4(g) + 2H_2O(g) \rightleftharpoons CO_2(g) + 4H_2(g)$$

14-42. Given that

$$C(s) + 2H_2O(g) \rightleftharpoons CO_2(g) + 2H_2(g) \qquad K_p = 3.85\ atm$$

and

$$H_2(g) + CO_2(g) \rightleftharpoons H_2O(g) + CO(g) \qquad K = 0.71$$

calculate K_p for the equation

$$C(s) + CO_2(g) \rightleftharpoons 2CO(g)$$

14-43. Given that at 973 K

$$MgCl_2(s) + \tfrac{1}{2}O_2(g) \rightleftharpoons MgO(s) + Cl_2(g) \qquad K_p = 2.95\ atm^{1/2}$$

$$MgCl_2(s) + H_2O(g) \rightleftharpoons MgO(s) + 2HCl(g) \qquad K_p = 8.40\ atm$$

determine the equilibrium constant at 973 K for the equation

$$2Cl_2(g) + 2H_2O(g) \rightleftharpoons 4HCl(g) + O_2(g)$$

14-44. Given the equilibrium constants at 1000 K for the following equations

$$CaCO_3(s) \rightleftharpoons CaO(s) + CO_2(g) \qquad K_1 = 0.039\ atm$$

$$C(s) + CO_2(g) \rightleftharpoons 2CO(g) \qquad K_2 = 1.9\ atm$$

determine the equilibrium constant at 1000 K for the equation

$$CaCO_3(s) + C(s) \rightleftharpoons CaO(s) + 2CO(g)$$

14-45. For the equation

$$PCl_5(g) \rightleftharpoons PCl_3(g) + Cl_2(g)$$

$\Delta H^\circ_{rxn} = +92.9\ kJ \cdot mol^{-1}$. The value of K_p is 1.78 atm at 250°C. Calculate K_p at 400°C.

14-46. The equilibrium constant for the equation

$$H_2(g) + I_2(g) \rightleftharpoons 2HI(g)$$

is 617 at 25°C and $\Delta H^\circ_{rxn} = -10.2\ kJ \cdot mol^{-1}$. Calculate K at 100°C.

14-47. The reaction of sulfur dioxide with oxygen

$$2SO_2(g) + O_2(g) \rightleftharpoons 2SO_3(g) \qquad \Delta H^\circ_{rxn} = -198\ kJ$$

occurs in the catalytic converter of an automobile. For this reaction $K_p = 0.14\ atm^{-1}$ at 627°C. Calculate the value of K_p at 1000°C.

14-48. A key component of photochemical smog is NO, which is produced by the reaction between N_2 and O_2 at the high temperatures that occur in the internal combustion engine. At 2000°C the equilibrium constant of the equation $N_2(g) + O_2(g) \rightleftharpoons 2NO(g)$ is 4×10^{-4}. The reaction takes place to a much lesser extent at low temperatures. Calculate the equilibrium constant at 25°C, using the value $\Delta H^\circ_{rxn} = 181$ kJ.

14-49. Given the following equilibrium-constant data for the Deacon process,

$$Cl_2(g) + H_2O(g) \rightleftharpoons 2HCl(g) + \tfrac{1}{2}O_2(g)$$

T/K	log K
723	−0.706
873	−0.002

calculate the value of ΔH°_{rxn}.

14-50. Carbon monoxide reacts with hydrogen to yield methanol according to

$$CO(g) + 2H_2(g) \rightleftharpoons CH_3OH(g)$$

For this equation, $\Delta H^\circ_{rxn} = -128$ kJ and $K_p = 2.25 \times 10^4\ atm^{-2}$ at 25°C. The reaction is normally run at 300°C. What is the value of the equilibrium constant at 300°C?

ADDITIONAL PROBLEMS

14-51. The value of K_p for the chemical equation

$$CuSO_4 \cdot 4NH_3(s) \rightleftharpoons CuSO_4 \cdot 2NH_3(s) + 2NH_3(g)$$

is $6.66 \times 10^{-3}\ atm^2$ at 20°C. Calculate the equilibrium pressure of ammonia at 20°C.

14-52. The equilibrium constant for the chemical equation

$$N_2(g) + 3H_2(g) \rightleftharpoons 2NH_3(g)$$

is $K_p = 0.10\ \text{atm}^{-2}$ at 227°C. Compute the value of K_c for the reaction at 227°C.

14-53. At 500°C, hydrogen iodide decomposes according to

$$2HI(g) \rightleftharpoons H_2(g) + I_2(g)$$

For HI heated to 500°C in a 1.00-L reaction vessel, chemical analysis gave the following concentrations at equilibrium: $[H_2] = 0.42$ M, $[I_2] = 0.42$ M, and $[HI] = 3.52$ M. If an additional mol of HI is introduced to the reaction vessel, what are the equilibrium concentrations after the new equilibrium has been reached?

14-54. The equilibrium constant for the methanol synthesis equation

$$2H_2(g) + CO(g) \rightleftharpoons CH_3OH(g)$$

is $K_p = 2.25 \times 10^4\ \text{atm}^{-2}$ at 25°C.

(a) Compute the value of P_{CH_3OH} at equilibrium when $P_{H_2} = 0.020$ atm and $P_{CO} = 0.010$ atm.
(b) Given that at equilibrium $P_{total} = 10.0$ atm and $P_{H_2} = 0.020$ atm, compute P_{CO} and P_{CH_3OH}.

14-55. Given

$$SO_2(g) + NO_2(g) \rightleftharpoons SO_3(g) + NO(g) \qquad \Delta H^\circ_{rxn} = -42\ \text{kJ}$$

Complete the following table:

Change	Effect on equilibrium
(a) decrease in total volume	
(b) increase in temperature	
(c) increase in partial pressure of $NO_2(g)$	
(d) decrease in partial pressure of products	

14-56. According to Table 14-1, $K_c = 0.20$ M at 100°C for the chemical equation

$$N_2O_4(g) \rightleftharpoons 2NO_2(g)$$

Calculate K_p at the same temperature.

14-57. Tin can be prepared by heating SnO_2 ore with hydrogen gas:

$$SnO_2(s) + 2H_2(g) \rightleftharpoons Sn(s) + 2H_2O(g)$$

When the reactants are heated to 500°C in a closed vessel, $[H_2O] = [H_2] = 0.25$ M at equilibrium. If more hydrogen is added so that its initial concentration becomes 0.50 M, what are the concentrations of H_2 and H_2O when equilibrium is restored?

14-58. The equilibrium constant for the chemical equation

$$N_2(g) + 3H_2(g) \rightleftharpoons 2NH_3(g)$$

is $K_p = 0.10\ \text{atm}^{-2}$ at 227°C.

(a) Given that at equilibrium $P_{N_2} = 1.00$ atm and $P_{H_2} = 3.00$ atm, compute P_{NH_3} at equilibrium.
(b) Given that at equilibrium the total pressure is 2.00 atm and also that the mole fraction of H_2, X_{H_2}, is 0.20, compute X_{NH_3}. (Note that $X_{N_2} + X_{H_2} + X_{NH_3} = 1$.)

14-59. The decomposition of ammonium carbamate, NH_2COONH_4, takes place according to

$$NH_2COONH_4(s) \rightleftharpoons 2NH_3(g) + CO_2(g)$$

Show that if all the NH_3 and CO_2 result from the decomposition of ammonium carbamate, then $K_p = (4/27)P^3$, where P is the total pressure at equilibrium.

14-60. The equilibrium constant for the chemical equation

$$SO_2(g) + NO_2(g) \rightleftharpoons SO_3(g) + NO(g)$$

is 3.0. Calculate the number of moles of NO_2 that must be added to 2.4 mol of SO_2 in order to form 1.2 mol of SO_3 at equilibrium.

14-61. Show that, for a reaction involving gaseous products and/or reactants,

$$K_p = K_c(RT)^{\Delta n}$$

where Δn is the number of moles of gaseous products minus the number of moles of gaseous reactants in the chemical equation as written.

14-62. Diatomic chlorine dissociates to chlorine atoms at elevated temperatures. For example, $K_p = 0.570$ atm at 2000°C. Calculate the fraction of chlorine molecules that are dissociated at 2000°C.

14-63. The value of the equilibrium constant for the equation

$$H_2(g) + I_2(g) \rightleftharpoons 2HI(g)$$

is $K = 85$ at 553 K.

(a) Is it possible at 553 K to have an equilibrium reaction mixture for which $P_{HI} = P_{H_2} = P_{I_2}$?
(b) Suppose a 5.0-g sample of $HI(g)$ is heated to 553 K in a 2.00-L vessel. Calculate the composition of the equilibrium reaction mixture.

14-64. The equilibrium constant at 1000 K for the chemical equation

$$CaCO_3(g) \rightleftharpoons CaO(s) + CO_2(g) \qquad \Delta H^\circ_{rxn} = 158\ \text{kJ}$$

is $K = 0.039$ atm. Compute the equilibrium pressure of $CO_2(g)$ at 1300 K.

14-65. The equilibrium constant at 823 K for the chemical equation

$$MgCl_2(s) + \frac{1}{2}O_2(g) \rightleftharpoons MgO(s) + Cl_2(g)$$

is $K_p = 1.75\ \text{atm}^{1/2}$. Suppose that 50 g of $MgCl_2(s)$ is placed in a reaction vessel with 2.00 L of oxygen at 25°C and 1.00 atm, and that the reaction vessel is sealed and heated to 823 K until equilibrium is attained. Compute P_{Cl_2} and P_{O_2} at equilibrium.

14-66. Consider the reaction equilibrium

$$COCl_2(g) \rightleftharpoons CO(g) + Cl_2(g)$$

If 2.00 mol of $COCl_2(g)$ is introduced into a 10.0-L flask at 1000°C, calculate the equilibrium concentrations of all species at this temperature. At 1000°C, $K_c = 0.329$ M.

14-67. Osmium dioxide occurs either as a black powder or as brown crystals. The density of the black powder form is 7.7 $g\cdot mL^{-1}$ and the density of the brown crystalline form is 11.4 $g\cdot mL^{-1}$. Which is the more stable form at high pressure?

14-68. Discuss the connection between the van't Hoff equation and the Clapeyron-Clausius equation.

14-69. Deoxygenated nitrogen is often prepared in the laboratory by passing tank nitrogen over hot copper gauze

$$2Cu(s) + \frac{1}{2}O_2(g) \rightleftharpoons Cu_2O(s) \qquad \Delta H^\circ_{rxn} = -167\text{kJ}$$

Given that the equilibrium constant for this reaction is $K = 5.50 \times 10^{25}\ \text{atm}^{-1/2}$ at 300 K, calculate K at 900 K. Also calculate the pressure of oxygen in the nitrogen at 900 K after passing over $Cu(s)$.

14-70. Given that $\Delta H^\circ_{rxn} = -297$ kJ for

$$S(s) + O_2(g) \rightleftharpoons SO_3(g)$$

and that the equilibrium constant at 25°C is $K = 2.02 \times 10^{52}$, calculate the value of K at 1000°C.

Nitrogen

A nitrogen production plant at Jay Field in western Florida. The nitrogen is obtained by the fractional distillation of liquid air.

The most significant property of elemental nitrogen, N_2, is its lack of chemical reactivity. Nitrogen, as N_2, does not take part in many chemical reactions. Although nitrogen compounds are essential nutrients for animals and plants, only a few microorganisms are able to utilize elemental nitrogen directly by converting it to water-soluble compounds of nitrogen. The conversion of nitrogen from the free element to nitrogen compounds is one of the most important problems of modern chemistry and is called **nitrogen fixation.** The principles of chemical equilibria discussed in Chapter 14 have played a key role in the development of commercial nitrogen-fixation processes.

I-1. NITROGEN RANKS SECOND AMONG CHEMICALS IN ANNUAL PRODUCTION

Nitrogen (atomic number 7, atomic mass 14.0067) is a colorless, odorless gas that exists as a diatomic molecule, N_2. The Lewis formula, :N≡N:, shows that N_2 is a triple-bonded molecule, which accounts for its high bond strength and consequent lack of chemical reactivity. The principal source of nitrogen is the atmosphere, which is about 78 percent N_2 by volume. Pure nitrogen is produced by the fractional distillation of liquid air. Nitrogen boils at −196°C, whereas oxygen, the other principal component of air, boils at −183°C.

In terms of U.S. industrial production, nitrogen is the second leading chemical. Over 40 billion pounds of nitrogen is produced from air each year. Nitrogen is also found in potassium nitrate, KNO_3 (saltpeter), and in sodium nitrate, $NaNO_3$ (Chile saltpeter). Vast deposits of these two nitrates are found in the arid northern region of Chile, where there is insufficient rainfall to wash away these soluble compounds. The Chilean nitrate deposits are about 200 miles long, 20 miles wide, and many feet thick. At one time the economy of Chile was based primarily upon the sale of nitrates for use as fertilizers. Nitrogen also occurs in all living organisms, both animal and vegetable. Proteins and nucleic acids, such as DNA and RNA, contain significant quantities of nitrogen.

Large quantities of nitrogen are stored and shipped as the liquid (Figure I-1) in insulated metal cylinders. Smaller quantities are shipped as the gas in heavy-walled steel cylinders. An alternative source is to heat an aqueous solution of ammonium nitrite, which thermally decomposes according to the equation

$$NH_4NO_2(aq) \rightarrow N_2(g) + 2H_2O(l)$$

Ammonium nitrite is a potentially explosive solid, and so the aqueous ammonium nitrite solution is made by adding ammonium chloride and sodium nitrite, both stable compounds, to water. Even so, the solution must be heated carefully.

Lithium is the only element that reacts with nitrogen at room temperature:

$$6Li(s) + N_2(g) \rightarrow 2Li_3N(s)$$

Other reactive metals, such as the other alkali metals, magnesium, calcium, and aluminum, react at high temperatures to form nitrides:

$$3Mg(s) + N_2(g) \xrightarrow{600°C} Mg_3N_2(s)$$

$$2Al(s) + N_2(g) \xrightarrow{800°C} 2AlN(s)$$

When magnesium is burned in air, most of it is converted to the white magnesium oxide, but small yellow flecks of magnesium nitride also can be seen. Metallic nitrides react with water to form ammonia, which is one of the most important compounds of nitrogen:

$$Mg_3N_2(s) + 6H_2O(l) \rightarrow 3Mg(OH)_2(s) + 2NH_3(g)$$

Figure I-1 Liquid nitrogen at its boiling point.

Ammonia is a colorless gas with a sharp, irritating odor. It is the active ingredient in some forms of "smelling salts." Over 30 billion pounds of ammonia is produced annually in the United States. Rated in terms of pounds produced per year, ammonia is the third-ranked U.S. industrial chemical, being surpassed only by sulfuric acid and nitrogen.

The ammonia molecule was the first complex molecule to be identifed in interstellar space. Ammonia occurs in the galactic dust clouds of the Milky Way and, in the solid form, constitutes part of the rings of Saturn.

Unlike nitrogen, which is sparingly soluble in water, ammonia is very soluble in water. Over 700 mL of ammonia at 0°C and 1 atm will dissolve in 1 mL of water. The solubility of ammonia in water can be demonstrated nicely by the **fountain effect.** This effect can be observed with the simple laboratory steup shown in Figure I-2. A dry flask is filled with anhydrous ("dry") ammonia gas at atmospheric pressure. When just a few drops of water is squirted into the flask from a syringe, some of the ammonia dissolves in the drops. The pressure of the ammonia falls below that of the atmosphere, and water is forced from the beaker, up the vertical glass tubing into the flask, producing a spectacular fountain. Household ammonia is an aqueous solution of ammonia (about 2 M) together with a detergent.

Figure I-2 An ammonia fountain. The inverted flask is filled initially with ammonia gas. The beaker contains water and the acid-base indicator phenolphthalein, which is colorless in acidic or neutral solution and red in basic solution. When a small amount of water is squirted into the flask, the dissolution of ammonia gas in water creates a pressure drop in the inverted flask, and the aqueous solution is driven up into the flask by the atmospheric pressure. The solution changes to a red color because ammonia is a base and phenolphthalein is red in basic solution. The high solubility of ammonia in water leads to a large pressure drop, which in turn produces a strong upsurge of solution, thus creating a fountain.

Ammonia does not burn in air but will burn in pure oxygen:

$$4NH_3(g) + 3O_2(g) \rightarrow 6H_2O(g) + 2N_2(g)$$

Certain compositions of ammonia and oxygen are explosive if sparked.

Ammonia reacts with various acids to form ammonium compounds, which contain the NH_4^+ ion. For example,

$$NH_3(aq) + HCl(aq) \rightarrow NH_4Cl(aq)$$

Many ammonium compounds are important commercially, particularly as fertilizers.

I-2. THE HABER PROCESS IS USED FOR THE INDUSTRIAL PRODUCTION OF AMMONIA

The large-scale demand for ammonia to make fertilizers and other nitrogen compounds requires an economical process for its production. The laboratory preparation of ammonia by reacting a metallic nitride with water is too expensive for large-scale use, as is the commonly used reaction of ammonium chloride with calcium hydroxide:

$$2NH_4Cl(s) + Ca(OH)_2(s) \rightarrow CaCl_2(s) + 2H_2O(l) + 2NH_3(g)$$

In this case, one of the starting materials (NH_4Cl) is produced from ammonia in the first place. Since around 1913, ammonia has been produced commercially by the **Haber process,** developed by the German chemist Fritz Haber, in which nitrogen reacts directly with hydrogen at high pressure and high temperature:

$$N_2(g) + 3H_2(g) \xrightleftharpoons[\text{Fe/Mo catalyst}]{500^\circ\text{C, 300 atm}} 2NH_3(g) \qquad \Delta H^\circ_{rxn} = -92 \text{ kJ}$$

As we found in Chapter 14, this reaction is favored by high pressure and low temperature, but it proceeds too slowly at low temperatures. Thus the conditions represent a compromise between equilibrium (yield) and rate. Originally, the hydrogen was obtained from the electrolysis of water, but now it is obtained either from the water-gas reaction (Interchapter D)

$$C(s) + H_2O(g) \rightarrow CO(g) + H_2(g)$$

or from the steam reforming of natural gas (Interchapter D)

$$CH_4(g) + H_2O(g) \rightarrow CO(g) + 3H_2(g)$$

Ammonia binds to many components of soil and is easily converted to usable plant food. Concentrated aqueous solutions of ammonia or pure liquid ammonia can be sprayed directly into the soil (Figure I-3). Ammonia is an inexpensive fertilizer that is high in nitrogen. The increased growth of plants when fertilized by ammonia is spectacular.

For some purposes it is more convenient to use a solid fertilizer instead of ammonia solutions. For example, ammonia combines directly with sulfuric acid to produce ammonium sulfate:

$$2NH_3(aq) + H_2SO_4(aq) \rightarrow (NH_4)_2SO_4(aq)$$

Ammonium sulfate is the most important solid fertilizer in the world. Its annual U.S. production exceeds 4 billion pounds.

The primary fertilizer nutrients are nitrogen, phosphorus, and potassium, and fertilizers are rated by how much of each they contain. For example, a 5-10-5 fertilizer has 5 percent by mass total available nitrogen, 10 percent by mass phosphorus (equivalent to the form P_2O_5), 5 percent by mass potassium (equivalent to the form K_2O), and 80 percent inert ingredients. The production of fertilizers is one of the largest and most important industries in the world. In 1980 over 114 million metric tons of fertilizer was utilized worldwide.

Figure I-3 This photo demonstrates the method of spraying ammonia into the soil. Liquid ammonia, called anhydrous ammonia, is used extensively as a fertilizer because it is cheap, high in nitrogen, and easy to apply.

I-3. NITRIC ACID IS PRODUCED COMMERCIALLY BY THE OSTWALD PROCESS

Shortly after the Haber process was put into large-scale production, World War I, with its great demand for munitions, started. Explosives and munitions are produced from nitric acid and nitrates, and at the beginning of the War the major source of nitrates was Chile. Germany recognized that this transoceanic supply was going to be difficult to maintain (one of the first naval battles of World War I was fought between Britain and Germany off the coast of Chile) and directed its attention to the production of nitric acid from ammonia.

In the early 1900s, the German chemist Wilhelm Ostwald showed that nitric acid could be produced from ammonia by a sequence of reactions that is now known as the **Ostwald process.** The first step in this process is the conversion of ammonia to nitrogen oxide:

$$(1) \qquad 4NH_3(g) + 5O_2(g) \xrightarrow[\text{Pt catalyst}]{800^\circ\text{C}} 4NO(g) + 6H_2O(g)$$

The second step involves the oxidation of NO to NO_2:

$$(2) \qquad 2NO(g) + O_2(g) \rightarrow 2NO_2(g)$$

In the final step, the NO_2 is reacted with water to yield nitric acid:

$$(3) \qquad 3NO_2(g) + H_2O(l) \rightarrow 2HNO_3(l) + NO(g)$$

DANGER!
CAUSES SEVERE BURNS.
VAPOR EXTREMELY HAZARDOUS.
MAY CAUSE NITROUS GAS POISONING.
MAY BE FATAL IF INHALED OR SWALLOWED.
SYMPTOMS OF LUNG INJURY MAY BE DELAYED.
STRONG OXIDIZER.
SPILLAGE MAY CAUSE FIRE OR LIBERATE DANGEROUS GAS.
Do not get in eyes, on skin, on clothing.
Do not breathe vapor.
Use only with adequate ventilation.
Wash thoroughly after handling.
Keep from contact with clothing and other combustible materials.
Do not store near combustible materials.
Store in tightly closed containers.
Keep out of reach of children.

POISON

Call a physician at once.
FIRST AID: In case of contact, immediately flush eyes or skin with plenty of water for at least 15 minutes while removing contaminated clothing and shoes. Wash clothing before reuse.
If inhaled, remove to fresh air. If not breathing give artificial respiration, preferably mouth-to-mouth. If breathing is difficult, give oxygen.
If swallowed, do not give emetics or baking soda. Give tap water, milk of magnesia or eggs beaten with water. Never give anything by mouth to an unconscious person.
In case of spill: Flush immediately with large volumes of water; after washing neutralize residue with soda ash or lime.
In case of fire: Use large quantities of water for extinguishing fire.

Exposure to light causes formation of colored oxides of nitrogen. Store in a cool, dark place.

Allied Corporation
Allied Chemical
Morristown, New Jersey 07960

Allied Chemical
An ALLIED Company

Nitric Acid

Code 108-002677
Semiconductor Low Mobile Ion Grade
Meets SEMI specifications.
For manufacturing use only.
Not for food or drug use.

HNO_3 F.W. 63.01
Assay (HNO_3) 70.0-71.0%
Density at 25° C (g/mL) .... 1.42 approx.

NET WT. 7 LBS. (3.18 kg)

108-002677-7-81

Figure I-4 Label from a bottle of concentrated nitric acid.

The NO(g) evolved is recycled back into step (2). The large-scale availability of ammonia from the Haber process enabled the Ostwald process to be run on an industrial scale.

Laboratory-grade nitric acid is approximately 70 percent HNO_3 by mass with a density of 1.42 $g \cdot mL^{-1}$ and a concentration of 16 M (Figure I-4). The U.S. annual production of nitric acid is over 16 billion pounds, which makes it the eleventh-ranked industrial chemical. The greatest use of nitric acid is in the production of ammonium nitrate for fertilizers. It is also used in the production of explosives such as trinitrotoluene (TNT), nitroglycerine, and nitrocellulose (gun cotton), and in etching and photoengraving processes to produce grooves in metal surfaces. For example, dilute nitric acid readily reacts with copper metal:

$$3Cu(s) + 8HNO_3(aq) \rightarrow 3Cu(NO_3)_2(aq) + 2NO(g) + 4H_2O(l)$$

In contrast, copper metal does not react directly with hydrochloric acid. Note that the above reaction does not involve the liberation of hydrogen gas, as in the reaction of, say, zinc metal with hydrochloric acid.

I-4. NITROGEN FORMS MANY IMPORTANT COMPOUNDS WITH HYDROGEN AND OXYGEN

▪ Household bleach should never be mixed with household ammonia because toxic and explosive chloramines, such as H_2NCl and $HNCl_2$, are produced as by-products.

The most important nitrogen-hydrogen compounds are ammonia, NH_3, hydrazine, N_2H_4, and hydrazoic acid, HN_3. Hydrazine is a colorless, fuming, reactive liquid. It is produced by the **Raschig synthesis,**in which ammonia is reacted with hypochlorite ion (household bleach is sodium hypochlorite in water) in basic solution:

$$2NH_3(aq) + ClO^-(aq) \xrightarrow{OH^-(aq)} N_2H_4(aq) + H_2O(l) + Cl^-(aq)$$

The reaction of hydrazine with oxygen,

$$N_2H_4(l) + O_2(g) \rightarrow N_2(g) + 2H_2O(g)$$

is accompanied by the release of a large amount of energy; hydrazine and some of its derivatives are used as rocket fuels (Interchapter E).

Nitrous acid, HNO_2, is prepared by reacting an equimolar mixture of nitrogen oxide and nitrogen dioxide with a basic solution (for example, NaOH):

$$NO(g) + NO_2(g) + 2NaOH(aq) \rightarrow 2NaNO_2(aq) + H_2O(l)$$

Addition of acid to the resulting solution yields nitrous acid:

$$NO_2^-(aq) + H^+(aq) \rightarrow HNO_2(aq)$$

Salts of nitrous acid are called nitrites. Sodium nitrite, $NaNO_2$, is used as a meat preservative. The nitrite ion combines with the hemoglobin in meat to produce a deep red color.

The reaction of nitrous acid with hydrazine in acidic solution yields hydrazoic acid:

$$N_2H_4(aq) + HNO_2(aq) \rightarrow HN_3(aq)2H_2O(l)$$

Hydrazoic acid is a colorless, toxic liquid and a dangerous explosive. Its lead and mercury salts, $Pb(N_3)_2$ and $Hg(N_3)_2$, which are called **azides,** are used in detonation caps; both compounds are dangerously explosive. Sodium azide, NaN_3, is used as the gas source in automobile air safety bags.

Dinitrogen oxide, which is more commonly called nitrous oxide, is prepared by gently heating liquid ammonium nitrate:

$$NH_4NO_3(l) \rightarrow N_2O(g) + 2H_2O(g)$$

The reaction is potentially explosive, with the evolution of nitrogen, oxygen, and water. Consequently, often an equimolar mixture of potassium nitrate and ammonium chloride is used:

$$KNO_3(s) + NH_4Cl(s) \rightarrow KCl(s) + 2H_2O(l) + N_2O(g)$$

Nitrous oxide is a colorless gas that is fairly unreactive. It is also called **laughing gas** and was once widely used as a general anesthetic in dentistry. Today it is used as a propellant in canned "whipped cream" products.

Nitrous oxide is one of several stable oxides of nitrogen. We have seen above that nitrogen oxide (NO) is produced by the reaction of a fairly unreactive metal like copper and dilute nitric acid. Nitrogen oxide is a colorless gas that is only slightly soluble in water. It reacts with oxygen to produce the red-brown gas nitrogen dioxide:

$$2NO(g) + O_2(g) \rightarrow 2NO_2(g)$$

Consequently, the reaction of copper with dilute nitric acid appears to produce NO_2 when it is run open to the atmosphere. Recall from Chapter 14 that nitrogen dioxide readily dimerizes to form the colorless oxide nitrogen tetroxide, N_2O_4.

Dinitrogen pentoxide is a reactive, colorless solid that is the acid anhydride of nitric acid. It can be prepared by carefully dehydrating pure nitric acid with $P_4O_{10}(s)$ at 0°C:

$$4HNO_3(l) + P_4O_{10}(s) \rightarrow 2N_2O_5(s) + 4HPO_3(l)$$

X-ray diffraction studies indicate that $N_2O_5(s)$ consists of an ionic array of the species NO_2^+ and NO_3^-. In the gas phase or in solutions with solvents such as CCl_4 and $CHCl_3$, however, the compound is molecular.

nitrogen fixation 460
fountain effect 462
Haber process 462
Ostwald process 463
Raschig synthesis 464
azide 465
laughing gas 465

QUESTIONS

I-1. Discuss what is meant by nitrogen fixation.

I-2. To produce small quantities of $N_2(g)$ from the thermal decomposition of $NH_4NO_2(aq)$, equimolar quantities of NH_4Cl and $NaNO_2$ are used instead of NH_4NO_2. Why?

I-3. Which is the only element that reacts with $N_2(g)$ at room temperature?

I-4. When magnesium is burned in air, yellow flecks can be found in the product. What are they?

I-5. Describe how an ammonia fountain works.

I-6. Briefly describe the Haber process.

I-7. Using balanced chemical equations, outline the Ostwald process.

I-8. What are azides? How are they made?

I-9. Describe the Raschig synthesis of hydrazine.

I-10. Why should you never mix household ammonia and bleach?

I-11. Outline a method for the preparation of DNO_3 in $D_2O(l)$.

I-12. What is "laughing gas"?

PROBLEMS

I-13. At 20°C, the solubility of ammonia in water is 33.1 percent by mass. Calculate the molarity of a saturated aqueous solution of ammonia at 20°C. The density of the solution is $0.890\ g\cdot mL^{-1}$.

I-14. How many metric tons of ammonium sulfate can be produced from one metric ton of ammonia?

I-15. How many metric tons of nitric acid can be produced from one metric ton of ammonia by the Ostwald process?

I-16. Laboratory-grade nitric acid is 70 percent HNO_3 by mass and has a density of $1.42\ g\cdot mL^{-1}$. Calculate both the molality and the molarity of laboratory grade nitric acid.

I-17. Ammonium hydrogen sulfide, $NH_4HS(s)$, decomposes into ammonia and hydrogen sulfide, $H_2S(g)$. If excess $NH_4HS(s)$ is placed in an evacuated reaction vessel at 25°C, the total pressure is 0.658 atm. Calculate the value of K_p and the value of K_c (see Problem 14-19).

I-18. Mercury reacts with hot nitric acid according to the equation

$$3Hg(l) + 8HNO_3(aq) \rightarrow 3Hg(NO_3)_2(aq) + 2NO(g) + 4H_2O(l)$$

Calculate how many grams of mercury will react with 40.0 mL of 2.00 M $HNO_3(aq)$.

I-19. A stock solution of nitric acid is 15.7 M. Describe how you would prepare 500 mL of 6.0 M nitric acid.

ACIDS AND BASES, I

Firefighters use an airport snowblower to blow sodium carbonate onto 20,000 gal of nitric acid spilled from a tankcar in Denver on April 3, 1983. The spill was brought under control (neutralized) in a few hours, and no serious injuries were caused by the accident. The neutralization reaction is $2HNO_3(aq) + Na_2CO_3(s) \rightarrow 2NaNO_3(aq) + H_2O(l) + CO_2(g)$.

A quantitative understanding of the chemistry of acids and bases is essential to an understanding of many chemical reactions and most biochemical reactions. In this chapter we first present a definition of acids and bases that incorporates the key role that water plays in acid-base chemistry. Then we introduce the concept of pH, which is a convenient measure of the strength of an acidic or a basic solution. We shall see that acids and bases can be classified as either strong or weak. A central, quantitative theme of this chapter is the calculation of the acidity, or pH, of an acidic or a basic solution as a function of the concentration of the acid or base. We shall learn that solutions of salts can be acidic, basic, or neutral.

15-1. AN ACID IS A PROTON DONOR AND A BASE IS A PROTON ACCEPTOR

In Chapter 2 we defined an acid as a substance that produces $H^+(aq)$ in aqueous solution and a base as a substance that produces $OH^-(aq)$ in aqueous solution. Acidic solutions taste sour (for example, vinegar and lemon juice), and basic solutions feel slippery and taste bitter (for example, soap).

The definition of acids and bases that we have used up to now is due to Arrhenius, and thus such substances are called **Arrhenius acids and bases.** In this chapter we use a more general definition of acids and bases, from the classification scheme proposed independently by the Danish chemist Johannes Brønsted and the English chemist Thomas Lowry. **Brønsted-Lowry acids** are defined as **proton donors,** and **Brønsted-Lowry bases** are defined as **proton acceptors.** For example, when $HCl(g)$ is dissolved in water, the acid HCl donates a proton to the base H_2O to produce a **hydronium ion,** $H_3O^+(aq)$, which is a hydrated proton:

$$HCl(g) + H_2O(l) \rightarrow H_3O^+(aq) + Cl^-(aq)$$

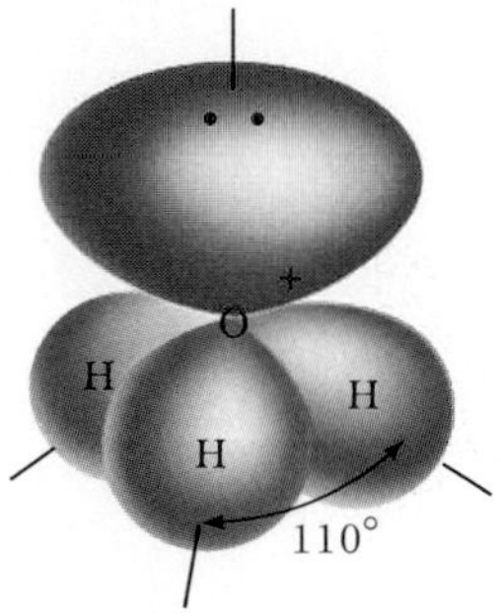

In water, a proton associates very strongly with a water molecule to form a hydronium ion, which has a trigonal pyramidal structure (Figure 15-1). The $H_3O^+(aq)$ notation tells us that the hydronium ion is further solvated by other water molecules, which are represented by (aq).

H O H H 110°

Figure 15-1 The hydronium ion has a tripod shape. All the oxygen-hydrogen bond lengths are identical (106 pm), and all the H—O—H angles are identical (110°). The Lewis formula is

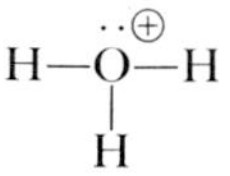

$$\mathrm{H{-}\ddot{O}^{\oplus}{-}H}\quad(\text{with a third H bonded below O})$$

and thus the ion has one lone pair of electrons.

Up to now we have designated a hydrated proton by $H^+(aq)$, but the designation $H_3O^+(aq)$ is more informative when we use the Brønsted-Lowry definition of acids and bases. For the dissociation of HCl, we write

$$HCl(aq) \rightarrow H^+(aq) + Cl^-(aq) \qquad \text{Arrhenius}$$

$$HCl(aq) + H_2O(l) \rightarrow H_3O^+(aq) + Cl^-(aq) \qquad \text{Brønsted-Lowry}$$

Note that the Brønsted-Lowry notation for the dissociation of HCl requires that we include $H_2O(l)$ explicitly to indicate the species to which HCl donates a proton.

A reaction involving the transfer of a proton from one molecule to another is called a **proton-transfer reaction** or a **protonation reaction.**

15-2. IN AN AQUEOUS SOLUTION THE ION CONCENTRATION PRODUCT $[H_3O^+][OH^-]$ IS A CONSTANT

Pure water contains a small number of hydronium ions and hydroxide ions, $H_3O^+(aq)$ and $OH^-(aq)$, that arise from the equilibrium

$$H_2O(l) + H_2O(l) \rightleftharpoons H_3O^+(aq) + OH^-(aq) \qquad (15\text{-}1)$$

■ Acid-base reactions are proton-transfer reactions.

In this reaction water molecules transfer protons to other water molecules. Note that water acts as both an acid (proton donor) and as a base (proton acceptor).

The equilibrium-constant expression for Equation (15-1) is

$$K_w = [H_3O^+][OH^-] \qquad (15\text{-}2)$$

where the subscript w refers to water. Note that the concentration of water does not appear in the K_w expression because $[H_2O]$ is effectively constant in aqueous solutions (Chapter 14). The quantity K_w is called the **ion-product constant of water.** At 25°C the experimental value of K_w is

$$K_w = [H_3O^+][OH^-] = 1.00 \times 10^{-14}\ M^2 \qquad (15\text{-}3)$$

▪ Equation (15-3) tells us that when $[H_3O^+]$ is large $[OH^-]$ is small, and when $[OH^-]$ is large $[H_3O^+]$ is small.

This small value of K_w means that in pure water the concentrations of $H_3O^+(aq)$ and $OH^-(aq)$ are low; that is, the equilibrium represented by Equation (15-1) lies far to the left. From the stoichiometry of Equation (15-1) we note that if we start with pure water, then $H_3O^+(aq)$ and $OH^-(aq)$ are produced on a one-for-one basis. Therefore, *in pure water* we have the equality

▪ The value of K_w depends on the temperature.

$$[H_3O^+] = [OH^-]$$

Using this equation to eliminate $[OH^-]$ in Equation (15-3) yields

$$[H_3O^+]^2 = 1.00 \times 10^{-14}\ M^2$$

Taking the square root of both sides yields

$$[H_3O^+] = 1.00 \times 10^{-7}\ M$$

Because $[H_3O^+] = [OH^-]$, we also have

$$[OH^-] = 1.00 \times 10^{-7}\ M$$

Thus both $[H_3O^+]$ and $[OH^-]$ are equal to 1.00×10^{-7} M in pure water at 25°C. Although $[H_3O^+] = [OH^-]$ for pure water, this is not necessarily true when substances are dissolved in water.

A **neutral** aqueous solution is defined as one in which

$$[H_3O^+] = [OH^-] \qquad \text{neutral solution}$$

An **acidic** aqueous solution is defined as one in which

$$[H_3O^+] > [OH^-] \qquad \text{acidic solution}$$

A **basic** aqueous solution is defined as one in which

$$[OH^-] > [H_3O^+] \qquad \text{basic solution}$$

15-3. STRONG ACIDS AND BASES ARE COMPLETELY DISSOCIATED IN AQUEOUS SOLUTIONS

Conductivity measurements on dilute $HCl(aq)$ solutions show that HCl in water is completely dissociated into $H_3O^+(aq)$ and $Cl^-(aq)$. There are essentially no undissociated HCl molecules in aqueous solution. Acids that are completely dissociated are referred to as **strong acids.** The term strong refers to the ability of such acids to donate protons to water molecules. Strong acids transfer all their dissociable protons to water molecules.

Example 15-1: Compute $[H_3O^+]$, $[Cl^-]$, and $[OH^-]$ in a 0.15 M aqueous solution of $HCl(aq)$.

Solution: Because HCl is a strong acid in water, it is completely dissociated, and thus

$$[H_3O^+] = 0.15 \text{ M} \qquad [Cl^-] = 0.15 \text{ M}$$

The corresponding value of $[OH^-]$ in this solution can be computed from the K_w expression:

$$K_w = [H_3O^+][OH^-] = 1.00 \times 10^{-14} \text{ M}^2$$

$$[OH^-] = \frac{1.00 \times 10^{-14} \text{ M}^2}{[H_3O^+]} = \frac{1.00 \times 10^{-14} \text{ M}^2}{0.15 \text{ M}} = 6.7 \times 10^{-14} \text{ M}$$

Because $[H_3O^+] \gg [OH^-]$, the solution is strongly acidic.

Note that we ignored the small contribution to $[H_3O^+]$ arising from the dissociation of water, which is roughly equal to $[OH^-]$ and has a value of about 6.7×10^{-14} M. This is indeed very small compared with 0.15 M.

Conductivity measurements show that sodium hydroxide in water is completely dissociated; that is, it exists as $Na^+(aq)$ and $OH^-(aq)$:

▪ Strong acids and strong bases are completely dissociated in solution.

$$NaOH(s) \xrightarrow[H_2O(l)]{} Na^+(aq) + OH^-(aq)$$

There is essentially no undissociated NaOH present in aqueous solution. Sodium hydroxide is a base because $OH^-(aq)$ is a proton acceptor:

$$H_3O^+(aq) + OH^-(aq) \rightarrow 2H_2O(l)$$

Completely dissociated bases are referred to as **strong bases.**

Example 15-2: Compute $[OH^-]$, $[Na^+]$, and $[H_3O^+]$ in a 0.15 M aqueous solution of $NaOH(aq)$.

Solution: Because NaOH is a strong base in water, it is completely dissociated, and thus

$$[OH^-] = 0.15 \text{ M} \qquad [Na^+] = 0.15 \text{ M}$$

The value of $[H_3O^+]$ can be computed from the K_w expression:

$$[H_3O^+] = \frac{1.00 \times 10^{-14} \text{ M}^2}{[OH^-]} = \frac{1.00 \times 10^{-14} \text{ M}^2}{0.15 \text{ M}} = 6.7 \times 10^{-14} \text{ M}$$

Because $[OH^-] \gg [H_3O^+]$, the solution is strongly basic.

There are only a few strong acids and bases in water. Most acids and bases, when dissolved in water, are only partly dissociated into their constituent ions. Acids that are incompletely dissociated are called **weak acids,** and bases that are incompletely dissociated are called **weak bases.**

▪ An acid not listed in Table 15-1 is a weak acid. A base not listed in Table 15-1 is a weak base. Most acids and bases are weak.

▪ The chemistry of $Tl^+(aq)$ is similar in many respects to that of an alkali metal ion.

The strong acids and bases are listed in Table 15-1. You should memorize their formulas because this information is essential in working problems in acid-base chemistry. Note that three of the six strong acids are halogen acids (HCl, HBr, and HI) and that five of the nine strong bases are alkali metal hydroxides (LiOH, NaOH, KOH, RbOH, and CsOH) and three are alkaline earth metal hydroxides ($Ca(OH)_2$, $Sr(OH)_2$, and $Ba(OH)_2$). In contrast to the other halogen acids, $HF(aq)$

Table 15-1 Strong acids and bases in water

Strong acids		Strong bases	
$HClO_4$	perchloric	$LiOH$	lithium hydroxide
HNO_3	nitric	$NaOH$	sodium hydroxide
H_2SO_4	sulfuric*	KOH	potassium hydroxide
HCl	hydrochloric	$RbOH$	rubidium hydroxide
HBr	hydrobromic	$CsOH$	cesium hydroxide
HI	hydroiodic	$TlOH$	thallium(I) hydroxide
		$Ca(OH)_2$	calcium hydroxide
		$Sr(OH)_2$	strontium hydroxide
		$Ba(OH)_2$	barium hydroxide

*First proton only.

is a weak acid. This is a result of the fact that HF has a much stronger bond than the H—X bonds in the other halogen acids.

Most organic acids are weak; the most common are **carboxylic acids,** which have the general formula RCOOH, where R is a hydrogen atom or an alkyl group such as methyl (CH_3—) or ethyl (CH_3CH_2—). The —COOH group is called the **carboxyl group.** The two simplest carboxylic acids are formic acid and acetic acid:

$$H{-}C(=O){-}OH \qquad CH_3{-}C(=O){-}OH$$

formic acid, $HCHO_2$ acetic acid, $HC_2H_3O_2$

Formic acid is one of the irritants in the bite of ants. Acetic acid is familiar in vinegar, which is a 5 percent aqueous solution of acetic acid.

The carboxyl group produces hydronium ions in water as shown by the equation for the acid-base reaction

$$\underset{\text{acetic acid}}{CH_3{-}C(=O){-}OH}(aq) + H_2O(l) \rightleftharpoons \underset{\text{hydronium ion}}{H_3O^+(aq)} + \underset{\text{acetate ion}}{CH_3{-}C(=O){-}O^{\ominus}}(aq)$$

Organic acids also react with bases such as sodium hydroxide to produce salts and water:

$$\underset{\text{formic acid}}{H{-}C(=O){-}OH}(aq) + NaOH(aq) \rightleftharpoons \underset{\text{sodium formate}}{Na^+H{-}C(=O){-}O^{\ominus}}(aq) + H_2O(l)$$

Example 15-3: Complete and balance the following chemical equation:

$$HC_2H_3O_2(aq) + Ca(OH)_2(aq) \rightarrow$$

and name the product of the reaction.

Solution: Each acetic acid molecule contributes one hydrogen ion, and so it requires 2 mol of acetic acid to neutralize completely 1 mol of calcium hydroxide. The balanced equation is

$$2HC_2H_3O_2(aq) + Ca(OH)_2(aq) \rightarrow Ca(C_2H_3O_2)_2(aq) + 2H_2O(l)$$

The product is a salt of calcium hydroxide and acetic acid. To name the anion of the salt, we change the *-ic* ending of the acid to *-ate* and drop the word acid. Thus we have, in this case, calcium acetate.

The anion formed by a carboxylic acid is called a **carboxylate ion** and has the general formula $RCOO^-$. Carboxylate ions are stabilized by charge delocalization. The carboxylate ion is described by the two resonance formulas

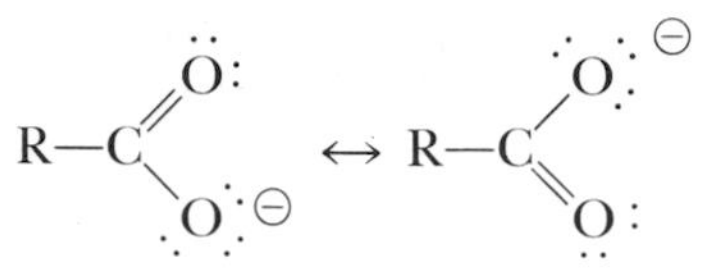

$$\text{R—C}(\text{=}\ddot{\text{O}}\text{:})(\text{—}\ddot{\text{O}}\text{:}^{\ominus}) \leftrightarrow \text{R—C}(\text{—}\ddot{\text{O}}\text{:}^{\ominus})(\text{=}\ddot{\text{O}}\text{:})$$

whose resonance hybrid is

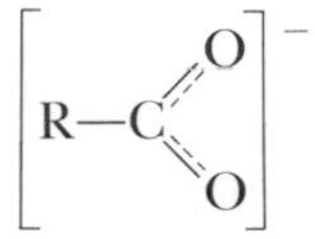

$$\left[\text{R—C}\begin{matrix}\text{O}\\ \text{O}\end{matrix}\right]^-$$

showing that the negative charge is distributed equally between the two oxygen atoms. The delocalization of the negative charge over the two oxygen atoms confers a degree of stability to a carboxylate ion. The two carbon-oxygen bonds in formic acid have different lengths, but in sodium formate the two carbon-oxygen bond lengths are identical and are intermediate between those of single and double carbon-oxygen bonds:

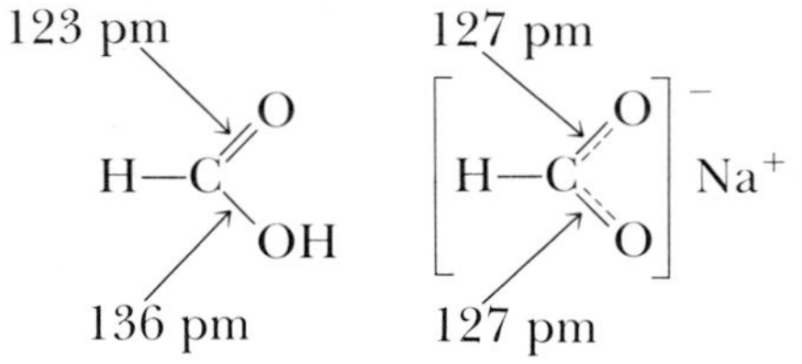

123 pm (C=O), 136 pm (C—OH): H—C(=O)OH
formic acid

127 pm, 127 pm: $[\text{H—C(O)O}]^-\ Na^+$
sodium formate

15-4. pH IS A MEASURE OF THE ACIDITY OF AN AQUEOUS SOLUTION

You will find throughout your study of chemistry that the rates of many chemical reactions depend upon the concentration of $H_3O^+(aq)$ in the reaction mixture even though $H_3O^+(aq)$ may not be one of the reactants or products. The rates of many reactions can be altered dramatically by the addition of a small amount of $H_3O^+(aq)$ (Section 13-6). As we shall see, concentrations of $H_3O^+(aq)$ often lie in the range from 1 M to 10^{-14} M. Such a wide range of concentrations makes it difficult to plot these values on ordinary graphs. Because of this, it is convenient to use a logarithmic scale and to define a quantity called **pH** as follows:

▪ pH is simply a numerically convenient measure of the acidity of a solution.

$$\text{pH} = -\log\,[H_3O^+] \qquad (15\text{-}4)$$

The properties of logarithms are reviewed in Appendix A-2. The following Example illustrates the use of Equation (15-4) to calculate the pH corresponding to various values of $[H_3O^+]$.

Example 15-4: Calculate the pH of a solution that has a $H_3O^+(aq)$ concentration of 5.0×10^{-10} M.

Solution: If you have a hand calculator, you enter 5.0×10^{-10} and press the log key and then the sign-change key to get

$$pH = -\log (5.0 \times 10^{-10}) = 9.30$$

If you use a table of logarithms instead of a calculator, you must use the relation

$$\log ab = \log a + \log b$$

Using this property of logarithms and also that $\log 5.0 = 0.70$, we have

$$\log (5.0 \times 10^{-10}) = \log 5.0 + \log 10^{-10}$$
$$= 0.70 - 10$$
$$= -9.30$$

or

$$pH = 9.30$$

Although it is easy to calculate pH using a calculator, you should also understand how to use a table of logarithms and remember the principal properties of logarithms.

▪ The number of significant figures after the decimal point in the logarithm of $(a \times 10^{-n})$ should be the same as the number of significant figures in the number a. Thus the 9 in pH = 9.30 only serves to position the decimal in 5.0×10^{-10}.

▪ A table of logarithms is given in Appendix C.

Example 15-5: Compute the pH of an aqueous solution prepared by dissolving 0.26 g of calcium hydroxide in water and diluting to a final volume of 0.500 L.

Solution: We first compute the number of moles of $Ca(OH)_2$. The formula mass of $Ca(OH)_2$ is 74.1, and so the number of moles is

$$(0.26 \text{ g } Ca(OH)_2)\left(\frac{1 \text{ mol } Ca(OH)_2}{74.1 \text{ g } Ca(OH)_2}\right) = 3.5 \times 10^{-3} \text{ mol}$$

The molarity of the solution is the number of moles per liter of solution:

$$\text{molarity} = \frac{\text{moles of solute}}{\text{liters of solution}} = \frac{3.5 \times 10^{-3} \text{ mol}}{0.500 \text{ L}} = 7.0 \times 10^{-3} \text{ M}$$

Calcium hydroxide is a strong base and yields two $OH^-(aq)$ per mole of $Ca(OH)_2(aq)$. Therefore, the molarity of the $OH^-(aq)$ is

$$[OH^-] = (2)(7.0 \times 10^{-3} \text{ M}) = 1.4 \times 10^{-2} \text{ M}$$

The value of $[H_3O^+]$ is calculated by using the ion-product constant of water:

$$[H_3O^+] = \frac{1.00 \times 10^{-14} \text{ M}^2}{[OH^-]} = \frac{1.00 \times 10^{-14} \text{ M}^2}{1.4 \times 10^{-2} \text{ M}} = 7.1 \times 10^{-13} \text{ M}$$

The pH of the solution is

$$pH = -\log [H_3O^+] = -\log (7.1 \times 10^{-13}) = 12.15$$

We can solve the problem in Example 15-5 a little more quickly in the following manner. First, we take the logarithm of Equation (15-3) to obtain

$$\begin{aligned}\log([H_3O^+][OH^-]) &= \log[H_3O^+] + \log[OH^-]\\ &= \log(1.00 \times 10^{-14}) = -14.00\end{aligned}$$

Next we multiply through by -1 and define the quantity **pOH** as follows:

$$\text{pOH} = -\log[OH^-] \tag{15-5}$$

Now, using the definitions of pH and pOH, we can write

$$\text{pH} + \text{pOH} = 14.00 \tag{15-6}$$

Equations (15-5) and (15-6) are often useful in solving problems involving basic solutions. In Example 15-5, we found that $[OH^-] = 1.4 \times 10^{-2}$ M. We substitute this value in Equation (15-5) to get pOH = 1.85, so the pH is 14.00 − 1.85 = 12.15. The pH values of some common aqueous solutions are given in Figure 15-2.

At 25°C, pure water has a hydronium-ion concentration of $[H_3O^+] = 1.00 \times 10^{-7}$ M, and thus the pH of a neutral aqueous solution at 25°C is

$$\text{pH} = -\log[H_3O^+] = -\log[1.00 \times 10^{-7}] = 7.00$$

At 25°C, acidic solutions have $[H_3O^+]$ values greater than 1.00×10^{-7} M, and thus acidic solutions have pH values less than 7.00. Basic solutions have a $[H_3O^+]$ less than 1.00×10^{-7} M and thus have pH values greater than 7.00. The pH scale is shown schematically as follows:

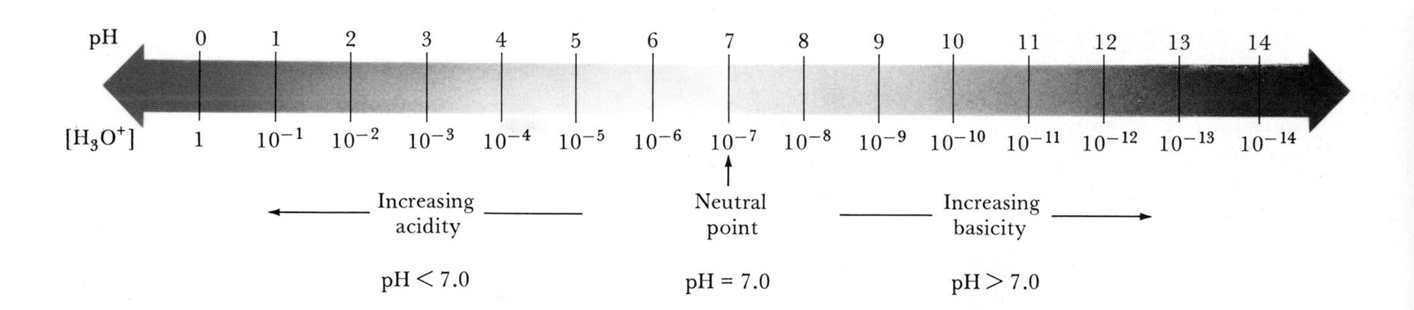

Note that a change in pH of one unit corresponds to a 10-fold change in $[H_3O^+]$.

The pH of a solution is conveniently measured in the laboratory with a **pH meter,** an electronic device that responds to the $[H_3O^+]$ of a solution (Figure 15-3). The meter scale or digital readout of the device is set up to display pH directly.

Examples 15-4 and 15-5 illustrate how to calculate pH from $[H_3O^+]$. It is often necessary to do the inverse calculation, that is, to calculate $[H_3O^+]$ from the pH, as shown in the following Example.

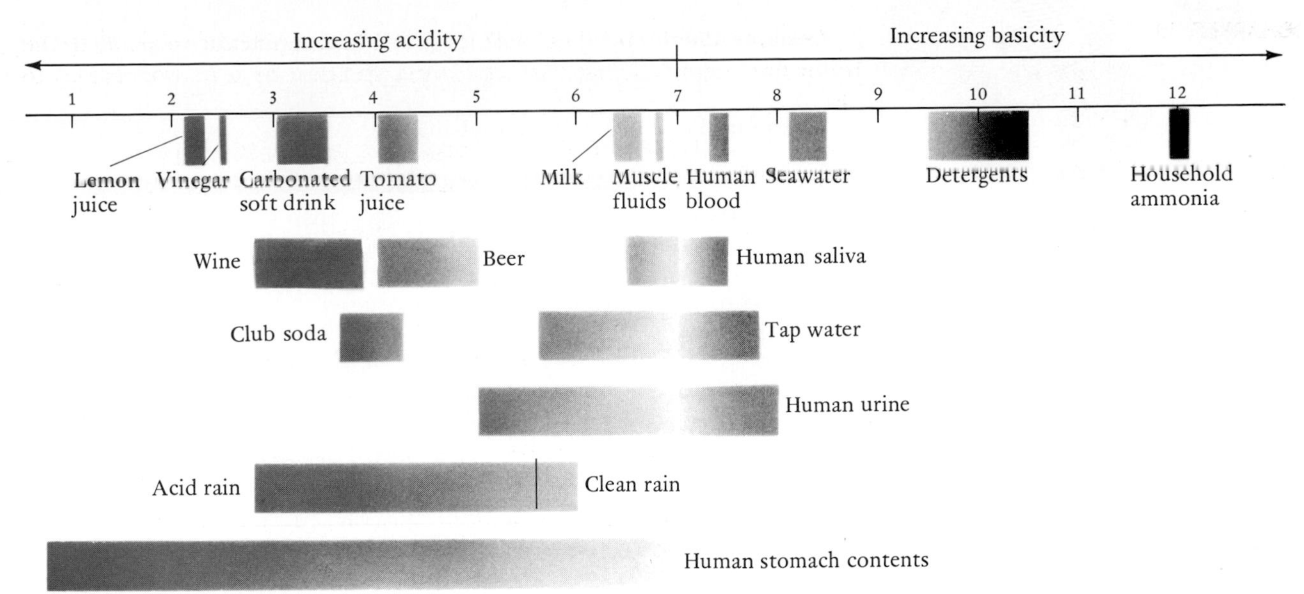

Figure 15-2 The range of pH values for some common aqueous solutions. The bars are spaced out to avoid overlapping.

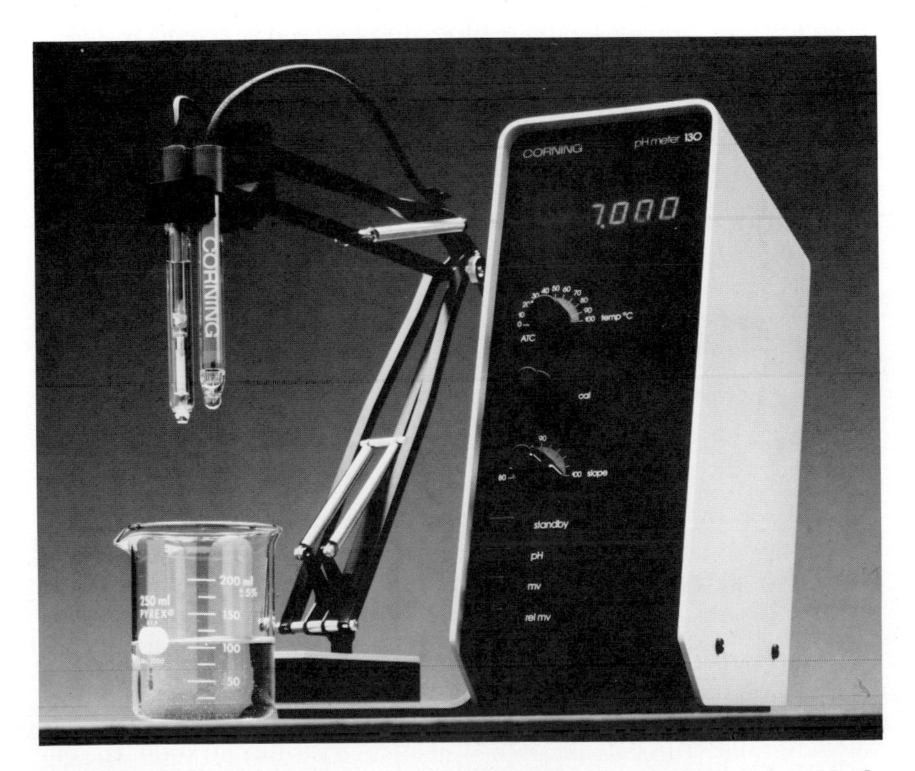

Figure 15-3 A pH meter and a pair of electrodes for measuring the pH of a solution. The electrodes are placed in the solution to measure the pH, which is displayed in digital form on the meter.

Example 15-6: The pH of milk is about 6.5. Compute the value of $[H_3O^+]$ for milk.

Solution: From the definition of pH we have

$$pH = -\log [H_3O^+]$$

A property of logarithms is that if $y = \log x$, then $x = 10^y$. Thus

$$[H_3O^+] = 10^{-pH}$$

The pH of milk is 6.5, and so

$$[H_3O^+] = 10^{-6.5}$$

The quantity $10^{-6.5}$ can be evaluated easily on your hand calculator by using the inverse logarithm operation (the 10^x key on some calculators):

$$[H_3O^+] = 10^{-6.5} = 3 \times 10^{-7}\ M$$

If you are using a table of logarithms instead of a hand calculator, then first you must write $10^{-6.5}$ as

$$10^{-6.5} = 10^{0.5} \times 10^{-7}$$

The antilogarithm of 0.5 is 3, and so once again we find that

$$[H_3O^+] = 3 \times 10^{-7}\ M$$

15-5. WEAK ACIDS AND WEAK BASES REACT ONLY PARTIALLY WITH WATER

If 0.10 mol of hydrogen chloride gas is dissolved in enough water to make 1.00 L of aqueous solution, then the observed pH of the resulting solution is 1.00. However, if 0.10 mol of hydrogen fluoride gas is dissolved in enough water to make 1.00 L of aqueous solution, then the observed pH of the resulting solution is 2.10. In each case we calculate the value of $[H_3O^+]$ from the pH using the relation (Example 15-6)

$$[H_3O^+] = 10^{-pH} \qquad (15\text{-}7)$$

The values that we obtain using Equation (15-7) are

0.10 M HCl(*aq*)	**0.10 M HF(*aq*)**
$pH = 1.00$	$pH = 2.10$
$[H_3O^+] = 10^{-1.00} = 0.10\ M$	$[H_3O^+] = 10^{-2.10} = 10^{0.90} \times 10^{-3}$
	$= 7.9 \times 10^{-3}\ M$

Comparison of these two values of $[H_3O^+]$ shows that, in contrast to hydrochloric acid, hydrofluoric acid is only partially dissociated in water. The **percent dissociation** of the hydrofluoric acid in the 0.10 M solution is only

$$\frac{[H_3O^+]}{[HF]_0} \times 100 = \frac{0.0079\ M}{0.10\ M} \times 100 = 7.9\%$$

where the subscript 0 denotes the stoichiometric concentration. In contrast, hydrochloric acid is completely dissociated:

$$\frac{[H_3O^+]}{[HCl]_0} \times 100 = \frac{0.10\ M}{0.10\ M} \times 100 = 100\%$$

Let's now examine a weak base in aqueous solution. Ammonia is a base in aqueous solution because of the reaction

$$NH_3(aq) + H_2O(l) \rightleftharpoons NH_4^+(aq) + OH^-(aq)$$

Note that the base $NH_3(aq)$ accepts a proton from the acid $H_2O(l)$, resulting in the production of $NH_4^+(aq)$ and $OH^-(aq)$. Ammonia is said to be a weak base because not all the ammonia molecules are protonated in aqueous solution.

Example 15-7: The pH of a 0.20 M $NH_3(aq)$ solution is 11.27. Calculate the percentage of ammonia molecules that are protonated in this solution.

Solution: The equation for the reaction is

$$NH_3(aq) + H_2O(l) \rightleftharpoons NH_4^+(aq) + OH^-(aq)$$

The percentage of ammonia molecules that are protonated is

$$\% \text{ protonated} = \frac{[NH_4^+]}{[NH_3]_0} \times 100$$

Because $[NH_4^+] = [OH^-]$, we have

$$\% \text{ protonated} = \frac{[OH^-]}{[NH_3]_0} \times 100$$

We need to determine $[OH^-]$. The pH is 11.27, and so

$$\text{pOH} = 14.00 - 11.27 = 2.73$$

The value of $[OH^-]$ can be calculated from the pOH:

$$[OH^-] = 10^{-\text{pOH}} = 10^{-2.73} = 1.9 \times 10^{-3}\ \text{M}$$

The percentage of ammonia molecules that are protonated in a 0.20 M $NH_3(aq)$ solution is

$$\% \text{ protonated} = \frac{1.9 \times 10^{-3}\ \text{M}}{0.20\ \text{M}} \times 100 = 0.95\%$$

Thus less than 1 percent of the ammonia molecules are protonated.

15-6. THE LARGER THE VALUE OF K_a, THE STRONGER IS THE ACID

The equilibrium constant expression for an **acid-dissociation reaction** is formulated according to the law of concentration action. Let's consider the weak acid acetic acid, $HC_2H_3O_2$, which produces $H_3O^+(aq)$ by the reaction with water

$$HC_2H_3O_2(aq) + H_2O(l) \rightleftharpoons H_3O^+(aq) + C_2H_3O_2^-(aq)$$

The equilibrium-constant expression for this equation is

$$K_a = \frac{[H_3O^+][C_2H_3O_2^-]}{[HC_2H_3O_2]}$$

▪ The law of concentration action is used to write K_a expressions.

where the subscript a on K reminds us that K_a is an **acid-dissociation constant.** Note that the $H_2O(l)$ concentration does not appear in the K_a

expression. The experimental value of K_a at 25°C for acetic acid is

$$K_a = 1.74 \times 10^{-5}\ \text{M} = \frac{[H_3O^+][C_2H_3O_2^-]}{[HC_2H_3O_2]} \tag{15-8}$$

For a given stoichiometric concentration, the percent dissociation of an acid depends on the value of K_a. The larger the value of K_a, the stronger is the acid. The small value of K_a reflects the fact that acetic acid is only slightly dissociated in aqueous solution. Table 15-2 gives the K_a values for a number of weak acids.

Because K_a values for aqueous acids range over many powers of 10 (Table 15-2), it is convenient to define a quantity **pK_a** as follows:

$$pK_a = -\log K_a \tag{15-9}$$

Note the similarity in the definitions of pH (Equation 15-4) and pK_a. The pK_a values at 25°C for several weak acids in water are given in Table 15-2. From the pK_a values given in Table 15-2, we see that the stronger the acid, the smaller is the value of pK_a.

Example 15-8: Given that $K_a = 1.74 \times 10^{-5}$ M for acetic acid in water at 25°C, calculate pK_a.

Solution: Using Equation (15-9) we have

$$\begin{aligned} pK_a &= -\log K_a \\ &= -\log(1.74 \times 10^{-5}) \\ &= 4.76 \end{aligned}$$

in agreement with the value in Table 15-2.

Let us now consider the problem of calculating the pH of a 0.050 M acetic acid solution. The concentration of acetic acid at equilibrium is

Table 15-2 The values of K_a and pK_a for some weak acids in water at 25°C

Acid	Formula	K_a/M	pK_a*
acetic	$HC_2H_3O_2$	1.74×10^{-5}	4.76
benzoic	$HC_7H_5O_2$	6.46×10^{-5}	4.19
chloroacetic	$HC_2H_2ClO_2$	1.35×10^{-3}	2.87
cyanic	HCNO	2.19×10^{-4}	3.66
formic	$HCHO_2$	1.78×10^{-4}	3.75
hydrazoic	HN_3	1.91×10^{-5}	4.72
hydrocyanic	HCN	4.79×10^{-10}	9.32
hydrofluoric	HF	6.76×10^{-4}	3.17
lactic	$HC_3H_5O_3$	1.40×10^{-4}	3.85
nitrous	HNO_2	4.47×10^{-4}	3.35
phenol	HC_6H_5O	1.0×10^{-10}	10.00

*pK_a is defined as pK_a = $-\log K_a$.

slightly less than 0.050 M because some of the $HC_2H_3O_2$ is converted to $C_2H_3O_2^-$ by the dissociation reaction. Thus we have the conservation condition

$$0.050\text{ M} = [HC_2H_3O_2] + [C_2H_3O_2^-]$$

Because we started with pure acetic acid, we know from the reaction stoichiometry that

$$[H_3O^+] \approx [C_2H_3O_2^-]$$

where we have neglected the contribution to $[H_3O^+]$ from the dissociation of water because K_w ($= 1.0 \times 10^{-14}\text{ M}^2$) is much smaller than K_a ($= 1.74 \times 10^{-5}$ M) for acetic acid. That is, $H_2O(l)$ is a much weaker acid than $HC_2H_3O_2(aq)$. We can set up a table for the initial and equilibrium concentrations of all the species in solution for the dissociation reaction:

	$HC_2H_3O_2(aq)$ +	$H_2O(l) \rightleftharpoons$	$H_3O^+(aq)$ +	$C_2H_3O_2^-(aq)$
initial concentration	0.050 M	—	≈0	0
equilibrium concentration	$0.050\text{ M} - [C_2H_3O_2^-]$	—	$[H_3O^+]$	$[C_2H_3O_2^-]$
or, using $[H_3O^+] = [C_2H_3O_2^-]$ from the reaction stoichiometry	$0.050\text{ M} - [H_3O^+]$	—	$[H_3O^+]$	$[H_3O^+]$

Note that we have set up this table so that the concentrations fall right under the species as they appear in the equation. Taking the expression for K_a

$$K_a = \frac{[H_3O^+][C_2H_3O_2^-]}{[HC_2H_3O_2]}$$

we substitute the experimental K_a value and the appropriate entries from the table to get

$$1.74 \times 10^{-5}\text{ M} = \frac{[H_3O^+]^2}{0.050\text{ M} - [H_3O^+]} \qquad (15\text{-}10)$$

Equation (15-10) can be written in the standard form of a quadratic equation:

▪ Recall that the solutions to the quadratic equation $ax^2 + bx + c = 0$ are

$$x = \frac{-b \pm \sqrt{b^2 - 4ac}}{2a}$$

$$[H_3O^+]^2 + (1.74 \times 10^{-5}\text{ M})[H_3O^+] - 8.70 \times 10^{-7}\text{ M}^2 = 0$$

The two solutions to this equation are

$$[H_3O^+] = 9.24 \times 10^{-4}\text{ M} \qquad \text{and} \qquad -1.86 \times 10^{-3}\text{ M}$$

We reject the physically unacceptable negative concentration, and thus

$$[H_3O^+] = 9.24 \times 10^{-4}\text{ M}$$

The pH of a 0.050 M acetic acid solution is

$$\begin{aligned} \text{pH} &= -\log[H_3O^+] \\ &= -\log[9.24 \times 10^{-4}] = 3.03 \end{aligned}$$

The percent dissociation of acetic acid in a 0.050 M solution of $HC_2H_3O_2(aq)$ is

$$\% \text{ dissociation} = \frac{[H_3O^+]}{[HC_2H_3O_2]_0} \times 100 = \frac{9.24 \times 10^{-4}\text{ M}}{0.050\text{ M}} \times 100 = 1.8\%$$

This result shows that over 98 percent of the acetic acid in a 0.050 M solution remains undissociated.

Figure 15-4 shows how the percent dissociation varies with the concentration of acetic acid. Notice that the percent dissociation increases as the acetic acid concentration decreases. This effect can be understood in terms of Le Châtelier's principle (Section 14-6). The addition of water increases the volume available to the reaction. Thus the reaction equilibrium shifts to the side with the greater number of moles of solute species; that is, the equilibrium shifts in the direction

$$HC_2H_3O_2(aq) + H_2O(l) \rightarrow H_3O^+(aq) + C_2H_3O_2^-(aq)$$

Note that the addition of solvent (dilution of the reaction mixture) increases the percent dissociation. As the initial concentration of the acid decreases, the percent dissociation increases, even though the value of $[H_3O^+]$, and thus the acidity of the solution, decreases.

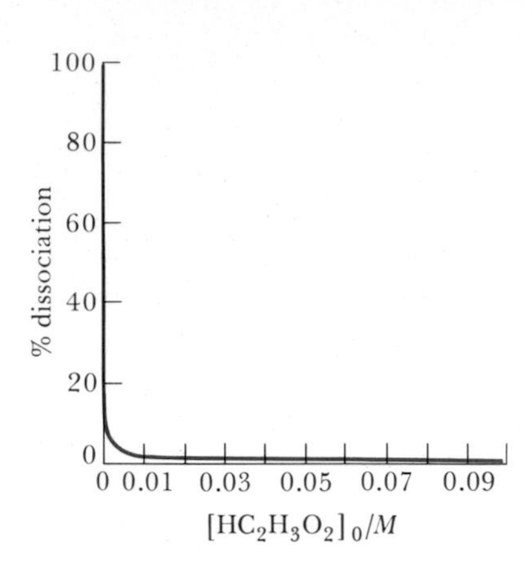

Figure 15-4 The percent dissociation of acetic acid as a function of the stoichiometric (initial) concentration of acid, $[HC_2H_3O_2]_0$. The calculations are carried out by solving the quadratic equation

$$\frac{[H_3O^+]^2}{[HC_2H_3O_2]_0 - [H_3O^+]} = 1.74 \times 10^{-5}\text{ M}$$

for various values of $[HC_2H_3O_2]_0$. The percent dissociation is then computed by using

$$\% \text{ dissociation} = \frac{[H_3O^+]100}{[HC_2H_3O_2]_0}$$

This procedure works down to $[HC_2H_3O_2]_0 \approx 10^{-5}$ M; below this concentration the contribution to $[H_3O^+]$ from the dissociation of water must be taken into account.

15-7. THE METHOD OF SUCCESSIVE APPROXIMATIONS IS OFTEN USED IN SOLVING ACID-BASE EQUILIBRIUM PROBLEMS

The use of the quadratic formula to solve Equation (15-10) is tedious. An alternate, and usually faster, method of solution of a quadratic equation is the **method of successive approximations.** Let's reconsider Equation (15-10):

$$1.74 \times 10^{-5}\text{ M} = \frac{[H_3O^+]^2}{0.050\text{ M} - [H_3O^+]}$$

Because acetic acid is a weak acid (as indicated by the K_a value), we expect that the percent dissociation of the acid is small. Furthermore, because essentially all the $H_3O^+(aq)$ arises from dissociation of the acid, we expect that $[H_3O^+]$ will be small relative to the initial concentration of the acid. We can express this algebraically as

$$0.050\text{ M} - [H_3O^+] \approx 0.050\text{ M} \qquad (15\text{-}11)$$

Using this approximation in Equation (15-10) yields for $[H_3O^+]$

$$[H_3O^+]^2 \approx (0.050\text{ M})(1.74 \times 10^{-5}\text{ M})$$

or

$$[H_3O^+] \approx \sqrt{(0.050\text{ M})(1.74 \times 10^{-5}\text{ M})} = 9.33 \times 10^{-4}\text{ M}$$

Note that this first approximation is close to the exact value of 9.24×10^{-4} M obtained by the solution of the full quadratic equation in the previous section.

We can now use this approximate value of $[H_3O^+]$ in the denominator of the right-hand side of Equation (15-10) to obtain a second, more accurate, approximation; that is, we take

$$1.74 \times 10^{-5}\ \text{M} \approx \frac{[\text{H}_3\text{O}^+]^2}{0.050\ \text{M} - 9.33 \times 10^{-4}\ \text{M}}$$

Thus, we calculate that

$$[\text{H}_3\text{O}^+] \approx \sqrt{(0.0491\ \text{M})(1.74 \times 10^{-5}\ \text{M})}$$
$$= 9.24 \times 10^{-4}\ \text{M}$$

which is the same as the exact solution. If we repeat the approximation procedure, using 9.24×10^{-4} M in the denominator, then we find no further change in the value of $[H_3O^+]$ to three significant figures. An unchanged value of $[H_3O^+]$ on successive approximations means that we have found the correct solution. This method of successive approximations is generally much faster and easier than the use of the full quadratic equation, particularly in using a hand calculator. In many cases, the first approximation, where c_0 is the initial concentration of the acid,

$$K_a \approx \frac{[\text{H}_3\text{O}^+]^2}{c_0}$$

or

$$[\text{H}_3\text{O}^+] \approx \sqrt{c_0 K_a}$$

gives a value for $[H_3O^+]$ that is accurate to within a few percent. Because of its simplicity and speed, we shall often use the method of successive approximations to solve equilibrium problems.

Example 15-9: The value of K_a for an aqueous solution of hypochlorous acid, HClO(*aq*), is 3.0×10^{-8} M. Calculate the pH of a 0.050 M HClO(*aq*) solution.

Solution: The equilibrium equation is

$$\text{HClO}(aq) + \text{H}_2\text{O}(l) \rightleftharpoons \text{H}_3\text{O}^+(aq) + \text{ClO}^-(aq)$$

We can set up the following table for initial and equilibrium concentrations of the species in solution:

	HClO(*aq*)	**+ $H_2O(l)$ ⇌**	**$H_3O^+(aq)$ +**	**$ClO^-(aq)$**
initial concentration	0.050 M	—	≈0	0
equilibrium concentration	0.050 M − $[ClO^-]$	—	$[H_3O^+]$	$[ClO^-]$
or, using $[H_3O^+] = [ClO^-]$ from the reaction stoichiometry	0.050 M − $[H_3O^+]$	—	$[H_3O^+]$	$[H_3O^+]$

If we combine the entries in this table with the expression for K_a, then we have

$$K_a = \frac{[\text{H}_3\text{O}^+][\text{ClO}^-]}{[\text{HClO}]} = \frac{[\text{H}_3\text{O}^+]^2}{0.050\ \text{M} - [\text{H}_3\text{O}^+]} = 3.0 \times 10^{-8}\ \text{M}$$

We use the method of successive approximations to solve this equation.

Neglecting $[H_3O^+]$ relative to 0.050 M, we obtain the first approximate solution:

$$[H_3O^+] \approx \sqrt{(0.050\text{ M})(3.0 \times 10^{-8}\text{ M})} = 3.9 \times 10^{-5}\text{ M}$$

Because $[H_3O^+]$ is very small relative to 0.050 M, the second approximation yields the same value of $[H_3O^+]$ as the first approximation; that is,

$$\frac{[H_3O^+]^2}{0.050\text{ M} - 3.9 \times 10^{-5}\text{ M}} = 3.0 \times 10^{-8}\text{ M}$$

and

$$[H_3O^+] \approx \sqrt{(0.050\text{ M} - 3.9 \times 10^{-5}\text{ M})(3.0 \times 10^{-8}\text{ M})} = 3.9 \times 10^{-5}\text{ M}$$

The pH of the solution is

$$\text{pH} = -\log [H_3O^+] = -\log (3.9 \times 10^{-5}) = 4.41$$

15-8. THE LARGER THE VALUE OF K_b, THE STRONGER IS THE BASE

Ammonia is a base in aqueous solution because it reacts with water to accept a proton and thereby liberates a hydroxide ion:

$$NH_3(aq) + H_2O(l) \rightleftharpoons NH_4^+(aq) + OH^-(aq)$$

In terms of Lewis formulas, the protonation of ammonia can be described by

$$H{-}\overset{H}{\underset{H}{\overset{|}{\underset{|}{N}}}}{:} + H_2\ddot{\underset{..}{O}} \rightleftharpoons H{-}\overset{H}{\underset{H}{\overset{|}{\underset{|}{N^{\oplus}}}}}{-}H + {:}\ddot{\underset{..}{O}}^{\ominus}{-}H$$

The lone pair of electrons on the nitrogen atom in the ammonia molecule is able to bond the proton donated by the water molecule. Ammonia is a weak base because the equilibrium constant for its reaction with water is 1.75×10^{-5} M, indicating that the reaction equilibrium lies far to the left. The most common weak bases are ammonia and some organic compounds that are similar to ammonia. For example, if we substitute the hydrogen atoms in ammonia with methyl groups, then we obtain the weak bases

$H{-}\ddot{N}{-}CH_3$ with H on N	$H_3C{-}\ddot{N}{-}CH_3$ with H on N	$H_3C{-}\ddot{N}{-}CH_3$ with CH_3 on N
methylamine	dimethylamine	trimethylamine

Some other organic bases are

$C_6H_5{-}\ddot{N}{-}H$ with H on N	$C_5H_5N{:}$
aniline	pyridine

Notice that the nitrogen atom in all these bases has a lone pair of electrons. Pyridine, for example, is basic in aqueous solution because of the reaction

$$\text{C}_5\text{H}_5\text{N:} + \text{H}_2\text{O}(l) \rightleftharpoons \text{C}_5\text{H}_5\overset{+}{\text{N}}\text{—H} + \text{OH}^-(aq)$$

pyridine   pyridinium ion

If we abbreviate pyridine by Py and the pyridinium ion by PyH^+, then the equilibrium-constant expression for this reaction is

$$K_b = \frac{[\text{PyH}^+][\text{OH}^-]}{[\text{Py}]}$$

The subscript b on K indicates that the reaction is a reaction of a weak base with water, that is, a base-protonation reaction. Thus K_b is called the **base-protonation constant.** The values of K_b and pK_b for some weak bases are given in Table 15-3. By analogy to pK_a, $\mathbf{pK_b}$ is defined as follows:

$$pK_b = -\log K_b \qquad (15\text{-}12)$$

The smaller the value of pK_b, the stronger is the base.

Example 15-10: Aniline is used in the manufacture of dyes and various pharmaceuticals. The solubility of aniline in water at 25°C is 1.00 g per 28.6 mL of solution. Calculate the pH of a saturated aqueous solution of aniline at 25°C.

Solution: The concentration of a saturated solution of aniline in water at 25°C is

$$\left(\frac{1.00\text{ g aniline}}{28.6\text{ mL soln}}\right)\left(\frac{1000\text{ mL}}{1\text{ L}}\right)\left(\frac{1\text{ mol aniline}}{93.13\text{ g aniline}}\right) = 0.375\text{ M}$$

Thus we must calculate the pH of a 0.375 M aqueous aniline solution. From Table 15-3, K_b is 4.17×10^{-10} M for the equilibrium

$$\text{C}_6\text{H}_5\text{—NH}_2(aq) + \text{H}_2\text{O}(l) \rightleftharpoons \text{C}_6\text{H}_5\text{—}\overset{+}{\text{N}}\text{H}_3(aq) + \text{OH}^-(aq)$$

If we let aniline be An and its protonated form be AnH^+, then

$$K_b = \frac{[\text{AnH}^+][\text{OH}^-]}{[\text{An}]} = 4.17 \times 10^{-10}\text{ M}$$

A table of initial and equilibrium concentrations has the form

	$An(aq)$	$+ H_2O(l) \rightleftharpoons$	$AnH^+(aq)$	$+ OH^-(aq)$
initial concentration	0.375 M	—	0	≈0
equilibrium concentration	$0.375\text{ M} - [\text{AnH}^+]$	—	$[\text{AnH}^+]$	$[\text{OH}^-]$
or, using $[\text{AnH}^+] = [\text{OH}^-]$ from the reaction stoichiometry	$0.375\text{ M} - [\text{OH}^-]$	—	$[\text{OH}^-]$	$[\text{OH}^-]$

Table 15-3 The values of K_b and pK_b for some weak bases in water at 25°C

Base	Formula	K_b/M	pK_b*
ammonia	$H-\ddot{N}(-H)-H$	1.75×10^{-5}	4.76
methylamine	$H-\ddot{N}(-H)-CH_3$	4.59×10^{-4}	3.34
dimethylamine	$H_3C-\ddot{N}(-H)-CH_3$	5.81×10^{-4}	3.24
trimethylamine	$H_3C-\ddot{N}(-CH_3)-CH_3$	6.11×10^{-5}	4.21
hydroxylamine	$H-\ddot{N}(-H)-OH$	1.07×10^{-8}	7.97
aniline	$C_6H_5-\ddot{N}(-H)-H$	4.17×10^{-10}	9.38
pyridine	C_5H_5N:	1.46×10^{-9}	8.84

*pK_b is defined as p$K_b = -\log K_b$.

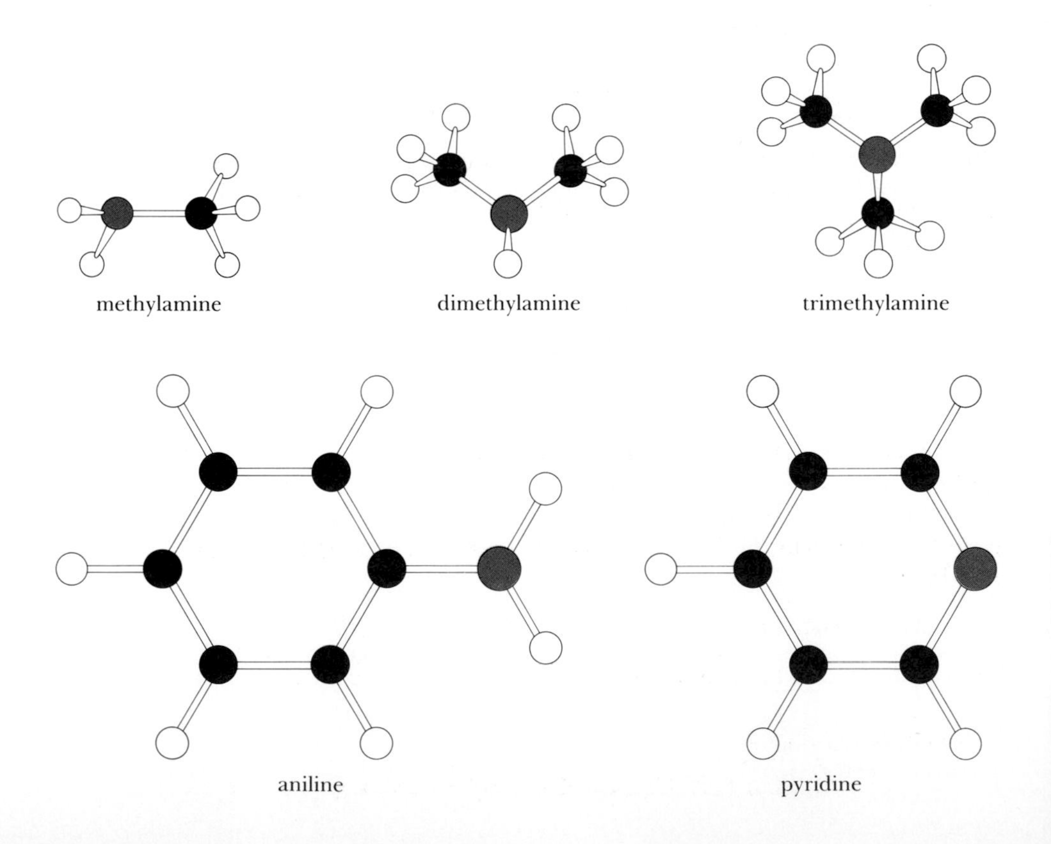

The expression for K_b is

$$K_b = \frac{[AnH^+][OH^-]}{[An]} = \frac{[OH^-]^2}{0.375\ M - [OH^-]} = 4.17 \times 10^{-10}\ M$$

Because K_b is so small, we expect $[OH^-]$ to be small also and thus negligible relative to 0.375 M. The K_b expression becomes

$$K_b = 4.17 \times 10^{-10}\ M \approx \frac{[OH^-]^2}{0.375\ M}$$

and

$$[OH^-] \approx 1.25 \times 10^{-5}\ M$$

We check to be sure that the value of $[OH^-]$ is negligible with respect to 0.375, and indeed it is. The pOH is

$$pOH = -\log(1.25 \times 10^{-5}\ M) = 4.90$$

and so the pH is

$$pH = 14.00 - pOH = 14.00 - 4.90 = 9.10$$

You probably noted the symmetry between weak-acid and weak-base equilibrium calculations. Denoting a weak acid by HB, we have

$$HB(aq) + H_2O(l) \rightleftharpoons H_3O^+(aq) + B^-(aq)$$

$$K_a = \frac{[H_3O^+][B^-]}{[HB]} = \frac{[H_3O^+]^2}{[HB]_0 - [H_3O^+]}$$

Denoting a weak base by B, we have

$$B(aq) + H_2O(l) \rightleftharpoons OH^-(aq) + HB^+(aq)$$

$$K_b = \frac{[OH^-][HB^+]}{[B]} = \frac{[OH^-]^2}{[B]_0 - [OH^-]}$$

Note the similarity in the algebraic forms of the K_a and K_b expressions. The methods of solving the K_a and K_b expressions are essentially the same. In the acid case we solve for $[H_3O^+]$ and in the base case we solve for $[OH^-]$.

15-9. POLYPROTIC ACIDS CAN DONATE MORE THAN ONE PROTON

Some acids have more than one dissociable proton and are called **polyprotic acids.** Examples are the **diprotic acids** sulfuric acid, H_2SO_4, and hydrosulfuric acid, H_2S, and the **triprotic acid** phosphoric acid, H_3PO_4. A 1-mol sample of H_2S is capable of neutralizing 2 mol of KOH:

$$H_2S(aq) + 2KOH(aq) \rightarrow K_2S(aq) + 2H_2O(l)$$

Hydrosulfuric acid has two distinct acid-dissociation equilibria:

1. dissociation of $H_2S(aq)$

$$H_2S(aq) + H_2O(l) \rightleftharpoons H_3O^+(aq) + HS^-(aq)$$

$$K_{a1} = \frac{[H_3O^+][HS^-]}{[H_2S]} = 9.10 \times 10^{-8}\ \text{M}$$

2. dissociation of $HS^-(aq)$

$$HS^-(aq) + H_2O(l) \rightleftharpoons H_3O^+(aq) + S^{2-}(aq)$$

$$K_{a2} = \frac{[H_3O^+][S^{2-}]}{[HS^-]} = 1.20 \times 10^{-13}\ \text{M}$$

As a rule of thumb, each successive acid-dissociation constant of a polyprotic acid is about 10^{-5} times the value of the preceding one. Each succeeding proton of a polyprotic acid is more difficult to remove, primarily because of the extra energy required to separate a (positively charged) proton from a negatively charged ion. The successive dissociation constants (as pK_a values) of some polyprotic acids are given in Table 15-4.

In general, the calculation of the pH of an aqueous solution of a polyprotic acid is more complicated than that of a monoprotic acid (an acid having only one dissociable proton). However, because the first acid-dissociation constant is usually much larger than the successive values, it is possible to obtain a very good approximate value of the pH of a solution of the acid by ignoring the second dissociation and by treating the polyprotic acid as though it were simply a monoprotic acid. As an example, let's calculate the pH of a 0.050 M $H_2S(aq)$ solution. We consider only the first acid-dissociation equilibrium:

$$H_2S(aq) + H_2O(l) \rightleftharpoons H_3O^+(aq) + HS^-(aq) \qquad K_{a1} = 9.10 \times 10^{-8}\ \text{M}$$

and solve this problem as we did for monoprotic acids in Section 15-6. We set up a table for initial and equilibrium concentrations of the major species in solution:

	$H_2S(aq)$	$+ H_2O(l) \rightleftharpoons$	$H_3O^+(aq)$	$+ HS^-(aq)$
initial concentration	0.05 M	—	≈ 0	0
equilibrium concentration	0.05 M − $[HS^-]$	—	$[H_3O^+]$	$[HS^-]$
or, using $[H_3O^+] = [HS^-]$ from the reaction stoichiometry	0.05 M − $[H_3O^+]$	—	$[H_3O^+]$	$[H_3O^+]$

If we combine the entries in this table with the expression for K_{a1}, then we have

$$K_{a1} = \frac{[H_3O^+][HS^-]}{[H_2S]} = \frac{[H_3O^+]^2}{0.050\ \text{M} - [H_3O^+]} = 9.10 \times 10^{-8}\ \text{M}$$

We solve for $[H_3O^+]$ and obtain 6.75×10^{-5} M, from which we find that

$$\text{pH} = -\log\,[H_3O^+] = -\log\,(6.75 \times 10^{-5}) = 4.17$$

Table 15-4 Successive acid-dissociation constants (as pK_a values) for some polyprotic acids in water at 25°C

Acid	pK_{a1}	pK_{a2}	pK_{a3}
Diprotic			
sulfuric H_2SO_4	strong	2.00	—
oxalic $H_2C_2O_4$	1.27	4.27	—
carbonic H_2CO_3	6.35	10.33	—
hydrogen sulfide H_2S	7.04	12.92	—
Triprotic			
phosphoric H_3PO_4	2.15	7.21	12.36

Because K_{a2} ($= 1.20 \times 10^{-13}$ M) is so much smaller than K_{a1} ($= 9.10 \times 10^{-8}$ M), it is a very good approximation to ignore the $H_3O^+(aq)$ produced by the second acid-dissociation equilibrium. The value of $[S^{2-}]$ is calculated from the K_{a2} expression using the relation $[H_3O^+] \approx [HS^-]$

$$[S^-] = \frac{K_{a2}[HS^-]}{[H_3O^+]} \approx K_{a2} = 1.2 \times 10^{-13}\text{ M}.$$

This small value of $[S^{2-}]$ relative to $[H_3O^+]$ confirms the assumption that the dissociation of $HS^-(aq)$ is a negligible source of $H_3O^+(aq)$.

15-10. THE ACID-BASE PAIR HB,B$^-$ IS CALLED A CONJUGATE ACID-BASE PAIR

When an acid such as acetic acid dissociates, water acts as a base by accepting a proton from the acid:

$$HC_2H_3O_2(aq) + H_2O(l) \rightleftharpoons H_3O^+(aq) + C_2H_3O_2^-(aq)$$

If we look at the reverse reaction, then we see that $H_3O^+(aq)$ donates a proton to $C_2H_3O_2^-(aq)$. In other words, $H_3O^+(aq)$ acts as an acid by donating a proton and $C_2H_3O_2^-$ acts as a base by accepting a proton. The base $C_2H_3O_2^-(aq)$ is called the **conjugate base** of the acid $HC_2H_3O_2(aq)$; the pair $HC_2H_3O_2(aq)$, $C_2H_3O_2^-(aq)$ is called a **conjugate acid-base pair.** Similarly, the hydronium ion, $H_3O^+(aq)$, is the **conjugate acid** of the base $H_2O(l)$. The conjugate acid-base pairs for the dissociation of acetic acid in water are

$$HC_2H_3O_2(aq) + H_2O(l) \rightleftharpoons H_3O^+(aq) + C_2H_3O_2^-(aq)$$

conjugate acid-base pair

conjugate acid-base pair

Note that a conjugate base has one less proton than its corresponding conjugate acid.

An acetate ion is basic because it is the anion of a weak acid, acetic acid. To see more explicitly that an acetate ion is a weak base, consider the equation for the reverse of the acid dissociation:

$$(1) \qquad H_3O^+(aq) + C_2H_3O_2^-(aq) \rightleftharpoons HC_2H_3O_2(aq) + H_2O(l)$$

Because this reaction is the reverse of the reaction associated with the K_a of the acetic acid, we have for the equilibrium constant K_1 of Equation (1)

$$K_1 = \frac{1}{K_a}$$

The equation for the dissociation of water is

$$(2) \qquad H_2O(l) + H_2O(l) \rightleftharpoons H_3O^+(aq) + OH^-(aq)$$

for which

$$K_2 = K_w$$

If we add Equations (1) and (2), then we obtain

$$(3) \qquad C_2H_3O_2^-(aq) + H_2O(l) \rightleftharpoons HC_2H_3O_2(aq) + OH^-(aq)$$

The equilibrium constant for this equation is the product of the equilibrium constants for Equations (1) and (2) (Section 14-4):

$$K_3 = K_1K_2 = \frac{K_w}{K_a}$$

Note that Equation (3) shows explicitly that an acetate ion accepts a proton from water to yield hydroxide ions. Equation (3) is the same form as the equation for the reaction between ammonia and water. We denote the equilibrium constant for Equation (3) by K_b instead of K_3:

$$C_2H_3O_2^-(aq) + H_2O(l) \rightleftharpoons HC_2H_3O_2(aq) + OH^-(aq)$$

$$K_b = \frac{K_w}{K_a} \qquad (15\text{-}13)$$

The value of K_a for acetic acid at 25°C is 1.74×10^{-5} M, and so at 25°C

$$K_b = \frac{1.00 \times 10^{-14}\ \text{M}^2}{1.74 \times 10^{-5}\ \text{M}} = 5.75 \times 10^{-10}\ \text{M}$$

The small value of K_b means that acetate ion is a weak base in water.

Generally, the conjugate base of any weak acid is itself a weak base. If we denote a weak acid by HB and its conjugate base by B^-, then we have the equations

$$HB(aq) + H_2O(l) \rightleftharpoons H_3O^+(aq) + B^-(aq)$$

with

$$K_a = \frac{[H_3O^+][B^-]}{[HB]} \qquad (15\text{-}14)$$

and

$$B^-(aq) + H_2O(l) \rightleftharpoons HB(aq) + OH^-(aq)$$

with

$$K_b = \frac{[HB][OH^-]}{[B^-]} \qquad (15\text{-}15)$$

The first equation explicitly shows HB as an acid, and the second equation explicitly shows its conjugate base B^- as a base. The relation between K_a and K_b for a conjugate acid-base pair is

$$K_aK_b = K_w \qquad (15\text{-}16)$$

as can be seen by multiplying Equations (15-14) and (15-15) together.

Example 15-11: Given that $K_a = 6.76 \times 10^{-4}$ M at 25°C for the acid HF(*aq*), compute K_b and pK_b for $F^-(aq)$.

Solution: The fluoride ion is the conjugate base of the weak acid HF(*aq*). The equilibrium equation associated with K_b is

$$F^-(aq) + H_2O(l) \rightleftharpoons HF(aq) + OH^-(aq)$$

where

$$K_b = \frac{[HF][OH^-]}{[F^-]}$$

Thus from Equation (15-16) we have at 25°C

$$K_b = \frac{K_w}{K_a} = \frac{1.00 \times 10^{-14}\ M^2}{6.76 \times 10^{-4}\ M}$$

$$= 1.48 \times 10^{-11}\ M \text{ at } 25°C$$

The value of pK_b for $F^-(aq)$ is obtained by using Equation (15-12):

$$pK_b = -\log K_b = -\log (1.48 \times 10^{-11}) = 10.83$$

The relation pH + pOH = 14.00 (Equation 15-6) for water is a special case of a more general relation for a conjugate acid-base pair. To see this, multiply the logarithm of Equation (15-16) by −1 to obtain

$$-\log (K_aK_b) = -\log K_a - \log K_b = -\log K_w = 14.00$$

at 25°C. Using the definitions of pK_a (Equation 15-9) and pK_b (Equation 15-12) yields the following equation for a conjugate acid-base pair:

$$pK_a + pK_b = 14.00 \qquad (15\text{-}17)$$

Thus the sum of the pK_a and pK_b values for a conjugate acid-base pair must equal 14.00 at 25°C.

Table 15-5 lists values of K_a, pK_a, K_b and pK_b for a number of conjugate acid-base pairs. Note that pK_a + pK_b = 14.00 in each case. Also note that the stronger the acid, the weaker is its conjugate base. This reciprocal relation between the strength of an acid and the strength of its conjugate base is given in Equation (15-16).

It is interesting to note that water can act as either an acid or a base:

$$H_2O(l) + H_2O(l) \rightleftharpoons H_3O^+(aq) + OH^-(aq)$$

conjugate acid-base pair

conjugate acid-base pair

It is this property of water that gives it such a central role in acid-base chemistry.

Table 15-5 Values of K_a, pK_a, K_b, and pK_b at 25°C for some conjugate acid-base pairs

Name	Acid	K_a/M	pK_a	Base	K_b/M	pK_b
sulfurous acid	H_2SO_3	1.54×10^{-2}	1.81	HSO_3^-	6.49×10^{-13}	12.19
hydrogen sulfate ion	HSO_4^-	1.20×10^{-2}	1.92	SO_4^{2-}	8.33×10^{-13}	12.08
phosphoric acid	H_3PO_4	5.93×10^{-3}	2.23	$H_2PO_4^-$	1.69×10^{-12}	11.77
chloroacetic acid	$HC_2H_2ClO_2$	1.35×10^{-3}	2.87	$C_2H_2ClO_2^-$	7.41×10^{-12}	11.13
hydrofluoric acid	HF	6.76×10^{-4}	3.17	F^-	1.48×10^{-11}	10.83
nitrous acid	HNO_2	4.47×10^{-4}	3.35	NO_2^-	2.24×10^{-11}	10.65
cyanic acid	HCNO	2.19×10^{-4}	3.66	CNO^-	4.57×10^{-11}	10.34
formic acid	$HCHO_2$	1.78×10^{-4}	3.75	CHO_2^-	5.62×10^{-11}	10.25
benzoic acid	$HC_7H_5O_2$	6.46×10^{-5}	4.19	$C_7H_5O_2^-$	1.55×10^{-10}	9.81
acetic acid	$HC_2H_3O_2$	1.74×10^{-5}	4.76	$C_2H_3O_2^-$	5.75×10^{-10}	9.24
pyridinium ion	$C_5H_5NH^+$	6.84×10^{-6}	5.16	C_5H_5N	1.46×10^{-9}	8.84
hydrosulfuric acid	H_2S	9.10×10^{-8}	7.04	HS^-	1.10×10^{-7}	6.96
dihydrogen phosphate ion	$H_2PO_4^-$	6.32×10^{-8}	7.20	HPO_4^{2-}	1.58×10^{-7}	6.80
hydrogen sulfite ion	HSO_3^-	6.21×10^{-8}	7.21	SO_3^{2-}	1.61×10^{-7}	6.79
hypochlorous acid	HClO	3.00×10^{-8}	7.52	ClO^-	3.33×10^{-7}	6.48
ammonium ion	NH_4^+	5.71×10^{-10}	9.24	NH_3	1.75×10^{-5}	4.76
boric acid	H_3BO_3	5.91×10^{-10}	9.23	$H_2BO_3^-$	1.69×10^{-5}	4.77
hydrocyanic acid	HCN	4.79×10^{-10}	9.32	CN^-	2.09×10^{-5}	4.68
phenol	HC_6H_5O	1.00×10^{-10}	10.00	$C_6H_5O^-$	1.00×10^{-4}	4.00
hydrogen carbonate ion	HCO_3^-	4.72×10^{-11}	10.33	CO_3^{2-}	2.19×10^{-4}	3.67
hydrogen phosphate ion	HPO_4^{2-}	4.84×10^{-13}	12.32	PO_4^{3-}	2.07×10^{-2}	1.68
hydrogen sulfide ion	HS^-	1.20×10^{-13}	12.92	S^{2-}	8.33×10^{-2}	1.08

↑ increasing acid strength (upward) — increasing base strength ↓ (downward)

For the acids: $HB(aq) + H_2O(l) \rightleftharpoons H_3O^+(aq) + B^-(aq)$ $\qquad K_a = \dfrac{[H_3O^+][B^-]}{[HB]}$

For the bases: $B^-(aq) + H_2O(l) \rightleftharpoons HB(aq) + OH^-(aq)$ $\qquad K_b = \dfrac{[HB][OH^-]}{[B^-]}$

$K_aK_b = K_w = 1.00 \times 10^{-14}$ M^2 and pK_a + pK_b = 14.00 for a conjugate-acid-base pair at 25°C.

pH + pOH = 14.00 at 25°C for any aqueous solution.

15-11. AQUEOUS SOLUTIONS OF MANY SALTS ARE EITHER ACIDIC OR BASIC

Suppose we dissolve sodium acetate, $NaC_2H_3O_2(s)$, in water. The resulting solution contains sodium ions, $Na^+(aq)$, and acetate ions, $C_2H_3O_2^-(aq)$. However, the acetate ion is the conjugate base of the weak acid $HC_2H_3O_2$. Consequently, a fraction of the $C_2H_3O_2^-$ ions are protonated according to

$$C_2H_3O_2^-(aq) + H_2O(l) \rightleftharpoons HC_2H_3O_2(aq) + OH^-(aq) \qquad (15\text{-}18)$$

The sodium ions do not react with water to produce hydronium ions or

Table 15-6 Acid-base properties of some common cations and anions in water

Cations

Acidic	**Neutral**		**Basic**
NH_4^+	Li^+	Mg^{2+}	none
Al^{3+}, Pb^{2+}, Sn^{2+}	Na^+	Ca^{2+}	
transition	K^+	Sr^{2+}	
metal ions	Rb^+	Ba^{2+}	
	Cs^+		

Anions

Acidic	**Neutral**		**Basic**	
HSO_4^-	Cl^-	ClO_4^-	F^-	CN^-
	Br^-	NO_3^-	$C_2H_3O_2^-$	S^{2-}
	I^-		NO_2^-	SO_4^{2-}
			HCO_3^-	HPO_4^{2-}
			CO_3^{2-}	PO_4^{3-}
			plus many others	

hydroxide ions, and so the net result is that the sodium acetate solution is basic because the protonation of acetate by water produces hydroxide ions.

Various ions react with water to produce hydronium ions or hydroxide ions. The acidic, neutral, or basic properties of a number of ions are given in Table 15-6, from which you should note the following:

1. The conjugate bases of strong monoprotic acids—for example, $Cl^-(aq)$—are **neutral anions;** they do not react with H_2O to produce $OH^-(aq)$. The conjugate bases of weak acids—for example, $C_2H_3O_2^-(aq)$—are **basic anions;** they react with H_2O to produce $OH^-(aq)$. The hydrogen sulfate ion, HSO_4^- is an **acidic anion** because of the second acid dissociation of H_2SO_4:

$$HSO_4^-(aq) + H_2O(l) \rightleftharpoons H_3O^+(aq) + SO_4^{2-}(aq) \qquad pK_a = 1.92$$

2. There are no basic cations; there are only **acidic cations** and **neutral cations.** The alkali and alkaline earth cations (except Be^{2+}) are all neutral. The conjugate acids of weak bases—$NH_4^+(aq)$, for example—are acidic.

3. Many metal ions are acidic in aqueous solution. These ions exist in aqueous solution with a certain number of water molecules bonded to them and are said to be solvated. An example of an acidic cation is $Fe(H_2O)_6^{3+}(aq)$. The equation for the acid-dissociation reaction of this ion is

$$Fe(H_2O)_6^{3+}(aq) + H_2O(l) \rightleftharpoons Fe(OH)(H_2O)_5^{2+}(aq) + H_3O^+(aq)$$
$$K_a = 1 \times 10^{-3}\ M$$

Note that the acidity of the ion $Fe(H_2O)_6^{3+}(aq)$ arises from the loss of a proton from an attached water molecule.

Salts of neutral cations and basic anions dissolved in water produce basic solutions. For example, when sodium nitrite is dissolved in water

$$NaNO_2(s) \xrightarrow[H_2O(l)]{} Na^+(aq) + NO_2^-(aq)$$

the protonation reaction of the weak base $NO_2^-(aq)$

$$NO_2^-(aq) + H_2O(l) \rightleftharpoons HNO_2(aq) + OH^-(aq)$$

produces $OH^-(aq)$ and thus yields a basic solution.

Salts of acidic cations and neutral anions produce acidic solutions when dissolved in water. For example, when aluminum nitrate is dissolved in water

$$Al(NO_3)_3(s) \xrightarrow[H_2O(l)]{} Al(H_2O)_6^{3+}(aq) + 3NO_3^-(aq)$$

the acid-dissociation reaction is

$$Al(H_2O)_6^{3+}(aq) + H_2O(l) \rightleftharpoons Al(OH)(H_2O)_5^{2+}(aq) + H_3O^+(aq)$$

which shows explicitly the production of $H_3O^+(aq)$.

If we have a salt with an acidic cation and a basic anion, for example, $NH_4CN(aq)$, then we need to know the K_a value of the cation and the K_b value of the anion to predict whether the salt solution is acidic, basic, or neutral.

Example 15-12: Predict whether the following salts produce acidic, neutral, or basic solutions when dissolved in water: KCl, NaCN, $AlBr_3$, and NH_4NO_3.

Solution:

KCl	neutral cation, neutral anion	neutral solution
NaCN	neutral cation, basic anion	basic solution
$AlBr_3$	acidic cation, neutral anion	acidic solution
NH_4NO_3	acidic cation, neutral anion	acidic solution

Example 15-13: Given that $K_a = 5.7 \times 10^{-10}$ M for $NH_4^+(aq)$ and that $K_b = 2.1 \times 10^{-5}$ M for $CN^-(aq)$, predict whether a solution of $NH_4CN(aq)$ is acidic, basic, or neutral.

Solution: For the salt solution we have $[NH_4^+]_0 = [CN^-]_0$. The acid-dissociation equilibrium is

$$NH_4^+(aq) + H_2O(l) \rightleftharpoons NH_3(aq) + H_3O^+(aq) \qquad K_a = 5.7 \times 10^{-10}\ M$$

The base-protonation equilibrium is

$$CN^-(aq) + H_2O(l) \rightleftharpoons HCN(aq) + OH^-(aq) \qquad K_b = 2.1 \times 10^{-5}\ M$$

Because $K_b \gg K_a$, the value of $[OH^-]$ (from the base protonation) at equilibrium is much greater than the value of $[H_3O^+]$ (from the acid dissociation) at equilibrium, and thus the $NH_4CN(aq)$ solution is basic.

In addition to predicting whether a salt solution is acidic or basic, we also can calculate its pH using the methods already developed. Let's calculate the pH of a 0.050 M NH_4Cl solution. A chloride ion is neutral

(Table 15-6). An ammonium ion is acidic because of the acid-dissociation equilibrium.

$$NH_4^+(aq) + H_2O(l) \rightleftharpoons H_3O^+(aq) + NH_3(aq)$$

From Table 15-5, we find that $K_a = 5.71 \times 10^{-10}$ M for this reaction. A table of the initial and equilibrium concentrations is

	$NH_4^+(aq)$	$+ H_2O(l) \rightleftharpoons$	$H_3O^+(aq)$	$+ NH_3(aq)$
initial concentration	0.050 M	—	≈0	0
equilibrium concentration	0.050 M − $[NH_3]$	—	$[H_3O^+]$	$[NH_3]$
or, because $[H_3O^+] \approx [NH_3]$ from the reaction stoichiometry	0.050 M − $[H_3O^+]$	—	$[H_3O^+]$	$[H_3O^+]$

The expression for K_a becomes

$$\frac{[H_3O^+]^2}{0.050\text{ M} - [H_3O^+]} = 5.71 \times 10^{-10}\text{ M}$$

We neglect $[H_3O^+]$ relative to 0.050 M in the denominator and obtain

$$[H_3O^+] = (0.050\text{ M} \times 5.71 \times 10^{-10}\text{ M})^{1/2} = 5.3 \times 10^{-6}\text{ M}$$

The method of successive approximations confirms this result for $[H_3O^+]$. The pH of the solution is

$$\text{pH} = -\log [H_3O^+] = -\log (5.3 \times 10^{-6}) = 5.28$$

Notice that the solution is acidic; NH_4Cl consists of an acidic cation and a neutral anion.

Example 15-14: Compute the pH of a 0.050 M $NaC_2H_3O_2(aq)$ solution.

Solution: Sodium acetate consists of a neutral cation and basic anion. The equation for the reaction of the anion with water is

$$C_2H_3O_2^-(aq) + H_2O(l) \rightleftharpoons HC_2H_3O_2(aq) + OH^-(aq)$$

From Table 15-5, we see that K_b for this reaction is 5.75×10^{-10} M. A table of the initial and equilibrium concentrations is

	$C_2H_3O_2^-(aq)$	$+ H_2O(l) \rightleftharpoons$	$HC_2H_3O_2(aq)$	$+ OH^-(aq)$
initial concentration	0.050 M	—	0	≈0
equilibrium concentration	0.050 M − $[HC_2H_3O_2]$	—	$[HC_2H_3O_2]$	$[OH^-]$
or, because $[OH^-] = [HC_2H_3O_2]$ from the reaction stoichiometry	0.050 M − $[OH^-]$	—	$[OH^-]$	$[OH^-]$

The expression for K_b is

$$5.75 \times 10^{-10}\ \text{M} = \frac{[\text{OH}^-]^2}{0.050\ \text{M} - [\text{OH}^-]}$$

Neglecting $[OH^-]$ relative to 0.050 M, we obtain

$$[\text{OH}^-] = (0.050\ \text{M} \times 5.75 \times 10^{-10}\ \text{M})^{1/2} = 5.4 \times 10^{-6}\ \text{M}$$

which is negligible compared with 0.050 M, as confirmed by a second approximation. The value of pOH can be obtained from the value of $[OH^-]$:

$$\text{pOH} = -\log [\text{OH}^-] = -\log (5.4 \times 10^{-6}) = 5.27$$

The pH of the solution is

$$\text{pH} = 14.00 - \text{pOH} = 8.73$$

which shows that the solution is basic ($pH > 7.0$).

The salt $NH_4Cl(aq)$ has an acidic cation and a neutral anion, whereas $NaC_2H_3O_2(aq)$ consists of a neutral cation and a basic anion. For a salt in which neither ion is neutral, such as $NH_4CN(aq)$, calculation of the pH is more involved.

15-12. A LEWIS ACID IS AN ELECTRON-PAIR ACCEPTOR

The acid-base classification system introduced by G. N. Lewis provides the most general definitions of acids and bases. **Lewis acids** are defined as **electron-pair acceptors,** and **Lewis bases** are defined as **electron-pair donors.** For example, consider the reaction between ammonia and boron trifluoride:

$$H_3N\!: \;+\; BF_3 \;\rightleftharpoons\; H_3N{-}BF_3$$

Lewis base — Lewis acid — donor-acceptor complex

The BF_3 reacts with the NH_3 by accepting the lone pair of electrons on nitrogen to form a nitrogen-boron bond. Note that BF_3 is not an acid in the Brønsted-Lowry sense because it has no protons to donate, but it is a Lewis acid because it acts as an electron-pair acceptor. In general, an electron-deficient species (Chapter 8) can act as a Lewis acid, and a species with a lone pair of electrons can act as a Lewis base.

An interesting example of a Lewis acid is boric acid. Boric acid is a moderately soluble monoprotic weak acid in water; its formula usually is written as $B(OH)_3$ rather than H_3BO_3 or $HBO(OH)_2$, because it acts as a Lewis acid by accepting a hydroxide ion rather than by donating a proton

$$\text{B(OH)}_3 + \text{H}_2\text{O}(l) \rightleftharpoons \text{B(OH)}_4^-(aq) + \text{H}^+(aq) \qquad \text{p}K_a = 9.3 \text{ at } 25^\circ\text{C}$$

Boric acid is used in mouth and eye washes.

Example 15-15: Classify the following species as either Lewis acids or Lewis bases: (a) $H^+(aq)$ (b) $OH^-(aq)$ (c) $AlCl_3(aq)$ (d) $Cl^-(aq)$

Solution: (a) A free proton is electron-deficient and combines with electron-pair donors:

$$H^+ + :\ddot{\underset{\cdot\cdot}{O}}H^- \rightarrow H_2O$$

Thus H^+ is a Lewis acid.

(b) As shown in (a), OH^- is a Lewis base (electron-pair donor).

(c) $AlCl_3$ is electron-deficient:

$$AlCl_3 \text{ (Al bonded to three } :\ddot{Cl}: \text{ atoms)}$$

and thus can act as a Lewis acid; for example,

$$\underset{\text{Lewis base}}{:\ddot{\underset{\cdot\cdot}{Cl}}:^-} + \underset{\text{Lewis acid}}{AlCl_3} \rightarrow AlCl_4^-$$

(d) As shown in (c), Cl^- is a Lewis base.

The Lewis acid-base classification scheme is especially useful in understanding the mechanisms of many organic chemical reactions.

SUMMARY

The Brønsted-Lowry definition of acids and bases is that an acid is a species that donates protons and a base is a species that accepts protons. Acid-base reactions are proton-transfer reactions. Water plays a central role in acid-base reactions because it can act either as an acid or as a base. Water undergoes a self-protonation reaction to produce $H_3O^+(aq)$ and $OH^-(aq)$. The product of the equilibrium concentrations of $H_3O^+(aq)$ and $OH^-(aq)$ ions is fixed by the value of the ion product constant of water, $K_w = [H_3O^+][OH^-]$. The value of K_w is $1.00 \times 10^{-14}\ M^2$ at 25°C.

If $[H_3O^+] > [OH^-]$, then the solution is acidic; if $[H_3O^+] < [OH^-]$, then the solution is basic. The value of $[H_3O^+]$ is presented conveniently on a pH scale, and pH, a measure of the acidity of a solution, is defined through the equation $pH = -\log[H_3O^+]$. At 25°C, if $pH < 7$, then the solution is acidic; if $pH > 7$, then the solution is basic; if $pH = 7$, then the solution is neutral. The pOH of a solution is related to the pH of the solution by $pOH = 14.00 - pH$ at 25°C.

Strong acids and strong bases are completely dissociated in solution; weak acids are only partially dissociated and weak bases are only partially protonated. The strength of a weak acid is governed by the value of K_a, its acid-dissociation constant; the strength of a weak base is governed by the value of K_b, its base-protonation constant. The larger the value of K_a or K_b, the stronger is the acid or the base. The treatment of weak acids and weak bases can be unified by the introduction of the idea of a conjugate acid-base pair. The values of K_a and K_b for a conjugate acid-base pair obey the relation $K_w = K_aK_b$. The stronger an acid, the weaker is its conjugate base. The pK_a value of an acid is given by $pK_a = -\log K_a$; the smaller the value of pK_a, the stronger is the acid. The pK_b value of a base is given by $pK_b = -\log K_b$; the smaller the value of pK_b, the stronger is the base. For any conjugate acid-base pair, $pK_a + pK_b = 14.00$ at 25°C.

The aqueous solutions of many salts are acidic or basic. Many cations and anions have acidic or basic properties. Salts that consist of these cations or anions may be acidic or basic.

A Lewis acid is defined as an electron-pair acceptor; a Lewis base is defined as an electron-pair donor.

TERMS YOU SHOULD KNOW

Arrhenius acids and bases 468
Brønsted-Lowry acids and bases 468
proton donor 468
proton acceptor 468
hydronium ion, $H_3O^+(aq)$ 468
proton-transfer reaction 468

EQUATIONS YOU SHOULD KNOW HOW TO USE

$K_w = [H_3O^+][OH^-] = 1.00 \times 10^{-14}\ M^2$	(15-3)	(ion product constant for water at 25°C)
$pH = -\log [H_3O^+]$	(15-4)	(definition of pH)
$pOH = -\log [OH^-]$	(15-5)	(definition of pOH)
$pH + pOH = 14.00$	(15-6)	(relation between pH and pOH at 25°C)
$[H_3O^+] = 10^{-pH}$	(15-7)	(calculation of $H_3O^+(aq)$ from pH)
$pK_a = -\log K_a$	(15-9)	(definition of pK_a)
$pK_b = -\log K_b$	(15-12)	(definition of pK_b)
$K_a = \dfrac{[H_3O^+][B^-]}{[HB]}$	(15-14)	(acid dissociation constant expression for the acid HB(*aq*))
$K_b = \dfrac{[HB][OH^-]}{[B^-]}$	(15-15)	(base protonation constant expression for the base $B^-(aq)$)
$K_w = K_a K_b$	(15-16)	(relation between K_a and K_b for a conjugate acid-base pair)
$pK_a + pK_b = 14.00$	(15-17)	(relation between pK_a and pK_b for a conjugate acid-base pair at 25°C)

PROBLEMS

STRONG ACIDS AND BASES

15-1. Calculate $[H_3O^+]$, $[ClO_4^-]$, and $[OH^-]$ in an aqueous solution that is 0.150 M in $HClO_4(aq)$. Is the solution acidic or basic?

15-2. Calculate $[OH^-]$, $[K^+]$, and $[H_3O^+]$ in an aqueous solution that is 0.25 M in KOH(*aq*). Is the solution acidic or basic?

15-3. Compute $[Tl^+]$, $[OH^-]$, and $[H_3O^+]$ for a solution that is prepared by dissolving 2.00 g of TlOH in enough water to make 500.0 mL of solution.

15-4. Compute $[Ca^{2+}]$, $[OH^-]$, and $[H_3O^+]$ for a solution that is prepared by dissolving 0.60 g of $Ca(OH)_2$ in enough water to make 1.00 L of solution.

pH CALCULATIONS

15-5. Calculate the pH of an aqueous solution that is 0.020 M in $HNO_3(aq)$. Is the solution acidic or basic?

15-6. Calculate the pH of an aqueous solution that is 0.20 M in $CsOH(aq)$. Is the solution acidic or basic?

15-7. Calculate the pH and the pOH of an aqueous solution that is 0.035 M in $HCl(aq)$ and 0.045 M in $HBr(aq)$.

15-8. Calculate the pOH and the pH of an aqueous solution that is 0.020 M in $Ba(OH)_2(aq)$.

15-9. Calculate the pOH and the pH of an aqueous solution prepared by dissolving 2.00 g of KOH pellets in water and diluting to a final volume of 0.500 L.

15-10. A solution of $NaOH(aq)$ contains 6.25 g of NaOH per 100 mL of solution. Calculate the pH and the pOH of the solution at 25°C.

CALCULATION OF $[H_3O^+]$ AND $[OH^-]$ FROM pH

15-11. The pH of human muscle fluids is 6.8. Compute the value of $[H_3O^+]$ in muscle fluid at 25°C.

15-12. The pH of household ammonia is about 12. Calculate the value of $[OH^-]$ for the ammonia solution.

15-13. The pH of the contents of the human stomach can be as low as 1.0. Estimate the value of $[H_3O^+]$ in the stomach when the pH = 1.0.

15-14. Normal rainwater has a pH of about 5.6, whereas what is called acid rain has been observed to have pH values as low as 3.0. Compute the ratio of $[H_3O^+]$ in pH = 3.0 acid rain to that in normal rain. What is the cause of the acidity of normal rain?

15-15. The pH of human blood is fairly constant at 7.4. Compute the hydronium ion concentration and the hydroxide ion concentration in human blood at 25°C.

15-16. The pH of the world's oceans is remarkably constant at 8.15. Compute the hydronium ion and hydroxide ion concentrations in the ocean. Assume a temperature of 25°C.

CALCULATION OF K_a FROM pH

15-17. The pH of a 0.050 M aqueous solution of propionic acid, $HC_3H_5O_2$, a component of milk, is found to be 3.09. Calculate K_a, the acid-dissociation constant, for propionic acid.

15-18. The pH of a 0.20 M aqueous solution of cyanic acid, HCNO, is found to be 2.19. Calculate K_a, the acid-dissociation constant, for cyanic acid.

15-19. We are given that the measured pH of a 1.00×10^{-2} M solution of $HC_2H_3O_2(aq)$ is 3.39 at 25°C. Compute the value of K_a for $HC_2H_3O_2(aq)$ at 25°C.

15-20. The pH of a 0.10 M aqueous solution of formic acid, $HCHO_2$, is 2.38. Calculate the value of K_a for formic acid.

15-21. In its undissociated acid form, acetylsalicylic acid, aspirin, can cross the stomach lining. Estimate the ratio of the concentration of anion to the undissociated acid when the pH of the stomach contents is 1.5. Take $K_a = 2.75 \times 10^{-5}$ M for acetylsalicylic acid at 37°C.

15-22. Uric acid is an end product of the metabolism of certain biological substances and is excreted from the body in urine. The acid dissociation constant of uric acid is $K_a = 4.0 \times 10^{-6}$ M. The pH of a urine sample is 6.0. Estimate the ratio of urate ion to uric acid in the urine.

CALCULATION OF pH FROM K_a

15-23. The value of K_a in water at 25°C for benzoic acid, $HC_7H_5O_2$, is 6.46×10^{-5} M. Calculate the pH of an aqueous solution with a total concentration of $HC_7H_5O_2$ of 0.020 M.

15-24. The value of K_a in water at 25°C for hypochlorous acid, HClO, is 3.00×10^{-8} M. Calculate the pH of an aqueous solution with a total concentration of HClO of 0.15 M.

15-25. The value of K_a in water at 25°C for trichloroacetic acid, $HC_2Cl_3O_2$, is 2.3×10^{-1} M. Calculate the pH of an aqueous solution with a total concentration of $HC_2Cl_3O_2$ of 0.030 M.

15-26. Calculate the pH of a 0.10 M aqueous solution of chloroacetic acid given that $K_a = 1.35 \times 10^{-3}$ M at 25°C.

15-27. Sulfamic acid, HO_3SNH_2, is used as a stabilizer for chlorine in swimming pools. Calculate the pH of a 0.040 M sulfamic acid solution given that $K_a = 0.10$ M at 25°C.

15-28. The compound sodium bisulfate, $NaHSO_4$, is used in cleaning metals and in leather treatment. Given that the K_a for $HSO_4^-(aq)$ is 0.012 M at 25°C, compute the pH of a $NaHSO_4(aq)$ solution prepared by dissolving 10.0 g of $NaHSO_4$ in sufficient water to make 100 mL of solution.

CALCULATIONS INVOLVING K_b

15-29. The measured pH of a 0.100 M solution of $NH_3(aq)$ at 25°C is 11.12. Compute K_b for $NH_3(aq)$ at 25°C.

15-30. The pH of a 0.50 M solution of the weak base ethyl amine, $C_2H_5NH_2$, is 12.20. Determine K_b for ethyl amine.

15-31. The organic solvent pyridine, C_5H_5N, is a base with a strong, irritating odor. Calculate the pH of a 0.300 M aqueous solution of pyridine.

15-32. The base hydroxylamine, $HONH_2$, is used to synthesize a variety of organic compounds. Calculate the pH of a 0.125 M aqueous solution of $HONH_2$.

15-33. Calculate the pH of a 0.060 M aqueous solution of dimethyl amine, $(CH_3)_2NH$.

15-34. Compute the pH of a household ammonia cleaning solution prepared by dissolving $NH_3(g)$ in water to yield a 0.20 M $NH_3(aq)$ solution.

LE CHATELIER'S PRINCIPLE

15-35. Use Le Châtelier's principle to predict what happens to the equilibrium

$$CO_2(aq) + H_2O(l) \rightleftharpoons H_3O^+(aq) + HCO_3^-(aq)$$

if

(a) $[CO_2]$ is decreased
(b) $[HCO_3^-]$ is decreased
(c) $[H_3O^+]$ is decreased
(d) the solution is diluted with water

15-36. Use LeChâtelier's principle to predict the direction in which the following acid-dissociation equilibrium shifts in response to the indicated change in conditions:

$$HCHO_2(aq) + H_2O(l) \rightleftharpoons H_3O^+(aq) + CHO_2^-(aq)$$

(a) addition of $NaOH(s)$
(b) addition of $NaCHO_2(s)$
(c) dilution of a 0.1 M solution to 0.01 M
(d) addition of $HCl(g)$

15-37. Use Le Châtelier's principle to predict the direction in which the following equilibrium shifts in response to the indicated change in conditions:

$$HC_7H_5O_2(aq) + H_2O(l) \rightleftharpoons H_3O^+(aq) + C_7H_5O_2^-(aq) \qquad \Delta H^\circ_{rxn} \approx 0$$

(a) evaporation of water from the solution at a fixed temperature
(b) decrease in the temperature of the solution
(c) addition of $KC_7H_5O_2(s)$
(d) addition of $NH_3(g)$
(e) addition of $HCl(g)$

15-38. Use Le Châtelier's principle to predict the direction in which the following equilibrium shifts in response to the indicated change in conditions:

$$HNO_2(aq) + H_2O(l) \rightleftharpoons H_3O^+(aq) + NO_2^-(aq) \qquad \Delta H^\circ_{rxn} < 0$$

(a) increase in the temperature of the solution
(b) dissolution of $NaNO_2(s)$
(c) dissolution of $NaOH(s)$
(d) removal of $NO_2^-(aq)$ as $AgNO_2(s)$ by addition of $AgNO_3(s)$

POLYPROTIC ACIDS

15-39. Calculate the volume of 0.10 M $NaOH(aq)$ required to neutralize completely 25.0 mL of 0.10 M oxalic acid, $H_2C_2O_4$.

15-40. Calculate the volume of 0.10 M $NaOH(aq)$ required to neutralize completely 100 mL of 0.10 M H_2S.

15-41. A 0.500-g sample of oxalic acid, a poisonous component of rhubarb leaves, is dissolved in 100 mL of water and titrated completely with 0.250 M $NaOH(aq)$. The volume of base is 44.4 mL. Calculate the molecular mass of oxalic acid, which has two dissociable protons per molecule.

15-42. A 1.20-g sample of fumaric acid, an essential component in the production of energy in living cells, is dissolved in water and titrated completely with 0.300 M $NaOH(aq)$. The volume of base is 69.0 mL. Calculate the molecular mass of fumaric acid, which contains two dissociable protons.

15-43. Calculate the pH of a 0.100 M $H_3AsO_4(aq)$ solution given that $pK_{a1} = 2.22$, $pK_{a2} = 6.96$, and $pK_{a3} = 11.40$.

15-44. Ascorbic acid (vitamin C) has the structure

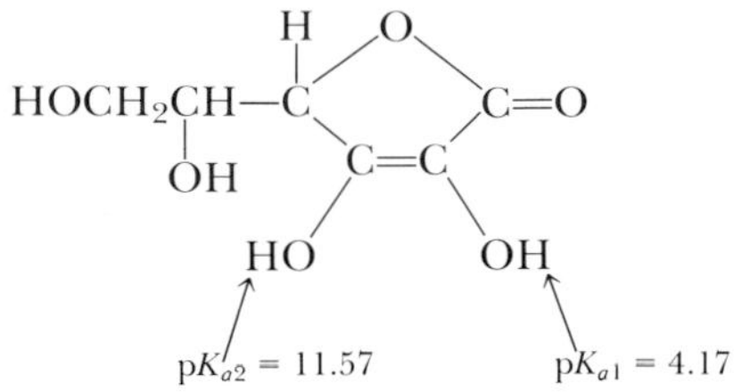

```
              H    O
              |  /   \
   HOCH2CH—C          C=O
         |    \      /
         OH    C==C
              /    \
            HO      OH
           ↑          ↑
   pKa2 = 11.57    pKa1 = 4.17
```

Compute the pH of an aqueous vitamin C solution obtained by dissolving 500 mg of vitamin C in enough water to make 1.00 L of solution.

CONJUGATE ACIDS AND BASES

15-45. Indicate the conjugate acid-base pairs in the following equations:

(a) $HC_7H_5O_2(aq) + H_2O(l) \rightleftharpoons H_3O^+(aq) + C_7H_5O_2^-(aq)$
(b) $CH_3NH_2(aq) + H_2O(l) \rightleftharpoons CH_3NH_3^+(aq) + OH^-(aq)$
(c) $HCHO_2(aq) + H_2O(l) \rightleftharpoons H_3O^+(aq) + CHO_2^-(aq)$

15-46. Indicate the conjugate acid-base pairs in the following reactions [(amm) denotes the solvent NH_3]:

(a) $NH_3(l) + NH_3(l) \rightleftharpoons NH_4^+(amm) + NH_2^-(amm)$
(b) $HNO_2(aq) + H_2O(l) \rightleftharpoons H_3O^+(aq) + NO_2^-(aq)$
(c) $C_5H_5N(aq) + H_2O(l) \rightleftharpoons C_5H_5NH^+(aq) + OH^-(aq)$

15-47. Give the conjugate base for each of the following acids:

(a) $HClO(aq)$ (b) $NH_4^+(aq)$
(c) $HN_3(aq)$ (d) $HS^-(aq)$

15-48. Give the conjugate base for each of the following acids:

(a) $HNO_3(aq)$ (b) $HCHO_2(aq)$
(c) $HC_6H_5O(aq)$ (d) $CH_3NH_3^+(aq)$

15-49. Identify which of the following species are Brønsted-Lowry acids and which are Brønsted-Lowry bases in water. In each case give the chemical formula for the conjugate member of the conjugate acid-base pair:

(a) $HCNO(aq)$ (b) $OBr^-(aq)$
(c) $HClO_3(aq)$ (d) $CH_3NH_3^+(aq)$
(e) $ClNH_2(aq)$ (f) $HONH_2(aq)$

15-50. Identify which of the following species are Brønsted-Lowry acids and which are Brønsted-Lowry bases in water. In each case give the chemical formula for the conjugate member of the conjugate acid-base pair.

(a) $HC_2H_2ClO_2(aq)$ (b) $NH_3(aq)$
(c) $ClO^-(aq)$ (d) $CHO_2^-(aq)$
(e) $HN_3(aq)$ (f) $NO_2^-(aq)$

15-51. Given the following acids and their dissociation constants, calculate K_b for the conjugate bases:

Acid	K_a/M
(a) $HC_3H_5O_2$, propionic acid	1.34×10^{-5}
(b) HF	6.76×10^{-4}
(c) NH_4^+	5.71×10^{-10}
(d) $H_2PO_4^-$	6.32×10^{-8}

15-52. Given the following bases and their values of K_b, calculate K_a for the conjugate acids:

Base	K_b/M
(a) C_5H_5N	1.46×10^{-9}
(b) CN^-	2.09×10^{-5}
(c) CNO^-	4.57×10^{-11}
(d) HS^-	1.10×10^{-7}

15-53. Use the data in Table 15-5 to compute the equilibrium constants at 25°C for the following equations:

(a) $HCNO(aq) + NO_2^-(aq) \rightleftharpoons HNO_2(aq) + CNO^-(aq)$
(b) $NH_4^+(aq) + HCO_3^-(aq) \rightleftharpoons NH_3(aq) + H_2CO_3(aq)$

15-54. Use the data in Table 15-5 to compute the equilibrium constants at 25°C for the following equations:

(a) $NH_4^+(aq) + C_2H_3O_2^-(aq) \rightleftharpoons NH_3(aq) + HC_2H_3O_2(aq)$
(b) $C_6H_5O^-(aq) + C_5H_5NH^+(aq) \rightleftharpoons HC_6H_5O(aq) + C_5H_5N(aq)$

15-55. Predict whether the following salts, when dissolved in water, produce acidic, basic, or neutral solutions:

(a) $Al(NO_3)_3$ (b) NH_4Br
(c) $NaHCO_3$ (d) LiCNO

15-56. Predict whether the following salts, when dissolved in water, produce acidic, basic, or neutral solutions:

(a) $CoBr_3$ (b) $NaNO_3$
(c) $KHSO_4$ (d) NaF

15-57. Predict whether the following salts, when dissolved in water, produce acidic, basic, or neutral solutions:

(a) Na_2CO_3 (b) $KClO_4$
(c) RbClO (d) $Al(ClO_4)_3$

15-58. Predict whether the following salts, when dissolved in water, produce acidic, basic, or neutral solutions:

(a) KCN (b) $Pb(NO_3)_2$
(c) $NaHSO_4$ (d) $CaCl_2$

15-59. Some soaps are produced by reacting caustic soda, NaOH, with animal fats, which contain a type of organic acid called a fatty acid. An example is stearic acid, $HC_{18}H_{35}O_2$. Fatty acids, like most organic acids, are weak acids. Is a soap solution acidic, basic, or neutral?

15-60. Various aluminum salts, such as $Al_2(SO_4)_3$ and $KAl(SO_4)_2$, are used as additives to increase the acidity of soils for "acid-loving" plants such as azaleas and tomatoes. Explain how these salts increase soil acidity.

pH CALCULATIONS OF SALT SOLUTIONS

15-61. Sodium hypochlorite, NaClO, is a bleaching agent. Calculate the pH and the concentration of HClO in an aqueous solution that is 0.030 M in NaClO.

15-62. Sodium propionate, $NaC_3H_7O_2$, is used as a food preservative. Calculate the pH of a 0.20 M solution of $NaC_3H_7O_2$, taking $K_a = 1.34 \times 10^{-5}$ M for propionic acid.

15-63. A solution of sodium cyanate, NaCNO, is prepared at a concentration of 0.20 M. Calculate the equilibrium concentrations of $OH^-(aq)$, $HCNO(aq)$, $CNO^-(aq)$, and $H_3O^+(aq)$ and the pH of solution.

15-64. Calculate the equilibrium concentrations of $OH^-(aq)$, $HNO_2(aq)$, $NO_2^-(aq)$, and $H_3O^+(aq)$ and the pH of a solution that is 0.25 M in $NaNO_2$.

15-65. Calculate the pH of a 0.30 M aqueous solution of pyridinium chloride, C_5H_6NCl.

15-66. Calculate the pH of a 0.15 M aqueous solution of potassium cyanide, KCN.

15-67. A saturated aqueous solution of ammonium perchlorate contains 23.7 g of NH_4ClO_4 per 100 mL of solution at 25°C. Estimate the pH of this solution at 25°C.

15-68. Calculate the pH of a solution that is prepared by dissolving 25.0 g of barium acetate in enough water to make exactly 1 L of solution.

15-69. The acid-dissociation constant at 25°C for the equilibrium

$$Fe(H_2O)_6^{3+}(aq) + H_2O(l) \rightleftharpoons H_3O^+(aq) + Fe(OH)(H_2O)_5^{2+}(aq)$$

is $K_a = 1.0 \times 10^{-3}$ M. Calculate the pH of a 0.20 M solution of $Fe(NO_3)_3(aq)$ at 25°C.

15-70. Calculate the pH at 25°C of a solution that is 0.10 M in $TlBr_3$. The acid-dissociation constant at 25°C for the equilibrium

$$Tl(H_2O)_6^{3+}(aq) + H_2O(l) \rightleftharpoons H_3O^+(aq) + Tl(OH)(H_2O)_5^{2+}(aq)$$

is $K_a = 6 \times 10^{-2}$ M

LEWIS ACIDS AND BASES

15-71. Determine whether each of the following substances is an Arrhenius acid, a Brønsted-Lowry acid, or a Lewis acid (it is possible for each to be of more than one type):

(a) HCl (b) $AlCl_3$ (c) BCl_3

15-72. Determine whether each of the following substances is an Arrhenius base, a Brønsted-Lowry base, or a Lewis base (it is possible for each to be of more than one type):

(a) NH_3 (b) Br^- (c) NaOH

15-73. Classify each of the following species as either a Lewis acid or a Lewis base:

(a) CH_3OCH_3 (b) $GaCl_3$ (c) H_2O

15-74. Classify each of the following species as either a Lewis acid or a Lewis base:

(a) $CH_3—\ddot{N}(CH_3)—CH_3$ (b) BCl_3 (c) BeF_2

ADDITIONAL PROBLEMS

15-75. The highly toxic compound 2,4-dinitrophenol is used in biological research to inhibit energy production in cells. In an experiment, a solution of 2,4-dinitrophenol was prepared with the pH adjusted to 7.4. Estimate the ratio of the concentrations of the anion to the undissociated acid:

2,4 dinitrophenol

[2,4-dinitrophenol (OH, NO_2, NO_2)] $(aq) + H_2O(l) \rightleftharpoons$ [2,4-dinitrophenolate (O^-, NO_2, NO_2)] $(aq) + H_3O^+(aq)$

undissociated acid — dissociated form

$$K_a = 1.1 \times 10^{-4}\ M$$

15-76. Nitrites, such as $NaNO_2$, are added to processed meats and hamburger both as a preservative and to give the meat a redder color by binding to hemoglobin in the red blood cells. When nitrite ion is ingested, it reacts with stomach acid to form nitrous acid, $HNO_2(aq)$, in the stomach. Given that $K_a = 4.47 \times 10^{-4}$ M for nitrous acid, compute the value of the ratio $[HNO_2]/[NO_2^-]$ in the stomach following ingestion of NO_2^- when $[H_3O^+]$ is 0.10 M.

15-77. A 6.15-g sample of benzoic acid, $HC_7H_5O_2$, is dissolved in enough water to make 600 mL of solution. Calculate the pH of the solution. The value of K_a for benzoic acid at 25°C is 6.46×10^{-5} M.

15-78. Given that K_w for water is $2.40 \times 10^{-14}\ M^2$ at 37°C, compute the pH of a neutral aqueous solution at 37°C, which is the normal human body temperature. Is a pH = 7.00 solution acidic or basic at 37°C?

15-79. A saturated solution of $Mg(OH)_2(aq)$ at 25°C has a pH of 10.52. Estimate the solubility of $Mg(OH)_2(s)$ in water at 25°C.

15-80. Sodium benzoate, $NaC_7H_5O_2$, is used as a food preservative because of its antimicrobial action. The K_a of benzoic acid is 6.46×10^{-5} M. Estimate the ratio of the concentration of benzoic acid to the concentration of benzoate in a food with a pH of 3.0.

15-81. Suppose two 5-grain (5 grains = 324 mg) aspirin tablets are dissolved in enough water to make 500 mL of solution at 25°C. Compute the pH of the resulting solution. Take $K_a = 2.75 \times 10^{-5}$ M at 25°C for aspirin (acetylsalicylic acid, molecular mass = 180.15).

15-82. Use Le Châtelier's principle to predict in which direction the equilibrium shifts if we

(a) add $HCl(g)$ to a 0.10 M NH_3 solution
(b) add $NaOH(s)$ to a 0.10 M NH_3 solution
(c) add $HCl(g)$ to a 0.10 M $HCHO_2$ solution
(d) add $NaOH(s)$ to a 0.10 M $HCHO_2$ solution

15-83. What is the pH of a 2.60×10^{-8} M solution of HCl at 25°C?

15-84. Self-protonation occurs in solvents other than water. For liquid ammonia, the self-protonation equilibrium is

$$2NH_3(l) \rightleftharpoons NH_4^+(amm) + NH_2^-(amm) \quad \text{with} \quad K \simeq 10^{-30}\ M^2 \quad \text{at } -50°C$$

where (*amm*) denotes a solute in liquid ammonia. Estimate the concentration of the ammonium ion in liquid ammonia at −50°C. How many molecules of ammonia are dissociated per mole of ammonia? Take the density of liquid ammonia as $0.77\ g \cdot mL^{-1}$ at −50°C.

15-85. The self-protonation equilibrium for the solvent liquid ammonia, $NH_3(l)$, is

$$2NH_3(l) \rightleftharpoons NH_4^+(amm) + NH_2^-(amm)$$

where (*amm*) denotes a species dissolved in $NH_3(l)$. What is the strongest acid capable of existing at appreciable concentrations in $NH_3(l)$? What is the strongest base capable of existing at appreciable concentrations in $NH_3(l)$?

15-86. The self-protonation constant for the solvent ethanol, $CH_3CH_2OH(l)$, at 25°C is $8 \times 10^{-20}\ M^2$ at 25°C. The self-protonation equilibrium is

$$2CH_3CH_2OH(l) \rightleftharpoons CH_3CH_2OH_2^+(alc) + CH_3CH_2O^-(alc)$$

where (*alc*) denotes alcohol solution.

(a) We define a pH scale in alcohol by the equation

$$pH = -\log_{10} [CH_3CH_2OH_2^+]$$

Compute the pH of a neutral alcohol solution at 25°C. (b) Compute the pH of a 0.010 M solution of sodium ethoxide, $CH_3CH_2O^-Na^+$, in alcohol at 25°C. Assume that $CH_3CH_2O^-Na^+$ is completely dissociated in alcohol.

15-87. Uric acid is an end product of the metabolism of certain biological compounds. Gout is a disease of the joints that is due to the precipitation of sodium urate crystals. Given that

$$\underset{\text{uric acid}}{C_5N_4O_3H_4(aq)} + H_2O(l) \rightleftharpoons \underset{\text{urate ion}}{C_5N_4O_3H_3^-(aq)} + H_3O^+(aq) \qquad K_a = 1.3 \times 10^{-4}\ M$$

determine the pH values for which [urate] > [uric acid].

15-88. The value of the ion-product constant for water, K_w, at 0°C is $0.12 \times 10^{-14}\ M^2$. Compute the pH of a neutral aqueous solution at 0°C. Is an aqueous solution with a pH = 7.25 acidic or basic at 0°C?

15-89. A saturated solution of $Sr(OH)_2(aq)$ at 25°C has a measured pH of 13.50. Estimate the solubility of $Sr(OH)_2(s)$ in water at 25°C.

Phosphorus

Red and white phosphorus. White phosphorus, one of the principal allotropes of solid phosphorus, is very reactive and must be handled with care because it produces severe burns when it comes in contact with skin. The sample shown here has a yellowish cast as a result of surface reactions with air. White phosphorus is usually stored under water. Red phosphorus, on the other hand, is much less reactive than white phosphorus and does not require special handling.

Phosphorus (atomic number 15, atomic mass 30.97376) was the first element whose discovery could be attributed to a specific person. It was discovered in 1669 by the German alchemist Hennig Brandt by the unsavory process of distilling putrefied urine. Although phosphorus constitutes less than 0.1 percent by mass of the earth's crust, all living organisms contain this element, and it is the sixth most abundant element in the human body. The energy requirements of essentially all biochemical reactions are supplied by phosphorus compounds. Plants require phosphorus as a nutrient, and most of the phosphorus compounds that are produced are used as fertilizers.

One of the most important compounds of phosphorus is phosphoric acid, H_3PO_4, a triprotic acid. Although the formulas of the other oxyacids of phosphorus are usually written as H_3PO_2 and H_3PO_3, we shall see that these acids are monoprotic and diprotic, respectively. Thus the three oxyacids of phosphorus illustrate the idea of monoprotic and polyprotic acids presented in Chapter 15.

J-1. THERE ARE TWO PRINCIPAL ALLOTROPES OF SOLID PHOSPHORUS

There are several allotropes of elemental solid phosphorus, the most important of which are **white phosphorus** and **red phosphorus.** White phosphorus is a white, transparent, waxy crystalline solid (Frontispiece) that often appears pale yellow because of impurities. It is insoluble in water and alcohol but soluble in carbon disulfide. A characteristic property of white phosphorus is its high chemical reactivity. It ignites spontaneously in air at about 25°C. White phosphorus is very poisonous, the lethal dose is 50 to 100 mg. White phosphorus should always be kept under water and handled with forceps.

▪ Recall that allotropes are forms of an element with different arrangements of the atoms.

When white phosphorus is heated above 400°C for several hours in the absence of air, a form called red phosphorus is produced. Red phosphorus is a red to violet powder that is less reactive than white phosphorus. The chemical reactions that the red form undergoes are the same as those of the white form, but they generally occur only at higher temperatures. For example, red phosphorus must be heated to 260°C before it burns in air. The toxicity of red phosphorus is much lower than that of white phosphorus.

White phosphorus consists of tetrahedral P_4 molecules (Figure J-1), whereas red phosphorus consists of large, random aggregates of phosphorus atoms. The structure of red phosphorus is called **amorphous,** which means that it has no definite shape. Butter is another example of an amorphous substance.

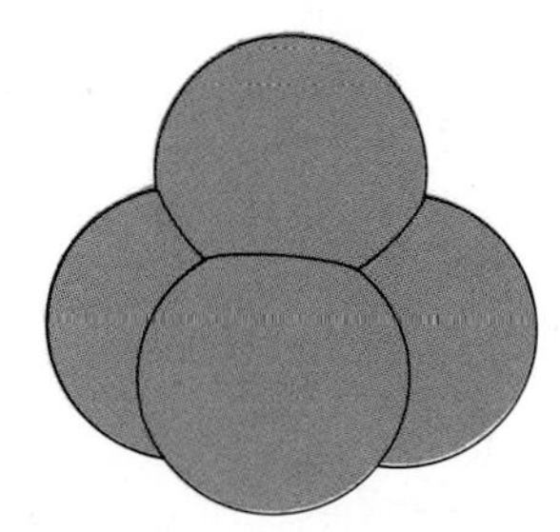

Figure J-1 White phosphorus consists of tetrahedral P_4 molecules.

Most of the phosphorus that is produced is used to make phosphoric acid or other phosphorus compounds. Elemental phosphorus, however, is used in the manufacture of pyrotechnics, matches, rat poisons, incendiary shells, smoke bombs, and tracer bullets.

Phosphorus is not found as the free element in nature. The principal sources are the **apatite ores** (Figure J-2):

Figure J-2 The apatite minerals. Left to right: hydroxyapatite, $Ca_{10}(OH)_2(PO_4)_6$; fluorapatite, $Ca_{10}F_2(PO_4)_6$; and chlorapatite, $Ca_{10}Cl_2(PO_4)_6$.

hydroxyapatite	$Ca_{10}(OH)_2(PO_4)_6$
fluorapatite	$Ca_{10}F_2(PO_4)_6$
chlorapatite	$Ca_{10}Cl_2(PO_4)_6$

These and other phosphate ores collectively are called **phosphate rock.** Large phosphate rock deposits occur in the Soviet Union, in Morocco, and in Florida, Tennessee, and Idaho. An electric furnace is used to obtain phosphorus from phosphate rock. The furnace is charged with powdered phosphate rock, sand (SiO_2), and carbon in the form of coke. The source of heat is an electric current that produces temperatures of over 1000°C. A simplified version of the overall reaction that takes place is

$$\underset{\text{phosphate rock}}{2Ca_3(PO_4)_2(s)} + \underset{\text{sand}}{6SiO_2(s)} + \underset{\text{coke}}{10C(s)} \rightarrow 6CaSiO_3(l) + 10CO(g) + P_4(g)$$

The liquid calcium silicate, $CaSiO_3(l)$, called slag, is tapped off from the bottom of the furnace, and the phosphorus vapor produced solidifies to the white solid when the mixture of $CO(g)$ and $P_4(g)$ is passed through water (carbon monoxide does not dissolve in water). The annual world production of elemental phosphorus is approximately one million tons.

Although some phosphate rock is used to make elemental phosphorus, most phosphate rock is used in the production of fertilizers. Phosphorus is a required nutrient of all plants, and phosphorus compounds have long been used as fertilizer. In spite of its great abundance, phosphate rock cannot be used as a fertilizer because, as the name implies, it is insoluble in water. Consequently, plants are not able to assimilate the phosphorus from phosphate rock. To produce a water-soluble source of phosphorus, phosphate rock is reacted with sulfuric acid to produce a water-soluble product called **superphosphate,** $Ca(H_2PO_4)_2$, one of the world's most important fertilizers.

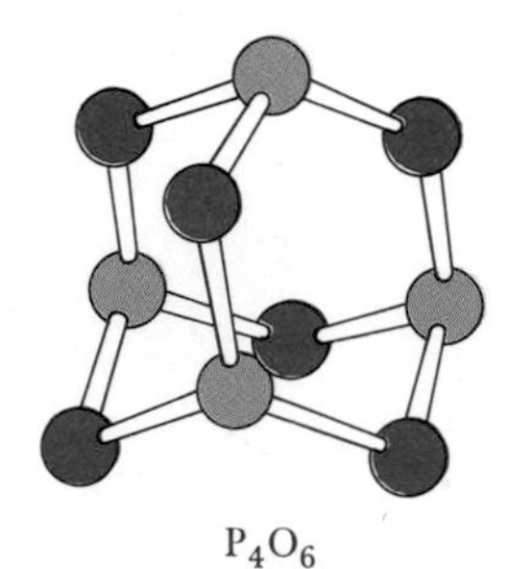

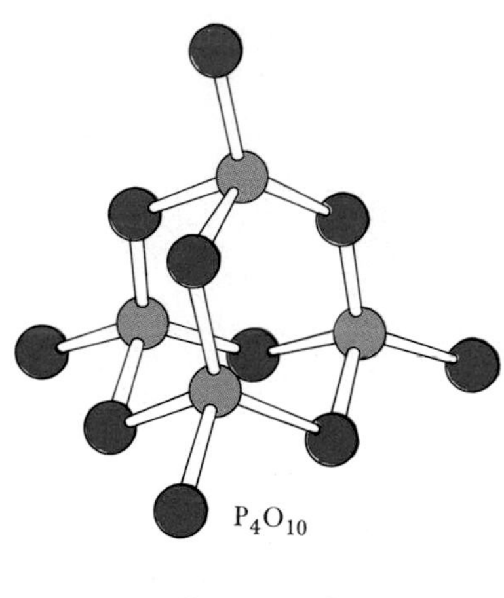

Figure J-3 Structure of P_4O_6 and P_4O_{10}. The P_4O_6 molecule can be viewed as arising from the tetrahedral P_4 molecule when an oxygen atom is inserted between each pair of adjacent phosphorus atoms. The P_4O_{10} molecule can be viewed as arising from P_4O_6 when an oxygen atom is attached to each of the four phosphorus atoms. Note that there are no phosphorus-phosphorus bonds in either P_4O_6 or P_4O_{10}.

J-2. PHOSPHORUS FORMS SEVERAL OXYACIDS

White phosphorus reacts directly with oxygen to produce the oxides P_4O_6 and P_4O_{10}. With excess phosphorus present, P_4O_6 is formed:

$$\underset{\text{excess}}{P_4(s)} + 3O_2(g) \rightarrow P_4O_6(s)$$

with excess oxygen present, P_4O_{10} is formed:

$$P_4(s) + \underset{\text{excess}}{5O_2(g)} \rightarrow P_4O_{10}(s)$$

In practice, a mixture of oxides is formed in each case, but one oxide can be greatly favored over the other by controlling the relative amounts of phosphorus and oxygen. The formulas for P_4O_6 and P_4O_{10} are often written P_2O_3 and P_2O_5, respectively. These obsolete (that is, now known to be incorrect) molecular formulas are the basis for the common names phosphorus *tri*oxide and phosphorus *pent*oxide.

It is interesting to compare the structures of P_4O_6 and P_4O_{10} (Figure J-3). The structure of P_4O_6 is obtained from that of P_4 by inserting an oxygen atom between each pair of adjacent phosphorus atoms; there are six edges on a tetrahedron, and thus a total of six oxygen atoms are required. The structure of P_4O_{10} is obtained from that of P_4O_6 by attaching an additional oxygen atom to each of the four phosphorus atoms.

The phosphorus oxides P_4O_6 and P_4O_{10} react with cold water to form the phosphorus oxyacids: phosphorous acid, H_3PO_3, and phosphoric acid, H_3PO_4:

$$P_4O_6(s) + 6H_2O(l) \rightarrow 4H_3PO_3(aq)$$

$$P_4O_{10}(s) + 6H_2O(l) \rightarrow 4H_3PO_4(aq)$$

The reaction of $P_4O_{10}(s)$ with cold water is quite vigorous and can be explosive.

Phosphorus pentoxide is a powerful dehydrating agent capable of removing water from concentrated sulfuric acid, which is itself a strong dehydrating agent, and in Interchapter I we saw that $N_2O_5(s)$ can be obtained by a similar reaction with nitric acid. The equations for the two reactions are

$$P_4O_{10}(s) + 6H_2SO_4(l) \rightarrow 6SO_3(g) + 4H_3PO_4(l)$$

$$P_4O_{10}(s) + 12HNO_3(l) \rightarrow 6N_2O_5(s) + 4H_3PO_4(l)$$

Phosphorus pentoxide is used as a drying agent in desiccators and dry boxes to remove water vapor.

Hypophosphorous acid, H_3PO_2, is prepared by reacting $P_4(g)$ with a warm aqueous solution of NaOH:

$$P_4(g) + 3OH^-(aq) + 3H_2O(l) \rightarrow 3H_2PO_2^-(aq) + PH_3(g)$$

$$H_2PO_2^-(aq) + H^+(aq) \rightarrow H_3PO_2(aq)$$

The Lewis formulas for the phosphate ion, PO_4^{3-}, the phosphite ion, HPO_3^{2-}, and the hypophosphite ion, $H_2PO_2^-$, are

phosphate ion — phosphite ion — hypophosphite ion

Using VSEPR theory, we predict that these ions are tetrahedral.

The hydrogen atoms attached to the phosphorus atom are not dissociable in aqueous solutions. Thus, phosphoric acid, H_3PO_4, is triprotic; phosphorous acid, $H_2(HPO_3)$, is diprotic; and hypophosphorous acid, $H(H_2PO_2)$, is monoprotic. The structures of these three acids are shown in Figure J-4. The pK_a values of these acids are given in Table J-1.

Phosphoric acid (Figure J-5) is the eighth-ranked industrial chemical and the second-ranked acid, about 22 billion pounds being produced annually in the United States. It is produced industrially by the reaction of phosphate rock and sulfuric acid. The 85 percent by mass (85 g of H_3PO_4 to 15 g of H_2O) laboratory-grade solution is a colorless, syrupy liquid. The 85% solution is equivalent to 15 M.

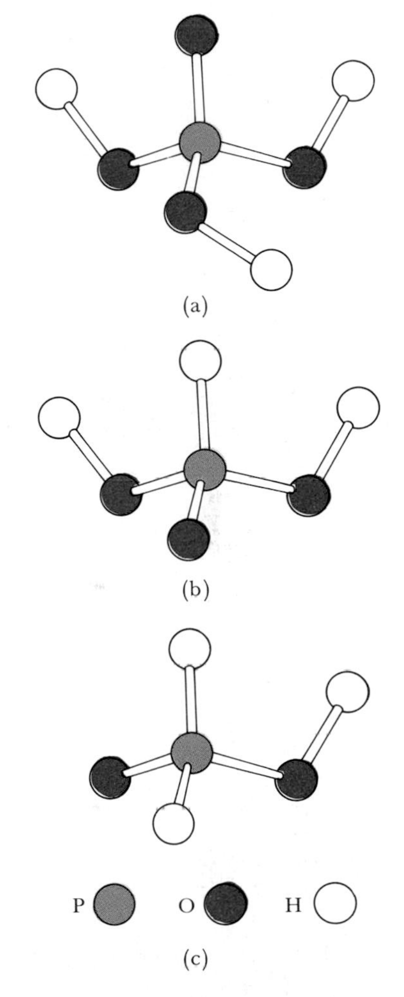

Figure J-4 The structures of phosphoric acid, phosphorous acid, and hypophosphorous acid. Note that all three hydrogen atoms of phosphoric acid are attached to oxygen atoms. One of the hydrogen atoms in phosphorous acid is attached directly to the phosphorus atom, and two of the hydrogen atoms in hypophosphorous acid are attached to the phosphorus atom. Only those hydrogen atoms attached to oxygen atoms are dissociable, and so phosphoric acid is triprotic, phosphorous acid is diprotic, and hypophosphorous acid is monoprotic.

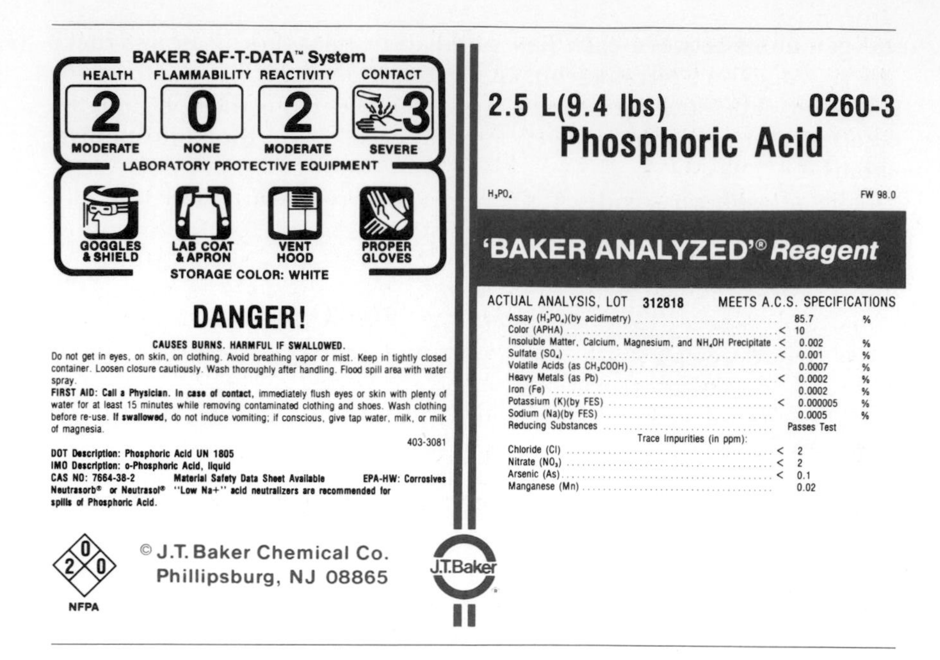

Figure J-5 Phosphoric acid label. Phosphoric acid is sold as a 15 M aqueous solution.

Table J-1 The pK_a values of phosphorus oxyacids in water at 25°C

Acid	pK_{a1}	pK_{a2}	pK_{a3}
H_3PO_4	2.2	7.2	12.4
H_3PO_3	1.8	~7	none
H_3PO_2	1.2	none	none

The principal use of phosphoric acid is in the manufacture of fertilizers. It is also used extensively in the production of soft drinks, and many of its salts are used in the food industry. For example, the monosodium salt, NaH_2PO_4, is used in a variety of foods to control acidity, and calcium dihydrogen phosphate, $Ca(H_2PO_4)_2$, is the acidic ingredient in baking powder. The evolution of carbon dioxide that takes place when baking powder is heated can be represented as

$$\underbrace{Ca(H_2PO_4)_2(s) + 2NaHCO_3(s)}_{\text{in baking powder}} \xrightarrow{300^\circ C} 2CO_2(g) + 2H_2O(g) + CaHPO_4(s) + Na_2HPO_4(s)$$

The slowly evolving $CO_2(g)$ gets trapped as small gas pockets and thereby causes the cake or bread to rise.

When phosphoric acid is heated gently, pyrophosphoric acid (*pyro*- means heat) is obtained as a result of the elimination of a water molecule from a pair of phosphoric acid molecules:

$$HO{-}\overset{\overset{O}{\|}}{\underset{\underset{OH}{|}}{P}}{-}\boxed{OH + H}O{-}\overset{\overset{O}{\|}}{\underset{\underset{OH}{|}}{P}}{-}OH \rightarrow HO{-}\overset{\overset{O}{\|}}{\underset{\underset{OH}{|}}{P}}{-}O{-}\overset{\overset{O}{\|}}{\underset{\underset{OH}{|}}{P}}{-}OH + H_2O$$

elimination of water

pyrophosphoric acid, $H_4P_2O_7$

Pyrophosphoric acid, which is also called diphosphoric acid, is a viscous, syrupy liquid that tends to solidify on long standing. In aqueous solution, it slowly reverts to phosphoric acid.

Longer chains of phosphate groups can be formed. The compound sodium triphosphate, $Na_5P_3O_{10}$, used to be the primary phosphate ingredient of detergents. Its role was to break up and suspend dirt and

$$\left[O{-}\overset{\overset{O}{|}}{\underset{\underset{O}{|}}{P}}{-}O{-}\overset{\overset{O}{|}}{\underset{\underset{O}{|}}{P}}{-}O{-}\overset{\overset{O}{|}}{\underset{\underset{O}{|}}{P}}{-}O\right]^{3-}$$

triphosphate ion

stains by forming water-soluble complexes with metal ions. (The formation of complexes is discussed in Chapter 22.) In the 1960s almost all detergents contained phosphates, sometimes as much as 50 percent by mass. It was discovered, however, that the phosphates led to a serious water pollution problem. The enormous quantity of phosphates discharged into rivers and lakes served as a nutrient for the rampant growth of algae and other organisms. When these organisms died, much of the oxygen dissolved in the water was consumed in the decay process, thus depleting the water's oxygen supply and destroying the ecological balance. This process is called **eutrophication.** As a result of legislation in the 1970s, phosphates have been eliminated from detergents, or at least their levels have been reduced markedly.

J-3. PHOSPHORUS FORMS A NUMBER OF BINARY COMPOUNDS

Phosphorus reacts directly with reactive metals, such as sodium and calcium, to form phosphides; for example,

$$12Na(s) + P_4(s) \rightarrow 4Na_3P(s)$$

Most metal phosphides react vigorously with water to produce phosphine, $PH_3(g)$:

$$Ca_3P_2(s) + 6H_2O(l) \rightarrow 2PH_3(g) + 3Ca(OH)_2(aq)$$

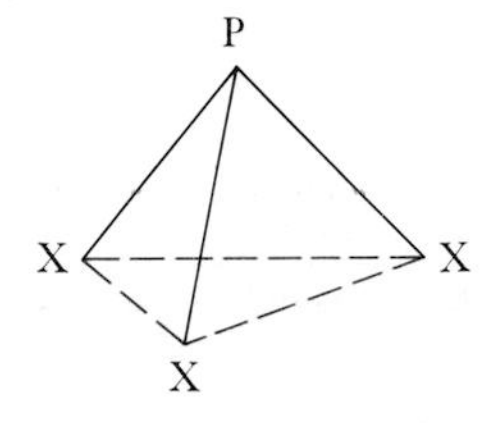

Figure J-6 The phosphorus trihalides, PX_3, have a trigonal pyramidal structure in the gas phase.

Phosphine has a trigonal pyramidal structure with an H—P—H bond angle of 93.7°. It is a colorless, extremely toxic gas with an offensive odor like that of rotten fish. Unlike ammonia, phosphine does not act as a base toward water, and few phosphonium (PH_4^+) salts are stable. Phosphine can also be prepared by the reaction of white phosphorus with a strong base. The equation for the reaction is

- Phosphine reacts violently with oxygen and the halogens.

$$P_4(s) + 3NaOH(aq) + 3H_2O(l) \rightarrow PH_3(g) + 3NaH_2PO_2(aq)$$

Phosphorus reacts directly with the halogens to form halides. If an excess of phosphorus is used, then the trihalide is formed. For example,

$$\underset{\text{excess}}{P_4(s)} + 6Cl_2(g) \rightarrow 4PCl_3(l)$$

Phosphorus trichloride reacts with chlorine to give phosphorus pentachloride:

$$PCl_3(l) + Cl_2(g) \rightarrow PCl_5(s)$$

Recall from Chapter 9 that phosphorus trihalide molecules in the gas phase have a trigonal pyramidal structure (Figure J-6) and that phosphorus pentahalide molecules in the gas phase have a trigonal bipyramidal structure (Figure J-7). In the solid phase, however, X-ray diffraction studies have shown that PCl_5 exists as PCl_4^+ and PCl_6^- ion pairs. Phosphorus halides react vigorously with water:

$$PCl_3(l) + 3H_2O(l) \rightarrow H_3PO_3(aq) + 3HCl(aq)$$

$$PCl_5(s) + 4H_2O(l) \rightarrow H_3PO_4(aq) + 5HCl(aq)$$

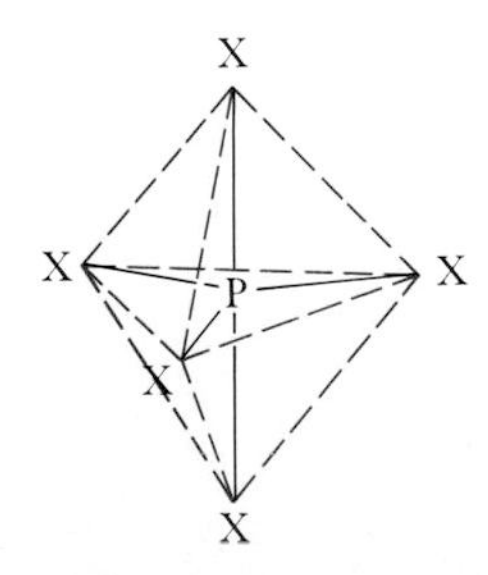

Figure J-7 The phosphorus pentahalides, PX_5, have a trigonal bipyramidal structure in the gas phase.

When phosphorus is heated with sulfur, the yellow crystalline compound tetraphosphorus trisulfide, P_4S_3, is formed. Matches that can be ignited by striking on any rough surface (Figure J-8) contain a tip composed of the yellow P_4S_3 on top of a red portion that contains lead dioxide, PbO_2, together with antimony trisulfide, Sb_2S_3. Friction causes the P_4S_3 to ignite in air, and the heat produced then initiates a reaction between antimony trisulfide and lead dioxide that produces a flame.

Safety matches consist of a mixture of potassium chlorate and antimony trisulfide. The match is ignited by striking on a special rough surface composed of a mixture of red phosphorus, glue, and abrasive. The red phosphorus is ignited by friction and in turn ignites the reaction mixture in the matchhead.

Figure J-8 Phosphorus, in the form of P_4S_3, is one of the principal components of the tips of "strike-anywhere" matches.

J-4. MANY PHOSPHORUS COMPOUNDS ARE IMPORTANT BIOLOGICALLY

Ordinary tooth enamel is hydroxyapatite, $Ca_{10}(OH)_2(PO_4)_6$. If low concentrations of fluoride ion are added to the diets of children, then a substantial amount of the tooth enamel formed will consist of fluorapatite, $Ca_{10}F_2(PO_4)_6$, which is much harder and less affected by acidic substances than hydroxyapatite. Consequently, fluorapatite is more resistant to tooth decay than is hydroxyapatite. Small quantities of fluoride are added to most municipal water supplies, and the incidence of tooth decay among children has decreased markedly over the past 30 years.

The energy requirements for all biochemical reactions are supplied by a substance called **adenosine triphosphate,** or simply **ATP.** (Figure J-9) Under physiological conditions, the reaction of 1 mol of ATP with water releases 31 kJ of energy. The energy released is used by all living species to drive biochemical reactions, in which typically the products are at a higher energy than the reactants. ATP is a biological fuel. The formation and consumption of ATP occur on the average within about 1 min of each other. The amount of ATP used by the human body is truly remarkable: at rest over a 24-h period about 40 kg of ATP is consumed. For strenuous exercise the rate of utilization of ATP can reach 5 kg in 10 min.

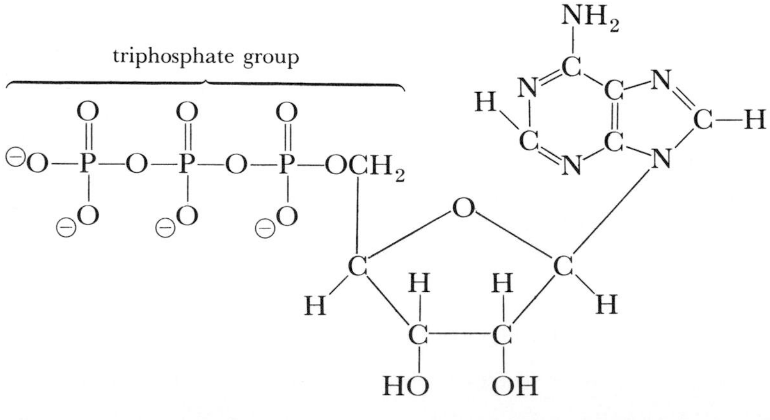

Figure J-9 A Lewis formula for ATP, the substances that is the energy source for all living species.

TERMS YOU SHOULD KNOW

allotrope 503
white phosphorus 503
red phosphorus 503
amorphous 503
apatite ores 503
phosphate rock 504
superphosphate 504
eutrophication 507
adenosine triphosphate (ATP) 508

QUESTIONS

J-1. Discuss the difference in reactivity between white phosphorus and red phosphorus.

J-2. What is the structure of P_4?

J-3. What is the general formula for an apatite mineral?

J-4. What is phosphate rock? What is its most important use?

J-5. Why can't phosphate rock be used directly as fertilizer?

J-6. Compare the structures of P_4O_6 and P_4O_{10}.

J-7. Compare the structures of H_3PO_2, H_3PO_3, and H_3PO_4. How many dissociable protons are there per mole of phosphorous acid? of hypophosphorous acid?

J-8. Describe the action of baking powder.

J-9. Discuss the process of eutrophication.

J-10. Compare ammonia and phosphine as bases.

J-11. Describe two ways to prepare phosphine.

J-12. Discuss the difference between safety matches and strike-anywhere matches.

PROBLEMS

J-13. Compare the pH values of 0.250 M solutions of $H_3PO_4(aq)$ and $H_3PO_3(aq)$.

J-14. What volume of 6.00 M $NaOH(aq)$ solution is required to react completely with 37.5 mL of 2.00 M $H_3PO_4(aq)$? with 37.5 mL of 2.00 M $H_3PO_3(aq)$?

J-15. The value of ΔH°_{rxn} for

$$PCl_5(g) \rightleftharpoons PCl_3(g) + Cl_2(g)$$

is 92.9 $kJ \cdot mol^{-1}$. Calculate the value of K_c at 500°C, given that $K_c = 1.8$ M at 250°C.

J-16. Given that K_c for the reaction

$$PCl_5(g) \rightleftharpoons PCl_3(g) + Cl_2(g)$$

is 1.8 M at 250°C, calculate the fraction of $PCl_5(g)$ that decomposes if $[PCl_5]_0 = 1.00$ M.

J-17. What is the maximum amount of $P_4(s)$ that can be obtained from 1 metric ton of $Ca_3(PO_4)_2(s)$?

J-18. Phosphorus oxychloride, $POCl_3(l)$, can be prepared by distilling $PCl_3(l)$ with $KClO_3(s)$:

$$3PCl_3(l) + KClO_3(s) \rightarrow 3POCl_3(l) + KCl(s)$$

How many grams of $POCl_3(l)$ can be prepared from 6.50 g of $PCl_3(l)$?

J-19. Deuterated phosphine, $PD_3(g)$, can be prepared by reacting calcium phosphide with heavy water. What volume of $PD_3(g)$ at 0°C and 700 torr can be prepared from 10.0 g of $D_2O(l)$? The atomic mass of deuterium is 2.016.

ACIDS AND BASES, II

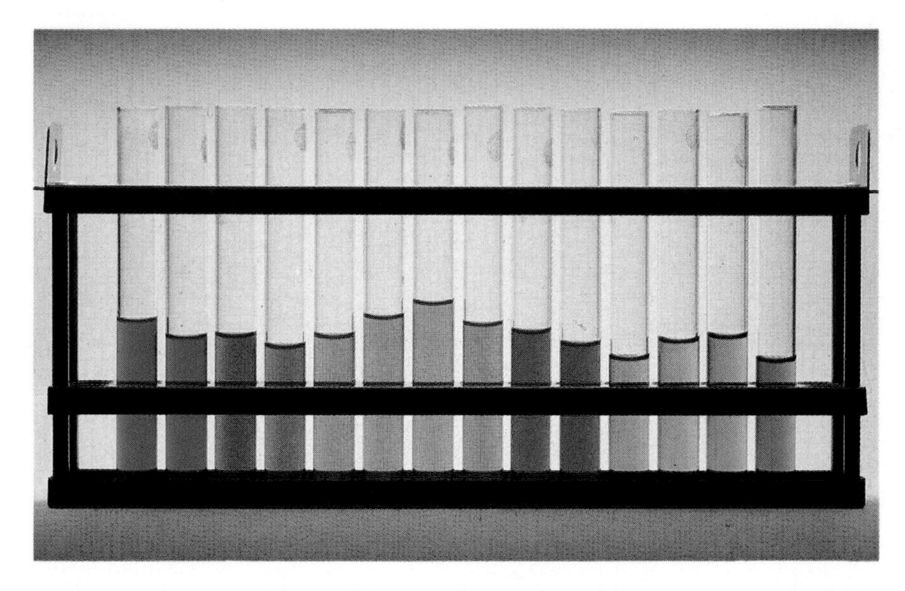

Aqueous solutions of the acid-base indicator thymol blue at pH values (left to right) of 1, 2, 3, 4, 5, 6, 7, 8, 9, 10, 11, 12, 13, and 14. Note that the indicator undergoes two color changes in the range 1 to 14.

This chapter continues our treatment of acid-base chemistry. We begin with a discussion of colored acid-base indicators and of how they can provide a qualitative measure of the acidity of a solution. We then discuss acid-base titration curves, in which the pH of a solution is plotted against the volume of an added base or acid. Acid-base titrations are used to determine the total acidity of a solution by titration with a base or the total basicity of a solution by titration with an acid. Acid-base indicators are used to signal the equivalence point in an acid-base titration.

Buffers, which are mixtures of conjugate acid-base pairs, are used to control pH in chemical and biochemical systems. The chemical basis of buffer action is discussed in terms of Le Châtelier's principle, and the Henderson-Hasselbalch equation is used to calculate the pH of buffer solutions.

16-1. AN INDICATOR IS A WEAK ORGANIC ACID WHOSE COLOR VARIES WITH pH

Numerous weak organic acids change color upon loss of a proton. These compounds are called **indicators** because they indicate the pH of a solution by their color. In Chapter 2 we discussed the use of litmus as an acid-base indicator. Another example of an acid-base indicator is the compound methyl orange, which we denote by HIn to emphasize that it is a weak acid and an indicator. The acid form of methyl orange (HIn) is red, and the conjugate base form (In^-) is yellow (Figure 16-1). We can represent the acid-base reaction of methyl orange by the equation

$$\underset{\text{red}}{HIn(aq)} + H_2O(l) \rightleftharpoons H_3O^+(aq) + \underset{\text{yellow}}{In^-(aq)} \qquad (16\text{-}1)$$

According to Le Châtelier's principle, the reaction equilibrium lies to the left if $[H_3O^+]$ is large (low pH) and to the right if $[H_3O^+]$ is small (high pH). Consequently, a solution containing methyl orange is red at low pH and yellow at high pH.

We can make this discussion quantitative by considering the acid-dissociation constant of methyl orange. The acid-dissociation constant for Equation (16-1) is

$$K_{ai} = \frac{[H_3O^+][In^-]}{[HIn]} \qquad (16\text{-}2)$$

Figure 16-1 Aqueous solutions of the acid-base indicator methyl orange at pH values of 2 (red), 4 (orange) and 6 (yellow). The orange color at pH = 4 is a mixture of the red and yellow forms of the indicator.

where K_{ai} denotes the acid-dissociation constant of an indicator. As we have noted, an aqueous solution containing methyl orange in the acid form (HIn) is red and an aqueous solution containing methyl orange in the basic form (In^-) is yellow. If both the HIn(*aq*) and In^-(*aq*) forms are present simultaneously in comparable concentrations, that is, if

$$[\mathrm{HIn}] \approx [\mathrm{In}^-]$$

then the solution is orange (red + yellow) (Figure 16-1). Substitution of the condition $[\mathrm{HIn}] \approx [\mathrm{In}^-]$ into Equation (16-2) yields

$$[\mathrm{H_3O^+}] \approx K_{ai}$$

By taking the negative of the logarithms of the terms in this expression and using the definitions of pH and pK_a, we obtain

$$\mathrm{pH} \approx \mathrm{p}K_{ai}$$

▪ The pH range over which an indicator changes color is approximately equal to $pK_{ai} \pm 1$.

In other words, if methyl orange is orange in a solution, then the pH of the solution is approximately equal to the pK_{ai} value for methyl orange. The value of pK_{ai} for methyl orange is 3.9, and so the pH of the orange solution is about 4.

If we place a few drops of a methyl orange solution into each of the following solutions:

0.010 M HCl (pH = 2) + methyl orange	0.00010 M HCl (pH = 4) + methyl orange	pure water (pH ≈ 7) + methyl orange
red solution	orange solution	yellow solution

Aqueous solutions containing the acid-base indicator methyl red at pH values of 10 (*yellow*), 5 (*orange*), and 2 (*red*).

the resulting solutions will have the colors indicated. This simple experiment shows that acid-base indicators can be used to determine qualitatively the acidity of solutions.

Because of the intense color of indicators, only a very small concentration is necessary to produce a visible color. The contribution of the indicator acid, HIn, to the total acidity of the solution is usually negligible. Several indicators, together with their colors at various pH values, are given in Figure 16-2. Using the indicators in Figure 16-2 we can estimate the pH of an aqueous solution to within about 0.5 pH unit. Of course, the solution must be colorless initially, for otherwise the color change of the indicator may be obscured.

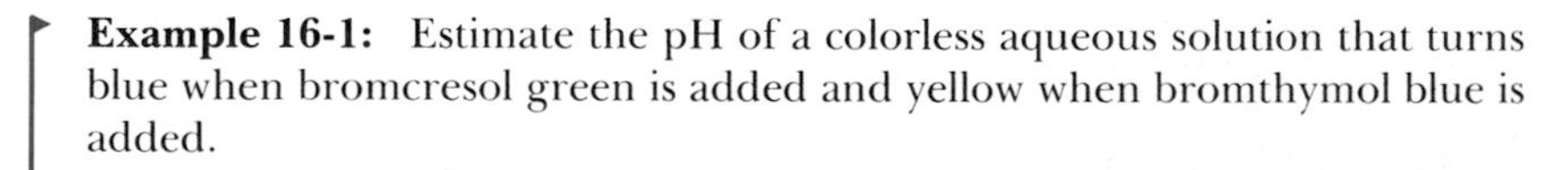

Example 16-1: Estimate the pH of a colorless aqueous solution that turns blue when bromcresol green is added and yellow when bromthymol blue is added.

Solution: From Figure 16-2 we note that the pH at which bromcresol green is blue is >5.6, and that bromthymol blue is yellow at pH < 6. Therefore the pH of the solution is about 5.8.

16-2. AT THE EQUIVALENCE POINT THE NUMBER OF MOLES OF ACID EQUALS THE NUMBER OF MOLES OF BASE

In Section 3-12 we discussed the stoichiometry of neutralization reactions between strong bases and strong acids. Recall that the neutraliza-

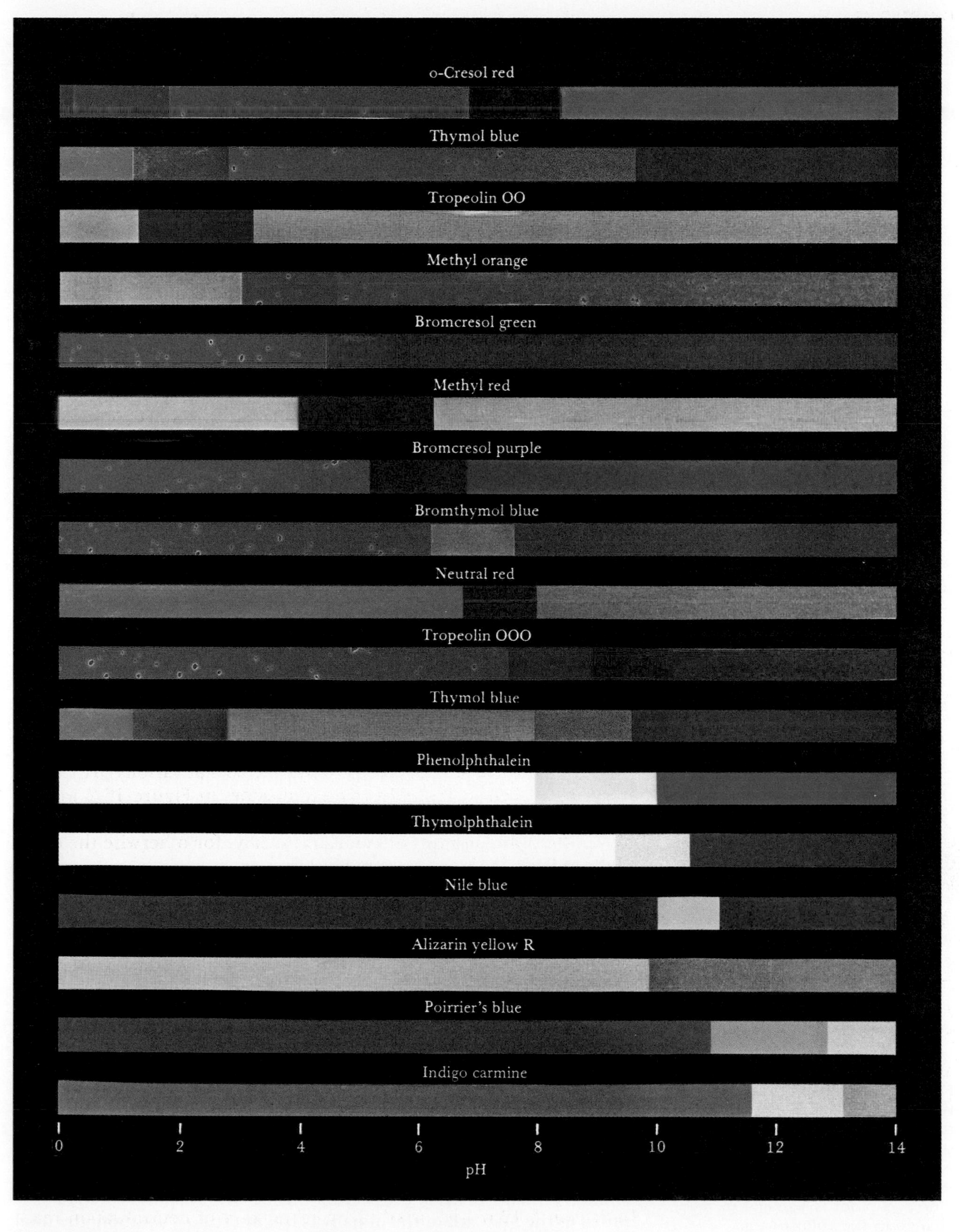

Figure 16-2 The colors of various indicators at different pH values.

tion reaction is complete when the number of moles of base reacted is equal to the number of moles of acid reacted; that is, when

$$M_{\text{base}}V_{\text{base}} = M_{\text{acid}}V_{\text{acid}}$$

Now suppose that an aqueous solution of a strong base, such as NaOH(*aq*), is added slowly to an aqueous solution of a strong acid, such as HCl(*aq*). This procedure is called **titration** (Figures 16-3 and 16-4), and a plot of the pH of the resulting solution as a function of the volume of added solution—that is, the **titrant**—is called a **titration curve.** The titration curve of a 50.0-mL sample of 0.100 M HCl(*aq*) titrated with 0.100 M NaOH(*aq*) is shown in Figure 16-5. The **equivalence point** of a titration is the point at which stoichiometrically equivalent amounts of acid and base have been brought together. At the equivalence point, the number of moles of base added is equal to the number of moles of acid initially present. The acid-base reaction is complete at the equivalence point.

Because HCl(*aq*) and NaOH(*aq*) are strong electrolytes, the net ionic equation for the titration is

$$H_3O^+(aq) + OH^-(aq) \rightarrow 2H_2O(l) \qquad (16\text{-}3)$$

At the equivalence point, the solution is simply an aqueous solution of NaCl. Because NaCl is a neutral salt, the pH is 7 at the equivalence point.

Indicators are used in titration solutions to signal the completion of the acid-base reaction. The point at which the indicator changes color is called the **end point** of the titration. The end point is the experimental estimate of the equivalence point.

0.100 M NaOH (*aq*)

0.100 M HCl (*aq*)

Magnetic stirring bar

Magnetic stirrer

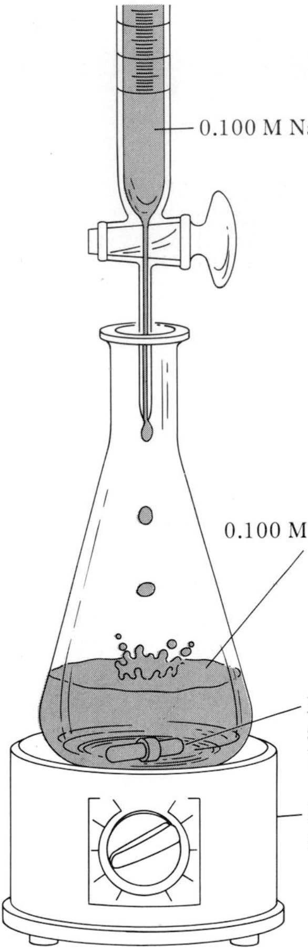

Figure 16-3 A diagrammatic representation of the setup used for titration. The solution is stirred constantly and the NaOH(*aq*) is added drop by drop as the end point is approached. The end point is signaled by a change in the color of the indicator.

Example 16-2: By referring to Figures 16-2 and 16-5, choose an indicator to determine the equivalence point shown in Figure 16-5.

Solution: From Figure 16-5 we note that the equivalence point occurs at pH = 7.0. The titration curve is very steep in the vicinity of the equivalence point, however, and thus an indicator with a color transition range lying between pH = 5 and pH = 9 would be suitable. Referring to Figure 16-2, we see that there are several possible choices, for example, bromthymol blue or phenolphthalein. We must choose our indicator such that the end point and the equivalence point are the same within the required accuracy of the titration.

The calculation of points on the HCl + NaOH titration curve is straightforward. The key factor to recognize is that the equilibrium constant for the titration reaction (Equation 16-3) is very large. Because it is the reverse of the water dissociation reaction, the equilibrium constant, K, for Equation (16-3) is equal to $1/K_w$, where $K_w = 1.0 \times 10^{-14}\ \text{M}^2$. Therefore, $K = 1.0 \times 10^{14}\ \text{M}^{-2}$. The large value of K for Equation (16-3) means that this reaction goes essentially to completion.

The concentration of $H_3O^+(aq)$ in the 0.100 M HCl(*aq*) solution before any base is added is $[H_3O^+] = 0.100$ M, and thus the pH of the solution initially is 1.00. After 10.0 mL of the 0.100 M NaOH(*aq*) is added to the 50.0 mL of the initially 0.100 M HCl(*aq*) solution, the total volume of the resulting solution is 50.0 mL + 10.0 mL = 60.0 mL. The

(a)

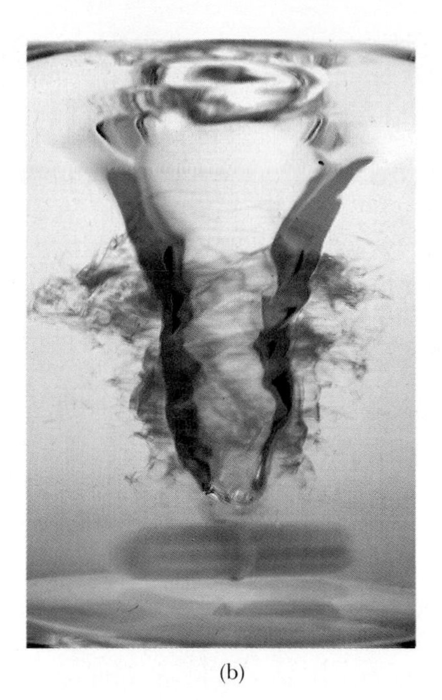

(b)

(c)

Figure 16-4 Experimental setup for the titration of HCl(*aq*) with NaOH(*aq*). A 50.0-mL sample of HCl (*aq*) is placed in the reaction flask together with a magnetic bar for stirring the reaction mixture during the titration. (a) Two or three drops of the acid-base indicator phenolphthalein is added to the HCl(*aq*), and then (b) NaOH(*aq*) is added from a buret. As the equivalence point is approached, the base is added dropwise until (c) a single drop turns the entire reaction mixture a pink color that does not revert to colorless with stirring.

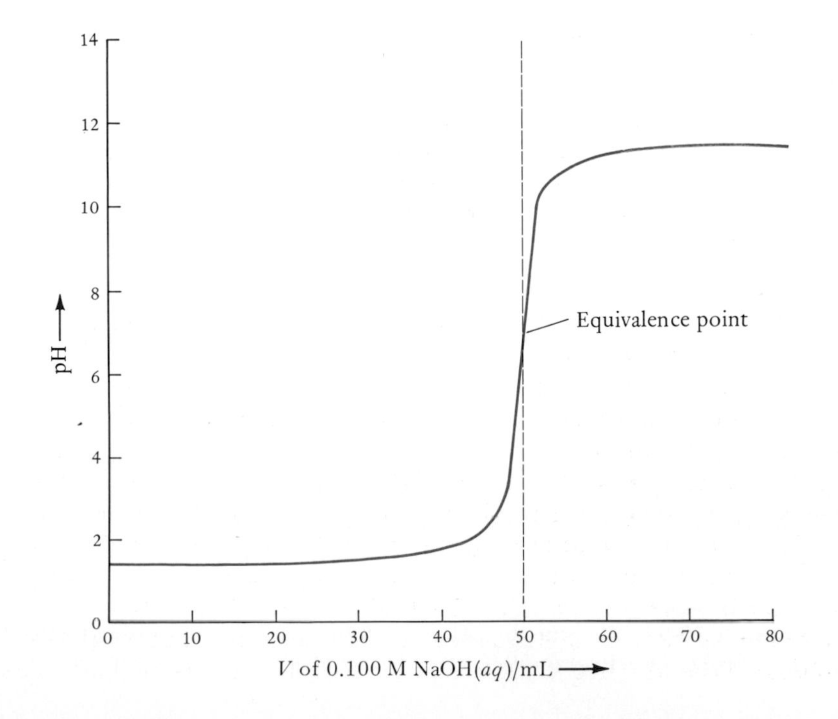

Figure 16-5 Titration curve for the titration of 50.0 mL of 0.100 M HCl(*aq*) with 0.100 M NaOH(*aq*). Note that the pH of the solution changes very slowly until the vicinity of the equivalence point is reached and then changes very rapidly in the vicinity of the equivalence point.

$OH^-(aq)$ from the $NaOH(aq)$ reacts with the $H_3O^+(aq)$ from the $HCl(aq)$ to produce water:

$$H_3O^+(aq) + OH^-(aq) \rightarrow 2H_2O(l)$$

Because the volumes in titration experiments are usually expressed in milliliters, it is convenient to carry out calculations in terms of millimoles and milliliters rather than moles and liters. The key relation is

$$\text{number of millimoles} = (\text{molarity}) \times (\text{number of milliliters}) \quad (16\text{-}4)$$

Thus, when 10.0 mL of 0.100 M $NaOH(aq)$ is added to 50.0 mL of 0.100 M $HCl(aq)$, the number of millimoles of $H_3O^+(aq)$ that reacts is equal to the number of millimoles of $OH^-(aq)$ added:

$$\begin{pmatrix}\text{mmol of}\\ OH^-(aq)\\ \text{added}\end{pmatrix} = (0.100\text{ M})(10.0\text{ mL})$$

$$= 1.00\text{ mmol} = \begin{pmatrix}\text{mmol of}\\ H_3O^+(aq)\\ \text{reacted}\end{pmatrix}$$

The total number of millimoles of $H_3O^+(aq)$ initially present in 50.0 mL of 0.100 M $HCl(aq)$ is

$$\begin{pmatrix}\text{initial}\\ \text{mmol of}\\ H_3O^+(aq)\end{pmatrix} = MV = (0.100\text{ M})(50.0\text{ mL}) = 5.00\text{ mmol}$$

▪ Note that

$$1\text{ M} = \frac{1\text{ mol}}{\text{L}} = \frac{1\text{ mmol}}{\text{mL}}$$

Thus

$$(1\text{ M})(1\text{ mL}) = 1\text{ mmol}$$

The concentration of $H_3O^+(aq)$ that remains after the addition of 10.0 mL of 0.100 M NaOH is equal to the number of millimoles of $H_3O^+(aq)$ that remains unreacted divided by the total volume in milliliters of resulting solution. The number of millimoles of $H_3O^+(aq)$ unreacted is

$$\begin{pmatrix}\text{mmol of}\\ H_3O^+(aq)\\ \text{unreacted}\end{pmatrix} = \begin{pmatrix}\text{initial}\\ \text{mmol of}\\ H_3O^+(aq)\end{pmatrix} - \begin{pmatrix}\text{mmol of}\\ H_3O^+(aq)\\ \text{reacted}\end{pmatrix}$$

$$= 5.00\text{ mmol} - 1.00\text{ mmol}$$

$$= 4.00\text{ mmol}$$

and the molarity of unreacted $H_3O^+(aq)$ is

$$\text{molarity} = \frac{4.00\text{ mmol}}{60.0\text{ mL}} = 6.67 \times 10^{-2}\text{ M}$$

Therefore, the pH of the solution after the addition of 10.0 mL of $NaOH(aq)$ is

$$\text{pH} = -\log[H_3O^+] = -\log(6.67 \times 10^{-2}) = 1.18$$

Proceeding in an analogous fashion, we can compute the pH of the resulting solution after the addition of other volumes of $NaOH(aq)$ (Table 16-1). Beyond the equivalence point, all the $H_3O^+(aq)$ has reacted and we simply have a diluted solution of $NaOH(aq)$.

Example 16-3. Compute the pH of a solution obtained by adding 60.0 mL of 0.100 M $NaOH(aq)$ to 50.0 mL of 0.100 M $HCl(aq)$.

Solution: The total number of millimoles of $H_3O^+(aq)$ in 50.0 mL of 0.100 M $HCl(aq)$ is

$$\begin{pmatrix}\text{mmol of}\\ H_3O^+(aq)\end{pmatrix} = MV = (0.100\ \text{M})(50.0\ \text{mL}) = 5.00\ \text{mmol}$$

The total number of millimoles of $OH^-(aq)$ in 60.0 mL of 0.100 M $NaOH(aq)$ is

$$\begin{pmatrix}\text{mmol of}\\ OH^-(aq)\\ \text{added}\end{pmatrix} = MV = (0.100\ \text{M})(60.0\ \text{mL}) = 6.00\ \text{mmol}$$

Note that the number of millimoles of $NaOH(aq)$ added (6.00 mmol) exceeds the number of millimoles of $H_3O^+(aq)$ present initially (5.00 mmol). The number of millimoles of $OH^-(aq)$ that remain unreacted after the NaOH addition is

$$\begin{pmatrix}\text{mmol of}\\ OH^-(aq)\\ \text{unreacted}\end{pmatrix} = 6.00\ \text{mmol} - 5.00\ \text{mmol} = 1.00\ \text{mmol}$$

The total volume of the solution is 50.0 mL + 60.0 mL = 110.0 mL, and thus the concentration of unreacted $OH^-(aq)$ is

$$[OH^-] = \frac{1.00\ \text{mmol}}{110.0\ \text{mL}} = 9.09 \times 10^{-3}\ \text{M}$$

▪ Note in Table 16-1 the rapid change in pH in the range 49.5 mL to 50.5 mL of added base.

Table 16-1 Calculation of various points for the titration of 50.0 mL of 0.100 M HCl(*aq*) with 0.100 M NaOH(*aq*)

Volume of 0.100 M NaOH(aq) added/mL	OH^-(aq) added/mmol	Unreacted H_3O^+(aq)/mmol	Total volume of solution/mL	Concentration of unreacted H_3O^+(aq)/M	pH
20.0	2.00	3.00	70.0	4.29×10^{-2}	1.37
30.0	3.00	2.00	80.0	2.50×10^{-2}	1.60
40.0	4.00	1.00	90.0	1.11×10^{-2}	1.95
45.0	4.50	0.50	95.0	5.26×10^{-3}	2.28
49.0	4.90	0.10	99.0	1.01×10^{-3}	3.00
49.5	4.95	0.05	99.5	5.03×10^{-4}	3.30
50.0 **equivalence point**	5.00	1.00×10^{-5} (from H_2O dissociation)	100.0	1.00×10^{-7}	7.00
Beyond the equivalence point:					
		Unreacted OH^-(aq)/mmol		**Concentration of unreacted OH^-(aq)/M**	
50.5	5.05	0.05	100.5	4.98×10^{-4}	10.70
51.0	5.10	0.10	101.0	9.90×10^{-4}	11.00
55.0	5.50	0.50	105.0	4.76×10^{-3}	11.68
60.0	6.00	1.00	110.0	9.09×10^{-3}	11.96
70.0	7.00	2.00	120.0	1.67×10^{-2}	12.22
80.0	8.00	3.00	130.0	2.31×10^{-2}	12.36
90.0	9.00	4.00	140.0	2.86×10^{-2}	12.46

The pOH of the solution is

$$\text{pOH} = -\log [\text{OH}^-] = -\log (9.09 \times 10^{-3}) = 2.04$$

and so the pH of the solution is

$$\text{pH} = 14.00 - \text{pOH} = 14.00 - 2.04 = 11.96$$

which is the result shown in Figure 16-5 and Table 16-1 for 60.0 mL of added base. You should reproduce some of the other values given in Table 16-1.

16-3. WEAK ACIDS CAN BE TITRATED WITH STRONG BASES

Figure 16-5 shows the titration curve of a strong acid with a strong base. The titration curve of a weak acid with a strong base looks somewhat different from Figure 16-5. The titration curve for 50.0 mL of 0.100 M acetic acid, $HC_2H_3O_2(aq)$, titrated with 0.100 M NaOH(aq) is shown in Figure 16-6. Note that the pH of the acetic acid solution is about 2.9 initially and increases slowly until the equivalence point is reached. Around the equivalence point the pH changes from 6 to 11. Note also that the equivalence point occurs not at pH = 7 but at pH = 9.

We can calculate the pH of the 0.100 M acetic acid solution before any sodium hydroxide is added by the method developed in Section 15-6. The equilibrium equation is

$$HC_2H_3O_2(aq) + H_2O(l) \rightleftharpoons H_3O^+(aq) + C_2H_3O_2^-(aq)$$

and thus

$$K_a = \frac{[H_3O^+][C_2H_3O_2^-]}{[HC_2H_3O_2]} = 1.74 \times 10^{-5}\ \text{M} \qquad (16\text{-}5)$$

Figure 16-6 Titration curve for the titration of 50.0 mL of 0.100 M $HC_2H_3O_2(aq)$ with 0.100 M NaOH(aq). The indicator used in this case is phenolphthalein, which changes from colorless to pink around pH = 9.

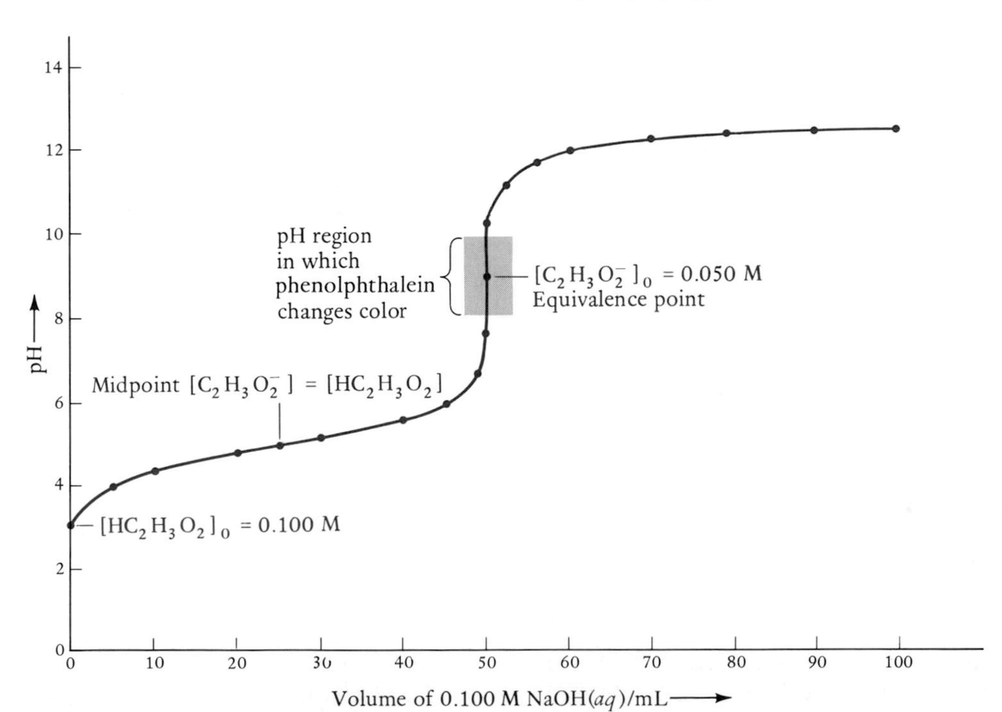

Before any $NaOH(aq)$ is added we have $[H_3O^+] = [C_2H_3O_2^-]$. Thus Equation (16-5) becomes

$$\frac{[H_3O^+]^2}{0.100\ M - [H_3O^+]} = 1.74 \times 10^{-5}\ M \qquad (16\text{-}6)$$

Solving Equation (16-6) by the method of successive approximations, or by using the quadratic formula, yields $[H_3O^+] = 1.31 \times 10^{-3}$ M. Thus we have

$$pH = -\log(1.31 \times 10^{-3}) = 2.88$$

in agreement with Figure 16-6.

To calculate the pH at the equivalence point, we must consider the principal species in the solution at the equivalence point. At the equivalence point, the number of moles of $NaOH(aq)$ added is equal to the number of moles of $HC_2H_3O_2(aq)$ initially present. The equation for the reaction is

$$HC_2H_3O_2(aq) + Na^+(aq) + OH^-(aq) \rightarrow$$
$$Na^+(aq) + C_2H_3O_2^-(aq) + H_2O(l) \qquad (16\text{-}7)$$

It is important to realize that the reaction described by Equation (16-7) goes essentially to completion even though $HC_2H_3O_2(aq)$ is a weak acid because $OH^-(aq)$ is a strong base. We can see this quantitatively, by calculating the equilibrium constant of Equation (16-7). The net ionic equation for the titration of $HC_2H_3O_2(aq)$ with $NaOH(aq)$ is

▪ The titration of a weak acid with a strong base goes essentially to completion because the value of the equilibrium constant is very large for the neutralization reaction.

$$HC_2H_3O_2(aq) + OH^-(aq) \rightleftharpoons C_2H_3O_2^-(aq) + H_2O(l) \qquad (16\text{-}8)$$

Equation (16-8) is simply the reverse of the base-protonation reaction for $C_2H_3O_2^-(aq)$. The value of K_b for $C_2H_3O_2^-(aq)$ at 25°C is 5.75×10^{-10} M (Table 15-5), so the value of K for Equation (16-8) is

$$K = \frac{1}{K_b} = \frac{1}{5.75 \times 10^{-10}\ M} = 1.74 \times 10^{9}\ M^{-1}$$

This large value of K indicates that the equilibrium described by Equation (16-8) lies far to the right, and thus $HC_2H_3O_2(aq)$ reacts essentially completely when a strong base is added to the solution.

Because the reaction described by Equation (16-8) goes essentially to completion, the solution at the equivalence point of the titration consists of a $NaC_2H_3O_2(aq)$ solution. To calculate the concentration of this $NaC_2H_3O_2(aq)$ solution, we use the fact that we started with 50.0 mL of 0.100 M $HC_2H_3O_2(aq)$. The initial number of millimoles of $HC_2H_3O_2(aq)$ is

$$\text{initial mmol of } HC_2H_3O_2(aq) = (0.100\ M)(50.0\ mL)$$
$$= 5.00\ mmol$$

According to Equation (16-7)

$$\begin{pmatrix}\text{mmol of } NaC_2H_3O_2(aq) \text{ at} \\ \text{the equivalence point}\end{pmatrix} = \begin{pmatrix}\text{initial mmol of} \\ HC_2H_3O_2(aq)\end{pmatrix}$$
$$= 5.00\ mmol$$

The total volume of the solution at the equivalence point is 100.0 mL, that is, 50.0 mL of $HC_2H_3O_2(aq)$ plus 50.0 mL of added $NaOH(aq)$, and so the concentration of the $NaC_2H_3O_2(aq)$ solution at the equivalence point is

$$M = \frac{5.00 \text{ mmol}}{100.0 \text{ mL}} = 0.0500\ M$$

We learned how to calculate the pH of a solution such as $NaC_2H_3O_2(aq)$ in Section 15-11. In fact, Example 15-14 involved the calculation of the pH of a 0.050 M $NaC_2H_3O_2(aq)$ solution, giving a value of pH = 8.73. Thus we see that the pH at the equivalence point in Figure 16-6 is 8.73.

Example 16-4: Which indicator would you use to signal the equivalence point of the titration of 50.0 mL of 0.100 M $HC_2H_3O_2(aq)$ with 0.100 M $NaOH(aq)$?

Solution: The pH at the equivalence point in the titration is 8.73 (Example 15-14). Referring to Figure 16-2, we see that phenolphthalein is a suitable indicator for the titration.

▪ The indicator is an acid and is titrated along with the acetic acid. The total moles of indicator, however, is usually negligible.

We have calculated the pH at only two points (the initial point and the equivalence point) on the titration curve of a weak acid with a strong base. There is another point that is easy to calculate. The **midpoint** of a titration of a weak acid with a strong base is the point at which $[HB]_0 = [B^-]_0$, where the subscript zeros denote stoichiometric concentrations. The acid-dissociation-constant expression

$$K_a = \frac{[H_3O^+][B^-]}{[HB]}$$

can be rearranged to the form

$$[H_3O^+] = K_a \frac{[HB]}{[B^-]}$$

At the midpoint, $[HB]_0 \approx [B^-]_0$ and $[HB] \approx [B^-]$; therefore

$$[H_3O^+] \approx K_a \qquad \text{at the midpoint}$$

Upon taking logarithms, we obtain

$$\text{pH} \approx \text{p}K_a \qquad \text{at the midpoint}$$

The value of $\text{p}K_a$ for acetic acid is 4.76. The midpoint for the titration of 50.0 mL of 0.100 M $HC_2H_3O_2(aq)$ with 0.100 M $NaOH(aq)$ occurs when 25.0 mL of $NaOH(aq)$ has been added, and so we find that the pH is 4.76 at this point (Figure 16-6).

The following examples illustrate the calculation of two additional points on the $HC_2H_3O_2$ + NaOH titration curve, one just before the equivalence point and one well beyond the equivalence point.

Example 16-5: Compute the pH of the solution that results when 49.0 mL of 0.100 M $NaOH(aq)$ is added to 50.0 mL of 0.100 M acetic acid.

Solution: The total number of millimoles of acetic acid present in the original solution is

$$\begin{pmatrix}\text{initial}\\ \text{mmol of}\\ HC_2H_3O_2(aq)\end{pmatrix} = (0.100\text{ M})(50.0\text{ mL}) = 5.00\text{ mmol}$$

The number of millimoles of $NaOH(aq)$ added is

$$\begin{pmatrix}\text{mmol of}\\ NaOH(aq)\\ \text{added}\end{pmatrix} = (0.100\text{ M})(49.0\text{ mL}) = 4.90\text{ mmol}$$

The number of millimoles of unreacted $HC_2H_3O_2(aq)$ is

$$\begin{pmatrix}\text{mmol of}\\ HC_2H_3O_2(aq)\\ \text{unreacted}\end{pmatrix} = \begin{pmatrix}\text{initial}\\ \text{mmol of}\\ HC_2H_3O_2(aq)\end{pmatrix} - \begin{pmatrix}\text{mmol of}\\ NaOH(aq)\\ \text{added}\end{pmatrix}$$

$$= 5.00\text{ mmol} - 4.90\text{ mmol} = 0.10\text{ mmol}$$

Because of Equation (16-8), the stoichiometric concentration of unreacted acetic acid is

$$[HC_2H_3O_2]_0 = \frac{\text{mmol of } HC_2H_3O_2}{\text{mL of solution}} = \frac{0.100\text{ mmol}}{99.0\text{ mL}}$$

$$= 0.00101\text{ M}$$

The stoichiometric concentration of $[C_2H_3O_2^-]_0$ in this solution is

$$[C_2H_3O_2^-]_0 = \frac{\text{mmol NaOH added}}{\text{mL of solution}} = \frac{4.90\text{ mmol}}{99.0\text{ mL}}$$

$$= 0.0495\text{ M}$$

The value of $[H_3O^+]$ is computed using the K_a expression for $HC_2H_3O_2(aq)$:

$$K_a = 1.74 \times 10^{-5}\text{ M} = \frac{[H_3O^+][C_2H_3O_2^-]}{[HC_2H_3O_2]} \tag{16-9}$$

The equation of the reaction that relates $[H_3O^+]$, $[HC_2H_3O_2]$, and $[C_2H_3O_2^-]$ is

$$HC_2H_3O_2(aq) + H_2O(l) \rightleftharpoons H_3O^+(aq) + C_2H_3O_2^-(aq)$$

From the stoichiometry of this equation, we see that

$$[HC_2H_3O_2] = [HC_2H_3O_2]_0 - [H_3O^+]$$

$$= 0.00101\text{ M} - [H_3O^+]$$

and that

$$[C_2H_3O_2^-] = [C_2H_3O_2^-]_0 + [H_3O^+]$$

$$= 0.0495\text{ M} + [H_3O^+]$$

If we substitute these results into Equation (16-9), then we obtain

$$1.74 \times 10^{-5}\text{ M} = \frac{[H_3O^+]\{0.0495\text{ M} + [H_3O^+]\}}{0.00101\text{ M} - [H_3O^+]} \tag{16-10}$$

A first approximation to the solution of Equation (16-10) is obtained by assuming that $[H_3O^+]$ is small relative to 0.00101 M and 0.0495 M. With this approximation we have

$$[H_3O^+] \approx \frac{(1.74 \times 10^{-5}\text{ M})(0.00101\text{ M})}{(0.0495\text{ M})}$$

$$= 3.55 \times 10^{-7}\text{ M}$$

Note that the result for $[H_3O^+]$ is indeed much less than 0.00101 M and thus is an accurate approximation, as is readily verified by using the method of successive approximations. The pH of the solution is

$$\text{pH} = -\log(3.55 \times 10^{-7}) = 6.45$$

Example 16-6: Compute the pH of the solution that results when 60.0 mL of 0.100 M NaOH(*aq*) is added to 50.0 mL of 0.100 M acetic acid.

Solution: The number of millimoles of NaOH added is

$$\begin{pmatrix}\text{mmol of}\\ \text{NaOH}(aq)\\ \text{added}\end{pmatrix} = (0.100\ \text{M})(60.0\ \text{mL}) = 6.00\ \text{mmol}$$

There are only 5.00 mmol of $HC_2H_3O_2(aq)$ in the original sample of acetic acid and thus we added an excess of NaOH(*aq*). The number of millimoles of unreacted NaOH(*aq*) is

$$\begin{pmatrix}\text{mmol of}\\ \text{NaOH}(aq)\\ \text{unreacted}\end{pmatrix} = \begin{pmatrix}\text{total mmol}\\ \text{of NaOH}(aq)\\ \text{added}\end{pmatrix} - \begin{pmatrix}\text{total mmol of}\\ \text{HC}_2\text{H}_3\text{O}_2(aq)\\ \text{available}\end{pmatrix}$$

$$= 6.00\ \text{mmol} - 5.00\ \text{mmol} = 1.00\ \text{mmol}$$

The stoichiometric unreacted (excess) concentration of $OH^-(aq)$ in the final solution is

$$[\text{OH}^-]_0 = \frac{1.00\ \text{mmol}}{110.0\ \text{mL}} = 9.09 \times 10^{-3}\ \text{M}$$

The stoichiometric concentration of acetate ion in the final solution is

$$[\text{C}_2\text{H}_3\text{O}_2^-]_0 = \frac{5.00\ \text{mmol}}{110.0\ \text{mL}} = 4.55 \times 10^{-2}\ \text{M}$$

The titration solution beyond the equivalence point consists of a mixture of sodium acetate and sodium hydroxide. Because acetate ion is a weak base, we have the equilibrium

$$\text{C}_2\text{H}_3\text{O}_2^-(aq) + \text{H}_2\text{O}(l) \rightleftharpoons \text{OH}^-(aq) + \text{HC}_2\text{H}_3\text{O}_2(aq) \qquad (16\text{-}11)$$

which is the equation for the base protonation of acetate ion. The equilibrium constant for Equation (16-11) is $K_b = 5.75 \times 10^{-10}$ M (Table 15-5). Because K_b is very small the equilibrium in Equation (16-11) lies far to the left. The equilibrium is driven even further to the left (Le Châtelier's principle) by the presence of excess $OH^-(aq)$ in the solution at a concentration of $[OH^-]_0 = 9.09 \times 10^{-3}$ M. Thus the reaction described by Equation (16-11) makes a negligible contribution to the total concentration of $OH^-(aq)$, and we have

$$[\text{OH}^-] \approx [\text{OH}^-]_0 = 9.09 \times 10^{-3}\ \text{M} \qquad (16\text{-}12)$$

The pOH of the solution is

$$\text{pOH} = -\log[\text{OH}^-] = -\log(9.09 \times 10^{-3}) = 2.04$$

and the pH of the solution is

$$\text{pH} = 14.00 - \text{pOH} = 14.00 - 2.04 = 11.96$$

in agreement with the point at 60.0 mL NaOH(*aq*) in Figure 16-6. The approximation given in Equation (16-12) can be checked by solving the K_b expression for $[HC_2H_3O_2]$ and noting that $[OH^-] = [OH^-]_0 +$

$[HC_2H_3O_2] \approx [OH^-]_0$. The detailed solution, which you should verify, confirms the approximation given by Equation (16-12).

16-4. WEAK BASES CAN BE TITRATED WITH STRONG ACIDS

We have considered the titration of a strong acid with a strong base and a weak acid with a strong base. Let's now discuss the titration of a weak base with a strong acid. The titration curve for 50.0 mL of 0.100 M $NH_3(aq)$ with 0.100 M $HCl(aq)$ is shown in Figure 16-7. Note that the initial pH is 11.1, which is the pH of a 0.100 M $NH_3(aq)$ solution, and that the pH is 5.3 at the equivalence point. The equation for the reaction is

$$NH_3(aq) + H_3O^+(aq) + Cl^-(aq) \rightleftharpoons NH_4^+(aq) + Cl^-(aq) + H_2O(l) \qquad (16\text{-}13)$$

As in the case of the reaction of a weak acid with a strong base, this reaction goes essentially to completion.

Example 16-7: Calculate K for Equation (16-13).

Solution: The reaction represented by Equation (16-13) is simply the reverse of the acid-dissociation reaction for $NH_4^+(aq)$, whose equation is

$$NH_4^+(aq) + H_2O(l) \rightleftharpoons NH_3(aq) + H_3O^+(aq) \qquad K_a = 5.71 \times 10^{-10}\ M$$

Thus the equilibrium constant for Equation (16-13) is

$$K = \frac{1}{K_a} = \frac{1}{5.71 \times 10^{-10}\ M} = 1.75 \times 10^9\ M^{-1}$$

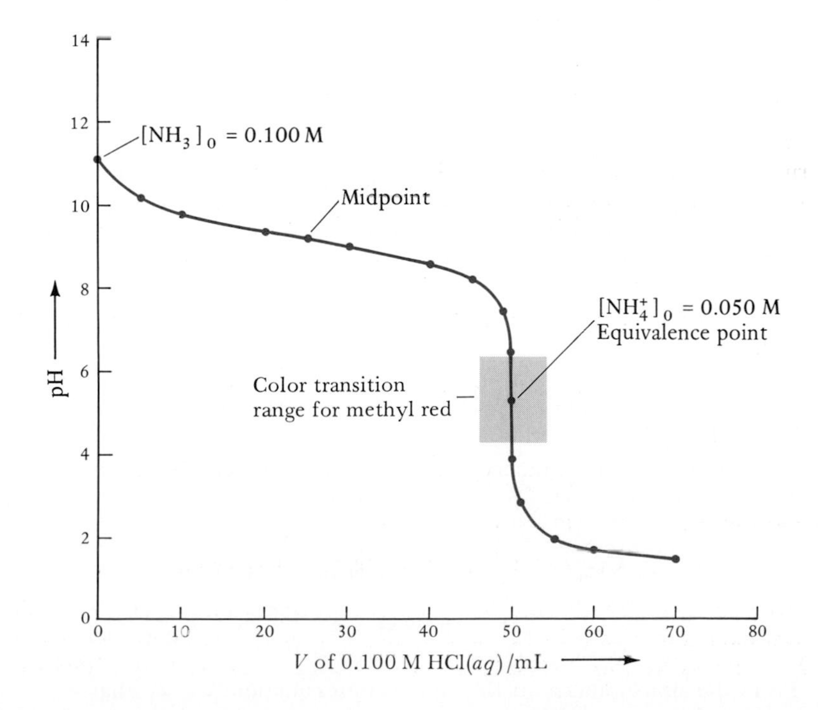

Figure 16-7 Titration curve for the titration of 50.0 mL of 0.100 M $NH_3(aq)$ with 0.100 M $HCl(aq)$. The pH at the equivalence point is acidic because at that point we have a 0.050 M solution of $NH_4Cl(aq)$, which is acidic because $NH_4^+(aq)$ is a weak acid.

This large value of K indicates that the reaction represented by Equation (16-13) goes essentially to completion.

The pH of a 0.100 M $NH_3(aq)$ solution ($K_b = 1.75 \times 10^{-5}$ M, Table 15-5) is calculated, in a manner analogous to that used for the weak base aniline in Example 15-10, to be pH = 11.12, in agreement with Figure 16-7.

By the same reasoning we used for the titration of 50.0 mL of 0.100 M $HC_2H_3O_2(aq)$ with 0.100 M $NaOH(aq)$, we find that the solution at the equivalence point in the titration of 50.0 mL of 0.100 M $NH_3(aq)$ with 0.100 M $HCl(aq)$ is a 0.050 M $NH_4Cl(aq)$ solution. In Section 15-11 we calculated the pH of a 0.050 M $NH_4Cl(aq)$ solution to be 5.28, in agreement with Figure 16-7. The pH at the midpoint of the titration is equal to pK_a for $NH_4^+(aq)$, which is 9.24, again in agreement with Figure 16-7. A suitable indicator for the titration is methyl red, which has a color-transition range centered at pH = 5 (see Figure 16-7).

Example 16-8: Calculate the pH of the solution obtained when 35.0 mL of 0.100 M $HCl(aq)$ is added to 50.0 mL of 0.100 M $NH_3(aq)$.

Solution: The number of millimoles of $NH_3(aq)$ present in the original solution is

$$\begin{pmatrix}\text{initial}\\ \text{mmol of}\\ NH_3(aq)\end{pmatrix} = (0.100\ \text{M})(50.0\ \text{mL}) = 5.00\ \text{mmol}$$

The number of millimoles of unreacted $NH_3(aq)$ after the addition of the $HCl(aq)$ is (see Equation 16-13)

$$\begin{pmatrix}\text{mmol of}\\ NH_3(aq)\\ \text{unreacted}\end{pmatrix} = \begin{pmatrix}\text{initial}\\ \text{mmol of}\\ NH_3(aq)\end{pmatrix} - \begin{pmatrix}\text{mmol of}\\ HCl(aq)\\ \text{added}\end{pmatrix}$$

$$= 5.00\ \text{mmol} - (0.100\ \text{M})(35.0\ \text{mL}) = 1.50\ \text{mmol}$$

The number of millimoles of $NH_4^+(aq)$ produced is equal to the number of millimoles of added $HCl(aq)$, that is, 3.50 mmol (see Equation 16-13). After the addition of the $HCl(aq)$ the solution contains both $NH_3(aq)$ and $NH_4^+(aq)$, as well as $H_3O^+(aq)$. The total volume of the solution is 85.0 mL, so the stoichiometric concentrations of $NH_4^+(aq)$ and $NH_3(aq)$ are

$$[NH_4^+]_0 = \frac{3.50\ \text{mmol}}{85.0\ \text{mL}} = 0.0412\ \text{M}$$

$$[NH_3]_0 = \frac{1.50\ \text{mmol}}{85.0\ \text{mL}} = 0.0176\ \text{M}$$

Because the solution contains appreciable stoichiometric concentrations of both $NH_4^+(aq)$ and $NH_3(aq)$, we use the K_a expression for $NH_4^+(aq)$ to compute $[H_3O^+]$:

$$NH_4^+(aq) + H_2O(l) \rightleftharpoons NH_3(aq) + H_3O^+(aq)$$

$$K_a = 5.71 \times 10^{-10}\ \text{M} = \frac{[NH_3][H_3O^+]}{[NH_4^+]}$$

From the stoichiometry of the equilibrium equation, we see that

$$[NH_4^+] = [NH_4^+]_0 - [H_3O^+]$$
$$= 0.0412\ M - [H_3O^+]$$

and that

$$[NH_3] = [NH_3]_0 + [H_3O^+]$$
$$= 0.0176\ M + [H_3O^+]$$

If we substitute these results into the K_a expression, then we obtain

$$5.71 \times 10^{-10}\ M = \frac{\{0.0176\ M + [H_3O^+]\}[H_3O^+]}{0.0412\ M - [H_3O^+]}$$

Because the solution is basic (not all the $NH_3(aq)$ has reacted), $[H_3O^+]$ is less than 10^{-7} M and consequently is negligible with respect to either 0.0176 M or 0.0412 M. Thus we have

$$5.71 \times 10^{-10}\ M \approx \frac{(0.0176\ M)[H_3O^+]}{(0.0412\ M)}$$

from which we compute

$$[H_3O^+] \approx 1.34 \times 10^{-9}\ M$$

Therefore the pH of the solution is

$$pH \approx -\log(1.34 \times 10^{-9}) = 8.87$$

in agreement with Figure 16-7.

We have considered three types of acid-base titrations. The common feature of the three types of titrations is that they all involve an acid-base reaction with a large equilibrium constant. In order for an acid-base titration to work, it is necessary that the reaction that occurs on addition of titrant go essentially to completion; that is, the equation for the titration reaction must have a large value of K.

16-5. THE pH OF A BUFFER SOLUTION CAN BE COMPUTED USING THE HENDERSON-HASSELBALCH EQUATION

A solution that contains both a weak acid and its conjugate base can resist a change in pH when either an acid or a base is added by reacting with the added acid or base. A solution that is resistant to changes in pH upon the addition of an acid or a base is called a **buffer.** As an example of a buffer solution, consider a solution that contains a mixture of $HC_2H_3O_2(aq)$ and its conjugate base, $C_2H_3O_2^-(aq)$. If an acid is added to this buffer solution, then the added H_3O^+ ions are removed from the solution by the reaction of $H_3O^+(aq)$ with $C_2H_3O_2^-(aq)$:

$$H_3O^+(aq) + C_2H_3O_2^-(aq) \rightleftharpoons HC_2H_3O_2(aq) + H_2O(l)$$

The value of K for this equation is the reciprocal of K_a for $HC_2H_3O_2(aq)$, or $K = 5.75 \times 10^4\ M^{-1}$. Thus the reaction goes essentially to completion and converts essentially all the added $H_3O^+(aq)$ into $HC_2H_3O_2(aq)$.

Suppose now that a base instead of an acid is added to the buffer solution. The added $OH^-(aq)$ ions are removed from solution by reaction with acetic acid:

$$HC_2H_3O_2(aq) + OH^-(aq) \rightleftharpoons C_2H_3O_2^-(aq) + H_2O(l)$$

We showed in Section 16-3 that $K = 1.74 \times 10^9\ M^{-1}$ for this equation, and so we see that essentially all the added $OH^-(aq)$ ions are removed from the solution by this reaction. Thus a solution containing both $HC_2H_3O_2(aq)$ and $C_2H_3O_2^-(aq)$ is resistant to changes in pH on the addition of either an acid or a base.

We know that acetic acid is a weak acid. Thus neither $[HC_2H_3O_2]$ nor $[C_2H_3O_2^-]$ differs significantly from the stoichiometric concentration, because the values of the equilibrium constants (Table 15-5) of the two equations

$$HC_2H_3O_2(aq) + H_2O(l) \rightleftharpoons H_3O^+(aq) + C_2H_3O_2^-(aq) \qquad K_a = 1.74 \times 10^{-5}\ M$$

$$C_2H_3O_2^-(aq) + H_2O(l) \rightleftharpoons HC_2H_3O_2(aq) + OH^-(aq) \qquad K_b = 5.75 \times 10^{-10}\ M$$

are small. Consequently, we can write

$$[HC_2H_3O_2] \approx [HC_2H_3O_2]_0$$

$$[C_2H_3O_2^-] \approx [C_2H_3O_2^-]_0$$

Because buffers usually consist of a solution of a mixture of a weak conjugate acid-base pair, we usually can assume that the stoichiometric and equilibrium concentrations are equal for both the weak acid and its conjugate base; that is,

$$[HB] \approx [HB]_0$$

$$[B^-] \approx [B^-]_0$$

The general form of a weak acid–conjugate base equilibrium is

$$\underset{\text{conjugate acid}}{HB(aq)} + H_2O(l) \rightleftharpoons H_3O^+(aq) + \underset{\text{conjugate base}}{B^-(aq)} \qquad (16\text{-}14)$$

Buffer action can be described quantitatively in terms of the acid-dissociation constant. For Equation (16-14), K_a is

$$K_a = \frac{[H_3O^+][B^-]}{[HB]} = \frac{[H_3O^+][\text{base}]}{[\text{acid}]} \qquad (16\text{-}15)$$

Taking the logarithm of both sides of Equation (16-15) yields

$$\log K_a = \log \frac{[H_3O^+][\text{base}]}{[\text{acid}]}$$

Using the fact that $\log ab = \log a + \log b$, we can rewrite this equation in the form

$$\log K_a = \log [H_3O^+] + \log \frac{[\text{base}]}{[\text{acid}]}$$

Multiplying by -1, and using $\text{pH} = -\log [H_3O^+]$ and $\text{p}K_a = -\log K_a$, we write

$$\text{pH} = \text{p}K_a + \log \frac{[\text{base}]}{[\text{acid}]} \qquad (16\text{-}16)$$

As noted previously, it is usually a good approximation to say that, in a buffer solution, the equilibrium concentration of the acid is equal to its stoichiometric concentration and the equilibrium concentration of the conjugate base is equal to its stoichiometric concentration. This approximation allows us to substitute

$$[\text{acid}] \approx [\text{acid}]_0 \qquad \text{and} \qquad [\text{base}] \approx [\text{base}]_0$$

where the subscript zeros denote stoichiometric concentrations, into Equation (16-16), and obtain

$$\text{pH} \approx \text{p}K_a + \log \frac{[\text{base}]_0}{[\text{acid}]_0} \tag{16-17}$$

Equation 16-17 is known as the **Henderson-Hasselbalch equation** and is useful in the analysis of buffer solutions. The advantage of the Henderson-Hasselbalch equation is that we use the stoichiometric concentration directly without the need for a stepwise analysis of the equilibrium. To apply Equation (16-17) we must meet the general conditions

$$[\text{base}]_0 \neq 0 \qquad \text{and} \qquad [\text{acid}]_0 \neq 0$$

and K_a and K_b for the conjugate acid-base pair both must be small, that is, less than about 10^{-3}.

▪ The Henderson-Hasselbalch equation is used extensively in biochemistry. The particular form of Equation (16-17) is remembered by the mnemonic "The pH increases when the concentration of *base goes up*." Thus when the ratio $[\text{base}]_0/[\text{acid}]_0$, in which the $[\text{base}]_0$ term is "up," or on top, in Equation (16-17) increases, the resulting value of the pH increases.

Example 16-9: Estimate the pH of a solution that is 0.20 M in $HC_2H_3O_2(aq)$ and 0.10 M in $NaC_2H_3O_2(aq)$. Take $\text{p}K_a = 4.76$ for acetic acid.

Solution: The stoichiometric concentrations of the acid and base forms are

$$[HC_2H_3O_2]_0 = [\text{acid}]_0 = 0.20\text{ M} \qquad [C_2H_3O_2^-]_0 = [\text{base}]_0 = 0.10\text{ M}$$

Because $K_a(=1.74 \times 10^{-5}\text{ M})$ for $HC_2H_3O_2(aq)$ and $K_b(=5.75 \times 10^{-10}\text{ M})$ for $C_2H_3O_2^-(aq)$ are small, we can use the Henderson-Hasselbalch equation instead of Equation (16-16) and write

$$\text{pH} = \text{p}K_a + \log \frac{[\text{base}]_0}{[\text{acid}]_0} = 4.76 + \log\left(\frac{0.10\text{ M}}{0.20\text{ M}}\right) = 4.76 - 0.30 = 4.46$$

Note that the buffer is acidic; in addition note that $\text{pH} < \text{p}K_a$, because $[\text{base}]_0 < [\text{acid}]_0$.

Example 16-10: Estimate the pH of a buffer solution that is 0.050 M in $NH_4Cl(aq)$ and 0.20 M in $NH_3(aq)$. Take $\text{p}K_a = 9.24$ for $NH_4^+(aq)$ (Table 15-5).

Solution: The stoichiometric concentrations of the conjugate acid and base are

$$[NH_4^+]_0 = [\text{acid}]_0 = 0.050\text{ M} \qquad [NH_3]_0 = [\text{base}]_0 = 0.20\text{ M}$$

Because $K_a(=5.71 \times 10^{-10}\text{ M})$ for $NH_4^+(aq)$ and $K_b(=1.75 \times 10^{-5}\text{ M})$ for $NH_3(aq)$ are small, we can use the Henderson-Hasselbalch equation and write

$$\text{pH} = \text{p}K_a + \log \frac{[\text{base}]_0}{[\text{acid}]_0} = 9.24 + \log\left(\frac{0.20\text{ M}}{0.050\text{ M}}\right) = 9.84$$

Note that the $NH_4^+(aq)$–$NH_3(aq)$ buffer is basic. In addition note that $[\text{base}]_0 > [\text{acid}]_0$ and thus $\text{pH} > \text{p}K_a$.

16-6. A BUFFER SOLUTION SUPPRESSES pH CHANGES WHEN ACID OR BASE IS ADDED

We now demonstrate the insensitivity of a buffer solution to the addition of small amounts of acid or base. Let the buffer be composed of a solution that is 0.10 M in $HC_2H_3O_2(aq)$ and 0.10 M in $C_2H_3O_2^-(aq)$. Suppose that 10 mL of 0.10 M $HCl(aq)$ is added to 100 mL of the buffer solution. The number of millimoles of acid in 10 mL of 0.10 M $HCl(aq)$ is

$$\begin{pmatrix}\text{mmol of}\\ H_3O^+\\ \text{added}\end{pmatrix} = (0.10\text{ M})(10\text{ mL}) = 1.0\text{ mmol}$$

The $H_3O^+(aq)$ reacts with $C_2H_3O_2^-(aq)$ in the buffer according to

$$H_3O^+(aq) + C_2H_3O_2^-(aq) \rightarrow HC_2H_3O_2(aq) + H_2O(l)$$

Before the HCl is added, the number of moles of acetate ion in the buffer solution is

$$\begin{pmatrix}\text{mmol of } C_2H_3O_2^-\\ \text{before HCl added}\end{pmatrix} = (0.10\text{ M})(100\text{ mL}) = 10\text{ mmol}$$

The number of millimoles of $C_2H_3O_2^-(aq)$ after the addition of 10 mL of 0.10 M $HCl(aq)$ is

$$\begin{pmatrix}\text{mmol of } C_2H_3O_2^-\\ \text{after HCl added}\end{pmatrix} = \begin{pmatrix}\text{mmol of } C_2H_3O_2^-\\ \text{before HCl added}\end{pmatrix} - \begin{pmatrix}\text{mmol of}\\ H_3O^+\text{ added}\end{pmatrix}$$
$$= 10\text{ mmol} - 1.0\text{ mmol} = 9\text{ mmol}$$

The number of millimoles of $HC_2H_3O_2(aq)$ after the addition of 1.0 mmol of $H_3O^+(aq)$ is

$$\begin{pmatrix}\text{mmol of } HC_2H_3O_2\\ \text{after HCl added}\end{pmatrix} = \begin{pmatrix}\text{mmol of } HC_2H_3O_2\\ \text{before HCl added}\end{pmatrix} + \begin{pmatrix}\text{mmol of}\\ H_3O^+\text{ added}\end{pmatrix}$$
$$= 10\text{ mmol} + 1.0\text{ mmol} = 11\text{ mmol}$$

The pH of the buffer solution before the addition of the $HCl(aq)$ is

$$\text{pH} \approx \text{p}K_a + \log\frac{[\text{base}]_0}{[\text{acid}]_0} = 4.76 + \log\left(\frac{0.10\text{ M}}{0.10\text{ M}}\right) = 4.76$$

and the pH of the buffer solution after the addition of the $HCl(aq)$ is

$$\text{pH} = 4.76 + \log\frac{(9\text{ mmol}/110\text{ mL})}{(11\text{ mmol}/110\text{ mL})} = 4.67$$

Thus the pH of the solution changes by less than 0.10 pH unit. Notice that the volumes (110 mL) cancel out in the log factor and thus we can work directly with the number of millimoles of the conjugate acid and base without the need to convert to molarities before taking the logarithm of the ratio.

In contrast to the above result, if we add 10 mL of 0.10 M $HCl(aq)$ to 100 mL of pure water, then the pH of the resulting solution is

$$[H_3O^+] = \frac{\text{mmol of } H_3O^+}{\text{mL of solution}} = \frac{(0.10\text{ M})(10\text{ mL})}{(110\text{ mL})} = 9.1 \times 10^{-3}\text{ M}$$

or $\quad pH = -\log [H_3O^+] = -\log (9.1 \times 10^{-3}) = 2.04$

The pH of pure water is 7.0, and so we see that the pH changes by 5.0 units (7.0 − 2.0 = 5.0). The value of $[H_3O^+]$ changes by a factor of 10^5, or 100,000, whereas in the buffer solution the pH changes by less than 0.1 pH unit.

Example 16-11: Compute the change in pH when 10 mL of 0.10 M NaOH(*aq*) is added to 100 mL of a buffer that is 0.10 M in $HC_2H_3O_2(aq)$ and 0.10 M in $C_2H_3O_2^-(aq)$. Take $pK_a = 4.76$ for $HC_2H_3O_2(aq)$.

▪ The buffering action of the $HC_2H_3O_2(aq)$–$C_2H_3O_2^-(aq)$ buffer can be seen in Figure 16-6. It is the region between 5 and 45 mL of added base, where the titration curve is relatively flat.

Solution: The number of millimoles of base in 10 mL of 0.10 M NaOH(*aq*) is

$$\begin{pmatrix}\text{mmol of OH}^-\\ \text{added}\end{pmatrix} = (0.10\ \text{M})(10\ \text{mL})$$

$$= 1.0\ \text{mmol}$$

The $OH^-(aq)$ reacts with $HC_2H_3O_2(aq)$ in the buffer via the reaction

$$OH^-(aq) + HC_2H_3O_2(aq) \rightarrow H_2O(l) + C_2H_3O_2^-(aq)$$

The number of millimoles of acetic acid in the buffer solution before the addition of NaOH(*aq*) is

$$\begin{pmatrix}\text{mmol of HC}_2\text{H}_3\text{O}_2\\ \text{before NaOH added}\end{pmatrix} = (0.10\ \text{M})(100\ \text{mL})$$

$$= 10\ \text{mmol}$$

The number of millimoles of acetic acid after the addition of 10 mL of 0.10 M NaOH(*aq*) is

$$\begin{pmatrix}\text{mmol of HC}_2\text{H}_3\text{O}_2\\ \text{after NaOH added}\end{pmatrix} = \begin{pmatrix}\text{mmol of HC}_2\text{H}_3\text{O}_2\\ \text{before NaOH added}\end{pmatrix} - \begin{pmatrix}\text{mmol of}\\ \text{OH}^-\ \text{added}\end{pmatrix}$$

$$= 10\ \text{mmol} - 1.0\ \text{mmol} = 9\ \text{mmol}$$

The number of millimoles of $C_2H_3O_2^-(aq)$ following the addition of 1.0 mmol of NaOH(*aq*) is

$$\begin{pmatrix}\text{mmol of C}_2\text{H}_3\text{O}_2^-\\ \text{after NaOH added}\end{pmatrix} = \begin{pmatrix}\text{mmol of C}_2\text{H}_3\text{O}_2^-\\ \text{before NaOH added}\end{pmatrix} + \begin{pmatrix}\text{mmol of}\\ \text{OH}^-\ \text{added}\end{pmatrix}$$

$$= 10\ \text{mmol} + 1\ \text{mmol} = 11\ \text{mmol}$$

The pH is

$$pH \approx pK_a + \log \frac{[\text{base}]_0}{[\text{acid}]_0} = 4.76 + \log \frac{11}{9} = 4.85$$

Once again, notice that the pH changes by less than 0.1 pH unit.

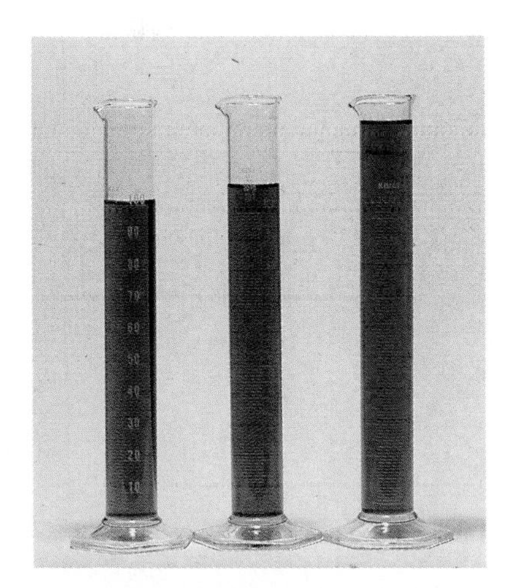

Figure 16-8 The cylinder on the left contains 100 mL of a sodium acetate and acetic acid buffer with bromcresol green indicator. The middle cylinder contains 100 mL of the $C_2H_3O_2^-\ (aq)/HC_2H_3O_2\ (aq)$ buffer plus 7 mL of 0.10 M NaOH. The right cylinder contains 100 mL of the $C_2H_3O_2^-\ (aq)/HC_2H_3O_2\ (aq)$ buffer plus 25 mL of 0.10 M NaOH. The change in the indicator color from green to blue shows that the buffer has been overwhelmed.

The capacity of a buffer to resist changes in pH is not unlimited. If sufficient acid (or base) is added to neutralize all the conjugate base (or all the conjugate acid), then the pH of the solution will change significantly. In such a case the buffer is simply overwhelmed (Figure 16-8).

Another property of buffers is their ability to resist changes in pH upon dilution with solvent. This property is readily understood from the Henderson-Hasselbalch equation. If we dilute a buffer solution by, say, a factor of 2, then the stoichiometric concentration of the base,

■ Buffer capacity is an indication of the amount of acid or base that can be added to a buffer before it is no longer able to resist pH change.

$[base]_0$, and the stoichiometric concentration of the acid, $[acid]_0$, both decrease by a factor of 2, but the ratio of the stoichiometric concentrations does not change:

$$\frac{[base]_0/2}{[acid]_0/2} = \frac{[base]_0}{[acid]_0}$$

and thus the pH of the buffer solution does not change on dilution. The effects of the addition of HCl(*aq*) and NaOH(*aq*) and of dilution on the pH of $HC_2H_3O_2(aq)$–$C_2H_3O_2^-(aq)$ and $NH_4^+(aq)$–$NH_3(aq)$ buffers are summarized in Table 16-2.

The principal utility of buffers is in the control of pH. Innumerable chemical and biochemical processes depend upon the value of $[H_3O^+]$, and the control of this key concentration is achieved by a buffer. The buffer must be chemically inert to the species in the system whose pH is to be controlled.

The activity of many enzymes depends upon pH, and numerous biological systems have natural buffers to control enzyme activities. Essentially constant pH values are required for maintaining the delicate balances in the complex sequences of biochemical reactions essential to the existence of life. For example, blood is buffered by a mixture of carbonates, phosphates, and proteins and exhibits a remarkably constant pH value of 7.4. At pH values lower than 7.3, the blood cannot efficiently remove CO_2 from the cells, and at pH values greater than 7.7 the blood cannot efficiently release CO_2 to the lungs. Blood pH values outside the range 7.0 to 7.8 cannot sustain life.

Table 16-2 Resistance of a buffer to change in pH upon addition of acid or base and upon dilution

Buffer	Initial pH	pH after addition of 10.0 mL of 0.10 M HCl to 100 mL of buffer	pH after addition of 10.0 mL of 0.10 M NaOH to 100 mL of buffer	pH after twofold dilution with water
0.10 M $HC_2H_3O_2(aq)$ plus 0.10 M $C_2H_3O_2^-(aq)$	4.76	4.67	4.85	4.76
0.10 M $NH_4^+(aq)$ plus 0.10 M $NH_3(aq)$	9.24	9.15	9.33	9.24

SUMMARY

A weak acid whose conjugate base has a different color can be used as a pH indicator. Indicators can be used to signal the equivalence point in an acid-base titration, that is, the point on the titration curve at which the number of moles of acid (or base) originally present is equal to the number of moles of base (or acid) added as titrant. The pH of a titration curve undergoes an especially rapid change in the vicinity of the equivalence point, and this rapid change in pH can be used to detect the equivalence point.

A buffer is a mixture of a weak acid and its conjugate base. A buffer suppresses the change in pH that would otherwise result from the addition of an acid or a base to a solution. The pH of a buffer solution can be calculated from the Henderson-Hasselbalch equation.

TERMS YOU SHOULD KNOW

indicator 511	titration curve 514	midpoint 520
titration 514	equivalence point 514	buffer 525
titrant 514	end point 514	Henderson-Hasselbalch equation 527

EQUATIONS YOU SHOULD KNOW HOW TO USE

number of millimoles = (molarity) × (number of milliliters) (16-4)

$$\text{pH} \approx \text{p}K_a + \log \frac{[\text{base}]_0}{[\text{acid}]_0}$$ (16-17) (Henderson-Hasselbalch equation)

PROBLEMS

INDICATORS

16-1. Estimate the pH of a colorless aqueous solution that turns blue when Nile blue is added and turns blue when thymol blue is added.

16-2. Estimate the pH of a colorless aqueous solution that turns yellow when methyl orange is added and yellow when bromcresol purple is added.

16-3. A colorless aqueous solution was obtained from a rain gauge. The following indicators were added to separate samples of the solution and the colors observed were

bromcresol purple:	yellow
bromcresol green:	green

Estimate the pH of the solution.

16-4. We wish to estimate the pH of a colorless aqueous solution. Litmus paper indicates that the solution is basic. The following indicators are added to separate samples of the solution, and the colors observed are

Nile blue:	red
Alizarin yellow R:	deep yellow

Estimate the pH of the solution.

16-5. A certain bacterium grows best in an acidic medium, and in an experiment with this bacterium the pH must be maintained at around 5. What indicator should be added to indicate a pH of 5 and to monitor changes in the pH of the medium?

16-6. The pH of the nutrient broth used to maintain cultures of tissue samples must be greater than 7 but should not rise above 8. What indicator can be added to the nutrient broth to monitor the pH?

16-7. The pH indicator bromcresol green changes from yellow to blue over the pH range 4 to 5. Estimate the K_{ai} for bromcresol green.

16-8. The pH indicator Nile blue changes from blue to pink over the pH range 10 to 11. Estimate the K_a of Nile blue.

TITRATIONS INVOLVING STRONG ACIDS AND STRONG BASES

16-9. Calculate the pH of the resulting solution if 20.0 mL of 0.200 M HCl(*aq*) is added to

(a) 25.0 mL of 0.200 M NaOH(*aq*)
(b) 30.0 mL of 0.350 M NaOH(*aq*)

16-10. Calculate the pH of the resulting solution if 20.0 mL of 0.200 M HCl(*aq*) is added to

(a) 40.0 mL of 0.100 M NaOH(*aq*)
(b) 20.0 mL of 0.150 M NaOH(*aq*)

16-11. Determine the equivalence point if a 0.100 M NaOH(*aq*) solution is used to titrate the following acidic solutions:

(a) 50.0 mL of 0.200 M HBr(*aq*)
(b) 30.0 mL of 0.150 M HNO_3(*aq*)

16-12. Determine the equivalence point if a 0.100 M NaOH(*aq*) solution is used to titrate the following acidic solutions:

(a) 30.0 mL of 0.600 M $HClO_4$(*aq*)
(b) 50.0 mL of 0.100 M H_2SO_4(*aq*)

16-13. Sketch, but do not calculate, the titration curve for the titration of 50.0 mL of 0.100 M NaOH(*aq*) with 0.100 M HCl(*aq*).

16-14. Sketch, but do not calculate, the titration curve for the titration of 50.0 mL of 0.25 M HNO_3(*aq*) with 0.50 M KOH(*aq*).

16-15. Calculate the molarity of an aqueous nitric acid solution if it requires 35.6 mL of 0.165 M NaOH(*aq*) to neutralize 25.0 mL of the nitric acid.

16-16. Calculate the molarity of an aqueous sodium hydroxide solution if it requires 34.7 mL of 0.125 M $HCl(aq)$ to neutralize 15.0 mL of the $NaOH(aq)$ solution.

TITRATIONS INVOLVING WEAK ACIDS OR WEAK BASES

16-17. Calculate the pH of the solution obtained by titrating 25.0 mL of a 0.100 M solution of the herbicide cacodylic acid, $HC_2H_6AsO_2$, with 0.095 M $NaOH(aq)$ to the equivalence point. Take $K_a = 5.4 \times 10^{-7}$ M. What indicator should be used to signal the equivalence point?

16-18. Calculate the pH of the solution obtained by titrating 50.0 mL of 0.100 M $HNO_2(aq)$ with 0.100 M $NaOH(aq)$ to the equivalence point. Take $K_a = 4.5 \times 10^{-4}$ M for $HNO_2(aq)$. What indicator should be used to signal the equivalence point?

16-19. Calculate the pH for the following cases in the titration of 25.00 mL of 0.200 M acetic acid, $HC_2H_3O_2$, with 0.200 M $NaOH(aq)$:

(a) before addition of any $NaOH(aq)$
(b) after addition of 5.00 mL of $NaOH(aq)$
(c) after addition of 12.50 mL of $NaOH(aq)$
(d) after addition of 25.00 mL of $NaOH(aq)$
(e) after addition of 26.00 mL of $NaOH(aq)$

16-20. Calculate the pH for each of the following cases in the titration of 50.0 mL of 0.150 M hypochlorous acid, $HClO(aq)$, with 0.150 M $KOH(aq)$:

(a) before addition of any $KOH(aq)$:
(b) after addition of 25.0 mL of $KOH(aq)$
(c) after addition of 50.0 mL of $KOH(aq)$
(d) after addition of 60.0 mL of $KOH(aq)$

16-21. Calculate the pH for each of the following cases in the titration of 25.0 mL of 0.150 M pyridine, $C_5H_5N(aq)$, with 0.150 M $HBr(aq)$:

(a) before addition of any $HBr(aq)$
(b) after addition of 10.0 mL of $HBr(aq)$
(c) after addition of 24.0 mL of $HBr(aq)$
(d) after addition of 25.0 mL of $HBr(aq)$
(e) after addition of 26.0 mL of $HBr(aq)$

16-22. Calculate the pH for each of the following cases in the titration of 35.0 mL of 0.200 M methylamine, $CH_3NH_2(aq)$, with 0.200 M $HCl(aq)$:

(a) before addition of any $HCl(aq)$
(b) after addition of 17.5 mL of $HCl(aq)$
(c) after addition of 34.9 mL of $HCl(aq)$
(d) after addition of 35.0 mL of $HCl(aq)$
(e) after addition of 35.1 mL of $HCl(aq)$

16-23. A 1.50-g sample of ascorbic acid, vitamin C, is dissolved in 100 mL of water and titrated with 0.250 M $NaOH(aq)$ to the equivalence point. The volume of base consumed is 34.1 mL. Calculate the molecular mass of vitamin C assuming one dissociable proton per molecule.

16-24. A 0.772-g sample of benzoic acid, which is found in most berries, is dissolved in 50.0 mL of water and titrated to the equivalence point with 0.250 M $NaOH(aq)$. The volume of base consumed is 25.3 mL. Calculate the molecular mass of benzoic acid, assuming one dissociable proton per molecule.

16-25. Vinegar is a dilute aqueous solution of acetic acid. A 21.0-mL sample of vinegar requires 38.5 mL of 0.400 M $NaOH(aq)$ to neutralize the $HC_2H_3O_2(aq)$. Given that the density of the vinegar is $1.060\ g{\cdot}mL^{-1}$, calculate the mass percentage of acetic acid in the vinegar.

16-26. A 2.00-g sample of acetylsalicylic acid, better known as aspirin, is dissolved in 100 mL of water and titrated with 0.200 M $NaOH(aq)$ to the equivalence point. The volume of base required is 55.5 mL. Calculate the molecular mass of the acetylsalicylic acid, which has one dissociable proton per molecule.

16-27. Calculate the pH at the equivalence point in the titration of 50.0 mL of 0.125 M $NH_3(aq)$ with 0.175 M $HCl(aq)$.

16-28. Calculate the pH at the equivalence point in the titration of 17.5 mL of 0.098 M pyridine with 0.117 M $HI(aq)$.

BUFFER CALCULATIONS

16-29. Estimate the pH of a solution that is 0.050 M in $HC_2H_3O_2(aq)$ and 0.050 M in $NaC_2H_3O_2(aq)$.

16-30. Estimate the pH of a solution that is 0.10 M in $HC_2H_3O_2(aq)$ and 0.20 M in $NaC_2H_3O_2(aq)$.

16-31. Estimate the pH of a solution that is 0.15 M in $HCHO_2(aq)$ and 0.25 M in $NaCHO_2(aq)$.

16-32. Estimate the pH of a solution that is 0.20 M in $HNO_2(aq)$ and 0.15 M in $NaNO_2(aq)$.

16-33. Calculate the pH of an aqueous solution that is 0.200 M in pyridine, C_5H_5N, and 0.250 M in pyridinium chloride, $C_5H_5NH^+Cl^-$.

16-34. Compute the pH of a $NH_4^+(aq)$–$NH_3(aq)$ buffer solution that is 0.40 M in $NH_4Cl(aq)$ and 0.20 M in $NH_3(aq)$.

16-35. A commonly used buffer in biological experiments is a phosphate buffer containing NaH_2PO_4 and Na_2HPO_4. Estimate the pH of an aqueous solution that is

(a) 0.050 M NaH_2PO_4 and 0.050 M Na_2HPO_4
(b) 0.050 M NaH_2PO_4 and 0.10 M Na_2HPO_4
(c) 0.10 M NaH_2PO_4 and 0.050 M Na_2HPO_4

The relevant equation is

$$H_2PO_4^-(aq) + H_2O(l) \rightleftharpoons H_3O^+(aq) + HPO_4^{2-}(aq) \qquad K_a = 6.2 \times 10^{-8}\ \mathrm{M}$$

16-36. Calculate the pH of a buffer solution obtained by dissolving 10.0 g of $KH_2PO_4(s)$ and 20.0 g of $Na_2HPO_4(s)$ in water and then diluting to 1.00 L. The relevant equation is $H_2PO_4^-(aq) + H_2O(l) \rightleftharpoons H_3O^+(aq) + HPO_4^{2-}(aq)$ with $pK_a = 7.21$.

16-37. Suppose you are performing an experiment during which the pH must be maintained at 3.70. What would be an appropriate buffer to use?

16-38. Suppose you are performing an experiment during which the pH must be maintained at 5.16. What would be an appropriate buffer to use?

ADDITION OF ACIDS AND BASES TO BUFFERS

16-39. The higher the concentration of acid and conjugate base in a buffer, the smaller is the pH change when acid or base is added. Compute the pH change in the following two buffers when 1.00 g of KOH(*s*) is added:

(a) 500 mL of a 0.10 M $NH_4Cl(aq)$–0.10 M $NH_3(aq)$ buffer
(b) 500 mL of a 1.00 M $NH_4Cl(aq)$–1.00 M $NH_3(aq)$ buffer

16-40. Calculate the change in pH when 5.00 mL of 0.100 M HCl(*aq*) is added to 100 mL of a buffer solution that is 0.100 M in $NH_3(aq)$ and 0.100 M in $NH_4Cl(aq)$. Calculate the change in pH when 5.00 mL of 0.100 M NaOH(*aq*) is added to the original buffer solution.

16-41. Calculate the change in pH if 2.0 mmol of HCl is added to 1.00 L of a buffer that is 0.0200 M in propionic acid, $HC_3H_5O_2$, and 0.0150 M in sodium propionate, $C_3H_5O_2^-$. The value of K_a for propionic acid is 1.30×10^{-5} M.

16-42. One liter of a buffer solution contains 0.100 mol of $HC_2H_3O_2$ and 0.100 mol of $NaC_2H_3O_2$. Calculate the pH of this buffer. Calculate the change in pH when 0.030 mol of NaOH is added to the buffer.

16-43. Calculate the pH of a solution that is prepared by mixing 2.16 g of propionic acid, $HC_3H_5O_2$, and 0.56 g of NaOH in enough water to make exactly 100 mL of solution. Take $pK_a = 4.89$ for propionic acid.

16-44. If 6.52 g of pyridine, C_5H_5N, is added to 30.0 mL of 0.950 M HCl(*aq*), then what will be the pH of the resulting solution? Take the final volume of the solution to be 36.0 mL.

16-45. A buffer is prepared such that $[HC_2H_3O_2]_0 = [C_2H_3O_2^-]_0 = 0.050$ M. Estimate the volume of 0.10 M NaOH that can be added to 100 mL of the solution before its buffering capacity is lost. Assume the buffer capacity is lost when $[\text{acid}]_0 \approx 0$.

16-46. A buffer solution is prepared such that $[H_3PO_4]_0 = [H_2PO_4^-]_0 = 0.20$ M. Estimate the volume of 0.010 M HCl that can be added to 200 mL of the solution before its buffering capacity is lost. Assume the buffer capacity is lost when $[H_2PO_4^-]_0 \approx 0$.

16-47. Calculate the mass of $NH_4Cl(s)$ that must be added to 1.00 L of 0.200 M $NH_3(aq)$ solution to obtain a solution of pH 9.50. Assume no change in volume.

16-48. Calculate the mass of NaOH(*s*) that must be added to 500.0 mL of 0.120 M $HC_2H_3O_2$ to yield a solution of pH = 4.52. Assume no change in volume.

ADDITIONAL PROBLEMS

16-49. Various antacid tablets contain water-insoluble metal hydroxides, such as $Mg(OH)_2(s)$. Given that stomach acid is about 0.10 M in HCl(*aq*), compute the number of milliliters of stomach acid that can be neutralized by 1.00 g of $Mg(OH)_2(s)$.

16-50. Given the following results for the titration of 30.0 mL of a 0.100 M HCl(*aq*) solution with a 0.100 M NaOH(*aq*) solution, plot the titration curve:

Volume of NaOH(aq) solution added/mL	pH
0.0	1.00
10.0	1.30
20.0	1.70
25.0	2.04
29.0	2.77
30.0	7.00
31.0	11.22
35.0	11.89
40.0	12.16
50.0	12.40

16-51. An unknown sample is thought to be either benzoic acid, $HC_7H_5O_2$, or chlorobenzoic acid, $HC_7H_4ClO_2$. When 1.89 g is dissolved in water, 15.49 mL of 1.00 M NaOH(*aq*) is required to reach the equivalence point. Which acid is the unknown sample?

16-52. The electronic meters used to measure pH are calibrated using standard buffer solutions of known pH. For example, the directions for preparing a certain buffer at 25°C are as follows. Dissolve 3.40 ± 0.01 g of $KH_2PO_4(s)$ and 3.55 ± 0.01 g of $Na_2HPO_4(s)$ in sufficient water to make 1.00 L of solution. What pH do you calculate for this solution?

16-53. It is important always to remember that the Henderson-Hasselbalch equation is based upon the assumption that [acid] ≈ $[\text{acid}]_0$ and [base] ≈ $[\text{base}]_0$. Calculate $[HC_2H_3O_2]$, $[HC_2H_3O_2]_0$, $[C_2H_3O_2^-]$, and $[C_2H_3O_2^-]_0$ for a solution that is 0.100 M in both $HC_2H_3O_2(aq)$ and $NaC_2H_3O_2(aq)$. Is $[HC_2H_3O_2] \approx [HC_2H_3O_2]_0$ and is $[C_2H_3O_2^-] \approx [C_2H_3O_2^-]_0$? Can you see from your calculation why this is so?

16-54. Calculate the change in pH when 20.0 mL of 0.200 M NaOH(*aq*) is added to 50.0 mL of a buffer solution that is 0.150 M in $HC_2H_3O_2(aq)$ and 0.150 M in $NaC_2H_3O_2(aq)$. Can you use the Henderson-Hasselbalch equation to do this problem?

16-55. The principal reaction that occurs when a salt containing an anion that can act as either an acid or a base, for example, $NaHCO_3$, is dissolved in water is of the type

$$2HCO_3^-(aq) \rightleftharpoons CO_3^{2-}(aq) + H_2CO_3(aq)$$

(a) Show that the $[H_3O^+]$ of the solution is given by

$$[H_3O^+] = (K_{a1}K_{a2})^{1/2}$$

where K_{a1} and K_{a2} are the first and second acid-dissociation constants of $H_2CO_3(aq)$. Note that the pH of the solution is independent of the salt concentration. Calculate the pH.

(b) Explain the buffer property of $NaHCO_3(aq)$ solutions.

16-56. A 2.500-g sample of oxalic acid, a diprotic acid, is dissolved in 250 mL of water and titrated with 1.000 M NaOH to the first equivalence point. The volume of base is 27.75 mL. Calculate the molecular mass of oxalic acid.

16-57. Alka-Seltzer contains sodium bicarbonate, $NaHCO_3(s)$, and the triprotic acid citric acid, $H_3C_6H_5O_7$, in addition to 324 mg of aspirin. Write the acid-base reaction that gives rise to the fizz (CO_2 evolution) when an Alka-Seltzer tablet is dissolved in water.

16-58. Suppose that 80.0 mL of a 0.200 M $NH_3(aq)$ solution is titrated with a 0.400 M HCl(*aq*) solution. Calculate the pH at 10.0-mL intervals of added HCl(*aq*), up to 70.0 mL added, and plot the titration curve.

16-59. Several types of commercial antacid tablets contain $Al(OH_3)(s)$ as the active ingredient. Given that stomach acid is about 0.10 M in HCl(*aq*), compute the number of milliliters of stomach acid that can be neutralized by 500 mg of $Al(OH)_3(s)$.

16-60. Suppose that you are titrating 25.0 mL of a 0.250 M HBr(*aq*) solution with a 0.250 M NaOH(*aq*) solution. The equivalence point occurs when 25.0 mL of the NaOH(*aq*) solution has been added. The pH at this point is 7.0. Calculate the pH if one more drop (0.05 mL) of the NaOH(*aq*) solution is added.

16-61. A 0.550-g sample of butyric acid is dissolved in 100 mL of water and titrated with 0.100 M NaOH(*aq*) to the equivalence point. The volume of base consumed is 62.4 mL. Calculate the molecular mass of butyric acid, which has one dissociable proton per molecule.

16-62. Indicate for which of the following solutions the Henderson-Hasselbalch equation cannot be used to calculate the pH:

(a) 0.15 M $HNO_2(aq)$ plus 0.20 M $NaNO_2(aq)$
(b) 0.15 M $HNO_2(aq)$
(c) 0.20 M $NaNO_2(aq)$
(d) 0.10 M $Na_2HPO_4(aq)$ plus 0.20 M $KH_2PO_4(aq)$

16-63. Suppose that you wish to determine whether a solution of unknown composition is buffered. Explain how you could do this with only two pH measurements.

16-64. The principal reaction when a salt composed of an acidic cation and a basic anion, for example, $NH_4C_2H_3O_2$, is dissolved in water is of the type

$$NH_4^+(aq) + C_2H_3O_2^-(aq) \rightleftharpoons NH_3(aq) + HC_2H_3O_2(aq)$$

(a) Show that the value of the equilibrium constant for this reaction is given by

$$K = \frac{K_{a,NH_4^+}}{K_{a,HC_2H_3O_2}}$$

(b) Given the above stoichiometry, show that the $[H_3O^+]$ of the solution is equal to

$$[H_3O^+] = (K_{a,NH_4^+}K_{a,HC_2H_3O_2})^{1/2}$$

Note that $[H_3O^+]$ and thus the pH of the solution are independent of the concentration of the salt.

(c) Is an $NH_4C_2H_3O_2(aq)$ solution a buffer? Explain.

16-65. A 1.20-g sample of an unknown acid is dissolved in water and titrated with 0.150 M NaOH(*aq*) to the equivalence point. The volume of base is 69.0 mL. Calculate the molecular mass of the acid. The titration curve shows only one dissociable proton per molecule.

16-66. Compute the number of grams of $NaHCO_3(s)$ required to neutralize 2.00 g of citric acid in water. Citric acid is a triprotic acid, $H_3C_6H_5O_7$.

16-67. Determine how many of the hydrogen atoms in citric acid, $C_6H_8O_7$, are acidic if it requires 147.2 mL of 0.135 M NaOH(*aq*) solution to titrate a 1.270-g sample of citric acid.

16-68. Calculate the pH of the solution that results when 2.00 g of $Mg(OH)_2(s)$ is dissolved in 850 mL of 0.160 M HCl(*aq*).

16-69. Calculate $[H_2C_2O_4]$, $[HC_2O_4^-]$, and $[C_2O_4^{2-}]$ in a 0.125 M $H_2C_2O_4$ solution that is buffered at a pH of 5.00 ($pK_{a1} = 1.27$, $pK_{a2} = 4.27$).

16-70. A solution is 0.0500 M in HCl(*aq*) and 0.060 M in $HC_2H_3O_2(aq)$. Calculate the pH of the resulting solution if 30.0 mL of 0.120 M NaOH(*aq*) is added to 50.0 mL of the original solution.

Natural Waters

Vents in the ocean floor through which hot, mineral-rich water enters the ocean. These minerals are the major source of dissolved solids in the oceans. The structure showing at the bottom of the photo is the bathyscope from which the picture was taken.

Approximately three fourths of the earth's surface is covered with water. The total amount of water on earth is estimated to be 1.4×10^{12} kg, of which 97.3 percent is seawater, 2.1 percent is ice, and only 0.6 percent is fresh water. The fresh water is located in rivers, lakes, and clouds and in the ground. Water is the solvent of life; the human body is about three-fourths water by mass. More chemical and biochemical reactions take place in water than in all other solvents combined. In the absence of water, no known form of life is possible.

K-1. SEAWATER CONTAINS MANY DISSOLVED SALTS

We can classify water according to the amount of dissolved minerals it contains (Table K-1). Seawater averages 3.5 percent by mass of dis-

Table K-1 Classification scheme for water

Type of water	Quantity of dissolved minerals/% by mass
fresh	0–0.1
brackish	0.1–1
salty	1–10
brine	>10

solved minerals, which puts it in the salty category. About 75 elements have been detected in seawater, but only 10 species (Table K-2) constitute over 99.9 percent of the mass of the various substances dissolved in seawater. Sodium ions plus chloride ions constitute about 86 percent by mass of the dissolved species in seawater.

Most of the ionic constituents of seawater enter the ocean in superheated (320°C), mineral-rich water that originates deep within the earth and flows through vent holes in the ocean floor (Frontispiece). **Manganese nodules** (Figure K-1), which are porous, roughly spherical chunks of metallic oxides ranging from 2 to 10 cm in diameter, form spontaneously in the vicinity of the vents. In addition to manganese, the nodules contain iron, cobalt, nickel, copper, zinc, chromium, vanadium, tungsten, and lead. Manganese nodules are a potentially rich source of scarce metals, such as cobalt and chromium, but they cannot be recovered economically at the present time.

Table K-2 Concentration of the ten principle ionic constituents of seawater

Ion	Concentration/M
Cl^-	0.55
Na^+	0.46
SO_4^{2-}	0.028
Mg^{2+}	0.054
Ca^{2+}	0.010
K^+	0.010
HCO_3^-	0.0023
CO_3^{2-}	0.0003
Br^-	0.00083
Sr^{2+}	9×10^{-5}

K-2. THE MAJOR NUTRIENTS IN THE OCEAN FOOD CHAIN ARE PHOSPHATE, NITRATE, CARBON DIOXIDE, AND OXYGEN

Temperature, oxygen concentration, carbon dioxide concentration, phosphate concentration as HPO_4^{2-}, nitrate concentration, and pH play key roles in the chemistry and biochemistry of the oceans.

The temperature of ocean water ranges from a high of about 32°C near the surface in some regions to a low of about −2°C near an ice shelf. The average surface temperature is around 22°C and decreases to about 2°C at a depth of 2 km; below 2 km the ocean temperature is fairly constant at 2°C.

The pH of ocean water is remarkably constant at a value of 8.2. The ocean pH is maintained at 8.2 by the buffering action of ocean sediments containing carbonates and phosphates.

Variations in the concentration of O_2, NO_3^-, and HPO_4^{2-} with depth are shown in Figure K-2. All these constituents are essential to the development of plants and marine organisms such as phytoplankton, which serve as food sources for higher forms of marine life.

The concentration of oxygen near the ocean surface is high relative to that of nitrate and hydrogen phosphate (Figure K-2) because of the dissolution of oxygen from the air and its production in photosynthesis by phytoplankton. The value of $[O_2]$ initially decreases with depth because of the consumption of oxygen during the decomposition of animal and plant matter.

The phosphates and nitrates rise to the surface from the mineral-rich deep water and are depleted near the surface as they are utilized in the production of marine life. The growth of phytoplankton in the ocean is limited by the amount of available phosphate, nitrate, and trace mineral nutrients. The carbon nutrient source is carbon dioxide from the atmosphere, and CO_2 is always present in stoichiometric excess amounts. There are certain places in the ocean where the deep water rises to the surface in large quantities, as a result of prevailing winds that blow away the warmer and thus lighter surface water. These regions are especially rich in animal life and constitute the great fishing areas. Examples are regions off the coasts of Newfoundland, Chile, and Peru. Areas where the nutrient-rich deep water rises to the surface

Figure K-1 Manganese nodules, which form in the vicinity of vents in the ocean floor, contain manganese, iron, cobalt, nickel, copper, zinc, chromium, vanadium, tungsten, and lead.

constitute only 0.1 percent of the ocean surface, but these areas supply over 50 percent of the total fish catch. Ninety percent of the ocean is a biological desert that yields only one percent of the total fish catch. Nutrient-poor tropical ocean waters, like those near Hawaii, for example, contain very little plant life and as a result are blue. Nutrient-rich ocean waters are teeming with microscopic plant life and as a result are greenish-brown.

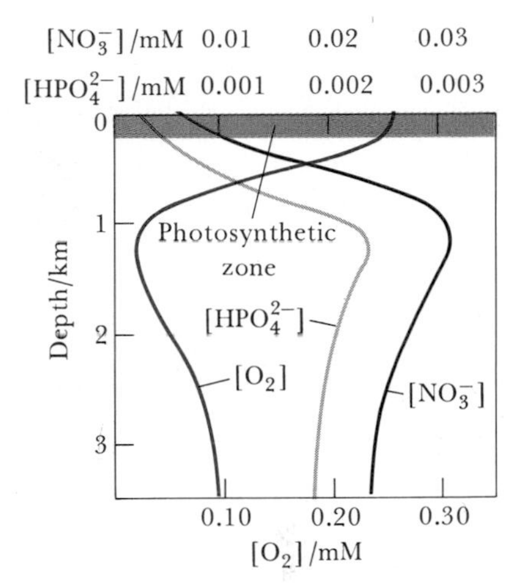

Figure K-2 Variation of $[O_2]$, $[HPO_4^{2-}]$, and $[NO_3^-]$ with ocean depth. Note that in the first kilometer the oxygen concentration decreases while the concentrations of nitrate and hydrogen phosphate increase. The value of the ratio $[NO_3^-]/[HPO_4^{2-}]$ is about 15, which is equal to the nitrogen-to-phosphorus ratio in phytoplankton.

K-3. FOUR CHEMICALS ARE OBTAINED COMMERCIALLY FROM SEAWATER

Seawater constitutes an enormous source of useful chemicals, but at the present time only four substances are obtained from seawater on a commercial scale: pure water, sodium chloride, bromine, and magnesium hydroxide.

Pure water is obtained from seawater on a commercial scale by various distillation techniques. The process of removing dissolved solids from seawater is called **desalination**. Economical methods of desalination are of paramount importance in the arid regions of the world.

About 40 million metric tons (1 metric ton = 1000 kg = 2200 lb) of sodium chloride is obtained from seawater each year. The process involves filtration of the seawater to remove particulate matter followed by natural evaporation of the water from storage ponds until the NaCl crystallizes from solution (Figure K-3).

Bromine is present in seawater as bromide ion at a concentration of 8.3×10^{-4} M; some oil-well brines have even higher concentrations of $Br^-(aq)$ than those in seawater. The economical recovery of bromine from seawater and brines depends upon the fact that elemental bromine, Br_2, is a volatile liquid. The pH of seawater is decreased from 8.2 to 3.5 by the addition of sulfuric acid and the bromide ion is then converted to bromine by a replacement reaction with chlorine:

$$2Br^-(aq) + Cl_2(g) \rightarrow Br_2(l) + 2Cl^-(aq)$$

Figure K-3 Evaporation ponds used to obtain salt from brines. The vivid coloration is caused by the natural growth of certain algae and bacteria in the brine as it gradually becomes more concentrated.

The Br_2 is removed from the seawater when a stream of air is passed through the solution to sweep out the bromine.

Magnesium ion is present in seawater at a concentration of 0.054 M. It is separated from seawater by the addition of lime, $CaO(s)$, which precipitates the $Mg^{2+}(aq)$ as the hydroxide, $Mg(OH)_2(s)$:

$$Mg^{2+}(aq) + CaO(s) + H_2O(l) \rightarrow Mg(OH)_2(s) + Ca^{2+}(aq)$$

The $Mg(OH)_2$ is collected by filtration and converted to magnesium chloride with hydrochloric acid. The magnesium chloride is recovered by evaporation of water from the solution until the $MgCl_2(s)$ crystallizes.

Table K-3 Comparison of concentrations of the major ionic constituents in seawater, fresh water, and human blood serum*

	Seawater	Fresh water (average values)	Human blood serum
$[Na^+]$	460	0.27	145
$[K^+]$	10	0.06	5.1
$[Ca^{2+}]$	10	0.38	2.5
$[Mg^{2+}]$	54	0.34	1.2
$[Cl^-]$	550	0.22	103
$[SO_4^{2-}]$	28	0.12	2.5
$[HCO_3^-]$	2.3	0.96	12
Total	1114	2.35	271

*All concentrations are expressed in millimolar (mM) units.

K-4. FRESH WATER CONTAINS A VARIETY OF DISSOLVED SUBSTANCES

The major source of fresh water is rain, which falls from clouds formed primarily from evaporated seawater. All rainwater contains dissolved oxygen, nitrogen, and carbon dioxide. Rain that has fallen through polluted air may contain a variety of contaminants, including sulfuric and sulfurous acids, carbon monoxide, oxides of nitrogen, dust, pollen, and salts of numerous trace metals, such as iron and lead. A beneficial feature of rain and snow is that they remove pollutants from the lower atmosphere. Rain that has fallen through clean air has a pH of about 5.6 because the dissolved carbon dioxide reacts with the water to produce $H^+(aq)$ and $HCO_3^-(aq)$ ions:

$$CO_2(aq) + H_2O(l) \rightleftharpoons H^+(aq) + HCO_3^-(aq)$$

Fresh water that has been in contact with the earth for some time contains a variety of anions and cations. Average values of the concentrations of the major ionic species in fresh water are given in Table K-3. **Hard water** contains appreciable amounts of divalent cations, primarily Ca^{2+}, Mg^{2+}, and Fe^{2+}. The major anions in hard water are the hydrogen carbonate ion (HCO_3^-) and the sulfate ion (SO_4^{2-}). These ions can arise in groundwater by reactions such as the interaction of water containing dissolved CO_2 with limestone ($CaCO_3$):

▪ The reverse of this reaction results in the formation of stalagtites and stalactites.

$$CaCO_3(s) + H_2O(l) + CO_2(aq) \rightleftharpoons Ca^{2+}(aq) + 2HCO_3^-(aq)$$

The result of this reaction is the dissolution of $CaCO_3(s)$. Magnesium ions in groundwater are the result of a similar reaction between $CO_2(aq)$ and **dolomite,** a mineral containing $CaCO_3$ and $MgCO_3$. The divalent cations in hard water form precipitates with soaps that appear as a scum in the wash water. Natural soaps are sodium salts of fatty acids, which are organic acids containing long hydrocarbon chains. A typical example is sodium stearate, $C_{17}H_{35}COO^-Na^+$:

$$\underbrace{CH_3CH_2CH_2CH_2CH_2CH_2CH_2CH_2CH_2CH_2CH_2CH_2CH_2CH_2CH_2CH_2CH_2}_{\text{hydrocarbon portion}}\overbrace{COO^-}^{\text{carboxyl portion}}\ Na^+$$

An example of scum formation is the reaction

$$Ca^{2+}(aq) + 2C_{17}H_{35}COO^-(aq) \rightarrow \underset{\text{calcium stearate (scum)}}{Ca(C_{17}H_{35}COO)_2(s)}$$

(where $2C_{17}H_{35}COO^-(aq)$ is the stearate ion (soap))

Hard water is classified as either temporary or permanent, depending on which anions it contains. Temporary hard water contains $HCO_3^-(aq)$ anions, along with $Ca^{2+}(aq)$ and/or $Mg^{2+}(aq)$. When temporary hard water is heated, calcium carbonate or magnesium carbonate precipitates because of the reaction

$$M^{2+}(aq) + 2HCO_3^-(aq) \rightarrow MCO_3(s) + H_2O(l) + CO_2(g)$$

where M stands for either metal. The metal carbonate can be seen as deposits called **boiler scale** in boilers, hot-water pipes, and tea kettles. Such deposits can clog pipes and, by acting as a heat insulator, increase the cost of heating water. In permanent hard water the primary anion is sulfate. Both calcium sulfate and magnesium sulfate are soluble in hot water and are not precipitated by heating. Thus, the water cannot be softened by heating.

Hard water can be converted by chemical means to **softened water,** in which divalent cations have been removed, or to **deionized water,** in which all cations and anions have been removed. The softening and deionization processes involve the use of water-insoluble **ion-exchange resins.** The chemical process for removal of $M^{2+}(aq)$ ions, for example, can be represented by the following equation:

$$\underset{\text{fixed resin group}}{\text{C}(COO^-Na^+)_2} + M^{2+}(aq) \rightarrow \underset{\text{fixed resin group}}{\text{C}(COO^-)_2 M^{2+}} + 2Na^+(aq)$$

This cation-exchange reaction is the chemical process that takes place in home water-softener systems. Note that the divalent ions, $M^{2+}(aq)$, in the hard water are replaced with sodium ions, $Na^+(aq)$, which do not cause the precipitation of scum in the presence of soap. The ion-exchange resin can be reactivated by running a concentrated salt solution (NaCl) through the ion-exchanger system, which reverses the cation-exchange reaction.

Deionization of water involves two ion-exchange reactions, which are described by the following equations:

1. metal cation removal (in the first ion exchanger)

$$\underset{\text{fixed resin group}}{\text{C}(COO^-H^+)_2} + M^{2+}(aq) \rightarrow \underset{\text{fixed resin group}}{\text{C}(COO^-)_2 M^{2+}} + 2H^+(aq)$$

2. anion removal (in the second ion exchanger)

$$\underset{\text{fixed resin group}}{-NH_3^+OH^-} + X^-(aq) \rightarrow \underset{\text{fixed resin group}}{-NH_3^+X^-} + OH^-(aq)$$

The OH^- released in the anion exchanger by the anion X^- reacts with the H^+ released in the cation exchanger to make water:

$$H^+(aq) + OH^-(aq) \rightarrow H_2O(l)$$

Ion exchangers are used to produce high-quality, essentially ion-free water.

K-5. SOAPS CLEAN BY FORMING MICELLES

The cleaning action of soap is a consequence of the dual affinity of soap molecules for grease and water. Realize that water and hydrocarbon compounds do not dissolve in each other because they have very different electrical properties. Water is a polar molecule, whereas hydrocarbon chains are nonpolar (the electronegativities of carbon and hydrogen are about the same). Water molecules interact much more strongly with each other than with nonpolar molecules. Nonpolar molecules have no regions of net charge that can interact with the charged regions in a water molecule. Consequently, nonpolar molecules are excluded from water and so are not soluble in it. Polar substances, on the other hand, can interact electrostatically with water molecules and so are soluble in water. As a rule of thumb, polar substances dissolve polar substances and nonpolar substances dissolve nonpolar substances. More succinctly, *like dissolves like.*

Soap is effective as a cleaning agent because the hydrocarbon portion of a soap molecule has a strong affinity for grease, whereas the charged portion of the molecule has a strong affinity for water. The anion of a soap molecule can be represented schematically as

$$\sim\sim\sim\sim\sim\sim\ominus$$

When soap molecules that are dissolved in water come into contact with grease, the hydrocarbon portions stick into the grease, leaving the anion portions at the grease-water interface. The penetration of the

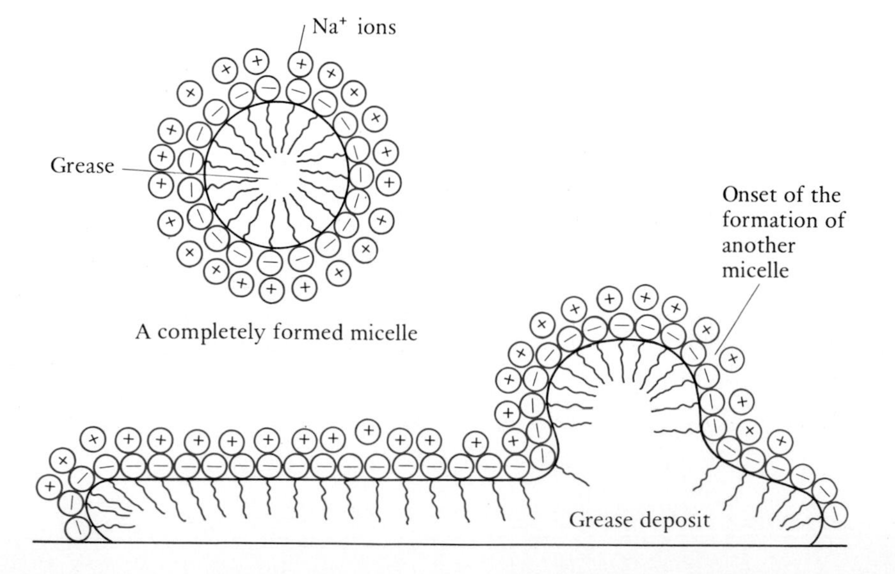

Figure K-4 The cleaning action of soap is due to the ability of soap molecules to form micelles that encapsulate grease and carry it away. The fatty-acid portions of the soap molecules dissolve in each other and in the grease particles, thus forming a water-soluble spherical particle with a charged group on its surface.

grease by the soap molecules is followed by a remarkable phenomenon—the formation of **micelles** (Figure K-4), which are small, spherical grease-soap droplets that are soluble in water as a result of the polar groups on their surface. Micelles do not combine into larger drops because their surfaces are all negatively charged. Thus the cleaning action of soap is a consequence of the specialized molecular properties of the soap molecules and their interaction with grease that leads to the formation of water-soluble micelles. The micelles encapsulate small grease particles and are subsequently rinsed away, leaving a clean region behind.

TERMS YOU SHOULD KNOW

manganese nodules	536	hard water	538	softened water	539	ion-exchange resin	539
desalination	537	boiler scale	539	deionized water	539	micelle	541

QUESTIONS

K-1. What are the principal ionic constituents in seawater?

K-2. What is the major source of dissolved salts in the oceans?

K-3. Name four substances that are obtained from seawater on a commercial scale.

K-4. What is desalination?

K-5. Describe how bromine is obtained from seawater.

K-6. Distinguish between fresh water, brackish water, salty water, and brine.

K-7. What are the principal constituents of hard water?

K-8. Describe how soap acts as a cleaning agent.

K-9. What is a micelle? What are the special properties of molecules that form micelles?

K-10. Describe the difference between temporary and permanent hard water.

K-11. What is the cause of the formation of soap scum?

K-12. What does it mean to soften water? What is the difference between softened and deionized water?

K-13. Describe how an ion-exchange resin works.

K-14. What are manganese nodules?

PROBLEMS

K-15. Verify that the data in Table K-2 are consistent with the charge-balance condition for an electrolyte solution to within the 1.0 percent accuracy of the data given. (Total positive charge per unit volume equals the total negative charge per unit volume.)

K-16. Compute the number of grams of magnesium and of bromine that can be obtained from 1.00 L of seawater.

K-17. Why is it that acid rain does not pose as serious a threat to living organisms in the oceans as it does to living organisms in lakes and ponds?

K-18. Given that seawater is 0.46 M in $NaCl(aq)$, calculate the number of liters of seawater that must be processed to produce 40 million metric tons of $NaCl(s)$. Assume 100 percent recovery.

K-19. Given that seawater is 8.3×10^{-4} M in $Br^-(aq)$, calculate the number of kilograms of $Cl_2(g)$ required to obtain the bromine from 1 billion liters of seawater.

CHAPTER • 17

SOLUBILITY AND PRECIPITATION REACTIONS

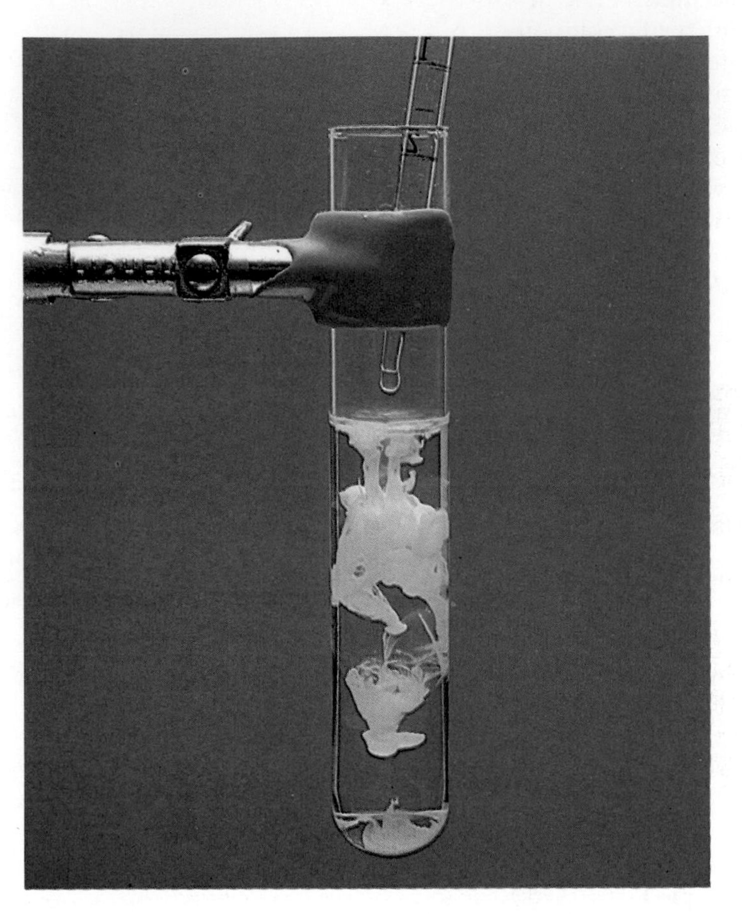

Precipitation of $Ag^+(aq)$ as $AgCl(s)$ by addition of $HCl(aq)$ to a solution containing $Ag^+(aq)$.

Many substances are insoluble or have a low solubility in water. In this chapter we discuss chemical equilibria involving slightly soluble substances. A saturated solution of a slightly soluble ionic compound involves an equilibrium between the solid and its constituent ions in solution. An equilibrium constant for a solubility equilibrium is defined according to the law of concentration action. The treatment of solubility as an equilibrium process allows us to calculate the solubility of a solid not only in pure water but in solutions of acids, bases, and salts as well. The solubility of many substances is shown to depend strongly upon pH and the presence of other salts.

17-1. THE DRIVING FORCE OF A DOUBLE-REPLACEMENT REACTION IS THE FORMATION OF AN INSOLUBLE PRODUCT

In Chapter 2 we learned that one possible driving force of a double-replacement reaction is the formation of a precipitate. For example, the reaction described by the chemical equation

$$BaCl_2(aq) + Na_2SO_4(aq) \rightarrow 2NaCl(aq) + BaSO_4(s)$$

occurs readily, with the formation of a white $BaSO_4$ precipitate. To predict whether a precipitate will form, it is necessary to know the **solubility** of a compound. Although all compounds have a characteristic solubility in water at a particular temperature, it is useful to set up a few general rules for characterizing solubility. We call any substance whose solubility is less than 0.01 $mol \cdot L^{-1}$ insoluble. If its solubility is greater than 0.1 $mol \cdot L^{-1}$, we call it soluble. If its solubility is between 0.01 and 0.1 $mol \cdot L^{-1}$, we say that it is slightly soluble. The following **solubility rules** can be used to determine solubilities in water:

▪ The designations soluble, slightly soluble, and insoluble given here are arbitrary.

Solubility Rules

▪ These solubility rules are neither infallible nor all-inclusive, but they are useful.

1. All sodium, potassium, and ammonium salts are soluble.
2. All nitrates, acetates, and perchlorates are soluble.
3. All silver, lead, and mercury(I) salts are insoluble.
4. All chlorides, bromides, and iodides are soluble.
5. All carbonates, sulfides, oxides, and hydroxides are insoluble.
6. All sulfates are soluble except calcium sulfate and barium sulfate.

The solubility rules must be applied in the order given; the rule with the lower number takes precedence in case of a conflict. For example, Na_2S is soluble because rule 1 states that all sodium salts are soluble and the rule concerning sulfide solubilities is rule 5. Similarly, $PbSO_4$ in insoluble because rule 3 states that all Pb^{2+} salts are insoluble.

Example 17-1: Predict the solubility of the following compounds in water:

(a) $(NH_4)_2SO_4$ (b) $CaCO_3$ (c) Al_2O_3 (d) $Pb(NO_3)_2$ (e) $PbCl_2$

Solution: Apply the solubility rules in the order given:

(a) soluble by rule 1: all ammonium salts are soluble
(b) insoluble by rule 5: all carbonates are insoluble
(c) insoluble by rule 5: all metal oxides are insoluble
(d) soluble by rule 2: all nitrates are soluble
(e) insoluble by rule 3: all lead salts are insoluble

These six solubility rules can be used to predict whether or not a double-replacement reaction occurs. Consider the reaction between barium chloride, $BaCl_2(aq)$, and potassium carbonate, $K_2CO_3(aq)$. The possible double-replacement products are KCl and $BaCO_3$. According to rule 5, $BaCO_3$ is insoluble, and so we have

$$BaCl_2(aq) + K_2CO_3(aq) \rightarrow 2KCl(aq) + BaCO_3(s)$$

The net ionic equation is

$$Ba^{2+}(aq) + CO_3^{2-}(aq) \rightarrow BaCO_3(s)$$

Example 17-2: Predict whether there is a reaction when solutions of $Cd(NO_3)_2$ and Na_2S are mixed. If there is a reaction, then write the complete equation and the net ionic equation that describe the reaction.

Solution: The reaction we are considering is

$$Na_2S(aq) + Cd(NO_3)_2(aq) \rightarrow ?$$

The possible double-replacement products are $NaNO_3$ and CdS. Although $NaNO_3$ is soluble, CdS is not (rule 5), and so we have

$$Na_2S(aq) + Cd(NO_3)_2(aq) \rightarrow 2NaNO_3(aq) + CdS(s)$$

The net ionic equation is

$$Cd^{2+}(aq) + S^{2-}(aq) \rightarrow CdS(s)$$

Cadmium sulfide is a yellow-orange solid. The addition of Na_2S to an unknown solution that may contain Cd^{2+} ions is a test for the presence of $Cd^{2+}(aq)$.

17-2. THE LAW OF CONCENTRATION ACTION GOVERNS THE EQUILIBRIUM BETWEEN AN IONIC SOLID AND ITS CONSTITUENT IONS IN SOLUTION

Consider the equilibrium between solid silver bromate and its constituent ions in water:

$$AgBrO_3(s) \rightleftharpoons Ag^+(aq) + BrO_3^-(aq) \qquad (17\text{-}1)$$

Application of the law of concentration action to Equation (17-1) yields the equilibrium-constant expression for this reaction:

$$K_{sp} = [Ag^+][BrO_3^-] \qquad (17\text{-}2)$$

The subscript *sp* stands for solubility product, and K_{sp} is called the **solubility-product constant.** Note that $AgBrO_3(s)$ does not appear in the K_{sp} expression because its concentration cannot vary. Recall from Chapter 14 that a pure solid does not appear in an equilibrium-constant expression.

The experimental value of K_{sp} at 25°C for Equation (17-1) is

$$K_{sp} = [Ag^+][BrO_3^-] = 5.8 \times 10^{-5}\ M^2 \qquad (17\text{-}3)$$

Equation (17-3) states that if $AgBrO_3(s)$ is in equilibrium with an aqueous solution of $AgBrO_3(aq)$ at 25°C, then the product of the concentrations of $Ag^+(aq)$ and $BrO_3^-(aq)$ at equilibrium must equal $5.8 \times 10^{-5}\ M^2$.

A K_{sp} expression can be used to estimate the solubility of a solid. For example, suppose that excess $AgBrO_3(s)$ is in contact with water at 25°C. Then at equilibrium

$$[Ag^+][BrO_3^-] = 5.8 \times 10^{-5}\ M^2$$

From the reaction stoichiometry, Equation (17-1), we have

$$[Ag^+] = [BrO_3^-]$$

because each formula unit of $AgBrO_3$ that dissolves produces one $Ag^+(aq)$ ion and one $BrO_3^-(aq)$ ion. Furthermore, if we denote the solubility of $AgBrO_3(s)$, in units of molarity, by s, then

$$s = \left(\begin{matrix}\text{solubility of } AgBrO_3(s)\\ \text{in water}\end{matrix}\right) = [Ag^+] = [BrO_3^-]$$

because the concentration of either $Ag^+(aq)$ or $BrO_3^-(aq)$ is equal to the number of moles of dissolved salt per liter of solution. Thus from the K_{sp} expression we have

$$K_{sp} = 5.8 \times 10^{-5}\ M^2 = [Ag^+][BrO_3^-] = s^2$$

and thus $$s = (5.8 \times 10^{-5}\ M^2)^{1/2} = 7.6 \times 10^{-3}\ M$$

The formula mass of $AgBrO_3(s)$ is 235.8, and thus the number of grams of $AgBrO_3(s)$ that dissolves in 1.00 L of solution at 25°C is

$$(7.6 \times 10^{-3}\ \text{mol}\cdot\text{L}^{-1})\left(\frac{235.8\ \text{g}\ AgBrO_3}{1\ \text{mol}\ AgBrO_3}\right) = 1.8\ \text{g}\cdot\text{L}^{-1}$$

Example 17-3: The K_{sp} for $PbCrO_4(s)$ in equilibrium with an aqueous solution of its constituent ions at 25°C is $2.8 \times 10^{-13}\ M^2$. Write the chemical equation that represents the solubility equilibrium for $PbCrO_4(s)$ and compute its solubility in water at 25°C.

Solution: The solubility equilibrium is

$$PbCrO_4(s) \rightleftharpoons Pb^{2+}(aq) + CrO_4^{2-}(aq)$$

The K_{sp} expression is

$$K_{sp} = [Pb^{2+}][CrO_4^{2-}] = 2.8 \times 10^{-13}\ M^2$$

If $PbCrO_4(s)$ is equilibrated with pure water, then, from the reaction stoichiometry, we have at equilibrium

$$[Pb^{2+}] = [CrO_4^{2-}] = s$$

where s is the solubility of $PbCrO_4(s)$ in pure water. Thus

$$K_{sp} = s^2 = 2.8 \times 10^{-13}\ M^2$$

and $$s = (2.8 \times 10^{-13}\ M^2)^{1/2} = 5.3 \times 10^{-7}\ M$$

When excess $AgBrO_3(s)$ is equilibrated with pure water, we have at equilibrium $[Ag^+] = [BrO_3^-]$, because each $AgBrO_3$ unit that dissolves yields one $Ag^+(aq)$ and one $BrO_3^-(aq)$. Consider the problem of calculating the solubility in water of copper(II) iodate, $Cu(IO_3)_2(s)$, which yields *two* $IO_3^-(aq)$ and one $Cu^{2+}(aq)$ for each $Cu(IO_3)_2$ unit that dissolves. The chemical equation that represents the solubility equilibrium of $Cu(IO_3)_2(s)$ in water is

$$Cu(IO_3)_2(s) \rightleftharpoons Cu^{2+}(aq) + 2IO_3^-(aq) \qquad (17\text{-}4)$$

According to the law of concentration action, the K_{sp} expression for Equation (17-4) is

$$K_{sp} = [Cu^{2+}][IO_3^-]^2 \qquad (17\text{-}5)$$

The experimental value of K_{sp} at 25°C is 7.4×10^{-8} M^3, and so we have at 25°C

$$K_{sp} = [Cu^{2+}][IO_3^-]^2 = 7.4 \times 10^{-8}\ \text{M}^3 \qquad (17\text{-}6)$$

Note that it is the *square* of the concentration of $IO_3^-(aq)$ that appears in the K_{sp} expression for $Cu(IO_3)_2(s)$ because each formula unit of $Cu(IO_3)_2$ that dissolves produces two iodate ions. Thus, when $Cu(IO_3)_2(s)$ is in equilibrium with pure water, the concentration of iodate ion is twice as great as the concentration of copper(II) ion:

$$[IO_3^-] = 2[Cu^{2+}]$$

The solubility of $Cu(IO_3)_2(s)$ in pure water is equal to $[Cu^{2+}]$ because each mole of $Cu(IO_3)_2$ that dissolves yields one mole of $Cu^{2+}(aq)$. If we denote the solubility of $Cu(IO_3)_2(s)$ in pure water by s, then

$$s = \begin{pmatrix}\text{solubility of } Cu(IO_3)_2(s) \\ \text{in water}\end{pmatrix} = [Cu^{2+}] = \frac{[IO_3^-]}{2}$$

It follows then that $[IO_3^-] = 2s$. Combining these results with the K_{sp} expression, Equation (17-6), yields

$$7.4 \times 10^{-8}\ \text{M}^3 = [Cu^{2+}][IO_3^-]^2 = (s)(2s)^2 = 4s^3$$

and therefore

$$s = \left(\frac{7.4 \times 10^{-8}\ \text{M}^3}{4}\right)^{1/3} = 2.6 \times 10^{-3}\ \text{M}$$

Note that $[Cu^{2+}] = s = 2.6 \times 10^{-3}$ M and also that $[IO_3^-] = 2s = 5.2 \times 10^{-3}$ M.

Example 17-4: The solubility-product constant for silver chromate in equilibrium with water at 25°C is 1.1×10^{-12} M^3. Compute the value of $[Ag^+]$ that results when pure water is saturated with $Ag_2CrO_4(s)$.

Solution: The $Ag_2CrO_4(s)$ solubility equilibrium is

$$Ag_2CrO_4(s) \rightleftharpoons 2Ag^+(aq) + CrO_4^{2-}(aq)$$

and the solubility-product expression is

$$K_{sp} = [Ag^+]^2[CrO_4^{2-}] = 1.1 \times 10^{-12}\ \text{M}^3$$

Each Ag_2CrO_4 unit that dissolves yields two $Ag^+(aq)$ and one $CrO_4^{2-}(aq)$; thus

$$s = \frac{[Ag^+]}{2} = [CrO_4^{2-}]$$

Substitution of this result into the K_{sp} expression for $Ag_2CrO_4(s)$ yields

$$K_{sp} = (2s)^2(s) = 1.1 \times 10^{-12}\ \text{M}^3$$

■ For a salt with the formula A_xB_y, the solubility is given by

$$s = \frac{[A]}{x} = \frac{[B]}{y}$$

and so

$$s^3 = \frac{1.1 \times 10^{-12} M^3}{4}$$

Solving for s yields

$$s = 6.5 \times 10^{-5} M$$

Example 17-5: The solubility of $PbCl_2$ in water at 25°C is 4.41 $g \cdot L^{-1}$. Calculate the K_{sp} of $PbCl_2$.

Solution: The solubility s of $PbCl_2$ in moles per liter is

$$s = (4.41\ g \cdot L^{-1}) \left(\frac{1\ mol\ PbCl_2}{278.1\ g\ PbCl_2} \right) = 1.59 \times 10^{-2}\ M$$

The chemical equation for the solubility equilibrium is

$$PbCl_2(s) \rightleftharpoons Pb^{2+}(aq) + 2Cl^-(aq)$$

The solubility s is related to $[Pb^{2+}]$ and $[Cl^-]$ by

$$s = [Pb^{2+}] \quad \text{and} \quad [Cl^-] = 2s$$

Therefore,

$$K_{sp} = [Pb^{2+}][Cl^-]^2 = (s)(2s)^2 = 4s^3 = 4(1.59 \times 10^{-2}\ M)^3$$
$$= 1.61 \times 10^{-5}\ M^3$$

Some solubility-product constants are given in Table 17-1.

17.3. THE SOLUBILITY OF AN IONIC SOLID IS DECREASED WHEN A COMMON ION IS PRESENT IN THE SOLUTION

Consider the problem of calculating the solubility of silver bromate in an aqueous solution at 25°C that is 0.10 M in sodium bromate. As we learned in Section 17-2, the solubility equilibrium of $AgBrO_3(s)$ is

$$AgBrO_3(s) \rightleftharpoons Ag^+(aq) + BrO_3^-(aq) \qquad (17\text{-}7)$$

and the solubility-product-constant expression is

$$K_{sp} = [Ag^+][BrO_3^-] = 5.8 \times 10^{-5}\ M^2 \qquad (17\text{-}8)$$

The $Na^+(aq)$ from the $NaBrO_3(aq)$ is simply a spectator ion and does not enter into any of our calculations.

▪ Recall that a spectator ion is an ion that is present but does not take a direct part in the reaction.

It is important to realize that Equation (17-8) fixes only the value of the product of the ionic concentrations, $[Ag^+][BrO_3^-]$. Any pair of values of $[Ag^+]$ and $[BrO_3^-]$ that when multiplied together equal $5.8 \times 10^{-5}\ M^2$ constitutes a pair of equilibrium ionic concentrations. It is important also to realize that the ionic concentrations in Equation (17-8) are the *total* ionic concentrations, *regardless of the source of each ionic species.* In the case of $AgBrO_3$ dissolved in a 0.10 M $NaBrO_3(aq)$ solution, the $Ag^+(aq)$ ions come only from the $AgBrO_3(s)$ that dissolves. The $BrO_3^-(aq)$ ions, on the other hand, come from two sources: from the 0.10 M $NaBrO_3(aq)$ initially present, which is completely dissociated into $Na^+(aq)$ and $BrO_3^-(aq)$, and from the $AgBrO_3(s)$ that dissolves. If we let s be the solubility of $AgBrO_3(s)$ in 0.10 M $NaBrO_3(aq)$, then

Table 17-1 Solubility-product constants for salts in water at 25°C

Halides	K_{sp}	Halides	K_{sp}
AgCl	1.8×10^{-10} M^2	Hg_2I_2*	4.5×10^{-29} M^3
AgBr	5.0×10^{-13} M^2	MgF_2	6.5×10^{-9} M^3
AgI	8.3×10^{-17} M^2	PbF_2	7.7×10^{-8} M^3
BaF_2	1.0×10^{-6} M^3	$PbCl_2$	1.6×10^{-5} M^3
CaF_2	5.3×10^{-9} M^3	$PbBr_2$	4.0×10^{-5} M^3
CuCl	1.2×10^{-6} M^2	PbI_2	7.1×10^{-9} M^3
CuBr	5.3×10^{-9} M^2	SrF_2	2.5×10^{-9} M^3
CuI	1.1×10^{-12} M^2	TlCl	1.7×10^{-4} M^2
Hg_2Cl_2*	1.3×10^{-18} M^3	TlBr	3.4×10^{-6} M^2
Hg_2Br_2*	5.6×10^{-23} M^3	TlI	6.5×10^{-8} M^2
Carbonates	**K_{sp}**	**Carbonates**	**K_{sp}**
Ag_2CO_3	8.1×10^{-12} M^3	$MgCO_3$	3.5×10^{-8} M^2
$BaCO_3$	5.1×10^{-9} M^2	$MnCO_3$	1.8×10^{-11} M^2
$CaCO_3$	2.8×10^{-9} M^2	$PbCO_3$	7.4×10^{-14} M^2
$CuCO_3$	1.4×10^{-10} M^2	$SrCO_3$	1.1×10^{-10} M^2
$FeCO_3$	3.2×10^{-11} M^2	$ZnCO_3$	1.4×10^{-11} M^2
Chromates	**K_{sp}**	**Oxalates**	**K_{sp}**
Ag_2CrO_4	1.1×10^{-12}M^3	$Ag_2C_2O_4$	3.4×10^{-11} M^3
$BaCrO_4$	1.2×10^{-10} M^2	CaC_2O_4	4×10^{-9} M^2
$PbCrO_4$	2.8×10^{-13} M^2	MgC_2O_4	7×10^{-7} M^2
Hydroxides	**K_{sp}**	**Hydroxides**	**K_{sp}**
$Al(OH)_3$	1.3×10^{-33}M^4	$Fe(OH)_3$	1.0×10^{-38} M^4
$Ca(OH)_2$	5.5×10^{-6} M^3	$Mg(OH)_2$	1.8×10^{-11} M^3
$Cd(OH)_2$	2.5×10^{-14} M^3	$Ni(OH)_2$	2.0×10^{-15} M^3
$Cr(OH)_3$	6.3×10^{-31} M^4	$Pb(OH)_2$	1.2×10^{-15} M^3
$Cu(OH)_2$	2.2×10^{-20} M^3	$Sn(OH)_2$	1.4×10^{-28} M^3
$Fe(OH)_2$	8.0×10^{-16} M^3	$Zn(OH)_2$	1.0×10^{-15} M^3
Sulfates	**K_{sp}**	**Sulfates**	**K_{sp}**
Ag_2SO_4	1.4×10^{-5} M^3	Hg_2SO_4	7.4×10^{-7} M^2
$BaSO_4$	1.1×10^{-10} M^2	$PbSO_4$	1.6×10^{-8} M^2
$CaSO_4$	9.1×10^{-6} M^2	$SrSO_4$	3.2×10^{-7} M^2
Sulfides	**K_{sp}**	**Sulfides**	**K_{sp}**
CdS	8.0×10^{-27} M^2	MnS	2.5×10^{-13} M^2
CuS	6.3×10^{-36} M^2	PbS	8.0×10^{-28} M^2
FeS	6.3×10^{-18} M^2	SnS	1.0×10^{-25} M^2
HgS	4×10^{-53} M^2	ZnS	1.6×10^{-24} M^2

*Recall that Hg(I) exists as Hg_2^{2+} (aq) in aqueous solution.

$$[Ag^+] = s$$

$$[BrO_3^-] = s + 0.10\ M$$

If we substitute these two expressions into Equation (17-8), then we obtain

$$s(s + 0.10\ M) = 5.8 \times 10^{-5}\ M^2 \qquad (17\text{-}9)$$

Because $AgBrO_3(s)$ is only sparingly soluble in pure water, we expect the value of s in Equation (17-9) to be small. In fact, the largest that s can be is the solubility of $AgBrO_3(s)$ in pure water, which we calculated in Section 17-2 to be 7.6×10^{-3} M. Therefore, we can neglect s relative to 0.10 M and write

$$s(0.10\ M) \approx 5.8 \times 10^{-5}\ M^2$$

$$s \approx 5.8 \times 10^{-4}\ M$$

The small value of $[Ag^+]$ justifies the assumption that $s = [Ag^+] \ll 0.10$ M. As a check, if we substitute $s = 5.8 \times 10^{-4}$ M into Equation (17-9), then we find that it is satisfied. Note that the common ion BrO_3^- decreases the solubility of $AgBrO_3(s)$. The solubility of $AgBrO_3(s)$ in pure water is 7.6×10^{-3} M, about 13 times greater than its solubility in a 0.10 M $NaBrO_3(aq)$ solution.

Example 17-6: Estimate the solubility of copper(II) iodate ($K_{sp} = 7.4 \times 10^{-8}\ M^3$) in an aqueous solution that is 0.20 M in copper(II) perchlorate.

Solution: The equilibrium expression that describes the solubility of $Cu(IO_3)_2$ in water is

$$Cu(IO_3)_2(s) \rightleftharpoons Cu^{2+}(aq) + 2IO_3^-(aq)$$

and the corresponding solubility-product expression is

$$K_{sp} = [Cu^{2+}][IO_3^-]^2 = 7.4 \times 10^{-8}\ M^3 \qquad (17\text{-}10)$$

The $ClO_4^-(aq)$ from the $Cu(ClO_4)_2(aq)$ is simply a spectator ion and does not enter into any of our calculations. The only source of $IO_3^-(aq)$ is from the $Cu(IO_3)_2(s)$ that dissolves. If we let s be the solubility of $Cu(IO_3)_2(s)$ in 0.20 M $Cu(ClO_4)_2(aq)$, then

$$[IO_3^-] = 2s$$

because two $IO_3^-(aq)$ ions are produced for each formula unit of $Cu(IO_3)_2(s)$ that dissolves. The $Cu^{2+}(aq)$ comes from the 0.20 M $Cu(ClO_4)_2(aq)$ and from the $Cu(IO_3)_2(s)$ that dissolves; thus

$$[Cu^{2+}] = 0.20\ M + s$$

If we substitute these expressions for $[Cu^{2+}]$ and $[IO_3^-]$ into Equation (17-10), then we obtain

$$(0.20\ M + s)(2s)^2 = 7.4 \times 10^{-8}\ M^3 \qquad (17\text{-}11)$$

Because $Cu(IO_3)_2(s)$ is a slightly soluble salt, we expect the value of s to be small. Therefore, we neglect s relative to 0.20 M in Equation (17-11) and write

$$(0.20\ M)(2s)^2 \approx 7.4 \times 10^{-8}\ M^3$$

Solving for s yields

$$s \approx \left(\frac{7.4 \times 10^{-8}\ M^3}{(0.20\ M)(4)} \right)^{1/2} = 3.0 \times 10^{-4}\ M$$

As a check, we note that the value of s that we obtain is indeed negligible relative to 0.20 M. The presence of the **common ion** $Cu^{2+}(aq)$ in the solution suppresses the solubility of $Cu(IO_3)_2(s)$. The solubility of $Cu(IO_3)_2(s)$ in pure water is 2.6×10^{-3} M, about 9 times greater than its solubility in a solution that is 0.20 M in $Cu^{2+}(aq)$.

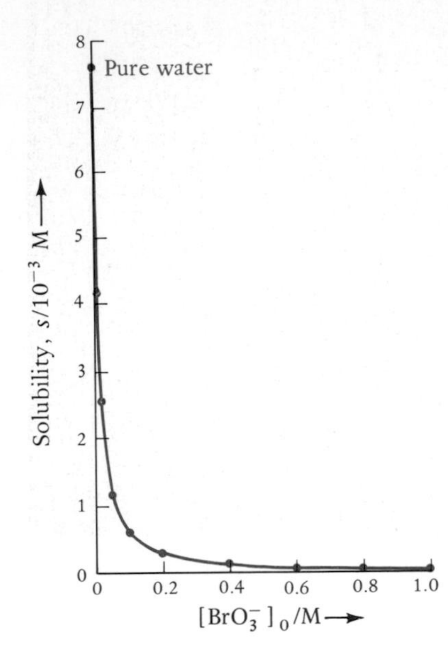

Figure 17-1 The solubility of $AgBrO_3(s)$ in water as a function of the bromate-ion concentration. The bromate-ion concentration can be controlled by adding a soluble salt such as $NaBrO_3$ to an aqueous solution of $AgBrO_3$. The plot illustrates the common-ion effect, whereby the solubility of $AgBrO_3(s)$ is decreased by the addition of $BrO_3^-(aq)$.

The decrease in the solubility of an ionic solid in the presence of one of its constituent ions is called the **common-ion effect.** This phenomenon is understood readily in terms of Le Châtelier's principle. Consider the equation for the solubility equilibrium for silver bromate

$$AgBrO_3(s) \rightleftharpoons Ag^+(aq) + BrO_3^-(aq)$$

for which

$$K_{sp} = [Ag^+][BrO_3^-] = 5.8 \times 10^{-5}\ M^2$$

An increase in the concentration of $BrO_3^-(aq)$—for example, by adding $NaBrO_3$, a strong electrolyte, to the solution—shifts the solubility equilibrium from right to left and thereby decreases the solubility of $AgBrO_3(s)$. The larger the value of $[BrO_3^-]$ at equilibrium, the smaller the value of $[Ag^+]$ because the product $[Ag^+][BrO_3^-]$ *must* equal $5.8 \times 10^{-5}\ M^2$ at equilibrium at 25°C. Similarly, an increase in the concentration of $Ag^+(aq)$—for example, by adding $AgNO_3$, which is water-soluble—also shifts the solubility equilibrium from right to left and thereby decreases the solubility of $AgBrO_3(s)$. The common-ion effect for $AgBrO_3(s)$ is illustrated in Figure 17-1.

17-4. THE SOLUBILITY OF A SOLID IS INCREASED BY THE FORMATION OF A SOLUBLE COMPLEX ION

In calculating the solubility of a solid from the K_{sp} expression, it is imperative to recognize the possibility of other important equilibria involving the ions. For example, $Ag^+(aq)$ reacts with $NH_3(aq)$ according to

$$Ag^+(aq) + 2NH_3(aq) \rightleftharpoons Ag(NH_3)_2^+(aq) \qquad K_{comp} = 2.0 \times 10^7\ M^{-2} \quad (17\text{-}12)$$

The product here is called a **complex ion,** which is a metal ion with small molecules or ions attached to it. The subscript *comp* on K denotes that the equilibrium involves the formation of a *comp*lex ion. (We shall study complex ions in Chapter 22.) If we add the equation for this **complexation reaction** to Equation (17-7), then we obtain

$$AgBrO_3(s) + 2NH_3(aq) \rightleftharpoons Ag(NH_3)_2^+(aq) + BrO_3^-(aq)$$

$$K = K_{sp}K_{comp} = 1.2 \times 10^3 \quad (17\text{-}13)$$

Note that the value of K for this equation is much larger than the value of K_{sp} for $AgBrO_3(s)$ and thus the equilibrium represented by Equation (17-13) will lie farther to the right than the K_{sp} equilibrium (Equation 17-7) for $AgBrO_3(s)$.

(a)

(b)

Figure 17-2 Silver bromate is only slightly soluble in water at 25°C. If $NH_3(aq)$ is added to a saturated solution of $AgBrO_3(aq)$, then the solubility of $AgBrO_3(s)$ is increased as a result of the formation of $Ag(NH_3)_2^+(aq)$. (a) Both test tubes initially contained 1.0 g of $AgBrO_3(s)$. (b) Then 20 mL of water was added to the left-hand tube and 20 mL of 6 M $NH_3(aq)$ was added to the right-hand tube.

If $NH_3(aq)$ is added to an aqueous solution in equilibrium with $AgBrO_3(s)$, then the solubility of $AgBrO_3(s)$ is enhanced, owing to the formation of the $Ag(NH_3)_2^+$ ion. The left-to-right solubility equilibrium shift in Equation (17-13) leads to an increased amount of dissolved $AgBrO_3(s)$. Figure 17-2 illustrates the increase in solubility of $AgBrO_3$ by the addition of $NH_3(aq)$. In the next section we present another example of the necessity of recognizing the possibility of other important equilibria in solubility calculations.

17.5. SALTS OF WEAK ACIDS ARE MORE SOLUBLE IN ACIDIC SOLUTIONS THAN IN NEUTRAL OR BASIC SOLUTIONS

Sodium benzoate, $NaC_7H_5O_2$, the salt of NaOH and the weak acid benzoic acid, has the formula

$$C_6H_5-C(=O)-O^{\ominus}Na$$

Sodium benzoate is a water-soluble food additive that functions as an antimicrobial agent in foods with pH values lower than about 4. Numerous acidic beverages, syrups, jams, jellies, and processed fruits contain about 0.1 percent sodium benzoate to prevent the growth of yeasts and harmful bacteria. Numerous benzoate salts are insoluble above $pH = 4$, and the formation of insoluble benzoate salts at these higher pH values removes benzoate ions from solution, rendering the added sodium benzoate solution ineffective as an antimicrobial agent. At $pH \leq 4$, an appreciable fraction of the benzoate exists as benzoic acid, which is the biologically active form.

Let us use silver benzoate, $AgC_7H_5O_2(s)$, to illustrate the effect of pH on the solubility of a salt of a weak acid. The equation for the solubility equilibrium of silver benzoate is

$$AgC_7H_5O_2(s) \rightleftharpoons Ag^+(aq) + C_7H_5O_2^-(aq) \quad (17\text{-}14)$$

The K_{sp} at 25°C for Equation (17-14) is

$$K_{sp} = 2.5 \times 10^{-5}\ M^2 = [Ag^+][C_7H_5O_2^-]$$

The solubility equilibrium fixes the value of the product $[Ag^+][C_7H_5O_2^-]$. However, we must recognize that the benzoate ion is the conjugate base of a weak acid, and that it reacts with $H_3O^+(aq)$ to form undissociated benzoic acid, as shown by the equation

$$C_7H_5O_2^-(aq) + H_3O^+(aq) \rightarrow HC_7H_5O_2(aq) + H_2O(l) \quad (17\text{-}15)$$

If we add Equations (17-14) and (17-15), then we obtain

$$AgC_7H_5O_2(s) + H_3O^+(aq) \rightleftharpoons Ag^+(aq) + HC_7H_5O_2(aq) + H_2O(l) \quad (17\text{-}16)$$

Thus, the addition of $H_3O^+(aq)$ to an aqueous solution in equilibrium with $AgC_7H_5O_2(s)$ shifts the solubility equilibrium, Equation (17-16), from left to right, thereby increasing the solubility of the silver benzoate. Figure 17-3 shows the solubility of silver benzoate in water as a function of pH. As Figure 17-3 shows, the solubility of a salt of a weak acid increases dramatically with decreasing pH. The variation in the solubility of salts of weak acids with pH is utilized in chemical analysis to separate metal ions, as we see in the next section.

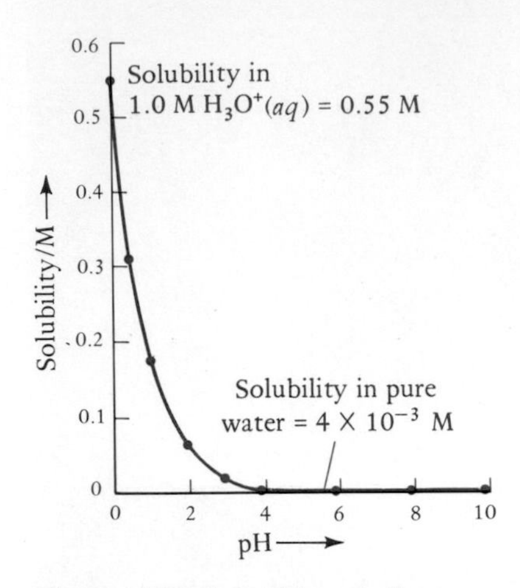

Figure 17-3 Solubility of silver benzoate in water as a function of pH. The addition of $H_3O^+(aq)$ shifts the solubility equilibrium to the right (increase in solubility) as $H_3O^+(aq)$ combines with $C_7H_5O_2^-(aq)$ to produce $HC_7H_5O_2(aq)$ and thereby decreases $[C_7H_5O_2^-]$.

17-6. INSOLUBLE SULFIDES ARE SEPARATED BY ADJUSTMENT OF SOLUTION pH

The solubility product values given in Table 17-1 shows that the solubilities of metal sulfides vary over an enormous range. We can selectively precipitate certain metals by controlling the concentration of $S^{2-}(aq)$. This can be done by controlling the pH because $S^{2-}(aq)$ is the conjugate base of the weak acid HS^-. The relevant equilibrium expressions are

$$H_2S(aq) + H_2O(l) \rightleftharpoons H_3O^+(aq) + HS^-(aq) \qquad K_{a1} = 9.1 \times 10^{-8}\ M$$

$$HS^-(aq) + H_2O(l) \rightleftharpoons H_3O^+(aq) + S^{2-}(aq) \qquad K_{a2} = 1.2 \times 10^{-13}\ M$$

The sum of these two equations is

$$H_2S(aq) + 2H_2O(l) \rightleftharpoons 2H_3O^+(aq) + S^{2-}(aq)$$

and the value of K for this equation is

$$K = K_{a1}K_{a2} = (9.1 \times 10^{-8}\ M)(1.2 \times 10^{-13}\ M) = 1.1 \times 10^{-20}\ M^2$$

$$= \frac{[H_3O^+]^2[S^{2-}]}{[H_2S]} \quad (17\text{-}17)$$

The concentration of $H_2S(aq)$ in a saturated aqueous solution at 25°C is 0.10 M. If we substitute $[H_2S] = 0.10$ M into Equation (17-17) and solve for $[S^{2-}]$, then we obtain

$$[S^{2-}] = (1.1 \times 10^{-20}\ M^2)\left(\frac{0.10\ M}{[H_3O^+]^2}\right)$$

or

$$[S^{2-}] = \frac{1.1 \times 10^{-21}\ M^3}{[H_3O^+]^2} \qquad (17\text{-}18)$$

▪ Equation (17-18) is valid only when $[H_2S] = 0.10$ M, which corresponds to a solution saturated with $H_2S(aq)$ at 25°C and 1 atm.

Equation (17-18) shows that the sulfide-ion concentration in a saturated $H_2S(aq)$ solution can be controlled by the hydrogen-ion concentration or, equivalently, by the pH.

To illustrate the use of Equation (17-18), we calculate the solubility of $ZnS(s)$ and $FeS(s)$ at pH = 2.0. According to Equation (17-18), at pH = 2.0, that is, at $[H_3O^+] = 1.0 \times 10^{-2}$ M,

$$[S^{2-}] = \frac{1.1 \times 10^{-21}\ M^3}{(1.0 \times 10^{-2}\ M)^2} = 1.1 \times 10^{-17}\ M$$

Using the K_{sp} values from Table 17-1, we have that

$$[Zn^{2+}][S^{2-}] = 1.6 \times 10^{-24}\ M^2$$

$$[Fe^{2+}][S^{2-}] = 6.3 \times 10^{-18}\ M^2$$

Therefore, the concentrations of $Zn^{2+}(aq)$ and $Fe^{2+}(aq)$ in a saturated $H_2S(aq)$ solution at pH = 2.0 are

$$[Zn^{2+}] = \frac{1.6 \times 10^{-24}\ M^2}{[S^{2-}]} = \frac{1.6 \times 10^{-24}\ M^2}{1.1 \times 10^{-17}\ M} = 1.5 \times 10^{-7}\ M$$

$$[Fe^{2+}] = \frac{6.3 \times 10^{-18}\ M^2}{[S^{2-}]} = \frac{6.3 \times 10^{-18}\ M^2}{1.1 \times 10^{-17}\ M} = 0.57\ M$$

If the pH is adjusted to 2.0, then, the $Zn^{2+}(aq)$ precipitates out as $ZnS(s)$ and the $Fe^{2+}(aq)$ remains in solution. Thus we achieve a separation of Zn^{2+} from Fe^{2+}.

17-7. SOME METAL CATIONS CAN BE SEPARATED FROM A MIXTURE BY THE FORMATION OF AN INSOLUBLE HYDROXIDE OF ONE OF THEM

It is possible to separate certain metal ions in aqueous solution by adjusting the pH of the solution such that the formation of an insoluble hydroxide of one of the ions is favored. For example, consider the hydroxide of $Zn^{2+}(aq)$, for which

$$Zn(OH)_2(s) \rightleftharpoons Zn^{2+}(aq) + 2OH^-(aq)$$

and

$$K_{sp} = 1.0 \times 10^{-15}\ M^3 = [Zn^{2+}][OH^-]^2$$

The solubility of $Zn(OH)_2(s)$ in water can be calculated from the K_{sp} expression:

$$s = [Zn^{2+}] = \frac{1.0 \times 10^{-15}\ M^3}{[OH^-]^2} \qquad (17\text{-}19)$$

The concentration of $OH^-(aq)$ can be related to $[H_3O^+]$ by using the ion-product-constant expression for water:

Table 17-2 Solubility of $Zn(OH)_2(s)$ in water at 25°C at various pH values

pH	$[H_3O^+]/M$	$[H_3O^+]^2/M^2$	$[Zn^{2+}]/M$
6.5	3.2×10^{-7}	1.0×10^{-13}	1.0
6.8	1.6×10^{-7}	2.5×10^{-14}	0.25
7.0	1.0×10^{-7}	1.0×10^{-14}	0.10
7.5	3.2×10^{-8}	1.0×10^{-15}	0.010
8.0	1.0×10^{-8}	1.0×10^{-16}	0.0010
8.5	3.2×10^{-9}	1.0×10^{-17}	0.00010

$$[OH^-] = \frac{K_w}{[H_3O^+]} = \frac{1.00 \times 10^{-14}\ M^2}{[H_3O^+]} \qquad (17\text{-}20)$$

Substitution of Equation (17-20) into Equation (17-19) yields

$$s = [Zn^{2+}] = \frac{1.0 \times 10^{-15}\ M^3}{(1.00 \times 10^{-14}\ M^2)^2}[H_3O^+]^2 \qquad (17\text{-}21)$$

$$= (1.0 \times 10^{13}\ M^{-1})[H_3O^+]^2$$

From this expression we can compute the solubility of $Zn(OH)_2(s)$, that is, $[Zn^{2+}]$, at various pH values, as shown in Table 17-2. These results are plotted in Figure 17-4, together with the analogous results for the solubility of $Fe(OH)_3(s)$. Note that $Fe^{3+}(aq)$ can be separated from $Zn^{2+}(aq)$ by adjusting the pH of the solution to about 5 with an acetic acid–acetate buffer. At pH ≈ 5, the $Fe(OH)_3(s)$ precipitates and the $Zn^{2+}(aq)$ remains in solution.

Example 17-7: Compute the solubilities of $Zn(OH)_2(s)$ and $Fe(OH)_3(s)$ in an aqueous solution buffered at pH = 6.8.

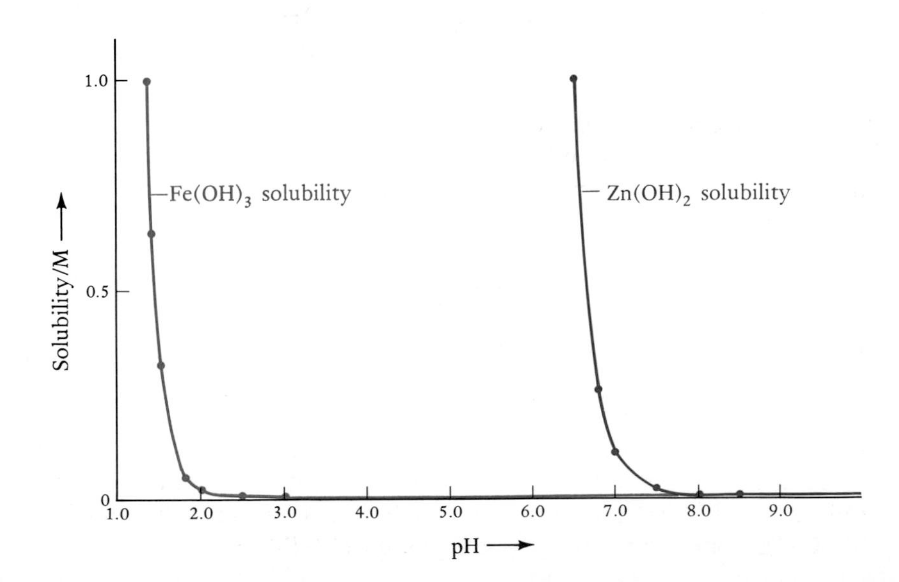

Figure 17-4 Solubilities of $Fe(OH)_3(s)$ and $Zn(OH)_2(s)$ as a function of pH. Note that a much lower solution pH is required to dissolve $Fe(OH)_3(s)$ than to dissolve $Zn(OH)_2(s)$. Therefore, at pH = 4.8, for example, $Fe(OH)_3(s)$ precipitates and $Zn^{2+}(aq)$ remains in solution. The $Fe(OH)_3(s)$ can be filtered off, thereby achieving a separation of $Fe^{3+}(aq)$ from $Zn^{2+}(aq)$.

Solution: The solubility of $Zn(OH)_2(s)$ in a solution buffered at pH = 6.8 can be computed from Equation (17-21):

$$s = [Zn^{2+}] = (1.0 \times 10^{13}\ M^{-1})[H_3O^+]^2$$

At pH = 6.8, we have $[H_3O^+] = 10^{-pH} = 10^{-6.8} = 1.58 \times 10^{-7}$ M, and thus

$$s = (1.0 \times 10^{13}\ M^{-1})(1.58 \times 10^{-7}\ M)^2 = 0.25\ M$$

The solubility of $Fe(OH)_3(s)$ ($K_{sp} = 1.0 \times 10^{-38}\ M^4$) (Table 17-1) is computed in a manner analogous to that just described for $Zn(OH)_2(s)$. Thus, we write

$$Fe(OH)_3(s) \rightleftharpoons Fe^{3+}(aq) + 3OH^-(aq)$$

and $$K_{sp} = [Fe^{3+}][OH^-]^3 = 1.0 \times 10^{-38}\ M^4$$

The solubility of $Fe(OH)_3(s)$ is equal to $Fe^{3+}(aq)$, and so

$$s = [Fe^{3+}] = \frac{1.0 \times 10^{-38}\ M^4}{[OH^-]^3} = \frac{1.0 \times 10^{-38}\ M^4[H_3O^+]^3}{K_w^3}$$

The final form here is obtained by using the relation $[OH^-] = K_w/[H_3O^+]$. Using the fact that $K_w = 1.0 \times 10^{-14}\ M^2$, we have

$$s = [Fe^{3+}] = \frac{1.0 \times 10^{-38}\ M^4[H_3O^+]^3}{(1.0 \times 10^{-14}\ M^2)^3} = (1.0 \times 10^4\ M^{-2})[H_3O^+]^3$$

At $[H_3O^+] = 1.58 \times 10^{-7}$ M, we calculate

$$s = [Fe^{3+}] = (1.0 \times 10^4\ M^{-2})(1.58 \times 10^{-7}\ M)^3 = 3.9 \times 10^{-17}\ M$$

At pH = 6.8, $Fe(OH)_3(s)$ is insoluble and $Zn(OH)_2$ is soluble (Figure 17-4).

17-8. AMPHOTERIC HYDROXIDES DISSOLVE IN BOTH HIGHLY ACIDIC AND HIGHLY BASIC SOLUTIONS

The separation of metal ions in aqueous solution by the formation of insoluble hydroxides is another example of the need to consider all possible competing reactions and their equilibrium expressions.

At high concentrations of $OH^-(aq)$, some hydroxides dissolve as the result of the formation of soluble hydroxy complex ions. For example,

$$Al(OH)_3(s) + OH^-(aq) \rightleftharpoons \underset{\text{tetrahydroxoaluminate ion}}{Al(OH)_4^-(aq)} \qquad (17\text{-}22)$$

Such hydroxides are said to be **amphoteric.** An **amphoteric metal hydroxide** is one that is soluble in both strong acids and strong bases but insoluble in neutral solutions. The equilibrium-constant expression at 25°C for Equation (17-22) is

$$K = \frac{[Al(OH)_4^-]}{[OH^-]} = 40 \qquad (17\text{-}23)$$

We can ignore the $Al(OH)_3(s)$ that dissolves in the reaction

$$Al(OH)_3(s) \rightleftharpoons Al^{3+}(aq) + 3OH^-(aq) \qquad (17\text{-}24)$$

because the value of K for Equation (17-22) is much greater than the value of K_{sp} for $Al(OH)_3(s)$ ($K_{sp} = 1.3 \times 10^{-33}\ M^3$). Therefore, the solubility of $Al(OH)_3(s)$ in a basic solution is essentially equal to $[Al(OH)_4^-]$, and so we can write Equation (17-23) as

$$K = \frac{s}{[OH^-]} = 40$$

or

$$s = 40[OH^-]$$

If $[OH^-] = 0.10$ M, then

$$s = 40(0.10 \text{ M}) = 4.0 \text{ M}$$

For Equation (17-24) we have

$$K_{sp} = 1.3 \times 10^{-33} \text{ M}^4 = [Al^{3+}][OH^-]^3$$

If Equation (17-22) did not occur, then the solubility of $Al(OH)_3(s)$ in a solution with $[OH^-] = 0.10$ M would be

$$s = [Al^{3+}] = \frac{1.3 \times 10^{-33} \text{ M}^4}{[OH^-]^3} = \frac{1.3 \times 10^{-33} \text{ M}^4}{(0.10 \text{ M})^3} = 1.3 \times 10^{-30} \text{ M}$$

which is very small indeed. If we computed the solubility of $Al(OH)_3(s)$ in 0.10 M $OH^-(aq)$ without considering Equation (17-22), then our result would be in error by a factor of 3×10^{30}.

At high pH values the solubility of $Al(OH)_3(s)$ is determined by Equation (17-22). In neutral solution, its solubility is determined by Equation (17-24). At low pH, the $Al(OH)_3(s)$ dissolves readily because of the acid-base neutralization reaction:

$$Al(OH)_3(s) + 3H_3O^+(aq) \rightleftharpoons Al^{3+}(aq) + 6H_2O(l) \qquad (17\text{-}25)$$

Figure 17-5 Aluminum hydroxide occurs as a white, flocculent precipitate that is used to clarify water.

Thus we see that $Al(OH)_3(s)$ dissolves in basic solutions (Equation 17-22) and in acidic solutions (Equation 17-25) but not in neutral solutions (Equation 17-24). The amphoteric behavior of $Al(OH)_3(s)$ is illustrated in Figure 17-6. Some other examples of amphoteric hydroxides are given in Table 17-3.

Example 17-8: Use the equilibrium-constant data for zinc hydroxide in Table 17-3 to estimate its solubility in a solution buffered at pH = 14.0.

Solution: For the equilibrium equation

$$Zn(OH)_2(s) + 2OH^-(aq) \rightleftharpoons Zn(OH)_4^{2-}(aq)$$

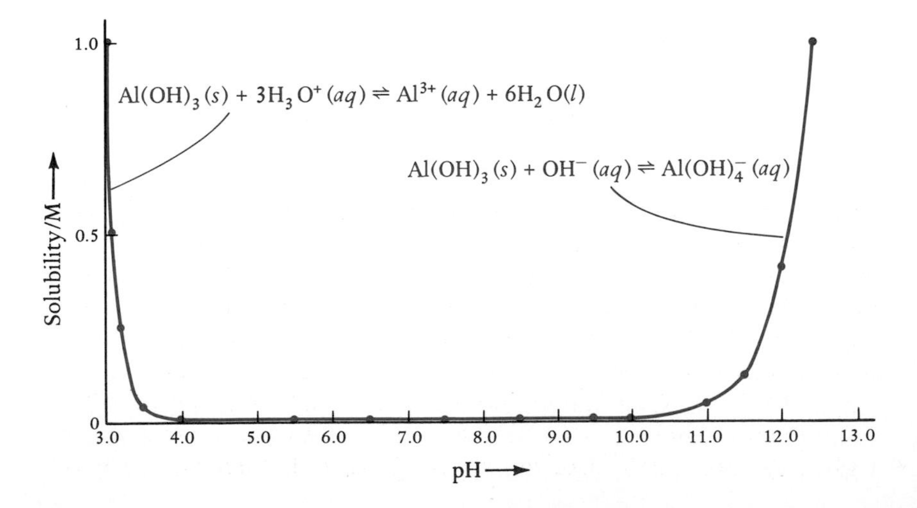

Figure 17-6 The solubility of $Al(OH)_3(s)$ as a function of pH. The amphoteric nature of $Al(OH)_3$ is clearly shown by its solubility in both highly acidic and highly basic solutions. Note that $Al(OH)_3(s)$ is essentially insoluble over the pH range 4 to 10.

Table 17-3 Some amphoteric hydroxides

Reaction	K
$Al(OH)_3(s) + OH^-(aq) \rightleftharpoons Al(OH)_4^-(aq)$	40
$Pb(OH)_2(s) + OH^-(aq) \rightleftharpoons Pb(OH)_3^-(aq)$	0.08
$Zn(OH)_2(s) + 2OH^-(aq) \rightleftharpoons Zn(OH)_4^{2-}(aq)$	$0.05\ M^{-1}$
$Cr(OH)_3(s) + OH^-(aq) \rightleftharpoons Cr(OH)_4^-(aq)$	0.04
$Sn(OH)_2(s) + OH^-(aq) \rightleftharpoons Sn(OH)_3^-(aq)$	0.01

we have

$$K = 0.05\ M^{-1} = \frac{[Zn(OH)_4^{2-}]}{[OH^-]^2} \approx \frac{s}{[OH^-]^2}$$

At pH = 14.0, $[H_3O^+] = 1.0 \times 10^{-14}$ M, and so

$$[OH^-] = \frac{K_w}{[H_3O^+]} = \frac{1.0 \times 10^{-14}\ M^2}{1.0 \times 10^{-14}\ M} = 1.0\ M$$

Substitution of this value for $[OH^-]$ into the expression for K yields for the solubility of $Zn(OH)_2(s)$

$$s = (0.05\ M^{-1})[OH^-]^2 = (0.05\ M^{-1})(1.0\ M)^2 = 0.05\ M$$

At pH = 14.0 the equilibrium

$$Zn(OH)_2(s) \rightleftharpoons Zn^{2+}(aq) + 2OH^-(aq) \qquad K_{sp} = 1.0 \times 10^{-15} M^3$$

makes a negligible contribution to the solubility of zinc hydroxide.

17-9. THE MAGNITUDE OF THE RATIO Q_{sp}/K_{sp} IS USED TO PREDICT WHETHER AN IONIC SOLID WILL PRECIPITATE

Consider the silver bromate solubility equilibrium

$$AgBrO_3(s) \rightleftharpoons Ag^+(aq) + BrO_3^-(aq) \qquad (17\text{-}26)$$

for which

$$K_{sp} = [Ag^+][BrO_3^-]$$

The concentration quotient, Q_{sp}, for Equation (17-26) is given by

$$Q_{sp} = [Ag^+]_0[BrO_3^-]_0 \qquad (17\text{-}27)$$

Recall from Section 14-8 that the distinction between K_{sp} and Q_{sp} is that only equilibrium values of $[Ag^+]$ and $[BrO_3^-]$ can be used in the K_{sp} expression, whereas arbitrary values of $[Ag^+]$ and $[BrO_3^-]$ can be used in the Q_{sp} expression. If we prepare a solution with arbitrary values of $[Ag^+]$ and $[BrO_3^-]$, then the criterion for whether or not precipitation occurs, that is, whether or not $AgBrO_3(s)$ separates out, is $Q_{sp}/K_{sp} > 1$. Recall that if $Q_{sp}/K_{sp} > 1$, then the reaction given by Equation (17-26) is spontaneous from right to left, which leads to the formation of a $AgBrO_3(s)$ precipitate. Conversely, if $Q_{sp}/K_{sp} < 1$, then no precipitate

Table 17-4 Criteria for the formation of a precipitate from a solution prepared with the constituent ions

For any arbitrary ion concentrations:

$\frac{Q_{sp}}{K_{sp}} > 1$	or $Q_{sp} > K_{sp}$	precipitate forms
$\frac{Q_{sp}}{K_{sp}} < 1$	or $Q_{sp} < K_{sp}$	no precipitate forms

When equilibrium is disturbed:

$\frac{Q_{sp}}{K_{sp}} > 1$	or $Q_{sp} > K_{sp}$	more precipitate forms until $Q_{sp} = K_{sp}$
$\frac{Q_{sp}}{K_{sp}} < 1$	or $Q_{sp} < K_{sp}$	precipitate dissolves either until $Q_{sp} = K_{sp}$ or until solid phase disappears completely

forms. If $Q_{sp} > K_{sp}$, then the precipitation continues until $Q_{sp} = K_{sp}$, that is, until equilibrium is established. If the equilibrium given by Equation (17-26) is disturbed in such a way that Q_{sp}/K_{sp} becomes less than unity, then additional $AgBrO_3(s)$ dissolves. These conditions are summarized in Table 17-4.

For example, suppose we mix 50.0 mL of 1.0 M $AgNO_3(aq)$ with 50.0 mL of 0.010 M $NaBrO_3(aq)$ at 25°C. Does $AgBrO_3(s)$ precipitate? The value of K_{sp} for $AgBrO_3(s)$ at 25°C is $5.8 \times 10^{-5}\ M^2$. The initial concentrations of $[Ag^+]$ and $[BrO_3^-]$ in the 100-mL mixture of the two solutions are

$$[Ag^+]_0 = \frac{(1.0\ M)(0.050\ L)}{(0.100\ L)} = 0.50\ M$$

$$[BrO_3^-]_0 = \frac{(0.010\ M)(0.050\ L)}{(0.100\ L)} = 0.0050\ M$$

The initial value of Q_{sp} for the mixture is

$$Q_{sp} = [Ag^+]_0[BrO_3^-]_0 = (0.50\ M)(0.0050\ M) = 2.5 \times 10^{-3}\ M^2$$

and

$$\frac{Q_{sp}}{K_{sp}} = \frac{2.5 \times 10^{-3}\ M^2}{5.8 \times 10^{-5}\ M^2} = 43 > 1$$

The fact that $Q_{sp}/K_{sp} > 1$ means that the precipitation of $AgBrO_3(s)$ from the mixture is a spontaneous process. Once started, the precipitation continues until $Q_{sp}/K_{sp} = 1$, that is, until equilibrium is attained.

Example 17-9: A 1.0×10^{-3} M $NaIO_3(aq)$ solution is made 0.010 M in $Cu^{2+}(aq)$ by dissolving the soluble salt $Cu(ClO_4)_2(s)$. Does $Cu(IO_3)_2(s)$ ($K_{sp} = 7.4 \times 10^{-8}\ M^3$) precipitate from the solution?

Solution: The value of Q_{sp} is

$$Q_{sp} = [Cu^{2+}]_0[IO_3^-]_0^2 = (1.0 \times 10^{-2}\ M)(1.0 \times 10^{-3}\ M)^2 = 1.0 \times 10^{-8}\ M^3$$

and

$$\frac{Q_{sp}}{K_{sp}} = \frac{1.0 \times 10^{-8}\ M^3}{7.4 \times 10^{-8}\ M^3} = 0.14 < 1$$

Hence no precipitate of $Cu(IO_3)_2(s)$ forms.

17-10. QUALITATIVE ANALYSIS IS THE IDENTIFICATION OF THE SPECIES PRESENT IN A SAMPLE

The laboratory work in most introductory chemistry courses involves **qualitative analysis.** The objective of a qualitative analysis scheme is the identification of various cations and anions in a solution or in a mixture of solids. In qualitative analysis we seek only to identify the ions, not to determine their concentration. The determination of the amounts present or the percentage compositions is called **quantitative analysis.**

The sample to be analyzed is called the unknown, but you begin your analysis knowing the list of *possible* cations and anions in the unknown. You also know, on the basis of the results of a set of preliminary experiments, what tests to perform for the presence or absence of the various possible species. Qualitative analysis schemes are included in laboratory programs because they illustrate the application of numerous chemical principles, such as solubility-product equilibria and buffer action. Furthermore, most students enjoy the challenge of identifying the ions in their unknowns.

The methods of qualitative analysis developed in general chemistry laboratories are not, in most cases, the same methods used by professional analytical chemists to identify the constituents of an unknown sample. An analytical chemist working in, say, a criminal investigations laboratory faces much more difficult challenges to his or her chemical ingenuity than those encountered in a general chemistry laboratory (Figure 17-7). The major differences are that, in the "real world," the

Figure 17-7 Although chemists now often use a variety of sophisticated electronic instruments in performing chemical analyses, they still employ "wet lab" chemical techniques in certain cases, as shown here at the Smithsonian Institution.

possible constituents of an unknown are essentially unlimited and often truly unknown. Furthermore, in many cases the available sample is very small and irreplaceable. Consequently, the practicing analytical chemist depends on techniques that are much more sophisticated than those you will use in your qualitative analysis experiments. Nonetheless, all the chemical principles you will use in your experiments are used by analytical chemists.

The basic approach used in the qualitative analysis of a mixture of cations is the separation of some cations from the mixture by addition of a reagent that precipitates certain cations but not others. In this way the analytical problem is simplified because the number of cations involved in each of the two resulting samples (precipitate and remaining solution) is smaller than in the original sample. It is essential in qualitative analysis to know the solubilities of the various possible salts that may be formed as a result of the addition of a precipitating agent to the solution.

As a simple example, consider a solution that may contain $KNO_3(aq)$ and/or $AgNO_3(aq)$. Because AgCl is insoluble and KCl is soluble, the formation of a white precipitate upon addition of 6 M $HCl(aq)$ indicates that $Ag^+(aq)$ is present:

$$Ag^+(aq) + Cl^-(aq) \rightarrow AgCl(s)$$

$$K^+(aq) + Cl^-(aq) \rightarrow \text{no reaction}$$

Addition of excess $HCl(aq)$ precipitates essentially all the $Ag^+(aq)$ as $AgCl(s)$. The settling of the $AgCl(s)$ is hastened by using a centrifuge, which spins the sample and thereby accelerates its settling. The resulting **supernatant** solution is then decanted (poured off) and tested separately for the presence of $K^+(aq)$. Water-insoluble silver chloride is soluble in 6 M $NH_3(aq)$ because of the formation of the $Ag(NH_3)_2^+(aq)$ ion:

$$AgCl(s) + 2NH_3(aq) \rightarrow Ag(NH_3)_2^+(aq) + Cl^-(aq)$$

The solubilization of the white $AgCl(s)$ in the $NH_3(aq)$ is a confirmatory test for $Ag^+(aq)$.

Almost all potassium salts are water-soluble. An exception is $K_3Co(NO_2)_6(s)$, potassium hexanitrocobaltate(III). Addition of $Na_3Co(NO_2)_6(aq)$, which is water-soluble, to a solution containing $K^+(aq)$ produces a pale yellow precipitate:

$$3K^+(aq) + Co(NO_2)_6^{3-}(aq) \rightarrow \underset{\text{pale yellow}}{K_3Co(NO_2)_6(s)}$$

▪ You will learn how to name compounds such as potassium hexanitritocobaltate(III) in Chapter 22.

Silver ion, as well as most cations other than sodium, also forms an insoluble salt with $Co(NO_2)_6^{3-}(aq)$:

$$3Ag^+(aq) + Co(NO_2)_6^{3-}(aq) \rightarrow Ag_3Co(NO_2)_6(s)$$

but recall that $Ag^+(aq)$ was removed from the unknown by precipitation as $AgCl(s)$.

Many of the common metal sulfides are colored, insoluble salts, and the different colors are useful in identifying cations. Metal sulfides are precipitated by adding H_2S to the solution. Many laboratories do not use H_2S directly because hydrogen sulfide is a foul smelling poisonous gas. In order to minimize H_2S exposure in the laboratory, H_2S is prepared in solution by the decomposition of thioacetamide on gentle heating in a laboratory hood:

$$CH_3\underset{\underset{\cdot\ddot{S}\cdot}{\|}}{C}NH_2(aq) + 2H_2O(l) \xrightarrow{60^\circ C} C_2H_3O_2^-(aq) + NH_4^+(aq) + H_2S(aq)$$

thioacetamide

Selective precipitation of metal sulfides is achieved by adjusting the value of $[H_3O^+]$ in the solution, because the value of $[H_3O^+]$ controls the value of $[S^{2-}]$ at a fixed value of $[H_2S]$ (Equation 17-18). The value of $[S^{2-}]$, in turn, controls the solubility of metal sulfides through the solubility equilibria (see Section 17-6).

An essential feature of a qualitative analysis scheme for a large group of ions is the *successive* removal of subgroups of the ions by precipitation reactions. It is essential to carry out the separation steps in a *systematic* fashion; otherwise ions that are presumed to have been removed may interfere with subsequent steps in the analytical scheme. If you are in doubt about how the test results look with certain mixtures of ions, then you should prepare a known mixture containing the ions in question and run the appropriate tests on the known for comparison to the unknown.

The keys to an accurate analysis of your unknown are careful observation and a thorough knowledge of the chemistry of the qualitative analysis scheme, such as solubilities at various pH values, possible complexation reactions, and so on. With a reasonable effort on your part to master the relevant chemistry, you will find your experience as a chemical sleuth both fun and informative.

SUMMARY

The solubility equilibrium between a salt and a solution saturated with its ions is characterized by a solubility-product constant, K_{sp}. The algebraic form of K_{sp} is given by the law of concentration action. The common-ion effect is the suppression of the solubility of a salt by the addition of one of its ions to the solution in contact with the salt. The ratio Q_{sp}/K_{sp} can be used to predict whether a salt dissolves in or precipitates out of a solution.

Slightly soluble salts of weak acids become more soluble as the pH of the solution is lowered. The increase in solubility is a consequence of the protonation of the conjugate base to form the weak acid; this protonation reaction shifts the solubility equilibrium.

The dependence of the solubility of salts of weak acids on pH, especially hydroxides and sulfides, can be used to achieve separations of various metal ions by successive adjustments of pH and addition of precipitating agents. Amphoteric hydroxides are soluble in both acidic and basic solutions; the solubility of amphoteric hydroxides in strong bases is a consequence of the formation of a soluble hydroxy ion of the metal ion.

Ions can be identified by their characteristic chemical reactions and the colors of their salts.

TERMS YOU SHOULD KNOW

solubility 543
solubility rules 543
solubility-product constant, K_{sp} 544
common ion 550
common-ion effect 550
complex ion 550

EQUATIONS YOU SHOULD KNOW HOW TO USE

For the salt A_xB_y:

$$K_{sp} = [A]^x[B]^y$$

$$s = \frac{[A]}{x} = \frac{[B]}{y}$$

$$[S^{2-}] = \frac{1.1 \times 10^{-21}\ M^3}{[H_3O^+]^2}$$

solution saturated with H_2S at 1 atm and 25°C (17-18)

For any arbitrary ion concentrations:

$\frac{Q_{sp}}{K_{sp}} > 1$ precipitate forms

$\frac{Q_{sp}}{K_{sp}} < 1$ no precipitate forms

When equilibrium is disturbed:

$\frac{Q_{sp}}{K_{sp}} > 1$ more precipitate forms until $Q_{sp} = K_{sp}$

$\frac{Q_{sp}}{K_{sp}} < 1$ precipitate dissolves either until $Q_{sp} = K_{sp}$ or until solid phase disappears completely

---

PROBLEMS

SOLUBILITY RULES

17-1. Use the solubility rules to predict whether the following compounds are soluble or insoluble in water:

(a) AgI (b) $Pb(ClO_4)_2$
(c) NH_4Br (d) K_2SO_4
(e) $SrCO_3$

17-2. Use the solubility rules to predict whether the following compounds are soluble or insoluble in water:

(a) Al_2O_3 (b) $CuCl_2$
(c) KNO_3 (d) Hg_2Br_2
(e) $PbCl_2$

17-3. Use the solubility rules to predict whether the following barium salts are soluble or insoluble in water:

(a) $BaCO_3$ (b) $Ba(ClO_4)_2$
(c) $BaCl_2$ (d) BaS
(e) $BaSO_4$

17-4. Use the solubility rules to predict whether the following silver salts are soluble or insoluble in water:

(a) AgBr (b) $AgNO_3$
(c) Ag_2S (d) $AgClO_4$
(e) Ag_2CO_3

17-5. Use the solubility rules to predict the products of the following reactions. In each case, complete and balance the equation and also write the corresponding net ionic equation. If no precipitate forms, then write no reaction.

(a) $CuCl_2(aq) + Na_2S(aq) \rightarrow$
(b) $MgBr_2(aq) + K_2CO_3(aq) \rightarrow$
(c) $BaCl_2(aq) + K_2SO_4(aq) \rightarrow$
(d) $Hg_2(NO_3)_2(aq) + KCl(aq) \rightarrow$

17-6. Use the solubility rules to predict the products of the following reactions. In each case complete and balance the equation and also write the net ionic equation. If no precipitate forms, then write no reaction.

(a) $H_2SO_4(aq) + Ca(ClO_4)_2(aq) \rightarrow$
(b) $AgNO_3(aq) + NaClO_4(aq) \rightarrow$
(c) $Hg_2(NO_3)_2(aq) + NaC_7H_5O_2(aq) \rightarrow$
(d) $Na_2SO_4(aq) + AgF(aq) \rightarrow$

17-7. In each of the following cases, the two solutions indicated are mixed. In each case for which a precipitate forms on mixing, write the complete equation. If no precipitate forms, then write "no reaction." Use the solubility rules and assume that all solutions before mixing are 0.20 M and that equal volumes of the two solutions are mixed.

(a) $Hg_2(ClO_4)_2(aq) + NaBr(aq) \rightarrow$
(b) $Fe(ClO_4)_3(aq) + NaOH(aq) \rightarrow$
(c) $Pb(NO_3)_2(aq) + LiIO_3(aq) \rightarrow$
(d) $H_2SO_4(aq) + Pb(NO_3)_2(aq) \rightarrow$

17-8. In each of the following cases, the two solutions indicated are mixed. In each case for which a precipitate forms on mixing, write the complete equation. If no precipitate forms, then write "no reaction." Use the solubility rules and assume that all solutions before mixing are 0.20 M and that equal volumes of the two solutions are mixed.

(a) $Hg_2(NO_3)_2(aq) + KCl(aq) \rightarrow$
(b) $Zn(ClO_4)_2(aq) + Na_2S(aq) \rightarrow$
(c) $CaCl_2(aq) + Na_2CO_3(aq) \rightarrow$
(d) $Cu(ClO_4)_2(aq) + LiOH(aq) \rightarrow$

17-9. Predict which of the following compounds are soluble in water:

(a) $(NH_4)_2CO_3$ (b) $Ag_2C_2O_4$
(c) $PbSO_4$ (d) CuO

17-10. Predict which of the following compounds are soluble in water:

(a) K_2CO_3 (b) $SnSO_4$
(c) $CaCl_2$ (d) ZnS

17-11. Which of the following compounds are soluble in water?

(a) $Zn(ClO_4)_2(s)$ (b) $AgBrO_3(s)$
(c) $CdSO_4(s)$ (d) $Fe(OH)_2(s)$
(e) $Mn(NO_3)_2(s)$

17-12. Which of the following compounds are soluble in water?

(a) $PbSO_4(s)$ (b) $AgNO_2(s)$
(c) $Cu(C_2H_3O_2)_2(s)$ (d) $NiI_2(s)$
(e) $Fe(NO_3)_3(s)$

K_{sp} CALCULATIONS

17-13. The K_{sp} for $PbCrO_4(s)$ in equilibrium with water at 25°C is 2.8×10^{-13} M^2. Write the chemical equation that represents the solubility equilibrium for $PbCrO_4(s)$ and calculate its solubility in water at 25°C.

17-14. The K_{sp} for $TlCl(s)$ in equilibrium with water at 25°C is 1.7×10^{-4} M^2. Write the chemical equation that represents the solubility equilibrium for $TlCl(s)$ and calculate its solubility in water at 25°C.

17-15. The solubility of silver bromide in pure water at 18°C is 1.33×10^{-4} $g \cdot L^{-1}$. Calculate the value of K_{sp} for silver bromide at 18°C.

17-16. The solubility of lead(II) iodate in pure water is 2.24×10^{-2} $g \cdot L^{-1}$ at 25°C. Calculate the value of K_{sp} for lead(II) iodate at 25°C.

17-17. The solubility product for $Mg(OH)_2(s)$ in equilibrium with water at 25°C is 1.8×10^{-11} M^3. Calculate the solubility of $Mg(OH)_2(s)$ in water at 25°C.

17-18. The solubility product for $PbBr_2(s)$ in equilibrium with water at 25°C is 4.0×10^{-5} M^3. Calculate the solubility of $PbBr_2(s)$ in water at 25°C.

17-19. Potassium perchlorate is soluble in water to the extent of 0.75 g per 100 mL at 0°C. Calculate the K_{sp} of $KClO_4(aq)$ at 0°C.

17-20. Lithium fluoride dissolves in water to the extent of 0.27 g per 100 mL at 18°C. Estimate its K_{sp} at 18°C.

17-21. The solubility product of zinc hydroxide is 1.0×10^{-15} M^3 at 25°C. Calculate the pH of a saturated $Zn(OH)_2(aq)$ solution at 25°C.

17-22. Given that the pH of a saturated $Ca(OH)_2(aq)$ solution is 12.45, compute the solubility of $Ca(OH)_2(s)$ in water at 25°C.

COMMON-ION EFFECT

17-23. Calculate the solubility (in $g \cdot L^{-1}$) of silver sulfate in a 0.55 M silver nitrate solution at 25°C.

17-24. Calculate the solubility (in $g \cdot L^{-1}$) of barium chromate in a 0.0553 M ammonium chromate solution at 25°C.

17-25. The solubility-product constant for TlCl at 25°C is $K_{sp} = 1.7 \times 10^{-4}$ M^2. Estimate its solubility in an aqueous solution that is 0.25 M in $NaCl(aq)$.

17-26. The solubility-product constant for $PbI_2(s)$ at 25°C is $K_{sp} = 7.1 \times 10^{-9}$ M^3. Compute its solubility in a 0.010 M $Pb(ClO_4)_2(aq)$ solution at 25°C.

17-27. Calculate the solubility of $AgI(s)$ in 0.20 M $CaI_2(aq)$ at 25°C.

17-28. Calculate the solubility of $CaSO_4(s)$ in 0.25 M $Na_2SO_4(aq)$ at 25°C.

17-29. The equilibrium constant for the equation

$$AgCl(s) + 2S_2O_3^{2-}(aq) \rightleftharpoons Ag(S_2O_3)_2^{3-}(aq) + Cl^-(aq)$$

is 5.20×10^3 at 25°C. Calculate the solubility of $AgCl(s)$ in a solution whose *equilibrium* concentration of $S_2O_3^{2-}(aq)$ is 0.010 M.

17-30. Copper(I) ions in aqueous solution react with NH_3 according to

$$Cu^+(aq) + 2NH_3(aq) \rightleftharpoons Cu(NH_3)_2^+(aq) \qquad K = 6.3 \times 10^{10}\ M^{-2}$$

Calculate the solubility of $CuBr(s)$ in a solution in which the equilibrium concentration of $NH_3(aq)$ is 0.15 M.

17-31. Consider the chemical equilibrium

$$AgBr(s) + 2S_2O_3^{2-}(aq) \rightleftharpoons Ag(S_2O_3)_2^{3-}(aq) + Br^-(aq)$$

Predict whether the solubility of $AgBr(s)$ is increased, decreased, or unchanged by

(a) an increase in the concentration of $Na_2S_2O_3(aq)$
(b) a decrease in the amount of $AgBr(s)$
(c) dissolution of $NaBr(s)$
(d) dissolution of $NaNO_3(s)$

17-32. Consider the chemical equilibrium

$$PbI_2(s) + 3OH^-(aq) \rightleftharpoons Pb(OH)_3^-(aq) + 2I^-(aq)$$

Predict whether the solubility of $PbI_2(s)$ is increased, decreased, or unaffected by

(a) an increase in the concentration of $OH^-(aq)$
(b) a decrease in the amount of $PbI_2(s)$
(c) a decrease in the concentration of $I^-(aq)$

SOLUBILITY AND pH

17-33. Lead fluoride is slightly soluble in water. Predict the effect on its solubility when

(a) the pH of the solution is decreased to 3
(b) $Pb(NO_3)_2(s)$ is added to the solution

17-34. Magnesium oxalate is sparingly soluble in water. Predict the effect on its solubility when

(a) the solution is made more acidic
(b) the solution is made more basic
(c) $Mg(NO_3)_2(s)$ is added to the solution

17-35. Indicate for which of the following compounds the solubility increases as the pH of the solution is lowered:

(a) $CaCO_3(s)$ (b) $CaF_2(s)$
(c) $PbSO_3(s)$ (d) $KClO_4(s)$
(e) $Fe(OH)_3(s)$ (f) $ZnS(s)$

17-36. Indicate for which of the following compounds the solubility increases as the pH of the solution is lowered:

(a) $FeS(s)$ (b) $ZnCO_3(s)$
(c) $PbCrO_4(s)$ (d) $Hg_2I_2(s)$
(e) $Ag_2C_2O_4(s)$ (f) $Ag_2O(s)$

17-37. Calculate the solubility of $Mg(OH)_2(s)$ in an aqueous solution buffered at pH = 8.5.

17-38. Calculate the solubility of $AgC_7H_5O_2(s)$ in an aqueous solution buffered at pH = 4.0. Take $K_a = 6.5 \times 10^{-5}$ M for $HC_7H_5O_2(aq)$ and $K_{sp} = 2.5 \times 10^{-5}\ M^2$ for $AgC_7H_5O_2(s)$.

17-39. Calculate the solubility of $Cu(OH)_2(s)$ in an aqueous solution buffered at pH = 7.0.

17-40. Calculate the solubility of $Cd(OH)_2(s)$ in an aqueous solution buffered at pH = 9.0.

17-41. Use Le Châtelier's principle to predict the effect on the solubility of

(a) $ZnS(s)$ when $HNO_3(aq)$ is added to a saturated $ZnS(aq)$ solution
(b) $AgI(s)$ when $NH_3(g)$ is added to a saturated $AgI(aq)$ solution

17-42. For the equilibrium

$$ZnS(s) + 2H_3O^+(aq) \rightleftharpoons Zn^{2+}(aq) + H_2S(aq) + 2H_2O(l)$$

predict the direction of shift in response to each of the following changes in conditions. (If the equilibrium is unaffected by the change, then write "no change.")

(a) bubbling in $HCl(g)$
(b) diluting the solution
(c) increasing the pH of the solution

SEPARATION OF CATIONS AS HYDROXIDES AND SULFIDES

17-43. Calculate the solubility of $Cr(OH)_3(s)$ and $Ni(OH)_2(s)$ in an aqueous solution buffered at pH = 5.0. Can $Cr(OH)_3$ be separated from $Ni(OH)_2$ at this pH?

17-44. Calculate the solubility of $Cu(OH)_2(s)$ and $Zn(OH)_2(s)$ in an aqueous solution buffered at pH = 4.0. Can $Cu(OH)_2$ be separated from $Zn(OH)_2$ at this pH?

17-45. Calculate the solubility of $CuS(s)$ in a solution buffered at pH = 2.0 and saturated with hydrogen sulfide so that $[H_2S]$ = 0.10 M.

17-46. Calculate the solubility of $SnS(s)$ in a solution buffered at pH = 2.0 and saturated with hydrogen sulfide so that $[H_2S]$ = 0.10 M.

17-47. What must the pH of a buffered solution saturated with H_2S ($[H_2S]$ = 0.10 M) be in order to precipitate PbS leaving $[Pb^{2+}] = 1 \times 10^{-6}$ M, without precipitating any MnS? The original solution is 0.025 M in both $Pb^{2+}(aq)$ and $Mn^{2+}(aq)$.

17-48. Iron(II) sulfide is used as the pigment in black paint. A sample of $FeS(s)$ is suspected of containing lead(II) sulfide, which can cause lead poisoning if ingested. Suggest a scheme based on pH for separating FeS from PbS.

17-49. Use the equilibrium-constant data in Table 17-3 to estimate the solubility of tin(II) hydroxide in a solution buffered at pH = 13.0.

17-50. Use the equilibrium-constant data in Table 17-3 to estimate the solubility of lead(II) hydroxide in a solution buffered at pH = 13.0.

17-51. The equilibrium constant for the equation

$$Al(OH)_3(s) + OH^-(aq) \rightleftharpoons Al(OH)_4^-(aq)$$

is $K = 40$ at 25°C. Compute the solubility of $Al(OH)_3(s)$ in a solution buffered at pH = 12.0 at 25°C.

17-52. The equilibrium constant for the equation

$$Zn(OH)_2(s) + 2OH^-(aq) \rightleftharpoons Zn(OH)_4^{2-}(aq)$$

is $K = 0.050\ M^{-1}$. Compute the solubility of $Zn(OH)_2(s)$ in a 0.10 M NaOH(*aq*) solution.

Q_{sp} CALCULATIONS

17-53. A 100-mL sample of water from a salt lake has a chloride ion concentration of 0.25 M. To the sample is added 5.0 mL of a 0.10 M aqueous solution of $AgNO_3$. Does AgCl(*s*) precipitate from solution?

17-54. Suppose we mix 50.0 mL of 0.20 M $AgNO_3(aq)$ with 150 mL of 0.10 M $H_2SO_4(aq)$. Does $Ag_2SO_4(s)$ precipitate from the solution?

17-55. If we mix 40.0 mL of 3.00 M $Pb(NO_3)_2(aq)$ with 20.0 mL of 2.00×10^{-3} M NaI(*aq*), does $PbI_2(s)$ precipitate from the solution? If yes, then compute how many moles of $PbI_2(s)$ precipitate and the values of $[Pb^{2+}]$, $[I^-]$, $[NO_3^-]$, and $[Na^+]$ at equilibrium.

17-56. Suppose we mix 50.0 mL of 0.50 M $AgNO_3(aq)$ with 50.0 mL of 1.00×10^{-4} M NaBr(*aq*). Does AgBr(*s*) precipitate from the solution? If yes, then compute how many moles of AgBr(*s*) precipitate and the values of $[Ag^+]$, $[Br^-]$, $[Na^+]$, and $[NO_3^-]$ at equilibrium.

17-57. Suppose we mix 100 mL of a 2.00 M NaCl(*aq*) solution with 100 mL of a 0.020 M $AgNO_3(aq)$ solution. Determine

(a) the number of grams of AgCl(*s*) that precipitate from the solution
(b) the concentration of $Ag^+(aq)$ at equilibrium following the precipitation of AgCl(*s*)

17-58. Suppose that 10.0 mL of a 0.30 M $Zn(NO_3)_2(aq)$ solution is added to 10.0 mL of a 2.00×10^{-4} M $Na_2S(aq)$ solution. Compute

(a) the number of milligrams of ZnS(*s*) that precipitate
(b) the concentrations of $Zn^{2+}(aq)$ and $S^{2-}(aq)$ at equilibrium

ADDITIONAL PROBLEMS

17-59. One treatment for poisoning by soluble lead compounds is to give $MgSO_4(aq)$ or $Na_2SO_4(aq)$ as soon as possible. Explain in chemical terms why this procedure is effective.

17-60. Calculate the solubility of HgS and CdS at pH = 3.0 and 6.0 for aqueous solutions that are saturated with H_2S ($[H_2S] = 0.10$ M).

17-61. It is observed that a precipitate forms when a 2.0 M NaOH(*aq*) solution is added dropwise to a 0.10 M $Pb(NO_3)_2(aq)$ solution and that, on further addition of NaOH(*aq*), the precipitate dissolves. Explain these observations using balanced chemical equations.

17-62. Insoluble $Pb(OH)_2$ and $Sn(OH)_2$ are formed when sodium hydroxide is added to a solution containing $Pb^{2+}(aq)$ and $Sn^{2+}(aq)$. At what pH can $Pb(OH)_2$ be separated from $Sn(OH)_2$? Assume that an effective separation requires a maximum concentration of the less soluble hydroxide of 1×10^{-6} M.

17-63. Oxalic acid and soluble oxalates can cause death if swallowed. The recommended treatment for oxalic acid or oxalate poisoning is to give, as soon as possible, a glassful of limewater (saturated solution of calcium hydroxide) or a 1 percent calcium chloride solution, followed by inducing vomiting several times. Then give 15 to 30 g of Epsom salt ($MgSO_4$) in water and do not induce vomiting. Explain in chemical terms why this procedure is effective.

17-64. A solution 0.30 M in $H_3O^+(aq)$ containing $Mn^{2+}(aq)$, $Cd^{2+}(aq)$, and $Fe^{2+}(aq)$ all at 0.010 M was saturated with $H_2S(g)$ at 25°C. Compute the equilibrium concentrations of $Mn^{2+}(aq)$, $Cd^{2+}(aq)$, and $Fe^{2+}(aq)$. Assume that the solution is continuously saturated with H_2S and that the pH remains constant.

17-65. It is observed that a precipitate forms when 2.0 M KOH(*aq*) solution is added dropwise to a 0.20 M $Zn(ClO_4)_2(aq)$ solution and that, on further addition of KOH(*aq*), the precipitate dissolves. Explain these observations using balanced chemical equations.

17-66. A deposit of limestone is analyzed for its calcium and magnesium content. A sample is dissolved, and then the calcium and magnesium are precipitated as $Ca(OH)_2$ and $Mg(OH)_2$. At what pH can $Ca(OH)_2$ be separated from $Mg(OH)_2$? Assume that an effective separation requires a maximum concentration of the less soluble hydroxide of 1×10^{-6} M.

17-67. A 2.000-g sample of a salt deposit was dissolved in an aqueous solution. A solution of $AgNO_3$ was added to precipitate all the chloride ions as AgCl. The precipitate was filtered, dried, and weighed. The amount of AgCl obtained was 4.188 g. Calculate the mass percentage of chloride ion in the sample.

17-68. Consider the following chemical equation:

$$PbCrO_4(s) + 3OH^-(aq) \rightleftharpoons Pb(OH)_3^-(aq) + CrO_4^{2-}(aq)$$

Predict whether the solubility of $PbCrO_4(s)$ is increased, is decreased or remains unchanged by

(a) a decrease in the concentration of $OH^-(aq)$
(b) an increase in the amount of $PbCrO_4(s)$
(c) an increase in the pH of the solution
(d) dissolution of $Na_2CrO_4(s)$
(e) addition of $H_2O(l)$ to the system
(f) addition of $HClO_4(aq)$

17-69. Given the equation

$$Ag^+(aq) + 2NH_3(aq) \rightleftharpoons Ag(NH_3)_2^+(aq) \qquad K_{comp} = 2.0 \times 10^7\ M^{-2}$$

determine the final concentration of $NH_3(aq)$ that is required to dissolve 250 mg of $AgCl(s)$ in 100 mL of solution.

17-70. Given the following data at 25°C

solubility of $I_2(s)$ in $H_2O(l)$: 0.00132 M
solubility of $I_2(s)$ in 0.1000 M $KI(aq)$: 0.05135 M

compute the equilibrium constants for the following set of equations:

$$I_2(s) \rightleftharpoons I_2(aq)$$
$$I_2(s) + I^-(aq) \rightleftharpoons I_3^-(aq)$$
$$I_2(aq) + I^-(aq) \rightleftharpoons I_3^-(aq)$$

17-71. Use the K_{sp} data in Table 17-1 to calculate the equilibrium constants for the following set of equations:

(1) $Ag_2CrO_4(s) + 2Br^-(aq) \rightleftharpoons 2AgBr(s) + CrO_4^{2-}(aq)$

(2) $PbCO_3(s) + Ca^{2+}(aq) \rightleftharpoons CaCO_3(s) + Pb^{2+}(aq)$

17-72. Compute the pH at which $Ca(OH)_2(s)$ will begin to precipitate from a solution that is 2.0×10^{-2} M in $Ca^{2+}(aq)$ at 25°C.

17-73. Suppose we have a solution containing $Pb^{2+}(aq)$ and $NO_3^-(aq)$. A solution of $NaCl(aq)$ is added slowly until no further precipitation occurs. The precipitate is collected by filtration, dried, and weighed. A total of 12.79 g of $PbCl_2(aq)$ is obtained from 200.0 mL of the original solution. Calculate the mass of $Pb(NO_3)_2$ present and the molarity of the solution.

17-74. Excess $HgI_2(s)$ was equilibrated with a solution of 0.10 M in $KI(aq)$. Calculate the solubility of $HgI_2(s)$ in this solution given

$$HgI_2(s) \rightleftharpoons Hg^{2+}(aq) + 2I^-(aq) \qquad K_{sp} = 2.0 \times 10^{-28}\ M^3$$

$$HgI_2(s) + 2I^-(aq) \rightleftharpoons HgI_4^{2-}(aq) \qquad K_{comp} = 0.79\ M^{-1}$$

OXIDATION-REDUCTION REACTIONS

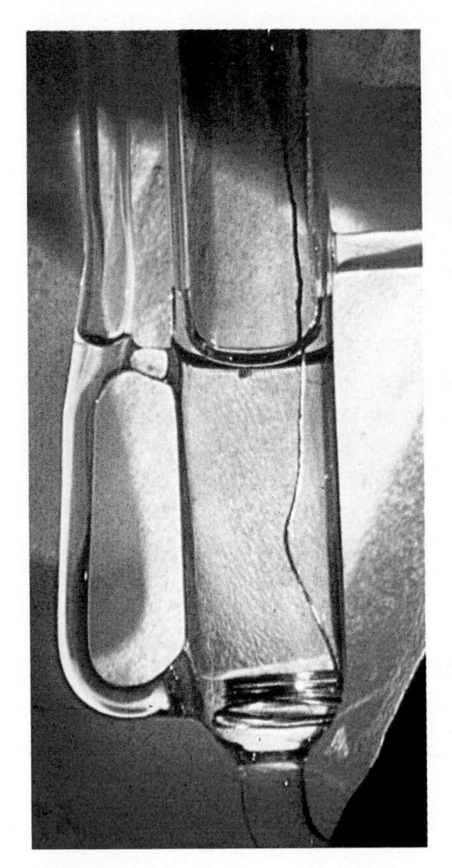

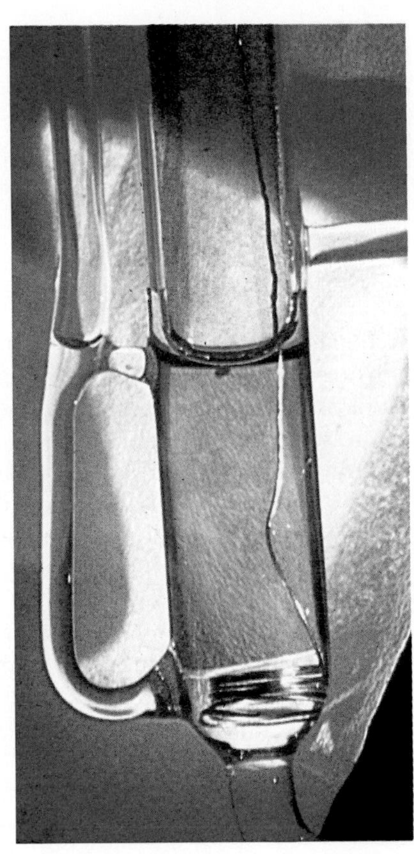

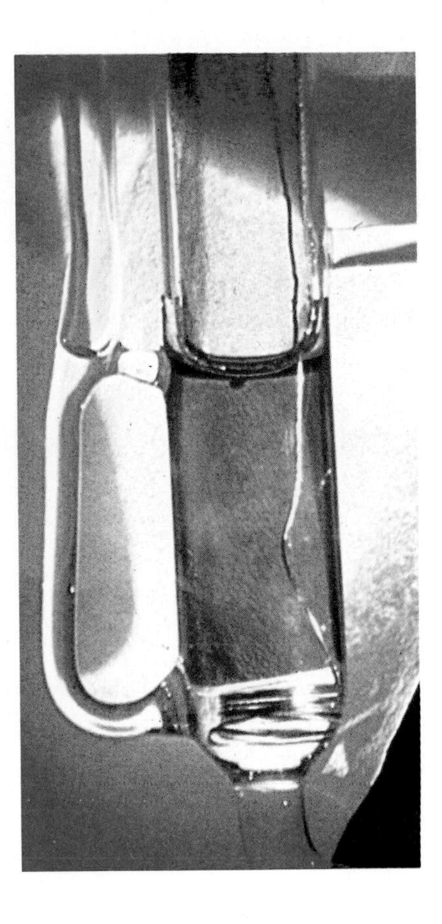

Photographs of an electrochemically active polymer coated on an optically transparent electrode (SnO_2). The color in the polymer is due to the ion RuL_3^{2+}, where L is a complex organic group. As the electrode potential is made progressively more negative, the oxidation state of the ruthenium complex in the film changes and so does its color. The colors and formal oxidation states of the ruthenium in the complex are as follows: pale orange, +2; blue, 0; and cherry red, −4.

All chemical reactions can be assigned to one of two classes: reactions in which electrons are transferred from one reactant to another and reactions in which electrons are not transferred. Reactions in which electrons are transferred from one reactant to another are called oxidation-reduction reactions or electron-transfer reactions. Most of the reactions that we have considered prior to this chapter (for example, acid-base reactions and precipitation reactions) do not involve electron transfer and thus are not oxidation-reduction reactions.

In this chapter we begin with a discussion of the concept of the oxidation states of elements in chemical species. Oxidation states are determined according to a set of rules and are used to balance equations for oxidation-reduction reactions in a systematic way. Much of this chapter is devoted to balancing equations for oxidation-reduction reactions. The final section of the chapter discusses the use of oxidation-reduction reactions in chemical analyses.

18-1. AN OXIDATION STATE CAN BE ASSIGNED TO EACH ATOM IN A CHEMICAL SPECIES

An **oxidation-reduction reaction,** commonly called a **redox reaction,** is one that involves the transfer of electrons from one chemical species to another. A simple example of an oxidation-reduction reaction is the reaction between zinc metal and a copper(II) salt in aqueous solution:

$$Zn(s) + Cu^{2+}(aq) \rightarrow Zn^{2+}(aq) + Cu(s) \qquad (18\text{-}1)$$

In Equation (18-1), zinc metal converts $Cu^{2+}(aq)$ to copper metal by transferring two electrons from each Zn atom to each $Cu^{2+}(aq)$ ion. The loss of two electrons from a Zn atom yields a $Zn^{2+}(aq)$ ion.

The study of oxidation-reduction reactions can be systematized by assigning an **oxidation state** to each atom in a chemical species according to a set of rules. The rules originate from a consideration of the number of electrons and of the electronegativities of the various elements in a species. In some cases, an assigned oxidation state is the actual charge on an atom, but in general this is not the case. Oxidation states are used to balance equations for oxidation-reduction reactions.

The general procedure for assigning oxidation states to elements in chemical species containing two or more atoms is given by the following set of rules, *which take priority in the order given:*

Rules for Assigning Oxidation States

1. Free elements are assigned an oxidation state of zero.
2. The sum of the oxidation states of all the atoms in a species must be equal to the net charge on the species.
3. The alkali metals (Li, Na, K, Rb, and Cs) in compounds are always assigned an oxidation state of +1.
4. Fluorine in its compounds is always assigned an oxidation state of −1.
5. The alkaline earth metals (Be, Mg, Ca, Sr, Ba, and Ra) and also Zn and Cd in compounds are always assigned an oxidation state of +2.
6. Aluminum and gallium are always assigned an oxidation state of +3 in their compounds.
7. Hydrogen in compounds is assigned an oxidation state of +1.
8. Oxygen in compounds is assigned an oxidation state of −2.

The +1 oxidation state of alkali metals (Group 1 metals) in compounds corresponds to the ionic charge of the alkali metal ions. The +1 state corresponds to the loss of an electron from the outermost *s* orbital in the neutral atoms. The +2 oxidation state of the alkaline

Table 18-1 Comparison of ionic charge and oxidation state for some metal ions

Group	Ionic charge	Oxidation state
alkali metal ions (Li^+, Na^+, K^+, Cs^+, Rb^+)	+1	+1
alkaline earth ions (Be^{2+}, Mg^{2+}, Ca^{2+}, Sr^{2+}, Ba^{2+})	+2	+2
Group 3 ions (Al^{3+}, Ga^{3+})	+3	+3

earth metals (Group 2 metals) corresponds to the ionic charge of the alkaline earth metal ions. The +2 state corresponds to a loss of two electrons from the outermost *s* orbital in the neutral atoms. The ionic charges of the metal ions discussed in Chapter 2 (Figure 2-14) correspond to the oxidation states of those elements.

Example 18-1: Determine the oxidation state of each atom in the following compounds:

(a) CsCl (b) NO_2 (c) $HClO_3$ (d) H_2O_2 (e) NaH

Solution: (a) We assign cesium an oxidation state of +1 (rule 3), and thus chlorine is assigned an oxidation state of −1 (rule 2) because CsCl is a neutral species and $+1 - 1 = 0$.

(b) We assign oxygen an oxidation state of −2 (rule 8). The oxidation state of nitrogen in NO_2, represented by x, is thus (rule 2)

$$x + 2(-2) = 0$$

$$x = +4$$

The oxidation state of nitrogen in NO_2 is +4.

(c) We assign hydrogen an oxidation state of +1 (rule 7) and oxygen an oxidation state of −2 (rule 8). Then the oxidation state x of chlorine is (rule 2)

$$+1 + x + 3(-2) = 0$$

$$x = +5$$

The oxidation state of chlorine in $HClO_3$ is +5.

(d) We assign hydrogen an oxidation state of +1 (rule 7), and the oxidation state x of oxygen is (rule 2)

$$2(+1) + 2x = 0$$

$$x = -1$$

Thus the oxidation state of oxygen in H_2O_2 is −1, which is characteristic of peroxides. This result does not contradict rule 8 because rules 2 and 7 take precedence.

▪ Peroxides involve an oxygen-oxygen single bond.

(e) We assign sodium an oxidation state of +1 (rule 3), and, according to rule 2, the oxidation state of hydrogen in NaH is −1, which is characteristic of hydrides. This result does not violate rule 7 because rules 2 and 3 take precedence.

Example 18-1 involves only neutral molecules, whose net charge must be zero. For ionic species, the sum of the oxidation states for each atom must equal the net charge on the ion.

■ The O_2^- ion is called the super-oxide ion.

Example 18-2: Determine the oxidation state of each atom in the following ions:

(a) CrO_4^{2-} (b) HS^- (c) NH_4^+ (d) O_2^-

Solution: (a) We assign oxygen an oxidation state of −2 (rule 8), and thus the oxidation state x of chromium is (rule 2)

$$x + 4(-2) = -2$$
$$x = +6$$

The oxidation state of chromium in CrO_4^{2-} is +6.

(b) We assign hydrogen an oxidation state of +1 (rule 7), and thus the oxidation state x of sulfur is (rule 2)

$$+1 + x = -1$$
$$x = -2$$

The oxidation state of S in HS^- is −2.

(c) We assign hydrogen an oxidation state of +1 (rule 7), and thus the oxidation state x of nitrogen is (rule 2)

$$x + 4(+1) = +1$$
$$x = -3$$

The oxidation state of N in NH_4^+ is −3.

(d) Using rule 2, we assign oxygen an oxidation state of $-\frac{1}{2}$ (since $2(-\frac{1}{2}) = -1$), which is characteristic of superoxides (Interchapter B).

All the atoms in Examples 18-1 and 18-2 can be assigned oxidation states using the rules listed previously. Although these rules do not cover all cases, they do cover many cases that arise in practice. Examples that are not covered by the rules can usually be solved by analogy with other elements in the periodic table, and by using clues provided by the names of the species involved. For example, consider phosphorus pentachloride, PCl_5. As the name implies, chlorine is more electronegative than phosphorus, and so chlorine is assigned a negative oxidation state. The only negative oxidation state of the halogens is −1 because the addition of one electron to a halogen atom yields a noble gas electron configuration. Thus, we have for PCl_5

$$x + 5(-1) = 0$$

or that the oxidation state of phosphorus in PCl_5 is +5. The following Example gives you more practice.

Example 18-3: Working by analogy with other elements in the periodic table and also using the clues provided by the names, assign oxidation states to each of the elements in the following compounds:

(a) As_2S_5 arsenic pentasulfide
(b) In_2Se_3 indium selenide

Solution: (a) The name arsenic pentasulfide tells us that sulfur is more electronegative than arsenic, and so the sulfur is assigned a negative oxidation state. Arsenic is below nitrogen, and sulfur is below oxygen in the periodic table; thus the analogous compound is N_2O_5. Therefore, we assign sulfur an oxidation state of -2 (analogous to rule 8 for oxygen), and the oxidation state x of arsenic is (rule 2)

$$2x + 5(-2) = 0$$

$$x = +5$$

(b) Note that Se is below S in the periodic table and recall that the oxidation state of sulfur in sulfides is -2. By analogy with sulfur we assign an oxidation state of -2 to selenium.

$$2x + 3(-2) = 0$$

$$x = +3$$

Note that indium is below aluminum in the periodic table and aluminum has a characteristic oxidation state of $+3$ in its compounds.

With a little experience (which you can get by doing Problems 18-1 to 18-10) you will become confident in assigning oxidation states.

18-2. OXIDATION-REDUCTION REACTIONS INVOLVE THE TRANSFER OF ELECTRONS FROM ONE REACTANT TO ANOTHER

Consider the oxidation-reduction reaction between zinc metal and $Cu^{2+}(aq)$ ions in aqueous solution:

$$Zn(s) + Cu^{2+}(aq) \rightarrow Cu(s) + Zn^{2+}(aq)$$

The $Cu^{2+}(aq)$ is said to be **reduced** to Cu(s) because the process involves a decrease (reduction) in the oxidation state of copper (from $+2$ to 0):

$$Cu^{2+}(aq) + 2e^- \rightarrow Cu(s) \qquad \textbf{reduction}$$

▪ Reduction is a gain of electrons.

The Zn(s) is said to be **oxidized** because the process involves an increase in the oxidation state of zinc (from 0 to $+2$):

$$Zn(s) \rightarrow Zn^{2+}(aq) + 2e^- \qquad \textbf{oxidation}$$

▪ Oxidation is a loss of electrons.

An essential feature of oxidation-reduction reactions is that in one reactant the oxidation state of an element increases, and in another reactant the oxidation state of an element decreases. Thus oxidation-reduction reactions involve a simultaneous oxidation and reduction. The simultaneous changes in oxidation states in oxidation-reduction reactions are the result of the transfer of electrons from one reactant to another, and thus oxidation-reduction reactions are also called **electron-transfer reactions.**

The reactant that contains the atom that is reduced is called the **oxidizing agent.** The reactant that contains the atom that is oxidized is called the **reducing agent.** The oxidizing agent accepts electrons from the atom that is oxidized and thus is an **electron acceptor.** The reducing agent donates electrons to the atom that is reduced and therefore is an **electron donor.** In the equation

$$\underset{\substack{\text{reducing agent}\\ \text{(oxidized)}}}{Zn(s)} + \underset{\substack{\text{oxidizing agent}\\ \text{(reduced)}}}{Cu^{2+}(aq)} \rightarrow Zn^{2+}(aq) + Cu(s)$$

the oxidizing agent (electron acceptor) is $Cu^{2+}(aq)$ because it accepts two electrons, and the reducing agent (electron donor) is $Zn(s)$ because it donates two electrons. In this reaction, as in all electron-transfer reactions, all the electrons donated by the reducing agent are accepted by the oxidizing agent.

Example 18-4: In the following chemical equation, identify the atom that is oxidized, the atom that is reduced, the oxidizing agent, and the reducing agent:

$$MnO_2(s) + 4HCl(aq) \rightarrow MnCl_2(aq) + Cl_2(g) + 2H_2O(l)$$

Solution: The oxidation state of Mn is +4 in MnO_2 and +2 in $MnCl_2$. The oxidation state of Cl is −1 in HCl and 0 in Cl_2. Therefore, Cl is oxidized and Mn is reduced in this reaction. The reactant that contains the atom that is reduced is MnO_2, and so MnO_2 is the oxidizing agent. The reactant that contains the atom that is oxidized is HCl, and so HCl is the reducing agent. Notice that two electrons are transferred in this reaction: one manganese atom accepts two electrons and each of two chlorine atoms donates one electron.

Table 18-2 Summary of oxidation-reduction reactions

The reducing agent:
contains the atom that is oxidized
contains the atom whose oxidation state increases
is the electron donor
The oxidizing agent:
contains the atom that is reduced
contains the atom whose oxidation state decreases
is the electron acceptor

Table 18-2 summarizes what we have learned so far about oxidation-reduction reactions.

18-3. ELECTRON-TRANSFER REACTIONS CAN BE SEPARATED INTO TWO HALF-REACTIONS

The electron-transfer reaction described by the equation

$$Zn(s) + Cu^{2+}(aq) \rightarrow Zn^{2+}(aq) + Cu(s) \quad (1)$$

can be separated into two **half-reactions** represented by the equations

$$Zn(s) \rightarrow Zn^{2+}(aq) + 2e^- \quad (2)$$

$$Cu^{2+}(aq) + 2e^- \rightarrow Cu(s) \quad (3)$$

We obtain Equation (1) if we add Equations (2) and (3). The half-reaction given by the equation in which electrons appear on the right-hand side (Equation 2) is called the **oxidation half-reaction** (recall that oxidation is a *loss* of electrons). The half-reaction given by the equation in which electrons appear on the left-hand side (Equation 3) is called the **reduction half-reaction** (recall that reduction is a *gain* of electrons).

The oxidation half-reaction supplies electrons to the reduction half-reaction.

Example 18-5: Identify the oxidizing and reducing agents and the oxidation and reduction half-reactions in the reaction described by the equation

$$Tl^{+}(aq) + 2Ce^{4+}(aq) \rightarrow 2Ce^{3+}(aq) + Tl^{3+}(aq)$$

Solution: The oxidation state of thallium increases from +1 in Tl^{+} to +3 in Tl^{3+}. Thus Tl^{+} is oxidized and acts as the reducing agent (electron donor). The oxidation state of cerium decreases from +4 in Ce^{4+} to +3 in Ce^{3+}, thus Ce^{4+} is reduced and acts as the oxidizing agent (electron acceptor).

We identify the two half-reactions by writing the equations for the oxidation and the reduction reactions separately:

$$Tl^{+}(aq) \rightarrow Tl^{3+}(aq) + 2e^{-} \qquad \text{oxidation half-reaction}$$

$$Ce^{4+}(aq) + e^{-} \rightarrow Ce^{3+}(aq) \qquad \text{reduction half-reaction}$$

Note that Tl^{+} is a two-electron reducing agent whereas Ce^{4+} is a one-electron oxidizing agent. Thus it requires 2 mol of Ce^{4+} to oxidize 1 mol of Tl^{+}. The number of electrons involved in the two half-reaction equations can be balanced by multiplying the equation for the cerium half-reaction by 2.

18-4. EQUATIONS FOR OXIDATION-REDUCTION REACTIONS CAN BE BALANCED BY BALANCING EACH HALF-REACTION SEPARATELY

Consider the equation for the reaction between iron metal and aqueous chlorine:

$$Fe(s) + Cl_2(aq) \rightarrow Fe^{3+}(aq) + Cl^{-}(aq)$$

This equation as it stands is not balanced. If we write

$$Fe(s) + Cl_2(aq) \rightarrow Fe^{3+}(aq) + 2Cl^{-}(aq)$$

then the equation is balanced with respect to the elements but not with respect to charge. The net charge on the left-hand side is zero, whereas the net charge on the right-hand side is $+3 + 2(-1) = +1$. The balanced equation is

$$2Fe(s) + 3Cl_2(aq) \rightarrow 2Fe^{3+}(aq) + 6Cl^{-}(aq)$$

The equation is now balanced both with respect to the elements and with respect to the charge; that is, the net charge is the same on both sides.

The balancing of equations for electron-transfer reactions must be done systematically because attempting to balance such equations by guessing the balancing coefficients can be a time-consuming and frustrating experience. The systematic procedure that we use is called the **method of half-reactions.** This method can be used to balance even the most complicated equation in a straightforward and systematic way. We illustrate the method of half-reactions by balancing the following equation:

$$Fe^{2+}(aq) + Cr_2O_7^{2-}(aq) \xrightarrow{H^{+}(aq)} Fe^{3+}(aq) + Cr^{3+}(aq) \qquad (18\text{-}2)$$

Note that the reaction occurs in acidic solution.

I. *Separate the equation into two equations representing the oxidation half-reaction and the reduction half-reaction.*

The oxidation state of iron changes from +2 to +3, and the oxidation state of chromium changes from +6 (in $Cr_2O_7^{2-}$) to +3. Thus the two half-reactions are

$$Fe^{2+} \rightarrow Fe^{3+} \qquad \text{oxidation}$$
$$Cr_2O_7^{2-} \rightarrow Cr^{3+} \qquad \text{reduction}$$

II. *Balance the equation for each half-reaction with respect to all elements other than oxygen and hydrogen.*

The equation for the iron half-reaction is already balanced with respect to iron (one Fe on each side); the equation for the chromium half-reaction is balanced with respect to chromium by placing a 2 in front of Cr^{3+}:

$$Fe^{2+} \rightarrow Fe^{3+}$$
$$Cr_2O_7^{2-} \rightarrow 2Cr^{3+}$$

Figure 18-1 Addition of a 0.10M solution of the oxidizing agent potassium dichromate, $K_2Cr_2O_7$ *(aq)* (orange solution), to a solution 0.10M in iron(II) sulfate, $FeSO_4$ *(aq)* (light green), and 0.10M in H_2SO_4 *(aq)*. The iron is oxidized from Fe(II) to Fe(III) and the chromium is reduced from Cr(VI) to Cr(III).

III. *Balance each half-reaction equation with respect to oxygen by adding the appropriate number of H_2O to the side deficient in oxygen.*

Only the chromium half-reaction involves oxygen. There are seven oxygen atoms on the left and none on the right. Therefore we balance the oxygen by adding seven H_2O to the right-hand side of the equation for the chromium half-reaction:

$$Fe^{2+} \rightarrow Fe^{3+}$$
$$Cr_2O_7^{2-} \rightarrow 2Cr^{3+} + 7H_2O$$

IV. *Balance each half-reaction equation with respect to hydrogen by adding the appropriate number of H^+ to the side deficient in hydrogen.*

Only the chromium half-reaction involves hydrogen. There are 14 hydrogens on the right and none on the left. Therefore we balance the hydrogen by adding 14 H^+ to the left-hand side of the equation for the chromium half-reaction:

$$Fe^{2+} \rightarrow Fe^{3+}$$
$$14H^+ + Cr_2O_7^{2-} \rightarrow 2Cr^{3+} + 7H_2O$$

The two half-reaction equations are now balanced with respect to atoms, but they are not balanced with respect to charge.

V. *Balance each half-reaction equation with respect to charge by adding the appropriate number of electrons to the side with the excess positive charge.*

▪ Recall that all the electrons consumed by the oxidizing agent must be supplied by the reducing agent (conservation of electrons).

The equation for the iron half-reaction has a charge of +2 on the left and +3 on the right. Thus we balance the charge by adding one electron to the right-hand side:

$$Fe^{2+} \rightarrow Fe^{3+} + e^- \qquad \text{oxidation}$$

The equation for the chromium half-reaction has a net charge of $+12\,[= 14(+1) + (-2)]$ on the left and $+6[= 2(+3)]$ on the right. Thus we balance the charge by adding six electrons to the left-hand side:

$$14H^+ + Cr_2O_7^{2-} + 6e^- \rightarrow 2Cr^{3+} + 7H_2O \qquad \text{reduction}$$

The two half-reaction equations are now balanced. Note that the iron half-reaction donates electrons (electrons on the right-hand side) and the chromium half-reaction accepts electrons (electrons on the left-hand side).

VI. *Multiply each half-reaction equation by integers that make the number of electrons supplied by the oxidation half-reaction equal to the number of electrons consumed by the reduction half-reaction.*

The iron half-reaction supplies one electron for each Fe^{2+} that is oxidized to Fe^{3+}, and the chromium half-reaction consumes six electrons for each $Cr_2O_7^{2-}$ that is reduced to Cr^{3+}. Therefore, we multiply the equation for the iron half-reaction through by 6:

$$6Fe^{2+} \rightarrow 6Fe^{3+} + 6e^-$$

$$14H^+ + Cr_2O_7^{2-} + 6e^- \rightarrow 2Cr^{3+} + 7H_2O$$

VII. *Obtain the complete balanced equation by adding the two balanced half-reaction equations and canceling any like terms.*

Adding the equations for the two half-reactions and canceling the $6e^-$ that appear on both sides yields

$$6Fe^{2+} \rightarrow 6Fe^{3+} + \cancel{6e^-}$$
$$14H^+ + Cr_2O_7^{2-} + \cancel{6e^-} \rightarrow 2Cr^{3+} + 7H_2O$$
$$\overline{6Fe^{2+} + 14H^+ + Cr_2O_7^{2-} \rightarrow 6Fe^{3+} + 2Cr^{3+} + 7H_2O}$$

Note that the electrons cancel. No electrons ever appear in the complete balanced equation. This fact serves as a nice intermediate check. You should also check your result by making sure that the final equation is balanced with respect to each element and with respect to charge. As a final step, we rewrite the balanced equation with phases indicated:

$$6Fe^{2+}(aq) + 14H^+(aq) + Cr_2O_7^{2-}(aq) \rightarrow 6Fe^{3+}(aq) + 2Cr^{3+}(aq) + 7H_2O(l)$$

Although the method of half-reactions involves numerous steps, it is actually simple to use and with a little practice becomes straightforward.

Example 18-6: Balance the following equation

$$CrO_4^{2-}(aq) + Cl^-(aq) \rightarrow Cr^{3+}(aq) + HClO_2(aq)$$

given that the reaction takes place in acidic solution.

Solution: The oxidation state of Cl changes from −1 (in Cl^-) to +3 (in $HClO_2$) and that of Cr changes from +6 (in CrO_4^{2-}) to +3 (in Cr^{3+}). Thus, the two half-reactions are

$$Cl^- \rightarrow HClO_2 \qquad \text{oxidation}$$

$$CrO_4^{2-} \rightarrow Cr^{3+} \qquad \text{reduction}$$

Let's balance each equation in turn. The oxidation half-reaction equation is balanced with respect to chlorine. To balance it with respect to oxygen, we add $2H_2O$ to the left-hand side:

$$Cl^- + 2H_2O \rightarrow HClO_2$$

We add $3H^+$ to the right-hand side to balance it with respect to hydrogen:

$$Cl^- + 2H_2O \rightarrow HClO_2 + 3H^+$$

and last, we add $4e^-$ to the right-hand side to balance it with respect to charge:

$$Cl^- + 2H_2O \rightarrow HClO_2 + 3H^+ + 4e^- \qquad \text{(oxidation)}$$

The reduction half-reaction equation is already balanced with respect to chromium. To balance it with respect to oxygen, we add $4H_2O$ to the right-hand side:

$$CrO_4^{2-} \rightarrow Cr^{3+} + 4H_2O$$

To balance it with respect to hydrogen, we add $8H^+$ to the left-hand side:

$$CrO_4^{2-} + 8H^+ \rightarrow Cr^{3+} + 4H_2O$$

and we add $3e^-$ to the left-hand side to balance it with respect to charge:

$$CrO_4^{2-} + 8H^+ + 3e^- \rightarrow Cr^{3+} + 4H_2O \qquad \text{(reduction)}$$

The oxidation half-reaction as written supplies four electrons and the reduction half-reaction as written consumes three electrons. If we multiply the oxidation half-reaction equation by 3 and the reduction half-reaction equation by 4, then both half-reaction equations will involve 12 electrons:

$$3Cl^- + 6H_2O \rightarrow 3HClO_2 + 9H^+ + 12e^-$$

$$4CrO_4^{2-} + 32H^+ + 12e^- \rightarrow 4Cr^{3+} + 16H_2O$$

Addition of these two half-reaction equations and cancellation of like terms yields

$$3Cl^- + 4CrO_4^{2-} + 23H^+ \rightarrow 3HClO_2 + 4Cr^{3+} + 10H_2O$$

Finally, we indicate the phases and write

$$3Cl^-(aq) + 4CrO_4^{2-}(aq) + 23H^+(aq) \rightarrow 3HClO_2(aq) + 4Cr^{3+}(aq) + 10H_2O(l)$$

Note that this equation is balanced with respect to each element and with respect to charge.

The following Example illustrates the balancing of an equation for a single half-reaction in acidic solution.

Example 18-7: Write the equation for the half-reaction for H_2O_2 acting as an oxidizing agent in acidic aqueous solution.

Solution: The oxidation state of oxygen in H_2O_2 is -1 (Example 18-1d). Because H_2O_2 is an oxidizing agent, oxygen must be reduced to an oxidation state of -2, which is the lowest possible oxidation state of oxygen, because the addition of two electrons to an oxygen atom yields a noble-gas electron configuration. Thus

$$H_2O_2 \rightarrow H_2O$$

Balancing the oxygen atoms by adding H_2O to the right-hand side yields

$$H_2O_2 \rightarrow 2H_2O$$

and balancing the hydrogen atoms by adding $2H^+$ to the left-hand side yields

$$2H^+ + H_2O_2 \rightarrow 2H_2O$$

We now balance the charge by adding $2e^-$ to the left-hand side and indicate phases to obtain

$$2H^+(aq) + H_2O_2(aq) + 2e^- \rightarrow 2H_2O(l)$$

18-5. CHEMICAL EQUATIONS FOR REACTIONS OCCURRING IN BASIC SOLUTION ARE BALANCED SLIGHTLY DIFFERENTLY THAN REACTIONS THAT TAKE PLACE IN ACIDIC SOLUTION

The reactions considered up to this stage have all taken place in acidic aqueous solution, where $H^+(aq)$ and H_2O are readily available and thus can be used in balancing the equations for the half-reactions. However, in basic solution $H^+(aq)$ is not available at significant concentrations to use in step IV (page 574) to balance the equations for the half-reactions with respect to hydrogen. Therefore, we must change step IV for reactions that take place in basic solution by using OH^- instead of H^+. However, if we simply add OH^- to balance with respect to hydrogen, then each OH^- adds one oxygen atom in addition to a hydrogen atom, and the oxygen-atom balance that was attained in step III will be lost. To preserve the oxygen-atom balance, we add one H_2O to the hydrogen-deficient side of the equation and one OH^- to the other side of the equation for each deficient hydrogen atom. Note that the net result is to add one hydrogen atom to the hydrogen-deficient side. This procedure is best illustrated by example. Let's balance the equation for the reduction half-reaction

$$ClO^-(aq) \rightarrow Cl^-(aq)$$

in basic solution. The half-reaction equation is already balanced with respect to chlorine. To balance it with respect to oxygen, we add H_2O to the right-hand side:

$$ClO^- \rightarrow Cl^- + H_2O$$

We now balance it with respect to hydrogen by adding $2H_2O$ to the left-hand side and $2OH^-$ to the right-hand side to obtain

$$ClO^- + 2H_2O \rightarrow Cl^- + H_2O + 2OH^-$$

or, upon canceling H_2O from each side,

$$ClO^- + H_2O \rightarrow Cl^- + 2OH^-$$

We now balance with respect to charge by adding $2e^-$ to the left-hand side:

$$ClO^- + H_2O + 2e^- \rightarrow Cl^- + 2OH^-$$

The balanced equation for the half-reaction with phases indicated is

$$ClO^-(aq) + H_2O(l) + 2e^- \rightarrow Cl^-(aq) + 2OH^-(aq)$$

Example 18-8: Given that the reaction takes place in basic solution, balance the following equation:

$$N_2H_4(aq) + Cu(OH)_2(s) \rightarrow N_2(g) + Cu(s)$$

Solution: The oxidation state of N changes from −2 (in N_2H_4) to 0 (in N_2) and that of Cu changes from +2 [in $Cu(OH)_2$] to 0 (in Cu). Thus, the two half-reactions we have

$$N_2H_4 \rightarrow N_2 \qquad \text{oxidation}$$

$$Cu(OH)_2(s) \rightarrow Cu(s) \qquad \text{reduction}$$

Both equations are balanced with respect to atoms other than oxygen and hydrogen. Let's finish balancing each one in turn. To balance the oxidation half-reaction equation with respect to hydrogen, we add $4H_2O$ to the right-hand side and $4OH^-$ to the left-hand side:

$$N_2H_4 + 4OH^- \rightarrow N_2 + 4H_2O$$

To balance with respect to charge, we add $4e^-$ to the right-hand side:

$$N_2H_4 + 4OH^- \rightarrow N_2 + 4H_2O + 4e^- \qquad \text{(oxidation)}$$

To balance the reduction half-reaction equation with respect to oxygen, we add $2H_2O$ to the right-hand side:

$$Cu(OH)_2 \rightarrow Cu + 2H_2O$$

The left-hand side is deficient by two hydrogen atoms. To balance with respect to hydrogen, we add $2H_2O$ to the left-hand side and $2OH^-$ to the right-hand side to obtain

$$Cu(OH)_2 + 2H_2O \rightarrow Cu + 2H_2O + 2OH^-$$

or, upon canceling $2H_2O$ from each side,

$$Cu(OH)_2 \rightarrow Cu + 2OH^-$$

To balance with respect to charge, we add $2e^-$ to the left-hand side:

$$Cu(OH)_2 + 2e^- \rightarrow Cu + 2OH^- \qquad \text{(reduction)}$$

To obtain a balanced equation, we multiply the equation for the reduction half-reaction by 2 and add the result to the equation for the oxidation half-reaction to obtain

$$N_2H_4 + 4OH^- + 2Cu(OH)_2 \rightarrow N_2 + 4H_2O + 2Cu + 4OH^-$$

Upon canceling like terms and indicating phases, we obtain

$$N_2H_4(aq) + 2Cu(OH)_2(s) \rightarrow N_2(g) + 2Cu(s) + 4H_2O(l)$$

▪ Note that the equation for the half-reaction

$$Cu(OH)_2 \rightarrow Cu$$

can be balanced, atomically simply by adding $2OH^-$ to the right-hand side. This is generally true for half-reactions involving metal hydroxides.

18-6. OXIDATION-REDUCTION REACTIONS ARE USED IN CHEMICAL ANALYSES

In this section we shall illustrate the use of oxidation-reduction reactions in chemical analysis. Suppose that a 3.532-g sample of iron ore is dissolved in $H_2SO_4(aq)$ and that all the iron(III) is reduced to $Fe^{2+}(aq)$ by adding powdered zinc to the solution. Now let's titrate the resulting filtered $Fe^{2+}(aq)$ solution with an oxidizing agent such as $KMnO_4(aq)$. The balanced equation for the reaction is

$$5Fe^{2+}(aq) + MnO_4^-(aq) + 8H^+(aq) \rightarrow 5Fe^{3+}(aq) + Mn^{2+}(aq) + 4H_2O(l)$$

The equilibrium constant of this equation at 25°C is very large ($K = 3 \times 10^{62}\ M^{-8}$), so essentially all the added $KMnO_4(aq)$ oxidizes the $Fe^{2+}(aq)$ to $Fe^{3+}(aq)$. Such a reaction is said to be a **quantitative reaction.** If it requires 34.58 mL of 0.1108 M $KMnO_4(aq)$ to oxidize all the

$Fe^{2+}(aq)$, then what is the mass percentage of iron in the ore?

The number of millimoles of $KMnO_4(aq)$ required is

$$\text{mmol } KMnO_4 = MV = (0.1108 \text{ mol·L}^{-1})(34.58 \text{ mL})$$
$$= 3.831 \text{ mmol}$$

From the balanced equation, we see that 5 mmol of $Fe^{2+}(aq)$ are oxidized for each millimole of $KMnO_4(aq)$ added, and so

$$\text{mmol } Fe^{2+} = (3.831 \text{ mmol } KMnO_4)\left(\frac{5 \text{ mmol } Fe^{2+}}{1 \text{ mmol } KMnO_4}\right)$$
$$= 19.16 \text{ mmol} = 0.01916 \text{ mol}$$

The atomic mass of iron is 55.85, and so

$$\text{g Fe} = \text{g } Fe^{2+} = (0.01916 \text{ mol})\left(\frac{55.85 \text{ g Fe}}{1 \text{ mol Fe}}\right)$$
$$= 1.070 \text{ g}$$

The mass percent of iron in the ore sample is

$$\text{mass \% Fe} = \left(\frac{1.070 \text{ g}}{3.523 \text{ g}}\right) \times 100 = 30.37\%$$

Example 18-9: The concentration of ozone in a sample of air can be determined by reacting the sample with a buffered $KI(aq)$ solution. The $O_3(g)$ oxidizes $I^-(aq)$ to $I_3^-(aq)$ according to

$$O_3(g) + 3I^-(aq) + 2H^+(aq) \rightarrow O_2(g) + I_3^-(aq) + H_2O(l)$$

The concentration of $I_3^-(aq)$ formed is then determined by titration with sodium thiosulfate, $Na_2S_2O_3(aq)$:

$$\underset{\text{thiosulfate}}{2S_2O_3^{2-}(aq)} + I_3^-(aq) \rightarrow \underset{\text{tetrathionate}}{S_4O_6^{2-}(aq)} + 3I^-(aq)$$

Given that 34.56 mL of 0.002475 M $Na_2S_2O_3(aq)$ is required to titrate the $I_3^-(aq)$ in a 50.00-mL sample of $KI(aq)$ that was reacted with a 43.15-g sample of air, calculate the mass percentage of ozone in the mixture.

Solution: The number of millimoles of $Na_2S_2O_3(aq)$ required is

$$\text{mmol } S_2O_3^{2-} = (0.002475 \text{ mol·L}^{-1})(34.56 \text{ mL}) = 8.554 \times 10^{-2} \text{ mmol}$$

The number of millimoles of $I_3^-(aq)$ reduced by the $S_2O_3^{2-}(aq)$ is

$$\text{mmol } I_3^-(aq) = (8.554 \times 10^{-2} \text{ mmol } S_2O_3^{2-})\left(\frac{1 \text{ mmol } I_3^-}{2 \text{ mmol } S_2O_3^{2-}}\right)$$
$$= 4.277 \times 10^{-2} \text{ mmol}$$

According to the equation for the O_3–I^- reaction,

$$\text{mmol } O_3 = \text{mmol } I_3^- = 4.277 \times 10^{-2} \text{ mmol}$$

The mass of ozone is

$$\text{g } O_3 = (4.277 \times 10^{-2} \text{ mmol})\left(\frac{1 \text{ mol}}{1000 \text{ mmol}}\right)\left(\frac{48.00 \text{ g } O_3}{1 \text{ mol } O_3}\right)$$
$$= 2.053 \times 10^{-3} \text{ g}$$

Figure 18-2 Addition of an aqueous iodine solution, I_2 (aq) (yellow), to an aqueous solution of sodium thiosulfate, $Na_2S_2O_3$ (aq) and starch indicator. The deep blue color is due to a starch-iodine complex. As the iodine reacts with the thiosulfate, the blue color disappears. The blue color will persist when all the thiosulfate is reacted with iodine.

and the mass percentage of ozone in the air sample is

$$\text{mass \% } O_3 = \left(\frac{\text{mass of ozone}}{\text{mass of sample}}\right) \times 100 = \left(\frac{2.053 \times 10^{-3}\text{ g}}{43.15\text{ g}}\right) \times 100$$

$$= 4.758 \times 10^{-3}\ \%$$

SUMMARY

In order to determine whether a reaction involves electron transfer, we first assign oxidation states to each element on both sides of the equation. In an electron-transfer reaction, the oxidation state of one element increases and the oxidation state of another element decreases.

Electron-transfer reactions can be separated into two half-reactions: the oxidation half-reaction (represented by the equation with electrons on the right) and the reduction half-reaction (represented by the equation with electrons on the left). The oxidation half-reaction supplies electrons to the reduction half-reaction. The equations for electron-transfer reactions can be balanced by a systematic procedure once the oxidation half-reaction and the reduction half-reaction have been identified by the assignment of oxidation states to the atoms involved in the reaction. The procedure for balancing equations for oxidation-reduction reactions in acidic solutions involves seven steps:

I. Separate the equation into an oxidation half-reaction equation and a reduction half-reaction equation.

II. Balance each half-reaction equation with respect to all elements other than oxygen and hydrogen.

III. Balance each half-reaction equation with respect to oxygen by adding the appropriate number of H_2O to the side deficient in oxygen.

IV. Balance each half-reaction equation with respect to hydrogen by adding the appropriate number of H^+ to the side deficient in hydrogen.

V. Balance each half-reaction equation with respect to charge by adding the appropriate number of electrons to the side with the excess positive charge.

VI. Multiply each half-reaction equation by an integer that makes the number of electrons supplied by the oxidation half-reaction equation equal to the number of electrons accepted by the reduction half-reaction equation.

VII. Obtain the complete balanced equation by adding the two half-reaction equations and canceling any like terms.

To balance the equation for a redox reaction that takes place in basic solution, step IV is changed to

IV. Balance each half-reaction equation with respect to hydrogen by adding one H_2O to the hydrogen-deficient side and one OH^- to the other side for each deficient hydrogen atom.

TERMS YOU SHOULD KNOW

oxidation-reduction reaction 568
redox reaction 568
oxidation state 568
reduced 571
reduction 571
oxidized 571
oxidation 571
electron-transfer reaction 571
oxidizing agent 572
reducing agent 572
electron acceptor 572
electron donor 572
half-reaction 572
oxidation half-reaction 572
reduction half-reaction 572
method of half-reactions 573
quantitative reaction 578

PROBLEMS

OXIDATION STATES

18-1. Assign an oxidation state to the metal in each of the following species:

(a) CaC_2 (b) Al_2O_3
(c) VO_2^+ (d) Co_3O_4

18-2. Assign an oxidation state to sulfur in each of the following species:

(a) S^{2-} (b) SO_3
(c) $S_2O_3^{2-}$ (d) SO_4^{2-}

18-3. Assign an oxidation state to the underlined element in each of the following compounds:

(a) $LiAl\underline{H}_4$ (b) $\underline{Cl}O_2$
(c) $Na\underline{Br}O_3$ (d) $H\underline{As}O_2$

18-4. Assign an oxidation state to the underlined element in each of the following compounds:

(a) $Na\underline{Cl}O$ (b) $\underline{Fe}_2O_3$
(c) $K\underline{O}_2$ (d) $K_2\underline{Mn}O_4$

18-5. Assign an oxidation state to nitrogen in each of the following species:

(a) NO_2 (b) N_2O
(c) N_2O_5 (d) N_2O_3

18-6. Determine the oxidation state of chlorine in each of the following chlorine oxides:

(a) Cl_2O (b) Cl_2O_3
(c) ClO_2 (d) Cl_2O_6

18-7. Determine the oxidation state of carbon in each of the following compounds:

(a) H_2CO (b) CH_4
(c) CH_3OH (d) $HCOOH$

18-8. Phosphorus forms a number of oxides. Assign an oxidation state to phosphorus in

(a) P_4O_6 (b) P_4O_7
(c) P_4O_8 (d) P_4O_{10}

18-9. Assign an oxidation state to antimony in each of the following species:

(a) $SbCl_3$ (b) Sb_4O_6
(c) SbF_5^{2-} (d) $SbCl_6^{3-}$

18-10. Assign an oxidation state to xenon in each of the following compounds:

(a) XeF_2 (b) XeF_4
(c) $XeOF_4$ (d) XeO_2F_2

OXIDIZING AGENTS AND REDUCING AGENTS

18-11. Identify the oxidizing and reducing agents in the equation

$$I_2(s) + 2Na_2S_2O_3(aq) \rightarrow 2NaI(aq) + Na_2S_4O_6(aq)$$

18-12. Sodium sulfide is manufactured by reacting sodium sulfate with carbon in the form of coke:

$$Na_2SO_4(s) + 4C(s) \rightarrow Na_2S(s) + 4CO(g)$$

Identify the oxidizing and reducing agents in this reaction.

18-13. Sodium nitrite, an important chemical in the dye industry, is manufactured by the reaction between sodium nitrate and lead:

$$NaNO_3(aq) + Pb(s) \rightarrow NaNO_2(aq) + PbO(s)$$

Identify the oxidizing and reducing agents in this equation.

18-14. Sodium chlorite, an industrial bleaching agent, is prepared as shown by the equation

$$4NaOH(aq) + Ca(OH)_2(aq) + C(s) + 4ClO_2(g) \rightarrow 4NaClO_2(aq) + CaCO_3(s) + 3H_2O(l)$$

Identify the oxidizing and reducing agents in this equation.

18-15. Identify the oxidizing and reducing agents and the oxidation and reduction half-reaction equations in the following equations:

(a) $2Fe^{3+}(aq) + 2I^-(aq) \rightarrow 2Fe^{2+}(aq) + I_2(s)$
(b) $2Ti^{2+}(aq) + Co^{2+}(aq) \rightarrow 2Ti^{3+}(aq) + Co(s)$

18-16. Identify the oxidizing and reducing agents and the oxidation and reduction half-reaction equations in the following equations:

(a) $H_2S(aq) + ClO^-(aq) \rightarrow S(s) + Cl^-(aq) + H_2O(l)$
(b) $In^+(aq) + 2Fe^{3+}(aq) \rightarrow 2Fe^{2+}(aq) + In^{3+}(aq)$

18-17. Potassium superoxide, KO_2, is a strong oxidizing agent. Explain why.

18-18. Lithium aluminum hydride, $LiAlH_4$, is a strong reducing agent. Explain why.

BALANCING OXIDATION-REDUCTION EQUATIONS

18-19. Balance the following equations for reactions that occur in acidic solution:

(a) $MnO(s) + PbO_2(s) \rightarrow MnO_4^-(aq) + Pb^{2+}(aq)$
(b) $As_2S_5(s) + NO_3^-(aq) \rightarrow H_3AsO_4(aq) + HSO_4^-(aq) + NO_2(g)$

For each of these reactions, identify the

electron donor reducing agent
electron acceptor species oxidized
oxidizing agent species reduced

18-20. Balance the following equations for reactions that occur in acidic solution:

(a) $ZnS(s) + NO_3^-(aq) \rightarrow Zn^{2+}(aq) + S(s) + NO(g)$
(b) $MnO_4^-(aq) + HNO_2(aq) \rightarrow NO_3^-(aq) + Mn^{2+}(aq)$

For each of these reactions, identify the

electron donor reducing agent
electron acceptor species oxidized
oxidizing agent species reduced

18-21. Complete and balance the following equations:

(a) $NH_4^+(aq) + NO_3^-(aq) \rightarrow N_2O(g)$ (acidic)
(b) $Fe(s) + O_2(g) \rightarrow Fe_2O_3 \cdot 3H_2O(s)$ (basic)

18-22. Complete and balance the following equations:

(a) $CoCl_2(s) + Na_2O_2(aq) \rightarrow Co(OH)_3(s) + Cl^-(aq) + Na^+(aq)$ (basic)
(b) $C_2O_4^{2-}(aq) + MnO_2(s) \rightarrow Mn^{2+}(aq) + CO_2(g)$ (acidic)

18-23. Complete and balance the following equations:

(a) $Fe(OH)_2(s) + O_2(g) \rightarrow Fe(OH)_3(s)$ (basic)
(b) $Cu(s) + NO_3^-(aq) \rightarrow Cu^{2+}(aq) + NO(g)$ (acidic)

18-24. Complete and balance the following equations:

(a) $Cr_2O_7^{2-}(aq) + I^-(aq) \rightarrow Cr^{3+}(aq) + I_2(s)$ (acidic)
(b) $CuS(s) + NO_3^-(aq) \rightarrow Cu^{2+}(aq) + S(s) + NO(g)$ (acidic)

18-25. Use the method of half-reactions to balance the following equations:

(a) $IO_4^-(aq) + I^-(aq) \rightarrow IO_3^-(aq) + I_3^-(aq)$ (basic)
(b) $H_2MoO_4(aq) + Cr^{2+}(aq) \rightarrow Mo(s) + Cr^{3+}(aq)$ (acidic)

18-26. Use the method of half-reactions to balance the following equations:

(a) $BrO_3^-(aq) + F_2(g) \rightarrow BrO_4^-(aq) + F^-(aq)$ (basic)
(b) $H_3AsO_3(aq) + I_2(aq) \rightarrow H_3AsO_4(aq) + I^-(aq)$ (acidic)

18-27. Use the method of half-reactions to balance the following equations:

(a) $CrO_4^{2-}(aq) + Cl^-(aq) \rightarrow Cr^{3+}(aq) + ClO_2^-(aq)$ (acidic)
(b) $Cu^{2+}(aq) + S_2O_3^{2-}(aq) \rightarrow Cu^+(aq) + S_4O_6^{2-}(aq)$ (acidic)

18-28. Use the method of half-reactions to balance the following equations:

(a) $Co(OH)_2(s) + SO_3^{2-}(aq) \rightarrow SO_4^{2-}(aq) + Co(s)$ (basic)
(b) $IO_3^-(aq) + I^-(aq) \rightarrow I_3^-(aq)$ (acidic)

18-29. For the strong of heart, balance

$CrI_3(s) + Cl_2(g) \rightarrow CrO_4^{2-}(aq) + IO_4^-(aq) + Cl^-(aq)$ (basic)

18-30. For the strong of heart, balance

$C_2H_5OH(aq) + I_3^-(aq) \rightarrow CO_2(g) + CHO_2^-(aq) + CHI_3(aq) + I^-(aq)$ (acidic)

18-31. Balance the following equations for half-reactions that occur in acidic solution:

(a) $Mo^{3+}(aq) \rightarrow MoO_2^{2+}(aq)$
(b) $P_4(s) \rightarrow H_3PO_4(aq)$
(c) $S_2O_8^{2-}(aq) \rightarrow HSO_4^-(aq)$

18-32. Balance the following equations for half-reactions that occur in acidic solution:

(a) $H_2BO_3^-(aq) \rightarrow BH_4^-(aq)$
(b) $ClO_3^-(aq) \rightarrow Cl_2(g)$
(c) $Cl_2(g) \rightarrow HClO(aq)$

18-33. Balance the following half-reaction equations:

(a) $WO_3(s) \rightarrow W_2O_5(s)$ (acidic)
(b) $U^{4+}(aq) \rightarrow UO_2^+(aq)$ (acidic)
(c) $Zn(s) \rightarrow Zn(OH)_4^{2-}(aq)$ (basic)

18-34. Balance the following half-reaction equations:

(a) $OsO_4(s) \rightarrow Os(s)$ (acidic)
(b) $S(s) \rightarrow SO_3^{2-}(aq)$ (basic)
(c) $Sn(s) \rightarrow HSnO_2^-(aq)$ (basic)

18-35. Balance the following equations for half-reactions that occur in basic solution:

(a) $SO_3^{2-}(aq) \rightarrow S_2O_4^{2-}(aq)$
(b) $Cu(OH)_2(s) \rightarrow Cu_2O(s)$
(c) $AgO(s) \rightarrow Ag_2O(s)$

18-36. Balance the following equations for half-reactions:

(a) $Au(CN)_2^-(aq) \rightarrow Au(s) + CN^-(aq)$ (acidic)
(b) $MnO_4^-(aq) \rightarrow MnO_2(s)$ (acidic)
(c) $Cr(OH)_3(s) \rightarrow CrO_4^{2-}(aq)$ (basic)

CALCULATIONS INVOLVING OXIDATION-REDUCTION EQUATIONS

18-37. The quantity of antimony in a sample can be determined by an oxidation-reduction titration with an oxidizing agent. A 9.62-g sample of stibnite, an ore of antimony, is dissolved in hot, concentrated HCl(*aq*) and passed over a reducing agent so that all the antimony is in the form Sb^{3+}. The $Sb^{3+}(aq)$ is completely oxidized by 43.7 mL of a 0.125 M solution of $KBrO_3$. The unbalanced equation for the reaction is

$$BrO_3^-(aq) + Sb^{3+}(aq) \rightarrow Br^-(aq) + Sb^{5+}(aq)$$

Calculate the amount of antimony in the sample and its percentage in the ore.

18-38. An ore is to be analyzed for its iron content by an oxidation-reduction titration with permanganate ion. A 4.23-g sample of the ore is dissolved in hydrochloric acid and passed over a reducing agent so that all the iron is in the form Fe^{2+}. The $Fe^{2+}(aq)$ is completely oxidized by 31.6 mL of a 0.0512 M solution of $KMnO_4$. The unbalanced equation for the reaction is

$$KMnO_4(aq) + HCl(aq) + FeCl_2(aq) \rightarrow MnCl_2(aq) + FeCl_3(aq) + H_2O(l) + KCl(aq)$$

Calculate the amount of iron in the sample and its mass percentage in the ore.

18-39. A rock sample is to be assayed for its tin content by an oxidation-reduction titration with $I_3^-(aq)$. A 10.0-g sample of the rock is crushed, dissolved in sulfuric acid, and passed over a reducing agent so that all the tin is in the form Sn^{2+}. The $Sn^{2+}(aq)$ is completely oxidized by 34.6 mL of a 0.556 M solution of NaI_3. The unbalanced equation for the reaction is

$$I_3^-(aq) + Sn^{2+}(aq) \rightarrow Sn^{4+}(aq) + I^-(aq)$$

Calculate the amount of tin in the sample and its mass percentage in the rock.

18-40. Sodium chlorite, $NaClO_2(s)$, is a powerful but stable oxidizing agent used in the paper industry, especially for the final whitening of paper. Sodium chlorite is capable of bleaching materials containing cellulose without oxidizing the cellulose. Sodium chlorite is made by the reaction

$$NaOH(aq) + Ca(OH)_2(s) + C(s) + ClO_2(g) \rightarrow NaClO_2(aq) + CaCO_3(s)$$

Balance the equation for the reaction and compute the number of kilograms of $ClO_2(g)$ required to make 1.00 metric ton of $NaClO_2$.

18-41. The amount of $I_3^-(aq)$ in a solution can be determined by titration with a solution containing a known concentration of $S_2O_3^{2-}(aq)$ (thiosulfate ion). The determination is based on the unbalanced equation

$$I_3^-(aq) + S_2O_3^{2-}(aq) \rightarrow I^-(aq) + S_4O_6^{2-}(aq)$$

Given that it requires 36.4 mL of 0.330 M $Na_2S_2O_3(aq)$ to titrate the $I_3^-(aq)$ in a 15.0-mL sample, compute the molarity of $I_3^-(aq)$ in the solution.

18-42. The amount of $Fe^{2+}(aq)$ in an $FeSO_4(aq)$ solution can be determined by titration with a solution containing a known concentration of $Ce^{4+}(aq)$. The determination is based on the reaction

$$Fe^{2+}(aq) + Ce^{4+}(aq) \rightarrow Fe^{3+}(aq) + Ce^{3+}(aq)$$

Given that it requires 37.5 mL of 0.0965 M $Ce^{4+}(aq)$ to oxidize the $Fe^{2+}(aq)$ in a 35.0-mL sample to $Fe^{3+}(aq)$, compute the molarity of $Fe^{3+}(aq)$ and the number of milligrams of iron in the sample.

18-43. Solid phosphorus reacts with $BaSO_4(s)$ under oxygen-free, anhydrous conditions to produce $P_4O_{10}(s)$ and $BaS(s)$; write a balanced equation for this process. How much phosphorus is required to react completely with 2.16 g $BaSO_4(s)$?

18-44. A solution of $I_3^-(aq)$ can be standardized by using it to titrate $As_4O_6(aq)$. The titration of 0.1021 g of $As_4O_6(s)$ dissolved in 30.00 mL of water requires 36.55 mL of $I_3^-(aq)$. Calculate the molarity of the $I_3^-(aq)$ solution. The unbalanced equation is

$$As_4O_6(s) + I_3^-(aq) \rightarrow As_4O_{10}(s) + I^-(aq)$$

ADDITIONAL PROBLEMS

18-45. Aqueous solutions of potassium permanganate decompose according to

$$MnO_4^-(aq) \rightarrow MnO_2(s) + O_2(g)$$

Balance the equation for this reaction under basic conditions.

18-46. Iodate, $IO_3^-(aq)$, can be used to titrate $Tl^+(aq)$ in a concentrated solution of HCl. Balance the equation

$$IO_3^-(aq) + Tl^+(aq) + Cl^-(aq) \rightarrow ICl_2^-(aq) + Tl^{3+}(aq)$$

18-47. Balance the following equations, which represent oxidations of $I^-(aq)$ to $I_3^-(aq)$:

(a) $Cr_2O_7^{2-}(aq) + I^-(aq) \rightarrow Cr^{3+}(aq) + I_3^-(aq)$ (acidic)
(b) $IO_4^-(aq) + I^-(aq) \rightarrow I_3^-(aq)$ (acidic)

18-48. Peroxydisulfate, $S_2O_8^{2-}$, is a strong oxidizing agent that in the presence of $Ag^+(aq)$ can oxidize Mn^{2+} to MnO_4^-, Cr^{3+} to $Cr_2O_7^{2-}$, and V(IV) to V(V). Excess reagent can be destroyed by boiling the solution after oxidation is complete. Balance the equation

$$S_2O_8^{2-}(aq) \rightarrow SO_4^{2-}(aq) + O_2(g) \qquad \text{(acidic)}$$

18-49. Silver(II) oxide dissolves in concentrated inorganic acids to produce $Ag^{2+}(aq)$, which is a powerful oxidizing agent. Excess Ag^{2+} can be removed by boiling the solution. Balance the equation

$$Ag^{2+}(aq) \rightarrow Ag^+(aq) + O_2(g) \qquad \text{(acidic)}$$

18-50. Potassium permanganate solutions can be standardized by titration with sodium oxalate, $Na_2C_2O_4$. Balance the equation

$$KMnO_4(aq) + Na_2C_2O_4(aq) \rightarrow Mn^{2+}(aq) + CO_2(g) \qquad \text{(acidic)}$$

18-51. Chromium(II), one of the most commonly used reducing agents, can be prepared by reducing $K_2Cr_2O_7$ with H_2O_2, followed by reduction of $Cr^{3+}(aq)$ to $Cr^{2+}(aq)$ by zinc. Balance the equations

(a) $Cr_2O_7^{2-}(aq) + H_2O_2(aq) \rightarrow Cr^{3+}(aq) + O_2(g)$ (acidic)
(b) $Cr^{3+}(aq) + Zn(s) \rightarrow Cr^{2+}(aq) + Zn^{2+}(aq)$

18-52. A 32.15-mL sample of a solution of $MoO_4^{2-}(aq)$ was passed through a Jones reductor (a column of zinc powder) in order to convert all the $MoO_4^{2-}(aq)$ to $Mo^{3+}(aq)$. The filtrate required 20.85 mL of 0.0955 M $KMnO_4(aq)$ for the reaction given by

$$MnO_4^-(aq) + Mo^{3+}(aq) \rightarrow Mn^{2+}(aq) + MoO_2^{2+}(aq)$$

Balance this equation and then calculate the concentration of the original $MoO_4^{2-}(aq)$ solution.

18-53. Hydrogen peroxide can act as either an oxidizing agent or a reducing agent, depending on the species present in solution. Write balanced half-reaction equations for each of the following:

(a) H_2O_2 acting as an oxidizing agent in an acidic solution
(b) H_2O_2 acting as a reducing agent in an acidic solution

Write a balanced equation for the reaction in which H_2O_2 oxidizes and reduces itself.

18-54. A 3.651-g sample of a lanthanum sulfate ore is

 dissolved in nitric acid, and the lanthanum precipitated as $La(IO_3)_3$ by the addition of 40.00 mL of 0.1105 M $KIO_3(aq)$. When an excess of KI is added to the acidified filtrate, the I_2 that results from the reaction

$$IO_3^-(aq) + 5I^-(aq) + 6H^+(aq) \rightarrow 3I_2(s) + 3H_2O(l)$$

requires 12.65 mL of 0.0650 M $Na_2S_2O_3(aq)$ to react according to

$$I_2(s) + 2Na_2S_2O_3(aq) \rightarrow Na_2S_4O_6(aq) + 2NaI(aq)$$

Calculate the mass percent of $La_2(SO_4)_3$ in the sample.

18-55. The concentration of $H_2S(g)$ in air can be determined by the following method. A sample of air is passed through a $Cd^{2+}(aq)$ solution, where the sulfur is precipitated as $CdS(s)$. This precipitate is then treated with an excess of $I_2(aq)$, which oxidizes the sulfide to elemental sulfur. The amount of excess $I_2(aq)$ is determined by titration with $Na_2S_2O_3(aq)$. Suppose that a 10.75-g sample of air is passed through a $Cd^{2+}(aq)$ solution. A 30.00-mL sample of 0.0115 M $I_2(aq)$ is then added to the $CdS(s)$. The unreacted $I_2(aq)$ required 7.65 mL of 0.0750 M $Na_2S_2O_3(aq)$ for reduction to $I^-(aq)$. Calculate the mass percent of $H_2S(s)$ in the air.

18-56. The concentration of ozone in oxygen-ozone mixtures can be determined by passing the gas mixture into a buffered $KI(aq)$ solution. The O_3 oxidizes $I^-(aq)$ to $I_3^-(aq)$:

$$O_3(g) + 3I^-(aq) + H_2O(l) \rightarrow O_2(g) + I_3^-(aq) + 2OH^-(aq)$$

The concentration of $I_3^-(aq)$ formed is then determined by titration with $Na_2S_2O_3(aq)$. Given that 22.50 mL of 0.0100 M $Na_2S_2O_3(aq)$ is required to titrate the $I_3^-(aq)$ in a 50.0-mL sample of $KI(aq)$ that was equilibrated with a 8.65-g sample of an O_2–O_3 mixture, compute the mass percent of O_3 in the sample.

18-57. Atmospheric SO_2 can be determined by reaction with $H_2O_2(aq)$:

$$H_2O_2(aq) + SO_2(g) \rightarrow H_2SO_4(aq)$$

followed by titration of the $H_2SO_4(aq)$ produced. Given that 18.50 mL of 0.00250 M NaOH was required to neutralize the $H_2SO_4(aq)$ in a 50.0-mL $H_2O_2(aq)$ sample that was equilibrated with a 812.1-g sample of air containing SO_2, compute the mass percent of SO_2 in the air sample.

18-58. Iodine pentoxide is a reagent for the quantitative determination of carbon monoxide. The equation for the reaction is

$$5CO(g) + I_2O_5(s) \rightarrow I_2(s) + 5CO_2(g)$$

The iodine produced is dissolved in Kl(*aq*) and then determined by reaction with $Na_2S_2O_3$:

$$2S_2O_3^{2-}(aq) + I_3^-(aq) \rightarrow 3I^-(aq) + S_4O_6^{2-}(aq)$$

Compute the mass percent of $CO(g)$ in a 56.04-g sample of air if the $CO(g)$ produces sufficient $I_3^-(aq)$ to react completely with the $S_2O_3^{2-}(aq)$ in 10.0 mL of 0.0350 M $Na_2S_2O_3(aq)$.

INTERCHAPTER • L

Sulfur

Elemental sulfur occurs in large underground deposits along the Gulf Coast of the United States. The sulfur shown here was mined by the Frasch process and is awaiting shipment at Pennzoil's terminal at Galveston, Texas.

Sulfur (atomic number 16, atomic mass 32.06) is a yellow, tasteless solid that is often found in nature as the free element. Sulfur is essentially insoluble in water but dissolves readily in carbon disulfide, CS_2. It does not react with dilute acids or bases, but is does react with many metals at elevated temperatures to form metal sulfides.

Sulfur, which constitutes only 0.05 percent of the earth's crust, is not one of the most prevalent elements. Yet it is one of the most commercially important ones because it is the starting material for the most important industrial chemical, sulfuric acid.

L-1. SULFUR IS RECOVERED FROM LARGE UNDERGROUND DEPOSITS BY THE FRASCH PROCESS

Io, one of the moons of Jupiter, appears yellow because of large deposits of sulfur from volcanic activity.

Prior to 1900, most of the world's supply of sulfur came from Sicily, where sulfur occurs at the surfaces around hot springs and volcanoes. In the early 1900s, however, large subsurface deposits of sulfur were found along the Gulf Coast of the United States. The sulfur occurs in limestone caves, more than 1000 feet beneath layers of rock, clay, and quicksand. The recovery of the sulfur from these deposits posed a great technological problem, which was solved by the engineer Herman Frasch. The **Frasch process** (Figure L-1) uses an arrangement of three concentric pipes (diameters of 1 in., 3 in., and 6 in.) placed in a bore hole that penetrates to the base of the sulfur-bearing calcite ($CaCO_3$) rock formation. Pressurized hot water (180°C) is forced down the space between the 6-in. and 3-in. pipes to melt the sulfur (melting point 119°C). The molten sulfur, which is twice as dense as water, sinks to the bottom of the deposit and then is forced up the space between the 3-in. and 1-in. pipes as a foam by the action of compressed air injected through the innermost pipe. The molten sulfur rises to the surface, where it is pumped into tank cars for shipment or into storage

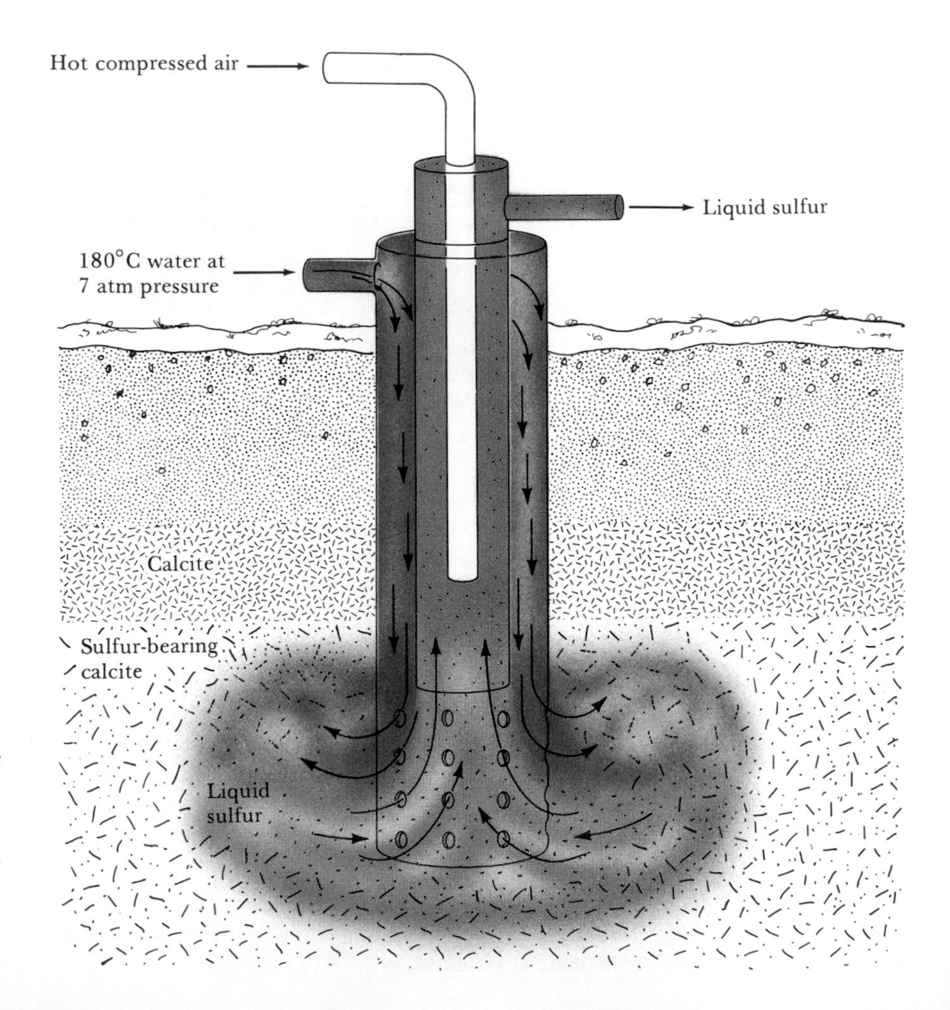

Figure L-1 The Frasch process for sulfur extraction. Three concentric pipes are sunk into sulfur-bearing calcite rock. Water at 180°C is forced down the outermost pipe to melt the sulfur. Hot compressed air is forced down the innermost pipe and mixes with the molten sulfur, forming a foam of water, air, and sulfur. The mixture rises to the surface through the middle pipe. The resulting dried sulfur has a purity of 99.5 percent.

Figure L-2 Many of the hot springs and geysers in the western United States have yellow deposits of sulfur around them.

areas. About 40 percent of the U.S. annual sulfur production of over 11 million metric tons is obtained by the Frasch process from the region around the Gulf of Mexico in Louisiana and Texas (see Frontispiece).

About 50 percent of U.S. and most Canadian sulfur is produced by the **Claus process,** in which sulfur is obtained from the hydrogen sulfide that occurs in some natural gas deposits and from the H_2S produced when sulfur is removed from petroleum. In the Claus process, hydrogen sulfide is burned in air to produce sulfur dioxide gas, which is then reacted with additional hydrogen sulfide to produce sulfur:

$$2H_2S(g) + 3O_2(g) \rightarrow 2SO_2(g) + 2H_2O(g)$$

$$SO_2(g) + 2H_2S(g) \rightarrow 3S(l) + 2H_2O(g)$$

These reactions also are thought to be responsible for the surface deposits of sulfur around hot springs and volcanoes (Figure L-2).

L-2. SULFIDE ORES ARE IMPORTANT SOURCES OF SEVERAL METALS

Deposits of metal sulfides are found in many regions and are valuable ores of the respective metals. Galena (PbS), cinnabar (HgS), and antimony sulfide (Sb_2S_3) are examples of metal sulfides that are ores (Figure L-3). In obtaining metals from sulfide ores, the ores usually are heated in an oxygen atmosphere. This process is called **roasting.** The chemical equation for the roasting of galena is

$$2PbS(s) + 3O_2(g) \rightarrow 2PbO(s) + 2SO_2(g)$$

The lead oxide then is reduced by heating with carbon in the form of coke:

$$PbO(s) + C(s) \rightarrow Pb(l) + CO(g)$$

Figure L-3 Many metal sulfides are valuable ores of the respective metals. Shown here are (*left to right*) galena, PbS; cinnabar, HgS; iron pyrite, FeS_2; and sphalerite, ZnS.

The sulfur dioxide produced by roasting is a serious atmospheric pollutant and should be recovered.

Sulfur is also found in nature in a few insoluble sulfates, such as gypsum, $CaSO_4 \cdot 2H_2O$ (calcium sulfate dihydrate) (Figure L-4), and barite, $BaSO_4$.

L-3. SOLID SULFUR CONSISTS OF RINGS OF EIGHT SULFUR ATOMS

Below 96°C sulfur exists as yellow, transparent **rhombic** crystals, shown in Figure L-5a. If rhombic sulfur is heated above 96°C, then it be-

Figure L-4 Large deposits of gypsum, $CaSO_4 \cdot 2H_2O$, an insoluble mineral, are found in many areas. Shown here are the dunes of White Sands National Monument in New Mexico, which are composed of gypsum.

(a)

(b)

Figure L-5 Sulfur occurs as (a) rhombic and (b) monoclinic crystals. Rhombic sulfur is the stable form below 96°C. From 96°C to 119°C (the normal melting point), monoclinic sulfur is the stable form. The terms rhombic and monoclinic are derived from the shape of the crystals.

comes opaque and the crystals expand into **monoclinic** crystals (Figure L-5b). The molecular units of the rhombic form are rings containing eight sulfur atoms, S_8 (Figure L-6). Monoclinic sulfur is the stable form from 96°C to the melting point. The molecular units of monoclinic sulfur are also S_8 rings, but the rings themselves are arranged differently in rhombic and monoclinic sulfur, which are allotropic forms of solid sulfur.

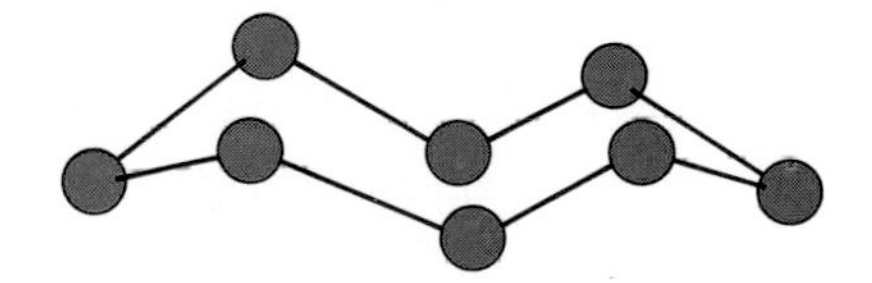

Figure L-6 Under most conditions, sulfur exists as eight-membered rings, S_8. The ring is not flat but puckered in such a way that four of the atoms lie in one plane and the other four lie in another plane. Each sulfur atom forms two sulfur-sulfur bonds and also has two lone pairs of electrons.

Monoclinic sulfur melts at 119°C to a thin yellow liquid (Figure L-7a) consisting of S_8 rings. Upon heating to about 150°C there is little change, but beyond 150°C the liquid sulfur begins to thicken and turns reddish brown. By 200°C, the liquid is so thick that is hardly pours (Figure L-7b). The molecular explanation for this behavior is simple. At about 150°C, thermal agitation causes the S_8 rings to begin to break apart and form chains of sulfur atoms:

$$S_8 \text{ ring} \longrightarrow \dot{S}-S-S-S-S-S-S-\dot{S}$$

These chains can then join together end to end to form longer chains, which become entangled in each other and cause the liquid to thicken. Above 250°C, the liquid begins to flow more easily because the thermal agitation is sufficient to begin to break the chains of sulfur atoms. At the boiling point (445°C), liquid sulfur pours freely. The vapor molecules consist mostly of S_8 rings.

If liquid sulfur at about 200°C is placed quickly in cold water—this process is called **quenching**—then a rubbery substance known as **plas-**

(a) 120°C

(b) 200°C

Figure L-7 Molten sulfur at (a) 120°C and (b) 200°C. The change in color and physical properties of liquid sulfur with increasing temperature (120° to about 250°C) is a result of the conversion of eight-membered rings to long chains of sulfur atoms. Above 250°C, the long chains begin to break up into smaller segments.

Figure L-8 If liquid sulfur at about 200°C is cooled quickly by pouring it into cold water, a rubbery substance called plastic sulfur is obtained.

tic sulfur is formed (Figure L-8). The material is rubbery because the long, coiled chains of sulfur atoms can straighten out somewhat if they are pulled. This molecular explanation of rubbery character (elasticity), is discussed in Chapter 24. As plastic sulfur cools, it slowly becomes hard again as it rearranges itself into the rhombic form.

L-4. SULFURIC ACID IS THE LEADING INDUSTRIAL CHEMICAL

By far the most important use of sulfur is in the manufacture of sulfuric acid. Most sulfuric acid is made by a process called the **contact process.** The sulfur is first burned in oxygen to produce sulfur dioxide:

$$S(s) + O_2(g) \rightarrow SO_2(g)$$

The sulfur dioxide is then converted to sulfur trioxide:

$$2SO_2(g) + O_2(g) \xrightarrow{V_2O_5(s)} 2SO_3(g)$$

The V_2O_5 over the arrow in this equation means that V_2O_5 (vanadium pentoxide) is a catalyst for the reaction. The sulfur trioxide is then absorbed into nearly pure liquid sulfuric acid (99 percent sulfuric acid plus 1 percent water) to form **fuming sulfuric acid,** also known as **oleum:**

$$H_2SO_4(l) + SO_3(g) \rightarrow \underset{\text{oleum}}{H_2S_2O_7} \ (35\% \text{ in } H_2SO_4)$$

The oleum is then added to water or aqueous sulfuric acid to produce the desired final concentration of aqueous sulfuric acid. Sulfur trioxide cannot be absorbed directly in water because the acid mist of H_2SO_4 that forms is very difficult to condense.

About 80 billion pounds of sulfuric acid is produced annually in the United States. Commercial-grade sulfuric acid is one of the least ex-

Table L-1 Annual U.S. industrial use of sulfuric acid (1980)

Use	Quantity used/millions of tons
manufacture of ammonium sulfate and phosphate fertilizers	20
manufacture of HNO_3, HCl, H_3PO_4, HF, explosives	10
purification of petroleum products	4
manufacture of paints and pigments	1
cleaning metal surfaces, metallurgy	1
other	4

pensive chemicals, costing less than 10 cents per pound in bulk quantities. Very large quantities of sulfuric acid are used in the production of fertilizers and numerous industrial chemicals, the petroleum industry, metallurgical processes, synthetic fiber production, and paints, pigments, and explosives manufacture. Table L-1 lists the quantities of sulfuric acid used in its various industrial applications.

Pure, anhydrous sulfuric acid is a colorless, syrupy liquid that freezes at 10°C and boils at 290°C. The standard laboratory acid is 98 percent H_2SO_4 by mass and is 18 M in H_2SO_4 (Figure L-9). Concentrated sulfuric acid is a powerful dehydrating agent. Gases are sometimes bubbled through it to remove traces of water vapor—provided, of course, that the gases do not react with the acid.

Sulfuric acid is such a strong dehydrating agent that it can remove water from carbohydrates, such as cellulose and sugar, even though these substances contain no free water. If concentrated sulfuric acid is

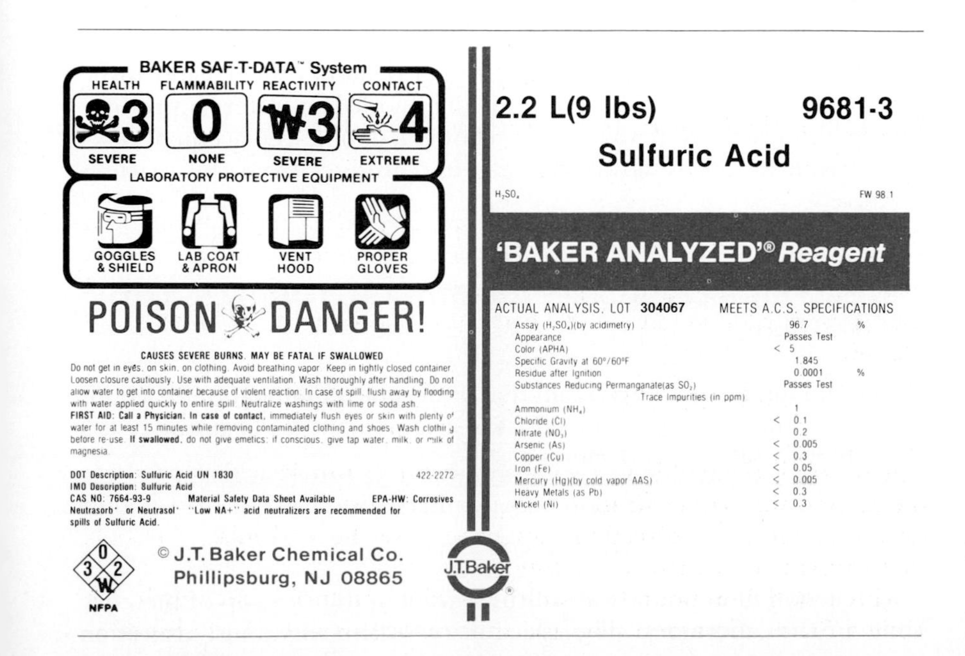

Figure L-9 Sulfuric acid is sold for laboratory use as an 18 M solution that is 98 percent sulfuric acid and 2 percent water by mass.

Figure L-10 Concentrated (98%) sulfuric acid is a powerful dehydrating agent capable of converting sucrose to carbon.

poured over sucrose, $C_{12}H_{22}O_{11}$, then we have the reaction described by the equation.

$$C_{12}H_{22}O_{11}(s) \xrightarrow[H_2SO_4\ (98\%)]{} 12C(s) + 11H_2O\ (\text{in } H_2SO_4)$$

This impressive reaction is shown in Figure L-10. Similar reactions are responsible for the destructive action of concentrated sulfuric acid on wood, paper, and skin.

The high boiling point and strength of sulfuric acid are the basis of its use in the production of other acids. For example, dry hydrogen chloride gas is produced by the reaction of sodium chloride with sulfuric acid:

$$2NaCl(s) + H_2SO_4(l) \rightarrow Na_2SO_4(s) + 2HCl(g)$$

The high boiling point of the sulfuric acid allows the $HCl(g)$ to be driven off by heating. The $HCl(g)$ is then added to water to produce hydrochloric acid. Note that this reaction is a double-replacement reaction driven by the removal of a gaseous product from the reaction mixture.

L-5. SULFUR REACTS DIRECTLY WITH MOST METALS AND NONMETALS

Sulfur is a fairly reactive element that undergoes combination reactions with most metals and nonmetals. For example, at 25°C sulfur reacts directly with lithium and mercury:

$$2Li(s) + S(s) \rightarrow Li_2S(s)$$

$$Hg(l) + S(s) \rightarrow HgS(s)$$

The reaction of mercury with sulfur is used in the cleanup of mercury spills in the laboratory. The reaction of sulfur with most transition metals requires a high temperature. With the exception of the alkali metal sulfides, most metal sulfides are insoluble in water. For example, at 25°C

$$FeS(s) \rightleftharpoons Fe^{2+}(aq) + S^{2-}(aq) \qquad K_{sp} = 6.3 \times 10^{-18} M^2$$

The insolubility of metal sulfides is utilized in the separation of metal ions in qualitative analysis (see Section 17-6). The addition of a solution of a strong acid to a metal sulfide yields hydrogen sulfide gas; for example, at 25°C

$$FeS(s) + 2H^+(aq) \rightleftharpoons Fe^{2+}(aq) + H_2S(g) \qquad K = 5.7 \times 10^3 \text{ atm} \cdot M^{-1}$$

This reaction is driven by the formation of the very weak diprotic hydrosulfuric acid, H_2S.

Hydrogen sulfide, H_2S, is a colorless gas with an offensive odor suggestive of rotten eggs. It is also very poisonous. Trace amounts of hydrogen sulfide occur naturally in the atmosphere owing to volcanic activity and the decay of organic matter. The presence of traces of hydrogen sulfide in the atmosphere is partly responsible for the tarnishing of silver. In the presence of oxygen, silver reacts with hydrogen sulfide according to

$$4Ag(s) + 2H_2S(g) + O_2(g) \rightarrow 2Ag_2S(s) + 2H_2O(l)$$

The silver sulfide formed by the reaction is a black, insoluble solid that appears as a dark tarnish on the surface of the silver (Figure L-11).

Organic compounds that contain an —SH group are called **mercaptans** and are notoriously foul-smelling. For example, the *Guinness Book of World Records* reports that ethyl mercaptan, CH_3CH_2SH, has the worst odor of any substance; many other mercaptans have comparably obnoxious odors. The odor of a skunk's spray is due to a mixture of mercaptans. Because pure CH_4 is odorless, trace amounts of methyl mercaptan, CH_3SH, are added to natural gas, which is mostly CH_4, so that the presence of a natural gas leak can be detected by smell.

Sulfur dissolves at 25°C in alkaline aqueous solutions containing sulfide ion to form the disulfide ion $S_2^{2-}(aq)$:

$$S(s) + S^{2-}(aq) \rightleftharpoons S_2^{2-}(aq) \qquad K = 2.3 \text{ (at 25°C)}$$

At higher temperatures, polysulfides such as S_3^{2-} and S_4^{2-}, form.

Sulfur burns in oxygen to form sulfur dioxide, a colorless gas with a characteristic choking odor. A pressure of 3 atm is sufficient to liquefy sulfur dioxide at 20°C. For this reason, SO_2 was once used in industrial refrigeration units, but the unpleasant odor and toxicity brought on its replacement by Freons. Most sulfur dioxide is used to make sulfuric acid, but some is used as a bleaching agent in the manufacture of paper products, oils, and starch, and as a food additive to inhibit browning. Large quantities are used in the wine industry as a fungicide for grapevines.

Sulfur dioxide is very soluble in water; at 25°C and 1 atm pressure of $SO_2(g)$ about 200 g of sulfur dioxide dissolves in one liter of water. Some of the sulfur dioxide reacts with the water to form sulfurous acid:

$$SO_2(g) + H_2O(l) \rightarrow H_2SO_3(aq)$$

but most of its exists in solution as $SO_2(aq)$. In contrast to sulfuric acid sulfurous acid is a weak acid in water ($pK_{a1} = 1.8$, $pK_{a2} = 7.2$ at 25°C).

Figure L-11 Silver sulfide is a black, insoluble solid that appears as a dark tarnish on the surface of silver.

The salts of sulfurous acid are called sulfites. For example, if sodium hydroxide is added to an aqueous solution of sulfur dioxide, then sodium sulfite is formed:

$$2NaOH(aq) + H_2SO_3(aq) \rightarrow \underset{\text{sodium sulfite}}{Na_2SO_3(aq)} + 2H_2O(l)$$

Sodium sulfite is used occasionally as a preservative, especially for dehydrated fruits. The sulfite ion acts as a fungicide; however, it imparts a characteristic sulfur dioxide odor and taste to the food.

The thiosulfate ion is produced when an aqueous solution of a metal sulfite, such as $Na_2SO_3(aq)$, is boiled in the presence of solid sulfur:

$$S(s) + SO_3^{2-}(aq) \rightarrow \underset{\text{thiosulfate}}{S_2O_3^{2-}(aq)}$$

The designation thio- denotes the replacement of an oxygen atom by a sulfur atom. Sodium thiosulfate is used extensively as "hypo" ($Na_2S_2O_3 \cdot 5H_2O$) in black-and-white photography, where it is used to dissolve silver salts via a complexation reaction,

$$AgBr(s) + 2S_2O_3^{2-}(aq) \rightarrow Ag(S_2O_3)_2^{3-}(aq) + Br^-(aq) \qquad K = 23 \text{ (at 25°C)}$$

L-6. SEVERAL SULFUR-OXYGEN SPECIES ARE USEFUL OXIDIZING OR REDUCING AGENTS

At 25°C and 1 M concentration, sulfuric acid is a poor oxidizing agent. However hot, concentrated sulfuric acid is a strong oxidizing agent that can oxidize copper metal:

$$Cu(s) + 2H_2SO_4(aq) \rightarrow CuSO_4(aq) + SO_2(g) + 2H_2O(l)$$

Electrochemical oxidation of aqueous sulfuric acid yields the peroxodisulfate ion, $S_2O_8^{2-}(aq)$, which has a peroxide bond (—O—O—) and consequently is a powerful oxidizing agent

$$S_2O_8^{2-}(aq) + 2e^- \rightleftharpoons 2SO_4^{2-}(aq)$$

that is capable of oxidizing Ag(I) to Ag(III).

The thiosulfate ion is a mild reducing agent that is used in analytical chemistry for the determination of iodine, which is produced by the action of many mild oxidizing agents; for example,

$$2S_2O_3^{2-}(aq) + I_3^-(aq) \rightarrow 3I^-(aq) + \underset{\text{tetrathionate}}{S_4O_6^{2-}(aq)}$$

The tetrathionate ion has a sulfur-sulfur bond, with a structure similar to that of $S_2O_8^{2-}$. The thiosulfate ion is unstable in acidic solutions and decomposes to sulfur and sulfur dioxide:

$$S_2O_3^{2-}(aq) + 2H^+(aq) \rightarrow S(s) + SO_2(aq) + H_2O(l)$$

The sulfur prepared in this way is unusual in that it consists of six-membered rings, S_6, of sulfur atoms.

Electrochemical reduction of the sulfite ion in aqueous solution yields the dithionite ion, $S_2O_4^{2-}$ (aq), which is a strong reducing agent in basic solution:

$$S_2O_4^{2-}(aq) + 4OH^-(aq) \rightleftharpoons 2SO_3^{2-}(aq) + 2H_2O(l) + 2e^-$$

The strong reducing power of dithionite is a result of the weak sulfur-sulfur bond. Zinc dithionite is extensively used to bleach paper and textiles.

TERMS YOU SHOULD KNOW

Frasch process 586
Claus process 587
roasting of an ore 587
rhombic sulfur 588
monoclinic sulfur 589
quenching 589
plastic sulfur 589
contact process 590
fuming sulfuric acid (oleum) 590
mercaptan 593

QUESTIONS

L-1. Describe the Frasch process.

L-2. Use balanced chemical equations to show how zinc can be obtained from the ore sphalerite (ZnS).

L-3. Give the chemical formula for each of the following substances:

(a) sulfur dioxide
(b) sulfur trioxide
(c) sulfuric acid
(d) sulfurous acid
(e) hydrogen sulfide
(f) iron pyrite
(g) gypsum
(h) cinnabar
(i) sodium thiosulfate pentahydrate

L-4. Describe what happens at various stages when sulfur (initially in the rhombic form) is heated slowly from 90°C to 450°C.

L-5. Describe, using balanced chemical equations, the contact process for the manufacture of sulfuric acid.

L-6. When gypsum, $CaSO_4 \cdot 2H_2O(s)$, is heated, part of the water of crystallization is driven off and plaster of Paris, $CaSO_4 \cdot \frac{1}{2}H_2O(s)$, is formed. Write a balanced chemical equation for this process. Addition of water to plaster of Paris gives back gypsum. Commercial plaster contains plaster of Paris mixed with fibrous material, such as animal hair, to provide structural strength.

L-7. Write a balanced chemical equation for the formation of the fertilizer, ammonium sulfate, from ammonia and sulfuric acid.

L-8. Write balanced chemical equations to explain the solubility of solid sulfur in $Na_2S(aq)$ and $Na_2SO_3(aq)$.

L-9. Describe the Claus process.

L-10. Give the chemical formula for each of the following substances:

(a) calcium thiosulfate
(b) sodium dithionite
(c) potassium peroxodisulfate
(d) lithium tetrathionate

PROBLEMS

L-11. Use VSEPR theory (Chapter 9) to predict the geometries of the following oxysulfur ions:

(a) SO_3^{2-} (b) SO_4^{2-} (c) $S_2O_3^{2-}$

L-12. Using data given in this interchapter, calculate the equilibrium constant at 25°C for the equation

$$SO_2(g) \rightleftharpoons SO_2(aq)$$

L-13. There are two known isomers of S_2F_2 with significantly different sulfur-sulfur bond lengths. Write Lewis formulas for the two isomers and use VSEPR theory (Chapter 9) to predict the structures of the two isomers.

L-14. In acid solutions $S_2O_8^{2-}(aq)$ decomposes to yield hydrogen peroxide, $H_2O_2(aq)$. Write a balanced equation for the decomposition reaction, and suggest a method for making $H_2O_2(aq)$ from sulfuric acid.

L-15. Using data from Table 15-5 calculate the pH of a 0.15 M $KHSO_4(aq)$ solution at 25°C.

L-16. Using data given in this Interchapter, calculate the density of laboratory sulfuric acid.

L-17. For the equation

$$SO_2(g) + \frac{1}{2}O_2(g) \rightleftharpoons SO_3(g)$$

$\Delta H^0_{rxn} = -190$ kJ. Given that $K_p = 0.65$ atm$^{-1/2}$ at 800°C, calculate K_p at 850°C.

L-18. Balance the following oxidation-reduction equations

(a) $S_2O_3^{2-}(aq) + H_2O_2(aq) + OH^-(aq) \rightarrow SO_4^{2-}(aq) + H_2O(l)$

(b) $SO_3^{2-}(aq) + ICl(aq) + H_2O(l) \rightarrow SO_4^{2-}(aq) + I_2(s) + HCl(aq)$

CHAPTER • 19

ENTROPY AND GIBBS FREE ENERGY

The spontaneous dispersal of ink throughout the entire volume of water is an example of an entropy-driven process.

It is natural to ask why some substances react with each other and others do not. When a reaction occurs on its own, we say that it is spontaneous, and the condition that must be met in order for a reaction to be spontaneous is called the criterion of spontaneity. The observation that all highly exothermic reactions are spontaneous led the French chemist Pierre Berthelot to put forth, in the 1860s, the hypothesis that all spontaneous reactions are exothermic. Berthelot's criterion of spontaneity for a chemical reaction was based on the sign of ΔH_{rxn}. According to Berthelot, if ΔH_{rxn} is negative, then a reaction is spontaneous.

Berthelot's criterion of reaction spontaneity was shown to be incorrect, and it was superseded by the criterion of reaction spontaneity developed by the American thermodynamicist J. Willard Gibbs, which is based on the second law of thermodynamics. Gibbs showed that reaction spontaneity is not just a matter of energetics. There is another property, called entropy, that also must be considered when determining whether a reaction is spontaneous or not. We shall see in this chapter that entropy is a measure of the randomness or disorder of a system, and we see how the spontaneity of a chemical reaction is governed by both the energy change and the entropy change. We shall also see that the Gibbs criterion of reaction spontaneity is given by the sign of ΔG_{rxn}, where G is the Gibbs free energy, which depends on both energy and entropy. If ΔG_{rxn} is negative, then the reaction is spontaneous.

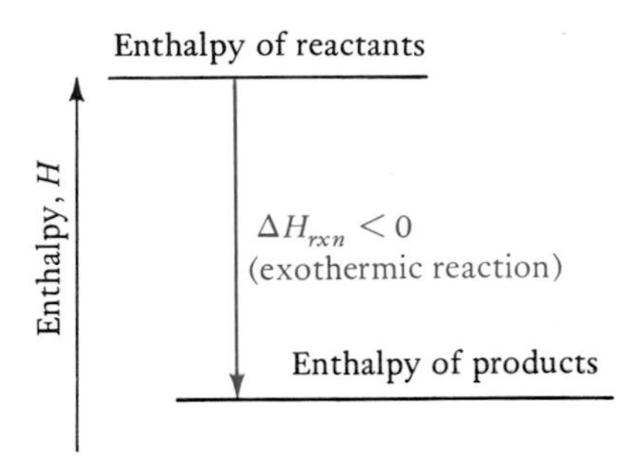

Figure 19-1 In an exothermic reaction, the total enthalpy of the products is less than the total enthalpy of the reactants.

19-1. NOT ALL SPONTANEOUS REACTIONS EVOLVE ENERGY

In Chapter 5 we discussed enthalpy changes for chemical reactions. Recall that an exothermic chemical reaction is one in which energy is evolved as heat. Exothermic reactions are energetically downhill in the sense that the total enthalpy of the products is less than the total enthalpy of the reactants (Figure 19-1). For an exothermic reaction, the **enthalpy change, ΔH_{rxn},** is negative. For an exothermic reaction run at standard conditions (all species at 1 atm pressure) we have

$$\Delta H^\circ_{rxn} = H^\circ(\text{products}) - H^\circ(\text{reactants}) < 0$$

where the superscript degree signs denote *standard* values for the thermodynamic quantities. Thus ΔH°_{rxn} is the **standard enthalpy change.** In Chapter 5 we learned to calculate ΔH°_{rxn} for a reaction by using a table of heats of formation (Table 5-1). Recall that the value of ΔH°_{rxn} is approximately equal to the value of the standard energy change, ΔU°_{rxn}, for a reaction; also, $\Delta H^\circ_{rxn} \approx \Delta H_{rxn}$, because the enthalpy is relatively independent of the pressure.

The natural tendency of simple mechanical systems is to undergo processes that lead to a decrease in energy of the system. For example, water flows spontaneously downhill without any help from us. Water at the bottom of a waterfall has a lower potential energy than water at the top. To get water back to the top of the waterfall, we have to use a pump, which requires energy. As another example, if we release the ball shown at the top of the hill in Figure 19-2, then it spontaneously rolls down the hill and eventually comes to rest at the bottom of the valley. The ball at the lowest point of the valley has the lowest possible potential energy for this system.

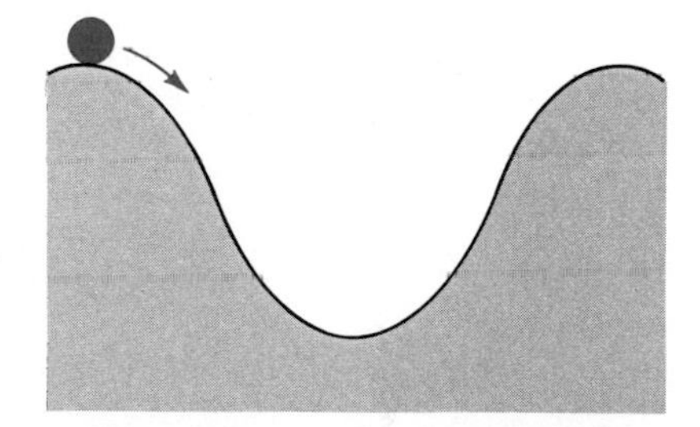

(a) Initial state

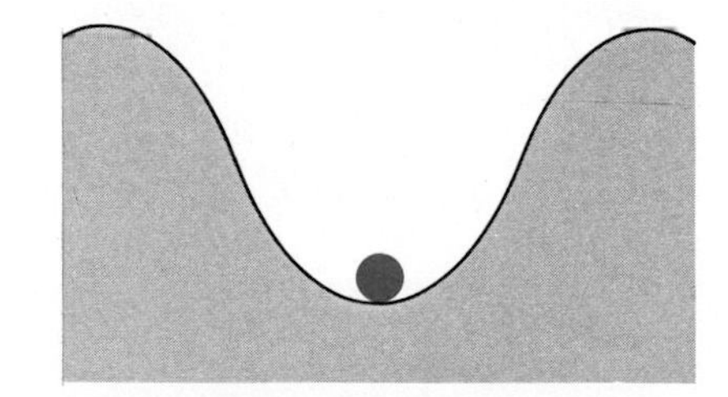

(b) Final state

Figure 19-2 A ball initially at the top of a hill will roll down the hill and eventually come to rest. The potential energy of the ball at the bottom of the hill is less than at the top. The ball spontaneously goes from a state of high potential energy to a state of low potential energy.

We have all observed a wide variety of spontaneous chemical processes. For example, natural gas, CH_4, once ignited, burns spontaneously in air to yield carbon dioxide and water:

$$CH_4(g) + 2O_2(g) \rightarrow CO_2(g) + 2H_2O(g) \qquad \Delta H^\circ_{rxn} = -802 \text{ kJ}$$

Iron, on exposure to air and moisture, spontaneously rusts:

$$4Fe(s) + 3O_2(g) \xrightarrow[H_2O(l)]{} 2Fe_2O_3(s) \qquad \Delta H^\circ_{rxn} = -1648 \text{ kJ}$$

Zinc metal reacts spontaneously with 1.0 M hydrochloric acid to yield hydrogen gas and $ZnCl_2(aq)$:

$$Zn(s) + 2HCl(aq) \rightarrow H_2(g) + ZnCl_2(aq) \qquad \Delta H^\circ_{rxn} = -150 \text{ kJ}$$

Furthermore, these reactions do not occur spontaneously in the reverse direction.

The three chemical reactions we have considered so far have been highly exothermic, and, by analogy with mechanical systems that move spontaneously to lower energy states, we predict (correctly) that they occur spontaneously. There are many spontaneous processes, however, that are not highly exothermic. Let's consider two gases, such as $N_2(g)$ and $I_2(g)$, occupying two separate containers, as shown in Figure 19-3. We all know that, if allowed to, the gases will mix spontaneously. Furthermore, simple gaseous mixtures do not separate spontaneously. It would be a disastrous occurrence if the air in a room suddenly separated so that part of the room contained pure oxygen and the rest contained nitrogen. Although the mixing of two gases is what we could call a spontaneous process, it turns out that $\Delta H^\circ_{rxn} \approx 0$ for such a process. Surely it is not the energy change that drives the process.

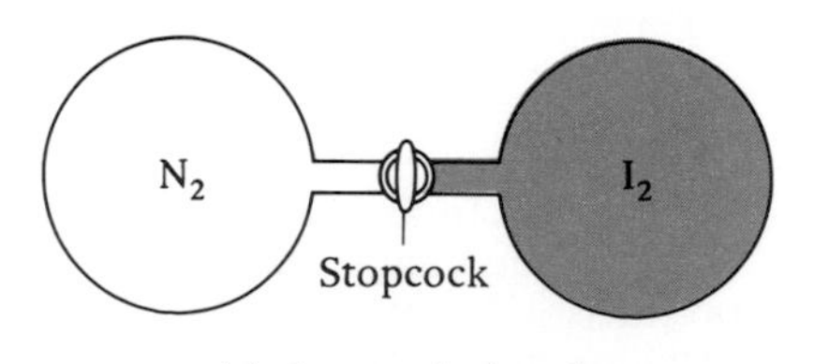

(a) Stopcock closed

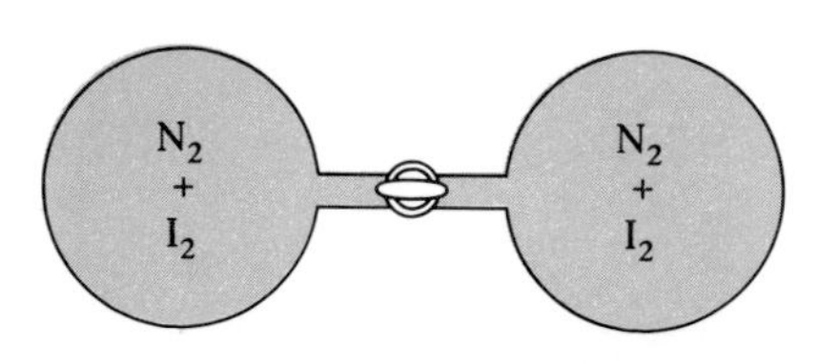

(b) After stopcock is opened

Figure 19-3 A simple example of a spontaneous process is the mixing of two gases.

Not only do some spontaneous processes have $\Delta H^\circ_{rxn} \approx 0$, but others are even endothermic. The energy of the products is *greater* than that of the reactants. For example, ice at any temperature greater than 0°C spontaneously melts:

$$H_2O(s) \rightarrow H_2O(l) \qquad \Delta H^\circ_{rxn} = +6.0 \text{ kJ}$$

and table salt spontaneously dissolves in water at 20°C:

$$NaCl(s) \xrightarrow[H_2O(l)]{} Na^+(aq) + Cl^-(aq) \qquad \Delta H^\circ_{rxn} = +6.4 \text{ kJ}$$

The fact that spontaneous processes can occur with negative, zero, or positive values of ΔH_{rxn} clearly indicates that heat evolution, that is, $\Delta H_{rxn} < 0$, is *not* a generally suitable criterion of reaction spontaneity. Reaction spontaneity is not determined solely by the energy change for a reaction. There is an additional factor involved. We shall see that this factor is the entropy change for the reaction.

19-2. THE SECOND LAW OF THERMODYNAMICS PLACES AN ADDITIONAL RESTRICTION ON ENERGY TRANSFERS

The only restriction placed on energy transfers by the first law of thermodynamics (Chapter 5) is that of conservation of energy. We know from every-day experience, however, that there are other restrictions on energy transfers. For example, energy as heat always flows spontaneously from a region of higher temperature to a region of lower temperature. A printed page of paper, once ignited, burns spontaneously in oxygen, but the reverse process, that is, the spontaneous recombination of the combustion products to the printed page and oxygen gas, has never been observed in nature. Nor is there any incantation that will cause a scrambled egg to reassemble itself. Innumerable other naturally occurring processes are unidirectional.

Understanding these additional restrictions on energy transfers involves a quantity called the **entropy,** denoted by S. We shall have more to say about the nature of entropy in subsequent sections; at this point it is sufficient to understand that entropy changes are closely related to the transfer of energy as heat. For example, when energy is transferred *only* as heat, q, under conditions of constant temperature, the entropy change of the system is given by

$$\Delta S = \frac{q}{T} \tag{19-1}$$

where ΔS is the **entropy change** and T is the absolute temperature. Note from Equation (19-1) that the units of entropy are joule per kelvin, that is, $J \cdot K^{-1}$.

The **second law of thermodynamics** states that the *total* entropy change for spontaneous processes must always be positive. By total entropy change, we mean not only the entropy of the system that we are studying but also that of its surroundings as well. Because the most general example of a system and its surroundings includes the entire universe, and because all naturally occurring processes are spontaneous, one statement of the second law of thermodynamics is that the entropy of the universe is constantly increasing.

In the next few sections we develop the concept of entropy in molecular terms.

19-3. ENTROPY IS A MEASURE OF THE AMOUNT OF DISORDER OR RANDOMNESS IN A SYSTEM

The entropy of a substance is a quantitative measure of the amount of disorder in the substance. The disorder is of two types: **positional disorder,** the distribution of the particles in space, and **thermal disorder,** the distribution of the available energy among the particles. Any process that produces a more random distribution of the particles in space or any constant-pressure process that increases the temperature of the particles gives rise to an increase in the total entropy of the substance.

Entropy probably is a new concept to you, and so to help you understand entropy, let's look at Figure 19-4, where the molar entropy of

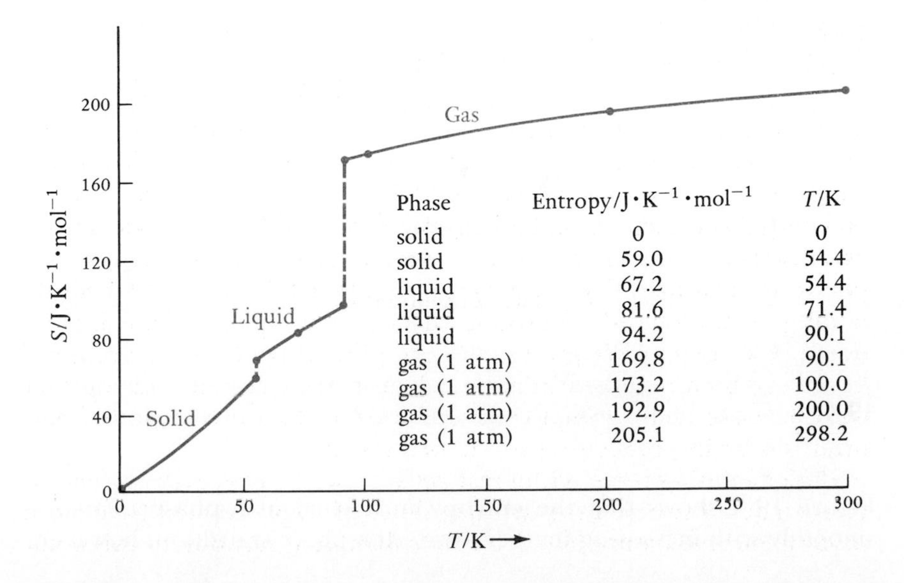

Phase	Entropy/$J \cdot K^{-1} \cdot mol^{-1}$	T/K
solid	0	0
solid	59.0	54.4
liquid	67.2	54.4
liquid	81.6	71.4
liquid	94.2	90.1
gas (1 atm)	169.8	90.1
gas (1 atm)	173.2	100.0
gas (1 atm)	192.9	200.0
gas (1 atm)	205.1	298.2

Figure 19-4 The molar entropy of oxygen as a function of temperature at 1 atm. Note that the entropy of each phase increases smoothly with increasing temperature. This gradual increase in entropy with increase in temperature is a result of the increase in thermal disorder. The jumps in entropy that occur when the solid melts and the liquid vaporizes are the result of an increase in positional disorder of the O_2 molecules.

oxygen is plotted against temperature from 0 K to 300 K. The entropy is zero at 0 K. At this temperature (absolute zero), there is neither positional disorder—the O_2 molecules in the crystal are perfectly arrayed in the lattice—nor thermal disorder—all the molecules are in the lowest possible energy state. Entropy is associated with disorder—no disorder, no entropy. The entropy of a (perfect) crystalline substance is zero at absolute zero.

- The statement that the entropy of a perfect crystalline substance is zero at absolute zero is known as the third law of thermodynamics.

As the temperature is increased from 0 K in Figure 19-4, the oxygen molecules begin to vibrate more freely about their lattice sites. The increase in temperature causes an increase in thermal disorder. When the temperature of a substance held at constant pressure increases, the entropy of the substance increases. The increase in entropy with an increase in temperature is associated with the greater amount of energy that must be distributed among the molecules of the substance. Molecules, like atoms, are restricted to discrete, quantized energy levels. The greater the amount of energy stored in a substance, the greater the number of ways in which the energy can be distributed and thus the greater the thermal disorder and the greater the entropy.

- When energy as heat is added to a system, the entropy of the system always increases (second law of thermodynamics).

When the volume of a gas is increased at a fixed temperature, the molar entropy of the gas increases, because the gas molecules have a larger available volume in which to move and thus have a higher degree of positional disorder. Similarly, the molar entropy of a solute increases when additional solvent is added to the solution, because the solute particles have more room in which to move and thus have a higher degree of positional disorder.

Example 19-1: Predict the sign of the entropy change, ΔS, for the following processes:

(a) $H_2O(g)$ (1 atm, 25°C) $\rightarrow$ $H_2O(g)$ (0.01 atm, 25°C)

(b) the dissolution of sodium chloride in water

(c) $NaCl(aq)$ (4.8 M, 20°C) $\rightarrow$ $NaCl(aq)$ (1.0 M, 20°C)

Solution: (a) A gas at a lower pressure, and thus with greater volume, is more disordered than a gas at a higher pressure at the same temperature; therefore

$$\Delta S_{gas} \text{ (constant-temperature expansion)} > 0$$

(b) Upon dissolving, the sodium and chloride ions are dispersed throughout the solvent and much more disordered than in the crystal, and so

$$\Delta S_{solute} \text{ (dissolution)} > 0$$

(c) Dilution with solvent increases the positional disorder of the solute, and so

$$\Delta S_{solute} \text{ (dilution)} > 0$$

19-4. THERE IS AN INCREASE IN ENTROPY ON MELTING AND VAPORIZATION

Figure 19-4 shows that the entropy in a particular phase increases smoothly with increasing temperature. At a phase transition, however, there is a jump in entropy. Let's investigate the origin of the two jumps in Figure 19-4. When solid oxygen melts to liquid oxygen (at 54.4 K), the ordered lattice of the solid breaks down into a liquid, where the molecules no longer are confined to lattice sites. Each molecule moves throughout the liquid volume, and so there is an increase in the positional disorder of the oxygen (Figure 19-5). This increase in positional disorder leads to an increase in entropy. Because all the entropy increase occurs at the melting point, ΔS of melting (fusion) appears as a vertical jump in Figure 19-4.

We can calculate the change in the molar entropy of a substance upon fusion. If we let ΔS_{fus} be the molar entropy change upon fusion, that is, the **molar entropy of fusion,** then using Equation (19-1) and the fact that $q = \Delta H_{fus}$, we have

$$\Delta S_{fus} = \frac{\Delta H_{fus}}{T_m} \tag{19-2}$$

where ΔH_{fus} is the molar enthalpy of fusion (Chapter 11) and T_m is the melting (fusion) point in kelvins. Using Equation (19-2), we find that the entropy increase of one mole of oxygen upon melting at 54.4 K is

$$\Delta S_{fus} = \frac{\Delta H_{fus}}{T_m} = \frac{440\ \mathrm{J{\cdot}mol^{-1}}}{54.4\ \mathrm{K}} = 8.09\ \mathrm{J{\cdot}K^{-1}{\cdot}mol^{-1}}$$

Note that ΔS_{fus} is positive, that is, the entropy of liquid oxygen is greater than the molar entropy of solid oxygen by $8.09\ \mathrm{J{\cdot}K^{-1}{\cdot}mol^{-1}}$ at 54.4 K.

When liquid oxygen is heated from its melting point to its boiling point, the entropy increases smoothly, owing to an increase in thermal disorder. At the boiling point, there is another jump in entropy. This jump occurs because the molecules in the gas phase can move throughout a much larger volume than the molecules in the liquid phase. Therefore, there is a sudden increase in positional disorder in going from the liquid phase to the gas phase (Figure 19-6).

We can calculate the change in the entropy of a substance upon vaporization by using a formula that is analogous to Equation (19-2). If we let ΔS_{vap} be the molar entropy change upon vaporization, that is, the **molar entropy of vaporization,** then using Equation (19-1) and the fact that $q = \Delta H_{vap}$, we have

$$\Delta S_{vap} = \frac{\Delta H_{vap}}{T_b} \tag{19-3}$$

where ΔH_{vap} is the molar enthalpy (heat) of vaporization (Chapter 11) and T_b is the boiling point in kelvins. Using Equation (19-3), we find that the change in the entropy of one mole of liquid oxygen upon being vaporized at 90.1 K is

$$\Delta S_{vap} = \frac{\Delta H_{vap}}{T_b} = \frac{6810\ \mathrm{J{\cdot}mol^{-1}}}{90.1\ \mathrm{K}} = 75.6\ \mathrm{J{\cdot}K^{-1}{\cdot}mol^{-1}}$$

Note that ΔS_{vap} is positive, that is, the molar entropy of gaseous oxygen is greater than the molar entropy of liquid oxygen.

Note also that $\Delta S_{vap} > \Delta S_{fus}$; this is a consequence of the fact that the molecules in the gas phase are much more disordered than the molecules in the liquid or the solid phase.

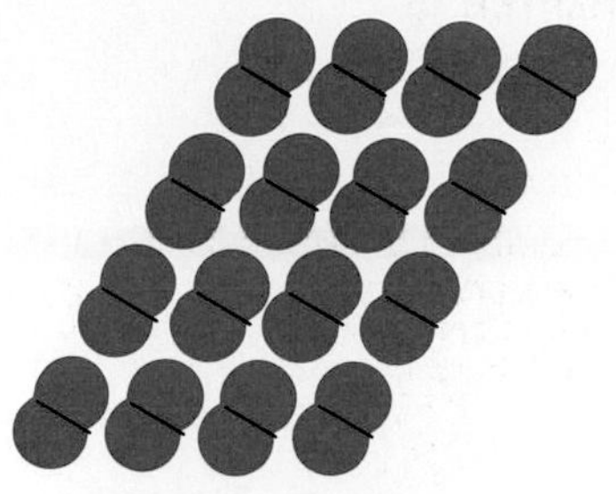

Figure 19-5 When a solid melts, the molecules become free to move throughout the volume of the liquid. This increased positional disorder in the liquid implies that, under the same conditions, the entropy of a liquid is always greater than the entropy of the solid.

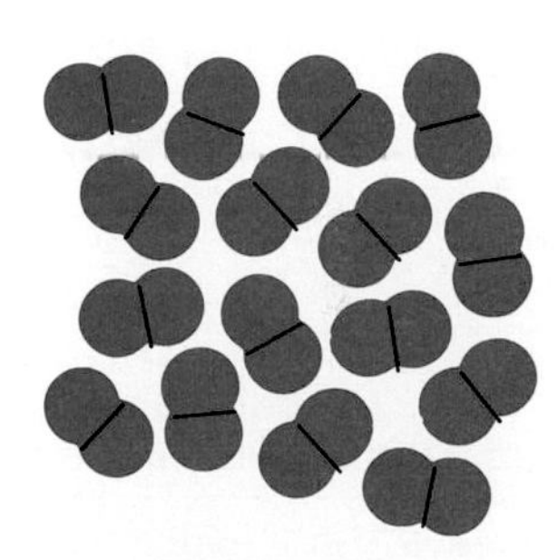

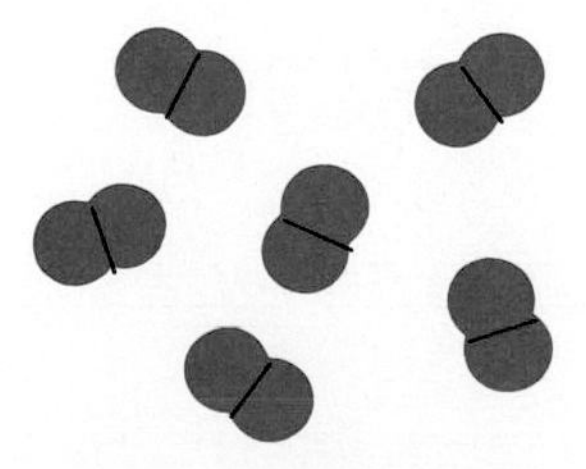

Figure 19-6 When a liquid vaporizes, there is a large increase in positional disorder and hence a large increase in entropy.

Example 19-2: The molar enthalpy of fusion of water is $\Delta H_{fus} = 6.02\ \text{kJ}\cdot\text{mol}^{-1}$ at 0°C, and the molar enthalpy of vaporization of water is $\Delta H_{vap} = 40.7\ \text{kJ}\cdot\text{mol}^{-1}$ at 100°C. Compute ΔS_{fus} and ΔS_{vap} for water.

Solution: The value of ΔS_{fus} is computed using Equation (19-2). The melting point of water is 0°C, or 273 K, and so we have

$$\Delta S_{fus} = \frac{\Delta H_{fus}}{T_m} = \frac{6.02 \times 10^3\ \text{J}\cdot\text{mol}^{-1}}{273\ \text{K}} = 22.1\ \text{J}\cdot\text{K}^{-1}\cdot\text{mol}^{-1}$$

The value of ΔS_{vap} is computed using Equation (19-3), with $T_b = 373$ K:

$$\Delta S_{vap} = \frac{\Delta H_{vap}}{T_b} = \frac{40.7 \times 10^3\ \text{J}\cdot\text{mol}^{-1}}{373\ \text{K}} = 109\ \text{J}\cdot\text{K}^{-1}\cdot\text{mol}^{-1}$$

Here, as in all cases, both ΔS_{fus} and ΔS_{vap} are positive, and $\Delta S_{vap} > \Delta S_{fus}$.

19-5. THE MOLAR ENTROPY DEPENDS ON MOLAR MASS AND MOLECULAR STRUCTURE

Remember that the superscript degree sign on a thermodynamic quantity tells us that the value given is the value of the quantity at standard conditions—that is, at 1 atm pressure. Thus we denote the **standard molar entropy** of a substance by $S°$. Table 19-1 lists the standard molar entropies of a variety of substances at 25°C. Notice that there is also a column headed ΔG_f° (the standard molar Gibbs free energy of formation) in Table 19-1. We shall learn about ΔG_f° in Section 19-10.

The molar entropy of a gas or a solute depends strongly on concentration because of the change in positional randomness with concentration. In this regard, the property of entropy differs sharply from enthalpy, which is relatively insensitive to pressure or concentration.

Let's look at the standard molar entropy values in Table 19-1 and try to determine some trends. First notice that the molar entropies of the

Table 19-1 Standard molar entropies ($S°$), heats of formation (ΔH_f°), and Gibbs free energies of formation (ΔG_f°) of various substances at 25°C and 1 atm

Substance	$S°/\text{J}\cdot\text{K}^{-1}\cdot\text{mol}^{-1}$	$\Delta H_f^\circ/\text{kJ}\cdot\text{mol}^{-1}$	$\Delta G_f^\circ/\text{kJ}\cdot\text{mol}^{-1}$	Substance	$S°/\text{J}\cdot\text{K}^{-1}\cdot\text{mol}^{-1}$	$\Delta H_f^\circ/\text{kJ}\cdot\text{mol}^{-1}$	$\Delta G_f^\circ/\text{kJ}\cdot\text{mol}^{-1}$
Ag(*s*)	42.6	0	0	H_2O_2(*l*)	110.0	−187.8	−120.4
AgBr(*s*)	107.1	−100.4	−96.90	HF(*g*)	173.6	−271.1	−273
AgCl(*s*)	96.2	−127.1	−109.8	HCl(*g*)	186.8	−92.31	−95.30
Al(*s*)	28.3	0	0	HBr(*g*)	198.6	−36.4	−53.43
Al_2O_3(*s*)	50.9	−1676	−1582	HI(*g*)	206.4	26.1	1.7
Ar(*g*)	154.7	0	0	H_2S(*g*)	205.7	−20.1	−33.0
Ba(*s*)	62.8	0	0	He(*g*)	126.0	0	0
$BaCO_3$(*s*)	112.2	−1216	−1138	I(*g*)	180.7	106.8	70.28
BaO(*s*)	70.2	−553.5	−525.1	I_2(*g*)	260.6	62.4	19.36
Br(*g*)	174.9	111.9	82.43	I_2(*s*)	116.5	0	0
Br_2(*g*)	245.4	30.91	3.14	Kr(*g*)	164.0	0	0

Table 19-1 (continued) Standard molar entropies ($S°$), heats of formation (ΔH_f°), and Gibbs free energies of formation (ΔG_f°) of various substances at 25°C and 1 atm

Substance	$S°$/J·K^{-1}·mol^{-1}	ΔH_f°/kJ·mol^{-1}	ΔG_f°/kJ·mol^{-1}	Substance	$S°$/J·K^{-1}·mol^{-1}	ΔH_f°/kJ·mol^{-1}	ΔG_f°/kJ·mol^{-1}
$Br_2(l)$	152.2	0	0	$N(g)$	153.2	472.6	455.5
$C(s$, diamond)	2.38	1.90	2.900	$N_2(g)$	191.5	0	0
$C(s$, graphite)	5.74	0	0	$NH_3(g)$	192.5	−46.19	−16.64
$CH_4(g)$	186.2	−74.86	−50.75	$N_2H_4(l)$	121.0	50.6	149
$C_2H_2(g)$	200.8	226.7	209.2	$NO(g)$	210.6	90.37	86.69
$C_2H_4(g)$	219.6	52.28	68.12	$NO_2(g)$	240.4	33.85	51.84
$C_2H_6(g)$	229.5	−84.68	−32.89	$N_2O(g)$	220.2	81.55	103.6
$C_3H_8(g)$	269.9	−103.8	−23.49	$N_2O_4(g)$	304.3	9.66	98.29
$C_6H_6(l)$	172.8	49.03	124.5	$N_2O_5(s)$	113.1	−41.8	134.2
$CH_3OH(l)$	126.9	−238.7	−166.3	$NOCl(g)$	262	51.9	66.1
$C_2H_5OH(l)$	160.8	−277.7	−174.8	$Na(g)$	153.6	107.1	77.30
$CH_3Cl(g)$	234.8	−80.83	−57.40	$Na(s)$	51.3	0	0
$CH_3Cl(l)$	145.3	−102	−51.5	$NaHCO_3(s)$	102	−947.7	−851.9
$CH_2Cl_2(g)$	270.4	−92.47	−65.90	$Na_2CO_3(s)$	138.8	−1131.1	−1048.2
$CH_2Cl_2(l)$	178.1	−121	−67.4	$Na_2O(s)$	75.06	−418.0	−379.1
$CHCl_3(g)$	294.9	−103.1	−70.37	$NaCl(s)$	72.13	−411.2	−384.0
$CHCl_3(l)$	202.6	−134.5	−73.72	$NaBr(s)$	86.82	−361.4	−349.3
$CCl_4(g)$	308.7	−103.0	−60.63	$NaI(s)$	98.49	−287.8	−284.6
$CCl_4(l)$	215.4	−135.4	−65.27	$Ne(g)$	146.2	0	0
$CO(g)$	197.8	−110.5	−137.2	$O(g)$	161.0	247.5	230.1
$CO_2(g)$	213.6	−393.5	−394.4	$O_2(g)$	205.0	0	0
$Ca(s)$	41.5	0	0	$O_3(g)$	238.8	142	163.4
$CaC_2(s)$	69.9	−59.8	−64.8	$P(s$, white)	41.1	0	0
$CaCO_3(s)$	92.9	−1207	−1129	$P(s$, red)	22.8	−18.4	−12.6
$CaO(s)$	39.7	−635.1	−604.0	$P_4O_{10}(s)$	228.9	−2984	−2698
$Cl(g)$	165.1	121.7	105.7	$POCl_3(g)$	325.3	−558.5	−513.0
$Cl_2(g)$	222.9	0	0	$POCl_3(l)$	222	−597.0	−520.9
$Cu(s)$	33.2	0	0	$PCl_3(g)$	311.7	−306.4	−286.3
$CuO(s)$	43.6	−157	−130	$PCl_5(g)$	354.5	−375.0	−305.0
$Cu_2O(s)$	93.1	−169	−146	$PH_3(g)$	210.1	5.4	13.1
$F(g)$	158.6	78.99	61.92	$S(s$, rhombic)	31.8	0	0
$F_2(g)$	203.7	0	0	$S(s$, monoclinic)	32.6	0.30	0.10
$Fe(s)$	27.3	0	0	$SO_2(g)$	248.4	−296.8	−300.2
$Fe_2O_3(s)$	87.9	−824.2	−742.2	$SO_3(g)$	256.3	−395.7	−371.1
$Fe_3O_4(s)$	146.3	−1118	−1015	$SF_6(g)$	291.7	−1209.3	−1105.2
$H(g)$	114.6	218.0	203.3	$Xe(g)$	169.6	0	0
$H_2(g)$	130.6	0	0	$Zn(s)$	41.63	0	0
$H_2O(g)$	188.7	−241.8	−228.6	$ZnO(s)$	43.64	−348.3	−318.3
$H_2O(l)$	69.9	−285.8	−237.2	$ZnS(s)$	57.7	−206.0	−210.3

gaseous substances are the largest and the molar entropies of the solid substances are the smallest. We have already discussed the reason for this.

Now consider the standard molar entropies of the noble gases, whose values in $J \cdot K^{-1} \cdot mol^{-1}$ at 25°C and 1 atm are: He(g), 126.0; Ne(g), 146.2; Ar(g), 154.7; Kr(g), 164.0; and Xe(g), 169.6. The increase in molar entropy of the noble gases is a consequence of their increasing mass as we move down the periodic table. The increased mass leads to an increased thermal disorder and thus to a greater entropy. The same trend can be seen by comparing the molar entropies of the gaseous halogens, which are, in $J \cdot mol^{-1} \cdot K^{-1}$: $F_2(g)$, 204; $Cl_2(g)$, 223; $Br_2(g)$, 245; and $I_2(g)$, 261.

Generally speaking, the more atoms of a given type there are in a molecule, the greater the capacity of the molecule to take up energy and thus the greater the entropy. This trend is illustrated by the series $C_2H_2(g)$, $C_2H_4(g)$, and $C_2H_6(g)$, whose Lewis formulas are

H—C≡C—H

acetylene (201)

$H_2C{=}CH_2$

ethylene (220)

$H_3C{-}CH_3$

ethane (230)

The standard molar entropies in $J \cdot K^{-1} \cdot mol^{-1}$ are shown in parentheses. For molecules with the same number of atoms, the standard molar entropy increases with increasing molecular mass.

Example 19-3: Without referring to Table 19-1, arrange the following molecules in order of increasing standard molar entropy:

$$CH_2Cl_2(g) \qquad CHCl_3(g) \qquad CH_3Cl(g)$$

Solution: The number of atoms is the same in each case, but chlorine has a greater mass than hydrogen. Thus we predict that

$$S°(CH_3Cl) < S°(CH_2Cl_2) < S°(CHCl_3)$$

The values of the standard molar entropies in $J \cdot K^{-1} \cdot mol^{-1}$ at 25°C are 234.8, 270.4, and 294.9, respectively.

Another interesting comparison involves the isomers acetone and trimethylene oxide. At 25°C their standard molar entropies are

$(H_3C)_2C{=}O$

acetone

$S° = 295\ J \cdot K^{-1} \cdot mol^{-1}$

$(CH_2)_3O$ (ring)

trimethylene oxide

$S° = 274\ J \cdot K^{-1} \cdot mol^{-1}$

▪ For molecules with approximately the same molecular masses, the more compact the molecule, the smaller is its entropy.

The entropy of acetone is higher than the entropy of trimethylene oxide because of the free rotation of the methyl groups about the carbon-carbon bonds in the acetone molecule. The relatively rigid ring structure of the trimethylene oxide molecule restricts the movement of the ring atoms. This restriction gives rise to a lower entropy because the capacity of the rigid isomer to take up energy is less than that of the more flexible acetone molecule.

19-6. ΔS°_{rxn} EQUALS THE ENTROPY OF THE PRODUCTS MINUS THE ENTROPY OF THE REACTANTS

We can use the values of the standard molar entropies in Table 19-1 to calculate standard entropy changes of reactions, ΔS°_{rxn}. Consider the reaction described by the general chemical equation

$$aA + bB \rightarrow yY + zZ \quad (19\text{-}4)$$

We define the **standard entropy change for a reaction,** ΔS°_{rxn}, as

$$\Delta S^\circ_{rxn} = S^\circ(\text{products}) - S^\circ(\text{reactants}) \quad (19\text{-}5)$$

Application of Equation (19-5) to Equation (19-4) yields

$$\Delta S^\circ_{rxn} = yS^\circ[Y] + zS^\circ[Z] - aS^\circ[A] - bS^\circ[B] \quad (19\text{-}6)$$

Let's use Equation (19-6) to compute ΔS°_{rxn} at 25°C for the equation

$$CO(g) + 3H_2(g) \rightarrow CH_4(g) + H_2O(g)$$

From Equation (19-6) we have

$$\Delta S^\circ_{rxn} = S^\circ[CH_4(g)] + S^\circ[H_2O(g)] - S^\circ[CO(g)] - 3S^\circ[H_2(g)]$$

The S°[compound] values are obtained from Table 19-1:

$$\begin{aligned}\Delta S^\circ_{rxn} &= (1\text{ mol})(186.2\text{ J}\cdot\text{K}^{-1}\cdot\text{mol}^{-1}) + (1\text{ mol})(188.7\text{ J}\cdot\text{K}^{-1}\cdot\text{mol}^{-1})\\ &\quad - (1\text{ mol})(197.8\text{ J}\cdot\text{K}^{-1}\cdot\text{mol}^{-1}) - (3\text{ mol})(130.6\text{ J}\cdot\text{K}^{-1}\cdot\text{mol}^{-1})\\ &= -214.7\text{ J}\cdot\text{K}^{-1}\end{aligned}$$

There is a large negative change in entropy for this reaction because there are only 2 mol of gaseous products but 4 mol of gaseous reactants. Under the same conditions, the volume of the products will be only about one-half the volume of the reactants. The product state is more ordered than the reactant state, and so the entropy of the products is less than that of the reactants.

By the same argument, we predict that the standard entropy change for the equation

$$2H_2O(l) \rightarrow 2H_2(g) + O_2(g)$$

is positive. The volume, and hence the disorder (entropy), of the products is much greater than the volume, and the entropy, of the reactants. Using Table 19-1, we obtain

$$\begin{aligned}\Delta S^\circ_{rxn} &= 2S^\circ[H_2(g)] + S^\circ[O_2(g)] - 2S^\circ[H_2O(l)]\\ &= (2\text{ mol})(130.6\text{ J}\cdot\text{K}^{-1}\cdot\text{mol}^{-1}) + (1\text{ mol})(205.0\text{ J}\cdot\text{K}^{-1}\cdot\text{mol}^{-1})\\ &\qquad - (2\text{ mol})(69.9\text{ J}\cdot\text{K}^{-1}\cdot\text{mol}^{-1})\\ &= +326.4\text{ J}\cdot\text{K}^{-1}\end{aligned}$$

We see that ΔS°_{rxn} is positive, just as we predicted.

▪ The greater the difference between the total number of moles of gaseous products and the total number of moles of gaseous reactants, the greater is the value of ΔS°_{rxn}.

Example 19-4: Predict the sign of ΔS°_{rxn} for the equation

$$N_2H_4(l) + O_2(g) \rightarrow 2H_2O(g) + N_2(g)$$

Use data in Table 19-1 to calculate ΔS°_{rxn} for the equation.

Solution: There are 3 mol of gaseous products and 1 mol of gaseous reactants. Therefore, we predict that the entropy of the products will be greater than the entropy of the reactants, and so $\Delta S^\circ_{rxn} > 0$. Using data in Table 19-1, we obtain

$$\begin{aligned}\Delta S^\circ_{rxn} &= 2S^\circ[H_2O(g)] + S^\circ[N_2(g)] - S^\circ[O_2(g)] - S^\circ[N_2H_4(l)] \\ &= (2\text{ mol})(188.7\text{ J}\cdot\text{K}^{-1}\cdot\text{mol}^{-1}) + (1\text{ mol})(191.5\text{ J}\cdot\text{K}^{-1}\cdot\text{mol}^{-1}) \\ &\quad -(1\text{ mol})(205.0\text{ J}\cdot\text{K}^{-1}\cdot\text{mol}^{-1}) - (1\text{ mol})(121.0\text{ J}\cdot\text{K}^{-1}\cdot\text{mol}^{-1}) \\ &= +242.9\text{ J}\cdot\text{K}^{-1}\end{aligned}$$

As we predicted, ΔS°_{rxn} is positive.

19-7. NATURE ACTS TO MINIMIZE THE ENERGY AND TO MAXIMIZE THE ENTROPY OF ALL PROCESSES

The ill-fated principle of Berthelot that a reaction has to be exothermic to be spontaneous was based on the notion that reactions that evolve energy are spontaneous because the products are energetically downhill from the reactants. However, we have seen that there are spontaneous reactions in which the products are either energetically uphill from the reactants or have essentially the same energy as the reactants. Such reactions are **entropy-driven reactions.** A reaction for which the total entropy of the products is greater than the total entropy of the reactants ($\Delta S_{rxn} > 0$) is an **entropy-favored reaction.**

A reaction for which the total energy of the products is less than the total energy of the reactants ($\Delta H_{rxn} < 0$) is an **energy-favored reaction.** If there is little or no change in entropy ($\Delta S_{rxn} \approx 0$), then a system will change in such a way that its energy is decreased. When the energy of the system is minimized, the system no longer changes and is at equilibrium. The simple mechanical examples of energy minimization for spontaneous processes that we discussed in Section 19-1 (for example, water flowing downhill) are processes for which there is little or no change in entropy.

If there is no change in energy ($\Delta H_{rxn} \approx 0$), then a system will change in such a way that its entropy is increased. When the entropy of the system is maximized, the system no longer changes and is at equilibrium. The mixing of two gases is a good example of this case. The gases mix spontaneously because the entropy of the final state (the mixture) is greater than the entropy of the initial state (separate gases). You may have had some idea that systems try to minimize energy, but the idea of entropy maximization may be new to you. Both ideas are equally important, however.

All processes that are *both* energy-favored and entropy-favored are spontaneous. That is, if $\Delta H_{rxn} < 0$ and $\Delta S_{rxn} > 0$, then the process is spontaneous. Conversely, if $\Delta H_{rxn} > 0$ *and* $\Delta S_{rxn} < 0$, then the process does not occur spontaneously. We must now consider processes in which the energy (enthalpy) and entropy factors oppose each other.

19-8. THE SIGN OF ΔG_{rxn} DETERMINES WHETHER OR NOT A REACTION IS SPONTANEOUS

Many reactions have values of ΔH_{rxn} and ΔS_{rxn} that oppose each other. For such a case we have either

$$\Delta H_{rxn} > 0 \quad \text{and} \quad \Delta S_{rxn} > 0$$

or

$$\Delta H_{rxn} < 0 \quad \text{and} \quad \Delta S_{rxn} < 0$$

The first case is entropy-favored but energy-disfavored, and the second case is energy-favored but entropy-disfavored. Such reactions may or may not be spontaneous, depending on the temperature and on the relative magnitudes of ΔH_{rxn} and ΔS_{rxn}. The spontaneity of such reactions was explained about 100 years ago by J. Willard Gibbs, one of America's greatest scientists. Gibbs introduced a quantity now called the **Gibbs free energy,** G, which, as we shall see, serves as a compromise function between the enthalpy change and the entropy change of a reaction. For a reaction run at a constant temperature, the **Gibbs free energy change,** ΔG_{rxn}, is given by

▪ The officially approved IUPAC (International Union of Pure and Applied Chemistry) name of G is Gibbs energy, but the common usage is Gibbs free energy.

$$\Delta G_{rxn} = \Delta H_{rxn} - T\,\Delta S_{rxn} \tag{19-7}$$

where T is the kelvin temperature.

Chemical reactions are forced by nature to seek a compromise between energy minimization and entropy maximization. For a reaction that occurs at a constant temperature and pressure, the nature of the compromise is given by the sign and magnitude of ΔG_{rxn}. The **Gibbs criteria of reaction spontaneity** are

$\Delta G_{rxn} < 0$	reaction is spontaneous and additional products can form
$\Delta G_{rxn} > 0$	reaction is not spontaneous and no additional products can form without energy input
$\Delta G_{rxn} = 0$	reaction is at equilibrium and no further net change occurs

The spontaneous mixing of two gases and the spreading of a drop of ink throughout a volume of water (see page 598) are examples of processes for which $\Delta H_{rxn} \approx 0$ and $\Delta S_{rxn} > 0$. For these processes,

$$\Delta G_{rxn} = \Delta H_{rxn} - T\,\Delta S_{rxn} \approx 0 - T\,\Delta S_{rxn} < 0$$

and so the Gibbs criteria predict that two gases mix spontaneously and that a drop of ink becomes uniformly dispersed throughout a volume of water. Furthermore, note that $\Delta G_{rxn} > 0$ for the reverse reactions. Therefore, the Gibbs criteria predict that we shall never see the reverse processes occur spontaneously.

Thus we see that, in general, a system will change in such a way that its Gibbs free energy is minimized. The Gibbs free energy of a substance consists of an enthalpy term and an entropy term. The enthalpy term is fairly independent of pressure (for a gaseous species) or concentration (for a species in solution), but as we have discussed in Section 19-3, the entropy term depends strongly on these quantities. In any spontaneous process, the composition of a system will change through the reaction of some species and the formation of others, thereby changing their pressures or concentrations so that the Gibbs free energy of the system is minimized.

In Chapter 14 we learned about another set of criteria that governed the direction in which a chemical reaction will proceed. These criteria involved the ratio of the reaction quotient Q to the equilibrium constant K. Recall that

if $\frac{Q}{K} > 1$ or $Q > K$ reaction proceeds spontaneously from right to left

if $\frac{Q}{K} < 1$ or $Q < K$ reaction proceeds spontaneously from left to right

if $\frac{Q}{K} = 1$ or $Q = K$ reaction is at equilibrium

Because the sign of ΔG_{rxn} also determines whether a reaction is spontaneous or not, we should expect a relation between ΔG_{rxn} and Q/K; the relation is

$$\Delta G_{rxn} = 2.30RT \log\left(\frac{Q}{K}\right) \quad (19\text{-}8)$$

▪ Note that because the units of Q and K are the same, the units cancel in the ratio Q/K.

Note that

if $\frac{Q}{K} > 1$ then $\Delta G_{rxn} = 2.30RT \log(>1) > 0$

if $\frac{Q}{K} < 1$ then $\Delta G_{rxn} = 2.30RT \log(<1) < 0$

if $\frac{Q}{K} = 1$ then $\Delta G_{rxn} = 2.30RT \log(1) = 0$

In applying any of these criteria to a reaction, it is worth remembering that, even though a reaction may be spontaneous under the prevailing conditions, it may not occur at a detectable rate. *Spontaneous is not synonymous with immediate.* On the other hand, if $\Delta G_{rxn} > 0$, then the reaction will not occur under the prevailing conditions. The *no* of thermodynamics is emphatic; the *yes* of thermodynamics is actually a *maybe*. For a reaction to occur, it is *absolutely necessary* that $\Delta G_{rxn} < 0$, but a negative value of ΔG_{rxn} is not sufficient to guarantee that the reaction will occur at a detectable rate.

The following Example illustrates the use of Equation (19-8).

Example 19-5: The reaction system

$$C(s) + H_2O(g) \rightleftharpoons CO(g) + H_2(g)$$

is prepared at 800°C with the initial concentration $[CO]_0 = 0.500$ M, $[H_2]_0 = 4.25 \times 10^{-2}$ M, and $[H_2O]_0 = 0.150$ M. Given that $K_c = 7.99 \times 10^{-2}$ M at 800°C, calculate ΔG_{rxn} for this reaction system and indicate in which direction the reaction will proceed spontaneously.

Solution: The value of Q for this reaction system is given by

$$Q_c = \frac{[CO]_0[H_2]_0}{[H_2O]_0} = \frac{(0.500\ M)(4.25 \times 10^{-2}\ M)}{0.150\ M} = 0.142\ M$$

The value of Q_c/K_c is $(0.142\ M / 7.99 \times 10^{-2}\ M) = 1.78$, and so

$$\Delta G_{rxn} = 2.30RT \log\left(\frac{Q_c}{K_c}\right)$$

$$= (2.30)(8.314\ J\cdot K^{-1})(1073\ K) \log (1.78)$$

$$= +5.14\ kJ$$

The positive value of ΔG_{rxn}, as well as the fact that $Q_c > K_c$, implies that the reaction system will evolve in such a way that the concentrations of $CO(g)$ and $H_2(g)$ will decrease and that of $H_2O(g)$ will increase, or that the reaction

$$C(s) + H_2O(g,\ 0.150\ M) \rightleftharpoons CO(g,\ 0.500\ M) + H_2(g,\ 0.0425\ M)$$

will proceed spontaneously from right to left.

At this point you may think that the introduction of ΔG_{rxn} has not given us anything new because we can determine the direction in which a reaction will proceed spontaneously by simply using the value of Q/K. But what if we are not given the value of the equilibrium constant? We shall learn in Section 19-10 that we can use Equation (19-8) and Table 19-1 to calculate the equilibrium constant for a given reaction and thus determine reaction spontaneity without being given the value of K. Another advantage of introducing ΔG_{rxn} is that the value of ΔG_{rxn} is related to the energy that can be obtained from a process. If ΔG_{rxn} is negative (spontaneous reaction), then its magnitude is equal to the maximum energy that can be obtained from the reaction. For example, ΔG_{rxn} at 25°C for the reaction

$$2H_2(g,\ 1\ atm) + O_2(g,\ 1\ atm) \rightarrow 2H_2O(l)$$

▪ The word "free" in Gibbs free energy is used in the sense of "available to perform work."

is -4.74×10^5 J. Thus this reaction can provide a maximum of 4.74×10^5 J of energy to an external device, such as an electric motor. The maximum amount of energy that we calculate for a reaction is an ideal value. In practice we would obtain less than the maximum amount.

If ΔG_{rxn} is positive, then its value is the minimum energy that must be supplied to the reaction in order to make it occur. For example, Example 19-5 shows that $\Delta G_{rxn} = +5.14$ kJ for the reaction

$$C(s) + H_2O(g,\ 0.150\ M) \rightarrow CO(g,\ 0.500\ M) + H_2(g,\ 0.0425\ M)$$

Thus we must supply at least 5.14 kJ to make the reaction occur. This energy could be supplied as heat.

19-9. IT IS THE SIGN OF ΔG_{rxn} AND NOT ΔG°_{rxn} THAT DETERMINES REACTION SPONTANEITY

We can use Equation (19-8) to express an equilibrium constant in terms of a standard Gibbs free energy change. Because the logarithm of a ratio is equal to the logarithm of the numerator minus the logarithm of the denominator, Equation (19-8) can be rewritten as

$$\Delta G_{rxn} = 2.30RT \log Q - 2.30RT \log K \qquad (19\text{-}9)$$

▪ $\log \frac{Q}{K} = \log Q - \log K$

When $Q = 1$, all reactants and products are at 1.00 M concentration for solution species or 1.00 atm pressure for gaseous species. Under these conditions, which we call the standard conditions, ΔG_{rxn} is equal to ΔG°_{rxn}, the **standard Gibbs free energy change.** Setting $Q = 1$ in Equation (19-9) thus yields

$$\Delta G^\circ_{rxn} = -2.30RT \log K \qquad (19\text{-}10)$$

Equation (19-10) shows that the magnitude of the standard Gibbs free energy change for a reaction at a particular temperature is determined by the magnitude of the equilibrium constant for the reaction at that temperature. If K is greater than unity, then $\log K$ is positive and ΔG°_{rxn} is negative. In other words, if $K > 1$, then the reaction is spontaneous *under standard conditions,* that is, for 1.00 M concentration for all solution species and 1.00 atm pressure for all gas species. If K is less than unity, then $\log K$ is negative, ΔG°_{rxn} is positive, and the reaction is not spontaneous under standard conditions.

Example 19-6: The equilibrium constant at 25°C for the chemical equation

$$HC_2H_3O_2(aq) + H_2O(l) \rightleftharpoons H_3O^+(aq) + C_2H_3O_2^-(aq)$$

is $K = 1.74 \times 10^{-5}$ M. Compute the value of ΔG°_{rxn} at 25°C.

Solution: The value of ΔG°_{rxn} can be computed from the value of K by using Equation (19-10):

▪ Note that we drop the units of K in this calculation because they cancel in Equation (19-8), which is used to derive Equation (19-10).

$$\begin{aligned}\Delta G^\circ_{rxn} &= -2.30RT \log K \\ &= -(2.30)(8.314\ \mathrm{J \cdot K^{-1}})(298\ \mathrm{K}) \log (1.74 \times 10^{-5}) \\ &= +2.71 \times 10^4\ \mathrm{J}\end{aligned}$$

The positive value of ΔG_{rxn} means that the reaction

$$HC_2H_3O_2(aq,\ 1\ \mathrm{M}) + H_2O(l) \rightleftharpoons H_3O^+(aq,\ 1\ \mathrm{M}) + C_2H_3O_2^-(aq,\ 1\ \mathrm{M})$$

will proceed spontaneously from right to left. We say that the reaction *as written* is not spontaneous from left to right under standard conditions.

It is important to remember the difference between ΔG_{rxn} and ΔG°_{rxn}. Using Equation (19-10), we can rewrite Equation (19-9) in the form

$$\Delta G_{rxn} = \Delta G^\circ_{rxn} + 2.30RT \log Q \qquad (19\text{-}11)$$

If all the reactants and products are at standard conditions, that is, if all concentrations are 1.00 M and all pressures are 1.00 atm, then $Q = 1$ in Equation (19-11) and

$$\Delta G_{rxn} = \Delta G^\circ_{rxn} \qquad \text{standard conditions}$$

We emphasize standard conditions here to point out that ΔG°_{rxn} is equal to ΔG_{rxn} only if $Q = 1$.

To see clearly the distinction between ΔG_{rxn} and ΔG°_{rxn}, consider the familiar reaction of the dissociation of the weak acid acetic acid (Example 19-6) at 25°C:

$$HC_2H_3O_2(aq) + H_2O(l) \rightleftharpoons H_3O^+(aq) + C_2H_3O_2^-(aq)$$
$$K_a = 1.74 \times 10^{-5}\ \text{M at } 25°\text{C}$$

where we calculated $\Delta G^\circ_{rxn} = +27.1$ kJ. The fact that ΔG°_{rxn} is positive does *not* mean that no acetic acid dissociates when we dissolve acetic acid in water at 25°C. Some acetic acid does dissociate because it is ΔG_{rxn}, not ΔG°_{rxn}, that determines whether or not a reaction occurs. The value of ΔG_{rxn} at 25°C for the dissociation of acetic acid in water is given by

$$\Delta G_{rxn} = \Delta G^\circ_{rxn} + 2.30RT \log Q$$
$$= 27.1\ \text{kJ} + (5.70\ \text{kJ}) \log \frac{[H_3O^+]_0[C_2H_3O_2^-]_0}{[HC_2H_3O_2]_0}$$

Let's consider a 0.10 M $HC_2H_3O_2(aq)$ solution. With $[HC_2H_3O_2]_0 =$ 0.10 M and $[H_3O]^+_0 = [C_2H_3O_2^-]_0 \approx 0$, the value of ΔG_{rxn} is very large and *negative,* because the logarithm of a very small number is a large negative number. Therefore, the dissociation of $HC_2H_3O_2(aq)$ takes place spontaneously. The concentration of $HC_2H_3O_2(aq)$ decreases and the concentrations of $H_3O^+(aq)$ and $C_2H_3O_2^-(aq)$ increase until equilibrium is reached. The equilibrium state is determined by the condition $\Delta G_{rxn} = 0$. If we set $\Delta G_{rxn} = 0$ in Equation (19-11), then we obtain

$$\Delta G^\circ_{rxn} = -2.30RT \log Q_{eq} \qquad \text{(at equilibrium)}$$

which, when compared with Equation (19-10), gives

$$Q_{eq} = K \qquad \text{equilibrium}$$

- At equilibrium, $\Delta G_{rxn} = 0$, but, in general, $\Delta G^\circ_{rxn} \neq 0$ at equilibrium.

Thus, initially ΔG_{rxn} has a large negative value, but this value decreases to zero as the reaction goes to equilibrium.

The value of ΔG_{rxn} depends upon the concentrations of the reactants and products through the quantity Q in Equation (19-11). The sign of ΔG_{rxn} depends upon the ratio Q/K. The value of ΔG°_{rxn}, on the other hand, is fixed at any given temperature and requires that all reactants and products be at standard conditions.

Example 19-7: The value of K_{sp} for AgCl(s) in equilibrium with water at 25°C is 1.8×10^{-10} M^2. Calculate ΔG_{rxn} at 25°C for the process

$$Ag^+(aq,\ 0.10\ \text{M}) + Cl^-(aq,\ 0.30\ \text{M}) \rightarrow AgCl(s)$$

Solution: This equation is the reverse of the K_{sp} equation for AgCl(s); thus we have for the equilibrium constant

$$K = \frac{1}{K_{sp}} = \frac{1}{1.8 \times 10^{-10}\ \text{M}^2} = 5.6 \times 10^9\ \text{M}^{-2}$$

Reaction spontaneity is governed by ΔG_{rxn}. From Equation (19-8) we have

$$\Delta G_{rxn} = 2.30RT \log (Q/K)$$

The value of Q is given by

$$Q = \frac{1}{[Ag^+]_0[Cl^-]_0} = \frac{1}{(0.10\ \text{M})(0.30\ \text{M})} = 33.3\ \text{M}^{-2}$$

and

$$\Delta G_{rxn} = (2.30)(8.314\ \text{J}\cdot\text{K}^{-1})(298\ \text{K}) \log \left(\frac{33.3\ \text{M}^{-2}}{5.6 \times 10^{9}\ \text{M}^{-2}}\right)$$

$$= -46.9 \times 10^{3}\ \text{J}$$

Thus, a $AgCl(s)$ precipitate will form if 0.10 M $Ag^{+}(aq)$ is added to 0.30 M $Cl^{-}(aq)$.

Let's redo the calculation in this Example, but with 2.0×10^{-6} M $Ag^{+}(aq)$ and 2.0×10^{-6} M $Cl^{-}(aq)$. In this case, $Q = 2.5 \times 10^{11}\ \text{M}^{-2}$ and $\Delta G_{rxn} = +9.40$ kJ, indicating that no $AgCl(s)$ precipitate will form if 2.0×10^{-6} M $Ag^{+}(aq)$ and 2.0×10^{-6} M $Cl^{-}(aq)$ are mixed.

19-10. ΔG°_{rxn} CAN BE CALCULATED FROM TABULATED ΔG°_{f} VALUES

In Chapter 5 we described the procedure for setting up a table of standard molar enthalpies of formation of compounds from their constituent elements. The procedure for setting up a table of **standard molar Gibbs free energies of formation** of compounds from their constituent elements is exactly the same as for ΔH°_{f} values. For the general chemical equation

$$aA + bB \rightarrow yY + zZ \qquad (19\text{-}12)$$

we have

$$\Delta G^{\circ}_{rxn} = y\Delta G^{\circ}_{f}[Y] + z\Delta G^{\circ}_{f}[Z] - a\Delta G^{\circ}_{f}[A] - b\Delta G^{\circ}_{f}[B] \qquad (19\text{-}13)$$

Table 19-1 lists ΔG°_{f} values for a variety of compounds. By convention, we take ΔG°_{f} of an element in its normal physical state at 25°C and 1 atm to be zero. This is similar to the convention that we introduced for ΔH°_{f} in Chapter 5. Let's consider equation

$$2H_2(g) + O_2(g) \rightarrow 2H_2O(l)$$

Using the ΔG°_{f} data in Table 19-1 yields

$$\Delta G^{\circ}_{rxn} = 2\Delta G^{\circ}_{f}[H_2O(l)] - 2\Delta G^{\circ}_{f}[H_2(g)] - \Delta G^{\circ}_{f}[O_2(g)]$$

$$= (2\ \text{mol})(-237.2\ \text{kJ}\cdot\text{mol}^{-1}) - (2\ \text{mol})(0) - (1\ \text{mol})(0)$$

$$= -474.4\ \text{kJ}\cdot\text{mol}^{-1}$$

With $P_{O_2} = P_{H_2} = 1$ atm, we have $Q = 1$ and $\Delta G_{rxn} = \Delta G^{\circ}_{rxn} \ll 0$. Thus the process described by the equation

$$2H_2(g,\ 1\ \text{atm}) + O_2(g,\ 1\ \text{atm}) \rightarrow 2H_2O(l)$$

is highly spontaneous, as we know from the explosive nature of hydrogen plus oxygen mixtures.

Example 19-8: Calculate ΔG°_{rxn} and K at 25°C for the equation

$$PCl_5(g) \rightleftharpoons PCl_3(g) + Cl_2(g)$$

Solution: Application of Equation (19-13) to this equation yields

$$\Delta G^{\circ}_{rxn} = \Delta G^{\circ}_{f}[PCl_3(g)] + \Delta G^{\circ}_{f}[Cl_2(g)] - \Delta G^{\circ}_{f}[PCl_5(g)]$$

Using data from Table 19-1 yields

$$\Delta G^\circ_{rxn} = (1\ \text{mol})(-286.3\ \text{kJ}\cdot\text{mol}^{-1}) + (1\ \text{mol})(0) - (1\ \text{mol})(-305.0\ \text{kJ}\cdot\text{mol}^{-1})$$
$$= +18.7\ \text{kJ}$$

The value of the equilibrium constant is calculated by using Equation (19-10):

$$\log K = -\frac{\Delta G^\circ_{rxn}}{2.30RT}$$
$$= \frac{-(18.7 \times 10^3\ \text{J})}{(2.30)(8.314\ \text{J}\cdot\text{K}^{-1})(298\ \text{K})} = -3.28$$

Thus

$$K = 10^{-3.28} = 5.2 \times 10^{-4}\ \text{atm}$$

Tables of ΔG°_f, ΔH°_f, and S° values are especially useful for the thermodynamic analysis of chemical reactions and, in particular, for the prediction of reaction spontaneity when combined with Equation (19-11).

Example 19-9: Will the reaction described by the equation

$$CO(g,\ 0.010\ \text{atm}) + 2H_2(g,\ 0.010\ \text{atm}) \rightarrow CH_3OH(l)$$

occur spontaneously at 25°C?

Solution: We first use Table 19-1 to calculate ΔG°_{rxn}:

$$\Delta G^\circ_{rxn} = \Delta G^\circ_f[CH_3OH(l)] - \Delta G^\circ_f[CO(g)] - 2\Delta G^\circ_f[H_2(g)]$$
$$= (1\ \text{mol})(-166.3\ \text{kJ}\cdot\text{mol}^{-1}) - (1\ \text{mol})(-137.2\ \text{kJ}\cdot\text{mol}^{-1}) - (2\ \text{mol})(0)$$
$$= -29.1\ \text{kJ}$$

The negative value here means that the reaction described by the equation

$$CO(g,\ 1\ \text{atm}) + 2H_2(g,\ 1\ \text{atm}) \rightarrow CH_3OH(l)$$

proceeds spontaneously at standard conditions but does not tell us whether the reaction will occur when the pressures of $CO(g)$ and $H_2(g)$ both are 0.010 atm, as originally stated. To find out, we must calculate ΔG_{rxn}:

$$\Delta G_{rxn} = \Delta G^\circ_{rxn} + 2.30RT \log Q$$
$$= -29.1\ \text{kJ} + (2.30)(8.314\ \text{J}\cdot\text{K}^{-1})(298\ \text{K}) \times \log\left(\frac{1}{(0.010)(0.010)^2}\right)$$
$$= -29.1\ \text{kJ} + 34.2\ \text{kJ} = +5.1\ \text{kJ}$$

Thus, the reaction will not occur spontaneously if the $CO(g)$ and $H_2(g)$ pressures are only 0.010 atm. If they are raised to 0.10 atm, then $\Delta G_{rxn} = -12.0$ kJ, and the reaction will occur.

19-11. THE VAN'T HOFF EQUATION GOVERNS THE TEMPERATURE DEPENDENCE OF EQUILIBRIUM CONSTANTS

Now we derive an equation that tells us how the value of an equilibrium constant varies with temperature. Equation (19-7) at standard conditions is

$$\Delta G^\circ_{rxn} = \Delta H^\circ_{rxn} - T\,\Delta S^\circ_{rxn}$$

The value of ΔG°_{rxn} is also given by Equation (19-10):

$$\Delta G^\circ_{rxn} = -2.30RT \log K$$

Equating these two expressions for ΔG°_{rxn} and solving for $\log K$ yields

$$\log K = -\frac{\Delta H^\circ_{rxn}}{2.30RT} + \frac{\Delta S^\circ_{rxn}}{2.30R} \tag{19-14}$$

Note that the first term on the right-hand side of Equation (19-14), $-\Delta H^\circ_{rxn}/2.30RT$, is inversely proportional to the absolute temperature, whereas the second term, $\Delta S^\circ_{rxn}/2.30R$, does not involve explicitly the absolute temperature. Because of the T in the denominator, the $-\Delta H^\circ_{rxn}/2.30RT$ term is of decreasing importance relative to the $\Delta S^\circ_{rxn}/2.30R$ term in determining the magnitude of K with increasing temperature. The entropy change for a chemical reaction is the dominant factor in determining the equilibrium distribution of species at high temperatures, whereas the energy change for a chemical reaction is the dominant factor in determining the equilibrium distribution of species at low temperatures. In general, at intermediate temperatures, the equilibrium distribution of species involves a compromise between minimization of energy and maximization of entropy.

Application of Equation (19-14) to a particular reaction that is run at two different temperatures, T_2 and T_1, yields

$$\log K_2 = -\frac{\Delta H^\circ_{rxn}}{2.30RT_2} + \frac{\Delta S^\circ_{rxn}}{2.30R}$$

$$\log K_1 = -\frac{\Delta H^\circ_{rxn}}{2.30RT_1} + \frac{\Delta S^\circ_{rxn}}{2.30R}$$

As usual, we have assumed that the values of ΔH°_{rxn} and ΔS°_{rxn} are the same at the temperatures T_1 and T_2. Subtracting the second equation from the first yields

$$\log K_2 - \log K_1 = -\frac{\Delta H^\circ_{rxn}}{2.30RT_2} + \frac{\Delta H^\circ_{rxn}}{2.30RT_1}$$

or

$$\log\left(\frac{K_2}{K_1}\right) = \frac{\Delta H^\circ_{rxn}}{2.30R}\left(\frac{1}{T_1} - \frac{1}{T_2}\right) = \frac{\Delta H^\circ_{rxn}}{2.30R}\left(\frac{T_2 - T_1}{T_1T_2}\right) \tag{19-15}$$

Equation (19-15), which describes the dependence of an equilibrium constant on the temperature, is called the **van't Hoff equation.** Note the similarity in mathematical form of the van't Hoff equation to the Clapeyron-Clausius equation (Equation 11-3). This similarity is a consequence of the fact that for a liquid-vapor equilibrium, which can be written as the general equation

$$A(l) \rightleftharpoons A(g) \qquad \Delta H^\circ_{rxn} \approx \Delta H_{vap}$$

we have $K = P_A$.

If $\Delta H^\circ_{rxn} > 0$ (endothermic reaction) and $T_2 > T_1$ in Equation (19-15), then $\log (K_2/K_1) > 0$, or $K_2 > K_1$. Thus we see that the value of the equilibrium constant for an endothermic reaction increases with increasing temperature, in accord with Le Châtelier's principle. Conversely, if $\Delta H^\circ_{rxn} < 0$ (exothermic reaction) and $T_2 > T_1$, then $\log (K_2/K_1) < 0$, or $K_2 < K_1$. The value of K for an exothermic reaction decreases with increasing temperature.

We shall illustrate the use of the van't Hoff equation by considering the **Fischer-Tropsch synthesis,** which involves reactions of hydrogen gas with carbon monoxide and carbon dioxide to produce straight-chain hydrocarbons and alcohols with up to ten carbon atoms. It is a practical method for converting coal to gasoline, diesel oil, and alcohols. Because the United States has massive coal supplies but limited petroleum deposits, the Fischer-Tropsch synthesis is expected to play a major role in the U.S. energy economy in the coming decades.

The first step in the overall process is the gasification of coal with steam:

$$C(s) + H_2O(g) \rightleftharpoons CO(g) + H_2(g)$$

This reaction is endothermic, with $\Delta H^\circ_{rxn} = 131$ kJ and with an equilibrium constant of 9.36×10^{-17} atm at 25°C. This small value of K_p means that the equilibrium lies far to the left at 25°C. Let's use the van't Hoff equation to calculate K_p for the reaction at 600°C. Using Equation (19-15), we have for K_p at 600°C

$$\log \left(\frac{K_p}{9.36 \times 10^{-17}\ \text{atm}}\right) = \left(\frac{131 \times 10^3\ \text{J}}{2.30 \times 8.314\ \text{J}\cdot\text{K}^{-1}}\right)\left(\frac{873\ \text{K} - 298\ \text{K}}{(298\ \text{K})(873\ \text{K})}\right)$$

$$= 15.14$$

and

$$K_p = (9.36 \times 10^{-17}\ \text{atm})(10^{15.14}) = 0.13\ \text{atm}$$

A value of $K_p = 0.13$ atm means that appreciable concentrations of $CO(g)$ and $H_2(g)$ exist in the equilibrium system at 600°C. A further increase in temperature will shift the equilibrium even farther to the right. The van't Hoff equation is invaluable in choosing the appropriate temperature for carrying out a reaction.

Example 19-10: A subsequent step in the Fischer-Tropsch process involves the reaction between hydrogen and carbon monoxide to produce methanol:

$$CO(g) + 2H_2(g) \rightleftharpoons CH_3OH(g) \qquad \Delta H^\circ_{rxn} = -90.2\ \text{kJ}$$

The equilibrium constant at 25°C for this chemical equation is $K_p = 2.23 \times 10^4\ \text{atm}^{-2}$. Compute the value of the equilibrium constant for this equation at 300°C.

Solution: Application of the van't Hoff equation (Equation 19-15) to the reaction, using the value of ΔH°_{rxn} given, yields

$$\log \left(\frac{K_2}{K_1}\right) = -\frac{90.2 \times 10^3 \text{J}}{2.30R}\left(\frac{T_2 - T_1}{T_1 T_2}\right)$$

At $T_1 = 298$ K, $K_1 = 2.23 \times 10^4$ atm^{-2}. Thus at $T_2 = 573$ K we have for K_2

$$\log\left(\frac{K_2}{2.23 \times 10^4 \text{ atm}^{-2}}\right) = \left(\frac{-90.2 \times 10^3 \text{ J}}{2.30 \times 8.314 \text{ J}\cdot\text{K}^{-1}}\right)\left(\frac{573 \text{ K} - 298 \text{ K}}{(573 \text{ K})(298 \text{ K})}\right)$$

and

$$K_2 = (2.23 \times 10^4 \text{ atm}^{-2})(2.53 \times 10^{-8})$$
$$= 5.64 \times 10^{-4} \text{ atm}^{-2}$$

The large decrease in K_p with increasing T for this chemical reaction is a consequence of the large negative ΔH°_{rxn} for the reaction.

The temperature dependence of an equilibrium constant can be presented in graphical form using the van't Hoff equation. Figure 19-7 shows a plot of $-\log K_w$ versus $1/T$ for the dissociation reaction of water.

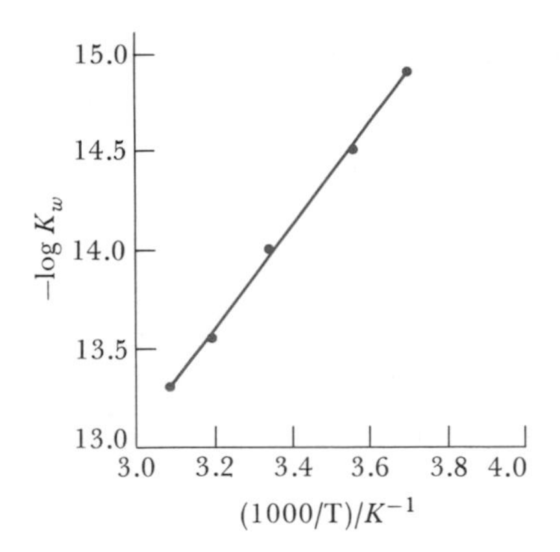

Figure 19-7 A plot of log K_w versus $1000/T$ for the equation

$$2H_2O(l) \rightleftharpoons H_3O^+(aq) + OH^-(aq)$$

Note that this plot is linear, which is consistent with Equation (19-15).

SUMMARY

Not all chemical reactions that evolve energy are spontaneous. Reaction spontaneity is determined by both energy and entropy changes. Entropy is a measure of the disorder, or randomness, of a system. In general, the position of equilibrium in a chemical reaction involves a compromise between a minimization of the energy and a maximization of the entropy.

The entropy of a compound increases with increasing temperature. Both fusion and vaporization processes lead to an increase in the entropy of a compound because molecules in a liquid are more disordered than the same molecules in the solid phase and molecules in a gas are more disordered than the same molecules in the liquid phase.

The entropy of a perfect crystalline substance is zero at 0 K. This property of crystals is the basis of the absolute entropy scale, on which the entropy of a compound at $T > 0$ K is always positive and increases with increasing temperature.

The Gibbs criterion for reaction spontaneity is that ΔG_{rxn} must be less than zero. The value of ΔG_{rxn} depends on the values of ΔH_{rxn} and ΔS_{rxn}. For a reaction with $\Delta G_{rxn} < 0$, the value of ΔG_{rxn} equals the maximum amount of work that can be obtained from the reaction under the stated conditions. The value of ΔG_{rxn} is related to the value of Q/K.

The criteria for reaction spontaneity are summarized as follows:

	Value of (Q/K)	**Gibbs free energy change**
spontaneous reaction	<1	$\Delta G_{rxn} < 0$
nonspontaneous reaction	>1	$\Delta G_{rxn} > 0$
reaction at equilibrium	=1	$\Delta G_{rxn} = 0$

Spontaneous is not synonymous with immediate. The fact that $\Delta G_{rxn} < 0$ is not sufficient to guarantee that a reaction proceeds toward equilibrium at a detectable rate.

The standard Gibbs free energy change ΔG°_{rxn} is equal to the value of ΔG_{rxn} when all products and reactants are at standard conditions (1 atm, 1 M). The value of ΔG°_{rxn} can be calculated from a table of standard molar Gibbs free energies of formation ΔG°_f (Table 19-1).

The van't Hoff equation describes the change in an equilibrium constant with temperature.

TERMS YOU SHOULD KNOW

enthalpy change, ΔH_{rxn} 597
standard enthalpy change, ΔH°_{rxn} 597
entropy, S 598
entropy change, ΔS 599
second law of thermodynamics 599
positional disorder 599
thermal disorder 599
molar entropy of fusion, ΔS_{fus} 601
molar entropy of vaporization, ΔS_{vap} 601
standard molar entropy, S° 602
standard entropy change for a reaction, ΔS°_{rxn} 605
entropy-driven reaction 606
entropy-favored reaction 606
energy-favored reaction 606
Gibbs free energy, G 607
Gibbs free energy change, ΔG_{rxn} 607
Gibbs criteria for reaction spontaneity 607
standard Gibbs free energy change, ΔG°_{rxn} 610
standard molar Gibbs free energy of formation, ΔG°_f 612
van't Hoff equation 614
Fischer-Tropsch synthesis 615

EQUATIONS YOU SHOULD KNOW HOW TO USE

$\Delta S = \dfrac{q}{T}$ (19-1) (from the second law of thermodynamics)

$\Delta S_{fus} = \dfrac{\Delta H_{fus}}{T_m}$ (19-2) (entropy change on fusion)

$\Delta S_{vap} = \dfrac{\Delta H_{vap}}{T_b}$ (19-3) (entropy change on vaporation)

$\Delta S^\circ_{rxn} = yS^\circ[\mathrm{Y}] + zS^\circ[\mathrm{Z}] - aS^\circ[\mathrm{A}] - bS^\circ[\mathrm{B}]$ (19-6) (standard entropy change for a reaction in terms of tabulated standard molar entropies)

$\Delta G_{rxn} = \Delta H_{rxn} - T\,\Delta S_{rxn}$ (19-7) (Gibbs free energy change for a reaction in terms of ΔH_{rxn} and ΔS_{rxn})

$\Delta G_{rxn} = 2.30RT \log (Q/K)$ (19-8) (relation between the Gibbs free energy change for a reaction and Q/K)

$\Delta G^\circ_{rxn} = -2.30RT \log K$ (19-10) (relation between the standard Gibbs free energy change for a reaction and the equilibrium constant)

$\Delta G_{rxn} = \Delta G^\circ_{rxn} + 2.30RT \log Q$ (19-11) (relation between the Gibbs free energy change for a reaction and the standard Gibbs free energy change for the reaction.)

$\Delta G^\circ_{rxn} = y\,\Delta G^\circ_f[\mathrm{Y}] + z\,\Delta G^\circ_f[\mathrm{Z}] - a\,\Delta G^\circ_f[\mathrm{A}] - b\,\Delta G^\circ_f[\mathrm{B}]$ (19-13) (relation between the standard Gibbs free energy change for a reaction and the standard Gibbs free energies of formation of the products and the reactants)

$\log\left(\dfrac{K_2}{K_1}\right) = \dfrac{\Delta H^\circ_{rxn}}{2.30R}\left(\dfrac{T_2 - T_1}{T_1 T_2}\right)$ (19-15) (the van't Hoff equation)

PROBLEMS

ENTROPIES OF FUSION AND VAPORIZATION

19-1. From the following data, calculate ΔS_{fus} and ΔS_{vap} for the compounds methane, ethane, and propane:

	t_m/°C	ΔH_{fus}/ kJ·mol^{-1}	t_b/°C	ΔH_{vap}/ kJ·mol^{-1}
CH_4	−182.5	0.9370	−164.0	8.907
C_2H_6	−183.3	2.859	−88.6	15.65
C_3H_8	−181.7	3.525	−42.1	20.13

19-2. From the following data, calculate ΔS_{fus} and ΔS_{vap} for hydrogen fluoride, hydrogen chloride, hydrogen bromide, and hydrogen iodide:

	t_m/°C	ΔH_{fus}/ kJ·mol^{-1}	t_b/°C	ΔH_{vap}/ kJ·mol^{-1}
HF	−83.11	4.577	19.54	25.18
HCl	−114.3	1.991	−84.9	17.53
HBr	−86.96	2.406	−67.0	19.27
HI	−50.91	2.871	−35.38	21.16

19-3. From the following data, calculate ΔS_{fus} and ΔS_{vap} for methyl alcohol, CH_3OH; ethyl alcohol, C_2H_5OH; and *n*-propyl alcohol, C_3H_7OH:

	t_m/°C	ΔH_{fus}/ kJ·mol^{-1}	t_b/°C	ΔH_{vap}/ kJ·mol^{-1}
CH_3OH	−97.8	3.177	64.96	37.57
C_2H_5OH	−114.5	5.021	78.7	40.48
C_3H_7OH	−126.1	5.196	97.4	43.60

19-4. From the following data, calculate ΔS_{fus} and ΔS_{vap} for lithium chloride, sodium chloride, and potassium chloride:

	t_m/°C	ΔH_{fus}/ kJ·mol^{-1}	t_b/°C	ΔH_{vap}/ kJ·mol^{-1}
LiCl	614	19.8	1325	161.4
NaCl	800	28.2	1413	181.7
KCl	770	26.8	1500	172.8

19-5. From the following data, calculate ΔS_{fus} and ΔS_{vap} for hydrogen sulfide:

	t_m/°C	ΔH_{fus}/ kJ·mol^{-1}	t_b/°C	ΔH_{vap}/ kJ·mol^{-1}
H_2S	−85.6	2.38	−60.7	18.7

Note the differences between these values and those for water (Example 19-2). Give a simple molecular interpretation for the differences.

19-6. Arrange the compounds NH_3, CH_4, and H_2O in order of increasing ΔS_{vap} values. Describe the reasoning that you used to reach your conclusions.

ENTROPY: MOLECULAR MASS, STRUCTURE, AND PHYSICAL STATE

19-7. In each case, predict which molecule of the pair has the greater molar entropy under the same conditions (assume gaseous species):

(a) H_2O D_2O

(b) CH_3CH_2OH (ethanol) $H_2C—CH_2$ bridged by O (ethylene oxide)

(c) $CH_3CH_2CH_2CH_2NH_2$ (butyl amine) $H_2C—CH_2$, H_2C CH_2, N–H ring (pyrrolidine)

19-8. In each case, predict which molecule of the pair has the greater molar entropy under the same conditions (assume gaseous species):

(a) CO CO_2

(b) $CH_3CH_2CH_3$ (propane) $H_2C—CH_2$ bridged by CH_2 (cyclopropane)

(c) $CH_3CH_2CH_2CH_2CH_3$ (*n*-pentane) $H_3C—C(CH_3)_2—CH_3$ (neopentane)

19-9. Arrange the following molecules in order of increasing values of total molar entropy at 1 atm:

$CH_3Cl(g)$ $CH_4(g)$ $CH_3OH(g)$

19-10. Rank the following compounds in order of increasing standard molar entropy:

$CH_4(g)$ $H_2O(g)$ $NH_3(g)$ $CH_3OH(g)$ $CH_3OD(g)$

19-11. Explain why the total molar entropy of $Fe_3O_4(s)$ is greater than that of $Fe_2O_3(s)$ at 25°C and 1 atm pressure.

19-12. Explain why the total molar entropy of $PCl_3(g)$ is less than that of $PCl_5(g)$ at 25°C and 1 atm pressure.

19-13. Explain why the molar entropy of $Br_2(g)$ is greater than that of $Br_2(l)$ when both are at 25°C and the same pressure.

19-14. Explain why the molar entropy of $H_2O(g)$ is greater than that of $H_2O(l)$ when both are at 100°C and 1 atm pressure.

19-15. Predict whether the entropy of the substance increases, decreases, or remains the same in the following processes:

(a) $Ar(l) \rightarrow Ar(g)$

(b) $O_2(g, 2.00\ \text{atm}, 300\ \text{K}) \rightarrow O_2(g, 1.00\ \text{atm}, 300\ \text{K})$

(c) $Cu(s, 300\ \text{K}) \rightarrow Cu(s, 800\ \text{K})$

(d) $CO_2(g) \rightarrow CO_2(s)$

19-16. Predict whether the entropy of the substance increases, decreases, or remains the same in the following processes:

(a) $H_2O(g$, 75 torr, 300 K$) \rightarrow H_2O(g$, 150 torr, 300 K$)$
(b) $Br_2(l$, 1 atm, 25°C$) \rightarrow Br_2(g$, 1 atm, 25°C$)$
(c) $I_2(g$, 1 atm, 125°C$) \rightarrow I_2(g$, 1 atm, 200°C$)$
(d) $Fe(s$, 250°C, 1 atm$) \rightarrow Fe(s$, 25°C, 1 atm$)$

VALUES OF ΔS°_{rxn}

19-17. Arrange the following reactions according to increasing ΔS°_{rxn} values (do not consult any references):

(a) $S(s) + O_2(g) \rightarrow SO_2(g)$
(b) $H_2(g) + O_2(g) \rightarrow H_2O_2(l)$
(c) $CO(g) + 3H_2(g) \rightarrow CH_4(g) + H_2O(g)$
(d) $C(s) + H_2O(g) \rightarrow CO(g) + H_2(g)$

19-18. Arrange the following reactions according to increasing ΔS°_{rxn} values (do not consult any references):

(a) $2H_2(g) + O_2(g) \rightarrow 2H_2O(l)$
(b) $NH_3(g) + HCl(g) \rightarrow NH_4Cl(s)$
(c) $K(s) + O_2(g) \rightarrow KO_2(s)$
(d) $N_2(g) + 3H_2(g) \rightarrow 2NH_3(g)$

19-19. Use the data in Table 19-1 to compute ΔS°_{rxn} values for the following reactions at 25°C:

(a) $4NH_3(g) + 7O_2(g) \rightarrow 4NO_2(g) + 6H_2O(g)$
(b) $CO(g) + 2H_2(g) \rightarrow CH_3OH(l)$
(c) $C(s$, graphite$) + H_2O(g) \rightarrow CO(g) + H_2(g)$
(d) $2CO(g) + O_2(g) \rightarrow 2CO_2(g)$

19-20. Use the data in Table 19-1 to compute ΔS°_{rxn} values for the following reactions at 25°C:

(a) $2H_2O_2(l) + N_2H_4(l) \rightarrow N_2(g) + 4H_2O(g)$
(b) $N_2(g) + O_2(g) \rightarrow 2NO(g)$
(c) $2CH_4(g) + O_2(g) \rightarrow 2CH_3OH(l)$
(d) $C_2H_4(g) + H_2(g) \rightarrow C_2H_6(g)$

19-21. Use the data in Table 19-1 to calculate ΔS°_{rxn} for the following reactions at 25°C:

(a) $C(s$, graphite$) + O_2(g) \rightarrow CO_2(g)$
(b) $2SO_2(g) + O_2(g) \rightarrow 2SO_3(g)$
(c) $CH_4(g) + 2O_2(g) \rightarrow CO_2(g) + 2H_2O(l)$
(d) $C_2H_2(g) + H_2(g) \rightarrow C_2H_4(g)$

19-22. Use the data in Table 19-1 to calculate ΔS°_{rxn} for the following reactions at 25°C:

(a) $I_2(s) \rightarrow I_2(g)$
(b) $BaCO_3(s) \rightarrow BaO(s) + CO_2(g)$
(c) $CH_4(g) + Cl_2(g) \rightarrow CH_3Cl(g) + HCl(g)$
(d) $2NaBr(s) + Cl_2(g) \rightarrow 2NaCl(s) + Br_2(l)$

SPONTANEITY AND ΔG_{rxn}

19-23. Water slowly evaporates at 25°C. Is the process described by the equation

$$H_2O(l) \rightarrow H_2O(g)$$

spontaneous? What are the signs of ΔG_{rxn}, ΔH_{rxn}, and $T\,\Delta S_{rxn}$ at 25°C? What drives the reaction?

19-24. Naphthalene, the active component of one variety of mothballs, sublimes at room temperature. Is the process described by the equation

$$\text{naphthalene}(s) \rightarrow \text{naphthalene}(g)$$

spontaneous? What are the signs of ΔG_{rxn}, ΔH_{rxn}, and $T\,\Delta S_{rxn}$ at 25°C? What drives the reaction?

19-25. For the equation

$$3C_2H_2(g) \rightleftharpoons C_6H_6(l) \qquad \Delta H^\circ_{rxn} = -631 \text{ kJ}$$

use the data in Table 19-1 to calculate ΔS°_{rxn} at 25°C. Combine your calculated value of ΔS°_{rxn} with the value of ΔH°_{rxn} and compute ΔG°_{rxn}. Indicate the direction in which the reaction is spontaneous at 1 atm pressure.

19-26. For the equation

$$C_2H_4(g) + H_2O(l) \rightleftharpoons C_2H_5OH(l) \qquad \Delta H^\circ_{rxn} = -35.4 \text{ kJ}$$

use the data in Table 19-1 to calculate ΔS°_{rxn} at 25°C. Combine your calculated value of ΔS°_{rxn} with the value of ΔH°_{rxn} and compute ΔG°_{rxn}. Indicate the direction in which the reaction is spontaneous at 1 atm pressure.

19-27. For the equation

$$C_2H_4(g) + 3O_2(g) \rightleftharpoons 2CO_2(g) + 2H_2O(g) \qquad \Delta H^\circ_{rxn} = -1323 \text{ kJ}$$

use the data in Table 19-1 to calculate ΔS°_{rxn} at 25°C. Combine your calculated value of ΔS°_{rxn} with the value of ΔH°_{rxn} and compute ΔG°_{rxn}. Indicate the direction in which the reaction is spontaneous when the pressures are as follows: $P_{C_2H_4} = 0.010$ atm; $P_{O_2} = 0.020$ atm; $P_{CO_2} = 20$ atm; and $P_{H_2O} = 0.010$ atm.

19-28. For the equation

$$C(s, \text{graphite}) + CO_2(g) \rightarrow 2CO(g) \qquad \Delta H^\circ_{rxn} = +172 \text{ kJ}$$

use the data in Table 19-1 to calculate ΔS°_{rxn} at 25°C. Combine your calculated value of ΔS°_{rxn} with the value of ΔH°_{rxn} and compute ΔG°_{rxn}. Indicate the direction in which the reaction is spontaneous when the pressures are as follows: $P_{CO} = 0.00050$ atm; $P_{CO_2} = 20$ atm.

19-29. A critical reaction in the production of energy to do work or drive chemical reactions in biological systems is the hydrolysis of adenosine triphosphate, ATP,

$$\text{ATP}(aq) + H_2O(l) \rightleftharpoons \text{ADP}(aq) + HPO_4^{2-}(aq)$$

for which $\Delta G^\circ_{rxn} = -30.5$ kJ at 37°C and pH 7.0. Calculate the value of ΔG_{rxn} in a biological cell in which [ATP] = 5.0 mM, [ADP] = 0.50 mM, and $[HPO_4^{2-}]$ = 5.0 mM. Is the hydrolysis of ATP spontaneous under these conditions?

19-30. The equilibrium constant for the equation

$$2SO_2(g) + O_2(g) \rightleftharpoons 2SO_3(g)$$

is $K = 10\ atm^{-1}$ at 900 K. Compute ΔG_{rxn} for this process at 900 K when it takes place with the indicated gas pressures:

$$2SO_2(1.0 \times 10^{-3}\ atm) + O_2(0.20\ atm) \rightarrow 2SO_3(1.0 \times 10^{-4}\ atm)$$

EQUILIBRIUM CONSTANTS AND ΔG°_{rxn}

19-31. The equilibrium constant at 250°C for the equation

$$PCl_5(g) \rightleftharpoons PCl_3(g) + Cl_2(g)$$

is $K_p = 4.5 \times 10^3$ atm. Calculate the value of ΔG°_{rxn} at 250°C. In which direction is the reaction spontaneous when PCl_3, Cl_2, and PCl_5 are at standard conditions? Calculate ΔG_{rxn} when $P_{PCl_3} = 0.20$ atm; $P_{Cl_2} = 0.80$ atm; and $P_{PCl_5} = 1.0 \times 10^{-6}$ atm; is the reaction to the right spontaneous?

19-32. The equilibrium constant at 527°C for the equation

$$COCl_2(g) \rightleftharpoons CO(g) + Cl_2(g)$$

is $K_c = 4.63 \times 10^{-3}$ M. Calculate the value of ΔG°_{rxn} at 527°C. In which direction is the reaction spontaneous when the concentrations of CO, Cl_2, and $COCl_2$ are 1.00 M? Calculate ΔG_{rxn} when [CO] = 0.010 M, $[Cl_2]$ = 0.010 M, and $[COCl_2]$ = 1.00 M. In which direction is the reaction spontaneous?

19-33. The equilibrium constant for the equation

$$HNO_2(aq) \rightleftharpoons H^+(aq) + NO_2^-(aq)$$

is $K_a = 4.5 \times 10^{-4}$ M at 25°C. Calculate the value of ΔG°_{rxn} at 25°C. Will nitrous acid spontaneously dissociate when $[NO_2^-] = [H^+] = [HNO_2] = 1.00$ M? when $[NO_2^-] = [H^+] = 1.0 \times 10^{-5}$ M and $[HNO_2] = 1.0$ M?

19-34. The equilibrium constant for the equation

$$HClO(aq) \rightleftharpoons H^+(aq) + ClO^-(aq)$$

is $K_a = 3.0 \times 10^{-8}$ M at 25°C. Calculate the value of ΔG°_{rxn} at 25°C. Will hypochlorous acid spontaneously dissociate when $[ClO^-] = [H^+] = [HClO] = 1.0$ M? When $[ClO^-] = [H^+] = 1.0 \times 10^{-6}$ M and $[HClO] = 0.10$ M?

19-35. The equilibrium constant for the equation

$$HC_2H_2ClO_2(aq) \rightleftharpoons H^+(aq) + C_2H_2ClO_2^-(aq)$$

is $K_a = 1.35 \times 10^{-3}$ M at 25°C. Calculate ΔG°_{rxn} at 25°C. Will chloroacetic acid spontaneously dissociate when $[C_2H_2ClO_2^-] = [H^+] = [HC_2H_2ClO_2] = 1.0$ M? when $[C_2H_2ClO_2^-] = 0.0010$ M, $[H^+] = 1.0 \times 10^{-5}$ M, and $[HC_2H_2ClO_2] = 0.10$ M?

19-36. The equilibrium constant for the equation

$$NH_3(aq) + H_2O(l) \rightleftharpoons NH_4^+(aq) + OH^-(aq)$$

is $K_b = 1.75 \times 10^{-5}$ M at 25°C. Calculate the value of ΔG°_{rxn} at 25°C. In which direction is the reaction spontaneous when $NH_3(aq)$, $NH_4^+(aq)$, and $OH^-(aq)$ are at standard conditions? Will ammonia react with water when both $[NH_4^+]$ and $[OH^-] = 1.0 \times 10^{-6}$ M and $[NH_3]$ = 0.050 M?

19-37. The equilibrium constant for the equation

$$AgCl(s) \xrightarrow[H_2O(l)]{} Ag^+(aq) + Cl^-(aq)$$

is the solubility-product constant, $K_{sp} = 1.78 \times 10^{-10}\ M^2$ at 25°C. Calculate the value of ΔG°_{rxn} at 25°C. Is it possible to prepare a solution that is 1.0 M in both $Ag^+(aq)$ and $Cl^-(aq)$?

19-38. The equilibrium constant for the equation

$$CaCO_3(s) \xrightarrow[H_2O(l)]{} Ca^{2+}(aq) + CO_3^{2-}(aq)$$

is the solubility-product constant, $K_{sp} = 2.8 \times 10^{-9}\ M^2$ at 25°C. Calculate the value of ΔG°_{rxn} at 25°C. What happens when a solution is prepared in which $[Ca^{2+}] = [CO_3^{2-}] = 1.0$ M?

19-39. The equilibrium constant at 25°C for the equation

$$Ag^+(aq) + 2NH_3(aq) \rightleftharpoons Ag(NH_3)_2^+(aq)$$

is $K_c = 2.5 \times 10^3\ M^{-2}$. Calculate the value of ΔG°_{rxn} at 25°C. In which direction is the reaction spontaneous when $Ag^+(aq)$, $NH_3(aq)$, and $Ag(NH_3)_2^+(aq)$ are at standard conditions? Calculate the value of ΔG_{rxn} when $[Ag^+] = 1.0 \times 10^{-3}$ M, $[NH_3] = 0.10$ M, and $[Ag(NH_3)_2^+] = 1.0 \times 10^{-3}$ M. In which direction is the reaction spontaneous under these conditions?

19-40. The equilibrium constant at 25°C for the equation

$$Co^{3+}(aq) + 6NH_3(aq) \rightleftharpoons Co(NH_3)_6^{3+}(aq)$$

is $K_c = 2.0 \times 10^7\ M^{-6}$. Calculate the value of ΔG°_{rxn} at 25°C. In which direction is the reaction spontaneous when $Co^{3+}(aq)$, $NH_3(aq)$, and $Co(NH_3)_6^{3+}(aq)$ are at standard conditions? Calculate the value of ΔG_{rxn} when $[Co^{3+}] = 0.0050$ M, $[NH_3] = 0.10$ M, and $[Co(NH_3)_6^{3+}] = 1.00$ M. In which direction is the reaction spontaneous under these conditions?

CALCULATION OF ΔG°_{rxn} FROM TABULATED DATA

19-41. Use the data in Table 19-1 to compute ΔG°_{rxn} and K at 25°C for the following equations:

(a) $CO(g) + 2H_2(g) \rightarrow CH_3OH(l)$
(b) $C(s) + H_2O(g) \rightarrow CO(g) + H_2(g)$
(c) $CO(g) + 3H_2(g) \rightarrow CH_4(g) + H_2O(g)$

19-42. Use the data in Table 19-1 to compute ΔG°_{rxn} and K at 25°C for the following equations:

(a) $2H_2O_2(l) + N_2H_4(l) \rightarrow N_2(g) + 4H_2O(g)$
(b) $N_2(g) + O_2(g) \rightarrow 2NO(g)$
(c) $2CH_4(g) + O_2(g) \rightarrow 2CH_3OH(l)$

19-43. Use the data in Table 19-1 to calculate ΔG°_{rxn} and ΔH°_{rxn} at 25°C for the equation

$$2HCl(g) + F_2(g) \rightleftharpoons 2HF(g) + Cl_2(g)$$

Calculate the equilibrium constant at 25°C for the equation.

19-44. Use the data in Table 19-1 to calculate ΔG°_{rxn} and ΔH°_{rxn} at 25°C for the reaction

$$Fe_3O_4(s) + 2C(s, \text{graphite}) \rightarrow 3Fe(s) + 2CO_2(g)$$

Calculate the equilibrium constant at 25°C for the equation.

19-45. The reaction described by the equation

$$2SO_2(g) + O_2(g) \rightleftharpoons 2SO_3(g)$$

is an important reaction in the manufacture of sulfuric acid. Use the data in Table 19-1 to calculate the values of ΔG°_{rxn} and ΔH°_{rxn} at 25°C for the reaction. Calculate the equilibrium constant for the reaction at 25°C.

19-46. Use the data in Table 19-1 to calculate ΔG°_{rxn} and ΔH°_{rxn} at 25°C for the equation

$$H_2(g) + I_2(g) \rightleftharpoons 2HI(g)$$

Calculate the equilibrium constant for the equation at 25°C.

19-47. Use the data in Table 19-1 to calculate the value of ΔG°_{rxn}, ΔH°_{rxn}, and ΔS°_{rxn} at 25°C for the reaction described by the equation

$$H_2(g) + CO_2(g) \rightleftharpoons H_2O(g) + CO(g)$$

What drives the reaction and in what direction at standard conditions?

19-48. Use the data in Table 19-1 for $CH_4(g)$, $Cl_2(g)$, and $HCl(g)$ to calculate ΔG°_f and ΔH°_f at 25°C for $CCl_4(l)$, given that $\Delta G^\circ_{rxn} = -395.7$ kJ and $\Delta H^\circ_{rxn} = -429.8$ kJ for the equation

$$CH_4(g) + 4Cl_2(g) \rightleftharpoons CCl_4(l) + 4HCl(g)$$

Compare your results for $CCl_4(l)$ with those given in Table 19-1.

19-49. Calculate the maximum amount of work that can be obtained from the combustion of 1.00 mol of ethane, $C_2H_6(g)$, at 25°C and standard conditions.

19-50. Calculate the maximum amount of work that can be obtained from the combustion of 1.00 mol of methane, $CH_4(g)$, at 25°C and standard conditions.

TEMPERATURE DEPENDENCE OF EQUILIBRIUM CONSTANTS

19-51. For the equation $N_2(g) + O_2(g) \rightleftharpoons 2NO(g)$, use the following data to calculate ΔH°_{rxn}:

T/K	$K_p/10^{-4}$
2000	4.08
2100	6.86
2200	11.0
2300	16.9
2400	25.1

19-52. For the dissociation of $Br_2(g)$ into $2Br(g)$, use the following data to calculate ΔH°_{rxn}:

t/°C	$K_p/10^{-3}$ atm
850	0.600
900	1.45
950	3.26
1000	6.88

19-53. For the equation

$$H_2(g) + CO_2(g) \rightleftharpoons CO(g) + H_2O(g)$$

use the following data to calculate ΔH°_{rxn}:

t/°C	K_p
600	0.39
700	0.64
800	0.96
900	1.34
1000	1.77

19-54. For the equation $2SO_2(g) + O_2(g) \rightleftharpoons 2SO_3(g)$, use the following data to calculate ΔH°_{rxn}:

T/K	K_p/atm^{-1}
900	43.1
1000	3.46
1100	0.44
1170	0.13

19-55. Use data in Table 19-1 to calculate the value of ΔH°_{rxn} for the equation

$$PCl_3(g) + Cl_2(g) \rightleftharpoons PCl_5(g)$$

Given that $K_p = 0.562$ atm^{-1} at 250°C, calculate the value of K_p at 400°C.

19-56. Use data in Table 19-1 to calculate the value of ΔH°_{rxn} for the equation

$$H_2(g) + I_2(g) \rightleftharpoons 2HI(g)$$

Given that $K = 58.0$ at 400°C, calculate K at 500°C.

19-57. Calculate ΔS_{fus} and ΔS_{vap} for the alkali metals:

Metal	T_m/K	ΔH_{fus}/ kJ·mol^{-1}	T_b/K	ΔH_{vap}/ kJ·mol^{-1}
Li	454	2.99	1615	134.7
Na	371	2.60	1156	89.6
K	336	2.33	1033	77.1
Rb	312	2.34	956	69
Cs	302	2.10	942	66

19-58. Suppose that you see an advertisement for a catalyst that decomposes water into hydrogen and oxygen at room temperature. Would you be skeptical of this claim? Explain.

19-59. Is it possible to have a reaction for which K is either zero or infinite? Explain in terms of ΔG°_{rxn}.

19-60. Given the following possibilities for ΔG°_{rxn}, what can you say in each case about the magnitude of the equilibrium constant for the reaction?

(a) $\Delta G^\circ_{rxn} > 0$ (b) $\Delta G^\circ_{rxn} = 0$ (c) $\Delta G^\circ_{rxn} < 0$?

19-61. Hydrogen peroxide can be prepared in several ways. One method is the reaction between hydrogen and oxygen:

$$H_2(g) + O_2(g) \rightleftharpoons H_2O_2(l)$$

Another method is the reaction between water and oxygen:

$$2H_2O(l) + O_2(g) \rightleftharpoons 2H_2O_2(l)$$

Calculate the value of ΔG°_{rxn} for both reactions. Predict which method requires less energy under standard conditions.

19-62. Given the following Gibbs free energies at 25°C

Substance	ΔG°_f/kJ·mol^{-1}
$Ag^+(aq)$	77.1
$Cl^-(aq)$	−131.2
AgCl(*s*)	−109.7
$Br^-(aq)$	−102.8
AgBr(*s*)	−96.8

calculate the solubility-product constant of (a) AgCl and (b) AgBr.

19-63. Plot the data in Problem 19-51 and show that $\log K$ versus $1/T$ is a straight line. Evaluate ΔH°_{rxn} from this plot.

19-64. Discuss the possible effects of a catalyst on the value of ΔG°_{rxn}.

19-65. Compute the value (in kilojoules) of the change in ΔG°_{rxn} that corresponds to a 10-fold change in K at 25°C.

19-66. Glucose is a primary fuel in the production of energy in biological systems. Given that $\Delta G^\circ_f = -916$ kJ·mol^{-1} for glucose, calculate the maximum amount of work that can be obtained from the complete combustion of 1.00 mol of glucose under standard conditions:

$$C_6H_{12}O_6(s) + 6O_2(g) \rightleftharpoons 6CO_2(g) + 6H_2O(l)$$

19-67. The solubility of gases in water decreases with increasing temperature. What does this tell you about the heats of solution of gases?

19-68. Estimate the heat of solution of AgCl(*s*) in water from the following data:

t/°C	K_{sp}/M^2
50.0	13.2×10^{-10}
100.0	2.15×10^{-8}

19-69. The variation of the Henry's law constant with temperature for the dissolution of CO_2 in water is

t/°C	K/atm·M^{-1}
0	13.2
25	29.4

Calculate ΔH°_{rxn} for the process

$$CO_2(aq) \rightleftharpoons CO_2(g)$$

19-70. The vapor pressure of water above mixtures of $CuCl_2\cdot H_2O(s)$ and $CuCl_2\cdot 2H_2O(s)$ is 3.72 torr at 18.0°C and 91.2 torr at 60.0°C. Calculate ΔH°_{rxn} for the equilibrium

$$CuCl_2\cdot 2H_2O(s) \rightleftharpoons CuCl_2\cdot H_2O(s) + H_2O(g)$$

19-71. Given the following data, compute ΔH°_{rxn}, ΔS°_{rxn}, and ΔG°_{rxn} at 298 K for the equilibrium

$$Mg(s) + 2HCl(aq) \rightleftharpoons H_2(g) + MgCl_2(aq)$$

Species	ΔH°_f/kJ·mol^{-1}	S°/J·K^{-1}·mol^{-1}
$H_2(g)$	0	130.6
HCl(*aq*)	−167.2	56.5
Mg(*s*)	0	32.6
$MgCl_2(aq)$	−801.2	−25.1

19-72. Given that $\Delta G^\circ_f = 3.142$ kJ·mol^{-1} for $Br_2(g)$ at 25°C, calculate the vapor pressure of bromine at 25°C.

ELECTROCHEMISTRY

A chlor-alkali plant, in which chlorine and sodium hydroxide are obtained from sodium chloride by electrolysis.

Electrochemistry is the study of the chemical processes involved when an electric current is passed through materials. The passage of an electric current through an electrolyte is called electrolysis. Chemical changes occur at the electrodes during electrolysis. In particular, many electrolytes can be decomposed by electrolysis. Many industrial processes involve electrolysis, and we shall discuss how to calculate how much electric current is required to produce a given quantity of product by electrolysis.

An electric current can be drawn from an oxidation-reduction, or electron-transfer, reaction by means of a device called an electrochemical cell. An electrochemical cell can be characterized by its voltage, which is a measure of how strongly an electric current is driven through a wire. The voltage of a cell depends upon the concentration of the various species in the cell reaction. In this chapter we shall learn how to calculate equilibrium constants from cell voltages, and to use cell voltages for chemical reactions to predict reaction spontaneity. The operation of batteries and fuel cells is based upon electrochemical cells, and we shall learn how lead storage batteries and nickel-cadmium rechargeable batteries work.

20-1. ELECTROLYSIS IS A CHEMICAL REACTION THAT OCCURS AS A RESULT OF THE PASSAGE OF AN ELECTRIC CURRENT THROUGH A SOLUTION

The science of **electrochemistry** began in 1791 when the Italian scientist Luigi Galvani showed that the contraction of a frog's leg produced an electric current. Galvani's discovery of "animal electricity" led his countryman Allessandro Volta to pursue its investigation with the objective of demonstrating that the phenomenon could be explained in purely physical and chemical terms, rather than by various mysterious forces invoked by scientists of that time. Volta constructed a device called a **voltaic pile** (Figure 20-1), in which alternate discs of dissimilar metals, such as zinc and copper, are separated by damp cloths that have been soaked in salt water. The most remarkable property of a voltaic pile is that a simple assembly of chemical substances can produce an electric current of considerable power capable of decomposing a wide variety of chemical substances. The voltage developed by a voltaic pile depends on the nature of the pair of metals used and is greater the larger the number of sets of metal discs. Voltaic piles played a major role in the work of the English scientist Humphrey Davy in the discovery of several reactive metals (Na, K, Mg, Ca, Sr, and Ba) in the early 1800s.

Michael Faraday, another English scientist, who was trained by Davy, used the voltaic pile to investigate the effect of the passage of an electric current through various electrolyte solutions. Faraday's primary observation was that under certain conditions, the passage of an electric current through a solution causes chemical reactions to occur that could not occur otherwise. For example, in this manner, water containing the salt Na_2SO_4 was decomposed into hydrogen and oxygen. A chemical reaction that results from the passage of an electric current through a solution is called an **electrolysis.** We can represent the electrolysis of water by

$$2H_2O(l) \xrightarrow{\text{electrolysis}} 2H_2(g) + O_2(g)$$

$Na_2SO_4(aq)$ does not appear in the net electrolysis equation because $Na^+(aq)$ and $SO_4^{2-}(aq)$ are not changed chemically by the passage of electric current through the solution. However, water is a poor conductor of electricity, and the ions $Na^+(aq)$ and $SO_4^{2-}(aq)$ are the current carriers in the water phase.

Faraday discovered that the metal ions of many salts are deposited as the metal when an electric current is passed through aqueous solutions of their salts. For example, silver metal is deposited from a solution of

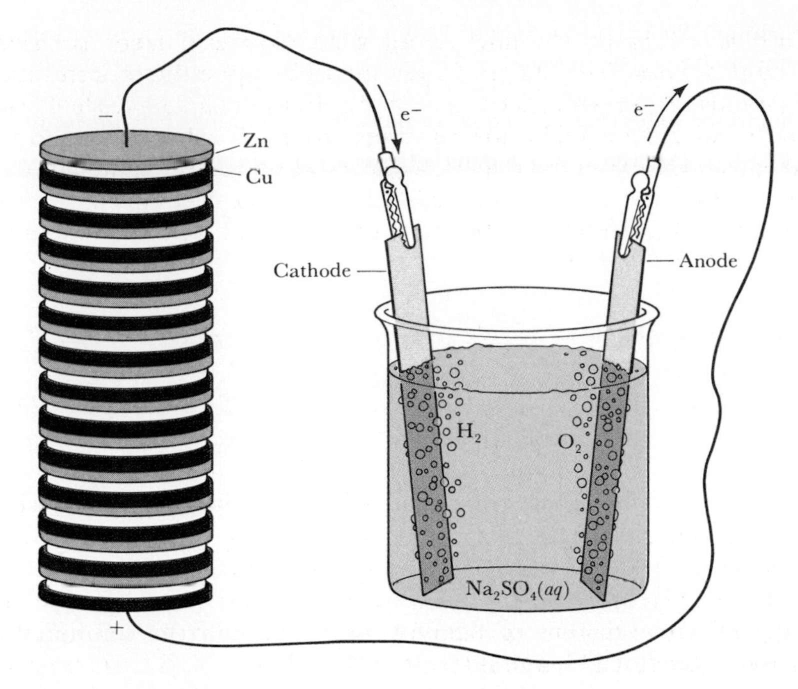

Figure 20-1 Electrolysis of water containing $Na_2SO_4(aq)$. The gas evolved from the platinum strip connected to the top zinc plate of the set of cells (a voltaic pile) is hydrogen. The gas evolved from the platinum strip connected to the bottom copper plate of the voltaic pile is oxygen. The current through the solution is carried by $Na^+(aq)$ and $SO_4^{2-}(aq)$ ions.

$AgNO_3(aq)$ and copper metal is deposited from $Cu(NO_3)_2(aq)$. The reactions that take place are

$$Ag^+(aq) + e^- \rightarrow Ag(s)$$

$$Cu^{2+}(aq) + 2e^- \rightarrow Cu(s)$$

Note that supplying 1 mol of electrons deposits 1 mol of silver from $Ag^+(aq)$ but 2 mol of electrons is needed to deposit 1 mol of copper from $Cu^{2+}(aq)$ (Figure 20-2).

The number of electrons supplied can be controlled by controlling the electric current through the solution. Electric currents are measured in **amperes** (A), and one ampere is the flow of one **coulomb** of charge per second. In an equation, we have

$$\text{current} = \text{charge/time}$$

The total charge that flows is

$$\text{charge} = \text{current} \times \text{time}$$

Using SI units, we have

$$\text{coulombs} = \text{amperes} \times \text{seconds}$$

Representing charge (in coulombs) by Z, current by I (in amperes), and time by t (in seconds), we have

$$Z = It \qquad (20\text{-}1)$$

Figure 20-2 Electrodeposition of silver (from $AgNO_3(aq)$) and copper (from $Cu(NO_3)_2(aq)$ species). The same quantity of electricity flows through the two solutions because they are placed in series. The number of moles of silver deposited after a given time is twice as great as the number of moles of copper deposited, because the reduction of 1 mol of Ag^+ requires 1 mol of electrons, whereas the reduction of 1 mol of Cu^{2+} requires 2 mol of electrons. The photo shows the electrodes removed from the solutions following the depositions. Finely divided copper metal is black.

Example 20-1: If a current of 1.50 A flows for 5.00 min, then what quantity of charge has flowed?

Solution: The charge is equal to the current multiplied by time, and so

$$\begin{aligned}\text{charge} &= (1.5\ \text{A})(5.00\ \text{min})\\ &= (1.50\ \text{C}\cdot\text{s}^{-1})(5.00\ \text{min})(60\ \text{s}\cdot\text{min}^{-1}) = 450\ \text{C}\end{aligned}$$

20-2. ELECTROLYSIS IS DESCRIBED QUANTITATIVELY BY FARADAY'S LAWS

Consider the passage of an electric current through a solution of $AgNO_3(aq)$. The charge that flows through the solution is directly related to the number of electrons that participate in the electrochemical reaction

$$Ag^+(aq) + e^- \rightarrow Ag(s)$$

Suppose that a current of 0.850 A flows through the solution for 20.0 min. The total charge is

$$\text{charge} = (0.850\ \text{C}\cdot\text{s}^{-1})(20.0\ \text{min})(60\ \text{s}\cdot\text{min}^{-1}) = 1020\ \text{C}$$

The number of moles of electrons that corresponds to 1020 C can be determined by using the fact that the charge on 1 mol of electrons has been determined to be 9.65×10^4 coulombs. The number of moles of electrons that corresponds to 1020 C is

$$\text{mol electrons} = \frac{1020\ \text{C}}{96{,}500\ \text{C}\cdot\text{mol}^{-1}} = 1.06 \times 10^{-2}\ \text{mol}$$

Because one electron deposits one atom of silver we have

$$\text{mol Ag deposited} = \text{mol electrons} = 1.06 \times 10^{-2}\ \text{mol}$$

The number of grams of silver deposited by the passage of 0.850 A through an $AgNO_3(aq)$ solution for 20.0 min is

$$\text{mass of Ag deposited} = (1.06 \times 10^{-2}\ \text{mol})\left(\frac{107.9\ \text{g}}{1\ \text{mol}}\right) = 1.14\ \text{g}$$

We can summarize these results by **Faraday's laws** of electrolysis:

First law: The extent of an electrochemical reaction depends solely upon the quantity of electricity that is passed through a solution.

Second law: The mass of a substance that is deposited as a metal, or evolved as a gas, by the passage of a given quantity of electricity is directly proportional to the molar mass of the substance divided by the number of electrons consumed or produced per formula unit.

We can determine the mathematical expression of Faraday's laws of electrolysis by the following argument. The mass of a metal deposited, or of a gas evolved, by a mole of electrons is equal to the molar mass of the metal or gas, A, divided by the number of electrons, n, required to produce from the starting material one formula unit of the substance deposited or evolved. Some examples of n and A are

Process	***n***	***A*/g·mol^{-1}**
$Cu^{2+}(aq) + 2e^- \rightarrow \underline{Cu}(s)$	2	63.55
$Ag^+(aq) + e^- \rightarrow \underline{Ag}(s)$	1	107.9
$2H_2O(l) \rightarrow \underline{O_2}(g) + 4H^+(aq) + 4e^-$	4	32.00

where the underlining identifies the species deposited or evolved. In terms of symbols we have

$$\begin{pmatrix}\text{mass deposited as metal}\\ \text{or evolved as gas per mole}\\ \text{of electrons passed through}\\ \text{the solution}\end{pmatrix} = \frac{A}{n} \tag{20-2}$$

The number of moles of electrons passed through the solution is equal to the total charge, Z, that is passed through the solution divided by the charge of one mole of electrons, which we denote by F. The quantity F is called **Faraday's constant** or, simply, a **faraday.** Its value is

$$\begin{aligned} F &= 96{,}500 \text{ coulombs per mole} \\ &= 9.65 \times 10^4 \text{ C·mol}^{-1} \end{aligned}$$

The number of moles of electrons passed through the solution is given by

$$\begin{pmatrix}\text{moles of electrons}\\ \text{passed through the solution}\end{pmatrix} = \frac{Z}{F} = \frac{It}{F} \tag{20-3}$$

where we have used Equation (20-1). We can now express Faraday's laws quantitatively by the equation

$$\begin{pmatrix}\text{mass deposited as}\\ \text{metal or evolved as gas}\end{pmatrix} = m = \left(\frac{It}{F}\right)\left(\frac{A}{n}\right) \tag{20-4}$$

Example 20-2: Magnesium is produced commercially by the electrolysis of molten $MgCl_2$ (melting point 650°C) in cells that have a capacity of about 8 metric tons of molten $MgCl_2$. The currents used are between 8.0×10^4 A and 1.00×10^5 A. Calculate how many metric tons of magnesium metal can be produced per day by the electrolysis of molten $MgCl_2$ using a current of 1.00×10^5 A.

Solution: The equation for the production of magnesium by electrolysis of molten $MgCl_2$ is

$$MgCl_2(l) \xrightarrow{710^\circ C} Mg(l) + Cl_2(g)$$

The magnesium deposition reaction is

$$Mg^{2+}(l) + 2e^- \rightarrow Mg(l)$$

Using Equation 20-4, we have for the mass in grams of magnesium deposited

$$\begin{aligned} m &= \left(\frac{It}{F}\right)\left(\frac{A}{n}\right) \\ &= \left[\frac{(1.00 \times 10^5 \text{ A})(24 \text{ h} \times 60 \text{ min·h}^{-1} \times 60 \text{ s·min}^{-1})}{9.65 \times 10^4 \text{ C·mol}^{-1}}\right]\left(\frac{24.31 \text{ g·mol}^{-1}}{2}\right) \end{aligned}$$

$$= 1.09 \times 10^6 \text{ g}$$

A metric ton is 1000 kilograms; thus

$$\begin{pmatrix}\text{number of}\\ \text{metric tons}\\ \text{of Mg}(s)\\ \text{deposited}\end{pmatrix} = (1.09 \times 10^6 \text{ g})\left(\frac{1 \text{ kg}}{1000 \text{ g}}\right)\left(\frac{1 \text{ metric ton}}{1000 \text{ kg}}\right)$$

$$= 1.09 \text{ metric tons}$$

The chlorine produced in the electrolysis also is collected and sold as a by-product.

The current through the metal sections of an electrolysis apparatus like that shown in Figure 20-1 is carried by electrons, but electrons do not pass through aqueous phases. The current through the solution is carried by cations and anions moving in opposite directions toward the two platinum strips, which are called **electrodes.** An electrode is a solid phase on the surface of which oxidation-reduction reactions occur. The electrode at which the reduction half-reaction occurs is called the **cathode,** and the electrode at which the oxidation half-reaction occurs is called the **anode.** These definitions are conveniently remembered with the aid of the mnemonic

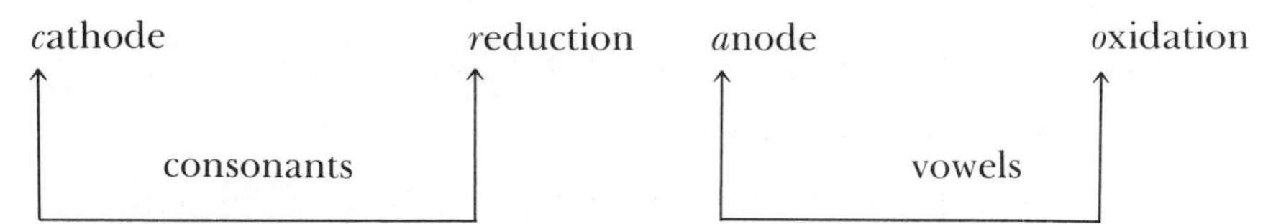

In electrolysis, the cations in the solution move toward the cathode, and the anions in the solution move toward the anode:

*cat*ions → *cat*hode   *an*ions → *an*ode

This is the origin of the names cathode and anode.

Let's return now to the electrolysis of water (Figure 20-1). The equation for the cathode half-reaction in the electrolysis of water is

$$4H^+(aq) + 4e^- \xrightarrow{\text{cathode}} 2H_2(g) \qquad \text{reduction, at cathode}$$

and the equation for the anode half-reaction is

$$2H_2O(l) \xrightarrow{\text{anode}} O_2(g) + 4H^+(aq) + 4e^- \qquad \text{oxidation, at anode}$$

The equation for the net reaction is the sum of the equations for the anode and cathode half-reactions:

$$2H_2O(l) \xrightarrow{\text{electrolysis}} 2H_2(g) + O_2(g)$$

The electrons that arrive at the cathode are consumed at the metal-solution interface by reaction with $H^+(aq)$ to produce H_2. An equal number of electrons is introduced back into the external circuit at the anode; these electrons are produced in the oxidation of H_2O to O_2, which occurs at the anode metal-solution interface.

Water electrolysis requires energy to break the oxygen-hydrogen bonds:

$$2H_2O(l) \rightarrow 2H_2(g) + O_2(g) \qquad \Delta G^\circ_{rxn} = +474.4 \text{ kJ}$$

The water-splitting reaction is driven uphill by the voltage applied across the electrodes. The electrochemical decomposition of water cannot occur unless the voltage is large enough to overcome the strength of the oxygen-hydrogen bonds. The minimum voltage necessary to electrochemically decompose a sample is called the **decomposition voltage.** The SI unit of voltage is a **volt,** denoted by V. A volt is defined as one joule per coulomb, $1 \text{ V} = 1 \text{ J} \cdot \text{C}^{-1}$. The minimum decomposition voltage for water at 25°C is about 1.2 V. In practice, more than 1.2 V is usually required in order to make the reaction proceed at an acceptable rate.

20-3. MANY CHEMICALS ARE PREPARED ON AN INDUSTRIAL SCALE BY ELECTROLYSIS

All the alkali and alkaline earth metals are prepared commercially by electrolysis (see Example 20-2). All the sodium hydroxide (about 20 billion pounds annually) and most of the chlorine produced in the United States (about 20 billion pounds annually) are made by the **chlor-alkali process,** which involves the electrolysis of concentrated aqueous solutions of sodium chloride. The overall electrolysis reaction is described by the equation

$$2NaCl(aq) + 2H_2O(l) \xrightarrow{\text{electrolysis}} 2NaOH(aq) + Cl_2(g) + H_2(g)$$

The two half-reaction equations are

$$2Na^+(aq) + 2H_2O(l) + 2e^- \rightarrow 2NaOH(aq) + H_2(g) \qquad \text{cathode}$$

and

$$2Cl^-(aq) \rightarrow Cl_2(g) + 2e^- \qquad \text{anode}$$

The electrolysis must be carried out in an apparatus in which the sodium hydroxide and hydrogen produced at the cathode and the chlorine produced at the anode are separated by some sort of diaphragm or membrane. If they were not separated, then the sodium hydroxide and chlorine would react to form sodium hypochlorite:

$$2NaOH(aq) + Cl_2(g) \rightarrow NaClO(aq) + NaCl(aq) + H_2O(l)$$

Sodium hypochlorite, which is a commonly used bleaching agent, is prepared in this manner. Figure 20-3 is a schematic drawing of a chlor-alkali membrane cell. The anode and cathode compartments are separated by a membrane (Figure 20-3b) that has a high internal negative charge, which excludes $Cl^-(aq)$ but allows $Na^+(aq)$ to pass through. A saturated sodium chloride solution enters the anode compartment, where the $Cl^-(aq)$ is oxidized to $Cl_2(g)$. The excess $Na^+(aq)$ migrates through the membrane to the cathode, where water is reduced to $H_2(g)$ and $OH^-(aq)$ according to the cathode half-reaction.

▪ Household bleach is about a 5 percent by mass aqueous solution of NaClO.

In an earlier version of the chlor-alkali process, the cathode was a pool of mercury at the bottom of the cell. In this type of cell, the $Na^+(aq)$ is reduced to metallic sodium, which immediately dissolves in the mercury, forming what is called an **amalgam.** The half-reaction equation is

▪ An amalgam is a solution of a metal in mercury.

(a)

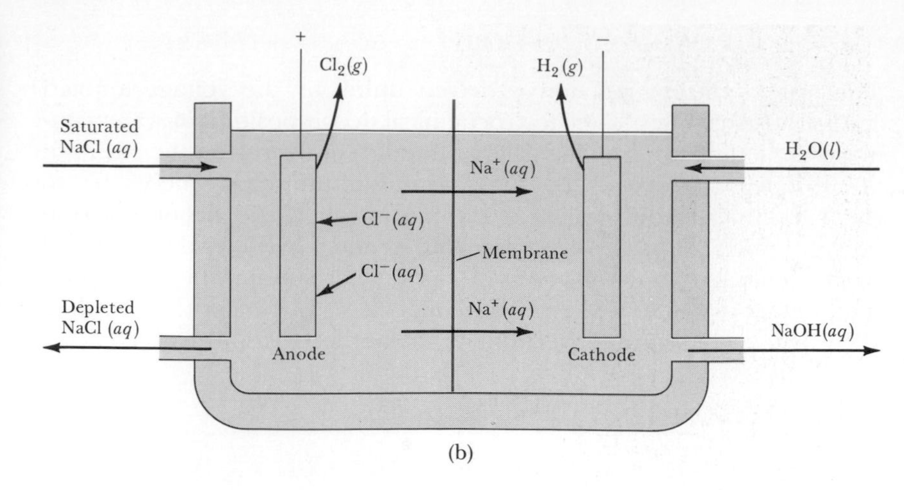

(b)

Figure 20-3 A chlor-alkali membrane cell. (a) A Nafion separator membrane. The membrane is a polymeric material with negatively charged groups ($-SO_3^-$) that permit the passage of Na^+ ions by migration from $-SO_3^-$ group to $-SO_3^-$ group in the membrane. The membrane is supported by a Teflon grid for additional mechanical strength. (b) A schematic view of the cell. Hydrogen gas and aqueous sodium hydroxide solution are produced at the cathode (−) and chlorine gas is produced at the anode (+). The migration of $Na^+(aq)$ through the membrane maintains equal numbers of positive and negative charges in the two cell solutions and also carries the current through the cell solutions.

$$Na^+(aq) + e^- \rightarrow Na(Hg) \qquad \text{mercury cathode}$$

The sodium amalgam is drawn off and then reacted with water in a specially prepared compartment:

$$2Na(Hg) + 2H_2O(l) \rightarrow 2NaOH(aq) + H_2(g)$$

Because of the poisonous nature of mercury, however, the mercury cells in older chlor-alkali apparatus are being replaced by membrane cells or their variants.

All aluminum metal is produced by the **Hall process,** which was patented by Charles Hall in 1889, when he was 26 years old. Hall conceived his process for the electrochemical production of aluminum while a student at Oberlin College in Ohio. A schematic of the electrolysis apparatus used in the Hall process, which is essentially the same as that used in the original process, is shown in Figure 20-4. The electroly-

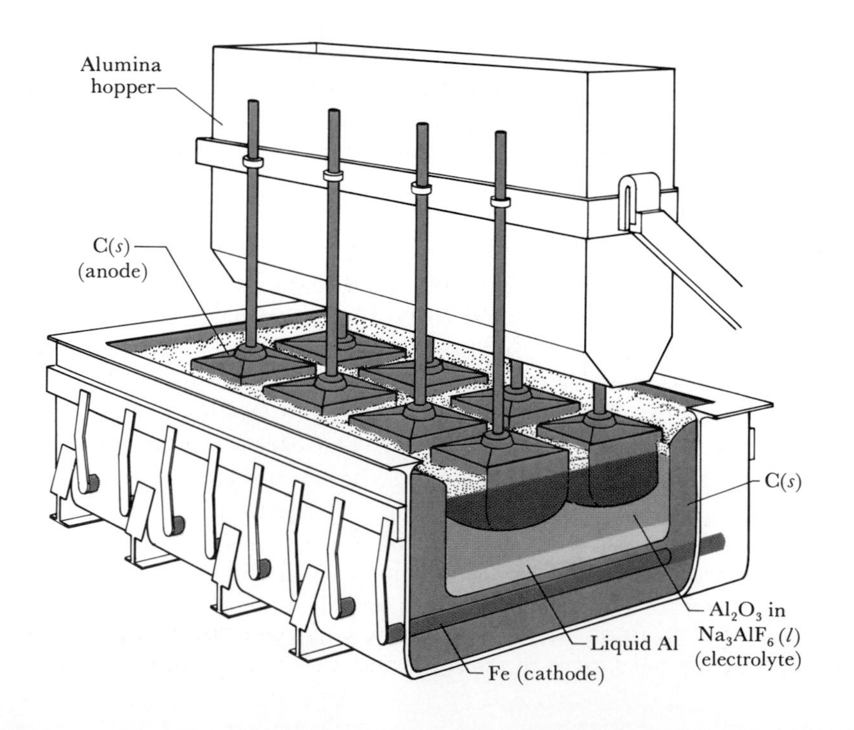

Figure 20-4 Hall process for the electrochemical production of aluminum. The cathodes are iron rods, and the anodes are carbon attached to metal rods. The electrolyte is alumina, Al_2O_3, dissolved in cryolite, Na_3AlF_6. The electrolysis is carried out at 980°C, where the aluminum is a liquid that can be drained from the cell.

Figure 20-5 The Hall process on an industrial scale. Shown here is an Alcoa potline (198 pots) at the Massena, New York, plant.

sis is carried out at about 980°C, where aluminum is a liquid and can be siphoned off from the cathode compartment. Electrical contact to the molten aluminum cathode is made through a steel shell that constitutes the bottom of the electrode compartment. The consumable anodes are composed of a petroleum coke that is obtained by heating to dryness the heavy petroleum fraction remaining from petroleum refining.

▪ The electrolytic production of aluminum consumes about 6 percent of the total U.S. production of electricity.

The key to the Hall process is the molten salt electrolyte consisting of powdered aluminum oxide dissolved in the mineral cryolite, Na_3AlF_6. Aluminum oxide dissolves in cryolite to form a conductive solution with a melting point low enough to allow the operation of the cell at 980°C. The equation for the overall electrochemical reaction is

$$2Al_2O_3(soln) + 3C(s) \rightarrow 4Al(l) + 3CO_2(g)$$

Aluminum oxide cannot be electrolyzed directly because its melting point is too high (2050°C). Over 5 million tons of 99.7 percent pure aluminum metal are produced each year in the United States by the Hall process (Figure 20-5).

Reactive metals can be protected from corrosive substances with a thin layer of a relatively nonreactive metal such as nickel, chromium, tin, silver or gold. The production of a layer of protective metal by electrochemical deposition is called **electroplating.** For example, gold electroplating is used to coat base metals with a layer of gold that functions both as a decorative and protective coating (Figure 20-6). Electroplating has innumerable industrial applications ranging from the manufacture of heavy machinery to the production of microcircuits.

Figure 20-6 Gold is purified by electroplating. The spongy gold deposits shown here were electroplated onto gold strips. The top of the original gold cathode is visible in the photo.

20.4 AN ELECTROCHEMICAL CELL PRODUCES ELECTRICITY DIRECTLY FROM A CHEMICAL REACTION

In the preceding sections we discussed electrolysis, which involves using an external source of current to drive a nonspontaneous chemi-

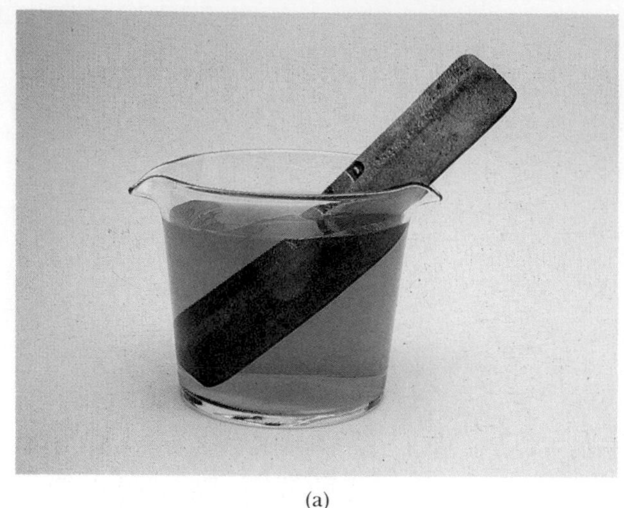

(a)

(b)

Figure 20-7 (a) When a zinc rod is placed in a copper sulfate solution, the zinc replaces the copper and (b) elemental copper forms.

cal reaction uphill in a Gibbs free energy sense. In this section we consider the reverse situation, in which a spontaneous chemical reaction is used to produce a current. The voltaic pile was one of the first examples of such a device.

Recall that electrons are transferred from one substance to another in oxidation-reduction (electron-transfer) reactions. Figure 20-7 shows that when a zinc rod is immersed in an aqueous solution of copper sulfate, the following reaction occurs spontaneously:

$$Zn(s) + CuSO_4(aq) \rightarrow Cu(s) + ZnSO_4(aq) \qquad (20\text{-}5)$$

It is possible to use redox reactions to produce an electric current. The basic idea is to keep the reactants (Zn and $CuSO_4$ in Equation 20-5) and the products (Cu and $ZnSO_4$) separated physically in such a way that the electron transfers take place via an external circuit. A setup in which an electric current is obtained from a chemical reaction is called an **electrochemical cell.** The simple electrochemical cell shown in Figure 20-8 has the remarkable property of being able to supply an electric current to power a light bulb or an electric motor. This cell consists of two different metals, Zn(*s*) and Cu(*s*), each immersed in an electrolyte solution containing their respective metal ions $Zn^{2+}(aq)$ and $Cu^{2+}(aq)$. Electrical contact between the two solutions is then made through a **salt bridge** consisting of a saturated KCl(*aq*) solution mixed with agar, a substance that forms a gel similar to Jell-O. The purpose of the gel is to hold the salt solution in the tube and thus prevent mixing. The salt bridge provides an ionic current path between the $ZnSO_4(aq)$ and the $CuSO_4(aq)$ solutions while simultaneously preventing their mixing. The electrodes are attached to metal leads (wires), which enable the cell to deliver electric current to the **external circuit.** Thus electricity is produced directly from a chemical reaction.

When the cell shown in Figure 20-8 is used as a source of electricity, the cell reaction when current is drawn from the cell is

$$Zn(s) + Cu^{2+}(aq) \xrightarrow{\text{discharge}} Zn^{2+}(aq) + Cu(s)$$

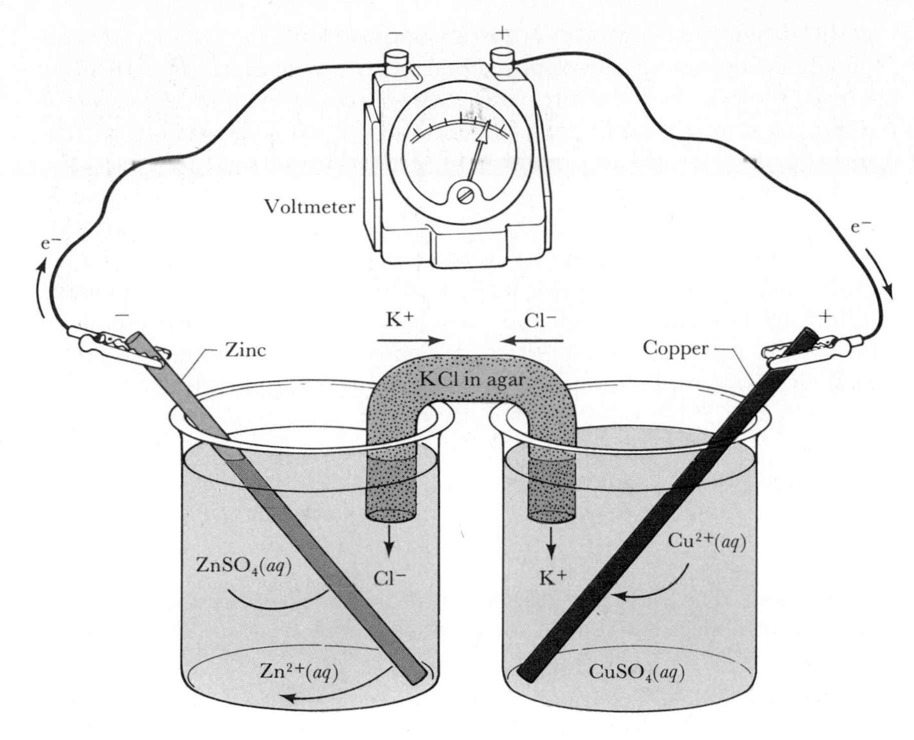

Figure 20-8 A zinc-copper electrochemical cell. The polarity of the cell (zinc electrode negative, copper electrode positive) is determined by experiment with a voltmeter. When the cell operates, electrons flow through the external circuit from the zinc electrode to the copper electrode.

where the word **discharge** indicates that current is drawn from the cell. The two electrode reactions on discharge are

$$Zn(s) \rightarrow Zn^{2+}(aq) + 2e^- \qquad \text{oxidation of } Zn(s)$$

$$Cu^{2+}(aq) + 2e^- \rightarrow Cu(s) \qquad \text{reduction of } Cu^{2+}(aq)$$

Note that the zinc-copper cell is based on the coupling of two half-reactions. When the external circuit is completed by an electrical connection between the Zn(*s*) and Cu(*s*) electrodes, electrons can flow in the external circuit, that is, through the wire leads and the voltmeter (Figure 20-8). The electrons produced at the zinc electrode by the oxidation of Zn(*s*) travel through the external circuit to the Cu(*s*) electrode, where they are consumed in the reduction of $Cu^{2+}(aq)$ to Cu(*s*).

The current in the electrolyte solutions and in the salt bridge is carried by ions and not by electrons. Because $Zn^{2+}(aq)$ ions are produced in the solution containing the Zn(*s*) electrode, negative ions—in this case, $Cl^-(aq)$—must enter the $ZnSO_4(aq)$ from the salt bridge in order to maintain electrical neutrality in the solution. Because $Cu^{2+}(aq)$ ions are removed from the solution containing the Cu(*s*) electrode, positive ions—in this case, $K^+(aq)$—must enter the $CuSO_4(aq)$ from the salt bridge in order to maintain electrical neutrality in the solution. Thus the current through the cell solutions is carried by moving ions, with negative ions moving toward the negative (zinc) electrode and positive ions moving toward the positive (copper) electrode.

The components of an electrochemical cell are set up in a manner that forces the electron-transfer reaction to proceed via the external circuit. Electrons are transferred from the reducing agent, Zn(*s*), to the oxidizing agent, $Cu^{2+}(aq)$, through the metal part (wire leads and voltmeter) of the circuit. If the two electrolyte solutions were mixed together, then the cell would not work because $Cu^{2+}(aq)$ would be spontaneously deposited as copper metal on the zinc electrode. In other words, the cell would be internally short-circuited. The reducing agent, Zn(*s*), must be physically separated from the oxidizing agent, $Cu^{2+}(aq)$, in order for the cell to operate.

A current of electrons flows spontaneously through a metal conductor from a region of negative voltage to a region of positive voltage. Thus, during the spontaneous discharge of the cell shown in Figure 20-8, the zinc electrode is negative and the copper electrode is positive. The total current through the cell electrolyte, which must equal the current in the external circuit, is carried by anions moving toward the zinc electrode and cations moving toward the copper electrode.

Example 20-3: Consider the electrochemical cell shown at the bottom of this page. Indicate the current flow in the external circuit and in the cell electrolytes.

Solution: The lead electrode is negative, and the silver electrode is positive. Therefore electrons flow in the external circuit from the lead electrode

$$Pb(s) \rightarrow Pb^{2+}(aq) + 2e^-$$

to the silver electrode

$$2Ag^+(aq) + 2e^- \rightarrow 2Ag(s)$$

Positive Pb^{2+} ions are produced at the lead electrode, and thus $NO_3^-(aq)$ ions flow from the salt bridge into the $Pb(NO_3)_2(aq)$ solution to maintain electrical neutrality. Positive Ag^+ ions are consumed at the silver electrode, and

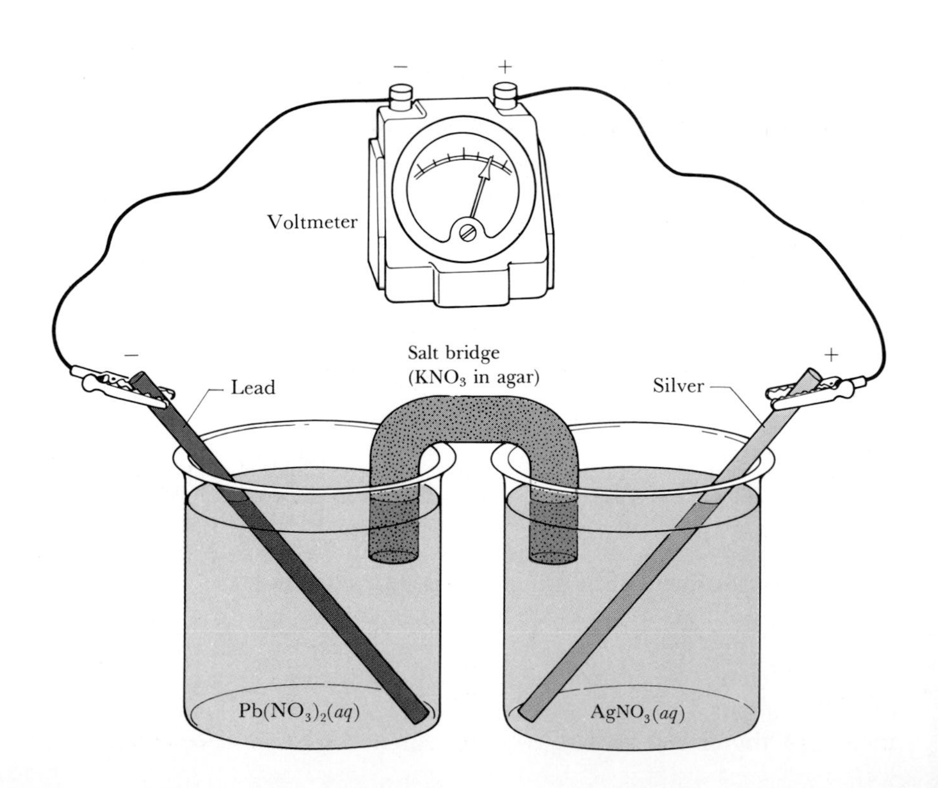

thus $K^+(aq)$ ions flow from the salt bridge into the $AgNO_3(aq)$ solution to maintain electrical neutrality. The equation for the overall cell reaction is

$$Pb(s) + 2Ag^+(aq) \rightarrow Pb^{2+}(aq) + 2Ag(s)$$

If we simply place a lead rod in $AgNO_3(aq)$, then the spontaneous reaction occurs directly (see Section 2-19) and $Ag(s)$ plates out on the $Pb(s)$ electrode. In contrast, the cell forces the reaction to proceed via the external circuit.

20-5. A CELL DIAGRAM IS USED TO REPRESENT AN ELECTROCHEMICAL CELL

The pictorial representation of an electrochemical cell can be made more compact by the use of a **cell diagram.** For example, the cell diagram for the cell shown in Figure 20-8 is:

$$Zn(s)|ZnSO_4(aq)||CuSO_4(aq)|Cu(s)$$

where the single vertical bars indicate boundaries of phases that are in contact and the double vertical bars indicate a salt bridge. Thus in the cell represented by this diagram, $Zn(s)$ and $ZnSO_4(aq)$ are separate phases in physical contact, as are $CuSO_4(aq)$ and $Cu(s)$, and a salt bridge separates the $ZnSO_4(aq)$ and $CuSO_4(aq)$ solutions.

The basic convention for writing equations for cell reactions is to *write the half-reaction of the left-hand electrode as an oxidation half-reaction equation and the half-reaction of the right-hand electrode as a reduction half-reaction equation.* This convention enables us to write the equation for the cell reaction unambiguously. For the above cell, then, we have, with oxidation occurring at the left-hand electrode,

$$Zn(s) \rightleftharpoons Zn^{2+}(aq) + 2e^- \qquad \text{oxidation at left-hand electrode}$$

If oxidation takes place at the left-hand electrode, then reduction must take place at the right-hand electrode:

$$Cu^{2+}(aq) + 2e^- \rightleftharpoons Cu(s) \qquad \text{reduction at right-hand electrode}$$

note *r* correspondence

The equation for the net cell reaction is given by the sum of the equations for the two electrode half-reactions:

$$Zn(s) + Cu^{2+}(aq) \rightleftharpoons Cu(s) + Zn^{2+}(aq)$$

Consider the redox reaction described by the equation

$$Zn(s) + 2H^+(aq) \rightleftharpoons H_2(g) + Zn^{2+}(aq)$$

Let's construct an electrochemical cell that utilizes this reaction. We first note that $Zn(s)$ is oxidized to $Zn^{2+}(aq)$, and thus the equation for the oxidation half-reaction of the cell is

$$Zn(s) \rightarrow Zn^{2+}(aq) + 2e^- \qquad \text{oxidation, left-hand electrode}$$

One of the electrode wires is connected to the zinc metal.

We next note that $H^+(aq)$ is reduced to hydrogen gas, and thus the equation for the reduction half-reaction of the cell is

$$2H^+(aq) + 2e^- \rightarrow H_2(g) \qquad \text{reduction, right-hand electrode}$$

This half-reaction requires a metal electrode that can provide a pathway for the electrons from the zinc electrode. The reduction of the $H^+(aq)$ to $H_2(g)$ takes place on the surface of this metal electrode, which acts as a source of electrons. The necessary electrode is provided by inserting a platinum coil into the cell compartment containing $H^+(aq)$ and $H_2(g)$. Platinum is a relatively unreactive metal that simply provides a metallic surface on which the reduction half-reaction occurs. An electrode involving a gaseous species is called a **gas electrode.** The cell diagram for the cell is

$$Zn(s)|Zn^{2+}(aq)||H^+(aq)|H_2(g)|Pt(s)$$

An electrochemical cell that incorporates the zinc oxidation and $H^+(aq)$ reduction half-reactions is shown in Figure 20-9. Hydrogen gas is bubbled continuously through the right-hand compartment in order to provide $H_2(g)$ at a known pressure. Because the cell is open to the atmosphere, the pressure of $H_2(g)$ in the cell is essentially equal to the atmospheric pressure.

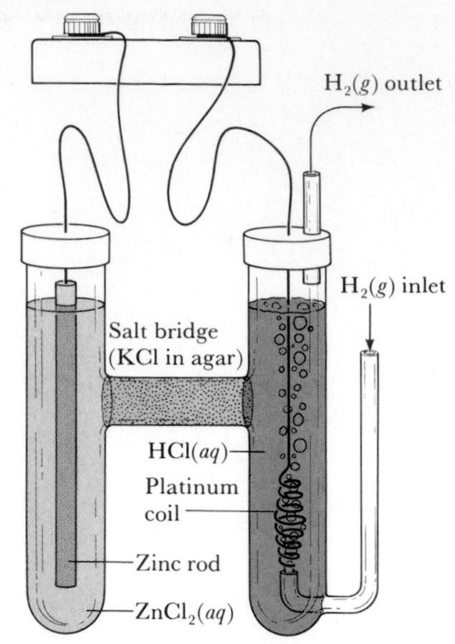

Figure 20-9 A zinc-hydrogen electrochemical cell. The cell diagram is

$$Zn(s)|ZnCl_2(aq)||HCl(aq)|H_2(g)|Pt(s)$$

and the equation for the cell reaction is

$$Zn(s) + 2H^+(aq) \rightleftharpoons H_2(g) + Zn^{2+}(aq)$$

Note the H-type geometry of the cell, which is used together with the salt bridge to separate the two electrolyte solutions.

Example 20-4: Write the equations for the electrode half-reactions and the net cell reaction for the electrochemical cell

$$Cd(s)|CdSO_4(aq)|Hg_2SO_4(s)|Hg(l)$$

Note that this cell does not have a salt bridge because it has only one electrolyte solution.

Solution: Oxidation is assumed to take place at the left-hand electrode. The oxidation of Cd(*s*) yields $Cd^{2+}(aq)$, and so we write

$$Cd(s) \rightarrow Cd^{2+}(aq) + 2e^- \qquad \text{oxidation}$$

Oxidation occurs at the left-hand electrode, and therefore reduction must occur at the right-hand electrode. The only element besides zinc that appears in two different oxidation states in the cell is mercury—zero in Hg(*l*) and +1 in $Hg_2SO_4(s)$. For the reduction at the mercury electrode, therefore, we write

$$Hg_2SO_4(s) \rightarrow 2Hg(l) \qquad \text{not balanced}$$

The balanced equation for the electrode half-reaction is

$$Hg_2SO_4(s) + 2e^- \rightarrow 2Hg(l) + SO_4^{2-}(aq) \qquad \text{reduction}$$

The sum of the equations for the oxidation and reduction half-reactions gives the equation for the net cell reaction:

$$Cd(s) + Hg_2SO_4(s) \rightarrow 2Hg(l) + Cd^{2+}(aq) + SO_4^{2-}(aq)$$

The cell described here is very similar to the Weston standard cell (see photo in margin), which is widely used as a voltage reference. The voltage of the cell is 1.018 V, and is essentially independent of temperature.

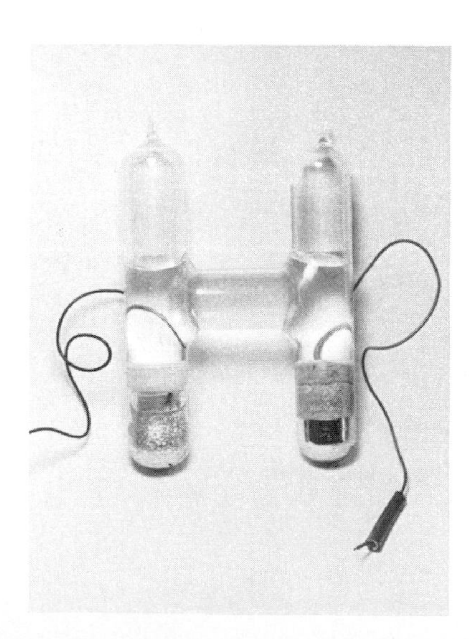

A Weston standard cell. The cell diagram is

$$Cd(Hg)|CdSO_4(aq)|Hg_2SO_4(s)|Hg(l)$$

The cell shown has two porcelain spacers with center holes that prevent the mixing of the cadmium amalgam, Cd(*Hg*), and the mercury. The grayish powder on the right is $Hg_2SO_4(s)$, which sits on top of the Hg(*l*).

20-6. CELL VOLTAGE DEPENDS ON THE CONCENTRATIONS OF THE REACTANTS AND PRODUCTS OF THE CELL REACTION

Walther Nernst, who received a Nobel Prize in chemistry for his pioneering work in electrochemistry, investigated the dependence of cell voltage on the concentration of the electrolytes, size of the electrodes, and other factors. He found, for example, that the voltage of the cell

$$Zn(s)|ZnSO_4(aq)||CuSO_4(aq)|Cu(s)$$

was independent of the size of the cell, the size of the electrodes, and the volume of the $ZnSO_4(aq)$ and $CuSO_4(aq)$ solutions. He also found that the voltage of the cell increased when the concentration of $Cu^{2+}(aq)$ increased and decreased when the concentration of $Zn^{2+}(aq)$ increased.

The effect of a change in reactant or product concentration on cell voltage is easily understood in qualitative terms by applying Le Châtelier's principle to the cell reaction, given that the **cell voltage** is a measure of the driving force of the reaction. We have seen that the reaction of the zinc-copper cell (Figure 20-8) is

$$Zn(s) + Cu^{2+}(aq) \rightleftharpoons Cu(s) + Zn^{2+}(aq)$$

An increase in the value of $[Cu^{2+}]$ drives the reaction from left to right, thereby increasing the cell voltage, whereas an increase in the value of $[Zn^{2+}]$ drives the reaction from right to left, thereby decreasing the cell voltage.

Example 20-5: Consider an electrochemical cell in which the cell reaction described by the equation

$$H_2(g) + 2AgCl(s) \rightleftharpoons 2Ag(s) + 2H^+(aq) + 2Cl^-(aq)$$

takes place. In this reaction $H_2(g)$ reduces $AgCl(s)$ to silver metal. Predict the effect of the following changes on the observed cell voltage:

(a) increase in the $H_2(g)$ pressure

(b) increase in the concentration of $H^+(aq)$

(c) increase in the amount of $AgCl(s)$

Solution: By using Le Châtelier's principle, we predict the following:

(a) An increase in P_{H_2} leads to an increase in the concentration of $H_2(g)$, by Henry's law, and this increases the reaction driving force from left to right. The cell voltage thus increases.

(b) An increase in $[H^+]$ increases the reaction driving force from right to left and thus decreases the cell voltage.

(c) An increase in the amount of $AgCl(s)$ has no effect on the cell voltage.

▪ A change in the amount of $AgCl(s)$ has no effect on the concentration of $AgCl(s)$ and thus has no effect on the voltage.

Recall from Chapter 14 that the value of Q/K, where Q is the concentration quotient and K is the equilibrium constant, can be used to predict the direction in which a reaction proceeds toward equilibrium. The

Nernst equation expresses the quantitative relationship between the cell voltage, E, and the value of Q/K:

$$E = -\frac{2.303\ RT}{nF}\log\left(\frac{Q}{K}\right) \qquad (20\text{-}6)$$

▪ The value of n is readily obtained as the number of electrons that cancel out from the two half-reaction equations when they are combined to give the balanced equation for the oxidation-reduction reaction.

where R is the gas constant, T is the Kelvin temperature, n is the number of electrons transferred from the reducing agent to the oxidizing agent for the cell reaction as written, and F is Faraday's constant.

Before we can use the Nernst equation to calculate the effect of concentration on cell voltage, we must discuss the units involved. As you may know from physics, a charged particle gains or loses energy when it is driven by a voltage difference. The energy change, the charge, and the voltage difference are related by the formula

$$\text{energy change} = (\text{electric charge}) \times (\text{voltage difference})$$

The SI units corresponding to this equation are

$$\text{joule} = \text{coulomb} \times \text{volt} \qquad (20\text{-}7)$$

or

$$1\ \text{J} = 1\ \text{C}\cdot\text{V}$$

▪ A volt is one joule per coulomb. Recall that mechanical work is force times displacement. Electrical work is voltage times charge. For this reason voltage is sometimes called "electromotive force" or "emf."

In Equation (20-6), RT has units of $\text{J}\cdot\text{mol}^{-1}$ and nF has units of $\text{C}\cdot\text{mol}^{-1}$, and so E has units of $\text{J}\cdot\text{C}^{-1}$, or V (volts).

We often apply the Nernst equation at 25.0°C, and so let's evaluate 2.303 RT/F at 25.0°C:

$$\frac{2.303\ RT}{F} = \frac{(2.303)(8.314\ \text{J}\cdot\text{K}^{-1}\cdot\text{mol}^{-1})(298.2\ \text{K})}{96{,}500\ \text{C}\cdot\text{mol}^{-1}}$$

$$= 0.0592\ \text{J}\cdot\text{C}^{-1} = 0.0592\ \text{V}$$

Thus at 25.0°C the Nernst equation becomes

$$E = -\left(\frac{0.0592\ \text{V}}{n}\right)\log\left(\frac{Q}{K}\right) \qquad (20\text{-}8)$$

Recall from Chapter 14 that, if $(Q/K) < 1$, then the reaction is spontaneous from left to right. If $(Q/K) < 1$, then $\log(Q/K)$ is negative because the logarithm of a number less than 1 is negative. Thus, if $(Q/K) < 1$, then Equation (20-8) shows that the cell voltage is positive. The various possibilities for the values of Q/K and the corresponding voltages are

$\frac{Q}{K} < 1$	$\log\left(\frac{Q}{K}\right) < 0$	$E > 0$	cell reaction is spontaneous from left to right
$\frac{Q}{K} = 1$	$\log\left(\frac{Q}{K}\right) = 0$	$E = 0$	cell reaction is at equilibrium
$\frac{Q}{K} > 1$	$\log\left(\frac{Q}{K}\right) > 0$	$E < 0$	cell reaction is spontaneous from right to left

▪ A reaction at equilibrium has no net driving force.

Thus we see that the sign of the cell voltage tells us the direction in which the cell reaction is spontaneous. Furthermore, the value of the cell voltage is a quantitative measure of the driving force of the cell reaction toward equilibrium. The larger the voltage, the greater is the reaction driving force. Note that, if the reaction is at equilibrium, then the corresponding cell voltage is zero.

Equation (20-8) can be written in a form that is often more convenient by using the relation

$$\log\left(\frac{Q}{K}\right) = \log Q - \log K \tag{20-9}$$

Combining this equation with Equation (20-8) yields

$$E = \left(\frac{0.0592\ \text{V}}{n}\right)\log K - \left(\frac{0.0592\ \text{V}}{n}\right)\log Q \tag{20-10}$$

If $Q = 1$ for the cell equation, that is, if all solution species involved in the reaction are at a concentration of 1.00 M and all gaseous species involved are at 1.00 atm, then the resulting cell voltage is called the **standard cell voltage** and is denoted by $E°$. Substituting $Q = 1$ into Equation (20-10) and noting that $\log 1 = 0$, we obtain

$$E\ (\text{at } Q = 1) = \underset{\substack{\uparrow \\ \text{standard} \\ \text{cell voltage}}}{E°} = \left(\frac{0.0592\ \text{V}}{n}\right)\log K \tag{20-11}$$

Equation (20-11) enables us to compute the value of the equilibrium constant, K, for the cell equation if the value of $E°$ is known and to compute the value of $E°$ if K is known.

Example 20-6: The value of $E°$ at 25.0°C for the equation

$$Zn(s) + Cu^{2+}(aq) \rightleftharpoons Cu(s) + Zn^{2+}(aq)$$

is 1.10 V. Compute the equilibrium constant at 25.0°C for this equation.

Solution: From Equation (20-11), we have

$$E° = \left(\frac{0.0592\ \text{V}}{n}\right)\log K$$

Solving this equation for $\log K$, we obtain

$$\log K = \left(\frac{nE°}{0.0592\ \text{V}}\right)$$

The value of n is 2 for the balanced equation, and thus

$$\log K = \frac{2(1.10\ \text{V})}{0.0592\ \text{V}} = 37.2$$

and

$$K = 10^{37.2} = 1.6 \times 10^{37} = \frac{[Zn^{2+}]_{eq}}{[Cu^{2+}]_{eq}}$$

The very large value of K means that at equilibrium the ratio of $[Zn^{2+}]$ to $[Cu^{2+}]$ is very large, or, in other words, that the value of $[Cu^{2+}]$ at equilibrium is very small indeed. (The subscript *eq* on a concentration term denotes an *eq*uilibrium value of that concentration.)

Substitution of Equation (20-11) into Equation (20-10) yields an especially useful form of the Nernst equation:

$$E = E^\circ - \left(\frac{0.0592\ \text{V}}{n}\right)\log Q \qquad (20\text{-}12)$$

where E and E° are in volts and all values are at 25.0°C. Equation (20-12) tells us that the cell voltage differs from the standard cell voltage when the reaction quotient is not equal to 1.00. If $Q < 1$, then $E > E^\circ$; if $Q > 1$, then $E < E^\circ$.

Example 20-7: The measured voltage at 25.0°C of a cell in which the reaction described by the equation

$$Zn(s) + Cu^{2+}(aq,\ 1.00\ \text{M}) \rightleftharpoons Cu(s) + Zn^{2+}(aq,\ 0.100\ \text{M})$$

occurs at the concentrations shown is 1.13 V. Compute E° for the cell equation.

Solution: From Equation (20-12) with $n = 2$, we have

$$E = E^\circ - \left(\frac{0.0592\ \text{V}}{n}\right)\log Q = E^\circ - \left(\frac{0.0592\ \text{V}}{2}\right)\log\frac{[Zn^{2+}]}{[Cu^{2+}]}$$

or

$$1.13\ \text{V} = E^\circ - \left(\frac{0.0592\ \text{V}}{2}\right)\log\left(\frac{0.100\ \text{M}}{1.00\ \text{M}}\right)$$

from which we compute

$$E^\circ = 1.13\ \text{V} + (0.0296\ \text{V})\log 0.100 = 1.10\ \text{V}$$

Note that the measurement of the cell voltage when the reactant and product concentrations are known yields the value of E° for the cell equation and hence, from Equation (20-11), the value of the equilibrium constant for the cell equation. Note also that $Zn(s)$ and $Cu(s)$ do not appear in the Q expression because they are both solids. Application of the Nernst equation at 25°C to the equation for the $Zn(s)$–$Cu^{2+}(aq)$ reaction at various values of Q yields the results shown in Figure 20-10.

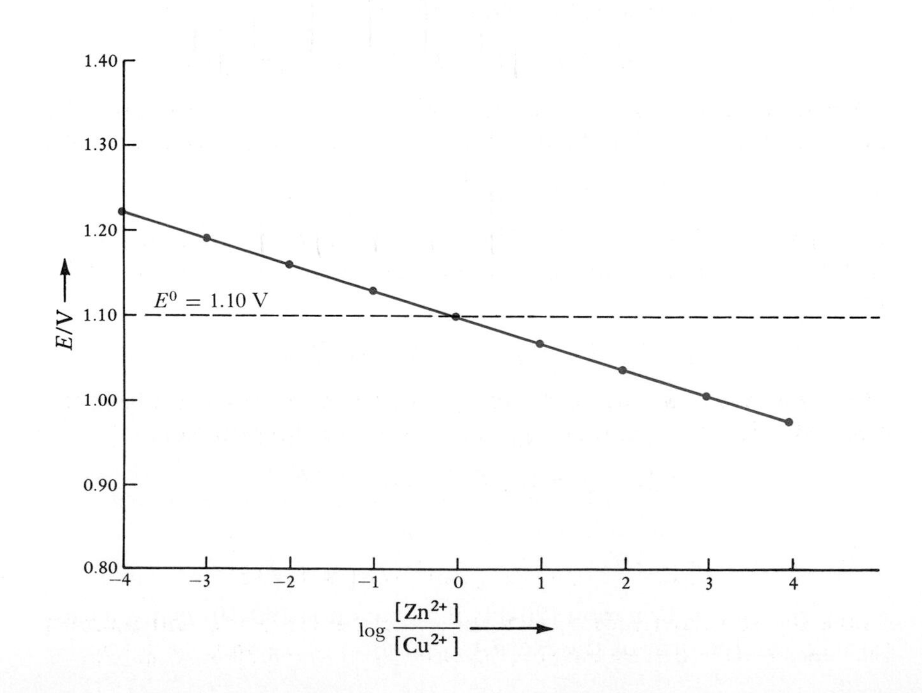

Figure 20-10 Plot of E (in volts) versus $\log Q$ for the cell equation

$$Zn(s) + Cu^{2+}(aq) \rightleftharpoons Cu(s) + Zn^{2+}(aq)$$

The plot is based on the Nernst equation applied to the above cell equation, that is,

$$E = 1.10\ \text{V} - (0.030\ \text{V})\log\frac{[Zn^{2+}]}{[Cu^{2+}]}$$

Note that a 10-fold increase in the value of $Q = [Zn^{2+}]/[Cu^{2+}]$ decreases the cell voltage by −0.030 V. The plot illustrates the linear dependence of cell voltage on $\log Q$.

20-7. $E°$ VALUES CAN BE ASSIGNED TO HALF-REACTION EQUATIONS

It is not possible to measure the voltage of a single electrode; only the difference in voltage between two electrodes can be measured. Recall that with our convention for the direction of current in a cell, *r*eduction occurs at the *r*ight electrode and oxidation occurs at the left electrode. Even though only differences in voltage can be measured, it is still possible to assign voltages to equations for half-reactions if we agree to choose a numerical value for the standard voltage of a particular electrode. The convention is to set the standard reduction voltage of the **hydrogen electrode** equal to zero; that is, we choose $E° = 0$ for the electrode half-reaction

$$2H^+(aq, 1\text{ M}) + 2e^- \rightarrow H_2(g, 1\text{ atm}) \qquad E° = 0 \text{ by convention} \qquad (20\text{-}13)$$

A reduction voltage denotes a voltage assigned to a half-reaction written as a reduction, that is, written with electrons on the left. A complete cell involves two half-reactions: a reduction half-reaction with voltage E_{red}, and an oxidation half-reaction with voltage E_{ox}. An oxidation half-reaction is a half-reaction written as an oxidation, that is, written with electrons on the right. We now write for a complete cell

$$E°_{\text{cell}} = E°_{red} + E°_{ox} \qquad (20\text{-}14)$$

Equation (20-14) can be used together with Equation (20-13) to set up a table of **standard reduction voltages,** that is, a table of $E°_{red}$ values. Note that for a particular half-reaction,

$$E°_{red} = -E°_{ox} \qquad \text{(same half-reaction)} \qquad (20\text{-}15)$$

Thus, if for the reduction half-reaction equation

$$Zn^{2+}(aq, 1\text{ M}) + 2e^- \rightarrow Zn(s) \qquad E°_{red} = -0.76\text{ V}$$

then for the oxidation half-reaction equation

$$Zn(s) \rightarrow Zn^{2+}(aq, 1\text{ M}) + 2e^- \qquad E°_{ox} = +0.76\text{ V}$$

We now show how standard reduction voltages are obtained. The experimental value for the standard voltage of the cell (Figure 20-9)

$$Zn(s)|ZnCl_2(aq)||HCl(aq)|H_2(g)|Pt(s)$$

is $E°_{\text{cell}} = 0.76\text{ V} = E°_{red} + E°_{ox}$. The right-hand electrode involves the H^+/H_2 pair, and the left-hand electrode involves the Zn^{2+}/Zn pair. Thus we write

$$E°_{\text{cell}} = E°_{red}[H^+/H_2] + E°_{ox}[Zn^{2+}/Zn] = 0.76\text{ V}$$

where the brackets enclose the oxidized and reduced species for each electrode. Substitution of $E°_{red}[H^+/H_2] = 0$ into this equation

$$E°_{\text{cell}} = 0 + E°_{ox}[Zn^{2+}/Zn] = 0.76\text{ V}$$

or

$$E°_{ox}[Zn^{2+}/Zn] = -E°_{red}[Zn^{2+}/Zn] = +0.76\text{ V}$$

Thus, the standard reduction voltage for the electrode half-reaction $Zn^{2+}(aq, 1\text{ M}) + 2e^- \rightarrow Zn(s)$ is $E°_{red}[Zn^{2+}/Zn] = -0.76\text{ V}$ at 25°C.

Once we have determined the standard reduction voltage of an electrode relative to the hydrogen electrode, we can use that electrode to determine the standard reduction voltage of other electrodes. This procedure is illustrated in the following Example.

Example 20-8: The standard voltage of the cell

$$Zn(s)|ZnSO_4(aq)||CuSO_4(aq)|Cu(s)$$

is $E^\circ_{cell} = 1.10$ V. Given that $E^\circ_{red} = -0.76$ V for the electrode half-reaction

$$Zn^{2+}(aq, 1\ M) + 2e^- \rightarrow Zn(s)$$

compute E°_{red} for the electrode half-reaction

$$Cu^{2+}(aq, 1\ M) + 2e^- \rightarrow Cu(s)$$

Solution: Assuming, as is our convention, that oxidation takes place at the left-hand electrode, we have for the half-reactions

$Zn(s) \rightarrow Zn^{2+}(aq) + 2e^-$ oxidation, left-hand electrode, $E^\circ_{ox}[Zn^{2+}/Zn]$

$Cu^{2+}(aq) + 2e^- \rightarrow Cu(s)$ reduction, right-hand electrode, $E^\circ_{red}[Cu^{2+}/Cu]$

Therefore, we have

$$E^\circ_{cell} = E^\circ_{red}[Cu^{2+}/Cu] + E^\circ_{ox}[Zn^{2+}/Zn] = 1.10\ V$$

because

$$E^\circ_{ox}[Zn^{2+}/Zn] = -E^\circ_{red}[Zn^{2+}/Zn] = +0.76\ V$$

we have

$$E^\circ_{cell} = E^\circ_{red}[Cu^{2+}/Cu] + 0.76\ V = 1.10\ V$$

and thus

$$E^\circ_{red}[Cu^{2+}/Cu] = 1.10\ V - 0.76\ V = +0.34\ V$$

The standard reduction voltage of the $Cu^{2+}(aq)/Cu(s)$ electrode is $E^\circ_{red} = +0.34$ V.

Thus we see that the standard reduction voltage of an electrode can be obtained from the standard cell voltage of a cell for which the standard reduction voltage of the other electrode is known.

Table 20-1 Standard reduction voltages at 25.0° for aqueous solutions

Electrode half-reaction	E°_{red}/V
Acidic solutions	
$F_2(g) + 2e^- \rightarrow 2F^-(aq)$	+2.87
$O_3(g) + 2H^+(aq) + 2e^- \rightarrow O_2(g) + H_2O(l)$	+2.07
$S_2O_8^{2-}(aq) + 2e^- \rightarrow 2SO_4^{2-}(aq)$	+2.01
$Co^{3+}(aq) + e^- \rightarrow Co^{2+}(aq)$	+1.81
$H_2O_2(aq) + 2H^+(aq) + 2e^- \rightarrow 2H_2O(l)$	+1.78
$PbO_2(s) + SO_4^{2-}(aq) + 4H^+(aq) + 2e^- \rightarrow PbSO_4(s) + 2H_2O(l)$	+1.68
$MnO_4^-(aq) + 8H^+(aq) + 5e^- \rightarrow Mn^{2+}(aq) + 4H_2O(l)$	+1.51
$Cl_2(g) + 2e^- \rightarrow 2Cl^-(aq)$	+1.36
$Cr_2O_7^{2-}(aq) + 14H^+(aq) + 6e^- \rightarrow 2Cr^{3+}(aq) + 7H_2O(l)$	+1.33
$O_2(g) + 4H^+(aq) + 4e^- \rightarrow 2H_2O(l)$	+1.23

Table 20-1 Continued

Electrode half-reaction	E°_{red}/V
$Br_2(l) + 2e^- \rightarrow 2Br^-(aq)$	+1.07
$Pd^{2+}(aq) + 2e^- \rightarrow Pd(s)$	+0.99
$Ag^+(aq) + e^- \rightarrow Ag(s)$	+0.80
$Hg_2^{2+}(aq) + 2e^- \rightarrow 2Hg(l)$	+0.79
$Fe^{3+}(aq) + e^- \rightarrow Fe^{2+}(aq)$	+0.77
$O_2(g) + 2H^+(aq) + 2e^- \rightarrow H_2O_2(aq)$	+0.68
$Hg_2SO_4(s) + 2e^- \rightarrow 2Hg(l) + SO_4^{2-}(aq)$	+0.62
$I_2(s) + 2e^- \rightarrow 2I^-(aq)$	+0.54
$Cu^+(aq) + e^- \rightarrow Cu(s)$	+0.52
$Fe(CN)_6^{3-}(aq) + e^- \rightarrow Fe(CN)_6^{4-}(aq)$	+0.36
$Cu^{2+}(aq) + 2e^- \rightarrow Cu(s)$	+0.34
$Hg_2Cl_2(s) + 2e^- \rightarrow 2Hg(l) + 2Cl^-(aq)$	+0.27
$AgCl(s) + e^- \rightarrow Ag(s) + Cl^-(aq)$	+0.22
$Cu^{2+}(aq) + e^- \rightarrow Cu^+(aq)$	+0.15
$2H^+(aq) + 2e^- \rightarrow H_2(g)$	0
$Pb^{2+}(aq) + 2e^- \rightarrow Pb(s)$	−0.13
$PbSO_4(s) + 2e^- \rightarrow Pb(s) + SO_4^{2-}(aq)$	−0.36
$Cd^{2+}(aq) + 2e^- \rightarrow Cd(s)$	−0.40
$Cr^{3+}(aq) + e^- \rightarrow Cr^{2+}(aq)$	−0.41
$Fe^{2+}(aq) + 2e^- \rightarrow Fe(s)$	−0.44
$Zn^{2+}(aq) + 2e^- \rightarrow Zn(s)$	−0.76
$Al^{3+}(aq) + 3e^- \rightarrow Al(s)$	−1.66
$H_2(g) + 2e^- \rightarrow 2H^-(aq)$	−2.25
$Mg^{2+}(aq) + 2e^- \rightarrow Mg(s)$	−2.36
$Na^+(aq) + e^- \rightarrow Na(s)$	−2.71
$Ca^{2+}(aq) + 2e^- \rightarrow Ca(s)$	−2.87
$K^+(aq) + e^- \rightarrow K(s)$	−2.93
$Li^+(aq) + e^- \rightarrow Li(s)$	−3.05
Basic solutions	
$O_3(g) + H_2O(l) + 2e^- \rightarrow O_2(g) + 2OH^-(aq)$	+1.24
$ClO^-(aq) + H_2O(l) + 2e^- \rightarrow Cl^-(aq) + 2OH^-(aq)$	+0.89
$O_2(g) + 2H_2O(l) + 4e^- \rightarrow 4OH^-(aq)$	+0.40
$HgO(s) + H_2O(l) + 2e^- \rightarrow Hg(l) + 2OH^-(aq)$	+0.10
$Cu(OH)_2(s) + 2e^- \rightarrow Cu(s) + 2OH^-(aq)$	−0.22
$O_2(g) + e^- \rightarrow O_2^-(aq)$	−0.56
$Fe(OH)_3(s) + e^- \rightarrow Fe(OH)_2(s) + OH^-(aq)$	−0.56
$2H_2O(l) + 2e^- \rightarrow H_2(g) + 2OH^-(aq)$	−0.83
$2SO_3^{2-}(aq) + 2H_2O(l) + 2e^- \rightarrow S_2O_4^{2-}(aq) + 4OH^-(aq)$	−1.12
$ZnS(s) + 2e^- \rightarrow Zn(s) + S^{2-}(aq)$	−1.41

Proceeding in the manner just outlined, we can construct a table of standard reduction voltages. In the arrangement used in Table 20-1, more positive E°_{red} values indicate more powerful oxidizing agents. Thus we can represent Table 20-1 in schematic form as follows:

increasing strength of oxidizing agents ↑		E°/V	increasing strength of reducing agents ↓
	$F_2(g) + 2e^- \rightarrow 2F^-(aq)$	+2.87	
	⋮	⋮	
	$H_2(g) + 2e^- \rightarrow 2H^+(aq)$	0	
	⋮	⋮	
	$Li^+(aq) + e^- \rightarrow Li(s)$	−3.05	

The more positive the E°_{red} value for a half-reaction, the stronger is the oxidizing agent (electron acceptor) in the half-reaction. The more negative the E°_{red} value for a half-reaction, the stronger is the reducing agent (electron donor) in the half-reaction. Thus we see that fluorine is the strongest oxidizing agent and that lithium is the strongest reducing agent in Table 20-1. Electrons appear on the left-hand side of the half-reactions in Table 20-1 because the values given refer to reduction half-reactions. The arrangement of half-reactions in order of the standard reduction voltages is sometimes called the **emf** (for *e*lectro*m*otive *f*orce) **series.**

Figure 20-11 A chromous bubbler, which is used to remove traces of oxygen from inert gases such as nitrogen or argon. The apparatus on the left is the chromous bubbler. The blue color of the solution is due to $Cr^{2+}(aq)$. The solid at the bottom of the tube is zinc amalgam (Zn + Hg). The green solution in the graduated cylinder is $Cr_2(SO_4)_3(aq)$, which is used to prepare the chromous bubbler. The green Cr(III) is reduced to the blue Cr(II) by the zinc amalgam. The gas to be deoxygenated enters the bubbler at the left inlet tube, bubbles up through the solution of Cr(II), and exits from the right tube. A glass frit at the bottom of the inlet tube breaks up the entering gas stream into small bubbles.

Table 20-1 contains a tremendous amount of information on the chemistry of aqueous solutions, which is extracted readily by using the Nernst equation. We can use the data in Table 20-1 to predict the thermodynamic stability of oxidizing and reducing agents under a wide variety of conditions. We also can predict the possible reactions between oxidizing and reducing agents. As an example, let's analyze the preparation and utilization of a **chromous bubbler** (Figure 20-11), which is a device used to remove traces of oxygen from gases. The chromous bubbler solution is prepared by reacting zinc metal with chromium(III) nitrate to form chromium(II):

$$Zn(s) + 2Cr^{3+}(aq) \rightarrow 2Cr^{2+}(aq) + Zn^{2+}(aq)$$

The resulting chromium(II) nitrate solution reduces oxygen to water in a gas mixture that is bubbled through the solution:

$$4Cr^{2+}(aq) + O_2(g) + 4H^+(aq) \rightarrow 4Cr^{3+}(aq) + 2H_2O(l)$$

The following Examples illustrate the calculations involved in assessing the effectiveness of a chromous bubbler in removing oxygen.

Example 20-9: A 1.0 M solution of $Cr(NO_3)_3(aq)$ is allowed to equilibrate with excess zinc metal at 25°C. Calculate the concentration of $Cr^{2+}(aq)$ at equilibrium. See Table 20-1 for the necessary data.

Solution: The equation for the reduction of $Cr^{3+}(aq)$ by $Zn(s)$ is

$$(1) \qquad Zn(s) + 2Cr^{3+}(aq) \rightleftharpoons 2Cr^{2+}(aq) + Zn^{2+}(aq)$$

In Table 20-1 we find

$$Cr^{3+}(aq) + e^- \rightarrow Cr^{2+}(aq) \qquad E^\circ_{red} = -0.41\ \text{V}$$

$$Zn^{2+}(aq) + 2e^- \rightarrow Zn(s) \qquad E^\circ_{red} = -0.76\ \text{V}$$

The E° value for Equation (1) is (see Equation 20-14)

$$E^\circ_{rxn} = E^\circ_{red}[Cr^{3+}/Cr^{2+}] + E^\circ_{ox}[Zn^{2+}/Zn]$$

$$= -0.41\ \text{V} + 0.76\ \text{V} = 0.35\ \text{V}$$

where $E^\circ_{rxn} = E^\circ_{\text{cell}}$ in Equation (20-14.) (We use E°_{rxn} for reactions that are not run in an electrochemical cell). The equilibrium constant for Equation (1) is calculated by using Equation (20-11):

$$\log K = \frac{nE^\circ_{rxn}}{0.0592\ \text{V}} = \frac{2 \times 0.35\ \text{V}}{0.0592\ \text{V}} = 11.82$$

and

$$K = 6.7 \times 10^{11}\ \text{M} = \frac{[Cr^{2+}]^2[Zn^{2+}]}{[Cr^{3+}]^2}$$

Because K is very large, the equilibrium lies far to the right and so the concentration of $Cr^{2+}(aq)$ will be very close to 1.00 M. Consequently, it is more convenient to solve for $[Cr^{3+}]$. Thus we have at equilibrium

$$[Cr^{2+}] = 1.00\ \text{M} - [Cr^{3+}] \qquad [Zn^{2+}] = \frac{1}{2}[Cr^{2+}] = \frac{1}{2}\{1.00\ \text{M} - [Cr^{3+}]\}$$

$$6.7 \times 10^{11}\ \text{M} = \frac{\{1.00\ \text{M} - [Cr^{3+}]\}^2\left(\frac{1}{2}\right)\{1.00\ \text{M} - [Cr^{3+}]\}}{[Cr^{3+}]^2}$$

Assuming $[Cr^{3+}] \ll 1.00$ M, we obtain

$$6.7 \times 10^{11}\ \text{M} \approx \frac{(1.00\ \text{M})^2(0.50\ \text{M})}{[Cr^{3+}]^2}$$

from which we compute

$$[Cr^{3+}] \approx \left(\frac{0.50\ \text{M}^3}{6.7 \times 10^{11}\ \text{M}}\right)^{1/2} = 8.6 \times 10^{-7}\ \text{M}$$

which confirms our assumption that $[Cr^{3+}] \ll 1.00$ M. Thus, at equilibrium, $[Cr^{2+}] = 1.00$ M, and essentially all the original $Cr^{3+}(aq)$ is reduced to $Cr^{2+}(aq)$.

Example 20-10: Oxygen reacts with $Cr^{2+}(aq)$ according to the equation

(1) $$4Cr^{2+}(aq) + O_2(g) + 4H^+(aq) \rightleftharpoons 4Cr^{3+}(aq) + 2H_2O(l)$$

Using data in Table 20-1, compute the equilibrium partial pressure of $O_2(g)$ over a solution that has

$$[Cr^{2+}] = 1.00\ \text{M} \qquad [Cr^{3+}] = 8.6 \times 10^{-7}\ \text{M} \qquad [H^+] = 0.10\ \text{M}$$

where the concentrations of $Cr^{2+}(aq)$ and $Cr^{3+}(aq)$ are maintained constant by the presence of excess $Zn(s)$ (see Example 20-10) and the value of $[H^+]$ is fixed by a buffer.

Solution: The equilibrium constant for Equation (1) is calculated by using the same method as in Example 20-10. Thus we have

$$\log K = \frac{nE^\circ_{rxn}}{0.0592\ \text{V}} = \frac{4(E^\circ_{red} + E^\circ_{ox})}{0.0592\ \text{V}}$$

$$= \frac{4\{E^\circ_{red}[O_2/H_2O] + E^\circ_{ox}[Cr^{3+}/Cr^{2+}]\}}{0.0592\ \text{V}}$$

$$= \frac{4(1.23\ \text{V} + 0.41\ \text{V})}{0.0592\ \text{V}} = 110.81$$

and $$K = 10^{110.81} = 6.5 \times 10^{110}\ \text{M}^{-4}\cdot\text{atm}^{-1}$$

For Equation (1) we have

$$(2) \qquad 6.5 \times 10^{110}\ \text{M}^{-4}\cdot\text{atm}^{-1} = \frac{[Cr^{3+}]^4}{[Cr^{2+}]^4[H^+]^4 P_{O_2}}$$

Substitution of the given values of $[Cr^{3+}]$, $[Cr^{2+}]$, and $[H^+]$ into Equation (2) yields

$$6.5 \times 10^{110}\ \text{M}^{-4}\cdot\text{atm}^{-1} = \frac{(8.6 \times 10^{-7}\ \text{M})^4}{(1.00\ \text{M})^4(0.10\ \text{M})^4 P_{O_2}}$$

from which we compute

$$P_{O_2} = 8.4 \times 10^{-132}\ \text{atm}$$

This value of P_{O_2} is infinitesimally small and means that $Cr^{2+}(aq)$ is capable of reducing the partial pressure of $O_2(g)$ to a value that is undetectably small.

20-8. ELECTROCHEMICAL CELLS ARE USED TO DETERMINE CONCENTRATIONS OF IONS

▪ Electrochemical cells are extensively used in analytical chemistry to measure the concentrations of ions in solution.

We have seen how the Nernst equation is used to calculate the voltage of an electrochemical cell when the concentrations of the species involved in the cell reaction are known. Conversely, a measured cell voltage can be used to determine the concentration of a species in solution. Consider the cell reaction given by

$$Zn(s) + Cu^{2+}(aq) \rightarrow Cu(s) + Zn^{2+}(aq) \qquad (20\text{-}16)$$

Application of the Nernst equation in the form of Equation (20-12) to Equation (20-16) at 25.0°C yields

$$E_{\text{cell}} = E^\circ_{\text{cell}} - \left(\frac{0.0592\ \text{V}}{2}\right)\log\frac{[Zn^{2+}]}{[Cu^{2+}]} \qquad (20\text{-}17)$$

The value of E° for Equation (20-16) is 1.10 V; thus we have from Equation (20-17)

$$E_{\text{cell}} = 1.10\ \text{V} - (0.0296\ \text{V})\log\frac{[Zn^{2+}]}{[Cu^{2+}]} \qquad (20\text{-}18)$$

We see from Equation (20-18) that if we measure E at a known value of, say, $[Cu^{2+}]$, then we can use the equation to compute the value of $[Zn^{2+}]$. For example, suppose that when $[Cu^{2+}] = 0.10$ M, we find that $E_{\text{cell}} = 1.20$ V. Substitution of these values for E and $[Cu^{2+}]$ into Equation (20-18) yields

$$1.20\text{ V} = 1.10\text{ V} - (0.0296\text{ V})\log\frac{[Zn^{2+}]}{0.10\text{ M}}$$

from which we compute

$$\log\frac{[Zn^{2+}]}{0.10\text{ M}} = \frac{1.10\text{ V} - 1.20\text{ V}}{0.0296\text{ V}} = -3.38$$

or

$$[Zn^{2+}] = (0.10\text{ M})(10^{-3.38}) = 4.2 \times 10^{-5}\text{ M}$$

The foregoing calculations show that electrochemical cells can be used to determine the concentrations of ions in solution.

Example 20-11: The measured voltage of the cell with the cell equation

$$H_2(g) + 2AgCl(s) \rightarrow 2H^+(aq) + 2Cl^-(aq,\ 1.0\text{ M}) + 2Ag(s)$$

is +0.34 V at 25°C when the pressure of $H_2(g)$ is 1.00 atm. Compute the pH of the solution.

Solution: Application of the Nernst equation to this equation, for which $n = 2$, yields

$$E_{cell} = E^\circ_{cell} - \left(\frac{0.0592\text{ V}}{2}\right)\log\left(\frac{[H^+]^2[Cl^-]^2}{P_{H_2}}\right)$$

The value of E°_{cell} is computed from the E° values for the half-reactions (Table 20-1) using Equation (20-14):

$$E^\circ_{cell} = E^\circ_{red}[AgCl/Ag] + E^\circ_{ox}[H^+/H_2]$$
$$= +0.22\text{ V} + 0 = +0.22\text{ V}$$

Substitution of the known values of E_{cell}, E°_{cell}, $[Cl^-]$, and P_{H_2} into the Nernst equation yields

$$+0.34\text{ V} = +0.22\text{ V} - (0.0296\text{ V})\log\left(\frac{[H^+]^2(1.0)^2}{1.00}\right)$$

Solving for $pH = -\log[H^+]$ gives

$$pH = -\log[H^+] = \frac{+0.34\text{ V} - 0.22\text{ V}}{2(0.0296\text{ V})} = +2.0$$

This Example shows how cells can be used to measure pH values, as is done routinely using pH meters.

20-9. THE VALUE OF ΔG_{rxn} IS EQUAL TO THE MAXIMUM AMOUNT OF WORK THAT CAN BE OBTAINED FROM THE REACTION

The Gibbs free energy change for an oxidation-reduction reaction is related to the corresponding cell voltage by the equation (see Equation 20-7)

Gibbs free energy change = charge transferred × voltage

or

$$\Delta G_{rxn} = -nFE_{rxn} \qquad (20\text{-}19)$$

where n is the number of moles of electrons transferred from the reducing agent to the oxidizing agent for the reaction as written and F is

Faraday's constant (96,500 C per mole of electrons). Recall from Equation (20-7) that charge × voltage has units of energy. If we use $F = 96{,}500\ \text{C·mol}^{-1}$ and E_{rxn} in volts in Equation (20-19), then the units of ΔG_{rxn} will be joules.

Combination of Equation (20-19) with Equation (19-8) yields

$$\Delta G_{rxn} = 2.30\,RT\log\left(\frac{Q}{K}\right) = -nFE_{rxn}$$

Solving for E_{rxn}, we obtain the Nernst equation:

$$E_{rxn} = -\frac{2.30\,RT}{nF}\log\left(\frac{Q}{K}\right)$$

The Nernst equation is thus seen to follow directly from the laws of thermodynamics.

Example 20-12: An electrochemical cell is set up in which the reaction described by the equation

$$2H_2(g) + O_2(g) \rightleftharpoons 2H_2O(l)$$

occurs. At 25°C the measured cell voltage is 1.23 V. Compute the value of ΔG_{rxn} for the reaction.

Solution: The value of ΔG_{rxn} can be computed using Equation (20-19), but we first must determine the value of n for the reaction. In the reaction 2 mol of oxygen *atoms* (1 mol of O_2) are reduced from an oxidation state of zero to an oxidation state of −2. Thus the reaction requires 4 mol of electrons per mole of O_2 and so $n = 4$ mol. Therefore

$$\begin{aligned}\Delta G_{rxn} &= -nFE_{rxn}\\ &= -(4\ \text{mol})(96{,}500\ \text{C·mol}^{-1})(1.23\ \text{V}) = -4.75\times10^{5}\ \text{C·V}\\ &= -4.75\times10^{5}\ \text{J}\end{aligned}$$

The negative value of ΔG_{rxn} indicates that the reaction is spontaneous, which we already know from the fact that mixtures of H_2 and O_2 can react explosively.

▪ The word free in Gibbs free energy is used in the sense of "available to perform work."

The value of ΔG_{rxn} is related to the energy that can be obtained from a process. If ΔG_{rxn} is negative (spontaneous reaction), then its magnitude is equal to the maximum energy that can be obtained from the reaction. In Example 20-12, the reaction of 2 mol of $H_2(g)$ with 1 mol of $O_2(g)$ can provide a maximum of 4.75×10^{5} J of energy to an external device, such as an electric motor. The maximum amount of energy that we calculate for a reaction is an ideal value. In practice we would obtain less than the maximum amount.

If ΔG_{rxn} is positive, then its value is the minimum energy that must be supplied to the reaction in order to make it occur. For example, Example 20-12 shows that $\Delta G_{rxn} = +4.75\times10^{5}$ J for the decomposition of water into $H_2(g)$ and $O_2(g)$ at 1 atm. Thus we must supply at least 4.75×10^{5} J to decompose 2 mol of water. This energy could be supplied by an electric current at an appropriate voltage, as is done in electrolysis.

20-10. A BATTERY IS AN ELECTROCHEMICAL CELL OR GROUP OF CELLS DESIGNED FOR USE AS A POWER SOURCE

A **battery** is a device that utilizes a chemical reaction to produce an electric current. There are many types of batteries. Some of the better-known examples are the lead storage battery, the mercury battery, the nickel-cadmium battery, the lithium battery, the dry cell, and the alkaline manganese cell. Batteries are especially useful as power sources when mobility is a prime consideration. They are adaptable to a wide range of power requirements, including the production of very high currents for short periods or stable voltages under low current drain for long periods. Batteries are classed as **primary batteries** if they are not rechargeable and as **secondary batteries** if they are rechargeable. Examples of primary batteries are the dry cell and the mercury battery, and examples of secondary batteries are the lead storage battery and the nickel-cadmium battery.

The 12-V **lead storage battery** consists of six of the following cells arranged in series:

▪ A 12-V lead storage automobile battery can provide over 300 amperes for short periods.

$$^{\ominus}Pb(s)|PbSO_4(s)|H_2SO_4(aq)|PbO_2(s),PbSO_4(s)|Pb(s)^{\oplus}$$

where the charges show the electrode polarities on discharge. The battery electrolyte is a concentrated $H_2SO_4(aq)$ solution (about 10 M). The equations for the electrode reactions on discharge are

$$Pb(s) + SO_4^{2-}(aq) \rightarrow PbSO_4(s) + 2e^- \qquad \text{oxidation}$$

$$2e^- + PbO_2(s) + 4H^+(aq) + SO_4^{2-}(aq) \rightarrow PbSO_4(s) + 2H_2O(l) \qquad \text{reduction}$$

The equation for the overall cell reaction is obtained by adding together the equations for the two half-reactions:

$$Pb(s) + PbO_2(s) + 2H_2SO_4(aq) \underset{\text{recharge}}{\overset{\text{discharge}}{\rightleftharpoons}} 2PbSO_4(s) + 2H_2O(l)$$

A single cell of this type develops about 2 V when fully charged, and the six cells in series in an ordinary 12-V automobile battery develop about 12 V. The cell voltage decreases on discharge because sulfuric acid is consumed and water is produced, and both of these reactions dilute the $H_2SO_4(aq)$ battery electrolyte.

On rapid recharge of a lead storage battery, appreciable amounts of $H_2(g)$ and $O_2(g)$ may be formed by the electrode reactions

$$2H^+(aq) + 2e^- \rightarrow H_2(g) \qquad \text{at the } Pb(s)|PbSO_4(s) \text{ electrode}$$

$$H_2O(l) \rightarrow \tfrac{1}{2}O_2(g) + 2H^+(aq) + 2e^- \qquad \text{at the } PbO_2(s), PbSO_4(s)|Pb(s) \text{ electrode}$$

There is danger of explosion because of the possible reaction

$$2H_2(g) + O_2(g) \xrightarrow{\text{spark}} 2H_2O(g)$$

Thus sparks and flames should not be brought near a lead storage battery.

Another rechargeable battery is the **nickel-cadmium** (nicad) **battery,** which is used, for example, in hand calculators, rechargeable flashlight batteries, and cordless electric shavers, drills, and toothbrushes. The equation for the cell reaction for the nicad battery is

$$2NiOOH(s) + Cd(s) + 2H_2O(l) \underset{\text{recharge}}{\overset{\text{discharge}}{\rightleftharpoons}} 2Ni(OH)_2(s) + Cd(OH)_2(s)$$

Note that the reaction consumes only water from the cell electrolyte, and thus the voltage on discharge is fairly constant. A completely sealed nicad battery is much more stable than a lead storage battery and can be left inactive for long periods without adverse effects, except for the need to recharge the battery before use.

Mercury batteries are used in heart pacemakers, hearing aids, some types of electric watches, and other electronic instruments. The main advantage of the mercury battery is a constant voltage (1.35 V) during discharge, which is a consequence of the fact that the cell reaction does not change the cell electrolyte composition. Mercury batteries are not safe to recharge because the container might rupture as a result of gas evolution in the sealed battery.

Example 20-13: The cell equation for the mercury battery is

$$Zn(s) + HgO(s) \rightarrow ZnO(s) + Hg(l)$$

Use the Nernst equation to explain why the voltage of the mercury battery remains constant during discharge.

Solution: The Nernst equation at 25°C is

$$E_{\text{cell}} = E^{\circ}_{\text{cell}} - \left(\frac{0.0592\ \text{V}}{n}\right) \log Q$$

All the reactants and products in the cell reaction have fixed concentrations and therefore do not appear in Q. Thus $Q = 1$ for the cell reaction and hence

$$E_{\text{cell}} = E^{\circ}_{\text{cell}} - \left(\frac{0.0592\ \text{V}}{2}\right) \log 1 = E^{\circ}_{\text{cell}}$$

Because E°_{cell} is a constant, E_{cell} is a constant. The voltage of a mercury battery remains at 1.35 V until all the $HgO(s)$ is used up (Zn is in excess), at which point the battery voltage drops to zero because the cell reaction can no longer occur.

The **dry cell** and the **alkaline manganese cell,** both of which are widely used in flashlights and battery-powered toys, are closely related. Both contain a zinc metal negative electrode and an inert positive electrode on which $MnO_2(s)$ is reduced to $Mn_2O_3(s)$. The positive electrode is a carbon rod in the dry cell and a steel rod in the alkaline manganese cell. The electrolyte in the dry cell is a paste of NH_4Cl; the electrolyte in the alkaline manganese cell is a concentrated solution of $NaOH(aq)$.

The cell reaction for the alkaline manganese cell is

$$Zn(s) + 2MnO_2(s) \rightarrow ZnO(s) + Mn_2O_3(s)$$

The alkaline manganese battery has twice the capacity, a steadier voltage under heavy current drain, and a higher available current than its major marketplace competitor, the dry cell. Its main disadvantage is that it costs roughly three times as much as the dry cell because of the more elaborate internal construction necessary to prevent leakage of the concentrated NaOH solution. No attempt should be made to recharge dry cells or alkaline manganese batteries because of the high explosion hazard arising from gas formation inside the sealed battery during recharge.

SUMMARY

Electrolysis is the process by which a chemical reaction is driven uphill on the energy scale by the application of a voltage across electrodes placed in a solution. The extent of the electrochemical reaction that occurs is proportional to the current that flows through the solution and is given quantitatively by Faraday's laws of electrolysis.

An electrochemical cell provides the means for obtaining electricity from an electron-transfer reaction. The cell consists of a pair of metal electrodes in contact with an electrolyte solution. The dependence of the cell voltage on the concentrations of the reactants and products of the cell reaction is described quantitatively by the Nernst equation.

The standard cell voltage, E°_{cell}, is the voltage of the cell when the reaction quotient $Q = 1$. The equilibrium constant of the cell reaction can be computed from the value of E°_{cell}. Electrode reactions can be arranged in a series of decreasing electrode reduction voltages. The assignment of standard reduction voltages to electrode reactions is achieved by setting $E^\circ = 0$ for the hydrogen gas electrode reaction. The E° values for electrode reactions given in Table 20-1 can be used to compute E°_{rxn} values for reactions, and to predict the thermodynamic stability of oxidizing agents and reducing agents. Electrochemical cells can be used to determine the concentration of an ion in solution.

Batteries are electrochemical cells designed to function as power sources. Their principal advantages are portability and the ability to supply large currents.

TERMS YOU SHOULD KNOW

electrochemistry 624
voltaic pile 624
electrolysis 624
ampere, A 625
coulomb, C 625
Faraday's laws 626
Faraday's constant, F 627
faraday 627
electrode 628
cathode 628
anode 628
decomposition voltage 629
volt, V 629
chlor-alkali process 629
amalgam 629
Hall process 630
electroplating 631
electrochemical cell 632
salt bridge 632
external circuit 632
discharge 633
cell diagram 635
gas electrode 636
cell voltage, E_{cell} 637
Nernst equation 638
standard cell voltage, E°_{cell} 639
hydrogen electrode 641
standard reduction voltage 641
emf series 644
chromous bubbler 644
battery 649
primary battery 649
secondary battery 649
lead storage battery 649
nickel-cadmium battery 650
mercury battery 650
dry cell 650
alkaline manganese cell 650

EQUATIONS YOU SHOULD KNOW HOW TO USE

1 coulomb = 1 ampere × 1 second

$Z = It$ (20-1) (definition of an ampere)

$$m = \left(\frac{It}{F}\right)\left(\frac{A}{n}\right) \quad \text{(20-4)} \quad \text{(Faraday's laws of electrolysis)}$$

1 joule = 1 volt × 1 coulomb (J = V·C) (20-7) (Definition of a volt)

$$E° = \left(\frac{0.0592 \text{ V}}{n}\right) \log K \text{ at } 25°\text{C} \quad \text{(20-11)} \quad \text{(Relation between } E° \text{ and the equilibrium constant)}$$

$$E = E° - \left(\frac{0.0592 \text{ V}}{n}\right) \log Q \quad \text{(20-12)} \quad \text{(Nernst equation at 25°C)}$$

$$E°_{\text{cell}} = E°_{red} + E°_{ox} \quad \text{(20-14)} \quad \text{(Cell voltage in terms of half-reaction voltages)}$$

$$\Delta G_{rxn} = -nFE_{rxn} \quad \text{(20-19)} \quad \text{(Relation between the cell voltage and the Gibbs free energy change)}$$

PROBLEMS

ELECTROLYSIS

20-1. How long will it take an electric current of 1.25 A to deposit all the copper from 500 mL of 0.150 M $CuSO_4(aq)$?

20-2. How much silver is deposited if an electric current of 0.150 A flows through a silver nitrate solution for 20.0 min?

20-3. Cesium metal is produced by the electrolysis of molten cesium cyanide. Calculate how much Cs(*s*) is deposited from CsCN(*l*) in 30 min by a current of 500 mA.

20-4. Beryllium occurs naturally in the form of beryl. The metal is produced from its ore by electrolysis after the ore has been converted to the oxide and then to the chloride. How much Be(*s*) is deposited from a $BeCl_2$ melt by a current of 5.0 A that flows for 1.0 h?

20-5. Fluorine is manufactured by the electrolysis of HF dissolved in molten KF. The equation is

$$2HF(KF) \rightarrow H_2(g) + F_2(g)$$

The KF acts as a solvent for HF and as the conductor of electricity. A commercial cell for producing fluorine operates at a current of 1500 A. How much F_2 can be produced per 24 h? Why isn't the electrolysis of liquid HF alone used?

20-6. Suppose that it is planned to electrodeposit 200 mg of gold onto the surface of a steel object via the process

$$Au(CN)_2^-(aq) + e^- \rightarrow Au(s) + 2CN^-(aq)$$

If the electric current in the circuit is set at 30 mA, for how long should the current be passed?

20-7. Gallium is produced by the electrolysis of an alkaline gallium oxide solution. Calculate the amount of Ga(*s*) deposited from a Ga(III) solution by a current of 0.50 A that flows for 30 minutes.

20-8. Sodium metal is produced commercially by the elctrolysis of molten sodium chloride. How much sodium will be produced if a current of 1.00×10^3 A is passed through NaCl(*l*) for 8.0 h? How many liters of chlorine at 10 atm and 20°C will be produced?

20-9. Hydrogen $H_2(g)$, and oxygen $O_2(g)$, can be produced by the electrolysis of water:

$$2H_2O(l) \xrightarrow{\text{electrolysis}} 2H_2(g) + O_2(g)$$

Compute the volume of $O_2(g)$ produced at 25°C and 1.00 atm when a current of 30.35 A is passed through a $K_2SO_4(aq)$ solution for 2.00 h.

20-10. Bauxite, the principle source of aluminum oxide, contains about 55 percent Al_2O_3 by mass. How much bauxite is required to produce the 5 million tons of aluminum metal produced each year by electrolysis?

CELL SETUPS

20-11. Consider a zinc-silver electrochemical cell. The negative electrode is a zinc rod immersed in a $ZnCl_2(aq)$ solution, and the positive electrode a silver rod in a $AgNO_3(aq)$ solution. The two solutions are connected by a KNO_3-agar salt bridge. Sketch a diagram of the cell, indicating the flow of electrons. Write the equation for the reaction that occurs at each electrode, and draw the cell diagram.

20-12. Consider a manganese-chromium electrochemical cell. The negative electrode is a manganese rod immersed in a $MnSO_4(aq)$ solution, and the positive electrode is a chromium rod in a $CrSO_4(aq)$ solution. The two solutions are connected by a salt bridge. Sketch a diagram of the cell, indicating the flow of electrons. Write the equation for the reaction that occurs at each electrode, and draw the cell diagram.

20-13. Consider a vanadium-copper electrochemical cell. The negative electrode is a vanadium rod immersed in a $VI_2(aq)$ solution, and the positive electrode is a copper rod in a $CuSO_4(aq)$ solution. The two solutions are connected by a salt bridge. Sketch a diagram of the cell, indicating the flow of electrons. Write the equation for the reaction that occurs at each electrode and draw the cell diagram.

20-14. Consider a cobalt-lead electrochemical cell. The negative electrode is a cobalt rod immersed in a $Co(NO_3)_2(aq)$ solution, and the positive electrode is a lead rod in a $Pb(NO_3)_2(aq)$ solution. The two solutions are connected by a salt bridge. Sketch a diagram of the cell, indicating the flow of electrons. Write the equation for the reaction that occurs at each electrode and draw the cell diagram.

CELL DIAGRAMS

20-15. Write the equations for the electrode reactions and the net cell reaction for the electrochemical cell

$$Pb(s)|PbI_2(s)|HI(aq)|H_2(g)|Pt(s)$$

20-16. Write the equations for the electrode reactions and the net cell reaction for the electrochemical cell

$$Cu(s)|Cu(ClO_4)_2(aq)||AgClO_4(aq)|Ag(s)$$

20-17. The cell diagram for an electrochemical cell is given as

$$In(s)|In(ClO_4)_3(aq)||ReCl_3(aq)|Re(s)$$

Write the equations for the half-reactions that occur at the electrodes. Make a sketch of the cell.

20-18. The cell diagram for an electrochemical cell is given as

$$Sn(s)|SnCl_2(aq)||AgNO_3(aq)|Ag(s)$$

Write the equations for the half-reactions that occur at the electrodes. Make a sketch of the cell.

20-19. Write the cell equation for the cell

$$Pt(s)|H_2(g)|HCl(aq)|Hg_2Cl_2(s)|Hg(l)$$

20-20. Write the cell equation for the cell

$$Pb(s)|PbSO_4(s)|K_2SO_4(aq)|Hg_2SO_4(s)|Hg(l)$$

ELECTROCHEMICAL CELLS AND LECHÂTELIER'S PRINCIPLE

20-21. Consider an electrochemical cell in which the reaction is described by the equation

$$2HCl(aq) + Ca(s) \rightarrow CaCl_2(aq) + H_2(g)$$

Predict the effect of the following changes on the cell voltage:

(a) decrease in amount of $Ca(s)$
(b) increase in pressure of $H_2(g)$
(c) increase in [HCl]
(d) dissolution of $Ca(NO_3)_2(s)$ in the $CaCl_2(aq)$ solution

20-22. Consider an electrochemical cell in which the reaction is described by the equation

$$Pb(s) + 2Ag^+(aq) + SO_4^{2-}(aq) \rightleftharpoons PbSO_4(s) + 2Ag(s)$$

Predict the effect of the following changes on the observed cell voltage:

(a) increase in $[Ag^+]$
(b) increase in amount of $PbSO_4(s)$
(c) increase in $[SO_4^{2-}]$

20-23. Given the following equation for an electrochemical cell reaction

$$H_2(g) + PbCl_2(s) \rightarrow Pb(s) + 2HCl(aq) \qquad \Delta H^\circ_{rxn} > 0$$

indicate whether the cell voltage will increase, not change, or decrease when the following changes in conditions are made:

(a) an increase in the amount of $PbCl_2(s)$
(b) dilution of the cell solution with $H_2O(l)$
(c) dissolution of $NaOH(s)$ in the cell solution
(d) dissolution of concentrated $HClO_4$ in the cell solution
(e) a decrease in P_{H_2}
(f) an increase in the temperature

20-24. Given the following equation for an electrochemical cell reaction

$$H_2(g) + PbSO_4(s) \rightleftharpoons 2H^+(aq) + SO_4^{2-}(aq) + Pb(s)$$

predict the effect of the following changes (increase, decrease, or no change) on the cell voltage:

(a) increase in pressure of $H_2(g)$
(b) increase in size of $Pb(s)$ electrode
(c) decrease in pH of cell electrolyte
(d) dilution of cell electrolyte with water
(e) dissolution of $Na_2SO_4(s)$ in cell electrolyte
(f) decrease in amount of $PbSO_4(s)$
(g) dissolution of a small amount of $NaOH(s)$ in the cell electrolyte

20-25. Consider the following equation for a reaction taking place in an electrochemical cell:

$$2Cr^{2+}(aq) + HClO(aq) + H^+(aq) \rightleftharpoons 2Cr^{3+}(aq) + Cl^-(aq) + H_2O(l)$$

Predict the effect of the following changes on the cell voltage:

(a) increase in [HClO]
(b) increase in size of inert electrodes
(c) increase in pH of cell solution
(d) dissolution of $KCl(s)$ in cell solution containing $Cl^-(aq)$

20-26. Given that the equation for the reaction in an electrochemical cell is

$$Fe^{2+}(aq) + Ag^+(aq) \rightleftharpoons Ag(s) + Fe^{3+}(aq)$$

 predict the effect of the following changes (increase, decrease, or no change) on the cell voltage:

(a) increase in $[Ag^+]$
(b) increase in $[Fe^{3+}]$
(c) twofold decrease in both $[Fe^{3+}]$ and $[Fe^{2+}]$
(d) decrease in amount of Ag(*s*)
(e) decrease in $[Fe^{2+}]$
(f) addition of NaCl(*aq*) to $Ag^+(aq)$ solution

NERNST EQUATION

20-27. Determine the value of n in the Nernst equation for the following equations:

(a) $CH_4(g) + 2O_2(g) \rightarrow CO_2(g) + 2H_2O(l)$
(b) $2Zn(s) + Ag_2O_2(s) + 2H_2O(l) + 4OH^-(aq) \rightarrow 2Ag(s) + 2Zn(OH)_4^{2-}(aq)$

20-28. Determine the value of n in the Nernst equation for the following equations:

(a) $Cu(s) + Mg^{2+}(aq) \rightarrow Cu^{2+}(aq) + Mg(s)$
(b) $2H_2O(l) + 2Na(s) \rightarrow 2Na^+(aq) + 2OH^-(aq) + H_2(g)$

20-29. The value of E°_{rxn} at 25°C for the equation

$$Pb(s) + Cu^{2+}(aq) \rightleftharpoons Pb^{2+}(aq) + Cu(s)$$

is 0.47 V. Calculate the equilibrium constant at 25°C for this equation.

20-30. The value of E°_{rxn} at 25°C for the equation

$$H_2(g) + 2AgCl(s) \rightleftharpoons 2Ag(s) + 2HCl(aq)$$

is +0.22 V. Calculate the equilibrium constant at 25°C for this equation.

20-31. The measured voltage at 25°C of a cell in which the reaction described by the equation

$$Cd(s) + Pb^{2+}(aq, 0.150\ M) \rightleftharpoons Pb(s) + Cd^{2+}(aq, 0.0250\ M)$$

takes place at the concentrations shown is 0.293 V. Calculate E°_{cell} and K, the equilibrium constant, for the cell equation.

20-32. The measured voltage at 25°C of a cell in which the reaction described by the equation

$$Co(s) + Sn^{2+}(aq, 0.18\ M) \rightleftharpoons Sn(s) + Co^{2+}(aq, 0.020\ M)$$

takes place at the concentrations shown is 0.168 V. Calculate E°_{cell} and K, the equilibrium constant, for the cell equation.

20-33. The measured voltage at 25°C of a cell in which the reaction described by the equation

$$Al(s) + Fe^{3+}(aq, 0.0050\ M) \rightleftharpoons Al^{3+}(aq, 0.250\ M) + Fe(s)$$

takes place at the concentrations shown is 1.59 V. Calculate E°_{cell} and K, the equilibrium constant, for the cell equation.

20-34. The measured voltage at 25°C of a cell in which the reaction described by the equation

$$Zn(s) + Hg_2^{2+}(aq, 0.30\ M) \rightleftharpoons 2Hg(l) + Zn^{2+}(aq, 0.50\ M)$$

takes place at the concentrations shown is 1.54 V. Calculate E°_{cell} and K, the equilibrium constant, for the cell equation.

20-35. Consider the electrochemical cell

$$Zn(s)|ZnCl_2(aq)|Hg_2Cl_2(s)|Hg(l)$$

(a) Determine the balanced equation for the cell reaction.
(b) The standard cell voltage for the cell at 25°C is $E^\circ_{cell} = +1.03$ V. Use the Nernst equation to compute the cell voltage when the concentration of $ZnCl_2(aq)$ is 0.040 M.

20-36. The standard voltage of the cell

$$Pt(s)|H_2(g)|H^+(aq)||Cd^{2+}(aq)|Cd(s)$$

is −0.40 V. Write the cell equation and calculate the cell voltage when $[H^+] = 0.10$ M, $P_{H_2} = 0.10$ atm, and $[Cd^{2+}] = 2.5 \times 10^{-3}$ M.

CONCENTRATIONS FROM CELL MEASUREMENTS

20-37. The reaction described by the equation

$$Zn(s) + Hg_2Cl_2(s) \rightleftharpoons 2Hg(l) + Zn^{2+}(aq) + 2Cl^-(aq)$$

is run in an electrochemical cell. The measured voltage of the cell at 25°C is 1.03 V when $Q = 1.00$. Suppose that in a cell solution with $[Cl^-] = 0.10$ M and $[Zn^{2+}]$ unknown, the measured cell voltage is 1.21 V. Compute the value of $[Zn^{2+}]$ in the cell solution.

20-38. Write out the cell equation for the cell

$$Zn(s)|Zn^{2+}(aq, 0.100\ M)||Zn^{2+}(aq, 2.00\ M)|Zn(s)$$

Calculate the voltage generated by this "concentration" cell at 25°C.

20-39. The measured voltage of the cell

$$Pt(s)|H_2(g, 1.00\ atm)|H^+(aq)||Ag^+(aq, 1.00\ M)|Ag(s)$$

is 0.900 V at 25°C. Given $E^\circ_{cell} = 0.800$ V, calculate the pH of the solution.

20-40. The measured voltage of the cell

$$Zn(s)|Zn^{2+}(aq, M)||Ag^+(aq, 0.100\ M)|Ag(s)$$

is 1.500 V at 25°C. Given $E^\circ_{cell} = 1.560$ V, calculate the value of $[Zn^{2+}]$.

USE OF TABULATED E° VALUES

20-41. Calculate E°_{rxn} for the following equations:

(a) $Cu(s) + Fe(CN)_6^{3-}(aq) \rightarrow Cu^+(aq) + Fe(CN)_6^{4-}(aq)$
(b) $Fe^{3+}(aq) + Ag(s) \rightarrow Fe^{2+}(aq) + Ag^+(aq)$
(c) $Zn(s) + F_2(g) \rightarrow Zn^{2+}(aq) + 2F^-(aq)$

20-42. Calculate E°_{rxn} for the following equations:

(a) $2Na(s) + 2H_2O(l) \rightarrow 2Na^+(aq) + 2OH^-(aq) + H_2(g)$
(b) $2H^+(aq) + Pd(s) \rightarrow Pd^{2+}(aq) + H_2(g)$
(c) $4Fe(OH)_2(s) + O_2(g) + 2H_2O(l) \rightarrow 4Fe(OH)_3(s)$

20-43. The standard voltage for the equation

$$HClO(aq) + H^+(aq) + 2Cr^{2+}(aq) \rightleftharpoons 2Cr^{3+}(aq) + Cl^-(aq) + H_2O(l)$$

is $E^{\circ}_{rxn} = 1.80$ V. Use data from Table 20-1 to calculate E°_{red} for the half-reaction equation

$$HClO(aq) + H^+(aq) + 2e^- \rightleftharpoons Cl^-(aq) + H_2O(l)$$

20-44. The standard voltage for the equation

$$NO_3^-(aq) + 2H^+(aq) + Cu^+(aq) \rightleftharpoons NO_2(g) + H_2O(l) + Cu^{2+}(aq)$$

is $E^{\circ}_{rxn} = 0.65$ V. Use data from Table 20-1 to calculate E°_{red} for the half-reaction equation

$$NO_3^-(aq) + 2H^+(aq) + e^- \rightleftharpoons NO_2(g) + H_2O(l)$$

20-45. Use data from Table 20-1 to calculate E°_{rxn} for the equation

$$Cd^{2+}(aq) + Zn(s) \rightleftharpoons Zn^{2+}(aq) + Cd(s)$$

Will zinc displace cadmium from the compound $Cd(NO_3)_2(aq)$ if $[Zn^{2+}] = [Cd^{2+}] = 1.00$ M?
Is the reaction spontaneous if $[Cd^{2+}] = 0.0010$ M and $[Zn^{2+}] = 1.00$ M?

20-46. Use data from Table 20-1 to calculate E°_{rxn} for the equation

$$S_2O_8^{2-}(aq) + 2H_2O(l) \rightleftharpoons H_2O_2(aq) + 2SO_4^{2-}(aq) + 2H^+(aq)$$

Is an aqueous solution of potassium peroxodisulfate ($K_2S_2O_8$) stable over a long period of time?

20-47. Compute the E°_{cell} value and the equilibrium constant at 25°C for the cell equation

$$V^{2+}(aq) + H^+(aq) \rightleftharpoons V^{3+}(aq) + \tfrac{1}{2}H_2(g)$$

The value of E°_{red} for the half-reaction equation

$$V^{3+}(aq) + e^- \rightarrow V^{2+}(aq)$$

is $E^{\circ}_{red} = -0.24$ V. Can $V^{2+}(aq)$ at 1.0 M liberate $H_2(g)$ at 1.00 atm from a solution with $[H^+] = 1.0$ M and $[V^{3+}] = 1.00 \times 10^{-4}$ M?

20-48. Compute the E°_{rxn} value and the equilibrium constant at 25°C for the cell reaction

$$Cu(s) + 2Ag^+(aq) \rightleftharpoons 2Ag(s) + Cu^{2+}(aq)$$

given that $[Ag^+] = 0.10$ M and $[Cu^{2+}] = 1.00 \times 10^{-4}$ M. Predict whether the reaction is spontaneous.

20-49. Calculate the voltage at 25°C of an electrochemical cell in which the reaction is described by the equation

$$2Zn(s) + O_2(g, 0.20\text{ atm}) + 4H^+(aq, 0.20\text{ M}) \rightleftharpoons 2Zn^{2+}(aq, 0.0010\text{ M}) + 2H_2O(l)$$

See Table 20-1 for the necessary E° data.

20-50. Calculate the voltage at 25°C of an electrochemical cell in which the reaction is described by the equation

$$Cd(s) + Pb(NO_3)_2(aq, 0.10\text{ M}) \rightleftharpoons Cd(NO_3)_2(aq, 0.010\text{ M}) + Pb(s)$$

See Table 20-1 for the necessary E° data.

20-51. The standard voltage for the equation

$$S_2O_3^{2-}(aq) + 2OH^-(aq) + O_2(g) \rightleftharpoons 2SO_3^{2-}(aq) + H_2O(l)$$

is $E^{\circ}_{rxn} = 0.98$ V. Write the two half-reaction equations and use data from Table 20-1 to determine the value of E°_{red} for the $SO_3^{2-}(aq)/S_2O_3^{2-}(aq)$ half-reaction in basic solution.

20-52. The standard voltage for the equation

$$BH_4^-(aq) + 8OH^-(aq) + 8O_2(g) \rightleftharpoons H_2BO_3^-(aq) + 5H_2O(l) + 8O_2^-(aq)$$

is $E^{\circ}_{rxn} = 0.68$ V. Write the two half-reaction equations and use data from Table 20-1 to determine the value of E°_{red} for the $H_2BO_3^-(aq)/BH_4^-(aq)$ half-reaction in basic solution.

CELLS AND ΔG_{rxn}

20-53. Suppose we have an aqueous solution at 25°C with $[Co^{3+}] = 0.20$ M, $[Co^{2+}] = 1.0 \times 10^{-4}$ M, and $[H^+] = 0.30$ M which is exposed to air ($P_{O_2} = 0.20$ atm). Use the Nernst equation to determine whether the oxidation of water by $Co^{3+}(aq)$ is spontaneous under the given conditions.

20-54. Using the data in Table 20-1, determine whether air ($P_{O_2} = 0.20$ atm) is capable of oxidizing $Fe^{2+}(aq)$ to $Fe^{3+}(aq)$ in a solution with $[Fe^{2+}] = [Fe^{3+}] = 0.10$ M and a pH = 2.0 at 25°C.

20-55. Using the data in Table 20-1, determine whether $Co^{3+}(aq, 0.010\text{ M})$ is capable of liberating $O_2(g)$ at 25°C and 1.0 atm from water with a pH = 1.0.

20-56. Use the data in Table 20-1 to calculate E°_{rxn} for the equation

$$O_3(g) + H_2O(l) + Cu(s) \rightleftharpoons Cu(OH)_2(s) + O_2(g)$$

Is the oxidation of copper by ozone a spontaneous process at 25°C under the conditions $P_{O_3} = 1.00 \times 10^{-4}$ atm and $P_{O_2} = 1.00$ atm?

20-57. An electrochemical cell is set up so that the reaction described by the equation

$$Zn(s) + Cu^{2+}(aq) \rightleftharpoons Zn^{2+}(aq) + Cu(s)$$

occurs. At 25°C the measured cell voltage is 1.05 V. Calculate the value of ΔG_{rxn} for the reaction.

20-58. An electrochemical cell is set up so that the reaction described by the equation

$$H_2O_2(aq) + Fe(s) + 2H^+(aq) \rightleftharpoons Fe^{2+}(aq) + 2H_2O(l)$$

occurs. At 25°C the measured cell voltage is 2.03 V. Calculate the value of ΔG_{rxn}.

20-59. An electrochemical cell is set up so that the reaction described by the equation

$$2NO_3^-(aq) + 4H^+(aq) + Cu(s) \rightleftharpoons 2NO_2(g) + 2H_2O(l) + Cu^{2+}(aq)$$

occurs. The standard voltage is $E° = 0.65$ V. Calculate the value of ΔG°_{rxn}.

20-60. An electrochemical cell is set up so that the reaction described by the equation

$$Cr_2O_7^{2-}(aq) + 14H^+(aq) + 6Fe^{2+}(aq) \rightleftharpoons 2Cr^{3+}(aq) + 6Fe^{3+}(aq) + 7H_2O(l)$$

occurs. At 25°C, the standard cell voltage is 0.56 V. Calculate the value of ΔG°_{rxn}.

20-61. Use the data in Table 20-1 to compute the ΔG°_{rxn} values for the following equations:

(a) $2Ag(s) + F_2(g) \rightleftharpoons 2Ag^+(aq) + 2F^-(aq)$
(b) $\frac{1}{2}H_2(g) + Fe^{3+}(aq) \rightleftharpoons Fe^{2+}(aq) + H^+(aq)$

20-62. Use the data in Table 20-1 to compute the ΔG°_{rxn} values for the following equations:

(a) $Zn(s) + Cu^{2+}(aq) \rightleftharpoons Zn^{2+}(aq) + Cu(s)$
(b) $Ag(s) + Fe^{3+}(aq) \rightleftharpoons Fe^{2+}(aq) + Ag^+(aq)$

20-63. For the following electrochemical cell

$$Zn(s)|Zn^{2+}(aq, 0.010\ M)||Cd^{2+}(aq, 0.050\ M)|Cd(s)$$

use the data in Table 20-1 to calculate the values at 25°C of $E°$, ΔG°_{rxn}, ΔG_{rxn}, and E for the cell equation. What is the equation for the cell reaction?

20-64. The standard voltage of the following cell at 25°C is $E° = 1.08$ V:

$$Co(s)|Co^{2+}(aq, 0.0155\ M)||Ag^+(aq, 1.50\ M)|Ag(s)$$

Calculate ΔG°_{rxn}, ΔG_{rxn} and E. What is the equation for the cell reaction?

BATTERIES

20-65. Calculate the standard voltage that can be obtained from an ethane-oxygen ($C_2H_6(g)$–$O_2(g)$) fuel cell at 25°C. (See Table 19-1 for data.) Take the cell reaction to be

$$C_2H_6(g) + \tfrac{7}{2}O_2(g) \rightleftharpoons 2CO_2(g) + 3H_2O(l)$$

20-66. Calculate the standard voltage that can be obtained from a methane-oxygen fuel cell at 25°C. (See Table 19-1 for data.)

20-67. In the Edison battery, an early version of the nickel-cadmium cell, iron is used in place of cadmium. Write the cell diagram for an Edison battery. In contrast to the nickel-cadmium cell, the Edison battery has heavy-duty industrial and railway uses.

20-68. The silver-zinc battery has the cell diagram

$$\ominus Zn(s)|K_2ZnO_2(s)|KOH(aq, 40\%)|Ag_2O_2(s)|Ag(s)\oplus$$

This battery is used in space satellites because of its exceptional compactness, high current capacity, and constant voltage during discharge. The high cost of silver precludes the use of this battery in routine applications, however. Write balanced chemical equations for

(a) the half-reaction that occurs at the left-hand electrode
(b) the half-reaction that occurs at the right-hand electrode
(c) the net cell reaction

ADDITIONAL PROBLEMS

20-69. Write the balanced chemical equation for the cell reaction in the following electrochemical cell:

$$Pt(s)|MnO_4^-(aq), Mn^{2+}(aq), H^+(aq)||IO_3^-(aq), I^-(aq), H^+(aq)|Pt(s)$$

20-70. A rechargeable silver oxide cell, involving a $Ag_2O_2(s)|Ag(s)$ cathode, in a LCD (liquid crystal display) calculator is estimated to last 1000 h while drawing a current of only 0.10 mA. What mass of silver will be produced over the lifetime of the cell?

20-71. The terminals of a lead storage battery corrode with the formation of a white powder. Given that oxygen and water are involved, write a balanced equation for the process.

20-72. Explain why there is a significant explosion hazard in the vicinity of a conventional lead storage battery that is being recharged.

20-73. Many metals can be refined electrolytically. The impure metal is used as the anode, and the cathode is made of the pure metal. The electrodes are placed in an electrolyte containing a salt of the metal being refined. When an electric current is passed between these electrodes, the metal leaves the impure anode and is deposited in a pure form on the cathode. How many ampere-hours of electricity are required to refine electrolytically 1 metric ton of copper? (Use $n = 2$.) An ampere-hour is an ampere times an hour.

20-74. Write a balanced equation for the cell reaction in the following electrochemical cell, and compute the cell voltage, E_{cell}:

$$Pt(s)|H_2(g, 0.50\ atm)|H_2SO_4(aq, 1.00\ M)|PbSO_4(s)|Pb(s)$$

20-75. Two electrolytic cells are placed in series. One cell contains a solution of $AgC_2H_3O_2(aq)$ and the other cell contains a solution of $Cd(C_2H_3O_2)_2(aq)$. An electric current is passed through the two cells until 0.876 g of Ag is deposited. How much Cd will be deposited?

20-76. Explain why $F_2(g)$ cannot be prepared by electrolysis of $NaF(aq)$.

20-77. Given that $E° = +0.728$ V at 25°C for the cell

$$Ag(s)|AgBr(s)|Br^-(aq)||Ag^+(aq)|Ag(s)$$

write the cell equation and determine the solubility-product constant of $AgBr(s)$ in water at 25°C.

20-78. The half-cell

$$Hg(l)|Hg_2Cl_2(s)|KCl(aq)$$

is called a calomel electrode. If the KCl solution is saturated, then $E^{\circ}_{red}(satd) = 0.2415$ V at 25°C. Explain how the cell

$$Pt(s)|H_2(g)|H^+(aq)||KCl(aq,\ satd)|Hg_2Cl_2(s)|Hg(l)$$

can be used to measure pH. Derive an equation for the voltage of this cell as a function of pH.

20-79. Show that the cell

$$Pt(s)|H_2(g)|HC_2H_3O_2(aq),NaC_2H_3O_2(aq)|| \quad KCl(aq,\ satd)|Hg_2Cl_2(s)|Hg(l)$$

can be used to determine K_a for $HC_2H_3O_2(aq)$.

20-80. The Weston standard cell is given by

$$^{\ominus}Cd(Hg)|CdSO_4(aq,\ satd)|Hg_2SO_4(s)|Hg(l)^{\oplus}$$
(12.5% Cd)

Write the equation that occurs in the cell. Ten Weston standard cells that use a saturated $CdSO_4(aq)$ solution are maintained at the U.S. Bureau of Standards as the official unit of voltage. The voltage of each cell is virtually constant at 1.01857 V. Explain why the voltage remains constant.

20-81. Suppose a zinc rod is dipped into a 1.0 M $CuSO_4(aq)$ solution containing a copper rod and the system is allowed to stand for several hours. What do you predict for the voltage measured between the Zn(s) and Cu(s) rods?

20-82. A battery that operates at −50°C was developed for the exploration of the moon and Mars. The electrodes are magnesium metal/magnesium chloride and silver chloride/silver. The electrolyte is potassium thiocyanate, KSCN, in liquid ammonia. Draw the cell diagram and write the equation for the reaction for the cell.

20-83. Suppose the leads of an electrochemical cell are connected together external to the cell and the cell is allowed to come to equilibrium. What will be the value of the cell voltage at equilibrium?

20-84. A battery that operates at 500°C was developed for the exploration of Venus. The electrodes are a magnesium metal anode and a mixture of copper(I) and copper(II) oxides in contact with an inert steel cathode. The electrolyte is a mixture of LiCl and KCl, which is melted to activate the cell. The MgO that is produced is sparingly soluble in the molten salt mixture and precipitates. Draw the cell diagram and write the equation for the reaction for the cell.

20-85. Electrolysis can be used to determine atomic masses. A current of 0.600 A deposits 2.42 g of a certain metal in exactly 1 h. Calculate the atomic mass of the metal if $n = 1$. What is the metal?

20-86. From 1882 to 1895 home electricity was provided as direct current rather than as alternating current, as is now the case. Thomas Edison invented a meter to measure the amount of electricity used by a consumer. A small amount of current was diverted to an electrolysis cell that consisted of zinc electrodes in a zinc sulfate solution. Once a month the cathode was removed, washed, dried, and weighed. The bill was figured in ampere-hours (Problem 20-73). In 1888 Boston Edison Company had 800 chemical meters in service. In one case, in one 30-day period, 65 g of zinc was deposited on the cathode. The meter used 11 percent of the current into the house. How many coulombs were used in the month? Calculate the current used in ampere-hours.

The Halogens

Bromine processing plant. The red color is produced by the bromine gas.

The elements fluorine, chlorine, bromine, iodine, and astatine are collectively called the **halogens.** At 25°C fluorine and chlorine are gases, bromine is a liquid, and iodine is a solid. There are no stable (nonradioactive) isotopes of astatine. These nonmetals, which occur in Group 7 of the periodic table, are all highly reactive diatomic molecules that are not found free in nature. They occur primarily as **halide** (F^-, Cl^-, Br^-, I^-) salts. The free elements are strong oxidizing agents and combine directly with almost all the other elements. The halogens have pungent, irritating odors and are poisonous.

Table M-1 shows that the properties of the halogens vary smoothly from the lightest, F_2, to the heaviest, I_2. The relative sizes of the diatomic halogen molecules and the halide ions are shown in Figure M-1. Note that the size increases with increasing atomic number. Halogen compounds have many uses and are produced commercially on a large scale.

M-1. FLUORINE IS THE MOST REACTIVE ELEMENT

Fluorine is a pale-yellow, corrosive gas that reacts directly, and in most cases vigorously, at room temperature with almost every element. It is the most reactive and the strongest oxidizing agent of all the elements. The extreme reactivity of fluorine is evidenced by its reactions with glass, ceramics, and most alloys. Water burns vigorously in fluorine:

$$2H_2O(l) + 2F_2(g) \rightarrow 4HF(g) + O_2(g)$$

and the light hydrocarbons react spontaneously; for example,

$$CH_4(g) + 4F_2(g) \rightarrow 4HF(g) + CF_4(g)$$

Even xenon and krypton, which were once thought to be completely inert, react with fluorine to form fluorides such as KrF_2, XeF_2, XeF_4, and XeF_6.

Because of its high electronegativity, fluorine is capable of stabilizing unusually high oxidation states of other elements. Some examples are

$$OF_2 \quad O(II) \qquad AgF_2 \quad Ag(II) \qquad IF_7 \quad I(VII)$$

In effect, fluorine stabilizes high oxidation states in other elements because its electronegativity is so high that even strong oxidizing agents, such as Ag(II), cannot remove electrons from it. Because its electronegativity is higher than that of any other element, fluorine does not occur with a positive oxidation state in any compound. All fluorine compounds are fluorides, where the oxidation state of fluorine is -1.

Fluorides are widely distributed in nature. The main natural sources of fluorine are the minerals fluorspar, CaF_2; cryolite, Na_3AlF_6; and fluorapatite, $Ca_{10}F_2(PO_4)_6$. Because of its extreme reactivity, elemental fluorine wasn't isolated until 1886. The French chemist Henri Moissan, who first isolated fluorine, received the 1906 Nobel Prize in chemistry for his work. Elemental fluorine is obtained by the electrolysis of hydrogen fluoride dissolved in molten potassium fluoride:

$$2HF \text{ (in KF melt)} \xrightarrow{\text{electrolysis}} H_2(g) + F_2(g)$$

The modern method of producing F_2 is essentially a variation of the method first used by Moissan. Prior to World War II, there was no

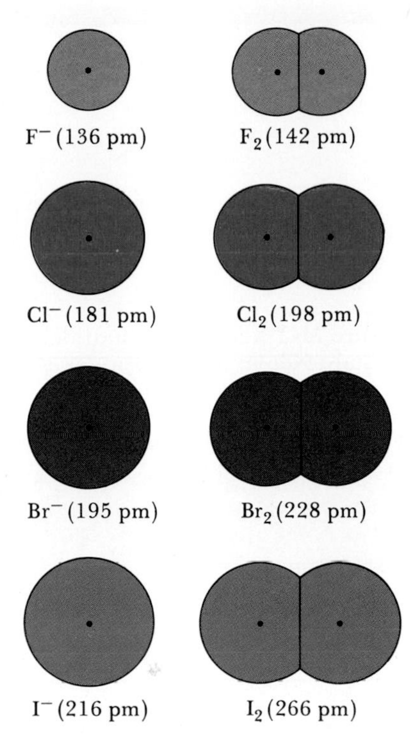

Figure M-1 Relative sizes of the halide ions *(circles)* and diatomic halogen molecules *(attached pairs)*. Distances are in picometers.

Table M-1 Physical properties of the halogens

Halogen	Molecular mass of diatomic species	Melting point/°C	Boiling point/°C	Bond length/pm	Bond energy/$kJ \cdot mol^{-1}$	Atomic electronegativity
fluorine, F_2	38.0	−220	−188	142	155	4.0
chlorine, Cl_2	70.9	−101	−35	198	243	3.2
bromine, Br_2	159.8	−7	59	228	192	3.0
iodine, I_2	253.8	114	184	266	150	2.7

Figure M-2 Xenon tetrafluoride crystals. Xenon tetrafluoride was first prepared in 1962 by the direct combination of $Xe(g)$ and $F_2(g)$ in a nickel reaction chamber at 6 atm and 400°C.

▪ The compound $^{235}UF_6$ is separated from $^{238}UF_6$ by gaseous effusion. The lighter $^{235}UF_6$ effuses more rapidly than does the heavier $^{238}UF_6$ (see Section 4-9).

commercial production of fluorine. The atomic bomb project required huge quantities of fluorine for the production of uranium hexafluoride, UF_6, a gaseous compound that is used in the separation of uranium-235 from uranium-238. It is uranium-235 that is used in nuclear devices. The production of uranium hexafluoride for the preparation of fuel for nuclear power plants is today a major commercial use of fluorine.

All the halogens form binary hydrogen compounds with the general formula HX. Hydrogen fluoride is a highly irritating, corrosive, colorless gas. It can be prepared by the reaction

$$CaF_2(s) + H_2SO_4(l) \rightarrow CaSO_4(s) + 2HF(g)$$

Hydrogen fluoride is used in petroleum refining and in the production of fluorocarbons such as Freons (refrigerants) and fluorocarbon polymers such as Teflon (Chapter 24). Unlike the other **hydrohalic acids,** hydrofluoric acid is a weak acid in aqueous solution ($pK_a = 3.17$). Fur-

Figure M-3 Samples of the minerals *(left to right)* fluorapatite, fluorite (fluorospar), and cryolite.

thermore, owing to the formation of strong hydrogen bonds, the equilibrium

$$2HF(aq) \rightleftharpoons H^+(aq) + \underset{\text{bifluoride ion}}{HF_2^-(aq)}$$

is important in the chemistry of HF(*aq*). For example, the reaction of hydrofluoric acid with potassium hydroxide produces potassium bifluoride, KHF_2:

$$2HF(aq) + KOH(aq) \rightarrow KHF_2(aq) + H_2O(l)$$

The other hydrohalic acids produce only normal salts such as KCl. Hydrofluoric acid reacts with glass and must be stored in plastic bottles. The reaction of hydrofluoric acid with glass is described by the equation

$$SiO_2(s) + 6HF(aq) \rightarrow H_2SiF_6(s) + 2H_2O(l)$$

This reaction can be used to etch, or "frost," glass for light bulbs and decorative glassware.

Various fluorides, such as tin(II) fluoride, SnF_2, and sodium monofluorophosphate, Na_2PO_3F, are used as toothpaste additives, and sodium fluoride is added to some municipal water supplies to aid in the prevention of tooth decay. The fluoride converts tooth enamel from hydroxyapatite, $Ca_{10}(OH)_2(PO_4)_6$, to fluorapatite, $Ca_{10}F_2(PO_4)_6$, which is harder and more resistant to acids than hydroxyapatite.

▪ Tin(II) fluoride is also known as stannous fluoride.

Many organofluoride compounds are used as refrigerants. Two common ones are dichlorodifluoromethane (Freon 12), CCl_2F_2, which is used in automobile air conditioners, and chlorodifluoromethane (Freon 21), $CHClF_2$, which is used in home air conditioners. Fluorocarbons have displaced refrigerants such as ammonia and sulfur dioxide in refrigerators because they are nontoxic, noncorrosive, and odorless. Oxygen is very soluble in numerous liquid fluorocarbons, and this unusual property of fluorocarbons has led to their study for potential use as artificial blood fluids. Figure M-4 shows a mouse totally submerged in dichlorofluoromethane saturated with oxygen. The mouse is able to breathe by absorbing oxygen from the oxygen-containing fluorocarbon that fills its lungs.

M-2. CHLORINE IS OBTAINED FROM CHLORIDES BY ELECTROLYSIS, AND BROMINE AND IODINE ARE OBTAINED BY THE OXIDATION OF BROMIDES AND IODIDES BY CHLORINE

Chlorine is the most abundant of the halogens. The major source of chlorine in nature is the chloride ion in the oceans and salt lakes. The major mineral sources of chloride are rock salt, NaCl, sylvite, KCl, and carnallite, $KMgCl_3{\cdot}6H_2O$.

Chlorine is prepared commercially by the electrolysis of either brines or molten rock salt:

$$\underset{\text{molten rock salt}}{2NaCl(l)} \xrightarrow{\text{electrolysis}} 2Na(l) + Cl_2(g)$$

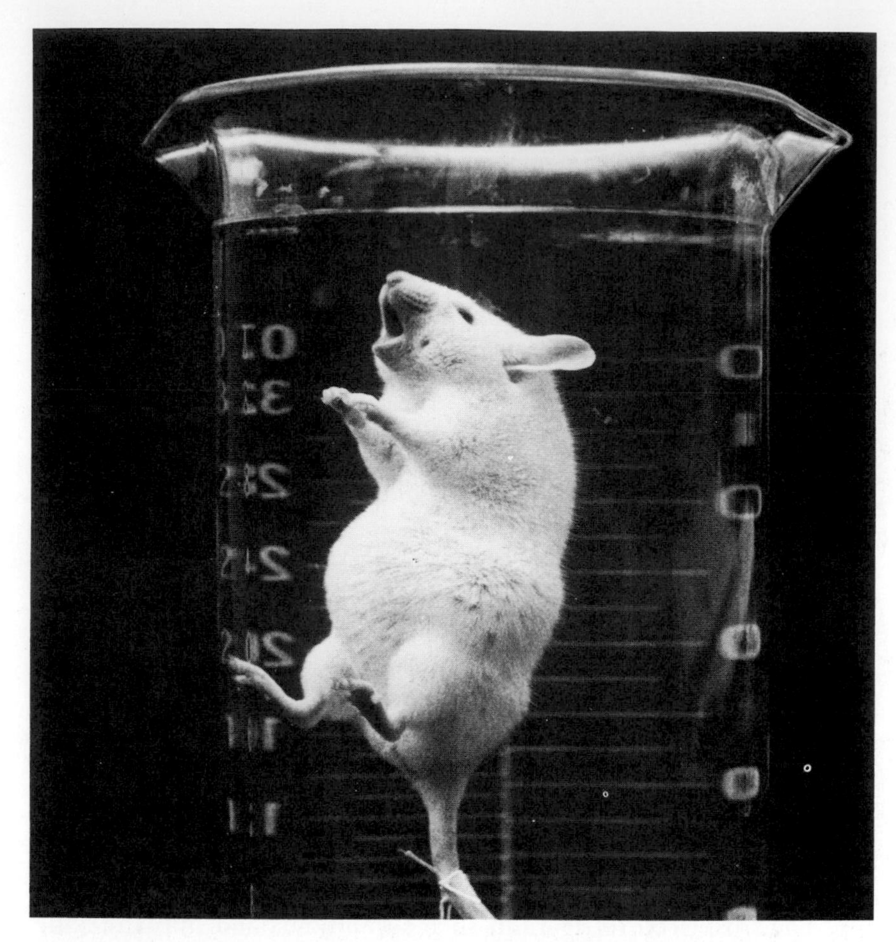

Figure M-4 This submerged mouse is breathing oxygen dissolved in a liquid fluorocarbon. The solubility of oxygen in this liquid is so great that the mouse is able to breath by absorbing oxygen from the oxygen-containing fluorocarbon that fills its lungs. When the mouse is removed from the liquid, the fluorocarbon vaporizes from its lungs and normal breathing resumes.

About 10 million metric tons of chlorine are produced annually in the United States, making it the ninth-ranked chemical in terms of production.

Most chlorine produced in the United States is used as a bleaching agent in the pulp and paper industry. It is also used extensively as a germicide in water purification and in the production of insecticides (DDT and chlordane) and herbicides (2,4-D).

The major source of bromine in the United States is from brines that contain bromide ions. The pH of the brine is adjusted to 3.5, and chlorine is added; the chlorine oxidizes bromide ion to bromine, which is swept out of the brine with a current of air:

$$2Br^-(aq) + Cl_2(aq) \xrightarrow{pH = 3.5} 2Cl^-(aq) + Br_2(aq)$$

About 1 kg of bromine can be obtained from 15,000 L of seawater. About 200,000 metric tons of bromine were produced in the United States during 1980. Bromine is a dense, red-brown liquid. Bromine vapor and solutions of bromine in nonpolar solvents are red. (Figure M-5).

Bromine is used to prepare a wide variety of metal bromide and organobromide compounds. Its major uses are in the production of dibromoethane, $BrCH_2CH_2Br$, which is added to leaded gasolines as a lead scavenger, and in the production of silver bromide emulsions for black-and-white photographic films. Bromine is also used as a fumigant and in the synthesis of fire retardants, dyes, and pharmaceuticals, especially sedatives.

Iodide ion is present in seawater and is assimilated and concentrated by many marine animals and by seaweed. Certain seaweeds are an especially rich source of iodine. The iodide ion in seaweed is converted to iodine by oxidation with chlorine. Iodine is also obtained from Chilean mineral deposits of sodium iodate, $NaIO_3$, and sodium periodate, $NaIO_4$. Iodine is the only halogen to occur naturally in a positive oxidation state. The free element is obtained by reduction of IO_3^- and IO_4^- with sodium hydrogen sulfite:

$$2IO_3^-(aq) + 5HSO_3^-(aq) \rightarrow I_2(aq) + 5SO_4^{2-}(aq) + 3H^+(aq) + H_2O(l)$$

Solid iodine is dark gray with a slight metallic luster. Iodine vapor and solutions of iodine in nonpolar solvents are purple. Solutions of iodine in water and alcohols are brown as a result of the specific polar interactions between molecular iodine and the oxygen-hydrogen bond. Iodine is only slightly soluble in pure water, but when iodide ion is present, the iodine forms a linear, colorless triiodide complex,

$$I_2(s) + I^-(aq) \rightleftharpoons I_3^-(aq) \qquad K_c = 720\ M^{-1}\ (25°C)$$

and the solubility of $I_2(s)$ is greatly enhanced. The presence of very low concentrations of aqueous triiodide can be detected by adding starch to the solution. The triiodide ion combines with starch to form a brilliant deep-blue species. (Figure M-6).

Iodide ion is essential for the proper functioning of the thyroid gland, which in humans is located in the base of the throat. Iodide

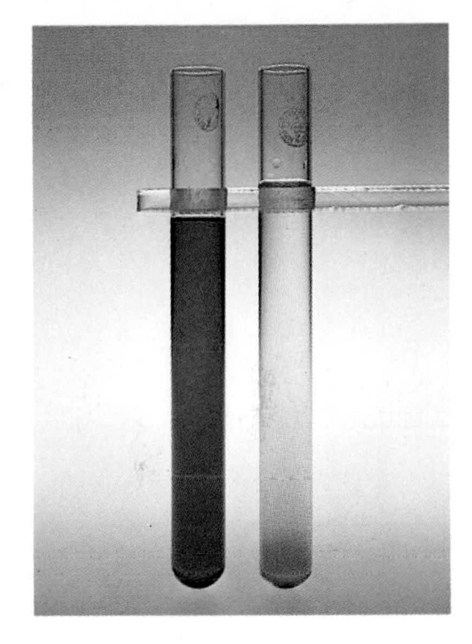

Figure M-5 Bromine dissolved in carbon tetrachloride *(left)* and in water *(right)*.

Figure M-6 Various solutions of iodine. *Left*, I_2 dissolved in CCl_4. *Center left*, I_2 dissolved in KI(*aq*). *Center right*, I_2 dissolved in water. *Right*, I_2 dissolved in KI(*aq*) with starch added to the solution.

deficiency disease is manifested as a goiter, which is a swelling in the neck caused by an enlargement of the thyroid gland. Potassium iodide is added to ordinary table salt, which is marketed as iodized salt. Iodized salt acts to prevent iodide deficiency. Alcohol solutions of iodine, known as tincture of iodine, were once used extensively as an antiseptic.

M-3. CHLORINE, BROMINE, AND IODINE HAVE VERY SIMILAR CHEMICAL PROPERTIES

Chlorine, bromine, and iodine have such similar chemical properties that it is convenient to consider them together. All three halogens can be prepared by the oxidation of a halide salt by manganese dioxide in concentrated sulfuric acid; for example,

$$2NaCl(s) + 4H^+(aq) + MnO_2(s) \rightarrow Mn^{2+}(aq) + 2Na^+(aq) + 2H_2O(l) + Cl_2(g)$$

This reaction was used by the Swedish chemist Karl Scheele in 1774 in the first laboratory preparation of chlorine, which was the first halogen to be isolated.

The standard reduction voltages

$$Cl_2(g) + 2e^- \rightarrow 2Cl^-(aq) \qquad E^\circ_{red} = +1.36\ V$$

$$Br_2(l) + 2e^- \rightarrow 2Br^-(aq) \qquad E^\circ_{red} = +1.07\ V$$

$$I_2(s) + 2e^- \rightarrow 2I^-(aq) \qquad E^\circ_{red} = +0.54\ V$$

indicate that the halogens act as oxidizing agents. For example, if bromine water is added to an aqueous solution of $FeSO_4(aq)$ in dilute sulfuric acid, the color of the bromine disappears and the yellow color of a solution of $Fe_2(SO_4)_3(aq)$ appears:

$$\underset{\text{pale green}}{2Fe^{2+}(aq)} + \underset{\text{reddish brown}}{Br_2(aq)} \rightarrow \underset{\text{yellow}}{2Fe^{3+}(aq)} + \underset{\text{colorless}}{2Br^-(aq)}$$

If $H_2S(g)$ is passed into an aqueous solution of chlorine, bromine, or iodine, then the halogen color disappears and a yellow precipitate of sulfur appears; for example,

$$H_2S(g) + I_2(aq) \rightarrow 2HI(aq) + S(s)$$

Chlorine reacts with most metals at moderate temperatures. Usually the metal is produced in its highest stable oxidation state. For example,

$$2Fe(s) + 3Cl_2(g) \rightarrow 2FeCl_3(s)$$

$$Sn(s) + 2Cl_2(g) \rightarrow SnCl_4(l)$$

Bromine and iodine react in a similar way but usually require higher temperatures.

The halogens also react with many nonmetallic elements. For example, we saw in Interchapter J that phosphorus forms a trichloride and a pentachloride with chlorine. Phosphorus tribromide results when carbon disulfide solutions of white phosphorus and bromine are mixed, and phosphorus triiodide can be made in a similar manner. Sulfur reacts with both chlorine and bromine to form a compound of the formula S_2X_2, but iodine does not. Except for the reaction between fluorine and oxygen, none of the halogens reacts directly with oxygen or nitrogen.

The most important binary compounds of the halogens and a nonmetal are the hydrogen halides. Chlorine reacts explosively with hydrogen in direct sunlight to produce hydrogen chloride according to

$$H_2(g) + Cl_2(g) \rightarrow 2HCl(g)$$

whereas bromine and hydrogen combine slowly in sunlight. Iodine combines with hydrogen even less readily and requires a catalyst for any appreciable reaction rate. The reaction is reversible:

$$H_2(g) + I_2(g) \rightleftharpoons 2HI(g)$$

with an equilibrium constant of 620 at 25°C. The corresponding values of the equilibrium constants for HBr(g) and HCl(g) are 1.6×10^{19} and 2.5×10^{38}, respectively.

Hydrogen chloride is a colorless, corrosive gas with a pungent, irritating odor. It is a major industrial chemical and is produced on a huge scale as a by-product in the chlorination of hydrocarbons, although the direct combination of hydrogen and chlorine is used if a high-purity product is desired. The annual world production of HCl(g) is well in excess of 10 billion pounds. Hydrochloric acid, which is also called **muriatic acid,** is an aqueous solution of hydrogen chloride. Its largest single use is in the treatment of steel and other metals to remove an adhering oxide coating, although it has many other industrial uses.

Hydrogen chloride is most readily prepared in the laboratory by the reaction

$$NaCl(s) + H_2SO_4(l) \rightarrow NaHSO_4(s) + HCl(g)$$

Hydrogen bromide and hydrogen iodide cannot be prepared in this manner because unlike HCl(g), they are oxidized by concentrated sulfuric acid. Hydrogen bromide and hydrogen iodide can be prepared by the reaction of the corresponding halogen with wet red phosphorus; for bromine, the equation for the reaction is

$$2P_4(s) + 6Br_2(l) + 12H_2O(l) \rightarrow 12HBr(g) + 4PBr_3(l) + 4H_3PO_3(l)$$

Like hydrochloric acid, hydrobromic acid and hydroiodic acid are strong acids. These acids become increasingly strong reducing agents in the order HCl(aq) to HBr(aq) to HI(aq). Although HCl(aq) is not regarded as a reducing agent, it can be oxidized to chlorine by strong oxidizing agents such as potassium permanganate or manganese dioxide:

$$2MnO_4^-(aq) + 10Cl^-(aq) + 16H^+(aq) \rightarrow 2Mn^{2+}(aq) + 8H_2O(l) + 5Cl_2(g)$$

$$MnO_2(s) + 4H^+(aq) + 2Cl^-(aq) \rightarrow Mn^{2+}(aq) + 2H_2O(l) + Cl_2(g)$$

Hydroiodic acid is a reducing agent and reacts quantitatively with many oxidizing agents; for example,

$$H_2O_2(aq) + 2H^+(aq) + 2I^-(aq) \rightarrow 2H_2O(l) + I_2(s)$$

Solutions of hydroiodic acid turn brown upon standing owing to air oxidation:

$$4HI(aq) + O_2(g) \rightarrow 2H_2O(l) + 2I_2(aq)$$

colorless   brown

The reducing properties of HBr(*aq*) are intermediate between those of HCl(*aq*) and HI(*aq*), but HBr(*aq*) is not a useful reducing agent.

M-4. THE HALOGENS FORM NUMEROUS OXYGEN-HALOGEN COMPOUNDS

The best-known and most important oxygen-halogen compounds are the halogen oxyacids. The halogens form a series of oxyacids in which the oxidation state of the halogen atom can be +1, +3, +5, or +7. For example, the oxyacids of chlorine are

HClO	hypochlorous acid	+1
$HClO_2$	chlorous acid	+3
$HClO_3$	chloric acid	+5
$HClO_4$	perchloric acid	+7

The numbers after the names give the oxidation state of chlorine in the acid. The Lewis formulas for these acids are

hypochlorous acid   chlorous acid   chloric acid   perchloric acid

Notice that in each case the hydrogen atom is attached to an oxygen atom. The anions of the chlorine oxyacids are

ClO^-	hypochlorite	ClO_3^-	chlorate
ClO_2^-	chlorite	ClO_4^-	perchlorate

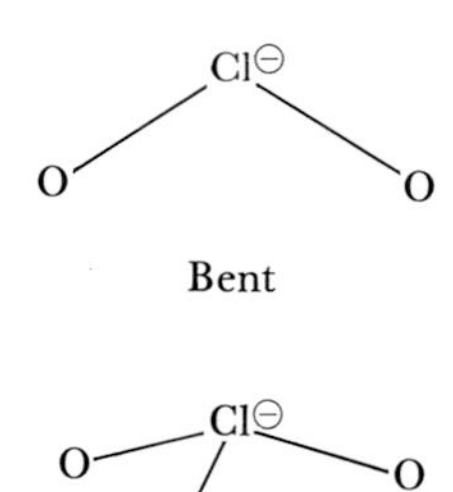

Tetrahedral

Figure M-7 The shapes of the oxyacid anions of chlorine.

The shapes of these ions are predicted correctly by VSEPR theory (Chapter 9) and are shown in Figure M-7. Table M-2 gives the known halogen oxyacids and their anions. Note that there are no oxyacids of fluorine; the only oxidation state of fluorine is −1.

The strength of an oxyacid depends strongly upon the number of oxygen atoms without attached hydrogen atoms that are attached to

Table M-2 The halogen oxyacids

Halogen oxidation state	Chlorine	Bromine	Iodine	Acid	Salt
+1	HClO	HBrO	HIO	hypohalous	hypohalite
+3	$HClO_2$	—	—	halous	halite
+5	$HClO_3$	$HBrO_3$	HIO_3	halic	halate
+7	$HClO_4$	$HBrO_4$	HIO_4	perhalic	perhalate

the central atom. Let this number of oxygen atoms be denoted by η_O. Thus, $\eta_O = 0$ for hypochlorous acid, HClO; $\eta_O = 1$ for nitrous acid, HNO_2, and phosphoric acid, H_3PO_4; $\eta_O = 2$ for sulfuric acid, H_2SO_4, and nitric acid, HNO_3; and $\eta_O = 3$ for perchloric acid, $HClO_4$. The following equation gives an approximate value of the pK_a of the oxyacid in terms of η_O:

$$pK_a \simeq 8 - 5.5\eta_O \tag{M-1}$$

Note that the strength of an oxyacid increases by a factor of about 10^5 for a unit increase in η_O. Equation (M-1) is valid to approximately ± 1 in the value of pK_a. Table M-3 lists the pK_a values for a number of oxyacids. The key point of Table M-3 is not the numerical accuracy of Equation (M-1) but the strength of the oxyacid as given by the value of η_O. Note that an oxyacid is very weak if $\eta_O = 0$; is weak if $\eta_O = 1$; is strong if $\eta_O = 2$; and is very strong if $\eta_O = 3$. In fact, $HClO_4$ is the strongest known oxyacid. Some of the oxyacids listed in Table M-3 are polyprotic acids. For polyprotic acids the successive acid-dissociation constants decrease by a factor of about 10^{-5}. Thus, for example, the three acid-dissociation constants of phosphoric acid are $K_{a1} = 7.1 \times 10^{-3}$ M, $K_{a2} = 6.2 \times 10^{-8}$ M, and $K_{a3} = 4.4 \times 10^{-13}$ M.

The various oxygen-halogen compounds, often called simply **oxyhalogens,** are prepared by the reactions of the halogens with water under various conditions. When Cl_2, for example, is dissolved in aqueous alkaline solution, the following reaction occurs:

(1) $Cl_2(g) + 2OH^-(aq) \rightleftharpoons Cl^-(aq) + ClO^-(aq) + H_2O(l)$

A solution of NaClO(*aq*) is a bleaching agent, and many household bleaches are a 5.25 percent aqueous solution of sodium hypochlorite. Commercially, solutions of NaClO are manufactured by the electrolysis of cold aqueous solutions of sodium chloride:

(2) $2Cl^-(aq) \rightarrow Cl_2(g) + 2e^-$ anode

(3) $2H_2O(l) + 2e^- \rightarrow H_2(g) + 2OH^-(aq)$ cathode

Table M-3 The pK_a values of various oxyacids*

$\eta_O = 0$		$\eta_O = 1$		$\eta_O = 2$		$\eta_O = 3$	
HClO	7.5	$HClO_2$	1.9	$HClO_3$	~−2.7	$HClO_4$	~−9
HBrO	8.7	H_2SO_3	1.9	$HBrO_3$	~−2		
HIO	10.6	H_2SeO_3	2.6	HIO_3	~−0.6		
H_6TeO_6	7.7	H_2TeO_3	2.5	H_2SO_4	~−3		
H_3AsO_3	9.2	HNO_2	3.2	H_2SeO_4	~−3		
H_4SiO_4	9.8	H_3PO_4	2.1	HNO_3	~−1		
		H_3PO_3	1.6				
		H_3PO_2	1.2				
		H_3AsO_4	2.3				

*Acids with pK_a values less than zero (for example, $HClO_4$ and HNO_3) are usually treated as completely dissociated ("strong") in aqueous solution.

The products of the two electrode reactions described by Equations (2) and (3) are allowed to mix, producing $ClO^-(aq)$ by the reaction described by Equation (1). Sodium hypochlorite is also employed as a disinfectant and deodorant in dairies, creameries, water supplies, and sewage disposals.

Hypohalite ions decompose in basic solution via reactions of the type

$$3IO^-(aq) \xrightarrow{OH^-} 2I^-(aq) + IO_3^-(aq)$$

The analogous reaction with $ClO^-(aq)$ is slow, which makes it possible to use hypochlorite as a bleach in basic solutions. The rate of decomposition of $ClO^-(aq)$, $IO^-(aq)$, and $BrO^-(aq)$ in hot alkaline aqueous solution is sufficiently fast that when Cl_2, Br_2, or I_2 is dissolved in basic solution and the resulting solution is heated to 60°C, the following type of reaction goes essentially to completion:

$$3Br_2(aq) + 6OH^-(aq) \xrightarrow{60°C} 5Br^-(aq) + BrO_3^-(aq) + 3H_2O(l)$$

Chlorates, bromates, and iodates also can be prepared by the reaction of the appropriate halogen with concentrated nitric acid or hydrogen peroxide or (commercially) by electrolysis of the halide. For example, the reaction for the oxidation of I_2 by H_2O_2 is

$$I_2(s) + 5H_2O_2(aq) \rightarrow 2IO_3^-(aq) + 4H_2O(l) + 2H^+(aq)$$

Perchlorate and periodate are prepared by the electrochemical oxidation of chlorate and iodate, respectively. For example,

$$ClO_3^-(aq) + H_2O(l) \xrightarrow{\text{electrolysis}} ClO_4^-(aq) + 2H^+(aq) + 2e^-$$

The perchlorate is obtained from the electrolyzed cell solution by adding potassium chloride in order to precipitate potassium perchlorate, which is only moderately soluble in water. Perchloric acid, $HClO_4$, also can be obtained from the electrolyzed solution containing perchlorate by adding sulfuric acid and then distilling. Concentrated perchloric acid should not be allowed to come into contact with reducing agents, such as organic matter, because of the extreme danger of a violent explosion. Solutions containing perchlorates should not be evaporated because of their treacherously explosive nature. Perchlorates are used in explosives, fireworks, and matches.

TERMS YOU SHOULD KNOW

halogen 658 | halide 658 | hydrohalic acid 660 | muriatic acid 665 | oxyhalogen 667

QUESTIONS

M-1. Explain why atomic electronegativity should decrease from fluorine to iodine.

M-2. Explain why the boiling points of the halogens increase in going from fluorine to iodine.

M-3. Given

$$HF(aq) + H_2O(l) \rightleftharpoons H_3O^+(aq) + F^-(aq) \qquad K = 6.8 \times 10^{-4}\ M$$

$$HF(aq) + F^-(aq) \rightleftharpoons HF_2^-(aq) \qquad K = 5.1\ M^{-1}$$

calculate the value of K for

$$2HF(aq) + H_2O(l) \rightleftharpoons H_3O^+(aq) + HF_2^-(aq)$$

M-4. Explain briefly why fluorine is capable of stabilizing unusually high oxidation states in many elements.

M-5. Describe how each of the halogens is prepared on a commercial scale.

M-6. Describe how the glass used in frosted light bulbs is etched.

M-7. Describe the role of fluoride in the prevention of tooth decay.

M-8. Iodic acid is usually prepared by heating iodine with concentrated nitric acid. Balance the equation

$$I_2(s) + HNO_3(aq, conc) \rightarrow HIO_3(aq) + NO_2(g) + H_2O(l)$$

M-9. When heated with concentrated hydrochloric acid, potassium chlorate oxidizes it to yield a dangerously explosive, bright-yellow mixture of chlorine and chlorine dioxide. Balance the equation

$$KClO_3(s) + HCl(aq, conc) \rightarrow KCl(aq) + Cl_2(g) + ClO_2(g) + H_2O(l)$$

M-10. Write a chemical equation for the laboratory preparation of bromine.

M-11. The rate of disproportionation of $I_2(aq)$ in basic solution is fast at all temperatures, and the reaction given by the following equation occurs rapidly and quantitatively:

$$I_2(s) + OH^-(aq) \rightarrow I^-(aq) + IO_3^-(aq) + H_2O(l)$$

Balance this equation.

M-12. What is household bleach?

M-13. Describe, using balanced chemical equations, how you would prepare $KIO_3(s)$ starting with $I_2(s)$.

M-14. Name the following oxyacids:

(a) $HBrO_2$ (b) HIO (c) $HBrO_4$ (d) HIO_3

M-15. Given the names of the following key oxyacids

H_2SO_4 sulfuric acid
H_3PO_4 phosphoric acid
HNO_3 nitric acid

(i) name the following oxyacids:

(a) HNO_2 (b) H_2SO_3 (c) H_3PO_2 (d) H_3PO_3
(e) $H_2N_2O_2$

(ii) and name the following salts:

(a) K_2SO_3 (b) $Ca(NO_2)_2$ (c) KIO_2 (d) $Mg(BrO)_2$

M-16. Use VSEPR theory to predict the shapes of I_3^-, IF_3, BrF_5, ICl_4^-, and BrF_6^+.

M-17. Give the oxidation state of the halogen in each of the following species:

(a) KIO_3 (b) IF_5
(c) $HClO$ (d) ClF
(e) $KBrO_2$ (f) $KClO_4$

PROBLEMS

M-18. Calculate the relative rates of effusion of $^{235}UF_6$ and $^{238}UF_6$. How many stages would have to be used in order to achieve a uranium-235 enrichment of 10 percent?

M-19. Using the values of the standard reduction voltages given in Table 20-1, determine whether $Fe^{2+}(aq)$ can be oxidized to $Fe^{3+}(aq)$ by (a) $Cl_2(g)$, (b) $Br_2(l)$, and (c) $I_2(s)$; in other words, calculate the equilibrium constant in each case for the chemical equation

$$2Fe^{2+}(aq) + X_2 \rightleftharpoons 2Fe^{3+}(aq) + 2X^-(aq)$$

where X_2 is the halogen.

M-20. Bromine is present in seawater as 8.3×10^{-4} M $Br^-(aq)$. Calculate how much elemental bromine can be obtained from 15,000 L of seawater.

M-21. Using the standard reduction voltages for

$$H^+(aq) + HBrO(aq) + e^- \rightarrow \tfrac{1}{2}Br_2(l) + H_2O(l) \qquad E^\circ_{red} = 1.590 \text{ V}$$

$$\tfrac{1}{2}Br_2(l) + e^- \rightarrow Br^-(aq) \qquad E^\circ_{red} = 1.065 \text{ V}$$

calculate the value of the equilibrium constant at 25°C for

$$Br_2(l) + H_2O(l) \rightleftharpoons H^+(aq) + Br^-(aq) + HBrO(aq)$$

M-22. Iodine pentoxide is a reagent for the quantitative determination of carbon monoxide. The reaction is

$$I_2O_5(s) + 5CO(g) \rightarrow I_2(s) + 5CO_2(g)$$

The iodine produced is dissolved in $KI(aq)$ and then determined by titration with $Na_2S_2O_3(aq)$:

$$2S_2O_3^{2-}(aq) + I_3^-(aq) \rightarrow 3I^-(aq) + S_4O_6^{2-}(aq)$$

Suppose that 10.0 L of air at 0°C and 1 atm is bubbled through a bed of $I_2O_5(s)$ and that it required 4.76 mL of 0.0100 M $Na_2S_2O_3(aq)$ to reduce the I_2 produced. Calculate the mass (in mg) of $CO(g)$ in the 10.0-L air sample.

M-23. The reaction betwen hypochlorite and iodide ions in acidic solution

$$OCl^-(aq) + 3I^-(aq) + 2H^+(aq) \rightarrow Cl^-(aq) + I_3^-(aq) + H_2O(l)$$

can be used to determine the strengths of bleaching solutions. A 3.285 g sample of household bleach is treated with excess $KI(aq)$, and then the resulting $I_3^-(aq)$ is acidified and titrated with $Na_2S_2O_3(aq)$ to react with the $I_3^-(aq)$. If it requires 18.55 mL of 0.2500 M $S_2O_3^{2-}(aq)$, then determine the mass percentage of NaOCl in the bleaching solution.

CHAPTER • 21

NUCLEAR AND RADIOCHEMISTRY

A sample of the radioactive substance, curium oxide, glowing from the heat released during radioactive decay.

Many nuclei are unstable and spontaneously emit subatomic particles. The process of nuclear decomposition is called radioactivity, which is the principal topic of this chapter. All elements with more than 83 protons are radioactive. In addition, many radioactive isotopes that do not occur in nature have been made in the laboratory. Radioactive isotopes are used in chemistry, physics, medicine, biology, agriculture, geology, and criminal investigations. Radioactivity is used to determine the age of rocks that are billions of years old as well as the age of archaeological findings a few thousand years old. As little as 10^{-12} g of an impurity in a sample can be measured by radiochemical methods. Although the emphasis of this chapter is on the chemical aspects of nuclear processes, we also study nuclear reactions, the source of enormous amounts of energy.

21-1. MANY NUCLEI SPONTANEOUSLY EMIT SMALL PARTICLES

Recall from Chapter 1 that nuclei consist of protons and neutrons. Collectively, protons and neutrons are called **nucleons.** The number of protons in a nucleus is the atomic number, Z, which specifies the element. The total number of nucleons (protons plus neutrons) in a nucleus is the mass number, A. The number of neutrons in two atoms of a given element need not be the same. Atoms that contain the same number of protons but different numbers of neutrons are called isotopes, which are denoted $^{A}_{Z}X$. For example, chlorine has two naturally occurring isotopes, $^{35}_{17}Cl$ and $^{37}_{17}Cl$, called chlorine-35 and chlorine-37. Both undergo the same chemical reactions.

▪ Isotopes are denoted by $^{A}_{Z}X$, where X is the symbol of the element, Z is its atomic number, and A is its mass number.

Most elements found in nature consist of mixtures of isotopes, as is shown in Table 1-9 for several elements. Example 1-6 illustrates the calculation of the atomic mass of chromium from the masses of the various chromium isotopes and their observed mass percentages.

Uranium-238 nuclei emit helium nuclei, $^{4}_{2}He$. When a $^{238}_{92}U$ nucleus emits a helium nucleus, the mass number of the uranium decreases by four and the atomic number decreases by two. The change in atomic number tells us that a different element has been produced. We can describe the process by a **nuclear equation:**

$$^{238}_{92}U \rightarrow \, ^{234}_{90}Th + \, ^{4}_{2}He$$

The product here is thorium because the atomic number of the resulting nucleus is 90. This nuclear equation is balanced: the total nuclear charge (number of protons) and the number of nucleons are the same on both sides. The spontaneous disintegration of a nucleus is called **radioactivity,** and such a nucleus is said to be **radioactive.** An isotope that is radioactive is called a **radioisotope.** Uranium-238 is an example of a radioisotope.

Radioactivity was discovered by the French scientist Henri Becquerel in 1896. He found that the compound potassium uranyl sulfate, $K_2UO_2(SO_4)_2$, emitted radiation and was able to show that the uranium was the source of the emissions. Becquerel referred to the radiation as α-rays (alpha rays) or **α-particles.** It was subsequently shown that α-particles are helium-4 nuclei. Uranium-238 is said to be an **α-emitter.**

Example 21-1: Radon-222 is an α-emitter. Write a balanced nuclear equation for the reaction.

▪ Nuclear equations must be balanced with respect to nuclear charge and number of nucleons.

Solution: The symbol for radon-222 is $^{222}_{86}Rn$. The mass number decreases by four and the atomic number decreases by two when an α-particle is emitted. The product has $Z = 84$ and $A = 218$. Polonium is the element that has $Z = 84$, and so the balanced nuclear equation is

$$^{222}_{86}Rn \rightarrow \, ^{218}_{84}Po + \, ^{4}_{2}He$$

Note that $86 = 84 + 2$ (charge balance) and $222 = 218 + 4$ (nucleon balance).

21-2. THERE ARE SEVERAL TYPES OF RADIOACTIVE EMISSIONS

Not all radioactive nuclei emit α-particles. Two other common types of emissions were found in early experiments on radioactivity; they were called **β-particle** (beta-particle) emission and **γ-ray** (gamma-ray) emission. It was discovered that β-particles are electrons. Electrons in nuclear equations are denoted by the symbol ${}_{-1}^{0}e$. The superscript zero refers to the small mass of an electron relative to that of a nucleon, and the subscript -1 refers to the negative charge on an electron. Two examples of β-emission reactions are

$$ {}_{49}^{116}\mathrm{In} \rightarrow {}_{50}^{116}\mathrm{Sn} + {}_{-1}^{0}\mathrm{e} $$

$$ {}_{90}^{234}\mathrm{Th} \rightarrow {}_{91}^{234}\mathrm{Pa} + {}_{-1}^{0}\mathrm{e} $$

Note that the mass numbers do not change but that the atomic number increases by one as a result of the emission of a β-particle.

There are no electrons in nuclei. The emission of a β-particle results from the conversion of a neutron to a proton within the nucleus, which can be represented as

$$ {}_{0}^{1}\mathrm{n} \rightarrow {}_{1}^{1}\mathrm{H} + {}_{-1}^{0}\mathrm{e} $$

Gamma rays are high-energy electromagnetic waves, similar to X-rays. The emission of γ-rays causes no change in either Z or A. The emission of an α-particle or a β-particle frequently leaves the product nucleus in an excited state. The excited nucleus can relax to its ground state by emitting a photon, whose frequency is given by the equation (Section 6-5)

$$ \Delta E = h\nu $$

where ΔE is the energy difference between the excited state and the ground state. Gamma-ray emission is analogous to an atom's emitting electromagnetic radiation when an electron falls from an excited state to a ground state. We often do not indicate γ-radiation in writing nuclear equations.

Several other types of radioactive emissions have been observed. For example, some nuclei emit a **positron,** which is a particle that has the same mass as an electron but a positive charge. The symbol for a positron in nuclear equations is ${}_{+1}^{0}e$. Two examples of positron emission are

$$ {}_{19}^{38}\mathrm{K} \rightarrow {}_{18}^{38}\mathrm{Ar} + {}_{+1}^{0}\mathrm{e} $$

$$ {}_{51}^{120}\mathrm{Sb} \rightarrow {}_{50}^{120}\mathrm{Sn} + {}_{+1}^{0}\mathrm{e} $$

The emission of a positron can be viewed as the conversion of a proton to a neutron:

$$ {}_{1}^{1}\mathrm{H} \rightarrow {}_{0}^{1}\mathrm{n} + {}_{+1}^{0}\mathrm{e} $$

Positrons exist for only a very short time. They combine with electrons in about 10^{-9} s and are converted to γ-radiation (Figure 21-1). Table 21-1 summarizes some of the types of radioactive processes.

When a radioactive nucleus emits a particle and transforms to another nucleus, we say that it decays to that nucleus. Thus, the expression **radioactive decay** refers to a radioactive process.

Figure 21-1 Not only do an electron and a positron annihilate each other to produce two γ-ray photons, but the reverse process also occurs; that is, a high-energy γ-ray photon can transform spontaneously into an electron and a positron, as seen in this photograph from a bubble chamber. Being uncharged, the photon leaves no track, but the electron and positron have clearly seen spiral tracks. An externally applied magnetic field causes the tracks to be spirals, and the opposite charges on the electron and positron cause them to spiral in opposite directions.

Table 21-1 The various particles emitted in radioactive processes

Emission	Symbol	Change in nucleus: Mass number	Change in nucleus: Atomic number	Example
α	${}^{4}_{2}He$	decreases by 4	decreases by 2	${}^{238}_{92}U \rightarrow {}^{234}_{90}Th + {}^{4}_{2}He$
β	${}^{0}_{-1}e$	no change	increases by 1	${}^{14}_{6}C \rightarrow {}^{14}_{7}N + {}^{0}_{-1}e$
γ	${}^{0}_{0}\gamma$	no change	no change	
positron	${}^{0}_{+1}e$	no change	decreases by 1	${}^{38}_{19}K \rightarrow {}^{38}_{18}Ar + {}^{0}_{+1}e$

Example 21-2: Fill in the missing symbols in the following nuclear equations:

(a) ${}^{214}_{82}Pb \rightarrow ? + {}^{214}_{83}Bi$ (b) ${}^{11}_{6}C \rightarrow {}^{0}_{+1}e + ?$

(c) $? \rightarrow {}^{0}_{-1}e + {}^{97}_{41}Nb$

Solution: (a) The missing particle has a charge of -1 ($82 = 83 - 1$), and A does not change; thus the particle is a β-particle, ${}^{0}_{-1}e$.

(b) The missing particle has $Z = 5$ and $A = 11$ and so is ${}^{11}_{5}B$.

(c) The missing particle has $Z = 40$ and $A = 97$. The element that has $Z = 40$ is zirconium, and so the nucleus that decays is ${}^{97}_{40}Zr$.

There is no general theory that we can use to predict the stability of nuclei, but there are a number of empirical observations that help. Figure 21-2 is a plot of all the known stable nuclei as a function of their

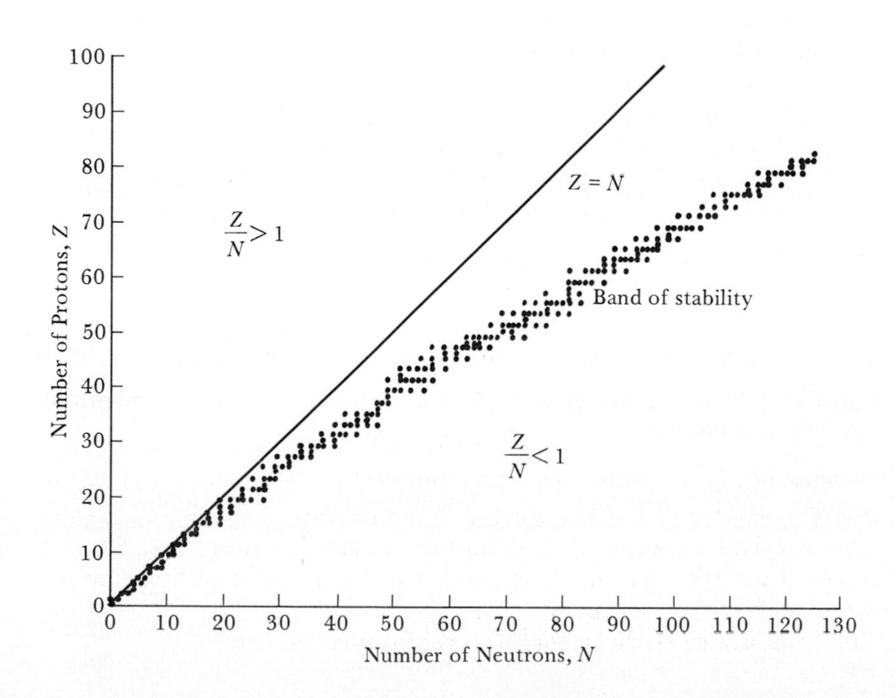

Figure 21-2 Known stable (nonradioactive) nuclei plotted on a graph of number of protons, Z, versus number of neutrons, N. The stable nuclei fall within the band of stability. The solid straight line is the line for which $Z = N$. Nuclei that lie above the band of stability may emit positrons because by doing so the product nucleus lies closer to the band of stability. Similarly, nuclei that lie below the band of stability emit β-particles. All nuclei with $Z > 83$ are radioactive.

number of protons, Z, and their number of neutrons, $N = A - Z$. It shows that the stable nuclei form what is called a **band of stability.** This band is determined by the ratio of protons to neutrons. For elements with an atomic number lower than 21, the proton-neutron ratio is unity or slightly less (the only exceptions are $^{1}_{1}H$ and $^{3}_{2}He$). Helium-4, carbon-12, oxygen-16, neon-20, and calcium-40 are examples of stable nuclei with equal numbers of neutrons and protons. As Z increases, the proton-to-neutron ratio decreases below unity and reaches a value of 0.7 at $^{209}_{83}Bi$, which is the stable nucleus with the largest value of Z. All nuclei with more than 83 protons are unstable.

Elements with an odd number of protons usually have only one or two stable isotopes, whereas elements with an even number of protons may have several stable isotopes (Table 21-2). Most stable nuclei contain an even number of protons or neutrons. In fact, there are only four stable nuclei that contain an odd number of both protons and neutrons ($^{2}_{1}H$, $^{6}_{3}Li$, $^{10}_{5}B$, $^{14}_{7}N$). Table 21-3 lists the numbers of stable nuclei with even and odd numbers of protons and neutrons. It can be seen from this table that nature seems to prefer nuclei with even numbers of protons and neutrons.

Nuclei that contain 2, 8, 20, 28, 50, or 82 protons or neutrons are particularly stable and abundant in nature. The numbers 2, 8, 20, 28, 50, and 82 are called **magic numbers,** and they have the same significance in nuclear structure theory that the numbers 2, 10, 18, 32, 54, and 86, the numbers of electrons in the noble gases, have in the theory of the electronic structure of atoms.

Table 21-2 Values of Z and numbers of stable isotopes of the elements Zr through Ba*

Z	Element	Number of stable isotopes
40	Zr	5
41	Nb	1
42	Mo	7
43	Tc	0
44	Ru	7
45	Rh	1
46	Pd	6
47	Ag	2
48	Cd	7
49	In	1
50	Sn	10
51	Sb	2
52	Te	7
53	I	1
54	Xe	9
55	Cs	1
56	Ba	7

*Notice that elements with odd values of Z usually have only one or two stable isotopes, and that elements with even values of Z have several stable isotopes.

Example 21-3: Predict whether or not the following nuclei are radioactive (unstable):

(a) $^{120}_{50}Sn$ (b) $^{213}_{87}Fr$ (c) $^{16}_{8}O$ (d) $^{15}_{8}O$

Solution: (a) Tin-120 has an even number of protons and neutrons and a magic number of protons; thus it is a stable isotope.

(b) Since there are no stable nuclei with more than 83 protons, francium-213 is an unstable nucleus.

(c) Oxygen-16 has a magic number (8) of both protons and neutrons and so is a stable isotope.

(d) Oxygen-15 has a magic number of protons, but its proton-to-neutron ratio (8/7) is greater than unity. We predict that oxygen-15 is a radioactive isotope. Because it lies above the band of stability in Figure 21-2, we predict (correctly) that it decays by positron emission.

Table 21-3 The number of stable nuclei with even and odd numbers of protons and neutrons

Number of protons	Number of neutrons	Number of stable nuclei	Examples
odd	odd	4	$^{2}_{1}H$ $^{6}_{3}Li$ $^{10}_{5}B$ $^{14}_{7}N$
even	odd	57	$^{13}_{6}C$ $^{25}_{12}Mg$ $^{47}_{22}Ti$
odd	even	50	$^{19}_{9}F$ $^{23}_{11}Na$ $^{27}_{13}Al$
even	even	168	$^{12}_{6}C$ $^{16}_{8}O$ $^{20}_{10}Ne$

21-3. THE RATE OF DECAY OF A RADIOACTIVE ISOTOPE IS A FIRST-ORDER PROCESS

Not all radioactive nuclei decay at the same rate. Some radioactive samples decay in a few millionths of a second; the same amount of another isotope may take billions of years to decay. An important aspect of radioactive decay is that its rate is independent of external factors, such as temperature and pressure, at least under normal conditions. At the present time, we are unable to alter the rate of radioactive decay processes. This is a serious problem in the disposal of radioactive waste.

Radioactive decay is a first-order rate process (Chapter 13). This means that the number of radioactive nuclei, N, remaining in a sample at time t—given that N_0 is the number of nuclei present initially—is given by (Section 13-3)

▪ Recall that

$$t_{1/2} = \frac{0.693}{k}$$

for a first-order reaction.

$$\log \frac{N_0}{N} = \frac{kt}{2.30} = \frac{0.301t}{t_{1/2}} \qquad (21\text{-}1)$$

where $t_{1/2}$ is the **half-life** for the decay of the radioactive nucleus. Recall from Chapter 13 that the half-life is the time required for one half of the reacting particles to undergo reaction. Thus the half-life for a radioactive isotope is the time it takes for one half of a sample of the isotope to undergo radioactive decay. Figure 21-3 shows a plot of N/N_0, which is the fraction of the nuclei remaining, versus time for the decay of sodium-25, which has a half-life of 1 min.

Different radioisotopes have different half-lives. The half-life is characteristic of a given radioisotope and is a direct measure of how rapidly it decays. Table 21-4 lists the half-lives of some common radioisotopes, and the following Examples illustrate the use of Equation (21-1).

▪ Recall from Chapter 13 that the fraction remaining, N/N_0 after n half-lives is given by

$$\frac{N}{N_0} = \left(\frac{1}{2}\right)^n$$

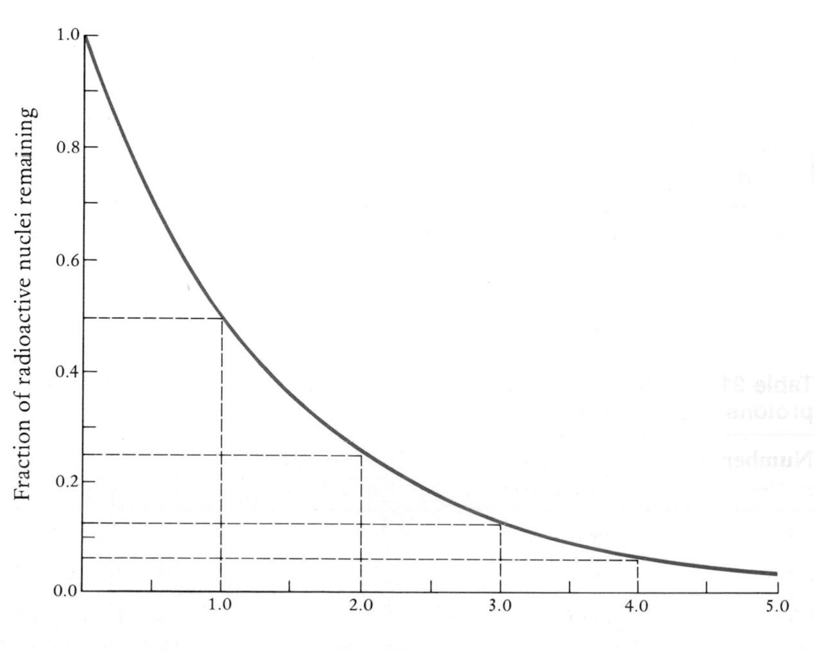

Figure 21-3 A graph of the decay of sodium-25 over time. The solid line is a plot of the fraction of sodium-25 nuclei present in a sample versus time. This curve is representative of a first-order kinetic process; the rate of decay is proportional to the number of sodium-25 nuclei present in the sample. The half-life ($t_{1/2}$) of sodium-25 is 1.0 min. One half of the sample decays in the first minute; one half of what remains decays in the next minute, and so on.

Table 21-4 Half-lives of some radioisotopes

Isotope	Half-life	Mode of decay
$^{3}_{1}H$	12.33 years	β
$^{14}_{6}C$	5730 years	β
$^{25}_{11}Na$	60 s	$\beta + \gamma$
$^{32}_{15}P$	14.28 days	β
$^{60}_{27}Co$	5.27 years	$\beta + \gamma$
$^{87}_{37}Rb$	4.9×10^{10} years	β
$^{90}_{38}Sr$	29 years	β
$^{131}_{53}I$	8.04 days	$\beta + \gamma$
$^{214}_{84}Po$	164 μs	α
$^{226}_{88}Ra$	1600 years	$\alpha + \gamma$
$^{238}_{92}U$	4.47×10^{9} years	$\alpha + \gamma$
$^{239}_{93}Pu$	2.41×10^{4} years	$\alpha + \gamma$

Example 21-4: The radioisotope cobalt-60 is used to destroy cancerous cells by directing γ-rays into the cancerous cell tissue. Calculate the fraction of a cobalt-60 sample left after 20.0 years.

Solution: From Table 21-4, we see that the half-life of cobalt-60 is 5.27 years. If we use Equation (21-1), then we have

$$\log \frac{N_0}{N} = \frac{0.301t}{t_{1/2}} = \frac{(0.301)(20.0 \text{ years})}{(5.27 \text{ years})} = 1.14$$

$$\frac{N_0}{N} = 10^{1.14} = 13.8$$

$$\frac{N}{N_0} = 0.0725$$

Thus, only 7.25 percent of the original sample remains after 20 years.

Example 21-5: A 2.00-mg sample of the radioisotope phosphorus-32 is found to contain 0.40 mg of phosphorus-32 after 33.3 days. Calculate the half-life of this isotope.

Solution: Solving Equation (21-1) for $t_{1/2}$, we have

$$t_{1/2} = \frac{0.301t}{\log (N_0/N)}$$

The value of N_0/N after 33.3 days is

$$\frac{N_0}{N} = \frac{(2.00 \times 10^{-3} \text{ g})\left(\dfrac{1 \text{ mol}}{32.0 \text{ g}}\right)(6.022 \times 10^{23} \text{ atom}\cdot\text{mol}^{-1})}{(0.40 \times 10^{-3} \text{ g})\left(\dfrac{1 \text{ mol}}{32.0 \text{ g}}\right)(6.022 \times 10^{23} \text{ atom}\cdot\text{mol}^{-1})}$$

$$= \frac{2.00}{0.400} = 5.00$$

Notice that we have just shown that

$$\frac{N_0}{N} = \frac{\text{initial mass of phosphorus-32}}{\text{present mass of phosphorus-32}}$$

Using the value of 5.00 for N_0/N, we compute for the half-life of phosphorus-32

$$t_{1/2} = \frac{(0.301)(33.3\text{ days})}{\log 5.00} = 14.3\text{ days}$$

which agrees with the value given in Table 21-4.

21-4. RADIOACTIVITY CAN BE USED TO DETERMINE THE AGE OF ROCKS

Naturally occurring radioactive uranium-238 decays through a series of processes to lead-206. The overall process is

$$^{238}_{92}\text{U} \rightarrow ^{206}_{82}\text{Pb} + 8\,^{4}_{2}\text{He} + 6\,^{0}_{-1}\text{e}$$

The half-life for this overall decay process is 4.51×10^9 years.

Now suppose that a sample of uranium ore is found to contain equal molar quantities of uranium-238 and lead-206 and that all the lead-206 arose from the decay of the uranium-238. Because half of the uranium-238 nuclei initially present have decayed since the rock was formed, we conclude that the rock is 4.51×10^9 years old. The following Example illustrates this **uranium-lead dating** method.

Example 21-6: A sample of uranium ore is found to contain 5.20 mg of uranium-238 and 1.85 mg of lead-206. Calculate the age of the rock.

Solution: If we solve Equation (21-1) for t, then we have

$$t = \frac{t_{1/2}}{0.301} \log \frac{N_0}{N}$$

The half-life of the conversion of uranium-238 to lead-206 is 4.51×10^9 years. In order to use this equation, we must now determine N_0/N. The initial amount of uranium-238 present in the ore is 5.20 mg plus whatever amount has been transformed into lead. The amount of uranium that has been transformed into lead is given by

$$\begin{pmatrix}\text{mg } ^{238}_{92}\text{U transformed} \\ \text{into } ^{206}_{82}\text{Pb}\end{pmatrix} =$$

$$(1.85\text{ mg } ^{206}_{82}\text{Pb})\left(\frac{1\text{ mol } ^{206}_{82}\text{Pb}}{206\text{ g } ^{206}_{82}\text{Pb}}\right)\left(\frac{1\text{ mol } ^{238}_{92}\text{U}}{1\text{ mol } ^{206}_{82}\text{Pb}}\right)\left(\frac{238\text{ g } ^{238}_{92}\text{U}}{1\text{ mol } ^{238}_{92}\text{U}}\right)$$

$$= 2.14\text{ mg } ^{238}_{92}\text{U}$$

Therefore, the initial amount of uranium-238 is

$$\text{initial mg } ^{238}_{92}\text{U} = 2.14\text{ mg} + 5.20\text{ mg} = 7.34\text{ mg}$$

As in Example 21-5, we can use the fact that

$$\frac{N_0}{N} = \frac{\text{initial mass of radioactive isotope}}{\text{present mass of radioactive isotope}} = \frac{7.34\text{ mg}}{5.20\text{ mg}} = 1.41$$

Using the equation we derived from Equation (21-1), we get

$$t = \left(\frac{4.51 \times 10^9 \text{ years}}{0.301}\right)(\log 1.41) = 2.24 \times 10^9 \text{ years}$$

The rock is over 2 billion years old. The oldest rocks analyzed by this method are 3.6 billion years old, which is about 1 billion years less than the present estimate of the age of the earth. Rocks that have been obtained from the moon in the Apollo program indicate that the moon is about the same age as the earth.

21-5. CARBON-14 CAN BE USED TO DATE CERTAIN ARCHAEOLOGICAL OBJECTS

▪ Willard Libby was a professor of chemistry at UCLA.

The radiodating methods that we have described for determining the age of rocks and ores are not useful for ages less than about a million years. The half-lives of the reactions are more than 1 billion years, and thus not enough decay occurs in less than 1 million years to be measured accurately. In the 1940s, Willard Libby developed a method of using carbon-14 to date carbon-containing objects derived from formerly living materials. The method is useful for dating objects that are less than about 30,000 years old. **Carbon-14 dating** has found wide use in archaeology, and Libby received the 1960 Nobel Prize in chemistry for his work.

The idea behind carbon-14 dating, also called **radiocarbon dating,** is as follows. The earth's upper atmosphere is being bombarded constantly by radiation from the sun and other parts of the universe. As a result, a small but fairly constant amount of carbon-14 is produced in the reaction between cosmic-ray neutrons and atmospheric nitrogen:

$${}^{14}_{7}\text{N} + {}^{1}_{0}\text{n} \rightarrow {}^{14}_{6}\text{C} + {}^{1}_{1}\text{H}$$

Carbon-14 is radioactive and decays by the reaction

$${}^{14}_{6}\text{C} \rightarrow {}^{14}_{7}\text{N} + {}^{0}_{-1}\text{e}$$

The half-life of carbon-14 is 5730 years. The carbon-14 occurs in $^{14}CO_2$. The $^{14}CO_2$ diffuses throughout the earth's atmosphere, and so a small but fairly constant fraction of atmospheric CO_2 contains carbon-14.

Living plants absorb CO_2 to build carbohydrates through photosynthesis, and so the carbon-14 is incorporated into plants and into the food chain of animals. As a result, all living plants and animals contain the same fraction of carbon-14 atoms as in atmospheric CO_2. The radiation due to the carbon-14 in all living organisms is 15.3 disintegrations per minute per gram of total carbon. When the organism dies, it no longer incorporates carbon-14, and so the quantity of carbon-14 decreases with a half-life of 5730 years.

The equation that we use to determine the age of an object is the one we used in Example 21-6:

$$t = \frac{t_{1/2}}{0.301} \log \frac{N_0}{N} \qquad (21\text{-}2)$$

The half-life of carbon-14 is 5730 years, and if we substitute this value into Equation (21-2), then we have

$$t = (1.90 \times 10^4 \text{ years}) \log \frac{N_0}{N}$$

We can convert the ratio N_0/N to a more convenient form by realizing that the 15.3 disintegrations per minute per gram of carbon is proportional to N_0, the initial number of carbon-14 nuclei per gram of carbon. We can write this as

$$N_0 \propto 15.3$$

Similarly, if R is the present disintegration rate per gram of carbon, then

$$N \propto R$$

The ratio N_0/N then is

$$\frac{N_0}{N} = \frac{15.3}{R} \qquad (21\text{-}3)$$

If we substitute Equation (21-3) into Equation (21-2), then we get

$$t = (1.90 \times 10^4 \text{ years}) \log \frac{15.3}{R} \qquad (21\text{-}4)$$

If we assume that the atmospheric level of carbon-14 is the same now as when the artifact was living matter, then the age of the object can be determined by measuring R. The following Example illustrates the use of Equation (21-4).

▪ The assumptions of carbon-14 dating have been tested extensively against other archeological dating techniques.

Example 21-7: Stonehenge is an ancient megalithic site in southern England. Some archaeologists believe that it was designed to make astronomical observations, but its actual purpose still is controversial. Charcoal samples taken from a series of holes at Stonehenge have a disintegration rate of 9.65 disintegrations per minute per gram of carbon. Calculate the age of the charcoal sample.

Solution: Using Equation (21-4), we have

$$t = (1.90 \times 10^4 \text{ years}) \log \frac{15.3}{9.65}$$

$$= (1.90 \times 10^4 \text{ years})(0.200)$$

$$= 3.80 \times 10^3 \text{ years}$$

Thus we estimate that the charcoal pits at Stonehenge are about 3800 years old (about 1800 B.C.). This result is in agreement with evidence based upon other archaeological data.

Radiocarbon dating has been used to date many archaeological objects. Figure 21-4 shows Equation (21-4) plotted as log R versus t, indicating a few of the many dates that have been determined by the carbon-14 method.

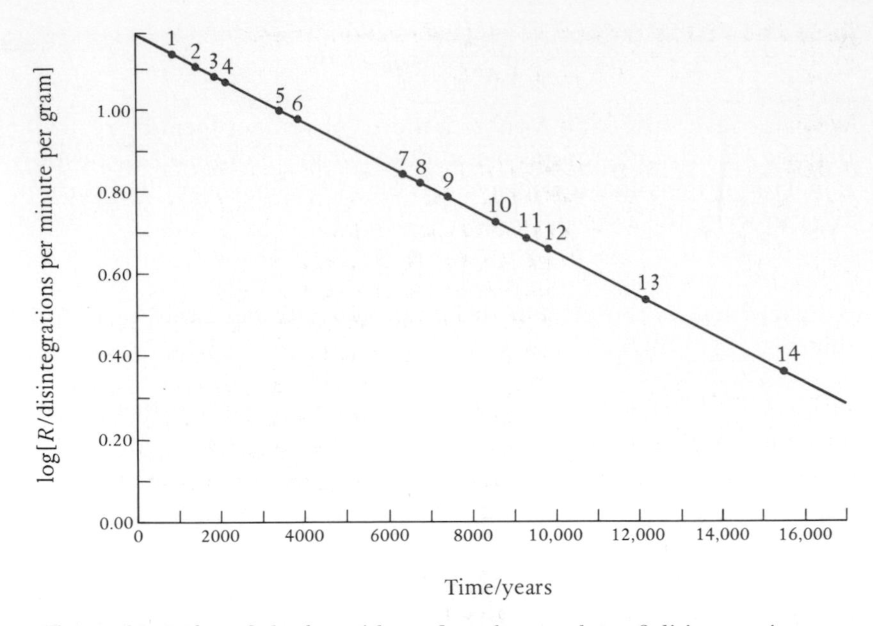

Figure 21-4 Plot of the logarithm of R, the number of disintegrations per minute per gram of carbon, versus time for the carbon-14 dating method. As Equation (21-4) indicates, this plot is a straight line. The numbers on the curve designate archaeological objects that have been dated by the carbon-14 method: (1) Charcoal from earliest Polynesian culture in Hawaii (946 ± 180 years). (2) Wooden lintels from a Mayan site in Tikal, Guatemala (1503 ± 110 years). (3) Linen wrappings from the Dead Sea Scrolls (1917 ± 200 years). (4) Wood from a coffin from the Egyptian Ptolemaic period (2190 ± 450 years). (5) Samples of oak from an ancient cooking place at Killeens, County Cork (3506 ± 230 years). (6) Charcoal sample from Stonehenge (3798 ± 275 years). (7) Charcoal from a tree destroyed by the explosion of Mount Mazama, the explosion that formed Crater Lake in Oregon (6453 ± 250 years). (8) Land-snail shells found at Jarmo, Iraq (6707 ± 320 years). (9) Charcoal from an archaeological site near Beer-Sheba, Israel (7420 ± 520 years). (10) Burned animal bones found near a site inhabited by humans in Palli Aike Cave in southern Chile (8639 ± 450 years). (11) Woven rope sandals found in Fork Rock Cave, Oregon (9053 ± 350 years). (12) Buried bison bone from Folsom Man site near Lubbock, Texas (9883 ± 350 years). (13) Glacial wood found near Skunk River, Iowa (12,200 ± 500 years). (14) Charcoal from the Lascaux cave in France, which contains many cave paintings (15,516 ± 900 years).

21-6. RADIOISOTOPES CAN BE PRODUCED IN THE LABORATORY

In 1919, the New Zealand scientist Ernest Rutherford was the first to carry out a nuclear reaction in the laboratory. He bombarded nitrogen with a beam of α-particles and was able to detect the reaction

$$^{14}_{7}\text{N} + ^{4}_{2}\text{He} \rightarrow ^{17}_{8}\text{O} + ^{1}_{1}\text{H}$$

This was the first laboratory synthesis of a nucleus, $^{17}_{8}\text{O}$. Since then, hundreds of different nuclear reactions have been achieved. The products of many of these reactions are radioactive isotopes that are not found in nature and so are called **artificial radioisotopes.** If these isotopes ever did exist in nature, they have long since disappeared because their half-lives are so much shorter than the age of the earth. For

Table 21-5 Some radioactive isotopes used in medicine

Isotope	Half-life		Use
	Biological*	Physical	
potassium-43	58 days	22.4 h	myocardial imaging
cobalt-58	9.5 days	71.4 days	Schilling test for pernicious anemia
gallium-67	4.8 days	78 h	imaging for Hodgkin's disease, lymphomas, malignant tumors
indium-111	10 h	2.8 days	spinal and cranial fluid imaging
indium-113	10 h	1.73 h	liver, lung, brain, blood pool scan
iodine-123	138 days	13.3 h	thyroid function, thyroid imaging
mercury-203	14.5 days	46.9 days	kidney scans

*Biological half-life is the time required for one half of an administered sample to be eliminated by the organism.

example, the element francium, which is the Group 1 metal with the largest atomic number, is not found in nature. Twenty-seven isotopes of francium have been made; the longest half-life among these isotopes is less than 20 min.

Many of the radioisotopes produced have applications in medicine, agriculture, insect control, oil exploration, and many other fields. For example, iodine-131, which has a half-life of 8 days, is used to measure the activity of the thyroid gland. A few of the many other radioisotopes that are used in medicine are sodium-24 (to follow blood circulation), technetium-99 (for brain, liver, and spleen scans) and phosphorus-32 (for treatment of chronic leukemia). Table 21-5 lists some other radioisotopes that are used in medicine.

Before the development of nuclear science, uranium lay at the end of the periodic table. Since the 1940s, elements with $Z = 93$ to 107 and 109 have been produced. Many of these have been produced at the Lawrence Radiation Laboratory in Berkeley, California, as the names

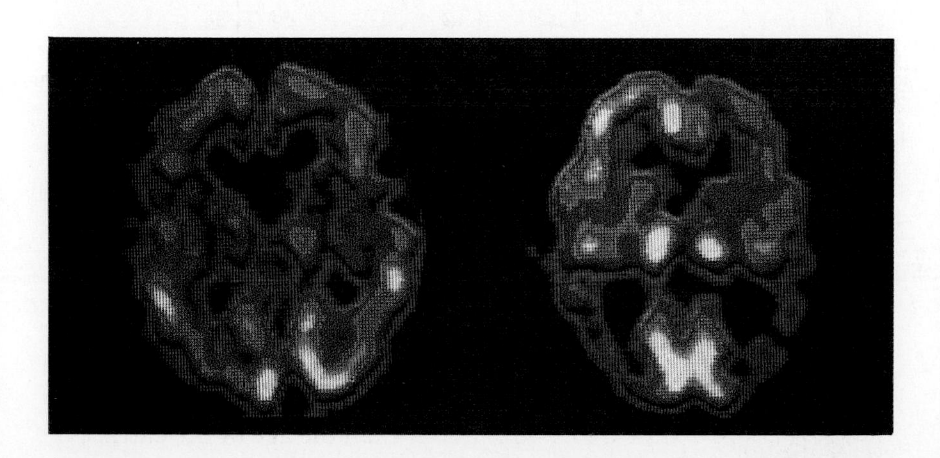

A molecule labeled with the radioisotope iodine-123 is used to study cerebral blood flow. The γ-ray photons emitted by iodine-123 are detected by probes placed around the patient's head. In the photos shown, a normal brain *(left)* and the brain of a patient with Alzheimer's disease are compared.

Figure 21-5 The elements beyond uranium in the periodic table are called transuranium elements. With the exception of recently discovered traces of plutonium, these elements do not occur in nature but have been created in the laboratory. Shown here is the first visible sample of americium produced, having a mass of only a few micrograms.

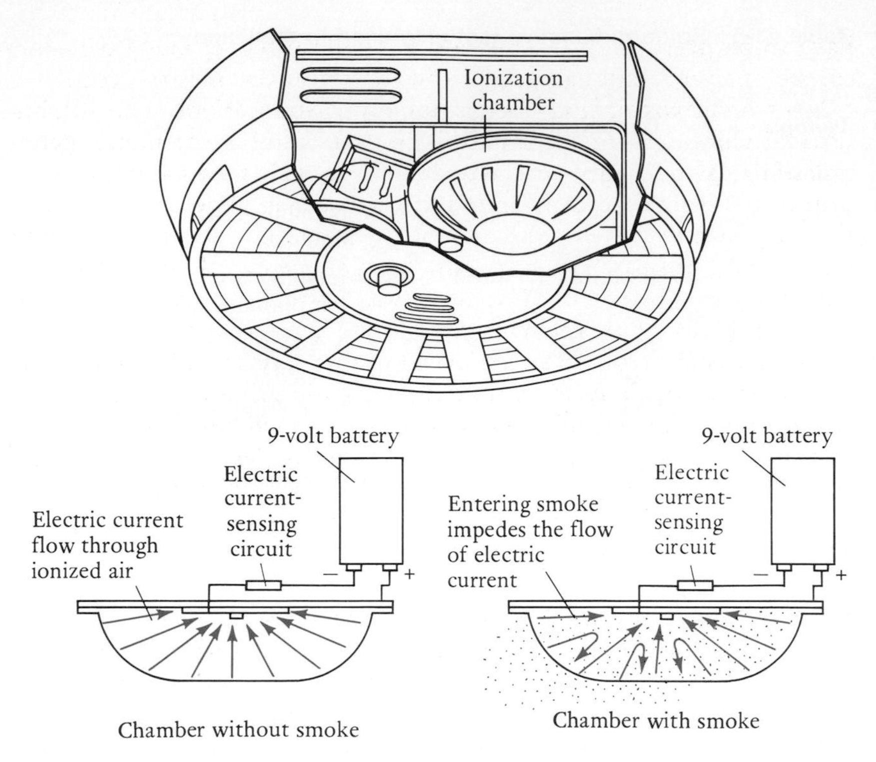

Figure 21-6 Diagram of a home smoke detector. A small quantity of americium-241 ($t_{1/2}$ = 432 years) ionizes the air in the ionization chamber. An electric voltage is applied across the ionization chamber, and the ions in the ionized air conduct an electric current, which is constantly monitored electronically. When smoke particles pass through the ionization chamber, they impede the flow of electricity, which is detected by electronic circuitry and signals an alarm.

americium (95), berkelium (97), and californium (98) indicate. Many of these **transuranium elements** can be made in commercial quantities. For example, americium-241 is used in many home smoke detectors (Figure 21-6).

21-7. NUCLEAR CHEMISTRY CAN BE USED TO DETECT EXTREMELY SMALL QUANTITIES OF THE ELEMENTS

One of the most important uses of radioisotopes in chemistry is in **neutron activation analysis,** which is one of the most sensitive analytical methods. It is capable of detecting as little as 10^{-12} g of some elements and is particularly useful for measuring trace quantities. In a neutron activation analysis, a sample is irradiated by a beam of neutrons. Various nuclei in the sample undergo nuclear reactions by absorbing neutrons. The product nuclei are usually radioactive and emit γ-rays. The energies of the γ-rays emitted by a radioisotope are characteristic of that radioisotope. Because each radioisotope emits γ-rays of only certain, well-defined energies, the frequencies of the γ-rays emitted by a sample after neutron irradiation can be used to identify the elements present. The advantages of neutron activation analysis over more conventional analytical techniques are that the sample does not have to be pretreated, the method is nondestructive, many elements can be analyzed simultaneously, and the sensitivity is very great.

Neutron activation analysis has numerous applications. The authenticity of paintings can be established by determining the mineral content of the paint used. In early times, each school of artists prepared its own paints from distinctive and individual recipes. The paint used in a painting in question is compared with the paint used in a painting known to be done by the artist. Another application of neutron activation analysis showed that Napoleon was poisoned by arsenic. When arsenic is ingested, it is concentrated in the hair. Analysis of several of Napoleon's hairs showed an abnormally large concentration of arsenic.

The results of neutron activation analysis in criminal investigations are at times dramatic. The basic idea is to match the distribution of elements in soil, paint, cosmetics, and so on found at the scene of a crime with those found with a criminal suspect. Neutron activation analysis of a wiping taken from a suspect's hand can reveal not only whether the person has fired a gun recently but also the type of ammunition used.

Another analytical method that involves nuclear techniques is **PIXE** (pronounced pixie), short for **particle-induced X-ray emission.** In the PIXE method, a sample is irradiated with a beam of protons, which strike the atoms in the sample and eject inner-shell electrons, such as K- and L-shell electrons. As electrons from the other shells fill these empty innermost shells, the atoms emit characteristic X-rays. The energies of the X-rays can be measured accurately and used to identify the atomic composition of the sample. A PIXE analysis can be carried out with less than 1 μg of the sample, an amount too small to be seen by the naked eye. One of the advantages of PIXE is that it requires only very small samples. It does not have the great sensitivity that neutron activation analysis has, however. PIXE and neutron activation analysis are just two of a number of analytical methods that use nuclear chemistry.

21-8. ENORMOUS AMOUNTS OF ENERGY ACCOMPANY NUCLEAR REACTIONS

Consider the nuclear reaction

$$\begin{array}{lcccc} & {}^{6}_{3}\text{Li} \;+ & {}^{1}_{1}\text{H} \;\rightarrow & {}^{3}_{2}\text{He} \;+ & {}^{4}_{2}\text{He} \\ \text{atomic mass:} & 6.01512 & 1.00782 & 3.01603 & 4.00260 \end{array}$$

The total mass of the reactants is 7.02294 amu, and the total mass of the products is 7.01863 amu. The mass is not the same on the two sides of the equation. The difference is

$$\begin{aligned} \Delta m &= 3.01603 + 4.00260 - (6.01512 + 1.00782) \\ &= -0.00431 \text{ amu} \end{aligned}$$

This number is far larger than the uncertainties of the masses of the reactants and products. Since $1 \text{ amu} = 1.66 \times 10^{-27}$ kg, -0.00431 amu is equal to

$$\begin{aligned} -0.00431 \text{ amu} &= (-0.00431 \text{ amu})(1.66 \times 10^{-27} \text{ kg}\cdot\text{amu}^{-1}) \\ &= -7.15 \times 10^{-30} \text{ kg} \end{aligned}$$

This amount of mass is lost when just one $^{6}_{3}Li$ nucleus reacts with one $^{1}_{1}H$ nucleus. If we calculate the mass lost on a per mole basis, then we get

$$\begin{aligned} -0.00431 \text{ amu·atom}^{-1} \\ &= (-7.15 \times 10^{-30} \text{ kg·atom}^{-1})(6.02 \times 10^{23} \text{ atom·mol}^{-1}) \\ &= -4.30 \times 10^{-6} \text{ kg·mol}^{-1} \end{aligned}$$

Up to this point, we have always stated that mass is conserved in reactions; yet in this nuclear reaction there is a mass discrepancy that lies far outside experimental error. The explanation is that the missing mass has been converted to energy. The relation between mass and energy is given by Einstein's famous formula,

■ The equation $E = mc^2$ comes from Einstein's theory of relativity.

$$E = mc^2 \tag{21-5}$$

where m is the mass lost and c is the speed of light. Because we are referring to energy changes and mass changes, we write Equation (21-5) in the form

$$\Delta E = c^2 \Delta m \tag{21-6}$$

Using the fact that the speed of light is 3.00×10^8 m·s^{-1}, we can write Equation (21-6) in the form

$$\Delta E = (9.00 \times 10^{16} \text{ m}^2\text{·s}^{-2})\Delta m \tag{21-7}$$

The mass lost in the reaction described by the equation

$$^{6}_{3}Li + ^{1}_{1}H \rightarrow ^{3}_{2}He + ^{4}_{2}He$$

is -4.30×10^{-6} kg·mol^{-1}. Therefore

■ $1 J = 1 kg·m^2·s^{-2}$

$$\begin{aligned} \Delta E &= c^2 \Delta m \\ &= (9.00 \times 10^{16} \text{ m}^2\text{·s}^{-2})(-4.30 \times 10^{-6} \text{ kg·mol}^{-1}) \\ &= -3.87 \times 10^{11} \text{ J·mol}^{-1} \end{aligned}$$

where we have used the fact that 1 joule is 1 kg·m^2·s^{-2} (Section 4-9). This value of ΔE is typical for nuclear reactions. Values of ΔE for ordinary chemical reactions are about 10^5 J·mol^{-1}, and so we see that nuclear reactions involve energies over 1 million times greater than conventional chemical reactions. This is why enormous amounts of energy are produced by nuclear explosions and nuclear reactors; a small amount of mass is converted to a large amount of energy.

Example 21-8: The value of ΔE for the reaction

$$H_2(g) + \tfrac{1}{2}O_2(g) \rightarrow H_2O(l)$$

is -282.1 kJ·mol^{-1} at 298 K and 1 atm. Calculate the loss in mass when 1 mol of liquid water is formed from hydrogen and oxygen at 298 K.

Solution: The negative value of ΔE means that energy is evolved and therefore that mass is lost. The mass corresponding to -282.1 kJ·mol^{-1} is

■ The value of ΔE for a chemical reaction is about 10^{-6} as large as the value of ΔE for a nuclear reaction.

$$\begin{aligned} \Delta m &= \frac{\Delta E}{c^2} = \frac{-282.1 \times 10^3 \text{ J·mol}^{-1}}{9.00 \times 10^{16} \text{ m}^2\text{·s}^{-2}} \\ &= -3.13 \times 10^{-12} \text{ kg·mol}^{-1} \end{aligned}$$

This is the mass lost when 1 mol of liquid water is formed. It is a very small mass and not directly measurable. The total energy evolved upon the formation of each molecule of water is equivalent to about 1 millionth of the mass of an electron. Mass changes in ordinary chemical reactions are negligible.

The energy required to break up a nucleus into its constituent protons and neutrons is called the **binding energy.** Let's calculate the binding energy of ${}^4_2He$. The reaction that we are considering is

$$\begin{array}{lccc} & {}^4_2He & \rightarrow 2\,{}^1_1H & + \; 2\,{}^1_0n \\ \text{mass/amu:} & 4.0026 & 1.0078 & 1.0087 \end{array}$$

The mass difference is

$$\begin{aligned}\Delta m &= 2(1.0078\text{ amu}) + 2(1.0087\text{ amu}) - 4.0026\text{ amu} \\ &= 4.0330\text{ amu} - 4.0026\text{ amu} \\ &= 0.0304\text{ amu}\end{aligned}$$

This value of Δm corresponds to an energy of

$$\begin{aligned}\Delta E &= c^2\Delta m \\ &= (9.00 \times 10^{16}\text{ m}^2\cdot\text{s}^{-2})(0.0304\text{ amu})(1.66 \times 10^{-27}\text{ kg}\cdot\text{amu}^{-1}) \\ &= 4.54 \times 10^{-12}\text{ J}\end{aligned}$$

Therefore, the binding energy of ${}^4_2He$ is 4.54×10^{-12} J.

Example 21-9: Given that the mass of ${}^{55}_{25}Mn$ is 54.9380 amu, calculate the binding energy and the binding energy per nucleon in ${}^{55}_{25}Mn$.

Solution: The list of physical constants on the inside back cover gives the following masses:

$$\begin{array}{ll} {}^1_1H & 1.0078\text{ amu} \\ {}^1_0n & 1.0087\text{ amu} \end{array}$$

The mass difference between ${}^{55}_{25}Mn$ and its constituent particles is

$$\begin{aligned}\Delta m &= (25 \times 1.0078\text{ amu}) + (30 \times 1.0087\text{ amu}) - 54.9380\text{ amu} \\ &= 55.4560\text{ amu} - 54.9380\text{ amu} \\ &= 0.518\text{ amu}\end{aligned}$$

which corresponds to an energy of

$$\begin{aligned}\Delta E = c^2\Delta m &= (9.00 \times 10^{16}\text{ m}^2\cdot\text{s}^{-2})(0.518\text{ amu})(1.66 \times 10^{-27}\text{ kg}\cdot\text{amu}^{-1}) \\ &= 7.74 \times 10^{-11}\text{ J}\end{aligned}$$

This is the binding energy of ${}^{55}_{25}Mn$. There are 55 nucleons in ${}^{55}_{25}Mn$, and so the binding energy per nucleon is

$$\begin{aligned}\text{binding energy per nucleon} &= \frac{7.74 \times 10^{-11}\text{ J}}{55\text{ nucleons}} \\ &= 1.41 \times 10^{-12}\text{ J}\cdot\text{nucleon}^{-1}\end{aligned}$$

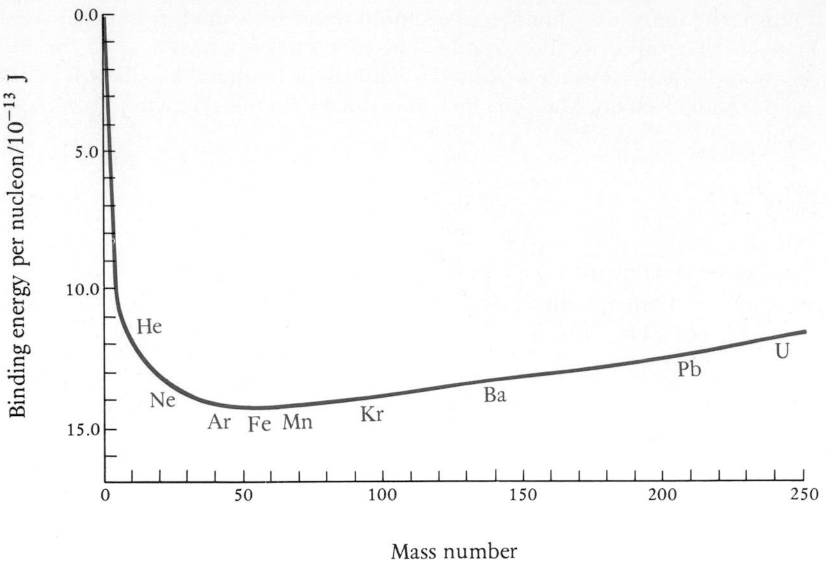

Figure 21-7 Binding energy per nucleon plotted versus mass number of a nucleus. This plot is called a binding energy curve.

Figure 21-7 is a plot of the binding energy per nucleon versus the number of nucleons (mass number). We use this **binding energy curve** in the next section.

21-9. SOME NUCLEI FRAGMENT IN NUCLEAR REACTIONS

Because neutrons are uncharged, they are relatively difficult to detect and so were not discovered until 1932. As soon as the neutron was discovered, nuclear scientists realized that it would be ideal to use to bombard nuclei. Being uncharged, neutrons can burrow into a nucleus more readily than can protons or α-particles. In the mid 1930s, neutrons were used to bombard uranium in an effort to produce elements beyond uranium in the periodic table. For example, the first element beyond uranium, neptunium (the planet Neptune lies beyond Uranus in the solar system), is produced by the reaction

$$^{238}_{92}\text{U} + ^{1}_{0}\text{n} \rightarrow ^{239}_{93}\text{Np} + ^{0}_{-1}\text{e}$$

Neptunium-239 has a half-life of about 2 days and decays to plutonium-239 (the planet Pluto lies beyond Neptune) by β-decay:

$$^{239}_{93}\text{Np} \rightarrow ^{239}_{94}\text{Pu} + ^{0}_{-1}\text{e}$$

Plutonium-239 has a half-life of about 24,000 years.

During the search for transuranium elements in the 1930s, it was discovered that when uranium-235 absorbs a neutron, it breaks into two fragments of roughly the same size (Figure 21-8). A typical reaction is

$$^{235}_{92}\text{U} + ^{1}_{0}\text{n} \rightarrow ^{92}_{36}\text{Kr} + ^{141}_{56}\text{Ba} + 3^{1}_{0}\text{n}$$

A nuclear reaction in which a nucleus splits into two smaller fragments is called **fission.** By referring to the binding energy curve in Figure 21-7, we see that barium and krypton are more stable than uranium.

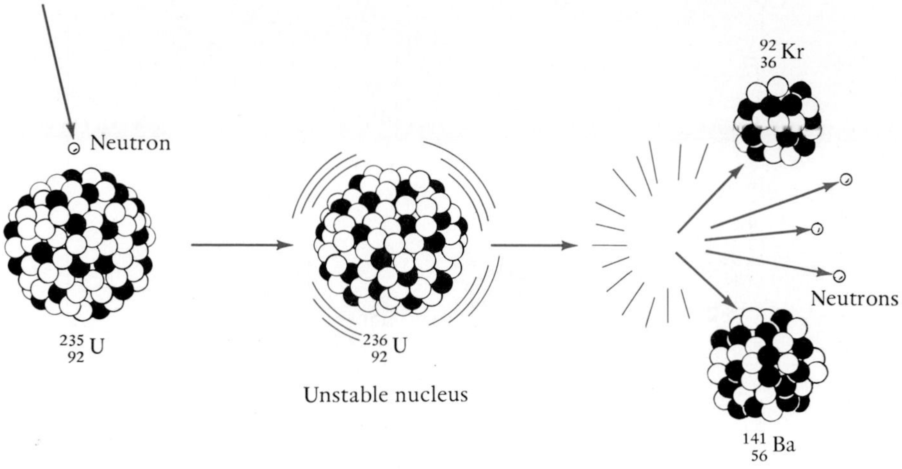

Figure 21-8 A fission reaction occurs when a uranium-235 nucleus absorbs a neutron. The fission products shown here are krypton-92 and barium-141, but in practice a distribution of fission products is obtained. The most important feature of the uranium-235 fission reaction is the simultaneous production of several neutrons.

Consequently, energy is released in the fission of uranium-235. The amount of energy released per mole of uranium is 2×10^{13} J. This is an enormous amount of energy, about 1 million times greater than that for an ordinary chemical reaction.

21-10. THE FISSION OF URANIUM-235 CAN INITIATE A CHAIN REACTION

Nuclear reactions often release great amounts of energy per mole, but usually so few nuclei are involved that the total energy output is small. The fission reaction of uranium-235 is special, however. Note that three neutrons are produced in the fission reaction given in the previous section. Each of these neutrons can cause another uranium-235 nucleus to undergo fission, producing nine neutrons. This process can continue, producing what is called a **chain reaction** (Figure 21-9).

There are several conditions that must be met in order for a chain reaction to occur. The level of impurities that absorb neutrons and do not produce any other neutrons must be kept very low. In addition, the quantity of fissionable material must be large enough to make the rate at which neutrons are lost through the surface of the material less than the rate at which neutrons are produced. The smallest mass for which more neutrons are produced than lost through the surface of the material is called the **critical mass.** A quantity of uranium-235 that is less than the critical mass will not support a chain reaction; a quantity that is greater than the critical mass will support a chain reaction. The critical mass of uranium-235 is just a few kilograms, about the size of a softball. Under the right conditions, a neutron chain reaction can occur extremely rapidly; it can be over in less than a microsecond. The energy is released as an enormous explosion, known to all of us.

The nuclear research that led to these discoveries was being carried out when the world was in turmoil and on the brink of World War II. A multinational effort went into using a neutron chain reaction to develop the atomic bomb. The theory behind the atomic bomb is very

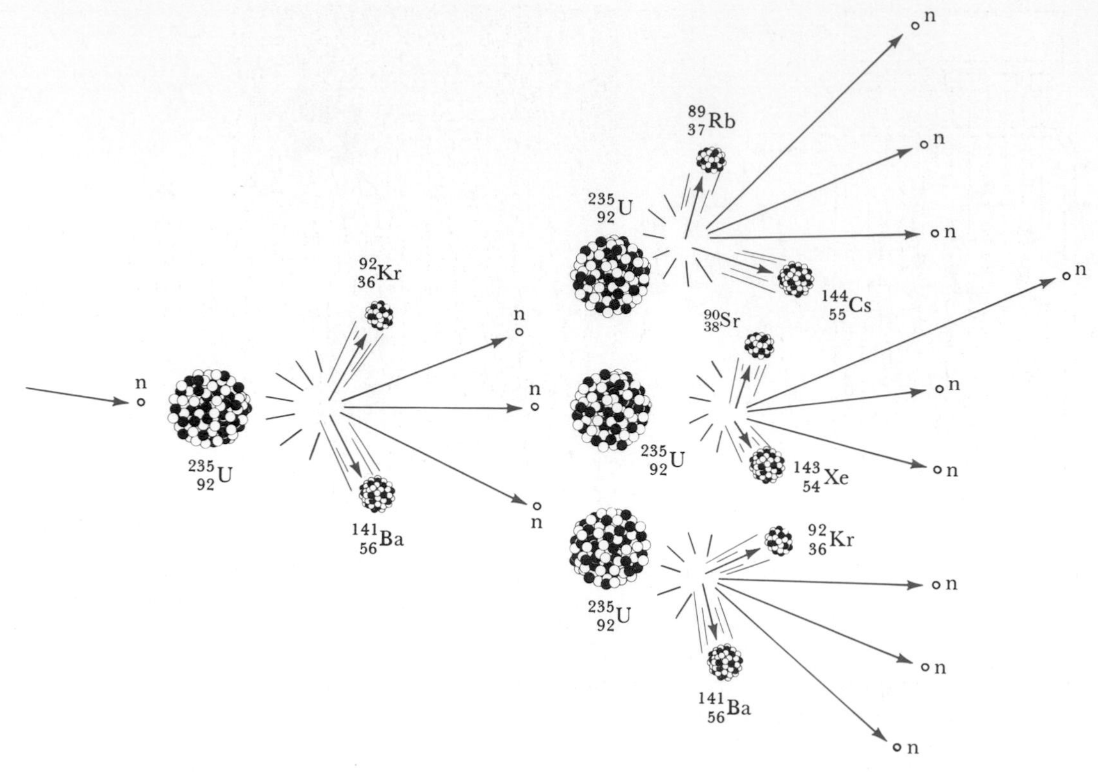

Figure 21-9 The chain reaction produced when uranium-235 undergoes fission after absorbing a neutron. The key point is that several (here, three) neutrons are produced in each reaction. Consequently, after the first step, three other nuclei absorb a neutron and undergo fission. These three nuclei produce 9 neutrons, which lead to the production of 27 neutrons in the third step, and so on. The number of fission reactions increases very rapidly with the number of steps, and the result can be an explosive release of energy.

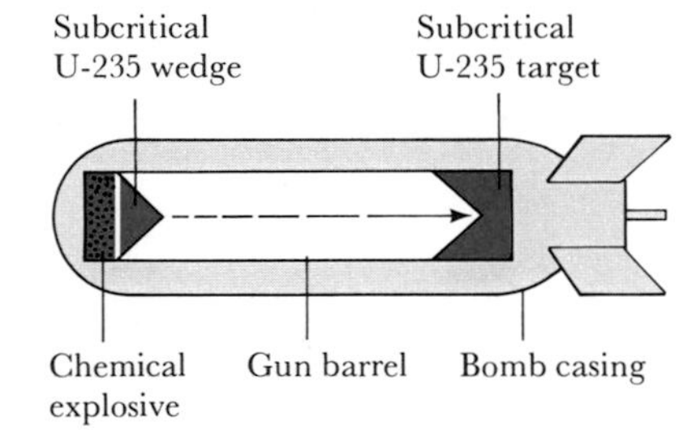

Figure 21-10 Simplified schematic of an atomic bomb. Two subcritical masses of a fissionable isotope, such as uranium-235, are located at the ends of the cylinder. A conventional chemical explosive charge is located behind one of the masses. When the charge is detonated, the left-hand subcritical mass is rapidly propelled toward the right-hand one. When the two collide, the total mass is supercritical and can sustain a neutron chain reaction and release the energy as an explosion.

simple. Two **subcritical masses** of uranium-235 are brought together very rapidly, producing a **supercritical mass,** which then explodes. A simplified diagram of an atomic bomb is shown in Figure 21-10. The actual production of the first atomic bomb was a monumental technological task.

2-11. A NUCLEAR REACTOR UTILIZES A CONTROLLED CHAIN REACTION

In a **nuclear reactor,** it is possible to tame a neutron chain reaction and produce the energy in a controlled manner. One way to do this is to insert a material that strongly absorbs neutrons (Figure 21-11) into the uranium core, which consists of a bank of **fuel rods** charged with uranium-235–enriched U_2O_3. Cadmium and boron both are strong absorbers of neutrons. The density of neutrons can be controlled by changing the height of the cadmium or boron rods, called **control rods.** In this manner, the uranium-235 chain reaction can be maintained at a steady, controlled rate, producing energy in the form of

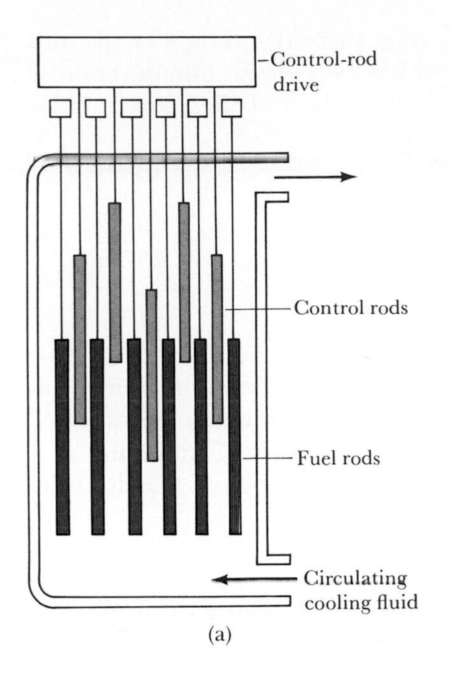

Figure 21-11 Diagram of the core of a nuclear reactor. (a) The fuel rods contain the fissionable material. The control rods consist of a material that is a good absorber of neutrons. By raising and lowering the control rods, the density of neutrons in the core, and thus the rate of production of energy, can be controlled. (b) Loading of fuel rods into a nuclear reactor core. The fuel rods are the long, thin, shiny metal tubes, which are filled with fissionable material.

heat. The heat produced is used to generate steam, which can run a turbine and produce electricity. A diagram of a nuclear reactor is shown in Figure 21-12.

One of several serious problems associated with the generation of electricity by nuclear reactors is the disposal of fission products and spent fuel. The high neutron densities generated in reactor cores produce large quantities of radioisotopes, which must be dealt with when

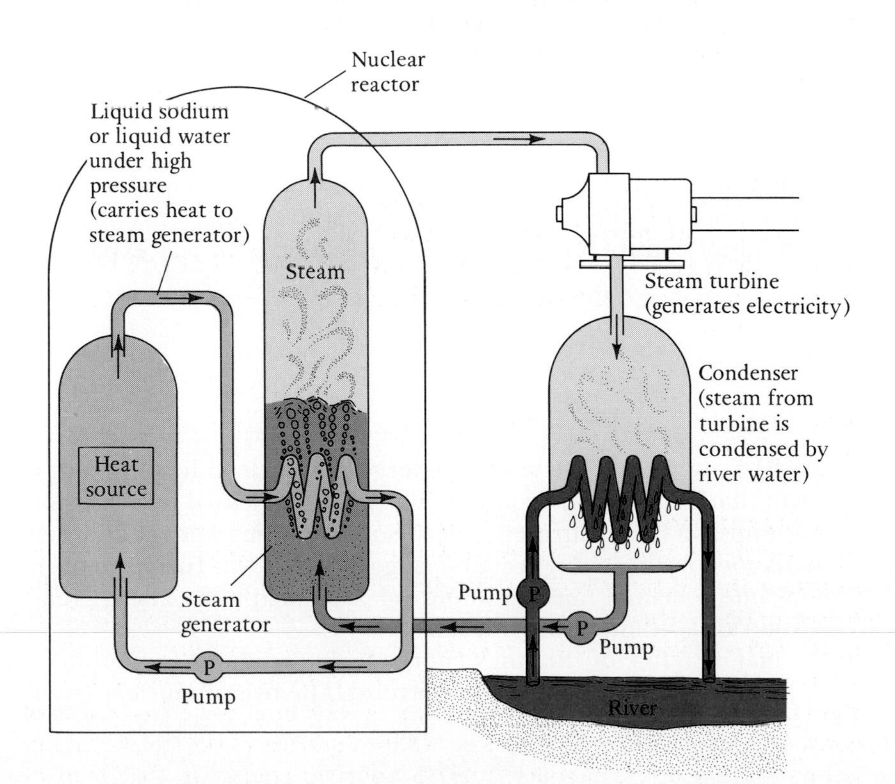

Figure 21-12 Diagram of a nuclear reactor. The heat source is the core of the reactor (Figure 21-11). The heat produced by the core is transferred by a closed loop of liquid sodium or liquid water to a steam generator. The steam produced runs a steam turbine, which produces electricity. The steam from the turbine is cooled by water from a nearby source, such as a river, or through a forced-flow cooling tower, and is pumped back into the steam generator.

the uranium fuel rods need replenishing. Just what to do with the tons of highly radioactive waste produced has become an intense political issue. At the present time, the spent fuel rods are being stored under water in vaults, usually at the reactor site.

21-12. A BREEDER REACTOR IS DESIGNED TO PRODUCE MORE FISSIONABLE MATERIAL THAN IT CONSUMES

It has been projected that the world's supply of uranium-235 will be exhausted sometime in the twenty-first century. Uranium-235 constitutes less than 1 percent of naturally occurring uranium, most of which (99.3 percent) is nonfissionable uranium-238. Some uranium-238 is converted to fissionable plutonium-239 by the series of reactions described by the equations

$$^{238}_{92}U + ^{1}_{0}n \rightarrow ^{239}_{92}U$$

$$^{239}_{92}U \rightarrow ^{239}_{93}Np + ^{0}_{-1}e \qquad t_{1/2} = 24 \text{ min}$$

$$^{239}_{93}Np \rightarrow ^{239}_{94}Pu + ^{0}_{-1}e \qquad t_{1/2} = 2.35 \text{ days}$$

The result of this process is plutonium-239, which has a half-life of 24,000 years. The high neutron flux in nuclear reactors provides the neutrons needed to initiate the $^{238}_{92}U \rightarrow ^{239}_{94}Pu$ process. If this process can be made to occur to such an extent that more than one plutonium-239 nucleus is produced for every uranium-235 nucleus that undergoes fission, then more fissionable material will be produced than consumed. A reactor that is designed to accomplish this task is called a **breeder reactor.** It would take about 20 breeders running for 1 year to make enough fuel to power one additional breeder for 1 year. Breeder reactor technology is inherently a very large-scale operation.

The design of breeder reactors has been fraught with technical and political difficulties. The units must be run at higher temperatures than those in conventional reactors, and controlling them is much more difficult. Furthermore, not only do they have the possibility of a core meltdown accident, but also, in contrast to ordinary fission reactors, an out-of-control breeder reactor can lead to a nuclear explosion.

21-13. FUSION REACTIONS RELEASE MORE ENERGY THAN FISSION REACTIONS

Figure 21-7 shows that the binding energy per nucleon for uranium is less than that of its fission products. The energy released in the fission of uranium-235 is the difference between its binding energy and that of its fission products. Figure 21-7 suggests that the **fusion** of four protons into a helium nucleus would release much more energy than the fission of a uranium-235 nucleus.

▪ The sun is powered by nuclear fusion.

The major source of the sun's energy is the fusion of four protons into a helium-4 nucleus and two positrons. The overall nuclear equation is

$$4\ ^{1}_{1}H \rightarrow ^{4}_{2}He + 2\ ^{0}_{+1}e$$

The energy released in this process is enormous. The sun produces about 10^{26} J·s^{-1}.

Fusion is the basis of the hydrogen bomb, in which lithium-6 deuteride, $^6Li^2H$, is surrounded by an ordinary fission bomb. Upon detonation, the fission bomb produces the temperatures of millions of degrees and the high compression that are necessary to initiate fusion reactions. The principal fusion reactions that occur are

$$^2_1H + ^2_1H \rightarrow ^3_2He + ^1_0n$$

$$^6_3Li + ^1_0n \rightarrow ^4_2He + ^3_1H$$

$$^2_1H + ^3_1H \rightarrow ^4_2He + ^1_0n$$

$$^2_1H + ^2_1H \rightarrow ^3_1H + ^1_1H$$

The quantity of energy released by 1 ton of $^6Li^2H$ is equivalent to that released by 60 million tons of TNT.

The control of fusion reactions for the generation of electricity is one of the most exciting, important, and difficult technological problems of our time. Excluding the use of the radioactive isotope tritium, the best possibilities seem to be the reactions

$$^2_1H + ^2_1H \rightarrow ^4_2He \qquad \text{and} \qquad ^1_1H + ^2_1H \rightarrow ^3_2He$$

Since hydrogen and deuterium are plentiful, the successful development of a fusion reactor would totally change the world we live in by allowing all countries to have an essentially inexhaustible supply of energy. The technological problems that bar the way are staggering, however. In addition to being heated to millions of degrees, the hydrogen and deuterium must be contained long enough for the reaction to occur. Of course, there is no structural material that can withstand such temperatures, but present research is aimed at confining the nuclei by using very strong magnetic and electric fields. The enormous temperatures required also can be achieved with high-power laser beams. Controlled fusion is the subject of intensive research. Some progress appears to have been made, but an operational fusion reactor is not yet close to realization.

21-14. EXPOSURE TO RADIATION DAMAGES CELLS, TISSUES, AND GENES

Example 21-4 notes that γ-radiation from cobalt-60 can be used to destroy cancerous cells. Actually, γ-radiation destroys healthy cells as well, but it is more destructive of cancerous cells because they grow and divide more rapidly than normal cells and so are more vulnerable. The effect of various kinds of radiation on living systems has received much study and has resulted in governments' placing limits on exposure to radiation.

One measure of the activity of a radioactive substance is its **specific activity,** which is the number of nuclei that disintegrate per second per gram of radioactive isotope. The specific activity of a radioisotope is related to its half-life, $t_{1/2}$, and atomic mass M by the equation

$$\text{specific activity} = \left(\frac{4.2 \times 10^{23}\ \text{disintegrations}\cdot\text{g}^{-1}}{Mt_{1/2}}\right) \qquad (21\text{-}8)$$

▪ Equation (21-8) is obtained as follows: the number of disintegrations per second is the rate of the decay process

$$\text{rate} = kN$$

The specific activity is the rate divided by the mass m.

$$\text{specific activity} = \frac{\text{rate}}{m} = \frac{kN}{m} = \frac{0.693}{t_{1/2}}\frac{N}{m} = \frac{0.693}{t_{1/2}}\frac{N_0}{M}$$

where N_0 is Avogadro's number.

Because specific activity is defined as the number of disintegrations per second per gram, the value of $t_{1/2}$ in Equation (21-8) should be in seconds. As an example, let's compute the specific activity of radium-226, which has a half-life of 1600 years. We first convert the half-life to seconds:

$$(1600\text{ yr})(365\text{ day}\cdot\text{yr}^{-1})(24\text{ h}\cdot\text{day}^{-1})(60\text{ min}\cdot\text{h}^{-1})(60\text{ s}\cdot\text{min}^{-1}) = 5.05 \times 10^{10}\text{ s}$$

The specific activity of radium-226 is thus

$$\text{specific activity} = \frac{4.2 \times 10^{23}\text{ disintegrations}\cdot\text{g}^{-1}}{(226)(5.05 \times 10^{10}\text{ s})} = 3.7 \times 10^{10}\text{ disintegrations}\cdot\text{s}^{-1}\cdot\text{g}^{-1}$$

The quantity 3.7×10^{10} disintegrations$\cdot$s^{-1} is called a **curie,** Ci, after Marie Curie, one of the pioneers in research in radioactivity (Figure 21-13). She received a Nobel Prize in physics in 1903 and one in chemistry in 1911. She is one of only four scientists to earn two Nobel prizes, and she is the only scientist to earn a Nobel prize in two different sciences.

Table 21-6 lists the specific activities of a number of important radioactive isotopes.

Figure 21-13 Marie Curie with her daughter, Irene. Both were pioneers in nuclear chemical research. Marie and her husband, Pierre, discovered radium and several other radioactive elements in the early 1900s. Irene and her husband, Frédéric Joliot, made the first artificial radioisotope, which has led to numerous applications in chemistry, biology, and medicine. Marie shared the 1903 Nobel Prize in physics with her husband and Henri Becquerel, and she received the Nobel Prize in chemistry singly in 1911. Irene Joliot-Curie shared the 1935 Nobel Prize in chemistry with her husband.

Example 21-10: In reading about the use of uranium-238 in radiodating rocks, you may have wondered how it is possible to determine a half-life as long as 4.47×10^9 years. The answer lies in our ability to count individual radioactive decay events, coupled with the fact that Avogadro's number is enormous. There are 6.02×10^{23} uranium-238 nuclei in a sample that contains 238 g of uranium-238. Compute the number of uranium-238 nuclei that disintegrate in 10 s in a sample that contains 2.0 mg of uranium-238.

Solution: Using Equation (21-8), we find that the specific activity of uranium-238, which has a half-life of 4.47×10^9 years, is

▪ The SI unit of the rate of radioactive decay is a becquerel (Bq), which is equal to one disintegration per second.

Table 21-6 Specific activities of some important radioactive isotopes

Isotope	Half-life	Specific activity/ disintegrations$\cdot$s$^{-1}\cdot$g^{-1}	Specific activity/ Ci$\cdot$g^{-1}
radium-226	1600 years	3.7×10^{10}	1.00
uranium-238	4.47×10^9 years	1.3×10^4	3.5×10^{-7}
plutonium-239	2.41×10^4 years	2.3×10^9	6.2×10^{-2}
iodine-131	8.04 days	4.6×10^{15}	1.2×10^5
cobalt-60	5.27 years	4.2×10^{13}	1.1×10^3
strontium-90	29 years	5.1×10^{12}	1.4×10^2
cesium-137	30.2 years	3.2×10^{12}	87
carbon-14	5730 years	1.7×10^{11}	4.6

$$\text{specific activity} = \frac{4.2 \times 10^{23}\ \text{disintegrations}\cdot\text{g}^{-1}}{(238)(4.47 \times 10^{9}\ \text{years})(3.15 \times 10^{7}\ \text{s}\cdot\text{year}^{-1})}$$

$$= 1.3 \times 10^{4}\ \text{disintegrations}\cdot\text{s}^{-1}\cdot\text{g}^{-1}$$

Thus the number of disintegrations in 2.0 mg in 10 s is

$$(1.3 \times 10^{4}\ \text{disintegrations}\cdot\text{s}^{-1}\cdot\text{g}^{-1})\ (2.0 \times 10^{-3}\ \text{g})(10\ \text{s}) = 260\ \text{disintegrations}$$

Thus, about 260 uranium-238 nuclei disintegrate in a 2.0-mg sample in only 10 s. It is easy with a modern radiocounting apparatus to measure such a large number of events in a 10-s interval. The half-life of uranium-238 can be measured by counting the number of disintegrations in a known mass of the radioisotope and then carrying out the reverse of the calculation just given with $t_{1/2}$ as the unknown, which is calculated from the measured activity level.

The damage that is produced by radiation depends on more than just the specific activity. As the radiation passes through tissue, it ionizes molecules and breaks chemical bonds, leaving behind a trail of molecular damage. The extent of the damage produced depends upon the energy and type of radiation. The different types of radiation vary in their ability to penetrate matter. Alpha particles can be stopped by a sheet of paper or by the skin. Beta particles penetrate deeper than α-particles, but β-particles of moderate energy are stopped by about 1 cm of water. Gamma rays are the most penetrating form of radiation, requiring walls of lead bricks to stop them.

SUMMARY

Many nuclei are radioactive, meaning that they spontaneously emit subatomic particles such as α-particles, β-particles, and positrons. The rate of decay of a radioactive substance is a first-order kinetic process. An important property of a first-order kinetic process is that its rate is characterized by a half-life, which is the time required for one half of the amount of a sample to react. The half-lives of radioisotopes range from less than 1 picosecond to billions of years. Rates of radioactive decay are used to determine the age of rocks and archaeological objects.

Many radioisotopes that are not found in nature can be produced by nuclear reactions in the laboratory. Such artificial radioisotopes have found wide application in chemistry, biology, forensic science, and medicine. All the elements beyond uranium in the periodic table have been artificially produced, allowing an extension of the table to atomic number 109.

The energies associated with nuclear reactions are about 1 million times greater than the energies associated with chemical reactions. Nuclear energy can be released in a controlled manner, as in nuclear reactors, or in an uncontrolled manner, as in atomic weapons. The future of nuclear energy as a commercial source of energy is clouded by uncertainty about the effect of radiation on living matter. It is well known that exposure to radiation causes damage to biological systems, and safe and acceptable levels of exposure (if any) have yet to be determined unequivocally.

TERMS YOU SHOULD KNOW

nucleon 671
nuclear equation 671
radioactivity 671
radioactive 671
radioisotope 671
α-particle 671
α-emitter 671
β-particle 672
γ-ray 672

EQUATIONS YOU SHOULD KNOW HOW TO USE

$$\log \frac{N_0}{N} = \frac{0.301t}{t_{1/2}}$$ (21-1) (number of radioactive nuclei that remain after time t)

$$t = (1.90 \times 10^4 \text{ years}) \log \frac{15.3}{R}$$ (21-4) (Carbon-14 dating)

$$\Delta E = c^2 \Delta m$$ (21-6) (relation between energy and mass)

$$\text{specific activity} = \left(\frac{4.2 \times 10^{23} \text{ disintegrations}\cdot\text{g}^{-1}}{M t_{1/2}} \right)$$ (21-8) (specific activity of a radioactive sample)

PROBLEMS

NUCLEAR EQUATIONS

21-1. Fill in the missing symbols in the following nuclear equations:

(a) ${}^{72}_{30}\text{Zn} \rightarrow {}^{0}_{-1}\text{e} + ?$
(b) ${}^{230}_{92}\text{U} \rightarrow {}^{4}_{2}\text{He} + ?$
(c) ${}^{136}_{57}\text{La} \rightarrow {}^{0}_{+1}\text{e} + ?$
(d) ${}^{14}_{7}\text{N} + ? \rightarrow {}^{1}_{1}\text{H} + {}^{14}_{6}\text{C}$

21-2. Fill in the missing symbols in the following nuclear equations:

(a) ${}^{238}_{92}\text{U} + {}^{16}_{8}\text{O} \rightarrow ? + 5\,{}^{1}_{0}\text{n}$
(b) ${}^{253}_{99}\text{Es} + {}^{4}_{2}\text{He} \rightarrow ? + {}^{1}_{0}\text{n}$
(c) ${}^{237}_{94}\text{Pu} \rightarrow {}^{96}_{40}\text{Zr} + ? + 3\,{}^{1}_{0}\text{n}$
(d) ${}^{12}_{6}\text{C} + {}^{12}_{6}\text{C} \rightarrow ? + {}^{1}_{1}\text{H}$

21-3. Equations for nuclear reactions are often written in a condensed form, for example, ${}^{14}_{7}\text{N}(\alpha, \text{p})$. This notation means that when ${}^{14}_{7}\text{N}$ is bombarded by an α-particle, a proton is produced. The complete equation is ${}^{14}_{7}\text{N} + {}^{4}_{2}\text{He} \rightarrow {}^{1}_{1}\text{H} + {}^{17}_{8}\text{O}$, where ${}^{17}_{8}\text{O}$ is deduced from the rules for balancing nuclear equations. Write out complete equations for the following processes (d stands for a deuteron, ${}^{2}_{1}\text{H}$):

(a) ${}^{25}_{12}\text{Mg}(\alpha, \text{p})$
(b) ${}^{27}_{13}\text{Al}(\text{n}, \alpha)$
(c) ${}^{17}_{8}\text{O}(\text{p}, \alpha)$
(d) ${}^{63}_{29}\text{Cu}(\text{p}, \text{n})$

21-4. As explained in Problem 21-3, equations for nuclear reactions are often written in a condensed notation. Write out complete nuclear equations for the following processes:

(a) ${}^{16}_{8}\text{O}(\text{p}, \gamma)$
(b) ${}^{59}_{27}\text{Co}(\text{d}, \text{p})$
(c) ${}^{22}_{11}\text{Na}(\text{p}, \alpha)$
(d) ${}^{18}_{8}\text{O}(\alpha, \text{n})$

STABILITY OF NUCLEI

21-5. Calcium-38 is an artificial radioisotope that has been produced in the laboratory. Predict its mode of decay.

21-6. The element technetium has no stable isotopes. Predict the mode of decay of the radioisotope technetium-91.

21-7. Predict the mode of decay of the following nuclei, all of which are produced in nuclear reactors:

(a) rubidium-76
(b) germanium-80
(c) chlorine-32
(d) iron-62

21-8. Predict the mode of decay of the following nuclei:

(a) xenon-142 (b) carbon-11
(c) hydrogen-3 (d) nitrogen-12

21-9. Which of the following nuclei would you predict to be radioactive?

(a) argon-35 (b) magnesium-24
(c) calcium-40 (d) neon-20

21-10. Which of the following nuclei would you predict to be stable?

(a) oxygen-16 (b) potassium-39
(c) boron-14 (d) germanium-72

21-11. Predict whether or not the following nuclei are radioactive:

(a) sodium-24 (b) astatine-215
(c) tin-118 (d) francium-224

21-12. Predict whether or not the following nuclei are radioactive:

(a) chlorine-38 (b) molybdenum-95
(c) tin-120 (d) francium-225

RATES OF NUCLEAR DECAY

21-13. The radioisotope argon-41 is used to measure the rate of the flow of gases from smokestacks. It is a γ-emitter with a half-life of 110 min. Calculate the fraction of an argon-41 sample that remains after one day.

21-14. The radioisotope bromine-82 is used as a tracer for organic materials in environmental studies. Its half-life is 36 h. Calculate the fraction of a sample of bromine-82 that remains after one day.

21-15. Mercury-203 is used to perform kidney scans. If a 0.200-mg sample of $^{203}_{80}Hg(NO_3)_2$ is purchased by a hospital, how much $^{203}_{80}Hg(NO_3)_2$ will remain after 6 months (182 days)? The half-life of mercury-203 is 46.9 days.

21-16. A sample of sodium-24 chloride containing 0.055 mg of sodium-24 is injected into an animal to study sodium balance. How much sodium-24 remains 6.0 h later? The half-life of sodium-24 is 15.0 h.

21-17. The radioisotope iron-59 is used in medicine in the diagnosis of anemia. Iron-59 emits γ-rays, which are detected and counted. Calculate the fraction of an iron-59 sample that will remain after one month (30 days). The half-life of iron-59 is 45 days.

21-18. Cesium-137 is produced in nuclear reactors. If this isotope has a half-life of 30.2 years, how many years will it take for it to decay to one tenth of a percent of its initial amount?

21-19. You order a sample of Na_3PO_4 containing the radioisotope phosphorus-32 ($t_{1/2}$ = 14.3 days). If the shipment is delayed in transit for two weeks, how much of the original activity will remain when you receive the sample?

21-20. Sulfur-31 is a positron emitter with a half-life of 2.54 s. Calculate how long it will take for 99 percent of the sulfur-31 to be converted to phosphorus-31.

21-21. The copper isotope $^{64}_{29}Cu$ ($t_{1/2}$ = 12.8 h) is used both in brain scans for tumors and in studies of Wilson's disease (a genetic disorder characterized by the inability to metabolize copper). Compute the number of days required for an administered dose of $^{64}_{29}Cu$ to drop to 0.10 percent of the initial value injected into a human. Assume no loss of $^{64}_{29}Cu$ except by radioactive decay.

21-22. The strontium isotope $^{90}_{38}Sr$ is a radioactive isotope that is produced in nuclear explosions. It decays by β-emission with a half-life of 29 years. Suppose that an infant ingests $^{90}_{38}Sr$ in mother's milk. Compute the fraction of the ingested $^{90}_{38}Sr$ that remains in the body when the infant reaches 74 years of age, assuming no loss of $^{90}_{38}Sr$ except by radioactive decay.

DATING

21-23. Burned animal bones found near a site inhabited by humans in Palli Aike Cave in southern Chile have a disintegration rate of 5.37 disintegrations$\cdot$min$^{-1}\cdot$g^{-1} of carbon. Estimate the age of the site.

21-24. Wood from a coffin from the Egyptian Ptolemic period has a disintegration rate of 11.7 disintegrations$\cdot$min$^{-1}\cdot$g^{-1} of carbon. Estimate the age of the coffin.

21-25. Charcoal from the Lascaux Cave in France (the cave with the remarkable prehistoric paintings) was found to have a $^{14}_{6}C$ content equal to 14.5 percent of that in living matter. Estimate the age of the charcoal.

21-26. The French explorer Fernand Navarra claims to have discovered Noah's Ark. In 1955 he discovered a log on Mt. Ararat in eastern Turkey, the legendary resting spot of the ark. Navarra claims that the log is a beam from the ark. Samples of the wood have a disintegration rate of 13.19 disintegrations$\cdot$min$^{-1}\cdot$g^{-1} carbon. Calculate the age of the log.

21-27. A sample of $CaCO_3$ from the shell of preserved ancient egg shell has a disintegration rate of 498 disintegrations per hour per gram of carbon. Estimate the age of the shell.

21-28. Samples of oak from an ancient Irish cooking site at Killeens, County Cork, have a carbon-14 content equal to 65 percent that in living matter. Estimate the age of the wood.

21-29. A sample of uranite is found to have a $^{206}_{82}Pb/^{238}_{92}U$ mass ratio of 0.395. Estimate the age of the uranite. The half-life of the conversion of uranium-238 to lead-206 is 4.51×10^9 years.

21-30. A sample of ocean sediment is found to contain 1.50 mg of uranium-238 and 0.460 mg of lead-206. Estimate the age of the sediment. The half-life for the conversion of uranium-238 to lead-206 is 4.51×10^9 years.

BINDING ENERGY

21-31. Calculate the binding energy per nucleon in $^{206}_{82}Pb$ (atomic mass = 205.97446 amu).

21-32. Iron-56 (atomic mass = 55.9346 amu) is the most stable nucleus. Calculate its binding energy and the binding energy per nucleon.

21-33. Given that the atomic mass of $^{35}_{17}Cl$ is 34.9689 amu, calculate the binding energy and the binding energy per nucleon.

21-34. Given that the atomic mass of $^{20}_{10}Ne$ is 19.9924 amu, calculate the binding energy and the binding energy per nucleon.

NUCLEAR REACTIONS AND ENERGY

21-35. Suppose that uranium-235 undergoes the fission reaction described by the equation

$$^{235}_{92}U + ^{1}_{0}n \rightarrow ^{141}_{56}Ba + ^{88}_{36}Kr + 7\,^{1}_{0}n$$

Given the atomic masses $^{235}_{92}U$ = 235.0439 amu, $^{141}_{56}Ba$ = 140.9137 amu, $^{88}_{36}Kr$ = 87.9142 amu, and $^{1}_{0}n$ = 1.0087 amu, calculate the energy released when one uranium-235 nucleus undergoes this reaction. Calculate the mass of uranium-235 that reacts in a 50-kiloton bomb. The designation 50-kiloton means that the bomb has the explosive equivalent of 50 kilotons of TNT. Assume that 2500 kJ is liberated per kilogram of TNT. For simplicity, assume that the tons here are metric tons, that is, 1000 kg.

21-36. Marie Curie observed that 1.0 g of pure radium-226 generates $148\ J\cdot h^{-1}$. Given that 1.0 g of radium-226 has a disintegration rate of 3.7×10^{10} $disintegrations\cdot s^{-1}$, calculate the energy released in each disintegration. Compare this with what you calculate given the equation

$$^{226}_{88}Ra \rightarrow ^{222}_{86}Rn + ^{4}_{2}He$$

and the atomic masses $^{226}_{88}Ra$ = 226.0254 amu, $^{222}_{86}Rn$ = 222.0154 amu, and $^{4}_{2}He$ = 4.0026 amu.

21-37. The half-life of a positron is very short. It reacts with an electron, and the masses of both are converted to two γ-rays (Figure 21-1):

$$^{0}_{+1}e + ^{0}_{-1}e \rightarrow 2\gamma$$

This reaction is called an annihilation reaction. Calculate the energy produced by the reaction between one electron and one positron. Assuming that the two γ-rays have the same frequency, calculate this frequency.

21-38. Calculate the difference in mass between the products and reactants for the equation

$$S(s) + O_2(g) \rightarrow SO_2(g) \qquad \Delta H^\circ_{rxn} = -297\ kJ\cdot mol^{-1}$$

21-39. Calculate the energy released in the fission reaction described by the equation

$$^{235}_{92}U + ^{1}_{0}n \rightarrow ^{134}_{51}Sb + ^{95}_{39}Y + ^{4}_{2}He + 3\,^{1}_{0}n$$

given the atomic masses $^{235}_{92}U$ = 235.0439 amu, $^{134}_{51}Sb$ = 133.8969 amu, $^{95}_{39}Y$ = 94.9125 amu, and $^{4}_{2}He$ = 4.0026 amu.

21-40. The fusion of 1.0 g of atomic hydrogen ($^{1}_{1}H$) to $^{4}_{2}He$ releases 6.4×10^{11} J of energy. Compute the number of metric tons of water that could supply the total annual U.S. energy requirement of 85×10^{15} kJ if all the hydrogen were converted to $^{4}_{2}He$. The percentage of $^{1}_{1}H$ atoms in ordinary hydrogen is 99.985 percent.

21-41. The world's energy consumption is estimated to be $3 \times 10^{17}\ kJ\cdot year^{-1}$. How much uranium-235 would be required to produce all this energy by fission? The amount of energy released by the fission of uranium-235 is $2 \times 10^{13}\ J\cdot mol^{-1}$. The world supply of uranium is estimated to be about 10^6 metric tons. About 0.7 percent of naturally occurring uranium is uranium-235. At a world energy consumption rate of $3 \times 10^{17}\ kJ\cdot year^{-1}$, how long will the supply of uranium last?

21-42. A nuclear power plant producing 1000 megawatts (1 watt = $1\ J\cdot s^{-1}$) of electricity produces 3000 megawatts of heat. Given that the available thermal energy per $^{235}_{92}U$ disintegration is 2.9×10^{-14} kJ, compute the number of moles of $^{235}_{92}U$ burned per month when the nuclear reactor operates continuously at maximum power.

SPECIFIC ACTIVITY

21-43. Fluorine-18 has a half-life of 110 min. Calculate the specific activity of fluorine-18 in units of curies per gram.

21-44. The radioisotope sulfur-35 is used extensively in biological and environmental studies. It is prepared in a cyclotron and is free of all other isotopes of sulfur. Its half-life is 87 days. Calculate the specific activity (in curies per gram) of a freshly prepared sample of sulfur-35.

21-45. The radioisotope gallium-67, which has a half-life of 78 h, is used as a diagnostic tool in tumor location. How many milligrams of $^{67}_{31}GaCl_3$ is equivalent to 350 mCi?

21-46. The radioisotope iodine-128 (half-life 25 min) is used as a diagnostic tool for thyroid imaging. A typical dose is 100 μCi. Compute the number of milligrams of $Na^{128}I$ equivalent to 100 μCi.

21-47. A sample of uranium-234 is observed to have an activity of $6.5 \times 10^4\ disintegrations\cdot min^{-1}$. What is the activity of the sample in μCi? Given that the half-life of uranium-234 is 2.44×10^5 years, how many micrograms of uranium-234 are present in the sample?

21-48. A 6.0-μCi sample of iodine-131 is observed using a Geiger counter with an efficiency of 40 percent. What disintegration rate will be measured?

ADDITIONAL PROBLEMS

21-49. Fill in the missing symbols in the following sequence of nuclear equations:

(a) ${}^{12}_{6}C + ? \rightarrow {}^{13}_{7}N + \gamma$
(b) $? \rightarrow {}^{13}_{6}C + {}^{0}_{+1}e$
(c) ${}^{13}_{6}C + {}^{1}_{1}H \rightarrow ? + \gamma$
(d) ${}^{14}_{7}N + ? \rightarrow {}^{15}_{8}O + \gamma$
(e) ${}^{15}_{8}O \rightarrow {}^{15}_{7}N + ?$
(f) ${}^{15}_{7}N + {}^{1}_{1}H \rightarrow {}^{12}_{6}C + ?$

This sequence of reactions is believed to power stars. What is the net reaction?

21-50. The radioisotope sulfur-38 can be incorporated into proteins to follow certain aspects of protein metabolism. If a protein sample initially has an activity of 10,000 disintegrations·min^{-1}, then what is the activity 6.0 h later? The half-life of sulfur-38 is 2.84 h.

21-51. The radioisotope hydrogen-3 is used in fusion reactors. It is a β-emitter with a half-life of 12.3 years. Calculate the fraction of a hydrogen-3 sample that will remain after 50.0 years.

21-52. The radioisotope phosphorus-32 can be incorporated into nucleic acids to follow certain aspects of their metabolism. If a nucleic acid sample initially has an activity of 40,000 disintegrations·min^{-1}, then what is the activity 220 h later? The half-life of phosphorus-32 is 14.28 days.

21-53. A sample of radioactive NaI (iodine-128 is the radioisotope here) is injected into a patient as part of radioiodine treatment of a thyroid condition. If the sample has an activity of 10,000 disintegrations·min^{-1} at 8 A.M., the time of injection, then what is the activity at 2 P.M. the same day? The half-life of iodine-128 is 25.0 min.

21-54. The body of an 80-kg human contains about 15 kg of carbon in various chemical compounds. Compute the number of ${}^{14}_{6}C$ disintegrations per second in the human body.

21-55. Calculate the energy released per mole of helium by the process

$$2\,{}^{2}_{1}H \rightarrow {}^{4}_{2}He$$

The atomic mass of ${}^{2}_{1}H$ is 2.0141 amu, and the atomic mass of ${}^{4}_{2}He$ is 4.0026 amu. How many kilograms of octane would have to be burned to produce as much energy as is produced by the formation of 1.0 g of helium in this reaction? For the combustion of octane, C_8H_{18}, $\Delta H^\circ_{rxn} = -5450\ \text{kJ·mol}^{-1}$.

21-56. Calculate the energy released per mole of lithium by the process

$${}^{7}_{3}Li + {}^{1}_{1}H \rightarrow 2\,{}^{4}_{2}He$$

The required atomic masses are ${}^{1}_{1}H = 1.0078$ amu, ${}^{7}_{3}Li = 7.0160$ amu, and ${}^{4}_{2}He = 4.0026$ amu. How many kilograms of octane would have to be burned to produce as much energy as is produced when 1.0 g of lithium-7 is consumed in this reaction? For the combustion of octane, C_8H_{18}, $\Delta H^\circ_{rxn} = -5450\ \text{kJ·mol}^{-1}$.

21-57. In one type of nuclear reactor, the uranium fuel is about 3 percent uranium-235. Only about a third of this uranium-235 can be utilized before released fission products eventually poison the fuel. If each fission reaction releases 2.9×10^{-14} kJ, calculate the energy released by the "burning" of 1 kg of the uranium fuel. Assuming that 30 percent of this energy is available for use, calculate the quantity of fuel required to operate a 1000-megawatt electric power plant at full power for 1 year. One watt is equal to one joule per second.

21-58. The only radioisotope of aluminum that is useful for tracer studies is aluminum-26, which has a half-life of 7.4×10^5 years. Calculate the specific activity (expressed in curies per gram) of a sample that contains only aluminum-26.

21-59. Radioactive decay is a first-order kinetic process, and so rate $= R = kN$. Substitute this expression into Equation (21-1) to obtain

$$\log\left(\frac{R_0}{R}\right) = \frac{0.301t}{t_{1/2}}$$

where R_0 is the initial decay rate and R is the decay rate at a later time t. The following data were obtained for chromium-51, one of the waste products in nuclear reactors:

Rate of decay/min^{-1}	Time/day
6000	0.0
4680	10.0
3650	20.0
2840	30.0
2200	40.0
1725	50.0

Calculate the half-life of chromium-51.

21-60. No stable isotope of the halogen astatine exists in nature. The radioisotope astatine-211 was made by bombarding bismuth-209 with α-particles. Plot the following data

Fraction of ^{211}At remaining	Time/h
0.909	1.0
0.825	2.0
0.681	4.0
0.464	8.0
0.215	16.0

and determine the half-life of astatine-211.

21-61. An organic acid reacts with an alcohol to produce an ester and water. For example

$$CH_3COOH(aq) + HOCH_3(aq) \rightarrow CH_3COOCH_3(aq) + H_2O(l)$$

In studies on the mechanism of this reaction, methanol, CH_3OH, was labeled with oxygen-18. All the oxygen-18 appeared in only the ester. What is the source of H_2O in the reaction? If acetic acid were labeled with oxygen-18, in which product or products would the oxygen-18 appear?

21-62. The atomic mass of calcium-40 is 39.9626 and the atomic mass of argon-36 is 35.9675. Do you think that the reaction described by the equation

$$^{40}_{20}Ca \rightarrow ^{36}_{18}Ar + ^{4}_{2}He$$

will occur? (The change in energy is much larger than the change in entropy for a nuclear reaction.)

21-63. An important thermonuclear reaction is

$$^{2}_{1}H + ^{3}_{1}H \rightarrow ^{4}_{2}He + ^{1}_{0}n$$

Given the atomic masses 2.0141, 3.0161, and 4.0026 for $^{2}_{1}H$, $^{3}_{1}H$, and $^{4}_{2}He$, respectively, calculate the energy released in this reaction.

21-64. Radium-226 (atomic mass = 226.025406) decays by α-emission to radon-222 (atomic mass = 222.017574). Calculate the energy released in the process.

21-65. In one theory of the synthesis of elements, the net reaction described by the equation

$$3\,^{4}_{2}He \rightarrow ^{12}_{6}C$$

plays a key role. Calculate the energy released in this process.

21-66. Beryllium-8 (atomic mass = 8.005305 amu) decays in 10^{-16} s into two α-particles. How much energy is released in this process?

21-67. Given that one out of 6700 hydrogen atoms is a deuterium atom, calculate the number of deuterium atoms in 1.00 kg of water. How much energy could be liberated by the fusion of all the deuterium into helium? The atomic mass of deuterium is 2.01410 amu.

21-68. The annual energy consumption in the United States is about 1×10^{17} kJ. What mass of water could supply this much energy utilizing the fusion of deuterium to helium? (See Problem 21-67.)

21-69. One of the most difficult problems involved in the utilization of nuclear fusion as an energy source is the requirement for very high temperatures in the reactant sample. Temperatures of the order of 10^8 to 10^{10} K are required. Why do you think such high temperatures are required for fusion?

21-70. A 10.0-g sample of SO_2 containing the radioisotope sulfur-35 has an activity of 3.23×10^{11} disintegrations$\cdot s^{-1}$. Given that sulfur-35 has a half-life of 87.2 days, calculate the fraction of radioactive sulfur atoms in the SO_2 sample.

21-71. A 1.00-mL sample of blood is withdrawn from an animal, and the red blood cells are labeled with phosphorus-32 ($t_{1/2}$ = 14.3 days). The activity of this sample is 50,000 disintegrations$\cdot min^{-1}$. The sample is then reinjected into the animal. A few hours later, another 1.00-mL sample is withdrawn, and its activity is 10.0 disintegrations$\cdot min^{-1}$. Determine the volume of blood in the animal. Assume that the phosphorus-32 is uniformly ditributed throughout the blood and that the activity due to phosphorus-32 remains constant during the experiment. For dogs, an approximately linear relationship has been established between blood volume and body weight. By using similar methods, it has been found that the human body contains about 75 mL of blood per kilogram of body weight.

21-72. The amount of oxygen dissolved in a sample of water can be measured by using the radioisotope thallium-204. Solid thallium reacts with oxygen according to the equation

$$4Tl(s) + O_2(aq) + 2H_2O(l) \rightarrow 4Tl^+(aq) + 4OH^-(aq)$$

The amount of oxygen can be determined by measuring the radioactivity due to thallium-204 in the water sample. In one experiment 10.0 mL of water is reacted with some thallium metal whose activity is 1.13×10^8 disintegrations$\cdot min^{-1}\cdot mol^{-1}$. The radioactivity of water is 563 disintegrations$\cdot min^{-1}$. Calculate the concentration of oxygen in the sample.

21-73. Low concentrations of sulfate ion in aqueous samples can be measured by precipitating $SO_4^{2-}(aq)$ with the radioisotope barium-131 in the form of $Ba^{2+}(aq)$. The radioactivity of the precipitate, $BaSO_4(s)$, is then measured. In one experiment, sulfate ion was precipitated with barium-131, whose specific activity is 7.6×10^7 disintegrations$\cdot min^{-1}\cdot g^{-1}$. The radioactivity of $BaSO_4(s)$ precipitated from a 10.0-mL sample of the aqueous solution was 3270 disintegrations$\cdot min^{-1}$. Calculate the concentration of SO_4^{2-} in the aqueous solution.

21-74. A 50.0-mL sample of a 0.075 M $Pb(NO_3)_2(aq)$ solution is mixed with 50.0 mL of a 0.15 M $NaI(aq)$ solution that has been labeled with the radioisotope iodine-131. The activity of the radioactive NaI solution is 20,000 disintegrations$\cdot$min$^{-1}\cdot$mL^{-1}. After precipitation, the mixture is filtered, and the supernatant liquid has an activity of 320 disintegrations$\cdot$min$^{-1}\cdot$mL^{-1}. Calculate the solubility-product constant of $PbI_2(s)$.

21-75. A 75-mL sample of a 0.010 M $Pb(NO_3)_2(aq)$ solution is mixed with 75 mL of a 0.010 M $Na_2SO_4(aq)$ solution that has been labeled with the radioisotope sulfur-35. The activity of the Na_2SO_4 solution is 14,000 disintegrations$\cdot$min$^{-1}\cdot$mL^{-1}. After precipitation, the mixture is filtered, and the supernatant liquid has an activity of 183 disintegrations$\cdot$min$^{-1}\cdot$mL^{-1}. Calculate the solubility-product constant of $PbSO_4(s)$.

The Transition Metals

Copper metal is purified by electrorefining. A current is passed between a pure copper metal cathode and an impure copper metal anode that are immersed in a solution containing Cu(II). Copper metal is oxidized to Cu(II) at the anode and Cu(II) is reduced to pure copper metal at the cathode. Note the refined copper electrodes that have been removed from a vat in the center of the photo.

In this interchapter we describe the sources, properties, and uses of the **transition metals.** Transition-metal alloys are the structural backbone of modern civilization. Human development progressed from the Stone Age, to the Bronze Age, and then to the Iron Age. The Industrial Revolution was powered by steam engines made from steels. The Space and Computer Age utilizes a truly remarkable variety of exotic alloys developed to meet a wide range of specialized requirements.

▪ Alloys are solid solutions of two or more elements. The major component of an alloy is always a metal.

Many of the transition metals are probably familiar to you. Iron, nickel, chromium, tungsten, and titanium are widely used in alloys for structural materials and play a key role in the world's technology. The precious metals—gold, platinum, and silver—are used as hard cur-

Figure N-1 Titanium has a relatively low density, high strength, excellent corrosion resistance and is easily machined.

rency, and to make jewelry and electrical components. Copper is the most widely used metal for electric wiring.

The transition metals vary greatly in abundance. Iron and titanium are the fourth and tenth most abundant elements in the earth's crust, whereas rhenium (Re) and hafnium (Hf) are unfamiliar even to many chemists because they are so rare. The characteristics of the transition metals vary from family to family, and yet they are all characterized by high densities and high melting points. The two metals with the greatest densities (iridium, Ir, 22.65 $g{\cdot}cm^{-3}$, and osmium, Os, 22.61 $g{\cdot}cm^{-3}$) and the metal with the highest melting point (tungsten, W, 3410°C) are transition metals.

The physical properties of the transition metals vary greatly. Iron, the most common and most important transition metal, is discussed in Section N-2. The second most common transition metal is titanium, which constitutes 0.6 percent of the earth's crust by mass. Pure titanium is a lustrous, white metal (Figure N-1). It is used to make lightweight alloys that are stable at high temperatures for use in missiles and high-performance aircraft. Titanium is as strong as most steels but 50 percent lighter. It is 60 percent heavier than aluminum but twice as strong. In addition, it has excellent resistance to corrosion.

Gold is a very dense, soft, yellow metal with a high luster (Figure N-2). It is found in nature as the free element and in tellurides. It occurs in veins and alluvial deposits and is often separated from rocks and other minerals by sluicing or panning. In many mining operations, about 5 g of gold is recovered from 1 ton of rock. Gold is very unreactive and has a remarkable resistance to corrosion. Pure gold is soft and often alloyed to make it harder. The amount of gold in an alloy is expressed in **karats.** Pure gold is 24 karat. Coinage gold is 22 karat, or $(22/24) \times 100 = 92$ percent. White gold, which is used in jewelry, is usually an alloy of gold and nickel. In addition to its use in jewelry and as a world monetary standard, gold is an excellent conductor of electricity and is used in microelectronic devices. It is also used extensively

Figure N-2 Gold bars from Echo Bay Mines in Edmonton, Alberta.

Table N-1 Properties of the 3*d* transition-series metals

Element	Density/$g \cdot cm^{-3}$	Melting point/°C	Principal sources	Uses
Sc	3.0	1541	thortveitite, $(Sc, Y)_2Si_2O_7$	no major industrial uses
Ti	4.5	1660	rutile, TiO_2	high-temperature, lightweight steel alloys, TiO_2 in white paints
V	6.0	1890	vanadinite, $(PbO)_9(V_2O_5)_3PbCl_2$	vanadium steels (rust-resistant)
Cr	7.2	1860	chromite, $FeCr_2O_4$	stainless steels, chrome plating
Mn	7.4	1244	pyrolusite, MnO_2 manganosite, MnO nodules on ocean floor	alloys
Fe	7.9	1535	hematite, Fe_2O_3 magnetite, Fe_3O_4	steels
Co	8.9	1490	cobaltite, $CoS_2 \cdot CoAs_2$ linnaetite, Co_3S_4	alloys, cobalt-60 medicine
Ni	8.9	1455	pentlandite, $(Fe, Ni)_9S_8$ pyrrhotite, $F_{0.8}S$	nickel plating, coins, magnets, catalysts
Cu	9.0	1083	chalcopyrite, $CuFeS_2$ cuprite, Cu_2O malachite, $Cu_2(CO_3)(OH)_2$	bronzes, brass, coins, electric conductors
Zn	7.1	420	zinc blende, ZnS smithsonite, $ZnCO_3$	galvanizing, bronze, brass, dry cells

in dentistry and medicine. The physical properties, principal sources, and major uses of the 3*d* transition-series metals are given in Table N-1.

N-1. THE MAXIMUM OXIDATION STATES OF SCANDIUM THROUGH MANGANESE ARE EQUAL TO THE TOTAL NUMBER OF 4*s* AND 3*d* ELECTRONS

The chemistry of even the first transition-metal series is especially rich owing to the several oxidation states available to many of the metals (Table N-2). In spite of the differences in the chemistries of the transition metals, certain trends do exist:

Table N-2 The common oxidation states of the 3*d* transition metals

Element	Atomic number	Oxidation states	Element	Atomic number	Oxidation states
Sc	21	+3	Fe	26	+2,+3
Ti	22	+3,+4	Co	27	+2,+3
V	23	+2,+3,+4,+5	Ni	28	+2
Cr	24	+2,+3,+6	Cu	29	+1,+2
Mn	25	+2,+4,+6,+7	Zn	30	+2

1. For scandium through manganese, the highest oxidation state is equal to the total number of 4*s* plus 3*d* electrons, and this oxidation state is achieved primarily in oxygen compounds or in fluorides and chlorides. Furthermore, the stability of the highest oxidation state decreases from scandium to manganese. Thus we have

Sc_2O_3	a stable oxide, Sc(III)
TiO_2	a stable, common ore of titanium, Ti(IV)
V_2O_5	mild oxidizing agent, V(V)
CrO_4^{2-}	strong oxidizing agent, Cr(VI)
MnO_4^-	strong oxidizing agent, Mn(VII)

2. Except for scandium and titanium, all the 3*d* transition metals form divalent ions in aqueous solution.

3. For a given metal, the halides become more covalent with increasing oxidation state and hydrolyze more readily.

Let's briefly discuss some of the chemical properties of the first transition-metal series. The ground-state electron configuration of scandium is $[Ar]4s^23d^1$, and it is somewhat similar to aluminum in its chemical properties. Scandium has a +3 oxidation state in almost all its compounds. It forms a very stable oxide, Sc_2O_3, and forms halides with the formula ScX_3. The addition of base to $Sc^{3+}(aq)$ produces a white, gelatinous precipitate with the formula $Sc_2O_3 \cdot nH_2O$. Like $Al(OH)_3$, this hydrated Sc_2O_3 is amphoteric. Scandium and its compounds have little technological importance.

The ground-state electron configuration of titanium is $[Ar]4s^23d^2$. Its most common and stable oxidation state is +4, as in the compounds TiO_2 and $TiCl_4$, which are covalently bonded. Pure titanium is difficult to prepare because the metal is very reactive at high temperatures.

The most important ore of titanium is rutile (Figure N-3), which is primarily TiO_2. Pure titanium metal is produced by first converting TiO_2 to $TiCl_4$, by heating TiO_2 to red heat in the presence of carbon and chlorine. The $TiCl_4$ is reduced to the metal by reacting it with magnesium in an inert atmosphere of argon. Most titanium is used in the production of titanium steels, but TiO_2, which is white when pure, is used as the white pigment in many paints. Titanium tetrachloride is also used to make smoke screens; when it is sprayed into the air it reacts with moisture to produce a dense and persistent white cloud of TiO_2:

$$TiCl_4(g) + 2H_2O(g) \rightarrow TiO_2(s) + 4HCl(g)$$

Titanium dioxide is the only transition-metal compound ranked (45th) in the top 50 industrial chemicals.

The ground-state electron configuration of vanadium is $[Ar]4s^23d^3$. Its maximum oxidation state is +5; the +2, +3, and +4 oxidation states are common, with the +2 state being the least common. Vanadium pentoxide is used as a catalyst in the oxidation of SO_2 to SO_3 in the production of sulfuric acid, as well as several other industrial processes. Except for V_2O_5, the compounds of vanadium have limited commercial importance, but vanadium itself is used in alloy steels, particularly ferrovanadium.

Figure N-3 The most important ore of titanium is rutile, which is principally TiO_2. Deposits of rutile are found in Georgia, Virginia, Australia, Brazil, Italy, and Mexico.

Chromium, with the ground-state electron configuration $[Ar]4s^1 3d^5$, has a maximum oxidation state of +6, although +2 and +3 are common oxidation states. Whereas +4 is the most common oxidation state for titanium and the +5 state of vanadium is only mildly oxidizing, the +6 oxidation state of chromium is strongly oxidizing. For example, the dichromate ion in acidic solution (Figure N-4) is a strong oxidizing agent:

Figure N-4 An acidic aqueous solution of sodium dichromate, $Na_2Cr_2O_7(aq)$, is a strong oxidizing agent.

$$14H^+(aq) + Cr_2O_7^{2-}(aq) + 6e^- \rightarrow 2Cr^{3+}(aq) + 7H_2O(l) \quad E^\circ_{red} = 1.33 \text{ V}$$

In contrast to chromium(VI) compounds, the chromium(II) ion is a fairly strong reducing agent:

$$Cr^{3+}(aq) + e^- \rightarrow Cr^{2+}(aq) \qquad E^\circ_{red} = -0.41 \text{ V}$$

and solutions containing $Cr^{2+}(aq)$ find use as reducing agents, as described for the chromous bubbler in Chapter 20.

The highest oxidation state of manganese is +7, which is best known in the strongly oxidizing permanganate ion, MnO_4^- (Figure N-5):

acidic solution:

$$MnO_4^-(aq) + 8H^+(aq) + 5e^- \rightarrow Mn^{2+}(aq) + 4H_2O(l) \qquad E^\circ_{red} = +1.51 \text{ V}$$

basic solution:

$$MnO_4^-(aq) + 2H_2O(l) + 3e^- \rightarrow MnO_2(s) + 4OH^-(aq) \qquad E^\circ_{red} = +1.23 \text{ V}$$

The most important permanganate is potassium permanganate, $KMnO_4$, which is used as an oxidizing agent in industry and medicine, as well as in many general chemistry laboratories. Freshly prepared solutions of potassium permanganate are deep purple but turn brown on long standing, because permanganate ion oxidizes water to oxygen and is thereby reduced to MnO_2, which is brown:

$$\underset{\text{purple}}{4MnO_4^-(aq)} + 2H_2O(l) \rightarrow \underset{\text{brown}}{4MnO_2(s)} + 3O_2(g) + 4OH^-(aq)$$

The reaction is catalyzed by $MnO_2(s)$ and is thus autocatalytic.

N-2. IRON IS THE MOST IMPORTANT METAL IN COMMERCE

With iron we no longer associate the highest oxidation state with the total number of 4*s* and 3*d* electrons. The highest-known oxidation state

Figure N-5 Potassium permanganate, $KMnO_4$, in water is a strong oxidizing agent. Freshly prepared $KMnO_4(aq)$ is purple; on standing, brown $MnO_2(s)$ precipitates as a result of the decomposition of $KMnO_4(aq)$.

of iron is +6, which is very rare. Only the +2 and +3 oxidation states of iron are common. Iron constitutes 4.7 percent by mass of the earth's crust. It is the cheapest metal and, in the form of steel, the most useful. Pure iron is a silvery-white, soft metal that rusts rapidly in moist air. It has little use as the pure element but is strengthened greatly by the addition of small amounts of carbon and of various other transition metals. It occurs in nature as hematite, Fe_2O_3; magnetite, Fe_3O_4; siderite, $FeCO_3$; and iron pyrite, FeS_2 (fool's gold) (Figure N-6).

Figure N-6 Iron pyrite, FeS_2, is known as fool's gold. Novice gold miners often mistake iron pyrite for gold because the two look so much alike. Gold is much more dense and is softer than iron pyrite.

Millions of tons of iron are produced annually in the United States by the reaction of Fe_2O_3 with coke, which is carried out in a **blast furnace.** A modern blast furnace is about 100 ft high and 25 ft wide and produces about 5000 tons of iron daily (Figure N-7). A mixture of iron ore, coke, and limestone ($CaCO_3$) is loaded into the top, and preheated compressed air and oxygen are blown in near the bottom. The reaction of the coke and the oxygen to produce carbon dioxide gives off a great deal of heat, and the temperature in the lower region of a blast furnace is around 1900°C. As the CO_2 rises, it reacts with more coke to produce hot carbon monoxide, which reduces the iron ore to iron. The molten iron metal is denser than the other substances and drops to the bottom, where it can be drained off to form ingots of what is called **pig iron.**

The function of the limestone is to remove the sand and gravel that normally occur with iron ore. The intense heat decomposes the limestone to CaO and CO_2. The CaO(*s*) combines with the sand and gravel (both of which are primarily silicon dioxide) to form calcium silicate:

$$\underset{\text{calcium oxide}}{CaO(s)} + \underset{\text{sand, gravel}}{SiO_2(s)} \rightarrow \underset{\text{molten calcium silicate}}{CaSiO_3(l)}$$

The molten calcium silicate, called **slag,** floats on top of the molten iron and is drained off periodically. It is used in building materials, such as cement and concrete aggregate, rock-wool insulation, and cinder block, and as railroad ballast.

Pig iron contains about 4 or 5 percent carbon together with lesser amounts of silicon, manganese, phosphorus, and sulfur. It is brittle, difficult to weld, and not strong enough for structural applications. To be useful, pig iron must be converted to steel, which is an alloy of iron with small but definite amounts of other metals and between 0.1 and 1.5 percent carbon. Steel is made from pig iron in several different processes, all of which use oxygen to oxidize most of the impurities. One such process is the **basic oxygen process,** in which hot, pure O_2 gas is blown through molten pig iron (Figure N-8). The oxidation of carbon and phosphorus is complete in less than 1 h. The desired carbon content of the steel is then achieved by adding high-carbon steel alloy.

There are two types of steels, carbon steels and alloy steels. Both types contain carbon, but carbon steels contain essentially no other metals besides iron. About 90 percent of all steel produced is carbon steel. Alloy steels contain other metals in small amounts. Different metals give different properties to steels. The alloy steels called stainless steels contain high percentages of chromium and nickel. Stainless steels resist corrosion and are used for cutlery, cooking pans, and hospital equipment. The most common stainless steel contains 18 percent chromium and 8 percent nickel.

▪ Steel additives:

Chromium improves hardness and resistance to corrosion.

Tungsten and molybdenum increase heat resistance.

Nickel adds toughness, as in armor plating.

Vanadium adds springiness.

Manganese improves resistance to wear.

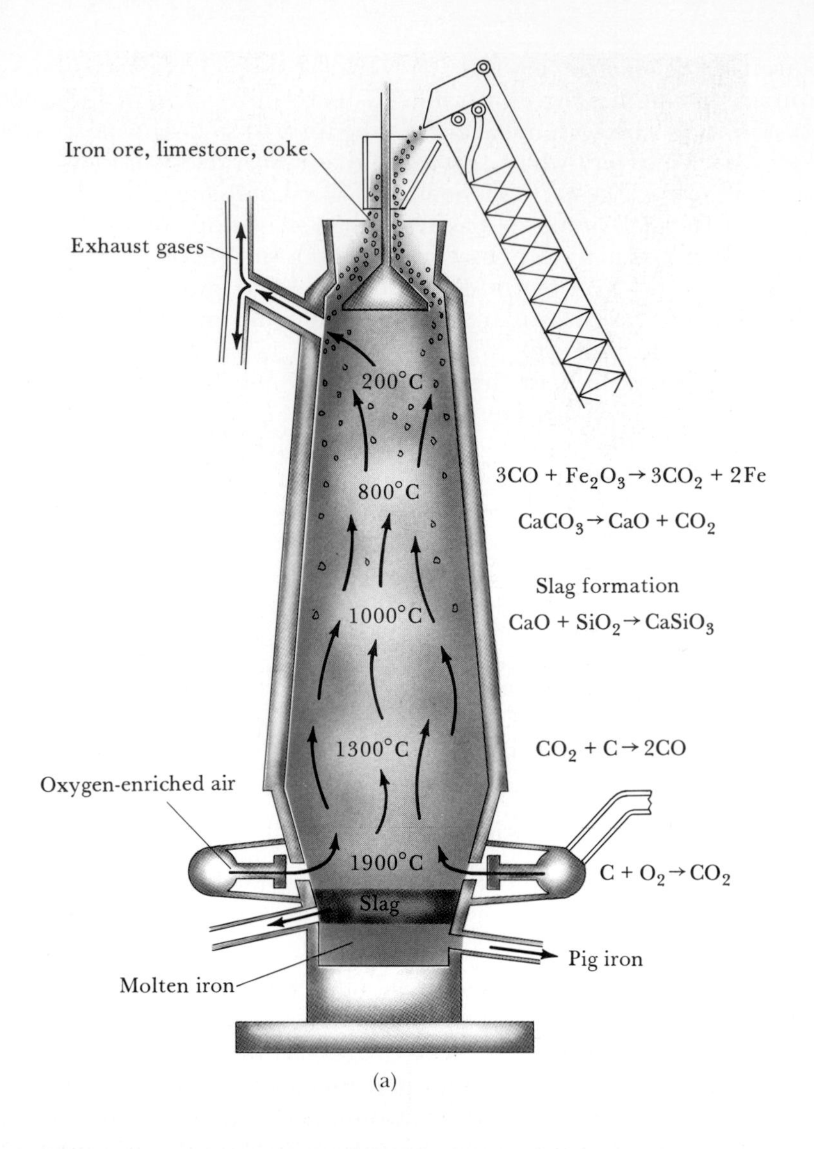

Figure N-7 Iron is produced in a blast furnace. (a) A diagram of a blast furnace. (b) A typical blast furnace like the one shown here runs continuously and consumes about 120 railroad cars of iron ore, 50 railroad cars of coke, and 40 railroad cars of limestone per day. The 5000 tons of iron produced requires about 75 railroad cars to transport it.

Figure N-8 Molten iron being charged into a basic oxygen furnace. Most steel is produced by a process called the basic oxygen process. A typical basic oxygen furnace is charged with about 200 tons of molten pig iron, 100 tons of scrap iron, and 20 tons of limestone (to form a slag). A stream of hot oxygen is blown through the molten mixture, where the impurities are oxidized and blown out of the iron. High-quality steel is produced in an hour or less.

N-3. THE +2 OXIDATION STATE IS THE MOST IMPORTANT OXIDATION STATE FOR COBALT, NICKEL, COPPER, AND ZINC

As we go from iron to zinc, there is an increasing prominence of the +2 oxidation state. Many compounds of cobalt and almost all compounds of nickel and zinc involve the metal in the +2 oxidation state. Of the 3*d* series metals, only copper has an important +1 oxidation state.

Cobalt is a fairly rare element, which ranks thirtieth in abundance in the earth's crust. It is usually found associated with ores of copper and nickel in nature. It is a hard, bluish-white metal that is used in the

Electrolytic refining facility for nickel at the Sudbury Nickel Mine (Inco) in Canada.

Figure N-9 The major copper ores. *Left to right:* chalcopyrite, $CuFeS_2$; malachite, $CuCO_3 \cdot Cu(OH)_2$; and chalcocite, Cu_2S.

manufacture of high-temperature alloys for aircraft engines and turbines and permanent magnets (Alnico). The pure metal is relatively unreactive and dissolves only slowly in dilute mineral acids. Most simple cobalt salts involve Co(II), and many of these are pink or red. The species $Co^{3+}(aq)$ is a strong oxidizing agent:

$$Co^{3+}(aq) + e^- \rightarrow Co^{2+}(aq) \qquad E^\circ_{red} = +1.84 \text{ V}$$

and oxidizes $H_2O(l)$ to oxygen.

Nickel is the twenty-second most abundant element in the earth's crust and occurs in a variety of sulfide ores, the most important deposit being found in the Sudbury basin of Ontario, Canada. Nickel is a silvery metal that takes a beautiful high polish. The bulk, pure metal is highly corrosion-resistant and is often used as a protective coating. It is used in a number of magnetic alloys, and in the alloy Monel, which is used to handle fluorine and reactive fluorine compounds. As a powder, nickel is used as a catalyst for the hydrogenation of vegetable oils. Nickel is more reactive than cobalt and dissolves readily in dilute acids. The aqueous solution chemistry of nickel involves primarily the species $Ni^{2+}(aq)$.

Copper is slightly less abundant than nickel. Copper generally occurs as various sulfides, although in some ores copper is present in the form of sulfates, carbonates, and other oxygen-containing compounds (Figure N-9). Deposits of the free metal are very rare, being found only in Michigan.

Brass is an alloy of copper with zinc, and **bronze** is an alloy of copper with tin. Brass and bronze are among the earliest-known alloys. Bronze usually contains from 5 to 10 percent tin and is very resistant to corrosion. It is used for casting, marine equipment, fine arts work, and spark-resistant tools. Yellow brasses contain about 35 percent zinc and have good ductility and high strength. Brass is used for piping, hose nozzles, marine equipment, and jewelry and in the fine arts.

Although copper is fairly unreactive, its surface turns green after long exposure to the atmosphere. The green patina (Figure N-10) is due to the surface formation of copper hydroxo carbonate and hydroxo sulfate. Copper does not replace hydrogen from dilute acids, because E°_{rxn} for the equation

$$Cu(s) + 2H^+(aq) \rightleftharpoons Cu^{2+}(aq) + H_2(g) \qquad E^\circ_{rxn} = -0.34\ V$$

is negative. Copper does react with oxidizing acids such as dilute nitric acid or hot concentrated sulfuric acid:

$$3Cu(s) + 8HNO_3(aq) \rightarrow 3Cu(NO_3)_2(aq) + 2NO(g) + 4H_2O(l)$$

$$Cu(s) + 2H_2SO_4(conc) \rightarrow CuSO_4(aq) + 2H_2O(l) + SO_2(g)$$

Most compounds of copper involve Cu(II). Copper (I) salts are often colorless and only slightly soluble in water. The $Cu^+(aq)$ ion is unstable and **disproportionates** (i.e., oxidizes and reduces itself) according to

$$2Cu^+(aq) \rightleftharpoons Cu(s) + Cu^{2+}(aq) \qquad K = 1.8 \times 10^6\ M^{-1}$$

Many copper(II) salts are blue or bluish-green. Copper(II) sulfate pentahydrate, $CuSO_4 \cdot 5H_2O$, which occurs as beautiful blue crystals, is the most common copper(II) salt. When the crystals are heated gently, the waters of hydration are driven off to produce anhydrous copper(II) sulfate, $CuSO_4$, which is a white powder.

Figure N-10 The statue of Liberty turned green as a result of the corrosion of copper metal. A major restoration project of the statue's surface ended in 1986.

Zinc is widely distributed in nature and is about as abundant as copper. Its principal ores are sphalerite (ZnS), or zinc blende, and smithsonite ($ZnCO_3$), from which zinc is obtained by roasting and reduction of the resultant ZnO with carbon. Zinc is a shiny white metal with a bluish-gray luster.

The 3*d* subshell of zinc is completely filled, and zinc behaves more like a Group 2 metal than like a transition metal. Metallic zinc is a strong reducing agent. It dissolves readily in dilute acids and combines with oxygen, sulfur, phosphorus, and the halogens upon being heated. The only important oxidation state of zinc is +2, and zinc(II) salts are colorless, unless color is imparted by the anion.

N-4. BILLIONS OF DOLLARS ARE SPENT EACH YEAR TO PROTECT METALS FROM CORROSION

We are all familiar with corrosion, the best-known example of which is the rusting of iron and steel. Brown rust (Figure N-11) is iron(III) oxide, and the corrosion of iron proceeds by air oxidation of the iron:

$$4Fe(s) + 3O_2(g) \xrightarrow{H_2O(l)} \underset{\text{rust}}{2Fe_2O_3(s)}$$

Most metals when exposed to air develop an oxide film. In some cases this film is very thin and protects the metal, and the metal maintains its luster. However, depending on the humidity, acidity and the presence of certain anions, corrosion can completely destroy a metal. For example, anions such as Cl^-, which is present in sea spray and on roads treated with rock salt, promote corrosion through the formation of chloro complexes. Certain gaseous species, such as the oxides of sulfur and nitrogen, combine with water to form acids that attack metal.

(a)

(b)

Figure N-11 The angular crystals seen in the scanning electron photomicrograph (a) are green rust, Fe(II) oxides, while the column-shaped and rounded crystals (b) are black rust, Fe_3O_4, and brown rust, Fe_2O_3, respectively. Because this type of microscope does not use optical wavelengths, the different rust colors are not actually seen in such pictures.

Corrosion is a major problem costing billions of dollars annually for replacements for corroded parts. Research is directed toward corrosion mechanisms because a detailed understanding of these mechanisms can provide important clues as to how to prevent the process.

Corrosion involves electron-transfer reactions between different sections of the same piece of metal or between two dissimilar metals in electrical contact with each other. One metal piece acts as the anode, and the other acts as the cathode. For example, iron in contact with air and moisture corrodes according to the mechanism sketched in Figure N-12. The anodic process is

$$2Fe(s) + 4OH^-(aq) \rightarrow 2Fe(OH)_2(s) + 4e^-$$

and the cathodic process is

$$2H_2O(l) + O_2(g) + 4e^- \rightarrow 4OH^-(aq)$$

where the $O_2(g)$ comes from the air. The iron(II) hydroxide formed is rapidly air-oxidized in the presence of water to iron(III) hydroxide:

$$4Fe(OH)_2(s) + O_2(g) + 2H_2O(l) \rightarrow 4Fe(OH)_3(s)$$

which in turn converts spontaneously to iron(III) oxide:

$$2Fe(OH)_3(s) \rightarrow Fe_2O_3 \cdot 3H_2O(s)$$

The corrosion of aluminum in air is not so pronounced as that of iron because the Al_2O_3 film that forms is tough and adherent and is

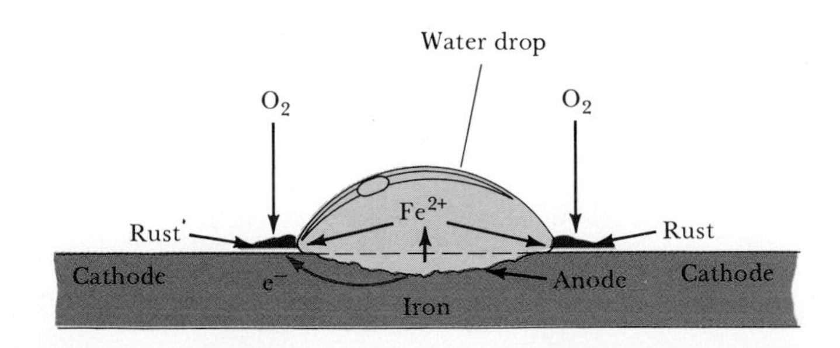

Figure N-12 Corrosion of iron. A drop of water on an iron surface can act as a corrosion center. The iron is oxidized by oxygen from the air. Moisture is necessary for corrosion because the mechanism involves the formation of dissolved $Fe^{2+}(aq)$ ions. Salt promotes the corrosion by enabling a larger current flow between anode and cathode.

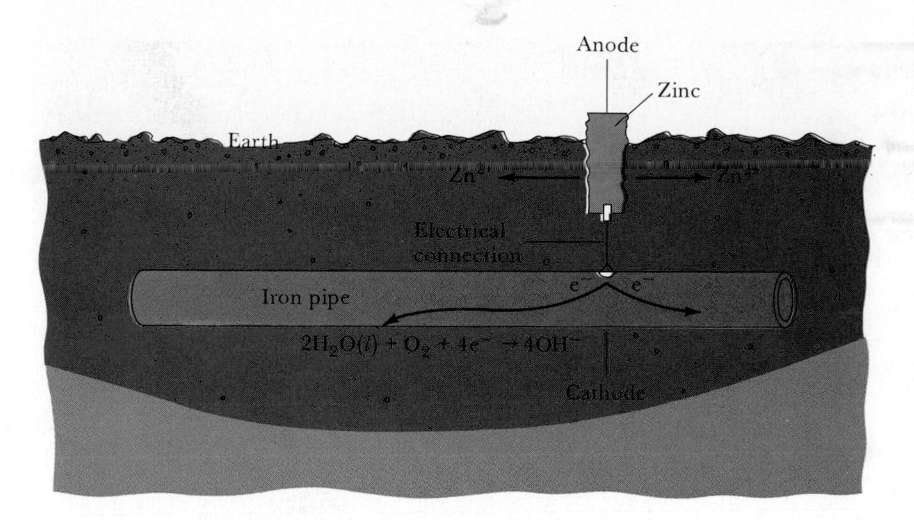

Figure N-13 Protection of an iron pipe from corrosion with a sacrificial zinc anode. Zinc is a stronger reducing agent than iron and thus is preferentially oxidized. The electrons produced in the oxidation flow to the iron pipe, on the surface of which O_2 is reduced to hydroxide ion. The net process is $2Zn(s) + O_2(aq) + 2H_2O(l) \rightarrow 2Zn(OH)_2(s)$, and the iron remains intact.

impervious to oxygen. The same is true for chromium and nickel.

The simplest method of corrosion prevention is to provide a protective layer of paint or of a corrosion-resistant metal, such as chromium or nickel. The weakness of such methods is that any scratch or crack in the protective layer exposes the metal surface. The exposed surface, even though small in area, can act as an anode in conjunction with other exposed metal parts, which act as cathodes. This combination then leads to corrosion of the metal under the no-longer-protective layer.

Another anticorrosion technique uses a replaceable **sacrificial anode,** which is a piece of metal electrically connected to a less active metal (Figure N-13). The more active metal is the stronger reducing agent and is thus preferentially oxidized; oxygen is reduced on the surface of the less active metal. Sacrificial anodes are used to protect water pipes and ship propellers. This method is also the electrochemical basis of galvanization, in which iron is protected from corrosion by a zinc coating, a process used in the manufacture of automobile bodies. A crack in the zinc coating does not affect the corrosion protection provided. Note that the less active metal (iron) promotes the corrosion of the more active metal (zinc).

TERMS YOU SHOULD KNOW

transition metal 700
karat 701
blast furnace 705
pig iron 705
slag 705
basic oxygen process 705
brass 708
bronze 708
disproportionate 709
sacrificial anode 711

QUESTIONS

N-1. How many *d* transition metals are there in each series?

N-2. Which metal has the highest melting point?

N-3. Which are the two most dense metals?

N-4. Which is the most abundant transition metal?

N-5. Describe how titanium metal is produced.

N-6. What is the percentage of gold in 14-karat gold?

N-7. Describe how iron is produced in a blast furnace.

N-8. What is pig iron?

N-9. What is slag? What is it used for?

N-10. Describe the basic oxygen process.

N-11. Discuss the differences between carbon steels and stainless steels.

N-12. Describe the reactions that occur in the corrosion of iron.

N-13. Why isn't the corrosion of aluminum as serious a problem as the corrosion of iron?

N-14. Describe how a sacrificial anode works.

PROBLEMS

N-15. Determine the oxidation state of manganese in each of the following species:

(a) MnO_4^-
(b) MnO_4^{2-}
(c) MnO_2
(d) $MnCl_2$
(e) $MnBr_3$
(f) Mn_2O_7

N-16. Determine the oxidation state of the transition metal in each of the following compounds:

(a) Fe_2O_3
(b) Fe_3O_4
(c) $Ni(OH)O$
(d) Ag_2O_2 (a peroxide)

N-17. Use VSEPR theory to predict the structures of the following species:

(a) $TiCl_4$
(b) MnO_4^-
(c) $AgCl_2^-$

N-18. Calculate the maximum number of grams of copper metal that can be obtained from 1.00 metric ton of a copper ore deposit that contains 3.5 percent by mass malachite.

N-19. For ocean water we have

$[Ca^{2+}] = 10.0\ \text{mM} \qquad [HCO_3^-] = 2.3\ \text{mM} \qquad \text{pH} = 8.15$

Determine in which direction the process

$$Ca^{2+}(aq) + CO_3^{2-}(aq) \rightleftharpoons CaCO_3(s)$$

is spontaneous in ocean water at 25°C. The K_{sp} for $CaCO_3(s)$ is $2.8 \times 10^{-9}\ \text{M}^2$, and K_a for $HCO_3^-(aq)$ is 4.27×10^{-11} M.

TRANSITION-METAL COMPLEXES

Crystals of the chlorides of the transition metals scandium (Sc) through zinc (Zn). Most compounds of the transition metals are colored.

Transition-metal complexes are species in which several anions or neutral molecules, called ligands, bond to a transition metal atom or ion. The chemistry of the transition metals is especially rich and interesting because of these complexes, which occur in a variety of geometries and oxidation states. Much of the chemistry of transition-metal complexes can be understood in terms of the electron occupancy of the *d* orbitals of the metal ion. In this chapter, we use *d*-orbital electron configurations to explain many of the spectral, magnetic, and structural properties of transition-metal complexes.

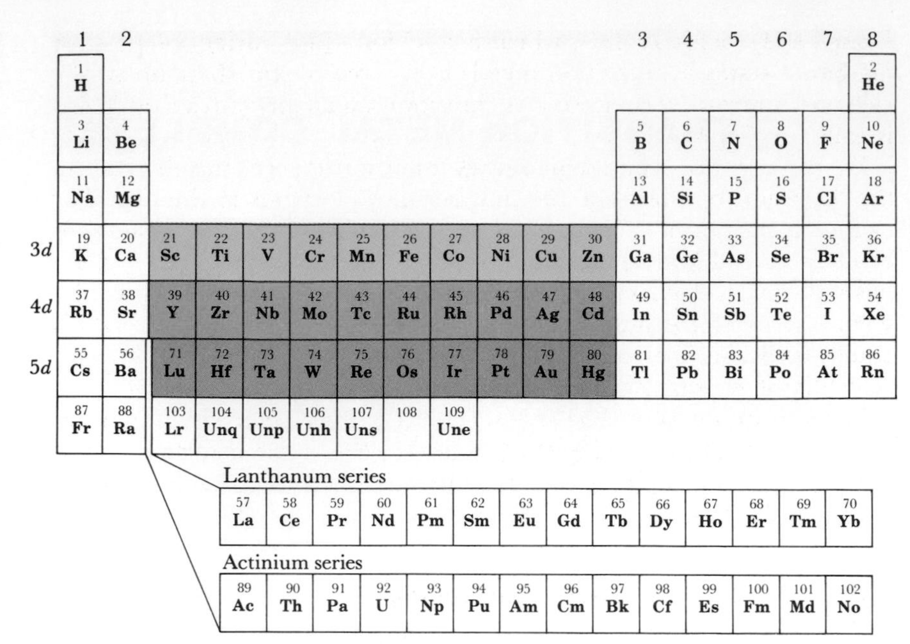

Figure 22-1 The periodic table with the 3*d* transition-metal series shown in blue, the 4*d* series shown in yellow and the 5*d* series shown in green.

22-1 THERE ARE 10 ELEMENTS IN EACH *d* TRANSITION-METAL SERIES

Recall from Chapter 6 that the angular momentum quantum number, l, for a d orbital is equal to 2. For $l = 2$, the magnetic quantum number, m_l, can be 2, 1, 0, -1, or -2. Thus there are five d orbitals for each value of the principal quantum number for $n \geqslant 3$. Each of the five d orbitals can hold a maximum of two electrons with opposite spins, giving a total of 10 electrons in a d subshell. For $n = 3$, we have the 3*d* series (Sc $\rightarrow$ Zn); for $n = 4$, we have the 4*d* series (Y $\rightarrow$ Cd); and for $n = 5$, we have the 5*d* series (Lu $\rightarrow$ Hg). The 3*d*, 4*d*, and 5*d* **transition-metal series** are shown as colored rows in Figure 22-1.

The shape and relative spatial orientation of the five ***d* orbitals** are shown in Figure 22-2. These orbitals are distinguished by x, y, and z subscripts that define the orientation of the orbitals with respect to the x, y, and z coordinate axes. Thus the d orbitals are named d_{xy}, d_{xz}, d_{yz}, $d_{x^2-y^2}$, and d_{z^2} **orbitals.** In the absence of any external electric or magnetic field, the energies of the five d orbitals for a given value of the principal quantum number are equal.

Recall from Chapter 7 that, for the transition-metal ions, the order of filling of the orbitals follows the arithmetic sequence: the $n = 1$ shell is filled first, then $n = 2$, then $n = 3$, and so on, until all the electrons are used up. Also recall that for the M(II) ions of the 3*d* series we have

d^1	d^2	d^3	d^4	d^5	d^6	d^7	d^8	d^9	d^{10}
Sc(II)	Ti(II)	V(II)	Cr(II)	Mn(II)	Fe(II)	Co(II)	Ni(II)	Cu(II)	Zn(II)
21	22	23	24	25	26	27	28	29	30

The M(III) ions of the 3*d* series have one fewer electron than the M(II) ions. Thus, for example, Fe(II) and Co(III) both have six d electrons; for this reason they are called d^6 ions. Removal of an electron from Fe(II) yields Fe(III), which is a d^5 ion. This variety of possible oxidation states in transition metals is one reason for their unusually rich and interesting chemistry. A transition-metal ion with x d electrons is called a d^x ion.

We can use the second digit of the atomic numbers of the elements in the 3d series to determine the number of d electrons in the ions. For example, scandium is element 21 and Sc(II) is a d^1 ion; manganese is element 25 and Mn(II) is a d^5 ion; iron is element 26 and Fe(III), which has one fewer electron than Fe(II), is a d^5 $(6 - 1 = 5)$ ion. Thus, the second digit of the atomic number of an element in the 3d series is equal to the number of d electrons in the M(II) ion of that element. Note, however, that Zn(II), element 30 (10th of the series), is a d^{10} ion.

For ions in the 4d and 5d series, we can quickly determine the number of d electrons by noting the position of the element relative to the 3d series. For example, rhodium is directly below cobalt (element 27) and thus Rh(II) is a d^7 ion.

▪ Ion	Atomic number	d^x
Sc(II)	21	d^1
Ti(II)	22	d^2
V(II)	23	d^3
Cr(II)	24	d^4
Mn(II)	25	d^5
Fe(II)	26	d^6
Co(II)	27	d^7
Ni(II)	28	d^8
Cu(II)	29	d^9
Zn(II)	30	d^{10}

Example 22-1: Determine the number of outer-shell d electrons in Ir(II), Pt(IV), and Mo(III).

Solution: Iridium is directly below cobalt ($Z = 27$) in the periodic table, and so Ir(II) is a d^7 ion. Platinum is below nickel ($Z = 28$), and so Pt(II) is a

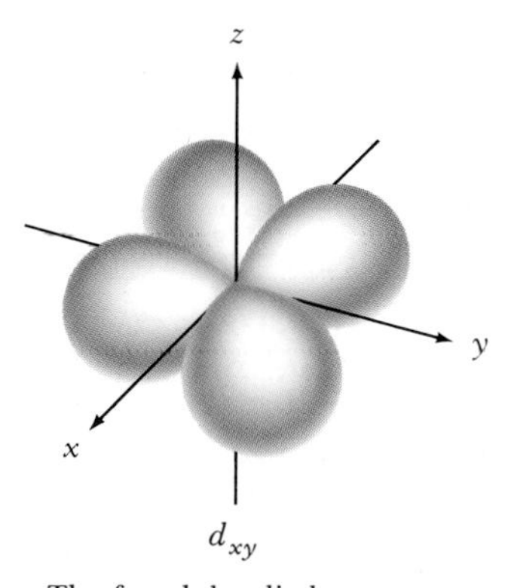

The four lobes lie between the x and y axes in the four quadrants on the xy plane.

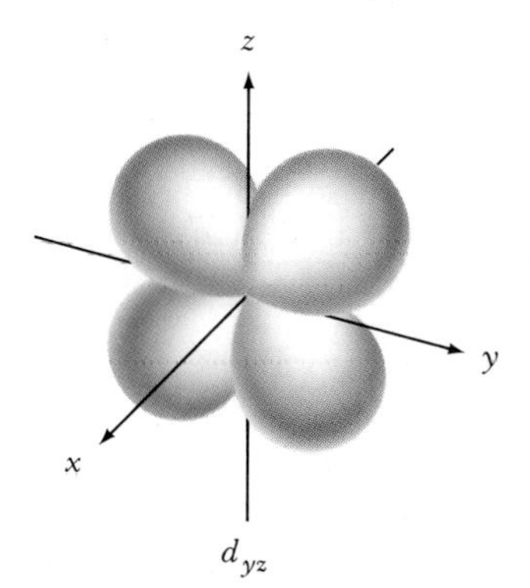

The four lobes lie between the y and z axes in the four quadrants on the yz plane.

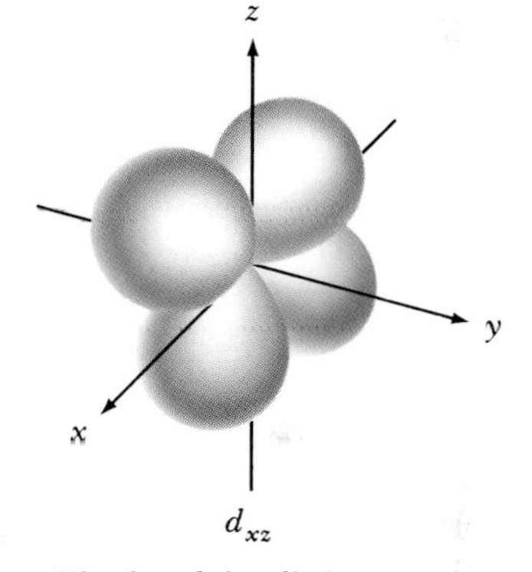

The four lobes lie between the x and z axes in the four quadrants on the xz plane.

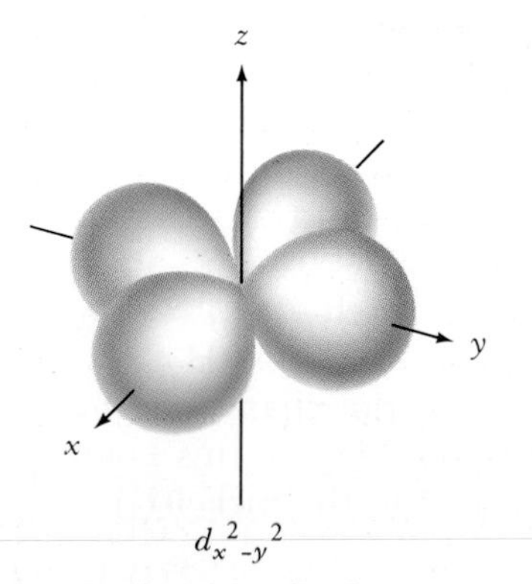

The four lobes lie along the x and y axes.

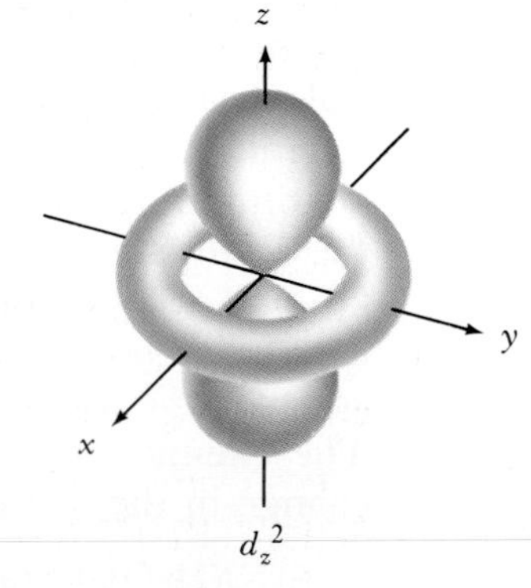

Two lobes are on the z axis and a donut-shaped lobe is symmetrically placed on the xy plane.

Figure 22-2 The shapes and relative orientations of the five d orbitals.

d^8 ion. Platinum (IV) has two electrons fewer than Pt(II), and so Pt(IV) is a d^6 ion. Molybdenum is below chromium ($Z = 24$), and so Mo(II) is a d^4 ion. Consequently, Mo(III), which has one electron fewer than Mo(II), is a d^3 ion.

Example 22-2: Give three examples of d^6 ions with an oxidation state of +3.

Solution: For an M(III) ion to be a d^6 ion, the corresponding M(II) ion must be a d^7 ion. The M(II) d^7 ions are Co(II), Rh(II), and Ir(II), and so the M(III) d^6 ions are Co(III), Rh(III), and Ir(III).

22-2. COMPLEXES CONSIST OF CENTRAL METAL ATOMS OR IONS THAT ARE BONDED TO LIGANDS

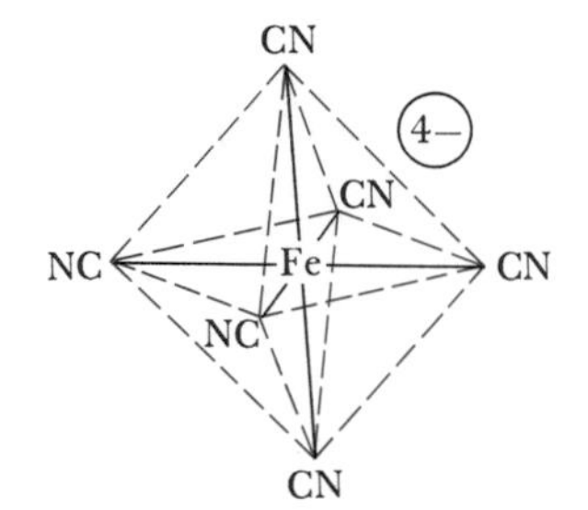

Figure 22-3 The yellow complex ion $[Fe(CN)_6]^{4-}$ is octahedral, with the six cyanide ions bonded to the central iron atom.

Simple cyanide salts, such as sodium cyanide, NaCN, are deadly poisons. A solution of NaCN contains the ions $Na^+(aq)$ and $CN^-(aq)$, and the solution is colorless. The toxicity is due to the $CN^-(aq)$ ions, which block the oxygen-carrying capacity of hemoglobin in red blood cells. If excess iron(II) nitrate, $Fe(NO_3)_2$, is added to an aqueous solution of NaCN, the solution turns yellow. Chemical tests show that $CN^-(aq)$ is no longer present and that the solution is no longer poisonous. What happens is that cyanide ions react with iron(II) to form the **complex ion** $[Fe(CN)_6]^{4-}(aq)$, in which six cyanide ions are bound directly to the iron in an octahedral structure (Figure 22-3). The $[Fe(CN)_6]^{4-}$ ion occurs in solution as a single ion. The charge on the $[Fe(CN)_6]^{4-}$ ion is −4, and its sodium salt has the formula $Na_4[Fe(CN)_6]$. If $Na_4[Fe(CN)_6]$ is dissolved in water, then the resulting solution contains $Na^+(aq)$ and $[Fe(CN)_6]^{4-}(aq)$ ions:

$$Na_4[Fe(CN)_6](s) \xrightarrow[H_2O(l)]{} 4Na^+(aq) + [Fe(CN)_6]^{4-}(aq)$$

Generally, a complex ion contains a central metal ion to which are attached anions or neutral molecules. A complex ion is a distinct chemical species with properties different from those of its constituent species. Transition-metal ions are capable of bonding to a wide variety of anions and neutral molecules to form complex ions. An anion or neutral molecule attached directly to a metal ion is called a **ligand.**

Transition-metal ions in aqueous solution form complex ions by bonding with the water molecules. For example, a nickel(II) ion forms the octahedral complex ion $[Ni(H_2O)_6]^{2+}$ in aqueous solution (Figure 22-4). A solution of nickel(II) perchlorate, $Ni(ClO_4)_2$, in water is brilliant green because of the presence of $[Ni(H_2O)_6]^{2+}$ ions. If $NH_3(aq)$ is added to the solution, then the color changes from green to blue-violet. The chemical reaction responsible for this color change involves a change in the ligands attached to the nickel(II) ion:

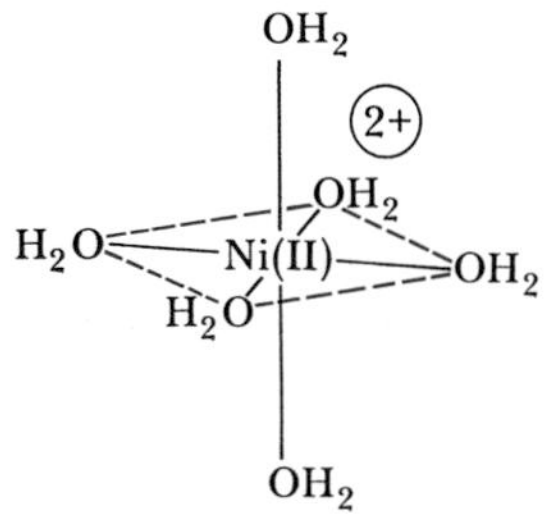

$$\underset{\text{green}}{[Ni(H_2O)_6]^{2+}(aq)} + \underset{\text{colorless}}{6NH_3(aq)} \rightleftharpoons \underset{\text{blue-violet}}{[Ni(NH_3)_6]^{2+}(aq)} + \underset{\text{colorless}}{6H_2O(l)}$$

OH2
2+
OH2
H2O
Ni(II)
OH2
H2O
OH2

Figure 22-4 Octahedral structure of the green complex ion $[Ni(H_2O)_6]^{2+}$. All six nickel-oxygen bonds are equivalent.

The H_2O ligands—that is, the water molecules attached to the Ni(II) ion—are displaced by the NH_3 molecules, which become the new li-

Figure 22-5 Aqueous solutions of Ni(II) complex ions: $[Ni(H_2O)_6]^{2+}$ (*green*), $[Ni(NH_3)_6]^{2+}$ (*violet*), and $[Ni(CN)_4]^{2-}$ (*orange-yellow*).

gands attached to the nickel(II) ion. The $[Ni(NH_3)_6]^{2+}$ ion is octahedral like $[Ni(H_2O)_6]^{2+}$, with NH_3 molecules instead of H_2O molecules surrounding the central nickel atom. A reaction involving a change in the ligands attached to the central metal ion in a complex ion is called a **ligand-substitution reaction.**

The change in ligands from H_2O to NH_3 around nickel(II) produces a modified electrical environment around the nickel(II) ion, and this changes the energies of the eight *d* electrons on the nickel(II) ion. This change in energy of the *d* electrons causes a change in the wavelength of the light absorbed by the ion and thus results in a color change. In general, changes in the energy of the *d* electrons give rise to the color of transition-metal complex ions.

▪ The colors of most transition-metal complexes are due to electronic transitions of their *d* electrons.

Not all complex ions are octahedral. For example, if we add $NaCN(aq)$ to a solution containing the blue-violet $[Ni(NH_3)_6]^{2+}(aq)$ ion, then cyanide ions displace the NH_3 ligands from the nickel(II) ion to form the bright-yellow $[Ni(CN)_4]^{2-}$ ion:

$$\underset{\text{blue-violet}}{[Ni(NH_3)_6]^{2+}(aq)} + \underset{\text{colorless}}{4CN^-(aq)} \rightleftharpoons \underset{\text{yellow}}{[Ni(CN)_4]^{2-}(aq)} + \underset{\text{colorless}}{6NH_3(aq)}$$

The structure of the $[Ni(CN)_4]^{2-}(aq)$ ion is square-planar; the four CN^- ligands are arranged in a plane at the four corners of an imaginary square around the nickel(II) ion.

It is the carbon end of the cyanide ion that is bonded to the nickel(II) ion. Note also that the overall charge on the $[Ni(CN)_4]^{2-}$ complex is -2 because the nickel(II) ion contributes a charge of $+2$ and the four CN^- ligands each contribute a charge of -1.

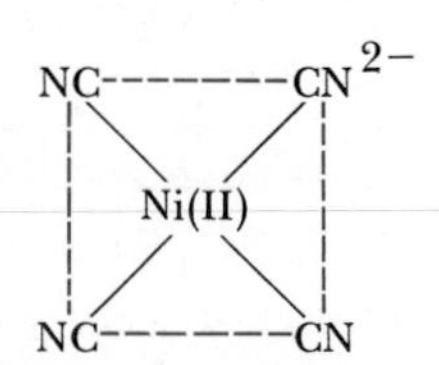

Figure 22-6 The $[Ni(CN)_4]^{2-}(aq)$ ion is square-planar.

Example 22-3: Determine the oxidation state of platinum in $[Pt(NH_3)_4Cl_2]^{2+}$.

Solution: The $[Pt(NH_3)_4Cl_2]^{2+}$ complex ion has two kinds of ligands around the central metal atom: four NH_3 ligands and two Cl^- ligands. The charge on the NH_3 ligands is zero, the charge on each Cl^- is -1, and the overall charge is $+2$. Denoting the charge on the Pt ion as x, we have

$$\underset{\text{charge on Pt}}{x} + \underset{4NH_3}{\underbrace{4(0)}} + \underset{2Cl^-}{\underbrace{2(-1)}} = \underset{\text{overall charge on ion}}{+2}$$

or

$$x + 2(-1) = +2$$

$$x = +4$$

Thus the oxidation state of platinum in the complex is +4.

Devices that indicate humidity levels by changing color are based on the ligand-substitution reaction

$$\underset{\text{pink}}{2[Co(H_2O)_6]Cl_2(s)} \rightleftharpoons \underset{\text{blue}}{Co[CoCl_4](s)} + 12H_2O(g)$$

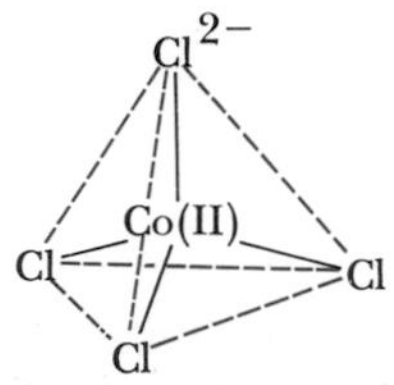

The $[Co(H_2O)_6]^{2+}$ ion is pink, and the $[CoCl_4]^{2-}$ ion is blue. When the humidity is high, $Co[CoCl_4]$ reacts with the water vapor in the air to form the pink $[Co(H_2O)_6]^{2+}$ ion. When the humidity is low, the equilibrium shifts from left to right, forming the blue $[CoCl_4]^{2-}$ ion. The $[CoCl_4]^{2-}$ ion is tetrahedral, with the cobalt(II) ion at the center of a tetrahedron formed by the four chloride ligands.

Figure 22-7 The $[CoCl_4]^{2-}(aq)$ ion is tetrahedral.

In many qualitative analysis schemes, $AgCl(s)$ is separated from other insoluble chlorides by the addition of $NH_3(aq)$ to form the soluble salt $[Ag(NH_3)_2]Cl$, which involves the complex $[Ag(NH_3)_2]^+$:

$$AgCl(s) + 2NH_3(aq) \rightarrow [Ag(NH_3)_2]^+(aq) + Cl^-(aq)$$

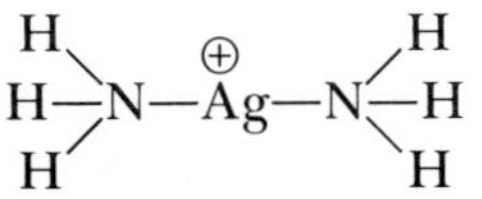

The $[Ag(NH_3)_2]^+$ ion is linear, with the N—Ag—N atoms arranged in a straight line (Figure 22-8).

Figure 22-8 The two nitrogen atoms and the silver atom in the ion $[Ag(NH_3)_2]^+$ are in a straight line.

The most common structures for transition-metal complexes are octahedral, tetrahedral, square-planar, and linear. Of these four geometries, octahedral is by far the most common. Examples of transition-metal complexes with these geometries are given in Table 22-1.

22-3. TRANSITION-METAL COMPLEXES HAVE A SYSTEMATIC NOMENCLATURE

The wide variety and large number of possible transition-metal complexes make a systematic procedure for naming them essential. An example of a systematic name for a transition-metal complex is

$[Ni(NH_3)_6](NO_3)_2$ hexaamminenickel(II) nitrate

Let's analyze the name of this compound. As with any salt, the cation is named first. Thus this name tells us that the compound consists of a hexaamminenickel(II) cation and nitrate anions. The Greek prefix hexa- denotes six, and ammine denotes the ligand NH_3. Thus the hexaammine- part of the name tells us that there are six NH_3 ligands in the cation. The roman numeral II tells us that the nickel is in the +2 oxidation state. Because ammonia is a neutral molecule and nickel is in a +2 oxidation state, the charge on the complex cation is +2. Its formula is $[Ni(NH_3)_6]^{2+}$.

A simplified set of nomenclature rules for complexes is as follows:

Table 22-1 Some examples of transition-metal complexes of various geometries

Octahedral	Tetrahedral	Square-planar	Linear
$[Fe(H_2O)_6]^{2+}$	$[Zn(NH_3)_4]^{2+}$	$[Pt(CN)_4]^{2-}$	$[AgCl_2]^-$
$[Co(NO_2)_6]^{3-}$	$[CoCl_4]^{2-}$	$[AuCl_4]^-$	$[CuI_2]^-$
$[Ru(NH_3)_6]^{3+}$	$[FeCl_4]^-$	$[Rh(CN)_4]^{3-}$	$[Ag(CN)_2]^-$
$[Pt(NH_3)_6]^{4+}$	$[Ni(CO)_4]$	$[Pt(NH_3)_4]^{2+}$	$[AuCl_2]^-$
$[Cr(CO)_6]$	$[HgI_4]^{2-}$	$[PdCl_4]^{2-}$	
		$[Cu(NH_3)_4]^{2+}$	

1. *Name the cation first and then the anion:* for example potassium tetracyanonickelate(II), $K_2[Ni(CN)_4]$.
2. *In any complex ion or neutral molecule, name the ligands first and then the metal:* for example, hexaamminenickel(II) or tetracyanonickelate(II). If there is more than one type of ligand in the complex, then name them in alphabetical order: for example, diamminedichloroplatinum(II), $[Pt(NH_3)_2Cl_2]$.
3. *End the names of negative ligands in the letter o, but give neutral ligands the name of the ligand molecule.* Some common neutral ligands have special names, such as aqua for H_2O, ammine for NH_3, and carbonyl for CO. Table 22-2 lists the names of a number of ligands.
4. *Denote the number of ligands of a particular type by a Greek prefix, such as di-, tri-, tetra-, penta-, or hexa-.*
5. *If the complex ion is a cation, then use the ordinary name for the metal; if the complex ion is an anion, then end the name of the metal in -ate:* for example, tetrachlorocobaltate(II), $[CoCl_4]^{2-}$, where the suffix -ate on the metal name tells us that the complex ion is an anion. Table 22-3 lists some exceptions to this rule.
6. *Denote the oxidation state of the metal by a roman numeral in parentheses following the name of the metal.*

The application of these rules is illustrated in the following Example.

Table 22-2 Names for some common ligands

Ligand*	Name as ligand
F^-	fluoro
Cl^-	chloro
Br^-	bromo
I^-	iodo
$\underline{C}N^-$	cyano
$\underline{O}H^-$	hydroxo
$\underline{N}O_2^-$	nitro
$\underline{C}O$	carbonyl
$H_2\underline{O}$	aqua
$\underline{N}H_3$	ammine

*For ligands with two or more different atoms, the underlined atom is the one bonded to the metal.

Example 22-4: The water-soluble yellow-orange compound $Na_3[Co(NO)_2)_6]$ is used in some qualitative analysis schemes to test for

Table 22-3 Some exceptions to rule 5 for naming complexes

Metal	Name in complex anion	Complex anion	Name of complex anion
silver	argentate	$[AgCl_2]^-$	dichloroargentate(I)
gold	aurate	$[Au(CN)_4]^-$	tetracyanoaurate(III)
copper	cuprate	$[CuCl_4]^{2-}$	tetrachlorocuprate(II)
iron	ferrate	$[Fe(CN)_6]^{3-}$	hexacyanoferrate(III)

▪ The names of the metals in Table 22-3 are derived from the Latin names.

$K^+(aq)$. Almost all potassium salts are water-soluble, but $K_3[Co(NO_2)_6]$ is only slightly soluble in water. The equation for the precipitation reaction is

$$3K^+(aq) + [Co(NO_2)_6]^{3-}(aq) \rightarrow K_3[Co(NO_2)_6](s)$$

Name the $[Co(NO_2)_6]^{3-}$ ion.

Solution: The oxidation state of cobalt in the complex ion is determined as follows. The overall charge on the ion is −3, and there are six nitrite ions in the complex, each with a charge of −1. Denoting the oxidation state of cobalt as x, we have

$$\underbrace{x}_{\text{Co}} + \underbrace{6(-1)}_{6NO_2^-} = \underbrace{-3}_{\text{net charge on complex ion}}$$

or $x = +3$. The ion is called hexanitrocobaltate(III), where the -ate ending tells us that the complex is an anion.

Some other examples of naming complexes are

$K_2[Ni(CN)_4]$ — potassium (cation) tetracyanonickelate(II) (complex anion); tetracyano: $4CN^-$ ligands; nickelate(II): Ni in +2 oxidation state

$[Cr(CO)_6]$ — hexacarbonylchromium(0) (neutral molecule); hexacarbonyl: 6CO ligands; chromium(0): Cr in zero oxidation state

$[Co(H_2O)_4Cl_2]Cl$ — tetraaquadichlorocobalt(III) (complex cation) chloride (anion); tetraaqua: $4H_2O$ ligands; dichloro: $2Cl^-$ ligands; cobalt(III): Co in +3 oxidation state

▪ If a transition-metal complex involves a metal in a zero oxidation state, as in $[Cr(CO)_6]$, all the outer electrons in the metal atom are placed in d orbitals. Thus, Cr in $[Cr(CO)_6]$ is a d^6 species. This occurs because the presence of the ligands raises the energy of the 4s orbital atom relative to the 3d orbital of the chromium atom.

The rules for writing a chemical formula from the name of a complex follow from the nomenclature rules. For example, the formula for the compound named potassium hexacyanoferrate(II) is determined as follows:

1. The cation is potassium, K^+.
2. The complex anion contains six CN^- (hexacyano) ions and an iron atom. The oxidation state of the iron is +2, as indicated by the roman numeral. The ending -ate tells us that the complex is an anion.

3. The charge on the complex anion is computed by adding up the charges on the metal ion and the ligands:

$$\underbrace{(+2)}_{\text{Fe(II)}} + \underbrace{6(-1)}_{6\text{CN}^-} = \underbrace{-4}_{\substack{\text{net charge}\\ \text{on complex}}}$$

4. The formula for a complex ion is enclosed in brackets, and so we write the formula for the anion as $[Fe(CN)_6]^{4-}$. The formula for the salt is $K_4[Fe(CN)_6]$ because four K^+ ions are required to balance the -4 charge on the anion.

Example 22-5: Give the formula for the compound

hexaamminecobalt(III) hexachlorocobaltate(III)

Solution: In this case both the cation and the anion are complex ions. The cation has six NH_3 ligands with zero charge and one cobalt in a +3 oxidation state. Therefore, the formula for the cation is

$$[Co(NH_3)_6]^{3+}$$

The anion has six Cl^- with a total ligand charge of -6 plus one cobalt in a $+3$ oxidation state. Therefore, the net charge on the complex anion is

$$6(-1) + (+3) = -3$$

and the formula is

$$[CoCl_6]^{3-}$$

The magnitudes of the charges on the cation and the anion are equal and thus appear in the formula for the salt on a one-to-one basis:

$$[Co(NH_3)_6][CoCl_6]$$

▪ Note that we usually write the formulas for complexes in brackets.

The nomenclature of transition-metal complexes may appear cumbersome at first because of the length of the names, but with a little practice you will find it straightforward and ultimately much simpler than memorizing a host of nonsystematic common names. Table 22-4

Table 22-4 Some examples of nomenclature for transition-metal compounds

Compound	Name
$[Co(NH_3)_6]Cl_3$	hexaamminecobalt(III) chloride
$K[AuCl_4]$	potassium tetrachloroaurate(III)
$Cu_2[Fe(CN)_6]$	copper(II) hexacyanoferrate(II)
$[Pt(NH_3)_6]Cl_4$	hexaammineplatinum(IV) chloride
$[Cu(NH_3)_4(H_2O)_2]Cl_2$	tetraamminediaquacopper(II) chloride
$[Cr(CO)_6]$	hexacarbonylchromium(0)
$K_3[CoF_6]$	potassium hexafluorocobaltate(III)

gives several additional examples of names of transition-metal complexes. You should try to name each one from the formula and write the formula from each name.

22-4. SOME OCTAHEDRAL AND SQUARE-PLANAR TRANSITION METAL COMPLEXES CAN EXIST IN ISOMERIC FORMS

Consider the compound

$[Pt(NH_3)_2Cl_2]$ diamminedichloroplatinum(II)

Platinum(II) complexes are invariably square-planar. As shown in Figure 22-9 (margin), there are two possible arrangements of the four ligands around the central platinum(II) ion, which give the **cis and trans isomers** of the compound. The cis and trans compounds are **geometric isomers.** The designation cis (the same) tells us that the identical ligands are placed adjacent to each other (on the same side) in the structure. The designation trans (opposite) tells us that the identical ligands are placed directly opposite each other in the structure. The cis and trans isomers of $[Pt(NH_3)_2Cl_2]$ are different compounds with different physical and chemical properties. For example, the *cis*-diamminedichloroplatinum(II) isomer is manufactured as the potent anticancer drug **cisplatin,** whereas the trans isomer does not exhibit anticancer activity. How the cis isomer destroys cancer cells is not understood, and this is an important area of chemical research.

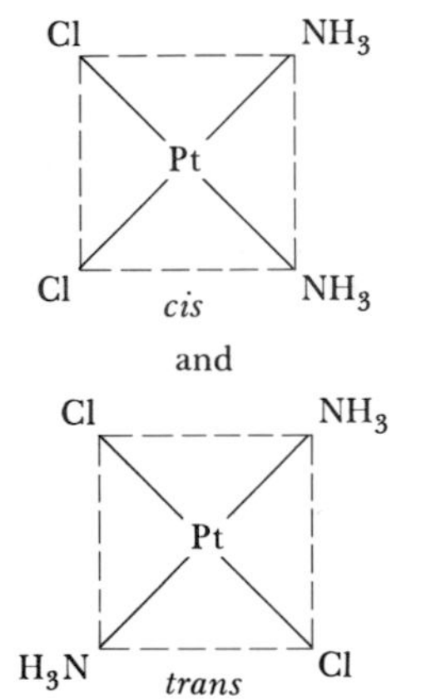

Figure 22-9 *Cis* and *trans* square-planar complexes of diamminedichloroplatinum(II).

Cis and trans isomers are also found in certain octahedral complexes. Consider the octahedral ion tetraamminedichlorocobalt(III), $[Co(NH_3)_4Cl_2]^+$. The two Cl^- ligands can be placed in adjacent (cis) or opposite (trans) positions around the central cobalt(III) ion, as shown below. Note that because the six **coordination positions** (that is, points

cis-$[Co(NH_3)_4Cl_2]$ (violet)

trans-$[Co(NH_3)_4Cl_2]$ (green)

of attachment) in an octahedral complex are equivalent, any other cis placement of the two Cl^- ligands around the cobalt(III) ion yields a structure identical to the cis structure shown here; this is also true for the trans placement of the two Cl^- ligands.

Example 22-6: The compound $[Co(NH_3)_3Cl_3]$ exists in two isomeric forms. Draw the structures of the two isomers.

Solution: The structures of the two isomers are shown in Figure 22-10. One compound is denoted cis,cis because each Cl^- is adjacent to the two others. The other compound is denoted cis,trans because one Cl^- is adjacent to and one Cl^- is opposite the other chloride.

The structure of transition-metal complexes was worked out by the Swiss chemist Alfred Werner in the late nineteenth and early twentieth centuries, without the aid of modern X-ray structure determination methods. In 1893, at the age of 26, he proposed a correct structural theory based on the number of different types of complexes, including isomers, that could be prepared for platinum(II), platinum(IV), and cobalt(III) amminechloro complexes. In 1913 Werner was awarded the Nobel Prize for his research in transition-metal chemistry.

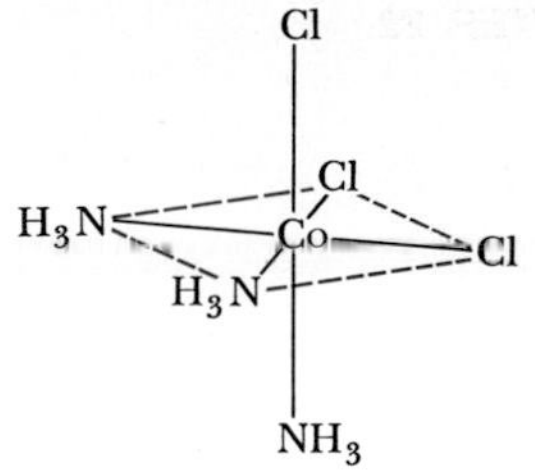

cis, cis-$[Co(NH_3)_3Cl_3]$

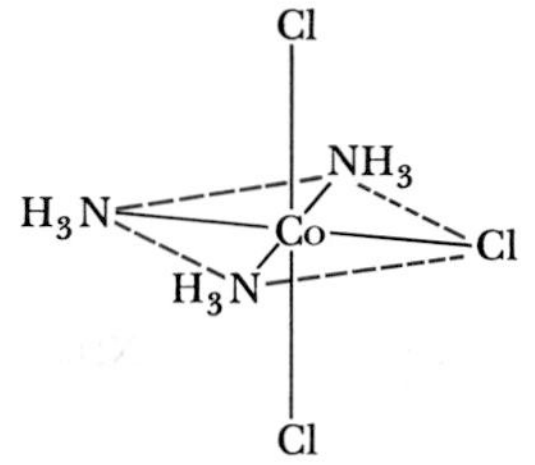

cis, trans-$[Co(NH_3)_3Cl_3]$

Figure 22-10 *Cis-cis* and *cis-trans* isomers of octahedral complexes.

22-5. POLYDENTATE LIGANDS BIND TO MORE THAN ONE COORDINATION POSITION AROUND THE METAL ION

Certain ligands can attach to a central metal cation at more than one coordination position. Examples of ligands that attach to two coordination positions are the oxalate ion (abbreviated ox) and ethylenediamine (abbreviated en):

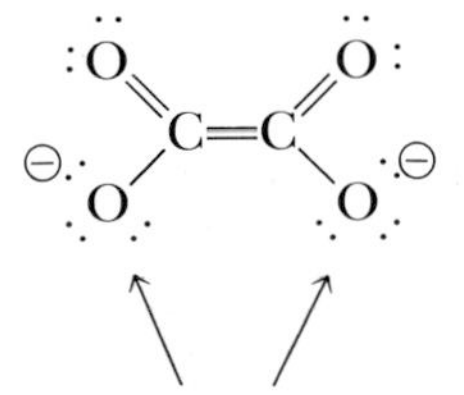

ligating atoms
oxalate ion (ox)

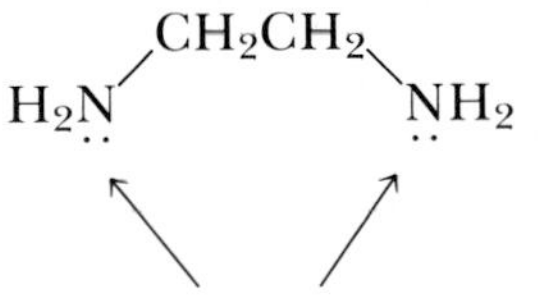

ligating atoms
ethylenediamine (en)

The atoms of the ligand that attach to the metal ion are called **ligating atoms.** Two complexes involving these two ligands are shown in the margin.

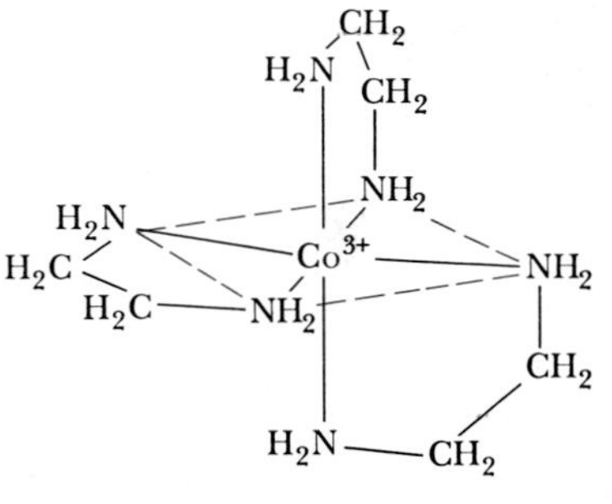

Ligands that attach to a metal ion at more than one coordination position are called **polydentate ligands** or **chelating ligands.** The resulting complex is called a **chelate,** which comes from the Greek word meaning claw. The attachment of a chelating ligand can be thought of as a grasping of the metal ion with molecular claws. A chelating ligand that attaches to two metal coordination positions is called **bidentate** (two teeth), one that attaches to three positions is **tridentate** (three teeth), and so on.

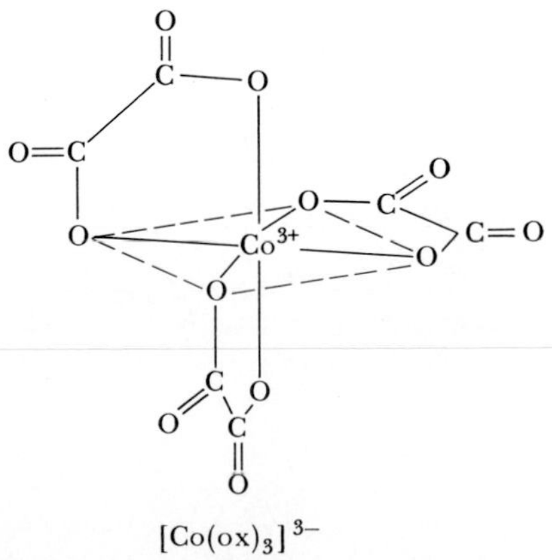

Ethylene-diaminetetraacetate ion, $EDTA^{4-}$, is the best known example of a hexadentate ligand:

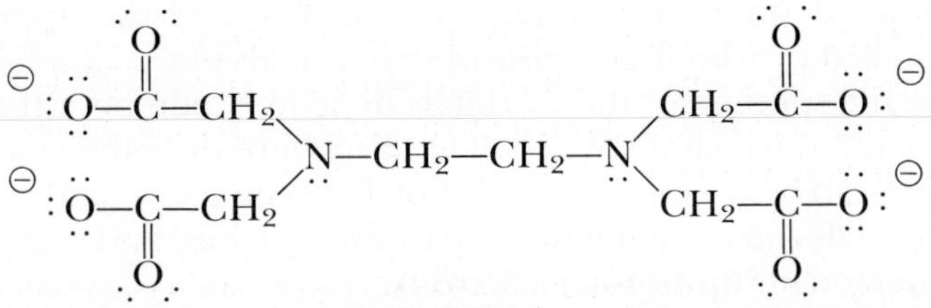

$EDTA^{4-}$

$[Co(en)_3]^{3+}$

$[Co(ox)_3]^{3-}$

The six ligating atoms are shown in color. EDTA binds strongly to a number of metal ions and has a great variety of uses. It is used as an antidote for poisoning by heavy metals such as lead and mercury; as a food preservative, where it complexes with and renders inactive metal ions that catalyze the reactions involved in the spoiling process; as an analytical reagent in the analysis of the hardness of water; in detergents, soaps and shampoos; to decontaminate radioactive surfaces; and many, many other applications.

The nomenclature for complex ions and molecules that have polydentate ligands follows the rules listed in Section 22-3, with one additional rule:

7. *If the ligand attached to the metal ion is a polydentate ligand, then enclose the ligand name in parentheses and use the prefix bis- for two ligands and tris- for three ligands:* for example, tris(ethylenediamine)cobalt(III), $[Co(H_2NCH_2CH_2NH_2)_3]^{3+}$.

Example 22-7: Give the chemical formula for ammonium tris(oxalato)ferrate(III).

Solution: The cation is the ammonium ion, NH_4^+. The anion is a complex ion with three (tris) oxalate ions, $C_2O_4^{2-}$, and an iron atom in the +3 oxidation state. The net charge on the complex anion is

$$\underbrace{(+3)}_{\text{Fe(III)}} + \underbrace{3(-2)}_{3C_2O_4^{2-}} = \underbrace{-3}_{\text{net charge}}$$

Therefore, the formula for the complex anion is $[Fe(C_2O_4)_3]^{3-}$ and the formula for ammonium tris(oxalato)ferrate(III) is

$$(NH_4)_3[Fe(C_2O_4)_3]$$

Note that we do not say triammonium because the number of NH_4^+ ions is umambiguously fixed by the net charge of −3 on the complex anion. There are four ions per formula unit in $(NH_4)_3[Fe(C_2O_4)_3]$.

22-6. THE FIVE *d* ORBITALS OF A TRANSITION-METAL ION IN AN OCTAHEDRAL COMPLEX ARE SPLIT INTO TWO GROUPS BY THE LIGANDS

The five *d* orbitals (Figure 22-2) in a gas-phase transition metal or metal ion without any attached ligands all have the same energy. However, when six identical ligands are attached to the transition-metal ion to form an octahedral complex, the *d* orbitals on the metal ion are split into two sets (see next page). The lower set of orbitals, called $\boldsymbol{t_{2g}}$ **orbitals,** consists of three orbitals (d_{xy}, d_{xz}, *and* d_{yz}), and the upper set of orbitals, called $\boldsymbol{e_g}$ **orbitals,** consists of two orbitals ($d_{x^2-y^2}$, d_{z^2}). The magnitude of the splitting of the *d* orbitals depends upon both the central metal ion and the ligands. We shall let $\boldsymbol{\Delta_o}$ (where the subscript *o* stands for octahedral) be the energy separation between the t_{2g} and e_g orbitals. The t_{2g} orbitals can accommodate up to six electrons, and the e_g orbitals can accommodate up to four electrons.

The reason for the particular *d*-orbital splitting pattern can be un-

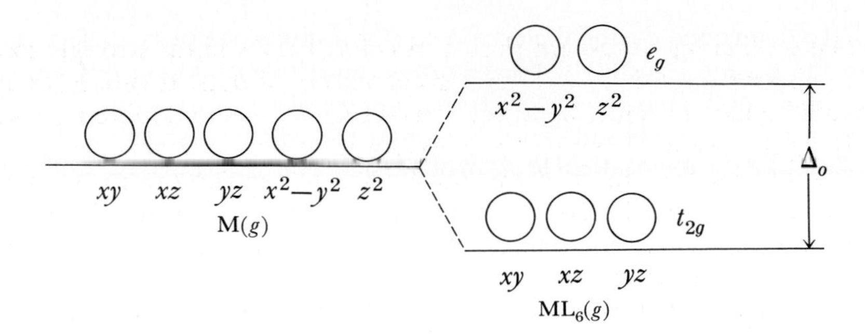

derstood from the placement of the six ligands relative to the d orbitals in the complex. Consider the octahedral transition-metal complex shown in Figure 22-11. The key point is the placement of the ligands relative to the lobes of the d orbitals. The ligand electrons, especially those in lone pairs directed at the central metal ion, provide an electric field. This electric field is not spherically symmetric around the central metal ion because the ligands are positioned in an octahedral array. Therefore, the energies of the metal d orbitals are changed by different amounts, depending on their positions with respect to the ligands. In the simplified picture described here, the ligand lone pairs remain with the ligand. Figure 22-11 shows that the lobes of the $d_{x^2-y^2}$ and d_{z^2} orbitals point directly at the ligands and that the lobes of the d_{xy}, d_{xz}, and d_{yz} orbitals all point between the ligands. Thus electrons placed in $d_{x^2-y^2}$ and d_{z^2} orbitals experience a greater electrostatic repulsion (like charges repel) and thus have a higher energy in the complex than electrons placed in the d_{xy}, d_{xz}, and d_{yz} orbitals. The d-orbital splitting pattern in an octahedral complex is thus seen to be a consequence of the positions of the ligands relative to the d orbitals.

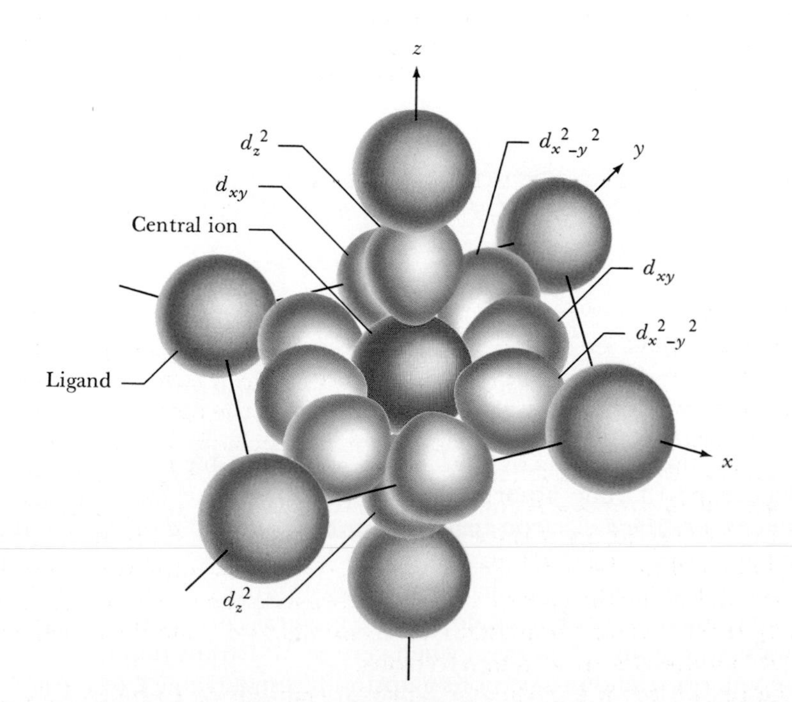

Figure 22-11 A regular octahedral complex, showing the orientation of the d_{z^2}, $d_{x^2-y^2}$, and d_{xy} orbitals relative to the ligands, which are brought in along the x, y, and z axes toward the central metal ion. For simplicity, the d_{xz} and d_{yz} orbitals are not shown. The d_{xy}, d_{xz}, and d_{yz} lobes point toward positions between the ligands, as shown here for d_{xy}.

To determine d-orbital electron configurations, we place d electrons in the t_{2g} and e_g orbitals. For example, an octahedral complex ion of chromium(III) has three d electrons, and we place them in the t_{2g} orbitals in accord with Hund's rule (Section 6-18), which says that we should place one electron in each t_{2g} orbital such that the spins are aligned. Thus, we write the ground-state d electron configuration of chromium(III), depicted in the margin, as $t_{2g}^3e_g^0$, or simply t_{2g}^3.

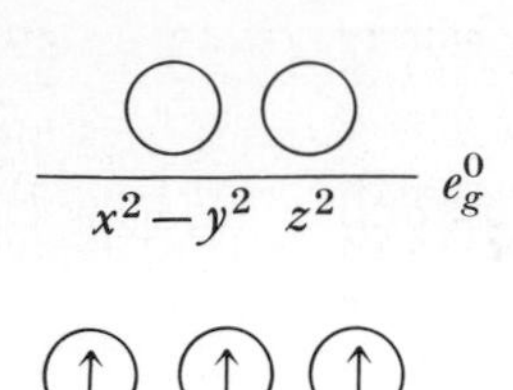

Example 22-8: Determine the ground-state d-electron configuration of nickel(II) in an octahedral complex.

Solution: Nickel(II) has eight d electrons. We place the eight d electrons in the t_{2g} and e_g orbitals in accord with the Pauli Exclusion Principle and Hund's rule. The ground-state d-electron configuration of nickel(II) is $t_{2g}^6e_g^2$.

The magnitude of the separation between the t_{2g} and e_g orbitals, Δ_o, depends upon the central metal ion and the ligands. For most cases Δ_o is such that the energy difference corresponds to the visible region of the electromagnetic spectrum. In other words, the frequency of the radiation that is absorbed, which obeys the relation

$$E = h\nu = \Delta_o$$

is in the visible region of the spectrum. Thus, many complexes absorb light in the visible region and are colored.

It is possible to understand the variety of colors of complex ions in terms of Δ_o and the electron occupancy of the t_{2g} and e_g orbitals. For example, consider the red-purple ion $[Ti(H_2O)_6]^{3+}(aq)$. Reference to Figure 22-1 shows that titanium is the second member of the $3d$ transition series, and thus Ti(II) has two d electrons. The $[Ti(H_2O)_6]^{3+}$ complex, which contains Ti(III), has one fewer d electron than Ti(II), and therefore $[Ti(H_2O)_6]^{3+}$ is a d^1 ion. The ground-state and excited-state d electron configurations of the d^1 $[Ti(H_2O)_6]^{3+}$ ion are

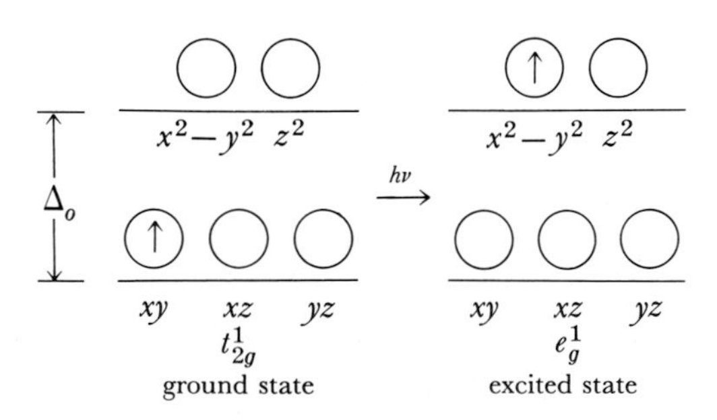

Figure 22-12 An aqueous solution of titanium(III) chloride. The d^1 ion $[Ti(H_2O)_6]^{3+}$ has a red-purple color.

The absorption spectrum of $[Ti(H_2O)_6]^{3+}$ in the visible region is shown in Figure 22-13. The absorption of a photon in the blue-to-orange region excites the d electron from the lower (t_{2g}) set of d orbitals to the upper (e_g) set.

The colors of many familiar gemstones (Figure 22-14) are due to transition-metal ions. For example, the mineral corundum, Al_2O_3, is colorless when pure. However, when certain M^{3+} transition metals are present in trace amounts, various gemstones result (Table 22-5). All the

Table 22-5 Transition-metal ion substitutions for various gemstones

Trace transition-metal	Color	Name
Corundum		
Cr^{3+}	brilliant red	ruby
Mn^{3+}	deep purple	amethyst
Fe^{3+}	yellow to reddish-yellow	topaz (yellow sapphire)
V^{3+} or Co^{3+}	rich blue	blue sapphire
Beryl		
Cr^{3+}	green	emerald
Fe^{3+}	pale blue to light greenish-blue	aquamarine
Silicates		
Fe^{3+}	red	garnet
Cr^{3+}	green	jade
Fe^{2+}	yellow-green	peridot

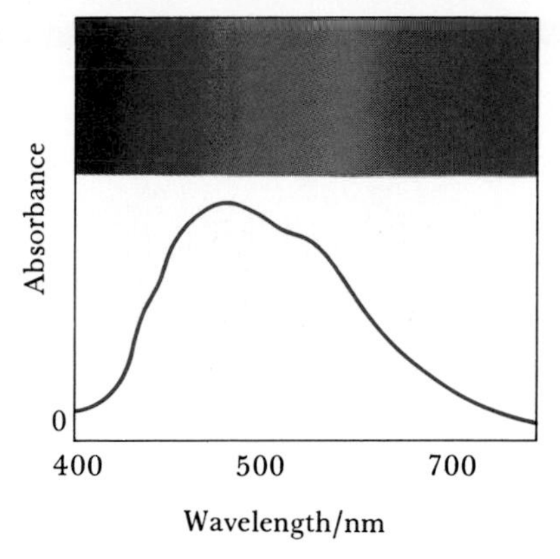

Figure 22-13 Absorption spectrum of the red-purple ion $[Ti(H_2O)_6]^{3+}$ in aqueous solution. The complex absorbs light in the blue, green, yellow, and orange regions, but most of the light in the red and purple regions passes through the sample and so is detected by the eye. Thus, the $[Ti(H_2O)_6]^{3+}$ complex is red-purple.

Figure 22-14 The colors of many gemstones are due to small quantities of transition-metal ions. Shown here are (1) kunzite, (2) garnet, (3) zircon, (4) aquamarine, (5) amethyst, (6) peridot, (7) morganite, (8) topaz, (9) ruby, (10) tourmaline (indicolite), (11) chrome tourmaline, (12) rose quartz, (13) rubellite tourmaline, (14) kyanite, (15) citrine, and (16) green tourmaline.

corundum and beryl gemstones in Table 22-5 can be made in the laboratory. Very large ruby rods are used in ruby lasers. The quality of synthetic rubies is at least as good as that of any natural ruby. Synthetic emeralds differ slightly in water content from natural emeralds, but the difference cannot be detected by eye. Synthetic emeralds can withstand much higher temperatures than natural emeralds without shattering. The monetary values of many types of gemstones are artificially maintained by limiting production of the synthetic ones.

22-7. *d*-ORBITAL ELECTRON CONFIGURATION IS THE KEY TO UNDERSTANDING MANY PROPERTIES OF THE *d* TRANSITION METAL IONS

Moving charges give rise to a magnetic field. Unpaired electrons act like tiny magnets as a result of their spinning about their axes. The magnetic fields from paired electrons, which have opposite spins, cancel out. Molecules with no unpaired electrons cannot be magnetized by an external magnetic field and are called **diamagnetic.** Molecules with unpaired electrons can be magnetized by an external field and are called **paramagnetic.** In paramagnetic molecules, an external magnetic field lines up the electron spins parallel to the applied field, and thus a paramagnetic substance behaves like a magnet and is drawn into an externally applied magnetic field (Figure 22-15). A diamagnetic substance is not drawn into an applied magnetic field and thus can be distinguished from a paramagnetic substance. Furthermore, in some cases, it is possible to determine the number of unpaired electrons by measuring the force with which the paramagnetic substance is drawn into the magnetic field.

Magnetic experiments have shown that the compound $K_4[Fe(CN)_6]$ is diamagnetic, whereas the compound $K_4[FeF_6]$ is paramagnetic. Furthermore, there are four unpaired electrons in $[FeF_6]^{4-}$. We can explain these observations in terms of *d*-orbital electron configurations.

Let's consider the ion $[FeF_6]^{4-}$. Iron(II) has six *d* electrons, and so there are two possibilities for the *d*-electron configuration of an octahedral iron(II) complex ion:

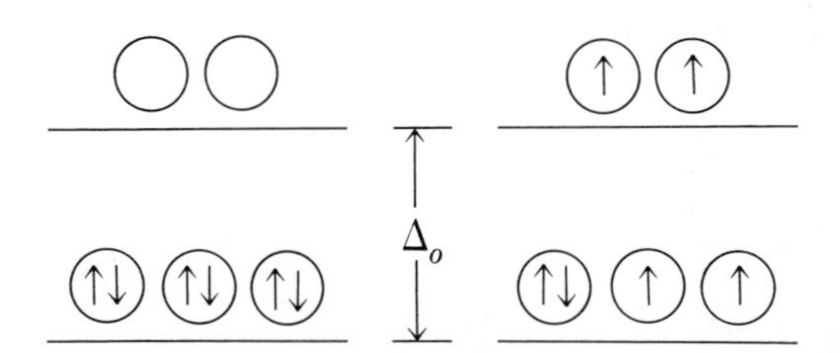

In one case the *d*-electron configuration is $t_{2g}^6e_g^0$, and in the other case it is $t_{2g}^4e_g^2$. In the t_{2g}^6 configuration, the spins of all the electrons are paired; in the $t_{2g}^4e_g^2$ configuration, the electrons are in different orbitals with four unpaired electrons, in accord with Hund's rule. The $t_{2g}^6e_g^0$ configuration is said to be a **low-spin configuration,** and the $t_{2g}^4e_g^2$ configuration is said to be a **high-spin configuration.**

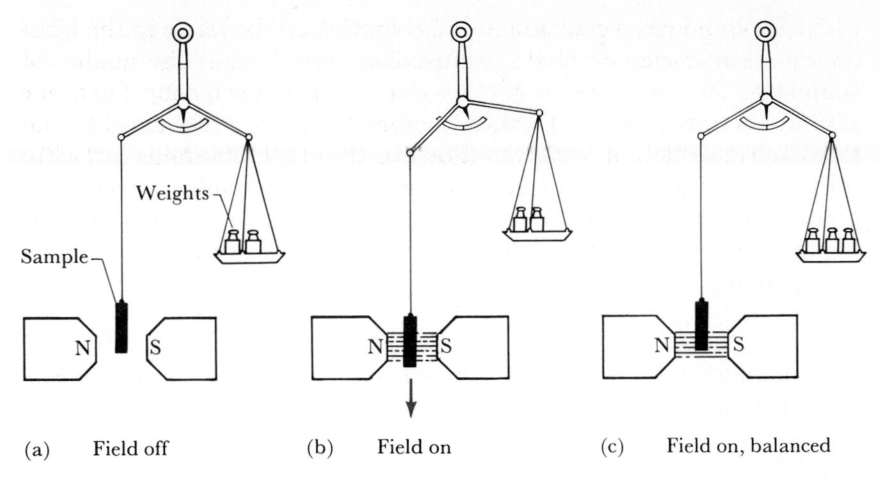

Figure 22-15 Attraction of a paramagnetic substance into a magnetic field. The magnetic pull on the sample makes it appear heavier. The number of unpaired electrons in the sample can be calculated from the apparent mass gain. Masses are added to the balance pan until balance is restored with the field on.

The value of Δ_o determines whether a d-electron configuration is low-spin or high-spin. If Δ_o is small, then the d electrons occupy the e_g orbitals before they pair up in the t_{2g} orbitals. If Δ_o is large, then the d electrons fill the t_{2g} orbitals completely before occupying the higher-energy e_g orbitals. For example, the d^6 ion $[FeF_6]^{4-}$ is high-spin and has four unpaired electrons:

$$e_g^2 \quad (\uparrow)(\uparrow)$$

$$\leftarrow \Delta_o(\mathrm{F}) \qquad [FeF_6]^{4-} \qquad t_{2g}^4 e_g^2$$

$$t_{2g}^4 \quad (\uparrow\downarrow)(\uparrow)(\uparrow)$$

This is because Δ_o is small relative to the energy required to pair up the electrons in the t_{2g} orbitals. The d^6 ion $[Fe(CN)_6]^{4-}$, on the other hand, is low-spin and has no unpaired electrons:

$$e_g^0 \quad (\ )(\ )$$

$$\leftarrow \Delta_o(\mathrm{CN}^-) \qquad [Fe(CN)_6]^{4-} \qquad t_{2g}^6$$

$$t_{2g}^6 \quad (\uparrow\downarrow)(\uparrow\downarrow)(\uparrow\downarrow)$$

This is because Δ_o is large relative to the energy required to pair up the electrons in the t_{2g} orbitals. Thus, we see that $[FeF_6]^{4-}$ is paramagnetic, with four unpaired electrons and that $[Fe(CN)_6]^{4-}$ is diamagnetic, with no unpaired electrons. The Δ_o values obtained from the spectra of the complexes show that the CN^- ligands interact much more strongly with iron(II) than do the F^- ligands; that is, $\Delta_o(CN^-) \gg \Delta_o(F^-)$. The increased *d*-orbital splitting energy for CN^- is sufficiently great to overcome the additional electron-electron repulsions that result from pairing up the electrons. A low-spin complex results whenever the energy difference between the t_{2g} and e_g orbitals is greater than the **electron-pairing energy.**

- Low-spin complex:
  $\Delta_o >$ pairing energy

  High-spin complex:
  $\Delta_o <$ pairing energy

Example 22-9: Give the *d*-electron configuration of the low-spin, complex $[Pt(NH_3)_6]^{4+}$.

Solution: Referring to Figure 22-1, we note that platinum is the eighth member of the 5*d* transition series. Therefore, platinum(II) is a d^8 ion. The platinum in $[Pt(NH_3)_6]^{4+}$ is platinum(IV), and thus platinum(IV) is a d^6 ion [two fewer *d* electrons than platinum(II)]. The *d*-electron configuration of a

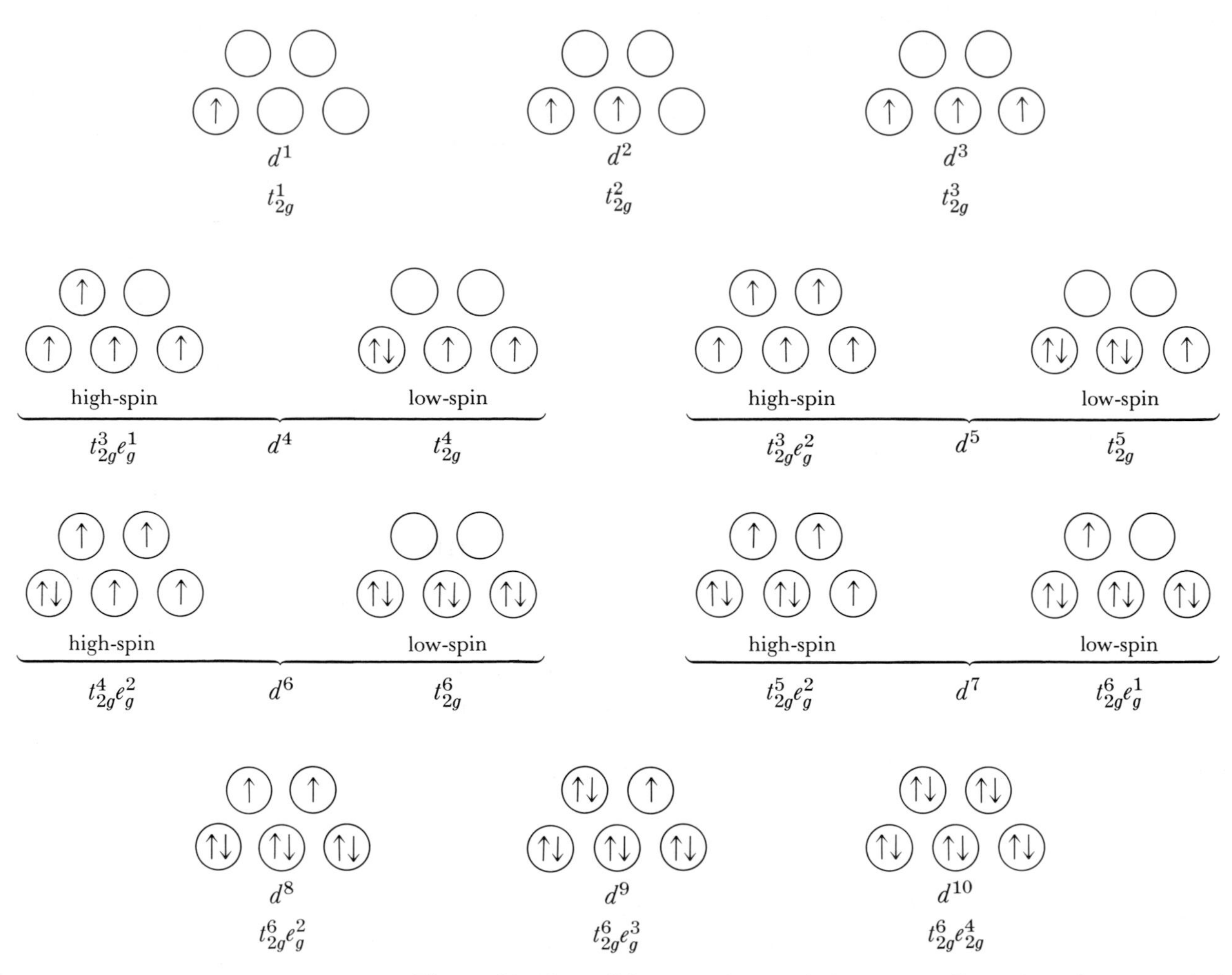

Figure 22-16 Possible ground-state *d*-electron configurations for octahedral d^x ions, where *x* is the total number of outer *d* electrons in the transition metal. Note that for d^4, d^5, d^6, and d^7 ions there are two possibilities: the high-spin and the low-spin configurations.

low-spin d^6 ion is $t_{2g}^6 e_g^0$ (see margin). The ion is diamagnetic because it has no unpaired electrons.

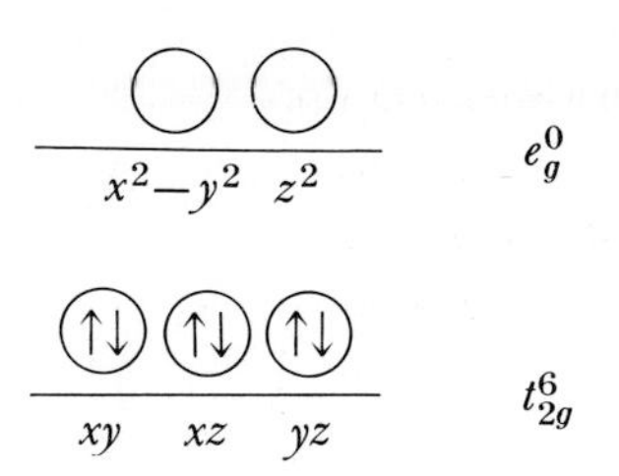

The various possible d-electron configurations for octahedral transition metal ions are shown in Figure 22-16. Note that for d^4, d^5, d^6, and d^7 octahedral complexes there are two possible d-electron configurations, namely, the high-spin (maximum possible number of unpaired d electrons) and the low-spin (minimum possible number of unpaired d electrons).

A particularly interesting case of a conversion of a transition-metal ion from a high-spin to a low-spin configuration occurs in hemoglobin. Hemoglobin is an iron-containing protein that is responsible for the color of red blood cells, which are about 35 percent hemoglobin by mass. Hemoglobin transports O_2 from the lungs to the various cells of the body, where it is used in the oxidation of foodstuffs, and transports CO_2, a waste product of the oxidation processes, back to the lungs for elimination. A hemoglobin molecule contains four so-called heme groups, each of which acts as a pentadentate (five-coordinate) ligand to an Fe^{2+} ion (Figure 22-17). The vacant, sixth coordinate position about the Fe^{2+} ion is taken up by an oxygen molecule, a hydrogen carbonate ion (in CO_2 transport) or a water molecule. The hemoglobin-oxygen complex, which is called oxyhemoglobin, is bright red and has a low-spin d-electron configuration. When oxygen is removed, the bluish-red high-spin deoxyhemoglobin is formed.

$$\underset{\text{red (low-spin)}}{\mathrm{Hmb}\cdot 4O_2} \rightleftharpoons \underset{\text{red-blue (high-spin)}}{\mathrm{Hmb}} + 4O_2$$

Δ_{red} Δ_{blue}

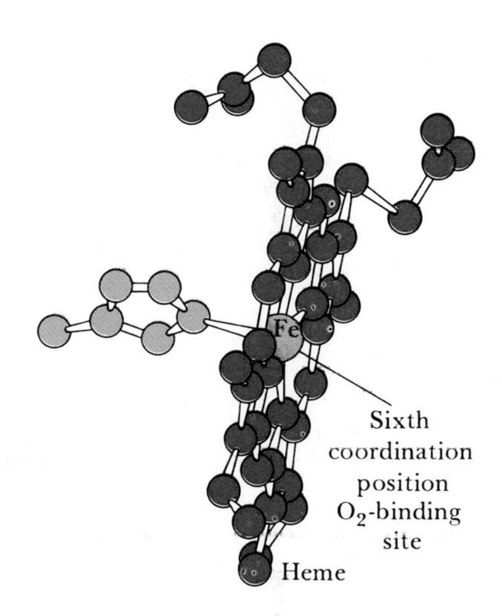

Figure 22-17 Model of the oxygen-binding site in a heme portion of hemoglobin. The iron atom is already attached to five atoms. The O_2 molecule attaches to the vacant sixth coordination position of the iron atom.

Arterial blood is red because it contains O_2 coordinated to iron(II); venous blood, which has passed through the capillaries and given up its oxygen to the tissues, is bluish-red (look at the bluish veins near the surface of your skin). Deprivation of oxygen, as occurs when the air passages to the lungs are blocked, produces a characteristic blue coloration in the affected person. The toxicity of carbon monoxide gas arises from the ability of this ligand to bind more tightly than O_2 to the iron(II) in hemoglobin and thereby block the uptake of O_2 from the lungs.

22-8. LIGANDS CAN BE ORDERED ACCORDING TO THEIR ABILITY TO SPLIT THE TRANSITION-METAL *d* ORBITALS

The Japanese chemist Ryutaro Tsuchida first noted that for octahedral complexes with the same metal oxidation state in the same transition series, the value of Δ_o depends primarily on the type of ligands and is

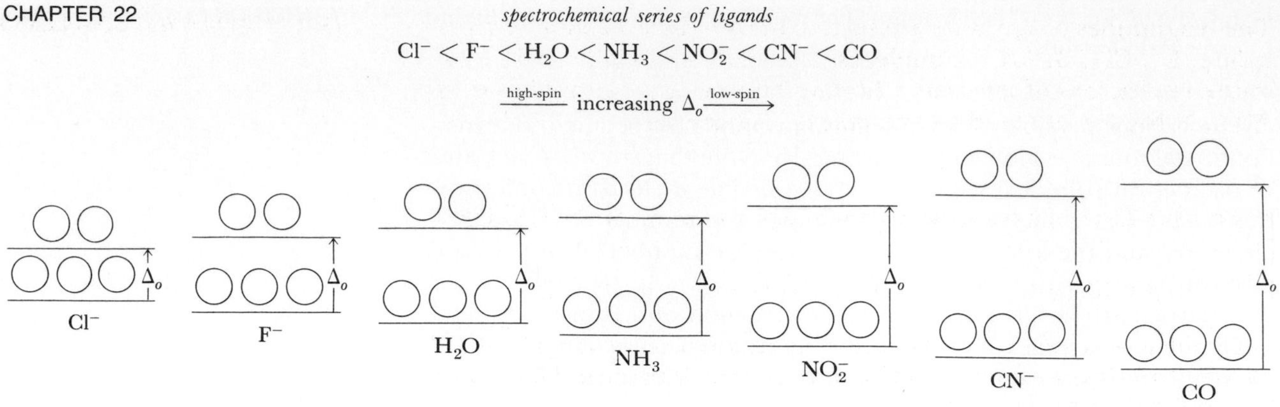

spectrochemical series of ligands

$Cl^- < F^- < H_2O < NH_3 < NO_2^- < CN^- < CO$

high-spin — increasing Δ_o — low-spin

Cl⁻ F⁻ H_2O NH_3 NO_2^- CN^- CO

Figure 22-18 The spectrochemical series for some common ligands.

independent of the metal ion. Tsuchida's observation enabled him to arrange common ligands in a **spectrochemical series** in order of increasing ability of ligands to split the metal *d* orbitals. An abbreviated version of Tsuchida's spectrochemical series is shown in Figure 22-18.

The ligands NO_2^-, CN^-, and CO cause large *d*-orbital splitting (large Δ_o), which results in low-spin complexes. The halide ions cause relatively small *d*-orbital splitting (small Δ_o), which results in high-spin complexes. Thus the chloride complex of the d^6 ion cobalt(III), $[CoCl_6]^{3-}$, is high-spin ($t_{2g}^4e_g^2$) and the cyanide complex, $[Co(CN)_6]^{3-}$, is low-spin ($t_{2g}^6e_g^0$).

Example 22-10: Predict the number of unpaired electrons in a hexachlorochromium(II) ion.

Solution: The chemical formula of the ion is $[CrCl_6]^{4-}$. Chromium(II) is a d^4 ion and the value of Δ_o for Cl^- is relatively small. Referring to Figure 22-16, we predict that $[CrCl_6]^{4-}$ is a high-spin complex with four unpaired electrons.

The ligands H_2O and NH_3 give rise to Δ_o values for M(II) 3*d* ions that are roughly equal to the pairing energy, and thus some H_2O and NH_3 complexes are high-spin and some are low-spin. For a given metal and ligand, the value of Δ_o increases as the oxidation state of the metal increases, because the higher ionic charge leads to a greater electrostatic interaction between the metal and the ligand. Spectroscopic measurements show that the value of Δ_o increases as we go from the 3*d* to the 4*d* to the 5*d* transition series for a particular number of *d* electrons. These trends are illustrated by the following data:

Complex	Δ_o/kJ·mol^{-1}	
$[Co(NH_3)_6]^{2+}$	121	increase in Δ_o due to increased oxidaton state of cobalt (121 → 274)
$[Co(NH_3)_6]^{3+}$	274	increase in Δ_o as we move down a column in the periodic table for a given oxidation state (274 → 408 → 478)
$[Rh(NH_3)_6]^{3+}$	408	
$[Ir(NH_3)_6]^{3+}$	478	

The magnitudes of Δ_o values are such that, with the exception of metal halide (F^-, Cl^-, Br^-, I^-) complexes, all M(III) and higher-oxidation-state complex ions of the 4*d* and 5*d* transition-metal series are low-spin. Also, all NO_2^-, CN^-, and CO complexes are low-spin.

Example 22-11: Arrange the following complexes in order of increasing Δ_o values:

$$[Co(NO_2)_6]^{3-} \quad [Fe(H_2O)_6]^{3+} \quad [Co(NH_3)_6]^{3+} \quad [Fe(H_2O)_6]^{2+}$$

Solution: The order of the ligands in the spectrochemical series is $H_2O < NH_3 < NO_2^-$. For a given ligand, Δ_o increases with increasing charge; thus, in order of increasing Δ_o values, we have

$$[Fe(H_2O)_6]^{2+} < [Fe(H_2O)_6]^{3+} < [Co(NH_3)_6]^{3+} < [Co(NO_2)_6]^{3-}$$

22-9. TRANSITION-METAL COMPLEXES ARE CLASSIFIED AS EITHER INERT OR LABILE

Many of the chemical properties of complex ions can be understood or predicted from the *d*-electron configuration. Here we consider the stability of complex ions. An **inert complex** is one that only slowly exchanges its ligands with other available ligands. A **labile complex** is one that rapidly exchanges its ligands with other available ligands. As a dramatic example of the contrasting ligand exchange rates of labile and inert complexes, consider the ligand-substitution reactions described by the equations

▪ The word labile means readily open to change.

$$(1)\ [Co(NH_3)_6]^{2+}(aq) + 6H_3O^+(aq) \rightleftharpoons [Co(H_2O)_6]^{2+}(aq) + 6NH_4^+(aq)$$

$$(2)\ [Co(NH_3)_6]^{3+}(aq) + 6H_3O^+(aq) \rightleftharpoons [Co(H_2O)_6]^{3+}(aq) + 6NH_4^+(aq)$$

The equilibrium for both reactions lies far to the right. Nonetheless, the reaction described by Equation (1), which involves the d^7 $[Co(NH_3)_6]^{2+}(aq)$ complex, attains equilibrium in 1 M $H^+(aq)$ in about 10 s, whereas the reaction described by Equation (2), which involves the d^6 $[Co(NH_3)_6]^{3+}(aq)$ complex, requires over a month to reach equilibrium at 25°C under the same conditions. In other words, a difference of only one *d* electron causes the reaction rates to differ by a factor of over 350,000.

Henry Taube, then at the University of Chicago, was the first to note that $t_{2g}^3e_g^0$, $t_{2g}^4e_g^0$, $t_{2g}^5e_g^0$, and $t_{2g}^6e_g^0$ octahedral complexes are inert and that all other octahedral complexes are labile. Thus $[Co(NH_3)_6]^{2+}(aq)$, an octahedral high-spin d^7 ion ($t_{2g}^5e_g^2$), is labile and $[Co(NH_3)_6]^{3+}(aq)$, an octahedral, low-spin d^6 ion ($t_{2g}^6e_g^0$), is inert.

$t_{2g}^6e_g^0$

e_g ○ ○

t_{2g} (↑↓) (↑↓) (↑↓)

$[Fe(CN)_6]^{4-}$

$t_{2g}^3e_g^0$

e_g ○ ○

t_{2g} (↑) (↑) (↑)

$[Cr(H_2O)_6]^{3+}$

$t_{2g}^2e_g^0$

e_g ○ ○

t_{2g} (↑) (↑) ○

$[V(H_2O)_6]^{3+}$

Example 22-12: Predict whether the following complex ions are labile or inert:

$$[Fe(CN)_6]^{4-} \quad [Cr(H_2O)_6]^{3+} \quad [V(H_2O)_6]^{3+}$$

Solution: The *d*-electron configurations of the three complexes are given in the margin. The d^6 low-spin $[Fe(CN)_6]^{4-}$ ion is $t_{2g}^6e_g^0$ and thus inert; the d^3 $[Cr(H_2O)_6]^{3+}$ ion is $t_{2g}^3e_g^0$ and thus inert; the d^2 $[V(H_2O)_6]^{3+}$ ion is $t_{2g}^2e_g^0$ and thus labile. These three predictions are confirmed by experiment.

Although Taube's rules can be given a theoretical foundation, we treat them here only as useful empirical rules because the theoretical interpretation is not simple.

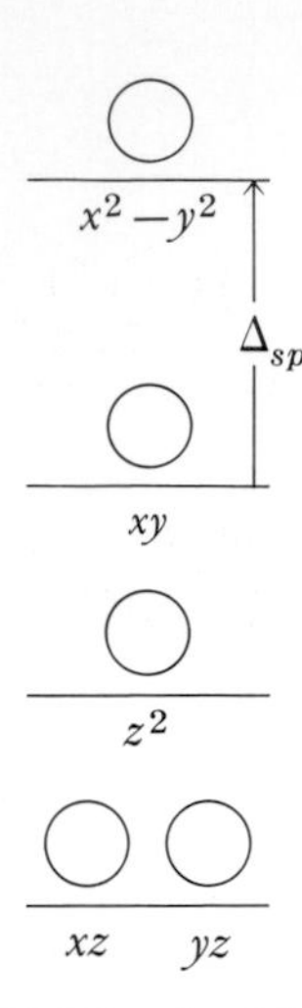

Figure 22-19 The *d*-orbital splitting pattern for a square-planar transition-metal complex. The value of Δ_{sp} (*sp* stands for square-planar) is the energy separation between the two highest-energy *d* orbitals.

22-10. THE *d*-ORBITAL SPLITTING PATTERNS IN SQUARE-PLANAR AND TETRAHEDRAL COMPLEXES ARE DIFFERENT FROM THOSE IN OCTAHEDRAL COMPLEXES

The splitting pattern of the *d* orbitals in an octahedral complex is explained in Figure 22-11. The splitting patterns in square-planar and tetrahedral complexes differ from the splitting pattern in octahedral complexes because the ligands are brought in from different directions toward the central metal ion.

The splitting pattern of a square-planar complex is shown in Figure 22-19. The closer the lobes of a particular *d* orbital to the ligands, the higher is the energy of that orbital. In a square-planar complex, the four ligands are placed on the *x* and *y* axes, and thus the lobes of the $d_{x^2-y^2}$ orbital point directly at the ligands and the $d_{x^2-y^2}$ orbital has the highest energy of the five *d* orbitals. The relative energies of the other *d* orbitals can be deduced by an analysis of their interaction with the ligands, as we did in Figure 22-11 for the octahedral case. The difference in energy between the $d_{x^2-y^2}$ orbital and the d_{xy} orbital in a square-planar complex is denoted by $\boldsymbol{\Delta_{sp}}$.

Example 22-13: Give the *d*-electron configuration for the square-planar complex $[Pd(CN)_4]^{2-}$.

Solution: Referring to Figure 22-1, we note that palladium is the eighth member of the 4*d* transition series. The oxidation state of palladium in $[Pd(CN)_4]^{2-}$ is Pd(II), and thus palladium(II) is a d^8 ion. Referring to Figure 22-19, we see that there are two possible configurations, which are given below.

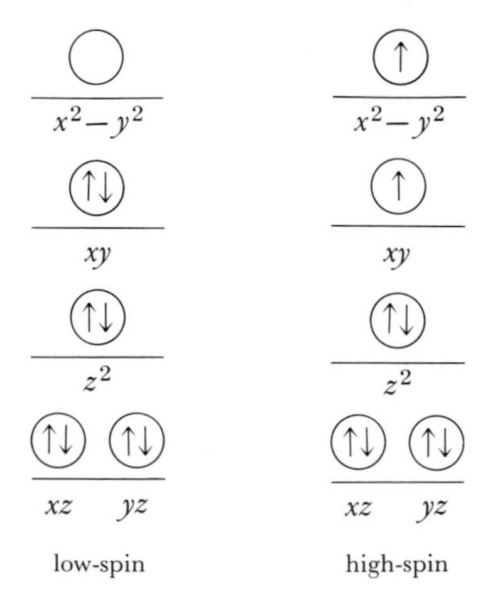

The actual configuration is the one with no unpaired electrons (low spin). The $d_{x^2-y^2}$ orbital is so high in energy relative to the d_{xy} orbital that all d^8 square-planar complexes are always low-spin.

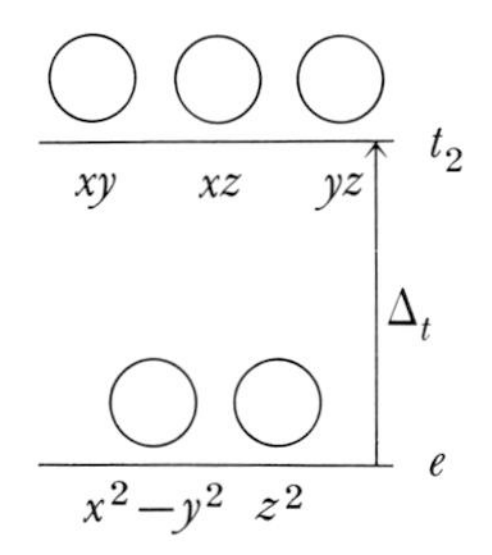

Figure 22-20 The *d*-orbital splitting pattern for a tetrahedral complex. Note that the pattern is the reverse of that for an octahedral complex. The difference in splitting patterns for tetrahedral and octahedral complexes is a direct consequence of the different placement of the ligands relative to the five *d* orbitals.

The d-orbital splitting pattern in a tetrahedral complex is shown in Figure 22-20. The two d orbitals that make up the lower set are called the ***e* orbitals,** and the three d orbitals that make up the upper set are called the **t_2 orbitals.** The value of Δ_t (the d-orbital splitting energy for a tetrahedral complex, where t denotes tetrahedral) is much smaller than the value of Δ_o (the d-orbital splitting energy for an octahedral complex) because none of the d-orbital lobes in tetrahedral complexes point directly at the ligands. A consequence of the relatively small value of Δ_t is that there are no low-spin tetrahedral complexes.

Example 22-14: Give the d-electron configuration in the tetrahedral complex $[CoCl_4]^{2-}$.

Solution: The oxidation state of cobalt in the complex is Co(II) because there are four Cl^- ligands, and the net charge on the complex is -2. Cobalt is the seventh member of the $3d$ transition series, and thus cobalt(II) is a d^7 ion. Referring to Figure 22-20, we have, for the d-electron configuration of a tetrahedral d^7 ion, $e^4t_2^3$, as is shown in the margin.

xy xz yz t_2^3

x^2-y^2 z^2 e^4

The d-orbital splitting patterns for octahedral, tetrahedral, and square-planar complexes are summarized in Figure 22-21.

The magnetic behavior of transition-metal complexes can, in some cases, be used for structure determination. For example, experiments show that the compound $K_2[NiBr_4]$ is paramagnetic and that the compound $K_2[Ni(CN)_4]$ is diamagnetic. Using this information, we can predict the structures of the $[NiBr_4]^{2-}$ and $[Ni(CN)_4]^{2-}$ ions. Both of these complexes contain nickel(II), a d^8 ion. Because there are four ligands, the two structure possibilities are tetrahedral and square-planar. The d^8 electron distributions for tetrahedral and square-planar complexes are given in the margin. A tetrahedral d^8 ion has two unpaired d electrons and is therefore paramagnetic. A square-planar d^8 ion has no unpaired d electrons and is therefore diamagnetic. Thus $[NiBr_4]^{2-}$, which is paramagnetic, must have a tetrahedral structure and $[Ni(CN)_4]^{2-}$, which is diamagnetic, must have a square-planar structure.

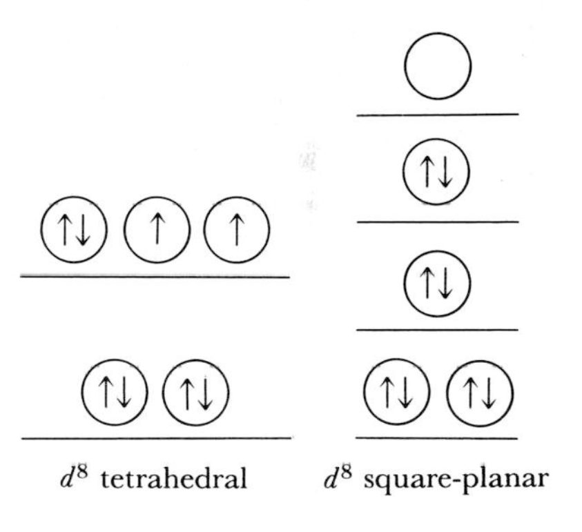

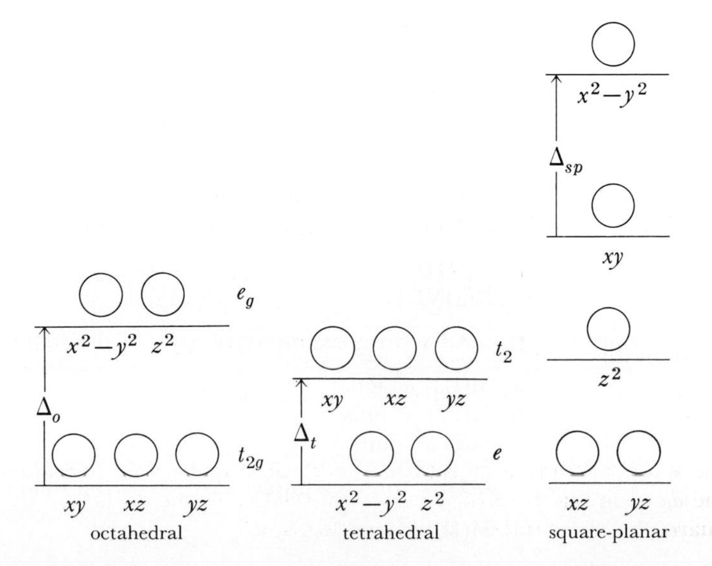

Figure 22-21 Comparison of the d-orbital splitting patterns in octahedral, tetrahedral, and square-planar complexes. The relative magnitudes of the d-orbital splitting energy for a particular metal ion with a particular ligand are $\Delta_{sp} \approx \Delta_o \approx 2\Delta_t$.

SUMMARY

There are three d transition-metal series ($3d$, $4d$, and $5d$), each having 10 members. The keys to understanding the chemistry of the d transition-series metals are the electron occupancy of the five d orbitals of the metal ion and the influence of the ligands on the relative energies and splitting patterns of the orbitals. Ligands are anions or neutral molecules that bind to metal ions to form complexes. Chelating or polydentate ligands are ligands that attach to two or more coordination positions on the metal ion. A transition-metal ion with x d electrons is called a d^x ion.

The structure of a complex may be octahedral (the most common), tetrahedral, square-planar, or linear. Certain octahedral and square-planar complexes exist as cis and trans isomers. The d-orbital splitting pattern is different for each geometry. The determining factor of the splitting of the d orbitals is the placement of the ligands in the complex ion relative to the positions of the d orbitals. The splitting of the d orbitals gives rise to the possibility of low-spin and high-spin complexes for octahedral d^4, d^5, d^6, and d^7 ions. The d-orbital splittings are denoted as Δ_o (octahedral), Δ_{sp} (square planar), and Δ_t (tetrahedral).

A paramagnetic complex ion has some electrons that are unpaired. Magnetic measurements can be used to detect the presence of unpaired electrons in a complex ion.

Transition-metal complexes are classified as either labile or inert, depending on whether they undergo ligand substitution reactions rapidly or slowly. Inert octahedral complexes are $t_{2g}^3e_g^0$, $t_{2g}^4e_g^0$, $t_{2g}^5e_g^0$, and $t_{2g}^6e_g^0$ ions.

TERMS YOU SHOULD KNOW

PROBLEMS

ELECTRON CONFIGURATIONS AND OXIDATION STATES

22-1. Write the ground-state electron configuration for each of the following transition-metal ions:

(a) Mn(III) (b) V(III)
(c) Ru(II) (d) Pt(IV)

22-2. Write the ground-state electron configuration for each of the following transition-metal ions:

(a) Co(III) (b) Ti(IV)
(c) Au(III) (d) Cu(I)

22-3. How many outer-shell d electrons are there in each of the following transition metal ions?

(a) Ag(I) (b) Pd(IV)
(c) Ir(III) (d) Co(II)

22-4. How many outer-shell d electrons are there in each of the following transition-metal ions?

(a) Re(III) (b) Sc(III)
(c) Ru(IV) (d) Hg(II)

22-5. Give three examples of each of the following:

(a) M(III) d^6 ions
(b) M(IV) d^4 ions
(c) M(I) d^{10} ions

22-6. Give three examples of each of the following:

(a) M(II) d^3 ions

(b) M(I) d^8 ions
(c) M(IV) d^0 ions

22-7. Give the oxidation state of the metal in each of the following complex ions:

(a) $[Os(NH_3)_4Cl_2]^+$ (b) $[CoCl_6]^{3-}$
(c) $[Fe(CN)_6]^{4-}$ (d) $[Nb(NO_2)_6]^{3-}$

22-8. Give the oxidation state of the metal in each of the following complex ions:

(a) $[Ir(H_2O)_6]^{3+}$ (b) $[Co(NH_3)_3(CO)_3]^{3+}$
(c) $[CuCl_4]^{2-}$ (d) $[Ni(CN)_4]^{2-}$

22-9. Give the oxidation state of the metal in each of the following complex ions:

(a) $[Cd(CN)_4]^{2-}$ (b) $[Pt(NH_3)_6]^{2+}$
(c) $[Pt(NH_3)_4Cl_2]$ (d) $[RhBr_6]^{3-}$

22-10. Give the oxidation state of the metal in each of the following complex ions:

(a) $[Mo(CO)_4Cl_2]^+$ (b) $[Ta(NO_2)_3Cl_3]^{3-}$
(c) $[Co(CN)_6]^{3-}$ (d) $[Ni(CO)_4]$

IONS FROM COMPLEX SALTS

22-11. Name the major species present when the following compounds are dissolved in water, and calculate the number of moles of each species present if 1.0 mol is dissolved:

(a) $K_3[Fe(CN)_6]$ (b) $[Ir(NH_3)_6](NO_3)_3$
(c) $[Pt(NH_3)_4Cl_2]Cl_2$ (d) $[Ru(NH_3)_6]Br_3$

22-12. Name the major species present when the following compounds are dissolved in water, and calculate the number of moles of each species present if 1.0 mol is dissolved:

(a) $[Cr(NH_3)_6]Br_3$ (b) $[Pt(NH_3)_3Cl_3]Cl$
(c) $[Mo(H_2O)_6]Br_3$ (d) $K_4[Cr(CN)_6]$

22-13. Some of the first complexes discovered by Werner in the 1890s had the empirical formulas given below. Also given are the number of chloride ions per formula unit precipitated by the addition of $Ag^+(aq)$. Explain these observations.

Empirical formula	Number of Cl^- per formula unit precipitated by $Ag^+(aq)$
$PtCl_4·6NH_3$	4
$PtCl_4·5NH_3$	3
$PtCl_4·4NH_3$	2
$PtCl_4·3NH_3$	1
$PtCl_4·2NH_3$	0

22-14. Some of the first complexes discovered by Werner in the 1890s had the empirical formulas given below. Also given are the number of chloride ions per formula unit precipitated by the addition of $Ag^+(aq)$. Explain these observations.

Empirical formula	Number of Cl^- per formula unit precipitated by $Ag^+(aq)$
$PtCl_2·4NH_3$	2
$PtCl_2·3NH_3$	1
$PtCl_2·2NH_3$	0

CHEMICAL FORMULAS AND NAMES

22-15. Give the systematic name for each of the following compounds:

(a) $K_3[Cr(CN)_6]$ (b) $[Cr(H_2O)_5Cl](ClO_4)_2$
(c) $[Co(CO)_4Cl_2]ClO_4$ (d) $[Pt(NH_3)_4Br_2]Cl_2$

22-16. Give the systematic name for each of the following compounds:

(a) $K_3[Fe(CN)_6]$ (b) $[Ni(CO)_4]$
(c) $[Ru(H_2O)_6]Cl_3$ (d) $Na[Al(OH)_4]$

22-17. Give the systematic name for each of the following compounds:

(a) $(NH_4)_3[Co(NO_2)_6]$ (b) $[Ir(NH_3)_4Br_2]Br$
(c) $K_2[CuCl_4]$ (d) $[Ru(CO)_5]$

22-18. Give the systematic name for each of the following compounds:

(a) $Na[Au(CN)_4]$ (b) $[Cr(H_2O)_6]Cl_3$
(c) $Na_3[Co(CN)_6]$ (d) $[Cu(NH_3)_6]Cl_2$

22-19. Give the chemical formula for each of the following compounds:

(a) sodium pentacyanocarbonylferrate(II)
(b) ammonium *trans*-dichlorodiiodoaurate(III)
(c) potassium hexacyanocobaltate(III)
(d) calcium hexanitrocobaltate(III)

22-20. Give the chemical formula for each of the following compounds:

(a) sodium bromochlorodicyanonickelate(II)
(b) rubidium tetranitrocobaltate(II)
(c) potassium hexachlorovanadate(III)
(d) pentaamminechlorochromium(III) acetate

22-21. Give the chemical formula for each of the following compounds:

(a) triamminechloroplatinum(II) nitrate
(b) sodium tetrafluorocuprate(II)
(c) lithium hexanitrocobaltate(II)
(d) barium hexacyanoferrate(II)

22-22. Give the chemical formula for each of the following compounds:

(a) calcium hexacyanomanganate(III)
(b) tetraamminepalladium(II) chlorate
(c) hexaaquarhenium(III) sulfate
(d) aluminum hexafluoroiridate(III)

22-23. Give the systematic name for each of the following compounds:

(a) $[Os(NH_3)_2(en)_2]Cl_3$ (b) $[Co(NH_3)_3(en)Cl]NO_3$
(c) $(NH_4)_2[Fe(EDTA)]$ (d) $K_3[Cr(C_2O_4)_3]$

22-24. Give the systematic name for each of the following compounds:

(a) $[Rh(en)_3]Br_3$ (b) $K_2[Zn(EDTA)]$
(c) $Li_3[Fe(C_2O_4)_3]$ (d) *trans*-$[CoCl_2(en)_2]ClO_4$

22-25. Give the chemical formula for each of the following compounds:

(a) potassium tris(oxalato)rhenate(III)
(b) tris(ethylenediamine)cobalt(II) acetate
(c) sodium (ethylenediaminetetraacetato)chromate(II)
(d) dichlorobis(ethylenediamine)palladium(IV) nitrate

22-26. Give the chemical formula for each of the following compounds:

(a) chlorohydroxobis(ethylenediamine)cobalt(III) nitrate
(b) bis(ethylenediamine)oxalatocadmium(II)
(c) lithium dinitrobis(oxalato)platinate(IV)
(d) bis(ethylenediamine)oxalatovanadium(III) acetate

GEOMETRIC ISOMERS

22-27. Draw all the geometric isomers for the following complexes:

(a) $[Co(en)_2Br_2]$ (b) $[RuCl_2Br_2(NO_2)_2]^{3-}$

22-28. Draw all the geometric isomers for the following complexes:

(a) $[Pd(C_2O_4)_2I_2]^{2-}$ (b) $[PtCl_3Br_3]^{2-}$

22-29. Draw the structure for each of the following complexes:

(a) *trans*-dichlorodibromoplatinum(IV) (square-planar)
(b) potassium *trans*-dichlorodiodoaurate(III) (square-planar)
(c) *cis,cis*-triamminetrichlorocobalt(III)
(d) *cis,trans*-triamminetrichloroplatinum(IV) chloride

22-30. Indicate whether each of the following complexes has geometric isomers:

(a) $[Cr(NH_3)_4Cl_2]^+$
(b) $[Cr(NH_3)_5Cl]^{2+}$
(c) $[Co(NH_3)_2Cl_2]^{2-}$ (tetrahedral)
(d) $[Pt(NH_3)_2Cl_2]$ (square-planar)

HIGH-SPIN AND LOW-SPIN COMPLEXES

22-31. Write the *d*-orbital electron configurations for the following octahedral complex ions:

(a) an Nb(III) complex
(b) an Mo(II) complex if Δ_o is greater than the electron-pairing energy
(c) an Mn(II) complex if Δ_o is less than the electron-pairing energy
(d) an Au(I) complex
(e) an Ir(III) complex if Δ_o is greater than the electron-pairing energy

22-32. Write the *d*-orbital electron configurations for the following octahedral complex ions:

(a) a high-spin Ni(II) complex
(b) a high-spin Mn(II) complex
(c) a low-spin Fe(III) complex
(d) a Ti(IV) complex
(e) a Ni(II) complex

22-33. Classify the following complex ions as high-spin or low-spin:

(a) $[Fe(CN)_6]^{4-}$ (no unpaired electrons)
(b) $[Fe(CN)_6]^{3-}$ (one unpaired electron)
(c) $[Co(NH_3)_6]^{2+}$ (three unpaired electrons)
(d) $[CoF_6]^{3-}$ (four unpaired electrons)
(e) $[Mn(H_2O)_6]^{2+}$ (five unpaired electrons)

22-34. Classify the following complex ions as high-spin or low-spin:

(a) $[Mn(NH_3)_6]^{3+}$ (two unpaired electrons)
(b) $[Rh(CN)_6]^{3-}$ (no unpaired electrons)
(c) $[Co(C_2O_4)_3]^{4-}$ (three unpaired electrons)
(d) $[IrBr_6]^{4-}$ (three unpaired electrons)
(e) $[Ru(NH_3)_6]^{3+}$ (one unpaired electron)

22-35. Classify the following complex ions as high-spin or low-spin and write the *d*-orbital electron configurations:

(a) $[Fe(CN)_6]^{4-}$ (b) $[MnF_6]^{4-}$
(c) $[Co(NO_2)_6]^{3-}$ (d) $[FeCl_4]^-$ (tetrahedral)

22-36. Classify the following complex ions as high-spin or low-spin and write the *d*-orbital electron configurations:

(a) $[Cr(NO_2)_6]^{4-}$ (b) $[CoF_6]^{3-}$
(c) $[Rh(CN)_6]^{3-}$ (d) $[MnCl_4]^{2-}$ (tetrahedral)

PARAMAGNETISM IN COMPLEX IONS

22-37. Predict the number of unpaired electrons in each of the following complexes:

(a) $[VCl_6]^{3-}$ (b) $[CoCl_4]^{2-}$ (tetrahedral)
(c) $[Cr(CO)_6]$ (d) $[Cr(CN)_6]^{4-}$

22-38. Predict the number of unpaired electrons in each of the following complex ions:

(a) $[Pd(NO_2)_4]^{2-}$ (square planar)
(b) $[Rh(NH_3)_6]^{3+}$
(c) $[Ir(H_2O)_6]^{3+}$
(d) $[FeF_6]^{3-}$

22-39. Indicate whether each of the following complexes is paramagnetic:

(a) $[Co(en)_3]^{3+}$ (low-spin)
(b) $[Fe(CN)_6]^{4-}$
(c) $[NiF_4]^{2-}$ (tetrahedral)
(d) $[CoBr_4]^{2-}$ (tetrahedral)

22-40. Indicate whether each of the following complexes is paramagnetic:

(a) $[Cu(NH_3)_6]^{2+}$
(b) $[CrF_6]^{3-}$
(c) $[CoCl_4]^{2-}$ (tetrahedral)
(d) $[Zn(H_2O)_6]^{2+}$

22-41. The complex $[NiF_4]^{2-}$ is paramagnetic but $[Ni(CN)_4]^{2-}$ is diamagnetic. Explain the difference.

22-42. The complex $[Fe(H_2O)_6]^{2+}$ is paramagnetic whereas $[Fe(CN)_6]^{4-}$ is diamagnetic. Explain the difference.

INERT AND LABILE COMPLEXES

22-43. Of the following complexes, which would be expected to be inert to ligand substitution?

(a) $[Ti(H_2O)_6]^{3+}$
(b) $[VF_6]^{3-}$
(c) $[Cr(NO_2)_6]^{3-}$
(d) $[CuCl_6]^{4-}$

22-44. Of the following complexes, which would be expected to be inert to ligand substitution?

(a) $[Mo(NO_2)_6]^{2-}$
(b) $[Fe(CN)_6]^{4-}$
(c) $[Fe(CN)_6]^{3-}$
(d) $[Rh(NH_3)_6]^{3+}$

22-45. Of the following complexes, which would be expected to be inert to ligand substitution?

(a) $[V(H_2O)_6]^{3+}$
(b) $[WF_6]^{2-}$
(c) $[Cr(CO)_6]$
(d) $[Zn(H_2O)_6]^{2+}$

22-46. Of the following complexes, which would be expected to be inert to ligand substitution?

(a) $[Re(NH_3)_6]^{2+}$
(b) $[NbCl_6]^{2-}$
(c) $[Rh(CN)_6]^{3-}$
(d) $[CoF_6]^{3-}$

ADDITIONAL PROBLEMS

22-47. Give the chemical formula for each of the following complex ions:

(a) hexanitrocobaltate(III)
(b) *trans*-dichlorobis(ethylenediamine)platinum(IV)
(c) pentacyanocarbonylferrate(II)
(d) *trans*-dichlorodiiodoaurate(III)

22-48. Write the chemical formula of the following octahedral complexes:

(a) The bromide complex of Rh(III)
(b) The oxalate complex of Co(III)
(c) The ammine complex of Cu(II)
(d) The EDTA complex of Cr(III)
(e) The ethylenediamine complex of Cr(II)

22-49. Give the chemical formulas corresponding to the following compounds:

(a) lithium tris(oxalato)ruthenate(III)
(b) potassium tetraiodomercurate(II)
(c) tetraamminedibromochromium(III) nitrate
(d) triaquatricarbonylmolybdenum(III) acetate

22-50. Write the formulas for each of the following compounds:
(a) hexaaquanickel(II) perchlorate
(b) triamminetrichloroplatinum(IV) bromide
(c) potassium chloropentacyanoferrate(III)
(d) strontium hexacyanoferrate(II)

22-51. Silver nitrate was added to solutions of the following octahedral complexes and AgCl was precipitated immediately in the mole ratios indicated:

Formula of the complex	(mol AgCl/mol complex)
$CoCl_3(NH_3)_6$	3
$CoCl_3(NH_3)_5$	2
$CoCl_3(NH_3)_4$ (purple)	1
$CoCl_3(NH_3)_4$ (green)	1

(a) Draw the structures expected for each of these complexes.
(b) Explain the fact that $CoCl_3(NH_3)_4$ can be purple or green but that both forms give 1 mol of AgCl per mol complex.

22-52. Arrange the following complexes in order of increasing values of Δ_o:

$[Cr(H_2O)_6]^{3+}$ $[Co(NH_3)_6]^{3+}$ $[CrF_6]^{3-}$
$[Cr(CN)_6]^{3-}$ $[Ru(CN)_6]^{3-}$

22-53. Comparing $[Co(CN)_6]^{3-}$ to $[CoCl_6]^{4-}$, indicate whether each of the following statements is true or false:

(a) $[Co(CN)_6]^{3-}$ has more d electrons than $[CoCl_6]^{4-}$.
(b) $[Co(CN)_6]^{3-}$ has the same number of d electrons as $[CoCl_6]^{4-}$.
(c) $[Co(CN)_6]^{3-}$ is paramagnetic while $[CoCl_6]^{4-}$ is diamagnetic.
(d) $[Co(CN)_6]^{3-}$ is diamagnetic while $[CoCl_6]^{4-}$ is paramagnetic.

22-54. Which of the following ions is inert to ligand substitution?

(a) $[Cr(C_2O_4)_3]^{3-}$
(b) $[Ti(H_2O)_6]^{3+}$
(c) $[Zn(H_2O)_6]^{2+}$
(d) $[Co(C_2O_4)_3]^{4-}$

22-55. How many unpaired electrons would you predict for the following complex ions?

(a) $[NiCl_4]^{2-}$ (tetrahedral)
(b) $[CoCl_4]^{2-}$ (tetrahedral)
(c) $[Co(CO)_6]^{3+}$
(d) $[Fe(CN)_6]^{3-}$

22-56. Why do you think most complexes of Zn(II) are colorless?

22-57. Use Le Châtelier's principle to predict the direction (if any) the following equilibrium will shift under the indicated perturbations:

$$AgCl(s) + 2CN^-(aq) \rightleftharpoons [Ag(CN)_2]^-(aq) + Cl^-(aq)$$

$$\Delta H^\circ_{rxn} = +21 \text{ kJ}$$

(a) addition of $NaCN(s)$
(b) addition of $AgCl(s)$
(c) a decrease in temperature
(d) addition of $HNO_3(l)$
(e) addition of $H_2O(l)$
(f) addition of $NaCl(s)$

22-58. Explain, using balanced chemical equations, the following observations:

(a) Dilute solutions of $CuSO_4(aq)$ are blue but become green on addition of 6 M HCl.
(b) Several commercially available rust removers contain sodium oxalate.
(c) Solid mercury(II) oxide is soluble in excess 2 M $KI(aq)$ solution.
(d) Dilute aqueous solutions of iron(III) nitrate are yellow in color. The yellow color is removed by addition of excess 2 M HNO_3 but not by excess 2 M HCl.

22-59. Excess $Pb_2[Fe(CN)_6](s)$ was equilibrated at 25°C with an aqueous solution of $NaI(aq)$. The equilibrium concentrations of $I^-(aq)$ and $[Fe(CN)_6]^{4-}(aq)$ were found by chemical analysis to be 0.57 M and 0.11 M, respectively. Estimate the K_{sp} of $Pb_2[Fe(CN)_6](s)$. See Table 17-1 for the K_{sp} of $PbI_2(s)$.

22-60. Which of the following complexes is labile?

(a) $[Cr(CN)_6]^{4-}$
(b) $[Pt(NH_3)_6]^{4+}$
(c) $[Cu(NH_3)_6]^{2+}$
(d) $[W(NH_3)_6]^{2+}$

ORGANIC CHEMISTRY

Natural gas burnoff at the Petro-Canada Terra Nova K-07 rig on the Grand Banks, off Newfoundland.

Organic chemistry is the chemistry of compounds that contain carbon atoms. The carbon atom is unusual in that it can bond covalently to other carbon atoms to form long chains and rings of carbon atoms. The vast majority of all compounds, and particularly almost all biologically important compounds, consist of molecules that contain many carbon atoms. Because it was once believed that all compounds containing carbon had their origins in living sources, compounds that contain carbon are called organic compounds. In this chapter we discuss a variety of organic compounds: saturated, unsaturated, and aromatic hydrocarbons, alcohols, aldehydes, ketones, amines, organic acids, and esters.

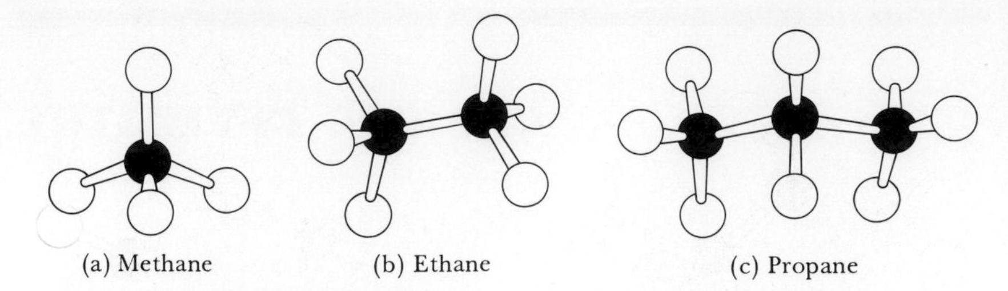

Figure 23-1 Ball-and-stick models of (a) methane, (b) ethane, and (c) propane. The hydrogen atoms in methane point to the vertices of a regular tetrahedron. Each of the carbon atoms in ethane and propane is surrounded by a tetrahedral array of atoms. The carbon-hydrogen bond length (110 pm) in ethane and propane is the same as in methane, and the H—C—H and H—C—C bond angles are equal to the tetrahedral bond angle (109.5°).

23-1. ALKANES ARE HYDROCARBONS THAT CONTAIN ONLY SINGLE BONDS

Methane, CH_4, ethane, C_2H_6, and propane, C_3H_8 (Figure 23-1), belong to the class of **organic compounds** called **hydrocarbons** because they consist of only hydrogen and carbon. Methane and ethane are the first two members of the series of hydrocarbons called the **alkanes.** Alkanes are hydrocarbons that contain only single carbon-carbon bonds, and their bonding is described in terms of sp^3 hybrid orbitals on the carbon atoms (Section 10-5).

The next member of the alkane series contains three carbon atoms in a row and is called propane, C_3H_8. Propane is a gas that is commonly used as fuel in areas not serviced by gas mains. The Lewis formula for the propane molecule is

```
     H   H   H
     |   |   |
 H — C — C — C — H
     |   |   |
     H   H   H
```

which can be written as the **structural formula** $CH_3CH_2CH_3$. Each of the carbon atoms in propane is surrounded by a tetrahedral array of atoms (Figure 23-1).

The fourth member of the alkane series, butane, C_4H_{10}, is interesting because there are two different types of butane molecules. (Figure 23-2). The Lewis formulas for the two forms of butane are

```
     H   H   H   H                       H     H     H
     |   |   |   |                       |     |     |
 H — C — C — C — C — H      and      H — C  —  C  —  C — H
     |   |   |   |                       |     |     |
     H   H   H   H                       H     |     H
                                           H — C — H
         n-butane                              |
                                               H

                                           isobutane
```

and their structural formulas are

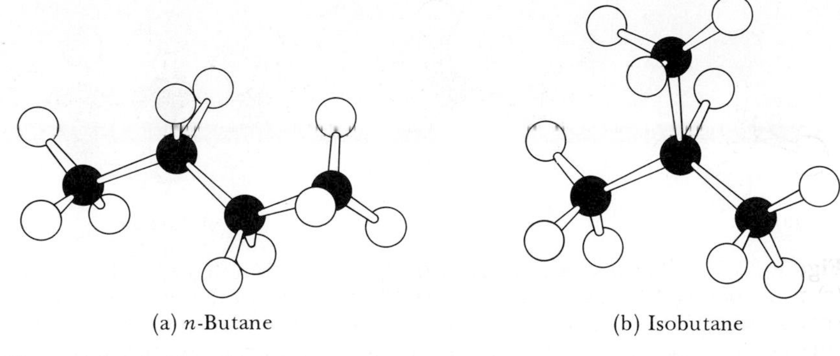

Figure 23-2 Ball-and-stick models of (a) n-butane and (b) isobutane. As in ethane and propane, the bonds from each carbon atom are tetrahedrally oriented.

$$\underset{n\text{-butane}}{CH_3CH_2CH_2CH_3} \qquad \text{and} \qquad \underset{\text{isobutane}}{\begin{array}{c}CH_3\overset{}{C}HCH_3\\ \quad|\\ \quad CH_3\end{array}}$$

▪ Note that the structural formulas for branched-chain molecules show the carbon-carbon bonds for branching carbon atoms.

The straight-chain molecule is called normal butane, written n-butane, and the branched molecule is called isobutane. It is important to realize that n-butane and isobutane are different compounds, with different physical and chemical properties. For example, the boiling point of n-butane is $-0.5°C$ and that of isobutane is $-10.2°C$. Compounds that have the same molecular formula but different structures are called **structural isomers.**

In alkanes, each of the four bonds from a carbon atom is connected to a different atom. The bonding about each carbon atom is saturated, and so alkanes are called **saturated hydrocarbons.**

Because the carbon-carbon bond in ethane is a σ bond (Figure 10-13), its energy is unaffected by rotation around the carbon-carbon axis. Thus one $—CH_3$ group in ethane can rotate relative to the other. All carbon-carbon single bonds are formed the same way as in ethane, and so all carbon-carbon single bonds are σ bonds. Consequently, we find that rotation can occur about carbon-carbon single bonds. Because there is rotation about carbon-carbon single bonds, the formula

$$\begin{array}{c}CH_3CH_2CH_2\\ \qquad\quad |\\ \qquad\quad CH_3\end{array}$$

does *not* represent a third isomer of butane. It simply represents four carbon atoms joined in a chain and is just a somewhat misleading way of writing the structural formula for n-butane.

After butane, the alkanes are assigned systematic names that indicate the number of carbon atoms in the molecule. For example, C_5H_{12} is called pentane. The prefix *pent-* indicates that there are five carbon atoms, and the ending, *-ane* denotes an alk*ane.* The names of the first 10 straight-chain alkanes are given in Table 23-1. The first four n-alkanes are gases at room temperature (20°C), whereas n-pentane through n-decane are liquids at room temperature. Higher alkanes are waxy solids at room temperature. The boiling points of the n-alkanes

Table 23-1 The first 10 straight-chain alkanes

Name	Molecular formula*
methane	CH_4
ethane	C_2H_6
propane	C_3H_8
n-butane	C_4H_{10}
n-pentane	C_5H_{12}
n-hexane	C_6H_{14}
n-heptane	C_7H_{16}
n-octane	C_8H_{18}
n-nonane	C_9H_{20}
n-decane	$C_{10}H_{22}$

*The molecular formulas for the alkanes fit the general formula C_nH_{2n+2}, where n represents the number of carbon atoms in the molecule.

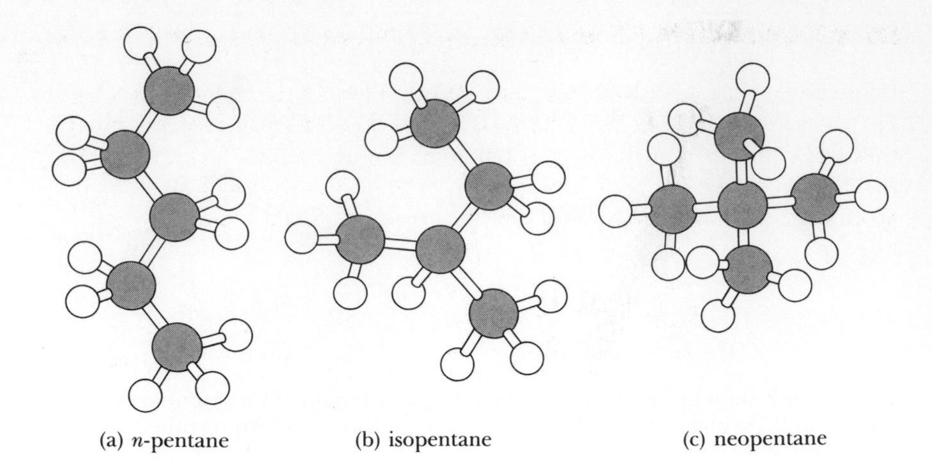

Figure 23-3 Ball-and-stick models of the three structural isomers of pentane: (a) *n*-pentane, (b) isopentane, and (c) neopentane.

increase with molecular mass, in accord with the fact that the total van der Waals force between molecules increases with the size of the molecules (Section 11-4).

The number of structural isomers increases with the number of carbon atoms in an alkane. There are three isomers of pentane (Figure 23-3):

$$CH_3CH_2CH_2CH_2CH_3 \qquad CH_3\underset{}{\overset{\overset{\displaystyle CH_3}{|}}{C}}HCH_2CH_3 \qquad CH_3\underset{\underset{\displaystyle CH_3}{|}}{\overset{\overset{\displaystyle CH_3}{|}}{C}}CH_3$$

n-pentane isopentane neopentane

Note that all three isomers have the molecular formula C_5H_{12}. The number of structural isomers increases very rapidly with the number of carbon atoms. For example, hexane has 5 structural isomers, heptane has 9, octane has 18, and nonane has 35.

▪ Petroleum is a complex mixture of hydrocarbons, organosulfur, and organonitrogen compounds.

All the normal straight-chain alkanes through $C_{33}H_{68}$, as well as many branched-chain hydrocarbons, have been isolated from petroleum. A few alkanes occur elsewhere in nature. For example, the skin of an apple contains the C_{27} and C_{29} *n*-alkanes. These are the waxes that produce the waxy feel of an apple when it is polished. Long-chain saturated hydrocarbons often form part of the protective coating on leaves and fruits. Similar hydrocarbons are found in beeswax. Apparently the major function of these waxes is to retard water loss.

Alkanes react with F_2, Cl_2, and Br_2 and burn in O_2 to form CO_2 and H_2O. These reactions are all highly exothermic. For example, ΔH°_{rxn} for the combustion of propane

$$C_3H_8(g) + 5O_2(g) \rightarrow 3CO_2(g) + 4H_2O(g)$$

is -2040 kJ. Reactions of this type constitute the basis for the use of hydrocarbons as heating fuels (see Interchapter E).

A mixture of an alkane and chlorine can be kept indefinitely in the dark. If such a mixture is heated or exposed to sunlight (ultraviolet radiation), however, a reaction occurs in which one or more of the hydrogen atoms in the alkane are replaced by chlorine atoms. Such a reaction is called a **substitution reaction** because the alkane hydrogen atoms are substituted by chlorine atoms. For example,

$$CH_4(g) + Cl_2(g) \xrightarrow{UV} CH_3Cl(g) + HCl(g)$$

(where UV denotes ultraviolet radiation). The product in this case can be considered to be a derivative of methane and so is named chloromethane. The function of the ultraviolet radiation is to break the bond in the Cl_2 molecule and produce free chlorine atoms. The chlorine atoms are free radicals—hence highly reactive—and react with methane:

$$:\ddot{\underset{..}{Cl}}-\ddot{\underset{..}{Cl}}: \xrightarrow{UV} 2:\ddot{\underset{..}{Cl}}\cdot$$

$$:\ddot{\underset{..}{Cl}}\cdot + H-\underset{H}{\overset{H}{\underset{|}{\overset{|}{C}}}}-H \rightarrow H-\ddot{\underset{..}{Cl}}: + H-\underset{H}{\overset{H}{\underset{|}{\overset{|}{C}}}}\cdot$$

$$H-\underset{H}{\overset{H}{\underset{|}{\overset{|}{C}}}}\cdot + :\ddot{\underset{..}{Cl}}-\ddot{\underset{..}{Cl}}: \rightarrow :\ddot{\underset{..}{Cl}}\cdot + H-\underset{H}{\overset{H}{\underset{|}{\overset{|}{C}}}}-\ddot{\underset{..}{Cl}}:$$

As the concentration of CH_3Cl builds up during the reaction, it reacts further with chlorine atoms to produce dichloromethane, CH_2Cl_2. The dichloromethane can react even further to produce trichloromethane, $CHCl_3$, and so on. Thus, this **free-radical reaction** leads to more than a single product. By varying the relative concentrations of CH_4 and Cl_2, however, it is possible to favor one product over another.

An alkane in which hydrogen atoms are replaced by halogen atoms is called an **alkyl halide** or a **haloalkane.** Alkyl halides are named by prefixing the name of the parent alkane to designate the attached halogen atoms. For example, the molecule

$$H-\underset{H}{\overset{H}{\underset{|}{\overset{|}{C}}}}-\underset{H}{\overset{H}{\underset{|}{\overset{|}{C}}}}-Cl$$

▪ For simplicity, we shall not always include lone pairs in Lewis formulas.

is called chloroethane.

▪ Chloroethane is also called ethyl chloride and is used as a topical anesthetic.

If two hydrogen atoms in ethane are replaced by chlorine atoms, then there are two distinct products:

$$H-\underset{Cl}{\overset{H}{\underset{|}{\overset{|}{\overset{1}{C}}}}}-\underset{Cl}{\overset{H}{\underset{|}{\overset{|}{\overset{2}{C}}}}}-H \quad \text{and} \quad H-\underset{Cl}{\overset{Cl}{\underset{|}{\overset{|}{\overset{1}{C}}}}}-\underset{H}{\overset{H}{\underset{|}{\overset{|}{\overset{2}{C}}}}}-H$$

1,2-dichloroethane — 1,1-dichoroethane

This is another example of structural isomerism. We must distinguish between these two dichloroethanes. We can do this by numbering the carbon atoms along the alkane chain and designating which carbon atoms have attached chlorine atoms. Here, the dichloroethane shown on the left is called 1,2-dichloroethane and the right-hand one is called 1,1-dichloroethane. Atoms or groups of atoms that replace a hydrogen atom bonded to a carbon atom are called **substituents.** Thus the two Cl atoms are the substituents in the **disubstituted** ethanes shown here.

▪ 1,2-Dichloroethane is extensively used as a dry-cleaning fluid because of its ability to dissolve oils and greases.

Although alkanes react with O_2 and Cl_2, they are relatively inert and are known for their lack of chemical reactivity. For example, some reactions that do *not* occur are

$$\left.\begin{array}{l} \text{alkane} + \text{strong acid (e.g., } H_2SO_4) \\ \text{alkane} + \text{strong base (e.g., NaOH)} \\ \text{alkane} + \text{oxidizing agent (e.g., } KMnO_4) \\ \text{alkane} + \text{reducing agent (e.g., NaH)} \end{array}\right\} \text{no reaction}$$

Alkyl halides are also fairly unreactive.

Example 23-1: Complete and balance the following equations. If there is no reaction, then write N.R.

(a) $C_4H_{10}(g) + O_2(g) \rightarrow$

(b) $C_2H_6(g) + Cl_2(g) \xrightarrow{\text{dark}}$

(c) $C_3H_8(g) + Cl_2(g) \xrightarrow{\text{UV}}$

Solution: (a) The equation for the combustion of butane is:

$$C_4H_{10}(g) + \frac{13}{2}O_2(g) \rightarrow 4CO_2(g) + 5H_2O(g)$$

(b) Chlorine and ethane do not react in the dark:

$$C_2H_6(g) + Cl_2(g) \xrightarrow{\text{dark}} \text{N.R.}$$

(c) Under ultraviolet radiation, we have

$$C_3H_8(g) + Cl_2(g) \xrightarrow{\text{UV}} C_3H_7Cl(g) + HCl(g)$$

Because this is a free radical reaction, dichloropropane and polychloropropanes also are formed.

23-2. ALKANES AND SUBSTITUTED ALKANES CAN BE NAMED SYSTEMATICALLY ACCORDING TO IUPAC RULES

Structural isomerism leads to an enormous number of alkanes and substituted alkanes. Consequently, it is necessary to have a systematic method of naming alkanes and their derivatives simply and unambiguously. A system of nomenclature for organic molecules has been devised and is used by chemists throughout the world. The system has been recommended by the International Union of Pure and Applied Chemistry (IUPAC).

The **IUPAC nomenclature** rules for alkanes and their derivatives are as follows:

1. For straight-chain alkanes, use the names in Table 23-1 *without* the *n* prefix. Thus, the straight-chain alkane containing eight carbon atoms is called octane.

2. To name a branched or a substituted alkane, first identify the longest chain (the main chain) of consecutive carbon atoms in the molecule. Name this main chain according to rule 1. For example, the main chain in the following molecule has five carbon atoms (shown in color):

$$\begin{array}{ccccccccc} & & \text{Cl} & & \text{CH}_3 & & & & \\ & & | & & | & & & & \\ \text{CH}_3 & - & \text{CH} & - & \text{CH} & - & \text{CH}_2 & - & \text{CH}_3 \end{array}$$

Because there are five carbon atoms in the main chain, this substituted alkane is named as a substituted pentane, even though the molecule has six carbon atoms in all.

3. Number the carbon atoms in the main chain consecutively, starting at the end that gives the *lowest* numbers to the carbon atoms that have attached groups. For our substituted pentane, we have

$$\begin{array}{ccccccccc} & & \text{Cl} & & \text{CH}_3 & & & & \\ {\scriptstyle 1} & & {\scriptstyle 2}| & & {\scriptstyle 3}| & & {\scriptstyle 4} & & {\scriptstyle 5} \\ \text{CH}_3 & - & \text{CH} & - & \text{CH} & - & \text{CH}_2 & - & \text{CH}_3 \end{array}$$

We number the carbon atoms from left to right so that the attached groups are on the lowest-numbered carbon atoms, 2 and 3, in this case. If we number the chain from right to left, then the carbon atoms with attached groups would be 3 and 4.

4. Name the groups attached to the main chain according to Table 23-2 and indicate their position along the chain by showing the number of the carbon atom to which they are attached. The substituted alkane we are using as our example is 2-chloro-3-methylpentane. Punctuation is important in assigning IUPAC names. Numbers are separated from letters by hyphens, and the name is written as one word.

5. When two or more different groups are attached to the main chain, list them in alphabetical order. For example, as we just saw,

$$\begin{array}{ccccccccc} & & \text{Cl} & & \text{CH}_3 & & & & \\ & & | & & | & & & & \\ \text{CH}_3 & - & \text{CH} & - & \text{CH} & - & \text{CH}_2 & - & \text{CH}_3 \end{array}$$

is called 2-chloro-3-methylpentane, whereas

$$\begin{array}{ccccccccc} & & \text{CH}_3 & & \text{Cl} & & & & \\ & & | & & | & & & & \\ \text{CH}_3 & - & \text{CH} & - & \text{CH} & - & \text{CH}_2 & - & \text{CH}_3 \end{array}$$

is called 3-chloro-2-methylpentane.

6. When two or more identical groups are attached to the main chain, use prefixes such as *di-*, *tri-*, or *tetra-*. For example,

$$\begin{array}{ccccccccc} & & \text{CH}_3 & & \text{CH}_3 & & & & \\ {\scriptstyle 1} & & {\scriptstyle 2}| & & {\scriptstyle 3}| & & {\scriptstyle 4} & & {\scriptstyle 5} \\ \text{CH}_3 & - & \text{CH} & - & \text{CH} & - & \text{CH}_2 & - & \text{CH}_3 \end{array}$$

is 2,3-dimethylpentane. Note that the numbers are separated by commas. Every attached group must be named and numbered, even if two identical groups are attached to the same carbon atom. For example, the IUPAC name for

$$\begin{array}{ccccccccc} & & \text{CH}_3 & & \text{CH}_3 & & & & \\ {\scriptstyle 1} & & {\scriptstyle 2}| & & {\scriptstyle 3}| & & {\scriptstyle 4} & & {\scriptstyle 5} \\ \text{CH}_3 & - & \text{C} & - & \text{CH} & - & \text{CH}_2 & - & \text{CH}_3 \\ & & | & & & & & & \\ & & \text{CH}_3 & & & & & & \end{array}$$

is 2,2,3-trimethylpentane.

The assignment of IUPAC names is best learned by example.

Table 23-2. Some common groups

Group	Name*
$—CH_3$	methyl
$—CH_2CH_3$	ethyl
$—CH_2CH_2CH_3$	propyl
$CH_3\underset{\vert}{C}HCH_3$	isopropyl
—F	fluoro
—Cl	chloro
—Br	bromo
—I	iodo
$—NH_2$	amino
$—NO_2$	nitro

*Groups that are derived from alkanes are called **alkyl groups.** The first four groups here are alkyl groups, and they are named by dropping the *-ane* ending from the name of the alkane and adding *-yl*.

Example 23-2: Assign a IUPAC name to neopentane.

Solution: The structural formula for neopentane is

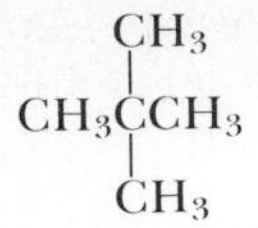

$$\begin{array}{c} CH_3 \\ | \\ CH_3CCH_3 \\ | \\ CH_3 \end{array}$$

Its main chain is

$$\overset{1}{C}H_3\text{—}\overset{2|}{\underset{|}{C}}\text{—}\overset{3}{C}H_3$$

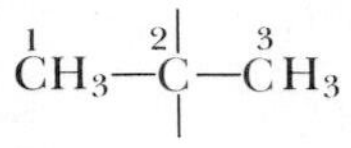

and so we name it as a derivative of propane. The IUPAC name is 2,2-dimethylpropane.

Example 23-3: Using Table 23-2, draw a structural formula for 1-chloro-2-methylbutane.

Solution: The parent alkane is butane. The name indicates that there is a chlorine atom bonded to the first carbon and a methyl group bonded to the second carbon atom in the butane chain. Thus, the structural formula is

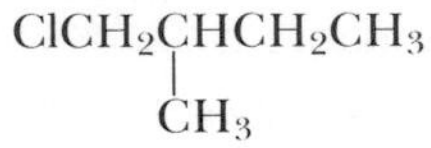

$$ClCH_2\underset{\underset{CH_3}{|}}{C}HCH_2CH_3$$

23-3. HYDROCARBONS THAT CONTAIN DOUBLE BONDS ARE CALLED ALKENES

All the hydrocarbons that we have discussed so far have been saturated. Each carbon atom has been bonded to four atoms. There is another class of hydrocarbons called **unsaturated hydrocarbons,** in which not all the carbon atoms are bonded to four other atoms. These molecules necessarily contain double or triple bonds. Unsaturated hydrocarbons that contain one or more double bonds are called **alkenes.** The simplest alkene is called ethene, or more commonly, ethylene:

▪ Ethylene is the starting material for about 40 percent of all organic substances produced commercially.

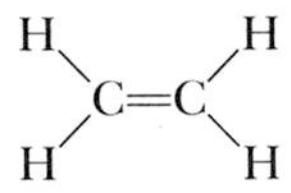

$$\begin{array}{ccc} H & & H \\ & C{=}C & \\ H & & H \end{array}$$

We learned in Section 10-8 that the bonding in ethene is described by sp^2 orbitals on each carbon atom (Figure 10-18). The double bond consists of a σ bond and a π bond. The σ bond results from the combination of two sp^2 orbitals, one from each carbon atom, and the π bond results from the combination of two p orbitals, also one from each carbon atom. The π orbital maintains the σ-bond framework in a planar shape and prevents rotation about the double bond. Consequently all six atoms in an ethene molecule lie in one plane and there are cis and trans isomers of 1,2-dichloroethene:

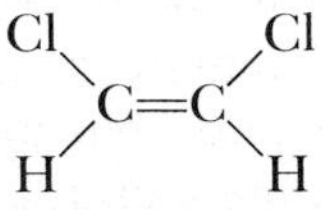

$$\begin{array}{ccc} Cl & & Cl \\ & C{=}C & \\ H & & H \end{array} \qquad \begin{array}{ccc} Cl & & H \\ & C{=}C & \\ H & & Cl \end{array}$$

cis 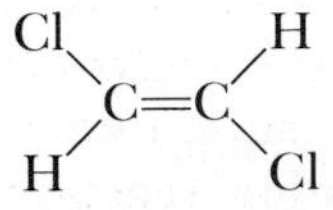 trans

The IUPAC nomenclature for alkenes uses the longest chain of consecutive carbon atoms *containing the double bond* as the parent compound. The parent compound is named by dropping the *-ane* and adding *-ene* to the name of the corresponding alkane and using a number to designate the carbon atom preceding the double bond. Thus we have

$$\underset{\text{ethene}}{H_2C{=}CH_2} \qquad \underset{\text{propene}}{H_2C{=}CH{-}CH_3}$$

There are two possible positions for the double bond in butene, and so we have

$$\underset{\text{1-butene}}{H_2C{=}CH{-}CH_2CH_3} \qquad \underset{\text{2-butene}}{H_3C{-}CH{=}CH{-}CH_3}$$

The planar $\rangle C{=}C\langle$ portion of each of these molecules is shaded. This emphasizes that the name 2-butene is ambiguous because of cis-trans isomerism. The cis-trans isomers of 2-butene are

$$\underset{\substack{\textit{cis}\text{-2-butene} \\ (\text{mp } -139°\text{C})}}{H_3C{-}CH{=}CH{-}CH_3} \qquad \underset{\substack{\textit{trans}\text{-2-butene} \\ (\text{mp } -106°\text{C})}}{H_3C{-}CH{=}CH{-}CH_3}$$

An important feature of a carbon-carbon double bond is the planar geometry it imposes in the region around it.

Example 23-4: Discuss the shape of a propene molecule, whose structural formula is

$$H_2C{=}CH{-}CH_3$$

Solution: The $\rangle C{=}C\langle$ portion of propene is planar, and so all three carbon atoms and the three hydrogen atoms attached to the double-bonded carbon atoms lie in one plane, as the shaded region shows:

$$H_2C{=}CH{-}CH_3$$

The methyl group can rotate about the carbon-carbon single bond.

Alkenes are more reactive than alkanes because the carbon-carbon double bond provides a reactive center in the molecule. In a sense, the double bond has "extra" electrons available for reaction. In addition to the combustion and substitution reactions that alkanes undergo, alkenes undergo **addition reactions.** Some examples of addition reactions are

1. Addition of hydrogen, **hydrogenation:**

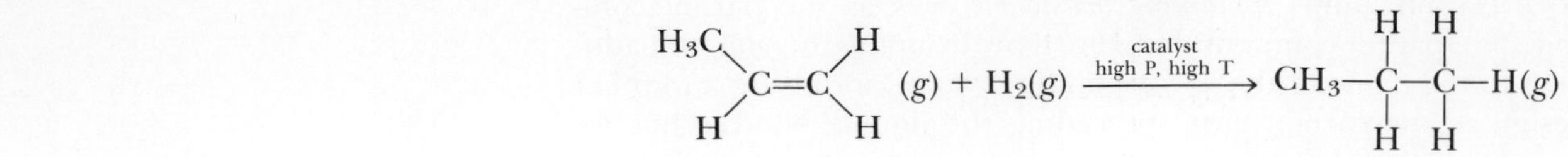

$$\underset{H}{\overset{H_3C}{}}\!\!>\!C{=}C\!<\!\underset{H}{\overset{H}{}}(g) + H_2(g) \xrightarrow[\text{high P, high T}]{\text{catalyst}} CH_3-\underset{H}{\overset{H}{C}}-\underset{H}{\overset{H}{C}}-H(g)$$

This reaction requires a catalyst and high pressure (P) and temperature (T). Usually powdered nickel or platinum is used as the catalyst. "Hydrogenated vegetable oils" are made by hydrogenating the double bonds in vegetable oils. This hydrogenation makes vegetable oils solid at room temperature.

2. Addition of chlorine or bromine:

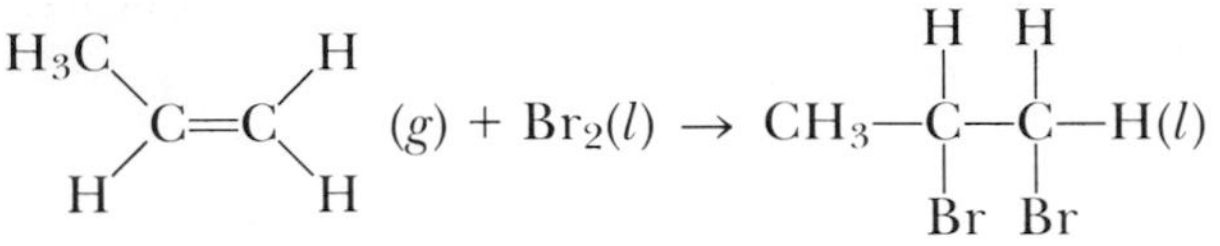

$$\underset{H}{\overset{H_3C}{}}\!\!>\!C{=}C\!<\!\underset{H}{\overset{H}{}}(g) + Br_2(l) \rightarrow CH_3-\underset{Br}{\overset{H}{C}}-\underset{Br}{\overset{H}{C}}-H(l)$$

This reaction can be carried out either with pure chlorine or bromine or by dissolving the halogen in some solvent, such as carbon tetrachloride. The addition reaction with bromine is a useful qualitative test for the presence of double bonds. A solution of bromine in carbon tetrachloride is red, whereas alkenes and bromoalkanes are usually colorless. As the bromine adds to the double bond, the red color disappears, giving a simple test for the presence of double bonds.

3. Addition of hydrogen chloride:

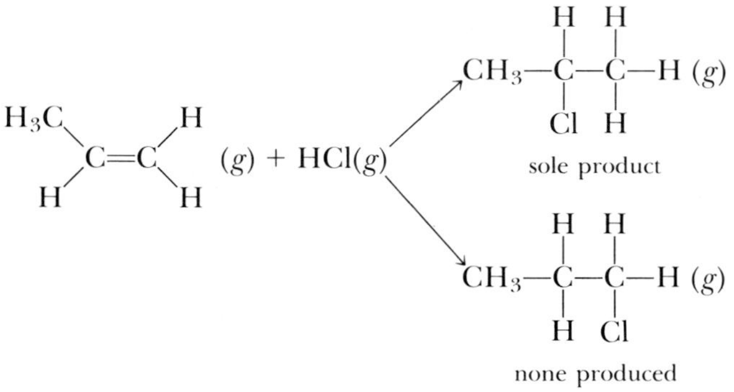

$$\underset{H}{\overset{H_3C}{}}\!\!>\!C{=}C\!<\!\underset{H}{\overset{H}{}}(g) + HCl(g) \nearrow CH_3-\underset{Cl}{\overset{H}{C}}-\underset{H}{\overset{H}{C}}-H\ (g)$$

sole product

$$\searrow CH_3-\underset{H}{\overset{H}{C}}-\underset{Cl}{\overset{H}{C}}-H\ (g)$$

none produced

Although two different products might seem possible in this reaction, only one is found. There is a simple rule for determining which product is produced. **Markovnikov's rule** states that, when HX adds to an alkene, the hydrogen atom attaches to the carbon atom in the double bond already bearing the larger number of hydrogen atoms. More succinctly, the hydrogen-rich get hydrogen-richer.

Example 23-5: Use Markovnikov's rule to predict the product of the reaction

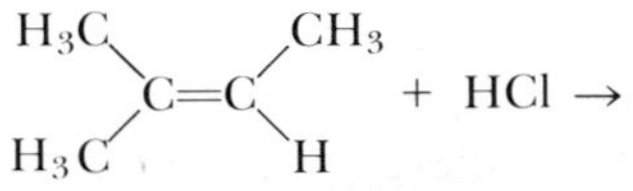

$$\underset{H_3C}{\overset{H_3C}{}}\!\!>\!C{=}C\!<\!\underset{H}{\overset{CH_3}{}} + HCl \rightarrow$$

Solution: According to Markovnikov's rule, the H of HCl will end up on the carbon atom of the double bond that already has the greater number of hydrogen atoms. This gives the sole product

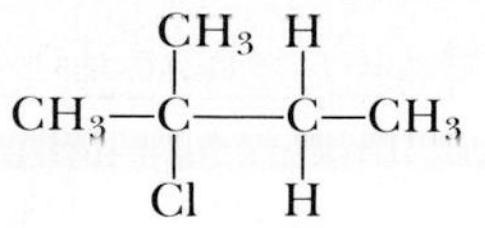

$$CH_3{-}\underset{\underset{Cl}{|}}{\overset{\overset{CH_3}{|}}{C}}{-}\underset{\underset{H}{|}}{\overset{\overset{H}{|}}{C}}{-}CH_3$$

4. Addition of water. In the presence of acid, which catalyzes the reaction, water adds to the more reactive alkenes:

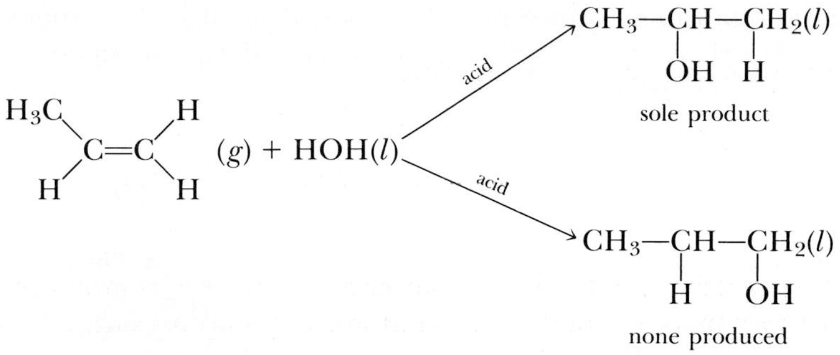

$$\underset{H}{\overset{H_3C}{}}\!C{=}C\!\underset{H}{\overset{H}{}}(g) + HOH(l) \xrightarrow{acid} CH_3{-}\underset{OH}{\underset{|}{CH}}{-}\underset{H}{\underset{|}{CH_2}}(l) \quad \text{sole product}$$

$$\xrightarrow{acid} CH_3{-}\underset{H}{\underset{|}{CH}}{-}\underset{OH}{\underset{|}{CH_2}}(l) \quad \text{none produced}$$

Note that the addition of water to an alkene obeys Markovnikov's rule. Simply picture the water as H—OH.

Example 23-6: Write structural formulas for and assign IUPAC names to the products of the following reactions:

(a) addition of Br_2 to *cis*-2-butene

(b) addition of HBr to 1-butene

Solution: (a) Bromine adds to the double bond:

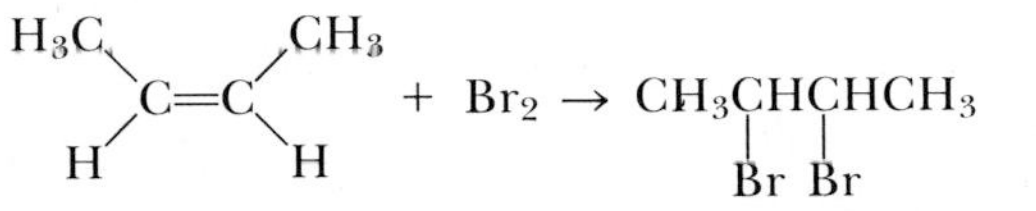

$$\underset{H}{\overset{H_3C}{}}\!C{=}C\!\underset{H}{\overset{CH_3}{}} + Br_2 \rightarrow CH_3\underset{Br}{\underset{|}{C}}H\underset{Br}{\underset{|}{C}}HCH_3$$

2,3-dibromobutane

(b) We must use Markovnikov's rule in this case:

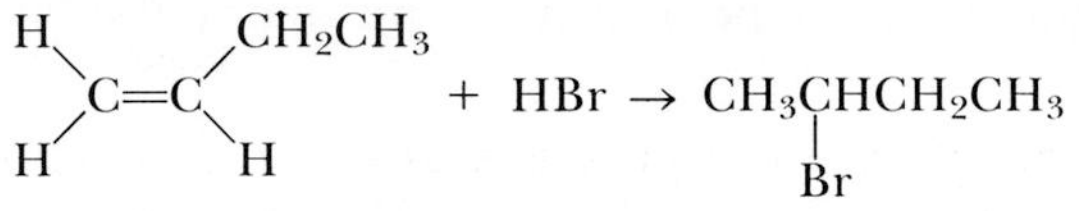

$$\underset{H}{\overset{H}{}}\!C{=}C\!\underset{H}{\overset{CH_2CH_3}{}} + HBr \rightarrow CH_3\underset{Br}{\underset{|}{C}}HCH_2CH_3$$

2-bromobutane

23-4. HYDROCARBONS THAT CONTAIN A TRIPLE BOND ARE CALLED ALKYNES

The simplest **alkyne** is acetylene, C_2H_2,

$$H{-}C{\equiv}C{-}H$$

Acetylene is a colorless gas with a penetrating odor. One of its most important uses is in oxyacetylene torches. We learned in Section 10-9 that acetylene is a linear molecule whose bonding can be described in terms of *sp* hybrid orbitals on the carbon atoms (Figure 10-21).

Figure 23-4 Chip Clark, spelunker, with a calcium carbide lamp on his helmet.

Acetylene can be produced by the reaction of calcium carbide and water:

$$\mathrm{CaC_2}(s) + 2\mathrm{H_2O}(l) \rightarrow \mathrm{Ca(OH)_2}(s) + \mathrm{C_2H_2}(g)$$

This reaction is used by spelunkers as a light source in caves. Acetylene is produced by allowing $H_2O(l)$ to drop slowly onto $CaC_2(s)$ in a canister. The $C_2H_2(g)$ pressure builds up, leaks out of the canister through a nozzle, and is burned in air:

$$2\mathrm{C_2H_2}(g) + 5\mathrm{O_2}(\text{air}) \rightarrow 4\mathrm{CO_2}(g) + 2\mathrm{H_2O}(g)$$

In addition to the combustion reaction that all hydrocarbons undergo, acetylene and related hydrocarbons undergo the same type of addition reactions as alkenes.

Example 23-7: Predict the product when HCl reacts with C_2H_2:

Solution: The reaction can be broken down into two steps. The first step is

$$\mathrm{H{-}C{\equiv}C{-}H} + \mathrm{HCl} \rightarrow \begin{matrix} \mathrm{Cl} & & \mathrm{H} \\ & \mathrm{C{=}C} & \\ \mathrm{H} & & \mathrm{H} \end{matrix}$$

The second step is

$$\begin{matrix} \mathrm{Cl} & & \mathrm{H} \\ & \mathrm{C{=}C} & \\ \mathrm{H} & & \mathrm{H} \end{matrix} + \mathrm{HCl} \rightarrow \begin{matrix} & \mathrm{Cl} & \mathrm{H} & \\ \mathrm{H}{-} & \mathrm{C} & {-}\mathrm{C} & {-}\mathrm{H} \\ & \mathrm{Cl} & \mathrm{H} & \end{matrix}$$

Note that we have used Markovnikov's rule to predict the product in the second step. The sole product is 1,1-dichloroethane.

The IUPAC nomenclature for alkynes uses the longest chain of consecutive carbon atoms *containing the triple bond* as the parent compound. The parent compound is named by dropping the *-ane* and adding *-yne* to the name of the corresponding alkane, and using a number to designate the carbon atom preceding the triple bond. (Compare with the nomenclature of alkenes given in Section 23-3.) For example, the IUPAC name of acetylene is ethyne and the IUPAC name of

$$\begin{matrix} & & \mathrm{CH} & & \\ \overset{1}{\mathrm{HC}} & {\equiv}\overset{2}{\mathrm{C}} & {-}\overset{3}{\mathrm{C}}\mathrm{H} & \overset{4}{\mathrm{C}}\mathrm{H_2} & \overset{5}{\mathrm{C}}\mathrm{H_3} \end{matrix}$$

is 3-methyl-1-pentyne. Some other examples are

$$\overset{4}{\mathrm{ClCH_2}}{-}\overset{3}{\mathrm{CH_2}}{-}\overset{2}{\mathrm{C}}{\equiv}\overset{1}{\mathrm{CH}} \qquad \text{4-chloro-1-butyne}$$

$$\overset{5}{\mathrm{CH_3}}{-}\overset{4}{\mathrm{CH_2}}{-}\overset{3}{\mathrm{C}}{\equiv}\overset{2}{\mathrm{C}}{-}\overset{1}{\mathrm{CH_2Cl}} \qquad \text{1-chloro-2-pentyne}$$

23-5. BENZENE BELONGS TO A CLASS OF HYDROCARBONS CALLED AROMATIC HYDROCARBONS

Benzene is a colorless, poisonous, flammable liquid (its boiling point is 80.1°C) with a characteristic odor. We learned in Section 10-10 that the

π electrons in benzene are delocalized (Figure 10-23). Recall that the two principal resonance forms of benzene are

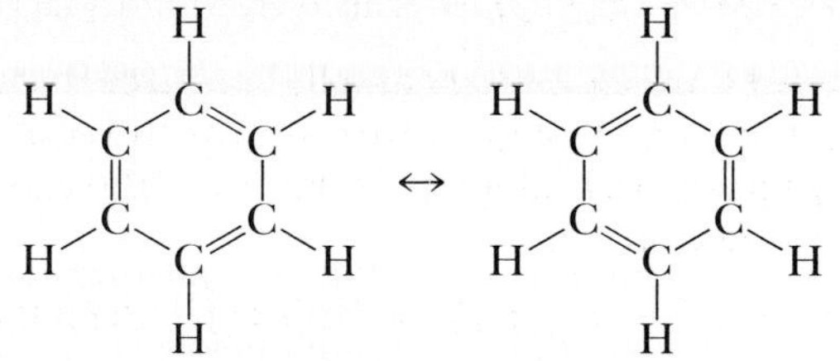

whose resonance hybrid is

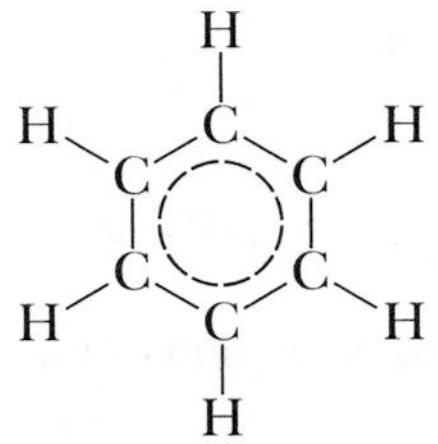

or, more compactly,

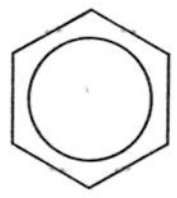

where a hydrogen atom is understood to be bonded to the carbon atom at each vertex in the benzene ring. Although each resonance form of benzene shows double bonds, the resonance hybrid does not. Benzene does not have double bonds and does not react as an unsaturated hydrocarbon. In fact, the π-electron delocalization causes the ring to be so stable that most of the reactions that benzene undergoes are substitution reactions in which the hydrogen atoms on the ring are replaced by other atoms or groups. For example, the usual reaction of benzene with bromine is a substitution reaction:

$$C_6H_6(l) + Br_2(l) \xrightarrow{FeBr_3} C_6H_5Br(l) + HBr(g)$$

The iron(III) bromide is a catalyst for this reaction. The fact that only one monobromobenzene has ever been isolated indicates that all the hydrogen atoms in benzene are equivalent, as is confirmed by X-ray and spectroscopic data.

Example 23-8: Draw structural formulas for all the isomers of dibromobenzene.

Solution: Because the benzene ring is a regular hexagon, there are three isomers of dibromobenzene:

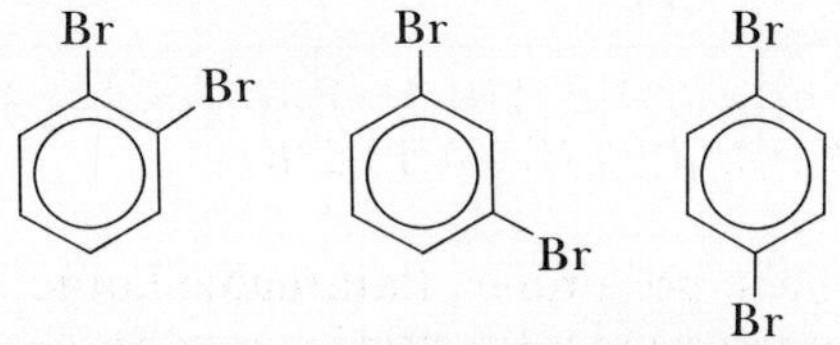

I II III

We can name these compounds by numbering the carbon atoms:

Therefore, we have

compound I	1,2-dibromobenzene
compound II	1,3-dibromobenzene
compound III	1,4-dibromobenzene

There is a less systematic but more common way to designate the positions of the bromine atoms in the three disubstituted benzenes in Example 23-8. Substituents at the 1,2 positions are designated **ortho-** (*o*-), those at the 1,3 positions are designated **meta-** (*m*-), and those at the 1,4 positions are designated **para-** (*p*-). Thus we write

o-dibromobenzene (ortho-dibromobenzene)   *m*-dibromobenzene (meta-dibromobenzene)   *p*-dibromobenzene (para-dibromobenzene)

Benzene belongs to the class of hydrocarbons called **aromatic hydrocarbons.** Aromatic hydrocarbons have relatively stable rings as a result of π-electron delocalization. Benzene and many other aromatic hydrocarbons are obtained from petroleum and coal tar. Almost 10 billion pounds of benzene are produced annually in the United States, making it the sixteenth-ranked industrial chemical in the United States. Benzene is used in the manufacture of medicinal chemicals, dyes, plastics, varnishes, lacquers, linoleum, and many other products. Some important derivatives of benzene are

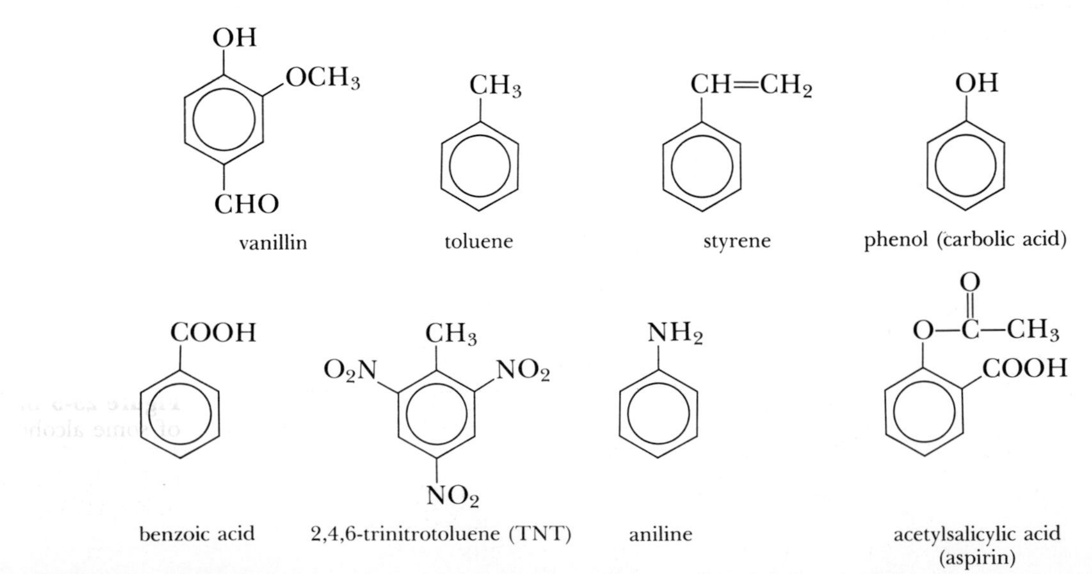

vanillin   toluene   styrene   phenol (carbolic acid)

benzoic acid   2,4,6-trinitrotoluene (TNT)   aniline   acetylsalicylic acid (aspirin)

23-6. ALCOHOLS ARE ORGANIC COMPOUNDS THAT CONTAIN AN —OH GROUP

The class of organic compounds called **alcohols** are characterized by an —OH group attached to a hydrocarbon chain. Some important common alcohols are

$$CH_3OH \qquad CH_3CH_2OH \qquad \begin{array}{c} CH_3CHCH_3 \\ | \\ OH \end{array}$$

methanol (methyl alcohol) ethanol (ethyl alcohol) 2-propanol (isopropyl alcohol)

These alcohols have both common names (given here in parentheses) and IUPAC names. The common names are formed by naming the alkyl group to which the —OH is attached and adding the word alcohol. The IUPAC name is formed by dropping the *-e* from the end of the alkane name of the longest chain of carbon atoms containing the —OH group and adding the suffix *-ol*. Thus, we have methanol, ethanol, and propanol. If the —OH group can be attached at more than one position, then its position is denoted by a number, as in

$$\overset{3}{C}H_3\overset{2}{C}H_2\overset{1}{C}H_2OH \qquad \begin{array}{c} \overset{3}{C}H_3\overset{2}{C}H\overset{1}{C}H_3 \\ | \\ OH \end{array}$$

1-propanol 2-propanol

Methanol sometimes is called wood alcohol because it can be produced by heating wood in the absence of oxygen. Over 8 billion pounds of methanol are produced annually in the United States by the reaction

$$CO(g) + 2H_2(g) \xrightarrow[\text{high P, high T}]{\text{catalyst}} CH_3OH(l)$$

Methanol is highly toxic and can cause blindness and death if taken internally. During Prohibition in the United States, many people died or became seriously ill from drinking methanol, either because they were not aware of the difference between methanol and ethanol or because the alcohol they purchased contained methanol as a major impurity.

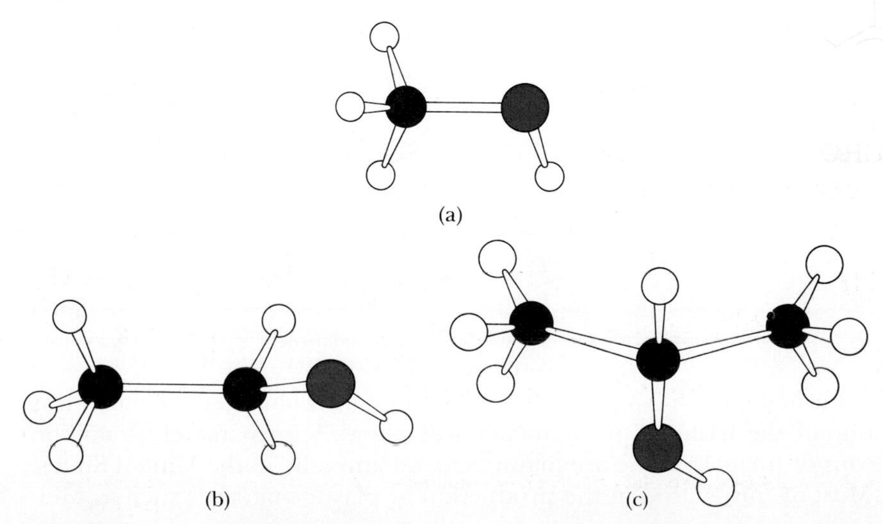

Figure 23-5 Ball-and-stick models of some alcohols: (a) methanol, CH_3OH; (b) ethanol, CH_3CH_2OH; and (c) 2-propanol, $\begin{array}{c} [CH_3CHCH_3. \\ | \\ OH \end{array}$

Ethanol is an ingredient in all fermented beverages. Various sugars and the starch in potatoes, grains, and similar substances can be used to produce ethanol by fermentation, a process in which yeast is used to convert sugar to ethanol. Although ethanol may be best known as an ingredient in alcoholic beverages, it is an important industrial chemical as well. An aqueous solution of 70 percent by volume of 2-propanol (isopropyl alcohol) is sold as rubbing alcohol.

It is sometimes convenient to view alcohols as derivatives of water, in which one of the hydrogen atoms is replaced by an alkyl group. This approach is most useful when the hydrocarbon portion of the alcohol is relatively small. The low-molecular-mass alcohols form hydrogen bonds and mix completely (i.e., are miscible) with water in all proportions. As the hydrocarbon portion becomes larger, however, the solubility of alcohols in water decreases and eventually approaches the low solubility of the comparable hydrocarbon. In addition, the low-molecular-mass alcohols have much higher boiling points than their comparable hydrocarbons. For example, whereas ethane (molecular mass 30) has a boiling point of $-88°C$, the boiling point of methyl alcohol (molecular mass 32) is 65°C. The relatively high boiling points of the low-molecular-mass alcohols is due to extensive hydrogen bonding in the liquids.

Alcohols undergo some of the reactions that water undergoes. For example, alcohols react with alkali metals to form alkoxides and liberate hydrogen:

$$2CH_3OH(l) + 2Na(s) \rightarrow \underset{\text{sodium methoxide}}{2NaOCH_3(alc)} + H_2(g)$$

Alkoxide ions are strong Lewis bases, that is, electron-pair donors (Section 15-12).

23-7. ALDEHYDES AND KETONES CONTAIN A CARBON-OXYGEN DOUBLE BOND

Alcohols can be oxidized to produce aldehydes and ketones. **Aldehydes** are compounds that have the general formula RCHO,

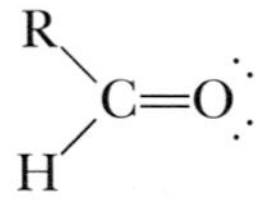

$$\begin{matrix} R \\ & \diagdown \\ & & C{=}\ddot{\underset{\cdot\cdot}{O}} \\ & \diagup \\ H \end{matrix}$$

where R is a hydrogen atom or an alkyl group. The —CHO group is called the **aldehyde group** and is characteristic of aldehydes. The two simplest aldehydes are

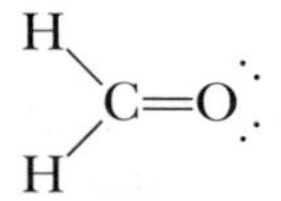

$$\begin{matrix} H \\ & \diagdown \\ & & C{=}\ddot{\underset{\cdot\cdot}{O}} \\ & \diagup \\ H \end{matrix} \qquad \begin{matrix} H_3C \\ & \diagdown \\ & & C{=}\ddot{\underset{\cdot\cdot}{O}} \\ & \diagup \\ H \end{matrix}$$

formaldehyde, HCHO 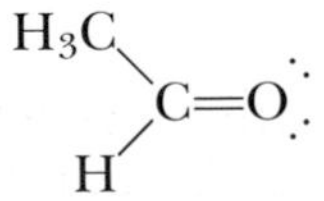 — acetaldehyde, CH_3CHO

Formaldehyde is a gas with an offensive and characteristic odor. It is one of the irritants in photochemical smog. Approximately 1 million tons of formaldehyde are manufactured annually in the United States. Most of this is used in the production of plastic materials such as Melmac and Bakelite. Melmac is used in plastic dinnerware, and Bakelite is used in the plastic casing of radios and telephones. A solution of formaldehyde in water is called formalin and is used to preserve biological specimens. However, this use is being phased out because formaldehyde is a carcinogen.

Acetaldehyde is a colorless, flammable liquid with a pungent, fruity odor. It is used in the manufacture of perfumes, flavors, plastics, synthetic rubbers, dyes, and a variety of other organic chemicals. Aldehydes are obtained from the oxidation of **primary alcohols,** which are alcohols in which the —OH-bearing carbon atom is bonded to only one other carbon atom. Examples of primary alcohols are

CH_3CH_2OH	ethanol
$CH_3CH_2CH_2OH$	1-propanol
$CH_3CH_2CH_2CH_2OH$	1-butanol

A commonly used oxidizing agent for the oxidation of primary alcohols is $K_2Cr_2O_7$ dissolved in an aqueous acidic solution:

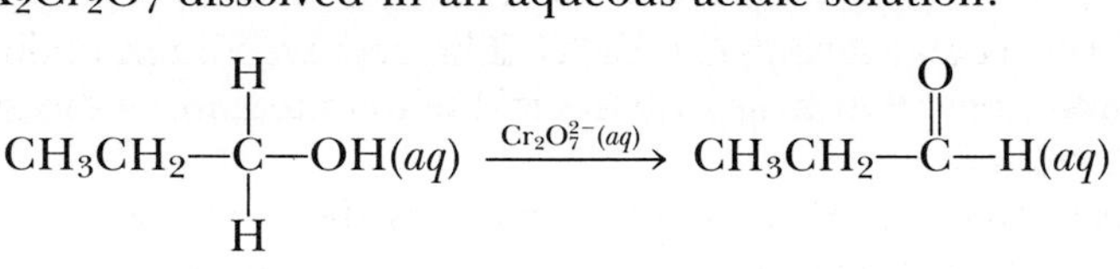

$$CH_3CH_2-\underset{\substack{|\\ H}}{\overset{\substack{H\\ |}}{C}}-OH(aq) \xrightarrow{Cr_2O_7^{2-}(aq)} CH_3CH_2-\overset{\substack{O\\ \|}}{C}-H(aq)$$

1-propanol → propanal

The systematic nomenclature for aldehydes is developed in Problem 23-37.

Secondary alcohols are alcohols in which the —OH-bearing carbon atom is bonded to two other carbon atoms. Examples of secondary alcohols are

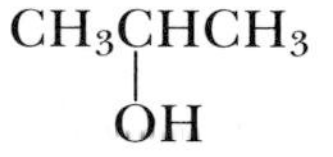

$CH_3\underset{\substack{|\\ OH}}{C}HCH_3$	2-propanol
$CH_3\underset{\substack{|\\ OH}}{C}HCH_2CH_3$	2-butanol
$CH_3CH_2\underset{\substack{|\\ OH}}{C}HCH_2CH_3$	3-pentanol

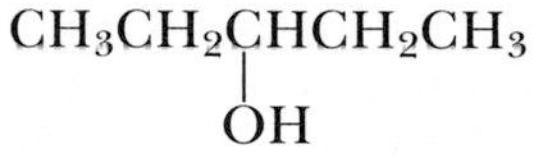

Secondary alcohols can be oxidized to compounds called **ketones,** whose general formula is

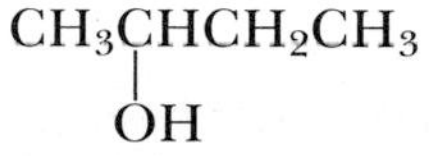

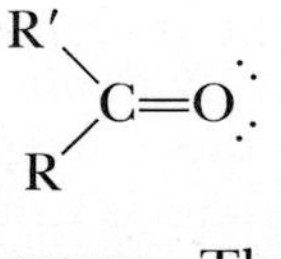

$$\begin{matrix} R' \diagdown & \\ & C{=}\ddot{\underset{..}{O}} \\ R \diagup & \end{matrix}$$

where R and R′ are alkyl groups. The simplest ketone, acetone, $(CH_3)_2CO$, is prepared from the oxidation of 2-propanol:

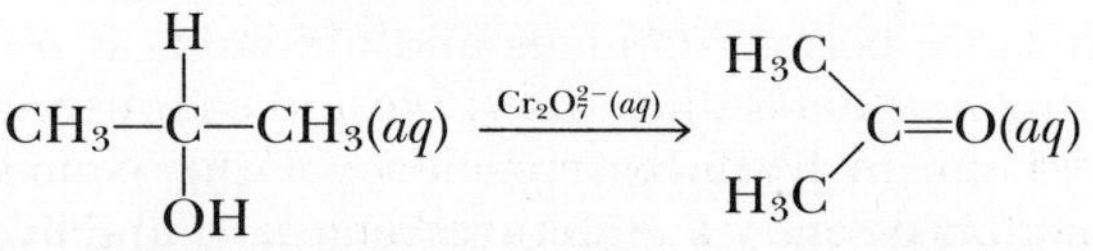

$$CH_3-\underset{\substack{|\\ OH}}{\overset{\substack{H\\ |}}{C}}-CH_3(aq) \xrightarrow{Cr_2O_7^{2-}(aq)} \begin{matrix} H_3C \diagdown & \\ & C{=}O(aq) \\ H_3C \diagup & \end{matrix}$$

2-propanol → acetone

Acetone is a colorless, volatile liquid with a sweet odor. About 2 billion pounds of acetone are produced annually in the United States. It is used extensively as a solvent because it dissolves many organic compounds, yet is completely miscible with water. It dissolves fats, oils, waxes, rubber, lacquers, varnishes, rubber cements, and coatings ranging from nail polish to exterior enamel paints.

Tertiary alcohols, which are alcohols in which the —OH-bearing carbon atom is attached to three other carbon atoms, are resistant to oxidation. The —OH-bearing carbon atom is already bonded to three other carbon atoms and so cannot form a double bond with an oxygen atom:

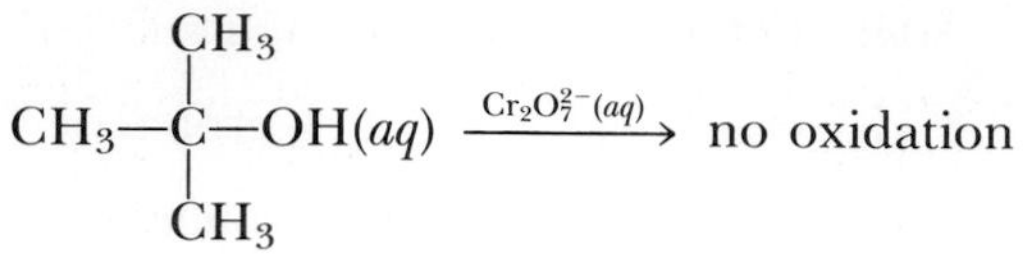

$$\begin{array}{c} CH_3 \\ | \\ CH_3\text{—}C\text{—}OH(aq) \\ | \\ CH_3 \end{array} \xrightarrow{Cr_2O_7^{2-}(aq)} \text{no oxidation}$$

tertiary alcohol:
2-methyl-2-propanol

Example 23-9: Which alcohol would you use to produce $CH_3CH_2\overset{\overset{\displaystyle O}{\|}}{C}CH_3$?

Solution: The product here is a ketone, and so we must oxidize a secondary alcohol. The secondary alcohol should have a methyl group and an ethyl group attached to the —OH-bearing carbon atom. Thus, we have

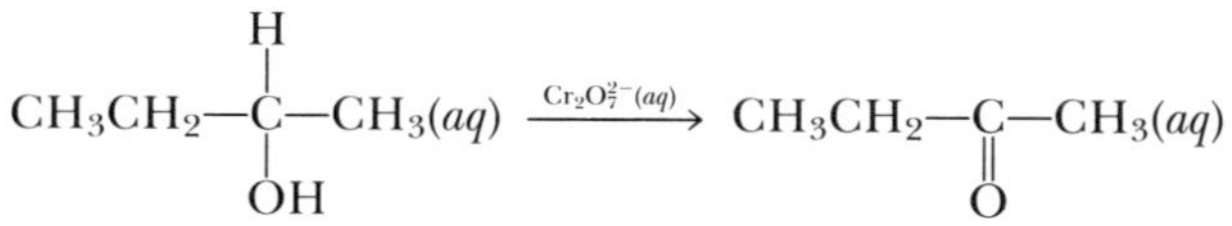

$$\begin{array}{c} H \\ | \\ CH_3CH_2\text{—}C\text{—}CH_3(aq) \\ | \\ OH \end{array} \xrightarrow{Cr_2O_7^{2-}(aq)} \begin{array}{c} CH_3CH_2\text{—}C\text{—}CH_3(aq) \\ \| \\ O \end{array}$$

2-butanol — methyl ethyl ketone

23-8. AMINES ARE ORGANIC DERIVATIVES OF AMMONIA

Amines are derivatives of ammonia in which one or more of the hydrogen atoms in NH_3 are replaced by hydrocarbon groups. Alkyl amines are frequently named by combining the names of the attached alkyl groups with the word *amine*. For example,

$$CH_3NH_2 \qquad (CH_3)_2NH \qquad \begin{array}{c} CH_3\text{—}N\text{—}CH_2CH_3 \\ | \\ CH_3 \end{array}$$

methylamine — dimethylamine — dimethylethylamine

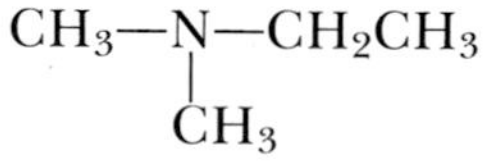

It is sometimes useful to classify amines according to the number of organic groups attached to the nitrogen atom. Thus, of the three amines shown here, methylamine is called a primary amine (with one attached group); dimethylamine is a secondary amine (with two attached groups); and dimethylethylamine is called a tertiary amine (with three attached groups).

The resemblance between amines and ammonia is much stronger than that between alcohols and water, but in both cases the similarity deteriorates as the hydrocarbon portions of the molecules become large. Primary and secondary amines have at least one hydrogen atom bonded to the central nitrogen atom, and so form hydrogen bonds in the liquid state. Thus, the boiling points of primary and secondary amines are higher than those of hydrocarbons of comparable molecular mass. The boiling points of tertiary amines, which have no hydrogen atom attached to the nitrogen atom, are closer to those of the comparable hydrocarbons; for example;

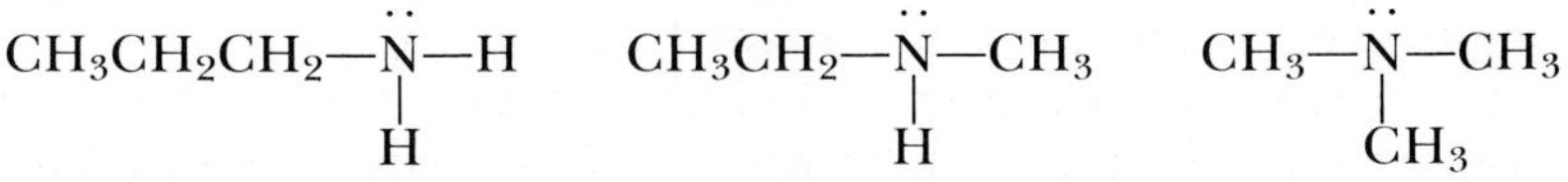

	$CH_3CH_2CH_2—\ddot{N}(H)—H$	$CH_3CH_2—\ddot{N}(H)—CH_3$	$CH_3—\ddot{N}(CH_3)—CH_3$
	propylamine	ethylmethylamine	trimethylamine
boiling point:	49°C	36°C	2.9°C

Notice that the molecular mass of each of these amines is 59.

Even tertiary amines can form hydrogen bonds with water molecules, and so the solubilities of amines in water are similar to those of alcohols of the same molecular mass. Amines having up to three or four carbon atoms are completely miscible with water, but the solubility decreases rapidly with an increasing number of carbon atoms.

Like ammonia, amines are weak bases (Table 15-3) and react with acids to form crystalline, nonvolatile salts that are very water-soluble. For example,

$$\underset{\text{dimethylamine}}{(CH_3)_2NH(aq)} + HBr(aq) \rightarrow \underset{\text{dimethylammonium ion}}{(CH_3)_2NH_2^+(aq)} + \underset{\text{bromide ion}}{Br^-(aq)}$$

Amines also behave as Lewis bases by reacting with alkyl halides, as in

$$\underset{\text{trimethylamine}}{(CH_3)_3N(aq)} + CH_3Cl(g) \rightarrow \underset{\text{tetramethylammonium ion}}{(CH_3)_4N^+(aq)} + \underset{\text{chloride ion}}{Cl^-(aq)}$$

The salts of amines are generally more water-soluble than the amines themselves.

Many alkyl amines have strong odors like that of decaying fish. The common names of some diamines suggest even worse odors:

$$\underset{\text{putrescine}}{H_2NCH_2CH_2CH_2CH_2NH_2} \qquad \underset{\text{cadaverine}}{H_2NCH_2CH_2CH_2CH_2CH_2NH_2}$$

23-9. THE REACTION OF A CARBOXYLIC ACID WITH AN ALCOHOL PRODUCES AN ESTER

One important use of aldehydes is in the production of **organic acids:**

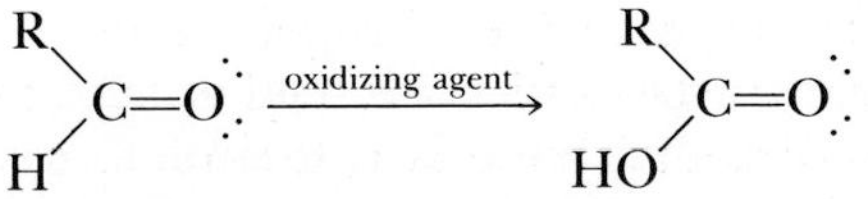

$$\underset{\text{aldehyde}}{R(H)C{=}\ddot{O}:} \xrightarrow{\text{oxidizing agent}} \underset{\text{organic acid}}{R(HO)C{=}\ddot{O}:}$$

For example,

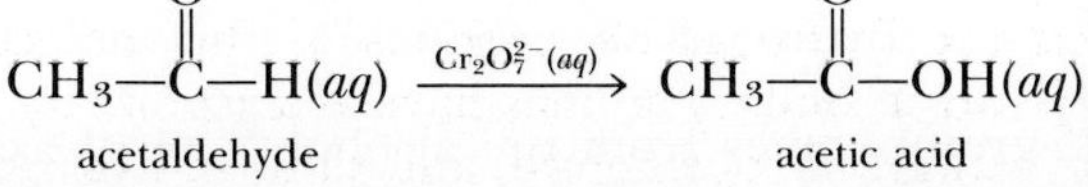

$$\underset{\text{acetaldehyde}}{CH_3—\overset{O}{\overset{\|}{C}}—H(aq)} \xrightarrow{Cr_2O_7^{2-}(aq)} \underset{\text{acetic acid}}{CH_3—\overset{O}{\overset{\|}{C}}—OH(aq)}$$

- alcohol: $R—OH$
- aldehyde: $R—C(H){=}O$
- ketone: $R(R')C{=}O$
- carboxylic acid: $R—C(OH){=}O$

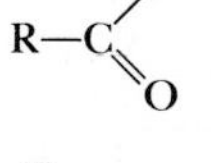

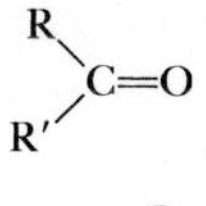

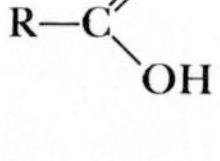

Note that the oxidizing agent inserts an oxygen atom between the carbon atom and the hydrogen atom in the aldehyde group.

We can also obtain organic acids by the direct oxidation of primary alcohols; in the oxidation, the aldehyde occurs as an intermediate species:

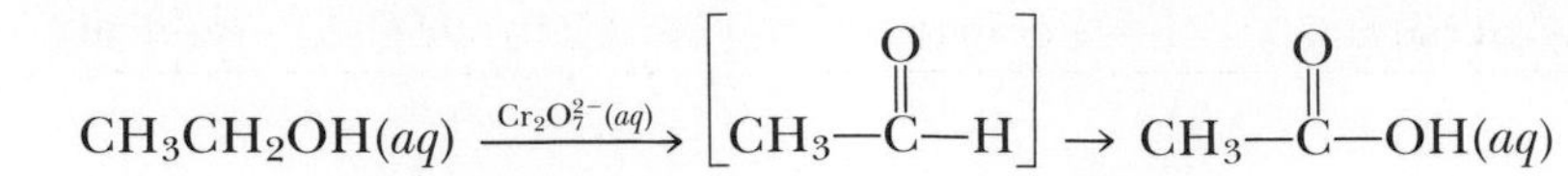

$$CH_3CH_2OH(aq) \xrightarrow{Cr_2O_7^{2-}(aq)} \left[CH_3{-}\overset{\overset{\displaystyle O}{\|}}{C}{-}H\right] \rightarrow CH_3{-}\overset{\overset{\displaystyle O}{\|}}{C}{-}OH(aq)$$

Carefully controlled conditions are required to isolate the aldehyde in good yield from this reaction because the aldehyde is readily oxidized to the acid.

The general formula for organic carboxylic acids is RCOOH. The —COOH group, called the **carboxyl group,** is characteristic of organic acids, or **carboxylic acids.** The two simplest carboxylic organic acids are formic acid and acetic acid (see Section 15-1):

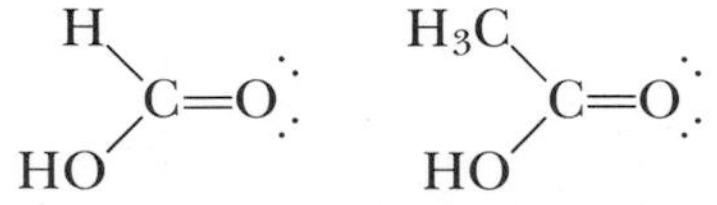

H, HO>C=Ö:	H_3C, HO>C=Ö:
formic acid, HCOOH	acetic acid, CH_3COOH

Formic acid is a colorless, fuming liquid that is very soluble in water. This acid is one of the irritants in the fluid secreted by ants when they bite. Acetic acid is a clear, colorless liquid with a pungent odor. The pure compound is called glacial acetic acid because it freezes at 18°C. Acetic acid is ranked thirty-fifth among industrial chemicals in terms of annual U.S. production. It is used in the manufacture of plastics, pharmaceuticals, dyes, insecticides, photographic chemicals, acetates, and many other organic chemicals.

Carboxylic acids react with alcohols in the presence of an acid catalyst; the equation for the reaction is:

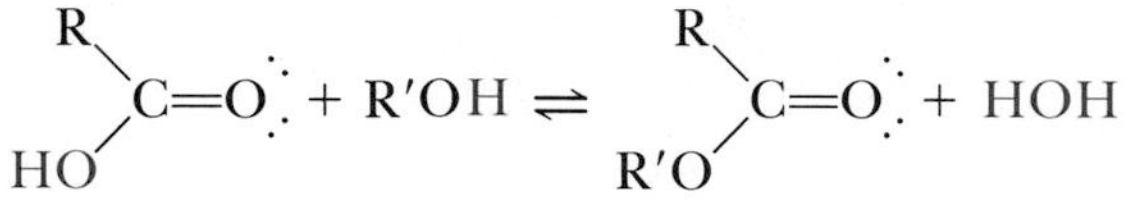

$$\begin{matrix} R \\ & C{=}\ddot{O}: \\ HO \end{matrix} + R'OH \rightleftharpoons \begin{matrix} R \\ & C{=}\ddot{O}: \\ R'O \end{matrix} + HOH$$

where R and R′ represent possibly different alkyl groups. The coloring used in this reaction emphasizes that the water is formed from the —OH group of the acid and the hydrogen atom of the alcohol. This is quite unlike an acid-base neutralization reaction, in which the water is formed by the reaction between a hydroxide ion from the base and a hydronium ion from the acid. Alcohols do not react like bases. The similarity in their chemical formulas, for example, CH_3OH versus NaOH, is superficial.

The product of the reaction between an organic acid and an alcohol is called an **ester.** As the carboxylic acid–alcohol reaction indicates, the general formula for an ester is

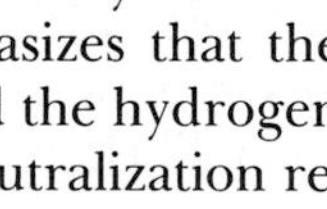

$$\begin{matrix} R \\ & C{=}\ddot{O}: \\ R'O \end{matrix}$$

where the R′ group comes from the alcohol. The pleasant odors of

Table 23-3 Some esters and their odors

Name	Odor	Name	Odor
ethyl formate	rum	methyl butyrate	apples
pentyl acetate	bananas	ethyl butyrate	pineapples
octyl acetate	oranges	pentyl butyrate	apricots

flowers and fruits are due to esters. Table 23-3 lists some naturally occurring esters and their odors. Esters are named by first naming the alkyl group from the alcohol and then designating the acid, with its *-ic* ending changed to *-ate*. For example,

$$\underset{\text{ethanol}}{C_2H_5OH} + \underset{\text{formic acid}}{H(HO)C{=}\ddot{O}:} \rightleftharpoons \underset{\text{ethyl formate}}{H(C_2H_5O)C{=}\ddot{O}:} + HOH$$

$$\underset{\text{1-pentanol}}{C_5H_{11}OH} + \underset{\text{acetic acid}}{CH_3(HO)C{=}\ddot{O}:} \rightleftharpoons \underset{\text{pentyl acetate}}{CH_3(C_5H_{11}O)C{=}\ddot{O}:} + HOH$$

Such reactions are called **esterification reactions.**

Example 23-10: Complete and balance the equation for the reaction between ethanol and acetic acid. Name the product.

Solution: The reaction between an organic acid and an alcohol yields an ester, so in this case we have

$$\underset{\text{ethanol}}{CH_3CH_2OH} + \underset{\text{acetic acid}}{H_3C(HO)C{=}\ddot{O}:} \rightleftharpoons \underset{\text{ethyl acetate}}{H_3C(CH_3CH_2O)C{=}\ddot{O}:} + H_2O$$

The equilibrium constants of equations for esterification reactions are usually not large, and so appreciable quantities of both reactants and products are present at equilibrium.

SUMMARY

Organic chemistry is the chemistry of compounds that contain carbon atoms. The simplest organic compounds are hydrocarbons, which are compounds that consist of only hydrogen and carbon. Hydrocarbons are divided into alkanes, alkenes, alkynes, and aromatic hydrocarbons. Alkanes have only single bonds, alkenes have one or more double bonds, alkynes have one or more triple bonds, and aromatic hydrocarbons have rings with delocalized electrons. Alkanes, which are also called saturated hydrocarbons, are relatively unreactive and undergo primarily combustion and substitution reactions. Alkenes and alkynes, which are called unsaturated hydrocarbons, undergo the same reactions as saturated hydrocarbons, as well as addition reactions, in which small molecules, such as H_2, Cl_2, Br_2, HCl, and H_2O, add across the double or triple bonds.

Other classes of organic compounds are alcohols (ROH), aldehydes (RCHO), ketones (RCOR′), carboxylic acids (RCOOH), and esters (RCOOR′). Compounds in each class undergo reactions that are characteristic of the class. For example, carboxylic acids and alcohols react to form esters, and aldehydes can be oxidized to carboxylic acids.

Because of the great diversity of organic compounds, a systematic nomenclature is necessary. An internationally recognized system (IUPAC) has been developed that allows each organic compound to be named unambiguously.

TERMS YOU SHOULD KNOW

organic compound 742
hydrocarbon 742
alkane 742
structural formula 742
structural isomer 743
saturated hydrocarbon 743
substitution reaction 744
free-radical reaction 745
alkyl halide (haloalkane) 745
substituent 745
disubstituted 745
IUPAC nomenclature 746
alkyl group 747
unsaturated hydrocarbon 748
alkene 748
addition reaction 749
hydrogenation 750
Markovnikov's rule 750
alkyne 751
ortho- 754
meta- 754
para- 754
aromatic hydrocarbon 754
alcohol 755
aldehyde 756
aldehyde group 756
primary alcohol 757
secondary alcohol 757
ketone 757
tertiary alcohol 758
amine 758
organic acid 759
carboxyl group 760
carboxylic acid 760
ester 760
esterification reaction 761

PROBLEMS

ALKANES

23-1. Complete and balance the following equations (if no reaction occurs, then write N.R.):

(a) $C_5H_{12}(g) + O_2(g) \rightarrow$
(b) $C_2H_6(g) + Cl_2(g) \xrightarrow{\text{dark}}$
(c) $C_4H_{10}(g) + H_2SO_4(aq) \rightarrow$
(d) $CH_4(g) + Cl_2(g) \xrightarrow{\text{UV}}$

23-2. Complete and balance the following equations (if no reaction occurs, then write N.R.):

(a) $C_2H_6(g) + KOH(aq) \rightarrow$
(b) $C_6H_{14}(l) + O_2(g) \rightarrow$
(c) $C_3H_8(g) + KMnO_4(aq) \rightarrow$
(d) $C_5H_{12}(g) + HCl(aq) \rightarrow$

23-3. Which of the following pairs of molecules are identical and which are different?

(a) $ClCH_2CH_2CH_3$ and $CH_3CH_2CH_2Cl$
(b) $CH_3CH_2\underset{\underset{CH_3}{|}}{C}HCl$ and $CH_3\underset{\underset{Cl}{|}}{C}HCH_2CH_3$
(c) $Cl—CH_2—\overset{\overset{CH_3}{|}}{C}H—CH_3$ and $CH_3—\overset{\overset{CH_3}{|}}{\underset{\underset{Cl}{|}}{C}}—CH_3$
(d) $CH_3—\overset{\overset{CH_2Cl}{|}}{C}H—CH_3$ and $ClCH_2—\overset{\overset{CH_3}{|}}{C}H—CH_3$

23-4. Which of the following pairs of molecules are identical and which are different?

(a) $CH_3CH_2\underset{\underset{Cl}{|}}{C}HCH_3$ and $CH_3\underset{\underset{Cl}{|}}{C}HCH_2CH_3$
(b) $H_2\overset{\overset{Cl}{|}}{C}CH_2CH_3$ and $CH_3CH_2CH_2Cl$
(c) $CH_3\overset{\overset{CH_3}{|}}{C}HCH_2CH_2CH_3$ and $CH_3CH_2\overset{\overset{CH_3}{|}}{C}HCH_2CH_3$
(d) $CH_3—\overset{\overset{H}{|}}{\underset{\underset{CH_2CH_3}{|}}{C}}—Cl$ and $CH_3—\overset{\overset{Cl}{|}}{\underset{\underset{H}{|}}{C}}—CH_2CH_3$

23-5. If you substitute a chlorine atom for a hydrogen atom in *n*-hexane, how many possible isomers do you get? Give IUPAC names for all these isomers.

23-6. Suppose that neopentane reacts with chlorine under UV radiation such that two hydrogen atoms are replaced by two chlorine atoms. Write the Lewis formulas for the possible isomers and name each isomer.

23-7. Give IUPAC names for the following compounds:

(a) 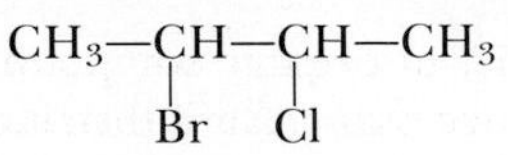 $CH_3{-}CH(Br){-}CH(Cl){-}CH_3$

(b) 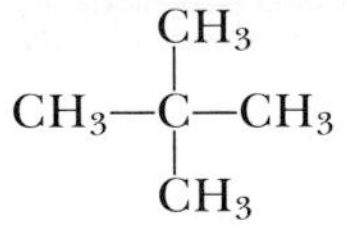 $CH_3{-}C(CH_3)_2{-}CH_3$

(c) $CH_3{-}C(CH_3)(NO_2){-}CH_3$

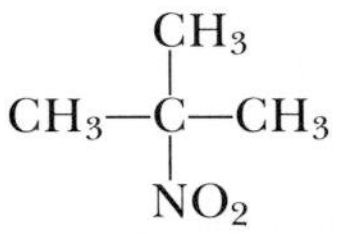

(d) 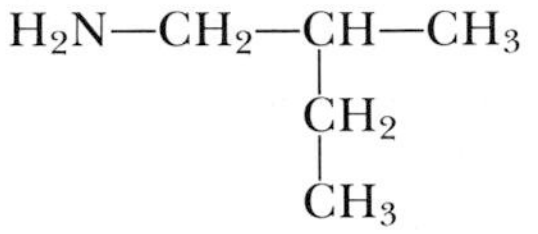 $H_2N{-}CH_2{-}CH(CH_2CH_3){-}CH_3$

23-8. Give IUPAC names for the following compounds:

(a) 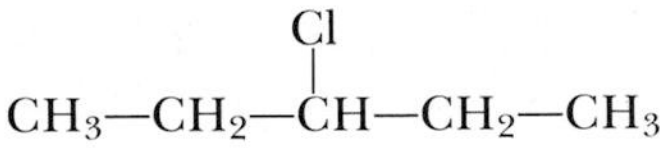 $CH_3{-}CH_2{-}CH(Cl){-}CH_2{-}CH_3$

(b) 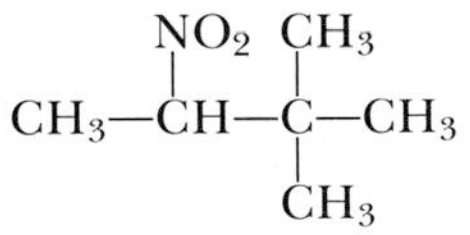 $CH_3{-}CH(NO_2){-}C(CH_3)_2{-}CH_3$

(c) $CH_3CH_2CH(Cl)CH_3$

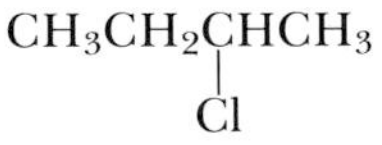

(d) 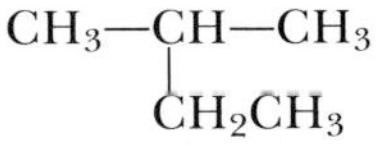 $CH_3{-}CH(CH_2CH_3){-}CH_3$

23-9. Explain why the following names are incorrect and give a correct IUPAC name in each case:

(a) 4-methylpentane
(b) 2-ethylbutane
(c) 2-propylhexane
(d) 2-dimethylpropane

23-10. Explain why the following names are incorrect and give a correct IUPAC name in each case:

(a) 2,3-dichloropropane
(b) 3-bromo-2-ethylpropane
(c) 1,1,3-trimethylpropane
(d) 2-dimethylbutane

23-11. Write the Lewis formula for and assign an IUPAC name to each of the following compounds:

(a) $CH_3CH(CH_3)CH_2CH_3$
(b) CH_2ClCH_2Br
(c) $CH_3CCl_2CCl_3$
(d) $CH_3C(CH_3)_2CH_3$

23-12. Write the Lewis formula for and assign an IUPAC name to each of the following compounds:

(a) $CH_3(CH_2)_4CH_3$
(b) $(CH_3)_2CHCH_2CH_2CH(CH_3)_2$
(c) $(CH_3)_4C$
(d) $CH_3CHClCCl_2CCl_3$

23-13. Write the structural formula for each of the following alkanes:

(a) 2,3-dimethylbutane
(b) 2,2,3-trimethylbutane
(c) 3,3-dimethyl-4-ethylhexane
(d) 4-isopropyloctane

23-14. Write the structural formula for each of the following trichloroalkanes:

(a) 1,1,2-trichlorobutane
(b) 1,1,1-trichloroethane
(c) 1,2,3-trichloropentane
(d) 2,2,4-trichlorohexane

23-15. Write the Lewis formulas for the following:

(a) 2,3-dimethylbutane
(b) 2-amino-3-methylbutane
(c) 3-chloro-3-ethylpentane
(d) 1,2,3-trichloropropane

ALKENES

23-16. Write a structural formula for each of the following alkenes:

(a) 2-methyl-3-hexene
(b) 2-methyl-1-pentene
(c) 2,3-dimethyl-2-butene
(d) 3-methyl-2-pentene

23-17. Name the product of the following reactions:

(a) addition of HCl to propene
(b) addition of HBr to 2-butene
(c) addition of HCl to 1-butene
(d) addition of HBr to 2-pentene

23-18. Write the structural formula for and assign an IUPAC name to the product when each of the following compounds reacts with 1 mol of bromine:

(a) 1-butene
(b) 2-butene
(c) 1,3-hexadiene (try this one)

23-19. Complete the following equations and name the products:

(a) 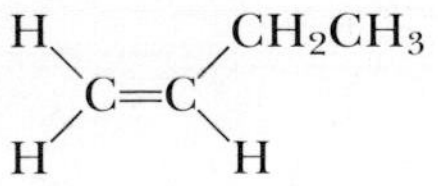 $(H)(H)C{=}C(CH_2CH_3)(H) + HCl \rightarrow$

(b) 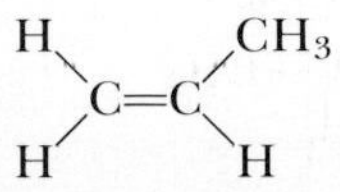 $(H)(H)C{=}C(CH_3)(H) + Cl_2 \rightarrow$

(c) 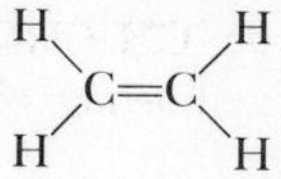 $H_2C{=}CH_2 + H_2O \xrightarrow{H^+(aq)}$

(d) 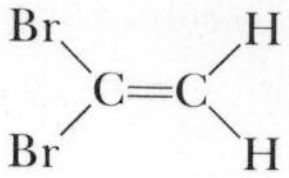 $Br_2C{=}CH_2 + HBr \rightarrow$

23-20. Complete and balance the following equations:

(a) 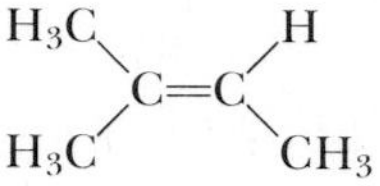 $(H_3C)_2C{=}CH(CH_3)(l) + Br_2(l) \rightarrow$

(b) 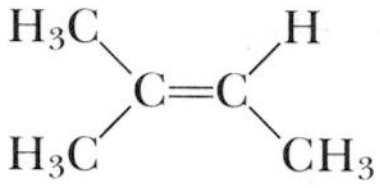 $(H_3C)_2C{=}CH(CH_3)(l) + HCl(g) \rightarrow$

(c) 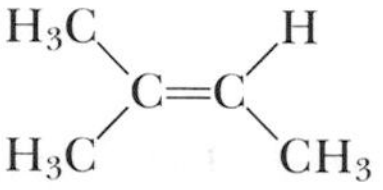 $(H_3C)_2C{=}CH(CH_3)(l) + H_2O(l) \xrightarrow{H^+(aq)}$

23-21. In each case, determine which unsaturated hydrocarbon reacts with what reagent to form the given product:

(a) 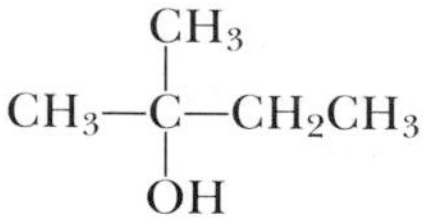 $CH_3{-}C(CH_3)(OH){-}CH_2CH_3$

(b) 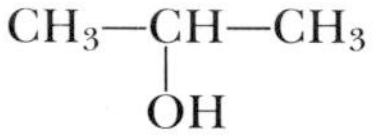 $CH_3{-}CH(OH){-}CH_3$

(c) $CH_3CH{=}CHCH_3$

(d) 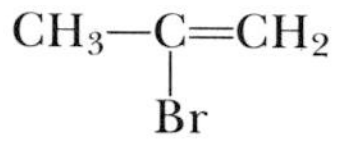 $CH_3{-}C(Br){=}CH_2$

23-22. In each case, determine which unsaturated hydrocarbon reacts with what reagent to form the given product:

(a) $CH_3CHBrCHBrCH_3$
(b) $CH_3CCl_2CCl_2CH_3$
(c) $(CH_3)_3COH$
(d) 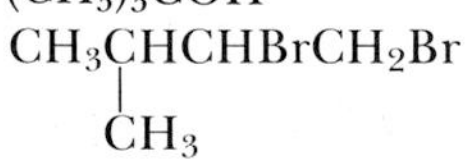 $CH_3CH(CH_3)CHBrCH_2Br$

ALKYNES

23-23. Assign IUPAC names to the following alkynes:

(a) $CH_3{-}C{\equiv}CH$
(b) $CH_3{-}C{\equiv}C{-}CH_3$
(c) 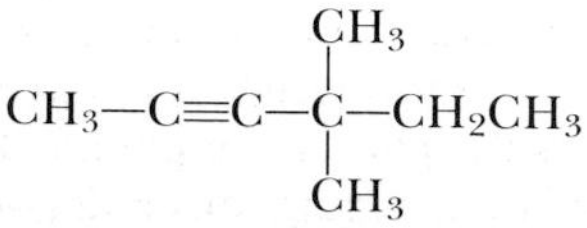 $CH_3{-}C{\equiv}C{-}C(CH_3)_2{-}CH_2CH_3$
(d) 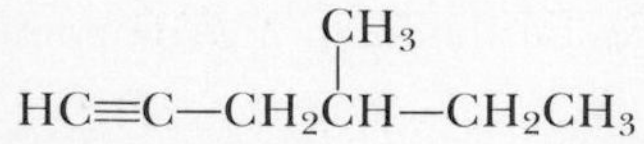 $HC{\equiv}C{-}CH_2CH(CH_3){-}CH_2CH_3$

23-24. Write a structural formula for each of the following alkynes:

(a) 2-pentyne
(b) 3-hexyne
(c) 5-ethyl-3-octyne
(d) 2,2-dimethyl-3-hexyne

23-25. Predict the product in the addition of 2 mol of HBr to 1 mol of propyne.

23-26. Predict the product in the addition of 2 mol of HBr to 1 mol of 1-butyne.

23-27. Complete and balance the following equations (assume complete saturation of the triple bonds):

(a) $CH_3C{\equiv}CH(g) + O_2(g) \xrightarrow{\text{combustion}}$
(b) $CH_3C{\equiv}CH(g) + HCl(g) \rightarrow$
(c) $CH_3C{\equiv}CH(g) + Br_2(l) \rightarrow$

23-28. Complete and balance the following equations (assume complete saturation of the triple bonds):

(a) $CH_3C{\equiv}CCH_3(g) + HCl(g) \rightarrow$
(b) $CH_3C{\equiv}CCH_3(g) + H_2(g) \xrightarrow{Ni(s)}$
(c) $CH_3C{\equiv}CCH_3(g) + Cl_2(g) \rightarrow$

BENZENE

23-29. Write the structural formula for each of the following benzene derivatives:

(a) ethylbenzene
(b) 1,3,5-trichlorobenzene
(c) 2-chloro-1-methylbenzene
(d) 1-chloro-3-bromobenzene

23-30. Name the following benzene derivatives:

(a) 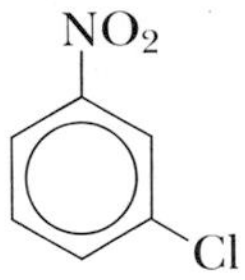 benzene ring with NO_2 and Cl (meta)

(b) 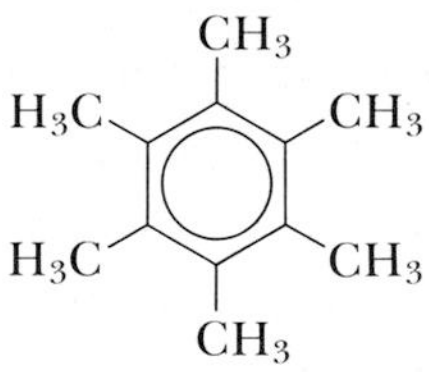 benzene ring with six CH_3 groups

(c) 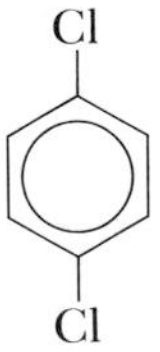 benzene ring with Cl and Cl (para)

(d) 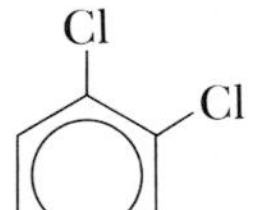 benzene ring with Cl and Cl (ortho)

ALCOHOLS

23-31. Assign IUPAC names to each of the following alcohols:

(a) $\begin{array}{c} CH_3\underset{|}{C}HOH \\ CH_3 \end{array}$ (b) $\begin{array}{c} CH_3 \\ | \\ CH_3CH_2CCH_2OH \\ | \\ CH_3 \end{array}$

(c) $\begin{array}{c} ClCH_2\underset{|}{C}HCH_2Cl \\ OH \end{array}$ (d) $\begin{array}{c} CH_3 \\ | \\ CH_3CH_2COH \\ | \\ H \end{array}$

23-32. Write a structural formula for each of the following alcohols:

(a) 2,3-dimethyl-1-butanol
(b) 2-methyl-4-hexanol
(c) 2-chloro-1-hexanol
(d) 1,2-dichloro-3-pentanol

23-33. Classify the following alcohols as primary, secondary, or tertiary:

(a) CH_3CH_2OH (b) $\begin{array}{c} H \\ | \\ CH_3CCH_3 \\ | \\ OH \end{array}$

(c) $\begin{array}{c} CH_3 \\ | \\ CH_3CH_2CH \\ | \\ OH \end{array}$ (d) $\begin{array}{c} CH_3 \\ | \\ CH_3COH \\ | \\ CH_3 \end{array}$

23-34. Classify the following alcohols as primary, secondary, or tertiary:

(a) $\begin{array}{c} CH_3 \\ | \\ CH_3CH_2COH \\ | \\ CH_3 \end{array}$ (b) $\begin{array}{c} OH \\ | \\ CH_3CH_2CHCH_3 \end{array}$

(c) $CH_3CH_2CH_2CH_2OH$ (d) $\begin{array}{c} CH_3 \\ | \\ CH_3CCH_2OH \\ | \\ CH_3 \end{array}$

23-35. Complete and balance the following equations:

(a) $CH_3CH_2OH(l) + Na(s) \rightarrow$
(b) $CH_3CH_2CH_2OH(l) + Na(s) \rightarrow$

23-36. Metal alkoxides (see Problem 23-35) are bases that are comparable in strength to sodium hydroxide. Write an equation that illustrates the basic character of alkoxides.

ALDEHYDES

23-37. The IUPAC nomenclature for aldehydes uses the longest chain of consecutive carbon atoms *containing the —CHO group* as the parent compound. The parent compound is named by dropping the terminal *-e* and adding *-al* to the name of the corresponding alkane (compare with the nomenclature for alcohols). For example, the IUPAC name for acetaldehyde, CH_3CHO, is ethanal. Assign IUPAC names to the following aldehydes:

(a) $CH_3CH_2CH_2CHO$ (b) $\begin{array}{c} CH_3 \\ | \\ CH_3CHCH_2CHO \end{array}$

(c) $HCHO$ (d) $\begin{array}{c} CH_3 \\ | \\ CH_3CHCHCH_2CHO \\ | \\ CH_3 \end{array}$

23-38. Using the IUPAC nomenclature for aldehydes presented in Problem 23-37, write the structural formula for each of the following aldehydes:

(a) propanal
(b) 2-methylpentanal
(c) 4-methylpentanal
(d) 3,3-dimethylhexanal

23-39. Determine which alcohol you would use to produce each of the following ketones:

(a) diethyl ketone
(b) methyl propyl ketone
(c) ethyl propyl ketone

23-40. Determine which alcohol you would use to produce each of the following aldehydes:

(a) ethanal
(b) 2-methylpropanal
(c) 2,2-dimethylpropanal

AMINES

23-41. Complete and balance the following equations:

(a) $C_2H_5NH_2(aq) + HBr(aq) \rightarrow$
(b) $(CH_3)_2NH(aq) + H_2SO_4(aq) \rightarrow$
(c) $C_6H_5NH_2$ (benzene ring with NH_2) $(aq) + HCl(aq) \rightarrow$
(d) $(C_2H_5)_3N(aq) + HCl(aq) \rightarrow$

23-42. Complete and balance the following equations:

(a) $H_2NCH_2CH_2NH_2(aq) + HI(aq) \rightarrow$
(b) $CH_3NH_2(aq) + H_2O(l) \rightarrow$
(c) $H_2NCH_2CH(NH_2)_2(aq) + HCl(aq) \rightarrow$

CARBOXYLIC ACIDS

23-43. The IUPAC nomenclature for carboxylic acids uses the longest chain of consecutive carbon atoms *containing the —COOH group* as the parent compound. The parent compound is named by changing the ending *-e* in the name of the corresponding alkane to *-oic acid.* For example, the IUPAC name for formic acid is methanoic acid and that for acetic acid is ethanoic acid. Write the structural formula for each of the following carboxylic acids:

(a) propanoic acid
(b) 2-methylpropanoic acid
(c) 3,3-dimethylbutanoic acid
(d) 3-methylpentanoic acid

23-44. Using the IUPAC nomenclature for carboxylic acids presented in Problem 23-43, name the following compounds:

(a)  CH_3CHCH_2COOH (with Cl attached to the second carbon)

(b) 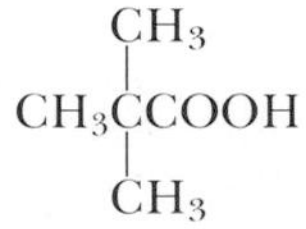 CH_3CCOOH (with CH_3 above and CH_3 below the second carbon)

(c) 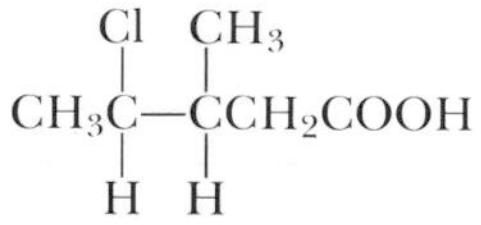 $CH_3C{-}CCH_2COOH$ (with Cl above and H below the second carbon; CH_3 above and H below the third carbon)

(d) 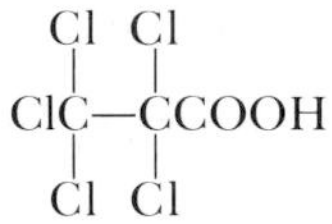 $ClC{-}CCOOH$ (with Cl above and Cl below each of the two carbons)

23-45. Complete and balance the following equations:

(a) $HCOOH(aq) + NaOH(aq) \rightarrow$
(b) $HCOOH(aq) + CH_3OH(aq) \xrightarrow{H^+(aq)}$
(c) $HCOOH(aq) + Ca(OH)_2(aq) \rightarrow$

23-46. Complete and balance the following equations:

(a) $CH_3CH_2COOH(aq) + NH_3(aq) \rightarrow$
(b) $CH_3CH_2COOH(aq) + CH_3OH(aq) \xrightarrow{H^+(aq)}$
(c) $CH_3CH_2COOH(aq) + CH_3CH_2OH(aq) \xrightarrow{H^+(aq)}$

23-47. To form the IUPAC name of anions of carboxylic acids, the ending *-ic acid* is replaced by *-ate*. For example, the IUPAC name for sodium acetate is sodium ethanoate. Complete and balance the following equations and name the salt in each case:

(a) $CH_3CH_2COOH(aq) + KOH(aq) \rightarrow$
(b) 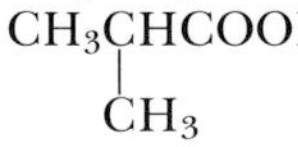 $CH_3CHCOOH(aq) + KOH(aq) \rightarrow$ (with CH_3 below the second carbon)
(c) $Cl_2CHCOOH(aq) + Ca(OH)_2(aq) \rightarrow$

23-48. Write the structural formula for each of the following salts (see Problem 23-47):

(a) sodium 2-chloropropanoate
(b) rubidium methanoate
(c) strontium 2,2-dimethylpropanoate
(d) lanthanum ethanoate

23-49. Complete and balance the following equations and name the products:

(a) 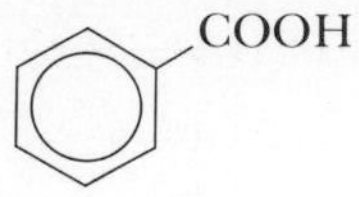 $C_6H_5COOH + CH_3CH_2OH \xrightarrow{H^+(aq)}$
benzoic acid

(b) $HOOC{-}COOH + CH_3CH_2CH_2OH \xrightarrow{H^+(aq)}$
oxalic acid

(c) $CH_3COOH + CH_3CHOHCH_3 \xrightarrow{H^+(aq)}$

(d) $CCl_3COOH + CH_3OH \xrightarrow{H^+(aq)}$
trichloroacetic acid

23-50. Complete and balance the following equations and name the products:

(a)  $COOH{-}COOH + KOH \rightarrow$
oxalic acid

(b) $CH_3(CH_2)_{16}COOH + NaOH \rightarrow$
stearic acid

(c) 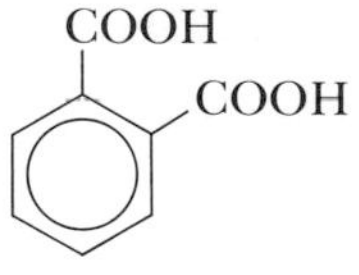 $C_6H_4(COOH)_2 + KOH \rightarrow$
phthalic acid

(d) 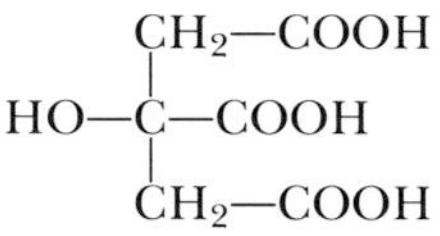 $HO{-}C(CH_2{-}COOH)_2{-}COOH + NaOH \rightarrow$
citric acid

ADDITIONAL PROBLEMS

23-51. In enumerating the monochlorosubstituted isomers of propane, we listed 1-chloropropane and 2-chloropropane but not 3-chloropropane. Why not?

23-52. An unknown cylinder is labeled "PENTANE." When the gas inside the cylinder is monochlorinated, five isomers of formula $C_5H_{11}Cl$ result. Was the gas pure *n*-pentane, pure isopentane, pure neopentane, or a mixture of two or all three of these?

23-53. Consider the following structures:

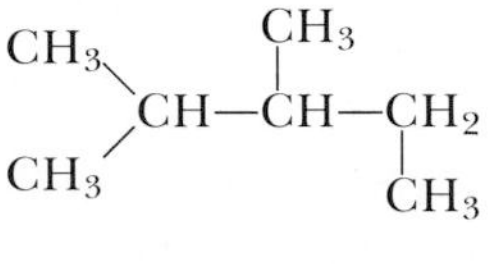

(1) $(CH_3)_2CH{-}CH(CH_3){-}CH_2{-}CH_3$ (drawn as CH_3, CH_3 on CH; CH_3 above the second CH; CH_3 below CH_2)

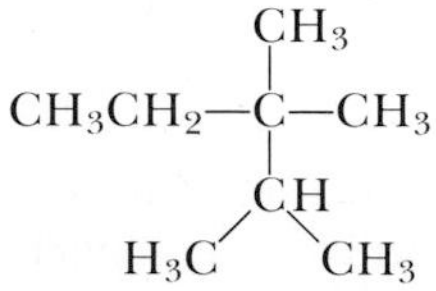

(2) $CH_3CH_2{-}C(CH_3)_2{-}CH(CH_3)_2$ (drawn as $CH_3CH_2{-}C{-}CH_3$ with CH_3 above and CH below, the CH bearing H_3C and CH_3)

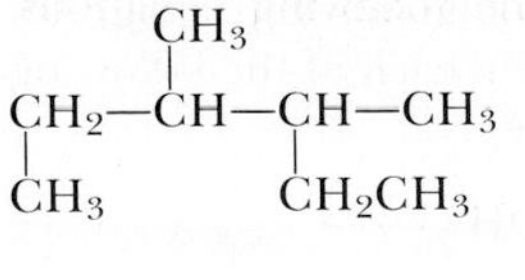

```
      CH3
      |
CH2—CH—CH—CH3
|         |
CH3       CH2CH3
         (3)
```

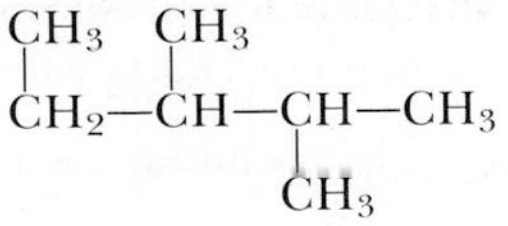

```
CH3   CH3
|     |
CH2—CH—CH—CH3
         |
         CH3
         (4)
```

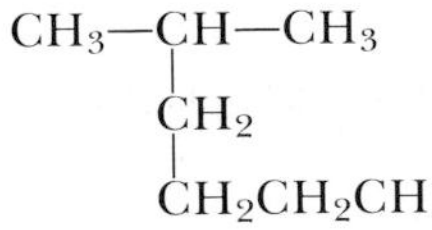

```
CH3—CH—CH3
    |
    CH2
    |
    CH2CH2CH3
      (5)
```

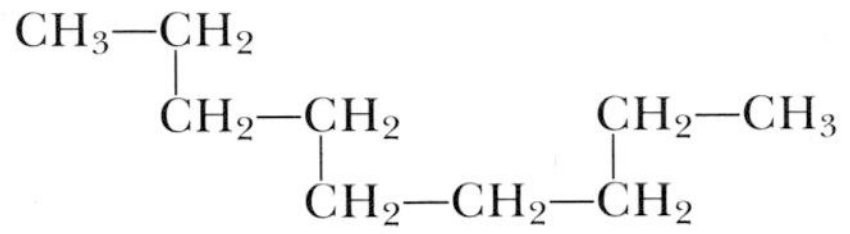

```
CH3—CH2
    |
    CH2—CH2          CH2—CH3
        |            |
        CH2—CH2—CH2
           (6)
```

Indicate which of these compounds represent(s)

(a) the same compound
(b) an isomer of octane
(c) a derivative of hexane
(d) the one with the most "methyl" groups

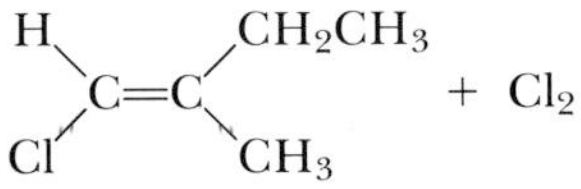

23-54. Name the products of the following addition reactions:

(a) 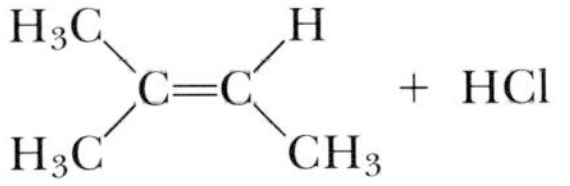 $(H_3C)_2C{=}C(H)(CH_3) + HCl$

(b) $(H)(Cl)C{=}C(CH_2CH_3)(CH_3) + Cl_2$

(c) $(H_3C)_2C{=}C(CH_2CH_3)_2 + H_2$

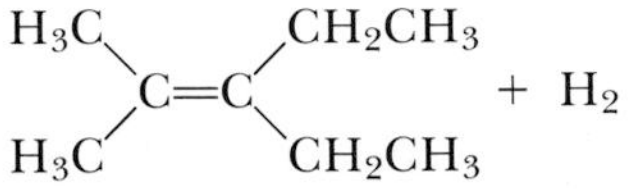

(d) $H{-}C{\equiv}C{-}CH_2Cl + 2Cl_2$

23-55. The early settlers of the United States used to make soap by heating animal fats with lye (NaOH). The chemical reaction is called saponification, and a typical saponification reaction is

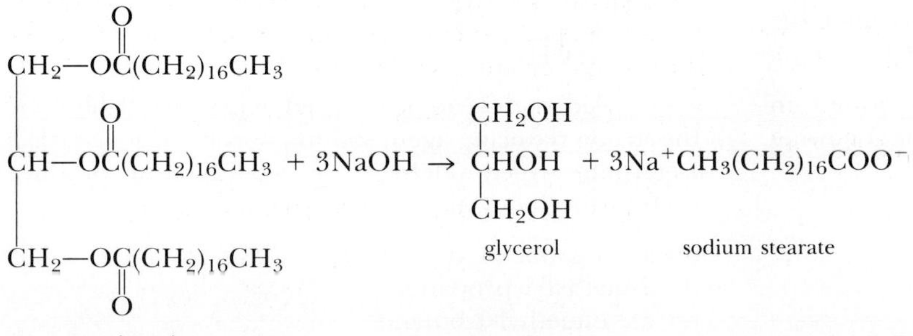

$$\begin{array}{l} CH_2{-}OC(O)(CH_2)_{16}CH_3 \\ | \\ CH{-}OC(O)(CH_2)_{16}CH_3 \\ | \\ CH_2{-}OC(O)(CH_2)_{16}CH_3 \\ \text{tristearin} \end{array} + 3NaOH \rightarrow \begin{array}{l} CH_2OH \\ | \\ CHOH \\ | \\ CH_2OH \\ \text{glycerol} \end{array} + \underset{\text{sodium stearate}}{3Na^+CH_3(CH_2)_{16}COO^-}$$

Explain why sodium stearate acts as a soap.

23-56. The boiling points and molecular masses of ethane, methylamine, and methanol are

	Boiling point/°C	Molecular mass
CH_3CH_3	−88	30
CH_3NH_2	−6.3	31
CH_3OH	65	32

Explain why their boiling points are so different even though they have similar molecular masses.

23-57. The boiling points and molecular masses of 1-butanol, diethyl ether, and pentane are

	Boiling point/°C	Molecular mass
$CH_3CH_2CH_2CH_2OH$	118	74
$CH_3CH_2OCH_2CH_3$	35	74
$CH_3CH_2CH_2CH_2CH_3$	36	72

Explain the relative values of these boiling points.

23-58. Alkoxides react with alkyl halides to produce ethers. For example,

$$Na^+CH_3O^- + CH_3Br \rightarrow CH_3OCH_3 + Na^+Br^-$$

The product here is called dimethyl ether. Ethers are compounds with the general formula ROR′, where R and R′ are alkyl groups. Devise a procedure to synthesize diethyl ether starting with ethylene and any inorganic substances. Diethyl ether was once used as an anesthetic.

23-59. An unknown gas was shown to have the molecular formula C_3H_6 from analysis of its combustion reaction. This gas reacts with HCl to give 2-chloropropane. What is the Lewis formula for the compound?

23-60. Animal fats and vegetable oils are esters of an alcohol called glycerol with long-chain acids called fatty acids. These esters are generally called glycerides. Write a Lewis formula for the glyceride product in the following reaction:

```
                         CH2OH
                         |
3CH3(CH2)14COOH +  CHOH   → 3H2O + a glyceride
                         |
                         CH2OH
    palmitic acid          glycerol
```

$$3CH_3(CH_2)_{14}COOH + \underset{\text{glycerol}}{CH_2OH{-}CHOH{-}CH_2OH} \rightarrow 3H_2O + \text{a glyceride}$$

palmitic acid   glycerol

23-61. Gasoline is a mixture of hydrocarbons, primarily containing five to ten carbon atoms. If we use octane as an example, calculate the volume of air at 20°C and 1 atm that is required to burn a tank of gasoline (say 75 liters). Air is 21 mole percent oxygen. If ΔH°_{rxn} for the combustion of octane is $-48\ \text{kJ}\cdot\text{g}^{-1}$, calculate the energy produced when 75 L of octane is burned. Take the density of octane to be $0.80\ \text{g}\cdot\text{mL}^{-1}$.

23-62. Crude oil is traded by the barrel, and one barrel is equivalent to 42 U.S. gallons. Given that the average heat of combustion of hydrocarbons is about $-47\ \text{kJ}\cdot\text{g}^{-1}$, and that their average density is about $0.80\ \text{g}\cdot\text{mL}^{-1}$, calculate the available energy in one barrel of crude oil (1 gallon = 4 quarts and 1.06 quarts = 1 liter).

23-63. A component of natural gas was completely burned in O_2 to give CO_2 and H_2O. From the analysis of the products, it was determined that the compound is 81.71 percent carbon and 18.29 percent hydrogen by mass. What is its empirical formula? If 0.75 g of the gas occupies 386 mL at 0°C and 750 torr, then determine the molecular formula of the gas. Name the gas.

23-64. Methyl ketones such as acetone, a widely used solvent of commercial importance, can be detected by the iodoform (CHI_3) test, in which the ketone is reacted with hypoiodite in aqueous base:

$$CH_3{-}\underset{\underset{O}{\|}}{C}{-}CH_3(aq) + 3IO^-(aq) \rightarrow$$

$$CH_3{-}C\begin{matrix}{/\!\!/}O \\ \diagdown O^-\end{matrix}\ (aq) \quad + \quad \underset{\text{yellow}}{CHI_3(s)} + 2OH^-(aq)$$

A 5.00 mL sample of acetone ($d = 0.792\ \text{g}\cdot\text{mL}^{-1}$) was allowed to react with excess hypoiodite under basic conditions. After filtration, washing, and drying, the iodoform product was found to weigh 15.6 g. Calculate the theoretical yield and the percentage yield (Chapter 3).

23-65. Write the structural formula for and assign an IUPAC name to the product when each of the following compounds reacts with 1-butene:

(a) Cl_2 (addition)
(b) HCl
(c) H_2O (acid catalyst)
(d) H_2 (platinum catalyst)

23-66. Write the structural formula for and assign an IUPAC name to the product when each of the following compounds reacts with 2-pentene:

(a) Cl_2 (addition)
(b) HCl
(c) H_2O (acid catalyst)
(d) H_2 (platinum catalyst)

23-67. Complete and balance the following equations (assume addition to all double bonds):

```
           H       H
            \     /
     H       C=C
      \     /     \
(a)    C=C         H     (g) + H2(g) --Pt-->
      /     \
     H       H

           H       H
            \     /
     H       C=C
      \     /     \
(b)    C=C         H     (g) + Cl2(g) →
      /     \
     H       H

           H       H
            \     /
     H       C=C
      \     /     \
(c)    C=C         H     (g) + HCl(g) →
      /     \
     H       H
```

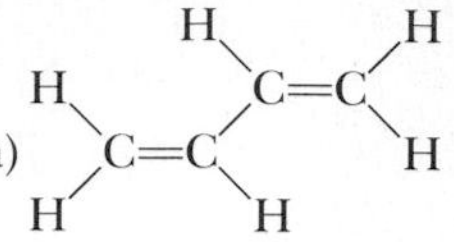

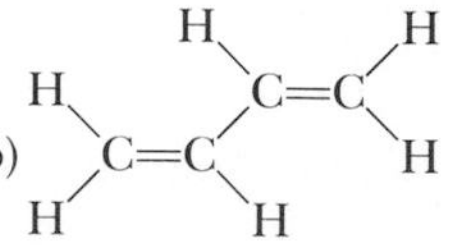

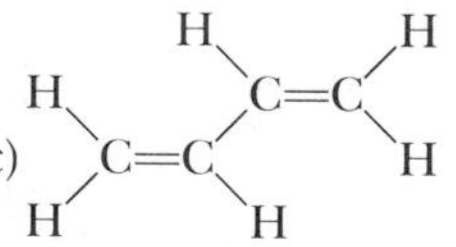

23-68. Write a structural formula for the alcohol formed when H_2O is added to each of the following alkenes:

(a) 3-methyl-1-butene
(b) 2-methyl-2-butene
(c) 2-pentene
(d) 2-butene

23-69. Write a structural formula for the product when HCl is added to each of the following alkenes:

(a) 3-chloro-1-butene
(b) 1-bromo-2-butene
(c) 2-methyl-1-propene
(d) 1-chloro-1-propene

23-70. Indicate whether each of the following alkenes shows cis-trans isomerism:

(a) $CH_2{=}CHCH_2CH_3$
(b) $CH_3CH{=}CHCH_2CH_3$
(c) $CH_2{=}\underset{\underset{CH_3}{|}}{C}CH_2CH_3$

23-71. Aldehydes can be reduced to primary alcohols by the strong reducing agent sodium borohydride, $NaBH_4$. Determine which aldehyde you would use to produce each of the following primary alcohols:

(a) 1-propanol
(b) 2-methyl-1-propanol
(c) 2,2-dimethyl-1-butanol

Synthetic and Natural Polymers

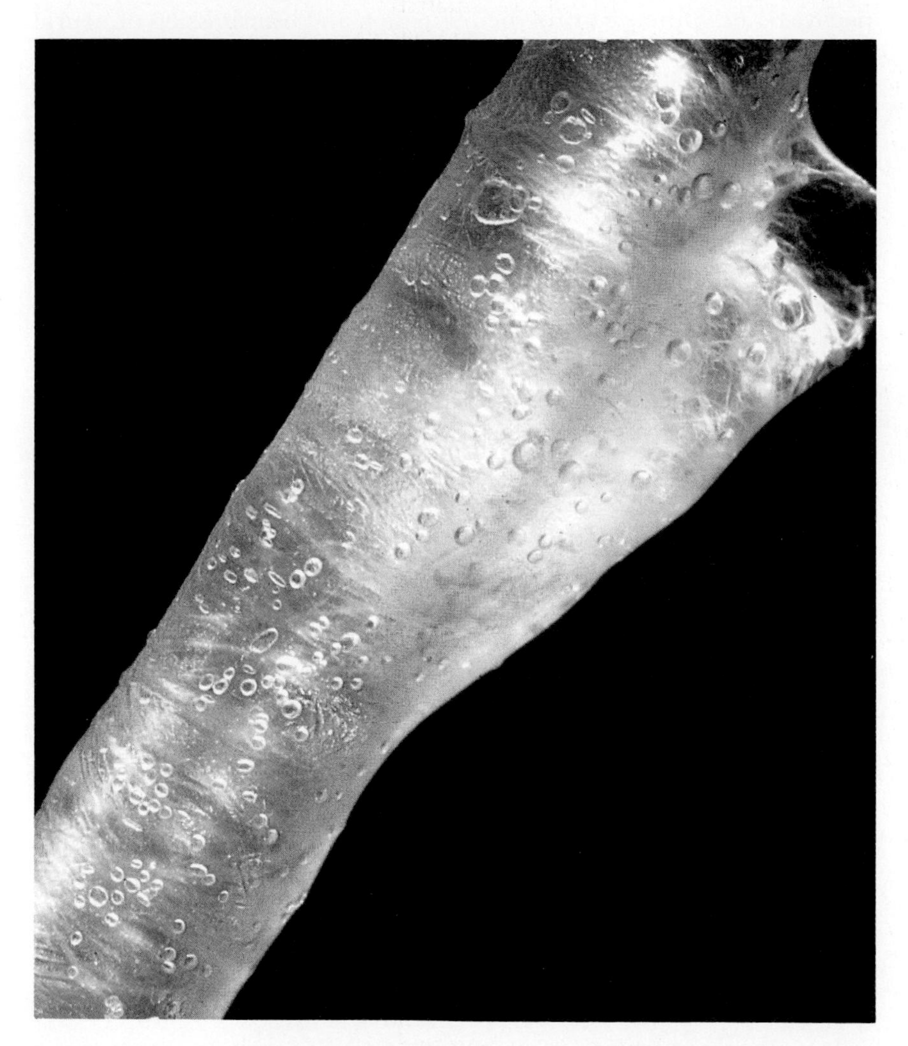

DNA that has been precipitated from a solution and then wound around a glass rod. The glistening bundle is composed of countless strands of human DNA.

Molecules composed of long chains of atoms are the basis of all synthetic and natural polymers. Synthetic polymers of incredibly diverse properties have been made by chemists over the last 50 years. Because of their superior physical and chemical properties, synthetic fibers and synthetic rubber have displaced natural materials for many purposes. Few of us have not heard the terms nylon, rayon, polyester, polyethylene, polystyrene, Teflon, Formica, and Saran, all of which are synthetic polymers. The technological impact of polymer chemistry on our daily lives is immense and continues to increase. There is no question that synthetic polymers have enriched the quality of life for most of us.

Many biologically important substances, such as proteins, starch, cellulose and nucleic acids, are polymers. Proteins are found in all cells. The variety of functions of proteins is staggering. Some proteins are major structural components of skin, tendons, muscles, hair, and connective tissue. Others control the regulation and transmission of neural impulses. Still others, such as enzymes, hormones, and gene regulators, direct and control the diverse chemical reactions that occur in the body. All biological reactions involve one or more protein catalysts called enzymes. The understanding of the structure and function of proteins is one of the great achievements of modern science and is still an active area of research.

The nucleic acids DNA and RNA are the molecules in which organisms store genetic information and through which they pass this information from generation to generation. DNA is the principal component of chromosomes, and RNA plays a variety of roles in the biosynthesis of proteins. We shall see at the end of the chapter how molecules are able to store and transmit genetic information.

24-1. POLYMERS ARE COMPOSED OF MANY MOLECULAR SUBUNITS JOINED END TO END

The simplest polymer is polyethylene which is formed by the repeated joining of ethylene molecules, $H_2C{=}CH_2$. This repeated addition of small molecules to form a long, continuous chain is called **polymerization,** and the resulting chain is called a **polymer.**

The polymerization of ethylene can be initiated by a free radical, such as $HO\cdot$ the hydroxyl radical. (Recall that a free radical is a species having one or more unpaired electrons.) The first step in the polymerization is

$$HO\cdot + H_2C{=}CH_2 \rightarrow HOCH_2CH_2\cdot$$

The product is a free radical that can react with another ethylene molecule to give

$$HOCH_2CH_2\cdot + H_2C{=}CH_2 \rightarrow HOCH_2CH_2CH_2CH_2\cdot$$

The product of this step is also a free radical that can react with another ethylene molecule:

$$HOCH_2CH_2CH_2CH_2\cdot + H_2C{=}CH_2 \rightarrow HOCH_2CH_2CH_2CH_2CH_2CH_2\cdot$$

The product here is a reactive chain that can grow longer by the sequential addition of more ethylene molecules. The chain continues to grow until some **termination reaction,** such as the combination of two free radicals, occurs. The polyethylene molecules formed in this manner typically contain thousands of carbon atoms. The **monomers** of polyethylene are ethylene.

The polymerization of ethylene can be written schematically as

$$\underset{\text{monomer}}{nH_2C{=}CH_2} \rightarrow \underset{\text{polymer}}{+\!\!\!(CH_2CH_2)\!\!\!+_n}$$

Table 24-1 Some common polymers prepared from substituted alkenes

Name	Structural unit	Name	Structural unit
polyethylene	$-\!\!(CH_2-CH_2)\!\!-_n$	Teflon	$(-CF_2-CF_2-)_n$
polypropylene	$(-CH(CH_3)-CH_2-)_n$		
polyvinylchloride (PVC)	$(-CH(Cl)-CH_2-)_n$	Orlon, Acrilan	$(-CH-CH(CN)-)_n$
Saran	$(-CH_2-CCl_2-)_n$	polystyrene	$(-CH_2-CH(C_6H_5)-)_n$

The notation $-\!\!(CH_2CH_2)\!\!-_n$ means that the group enclosed in the parentheses is repeated n times. The free-radical initiator is not indicated because n is large and thus the end group constitutes only a trivial fraction of the macromolecule. The precise number of monomer molecules incorporated into a polymer molecule is not important for typically large values of n. It makes little difference whether a polyethylene molecule consists of 5000 or 5100 monomer units, for example.

Polyethylene is a tough, flexible plastic that is used in the manufacture of packaging films and sheets, wire and cable insulation, ice cube trays, refrigerator dishes, squeeze bottles, bags for foods and clothes, trash bags, and many other articles. Other well-known polymeric materials are made from other monomers. For example, Teflon is produced from the monomer tetrafluoroethylene:

$$n\ F_2C{=}CF_2 \rightarrow (-CF_2-CF_2-)_n$$

tetrafluoroethylene → Teflon

An example of a thermally stable, heat-resistant plastic material.

Teflon is a tough, nonflammable, and exceptionally inert polymer that is used for nonstick surfaces in pots and pans, electrical insulation, plastic pipes, and cryogenic bearings. Some other polymers that are produced from alkenes are polypropylene (indoor-outdoor carpeting, pipes, valves); polyvinyl chloride, PVC (pipes, floor tiles, records), Saran (food packaging, fibers); Orlon and Acrilan (fabrics); and polystyrene (Table 24-1).

24-2. NYLON AND DACRON ARE MADE BY CONDENSATION REACTIONS

The polymerization reaction of ethylene is called an **addition polymerization reaction** because it involves the direct addition of monomer molecules. Another type of polymerization reaction is a **condensation polymerization reaction.** In a condensation polymerization reaction, a small molecule, such as water, is split out as each monomer is added to the polymer chain. The formation of nylon is an example of a condensation polymerization reaction. Nylon is formed by the reaction of a diamino compound, such as

$$\mathrm{H{-}\underset{\displaystyle H}{\underset{|}{N}}{-}CH_2CH_2CH_2CH_2CH_2CH_2{-}\underset{\displaystyle H}{\underset{|}{N}}{-}H}$$

and a dicarboxylic acid, such as

$$\mathrm{HO{-}\underset{\displaystyle \ddot{O}}{\underset{\|}{C}}{-}CH_2CH_2CH_2CH_2{-}\underset{\displaystyle \ddot{O}}{\underset{\|}{C}}{-}OH}$$

These two molecules can be linked by the reaction

$$\mathrm{H{-}\underset{\displaystyle H}{\underset{|}{N}}{-}CH_2CH_2CH_2CH_2CH_2CH_2{-}\underset{\displaystyle H}{\underset{|}{N}}{-}\boxed{H + HO}{-}\underset{\displaystyle \ddot{O}}{\underset{\|}{C}}{-}CH_2CH_2CH_2CH_2{-}\underset{\displaystyle \ddot{O}}{\underset{\|}{C}}{-}OH} \rightarrow$$

$$\mathrm{H{-}\underset{\displaystyle H}{\underset{|}{N}}{-}CH_2CH_2CH_2CH_2CH_2CH_2{-}\underset{\displaystyle H}{\underset{|}{N}}{-}\underset{\displaystyle \ddot{O}}{\underset{\|}{C}}{-}CH_2CH_2CH_2CH_2{-}\underset{\displaystyle \ddot{O}}{\underset{\|}{C}}{-}OH + HOH}$$

The product in this reaction is called a **dimer.** Note that one end of the dimer is an amino group and the other end is a carboxyl group. The dimer can grow by the reaction of its amino end with a dicarboxylic acid monomer or by the reaction of its carboxyl end with a diamine monomer. This process can continue to produce nylon (Figure 24-1), whose general formula is

$$\left(\mathrm{\underset{\displaystyle H}{\underset{|}{N}}{-}CH_2CH_2CH_2CH_2CH_2CH_2{-}\underset{\displaystyle H}{\underset{|}{N}}{-}\underset{\displaystyle O}{\underset{\|}{C}}{-}CH_2CH_2CH_2CH_2{-}\underset{\displaystyle O}{\underset{\|}{C}}}\right)_n$$

Figure 24-1 The formation of nylon by a condensation polymerization reaction at the interface of two immiscible solvents, one of which (water) contains the diamine monomer $H_2N(CH_2)_6NH_2$ plus NaOH(aq) and the other of which (hexane) contains the compound $\mathrm{Cl\underset{\displaystyle O}{\underset{\|}{C}}(CH_2)_4\underset{\displaystyle O}{\underset{\|}{C}}Cl}$. The small molecule that splits out in the reaction is HCl. The polymer forms at the interface between the two solutions and is pulled out by the glass rod as a strand of polymer.

Note that a condensation polymerization reaction involves two different monomeric molecules that combine to form the basic repeating unit of the polymer.

Over 2 billion pounds of nylon are produced annually in the United States. It is used to make strong, long-wearing fibers that find extensive use in rugs and in hosiery, sweaters, and other clothing. Nylon resembles silk in many of its properties but is cheaper to produce. The resemblance to silk is not at all coincidental: silk is a protein, and proteins are polymers in which the monomers are linked together by amide bonds (see margin) in the same manner as in nylon (Section 24-5).

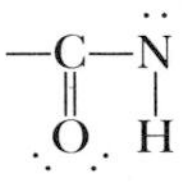

Dacron is a condensation polymer formed from the monomers ethylene glycol (a dialcohol), $HOCH_2CH_2OH$, and *para*-terephthalic acid (a dicarboxylic acid),

$$HO-\underset{\underset{O}{\|}}{C}-C_6H_4-\underset{\underset{O}{\|}}{C}-OH$$

The reaction proceeds by the formation of an ester linkage between a carboxyl group and an —OH group of the dialcohol. The basic polymer unit in Dacron is

$$\left(OCH_2CH_2-O-\underset{\underset{O}{\|}}{C}-C_6H_4-\underset{\underset{O}{\|}}{C}\right)_n$$

Dacron is called a polyester because the monomers are joined together through ester linkages.

$$-O-\underset{\underset{O}{\|}}{C}-$$

ester linkage

Example 24-1: Write the equation for the reaction that produces the ester linkage in Dacron.

Solution: The ester linkage results from the esterification reaction (Section 23-9) of one end of ethylene glycol with one end of *para*-terephthalic acid:

$$HO-CH_2CH_2-OH + HO-\underset{\underset{O}{\|}}{C}-C_6H_4-\underset{\underset{O}{\|}}{C}-OH \rightarrow$$

$$HO-CH_2CH_2-O-\underset{\underset{O}{\|}}{C}-C_6H_4-\underset{\underset{O}{\|}}{C}-OH + HOH$$

ester linkage

This dimer can grow through the reaction of its left-hand end with another *para*-terephthalic acid molecule and the reaction of its right-hand end with an ethylene glycol molecule.

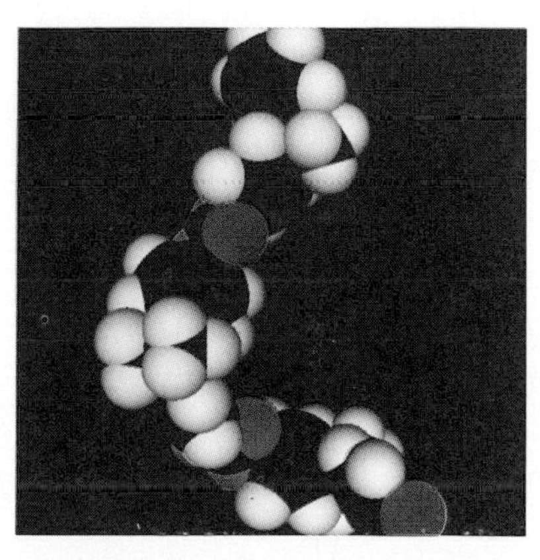

This three-dimensional space-filling model of a polymer molecule was generated by a computer as part of an exploratory research project conducted by Dr. John Bendler, a polymer physicist at the General Electric Research and Development Center in Schenectady, New York. The blue spheres represent the carbon "backbone" of a LEXAN polycarbonate polymer chain; the white spheres indicate the hydrogen atoms; and the red spheres mark the location of oxygen atoms.

Dacron, which is light and tough, is used to make clear films, skis, boat and aircraft components, surgical components, and permanent-press clothing. When used for clothing, it is usually blended in a roughly two-to-one ratio with cotton because the resulting blended fibers are softer and pass moisture more readily than pure Dacron.

In addition to nylon and Dacron, there are many other condensation polymers, including polycarbonates (safety helmets, lenses, electrical components, photographic film), polyurethanes (insulation, furniture), and phenolics (brake linings, structural components). The essential feature used to synthesize condensation polymers is the presence of reactive groups at both ends of the monomers.

Polymer	Annual production/ millions of pounds
polyethylene	12388
polyvinylchloride	5707
polyester	4176
polypropylene	3955
polystyrene	3621
phenolics	2333
nylon	2332
polyesters	997

24-3. POLYMERS WITH CROSS-LINKED CHAINS ARE ELASTIC

Depending on the length of the polymer chains and on the temperature, a particular type of polymer may occur as a viscous liquid, a rubbery solid, a glass, or a partially crystalline solid. Generally, the longer the average length of the polymer chains, the less liquid-like the polymer at a given temperature. For chain lengths involving more than 10,000 bonds, liquid flow is negligible at normal temperatures but elastic-like deformations are possible.

The elasticity of polymers can be explained in terms of structure. The polymer chains in a sample are coiled and intertangled with each other. If the polymer is stretched, then the chains slowly untangle and the sample appears to flow. The relative movement of polymer chains that occurs when the sample is stretched can be decreased by connecting the polymer chains through chemical bonds called cross-links (Figure 24-2a). When the cross-linked network is stretched, the coils become elongated (Figure 24-2b), but when the stress is released, the polymer network returns to its original coiled state. A cross-linked polymer that exhibits elastic behavior is called an **elastomer.** If the cross-links occur at average intervals of about 100 to 1000 bonds along the chain, then the polymer may be stretched to several times its unstretched length without breaking. The stretched polymer returns to its initial length when the force is released. The resistance of an elastomer to stretching can be increased by increasing the number of cross-links between chains. High elasticity is found in substances composed of long polymer chains joined by sparsely distributed cross-links.

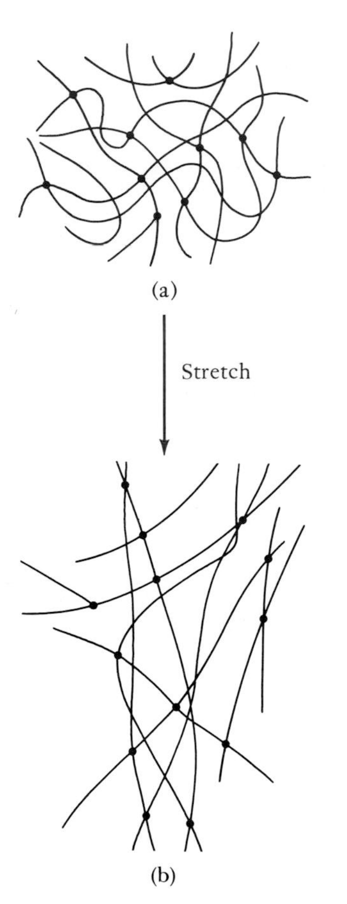

Figure 24-2 Schematic representation of a cross-linked polymer network. The cross-links between chains are represented by dots. (a) In the natural (unstretched) state, the polymer chains are coiled. (b) When the polymer is stretched, the chains become elongated.

Natural rubber is composed of chains of *cis*-1,4-isoprene units (see margin of next page) with an average chain length of 5000 isoprene units. The major problem associated with natural rubber is tackiness. This problem was solved in 1839 by Charles Goodyear, who discovered how to vulcanize natural rubber with sulfur. The **vulcanization** of natural rubber involves the formation of —S—S— cross-links between the polyisoprene chains (see margin of next page).

The U.S. sources of natural rubber were cut off by Japanese territorial expansions in 1941. As a result, research in the United States directed toward the production of synthetic rubber underwent a major expansion that culminated in the development of several varieties of synthetic rubber, including a product essentially identical to natural rubber. As with all rubbers, vulcanization is used to produce the cross-links that give rise to the desired degree of elasticity. The annual production of synthetic rubber in the United States exceeds 2,000,000 metric tons.

Example 24-2: An example of a synthetic rubber is polybutadiene. Draw two polybutadiene chain segments that are cross-linked by vulcanization.

Solution: As in the case of the vulcanization of isoprene on page 774, the —S—S— cross-links result from carbon-carbon double bonds between adjacent chains:

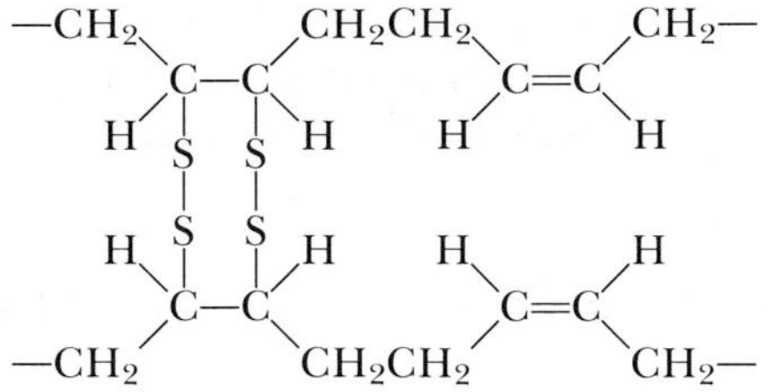

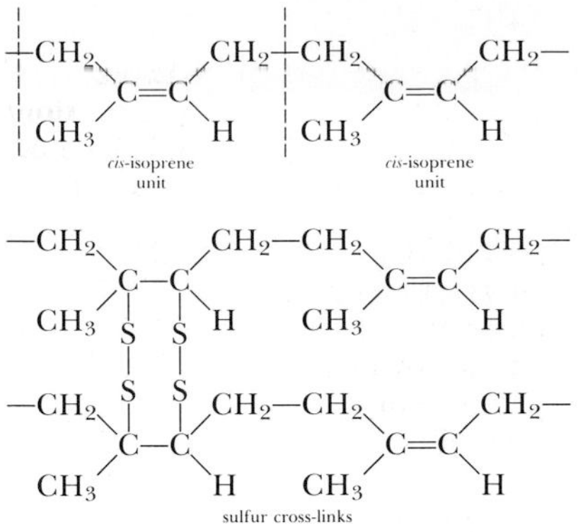

$$
\begin{array}{l}
-CH_2-CH(-S-)-CH(-S-)-CH_2CH_2-CH{=}CH-CH_2- \\
\quad\quad\quad\quad\; | \quad\quad\;\; | \\
-CH_2-CH(-S-)-CH(-S-)-CH_2CH_2-CH{=}CH-CH_2-
\end{array}
$$

cis-isoprene unit   *cis*-isoprene unit

sulfur cross-links

24-4. AMINO ACIDS HAVE AN AMINO GROUP AND A CARBOXYL GROUP ATTACHED TO A CENTRAL CARBON ATOM

Proteins are polymers whose monomer units are **amino acids.** The general formula of an amino acid is

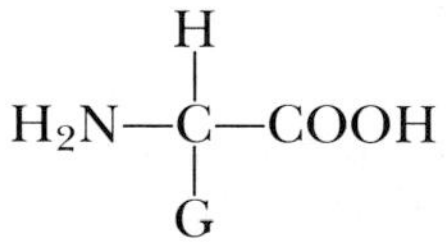

$$H_2N-\underset{G}{\overset{H}{C}}-COOH$$

▪ The word *protein* was coined by the Swedish chemist Jöns Berzelius in 1838. The word is derived from the Greek word *proteios*, which means "of the first rank."

Amino acids differ from each other only in the **side group,** G, attached to the central carbon atom. These compounds are called amino acids because they contain both an amino group, $—NH_2$, and an acidic group, —COOH. There is a total of 20 amino acids commonly found in proteins. This set of 20 amino acids occurs in proteins at all levels of life, from the simplest bacteria to humans.

Except for glycine, which is the simplest amino acid,

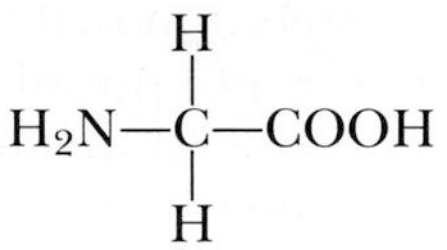

$$H_2N-\underset{H}{\overset{H}{C}}-COOH$$

glycine

all the amino acids have four different groups attached to a central carbon atom. For example, the structural formula for the amino acid alanine is

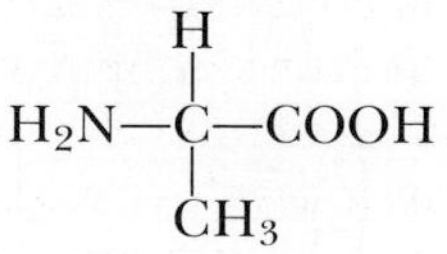

$$H_2N-\underset{CH_3}{\overset{H}{C}}-COOH$$

alanine

Some side groups and the names of the corresponding amino acids are shown on page 776.

Side group	Amino acid	Side group	Amino acid
Nonpolar side groups		Acidic side groups	
$—CH_3$	alanine (ala)	$—CH_2C(=O)OH$	aspartic acid (asp)
$—CH(CH_3)CH_3$	valine (val)		
Uncharged polar side groups		$—CH_2CH_2C(=O)OH$	glutamic acid (glu)
$—CH_2OH$	serine (ser)		
$—CH_2SH$	cysteine (cys)		
$—CH_2—C_6H_4—OH$	tyrosine (tyr)	Basic side groups	
		$—CH_2CH_2CH_2CH_2NH_2$	lysine (lys)

The four bonds about the central carbon atom in an amino acid are tetrahedrally oriented, which can be represented as

$$\begin{array}{c} H \\ \vdots \\ H_2N\blacktriangleright C \blacktriangleleft COOH \\ \vdots \\ G \end{array}$$

The dashed bonds indicate that the —H and —G groups lie below the page, and the dark, wedge-shaped bonds indicate that the $—NH_2$ and —COOH groups lie above the page.

The amino acids display a type of isomerism that we have not studied so far in this text. They exist as **optical isomers,** which are nonsuperimposable isomers that are mirror images of each other. To get an idea of what a nonsuperimposable mirror image means, consider the mirror images of the words MOM and DAD (Figure 24-3a). The mirror image of the word MOM is superimposable on the word MOM, whereas the mirror image of the word DAD, that is, ᗡAᗡ, is not superimposable on the word DAD. Similarly, your right hand is a mirror image of your left hand, and the two cannot be superimposed (Figure 24-3b). A molecule can exist in optically isomeric forms if the mirror image of the molecule cannot be superimposed onto itself. If we let the four groups attached

Figure 24-3 Mirror images may be superimposable or nonsuperimposable. (a) Mirror images of the words MOM and DAD. The mirror image of MOM is superimposable on the original, but the mirror image of DAD is not superimposable on the original. (b) Your two hands are an excellent example of nonsuperimposable mirror images. Your right hand is not superimposable on your left hand.

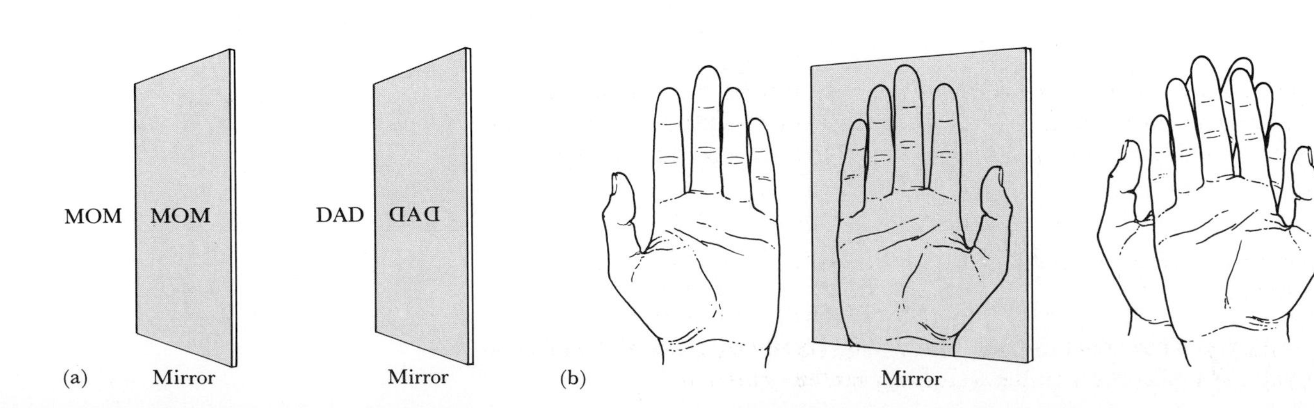

to the central carbon atom in alanine be H, X, Y, and Z for simplicity, then we see

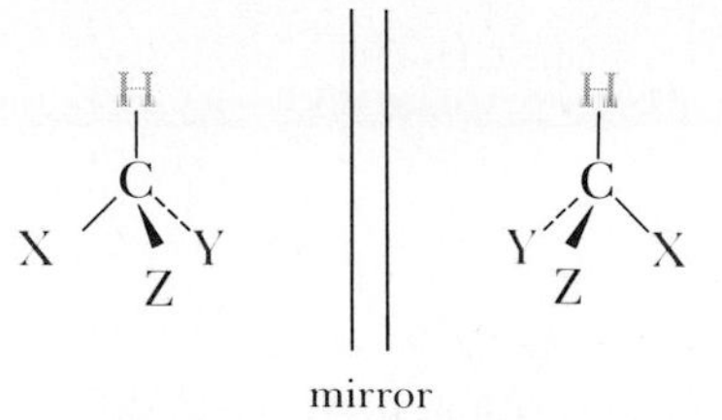

H H
C | C
X Y | Y X
Z | Z

mirror

These two mirror images cannot be superimposed, just as a right hand cannot be superimposed onto a left hand. Two optical isomers are distinguished from each other by a D or L placed in front of the name of the amino acid:

▪ The D and the L are derived from *dextro-* (right) and *levo-* (left).

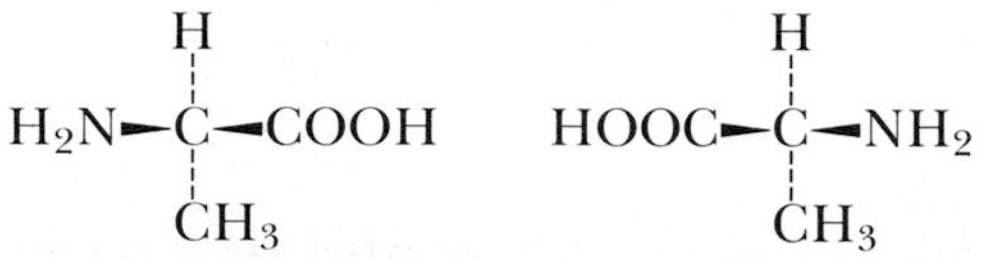

$$H_2N-\overset{H}{\underset{CH_3}{C}}-COOH \qquad HOOC-\overset{H}{\underset{CH_3}{C}}-NH_2$$

D-alanine L-alanine

Optical isomers ordinarily display the same chemical properties, but, with very few exceptions, only the L isomers of the amino acids occur in biological systems. Biochemical reactions are exceptionally **stereospecific;** that is, they are exceptionally dependent upon the shape of the reactants. Apparently life on earth originated from L amino acids, and once the process started, it continued to use only L isomers.

24-5. AMINO ACIDS ARE THE MONOMERS OF PROTEINS

Proteins are formed by condensation reactions similar to the reaction of the formation of nylon. The carboxyl group on one amino acid reacts with the amino group on another; for example,

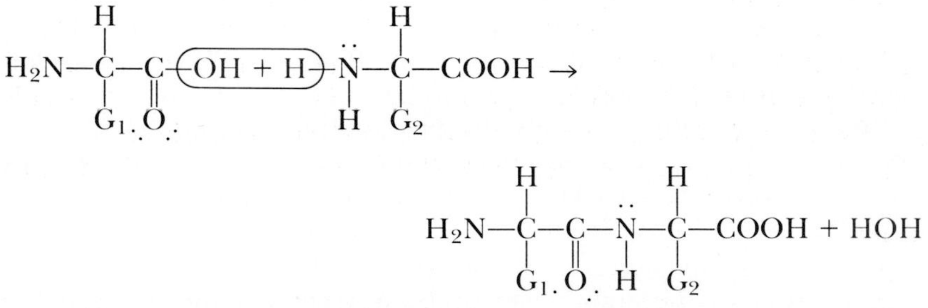

$$H_2N-\overset{H}{\underset{G_1}{C}}-\underset{O}{\overset{\|}{C}}-(OH + H)-\underset{H}{N}-\overset{H}{\underset{G_2}{C}}-COOH \rightarrow$$

$$H_2N-\overset{H}{\underset{G_1}{C}}-\underset{O}{\overset{\|}{C}}-\underset{H}{N}-\overset{H}{\underset{G_2}{C}}-COOH + HOH$$

▪ The two amino-acid monomers are linked by a peptide bond:

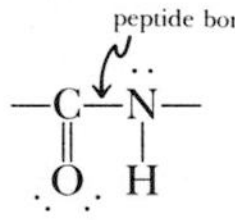

peptide bond

$$-\underset{O}{\overset{\|}{C}}-\underset{H}{N}-$$

The product of the reaction is called a **dipeptide** because it contains two amino-acid residues. An amino-acid residue is the portion of the amino acid that remains in the chain after the water molecule is split out.

Example 24-3: Write an equation for the reaction between alanine and serine.

Solution: The side groups in alanine and serine are given in the table on page 776. The reaction between them can be written as

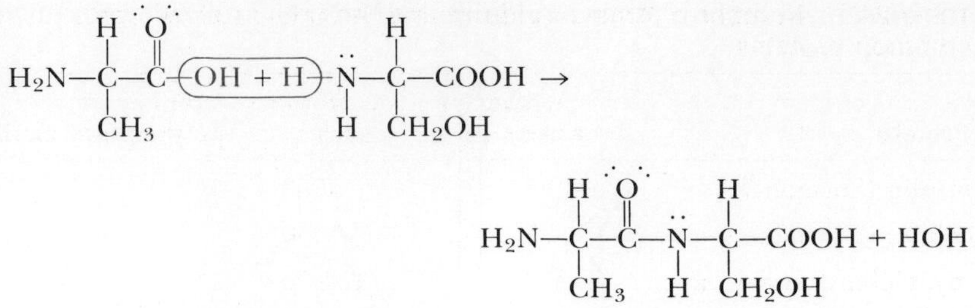

$$H_2N-\underset{CH_3}{\overset{H}{C}}-\overset{O}{\overset{\|}{C}}-OH + H-\underset{H}{\ddot{N}}-\underset{CH_2OH}{\overset{H}{C}}-COOH \rightarrow$$

$$H_2N-\underset{CH_3}{\overset{H}{C}}-\overset{O}{\overset{\|}{C}}-\underset{H}{\ddot{N}}-\underset{CH_2OH}{\overset{H}{C}}-COOH + HOH$$

This is not the only possible result, however. A different dipeptide is formed when the carboxyl group on serine reacts with the amino group on alanine:

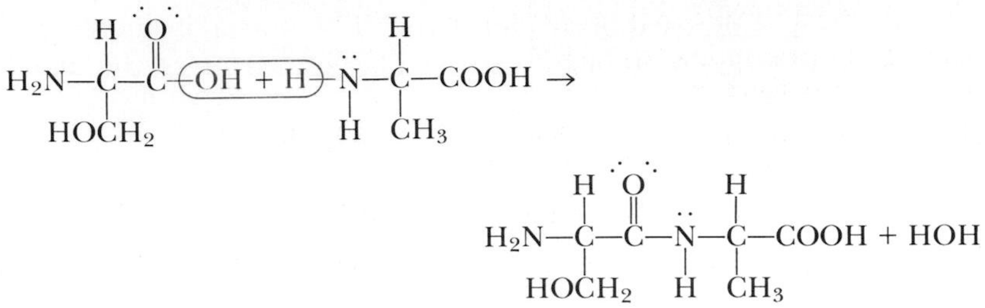

$$H_2N-\underset{HOCH_2}{\overset{H}{C}}-\overset{O}{\overset{\|}{C}}-OH + H-\underset{H}{\ddot{N}}-\underset{CH_3}{\overset{H}{C}}-COOH \rightarrow$$

$$H_2N-\underset{HOCH_2}{\overset{H}{C}}-\overset{O}{\overset{\|}{C}}-\underset{H}{\ddot{N}}-\underset{CH_3}{\overset{H}{C}}-COOH + HOH$$

Thus we see that it is necessary to specify the order of the amino acids in a dipeptide.

Further condensation reactions of a dipeptide with additional amino acid molecules produce a **polypeptide,** which is a polymer having amino acids as monomers. For example,

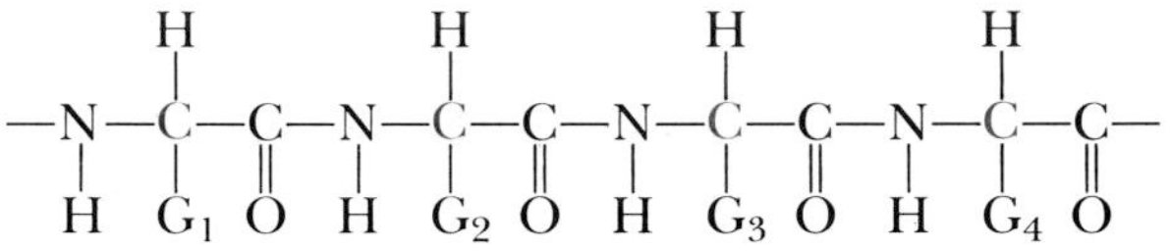

$$-\underset{H}{N}-\underset{G_1}{\overset{H}{C}}-\underset{O}{\underset{\|}{C}}-\underset{H}{N}-\underset{G_2}{\overset{H}{C}}-\underset{O}{\underset{\|}{C}}-\underset{H}{N}-\underset{G_3}{\overset{H}{C}}-\underset{O}{\underset{\|}{C}}-\underset{H}{N}-\underset{G_4}{\overset{H}{C}}-\underset{O}{\underset{\|}{C}}-$$

a portion of a polypeptide

where the carbon atoms that are bonded to the amino acid side groups are shown in color and the peptide bonds are shown in black. Polypeptides are long chains of amino acids joined together by peptide bonds. The chain to which the amino acid side groups are attached is called the **polypeptide backbone.**

Proteins are naturally occurring polypeptides. Each protein has a specific number of amino acid units and a specific order of these units along the polypeptide backbone. The number of amino acid units in a protein varies, ranging from a few to hundreds. Table 24-2 lists some important proteins and the number of amino acid units in each.

The order of the amino acid units in a polypeptide is called the **primary structure** of the polypeptide. The primary structure of a protein characterizes the protein uniquely. The primary structures of hundreds of proteins have been determined since the 1950s. Figure 24-4 shows the primary structure of the protein beef insulin, which is a

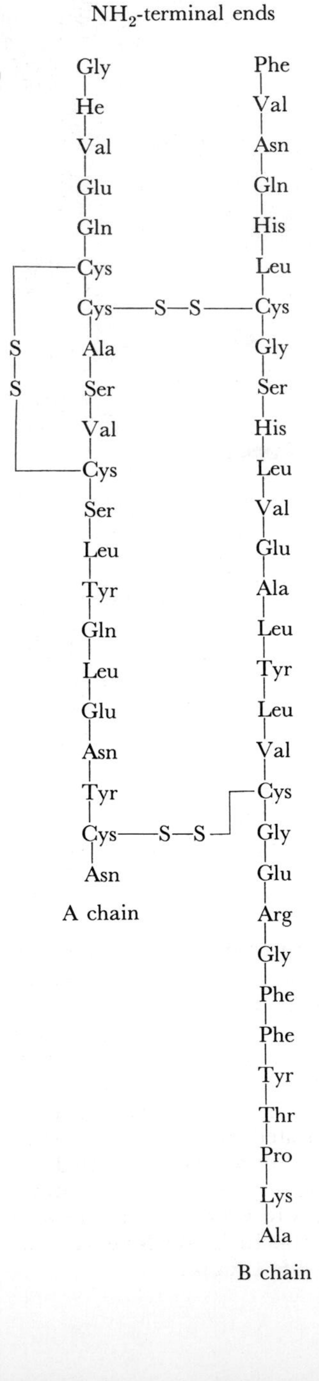

NH_2-terminal ends

A chain: Gly–Ile–Val–Glu–Gln–Cys–Cys–Ala–Ser–Val–Cys–Ser–Leu–Tyr–Gln–Leu–Glu–Asn–Tyr–Cys–Asn

B chain: Phe–Val–Asn–Gln–His–Leu–Cys–Gly–Ser–His–Leu–Val–Glu–Ala–Leu–Tyr–Leu–Val–Cys–Gly–Glu–Arg–Gly–Phe–Phe–Tyr–Thr–Pro–Lys–Ala

(Disulfide bridges: A-chain Cys 6 —S—S— A-chain Cys 11; A-chain Cys 7 —S—S— B-chain Cys 7; A-chain Cys 20 —S—S— B-chain Cys 19)

Figure 24-4 The primary structure of the protein beef insulin. The amino acids are designated by standard three-letter abbreviations. The determination of the primary structure of a protein is like a complicated chemical jigsaw puzzle. The protein is hydrolyzed into shorter chains, which are separated and analyzed separately. The first primary structure determination was completed by the British chemist Frederick Sanger in 1953. Sanger received the 1958 Nobel Prize in chemistry for this work.

Table 24-2 Number of amino acids in and formula mass of some common proteins

Protein	Number of amino acids	Formula mass	Number of polypeptide chains
insulin (hormone)	51	5700	2
cobratoxin (snake toxin)	62	7000	1
myoglobin (carries oxygen in muscles)	153	16,900	1
keratin (wool protein)	204	21,000	1
actin (muscle protein)	410	46,000	1
hemoglobin (transports oxygen in bloodstream)	574	64,500	4
alcohol dehydrogenase (metabolism of alcohol)	748	80,000	2
γ-globulin (antibody)	1250	150,000	4
collagen (skin, tendons, cartilage)	3000	300,000	3

polypeptide hormone that regulates carbohydrate metabolism. A deficiency of insulin leads to diabetes mellitus.

24-6. THE SHAPE OF A PROTEIN MOLECULE IS CALLED ITS TERTIARY STRUCTURE

A first step in understanding how a particular protein functions is a determination of its shape. Because many proteins are such large molecules, this is a very complex task. The ultimate step in the determination of a protein's structure is X-ray crystallography. We saw in Chapter 11 that X-ray patterns can be used to determine the arrangement of atoms in crystalline solids. The X-ray patterns obtained from proteins are more difficult to analyze and interpret because there are so many atoms involved (Figure 24-5).

In the 1950s two American chemists, Linus Pauling and R. B. Corey, were able to interpret X-ray patterns of proteins to show that many proteins have regions in which the chain twists into a helix (which is the shape of a spiral staircase). Pauling and Corey called the helix an ***α*-helix** (Figure 24-6). The helical shape results from the formation of

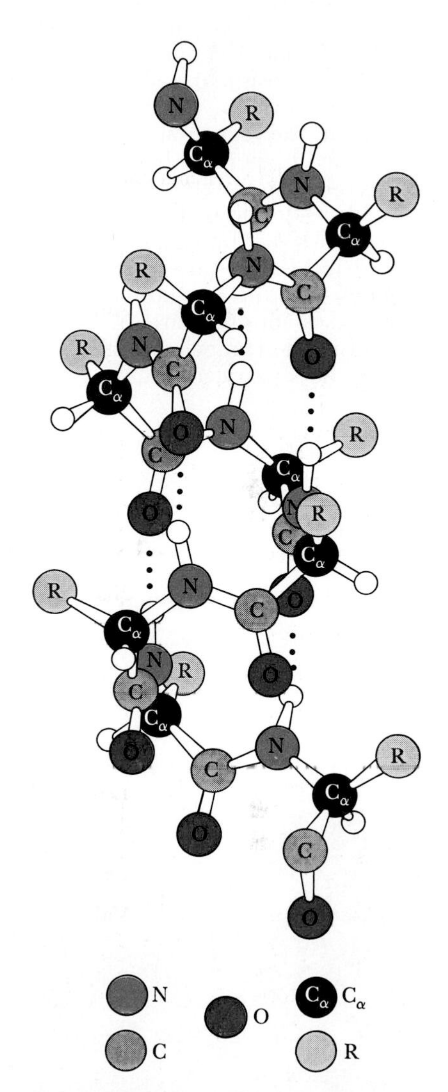

Figure 24-6 A segment of an α-helical region along a polypeptide chain. The chain is held in a helical shape by hydrogen bonds (dotted lines), which are formed between hydrogen and oxygen atoms in peptide bonds that are separated by three other peptide bonds along the polypeptide chain. C_α refers to the carbon atoms to which the side groups are attached, and R represents the side groups.

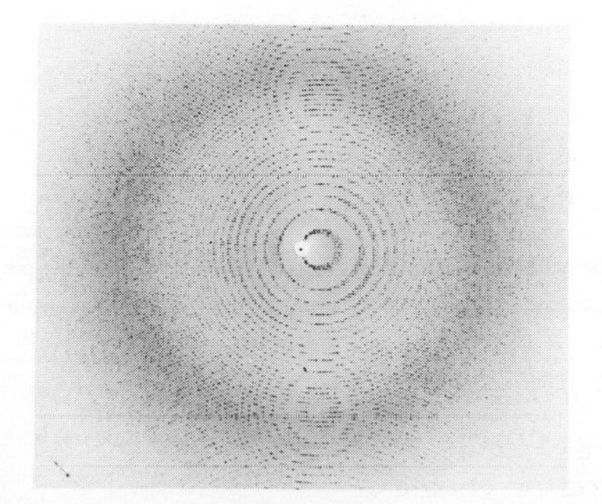

Figure 24-5 X-ray diffraction patterns like this one from a polio virus can be used to determine the structures of proteins and other biological molecules.

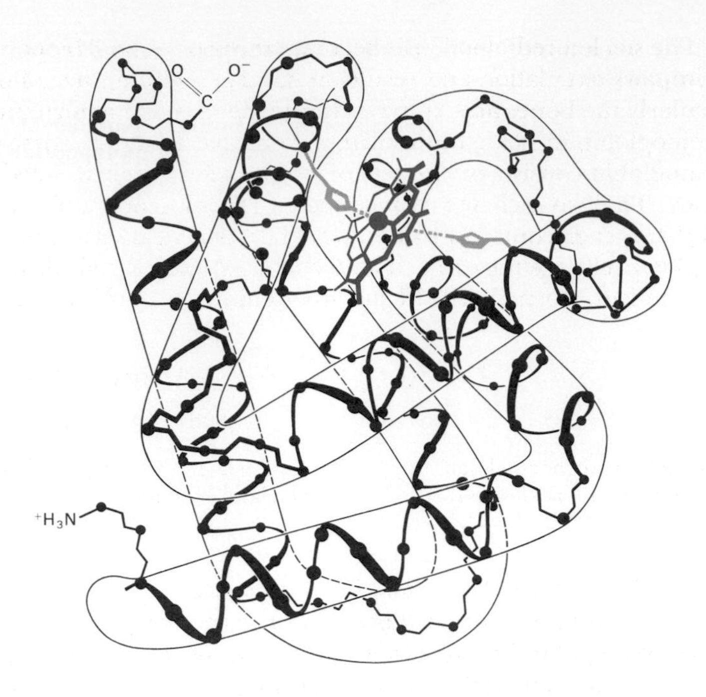

Figure 24-7 The tertiary structure—that is, the overall three-dimensional shape—of the protein myoglobin, which stores and transports oxygen in muscles. Myoglobin has 153 amino-acid units and consists of one polypeptide chain. The black dots represent the carbon atoms that are linked by peptide bonds. The oxygen-binding site, which is buried inside the molecule, is shown in red. The straight, cylindrical regions are α-helical regions. It is these regions that are packed into the three-dimensional arrangement that forms the tertiary structure.

hydrogen bonds between peptide linkages in the peptide chain. Individually, these hydrogen bonds are relatively weak, but collectively they combine to bend the protein chain into an α-helix. This coiled, helical shape in different regions of a protein chain is called **secondary structure.**

The overall shape of a protein molecule in water results from a complicated interplay between the amino acid side groups along the protein chain and the solvent, water. This interplay causes the protein to coil, fold, and bend into a three-dimensional overall shape called the **tertiary structure** (Figure 24-7). The tertiary structure of a protein is obtained from X-ray analysis.

Myoglobin (Figure 24-7) is an oxygen-binding protein that stores and transports oxygen in muscles. It appears to be tailor-made by nature to bind, transport, and release oxygen molecules. The tertiary structure of a protein depends upon the nature and order of the amino acid units making up the protein chain. Nature can build proteins to perform specific tasks by incorporating the appropriate amino acids. These cause the protein to take on a specific shape that is suited to the function of the protein.

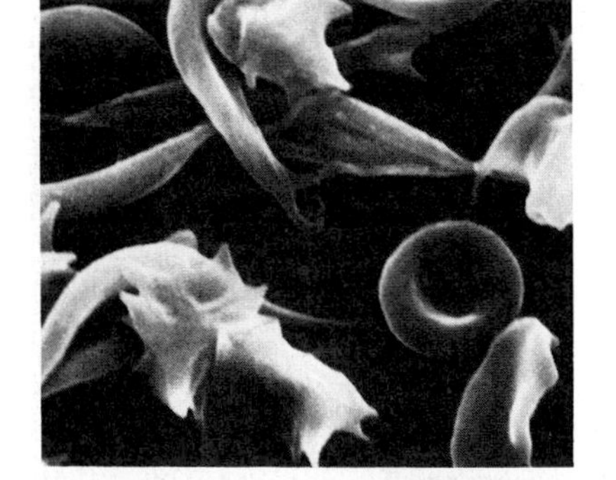

Figure 24-8 A small variation in the chemical composition of a hemoglobin molecule is sufficient to change the shape of red blood cells from normal (circular) to sickled (irregular). The difference in shape is caused by a difference of only 2 amino acids out of the 574 in a hemoglobin molecule.

24-7. NORMAL HEMOGLOBIN AND SICKLE-CELL HEMOGLOBIN DIFFER BY 2 OUT OF 574 AMINO ACIDS

Hemoglobin, the protein that transports oxygen through the bloodstream from the lungs to the various tissues in the body, serves as a striking example of how the shape, and hence the function, of a protein depend in exacting detail upon its amino acid sequence. Figure 24-8 shows an electron micrograph of normal and sickled red blood cells. The sickled red blood cells become trapped in small blood vessels. This impairs circulation and results in damage to a number of organs, particularly the bones and kidneys. In the 1950s it was determined that the hemoglobin in sickled cells differs from that found in normal cells.

Hemoglobin consists of two sets of two identical chains, called α and β chains. The two α chains each contain 141 amino acid units, and the two β chains each contain 146 amino acid units. The α chains of normal and sickled cells are the same. The β chains of sickled cells differ from the β chains of normal cells by just one amino acid. The difference is

	Position in β chain 1 2 3 4 5 6 7 8
hemoglobin from normal cell	H_2N-val-his-leu-thr-pro-<u>glu</u>-glu-lys-
hemoglobin from sickled cell	H_2N-val-his-leu-thr-pro-<u>val</u>-glu-lys-

Only 2 amino acids out of a total of 574 in the hemoglobins differ. The difference lies in the sixth position from the amino end in the β chains. Normal hemoglobin has a glutamic acid unit there, and sickle-cell hemoglobin has a valine unit. A change from glutamic acid to valine (see margin) is a change from a hydrophilic group to one that is hydrophobic. This is apparently enough to alter the shape of the protein molecule in a profound way and is responsible for the sickle shape of the red blood cells that contain the abnormal hemoglobin. Sickle-cell anemia is a molecular disease, as was first shown by Linus Pauling. The shape versus function of proteins is an illustration of the importance of molecular structure. An understanding of the underlying causes of a disease is a first step in its eradication.

$$-CH_2CH_2-C(=O)OH$$

glutamic acid side group

$$-CH(CH_3)-CH_3$$

valine side group

24-8. DNA IS A DOUBLE HELIX

The final class of biopolymers that we study in this chapter are the **polynucleotides.** The two most important polynucleotides are **DNA** (deoxyribonucleic acid) and **RNA** (ribonucleic acid). DNA occurs in the nuclei of cells and is the principal component of chromosomes. The genetic information that is passed from one generation to another is stored in DNA molecules. The discovery of just how this is done was made in 1953 and has led to a revolution in biology that is as profound and far-reaching as the harnessing of nuclear energy in the 1940s and 1950s. The study of heredity, reproduction, and aging at the molecular level has produced an entirely new field of science, molecular biology, which has given birth to genetic engineering, with its awesome possibilities. In order to see how DNA can store and pass on information, we must learn about its molecular structure.

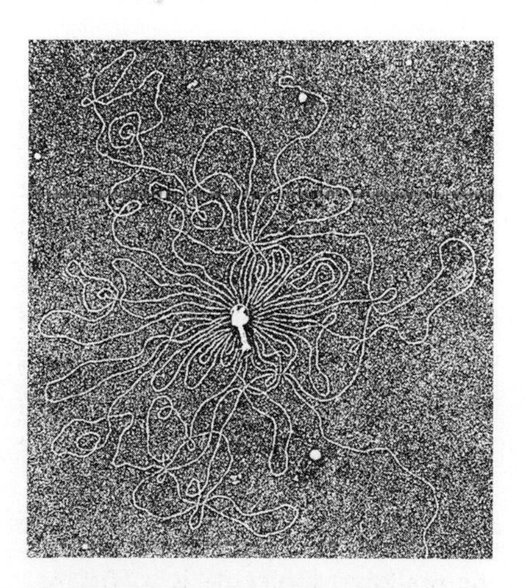

Figure 24-9 An electron micrograph of a virus particle that has burst and released strands of DNA. The long, cylindrical molecule is revealed beautifully in the photo.

DNA is a polynucleotide, which implies that it is a polymer made up of nucleotides. **Nucleotides,** the monomers of DNA and RNA, consist of three parts: a sugar (carbohydrate), a phosphate group, and a nitrogen-containing ring compound called a **base.** The sugar in DNA is 2-deoxyribose and that in RNA is ribose:

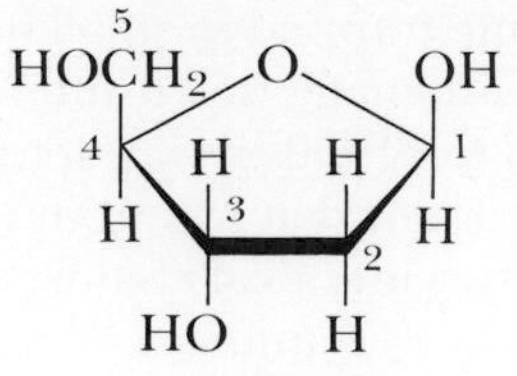

2-deoxyribose 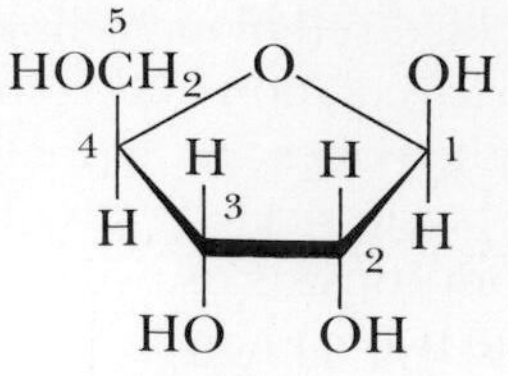 ribose

Carbon atoms, which are understood to constitute the vertices of these rings, are numbered 1 to 4 in the formulas. Notice that the difference between 2-deoxyribose and ribose is that 2-deoxyribose is lacking an oxygen atom at the number 2 carbon atom.

■ The phosphate group is bonded to the sugar molecule by the condensation reaction

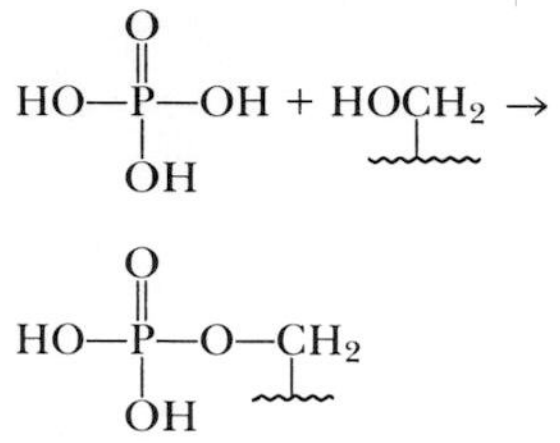

where the wavy line is shorthand for the rest of the sugar molecule.

In both DNA and RNA the phosphate group is attached to the number 5 carbon atom in the sugar:

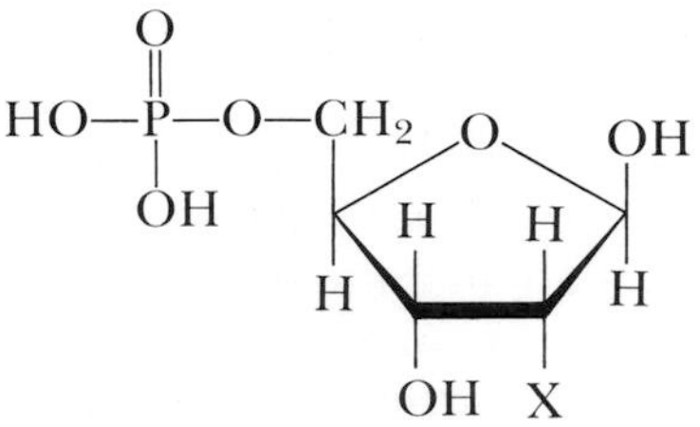

The group labeled X is —OH in ribose and —H in 2-deoxyribose.

The five bases that occur in DNA and RNA are given in the margin. The bases are bonded to the ribose or deoxyribose rings by condensation reactions involving the hydrogen atoms shown in red on the bases in the margin and the —OH group on the number 1 carbon atom in the ribose and deoxyribose rings. For example, we have

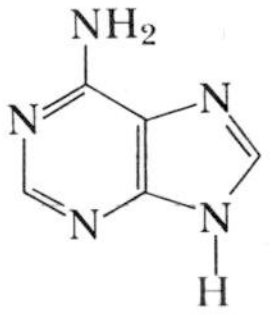

adenine 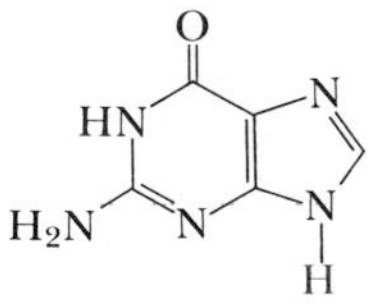 guanine

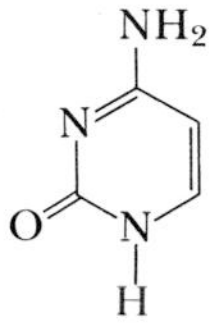

cytosine

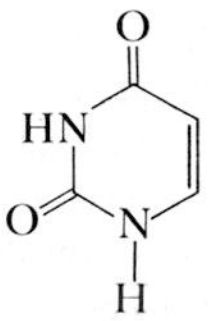

uracil 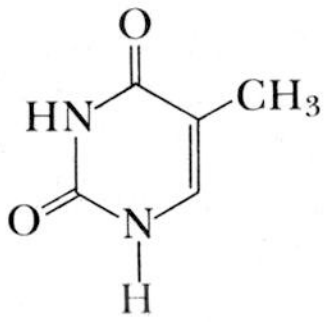 thymine

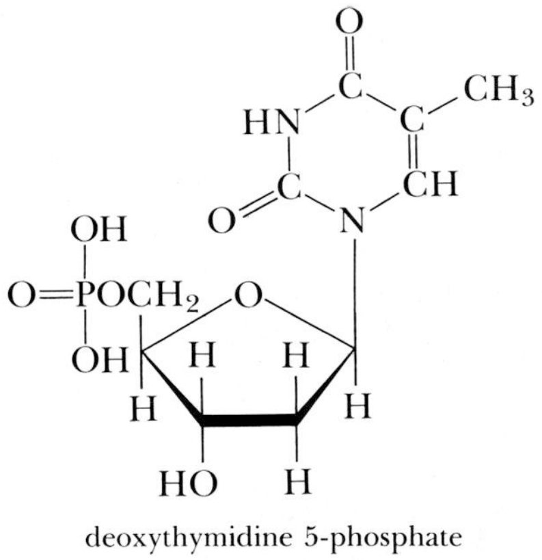

deoxythymidine 5-phosphate 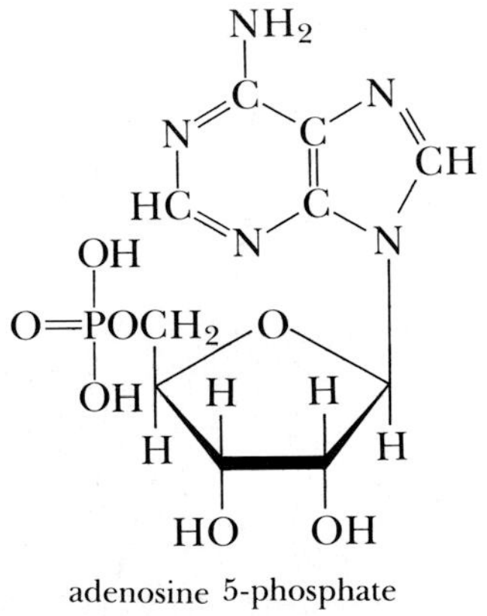 adenosine 5-phosphate

Both of these molecules are nucleotides. Deoxythymidine 5-phosphate is one of four monomers of DNA, and adenosine 5-phosphate is one of four monomers of RNA. DNA contains only the four bases adenine (A), guanine (G), cytosine (C), and thymine (T), and RNA contains adenine (A), guanine (G), cytosine (C), and uracil (U).

Nucleotides can be joined together by a condensation reaction between the phosphate group of one nucleotide and the 3-hydroxyl group of another. The result is a polynucleotide, part of which might look like this:

Thus we see that DNA and RNA consist of a **sugar-phosphate backbone** (shown in black) with bases attached at intervals (shown in red). Let's see now how a molecule like DNA can store and pass on genetic information.

The key to understanding how DNA works lies in its three-dimensional structure. In 1953, James Watson and Francis Crick (Figure 24-10) proposed that DNA consists of two polynucleotide chains inter-

Figure 24-10 In the early 1950s, James Watson (*left*), who had recently received his Ph.D. in zoology from Indiana University, went to Cambridge University on a postdoctoral research fellowship. He and the British physicist Francis Crick (*right*) worked together on the molecular structure of DNA. In 1953 they proposed the double helical model of DNA, which explains beautifully how DNA can store and transmit genetic information. Their proposal is touted as one of the most important scientific breakthroughs of modern times. Watson and Crick were awarded the Nobel Prize in Physiology and Medicine in 1962. The details of their discovery are given by Watson in his book *The Double Helix.*

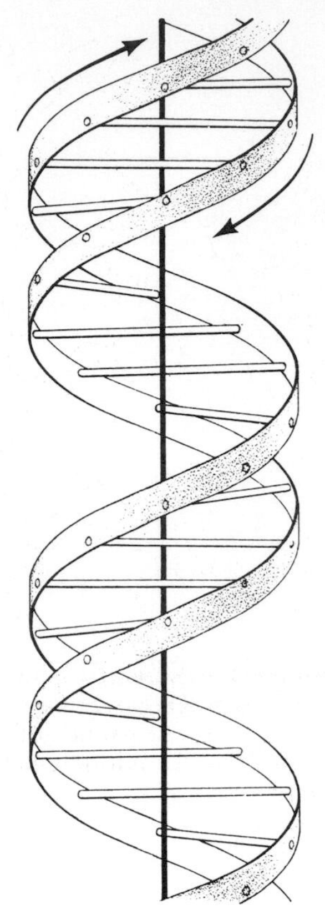

Figure 24-11 The double-helical structure of DNA consists of two polynucleotide strands twisted about each other.

twined in a **double helix** (Figure 24-11). Their proposal of such a structure was based on two principal observations: X-ray data indicated that DNA is helical, and chemical analysis revealed that, regardless of the source of DNA, be it a simple bacterium or the higher vertebrates, the amount of guanine is always equal to the amount of cytosine and the amount of adenine is always equal to the amount of thymine.

Watson and Crick realized that the bases in DNA must somehow be paired. Working with molecular models, they discovered that A and T were of the right shape and size to form two hydrogen bonds:

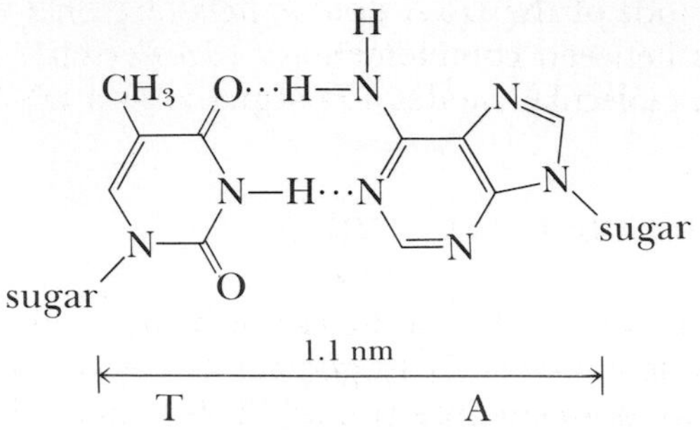

1.1 nm

T A

Similarly, G and C fit together at the same distance, forming three hydrogen bonds:

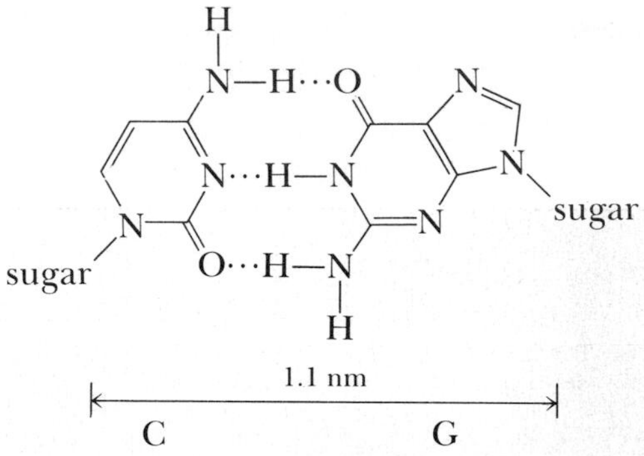

1.1 nm

C G

Notice that both base-pairing schemes allow the two strands of the double helix to have a separation of 1.1 nm. Other possible base pairs, such as C—C, T—T, and C—T, are too small, and A—A, G—G, and A—G are too large. Others that are the right size (A—C and G—T) cannot pair because their atoms are not in suitable positions to form hydrogen bonds:

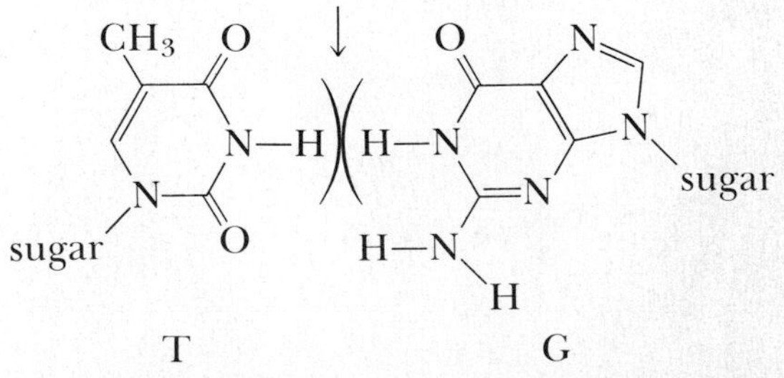

unfavorable for hydrogen bonding

T G

Thus only A—T and G—C base pairs can form, and this fact accounts for the structure of DNA. The two strands of the double helix are **complementary** to each other.

Example 24-4: If the base sequence along a portion of one strand of a double helix is . . . AGCCTCG . . . , what must be the corresponding sequence on the other strand?

Solution: The two sequences must be complementary to each other, meaning that a T and an A must be opposite each other and a G and a C must be opposite each other. Thus, the other sequence must have the base sequence . . . TCGGAGC . . . .

The two strands of the DNA double helix are held together by hydrogen bonds between complementary bases (Figure 24-12). Figure 24-13 shows a molecular model of a segment of a DNA double helix.

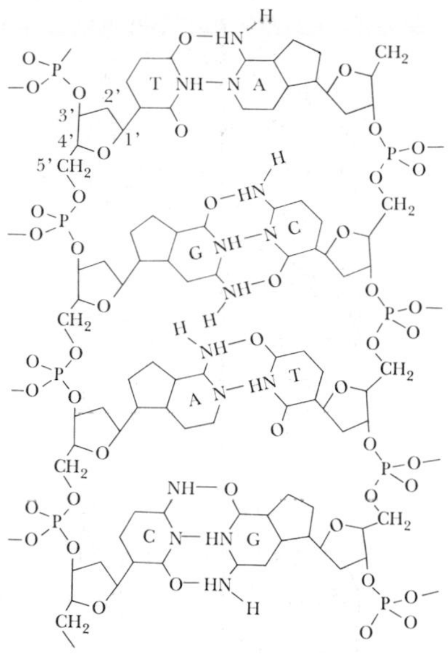

Figure 24-12 Hydrogen bonds between complementary base pairs hold the two strands of DNA together in a double-helical configuration.

24-9. DNA CAN DUPLICATE ITSELF

The two strands of the double helix are held together by hydrogen bonds, not covalent bonds. Hydrogen bonds are weak enough to allow the double helix to uncoil into two separate strands. Each strand can then act as a template for building a complementary strand, and the result will be two double helices that are identical to the first. Thus, the Watson-Crick model for DNA explains DNA replication. Example 24-5 illustrates this process.

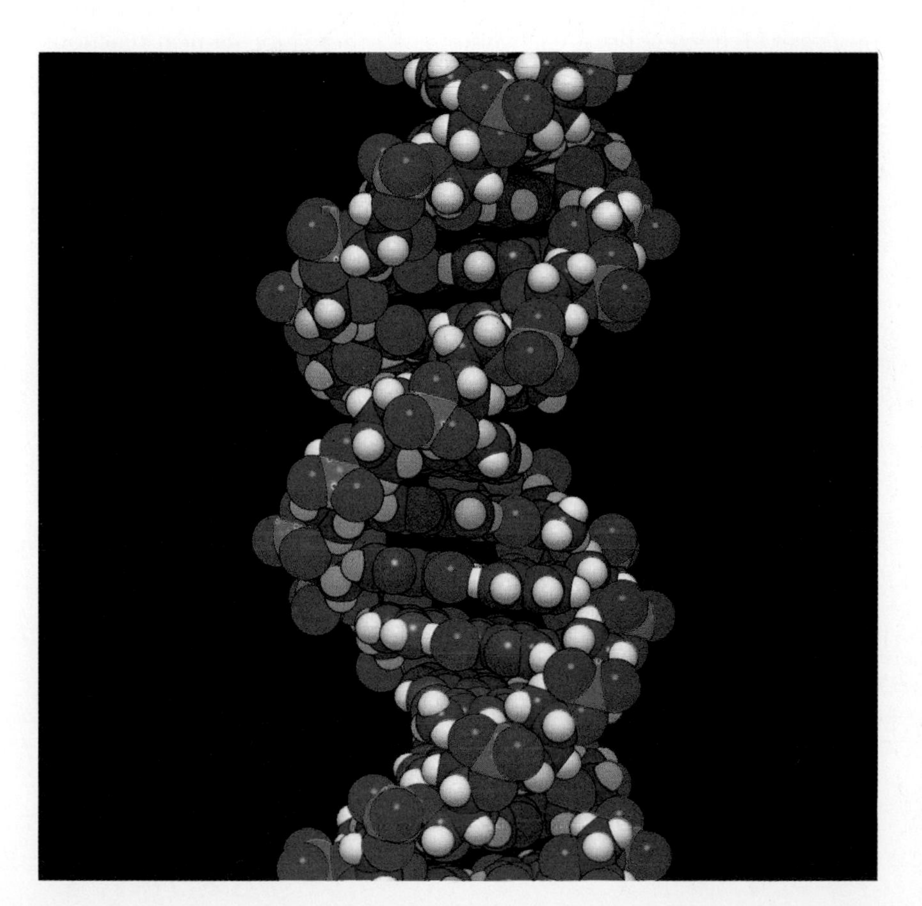

Figure 24-13 A computer-drawn simulation of a space-filling molecular model of DNA that includes 20 base pairs. The base pairs are roughly perpendicular to and interior to the long axis of the helix. The different atoms are shown in color: hydrogen *(white)*, nitrogen *(blue)*, oxygen *(red)*, phosphorus *(yellow)*, and carbon *(black)*.

Example 24-5: Suppose a segment along a double helix is

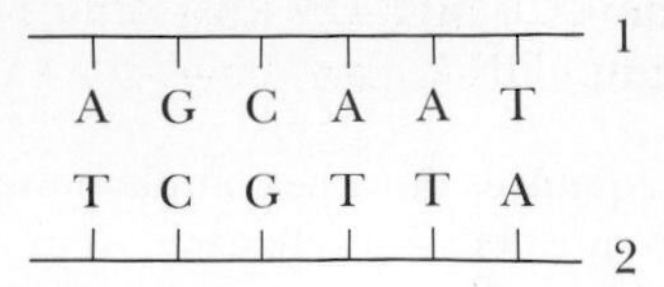

Show that if the double helix uncoils, then each separate strand can act as a template to build two identical double helices.

Solution: The two strands come apart to give

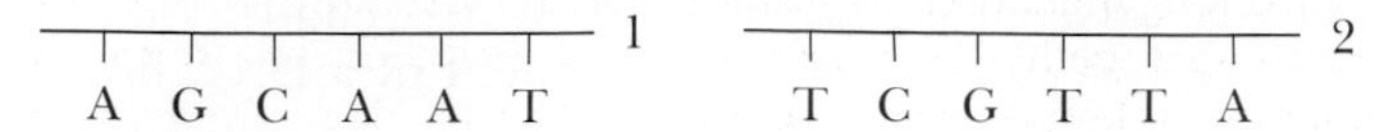

The two strands act as templates to form complementary strands, yielding

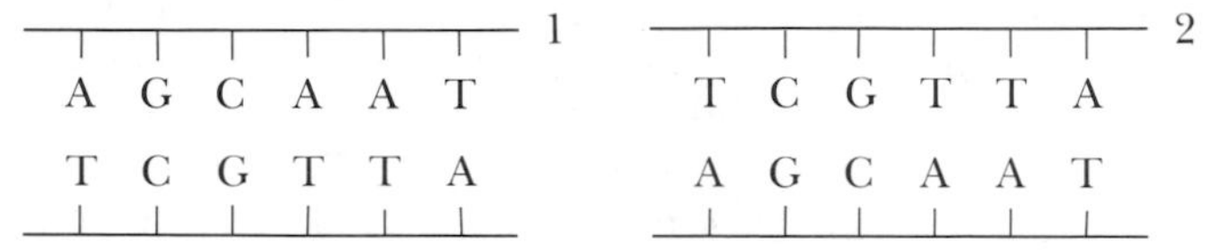

These two double helices are identical to each other and to the first double helix. Thus, the original double helix has replicated itself.

Living systems differ from one another by the myriad biochemical processes characteristic of each system. All these biochemical reactions are controlled by enzymes, and many of them involve other proteins as well. In a sense, each living system is a reflection of its various proteins. What we mean by genetic information is the information that calls for the production of all the proteins characteristic of a given organism. This is the information that is stored in DNA.

It was discovered in the 1950s that each series of three bases along a DNA segment is a code that leads to a particular amino acid. For example, the triplet AAA is a DNA code for phenylalanine and TTC is a DNA code for lysine. Thus the segment AAATTC in DNA would give rise to a segment phe-lys in a protein. Because a sequence of three bases is the code for one particular amino acid, the code is called the **triplet code.**

The chemical reactions that are involved in transcribing the base sequence along a DNA strand into a protein molecule are complicated but fairly well understood. They involve several types of RNA and numerous enzymes and other proteins. If you go on to take a course in biochemistry or biology, you will study the DNA-protein pathway in detail.

In summary, then, DNA is a double helix that is held together by hydrogen bonds between complementary bases. During replication, the two strands come apart and act as templates for the production of two identical double helices. It is in this way that genetic information is transmitted. The genetic information is stored in the form of the sequence of bases along each strand of the double helix. Each sequence of three bases codes a particular amino acid during protein synthesis. Because the nature and the order of the bases are equivalent to genetic information, the bases are in the interior of the double helix for protection. A **gene** is a segment along a DNA molecule that codes the synthesis of one polypeptide. DNA can have a molecular mass of over 100,000,000.

SUMMARY

Polymers are long, chainlike molecules that are formed by the bonding together of relatively small molecules called monomers. Monomers are the subunits, or links, of the molecular chains. Synthetic polymers of amazing diversity have been made by chemists in the last 50 years. All plastics, rubbers, synthetic fibers, and many other common materials are synthetic polymers. Proteins are naturally occurring polymers whose monomers are the 20 amino acids that are common to all living species. The order of the amino acids in a protein chain is called the primary structure of the chain and specifies the protein uniquely. A protein chain can be pictured as a polypeptide backbone with amino acid side groups attached at intervals. The side groups interact with each other and with solvent to effect a three-dimensional shape that is characteristic of the protein and called its tertiary structure. The function and efficacy of a protein are extraordinarily sensitive to its tertiary structure. Another class of biopolymers are polynucleotides. The polynucleotide DNA stores and transmits genetic information. The Watson and Crick double helical model pictures DNA as two complementary strands joined together by hydrogen bonds. This model explains how DNA can replicate itself and transmit genetic information from generation to generation. The genetic information is stored in DNA as a triplet code.

TERMS YOU SHOULD KNOW

polymerization 770
polymer 770
termination reaction 770
monomer 770
addition polymerization reaction 771
condensation polymerization reaction 772
dimer 772
elastomer 774
vulcanization 774
protein 775, 778
amino acid 775
side group 775
optical isomer 776
stereospecific 777
peptide bond 777
dipeptide 777
polypeptide 778
polypeptide backbone 778
primary structure 778
α-helix 779
secondary structure 780
tertiary structure 780
polynucleotide 781
DNA 781
RNA 781
nucleotide 781
base 781
sugar-phosphate backbone 783
double helix 784
complementary 784
triplet code 786
gene 787

PROBLEMS

POLYMERS

24-1. What is the difference between an addition polymerization and a condensation polymerization? Give an example of each.

24-2. Sketch the equation for the reaction in which Orlon is produced. Is it an addition polymerization reaction or a condensation polymerization reaction?

24-3. Why do you think the nylon discussed in Section 24-2 is called nylon 66?

24-4. Polystyrene, which has many uses including the manufacture of Styrofoam, has the formula

$-\!\!(CH_2-CH)\!\!-_n$ (with a benzene ring attached to the CH)

Write the formula for the monomer.

OPTICAL ISOMERS

24-5. Indicate whether each of the following compounds can exist as optical isomers:

(a) CH_3Cl

(b) H_2NCHCH_2OH (with COOH on the CH)

(c) $Cl-C-COOH$ (with H above and Br below the C)

(d) $CH_3CH_2-Si-Cl$ (with CH_3 above and Br below the Si)

24-6. Indicate whether each of the following compounds can exist as optical isomers:

(a) $Br-C-COOH$ (with NH_2 above and H below the C)

(b) CH_2Cl_2

(c) $(CH_3)_2SiCl_2$

(d) $F-C-H$ (with Cl above and Br below the C)

POLYPEPTIDES

24-7. Write the equations for the reactions between tyrosine and valine.

24-8. Write the equations for the reactions between lysine and serine.

24-9. Draw the structural formulas for the two possible dipeptides that can be formed from the reaction between glycine and alanine.

24-10. Draw the structural formulas for the two possible dipeptides that can be formed from the reaction between valine and aspartic acid.

24-11. How many different tripeptides can be formed from two different amino acids? Draw the structural formula for each of your tripeptides.

24-12. How many different tripeptides can be formed from three different amino acids? Draw the structural formula for each of your tripeptides.

24-13. Draw the structural formula for the tripeptide glu-asp-tyr.

24-14. Draw the structural formula for the tripeptide glu-val-cys.

24-15. Identify the atom(s), if any, in the following amino-acid side groups that can form a hydrogen bond with water:

(a) $-CH_2SH$

(b) $-CHCH_3$ (with CH_3 on the CH)

(c) $-CH_2CH_2CH_2CH_2NH_2$

(d) $-CH_2CH_2C(=O)O^{\ominus}$

24-16. Identify the atom(s), if any, in the following amino-acid side groups that can form a hydrogen bond with water:

(a) $-CH_2OH$

(b) $-CH_2-C_6H_4-OH$ (benzene ring)

(c) $-CH_2C(=O)OH$

(d) $-CH_3$

24-17. What is the difference between the primary, secondary, and tertiary structures of proteins?

24-18. What are the factors that govern the secondary and tertiary structures of proteins?

POLYNUCLEOTIDES

24-19. Draw the structural formula for the DNA triplet GAT.

24-20. Draw the structural formula for the DNA triplet ATC.

24-21. Draw the structural formula for the RNA triplet UCU.

24-22. Draw the structural formula for the RNA triplet CUG.

24-23. If the base sequence along a portion of one strand of a double helix is AAGTCTCGA, what must the corresponding sequence on the other strand be?

24-24. If the base sequence along a portion of one strand of a double helix is CATGGCTAA, what must the corresponding sequence on the other strand be?

24-25. Determine the complementary base sequence that corresponds to the following sequence of DNA bases:

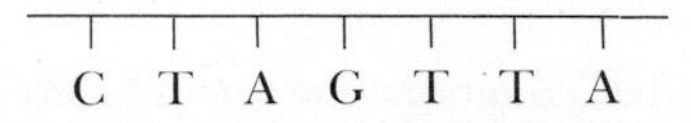

24-26. Determine the complementary base sequence that corresponds to the following sequence of DNA bases:

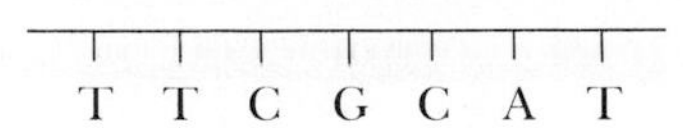

T T C G C A T

24-27. Suppose a segment along a double helix is

G C T T A C G — 1

C G A A T G C — 2

Draw the segments obtained when the DNA duplicates itself.

24-28. Suppose a segment along a double helix is

T C G T A C G — 1

A G C A T G C — 2

Draw the segments obtained when the DNA duplicates itself.

24-29. Determine the number of hydrogen bonds that must be broken to separate the strands of the DNA segment shown in Example 24-5.

24-30. Determine the number of hydrogen bonds that must be broken to separate the strands of the DNA segment shown in Problem 24-28.

ENERGY AND BIOCHEMICAL REACTIONS

24-31. An important source of energy for cells is the combustion of glucose:

$$C_6H_{12}O_6(aq) + 6O_2(g) \rightarrow 6CO_2(g) + 6H_2O(l)$$

$$\Delta G^\circ_{rxn} = -2.87 \times 10^3 \text{ kJ at pH} = 7.0,\ 25°\text{C}$$

Calculate the maximum amount of work that can be obtained from 1.0 g of glucose under standard conditions and pH = 7.0.

24-32. Plants synthesize carbohydrates from CO_2 and H_2O by the process of photosynthesis. For example,

$$6CO_2(g) + 6H_2O(l) \rightarrow \underset{\text{glucose}}{C_6H_{12}O_6(aq)} + 6O_2(g)$$

Calculate ΔG°_{rxn} for this reaction (see Problem 24-31). What is the equilibrium constant at 25°C for the chemical equation?

24-33. Given that at 25°C and pH = 7.0

$$\text{glucose}(aq) + 6O_2(g) \rightarrow 6CO_2(g) + 6H_2O(l) \qquad \Delta G^\circ_{rxn} = -2.87 \times 10^3 \text{ kJ}$$

$$\text{sucrose}(aq) + H_2O(l) \rightarrow \text{glucose}(aq) + \text{fructose}(aq) \qquad \Delta G^\circ_{rxn} = -29.3 \text{ kJ}$$

$$\text{fructose}(aq) \rightarrow \text{glucose}(aq) \qquad \Delta G^\circ_{rxn} = -1.6 \text{ kJ}$$

calculate the value of ΔG°_{rxn} for the combustion of sucrose. Calculate the maximum amount of work that can be obtained from the combustion of 1.0 g of sucrose under standard conditions.

24-34. An important biochemical reaction is the hydrolysis of adenosine triphosphate, ATP:

$$\text{ATP}(aq) + H_2O(l) \rightarrow \text{ADP}(aq) + HPO_4^{2-}(aq)$$

$$\Delta G^\circ_{rxn} = -29 \text{ kJ at pH} = 7.0 \text{ and } 25°\text{C}$$

Calculate the value of ΔG°_{rxn} for the equation

$$\text{glucose}(aq) + 38HPO_4^{2-}(aq) + 38\text{ADP}(aq) + 6O_2(g) \rightarrow 6CO_2(g) + 44H_2O(l) + 38\text{ATP}(aq)$$

Calculate the equilibrium constant at 25°C for the equation. (See Problem 24-31.)

24-35. Much of the energy from the combustion of glucose is used to synthesize adenosine triphosphate, ATP. The hydrolysis of ATP provides the energy to drive unfavorable reactions, perform work, and carry out other functions of the cells. The equation is

$$\text{ATP}(aq) + H_2O(l) \rightarrow \text{ADP}(aq) + HPO_4^{2-}(aq)$$

$$\Delta G^\circ_{rxn} = -31 \text{ kJ (cellular conditions, 37°C)}$$

Calculate the equilibrium constant at 37°C for this equation.

24-36. The hydrolysis of ATP provides the energy needed for the contraction of muscles. In a resting muscle, the concentration of ATP is 50 mM, the concentration of ADP is 0.5 mM, and the concentration of HPO_4^{2-} is 1.0 mM. Calculate the value of ΔG_{rxn} for the hydrolysis of ATP under these conditions. Take $\Delta G^\circ_{rxn} = -31$ kJ and a temperature of 37°C. (See Problem 24-35).

24-37. Another important source of energy in biological systems is glycolysis, the process by which glucose is broken down to lactic acid:

$$C_6H_{12}O_6(aq) \rightarrow 2CH_3CHOHCOOH(aq)$$

$$\Delta G^\circ_{rxn} = -200 \text{ kJ at pH} = 7.0 \text{ and } 25°\text{C}$$

Calculate the equilibrium constant at 25°C for the chemical equation.

24-38. The source of energy in human erythrocytes (red blood cells) is glycolysis. In erythrocytes the concentration of glucose is 5.0 mM and the concentration of lactic acid is 2.9 mM. Calculate the value of ΔG_{rxn} under these conditions at 25°C. See Problem 24-37 for the necessary data.

A MATHEMATICAL REVIEW

A1. SCIENTIFIC NOTATION AND EXPONENTS

The numbers encountered in chemistry are often extremely large (such as Avogadro's number) or extremely small (such as the mass of an electron in kilograms). When working with such numbers, it is convenient to express them in **scientific notation,** where we write the number as a number between 1 and 10 multiplied by 10 raised to the appropriate power. For example, the number 171.3 is $1.713 \times 100 = 1.713 \times 10^2$ in scientific notation. Some other examples are

$$7320 = 7.32 \times 10^3$$

$$1{,}624{,}000 = 1.624 \times 10^6$$

The zeros in these numbers are not regarded as significant figures and are dropped in scientific notation. Notice that in each case the power of 10 is the number of places that the decimal point has been moved to the left:

7320. (3 places) 1624000. (6 places)

When numbers that are smaller than 1 are expressed in scientific notation, the 10 is raised to a negative power. For example, 0.614 becomes 6.14×10^{-1}. Recall that a negative exponent is governed by the relation

$$10^{-n} = \frac{1}{10^n} \tag{A1-1}$$

Some other examples are

$$0.0005 = 5 \times 10^{-4}$$

$$0.000000000446 = 4.46 \times 10^{-10}$$

Notice that the power of 10 in each case is the number of places that the decimal point has been moved to the right:

0.0005 (4 places) 0.000000000446 (10 places)

It is necessary to be able to work with numbers in scientific notation. To add or subtract two or more numbers expressed in scientific notation, the power of 10 must be the same in both. For example, consider the sum

$$5.127 \times 10^4 + 1.073 \times 10^3$$

We rewrite the first number so that we have 10^3:

$$5.127 \times 10^4 = 51.27 \times 10^3$$

■ To change a number such as 51.27×10^3 to 5.127×10^4, we make the number in front one factor of 10 smaller, and so we must make 10^3 one factor of 10 larger.

Note that we have changed the 10^4 factor to 10^3, and so we must make the factor in front of 10^3 one power of 10 larger. Thus we have

$$\begin{aligned} 5.127 \times 10^4 + 1.073 \times 10^3 &= (51.27 + 1.073) \times 10^3 \\ &= 52.34 \times 10^3 \\ &= 5.234 \times 10^4 \end{aligned}$$

■ Note that in changing 2.156×10^{-7} to 0.2156×10^{-6}, we make 2.156 one factor of 10 smaller and 10^{-7} one factor of 10 larger.

Similarly, we have

$$\begin{aligned} (4.728 \times 10^{-6}) - (2.156 \times 10^{-7}) &= (4.728 - 0.2156) \times 10^{-6} \\ &= 4.512 \times 10^{-6} \end{aligned}$$

When multiplying two numbers, we add the powers of 10 because of the relation

$$(10^x)(10^y) = 10^{x+y} \tag{A1-2}$$

For example,

$$\begin{aligned} (5.00 \times 10^2)(4.00 \times 10^3) &= (5.00)(4.00) \times 10^5 \\ &= 20.0 \times 10^5 \\ &= 2.00 \times 10^6 \end{aligned}$$

$$\begin{aligned} (3.014 \times 10^3)(8.217 \times 10^{-6}) &= (3.014)(8.217) \times 10^{-3} \\ &= 24.77 \times 10^{-3} \\ &= 2.477 \times 10^{-2} \end{aligned}$$

To divide, we subtract the power of 10 of the number in the denominator from the power of 10 of the number in the numerator because of the relation

$$\frac{10^x}{10^y} = 10^{x-y} \tag{A1-3}$$

For example,

$$\frac{4.0 \times 10^{12}}{8.0 \times 10^{23}} = \left(\frac{4.0}{8.0}\right) \times 10^{12-23}$$
$$= 0.50 \times 10^{-11}$$
$$= 5.0 \times 10^{-12}$$

$$\frac{2.80 \times 10^{-4}}{4.73 \times 10^{-5}} = \left(\frac{2.80}{4.73}\right) \times 10^{-4+5}$$
$$= 0.592 \times 10^{1}$$
$$= 5.92$$

To raise a number to a power, we use the fact that

$$(10^x)^n = 10^{nx} \qquad \text{(A1-4)}$$

For example,

$$(2.187 \times 10^2)^3 = (2.187)^3 \times 10^6$$
$$= 10.46 \times 10^6$$
$$= 1.046 \times 10^7$$

To take a root of a number, we use the relation

$$\sqrt[n]{10^x} = (10^x)^{1/n} = 10^{x/n} \qquad \text{(A1-5)}$$

Thus, the power of 10 must be written such that it is divisible by the root. For example,

$$\sqrt[3]{2.70 \times 10^{10}} = (2.70 \times 10^{10})^{1/3} = (27.0 \times 10^9)^{1/3}$$
$$= (27.0)^{1/3} \times 10^3 = 3.00 \times 10^3$$

and

$$\sqrt{6.40 \times 10^5} = (6.40 \times 10^5)^{1/2} = (64.0 \times 10^4)^{1/2}$$
$$= (64.0)^{1/2} \times 10^2 = 8.00 \times 10^2$$

A2. LOGARITHMS

You know that $100 = 10^2$, $1000 = 10^3$, and so on. You also know that

$$\sqrt{10} = 10^{1/2} = 10^{0.50} = 3.16$$

By taking the square root of both sides of

$$10^{0.50} = 3.16$$

we find that

$$\sqrt{10^{0.50}} = 10^{(1/2)(0.50)} = 10^{0.25} = \sqrt{3.16} = 1.78$$

Furthermore, because

$$(10^x)(10^y) = 10^{x+y}$$

we can write

$$10^{0.25} \times 10^{0.50} = 10^{0.75} = (3.16)(1.78) = 5.62$$

By continuing this process, we can express any number y as

$$y = 10^x \qquad \text{(A2-1)}$$

The number x to which 10 must be raised to get y is called the **logarithm** of y and is written as

$$x = \log y \qquad \text{(A2-2)}$$

Equations (A2-1) and (A2-2) are equivalent. For example, we have shown that

$$\log 1.78 = 0.25$$
$$\log 3.16 = 0.50$$
$$\log 5.62 = 0.75$$
$$\log 10.00 = 1.00$$

Logarithms of other numbers may be obtained from tables (Appendix C) or, more conveniently, from a hand calculator. If you use tables, you must always write the number y in standard scientific notation. Thus, for example, you must write 782 as 7.82×10^2 and 0.000465 as 4.65×10^{-4}. To take the logarithm of such numbers, we use the fact that

$$\log ab = \log a + \log b \tag{A2-3}$$

Thus we write

$$\begin{aligned}\log 782 &= \log (7.82 \times 10^2) = \log 7.82 + \log 10^2 \\ &= \log 7.82 + 2.000\end{aligned}$$

Logarithm tables are set up such that the number a in $\log a$ is between 1 and 10 and the numbers in the tables are between 0 and 1. Thus from Appendix C we find, for example, that

$$\log 4.12 = 0.6149$$
$$\log 8.37 = 0.9227$$

and so on. If we look up log 7.82 in Appendix C, then we find that it is equal to 0.8932. Therefore,

$$\begin{aligned}\log 782 = \log (7.82 \times 10^2) &= \log 7.82 + \log 10^2 \\ &= 0.8932 + 2.000 = 2.8932\end{aligned}$$

If you use your calculator, you simply enter 782 and push a log key to get 2.8932 directly.

To find log 0.000465, we write

$$\begin{aligned}\log 0.000465 &= \log (4.65 \times 10^{-4}) \\ &= \log 4.65 + \log 10^{-4} \\ &= \log 4.65 - 4.000\end{aligned}$$

We find $\log 4.65 = 0.6675$ from Appendix C, and so

$$\begin{aligned}\log 0.000465 &= 0.6675 - 4.000 \\ &= -3.3325\end{aligned}$$

If you use your calculator, simply enter 0.000465 and push the log key to get -3.3325 directly. Although a hand calculator is much more convenient than a table of logarithms, you should be able to handle logarithms by either method.

Because logarithms are exponents ($y = 10^x$), they have certain special properties, such as

$$\log ab = \log a + \log b \tag{A2-3}$$

$$\log \frac{a}{b} = \log a - \log b \tag{A2-4}$$

$$\log a^n = n \log a \tag{A2-5}$$

$$\log \sqrt[n]{a} = \log a^{1/n} = \frac{1}{n} \log a \tag{A2-6}$$

If we let $a = 1$ in Equation (A2-4), then we have

$$\log \frac{1}{b} = \log 1 - \log b$$

or, because $\log 1 = 0$,

$$\log \frac{1}{b} = -\log b \qquad \text{(A2-7)}$$

Thus we change the sign of a logarithm by taking the reciprocal of its argument. Notice that because $\log 1 = 0$,

$$\log y > 0 \quad \text{if} \quad y > 1$$

$$\log y < 0 \quad \text{if} \quad y < 1$$

Up to this point we have found the value of x in

$$y = 10^x$$

when y is given. It is often necessary to find the value of y when x is given. Because x is called the logarithm of y, y is called the **antilogarithm** of x. For example, suppose that $x = 6.1303$ and we wish to find y. We write

$$y = 10^{6.1303} = 10^{0.1303} \times 10^6$$

From the logarithm table, we see that the number whose logarithm is 0.1303 is 1.35. Thus we find that

$$10^{6.1303} = 1.35 \times 10^6$$

You can obtain this result directly from your calculator. On a calculator having inv and log keys, for example, enter 6.1303, press the inv key (for inverse) and then the log key.

To obtain the antilogarithm of y using logarithm tables, you must express y as

$$y = 10^a \times 10^n \qquad \text{(A2-8)}$$

where n is a positive or negative integer and a is between 0 and 1. The quantity a is found in the table and the antilog of a is then read.

As another example, let's find the antilog of 1.9509. We write

$$y = 10^{1.9509} = 10^{0.9509} \times 10^1$$

We find the value 0.9509 in the log table and see that its antilog is 8.93. Thus we have

$$y = 10^{1.9509} = 8.93 \times 10^1 = 89.3$$

You should be able to obtain this result directly from your calculator. If your calculator has a 10^x key, then you can obtain the antilog of 1.9509 by entering 1.9509 and pressing the 10^x key. This operation is equivalent to using the inv key followed by the log key.

In many problems, it is necessary to find the antilogarithm of negative numbers. For example, let's find the antilogarithm of -4.167, or the value of y in

$$y = 10^{-4.167}$$

Even though the exponent is negative, we still must express y in the form of Equation (A2-8). To do this, we write $-4.167 = 0.833 - 5.000$ so that

$$y = 10^{0.833} \times 10^{-5}$$

Now we find 0.833 in a logarithm table and see that its antilogarithm is 6.81. Thus

$$y = 6.81 \times 10^{-5}$$

You should be able to obtain this same result from your calculator by entering -4.167 and finding the inverse logarithm directly.

A3. THE QUADRATIC FORMULA

The standard form for a quadratic equation in x is

$$ax^2 + bx + c = 0 \tag{A3-1}$$

where a, b, and c are constants. The two solutions to the quadratic equation are

$$x = \frac{-b \pm \sqrt{b^2 - 4ac}}{2a} \tag{A3-2}$$

Equation (A3-2) is called the **quadratic formula** and is used to obtain the solutions to a quadratic equation expressed in the standard form.

For example, let's find the solutions to the quadratic equation

$$2x^2 - 3x - 1 = 0$$

In this case, $a = 2$, $b = -3$, and $c = -1$ and Equation (A3-2) gives

$$\begin{aligned} x &= \frac{3 \pm \sqrt{(-3)^2 - 4(2)(-1)}}{(2)(2)} \\ &= \frac{3 \pm 4.123}{4} \\ &= 1.781 \quad \text{and} \quad -0.2808 \end{aligned}$$

To use the quadratic formula, it is first necessary to put the quadratic equation in the standard form so that we know the values of the constants a, b, and c. For example, consider the problem of solving for x in the quadratic equation

$$\frac{x^2}{0.50 - x} = 0.040$$

To identify the constants a, b, and c, we must write this equation in the standard quadratic form. Multiplying both sides by $0.50 - x$ yields

$$\begin{aligned} x^2 &= (0.50 - x)0.040 \\ &= 0.020 - 0.040x \end{aligned}$$

Rearrangement to the standard quadratic form yields

$$x^2 + 0.040x - 0.020 = 0$$

Thus $a = 1$, $b = 0.040$, and $c = -0.020$. Using Equation (A3-2), we have

$$x = \frac{-0.040 \pm \sqrt{(0.040)^2 - 4(1)(-0.020)}}{2(1)}$$

from which we compute

$$\begin{aligned} x &= \frac{-0.040 \pm \sqrt{0.0816}}{2} \\ &= \frac{-0.040 \pm 0.286}{2} \end{aligned}$$

Thus the solutions for x are

$$x = \frac{-0.040 + 0.286}{2} = 0.123$$

and $$x = \frac{-0.040 - 0.286}{2} = -0.163$$

If x represents, say, a concentration or gas pressure, then the only physically possible value is $+0.123$ because concentrations and pressures cannot have negative values.

Many problems involving chemical equilibria lead to a quadratic equation of the form

$$\frac{x^2}{M_0 - x} = K \tag{A4-1}$$

where x is the concentration of a particular species, M_0 is an initial concentration, and K is an equilibrium constant. For example, the equation

$$\frac{[H_3O^+]^2}{0.100 - [H_3O^+]} = 2.19 \times 10^{-4}\ \text{M} \tag{A4-2}$$

arises if we wish to calculate $[H_3O^+]$ and the pH of a 0.100 M HCNO(*aq*) solution. If the value of K is small, then it is much more convenient to solve an equation like Equation (A4-1) by the **method of successive approximations** than by using the quadratic equation.

The first step in the method of successive approximations is to neglect the unknown in the denominator on the left-hand side of the equation. This step allows the unknown to be found by simply multiplying through by the initial concentration and taking the square root of both sides:

$$[H_3O^+]_1 \simeq [(0.100\ \text{M})(2.19 \times 10^{-4}\ \text{M})]^{1/2} = 4.68 \times 10^{-3}\ \text{M} \tag{A4-3}$$

We have subscripted $[H_3O^+]$ with a 1 in this result because it represents a first approximation to $[H_3O^+]$. To obtain a second approximation, use the value of $[H_3O^+]_1$ in the denominator of the left-hand side of Equation (A4-2), multiply both sides by the result in the denominator, and then take the square root:

$$\begin{aligned}[H_3O^+]_2 &\simeq [(0.100\ \text{M} - 4.68 \times 10^{-3}\ \text{M})(2.19 \times 10^{-4}\ \text{M})]^{1/2} \\ &= 4.57 \times 10^{-3}\ \text{M}\end{aligned}$$

We now carry out the cycle, called an **iteration,** over again to obtain a third approximation:

$$\begin{aligned}[H_3O^+]_3 &\simeq [(0.100\ \text{M} - 4.57 \times 10^{-3}\ \text{M})(2.19\ 10^{-4}\ \text{M})]^{1/2} \\ &= 4.57 \times 10^{-3}\ \text{M}\end{aligned}$$

Note that $[H_3O^+]_3 \simeq [H_3O^+]_2$; when this occurs, we say that the procedure has converged. After convergence is achieved, the same result will occur in any subsequent iteration, and the value obtained is the solution to the original equation because the equation is satisfied by the same value of $[H_3O^+]$ in the numerator as in the denominator.

The method of successive approximations is particularly convenient when you are using a hand calculator. Although it is usually necessary to carry out several iterations to obtain the solution, each cycle is easy to perform on a calculator, and the total effort involved usually is less than using the quadratic formula. In fact, as you use this method, think about the sequence of steps that you use on your calculator. Consider the equation

$$\frac{x^2}{0.250 - x} = 7.63 \times 10^{-4}$$

First, neglect x compared to 0.250, multiply through by 0.250, and take the square root to obtain

$$x_1 = 1.38 \times 10^{-2}$$

Now subtract x_1 from 0.250, multiply the result by 7.63×10^{-4}, and take the square root to obtain

$$x_2 = 1.34 \times 10^{-2}$$

One more iteration gives $x_3 = x_2$, and we say that the method has converged in three iterations.

Usually you will obtain convergence after only a few iterations. If you don't see the successive iterations approaching some value after just a few iterations, then probably it is more convenient to use the quadratic formula.

Here are some examples to practice with:

1. $\dfrac{x^2}{0.500 - x} = 1.07 \times 10^{-3}$ ($x_1 = 2.31 \times 10^{-2}$, $x_2 = 2.26 \times 10^{-2}$, $x_3 = 2.26 \times 10^{-2}$)

2. $\dfrac{x^2}{0.0100 - x} = 6.80 \times 10^{-4}$ ($x_1 = 2.61 \times 10^{-3}$, $x_2 = 2.24 \times 10^{-3}$, $x_3 = 2.30 \times 10^{-3}$, $x_4 = 2.29 \times 10^{-3}$, $x_5 = 2.29 \times 10^{-3}$)

3. $\dfrac{x_2}{0.150 - x} = 0.0360$ ($x_1 = 7.35 \times 10^{-2}$, $x_2 = 5.25 \times 10^{-2}$, $x_3 = 5.92 \times 10^{-2}$, $x_4 = 5.72 \times 10^{-2}$, $x_5 = 5.78 \times 10^{-2}$, $x_6 = 5.76 \times 10^{-2}$, $x_7 = 5.77 \times 10^{-2}$, $x_8 = 5.77 \times 10^{-2}$)

Even in this last case, which requires eight iterations, the method of successive approximations is easier than using the quadratic formula.

A5. PLOTTING DATA

The human eye and brain are quite sensitive to recognizing straight lines, and so it is always desirable to plot equations or experimental data such that a straight line is obtained. The mathematical equation for a straight line is of the form

$$y = mx + b \tag{A5-1}$$

In this equation, m and b are constants: m is the **slope** of the line and b is its **intercept** with the y axis. The slope of a straight line is a measure of its steepness; it is defined as the ratio of its vertical rise to the corresponding horizontal distance.

Let's plot the two straight lines

I $y = x + 1$

II $y = 2x - 2$

We first make a table of values of x and y:

I		II	
x	y	x	y
−3	−2	−3	−8
−2	−1	−2	−6
−1	0	−1	−4
0	1	0	−2
1	2	1	0
2	3	2	2
3	4	3	4
4	5	4	6
5	6	5	8

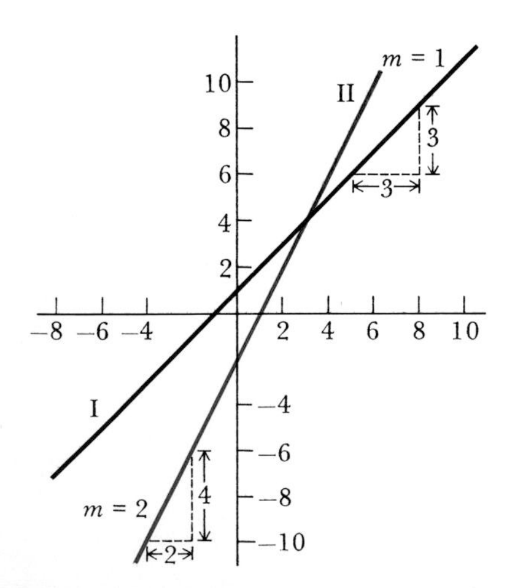

Figure A1 Plots of the equations (I) $y = x + 1$ and (II) $y = 2x - 2$.

These results are plotted in Figure A1. Note that curve I intersects the y axis at $y = 1$ ($b = 1$) and has a slope of 1 ($m = 1$). Curve II intersects the y axis at $y = -2$ ($b = -2$) and has a slope of 2 ($m = 2$).

Usually the equation to be plotted will not appear to be of the form of Equation (A5-1) at first. For example, consider the Boyle's law (Chapter 4) relation between the volume of a gas and its pressure:

$$V = \frac{c}{P} \qquad \text{constant temperature} \qquad \text{(A5-2)}$$

where c is a proportionality constant whose value depends upon the temperature and the mass of a given sample. For example, for a 0.29-g sample of air, $c = 0.244$ L·atm at 25°C. Some results for such a sample are presented in Table A1, and the data in Table A1 are plotted as volume versus pressure in Figure A2.

Table A1 Pressure-volume data for 0.29 g of air at 25°C

P/atm	V/L	$\frac{1}{P}$/atm^{-1}
0.26	0.938	3.85
0.41	0.595	2.44
0.83	0.294	1.20
1.20	0.203	0.83
2.10	0.116	0.48
2.63	0.093	0.38
3.14	0.078	0.32

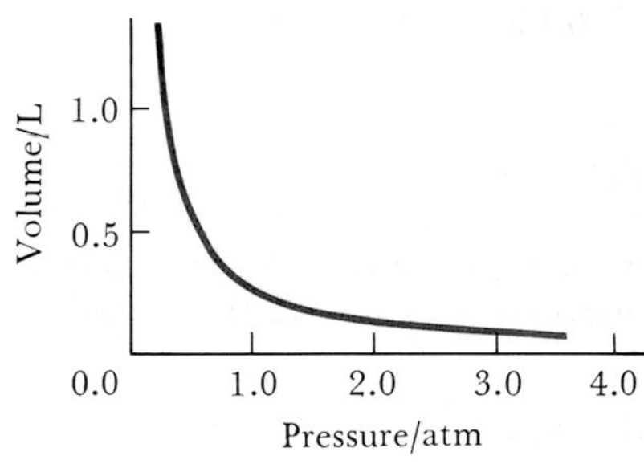

Figure A2 The volume of 0.29 g of air plotted versus pressure at 25°C. The data are given in Table A-1. The curve in this figure is from the Boyle's law equation

$$V = \frac{0.244\text{L·atm}}{P}$$

It may appear at first sight that Equation (A5-2) is not of the form $y = mx$. However, if we let $V = y$ and $1/P = x$, then Equation (A5-2) becomes

$$y = cx$$

Thus, if we plot V versus $1/P$ instead of versus P, then we get a straight line. The data in Table A1 are plotted as V versus $1/P$ in Figure A3. Note that a straight line is obtained.

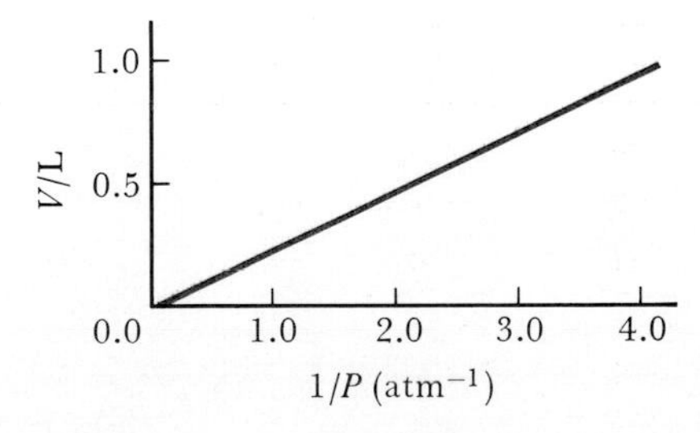

Figure A3 The volume of 0.29 g of air plotted versus the reciprocal of the pressure ($1/P$) at 25°C. If we compare this curve with Figure A2, we see that a straight line results when V is plotted versus $1/P$ instead of versus P. Straight lines are much easier to work from than other curves, and so it is usually desirable to plot equations and data in the form of a straight line.

SI UNITS AND CONVERSION FACTORS

Measurements and physical quantities in the sciences are expressed in the **metric system,** which is a system of units that was formalized by the French National Academy in 1790. There are several subsystems of units in the metric system, and in an international effort to achieve uniformity, the International System of Units (abbreviated SI from the French Système Internationale d'Unités) was adopted by the General Conference of Weights and Measures in 1960 as *the* recommended units for science and technology. The SI is constructed from the seven basic units given in Table B1. The first five units in Table B1 are used frequently in general chemistry. Each has a technical definition that serves to define the unit in an unambiguous, reproducible way, but here we simply relate the SI units to the English system.

Table B1 The seven SI basic units

Physical quantity	Name of unit	Symbol
length	meter	m
mass	kilogram	kg
time	second	s
temperature	kelvin	K
amount of substance	mole	mol
electric current	ampere	A
luminous intensity	candela	cd

▪ In 1983, one meter was redefined as the distance that light travels through space in 1/299,792,458 s.

1. Length: One meter is equivalent to 1.0936 yards, or to 39.370 in. Thus, a meter stick is about 3 in. longer than a yardstick.
2. Mass: One kilogram is equivalent to 2.2046 lb. The mass of a substance is determined by balancing it against a set of standard masses using a balance.
3. Temperature: The Kelvin temperature scale is related to the Celsius, or centigrade, temperature scale. On the Celsius scale, the normal freezing point of water is set at 0°C and its normal boiling point at sea level is set at 100°C. The Kelvin and Celsius scales are related by the equation (Chapter 4)

$$K = {}^\circ C + 273.15$$

Recall that the normal freezing point of water is 32°F and its normal boiling point is 212°F on the Fahrenheit scale. The relation between the Celsius and Fahrenheit scales is given by

$$^\circ C = \frac{5}{9}(^\circ F - 32) \qquad \text{(B-1)}$$

Thus, for example, 50°F corresponds to 10°C and 86°F corresponds to 30°C. Note that the symbol for kelvin is K and not °K.

4. Amount of substance: One mole is the amount of substance that contains as many elementary entities as there are atoms in exactly 0.012 kg of carbon-12 (Chapter 3).

▪ Temperature is popularly recorded two ways in the United States. On the Fahrenheit scale, named after the German physicist Gabriel Daniel Fahrenheit (1686–1736), 0° is the temperature of a mixture of equal weights of snow and sodium chloride, the freezing point of water is 32° and its boiling point is 212°. On the Celsius scale, named after the Swedish astronomer Anders Celsius (1701–44), the freezing point of water is 0° and its boiling point is 100°.

An important feature of the metric system and the SI is the use of prefixes to designate multiples of the basic units (Table B2).

The units of all quantities not listed in Table B1 involve combinations of the basic SI units and are called **derived units.** The derived units frequently used in general chemistry are given in Table B3. Many of these units may not be familiar to you unless you have had a course in physics. For example, the SI unit of force is a **newton,** which is defined as the force required to give a 1-kg body a speed of 1 $m \cdot s^{-1}$ when the force is applied for 1 s. The SI unit of pressure is the **pascal.** Pressure is force per area, and a pascal is defined as the pressure produced by a force of 1 N acting on an area of 1 m^2. The SI unit of energy is the **joule,** which is the energy that a 2-kg mass has when it is traveling at a speed of 1 $m \cdot s^{-1}$. A joule is also the energy that a mass gains when it is acted upon by a force of 1 N through a distance of 1 m. Thus, we have J = N·m.

Although the SI is gradually becoming the universally accepted system of units, there are a number of older units that are used frequently (Table B4). For example, volume is usually expressed in **liters,** L. A liter is defined as a cubic decimeter and is slightly larger than a quart, being equivalent to 1.0567 qt. The glassware in your laboratory is measured in milliliters (mL). One milliliter is equivalent to one cubic centimeter (cm^3).

Table B2 Prefixes used for multiples and fractions of SI units

Prefix	Symbol	Multiple	Example
tera-	T	10^{12}	terawatt, 1 TW = 10^{12} W
giga-	G	10^{9}	gigavolt, 1 GV = 10^{9} V
mega-	M	10^{6}	megawatt, 1 MW = 10^{6} W
kilo-	k	10^{3}	kilometer, 1 km = 10^{3} m
deci-	d	10^{-1}	decimeter, 1 dm = 10^{-1} m
centi-	c	10^{-2}	centimeter, 1 cm = 10^{-2} m
milli-	m	10^{-3}	millisecond, 1 ms = 10^{-3} s
micro-	μ*	10^{-6}	microsecond, μs = 10^{-6} s
nano-	n	10^{-9}	nanosecond, 1 ns = 10^{-9} s
pico-	p	10^{-12}	picometer, 1 pm = 10^{-12} m
femto-	f	10^{-15}	femtometer, 1 fm = 10^{-15} m
atto-	a	10^{-18}	attomole, 1 amol = 10^{-18} mol

*This is the Greek letter mu, pronounced mew.

▪ Chemists can now measure processes that occur in one picosecond.

Table B3 Names and symbols for some SI derived units

Quantity	Unit	Symbol	Definition
area	square meter	m^2	
volume	cubic meter	m^3	
density	kilogram per cubic meter	$kg \cdot m^{-3}$	
speed	meter per second	$m \cdot s^{-1}$	
frequency	hertz	Hz	s^{-1} (cycles per second)
force	newton	N	$kg \cdot m \cdot s^{-2}$
pressure	pascal	Pa	$N \cdot m^{-2} = kg \cdot m^{-1} \cdot s^{-2}$
energy	joule	J	$kg \cdot m^2 \cdot s^{-2} = N \cdot m$
electric charge	coulomb	C	$A \cdot s$
electric potential difference	volt	V	$J \cdot A^{-1} \cdot s^{-1} = kg \cdot m^2 \cdot s^{-3} \cdot A^{-1}$

Table B4 Some commonly used non-SI units

Quantity	Unit	Symbol	SI definition
length	angstrom	Å	10^{-10} m
length	micron	μ	10^{-6} m = 1 μm
volume	liter	L	10^{-3} m^3
energy	calorie	cal	4.184 J
pressure	atmosphere	atm	101.325 kPa
pressure	torr	torr	133.322 Pa
pressure	bar	bar	10^5 Pa

The SI unit of pressure, the pascal, is rarely used in the United States. The most commonly used units of pressure are the **atmosphere** (atm) and the **torr** (Chapter 4).

We can use the relation between atmospheres and pascals to derive a relation between liter-atmospheres and joules. We use this relationship in Chapter 4. We start with

$$1 \text{ atm} = 101.32 \text{ kPa} = 1.0132 \times 10^5 \text{ Pa}$$

and multiply by L:

$$1 \text{ L·atm} = 1.0132 \times 10^5 \text{ L·Pa}$$

Using the relations

$$\text{Pa} = \text{N·m}^{-2} \qquad \text{J} = \text{N·m} \qquad \text{L} = \text{dm}^3 = 10^{-3} \text{ m}^3$$

we obtain

$$\begin{aligned} 1 \text{ L·atm} &= (1.0132 \times 10^5 \text{ N·m}^{-2})(10^{-3} \text{ m}^3) \\ &= 101.32 \text{ N·m} = 101.32 \text{ J} \end{aligned}$$

or writing this result as a unit conversion factor

$$\frac{101.32 \text{ J}}{\text{L·atm}} = 1$$

In particular, in Chapter 4 we need the relation

$$0.08206 \text{ L·atm} = (0.08206 \text{ L·atm})(101.32 \text{ J} \cdot \text{L}^{-1}\text{·atm}^{-1}) = 8.314 \text{ J}$$

The SI units and their conversion factors are given on the inside back cover of this book.

FOUR-PLACE LOGARITHMS

N	0	1	2	3	4	5	6	7	8	9
10	0000	0043	0086	0128	0170	0212	0253	0294	0334	0374
11	0414	0453	0492	0531	0569	0607	0645	0682	0719	0755
12	0792	0828	0864	0899	0934	0969	1004	1038	1072	1106
13	1139	1173	1206	1239	1271	1303	1335	1367	1399	1430
14	1461	1492	1523	1553	1584	1614	1644	1673	1703	1732
15	1761	1790	1818	1847	1875	1903	1931	1959	1987	2014
16	2041	2068	2095	2122	2148	2175	2201	2227	2253	2279

N	**0**	**1**	**2**	**3**	**4**	**5**	**6**	**7**	**8**	**9**
17	2304	2330	2355	2380	2405	2430	2455	2480	2504	2529
18	2533	2577	2601	2625	2648	2672	2695	2718	2742	2765
19	2788	2810	2833	2856	2878	2900	2923	2945	2967	2989
20	3010	3032	3054	3075	3096	3118	3139	3160	3181	3201
21	3222	3243	3263	3284	3304	3324	3345	3365	3385	3404
22	3424	3444	3464	3483	3502	3522	3541	3560	3579	3598
23	3617	3636	3655	3674	3692	3711	3729	3747	3766	3784
24	3802	3820	3838	3856	3874	3892	3909	3927	3945	3962
25	3979	3997	4014	4031	4048	4065	4082	4099	4116	4133
26	4150	4166	4183	4200	4216	4232	4249	4265	4281	4298
27	4314	4330	4346	4362	4378	4393	4409	4425	4440	4456
28	4472	4487	4502	4518	4533	4548	4564	4579	4594	4609
29	4624	4639	4654	4669	4683	4698	4713	4728	4742	4757
30	4771	4786	4800	4814	4829	4843	4857	4871	4886	4900
31	4914	4928	4942	4955	4969	4983	4997	5011	5024	5038
32	5051	5065	5079	5092	5105	5119	5132	5145	5159	5172
33	5185	5198	5211	5224	5237	5250	5263	5276	5289	5302
34	5315	5328	5340	5353	5366	5378	5391	5403	5416	5428
35	5441	5453	5465	5478	5490	5502	5514	5527	5539	5551
36	5563	5575	5587	5599	5611	5623	5635	5647	5658	5670
37	5682	5694	5705	5717	5729	5740	5752	5763	5775	5786
38	5798	5809	5821	5832	5843	5855	5866	5877	5888	5899
39	5911	5922	5933	5944	5955	5966	5977	5988	5999	6010
40	6021	6031	6042	6053	6064	6075	6085	6096	6107	6117
41	6128	6138	6149	6160	6170	6180	6191	6201	6212	6222
42	6232	6243	6253	6263	6274	6284	6294	6304	6314	6325
43	6335	6345	6355	6365	6375	6385	6395	6405	6415	6425
44	6435	6444	6454	6464	6474	6484	6493	6503	6513	6522
45	6532	6542	6551	6561	6571	6580	6590	6599	6609	6618
46	6628	6637	6646	6656	6665	6675	6684	6693	6702	6712
47	6721	6730	6739	6749	6758	6767	6776	6785	6794	6803
48	6812	6821	6830	6839	6848	6857	6866	6875	6884	6893
49	6902	6911	6920	6928	6937	6946	6955	6964	6972	6981
50	6990	6998	7007	7016	7024	7033	7042	7050	7059	7067
51	7076	7084	7093	7101	7110	7118	7126	7135	7143	7152
52	7160	7168	7177	7185	7193	7202	7210	7218	7226	7235
53	7243	7251	7259	7267	7275	7284	7292	7300	7308	7316
54	7324	7332	7340	7348	7356	7364	7372	7380	7388	7396
55	7404	7412	7419	7427	7435	7443	7451	7459	7466	7474
56	7482	7490	7497	7505	7513	7520	7528	7536	7543	7551
57	7559	7566	7574	7582	7589	7597	7604	7612	7619	7627
58	7634	7642	7649	7657	7664	7672	7679	7686	7694	7701
59	7709	7716	7723	7731	7738	7745	7752	7760	7767	7774

N	0	1	2	3	4	5	6	7	8	9
60	7782	7789	7796	7803	7810	7818	7825	7832	7839	7846
61	7853	7860	7868	7875	7882	7889	7896	7903	7910	7917
62	7924	7931	7938	7945	7952	7959	7966	7973	7980	7987
63	7993	8000	8007	8014	8021	8028	8035	8041	8048	8055
64	8062	8069	8075	8082	8089	8096	8102	8109	8116	8122
65	8129	8136	8142	8149	8156	8162	8169	8176	8182	8189
66	8195	8202	8209	8215	8222	8228	8235	8241	8248	8254
67	8261	8267	8274	8280	8287	8293	8299	8306	8312	8319
68	8325	8331	8338	8344	8351	8357	8363	8370	8376	8382
69	8388	8395	8401	8407	8414	8420	8426	8432	8439	8445
70	8451	8457	8463	8470	8476	8482	8488	8494	8500	8506
71	8513	8519	8525	8531	8537	8543	8549	8555	8561	8567
72	8573	8579	8585	8591	8597	8603	8609	8615	8621	8627
73	8633	8639	8645	8651	8657	8663	8669	8675	8681	8686
74	8692	8698	8704	8710	8716	8722	8727	8733	8739	8745
75	8751	8756	8762	8768	8774	8779	8785	8791	8797	8802
76	8808	8814	8820	8825	8831	8837	8842	8848	8854	8859
77	8865	8871	8876	8882	8887	8893	8899	8904	8910	8915
78	8921	8927	8932	8938	8943	8949	8954	8960	8965	8971
79	8976	8982	8987	8993	8998	9004	9009	9015	9020	9025
80	9031	9036	9042	9047	9053	9058	9063	9069	9074	9079
81	9085	9090	9096	9101	9106	9112	9117	9122	9128	9133
82	9138	9143	9149	9154	9159	9165	9170	9175	9180	9186
83	9191	9196	9201	9206	9212	9217	9222	9227	9232	9238
84	9243	9248	9253	9258	9263	9269	9274	9279	9284	9289
85	9294	9299	9304	9309	9315	9320	9325	9330	9335	9340
86	9345	9350	9355	9360	9365	9370	9375	9380	9385	9390
87	9395	9400	9405	9410	9415	9420	9425	9430	9435	9440
88	9445	9450	9455	9460	9465	9469	9474	9479	9484	9489
89	9494	9499	9504	9509	9513	9518	9523	9528	9533	9538
90	9542	9547	9552	9557	9562	9566	9571	9576	9581	9586
91	9590	9595	9600	9605	9609	9614	9619	9624	9628	9633
92	9638	9643	9647	9652	9657	9661	9666	9671	9675	9680
93	9685	9689	9694	9699	9703	9708	9713	9717	9722	9727
94	9731	9736	9741	9745	9750	9754	9759	9763	9768	9773
95	9777	9782	9786	9791	9795	9800	9805	9809	9814	9818
96	9823	9827	9832	9836	9841	9845	9850	9854	9859	9863
97	9868	9872	9877	9881	9886	9890	9894	9899	9903	9908
98	9912	9917	9921	9926	9930	9934	9939	9943	9948	9952
99	9956	9961	9965	9969	9974	9978	9983	9987	9991	9996

INSTRUCTIONS FOR BUILDING A TETRAHEDRON AND AN OCTAHEDRON

Tetrahedron: To construct a tetrahedron, trace out the following figure composed of four equilateral triangles on a piece of light cardboard:

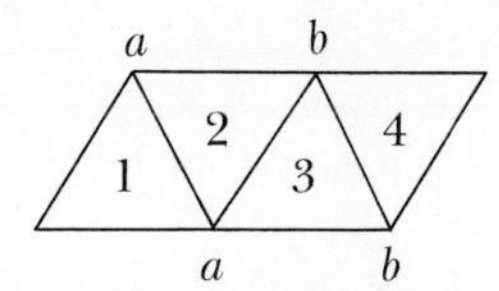

Bend face 1 upward about the line *aa*, and bend face 4 upward about the line *bb*:

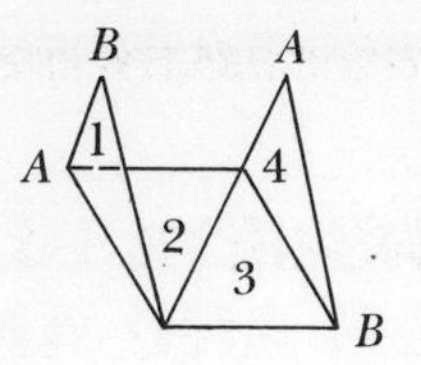

Now bend face 3 upward about the line between faces 2 and 3 and connect points *A* to *A* and *B* to *B* to get

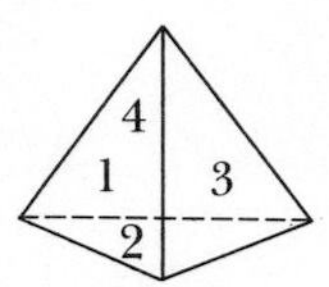

Octahedron: To construct an octahedron, trace out the following figure composed of eight equilateral triangles on a piece of light cardboard:

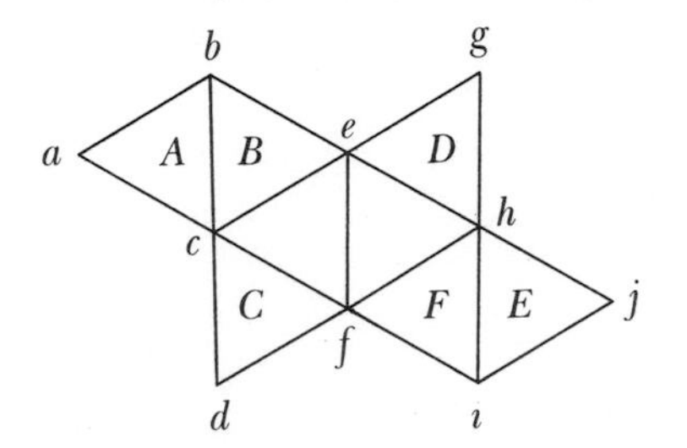

Bend face *A* upward about the line *bc* and face *C* upward about the line *cf*. Then bend face *B* upward about the line *ce* and join points *a* and *d*. Lines *ac* and *cd* are now aligned and should be taped together. Now bend face *D* upward about the line *eh* and face *E* upward about the line *hi*. Then bend face *F* upward about the line *fh* and join points *g* and *j*. Lines *gh* and *hj* are now aligned and should be taped in place. Last, bend both sides about the line *ef* and tape the resulting octahedron together to get

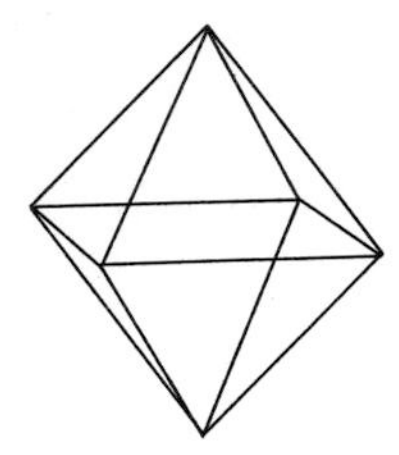

TABLE OF ATOMIC MASSES

Element	Symbol	Atomic number	Atomic mass[a]
actinium	Ac	89	(227)
aluminum	Al	13	26.98154
americium	Am	95	(243)
antimony	Sb	51	121.75
argon	Ar	18	39.948
arsenic	As	33	74.9216
astatine	At	85	(210)

Element	Symbol	Atomic number	Atomic mass[a]
barium	Ba	56	137.33
berkelium	Bk	97	(247)
beryllium	Be	4	9.01218
bismuth	Bi	83	208.9804
boron	B	5	10.81
bromine	Br	35	79.904
cadmium	Cd	48	112.41
calcium	Ca	20	40.08
californium	Cf	98	(251)
carbon	C	6	12.011
cerium	Ce	58	140.12
cesium	Cs	55	132.9054
chlorine	Cl	17	35.453
chromium	Cr	24	51.996
cobalt	Co	27	58.9332
copper	Cu	29	63.546
curium	Cm	96	(247)
dysprosium	Dy	66	162.50
einsteinium	Es	99	(252)
erbium	Er	68	167.26
europium	Eu	63	151.96
fermium	Fm	100	(257)
fluorine	F	9	18.998403
francium	Fr	87	(223)
gadolinium	Gd	64	157.25
gallium	Ga	31	69.72
germanium	Ge	32	72.59
gold	Au	79	196.9665
hafnium	Hf	72	178.49
helium	He	2	4.00260
holmium	Ho	67	164.9304
hydrogen	H	1	1.0079
indium	In	49	114.82
iodine	I	53	126.9045
iridium	Ir	77	192.22
iron	Fe	26	55.847
krypton	Kr	36	83.80
lanthanum	La	57	138.9055
lawrencium	Lr	103	(260)
lead	Pb	82	207.2
lithium	Li	3	6.941
lutetium	Lu	71	174.967
magnesium	Mg	12	24.305

[a]A value given in parentheses denotes the mass number of the longest-lived isotope.

Element	Symbol	Atomic number	Atomic mass[a]
manganese	Mn	25	54.9380
mendelevium	Md	101	(258)
mercury	Hg	80	200.59
molybdenum	Mo	42	95.94
neodymium	Nd	60	144.24
neon	Ne	10	20.179
neptunium	Np	93	(237)
nickel	Ni	28	58.70
niobium	Nb	41	92.9064
nitrogen	N	7	14.0067
nobelium	No	102	(259)
osmium	Os	76	190.2
oxygen	O	8	15.9994
palladium	Pd	46	106.4
phosphorus	P	15	30.97376
platinum	Pt	78	195.09
plutonium	Pu	94	(244)
polonium	Po	84	(209)
potassium	K	19	39.0983
praseodymium	Pr	59	140.9077
promethium	Pm	61	(145)
protactinium	Pa	91	(231)
radium	Ra	88	226.0254
radon	Rn	86	(222)
rhenium	Re	75	186.207
rhodium	Rh	45	102.9055
rubidium	Rb	37	85.4678
ruthenium	Ru	44	101.07
samarium	Sm	62	150.4
scandium	Sc	21	44.9559
selenium	Se	34	78.96
silicon	Si	14	28.0855
silver	Ag	47	107.868
sodium	Na	11	22.98977
strontium	Sr	38	87.62
sulfur	S	16	32.06
tantalum	Ta	73	180.9479
technetium	Tc	43	(98)
tellurium	Te	52	127.60
terbium	Tb	65	158.9254
thallium	Tl	81	204.37
thorium	Th	90	232.0381
thulium	Tm	69	168.9342

[a]A value given in parentheses denotes the mass number of the longest-lived isotope.

Element	Symbol	Atomic number	Atomic mass[a]
tin	Sn	50	118.69
titanium	Ti	22	47.90
tungsten	W	74	183.85
uranium	U	92	238.029
vanadium	V	23	50.9415
xenon	Xe	54	131.30
ytterbium	Yb	70	173.04
yttrium	Y	39	88.9059
zinc	Zn	30	65.38
zirconium	Zr	40	91.22

[a]A value given in parentheses denotes the mass number of the longest-lived isotope.

ANSWERS TO ODD-NUMBERED PROBLEMS*

CHAPTER 1

1-1. (a) Se (b) In (c) Mn (d) Tm (e) Hg (f) Kr (g) Pd (h) Tl (i) U (j) W

1-3. (a) germanium (b) scandium (c) iridium (d) cesium (e) strontium (f) americium (g) molybdenum (h) indium (i) plutonium (j) xenon

1-5. 59.0% Na, 41.0% O

1-7. 80.0% Cu, 20% S

*Interchapter answers follow those for Chapter 24.

1-9. 60.0% K, 18.4% C, 21.5% N

1-11. (a) lithium sulfide (b) barium oxide (c) magnesium phosphide (d) cesium bromide

1-13. (a) calcium carbide (b) gallium phosphide (c) aluminum oxide (d) beryllium chloride

1-15. (a) chlorine trifluoride, chlorine pentafluoride (b) sulfur tetrafluoride, sulfur hexafluoride (c) krypton difluoride, krypton tetrafluoride (d) bromine oxide, bromine dioxide

1-17. (a) 79.90 (b) 159.70 (c) 181.88 (d) 283.88

1-19. (a) 121.93 (b) 197.9 (c) 120.91 (d) 389.91

1-21. 45.68% Br, 54.32% F

1-23. 42.10% C, 6.479% H, 51.42% O

1-25. 1.267 g

1-27. (a) 53, 53, 78 (b) 27, 27, 33 (c) 19, 19, 24 (d) 49, 49, 64

1-29. $^{14}_{6}C$ (6, 8, 14), $^{241}_{95}Am$ (95, 146, 241), $^{123}_{53}I$ (53, 70, 123), $^{18}_{8}O$ (8, 10, 18)

1-31. $^{67}_{31}Ga$ (31, 36, 67), $^{15}_{7}N$ (7, 8, 15), $^{58}_{27}Co$ (27, 31, 58), $^{133}_{54}Xe$ (54, 79, 133)

1-33. 1.0080

1-35. 20.18

1-37. 49.35% $^{81}_{35}Br$

1-39. 0.36%

1-41. (a) 54 (b) 54 (c) 36 (d) 10

1-43. (a) 54 (b) 78 (c) 24 (d) 18

1-45. (a) Ca^{2+}, Cl^-, S^{2-} (b) Rb^+, Sr^{2+}, Br^-, Se^{2-} (c) Na^+, Mg^{2+}, F^-, O^{2-} (d) Cs^+, Ba^{2+}, Te^{2-}

1-47. (a) 3 (b) 4 (c) 4 (d) 3 (e) exact

1-49. (a) 18.0152 (b) 95.211 (c) 407.6950 (d) 258

1-51. (a) 2 (b) 2.08×10^4 (c) 2.8 (d) 3.4×10^{22}

1-53. (a) 1.06 quarts (b) 2.99×10^8 $m \cdot s^{-1}$ (c) 1.987 $cal \cdot K^{-1} \cdot mol^{-1}$

1-55. 9.46×10^{15} m, 5.88×10^{12} miles

1-57. the two-liter bottle

1-59. 3000 m

1-61. (c) 82.27%

1-63. 3.23×10^4 g

1-65. 63 mL

1-67. 4.67% $^{29}_{14}Si$, 3.10% $^{30}_{14}Si$

1-69. 172 mL

CHAPTER 2

2-1. (a) $2P(s) + 3Br_2(l) \rightarrow 2PBr_3(l)$
(b) $2H_2O_2(l) \rightarrow 2H_2O(l) + O_2(g)$
(c) $4CoO(s) + O_2(g) \rightarrow 2Co_2O_3(s)$
(d) $PCl_5(s) + 4H_2O(l) \rightarrow H_3PO_4(l) + 5HCl(g)$

2-3. (a) $CaH_2(s) + 2H_2O(l) \rightarrow Ca(OH)_2(aq) + 2H_2(g)$
(b) $CaCO_3(s) + 2HCl(aq) \rightarrow CaCl_2(aq) + CO_2(g) + H_2O(l)$
(c) $C_6H_{12}O_2(aq) + 8O_2(g) \rightarrow 6CO_2(g) + 6H_2O(l)$
(d) $2Li(s) + 2CO_2(g) + 2H_2O(g) \rightarrow 2LiHCO_3(s) + H_2(g)$

2-5. (a) $NaH(s) + H_2O(l) \rightarrow NaOH(aq) + H_2(g)$
sodium hydride; water; sodium hydroxide; hydrogen
(b) $2SO_2(g) + O_2(g) \rightarrow 2SO_3(g)$
sulfur dioxide; oxygen; sulfur trioxide
(c) $H_2S(g) + 2LiOH(aq) \rightarrow Li_2S(aq) + 2H_2O(l)$
hydrogen sulfide; lithium hydroxide; lithium sulfide; water
(d) $ZnO(s) + CO(g) \rightarrow Zn(s) + CO_2(g)$
zinc oxide; carbon monoxide; zinc; carbon dioxide

2-7. (a) $2Na(s) + I_2(s) \rightarrow 2NaI(s)$
sodium iodide
(b) $Sr(s) + H_2(g) \rightarrow SrH_2(s)$
strontium hydride
(c) $3Ca(s) + N_2(g) \rightarrow Ca_3N_2(s)$
calcium nitride
(d) $2Mg(s) + O_2(g) \rightarrow 2MgO(s)$
magnesium oxide

2-9. (a) solid (b) NaAt (c) white (d) At_2 (e) black

2-11. Tl, main-group (5) metal; Eu, inner transition metal; Xe, main-group (8) nonmetal; Hf, transition metal; Ru, transition metal; Am, inner transition metal; B, main-group (3) semimetal

2-13. (a) $2Ra(s) + O_2(g) \rightarrow 2RaO(s)$
(b) $Ra(s) + Cl_2(g) \rightarrow RaCl_2(s)$
(c) $Ra(s) + 2HCl(g) \rightarrow RaCl_2(s) + H_2(g)$
(d) $Ra(s) + H_2(g) \rightarrow RaH_2(s)$
(e) $Ra(s) + S(s) \rightarrow RaS(s)$

2-15. (a) Xe (c) Ar (d) Ar (e) Ne

2-17. (a) Mg^{2+}, S^{2-}; magnesium sulfide (b) Al^{3+}, P^{3-}; aluminum phosphide (c) Ba^{2+}, F^-; barium fluoride (d) Ga^{3+}, O^{2-}; gallium oxide

2-19. (a) Ga_2S_3 (b) AlP (c) KI (d) SrF_2

2-21. (a) Li_3N (b) Ga_2Te_3 (c) Ba_3N_2 (d) $MgBr_2$

2-23. (a) Fe_2O_3 (b) CdS (c) RuF_3 (d) Tl_2S

2-25. (a) calcium cyanide (b) silver perchlorate (c) potassium permanganate (d) strontium chromate

2-27. (a) ammonium sulfate (b) ammonium phosphate (c) calcium phosphate (d) potassium phosphate

2-29. (a) $Na_2S_2O_3$ (b) $KHCO_3$ (c) NaClO (d) $CaSO_3$

2-31. (a) Na_2SO_3 (b) K_3PO_4 (c) Ag_2SO_4 (d) NH_4NO_3

2-33. (a) mercury(I) chloride (b) chromium(III) nitrate (c) cobalt(II) bromide (d) copper(II) carbonate

2-35. (a) Cr_2O_3 (b) $Sn(OH)_2$ (c) $Cu(C_2H_3O_2)_2$ (d) $Co_2(SO_4)_3$

2-37. (a) decomposition (b) combination (c) single-replacement (d) double-replacement

2-39. (a) decomposition, equation is balanced (b) combination, $4Fe(s) + 3O_2(g) \rightarrow 2Fe_2O_3(s)$
(c) single-replacement, $2Al(s) + Mn_2O_3(s) \rightarrow 2Mn(s) + Al_2O_3(s)$
(d) double-replacement,
$2AgNO_3(aq) + H_2SO_4(aq) \rightarrow Ag_2SO_4(s) + 2HNO_3(aq)$
(e) double-replacement, $Ca(OH)_2(aq) + 2HBr(aq) \rightarrow CaBr_2(aq) + 2H_2O(l)$
(f) single-replacement, $Cd(s) + 2HCl(aq) \rightarrow CdCl_2(aq) + H_2(g)$

2-41. (a) $3Mg(s) + N_2(g) \rightarrow Mg_3N_2(s)$
(b) $H_2(g) + S(s) \rightarrow H_2S(g)$
(c) $2K(s) + Br_2(l) \rightarrow 2KBr(s)$
(d) $4Al(s) + 3O_2(g) \rightarrow 2Al_2O_3(s)$
(e) $MgO(s) + SO_2(g) \rightarrow MgSO_3(s)$

2-43. (a) $Zn(s) + 2HBr(aq) \rightarrow ZnBr_2(aq) + H_2(g)$
(b) $2Al(s) + Fe_2O_3(s) \rightarrow 2Fe(s) + Al_2O_3(s)$
(c) $Pb(s) + Cu(NO_3)_2(aq) \rightarrow Cu(s) + Pb(NO_3)_2(aq)$
(d) $Br_2(l) + 2NaI(aq) \rightarrow 2NaBr(aq) + I_2(s)$

2-45. (a) calcium is oxidized, chlorine is reduced (b) aluminum is oxidized, oxygen is reduced (c) rubidium is oxidized, bromine is reduced (d) sodium is oxidized, sulfur is reduced

2-47. (a) 2 (b) 6 (c) 1 (d) 2

2-49. (a) $2H^+(aq) + S^{2-}(aq) \rightarrow H_2S(g)$
(b) $Pb^{2+}(aq) + S^{2-}(aq) \rightarrow PbS(s)$
(c) $H^+(aq) + OH^-(aq) \rightarrow H_2O(l)$
(d) $OH^-(aq) + H^+(aq) \rightarrow H_2O(l)$
(e) $NH_3(aq) + H^+(aq) \rightarrow NH_4^+(aq)$

2-51. (a) $Fe(NO_3)_3(aq) + 3NaOH(aq) \rightarrow Fe(OH)_3(s) + 3NaNO_3(aq)$
$Fe^{3+}(aq) + 3OH^-(aq) \rightarrow Fe(OH)_3(s)$
(b) $Zn(ClO_4)_2(aq) + K_2S(aq) \rightarrow ZnS(s) + 2KClO_4(aq)$
$Zn^{2+}(aq) + S^{2-}(aq) \rightarrow ZnS(s)$
(c) $Pb(NO_3)_2(aq) + 2KOH(aq) \rightarrow Pb(OH)_2(s) + 2KNO_3(aq)$
$Pb^{2+}(aq) + 2OH^-(aq) \rightarrow Pb(OH)_2(s)$
(d) $Zn(NO_3)_2(aq) + Na_2CO_3(aq) \rightarrow ZnCO_3(s) + 2NaNO_3(aq)$
$Zn^{2+}(aq) + CO_3^{2-}(aq) \rightarrow ZnCO_3(s)$
(e) $Cu(ClO_4)_2(aq) + Na_2CO_3(aq) \rightarrow CuCO_3(s) + 2NaClO_4(aq)$
$Cu^{2+}(aq) + CO_3^{2-}(aq) \rightarrow CuCO_3(s)$

2-53. (a) acidic (b) acidic (c) basic (d) acidic (e) basic

2-55. (a) $2HClO_3(aq) + Ba(OH)_2(aq) \rightarrow Ba(ClO_3)_2(aq) + 2H_2O(l)$
barium chlorate
(b) $HC_2H_3O_2(aq) + KOH(aq) \rightarrow KC_2H_3O_2(aq) + H_2O(l)$
potassium acetate
(c) $2HI(aq) + Mg(OH)_2(s) \rightarrow MgI_2(aq) + 2H_2O(l)$
magnesium iodide
(d) $H_2SO_4(aq) + 2RbOH(aq) \rightarrow Rb_2SO_4(aq) + 2H_2O(l)$
rubidium sulfate

2-57. (a) $HCl(aq) + KCN(aq) \rightarrow HCN(g) + KCl(aq)$
(b) $2K(s) + 2H_2O(l) \rightarrow H_2(g) + 2KOH(aq)$

(c) $2H_2O_2(aq) \rightarrow O_2(g) + 2H_2O(l)$
(d) $H_2(g) + Br_2(l) \rightarrow 2HBr(g)$

2-59. $2Pb(l) + O_2(g) \rightarrow 2PbO(s)$
$Ag(l) + O_2(g) \rightarrow$ N.R.

2-61. (a) $2Na(s) + S(s) \rightarrow Na_2S(s)$
(b) $Ca(s) + Br_2(l) \rightarrow CaBr_2(s)$
(c) $2Ba(s) + O_2(g) \rightarrow 2BaO(s)$
(d) $2SO_2(g) + O_2(g) \rightarrow 2SO_3(g)$
(e) $3Mg(s) + N_2(g) \rightarrow Mg_3N_2(s)$

2-63. $HgS(s) + O_2(g) \xrightarrow{\text{heat}} Hg(g) + SO_2(g) \xrightarrow{\text{cool}} Hg(l) + SO_2(g)$

2-65. Na, Fe, Sn, Au

2-67. (a) $2Na(s) + H_2(g) \rightarrow 2NaH(s)$
(b) $2Al(s) + 3S(s) \rightarrow Al_2S_3(s)$
(c) $H_2O(g) + C(s) \rightarrow CO(g) + H_2(g)$
(d) $C(s) + 2H_2(g) \rightarrow CH_4(g)$
(e) $PCl_3(l) + Cl_2(g) \rightarrow PCl_5(s)$

2-69. $2CuO(s) + C(s) \xrightarrow{\text{high T}} 2Cu(s) + CO_2(g)$

$SnO_2(s) + C(s) \xrightarrow{\text{high T}} Sn(s) + CO_2(g)$

$2Fe_2O_3(s) + 3C(s) \xrightarrow{\text{high T}} 4Fe(s) + 3CO_2(g)$

CHAPTER 3

3-1. (a) 1.55 mol (b) 0.0167 mol (c) 7.77 mol (d) 1.78×10^4 mol

3-3. (a) 3.03 mol (b) 0.219 mol (c) 3.20×10^{-4} mol
(d) 9.368×10^{-3} mol

3-5. 8.55×10^{25} g, 1.4×10^{-2} of earth mass

3-7. (a) 7.308×10^{-23} g (b) 2.992×10^{-22} g (c) 1.843×10^{-22} g

3-9. (a) 1.855×10^{-20} g (b) 3.0×10^{-7} g (c) 2.7×10^{-17} g
(d) 5.3×10^{-17} g

3-11. 2.78 mol, 1.671×10^{24} molecules, 5.01×10^{24} atoms

3-13. (a) 3.54×10^{24} molecules (b) 3.52×10^{23} molecules
(c) 2.46×10^{24} molecules (d) 6.74×10^{19} molecules

3-15. CaC_2

3-17. $CuCl_2$

3-19. $CoCl_2$

3-21. CCl_2F_2

3-23. (a) Li_2O (b) Li_3N (c) LiN_3 (d) $CaCl_2$

3-25. 144.3, Nd

3-27. atomic mass 137, Ba

3-29. $C_3H_4O_3$

3-31. $C_4H_{10}O$

3-33. C_5H_5N

3-35. C_3H_6O

3-37. $Na_6P_6O_{18}$

3-39. 36.3 g

3-41. 81.6 g

3-43. 599 mg

3-45. 21.7 g

3-47. 1.34×10^5 kg

3-49. 1.58×10^5 kg

3-51. 67.80 kg of KNO_3, 23.80 kg of Cl_2

3-53. 29.0 g

3-55. 46.3%

3-57. 63.0%

3-59. 0.0250 M

3-61. 14.3 M

3-63. Dissolve 42.8 g of sucrose in about 250 mL of water in a 500-mL volumetric flask and then dilute the solution to the 500-mL mark.

3-65. (a) 3.195×10^{-3} mol (b) 1.0×10^{-6} mol

3-67. Add 25 mL of 1.0 M $NaH_2PO_4(aq)$ to a 500-mL volumetric flask half-filled with water, swirl the solution, and dilute with water to the 500-mL mark on the flask.

3-69. 8.3 mL

3-71. 39.0 mL

3-73. 1060 g

3-75. 0.171 M

3-77. (a) 2.8 μL (b) 8.33 mL

3-79. 0.114 M

3-81. 52.1%

3-83. 60.1

3-85. 1280 g Na_2CS_3, 442 g Na_2CO_3, 225 g H_2O

3-87. (a) 14.8 g (b) 66.4%

3-89. 10.0%

3-91. 71.4%

3-93. 13.2 M

3-95. 2.77 g

CHAPTER 4

4-1. (a) 5.7×10^4 torr, 76 bar (b) 0.763 atm, 773 mbar (c) 5.3×10^5 Pa, 530 kPa (d) 1.23×10^5 Pa, 1.21 atm

4-3. 2.25 mL, 1.07 cm, 1.63 cm

4-5. 6.0 atm

4-7. (a) 310 K (b) 293 K (c) 14 K (d) 472 K

4-9. 20.1 mL

4-11. 50 L

4-13. 21 L

4-15. 1.7 atm, 18 mL

4-17. 2.7×10^{22} molecules

4-19. 4.1×10^{12} molecules

4-21. 780 K

4-23. 35.0 L, 50.3 L

4-25. 0.848 L

4-27. 1.75 g

4-29. The ratio of densities is 1630.

4-31. $121\ g{\cdot}mol^{-1}$

4-33. C_2H_4

4-35. C_2H_4

4-37. $P_{H_2} = 0.610$ atm, $P_{N_2} = 1.38$ atm

4-39. $P_{N_2} = 518$ torr, $P_{O_2} = 222$ torr

4-41. 5.1 L, 10 atm

4-43. $442\ m{\cdot}s^{-1}$

4-45. by a factor of $\sqrt{2}$

4-47. ${}^{238}_{92}UF_6 < {}^{235}_{92}UF_6 < NO_2 < CO_2 < O_2 < N_2 < H_2O$

4-49. 3.5×10^{-7} m, 1.4×10^{9} $collisions{\cdot}s^{-1}$

4-51. 2.2×10^{31} pm

4-53. 1.6×10^{10} $collisions{\cdot}s^{-1}$

4-55. $200\ mL{\cdot}h^{-1}$

4-57. 70.5

4-59. 15.0 atm (van der Waals), 16.5 atm (ideal)

4-61. 2.49×10^{19} molecules

4-63. 5.67 μL, 56.2 μg

4-65. $17.7\ g{\cdot}m^{-3}$

4-67. ——

4-69. A plot of V versus $1/P$ is a straight line and $PV = 0.244$ L·atm.

4-71. 1.2 L

4-73. 8600 gal

4-75. ——

4-77. 62.8%

4-79. $1.20\ g{\cdot}L^{-1}$, 28.9

4-81. 21.2 mL

4-83. 0.55 atm

4-85. 1.4 lb O_2

4-87. 1764 K

4-89. 337 $m \cdot s^{-1}$ (sound), 493 $m \cdot s^{-1}$ (molecules)

CHAPTER 5

5-1. 804 $kJ \cdot mol^{-1}$

5-3. 89.3 $kJ \cdot mol^{-1}$

5-5. 145.4 kJ

5-7. 88.6 kJ

5-9. +229.27 kJ

5-11. −4.2 kJ

5-13. 180 kJ

5-15. −106 kJ

5-17. (a) −534.2 kJ, exothermic (b) −44.2 kJ, exothermic (c) −429.8 kJ, exothermic

5-19. (a) −1366.7 kJ and −29.67 $kJ \cdot g^{-1}$ (b) −1559.7 kJ and −51.87 $kJ \cdot g^{-1}$

5-21. −1249.1 $kJ \cdot mol^{-1}$

5-23. (a) 472.6 $kJ \cdot mol^{-1}$ (b) 79.0 $kJ \cdot mol^{-1}$ (c) 218.0 $kJ \cdot mol^{-1}$ (d) 121.7 $kJ \cdot mol^{-1}$ N_2 has the greatest bond strength.

5-25. 32.4 kJ

5-27. 171 $kJ \cdot mol^{-1}$

5-29. −608 kJ

5-31. 502 $kJ \cdot mol^{-1}$

5-33. 414 $kJ \cdot mol^{-1}$

5-35. 113 $J \cdot K^{-1} \cdot mol^{-1}$

5-37. 3.34×10^6 J

5-39. 21.6°C

5-41. 353°C

5-43. −56.2 $kJ \cdot mol^{-1}$

5-45. -1.48×10^3 kJ

5-47. 17.2 $kJ \cdot mol^{-1}$

5-49. −46.3 $kJ \cdot g^{-1}$, −2040 $kJ \cdot mol^{-1}$

5-51. −246 $kJ \cdot mol^{-1}$, −827 $kJ \cdot mol^{-1}$

5-53. 150 kJ, 3.8 g, 310 g of ice

5-55. TlCl

5-57. ZnSe

5-59. 120 miles

5-61. (a) −565 kJ (b) −540 kJ (c) −180 kJ. In each case $\Delta H^\circ_{rxn} \simeq \Delta U^\circ_{rxn}$.

5-63. −1049 kJ

5-65. 49 kJ

CHAPTER 6

6-1. Kr, Ne, He, Be^+

6-3. two electrons in an inner shell and three in an outer shell

6-5. $\mathrm{Li}\cdot$, $\mathrm{Na}\cdot$, $\mathrm{K}\cdot$, $\mathrm{Rb}\cdot$, $\mathrm{Cs}\cdot$ and $:\ddot{\underset{\cdot\cdot}{\mathrm{F}}}\cdot$, $:\ddot{\underset{\cdot\cdot}{\mathrm{Cl}}}\cdot$, $:\ddot{\underset{\cdot\cdot}{\mathrm{Br}}}\cdot$, $:\ddot{\underset{\cdot\cdot}{\mathrm{I}}}\cdot$

All the alkali metal atoms have one outer electron and all the halogen atoms have seven outer electrons.

6-7. $:\ddot{\underset{\cdot\cdot}{\mathrm{Ar}}}:$, $\cdot\ddot{\underset{\cdot\cdot}{\mathrm{S}}}\cdot$, $:\ddot{\underset{\cdot\cdot}{\mathrm{S}}}:^{2-}$, Al^{3+}, $:\ddot{\underset{\cdot\cdot}{\mathrm{Cl}}}:^-$

6-9. $4.74 \times 10^{14}\ s^{-1}$

6-11. 286 nm

6-13. 6 photons

6-15. Electrons will be ejected.

6-17. 2.07×10^{-19} J

6-19. 3.97 pm

6-21. 99.0 pm

6-23. 3.03×10^{-19} J, 656 nm

6-25. $n_i = 4$, $n_f = 2$

6-27. He^+: 5.25 $MJ\cdot mol^{-1}$; Li^{2+}: 11.8 $MJ\cdot mol^{-1}$; Be^{3+}: 21.0 $MJ\cdot mol^{-1}$. These values are in excellent agreement with those in Table 6-1.

6-29. (a), (c), (e) are possible

6-31. (a) $4p$ (b) $3d$ (c) $4d$ (d) $2s$

6-33. n must be at least 3; l must be at least 3

6-35. $n = 3$; $l = 2$; $m_l = -2, -1, 0, +1, +2$ and $m_s = +\frac{1}{2}$ or $-\frac{1}{2}$ for each value of m_l

6-37. 2, 6, 10, 14

6-39. Because each of the five d orbitals can be occupied by a maximum of 2 electrons and $5 \times 2 = 10$.

6-41. (a), (b), (c) are ruled out.

6-43. (a) $1s^22s^22p^63s^23p^2$, silicon
(b) $1s^22s^22p^63s^23p^64s^13d^5$, chromium
(c) $1s^22s^22p^63s^23p^64s^23d^{10}4p^2$, germanium
(d) $1s^22s^22p^63s^23p^64s^23d^{10}4p^5$, bromine
(e) $1s^22s^22p^1$, boron

6-45. (a) $[Ar]4s^23d^2$ (b) $[Ar]4s^1$ (c) $[Ar]4s^23d^6$ (d) $[Ar]4s^23d^{10}4p^3$

6-47. (a) $[Ar]4s^2$ (b) $[Ar]4s^23d^{10}4p^5$ (c) $[Kr]5s^14d^{10}$ (d) $[Ar]4s^23d^{10}$

6-49. (a) Groups 1 and 2 (b) Groups 3, 4, 5, 6, 7, 8 (c) the transition metals (d) the lanthanides and the actinides

6-51. (a) 2 (b) 2 (c) 3 (d) 4

6-53. (a) $\cdot\mathrm{H}\cdot^-$, helium (b) $:\ddot{\underset{\cdot\cdot}{\mathrm{O}}}:^{2-}$, neon

(c) $:\ddot{\underset{..}{C}}:^{4-}$, neon (d) $:\ddot{\underset{..}{S}}:^{2-}$, argon

6-55. (a) [Ar] (b) [Kr] (c) [Kr] (d) [Xe] They all have a noble-gas electron configuration.

6-57. (a) 2 (b) 0 (c) 0 (d) 1

6-59. (a) $O(1s^22s^22p^4) + 2e^- \rightarrow O^{2-}(1s^22s^22p^6)$
(b) $Ca([Ar]4s^2) + Sr^{2+}([Kr]) \rightarrow Ca^{2+}([Ar]) + Sr([Kr])5s^2)$

6-61. (a) $1s^12s^1$ (b) $2s^1$ (c) $1s^22s^22p^53s^1$ (d) $1s^22s^22p^53s^1$

6-63. (a) P (b) P (c) S (d) Kr

6-65. (a) $Li < Na < Rb < Cs$ (b) $P < Al < Mg < Na$ (c) $Ca < Sr < Ba$

6-67. The core electrons partially screen the nuclear charge, and the farther the electrons from the nucleus, the lower is the ionization energy.

6-69. 1.89×10^{22} photons

6-71. The plot gives a straight line whose slope is Planck's constant (6.63×10^{-34} J·s). The intercept of the straight line with the horizontal axis gives the threshold frequency ($1.1 \times 10^{15}\ s^{-1}$).

6-73. 1.88×10^{-20} J, 5×10^{19} photons

6-75. 6.02×10^{-20} J

6-77. A plot of $\nu_{n\rightarrow 1}$ versus $1/n^2$ is a straight line.

6-79. (a) 20 kJ (b) −710 kJ (c) −5.14 MJ

6-81. (a) 8, 34, 52, 84 (b) 1, 3, 19, 37, 55, 87

6-83. (a) 0, 9, 3 (b) 6, 12, 4 (c) 10, 14, 2

6-85. The ground-state outer-electron configurations of the second-row elements would be $2s^1$, $2s^12p_x^1$, $2s^12p_x^12p_y^1$, $2s^12p_x^12p_y^12p_z^1$, $2s^22p_x^12p_y^12p_z^1$, $2s^22p_x^22p_y^12p_z^1$, $2s^22p_x^22p_y^22p_z^1$, and $2s^22p_x^22p_y^22p_z^2$.

CHAPTER 7

7-1. (a) $Ca([Ar]4s^2) + 2F([He]2s^22p^5) \rightarrow$
$Ca^{2+}([Ar]) + 2F^-([Ne]) \rightarrow CaF_2(g)$
(b) $Sr([Kr]5s^2) + 2Br([Ar]4s^23d^{10}4p^5) \rightarrow$
$Sr^{2+}([Kr]) + 2Br^-([Kr]) \rightarrow SrBr_2(g)$
(c) $2Al([Ne]3s^23p^1) + 3O([He]2s^22p^4) \rightarrow$
$2Al^{3+}([Ne]) + 3O^{2-}([Ne]) \rightarrow Al_2O_3(g)$

7-3. (a) $3Li\cdot + \cdot\ddot{\underset{\cdot}{N}}\cdot \rightarrow 3Li^+ + :\ddot{\underset{..}{N}}:^{3-}$

(b) $Na\cdot + H\cdot \rightarrow Na^+ + H:^-$

(c) $\cdot\dot{\underset{\cdot}{Al}}\cdot + 3:\ddot{\underset{.}{I}}: \rightarrow Al^{3+} + 3:\ddot{\underset{..}{I}}:^-$

7-5. (a) $Cr^{2+}([Ar]3d^4)$ (b) $Cu^{2+}([Ar]3d^9)$ (c) $Co^{3+}([Ar]3d^6)$
(d) $Mn^{2+}([Ar]3d^5)$

7-7. (a) Fe, Ru, Os (b) Zn, Cd, Hg (c) Sc, Y, Lu (d) Mn, Tc, Re

7-9. (a) $Cd^{2+}([Kr]4d^{10})$ (b) $In^{3+}([Kr]4d^{10})$ (c) $Tl^{3+}([Xe]4f^{14}5d^{10})$
(d) $Zn^{2+}([Ar]3d^{10})$

7-11. (b), (d), (e), (f)

7-13. (a) Y_2S_3 (b) $LaBr_3$ (c) MgTe (d) Rb_3N (e) Al_2Se_3 (f) CaO

7-15. (a) Cl^- (b) Ag^+ (c) Cu^+ (d) O^{2-}

7-17. Cl > Br > I > H

7-19. $-270\ \text{kJ·mol}^{-1}$

7-21. (a) $196\ \text{kJ·mol}^{-1}$ (b) $-53\ \text{kJ·mol}^{-1}$ (c) $428\ \text{kJ·mol}^{-1}$

7-23. $-4.32 \times 10^{-18}\ \text{J}$

7-25. $-329\ \text{kJ·mol}^{-1}$

7-27. $-130\ \text{kJ·mol}^{-1}$

7-29. $\text{Li}(g) + \text{F}(g) \rightarrow \text{Li}^+\text{F}^-(g)$, $-523\ \text{kJ·mol}^{-1}$; $\text{Li}(g) + \text{F}(g) \rightarrow \text{Li}^+(g) + \text{F}^-(g)$, $187\ \text{kJ·mol}^{-1}$; $\text{Li}^+(g) + \text{F}^-(g) \rightarrow \text{Li}^+\text{F}^-(g)$, $-710\ \text{kJ·mol}^{-1}$

7-31. $-581\ \text{kJ·mol}^{-1}$

7-33. $-299\ \text{kJ·mol}^{-1}$

7-35. $-795\ \text{kJ·mol}^{-1}$

7-37. N^{3-}, O^{2-}, Na^+, Mg^{2+}, Al^{3+}

7-39. The nuclear charge of Cu^+ is greater than that of K^+.

7-41. $-348\ \text{kJ·mol}^{-1}$

7-43. 4 in Fe^{2+}; 0 in Zn^{2+}

7-45. The magnitude of the electron affinity decreases and the ionic size of X^- increases.

7-47. Ti^+, Zr^+, Hf^+

7-49. (a) strong (b) weak (c) nonelectrolyte (d) nonelectrolyte

CHAPTER 8

8-1.

```
                                      ..
                                     :Cl:
      ..   ..   ..              ..    |    ..
(a) :Cl—S—Cl:            (b) :Cl—Ge—Cl:
      ..   ..   ..              ..    |    ..
                                     :Cl:
                                      ..

      ..   ..   ..               ..
(c) :Br—As—Br:           (d) H—P—H
      ..   |    ..               |
          :Br:                   H
           ..
```

8-3.

```
   ..  ..
H—O—O—H
   ..  ..
```

8-5.

```
        H               H
        |               |  ..
(a) H—C—H       (b) H—C—F:
        |               |  ..
        H               H

        H
        |  ..
(c) H—C—N—H
        |  |
        H  H
```

8-7.

```
        H  H  H              H  H  H  H
        |  |  |              |  |  |  |
(a) H—C—C—C—H       (b) H—C—C—C—C—H
        |  |  |              |  |  |  |
        H  H  H              H  H  H  H

        H  H  H  H  H  H  H  H
        |  |  |  |  |  |  |  |
(c) H—C—C—C—C—C—C—C—C—H
        |  |  |  |  |  |  |  |
        H  H  H  H  H  H  H  H
```

8-9. (a) 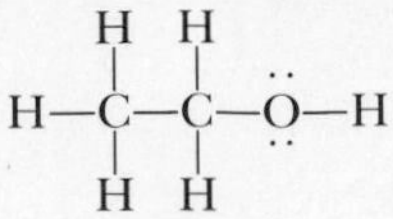 H—C(H)(H)—C(H)(H)—Ö—H (b) H—C(H)(H)—C(H)(H)—C(H)(H)—Ö—H

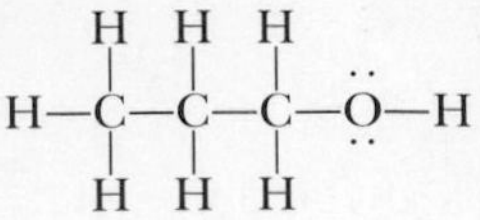

(c) H—C(H)(H)—C(H)(:Ö—H)—C(H)(H)—H

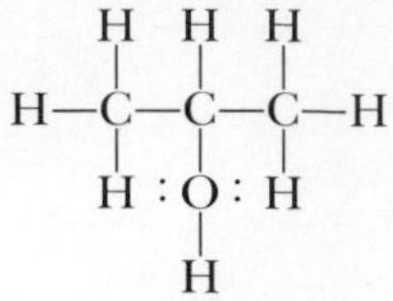

8-11. H—N̈⊖—H, sodium amide, barium amide

8-13. (a) H—C≡C—H (b) H—N̈=N̈—H

(c) 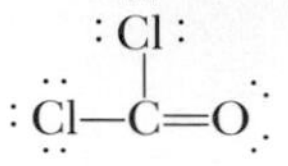 :C̈l—C(:C̈l:)=Ö:

8-15. H—C(=Ö:)—Ö—H

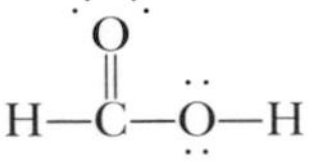

8-17. H—C(H)=C(H)—C̈l:

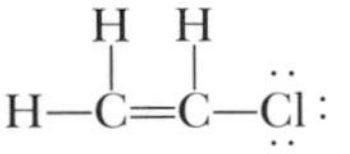

8-19. :F̈—N̈(:F̈:)—F̈:

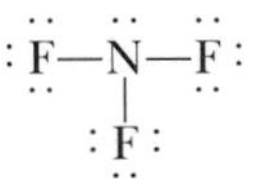

8-21. NNO

8-23. H—C(=Ö:)—Ö:⊖ ↔ H—C(:Ö:⊖)=Ö:

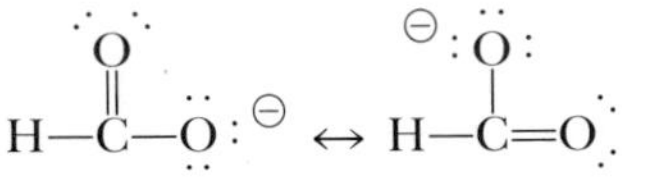

The two carbon-oxygen bonds have the same bond length and bond energy.

8-25. C(=Ö:)(⊖:Ö:)(:Ö:⊖) ↔ C(:Ö:⊖)(⊖:Ö:)(=Ö:) ↔ C(:Ö:⊖)(=Ö:)(:Ö:⊖)

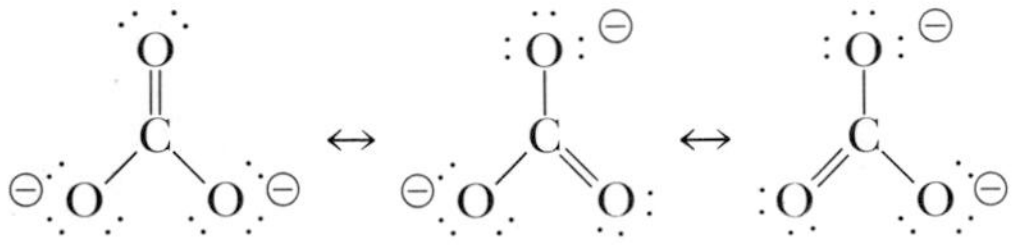

The three carbon-oxygen bonds have the same bond length and bond energy.

8-27. (a) 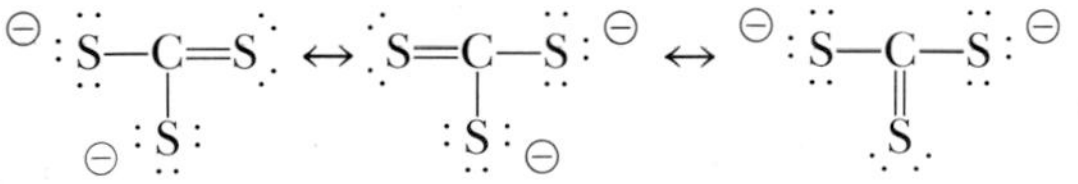 ⊖:S̈—C(:S̈:⊖)=S̈: ↔ :S̈=C(:S̈:⊖)—S̈:⊖ ↔ ⊖:S̈—C(=S̈:)—S̈:⊖

(b) :Ö=C(:Ö:)—C(:Ö:⊖)=Ö: ↔ ⊖:Ö—C(=Ö:)—C(:Ö:⊖)=Ö: ↔

:Ö=C(:Ö:⊖)—C(=Ö:)—Ö:⊖ ↔ ⊖:Ö—C(=Ö:)—C(=Ö:)—Ö:⊖

(c) :N≡C—S̈:⊖ ↔ ⊖:N̈=C=S̈: ↔ (2−):N̈—C≡S:⊕

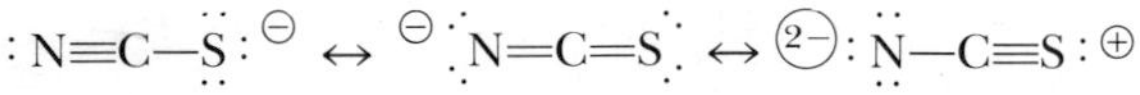

(d)  ⊖:Ö—Ö· ↔ ·Ö—Ö:⊖

8-29. (a) 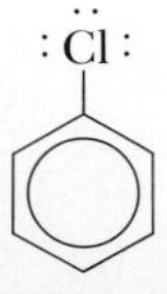 :C̈l: (on benzene ring) (b) 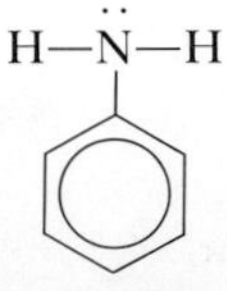 H—N̈—H (on benzene ring)

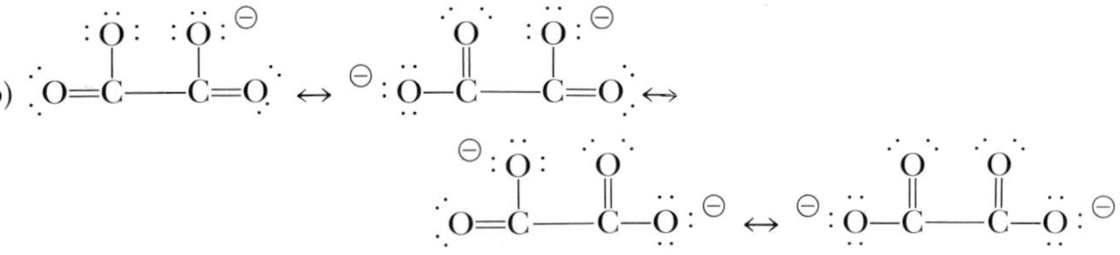

(c) 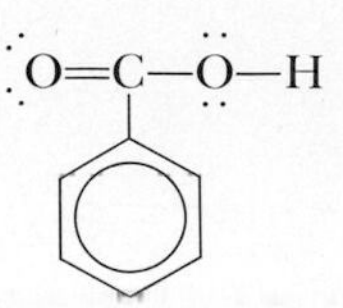 :O=C—O—H (d) 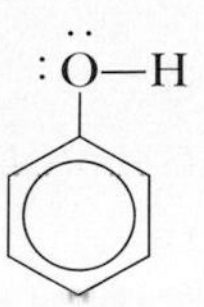 :O—H (each attached to a benzene ring)

8-31. (a) 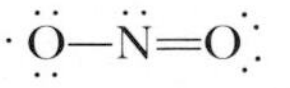 ·O—N=O: (c) 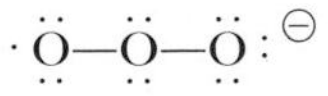 ·O—O—O: ⊖

(d) 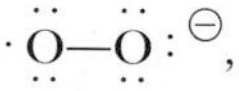 ·O—O: ⊖, plus other resonance forms in each case

8-33.

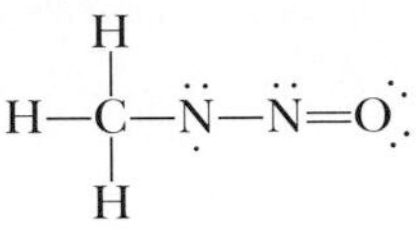

```
     H
     |
H—C—N—N=O:     (free radical)
     |
     H
```

8-35. (a)

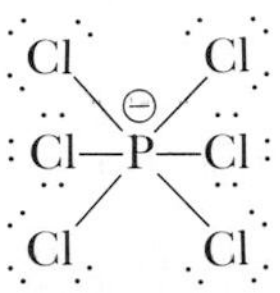

```
  :Cl:    :Cl:
      \ ⊖ /
:Cl—P—Cl:
      /   \
  :Cl:    :Cl:
```

(b) 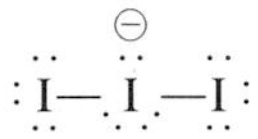 :I—I—I: (⊖ on central I)

(c)

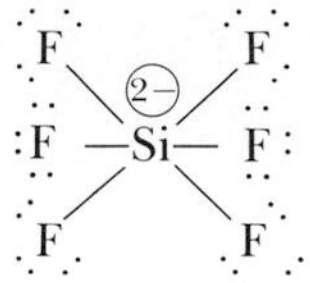

```
  :F:      :F:
      \ (2−) /
:F — Si — F:
      /     \
  :F:      :F:
```

8-37. :Br—Cl: (dipole arrow ⟶ pointing from Br toward Cl)

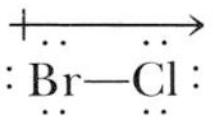

8-39. (a)

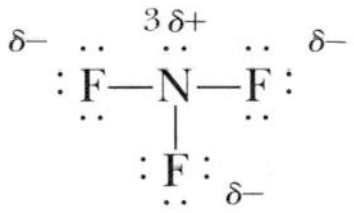

```
δ−      3δ+      δ−
:F — N — F:
         |
        :F:  δ−
```

(b)

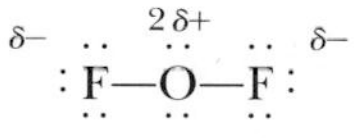

```
δ−     2δ+     δ−
:F — O — F:
```

(c)

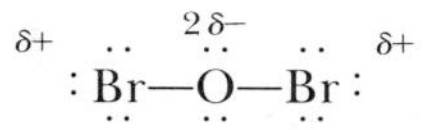

```
δ+      2δ−      δ+
:Br — O — Br:
```

8-41. (a)

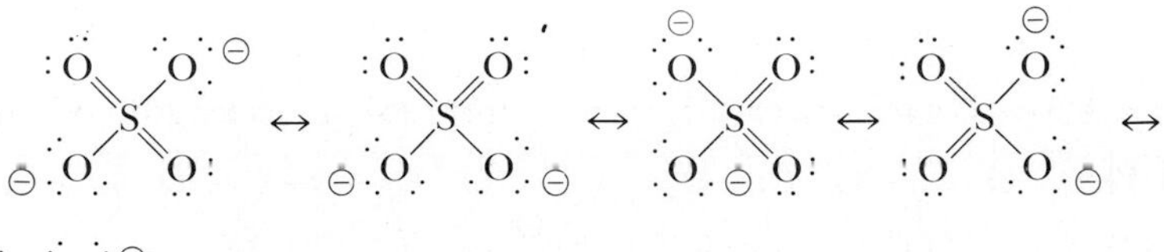

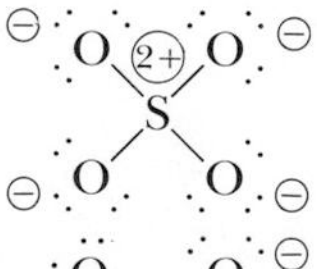

```
:O    O:⊖     :O    O:     ⊖O    O:     :O    O⊖
  \\ /          \\ //         \ //        \\ /
   S      ↔      S      ↔      S      ↔     S      ↔
  / \\          /  \          / \\         /  \
⊖O    O:     ⊖O    O⊖      O⊖   O:     :O    O⊖

⊖O    O⊖
   \(2+)/
    S
   /  \
⊖O    O⊖
```

All four bonds are equivalent.

(b)

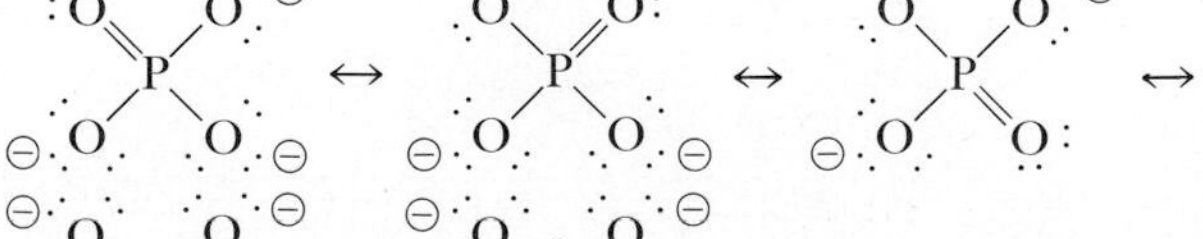

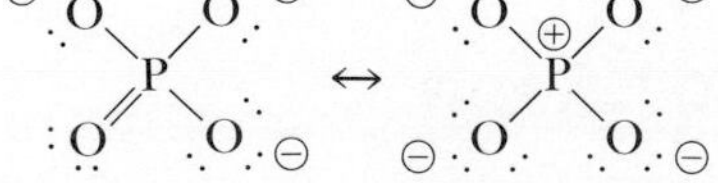

```
:O    O⊖       ⊖O    O:       ⊖O    O⊖
  \\ /           \ //           \ /
   P      ↔       P      ↔       P      ↔
  /  \           /  \           / \\
⊖O    O⊖      ⊖O    O⊖      ⊖O    O:

⊖O    O⊖       ⊖O    O⊖
   \ /            \(+)/
    P      ↔       P
  //  \           /  \
:O    O⊖      ⊖O    O⊖
```

All four bonds are equivalent.

(c)

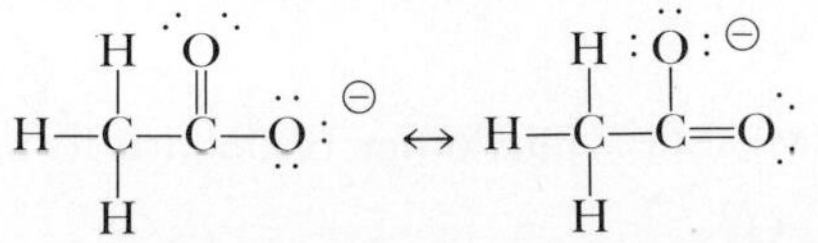

```
     H  :O:                    H  :O:⊖
     |   ||                    |   |
H—C—C—O:⊖       ↔     H—C—C=O:
     |                         |
     H                         H
```

The two carbon-oxygen bonds are equivalent.

8-43. (a) 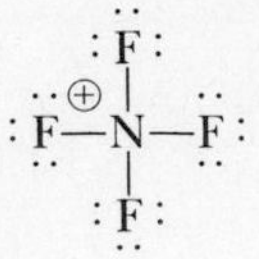 F—N—F with F above and below N; $N^{\oplus}$ (b) 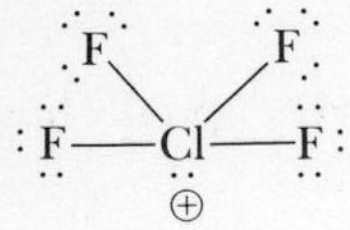 F—Cl—F with two F on Cl; $Cl^{\oplus}$

(c) 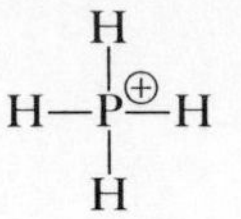 H—P—H with H above and below P; $P^{\oplus}$ (d) 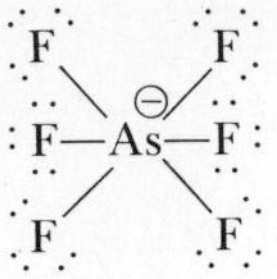 F—As—F with four F on As; $As^{\ominus}$

(e) 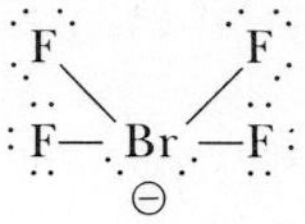 F—Br—F with two F on Br; $Br^{\ominus}$

8-45. (a) 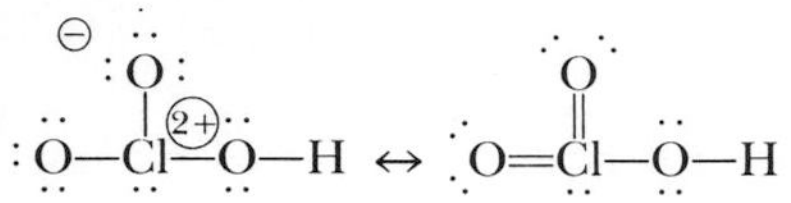 $^{\ominus}$O—Cl^{2+}—O—H with $^{\ominus}$O above Cl ↔ O=Cl—O—H with O above (=) Cl plus other resonance forms

(b) 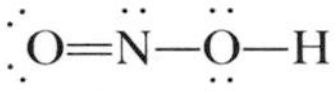 O=N—O—H

(c) 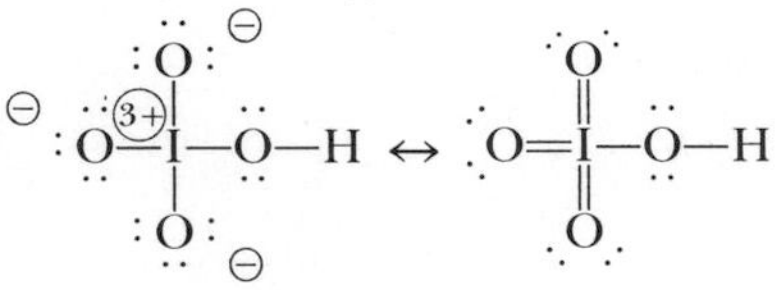 $^{\ominus}$O—I^{3+}—O—H with $^{\ominus}$O above and below I ↔ O=I—O—H with O above (=) and below (=) I plus other resonance forms

(d) 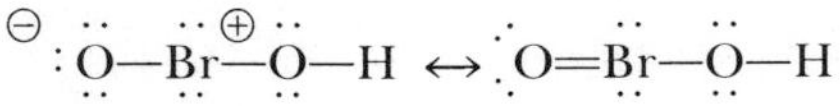 $^{\ominus}$O—Br$^{\oplus}$—O—H ↔ O=Br—O—H

8-47. (a) 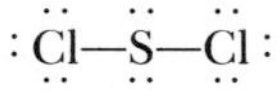 Cl—S—Cl (b) 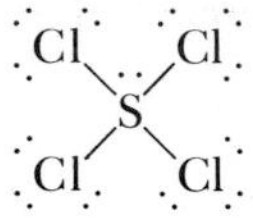 S with four Cl

(c) 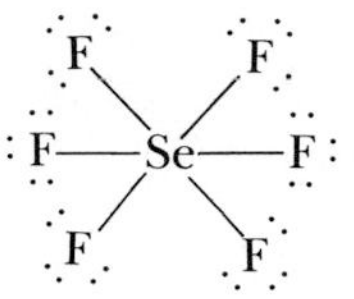 F—Se—F with four more F on Se (d) 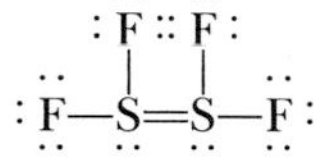 F—S=S—F with one F above each S

8-49. $^{\ominus}$O—Cr^{2+}—O—Cr^{2+}—O$^{\ominus}$ with $^{\ominus}$O above and below each Cr ↔ $^{\ominus}$O—Cr—O—Cr—O$^{\ominus}$ with O above (=) and below (=) each Cr

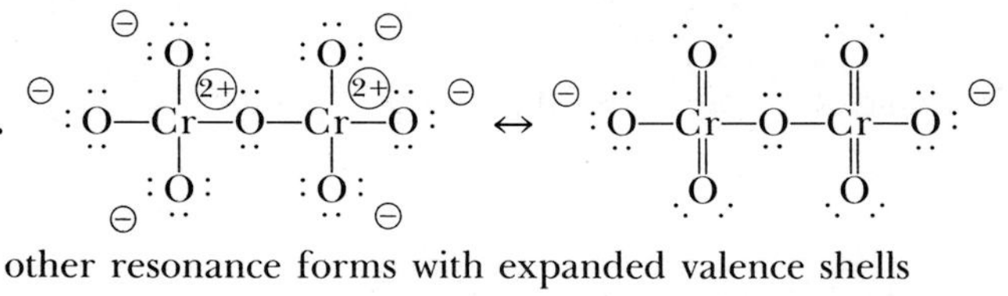

plus other resonance forms with expanded valence shells

8-51. (a) and (c)

8-53. Eight S atoms in a ring: S—S—S—S—S—S—S—S

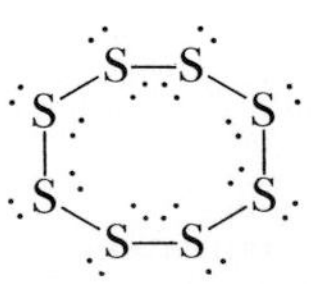

8-55. (a) 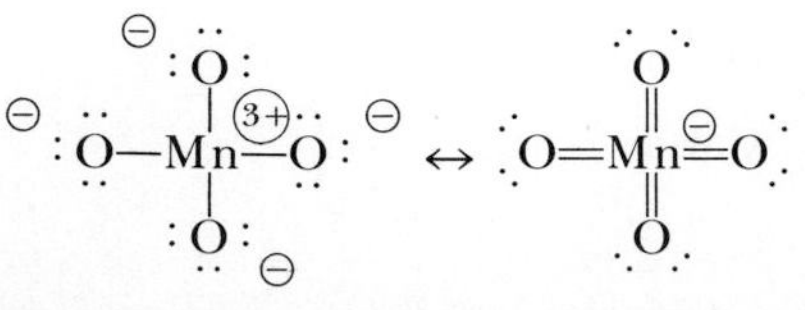 $^{\ominus}$O—Mn^{3+}—O$^{\ominus}$ with $^{\ominus}$O above and below Mn ↔ O=Mn=O with O above (=) and below (=) Mn; $Mn^{\ominus}$ plus other resonance forms

(b) 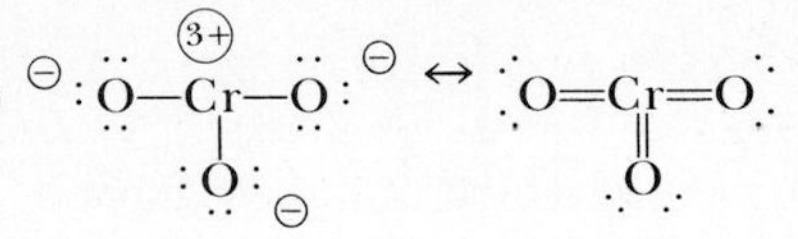 $^{\ominus}\!:\ddot{O}-\overset{(3+)}{Cr}(-\ddot{O}:^{\ominus})-\ddot{O}:^{\ominus} \leftrightarrow \; O{=}Cr({=}O){=}O$ plus other resonance forms

(c) 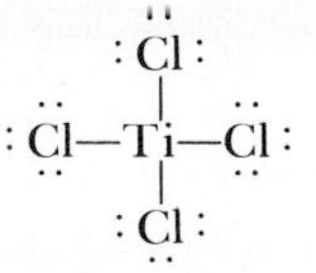 Cl_4Ti: :Cl—Ti(—Cl)(—Cl)—Cl: (d) 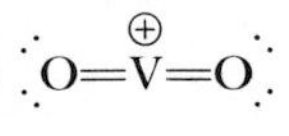 $O{=}\overset{\oplus}{V}{=}O$

8-57. The arrangement FNNF has the lower formal charges and so we predict that the structure of N_2F_2 is FNNF.

8-59. The arrangement H_2CO has the lowest formal charges and so we predict that the structure of formaldehyde is H_2CO.

CHAPTER 9

9-1. (a), (b), (d)

9-3. (c), (d)

9-5. (a) bent (b) bent (c) linear (d) bent

9-7. (c)

9-9. (b), (c)

9-11. 90°: (a), (b), (c); 120°: (b), (d)

9-13. 90°: (b), (c); 109.5°: (a), (d)

9-15. IF, linear; IF_3, BrF_3, ClF_3, T-shaped; IF_5, BrF_5, ClF_5, square pyramidal

9-17. (a) trigonal pyramidal, $:\ddot{Cl}-\overset{\oplus}{Se}(-\ddot{O}:^{\ominus})-\ddot{Cl}:$ + other resonance forms

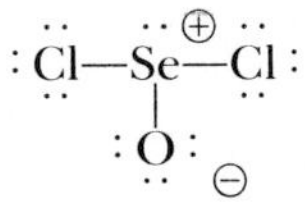

(b) tetrahedral, $^{\ominus}:\ddot{O}-\overset{(2+)}{S}(Cl)(Cl)-\ddot{O}:^{\ominus}$ + other resonance forms

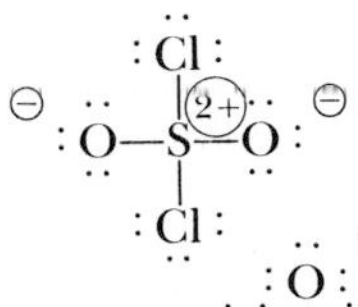

(c) trigonal bipyramidal, $\overset{\oplus}{S}F_4(-\ddot{O}:^{\ominus})$ + other resonance forms

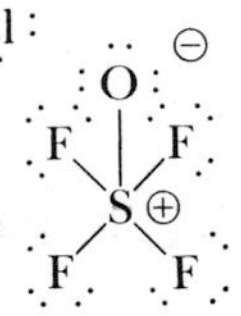

(d) tetrahedral, $^{\ominus}:\ddot{O}-\overset{(3+)}{Cl}(F)(-\ddot{O}:^{\ominus})-\ddot{O}:^{\ominus}$ + other resonance forms

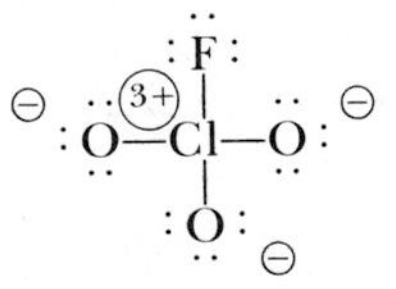

9-19. (a) trigonal planar, $Cl_2C{=}O$

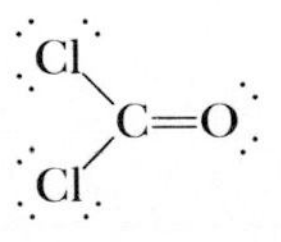

(b) tetrahedral, $:\ddot{F}-\overset{\oplus}{N}(-\ddot{S}:^{\ominus})(-\ddot{F}:)-\ddot{F}:$

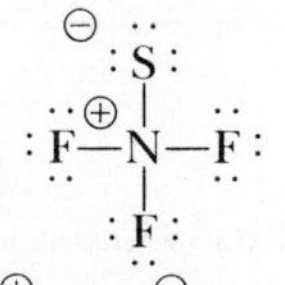

(c) linear, $\overset{\ominus}{N}{=}\overset{\oplus}{N}{=}\overset{\ominus}{N}$

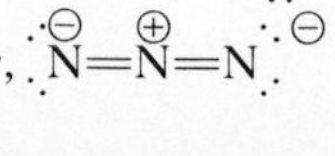

(d) bent, Cl—Sb=O

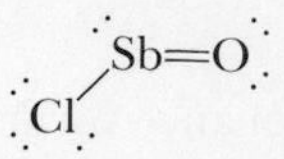

9-21. (a) bent (b) square pyramidal (c) tetrahedral (d) trigonal pyramidal

9-23. (a) octahedral, XeO_2F_4 (Xe 2+, each O 1−) + other resonance forms

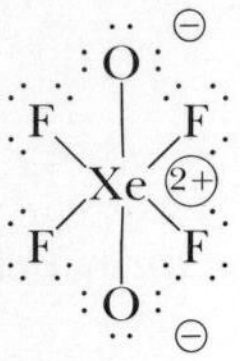

(b) trigonal bipyramidal, IO_2F_3 (I 2+, each O 1−) + other resonance forms

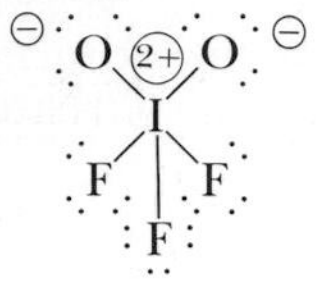

(c) trigonal pyramidal, O—I—O with F on I (I 2+, each O 1−) + other resonance forms

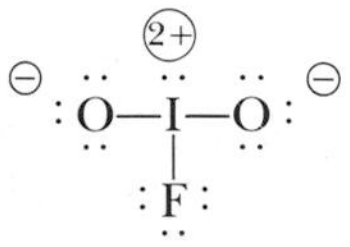

(d) tetrahedral, IO_3F (I 3+, each O 1−) + other resonance forms

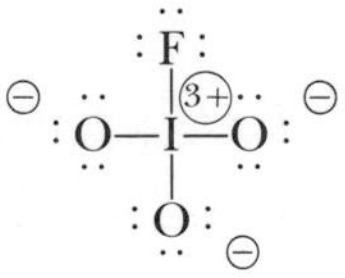

9-25. O=N=O (N ⊕) linear; ⊖O—N=O bent

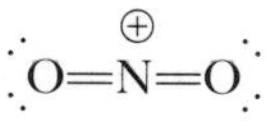

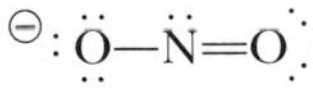

9-27. (a) 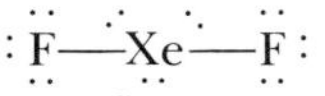 F—Xe—F linear (no dipole moment) (b) 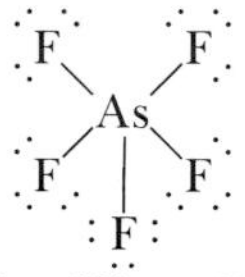 AsF_5 trigonal bipyramidal (no dipole moment)

(c) 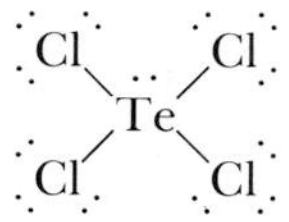 $TeCl_4$ seesaw (dipole moment) (d) 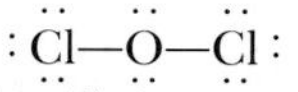 Cl—O—Cl bent (dipole moment)

9-29. (a) 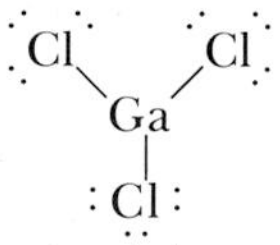 $GaCl_3$ trigonal planar (no dipole moment) (b) 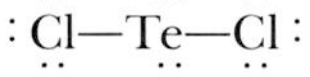 Cl—Te—Cl bent (dipole moment)

(c) 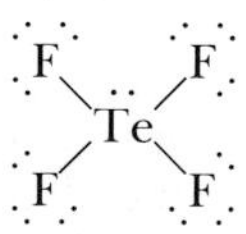 TeF_4 seesaw (dipole moment) (d) 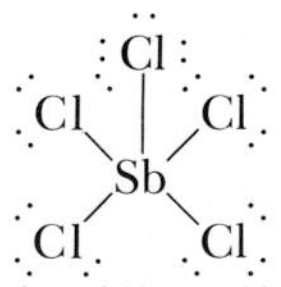 $SbCl_5$ trigonal bipyramidal (no dipole moment)

9-31. (b), (d)

9-33. (a) 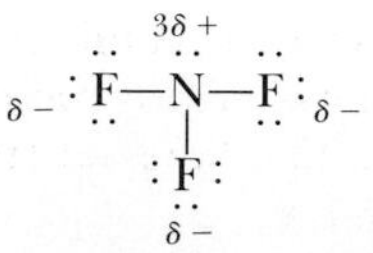 NF_3: N $3\delta+$, each F $\delta-$ (b) 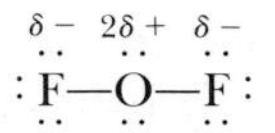 F—O—F: F $\delta-$, O $2\delta+$, F $\delta-$

(c) $\overset{\delta+}{:\ddot{\text{Br}}}-\overset{2\delta-}{\ddot{\text{O}}}-\overset{\delta+}{\ddot{\text{Br}}:}$

9-35. (a), (b), and (c); no isomers possible

(d) two isomers (*trans* and *cis*):

```
   Y                X
   |                |
X—A—X    and    Y—A—X
   |                |
   Y                Y
```

9-37. (a) only one isomer (b) two isomers, Ys at opposite vertices or at adjacent vertices (c) two isomers, 3 Ys on the vertices of a square and 3 Ys on a face

9-39. (a) 1 (b) 1 (c) 2 (d) 2

9-41. (a) seesaw (b) square planar (c) tetrahedral (d) tetrahedral

9-43. (a) H_2O (b) NH_2^- (c) AlH_4^- (d) SF_6

9-45. (a) trigonal pyramidal (b) tetrahedral (c) seesaw (d) octahedral

9-47. (a) bent (b) trigonal pyramidal (c) tetrahedral

9-49. (a) $HF > HCl > HBr > HI$ (b) $NH_3 > PH_3 > AsH_3$ (c) $IF_3 > BrF_3 > ClF_3$ (d) $H_2O > H_2S > H_2Se > H_2Te$

9-51. (a) XeF_4

9-53. (a) octahedral (b) linear (c) tetrahedral (d) tetrahedral

9-55. (a) trigonal bipyramidal (b) octahedral (c) octahedral (d) tetrahedral

9-57. (a) trigonal bipyramidal (b) trigonal pyramidal (c) tetrahedral

CHAPTER 10

10-1. There are 10 localized bonds in propane. A total of 20 valence electrons occupy the 10 localized bond orbitals.

10-3. (a) 14 valence electrons; 7 localized bonds; 0 lone pairs (b) 8, 3, 1 (c) 14, 5, 2 (d) 20, 2, 8

10-5. Six valence electrons; we use sp^2 orbitals on the boron atom and form three localized bond orbitals by combining each boron sp^2 orbital with a hydrogen $1s$ orbital. Two valence electrons occupy each of the three bond orbitals.

10-7. Thirty-two valence electrons; we use sp^3 orbitals on the carbon atom and combine each carbon sp^3 orbital with a fluorine $2p$ orbital to form four localized bond orbitals that point toward the vertices of a regular tetrahedron. Two valence electrons occupy each of the four bond orbitals. The remaining 24 valence electrons occupy the remaining s and p orbitals (as lone pairs) on the fluorine atoms.

10-9. The three aluminum-chlorine bonds in $AlCl_3$ are σ bonds. Each σ bond orbital is formed by combining an sp^2 hybrid orbital on the aluminum atom with a $3p$ chlorine orbital. Of the 24 valence electrons, 6 occupy the three bond orbitals and 18 occupy the remaining s and p orbitals (as lone pairs) on the chlorine atoms.

10-11. (a) Each of the three σ bond orbitals is formed by combining an sp^3 orbital on the oxygen atom with a $1s$ hydrogen orbital. Six of the eight valence electrons occupy these three bond orbitals. The lone electron pair occupies the remaining sp^3 orbital on the oxygen atom.

10-13. Six of the 26 valence electrons occupy the three bond orbitals formed by combining an sp^3 orbital on the nitrogen atom with a $2p$ orbital on a fluorine atom. A lone electron pair on the nitrogen atom occupies the remaining sp^3 orbital on the nitrogen atom. The remaining 18 valence electrons occupy lone pair orbitals on the fluorine atoms.

10-15. Six of the twenty-six valence electrons occupy the three σ bond orbitals formed by combining sp^3 orbitals on the phosphorus atom with a chlorine $3p$ orbital. A lone electron pair occupies the remaining sp^3 orbital on the phosphorus atom. The remaining 18 electrons occupy lone pair orbitals on the chlorine atoms.

10-17. The three σ bonds on each nitrogen atom are formed by combining two of the sp^3 orbitals on the nitrogen atom with two $1s$ hydrogen orbitals, and one of the sp^3 nitrogen orbitals with an sp^3 orbital on the other nitrogen atom. Ten of the 14 valence electrons occupy these five bond orbitals. The remaining four valence electrons occupy the remaining sp^3 orbital on each nitrogen atom as lone pairs.

10-19. We use sp^3 hybrid orbitals on both the carbon atom and the nitrogen atom. Twelve of the 14 valence electrons occupy the six σ bond orbitals formed by combining a carbon sp^3 orbital with a nitrogen sp^3 orbital and by combining the other carbon and nitrogen sp^3 orbitals with hydrogen $1s$ orbitals. The remaining two valence electrons occupy a nitrogen sp^3 orbital as a lone pair.

10-21. We use sp^3 orbitals on the oxygen atom and the two carbon atoms. We have six $C(sp^3) + H(1s)$ σ bond orbitals and two $O(sp^3) + C(sp^3)$ σ bond orbitals. Two lone electron pairs occupy the oxygen sp^3 orbitals.

10-23. (a) five σ, one π (b) nine σ, two π (c) seven σ, one π (d) fourteen σ, two π

10-25. nine σ, two π, 22 valence electrons

10-27. The σ bond between the carbon and oxygen atoms is formed by combining an sp orbital on the carbon atom and an sp orbital on the oxygen atom. The two π bonds are formed by combining the $2p$ orbitals on the carbon atom and the $2p$ orbitals on the oxygen atom. One lone pair occupies an sp orbital on the carbon atom; the other pair occupies an sp orbital on the oxygen atom.

10-29. There are 13 σ bonds: six $C(sp^2) + C(sp^2)$ bonds; five $C(sp^2) + H(1s)$ bonds; one $C(sp^2) + O(sp^3)$ bond; and, one $O(sp^3) + H(1s)$ bond. There are six π bond orbitals, the three of lowest energy being occupied by a total of six valence electrons that are delocalized over the entire ring. The remaining four valence electrons (out of 36 total) constitute two lone pairs on the oxygen atom.

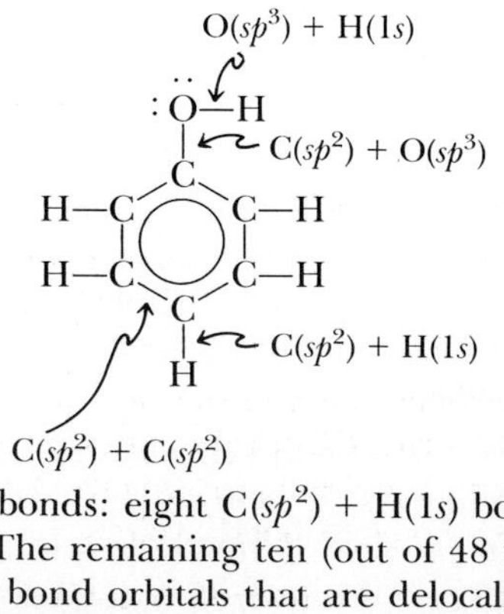

10-31. There are 19 σ bonds: eight $C(sp^2) + H(1s)$ bonds; and eleven $C(sp^2) + C(sp^2)$ bonds. The remaining ten (out of 48 total) valence electrons are in five of the ten π bond orbitals that are delocalized over the entire two rings, as indicated by the circles in the Lewis formula.

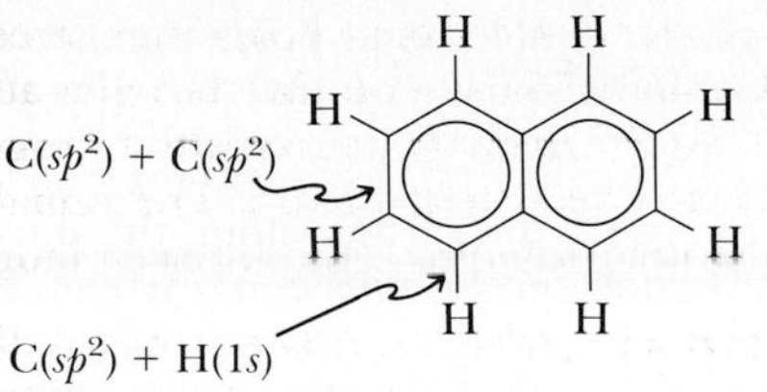

$C(sp^2) + C(sp^2)$

$C(sp^2) + H(1s)$

10-33. There are 24 valence electrons. We use sp^2 orbitals on each of the four atoms and form three localized σ bond orbitals that lie in the plane of the ion. Six valence electrons occupy the three σ orbitals. The remaining p orbital on each atom combines with the others to form four delocalized π orbitals. Six valence electrons occupy the delocalized π orbitals. The remaining 12 valence electrons occupy six sp^2 orbitals (2 on each oxygen) as lone pairs.

10-35. There are eight electrons in diatomic beryllium. There are four electrons in bonding orbitals and four electrons in antibonding orbitals, and so Be_2 has no net bonding.

10-37. We find that the bond order of N_2 is 3, whereas the bond order of N_2^+ is $2\frac{1}{2}$. The bond energy increases as the bond order increases; therefore, the bond energy of N_2 is greater than that of N_2^+. However, we find that the bond order of O_2 is 2, whereas the bond order of O_2^+ is $2\frac{1}{2}$. Therefore, the bond energy of O_2 is less than that of O_2^+.

10-39. Because the bond order of C_2^{2-} is greater than that of C_2, we predict that C_2^{2-} has a larger bond energy and a shorter bond length than C_2.

10-41. The bond order of CO is 3. Both molecular orbital theory and the Lewis formula predict that there is a triple bond in CO.

10-43. (a) $(1\sigma)^2(1\sigma^*)^2(2\sigma)^2(2\sigma^*)^2(1\pi)^2(1\pi)^2(3\sigma)^2(1\pi^*)^2(1\pi^*)^2$, 1
(b) $(1\sigma)^2(1\sigma^*)^2(2\sigma)^2(2\sigma^*)^2(1\pi)^2(1\pi)^1$, $1\frac{1}{2}$
(c) $(1\sigma)^2(1\sigma^*)^2(2\sigma)^2(2\sigma^*)^1$, $\frac{1}{2}$
(d) $(1\sigma)^2(1\sigma^*)^2(2\sigma)^2(2\sigma^*)^2(1\pi)^2(1\pi)^2(3\sigma)^2(1\pi^*)^2(1\pi^*)^2(3\sigma^*)^1$, $\frac{1}{2}$

10-45. If the additional electron occupies a bonding orbital, then a stronger net bonding will result; e.g., B_2 versus B_2^-.

10-47. B_2, C_2^{2+} and F_2^{2+}.

10-49. Twelve of the 18 valence electrons occupy the six σ bond orbitals and two of them occupy the π orbital. The two lone electron pairs on the oxygen atom occupy the other two oxygen sp^2 orbitals. The shape of the acetaldehyde is trigonal planar around the central carbon atom.

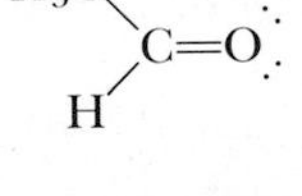

10-51. The σ bond framework is

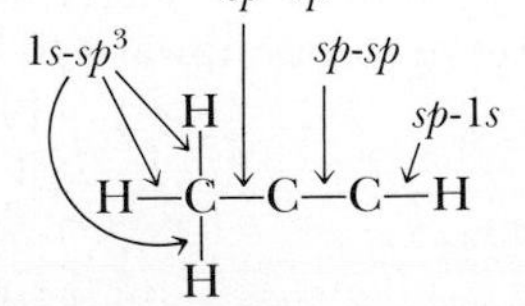

$1s$-sp^3 sp^3-sp sp-sp sp-$1s$

The remaining bond orbitals between the two carbon atoms on the right are two π bond orbitals formed from the $2p$ orbitals on each atom. There are six σ bond orbitals and two π bond orbitals. There are 16 valence electrons, which occupy the eight bond orbitals.

10-53. Use the $1s$ orbitals on the hydrogen atom and two of the $5p$ orbitals on the tellurium atom to form the two σ bond orbitals.

10-55. The π bond of the double bond holds the atoms in place about a double bond. Any rotation about the double bond would require rupture of the π bond.

10-57. We use sp^2 orbitals on the carbon atom, the nitrogen atom and the oxygen atom. The p orbitals on the nitrogen, carbon, and oxygen atoms combine to form three delocalized π orbitals, which fix the molecule into a planar configuration. Ten of the 16 valence electrons occupy the five localized σ orbitals, four of them occupy the remaining two oxygen sp^2 orbitals as two lone pairs, and the remaining four valence electrons occupy two of the delocalized π orbitals.

CHAPTER 11

11-1. 6880 kJ

11-3. 0.410 kJ

11-5. 418 J

11-7. 57.3 kJ

11-9. Time to heat Hg from 200 K to its melting point is 20.8 s; time to melt Hg is 51.6 s; time to heat Hg from 234 K to 630 K is 245 s; time to vaporize Hg is 1330 s; time to heat mercury from 630 K to 800 K is 79.3 s.

11-11. $29.2\ \text{kJ}\cdot\text{mol}^{-1}$

11-13. ClF and NF_3

11-15. $T_b[\text{He}] < T_b[C_2H_6] < T_b[C_2H_5OH] < T_b[\text{KBr}]$

11-17. $\Delta H_{vap}[CH_4] < \Delta H_{vap}[C_2H_6] < \Delta H_{vap}[CH_3OH] < \Delta H_{vap}[C_2H_5OH]$

11-19. between 45 and 50 torr

11-21. Liquid will be present.

11-23. condensation at about 60°C

11-25. the 20°C day

11-27. about 23°C

11-29. 179 torr

11-31. 80°C

11-33. $31.8\ \text{kJ}\cdot\text{mol}^{-1}$

11-35. 1.32×10^6

11-37. (a) gas (b) solid (c) liquid (d) liquid

11-39. Solid oxygen does not melt under an applied pressure, because the melting-point curve slopes upward to the right.

11-41. 2

11-43. 408.7 pm

11-45. $9.26\ \text{g}\cdot\text{cm}^{-3}$

11-47. $6.02 \times 10^{23}\ \text{atom}\cdot\text{mol}^{-1}$

11-49. 537.7 pm, 268.9 pm

11-51. Your body uses energy to melt the snow. It requires 0.334 kJ to melt one gram of snow if the temperature of the snow is 0°C.

11-53. Water is strongly hydrogen-bonded. Therefore, Trouton's rule does not apply to water.

11-55. $\Delta H_{sub} = \Delta H_{fus} + \Delta H_{vap}$

11-57. They both form a diamond-like covalent crystal network.

11-59. The boiling point of water was used to determine the normal atmospheric pressure, from which the altitude can be estimated.

11-61. liquid

11-63. 172 K, 10.7 torr

11-65. 76.9 $kJ \cdot mol^{-1}$

11-67. four formula units, NaCl-type unit cell

11-69. 3.995 $g \cdot cm^{-3}$

11-71. The normal boiling point is slightly greater than 78°C. $\Delta H_{vap} =$ 32.4 $kJ \cdot mol^{-1}$

CHAPTER 12

12-1. 4.9×10^{-4} M

12-3. 0.069 M

12-5. 130 ft

12-7. $X_{H_2O} = 0.911$, $X_{C_2H_5OH} = 0.0891$

12-9. Mix 431 g of acetone with 569 g of water.

12-11. 0.354

12-13. 46.9 torr, 0.2 torr

12-15. 23.64 torr, 0.12 torr

12-17. 51.3 torr, 7.9 torr

12-19. Dissolve 115 g of formic acid in 1000 g of acetone.

12-21. 0.1026 m

12-23. (a) 2.0 m_c (b) 3.0 m_c (c) 1.0 m_c (d) 5.0 m_c

12-25. (a) 0.16 torr (b) 0.23 torr (c) 0.08 torr (d) 0.31 torr

12-27. 104.2°C

12-29. 81.9°C

12-31. 212.3°C

12-33. −3.11°C

12-35. 4.59°C

12-37. 3.18°C

12-39. 450

12-41. 10 m

12-43. $HgCl_2$ is essentially undissociated.

12-45. 6.4 atm

12-47. about 5750

12-49. 13.6 L

12-51. (a) 412 torr (b) 0.466

12-53. about 150,000

12-55. Ethyl alcohol is a temporary antifreeze because its equilibrium vapor pressure is much greater than 1 atm at 100°C. Thus ethyl alcohol is much more readily lost by evaporation from the coolant system than is a liquid with a relatively high boiling point like ethylene glycol.

12-57. 0.11 m

12-59. 19 m; 110°C

12-61. $X_{met} = 0.0167$, $X_{eth} = 0.0104$
$m_{met} = 0.952$ m, $m_{eth} = 0.597$ m

12-63. (a) no net flow (b) no net flow (c) to the 0.50 M solution.

12-65. 7.23 m, 4.78 M; molality is independent of temperature.

12-67. (a) 0.115 m, 0.345 m_c, −0.642°C, 100.18°C (b) 0.434 m, 0.434 m_c, −0.807°C, 100.23°C (c) 0.215 m, 0.215 m_c, 171.2°C; 209.3°C

12-69. P_4

12-71. 2.26 M

12-73. 108.4°C

12-75. 18 M

CHAPTER 13

13-1. (a) Use a spectrophotometer to measure the decrease in the yellow color due to $I_2(aq)$ as a function of time. (b) Measure the total pressure in the reaction vessel as a function of time.

13-3. 7.2×10^{-2} M·s^{-1}

13-5. 7.5×10^{-4} atm·s^{-1}

13-7. From 0 s to 175 s, rate = 4.6×10^{-4} M·s^{-1}. From 175 s to 506 s, rate = 3.8×10^{-4} M·s^{-1}. From 506 s to 845 s, rate = 2.9×10^{-4} M·s^{-1}.

13-9. first-order

13-11. rate = $(7.3 \times 10^{-30}$ s$^{-1})[C_2H_5Cl]$

13-13. rate = $(2.8 \times 10^{-5}$ M^{-1}·s$^{-1})[NOCl]^2$

13-15. rate = $(2.0 \times 10^{-6}$ M^{-1}·s$^{-1})[Cr(H_2O)_6^{3+}][SCN^-]$

13-17. rate = $(5.0 \times 10^4$ M^{-1}·s$^{-1})[NO_2][O_3]$

13-19. 0.67

13-21. 0.24

13-23. rate = (0.041 min^{-1})$[S_2O_8^{2-}]$

13-25. (a) rate = $k[N_2O][O]$ (b) rate = $k[O][O_3]$ (c) rate = $k[ClCO][Cl_2]$

13-27. Yes. The rate law of the overall reaction is determined by the slow step.

13-29. 100 kJ·mol^{-1}

13-31. 3.11×10^{-2} s^{-1}

13-33. 0.017 h

13-35. No. A catalyst affects only the reaction rate, not the equilibrium concentrations.

13-37. (a) $H^+(aq)$ and $Br^-(aq)$ (b) third order (c) $[H^+]$ and $[Br^-]$ do not change with time because $H^+(aq)$ and $Br^-(aq)$ are catalysts. For $[H^+]$ fixed at 1.00×10^{-3} M and $[Br^-]$ fixed at 1.00×10^{-3} M, $[H_2O_2]$ is governed by a first-order rate law with $k = 1.0 \times 10^{-3}\ s^{-1}$.

13-39. Change the surface area of the wall material to see if the rate of the reaction changes.

13-41. The surface sites for adsorption of O_2 are occupied completely except at very low O_2 pressures. The rate-determining step is the dissociation of the adsorbed O_2 molecules to O atoms.

13-43. $4.8 \times 10^{-6}\ M^{-1}{\cdot}s^{-1}$

13-45. rate $= k_r[H_2][I_2]$

13-47. reverse rate $= \dfrac{k_r[N_2O_5][O_2]}{[NO_2]}$

13-49. $2N_2O_5(g) \rightleftharpoons 4NO_2(g) + O_2(g)$, rate $= k_1[N_2O_5]$

13-51. Take rate $= k[ClCO][Cl_2]$ and use equilibrium conditions in the first two steps to eliminate [ClCO] and then [Cl].

13-53. rate $= k'[NO]^2[O_2]$

13-55. rate $= (5.2 \times 10^{-7}\ torr^{-1}{\cdot}s^{-1})P^2_{CO}$

13-57. $3.6 \times 10^{-4}\ M{\cdot}s^{-1}$, $1.3\ mol \cdot L^{-1}$

13-59. First order; there will be 25,600 bacteria after 2 hours; the rate constant is $0.046\ min^{-1}$.

13-61. 230 days

13-63. CH_3 is a free radical, and two CH_3 radicals can join to form a stable (C_2H_6) molecule without the need to break any chemical bonds. The activation energy for the process is essentially zero.

13-65. $k = (2.2 \times 10^{-5}\ s^{-1})$

13-67. Run (1): $2.0 \times 10^{-4}\ M{\cdot}s^{-1}$; run (2): 0.40 M; run (3): 0.80 M

13-69. 1.4×10^6 bacteria·mL^{-1}

13-71. (a) rate $= 2 \times 10^5$ neurons·day^{-1} (b) 85 years

13-73. $k = 6 \times 10^{-4}\ s^{-1}$; $105\ kJ{\cdot}mol^{-1}$

13-75. 0.927

13-77. ——

13-79. 7.7×10^{-9} s

CHAPTER 14

14-1. (a) $\dfrac{[CO_2]}{[CO]}$ (b) $\dfrac{[O_2]^3}{[O_3]^2}$ (c) $\dfrac{[C_{10}H_{12}]}{[C_5H_6]^2}$ (d) $\dfrac{[NO_2]^4[O_2]}{[N_2O_5]^2}$

14-3. (a) $\dfrac{[SO_2][Cl_2]}{[SO_2Cl_2]}$ (b) $\dfrac{[O_2]}{[H_2O_2]^2}$ (c) $\dfrac{1}{[H_2O]^3}$

14-5. (a) $\dfrac{P_{SO_2}P_{Cl_2}}{P_{SO_2Cl_2}}$ (b) $\dfrac{P_{O_2}}{P^2_{H_2O}}$ (c) $\dfrac{1}{P^3_{H_2O}}$

14-7. $K_c = 4.7 \times 10^{-3}$ M

14-9. $6.7 \times 10^{-3}\ atm^2$

14-11. 50.3

14-13. 0.064 M

14-15. $[PCl_5] = 0.03$ M; $[PCl_3] = 0.22$ M

14-17. 0.98 mol

14-19. 0.658 atm

14-21. $[ICl] = 0.34$ M, $[I_2] = 0.26$ M, $[Cl_2] = 0.05$ M

14-23. $P_{N_2O_4} = 0.28$ atm, $P_{NO_2} = 1.17$ atm

14-25. $P_{CO_2} = 1.80$ atm, $P_{CO} = 3.00 \times 10^{-3}$ atm

14-27. (a) equilibrium will shift from right to left, P_{CO} decreases and P_{CO_2} increases. (b) no change

14-29. (a) to the right (b) to the right (c) to the left (d) to the left (e) no change

14-31. (a) to the left (b) to the left (c) to the left (d) to the right

14-33. (1) (a) $K_c = \frac{[CO][H_2]}{[H_2O]}$ (b) (i) to the right, (ii) to the left

(2) (a) $K_c = \frac{[CO_2][H_2]}{[CO][H_2O]}$ (b) (i) to the left, (ii) no effect

(3) (a) $K_c = \frac{[H_2O][CH_4]}{[CO][H_2]^3}$ (b) (i) to the left, (ii) to the right

14-35. (a) to the right (b) to the left

14-37. The reaction is not at equilibrium and will proceed to the right.

14-39. (a) 0.89 atm^{-1} (b) to the left

14-41. 36.9 atm^2

14-43. 812 atm

14-45. 209 atm

14-47. 6.0×10^{-5} atm^{-1}

14-49. 56.6 $kJ \cdot mol^{-1}$

14-51. 8.16×10^{-2} atm

14-53. $[H_2] = [I_2] = 0.52$ M, $[HI] = 4.33$ M

14-55. (a) no change (b) to the left (c) to the right (d) to the right

14-57. $[H_2] = [H_2O] = 0.38$ M

14-59. Start from $K_p = P^2_{NH_3}P_{CO_2}$ and use $P_{total} = P_{NH_3} + P_{CO_2}$ and the reaction stoichiometry.

14-61. Use $P_g = [\text{gas}]RT$ for each gaseous reactant and product species in the general equation $a\text{A}(g) + b\text{B}(g) \rightleftharpoons x\text{X}(g) + y\text{Y}(g)$.

14-63. (a) not possible (b) $[H_2] = [I_2] = 1.74 \times 10^{-3}$ M, $[HI] = 0.0160$ M

14-65. $P_{Cl_2} = 2.24$ atm, $P_{O_2} = 1.64$ atm

14-67. The brown form is more stable.

14-69. 2.14×10^6 $atm^{-1/2}$; 2.18×10^{-13} atm

CHAPTER 15

15-1. $[H_3O^+] = [ClO_4^-] = 0.150$ M, $[OH^-] = 6.67 \times 10^{-14}$ M, acidic

15-3. $[Tl^+] = [OH^-] = 1.81 \times 10^{-2}$ M, $[H_3O^+] = 5.52 \times 10^{-13}$ M, basic

15-5. 1.70, acidic

15-7. pH = 1.10, pOH = 12.90

15-9. pOH = 1.15, pH = 12.85

15-11. 1.6×10^{-7} M

15-13. 0.10 M

15-15. $[H_3O^+] = 4.0 \times 10^{-8}$ M, $[OH^-] = 2.5 \times 10^{-7}$ M

15-17. 1.3×10^{-5} M

15-19. 1.7×10^{-5} M

15-21. 0.086%

15-23. 2.96

15-25. 1.6

15-27. 1.5

15-29. 1.8×10^{-5} M

15-31. 9.32

15-33. 11.75

15-35. (a) shifts to the left (b) shifts to the right (c) shifts to the right (d) shifts to the right

15-37. (a) shifts to the left (b) no change (c) shifts to the left (d) shifts to the right (e) shifts to the left

15-39. 50.0 mL

15-41. 90.1

15-43. 1.66

15-45. (a) $HC_7H_5O_2(aq)$–$C_7H_5O_2^-(aq)$ and $H_3O^+(aq)$–$H_2O(l)$ (b) $CH_3NH_3^+(aq)$–$CH_3NH_2(aq)$ and $H_2O(l)$–$OH^-(aq)$ (c) $HCHO_2(aq)$–$CHO_2^-(aq)$ and $H_3O^+(aq)$–$H_2O(l)$

15-47. (a) $ClO^-(aq)$ (b) $NH_3(aq)$ (c) $N_3^-(aq)$ (d) $S^{2-}(aq)$

15-49. (a) acid, $CNO^-(aq)$ (b) base, $HOBr(aq)$ (c) acid, $ClO_3^-(aq)$ (d) acid, $CH_3NH_2(aq)$ (e) base, $ClNH_3^+(aq)$ (f) base, $HONH_3^+(aq)$

15-51. (a) 7.46×10^{-10} M (b) 1.48×10^{-11} M (c) 1.75×10^{-5} M (d) 1.58×10^{-7} M

15-53. (a) 0.490 (b) 1.28×10^{-3}

15-55. (a) acidic (b) acidic (c) basic (d) basic

15-57. (a) basic (b) neutral (c) basic (d) acidic

15-59. basic

15-61. pH = 10.00; $[HClO] = 9.98 \times 10^{-5}$ M

15-63. $[OH^-] = [HCNO] = 3.02 \times 10^{-6}$ M, $[CNO^-] = 0.20$ M, $[H_3O^+] = 3.31 \times 10^{-9}$ M pH = 8.48

15-65. 2.84

15-67. 4.47

15-69. 1.87

15-71. (a) Arrhenius acid and a Brønsted-Lowry acid (b) Arrhenius acid, a Brønsted-Lowry acid, and a Lewis acid (c) Lewis acid

15-73. (a) Lewis base (b) Lewis acid (c) Lewis base

15-75. 2.8×10^3

15-77. 2.64

15-79. 9.66×10^{-4} gram per 100 mL of solution

15-81. 3.37

15-83. 6.94

15-85. $NH_4^+(amm)$, $NH_2^-(amm)$

15-87. at $pH > 3.9$

15-89. 0.16 M or $19\ g\cdot L^{-1}$

CHAPTER 16

16-1. 9.5 ± 0.5

16-3. 4.5 ± 0.5

16-5. bromcresol green or methyl red

16-7. 3×10^{-5} M

16-9. (a) 12.35 (b) 13.11

16-11. (a) 100 mL (b) 45.0 mL

16-13. Before any acid is added, pH = 13.00; at the equivalence point (100.0 mL total volume), pH = 7.00; after 100.0 mL of acid is added, pH = 1.48.

16-15. 0.235 M

16-17. 9.48; thymolphthalein or phenolphthalein

16-19. (a) 2.73 (b) 4.16 (c) 4.76 (d) 8.88 (e) 11.59

16-21. (a) 9.17 (b) 5.34 (c) 3.78 (d) 3.15 (e) 2.51

16-23. 176

16-25. 4.15%

16-27. 5.19

16-29. 4.76

16-31. 3.97

16-33. 5.06

16-35. (a) 7.20 (b) 7.50 (c) 6.90

16-37. A solution of equal concentrations of $HCHO_2(aq)$ and $NaCHO_2(aq)$.

16-39. (a) 9.57 to 9.24 (b) 9.27 to 9.24

16-41. −0.11

16-43. 4.85

16-45. 50 mL

16-47. 5.88 g

16-49. 340 mL

16-51. benzoic acid

16-53. $[HC_2H_3O_2] \approx [HC_2H_3O_2]_0$ and $[C_2H_3O_2^-] \approx [C_2H_3O_2^-]_0$ because $[H_3O^+]$ is negligible compared to $[HC_2H_3O_2]_0$ and $[C_2H_3O_2^-]_0$

16-55. (a) pH = 8.34 (b) A $NaHCO_3(aq)$ solution acts as a buffer as shown by the equations $HCO_3^-(aq) + H_3O^+(aq) \rightleftharpoons H_2CO_3(aq) + H_2O(l)$ and $HCO_3^-(aq) + OH^-(aq) \rightleftharpoons CO_3^{2-}(aq) + H_2O(l)$

16-57. $3HCO_3^-(aq) + H_3C_6H_5O_7(aq) \rightleftharpoons 3CO_2(g) + C_6H_5O_7^{3-}(aq) + 3H_2O(l)$

16-59. 190 mL

16-61. 88.1

16-63. Measure the pH of the solution before and after dilution; the pH of a buffer is unaffected by dilution.

16-65. 116

16-67. There are three dissociable protons.

16-69. $[H_2C_2O_4] = 3.7 \times 10^{-6}$ M, $[HC_2O_4^-] = 0.020$ M, $[C_2O_4^{2-}] = 0.105$ M

CHAPTER 17

17-1. (a) insoluble (b) soluble (c) soluble (d) soluble (e) insoluble

17-3. (a), (d), (e)

17-5. (a) $CuCl_2(aq) + Na_2S(aq) \rightarrow CuS(s) + 2NaCl(aq)$
$Cu^{2+}(aq) + S^{2-}(aq) \rightarrow CuS(s)$
(b) $MgBr_2(aq) + K_2CO_3(aq) \rightarrow MgCO_3(s) + 2KBr(aq)$
$Mg^{2+}(aq) + CO_3^{2-}(aq) \rightarrow MgCO_3(s)$
(c) $BaCl_2(aq) + K_2SO_4(aq) \rightarrow BaSO_4(s) + 2KCl(aq)$
$Ba^{2+}(aq) + SO_4^{2-}(aq) \rightarrow BaSO_4(s)$
(d) $Hg_2(NO_3)_2(aq) + 2KCl(aq) \rightarrow Hg_2Cl_2(s) + 2KNO_3(aq)$
$Hg_2^{2+}(aq) + 2Cl^-(aq) \rightarrow Hg_2Cl_2(s)$

17-7. (a) $Hg_2(ClO_4)_2(aq) + 2NaBr(aq) \rightarrow 2NaClO_4(aq) + Hg_2Br_2(s)$
(b) $Fe(ClO_4)_3(aq) + 3NaOH(aq) \rightarrow 3NaClO_4(aq) + Fe(OH)_3(s)$
(c) $Pb(NO_3)_2(aq) + 2LiIO_3(aq) \rightarrow 2LiNO_3(aq) + Pb(IO_3)_2(s)$
(d) $H_2SO_4(aq) + Pb(NO_3)_2(aq) \rightarrow 2HNO_3(aq) + PbSO_4(s)$

17-9. (a)

17-11. (a), (c), (e)

17-13. $PbCrO_4(s) \rightleftharpoons Pb^{2+}(aq) + CrO_4^{2-}(aq)$; $s = 5.3 \times 10^{-7}$ M

17-15. $5.01 \times 10^{-13}\ M^2$

17-17. 1.7×10^{-4} M

17-19. $2.9 \times 10^{-3}\ M^2$

17-21. 9.10

17-23. $1.4 \times 10^{-2}\ g \cdot L^{-1}$

17-25. 6.8×10^{-4} M, $0.16\ g \cdot L^{-1}$

17-27. 2.1×10^{-16} M, $4.7 \times 10^{-14}\ g \cdot L^{-1}$

17-29. 0.72 M

17-31. (a) increased (b) unchanged (c) decreased (d) unchanged

17-33. (a) increases (b) decreases

17-35. (a), (b), (c), (e), (f)

17-37. 1.9 M

17-39. 2.2×10^{-6} M

17-41. (a) increases (b) increases

17-43. $[Cr(OH)_3] = 6.3 \times 10^{-4}$ M; $[Ni(OH)_2] = 2.0 \times 10^3$ M (very soluble). Yes, $Cr(OH)_3$ can be separated from $Ni(OH)_2$ at this pH.

17-45. 5.7×10^{-19} M

17-47. pH ≈ -0.08

17-49. 1×10^{-3} M

17-51. 0.40 M

17-53. AgCl(*s*) precipitates.

17-55. Yes; 1.82×10^{-5} mol of $PbI_2(s)$ precipitates; $[Pb^{2+}] = 2.00$ M, $[I^-] = 6.0 \times 10^{-5}$ M, $[NO_3^-] = 4.00$ M, $[Na^+] = 6.67 \times 10^{-4}$ M.

17-57. (a) 0.29 g (b) 1.8×10^{-10} M

17-59. The lead precipitates out of solution as $PbSO_4(s)$ and thus is no longer physiologically active.

17-61. $Pb(NO_3)_2(aq) + 2NaOH(aq) \rightleftharpoons 2NaNO_3(aq) + Pb(OH)_2(s)$
$Pb(OH)_2(s) + OH^-(aq) \rightleftharpoons Pb(OH)_3^-(aq)$

17-63. Calcium ion forms an insoluble oxalate, $CaC_2O_4(s)$. The excess Ca^{2+} is removed by adding $MgSO_4(aq)$ to form $CaSO_4(s)$, which is insoluble in water and in stomach acid. Vomiting of the $CaC_2O_4(s)$ is necessary because the solubility of $CaC_2O_4(s)$ in stomach acid is sufficiently high to permit toxic levels of oxalic acid (a weak acid) to pass through the stomach walls into the bloodstream.

17-65. $Zn(ClO_4)_2(aq) + 2KOH(aq) \rightleftharpoons Zn(OH)_2(s) + 2KClO_4(aq)$
$Zn(OH)_2(s) + 2OH^-(aq) \rightleftharpoons Zn(OH)_4^{2-}(aq)$

17-67. 51.75%

17-69. 0.29 M

17-71. (a) 4.4×10^{12} M^{-1} (b) 2.6×10^{-5}

17-73. 15.23 g, 0.230 M

CHAPTER 18

18-1. (a) +2 (b) +3 (c) +5 (d) +8/3

18-3. (a) −1 (b) +4 (c) +5 (d) +3

18-5. (a) +4 (b) +1 (c) +5 (d) +3

18-7. (a) 0 (b) −4 (c) −2 (d) +2

18-9. (a) +3 (b) +3 (c) +3 (d) +3

18-11. I_2 is the oxidizing agent and $Na_2S_2O_3$ is the reducing agent.

18-13. $NaNO_3$ is the oxidizing agent and Pb is the reducing agent.

18-15. (a) Fe^{3+} is the oxidizing agent and I^- is the reducing agent: $2I^-(aq) \rightarrow I_2(s) + 2e^-$ (oxidation); $Fe^{3+}(aq) + e^- \rightarrow Fe^{2+}(aq)$ (reduction).

(b) $Co^{2+}(aq)$ is the oxidizing agent and Ti^{2+} is the reducing agent: $Ti^{2+}(aq) \rightarrow Ti^{3+}(aq) + e^-$ (oxidation); $Co^{2+}(aq) + 2e^- \rightarrow Co(s)$ (reduction).

18-17. Oxygen in superoxides is in an unusual oxidation state of $-\frac{1}{2}$ and readily converts to its more normal (oxide) oxidation state of -2.

18-19. (a) $2MnO(s) + 5PbO_2(s) + 8H^+(aq) \rightarrow 2MnO_4^-(aq) + 5Pb^{2+}(aq) + 4H_2O(l)$; electron donor and reducing agent, MnO; electron acceptor and oxidizing agent, PbO_2; element oxidized, Mn; element reduced, Pb
(b) $As_2S_5(s) + 40NO_3^-(aq) + 35H^+(aq) \rightarrow 5HSO_4^-(aq) + 2H_3AsO_4(aq) + 40NO_2(g) + 12H_2O(l)$; electron donor and reducing agent, As_2S_5; electron acceptor and oxidizing agent, NO_3^-; element oxidized, S; element reduced, N

18-21. (a) $NH_4^+(aq) + NO_3^-(aq) \rightarrow N_2O(g) + 2H_2O(l)$
(b) $4Fe(s) + 3O_2(g) + 6H_2O(l) \rightarrow 2Fe_2O_3 \cdot 3H_2O(s)$

18-23. (a) $4Fe(OH)_2(s) + O_2(g) + 2H_2O(l) \rightarrow 4Fe(OH)_3(s)$
(b) $3Cu(s) + 2NO_3^-(aq) + 8H^+(aq) \rightarrow 2NO(g) + 3Cu^{2+}(aq) + 4H_2O(l)$

18-25. (a) $IO_4^-(aq) + 3I^-(aq) + H_2O(l) \rightarrow IO_3^-(aq) + I_3^-(aq) + 2OH^-(aq)$
(b) $H_2MoO_4(aq) + 6Cr^{2+}(aq) + 6H^+(aq) \rightarrow Mo(s) + 6Cr^{3+}(aq) + 4H_2O(l)$

18-27. (a) $4CrO_4^{2-}(aq) + 3Cl^-(aq) + 20H^+(aq) \rightarrow 3ClO_2^-(aq) + 4Cr^{3+}(aq) + 10H_2O(l)$
(b) $2S_2O_3^{2-}(aq) + 2Cu^{2+}(aq) \rightarrow S_4O_6^{2-}(aq) + 2Cu^+(aq)$

18-29. $2CrI_3(s) + 27Cl_2(g) + 64OH^-(aq) \rightarrow 2CrO_4^{2-}(aq) + 6IO_4^-(aq) + 54Cl^-(aq) + 32H_2O(l)$

18-31. (a) $Mo^{3+}(aq) + 2H_2O(l) \rightarrow MoO_2^{2+}(aq) + 4H^+(aq) + 3e^-$
(b) $P_4(s) + 16H_2O(l) \rightarrow 4H_3PO_4(aq) + 20H^+(aq) + 20e^-$
(c) $S_2O_8^{2-}(aq) + 2H^+(aq) + 2e^- \rightarrow 2HSO_4^-(aq)$

18-33. (a) $2WO_3(s) + 2H^+(aq) + 2e^- \rightarrow W_2O_5(s) + H_2O(l)$
(b) $U^{4+}(aq) + 2H_2O(l) \rightarrow UO_2^+(aq) + 4H^+(aq) + e^-$
(c) $Zn(s) + 4OH^-(aq) \rightarrow Zn(OH)_4^{2-}(aq) + 2e^-$

18-35. (a) $2SO_3^{2-}(aq) + 2H_2O(l) + 2e^- \rightarrow S_2O_4^{2-}(aq) + 4OH^-(aq)$
(b) $2Cu(OH)_2(s) + 2e^- \rightarrow Cu_2O(s) + 2OH^-(aq) + H_2O(l)$
(c) $2AgO(s) + H_2O(l) + 2e^- \rightarrow Ag_2O(s) + 2OH^-(aq)$

18-37. 2.00 g, 20.8%

18-39. 2.28 g, 22.8%

18-41. 0.400 M

18-43. $2P_4(s) + 5BaSO_4(s) \rightarrow 2P_4O_{10}(s) + 5BaS(s)$, 0.459 g

18-45. $4MnO_4^-(aq) + 2H_2O(l) \rightarrow 4MnO_2(s) + 3O_2(g) + 4OH^-(aq)$

18-47. (a) $Cr_2O_7^{2-}(aq) + 9I^-(aq) + 14H^+(aq) \rightarrow 2Cr^{3+}(aq) + 3I_3^-(aq) + 7H_2O(l)$
(b) $3IO_4^-(aq) + 33I^-(aq) + 24H^+(aq) \rightarrow 12I_3^-(aq) + 12H_2O(l)$

18-49. $4Ag^{2+}(aq) + 2H_2O(l) \rightarrow 4Ag^+(aq) + O_2(g) + 4H^+(aq)$

18-51. (a) $Cr_2O_7^{2-}(aq) + 8H^+(aq) + 3H_2O_2(aq) \rightarrow 2Cr^{3+}(aq) + 3O_2(g) + 7H_2O(l)$
(b) $2Cr^{3+}(aq) + Zn(s) \rightarrow 2Cr^{2+}(aq) + Zn^{2+}(aq)$

18-53. (a) $H_2O_2(aq) + 2H^+(aq) + 2e^- \rightarrow 2H_2O(l)$
(b) $H_2O_2(aq) \rightarrow O_2(g) + 2H^+(aq) + 2e^-$;
$2H_2O_2(aq) \rightarrow 2H_2O(l) + O_2(g)$

18-55. 0.018% **18-57.** 1.82×10^{-4} %

CHAPTER 19

19-1. For CH_4: $\Delta S_{fus} = 10.3\ J \cdot K^{-1} \cdot mol^{-1}$, $\Delta S_{vap} = 81.57\ J \cdot K^{-1} \cdot mol^{-1}$
For C_2H_6: $\Delta S_{fus} = 31.8\ J \cdot K^{-1} \cdot mol^{-1}$, $\Delta S_{vap} = 84.78\ J \cdot K^{-1} \cdot mol^{-1}$
For C_3H_8: $\Delta S_{fus} = 38.5\ J \cdot K^{-1} \cdot mol^{-1}$, $\Delta S_{vap} = 87.11\ J \cdot K^{-1} \cdot mol^{-1}$

19-3. For CH_3OH: $\Delta S_{fus} = 18.11\ J \cdot K^{-1} \cdot mol^{-1}$, $\Delta S_{vap} = 111.1\ J \cdot K^{-1} \cdot mol^{-1}$
For C_2H_5OH: $\Delta S_{fus} = 31.64\ J \cdot K^{-1} \cdot mol^{-1}$, $\Delta S_{vap} = 115.1\ J \cdot K^{-1} \cdot mol^{-1}$
For C_3H_7OH: $\Delta S_{fus} = 35.32\ J \cdot K^{-1} \cdot mol^{-1}$, $\Delta S_{vap} = 117.7\ J \cdot K^{-1} \cdot mol^{-1}$

19-5. For H_2S: $\Delta S_{fus} = 12.7\ J \cdot K^{-1} \cdot mol^{-1}$; $\Delta S_{vap} = 88.0\ J \cdot K^{-1} \cdot mol^{-1}$
The corresponding values for H_2O are larger because of the breaking of hydrogen bonds.

19-7. (a) D_2O (b) ethanol (c) butyl amine

19-9. $S°(CH_4) < S°(CH_3OH) < S°(CH_3Cl)$

19-11. $Fe_3O_4(s)$ contains more atoms and has a greater mass than $Fe_2O_3(s)$.

19-13. The positional disorder in the gaseous state is greater than in the liquid state.

19-15. (a) increases (b) increases (c) increases (d) decreases

19-17. $\Delta S^\circ_{rxn}(c) \approx \Delta S^\circ_{rxn}(b) < \Delta S^\circ_{rxn}(a) < \Delta S^\circ_{rxn}(d)$

19-19. (a) $-111.2\ J \cdot K^{-1}$ (b) $-332.1\ J \cdot K^{-1}$ (c) $134.0\ J \cdot K^{-1}$
(d) $-173.4\ J \cdot K^{-1}$

19-21. (a) $2.9\ J \cdot K^{-1}$ (b) $-189.2\ J \cdot K^{-1}$ (c) $-242.8\ J \cdot K^{-1}$
(d) $-111.8\ J \cdot K^{-1}$

19-23. The reaction is spontaneous. $\Delta G_{rxn} < 0$, $\Delta H_{rxn} > 0$, $T\Delta S_{rxn} > 0$. The reaction is entropy-driven.

19-25. $\Delta S^\circ_{rxn} = -429.6\ J \cdot K^{-1}$, $\Delta G^\circ_{rxn} = -503$ kJ, spontaneous from left to right when both $C_2H_2(g)$ and $C_6H_6(l)$ are at standard conditions.

19-27. $\Delta S^\circ_{rxn} = -30.0\ J \cdot K^{-1}$, $\Delta G^\circ_{rxn} = -1314$ kJ, spontaneous from left to right when all the reactants are products are at standard conditions.

19-29. -50.1 kJ, yes

19-31. $\Delta G^\circ_{rxn} = -36.5 \times 10^3$ kJ, spontaneous from left to right when PCl_5, PCl_3 and Cl_2 are at standard conditions; $\Delta G_{rxn} = +15.5$ kJ, spontaneous from right to left

19-33. $+19.1$ kJ. Nitrous acid will not dissociate spontaneously under standard conditions. It will dissociate when $[H^+] = [NO_2^-] = 1.0 \times 10^{-5}$ M and $[HNO_2] = 1.0$ M.

19-35. $+16.4$ kJ. Chloroacetic acid will not dissociate spontaneously under standard conditions. It will dissociate spontaneously when $[H^+] = 1.0 \times 10^{-5}$ M, $[C_2H_2ClO_2^-] = 0.0010$ M, and $[HC_2H_2ClO_2] = 0.10$ M.

19-37. $+55.6$ kJ, no

19-39. -19.4 kJ, left to right; -7.97 kJ, left to right.

19-41. (a) -29.1 kJ, $1.29 \times 10^5\ atm^{-3}$ (b) 91.4 kJ, 9.12×10^{-17} atm
(c) -142.2 kJ, $8.99 \times 10^{24}\ atm^{-2}$

19-43. $\Delta G^\circ_{rxn} = -355$ kJ, $\Delta H^\circ_{rxn} = -357.6$ kJ, $K = 2.0 \times 10^{62}$

19-45. $\Delta G^\circ_{rxn} = -141.8$ kJ, $\Delta H^\circ_{rxn} = -197.8$ kJ, $K = 7.66 \times 10^{24}$ atm^{-1}, $K = 3.01 \times 10^5$ atm^{-1} at 400°C.

19-47. $\Delta G^\circ_{rxn} = 28.6$ kJ, $\Delta H^\circ_{rxn} = 41.2$ kJ, $\Delta S^\circ_{rxn} = 42.3$ J·K^{-1}. The reaction is enthalpy-driven to the left.

19-49. 1468 kJ

19-51. 181 kJ

19-53. 34.5 kJ

19-55. $\Delta H^\circ_{rxn} = -68.6$ kJ, $K_p = 0.0166$ atm^{-1}

19-57. Li: $\Delta S_{fus} = 6.59$ J·K^{-1}·mol^{-1},
$\Delta S_{vap} = 83.41$ J·K^{-1}·mol^{-1}
Na: $\Delta S_{fus} = 7.01$ J·K^{-1}·mol^{-1},
$\Delta S_{vap} = 77.5$ J·K^{-1}·mol^{-1}
K: $\Delta S_{fus} = 6.93$ J·K^{-1}·mol^{-1},
$\Delta S_{vap} = 74.6$ J·K^{-1}·mol^{-1}
Rb: $\Delta S_{fus} = 7.50$ J·K^{-1}·mol^{-1},
$\Delta S_{vap} = 72$ J·K^{-1}·mol^{-1}
Cs: $\Delta S_{fus} = 6.95$ J·K^{-1}·mol^{-1},
$\Delta S_{vap} = 70$ J·K^{-1}·mol^{-1}

19-59. No, because then ΔG°_{rxn} is either negative infinite or positive infinite, which is not possible.

19-61. −120.4 kJ, 116.8 kJ; the first method is more energy efficient.

19-63. +181 kJ

19-65. −5.70 kJ

19-67. ΔH°_{rxn} is negative.

19-69. +21.6 kJ

19-71. $\Delta H^\circ_{rxn} = -466.8$ kJ, $\Delta S^\circ_{rxn} = -40.4$ J·K^{-1}, $\Delta G^\circ_{rxn} = -454.8$ kJ

CHAPTER 20

20-1. 1.16×10^4 s

20-3. 1.24 g

20-5. 2.55×10^4 g; liquid HF is a covalent compound and thus is a poor conductor of an electric current.

20-7. 0.217 g

20-9. 13.8 L

20-11. $Zn(s) \rightarrow Zn^{2+}(aq) + 2e^-$ (negative electrode),
$2Ag^+(aq) + 2e^- \rightarrow 2Ag(s)$ (positive electrode), $Zn(s)|ZnCl_2(aq)||AgNO_3(aq)|Ag(s)$

20-13. $V(s) \rightarrow V^{2+}(aq) + 2e^-$ (negative electrode),
$Cu^{2+}(aq) + 2e^- \rightarrow Cu(s)$ (positive electrode), $V(s)|VI_2(aq)||CuSO_4(aq)|Cu(s)$

20-15. $Pb(s) + 2I^-(aq) \rightarrow PbI_2(s) + 2e^-$ (left electrode),
$2H^+(aq) + 2e^- \rightarrow H_2(g)$ (right electrode), $Pb(s) + 2HI(aq) \rightarrow PbI_2(s) + H_2(g)$

20-17. $In(s) \rightarrow In^{3+}(aq) + 3e^-$ (left electrode),
$Re^{3+}(aq) + 3e^- \rightarrow Re(s)$ (right electrode)

20-19. $H_2(g) + Hg_2Cl_2(s) \rightarrow 2Hg(l) + 2H^+(aq) + 2Cl^-(aq)$

20-21. (a) no effect (b) decreases (c) increases (d) decreases

20-23. (a) no effect (b) increases (c) increases (d) decreases (e) decreases (f) increases

20-25. (a) increases (b) no effect (c) decreases (d) decreases

20-27. (a) 8 (b) 4

20-29. 7.6×10^{15}

20-31. $E° = 0.270$ V, $K = 1.3 \times 10^{9}$

20-33. $E° = 1.62$ V, $K = 1.2 \times 10^{82}$

20-35. (a) $Zn(s) + Hg_2Cl_2(s) \rightarrow 2Hg(l) + Zn^{2+}(aq) + 2Cl^-(aq)$
(b) 1.14 V

20-37. 8.3×10^{-5} M

20-39. 1.69

20-41. (a) −0.16 V (b) −0.03 V (c) 3.63 V

20-43. 1.39 V

20-45. The reaction is spontaneous under either of the conditions given.

20-47. $E° = +0.24$ V, $K = 1.12 \times 10^4$ atm$^{1/2}$M^{-1}, yes

20-49. 2.03 V

20-51. Oxidation half-reaction in basic solution:

$$S_2O_3^{2-}(aq) + 6OH^-(aq) \rightarrow 2SO_3^{2-}(aq) + 3H_2O(l) + 4e^-$$

reduction half-reaction in basic solution:

$$O_2(g) + 2H_2O(l) + 4e^- \rightarrow 4OH^-(aq) \quad E°[SO_3^{2-}/S_2O_3^{2-}] = -0.58\ V$$

20-53. spontaneous

20-55. yes

20-57. −203 kJ

20-59. −130 kJ

20-61. (a) −400 kJ (b) −74 kJ

20-63. $E° = +0.36$ V, $\Delta G°_{rxn} = -69$ kJ, $\Delta G_{rxn} = -73$ kJ, $E = +0.38$ V; $Zn(s) + Cd^{2+}(aq) \rightarrow Zn^{2+}(aq) + Cd(s)$

20-65. 1.086 V

20-67. $\ominus$steel|Fe(s)|Fe(OH)$_2$(s)|LiOH(aq)|NiOOH(s), Ni(OH)$_2$(s)|steel$\oplus$

20-69. $9H_2O(l) + 6Mn^{2+}(aq) + 5IO_3^-(aq) \rightarrow$

$$18H^+(aq) + 6MnO_4^-(aq) + 5I^-(aq)$$

20-71. Oxidation of lead to $Pb(OH)_2$ (white) occurs spontaneously at the negative electrode.

20-73. 8.44×10^5 A·h

20-75. 0.456 g

20-77. $Ag^+(aq) + Br^-(aq) \rightleftharpoons AgBr(s)$, $K_{sp} = 5.0 \times 10^{-13}$ M^2

20-79. $E_{cell} = 0.2415\ V - (0.0592\ V) \log [H^+]$

$$\approx 0.2415\ V - (0.0592\ V) \log \left\{ \frac{K_a[HC_2H_3O_2]_0}{[C_2H_3O_2^-]_0} \right\}$$

20-81. Zero, because the cell is short-circuited.

20-83. 0 V

20-85. 108, silver

21-1. (a) $^{72}_{31}Ga$ (b) $^{226}_{90}Th$ (c) $^{136}_{56}Ba$ (d) $^{1}_{0}n$

21-3. (a) $^{25}_{12}Mg + ^{4}_{2}He \rightarrow ^{28}_{13}Al + ^{1}_{1}H$
(b) $^{27}_{13}Al + ^{1}_{0}n \rightarrow ^{4}_{2}He + ^{24}_{11}Na$
(c) $^{17}_{8}O + ^{1}_{1}H \rightarrow ^{4}_{2}He + ^{14}_{7}N$
(d) $^{63}_{29}Cu + ^{1}_{1}H \rightarrow ^{63}_{30}Zn + ^{1}_{0}n$

21-5. positron emitter

21-7. (a) positron emitter (b) β-emitter (c) positron emitter (d) β-emitter

21-9. (a)

21-11. (a) radioactive (b) radioactive (c) not radioactive
(d) radioactive

21-13. 1.15×10^{-4}

21-15. 0.0135 mg

21-17. 0.63

21-19. 0.507

21-21. 5.33 days

21-23. 8640 years

21-25. 15,900 years

21-27. 5050 years

21-29. 2.44×10^{9} years

21-31. 1.26×10^{-12} J·nucleon^{-1}

21-33. 4.79×10^{-11} J, 1.37×10^{-12} J·nucleon^{-1}

21-35. 2.45×10^{-11} J·atom^{-1}, 1.99 kg

21-37. 1.64×10^{-13} J, 1.24×10^{20} Hz

21-39. -1.93×10^{13} J·mol^{-1}

21-41. 3.5×10^{3} metric tons, 2 years

21-43. 9.6×10^{7} Ci·g^{-1}

21-45. 1500 mg

21-47. 0.029 μCi, 4.6 μg

21-49. (a) $^{1}_{1}H$ (b) $^{13}_{7}N$ (c) $^{14}_{7}N$ (d) $^{1}_{1}H$ (e) $^{0}_{+1}e$ (f) $^{4}_{2}He$
$4\,^{1}_{1}H \rightarrow ^{4}_{2}He + 2\,^{0}_{+1}e + 3\gamma$

21-51. 0.0602

21-53. 0.46 disintegration·min^{-1}

21-55. 2.30×10^{12} J·mol^{-1}, 1.21×10^{4} kg

21-57. 7.4×10^{11} J, 1.4×10^{5} kg

21-59. 27.8 days

21-61. $CH_3-C(=\ddot{O}:)-\ddot{O}-H + H-\ddot{O}^{*}-CH_3 \rightarrow CH_3-C(=\ddot{O})-\ddot{O}^{*}-CH_3 + H_2O$

When the oxygen-18 label is in CH_3OH, all the oxygen-18 shows up in the ester. When the oxygen-18 label is in CH_3COOH, the oxygen-18 shows up in both H_2O and the ester.

21-63. 1.70×10^{12} J·mol^{-1}

21-65. 7.0×10^{11} J·mol^{-1}

21-67. 9.97×10^{21} D atoms, 1.91×10^{10} J·kg^{-1}

21-69. It requires large kinetic energies to overcome the coulombic repulsion involved in a fusion nuclear reaction.

21-71. 5.0 L

21-73. 3.3×10^{-5} M

21-75. 1.7×10^{-8} M^2

CHAPTER 22

22-1. (a) [Ar]$3d^4$ (b) [Ar]$3d^2$ (c) [Kr]$4d^6$ (d) [Xe]$4f^{14}5d^6$

22-3. (a) 10 (b) 6 (c) 6 (d) 7

22-5. (a) Co(III), Rh(III), Ir(III) (b) Fe(IV), Ru(IV), Os(IV) (c) Cu(I), Ag(I), Au(I)

22-7. (a) +3 (b) +3 (c) +2 (d) +3

22-9. (a) +2 (b) +2 (c) +2 (d) +3

22-11. (a) 3 mol $K^+(aq)$ and 1 mol $[Fe(CN)_6]^{3-}(aq)$
(b) 1 mol $[Ir(NH_3)_6]^{3+}(aq)$ and 3 mol $NO_3^-(aq)$
(c) 1 mol $[Pt(NH_3)_4Cl_2]^{2+}(aq)$ and 2 mol $Cl^-(aq)$
(d) 1 mol $[Ru(NH_3)_6]^{3+}(aq)$ and 3 mol $Br^-(aq)$

22-13. The given empirical formulas correspond to $[Pt(NH_3)_6]Cl_4$, $[Pt(NH_3)_5Cl]Cl_3$, $[Pt(NH_3)_4Cl_2]Cl_2$, $[Pt(NH_3)_3Cl_3]Cl$, and $[Pt(NH_3)_2Cl_4]$.

22-15. (a) +3, potassium hexacyanochromate(III) (b) +3, pentaaquachlorochromium(III) perchlorate (c) +3, tetracarbonyldichlorocobalt(III) perchlorate (d) +4, tetraamminedibromoplatinum(IV) chloride

22-17. (a) +3, ammonium hexanitrocobaltate(III) (b) +3, tetraamminedibromoiridium(III) bromide (c) +2, potassium tetrachlorocuprate(II) (d) 0, pentacarbonylruthenium(0)

22-19. (a) −3, $Na_3[Fe(CN)_5CO]$ (b) −1, *trans*-$NH_4[AuCl_2I_2]$ (c) −3, $K_3[Co(CN)_6]$ (d) −3, $Ca_3[Co(NO_2)_6]_2$

22-21. (a) $[Pt(NH_3)_3Cl]NO_3$ (b) $Na_2[CuF_4]$ (c) $Li_4[Co(NO_2)_6]$ (d) $Ba_2[Fe(CN)_6]$

22-23. (a) diammine-*bis*(ethylenediamine)osmium(III) chloride (b) triamminechloroethylenediaminecobalt(II) nitrate (c) ammonium ethylenediaminetetraacetatoferrate(II) (d) potassium *tris*(oxalato)chromate(III)

22-25. (a) $K_3[Re(C_2O_4)_3]$ (b) $[Co(en)_3](C_2H_3O_2)_2$ (c) $Na_2[Cr(EDTA)]$ (d) $[PdCl_2(en)_2](NO_3)_2$

22-27. (a)

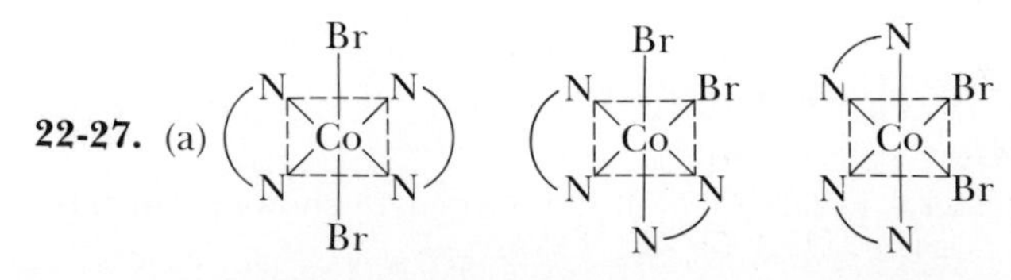

(b) 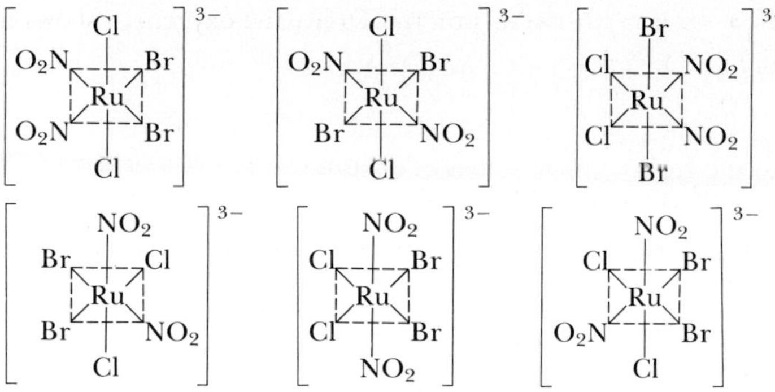

22-29. (a) 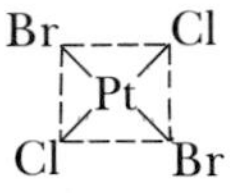

trans

(b) 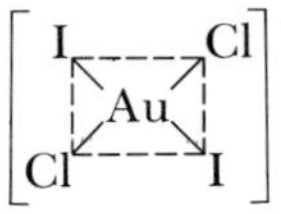

trans

(c) 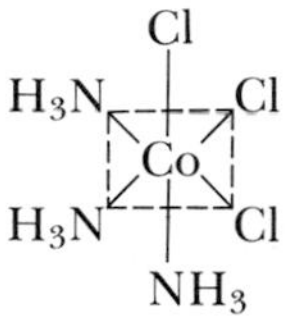

cis,cis

(d) 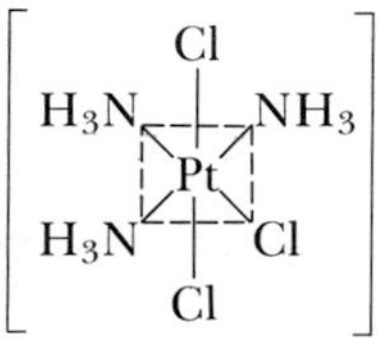

cis,trans

22-31. (a) $t_{2g}^2e_g^0$ (b) $t_{2g}^4e_g^0$ (c) $t_{2g}^3e_g^2$ (d) $t_{2g}^6e_g^4$ (e) $t_{2g}^6e_g^0$

22-33. (a) low-spin (b) low-spin (c) high-spin (d) high-spin (e) high-spin

22-35. (a) low-spin, $t_{2g}^6e_g^0$ (b) high-spin, $t_{2g}^3e_g^2$ (c) low-spin, $t_{2g}^6e_g^0$ (d) high-spin, $e^2t_2^3$

22-37. (a) 2 (b) 3 (c) 0 (d) 2

22-39. (c), (d)

22-41. $[NiF_4]^{2-}$ is tetrahedral with an $e^4t_2^4$ *d* electron configuration, and $[Ni(CN)_4]^{2-}$ is square planar with no unpaired electrons.

22-43. (c)

22-45. (c)

22-47. (a) $[Co(NO_2)_6]^{3-}$ (b) *trans*-$[PtCl_2(en)_2]^{2+}$ (c) $[Fe(CN)_5CO]^{3-}$ (d) *trans*-$[AuCl_2I_2]^-$

22-49. (a) $Li_3[Ru(C_2O_4)_3]$ (b) $K_2[HgI_4]$ (c) $[Cr(NH_3)_4Br_2]NO_3$ (d) $[Mo(H_2O)_3(CO)_3](C_2H_3O_2)_3$

22-51. (a) 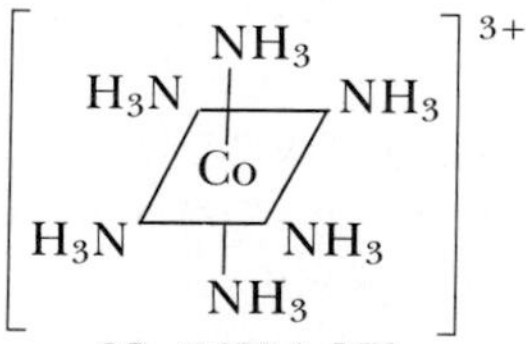

$[Co(NH_3)_6]Cl_3$

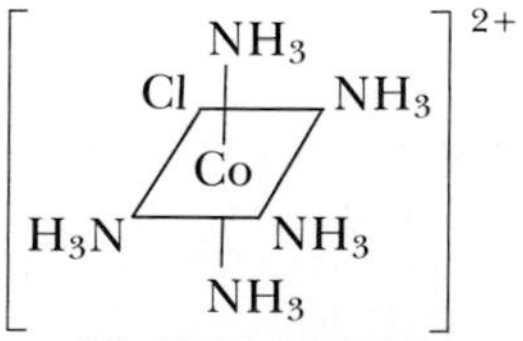

$[Co(NH_3)_5Cl]Cl_2$

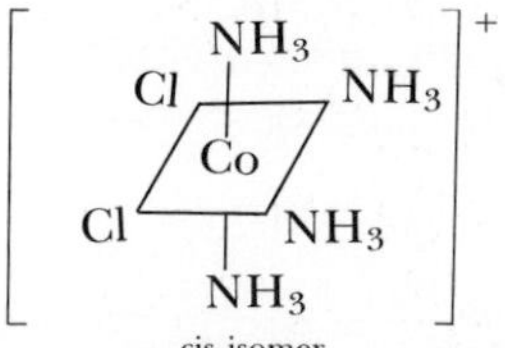

cis isomer

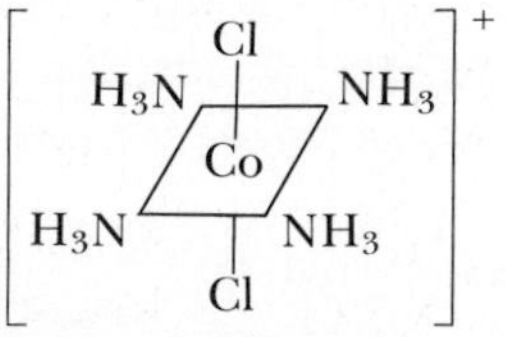

trans isomer

$[Co(NH_3)_4Cl_2]Cl$

(b) There are cis and trans isomers of $[Co(NH_3)_4Cl_2]^+$.

22-53. (a) false, (b) false, (c) false, (d) true

22-55. (a) 2 (b) 3 (c) 0 (d) 1

22-57. (a) to the right (b) no effect (c) to the left (d) to the left
(e) no effect (f) to the left

22-59. $5.3 \times 10^{-17}\ M^3$

CHAPTER 23

23-1. (a) $C_5H_{12}(g) + 8O_2(g) \rightarrow 5CO_2(g) + 6H_2O(l)$ (b) N.R. (c) N.R.
(d) $CH_4(g) + Cl_2(g) \xrightarrow{UV} CH_3Cl(g) + HCl(g)$ plus other chloromethanes

23-3. (a) identical (b) identical (c) different (d) identical

23-5. three: 1-chlorohexane, 2-chlorohexane, 3-chlorohexane

23-7. (a) 2-bromo-3-chlorobutane or 2-chloro-3-bromobutane
(b) 2, 2-dimethylpropane (c) 2-methyl-2-nitropropane
(d) 1-amino-2-methylbutane

23-9. (a) violates rule 3, 2-methylpentane (b) violates rule 2, 3-methylpentane (c) violates rule 2, 4-methyloctane (d) violates rule 6, 2,2-dimethylpropane

23-11. (a)
$$\begin{array}{l} CH_3-\underset{\displaystyle CH_3}{\underset{|}{CH}}-CH_2-CH_3 \end{array}$$
2-methylbutane

(b) $Cl-CH_2-CH_2-Br$

1-bromo-2-chloroethane
2-bromo-1-chloroethane

(c)
$$CH_3-\overset{\displaystyle Cl}{\overset{|}{\underset{\displaystyle Cl}{\underset{|}{C}}}}-\overset{\displaystyle Cl}{\overset{|}{\underset{\displaystyle Cl}{\underset{|}{C}}}}-Cl$$
1,1,1,2,2-pentachloropropane

(d)
$$CH_3-\overset{\displaystyle CH_3}{\overset{|}{\underset{\displaystyle CH_3}{\underset{|}{C}}}}-CH_3$$
2,2-dimethylpropane

23-13. (a)
$$CH_3-\underset{\displaystyle CH_3}{\underset{|}{CH}}-\underset{\displaystyle CH_3}{\underset{|}{CH}}-CH_3$$

(b)
$$CH_3-\overset{\displaystyle CH_3}{\overset{|}{\underset{\displaystyle CH_3}{\underset{|}{C}}}}-\underset{\displaystyle CH_3}{\underset{|}{CH}}-CH_3$$

(c)
$$CH_3-CH_2-\overset{\displaystyle CH_3}{\overset{|}{\underset{\displaystyle CH_3}{\underset{|}{C}}}}-\underset{\underset{\displaystyle CH_3}{\underset{|}{\displaystyle CH_2}}}{\underset{|}{CH}}-CH_2-CH_3$$

(d)
$$CH_3-CH_2-CH_2-\underset{\displaystyle CH_3-CH-CH_3}{\underset{|}{CH}}-CH_2-CH_2-CH_2-CH_3$$

23-15. (a)
$$H_3C-\overset{\displaystyle CH_3}{\overset{|}{\underset{\displaystyle H}{\underset{|}{C}}}}-\overset{\displaystyle CH_3}{\overset{|}{\underset{\displaystyle H}{\underset{|}{C}}}}-CH_3$$

(b)
$$H_3C-\overset{\displaystyle NH_2}{\overset{|}{\underset{\displaystyle H}{\underset{|}{C}}}}-\overset{\displaystyle CH_3}{\overset{|}{\underset{\displaystyle H}{\underset{|}{C}}}}-CH_3$$

(c)
$$H_3C-\overset{\displaystyle H}{\overset{|}{\underset{\displaystyle H}{\underset{|}{C}}}}-\overset{\displaystyle Cl}{\overset{|}{\underset{\underset{\displaystyle CH_3}{\underset{|}{\displaystyle CH_2}}}{\underset{|}{C}}}}-\overset{\displaystyle H}{\overset{|}{\underset{\displaystyle H}{\underset{|}{C}}}}-CH_3$$

(d)
$$H-\overset{\displaystyle Cl}{\overset{|}{\underset{\displaystyle H}{\underset{|}{C}}}}-\overset{\displaystyle Cl}{\overset{|}{\underset{\displaystyle H}{\underset{|}{C}}}}-\overset{\displaystyle Cl}{\overset{|}{\underset{\displaystyle H}{\underset{|}{C}}}}-H$$

23-17. (a) 2-chloropropane (b) 2-bromobutane (c) 2-chlorobutane
(d) 3-bromopentane and 2-bromopentane

23-19. (a) 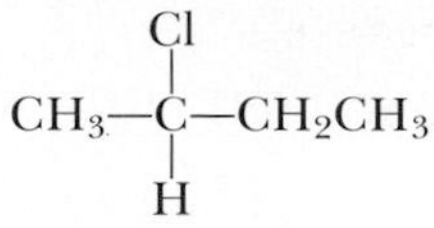 $CH_3{-}CCl(H){-}CH_2CH_3$ 2-chlorobutane (b) 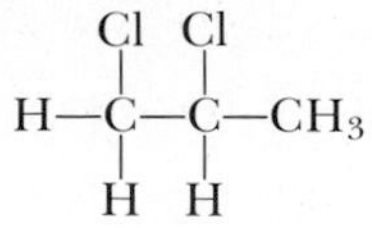 $H{-}CHCl{-}CHCl{-}CH_3$ 1,2-dichloropropane

(c) CH_3CH_2OH ethanol (d) 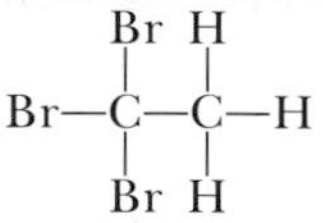 $Br{-}CBr_2{-}CH_2{-}H$ 1,1,1-tribromoethane

23-21. (a) 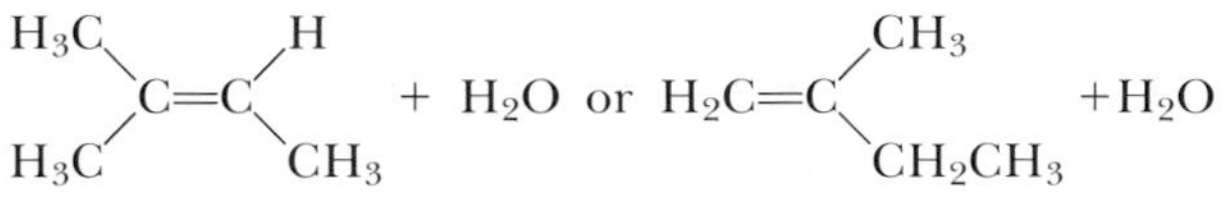 $(H_3C)_2C{=}CH(CH_3) + H_2O$ or $H_2C{=}C(CH_3)(CH_2CH_3) + H_2O$

(b) 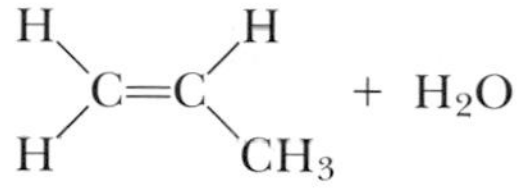 $H_2C{=}CH(CH_3) + H_2O$

(c) $CH_3{-}C{\equiv}C{-}CH_3 + H_2$

(d) $CH_3{-}C{\equiv}CH + HBr$

23-23. (a) propyne (b) 2-butyne (c) 4,4-dimethyl-2-hexyne (d) 4-methyl-1-hexyne

23-25. 2,2-dibromopropane

23-27. (a) $CH_3C{\equiv}CH(g) + 4O_2(g) \rightarrow 3CO_2(g) + 2H_2O(l)$
(b) $CH_3C{\equiv}CH(g) + 2HCl(g) \rightarrow CH_3CCl_2CH_3(l)$
(c) $CH_3C{\equiv}CH(g) + 2Br_2(l) \rightarrow CH_3CBr_2CHBr_2(l)$

23-29. (a) 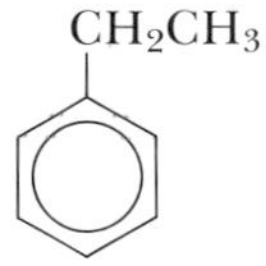 benzene ring with CH_2CH_3 (b) 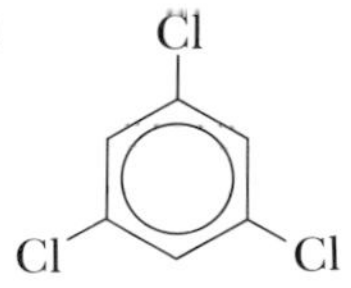 benzene ring with Cl, Cl, Cl (1,3,5 positions)

(c) 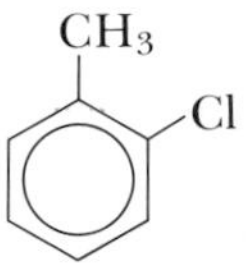 benzene ring with CH_3 and Cl (ortho) (d) 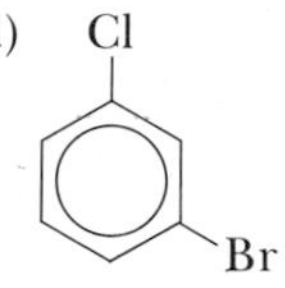 benzene ring with Cl and Br (meta)

23-31. (a) 2-propanol (b) 2,2-dimethyl-1-butanol (c) 1,3-dichloro-2-propanol (d) 2-butanol

23-33. (a) primary (b) secondary (c) secondary (d) tertiary

23-35. $2CH_3CH_2OH(l) + 2Na(s) \rightarrow 2Na^+CH_3CH_2O^-(s) + H_2(g)$
$2CH_3CH_2CH_2OH(l) + 2Na(s) \rightarrow 2Na^+CH_3CH_2CH_2O^-(s) + H_2(g)$

23-37. (a) butanal (b) 3-methylbutanal (c) methanal (d) 3,4-dimethylpentanal

23-39. (a) 3-pentanol (b) 2-pentanol (c) 3-hexanol

23-41. (a) $C_2H_5NH_2(aq) + HBr(aq) \rightarrow C_2H_5NH_3^+Br^-(aq)$
(b) $2(CH_3)_2NH(aq) + H_2SO_4(aq) \rightarrow [(CH_3)_2NH_2^+]_2SO_4^{2-}(aq)$
(c) 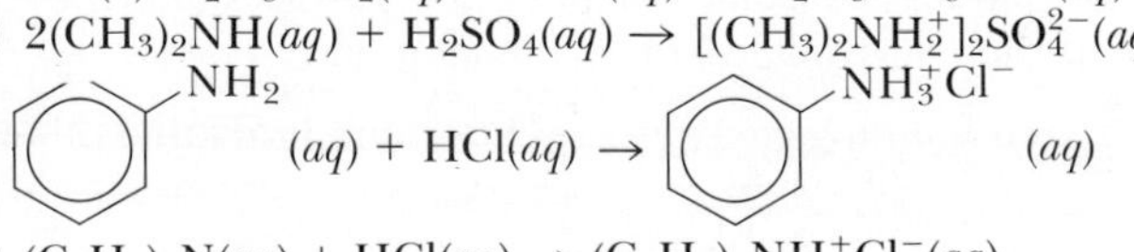 $C_6H_5NH_2(aq) + HCl(aq) \rightarrow C_6H_5NH_3^+Cl^-(aq)$
(d) $(C_2H_5)_3N(aq) + HCl(aq) \rightarrow (C_2H_5)_3NH^+Cl^-(aq)$

23-43. (a) 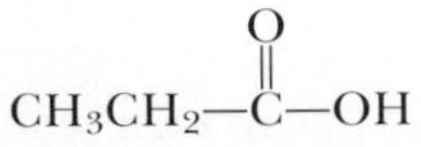 $CH_3CH_2{-}\overset{O}{\overset{\|}{C}}{-}OH$ (b) 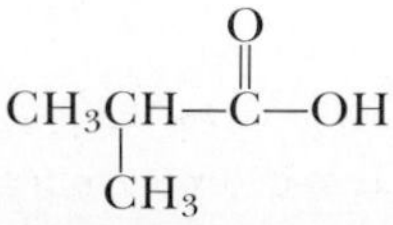 $CH_3CH(CH_3){-}\overset{O}{\overset{\|}{C}}{-}OH$

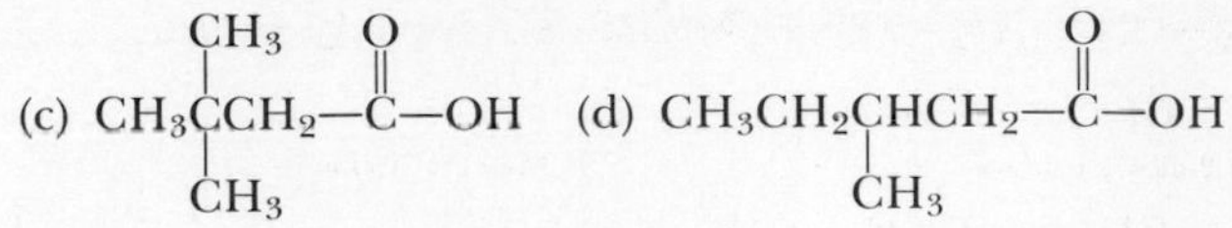

(c) $CH_3C(CH_3)_2CH_2-C(=O)-OH$ (d) $CH_3CH_2CH(CH_3)CH_2-C(=O)-OH$

23-45. (a) $HCOOH(aq) + NaOH(aq) \rightarrow Na^+HCOO^-(aq) + H_2O(l)$
(b) $HCOOH(aq) + CH_3OH(aq) \xrightarrow{H^+(aq)} HCOOCH_3(aq) + H_2O(l)$
(c) $2HCOOH(aq) + Ca(OH)_2(aq) \rightarrow Ca(OOCH)_2(aq) + 2H_2O(l)$

23-47. (a) $CH_3CH_2COOH(aq) + KOH(aq) \rightarrow K^+CH_3CH_2COO^-(aq) + H_2O(l)$
potassium propanoate

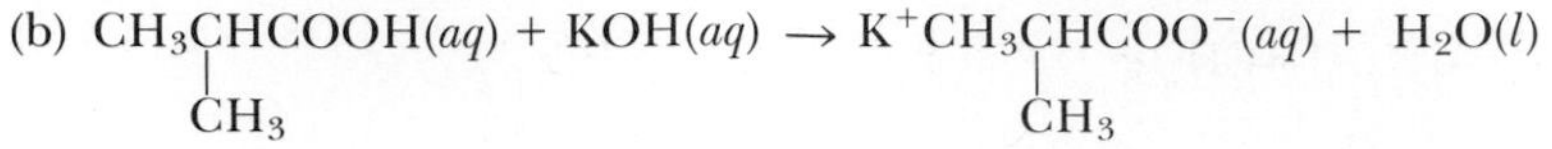

(b) $CH_3CH(CH_3)COOH(aq) + KOH(aq) \rightarrow K^+CH_3CH(CH_3)COO^-(aq) + H_2O(l)$
potassium 2-methylpropanoate

(c) $2Cl_2CHCOOH(aq) + Ca(OH)_2(aq) \rightarrow Ca(Cl_2CHCOO)_2(aq) + 2H_2O(l)$
calcium dichloroethanoate

23-49. (a) $C_6H_5C(=O)OH + HOCH_2CH_3 \xrightarrow{H^+(aq)} C_6H_5C(=O)OCH_2CH_3 + H_2O(l)$
benzoic acid ethanol ethyl benzoate

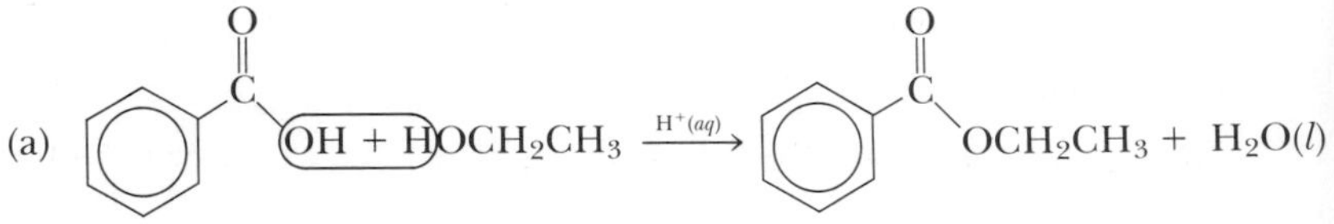

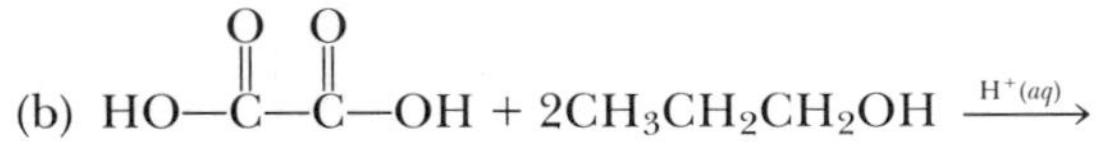

(b) $HO-C(=O)-C(=O)-OH + 2CH_3CH_2CH_2OH \xrightarrow{H^+(aq)}$
oxalic acid 1-propanol

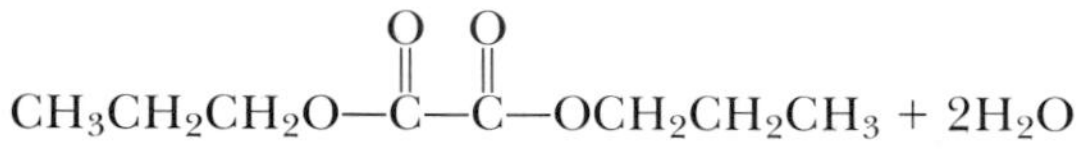

$CH_3CH_2CH_2O-C(=O)-C(=O)-OCH_2CH_2CH_3 + 2H_2O$
propyl oxalate

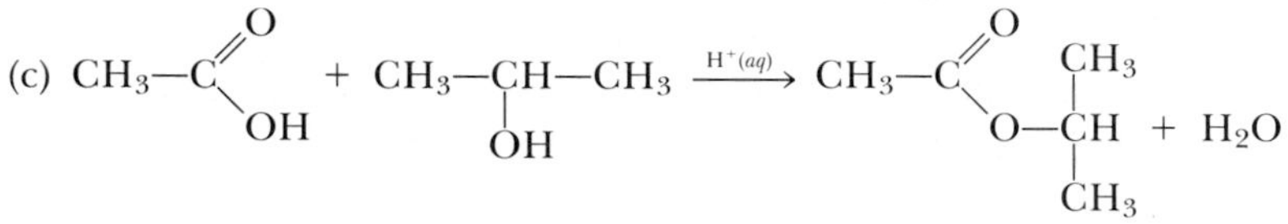

(c) $CH_3-C(=O)OH + CH_3-CH(OH)-CH_3 \xrightarrow{H^+(aq)} CH_3-C(=O)O-CH(CH_3)_2 + H_2O$
acetic acid 2-propanol (isopropyl alcohol) isopropyl acetate

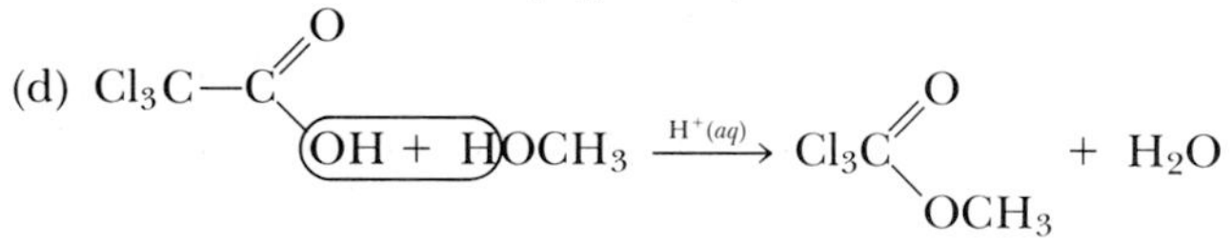

(d) $Cl_3C-C(=O)OH + HOCH_3 \xrightarrow{H^+(aq)} Cl_3C-C(=O)OCH_3 + H_2O$
trichloroacetic acid methanol methyl trichloroacetate

23-51. 3-chloropropane is 1-chloropropane rotated 180°.

23-53. (a) 1 and 4 (b) 2 and 3 (c) 3 and 5 (d) 2

23-55. The stearate ion consists of a hydrophobic part and a hydrophilic part. The hydrophobic parts encapsulate small grease particles and the hydrophilic parts keep the encapsulated entity in solution.

23-57. There is hydrogen bonding in 1-butanol but none in diethyl ether or pentane.

23-59. propene

23-61. 7.5×10^5 L, 2.9×10^6 kJ

23-63. C_3H_8, propane

23-65. (a) $Cl-CH_2-CH-CH_2-CH_3$ (with Cl on CH) 1,2-dichlorobutane (b) $CH_3-CH-CH_2-CH_3$ (with Cl on CH) 2-chlorobutane

(c) $CH_3-CH-CH_2-CH_3$ (with OH on CH) 2-butanol (d) $CH_3-CH_2-CH_2-CH_3$ butane

23-67. (a) $CH_2{=}CHCH{=}CH_2(g) + 2H_2(g) \xrightarrow{Pt} CH_3CH_2CH_2CH_3(g)$

(b) $CH_2{=}CHCH{=}CH_2(g) + 2Cl_2(g) \rightarrow ClCH_2CHCHCH_2Cl(l)$ (Cl, Cl on the two CH)

(c) $CH_2{=}CHCH{=}CH_2(g) + 2HCl(g) \rightarrow CH_3CHCHCH_3(l)$ (Cl, Cl on the two CH)

23-69. (a) $CH_3CHCHCH_3$ (Cl, Cl) (b) $BrCH_2CHCH_2CH_3$ (Cl) and $BrCH_2CH_2CHCH_3$ (Cl)

(c) CH_3CCH_3 (Cl above, CH_3 below) (d) $ClCH_2CHCH_3$ (Cl) and $Cl_2CHCH_2CH_3$

23-71. (a) propanal (b) 2-methylpropanal (c) 2,2-dimethylbutanal

CHAPTER 24

24-1. In addition polymerization, monomers are joined to each other directly. In condensation polymerization, monomers are joined together with the formation of small molecules as joint products.

24-3. Both monomers contain six carbon atoms.

24-5. (b), (c), (d)

24-7. (a) $H_2N-C(H)(CH_2C_6H_4OH)-COOH + H_2N-C(H)(CH(CH_3)_2)-COOH \rightarrow H_2N-C(H)(CH_2C_6H_4OH)-C(=O)-N(H)-C(H)(CH(CH_3)_2)-COOH$

or $H_2N-C(H)(CH(CH_3)_2)-COOH + H_2N-C(H)(CH_2C_6H_4OH)-COOH \rightarrow H_2N-C(H)(CH(CH_3)_2)-C(=O)-N(H)-C(H)(CH_2C_6H_4OH)-COOH + H_2O$

24-9. $H_2N-CH_2-C(=O)-NH-CH(CH_3)-COOH$ (gly, ala) OR $H_2N-CH(CH_3)-C(=O)-NH-CH_2-COOH$ (ala, gly)

24-11. 6; $H_2N-CH(G_1)-C(=O)-NH-CH(G_1)-C(=O)-NH-CH(G_2)-COOH$ plus five others with different sequences of G_1 and G_2

24-13. $H_2N-CH(CH_2CH_2COOH)-C(=O)-NH-CH(CH_2COOH)-C(=O)-NH-CH(CH_2C_6H_4OH)-COOH$

24-15. (a) S (b) none (c) N (d) both O

24-17. The primary structure of a protein is the order of amino acids along the polypeptide backbone. The secondary structure is the structure within sections of the protein. The tertiary structure is the overall three-dimensional shape.

24-19.

24-21.

24-23. TTCAGAGCT

24-25. G A T C A A T

24-27.

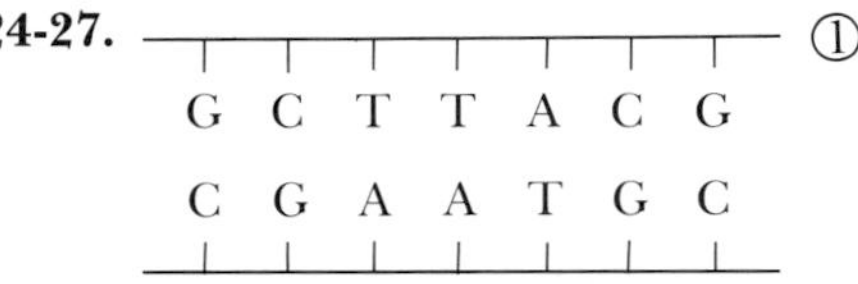

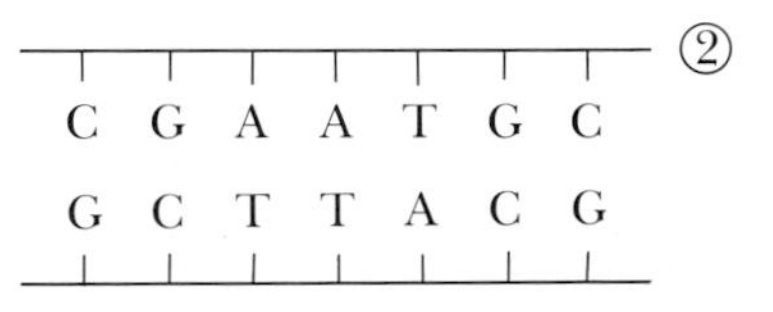

24-29. 14

24-31. 15.9 kJ

24-33. $-16.9\ \text{kJ}\cdot\text{g}^{-1}$, $16.9\ \text{kJ}\cdot\text{g}^{-1}$

24-35. 1.7×10^{5} M

24-37. 1.3×10^{35} M

INTERCHAPTERS

INTERCHAPTER B

B-11. (a) $2Na(s) + O_2(g) \rightarrow Na_2O_2(s)$
(b) $2Na(s) + 2H_2O(l) \rightarrow 2NaOH(aq) + H_2(g)$
(c) $6Li(s) + N_2(g) \rightarrow 2Li_3N(s)$

(d) $NaH(s) + H_2O(l) \rightarrow NaOH(aq) + H_2(g)$
(e) $Li_3N(s) + 3H_2O(l) \rightarrow 3LiOH(aq) + NH_3(g)$

B-13. (a) $NaHCO_3$ (b) KOH (c) $Ca(OH)_2$ (d) CsH

INTERCHAPTER C

C-11. 113 g Ca; 158 g CaO

C-13. (a) 0.866 kg (b) 0.683 kg (c) 0.775 kg

C-15. 7200 mL

INTERCHAPTER D

D-19. 0.421 L

D-21. 1.03 L

INTERCHAPTER E

E-11. 7.8×10^6 barrels per day

E-13. 2.4×10^4 kJ

E-15. -1234 kJ

E-17. 30 $kW \cdot h \cdot day^{-1}$

INTERCHAPTER F

F-13. 4.99×10^5 $J \cdot mol^{-1}$

F-15. 149 kg

F-17. 0.875 metric ton

INTERCHAPTER H

H-11. (a) Zircon contains SiO_4^{4-} ions. (b) Enstatite is a silicate that contains long, straight-chain silicate polyanions. (c) Talc, like mica, contains two-dimensional, polymeric silicate sheets.

H-13. For an *n*-type semiconductor, we can add small quantities of arsenic or antimony to silicon. For a *p*-type semiconductor, we can add gallium or indium.

H-15. 4.67% silicon-29 and 3.10% silicon-30.

H-17. 176 tons

INTERCHAPTER I

I-13. 17.3 M

I-15. 3.70 metric tons

I-17. $K_p = 0.108\ atm^2$, $K_c = 1.80 \times 10^{-4}\ M^2$

I-19. Add 191 mL of 15.7 M HNO_3 to enough water to make 500 mL of final solution

J-13. H_3PO_4: pH = 1.4; H_3PO_3: pH = 1.3

J-15. 1.8×10^3 M

J-17. 0.200 metric ton

J-19. 4.04 L

INTERCHAPTER K

K-15. ——

K-17. The ocean is buffered at a pH = 8.15. By contrast, most lakes and ponds are not buffered and thus they undergo a drop in pH when acid rain falls into the lake or pond.

K-19. 29.4 metric tons of chlorine

INTERCHAPTER L

L-11. SO_3^{2-}: trigonal pyramidal; SO_4^{2-}: tetrahedral; $S_2O_3^{2-}$: tetrahedral

L-13. S=S(F)F — trigonal pyramidal; F—S—S—F — bent

L-15. pH = 1.43

L-17. 0.25 $atm^{-1/2}$

INTERCHAPTER M

M-19. $Fe^{2+}(aq)$ can be oxidized to $Fe^{3+}(aq)$ by $Cl_2(g)$ and $Br_2(l)$, but not by $I_2(s)$.

M-21. $1.35 \times 10^{-9}\ M^3$

M-23. 5.254%

INTERCHAPTER N

N-15. (a) +7 (b) +6 (c) +4 (d) +2 (e) +3 (f) +7

N-17. (a) tetrahedral (b) tetrahedral (c) linear

N-19. from left to right

ILLUSTRATION CREDITS

All photos are by Chip Clark except the following:

p. 1, IBM Zurich Research Laboratory; p. 7 (Fig. 1-2), Metropolitan Museum of Art; p. 8 (Fig. 1-3), Burndy Library of Medicine; p. 9 (Fig. 1-4), Division of Mineral Sciences, Smithsonian Institution; p. 9 bottom, The Science Museum, London; p. 21 (Fig. 1-10), Lawrence Berkeley Laboratory, University of California; p. 22 (Fig. 1-11), Mettler Instrument Corporation; p. 36, Texasgulf; p. 41 (Fig. A-6), Chromatonix; p. 47 (Fig. 2-3a and b), Alexander Boden; p. 67 (Fig. 2-20a and b), W. H. Fahrenbach; p. 84 (Fig. B-2), Bill Tronca/Tom Stack and Assoc.; p. 85 (Fig. B-5), Bob Hammer/James Dye, Michigan State University; p. 86 (Fig. B-6), Tenneco Inc.; p. 120 (Fig. C-1), Division of Mineral Sciences, Smithsonian Institution; p. 123 (Fig. C-5), Alcoa; p. 126, Travis

Amos; p. 129 (Fig. 4-4), National Maritime Museum, Greenwich, England; p. 145 (Fig. 4-15), Oak Ridge National Laboratory; p. 156, New York Public Library; p. 161 (Fig. D-6), Ross Chapple; p. 163, NASA; p. 167 (Fig. 5-4), Tenneco Inc.; p. 178 (Fig. 5-8a), Parr Instrument Company; p. 187, U.S. Air Force; p. 188, Dennis Harding, Chevron Corp.; p. 191 (Fig. E-1), NASA; p. 191 (Fig. E-2), Department of the Navy; p. 192 (Fig. E-3), Peter Menzel; p. 194, Sandia National Laboratories; p. 202 (Fig. 6-6), Culver Pictures Inc.; p. 203 (Fig. 6-8), Yerkes Observatory; p. 204 (Fig. 6-9), American Institute of Physics, Center for History of Physics; p. 205 (Fig. 6-10a and b), Education Development Center, Inc., Newton, Mass.; p. 206 (Fig. 6-11), The Neils Bohr Institute; p. 210 (Fig. 6-17), University of Hamburg, from *From X-rays to Quarks,* by Emilio Segre, W. H. Freeman and Co., 1980; p. 211 (Fig. 6-18), A. K. Kleinschmidt/American Institute of Physics; p. 217 (Fig. 6-26), CERN/American Institute of Physics; p. 238, NASA; p. 243 (Fig. F-4a and b), Westfälisches Amt für Denkmalpflege; p. 262, The University of California Archives, The Bancroft Library; p. 288, Molenmuseum De Valk, Leiden; p. 308, General Electric Research and Development Center; p. 316, Joe McNally/Wheeler Pictures; p. 361 (Fig. 11-16), Håkon Hope; p. 374, General Electric Research and Development Center; p. 379 (Fig. H-8a and b), Steven Smale; p. 379 (Fig. H-9), Corning Glass Works; p. 381, Kenrick Day; p. 395 (Fig. 12-13), from "Biological membranes as bilayer couples," by M. Sheetz, R. Painter, and S. Singer, *J. of Cell Biology,* 1976, 70:193; p. 404, Engelhard Corporation; p. 446, Smithsonian Institution; p. 448 (Fig. 14-3), General Electric; p. 460, Air Products and Chemicals, Inc.; p. 463 (Fig. I-3), Grant Heilman/Grant Heilman; p. 467, Bill Wunsch/Denver Post; p. 475 (Fig. 15-3), Corning Glass Works; p. 535, Dudley Foster/Woods Hole Oceanographic Institution; p. 536 (Fig. K-1), Deepsea Ventures, Inc.; p. 537 (Fig. K-3), Freeberg, USGS; p. 559 (Fig. 17-7), Kjell Sandved; p. 567, Elliot Morris, Colorado State University; p. 585, Pennzoil Sulfur Co.; p. 586, NASA; p. 587 (Fig. L-2), Grant Heilman/Grant Heilman; p. 588 (Fig. L-4, right), Alan Pitcairn/Grant Heilman; p. 589 (Fig. L-5a), Division of Mineral Sciences, Smithsonian Institution; p. 589 (Fig. L-5b), Vulcain Explorer/Science Source/Photo Researchers; p. 623, Vulcan Materials Company; p. 630 (Fig. 20-3a), E. I. du Pont de Nemours & Co.; p. 631 (Fig. 20-5), Alcoa; p. 632 (Fig. 20-6), AMAX/Ray Manley Photography; p. 636, Travis Amos; p. 658, Ethyl Corp.; p. 660 (Fig. M-2), Argonne National Laboratory; p. 662 (Fig. M-4), Leland C. Clark; p. 670, Oak Ridge National Laboratory; p. 672 (Fig. 21-1), Lawrence Berkeley Laboratory, University of California; p. 681 (Fig. 21-5), Lawrence Berkeley Laboratory, University of California; p. 681, bottom, B. Leonard Holman and Thomas C. Hill, Harvard Medical School; p. 682, margin, Los Alamos National Laboratory; p. 689 (Fig. 21-11b), Pacific Gas and Electric; p. 692 (Fig. 21-13), Archives of the Institute of Radium, from *From X-rays to Quarks,* by E. Segre, W. H. Freeman and Company, 1980; p. 700, Ray Manley Photography; p. 701 (Fig. N-1), Martin Marietta; p. 701 (Fig. N-2), Echo Bay Mines; p. 705 (Fig. N-6), Division of Mineral Sciences, Smithsonian Institution; p. 706 (Fig. N-7b), Nippon Kokan K.K.; p. 707 (Fig. N-8), Bethlehem Steel Corp.; p. 707, bottom, Inco; p. 709 (Fig. N-10), Peter Arnold/Peter Arnold; p. 710 (Fig. N-11), Rodney Cotterill and Flemming Kragh, The Technical University of Denmark; p. 741, Petro-Canada; p. 769, Roger Tully © *Discover* Magazine, July 1983, Time Inc.; p. 770, General Electric Research and Development Center; p. 773, General Electric; p. 779 (Fig. 24-5), J. M. Hogle, Department of Molecular Biology, Research Institute of Scripps Clinic; p. 780 (Fig. 24-8), Wallace Jensen and Panpit Klug, The University of Miami Medical School; p. 781 (Fig. 24-9), A. K. Kleinschmidt, Elsevier; p. 783 (Fig. 24-11), A. C. Barrington Brown, from *The Double Helix,* by J. D. Watson, Atheneum, N.Y., p. 215. Copyright 1968; p. 785 (Fig. 24-13), Nelson Max, Lawrence Livermore Laboratory.

INDEX

Physical Constants

Constant	*Symbol*	*Value*
atomic mass unit	amu	1.66056×10^{-27} kg
Avogadro's number	N	6.02205×10^{23} mol^{-1}
Bohr radius	a_0	5.292×10^{-11} m
Boltzmann constant	k	1.38066×10^{-23} J·K^{-1}
charge of a proton	e	1.60219×10^{-19} C
Faraday constant	F	96,485 C·mol^{-1}
gas constant	R	8.31441 J·K^{-1}·mol^{-1}
		0.08206 L·atm·K^{-1}·mol^{-1}
mass of an electron	m_e	9.10953×10^{-31} kg
		5.48580×10^{-4} amu
mass of a neutron	m_n	1.67495×10^{-27} kg
		1.00866 amu
mass of a proton	m_p	1.67265×10^{-27} kg
		1.00728 amu
Planck's constant	h	6.62618×10^{-34} J·s
speed of light	c	2.997925×10^{8} m·s^{-1}

SI Prefixes

Prefix	*Multiple*	*Symbol*	*Prefix*	*Multiple*	*Symbol*
tera	10^{12}	T	deci	10^{-1}	d
giga	10^{9}	G	centi	10^{-2}	c
mega	10^{6}	M	milli	10^{-3}	m
kilo	10^{3}	k	micro	10^{-6}	μ
			nano	10^{-9}	n
			pico	10^{-12}	p
			femto	10^{-15}	f
			atto	10^{-18}	a